8th International Scientific Conference 'Actual Problems of Engineering Mechanics' (APEM 2021)

AA003681

IOP Conference Series: Materials Science and Engineering
Volume 1164

Online
11-14 May 2021

ISBN: 978-1-7138-4627-7
ISSN: 1757-8981

Printed from e-media with permission by:

Curran Associates, Inc.
57 Morehouse Lane
Red Hook, NY 12571

Some format issues inherent in the e-media version may also appear in this print version.

This work is licensed under a Creative Commons Attribution 3.0 International Licence.
Licence details: http://creativecommons.org/licenses/by/3.0/.

No changes have been made to the content of these proceedings. There may be changes to pagination and minor adjustments for aesthetics.

Printed with permission by Curran Associates, Inc. (2022)

For permission requests, please contact the Institute of Physics at the address below.

Institute of Physics
Dirac House, Temple Back
Bristol BS1 6BE UK

Phone: 44 1 17 929 7481
Fax: 44 1 17 920 0979

techtracking@iop.org

Additional copies of this publication are available from:

Curran Associates, Inc.
57 Morehouse Lane
Red Hook, NY 12571 USA
Phone: 845-758-0400
Fax: 845-758-2633
Email: curran@proceedings.com
Web: www.proceedings.com

TABLE OF CONTENTS

IOP Conf. Series: Materials Science and Engineering 1164 (2021) 011001 doi:10.1088/1757-899X/1164/1/011001

Preface

The annual International Conference "Actual Problems of Engineering Mechanics" is held for the eighth time. The conference is organized by the Odessa State Academy of Civil Engineering and Architecture (Ukraine, Odessa). As partners involved colleagues from Ukraine (G.S. Pisarenko Institute for Problems of Strength of the National Academy of Sciences of Ukraine), Greece (University of West Attica) and Croatia (University «North»).

Unfortunately, because of COVID-19, the conference participants were unable to get together, as in previous years, and were forced to limit themselves to online communication. Nevertheless, the conference in this format of the meeting aroused great interest.

The scientific program of the conference included plenary and sectional reports in the following areas:

1. Fundamental research in mechanics.
2. Computer mechanics.
3. Building structures.
4. Machine building structures.
5. Construction materials.

More than 500 participants were registered, 140 reports were submitted. The conference was attended by scientists from 14 countries of the world - Ukraine, Greece, Spain, USA, France, Vietnam, Thailand, China, Japan, Croatia, Poland, Belarus, Russia, Uzbekistan.

Conference materials are available in pdf format.

List of committees:

Content from this work may be used under the terms of the Creative Commons Attribution 3.0 licence. Any further distribution of this work must maintain attribution to the author(s) and the title of the work, journal citation and DOI.
Published under licence by IOP Publishing Ltd

Antoniuk N.R. – technical editor of «OSACA bulletin» journal, PhD, Associate Professor.

Balduk P.H. – conference secretary, PhD, Professor of Department of Structural mechanics of Odesa State Academy of Civil Engineering and Architecture.

Zinkovskyi A.P. – Deputy Research Director of G.S. Pisarenko for Problems of Strength, Dr. Tech. Sc., Professor.

Klymenko Y.V. – Dr. Tech. Sc., Professor of Odesa State Academy of Civil Engineering and Architecture.

Kovrov A.V. – Chairman of the Conference Organizing Committee, Rector of Odesa State Academy of Civil Engineering and Architecture, PhD, Professor.

Krutii Y.S. – vice-rector of Odesa State Academy of Civil Engineering and Architecture, Dr. Tech. Sc., professor.

Surianinov N.G. – Coordinator, Deputy Chairman of the Conference Organizing Committee, Chairman of Department of Structural Mechanics of Odesa State Academy of Civil Engineering and Architecture, Dr. Tech. Sc., Professor, sng@ogasa.org.ua.

Kharchenko V.V. – Director of G.S. Pisarenko for Problems of Strength, Academician of NAS Ukraine, Dr. Tech. Sc., Professor.

Shvabiuk V.I. – Lutsk National Technical University, Dr. Tech. Sc., Professor.

Hendrik Doss – Professor, Mainz Universsity of Applied Sciences (Germany).

Kyriazopoulos A. – Professor, University of West Attica.

Demakos K. – Professor, University of West Attica.

Pnevmatikos N. – Associate Professor, University of West Attica.

Milkovich Marin – rector of the University «Sjever», professor.

Peer review declaration

All papers published in this volume of *IOP Conference Series: Materials Science and Engineering* have been peer reviewed through processes administered by the Editors. Reviews were conducted by expert referees to the professional and scientific standards expected of a proceedings journal published by IOP Publishing.

- **Type of peer review:**

For each article, two reviews are presented (the first is internal, the second is external). The peer review is open.

- **Conference submission management system:** e-mail
- **Number of submissions received:** 105 manuscripts
- **Number of submissions sent for review:** 105 manuscripts
- **Number of submissions accepted:** 96 manuscripts
- **Acceptance Rate (Number of Submissions Accepted / Number of Submissions Received X 100):** 91,4%
- **Average number of reviews per paper:** 2 per paper
- **Total number of reviewers involved:** 124 reviewers
- **Any additional info on review process:**

Manuscripts were sent to the conference organizing committee. Then the members of the organizing committee sent these manuscripts to the reviewers. Each manuscript was reviewed by two reviewers - candidates of technical sciences and doctors of technical sciences. The following requirements were imposed on the articles: clarity of presentation, relevance of the topic, originality and novelty of the idea, correspondence of the material to the declared topic of the conference, transparency of the result obtained. If the reviewers accepted the articles or pointed out deficiencies requiring changes, the organizing committee sent the manuscript back to the authors so that the authors could make the necessary corrections and changes. After making corrections to the article, the authors sent the final version of the manuscripts to the organizing committee. After re-examination, the editors completed proofreading of the manuscripts to fully comply with IOP Publishing requirements.

- **Contact person for queries:** Mykola Surianinov, sng@ogasa.org.ua

Content from this work may be used under the terms of the Creative Commons Attribution 3.0 licence. Any further distribution of this work must maintain attribution to the author(s) and the title of the work, journal citation and DOI.
Published under licence by IOP Publishing Ltd

Reducing the environmental impact of mining production through the use of man-made waste

N Gorbatyuk[1*], A Adigamov[2], O Kobelev[2] and A Pashkov[2]

[1]V I Vernadsky Crimean Federal University, Prospekt Vernadskogo, 4 295007, Simferopol, Russia
[2]National University of Science and Technology «MISiS», Leninsky av., 4, 119991, Moscow, Russia

*E-mail: gorbatyuk54@bk.ru

Abstract. The admissibility of replacing traditional cement, used as a binder with magnesium-containing man-made waste in the preparation of a construction and fill mixture is justified. The possibility of replacing the traditional specially mined aggregate in construction and fill mixtures is proved. The results of testing samples after cementing and a construction-fill mixture prepared on the basis of man-made waste are presented. The positive effect of mechanical activation of the backfill mixture components on its rheological properties and strength characteristics of the fill mass is studied. The use of lignosulfonate as an additive was confirmed to improve the binding and rheological properties of the backfill mixture. The involvement of man-made waste in the closed cycle of the main and auxiliary production reduces the volume of man-made massifs, which significantly reduces the impact of mining production on the environment.

1. Referat

Every year there is an increase in the volume of consumption of georesources, which leads to an intensification of their extraction from the subsurface. This causes an increase in the accumulation of man-made waste stored on the surface, which has a significant impact on the environment. The term of existence of man-made massifs is unlimited, therefore, the environmental damage can subsequently significantly exceed the economic benefit from the extracted resources.

At the same time, it is necessary to consider that in man-made waste, useful components remain in significant volumes and the waste itself can be raw materials for secondary use. In this regard, it is necessary to introduce the concept of non-waste (low-waste) production.

The creation of closed-loop production using secondary products is a paradigm for the development of mineral resources. This will significantly reduce the impact on the environment. Industrial waste is perceived as a" by-product" of mining production. They must be qualified as a secondary (intermediate) product used for subsequent use or processing.

The main direction of using industrial waste can be the replacement of traditional, specially extracted aggregate when creating a backfill or construction mixture.

The re-use of man-made waste in a non-waste technological chain, in addition to their disposal, significantly reduces the risk of environmental impact of mining and processing production.

Before using man-made waste in construction and backfill mixes, it is necessary to consider that they have useful components in their composition, although not in a significant amount. In this regard, it should be noted that before using man-made waste in mixtures, it is necessary to extract useful components by deep processing.

When choosing the compositions of building and laying mixes, it is necessary to consider their features. It should be noted that the mixtures used to produce laying works in the extraction of minerals have a higher water-solid ratio in comparison with the solutions used in construction.

The use of man-made waste is hindered by the presence of harmful components in the waste that have a negative impact on monolithic structures, or the artificial mass being created. Neutralization of

Content from this work may be used under the terms of the Creative Commons Attribution 3.0 licence. Any further distribution of this work must maintain attribution to the author(s) and the title of the work, journal citation and DOI.
Published under licence by IOP Publishing Ltd

the negative impact of the components remaining in man-made waste is possible as a result of their deep processing. This will not only neutralize the impact of harmful components, but also introduce resource-renewable technologies.

The maximum ecological and economic effect of the use of man-made waste in construction and backfill mixtures can be achieved only after their deep processing and additional extraction of useful components and neutralization of the negative effect of the remaining ones. At the same time, the involvement of mining and processing industries in deep processing creates prerequisites for the creation of a new material and raw material base of the mining complex and eliminates the costs of exploration and development of new deposits.

2. Introduction

Russia is one of the world leaders in the extraction of mineral resources and the production of products from extracted raw materials. Russia accounts for up to 5% of the world's iron ore production, about 25% of potash raw materials, and in total, Russia produces about 10% of the total world production of the mining industry [1].

But at the same time, it should be noted that Russia lags far behind European countries and Japan in terms of ensuring environmental safety [2]. The Russian mining and processing industry is not at a sufficiently high level of ensuring environmental safety, processing and disposal of man-made waste from mining and processing industries and their use in the further technological scheme or production of products that are not related to the main activity of the mining and processing enterprise [3].

Every year, the useful content of recoverable reserves decreases, which leads to an increase in processing waste [4]. Developments go to great depths, which leads to an increase in the volume of rocks from overburden and tunneling operations. All these industrial wastes are stored on the surface, forming huge massifs [5]. Waste storage in industrial massifs is a form of anthropogenic impact on the environment, which causes qualitative and quantitative changes in its subjects [6].

About 30 tons of water-soluble industrial waste is generated during the production of 1 ton of potash fertilizers. Approximately 5 - 8% of the production cost is spent on warehousing, storage, construction of storage facilities for man-made waste, environmental deductions, etc. [7]. At the same time, it should be noted that the lifetime of man-made massifs is unlimited, therefore, the multiplicative damage from their placement will subsequently exceed the value of the extracted raw materials [8].

Until now, when reused, up to 10% of the extracted rocks from sinking and overburden operations are disposed of [9] and no more than 20% of water-soluble industrial waste [10]. According to the Ministry of Natural Resources and Ecology, about 45 billion tons of man-made waste from the mining and processing sector of various hazard classes are stored in industrial massifs in Russia [11].

All over the world, including Russia, there is an annual increase in the level of pollution of the environment by man-made waste. Since the beginning of the XXI century, the annual volume of expansion of man-made massifs with waste rock increases by 30% and amounts to 210 million m^3 / year, and waste from processing production amounts to 140 million m^3 / year [12]. The formed industrial massifs disrupt the ecological balance, withdraw fertile agricultural land from circulation [13].

Often, it is proposed to preserve the pristine nature and mineral diversity of the Earth by transferring mining production to cosmic bodies [14]. But the lack of technical possibilities for implementation [15] and unresolved legal issues [16] postpone the implementation of this idea for the distant future.

3. The use of man-made waste in construction and backfill mixes

The development of the society leads to an increase in the extraction of mineral resources and the construction of both industrial facilities [17] and civil buildings [18]. About 35% of enterprises in the mining sector use geotechnology with the use of developed stopes [19] and their share is constantly increasing.

In the production of construction and backfill mixes, cement or its derivatives are used as a binder, and specially extracted materials are used as an inert component: sand, gravel, crushed stone. Considering the increase in the volume of construction of buildings and structures and the fairly active introduction of geotenology with a backfill mass, the issue of replacing the components of the mixture with cheaper and more affordable ingredients is acute. Therefore, the use of man-made waste from mining and processing production is seen as a good option [20].

The idea of using man-made waste to replace the traditional, specially extracted aggregate when creating a backfill mixture is not new [21]. At the same time, it should be taken into account that the creation of building mixes has similar technological aspects to the production and design of backfill mixes [22]. The use of waste water-soluble ores formed after their enrichment as an aggregate in construction and backfill mixtures will allow the introduction of waste-free (low-waste) production [23]. The re-use of man-made waste in a non-waste technological chain, in addition to their disposal, significantly reduces the risk of environmental impact of mining and processing production [24].

Before using man-made waste in construction and backfill mixes, it is necessary to take into account that they have useful components in their composition, although not in significant quantities (Table 1). In this regard, it should be noted that before using man-made waste in mixtures, it is necessary to extract useful components by deep processing.

In addition, the use of man-made waste in construction and backfill mixes is hindered by the presence of harmful components in the waste that have a negative effect (softening, swelling, increasing porosity, etc.) on monolithic structures or the artificial mass created after the mixture solidifies [25]. Similarly, the harmful components of man-made waste can remain in a conditional equilibrium state for a long time [26]. But after exposure to the environment or when interacting with other components, they pass into a mobile state, are washed out of structures or artificial massifs, which leads to a violation of the ecological balance [27].

Currently, technologies for deep processing of man-made waste have been developed, which allow for additional extraction of useful components [28]. The technology of deep processing will also allow to neutralize the negative impact of harmful components on the created structures or mass and the environment [29]. This technology creates prerequisites for the introduction of resource-renewable technologies, contributes to the increase in the mineral resource base of the mining enterprise and increases its working life [30].

The most effective technology of deep processing at the present time is the mechanochemical technology. This technology combines chemical irrigation of industrial waste and their mechanical treatment in a disintegrator [31].

The idea of using water-soluble ore waste in mixtures to replace an aggregate is not new. The replacement of the traditional aggregate with the waste of water-soluble ores was previously carried out without their additional activation treatment [32]. At the same time, it should be noted that earlier studies have shown a positive effect of the activation treatment of industrial waste [33].

It should be noted that the mixtures used to produce laying works in the extraction of minerals have a higher water-solid ratio in comparison with the solutions used in construction.

The increased volume of water in the backfill mixtures is necessary in order to improve its transportability from the place of production to the place of laying and spreadability in the waste chamber. Strength and transportability have opposite dependencies on the amount of liquid in the mixture.

Since the main cost in the construction and backfill mix is cement, the search for its replacement always remains relevant. Earlier experiments have shown the possibility of using metallurgical slags after activation treatment as a binder [34].

Research work carried out earlier in the field in the creation of mixtures using water-soluble ore waste as an aggregate has demonstrated the advantage of binders containing magnesium.

Table 1. Chemical composition of industrial waste from water-soluble ore dressing.

Waste enrichment		Components	KCl	NaCl	$MgCl_2$	$CaSO_4$	Insoluble residue	Br^-	H_2O_{cryst}
Halurgic	Stale	Mass fraction %	3.35	92.66	0,07	1.91	1.9	0.03	0.08
	Current		1.91	94.3	0.07	1.914	1.7	0.026	0.08
Flotation	Stale		9.96	82.64	1.1	4.488	1.7	0.032	0.08
	Current		4.88	87.78	1.1	4.331	1.8	0.029	0.08

Therefore, the purpose of this research work is to develop a cementless construction-laying mixture based on waste water-soluble ores with specified rheological properties and intended for the formation of a monolith or artificial massif with specified strength characteristics.

When developing the composition of the mixture, waste from water-soluble ores of Uralkali was used as an aggregate, and slags from the Chusovsky Metallurgical Combine were used as a binder. All components were applied after activation treatment. The chemical and granulometric compositions of waste from the water-soluble ore processing plant are given in Tables 1 and 2.

Table 2. Granulometric composition of industrial waste from water-soluble ore dressing.

Waste	Particle size, mm								Average size
	+7	7-5	5-3	3-2	2-1	1-0,5	0.5-0.25	-0.25	
	7.4	7.3	17	16.3	20.9	19,5	8.6	3.0	2.54

4. Creation of an artificial mass based on man-made waste of water-soluble ores

As can be seen from Table 1, industrial waste of water-soluble ores consists of 94-98% of salts, therefore, it is easily soluble in water. To exclude their dissolution in the construction-laying mixtures, saturated salt solutions were used as a recluse. Previous studies have proven the effectiveness of saturated solutions.

Magnesia binders, including magnesia cement, are able to form sufficiently strong crystal bonds in small quantities and combine a large amount of aggregate into a monolith. In industrial applications, it is not necessary to use derivatives of magnesium-containing cement. In order to reduce the cost of the construction and backfill mixture, magnesia cement or caustic magnesite can be replaced with slags that have an increased magnesium content in their composition. Due to the large volume of construction and monolithic works on construction sites and laying works at mining enterprises, the use of magnesium-containing slags as a substitute for binder is very promising.

It is possible to increase the ability of slags to solidification by adding a small amount of cement or an activating additive to the mixture. Earlier studies of the mechanical activation treatment of the components of the backfill mixture showed positive results [35]. In this regard, in this study, the activation treatment of slags and waste of water-soluble ores in disintegrators was carried out.

When conducting experimental studies, they were based on methodological guidelines and instructions.

Waste from the processing of water-soluble ores and magnesia slags in the construction-backfill mixture were mixed in the proportion of 51% and 25% of the total volume of the composite, respectively. Before mixing, separate activation treatment of the components of the construction-laying mixture was performed in the DESI-11 disintegrator. In the activation unit, the magnesium – containing slags were crushed to a class content of less than 0.071 mm of at least 68-72%, and the waste of water-soluble ores-at least 88-92%.

The rheological properties of the mixture are important for quenching operations, since it often has to be transported over long distances from the production site to the place of laying in the stopes. Therefore, the rheological characteristics for the backfill mixture were carried out according to the established standard methods: the depth of shrinkage of the cone; the spreadability of the mixture on the viscometer.

The samples were solidified under the conditions specified in the procedure (T=20±2°C; W=95±5%) and their compression test was performed after the specified time specified in the procedure: 7; 28; 60 and 90 days. Samples for uniaxial compression were tested on a PI-2000-A press. The results of the experiments are listed in Table 3.

Table 3. Test results of backfill mixtures.

№	Backfill components: Magnesian slag: masses.%	Backfill components: Magnesian slag: Fraction content less than 0.071 mm	Backfill components: Enrichment waste of water-soluble ores: masses.%	Backfill components: Enrichment waste of water-soluble ores: Fraction content less than 0.071 mm,%	Lingosulfanate % of solid	Water, mass. %	Spreading, mm	Cone draft, cm	Uniaxial compression strength, MPa: Duration of hardening, days: 7	28	60	90
1	25.0	68 – 72	51.0	88 – 92	-	24.0	105	10	0.2	1.5	1.9	2.2
2	25.0	88 – 92	51.0	88 – 92	-	24,0	95	10	0.15	1.6	2.2	2.5
3	25.0	88 – 92	51.0	68 – 72	-	24.0	120	14	0.1	1.2	1.7	1.8
4	25.0	88 – 92	51,0	88 – 92	1	23.25	165	16	0.25	2.0	2.75	3.1

From the comparative analysis, it follows that the first composition of the construction-laying mixture demonstrated low strength characteristics when tested for uniaxial compression after solidification and its low mobility (composition No. 1 in Table 3).

To increase the hydraulic activity, the binder, in our case magnesia slag, was machined in a disintegrator to a fineness of 88-92% of the class less than 0.071 mm. The size of the aggregate was left unchanged. Testing of the samples after solidification demonstrated a significant improvement in strength parameters. But at the same time, the rheological characteristics were unsatisfactory, which indicates poor transportability of such a mixture (composition No. 2 in Table 3).

To increase the spreadability of the construction-backfill mixture and, in general, its rheological properties, the fraction fraction of less than 0.071 mm of the aggregate was reduced. Testing of samples after solidification for uniaxial compression showed a sharp drop in strength characteristics (composition No. 3 in Table 3).

The increase in the water-solid ratio, which makes it possible to change the rheololgic characteristics of the mixture, was immediately abandoned, due to the inexpediency of this experiment. Previous studies have shown that the addition of water to the mixture reduces the strength of the monolith, structures and the mass and increases the volume of drainage.

From the analysis of previous studies, it was concluded that the use of activating additives makes it possible to create with stronger structural-crystal bonds [36], which significantly increases the strength characteristics of monoliths and artificial massifs [37]. In addition, chemical additives have positive changes on the rheological characteristics of the created mixture. Such additives include lignosulfanate-waste from the pulp and paper industry obtained by processing spent sulfite and bisulfite liquors [38, 39].

The introduction of lingosulfanate into the composition of the construction-laying mixture allowed us to obtain samples with increased strength characteristics. At the same time, the rheological characteristics of this mixture fully meet the requirements of its transportability through pipelines (composition No. 4 in Table 3).

When creating monolithic structures in construction and artificial massifs in the mining industry, it is necessary to strictly take into account the long-term influence of the components of the mixture and selectively approach their choice. The processes that occur in monolithic structures and in the mass after their complete solidification are difficult to predict and model. Methods and methods of control of monolithic structures and artificial massifs after their solidification are sufficiently fully investigated in the works.

The use of man-made waste in the preparation of construction and backfill mixture without additional, deep processing is contrary to environmental safety and economic feasibility. Resource-renewable technologies for deep processing of man-made waste have been developed, which make it possible to additionally extract useful components from them.

With the introduction of innovative enrichment technologies, there is an improvement in qualitative and quantitative indicators, which is proved by the analysis of the composition of the tailings of the processing plant (Table 1).

Therefore, it is recommended to use the current tailings of the galurgical method of enrichment as part of the construction-backfill mixture without additional processing. When using the stale tailings of the galurgic method of enrichment and the tailings of the flotation method of enrichment, it is necessary to make their additional enrichment. Additional enrichment can be carried out at the processing plant facilities that were released as a result of the reduction in production volumes of the enterprise due to the decommissioning of "Uralkali" Berezniki mines.

5. Conclusions

Replacing the traditional cement binder with magnesium-containing slags in the construction-laying mixture allows us to develop a composition that meets the strength characteristics of artificial massifs or building structures that do not carry a high load. The use of man-made waste in the construction-laying mixture is possible after their mechanical activation treatment in disintegrators, which increases the strength characteristics of structures and artificial massifs.

The maximum ecological and economic effect of the use of man-made waste in construction and backfill mixes can be achieved only after their deep processing and additional extraction of useful components and neutralization of the negative effect of the remaining ones. At the same time, the involvement in deep processing of waste from mining and processing industries creates prerequisites for the creation of a new material and raw material base of the mining complex and eliminates the costs of exploration and development of new deposits.

The involvement of man-made waste in the closed cycle of the main and auxiliary production reduces the volume of man-made massifs, which significantly reduces the impact of mining production on the environment.

References

[1] Kobelev O *et al* 1995 Electrocapillary curve of copper in boroaluminate melt *Rasplavy* **1** pp 39-45

[2] Kouziyev D *et al* 2019 Sensing of dynamic loads in the open-cast mine combine *E3S Web of Conf.* **105** 03014 doi: 10.1051/e3sconf/201910503014

[3] Kongar-Syuryun Ch *et al* 2021 Geotechnology using composite materials from man-made waste is a paradigm of sustainable development *Materials Today: Proceedings* **38** pp 2078-2082 doi: 10.1016/j.matpr.2020.10.145.

[4] Niemiec D *et al* 2017 Selected black-coal mine waste dumps in the ostrava karviná region: an analysis of their potential use *IOP Conf. Series: Earth Environ. Sci.* **95** 042061 doi: 10.1088/1755-1315/95/4/042061

[5] Rybak J *et al* 2021 Utilization of mineral waste: a method for expanding the mineral resource base of a mining and smelting company *Metallurgist* **64** pp 851-861 doi: 10.1007/s11015-021-01065-5

[6] Kobelev O *et al* 1998 Analysis of the electrocapillary curves and adsorption process models in copper || Boron + Aluminium oxide melt *Zeitschrift fur Physikalische Chemie* **203** pp 143-158 doi: 10.1524/zpch.1998.203.Part_1_2.143

[7] Kowalik T *et al* 2019 Chemical hazards in construction industry *E3S Web of Conf.* **97** 03032 doi: 10.1051/e3sconf/20199703032

[8] Kobelev O *et al* 2009 Effective production of large pipe blanks *Steel in Translation* **39 (6)** pp 501-505 doi: 10.3103/S0967091209060163
[9] Kongar-Syuryun Ch *et al* 2020 Industrial waste in concrete mixtures for construction of underground structures and minerals extraction *IOP Conf. Series: Mater. Sci. Eng.* **869** 032004 doi: 10.1088/1757-899X/869/3/032004
[10] Khayrutdinov M *et al* 2020 Cementless backfill mixtures based on water-soluble manmade waste *Bulletin of the Tomsk Polytechnic University, Geo Assets Engineering* **331(11)** pp 30–36 doi: 10.18799/24131830/2020/11/2883
[11] Albagachiev A e*t al* 2020 Determination of rational friction temperature in lengthwise rolling *CIS Iron and Steel Review* **19** 33-36 doi: 10.17580/cisisr.2020.01.07
[12] Pimenov G *et al* 1991 Making of large-sized solid-forged blanks *Tyazheloe Mashinostroenie* **9** pp 21-24
[13] Zástěrová P *et al* 2016 Analysis the purposes of land use planning on the hard coal tailing dumps *IOP Conf. Series: Earth Environ. Sci.* **44** 022034 doi: 10.1088/1755-1315/44/2/022034
[14] Khayrutdinov M *et al* 2020 Planetary technology. prerequisites for the formation of a new scientific discipline *Gornaya Promyshlennost* **3** pp 113–120 doi: 10.30686/1609-91922020-3-113-120
[15] Khayrutdinov A 2020 Current issues of mining activities on celestial bodies: International law aspects *Advances in the Astronautical Sciences* **170** pp 895–902
[16] Gugunskiy D *et al* 2020 Legal models for activities on the exploration and utilization of space resources: towards the «Space-2030» Agenda *Advances in Intelligent Systems and Computing* **1100** 657-664 doi: 10.1007/978-3-030-39319-9_73
[17] Kawalec J and Warchał T 2015 Dynamic replacement columns with aggregate transition zone stabilized by geosynthetics for embankment foundation over weak deposits *XVI ECSMGE Geotechnical Engineering for Infrastructure and Development* 1511-1516 doi: 10.1680/ecsmge.60678.vol3.221
[18] Kawalec J *et al* 2017 Application of crushed concrete in geotechnical engineering - selected issues *IOP Conf. Series: Earth Environ. Sci.* **95** 022057 doi: 10.1088/1755-1315/95/2/022057
[19] Khayrutdinov A *et al* 2020 Stress-strain behavior control in rock mass using different-stregth backfill *Mining Informational and Analytical Bulletin* **10** 42-55 doi: 10.25018/0236-1493-2020-10-0-42-55
[20] Krivenko A and Khanh Z 2020 Influence of power fluid temperature in hydraulics on operating efficiency of hydraulic mining excavators *Gornyi Zhurnal* **12** pp 78-81 doi: 10.17580/gzh.2020.12.18
[21] Khayrutdinov M *et al* 2021 Improving safety when extracting water-soluble ores by optimizing the parameters of the backfill mass *Bezopasnost' Truda v Promyshlennosti* **1** pp 53-59 doi: 10.24000/0409-2961-2021-01-53-59
[22] Keropyan A *et al* 2020 Process research of wheel-rail mining machines *Traction lecture notes in mechanical engineering* 703-709 doi: 10.1007/978-3-030-22063-1_75
[23] Jiawen L *et al* 2019 Modeling pressure pulsation fields in a screw centrifugal pump *Proceedings of the 26th International Congress on Sound and Vibration ICSV26* **697**
[24] Kongar-Syuryun Ch *et al* 2021 Models and algorithms of choice of development technology of deposits when selecting the composition of the backfilling mixture *IOP Conf. Series: Earth Environ. Sci.* **684** 012008. doi:10.1088/1755-1315/684/1/012008
[25] Shakhov S and Vdovin K 2019 Electromagnetic Stirring in Continuous-Casting Machines *Steel in Translation* **49 (4)** pp 261-264 doi: 10.3103/S0967091219040120/
[26] Sivak B *et al* 2020 Development of a system for electromagnetic stirring of liquid steel in molds of billet and bloom CCMS *Metallurgist* **63(9-10)** pp 910-914 doi: 10.1007/s11015-020-00909-w

[27] Herbut A *et al* 2019 The surface wave attenuation as the effect of vibratory compaction of building embankments *IOP Conf. Series: Earth Environ. Sci.* **362** 012131 doi: 10.1088/1755-1315/362/1/012131
[28] Ivannikov A *et al* 2020 Express determination of the grain size of nickel-containing minerals in ore material. *Materials Today: Proceedings* **38** 2059-2062 https://doi.org/10.1016/j.matpr.2020.10.141
[29] Chumakov A *et al* 2021 Improving the control system of concentration plants based on express control of dissemination of magnetic minerals *IOP Conf. Series: Earth Environ. Sci.* **684** 012005 doi:10.1088/1755-1315/684/1/012005
[30] Valiullin I *et al* 2021 Application of a discrete double-row scraper cleaner as an efficient procedure of cleaning of belt conveyors aimed at transporting bulk materials *Metallurgist* **64** 1340-1346 doi: 10.1007/s11015-021-01124-x
[31] Khayrutdinov A *et al* 2020 Improvement of the backfilling characteristics by activation of halite waste for non-waste geotechnology *IOP Conf. Series: Mater. Sci. Eng.* **867** 012018 doi: 10.1088/1757-899X/867/1/012018
[32] Gorbatyuk S *et al* 2020 Technologies for applying Ni-Au coatings to heat sinks of SiC-Al metal matrix composite material *Materials Today: Proceeding* **38** 1889-1893 doi: 10.1016/j.matpr.2020.08.581
[33] Kongar-Syuryun Ch *et al* 2021 Effect of activating treatment of halite flotation waste in backfill mixture preparation *Mining Informational and Analytical Bulletin* **1** pp 43–57 doi: 10.25018/0236-1493-2021-1-0-43-57
[34] Rybak J *et al* 2021 Geomechanical substantiation of parameters of technology for mining salt deposits with a backfill *Mining Science* **28** 19-32 doi: 10.37190/msc212802
[35] Kongar-Syuryun Ch *et al* 2021 Research of rheological characteristics of the mixture as a way to create a new backfill material with specified characteristics *Materials Today: Proceedings* **38** 2052-2054 doi: 10.1016/j.matpr.2020.10.139
[36] Rakhutin M and Boyko P 2017 Ways to improve assessment methods of the main characteristics of grinding balls *Russian Coal Journal* **12** pp 49-52 doi: 10.18796/0041-5790-2017-12-49-52
[37] Boyko P *et al* 2019 Provision of operational life-time period of crushers liners by applying new technologies of their manufacturing and wear-out diagnosing *Equipment and technologies for oil and gas complex* **5** pp 42-47 doi: 10.33285/1999-6934-2019-5(113)-42-47
[38] Kobelev O and Tyurin V 2007 Production of large plates Steel in Translation **37 (9)** pp 727-729 doi: 10.3103/S096709120709001X
[39] Kulikova E and Ivannikov A 2020 The terms of soils removal from the defects of the underground structures' lining *J. Phys.: Conf. Series* **1425** 012062 doi.org/10.1088/1742-6596/1425/1/012062

Influence of fillers on the properties of under-rail gaskets

A I Adylkhodzhaev[1], V M Tsoy[1] and V I Kondrashchenko[2]

[1] Department of Construction of Buildings and Structures of the Tashkent Transport University, 100067, Tashkent, Adilkhodjaev st., 1, Uzbekistan
[2] Department of Construction Materials and Technologies of Russian University of Transport, 127994, Moscow, Obrazcova st, 15, Russia

Abstract. Using the criterion of the reduced hydration activity of inorganic materials, the choice of magnetic separation waste as a mineral filler for the production of under-rail gaskets is justified. By the methods of experimental planning, mathematical models were obtained that adequately describe the hardness (Shore A) and the conditional tensile strength of the gasket material, using which the composition of the material was developed that exceeds the physical and technical characteristics of the production composition.

1. Introduction

Consumers of rubber products (RTP) of the industry are almost all branches of production: motor vehicle engineering (automotive, tractor construction, agricultural and railway industries, aircraft construction, shipbuilding, etc.), construction, oil and gas production, production of consumer goods, medical devices, etc.

As you know, such components of RTP as rubber, technical soot, and various chemical modifiers are imported to the Republic of Uzbekistan from countries far and near abroad, which significantly increases the cost of manufactured products and is a deterrent to expanding its range and increasing production volumes.

The article presents the results of research on the development of composite sub-rail linings, the main purpose of which is to reduce the negative impact of vibrodynamic loads transmitted to the elements of the upper structure of the railway track from the rolling stock. In view of their functional purpose and large-capacity use in railway transport, when developing effective and competitive linings, it is important to develop such compositions that would provide the required indicators of physical and technical properties [1, 2].

One of the ways to implement this task is to develop RTP based on the integrated use of chemical modifiers and mineral dispersed fillers of natural and man-made origin of a wide range presented on the market of the Republic of Uzbekistan [3, 4].

2. Methods and techniques used in research

When conducting experimental studies, the generally accepted methods of conducting laboratory tests of materials, set out in the relevant standards, in particular, GOST 34078-2017 "Laying of rail fasteners of a railway track, were used. Technical conditions", as well as non-standard methods developed by specialists of leading foreign research institutes.

3. Research results

To achieve this goal, the authors conducted studies on the quantitative assessment of the activity of dispersed mineral fillers of various nature. As a result, a criterion was proposed – the reduced hydration activity of mineral additives P_{pga}, which characterizes the acid-base properties of the surface of mineral fillers depending on the acidity constant pK_a [5-13], calculated by the formula:

$$P_{pga} = P_{kb} + P_{kl} + 0.33P_{ol} - 0.1P_{ob}, \quad (1)$$

Content from this work may be used under the terms of the Creative Commons Attribution 3.0 licence. Any further distribution of this work must maintain attribution to the author(s) and the title of the work, journal citation and DOI.
Published under licence by IOP Publishing Ltd

where P_{kb}, P_{kl}, P_{ol}, P_{ob} - are the number of adsorption centers, respectively, in the regions $0 \le pK_a < 7$; $pK_a \ge 13$; $-4 < pK_a < 0$; $7 \le pK_a < 13$ in 10^{-3} mg-eкv/g.

Based on the P_{pga} criterion, the reaction activity of mineral fillers was evaluated, as shown in Table 1.

Table 1. Values of the P pga criterion for the reactivity of mineral fillers.

Mineral filler	pKa values, rel. units.				Values of the P_{pga} criterion, rel. units.
	-4…0	0…7	7…13	> 13	
	P_{ol}	P_{kb}	P_{ob}	P_{kl}	
Quartz sand	8.04	9.11	8.75	1.88	12.77
Dune sand	4.12	7.08	9.95	1.07	8.52
Gliezh	13.22	16.47	10.08	2.87	22.39
Basalt	23.4	22.15	11.16	1.96	30.71
ESP	41.2	5.48	9.34	1.14	19.28
CSW	6.61	23.88	16.37	4.32	28.74
Fly ash	43.1	27.61	11.77	5.32	46.68
ZR	102.1	24.88	12.62	2.14	59.44
MSW	112.03	27.82	13.51	2.45	67.53

Note: ESP – waste of electric smelting production; CSW -waste of copper smelting production; ZR – zeolite-containing rock; MSW - waste of magnetic separation

Based on the maximum value of the Ppga coefficient determined by equation (1), as well as the world experience of leading companies in the field of RTP, such as Hankook, Michelin, Goodyear and others, which use mainly "silica" SiO2 as a mineral filler, a regenerate – magnetic separation waste (MSW) containing 97% silicon oxide SiO2 was adopted for testing as a mineral powder.

Magnetic separation waste is a technogenic fine powder of dark gray color obtained as a result of the production of glass and glass products, consisting mainly of unrolled quartz particles and inclusions of metal oxides, carbonates, hematite and their aggregates. The particle size modulus of the powder is significantly lower than 1, and the particle content of less than 0.074 mm is 80-98%.

In the course of the conducted studies, using the planning of the experiment, the composition of the rubber mixture for the production of under-rail gaskets was selected. The variable factors were: X_1 – stearic acid, kg; X_2 – regenerate, %; X_3 – synthetic fibers with a length of 2 mm (a product of recycling of decommissioned tires), kg. The values of the variable factors are given in Table 2.

Table 2. Values of variable factors.

Variable factors	Levels of variation in the coded x scale		
	$x_{min} = -1$	$x_0 = 0$	$x_{max} = +1$
Consumption of stearic acid X_1, kg	0.2	0.5	0.8
Regenerate consumption X_2, %	2	5	8
Fiber consumption X_3, kg	1.0	1.5	2.0

The response functions were such properties of the gasket material as the hardness T (Shore A), standard units, and the conditional tensile strength Rp, MPa. As an experimental plan, we used plan B3, which is close to the D-optimal plan. The matrices of the plan and the results of the active experiment are shown in Table 3.

Table 3. Matrix and arithmetic mean values of the response functions at the points of the experiment plan.

Experienc e number	Values of variable factors on a coded scale			Hardness T, standard units	Strength Rp, MPa
	X_1	X_2	X_3		

1	+1	+1	+1	40.2	24
2	-1	+1	+1	48.7	20
3	+1	-1	+1	60.4	16
4	-1	-1	+1	70.4	25
5	+1	+1	-1	65.7	26
6	-1	+1	-1	84.5	13
7	+1	-1	-1	44.5	15
8	-1	-1	-1	59.2	24
9	0	0	-1	55.7	13
10	0	0	+1	75.4	27
11	0	-1	0	64.8	26
12	0	+1	0	80.4	24
13	+1	0	0	71.5	17
14	-1	0	0	46.5	20
15	0	0	0	75.4	16

As a result of processing the experimental data, polynomial mathematical models were obtained that adequately describe the hardness T (Shore A), standard units, and the conditional tensile strength Rp, MPa, of the gaskets:

$$T = 34.45 + 19.5X_1 - 0.126X_1^2 - 12.81X_2 + 1.03X_2^2 + 10.32X_3 + 0.30X_3^2, \quad (2)$$

$$R_p = 13.4 + 1.8X_1 - 0.41X_1X_2 + 12.81X_2 - 1.03X_2^2 + 8.32X_3 + 0.30X_3^2. \quad (3)$$

The analysis of the obtained regression equations showed that the chemical modifier-steoric acid- has the main influence on the properties of the under-rail gaskets among the variable factors. By varying the content of the variable factors X_i (i = 1, 2, 3), it is possible to obtain gaskets with a wide range of response function values that differ more than twice in both Shore A hardness (from 40.2 us units in Experiment 1 to 84.5 us units in Experiment 6) and conditional tensile strength (from 13 MPa in experiments 6, 9 to 27 MPa in experiment 10).

Using the obtained mathematical models (2) and (3), rational compositions of under-rail gaskets were established that provide a Shore A hardness of not less than 65 US units (permissible range according to GOST 34078-2017 from 65 to 90 US units) and a conditional tensile strength of not less than 15 MPa, which corresponds to gaskets of increased durability (gaskets of category II, version "PD"), providing 1.1 billion tons of gross cargo passed and operated in the air in the temperature range from minus 60°C to plus 60°C, not only in sleepers, but also in the bars of switches.

The compositions of experimental gaskets were tested at the Toshkent Rubber Products company. As a mineral filler, magnetic separation waste with a SiO_2 content of 92% was selected, modified by surfactants during grinding in a ball mill to a specific surface area of Sud = 3000 g/cm^2. The production and developed compositions for the production of pilot batches of gaskets are shown in Table 4.

Table 4. Production composition and compositions of pilot batches of gaskets.

Name of the components	Consumption of the components in the compositions				
	PS	№1	№2	№3	№4
Rubber, кg	15.0	13.5	12	13.5	12.0
CB carbon black 803, кg	15	13	13	13	13
Technical oil, L	-	6	6	6	6
Chalk, кг	10	-	-	-	-
Sulfur, kg	0.50	0.65	0.65	0.65	0.65

Tiuran, kg	0.20	0.25	0.25	0.25	0.25
Regenerate, %	-	10	15	15	15
Fiber size of 2 mm, kg	0.5	0.5	0.5	0.5	0.5
Stearic acid, kg	0.2	1.2	0.8	0.6	0.4
Zinc oxide, kg	0.20	0.20	0.25	0.25	0.25
MSW + superficially active properties, %	-	10	15	20	25

Note: the "*" icon indicates properties regulated by GOST 34078-2017.

The results of physical and technical tests shown in Table 5 show that the proposed rational compositions of the material of under-rail gaskets according to the properties normalized by GOST 34078-2017 have higher indicators compared to the production composition: in terms of conditional tensile strength by 1.9 times (composition No. 4), elongation at break by 1.6 times (composition No. 4), abrasion by 25% (composition No. 4), at the same time providing less rigidity of the railway track-the hardness of gaskets on Shore A is lower to 31% (composition #4). At the same time, other properties in comparison with the production composition differ slightly (plasticity, compositions no.2 and No. 4; viscosity, composition No. 1; elasticity at 20°C for composition No. 4 and at 1000C for composition No. 1; coefficient of temperature resistance at 1000C) or a significant improvement is achieved (for composition No. 4, resistance to sub-vulcanization increased by 16%, conditional stress at fixed elongation by 52%, tear resistance by 17%, resistance to crack growth by 33%, coefficient of thermal aging in strength by 25% and elongation by 20%, multiple stretching at 70°C by 1.4 times).

The analysis of rational compositions No. 1-No. 4 according to the complex of physical and technical indicators achieved by them allowed us to recommend composition No. 4 as a composition that provides the best normalized GOST 34078-2017 and production indicators. With the introduction of this composition in the production of under-rail gaskets, a number of technological parameters were improved, in particular, reducing the temperature and rolling time by 200C (from 80 to 600C) and 25 minutes, respectively, increasing productivity by 17% and reducing the cost of production due to the use of local man-made raw materials.

4. Conclusions

Based on the criterion of the reduced hydration activity of inorganic Ppga materials developed by the authors of the article, the choice of magnetic separation waste in the production of glass products as a mineral filler for the manufacture of under-rail gaskets is scientifically justified. Using the methods of planning active experiments, rational compositions of the material of under-rail gaskets were established, which provide an improvement in their physical and technical properties in comparison with the composition adopted in production. At the same time, the introduction of composition No. 4 in production has improved the technological parameters, increased productivity and reduced the cost of production due to the use of local man-made raw materials.

References

[1] Bekin N G and Shanin N G 1996 *Equipment for rubber industry factories* (L.: Chemistry) 376
[2] Barskova D M 1975 *Machines and apparatus for rubber production* (M.: Chemistry) 600
[3] Lepetov V A 1976 *Rubber technical products* 3-rd ed. rev. (L.: Chemistry) 440
[4] Minulenko L I Denisova O I and Strubel'skaya E M 2002 Vulcanization accelerators *Raw materials and materials for the production of rubber goods* **1** pp 8-11
[5] Adilxodjaev A I and Tsoy V M 2016 Impact of the technological factors on strength of the concrete with new generation chemical additive *European Science Review* (Austria) **3-4** pp 310-312
[6] Adilxodjaev A I and Tsoy V M 2016 Role of the acid-base nature of interphase interactions in structurization of composite construction materials *European Science Review* (Austria) **1-2** pp 163-165

[7] Adilxodjaev A I, Tsoy V M and Makhamataliev I M 2016 On a new methodological approach to the study of surface active properties of mineral fillers in cement systems *Architecture and construction of Uzbekistan*. **4-5** pp 79-82
[8] Adilxodjaev A I, Tsoy V M and Makhamataliev I M 2015 Assessment of reinforcement corrosion in high-filled ash-containing concrete designed for railway buildings and structures *Transport Problems* (Poland) pp 14-18
[9] Adilxodjaev A I 2017 Mathematical modeling of the properties of effective ceramic brick with burn-out additives from agricultural wastes *Internauka Nauchnyj zhurnal* **23 (57)** pp 37-40
[10] Adilxodjaev A I Makhamataliev I M Tsoy V M Ruzmetov F Sh and Umarov K S 2020 About ways of application of mineral fillers in the composition of cement concretes, taking into account the character of distribution of adsorption centers on their surface *International Journal of Advanced Science and Technology* vol **29** no 5 pp 1894-1900
[11] Adilxodjaev A, Tsoy V, Shaumarov S and Khodzhaev S 2020 Research of the influence of silicon-organic hydrophobizer on the basic properties of cement stone and mortar *International Journal of Advanced Science and Technology* vol **29** no 5 pp 1918-21

Investigation of the influence of elastic-mass characteristics of the axle test stand links on its own oscillation frequencies

S Raksha[1,2], P Anofriev[1,2,*], O Kuropiatnyk[1,2] and S Plitchenko[1,2]

[1]Department of Applied Mechanics and Materials Science
[2]Dnipro National University of Railway Transport named after Academician V. Lazaryan, Dnipro, Ukraine

*E-mail: anofrievp@ukr.net

Abstract. The reliability of work according to the criteria of vibration stability of the stand for testing the axles of wheelsets for endurance is largely determined by the natural vibration frequencies of the stand. The work considers a stand with a lever structure for loading the tested axles. In order to determine the influence of the parameters of the cross-section of the links on the natural frequencies of the vibration stand, the frequency characteristics were obtained by the method of mathematical modeling for two variants of the design schemes of the stand. Studies have shown that increasing the cross-sectional area of the levers without measuring other linear dimensions leads to a simultaneous increase in their mass and cruelty. The rigidity of the levers is defined as for beams on two supports with a load of a transverse concentrated force. As a result, the calculated values of the natural frequencies of the stand vibrations change insignificantly. A deeper analysis of the Eigen frequency characteristics of the stand with the tools of the Matlab system made it possible to obtain linear regression equations with the coefficients of determination close to one. The obtained research results gave reason to expect a decrease in stresses in the metal structures of the stand with an increase in the cross-sectional area of the levers, however, the values of the natural frequencies of the stand vibrations vary within a few percent. The way to control the natural frequencies of the stand vibrations only by changing the profile parameters is ineffective.

1. Introduction

The organization of the safe operation of transport is a complex task. This task is subordinated to the issues of safe operation, repair and modernization of rolling stock and, first of all, locomotives. [1 – 3].

The operational safety of a railway transportation vehicle is directly related to the strength of wheelset basic part, i.e. its axle. Therefore, latterly numerical studies of theoretic and experimental nature are conducted, both for railway transport vehicles [4 – 8]. A variety of special stands are used to test parts of wheelsets of railway rolling stock. [9, 10]. On the stands, geometrical dimensions, electrical parameters and mechanical characteristics are checked. Axle fatigue test stands create symmetrical or pulsating loads.

Symmetrical loads are more often created by inertial loading devices - vibrators [11]. Pulsating loads during axle testing are implemented by hydraulic and hydromechanical mechanisms. Such mechanisms contain levers necessary for design reasons and to increase the value of the transfer function of force. The critical frequencies of the stand with a lever loading mechanism are determined as a result of mathematical and simulation modeling of oscillations of the stand links [12]. Modeling significantly speeds up the process of choosing a rational scheme and design parameters of the stand according to the reliability requirements, as well as according to the criteria of strength, stiffness and vibration stability [13, 14].

2. Purpose and problem statement

The aim of the work is to assess the influence of the dimensions of the cross-sections of the arms of the stand for testing the axles of railway wheelsets on its natural frequencies.

Content from this work may be used under the terms of the Creative Commons Attribution 3.0 licence. Any further distribution of this work must maintain attribution to the author(s) and the title of the work, journal citation and DOI.
Published under licence by IOP Publishing Ltd

To achieve this goal, the following tasks were solved:

1) preparation of the initial data necessary to determine the natural frequencies of the stand using the previously developed mathematical models of its oscillations;

2) obtaining the dependences of the natural frequencies of the stand on the total mass of its levers;

3) evaluation of the influence of the dimensions of the sections of the levers of the stand on its natural frequencies.

The methodology for performing the tasks of the work is based on fundamental studies of the dynamics of mechanical systems with several degrees of freedom. The Mathcad engineering calculation system was chosen as a tool for performing studies of natural vibration frequencies.

3. The construction of the test bench

A stand for testing railway axles for fatigue strength (Figure 1) is similar in design to a stand for testing wheels [15]. It consists of an electric motor with a gearbox (stand drive) 1, a spring damper 2, the pressure force of which is controlled by a force sensor, an intermediate lever 3, which rests on a rack 4, a pusher 5 and a main lever 6 with racks 7. The tested axle 8 is mounted on two supports struts 9 and is loaded by the console of the main lever 6. The struts have the ability to move to change the axle load.

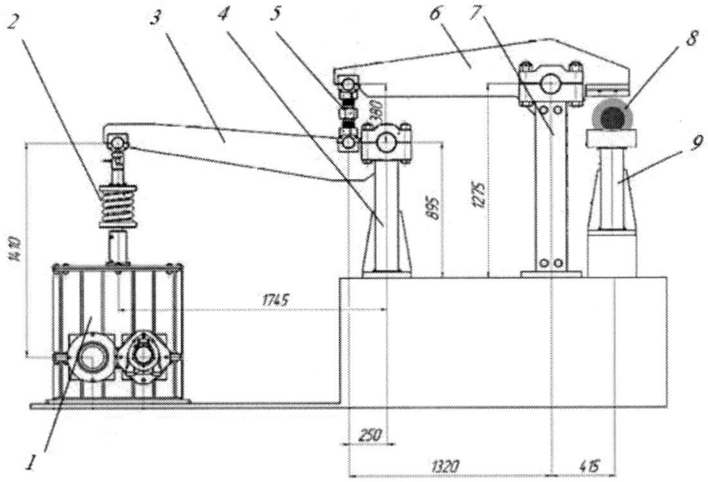

Figure 1. Test bench for axles of railway wheelsets [15].

The lever system of the stand allows increasing the force generated by the electric motor with gearbox 1 by about 20 times. Each arm of the stand can be viewed as a beam on two supports, loaded with a concentrated force. The cross-sections of the levers can be rectangular box-shaped; also the levers can have the shape of a beam of equal resistance. Therefore, the elastic-mass characteristics of the levers will be different and, therefore, the natural frequencies of the stand will also differ.

4. Mathematical model of stand vibrations

4.1 Kinematic scheme of the stand lever system

By its construction, the stand is a spatial mechanism. In order to simplify the calculations, the kinematic scheme of the stand was idealized (Figure 2) and was considered as a flat system of the type of a double rod pendulum with a uniformly distributed mass along the rods and an elastic connection between them due to their flexibility [12]. By the type of relative mobility, the beam 1, interacting with the stand drive, is a pendulum, and the beam 3 loads the tested axle of the wheelset and acts as a lever. The beams are interconnected by a short link - pusher 2.

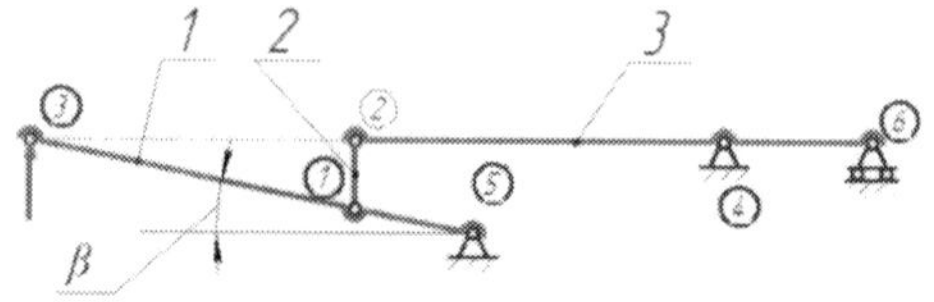

Figure 2. Idealized kinematic scheme of the stand.

4.2 Equations of motion of the links of the stand

In the first approximation, the construction of a mathematical model of the stand vibrations is carried out as for a simple oscillatory system with levers of constant cross-section and concentrated masses m_1 and m_2 (Figure 3). Linear displacements of the centers of mass of the levers of the stand corresponded to the generalized coordinates q_1 and q_2.

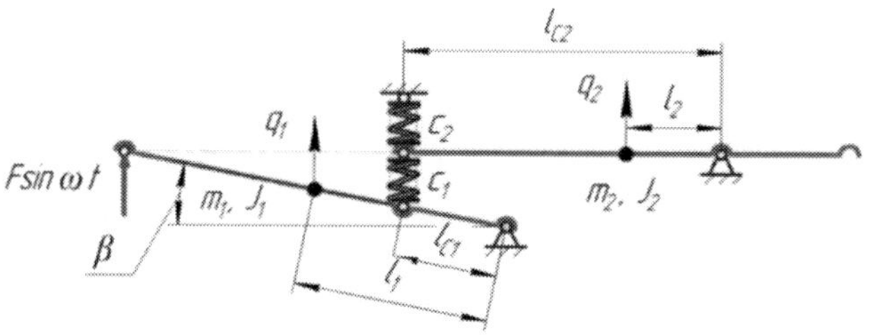

Figure 3. Calculated scheme of the stand.

Small oscillations of such a conservative dynamical system are described by a pair of differential equations [11]

$$\begin{cases} a_{11}\ddot{q}_1 + c_{11}q_1 + c_{12}q_2 = F'\sin\omega t; \\ a_{22}\ddot{q}_2 + c_{21}q_1 + c_{22}q_2 = 0, \end{cases} \tag{1}$$

where $a_{11} = m_1 + J_1/(l_1\cos\beta)^2$, $a_{22} = m_2 + J_2/l_2^2$ – inertial coefficients of the system; $c_{11} = c_1(l_{C1}\cos\beta/l_1)^2$, $c_{12} = -c_1(l_{C1}l_{C2}\cos\beta/l_1l_2)$, $c_{21} = c_{12}$; $c_{22} = (c_1 + c_2)(l_{C2}/l_2)^2$ – elastic coefficients of the system; m_1, m_2 – stand levers masses; J_1, J_2 – moments of inertia of the drive levers relative to their axes of rotation; c_1, c_2 – rigidity of the stand levers relative to their axes of rotation; l_1, l_2 – distance from the axes of rotation of the levers to their centers of mass; l_{C1}, l_{C2} – distance from the axes of rotation of the levers to the elastic elements; β – angle of inclination of the lever 1 of the stand; $F' = 2F$ – the amplitude of the loading force reduced to the center of mass of the lever 1; ω – system loading frequency; t – time.

Frequency equation of the system (1)

$$Ap^2 - Bp^2 + C = 0, \tag{2}$$

where $A = a_{11}a_{22}$, $B = a_{11}c_{22} + a_{22}c_{11}$, $C = c_{11}c_{22} - c_{12}^2$.

The second approximation, when constructing a mathematical model, takes into account the actual deviations and rotation of the centers of mass of the levers from the equilibrium position (Figure 4).

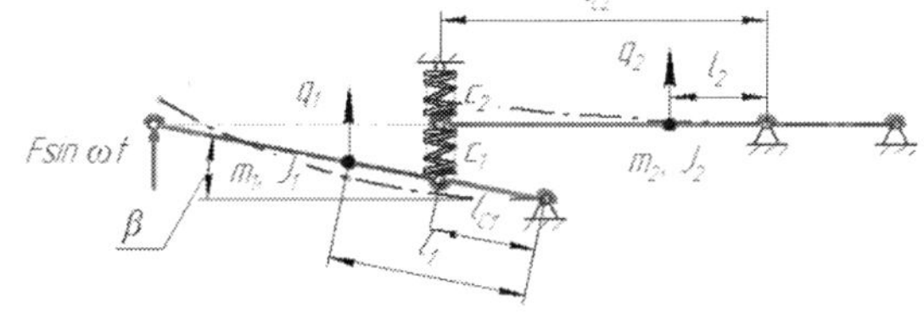

Figure 4. Calculated scheme of the stand, taking into account the actual deviations and rotation of the centers of mass of the levers from the equilibrium position.

For this model, the general view of the system of equations describing the movement of the links of the mechanism is the same as for the previous design scheme (Figure 3), but its inertial and elastic coefficients are different:

$$\begin{cases} a'_{11}\ddot{q}_1 + c'_{11}q_1 + c'_{12}q_2 = F' \sin \omega t; \\ a'_{22}\ddot{q}_1 + c'_{21}q_1 + c'_{22}q_2 = 0, \end{cases} \tag{3}$$

where $a'_{11} = m_1 + J_1 k^2_{\theta m1}$, $a'_{22} = m_2 + J_2 k^2_{\theta m2}$ – system inertial coefficients;

$c'_{11} = c_1 (k_{F1}/\cos\beta)^2$, $c'_{12} = -c_1 (k_{F1} k_{F2}/\cos\beta)$, $c'_{21} = c'_{12}$, $c'_{22} = (c_1 + c_2) k^2_{F2}$ – elastic coefficients of the system; $k_{\theta m1} = (\theta_1/q_1)\ \mathrm{m}^{-1}$, $k_{\theta m2} = (\theta_2/q_2)\ \mathrm{m}^{-1}$ – the coefficients of influence of the displacements of the centers of mass on the angle of their rotation;

$k_{F1} = \Delta_{C1}/q_1$, $k_{F1} = \Delta_{C2}/q_2$ – the coefficients of the influence of displacements of the centers of mass on the deformation of the elastic elements of the system.

The form of the frequency equation for system (3) is the same as for the system (1):

$$A'p^4 - B'p^2 + C' = 0, \tag{4}$$

where $A' = a'_{11}a'_{22}$, $B' = a'_{11}c'_{22} + a'_{22}c'_{11}$, $C' = c'_{11}c'_{22} - c'^2_{12}$.

5. Influence of the dimensions of the cross-sections of the levers on the natural frequencies of the stand

5.1 The construction of the sections of the levers of the stand

We will assume that the sections of the stand levers have a rectangular profile of a constant length, formed by welded steel sheets (Figure 5, Figure 6). The dimensions of the arm profiles are determined by design considerations and strength criteria [16].

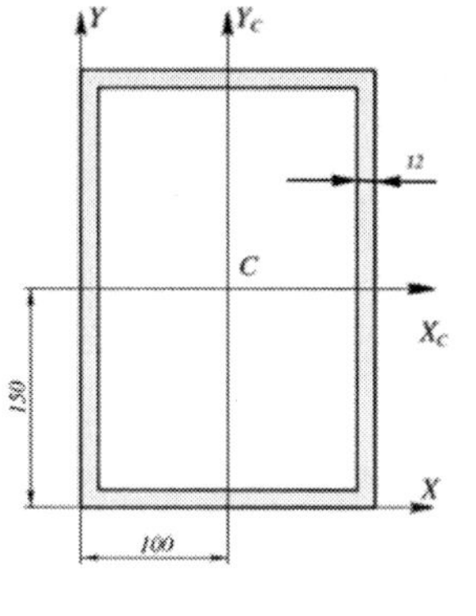

Figure 5. Profile lever 1.

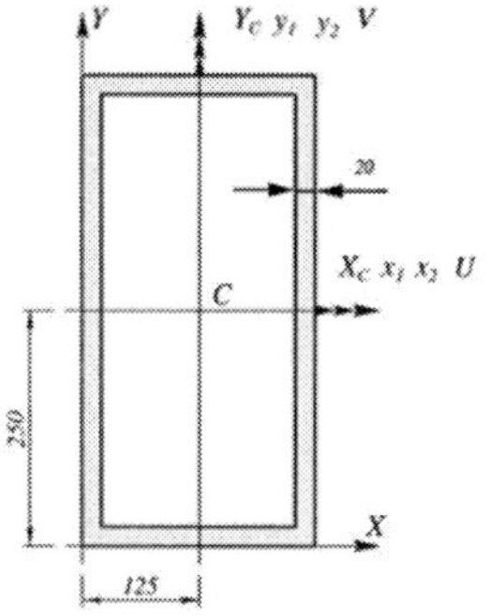

Figure 6. Profile lever 3.

5.2 Elastic-mass characteristics of the levers

The calculation of the parameters of the levers of the stand is made without changing their length. The masses of the levers of the stand changed due to a proportional increase in the thickness of their walls. Such changes in cross-sections led to a change in the masses of the levers and their stiffness's.

The rigidity of lever 1 (see Figure 2) was determined as the rigidity of the beam on two supports, loaded with a concentrated force *F* (Figure 7):

$$c_1 = \frac{F}{y_F}, \tag{5}$$

where $y_F = Fa^2b^2/3EIl$ - deflection of a beam under force *F* [16].

The rigidity of lever 3 was determined as the rigidity of a beam on two supports with a console loaded with a concentrated force F_1 (Figure 8):

$$c_2 = \frac{F_1}{y_F}, \tag{6}$$

where $y_C = F_1c^2(l+c)/3EI$ - deflection of a beam under force F_1.

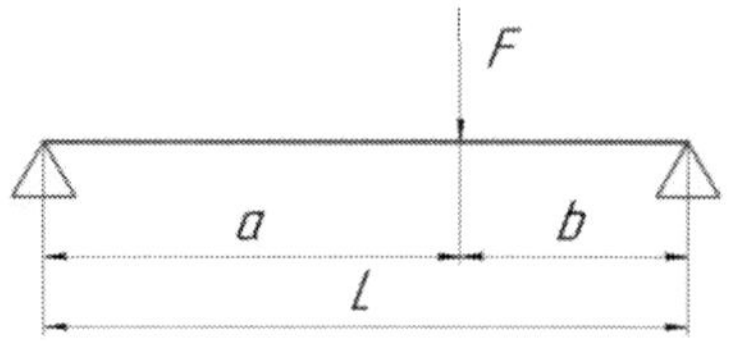

Figure 7. Calculated loading scheme for lever 1.

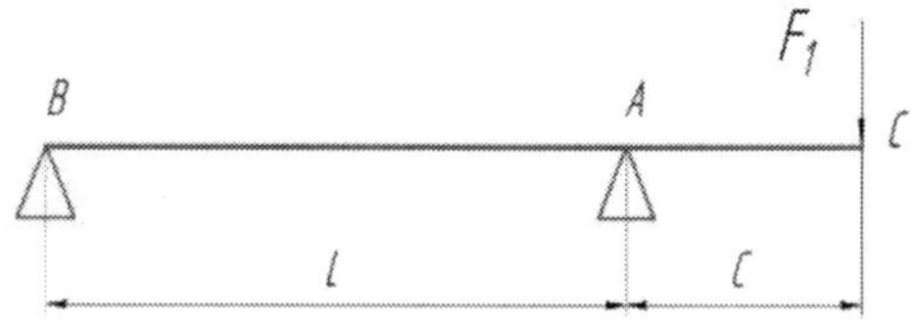

Figure 8. Calculated loading scheme for lever 3.

The results of determining the parameters of levers with different wall thicknesses, carried out using a PC, are summarized in Table 1.

Table 1. Elastic-mass characteristics of the levers.

№	Lever weight, kg		Lever moment of inertia, kg · m²		Lever stiffness, MN / m		Lever wall thickness, mm		Lever cross-sectional area, cm²	
	L1	*L3*	*L1*	*L3*	*L1*	*L3*	*L1*	*L3*	*L1*	*L3*
1	136.5	388	26.6	98.9	639	204	12	20	114	284
2	168.3	478	32.8	122	772	245	15	25	141	350
3	188.6	531	36.8	136	856	269	17	28	158	389
4	219.6	616	42.8	157	973	305	20	33	184	451
5	239.9	666	46.8	170	1046	326	22	36	201	488

Note: *L1* – lever 1; *L3* – lever 3

5.3 Own stand frequencies

The roots of the biquadratic frequency equations (2) and (4) are the own frequencies of the stand. After substituting the characteristics of the levers in (2), (4) and solving the equations in the Mathcad package, a tabular dependence of the own frequencies of the stand on the thickness of the walls of the levers was obtained (Table 2).

According to the strength conditions, the levers have different initial wall thicknesses, therefore, it is more convenient to represent the values of the natural vibration frequencies of the stand as a dependence on the total mass of the levers. In order to automate the statistical processing of the results obtained and plotting graphs, the data of Table 2 were loaded into the Matlab system.

Table 2. Own oscillation frequencies of the stand.

№	Lever wall thickness, mm		Total weight of levers, kg	Scheme Figure 3		Scheme Figure 4	
	L1	*L3*		P_{01}	P_{02}	P_{01}	P_{02}
1	12	20	524.5	460	3053	836	4234
2	15	25	646.3	455	3022	826	4190
3	17	28	719.6	450	3013	818	4177
4	20	33	835.6	445	2983	808	4135
5	22	36	905.9	440	2970	800	4116

Note: *L1* – lever 1; *L3* – lever 3; P_{01}, P_{02} – first and second own frequencies

An increase in the thickness of the profile of the levers of the stand is accompanied by an increase in the moments of resistance of the sections. Therefore, the stresses in the levers of the stand will decrease.

The dependence of the natural frequencies of the stand vibrations on the total mass of the levers (section sizes), firstly, has an almost linear character; secondly - they practically remain unchanged with an increase in the mass of the levers (Figure 9, Figure 10).

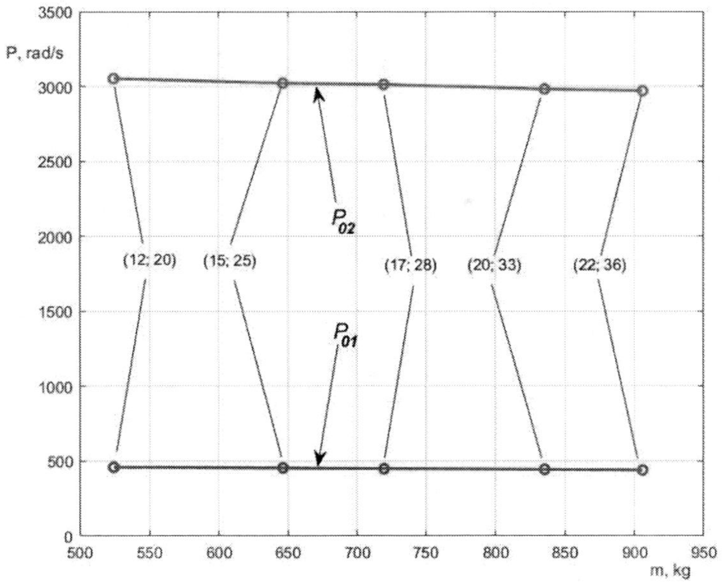

Figure 9. Dependence of the own frequencies of the stand vibrations on the total mass levers (section sizes) with a simplified design scheme (Figure 3).

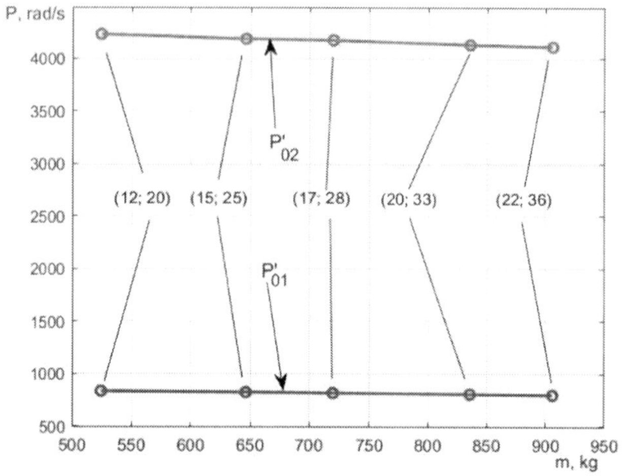

Figure 10. Dependence of the own frequencies of the stand vibrations on the total mass levers (section sizes) with a refined design scheme (Figure 4).

5.4 Analysis of the dependences of the own frequencies of the stand vibrations

A detailed analysis of the dependences of the own frequencies of the stand vibrations on the total mass of the levers (Figure 11, Figure12) using the Matlab system tools showed a slight decrease in frequencies with an increase in the total mass of the levers.

The linear regression equations reflecting the dependence of the own frequencies of the stand on the total mass of the levers have the form:

$$P_{01} = 487{,}8 - 0{,}0521m\,; \quad (7)$$

$$P_{02} = 3166 - 0{,}2178m\,, \quad (8)$$

where m – total weight of stand levers.

The coefficient of determination of equations (7), (8) is 0.99, which indicates a high accuracy of the selection of equations.

An analysis of expressions (7) and (8) showed that with an increase in the mass of the levers by 72.7%, the first natural frequency decreases by 4.3%, and the second decreases only by 2.7%.

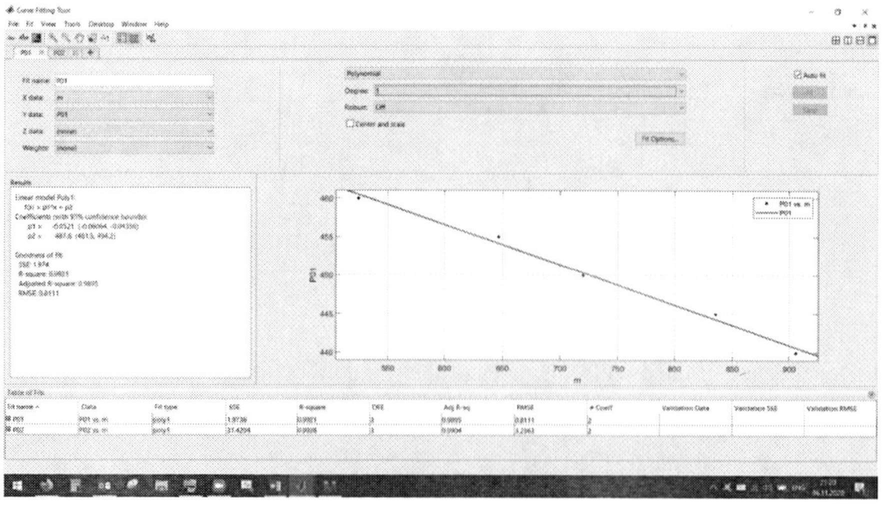

Figure 11. Analysis window for the dependence of the first ownvibration frequency of the stand on the mass of the levers.

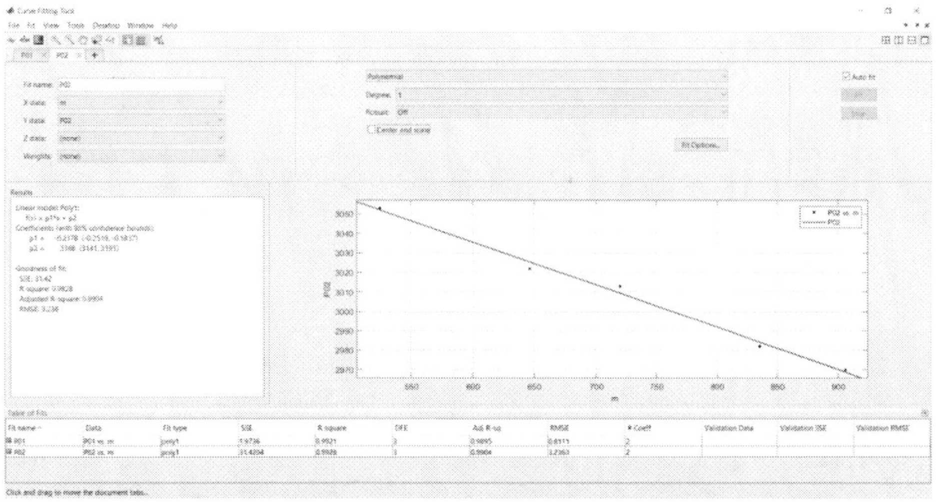

Figure 12. Analysis window for the dependence of the second own vibration frequency of the stand on the mass of the levers.

6 Conclusions

1. The results obtained make it possible to assert that a change in the thickness of the walls of the stand arms has practically no effect on the values of the own frequencies of the stand. This is due to the simultaneous increase in mass and lateral stiffness of the levers.
2. The dependence of the own frequencies of the stand vibrations on the mass of the levers is linear: with an increase in the total mass of the levers by 72.7%, the first natural frequency decreases by 4.3%, and the second - by 2.7%.

References

[1] Bannikov D and Radkevich A 2019 Analytical method for compiling and applying a ballast map for the traction unit PE2U *Eastern-European Journal of Enterprise Technologies* **2 1 (98)** pp 6-14

[2] Bannikov D and Yakovliev S 2020 Development of dynamic integral evaluation method of technical state of one-section electric locomotive body *Eastern-European Journal of Enterprise Technologies* **1 7** (**103**) pp 57-64

[3] Bannikov D, Radkevich A and Muntian A 2020 Modernization of the buffer beam of PE2U traction unit electric locomotive *IOP Conference Series: Materials Science and Engineering* **985** 012035

[4] Novosad M, Fajkoš R, Řeha B and Řezníček R 2010 Fatigue tests of railway axles *Procedia Engineering* **2** 2259–2268 doi: https://doi.org/10.1016/j.proeng.2010.03.242

[5] Klimenko I, Kalivoda J and Neduzha L 2018 Parameter optimization of the locomotive running gear *Proc. of 22-nd Intern. Scientific Conf. «Transport Means* pp 1095-1098

[6] Filippini M, Luke M, Varfolomeev I, Regazzi D and Beretta S 2017 Fatigue strength assessment of railway axles considering small-scale tests and damage calculations *Procedia Structural Integrity* **4** pp 11–18. doi: https://doi.org/10.1016/j.prostr.2017.07.013

[7] Răduţă A, Locovei C, Nicoară M and Cucuruz L (2010) On the influence of residual stresses on fatigue fracture of railway axles. *WSEAS Transactions on Applied and Theoretical Mechanics* **5** pp 197–207

[8] Raksha S, Anofriev P and Kuropiatnyk O 2021 Justification of the parameters of the life-test bench for railway wheelsets *IOP Conf. Series: Materials Science and Engineering* **1021** 012047 doi: https://doi.org/10.1088/1757-899X/1021/1/012047

[9] Kochetkov Ye V, Knyazev D A and Trepacheva T V 2018 *Stend dlya ispytaniya koles i osey kolesnykh par na soprotivlenie ustalosti* Patent No. 2650327

[10] Kochetkov Ye V, Ogan'yan E S, Bubnov A A and Trepacheva T V 2019 *Stend i sposob ispytaniya na stende koles i osey kolesnykh par na soprotivlenie ustalosti i issledovaniya povedeniya metalla v zone kontaktnogo vzaimodeystviya kolesa s relsom* Patent No. 2665358

[11] Raksha S, Kuropiatnyk O and Anofriev P 2019 Simulation modelling of the rolling stock axle test-bench *E3S Web of Conferences 2019* **123** *Ukrainian School of Mining Engineering 01032* doi: https://doi.org/10.1051/e3sconf/201912301032.

[12] Raksha S, Anofriev P and Kuropiatnyk O 2020 Stand for accelerated tests of rail vehicles wheelsets *IOP Conf. Series: Materials Science and Engineering* **985** 012030 doi: https://doi.org/10.1088/1757-899X/985/1/012030.

[13] Raksha S, Anofriev P and Kuropiatnyk O 2020 Stand for testing railway wheels for contact strength *Science, engineering and technology: global trends, problems and solutions Conference proceedings*, September 25–26, 2020. (Prague: Izdevnieciba Baltija Publishing) pp 173–177 doi: https://doi.org/10.30525/978-9934-588-79-2-2.42 .

[14] Raksha S, Anofriev P and Kuropiatnyk O 2020 Determination and analysis of technical parameters of the stand for complex tests of railway wheels *Science and transport progress. Bulletin of Dnipropetrovsk National University of Railway Transport* **5 (89)** pp 134-141 doi: https://doi.org/10.15802/stp2020/217771

[15] *The stand for tests of railway wheels on fatigue strength* 2019: pat. 136718 Ukraine / Kebal I. Yu., Zgrebna SM, Tyokotev OM, Tarasyuk M. Yu., Raksha SV, Kuropyatnik OS; G01M 17/1 (2006.01); declared March 29, 2019; publ. 27.08.2019 **16** 5

[16] Pisarenko G S, Agarev V A, Kvitka A L, Popkov V G and Umanskiy E S 1979 *Soprotivlenie materialov* (Kiev: Naukova dumka) 696

Assessment of the impact of modifier and filler on the physical and mechanical and operational properties of plastering mortars

S Savchenko[1,*] **and N Antoniuk**[1]

[1]Odessa State Academy of Civil Engineering and Architecture, 4, Didrichson str., Odessa, 65029, Ukraine

*E-mail: koval_sv@ukr.net

Abstract. To assessment of the impact of the addition of microsilica on the mechanical properties of the plaster mortar, experimental-statistical modeling was used. A complex of engineering problems based on models describing the change in the technological and physical and mechanical properties of the plaster mortar during the control of the dosages of microsilica and superplasticizing admix was solved. The problem of choosing the optimal dosage of the filler was solved under the condition of the equality of the water-cement ratio and the mobility of the compositions without and with microsilica additive, which was achieved by appropriate adjusting the dosage of the superplasticizing admix on the consumption of microsilica, i.e. the concentration of the superplasticizing admix was dependent on the consumption of microsilica. Analysis of the effectiveness of the addition of microsilica in the entire factor space using the model of relative efficiency was carried out. The areas of optimal compositions of plaster mortars, in which the increased technological and mechanical parameters are performed, have been determined. An assessment of the effect of ultrafine filler on the operating ability of plaster mortars has been carried out. The main indicators were: the value of adhesion of the plaster layer to the concrete base; the strength of the concrete base ("old" concrete); strength of the plaster layer; indicator of water absorption of the plaster layer. The assessment of the frost resistance of the modified plaster mortar was carried out after 20, 40 and 60 cycles of freezing and thawing. It was found that the proposed modification made it possible to increase the adhesion strength of the plaster layer to the base, save the characteristics of the concrete base after exposure to an aggressive environment, increase the strength of the plaster layer and make it less permeable due to the compaction of the structure, and also to reduce the thickness of the plaster layer without losing its protective properties.

1. Introduction

In the modern construction industry, ready-made dry plaster mixes are the main material for external and internal finishing works. For building facades, cement-based dry plaster mixes are optimal. In respect that the aggressiveness of the environment, the variability of weather conditions, special requirements are imposed on the plaster mixture.

The most important properties of plaster mortars are: the uniformity of the composition achieved due to well mixing of the initial materials, which excludes the tendency of the mixture to segregation; the ability to actively adhere to the base under the coating; provided by the composition of the water-retaining ability; the course of the hardening process, using a hardening process in accordance with construction technology, taking into account weather conditions. The above-mentioned properties are ensured through the use of optimal formulations of plaster compositions, which include chemical additives and fillers that increase adhesion, strength, and water resistance. Facade mixtures can be used not only as the final stage of finishing, but also as a basis for subsequent decorative cladding.

The system becomes more complex if polyfunctional modifiers are used, which can especially strongly affect the structure and properties of composites. These modifiers include superplasticizing admix (SP) and ultrafine fillers such as technogenic microsilica (MS). Microsilica – a waste of

Content from this work may be used under the terms of the Creative Commons Attribution 3.0 licence. Any further distribution of this work must maintain attribution to the author(s) and the title of the work, journal citation and DOI.
Published under licence by IOP Publishing Ltd

ferroalloy production is distinguished by a uncharacteristic specific surface area for most used fillers (ash, slags, organic mineral additives) and an increased content of a silica component.

2. Analysis of recent research and publications

Analysis of literature data [1-3] showed that using microsilica in combination with a superplasticizing admix can increase the strength of cement stone by 1.5-2 times with ordinary materials. Plasters can be prepared with high physical-mechanical and operational characteristics if select optimal recipe-technological conditions. The complex use of MS and SP also saves resources by reducing the consumption of the binder.

The improvement of the structure of the cement stone occurs due to the fact that microsilica, interacting with the products of cement hydration, forms an additional amount of hydrated neoplasms, thus contributing to an increase in the strength and density of the solution [4-6]. At the same time, microsilica contributes to the acceleration of carbonization, which significantly reduces destructive processes, increases the chloride and sulfate resistance of cement systems [7].

The introduction of MS promotes an increase in pore volume from 37 to 0.175 μm and, while maintaining the integral porosity, a decrease in the average diameter of air pores and the formation of a microporous structure of a cement stone with a reduced capillary porosity and which causes an increased impermeability of the cement slurry.

Other factors that have a significant effect on the durability of plaster mortars are the composition of the mortar in the pores. During the introduction of microsilica, the pH value of the solution in the pores decreases.

3. Purpose of the study

The purpose of this study is to determine the optimal concentrations of microsilica and superplasticizer to improve the physical, mechanical and operational properties of plaster mortar.

4. Objects and research methods

To find optimal solutions, mathematical models were used, i.e. mathematical description of specific recipe-technological situations [8-11]. A complex of engineering tasks was solved using models describing the change in the technological and mechanical properties of the plaster mortar during controlling the dosage of microsilica $X_1 = 10\pm10\%$ (from the mass of the binder), the ratio between sand and cement $S:C = X_2 = 2.5\pm0.5$ with this normalization: $x_1 = -1$ corresponds to no additive composition, $x_2 = -1$ – high cement consumption. The mobility (M) of the mortar mixtures corresponded to $M = X_3 = 5\pm2$ cm by the immersion of the cone.

The basis for taking these factors into account in the model is the following prerequisites. The addition of the MS – the main regulating factor. An increase in the MS dosage over 20% can cause increased plastic shrinkage of the layer and lead to unreasonable consumption of the superplasticizing admix.

A standard C-3 plasticizer with a density of 1.17 g/cm^3 at a concentration of 38.4% was used as a superplasticizing admix. In the experiments, no additives Portland cement with an activity of 375 kg/cm^2 of the Olshansk plant was used as a binder. River sand with $M_{cr} = 2.1$ was used as a fine aggregate, which meets the requirements for plaster mortar. To determine the mechanical properties, samples- beams with size 4×4×16 cm were tested.

Taking into account the possible operating conditions of the plaster mortar, its properties were studied (Figure 1, c) after 20, 40 and 60 cycles of freezing (at T = -20°C) and thawing under capillary suction conditions with a plaster mortar of salt solution (at T = +20°C).

The method of manufacturing and testing samples consisted of the following operations:

1. Samples of "old" concrete (beams with a section of 40×40 mm and a length of 120-140 mm – depending on the expected thickness of the plaster mortar) on the ends of which a plaster mortar was applied (Figure 1, b) and placed in special special form (Figure 1, a). Thus, filling completely one cell, depending on the length of the "old" concrete beam, a layer of plaster mortar of a given thickness was

formed (2, 3 and 4 cm).

2. Preparation of samples of "old" concrete before applying the plastering solution included treatment with an abrasive material and wetting the surface of the application of the plastering solution. After application, the mortar sags were carefully cut away without damaging the freshly formed layer.

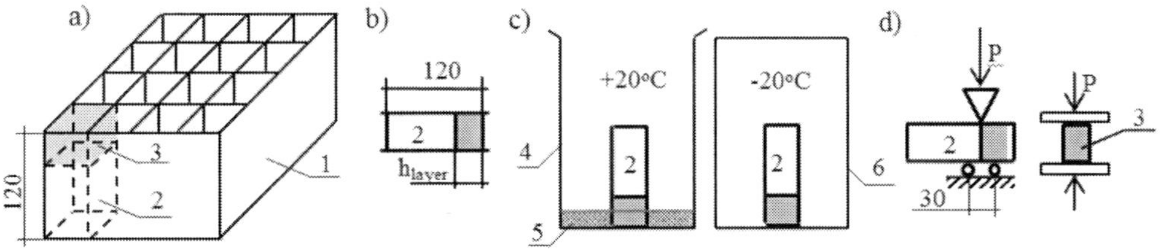

Figure 1. Method of manufacturing and testing samples of the plaster layer: a) special form for applying a layer of plaster on samples; b) sample with a layer of solution; c) holding samples during alternate freezing and thawing; d) the scheme of testing the plaster layer for spalling and compression; 1 – special form; 2 – "old" concrete; 3 – a layer of plaster mortar; 4 – a collecting gutter for thawing; 5 – salt solution; 6 – refrigerator.

3. After a 14-day exposure in normal humidity conditions, the samples were sent for freezing for 12 hours and thawing in a salt solution also for 12 hours (Figure 1, c). The salt solution level was 1 cm and monitored every cycle. The temperature of the salt solution was automatically maintained at +20°C. Freezing was carried out in refrigerator with an accuracy of maintaining the temperature of ±2°C. Thus, without taking into account the 1 cm layer, which was constantly in a liquid medium during thawing, the working layer thickness varied from 2 to 4 cm.

4. After 60 cycles of freezing and thawing, the plaster mortar layer was tested for spall relative to the "old" concrete and the strength of the plaster mortar layer (Figure 1, d).

5. The results of research

The selection of the optimal dosage of the filler was performed under the condition of the equality of the water-cement ratio and the mobility of non-additive compositions and with microsilica, which was achieved by adjustment of the dosage of the superplasticizing admix as the MS consumption increased, i.e. the concentration of the superplasticizing admix was dependent on the consumption of microsilica.

The experiment was carried out according to the B_3 plan [8] with one central point. The dosages of the superplasticizing admix required to obtain a predetermined mobility of the mixture with an increase in the amount of filler from 0 to 20% (the water content of the mixture was constant) were preliminarily determined.

The dependence between the concentration of the superplasticizing admix and the factors under study had a nonlinear character (Figure 2, a):

$$\begin{aligned} D_{SP} = 0.77 &+ 0.62x_1 - 0.09x_1^2 - 0.15x_1x_2 \\ &- 0.14x_2 - 0.08x_2^2 + 0.18x_1x_3 \\ &+ 0.17x_3 \qquad\qquad + 0.15x_2x_3 \end{aligned} \quad (1)$$

from which it follows that to maintain the specified mobility (5 cm) and water content, it is necessary to introduce 0.3-0.4% of the superplasticizing admix for every 5% of the filler. The use of SP in concentrations above 1.0%, although it allows the use of increased (more than 15-20%) dosages of MS, isn't economically justified, since it increases the cost of the material. Isoline D_{SP}= 1.0 % on the diagram D_{SP}= φ (x_1, x_2, x_3) limits the area of economically justified SP dosages with a mixture mobility of 5 cm (dashed line in Figure 2, a).

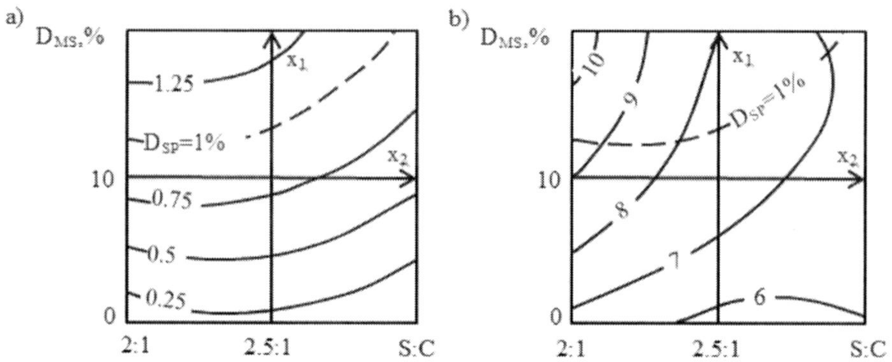

Figure 2. Influence of microsilica on the main physical and mechanical properties of the plaster mortar: a) change in the concentration of SP; b) flexural strength (R_b, MPa).

The nomogram of the strength R_b of the plaster solution depending on the concentration of MS and the S:C ratio is shown in Figure 2, b. The coefficients of the models used to plot the strength nomograms are given in Table 1.

Table 1. Coefficients of the strength models of solutions on the 28th day after the normal hardening regime.

Strength, MPa	Model coefficients								
	b_0	b_1	b_2	b_3	b_{11}	b_{22}	b_{12}	b_{13}	b_{23}
R_b	7.3	1.1	-1.1	-0.3	-0.5	0.6	-0.7	0.2	–
R_c	26.4	5.0	-4.5	-2.4	–	1.5	2.9	1.3	–

The positive effect of the ultrafine filler on the hardening of the solution is due to the increase in the number of contacts in the cement system and the intensive accumulation of an additional amount of hydrated neoplasms as a result of the interaction of MS with the products of cement hydration. Due to the introduction of microsilica fractions, a well compaction of the cement system is achieved and about a third of the space between the cement grains is filled with filler fractions. This explains the fact that the compressive strength R_c of plaster compositions with the addition of microsilica (curve 2 in Figure 3, a) is 40-50% higher than the strength of the non-additive composition (æ{R^c} = 100%). However, these data refer to one of the S:C = 2.5:1 ratios. If we consider the change in strength R_c with varying two factors – MS concentration and S:C ratio, it can be seen that the strength for mixtures with low S:C increases significantly more than for mixtures with high S:C (Figure 3, b) which requires careful selection of the composition of the plaster solution.

The analysis shows that by increasing the D_{MS} from 0 to 18%, the consumption of the binder can be reduced to 1.5 times (for example, when obtaining the grade strength R_c^{28} = 30 MPa mixtures with S:C = 3:1 can be used instead of S:C = 2:1). In this case, the effective concentration of the superplasticizer can be reduced (D_{SP}<1.0 %).

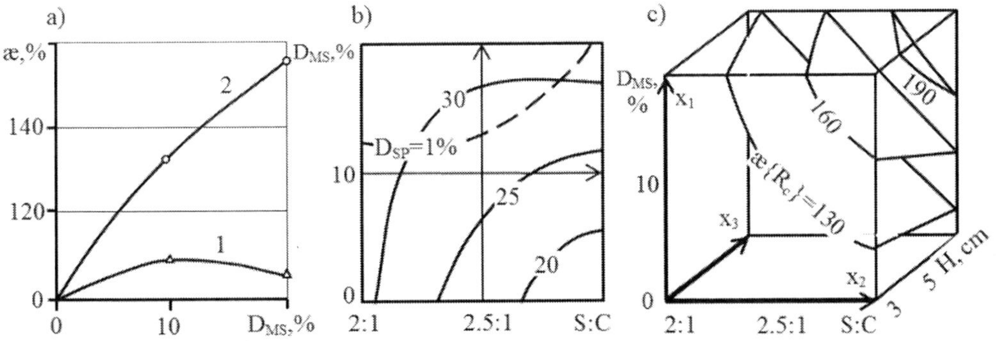

Figure 3. Influence of microsilica addition on the compressive strength of the plaster mortar: a) increase in strength (1 – initial mortar S:C = 2.5:1, S = 5; 2 – modified; b) nomogram of compressive strength; c) the relative efficiency of MS in the entire recipe-technological space.

This conclusion is confirmed by the analysis of the effectiveness of the addition of microsilica in the entire factor space, carried out using the relative efficiency model:

$$æ\{R\} = R\{D_{MS}\} : R\{D_{MS} = 0\},\ \% \tag{2}$$

The main advantage of a model of type (2) is that it can be used to compare the results of testing qualitatively different additives and various factors, since on the base side of the square (in particular, shown in Figure 3, c) there is always $æ\{x_i\} = 1$.

A function of type (2) was built for a 3-dimensional space, and the two-factor model of no additive plaster mortar $R = \varphi\ (X_1, X_2)$, was taken as the base one, which was provided for organizing this experiment. The analysis of the $æ\{R_c\}$ model indicated the peculiarities of the effect of microsilica on the strength of the plaster solution under compression. In the recipe space, there is a large area, within which the strength of plaster mortars with filler is practically equal to the strength of concrete at $D_{MS} = 0\%$. It was noted that the loss of strength of mortars with a decrease in the binder content becomes less noticeable in the presence of microsilica.

With an increase in the water content of the mixtures, the increase in the strength of the solutions on the filled binder increases, but the zone of the most effective effect of microsilica, in contrast to $æ\{R_b\}$, moves towards compositions with a low binder content. The combined positive effect of the plasticity index of mixtures and the amount of ultrafine filler can be explained by an increase in the cohesion of solutions, while for a non-additive mortar, prevails the negative effect of an increase in W/C on strength.

For solutions from plastic mixtures (7 cm) with S:C = 3:1, containing 20% microsilica additive, the compressive strength is 2.3 times higher than the strength of no additives composition. At the next stage, the task of obtaining a composition of a plaster mortar with improved technological and mechanical parameters was set. For this purpose, the models describing the influence of the investigated technology parameters on the quality indicators of the plaster mortar were analyzed in one factor space:

– concentration of SP;
– the relative increase in tensile strength in bending;
– the relative increase in compressive strength.

The need for a compromise solution is due to the fact that microsilica acts ambiguously on these indicators and its effectiveness depends on the levels of other factors, in particular, S:C ratio and the mobility of the mixture. Thus, the increment in strength R_b increases for a mixture with a high S:C, but under the same conditions the effect of MS on the increment in the brand strength in compression may be insignificant. The compromise task was solved under the following restrictions: a) concentration of SP not more than 1.0% (economy of superplasticizing admix); b) an increment in tensile strength in bending of at least 20%; c) increment in compressive strength æ{R_c} not less than 25%. For this purpose, the diagram combined the corresponding isolines of models of quality and efficiency indicators. The graphical-analytical analysis of the task posed is presented in the form of a field bounded by isolines D_{SP} = 1.0%, æ{R_b} = 120% and æ{R_c} = 125% (Figure 4, a).

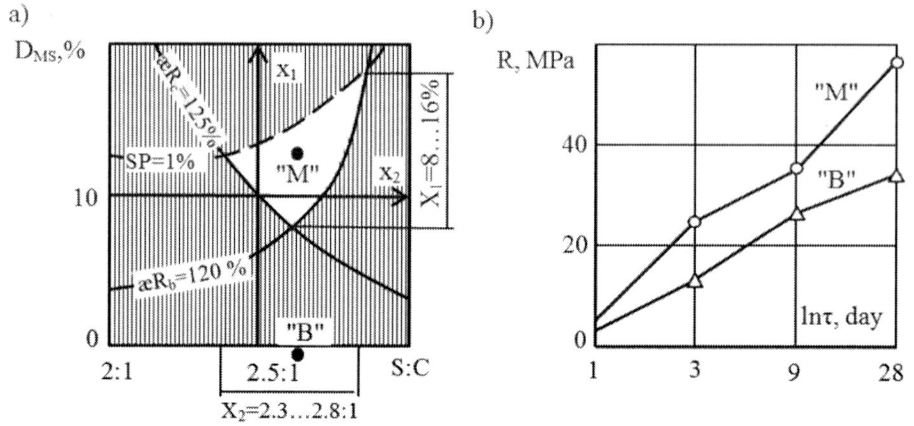

Figure 4. The field of effective compositions of plaster mortars:
a) search for a compromise solution (p. "B" – no additives composition, p. "M" – modified);
b) the kinetics of changes in the strength of the optimal composition.

In the field of effective solutions, the requirements for the plaster mortar are fulfilled (unshaded area). It is characteristic that they cannot be performed for additive-free compositions, but only for modified compositions containing 8-16% microsilica additives and 0.7-1.0% superplasticizing admix. In this case, the rational ratio of sand and cement is 2.3-2.8:1, i.e. the mixtures should contain the average values of the binder consumption for the given experiment. Thus, microsilica can be used both as a quality regulator and as a factor contributing to the reduction of the clinker component of cement.

Further, the properties of the modified compositions were analyzed in comparison with those without additives. In particular, in Figure 4, b shows the change in the kinetics of the strength gain of the modified composition at point "M", selected from the field of effective formulating solutions (D_{MS} = 13%, D_{SP} = 1.0%) and no additive at point "B" at the same S:C = 2.5:1. The analysis of changes in the strength of plaster solutions showed that at all control periods the strength of the modified plaster of the optimal composition "M" is higher than the strength of the non-additive composition "B" and at the brand age the increment in strength is at least 40%.

Among the main factors affecting the frost resistance of the mortar, the following are highlighted: the mineralogical composition of cement, the water content and structure of the cement stone, as well as the effect of additives.

The structure of the cement stone and its pore space depends on the mineralogical composition of the clinker, the content of alkalis, the type and amount of additives [12-14]. Consequently, these factors will also effect on frost resistance. Researches by various authors, including [13, 15, 16] have shown that the properties of the main minerals – hydrosilicates and hydroaluminates are very different and determine the difference in strength and rate of its formation. At the same time, calcium hydroaluminate, which is the main mineral of the aluminate phases ($3CaO{\cdot}Al_2O_3{\cdot}6H_2O$), has low water resistance and can lose strength in water [15]. This gives reason to assert that the reduction of tricalcium aluminate in the clinker will contribute to increase the hardiness experimentally confirmed that during using the active mineral additives capable of binding C_3A, the greatest effect is achieved when the complex used in conjunction with an air-entraining and plasticizing additives.

The high activity of microsilica, the ability to absorb calcium hydroxide, the compacting effect on the structure of the cement stone and the increase in strength will probably determine the increased frost resistance of the mortar.

In order to assess the frost resistance (F) of the modified plaster, special researches were carried out, reflecting the working conditions of the surface. The hardened plaster was evaluated for various quality parameters, which works in conjunction with a base: compressive and flexural tensile strength, adhesion strength to strip-off, shear and spall from the surface of the base, water permeability. In any case, those indicators that will be decisive for plaster solutions in real-life conditions are used as an assessment criterion. Taking into account the peculiarities of plastering work and possible difficult working conditions of the plaster layer (up and down temperature changes, especially in winter, saline waters, etc.), a method was developed for assessing the plaster layer after testing for frost resistance. The layer of plaster mortar is exposed to the one-sided effect of the external environment, therefore, moistening was carried out with capillary suction (Figure 1, c) in a salt solution (freezing at T=-20°C). Since the most important indicator that determines the intensity of the destructive effect of freezing is the degree of water saturation of the mortar, freezing with capillary suction of water and with full immersion in a liquid medium turns out to be more destructive than freezing in air.

Determination of the properties of the plaster layer was carried out after 20, 40 and 60 cycles of freezing and thawing. The main indicators were considered:

– the amount of adhesion of the plaster layer R_{ad} to the concrete base;
– strength of the concrete base ("old" concrete);
– the strength of the plaster layer;
– indicator of water absorption of the plaster layer.

In Figure 5, a shows the curves of the adhesion strength to the cleavage of the non-additive plaster layer "B" and the modified "M". The modification allows to increase the adhesion strength of the plaster layer to the base ("old" concrete) by 10-15%, depending on the number of freezing and thawing cycles. This fact can, in particular, be explained by an increase in the adhesion forces of the modified composition.

It is known [17] that the seam (contact zone) is the boundary of the change in the directed shrinkage deformations, so the contact zone becomes "prestressed" tensile force. As shown by the results of experiments for various S:C, with an increase in the amount of cement, the adhesion strength increases by 5-7%. With an increase in the amount of microsilica up to 10%, the adhesion value increases by 10-15%, however, a further increase in the concentration of the ultrafine filler is ineffective, especially for compositions with a reduced binder consumption. This confirms the conclusions about the need to use low dosages of microsilica in plaster mortars. Limitations on the dosage of MS are also caused by a certain tendency of such mortars to plastic shrinkage, which requires the use of special measures to protect against moisture loss.

To assessment the possibility of reducing material consumption, a plaster layer was obtained not only with a thickness of h = 4 cm, but also h = 2 cm. The results of the experiments are presented in the form of a graph in Figure 5, b. The strength of the plaster layer with ultrafine filler and SP ("M") is higher than the strength of the no-additive plaster mortar "B" at all times of frost resistance tests.

Comparison of curves "M_2" and "B_1" shows that the strength of the concrete base ("old" concrete) on which no additive plaster mortar with a thickness of h = 4 cm and a layer of modified plaster mortar with a thickness of h = 2 cm were applied are close in their values. Thus, due to the use of a modified plaster mortar, not only the operational characteristics of the structure can be improved, but also the thickness of the plaster layer itself can be reduced without changing the quality of the base.

The strength of the modified plaster mortar in all test periods is higher than the strength of the unadded one by 20-30% (Figure 5, c), which is explained by an increase in its density and an improvement in the pore structure of the cement stone.

Changes in density can be characterized by water absorption curves for compositions "M" and "B", which show that the density of the modified composition changes insignificantly with an increase in the effect of the destructive medium and, in general, becomes smaller with the introduction of modifiers by 1.3-1.5 times.

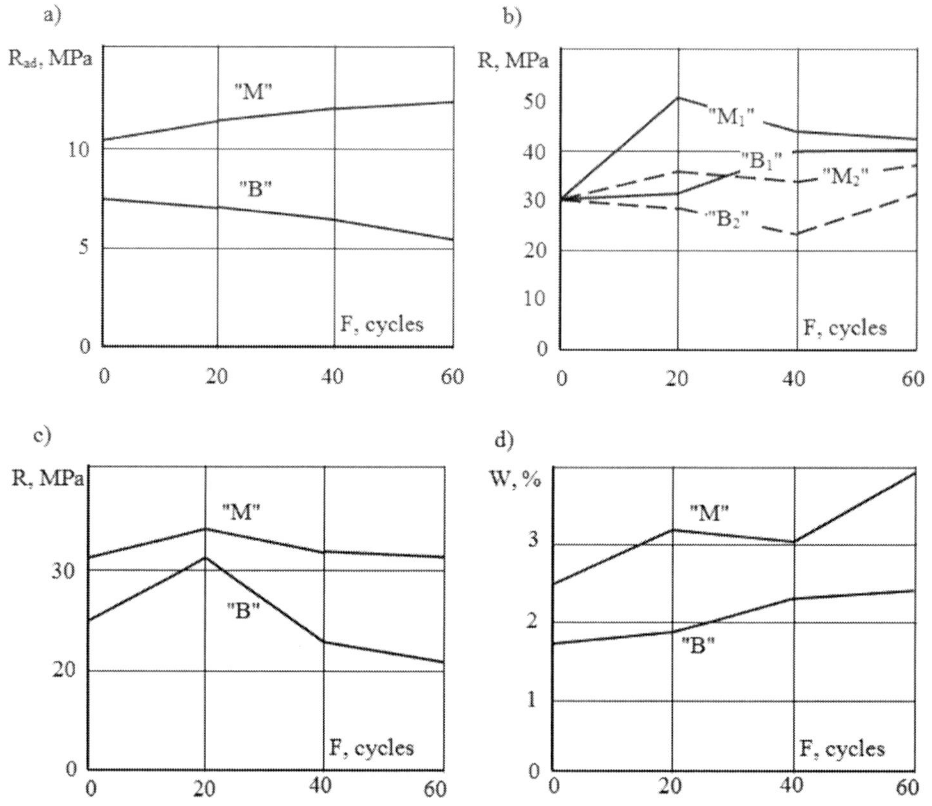

Figure 5. Assessment of the properties of the plaster layer after frost resistance tests (F): a) the strength of adhesion of the plaster layer to the base ("B" – no additives composite, "M" – modified, S:C = 2.5:1); b) the strength of the base – "old" concrete with different thickness of the plaster layer ("B_1", "M_1" – 4 cm, "B_2", "M_2" – 2 cm); c) the strength of the plaster layer; d) the density of the structure, assessment by water absorption (W).

According to [14, 18, 19], the micro-filler additives are characterized by a densifying effect. These additives contribute to the creation of a fine-grained structure characterized by evenly distributed and difficult-to-pass pores for the filtering liquid. At the same time, "microskeletons" are created in the intergranular cavities, which impede the flow of contraction and sedimentation processes, as well as the development of a system of filter pores and capillaries.

6. Conclusions

The use of a modified plaster solution improves its technological and mechanical characteristics. At the same time, in order to maintain a predefined mobility of plaster mortars (without changing their water content), it is necessary to introduce, as a rule, 0.3-0.4% of a superplasticizing admix for every 5% filler. The use of micro-filler makes it possible to increase the tensile strength in bending in 15-20%, and the compressive strength by more than 1.5 times, and the efficiency of microsilica increases with a decrease in the cement-sand ratio and an increase in the mobility of the mixture. With optimal ratios of all the factors under investigation, the economy of the superplasticizing admix can be ensured, as well as a noticeable increase in the tensile strength in bending and compression.

The modification made it possible to increase the adhesion strength of the plaster layer to the base by 10-15%, retain the characteristics of the base concrete after impact of an aggressive environment, increase the strength of the plaster layer itself by 20-30% and make it less permeable due to the compaction of the structure, and also reduce the thickness of the plaster layer almost 2 times while maintaining its protective properties.

References

[1] Dhagat A and Mittal M 2013 Effect of Microsilica and fly ash on the strength of concrete *IJSER* **4 8** pp 1399–1402
[2] Moskalenko O A 2020 Investigation of the influence of the organo-mineral complex produced by MC Bauchemi LLC on the physical and mechanical properties of slag-containing cements *Bulletin of the Odessa State Academy of Civil Engineering and Architecture* **80** pp 75–83
[3] Savchenko S V, Antoniuk N R and Bachynckyi V V 2021 Assessment of the impact of modifier and filler on the rheological and physical and mechanical properties of plaster solutions *Bulletin of Odessa State Academy of Civil Engineering and Architecture* **82** pp 105–113
[4] Pushkareva K K, Gonchar O A and Kaverin K O 2015 Influence of organo-mineral additives on rheological properties of cement compositions and their physical and mechanical characteristics *Coll. Science. works. of the Ukrainian state University of Railway Transport* **155** pp 124–128
[5] Pushkarova K K, Kaverin K O and Kalantaevsky D O 2015 Research of high-strength cement compositions modified by complex organic-silica additives *Eastern-European Journal of Enterprise Technologies* **5 5 (77)** pp 42–51
[6] Kroviakov S, Mishutin A and Pishev O 2018 Management of the properties of shipbuilding expanded clay lightweight concrete *International Journal of Engineering & Technology* **7 3.2** pp 245–249
[7] Frolov Yu G 1981 Colloidal chemistry course (surface phenomena and dispersed systems) 220
[8] Lyashenko T V and Voznesensky V A 2017 *Composition-process fields methodology in computational building materials science* 168
[9] Lyashenko T V 2015 Composition-process fields methodology for design of composites structure and properties *Brittle Matrix Composites* **11** pp 289–298
[10] Lyashenko T and Kryukovskaya S 2012 Modelling the influence of composition on rheological parameters and mechanical properties of fibre reinforced polymer-cement mortars *Brittle Matrix Composites* **10** pp 169–178
[11] Lyashenko T 1991 Brittle matrix composites optimisation on the base of structured experimental-statistical models *Brittle matrix composites* **3** pp 448–457
[12] Bazhenov Yu M 1987 *Concrete technology* 415
[13] Kramar L Ya, Trofimov B Ya, Talisman L S and other 1989 Influence of microsilica addition on alite hydration and sulfate resistance of cement stone *Cement* **6** pp 14–17
[14] Volzhensky A V, Burov Yu S and Kolokolnikov V D 1979 *Mineral binders* 476
[15] Alekseev S N, Ivanov F M, Modry S and Shisel P 1990 *Durability of reinforced concrete in aggressive environments* 520
[16] Ramachandran V, Feldman R and Baudouin J 1986 Concrete Science. *Physico-chemical Concrete Science* 278
[17] Khayutin Yu G 1981 *Monolithic concrete* 448
[18] Batrakov V G 1998 *Modified concrete. Theory and practice* 768
[19] Koval S V 2012 *Modeling and optimization of the composition and properties of modified concrete* 424

Use of the dactyling effect to obtain a compositional structure in white iron during forming

T Myronova[1], A Ashkelianets[2,*], V Chukhlib[2], N Biba[3], S Gubskii[2] and A Okun[2]

[1]Department of material science the named after U N Taran-Zhovnir, National Metallurgical Academy of Ukraine, Gagarina str., 4, Dnipro, Ukraine, 49000
[2]Department of Computer Modelling and Integrated Forming Technologies, National Technical University "Kharkiv Polytechnic Institute", Kyrpycheva str., 2, Kharkiv, Ukraine, 61002
[3]MICAS Simulations Ltd., 107 Oxford Road, Oxford, OX4ER, UK

*E-mail: ashkelianets@gmail.com

Abstract. Cast iron is traditionally used as a casting alloy for the manufacture of products operating in various industries. White cast irons are used for parts with high wear resistance due to the presence in their composite structure a large number of eutectic carbides, which are located around the dendrites of the solid solution and provide a high level of hardness and strength. The location of eutectic colonies in the form of a mesh causes a low level of ability to withstand impact loading, facilitates the chipping of individual "grains" in abrasive wear and limits the use of these alloys. The use of forming process allows us to grind the eutectic mesh in the structure of white cast iron, which significantly improves the complex of mechanical and operational properties. But the low level of plasticity of white cast iron in the cast state does not allow widespread using white cast iron plastic deformation. The possibility of hot deformation of cast iron workpieces with different speeds and schemes was investigated. The deformation rate of alloys was varied due to the use of different equipment for mass and small-series production using different types of metal forming.

1. Introduction

In modern mechanical engineering and metallurgy, cast iron continues to be used as one of the main foundry materials, retaining its primacy in the future, despite the growing interest in new structural and tool materials, including composite. Increasingly, cast irons of various grades are used for parts that require high structural strength and other special properties [1-3].

White cast irons have high hardness and wear resistance due to the presence a large number of carbide phases in the structure. However, low plasticity and toughness significantly limit the possibilities of their application from a technological point of view. The mechanical properties of eutectic-type alloys, which include cast iron and carbide steels, in the cast state are determined not only by the number but also by the morphology and distribution of the eutectic in the structure.

Then the directional influence on the structure of cast materials, including cast iron, is of great importance, which leads to its transformation and replacement by the reverse relative position of the components, and provides inversion of the microstructure [1, 2].

Nowadays, there are various methods of such focused influence on the formation of the structure of eutectic type alloys as follows: 1) the creation of favorable thermokinetic conditions of structure formation during solidification; 2) the use of alloying and modification; 3) heat treatment; 4) plastic deformation; 5) a combination of different methods [4, 5, 7].

One of the most transforming, both the shape of products from eutectic type alloys, and their microstructure, is hot metal forming in which there is a grinding of eutectic mesh.

The key factor is the reduced plasticity of eutectic carbides, especially iron carbide, which is the matrix phase in the eutectic of white low-alloy cast irons. The formation of cracks in the locations of ledeburite colonies, precedes the dissection of the carbide mesh and leads to the workpiece destruction

Content from this work may be used under the terms of the Creative Commons Attribution 3.0 licence. Any further distribution of this work must maintain attribution to the author(s) and the title of the work, journal citation and DOI.
Published under licence by IOP Publishing Ltd

during plastic deformation. This problem was solved by creating a new class of alloys – dactylated cast iron [5]. The "dactylating" effect is the plasticization of cementite directly in the process of hot deformation due to its structuring during the development of phase transformations, which is observed in white cast iron alloyed with carbide-forming elements, especially vanadium [7-10].

Due to this effect in the deformation process ledeburite colonies change their original structure and they stretched, bent and crushed.

The aim of this work is to develop effective methods and deformation modes that facilitate the formation of a composite structure in cast iron workpieces, which is characterized by the presence of fibers from the solid carbide phase, surrounded by a more viscous matrix. Moreover, the structure of this matrix can be changed due to additional heat treatment.

The study was performed on alloys containing about 1.5-3.2 % by weight of vanadium and 2.35-3.02 % of C as shown on Table 1.

Table 1. Chemical composition of experimental alloys.

Number of an alloy	Chemical composition by weight, %					
	C	Cr	V	Si	S	P
1	2.35	0.89	1.50	0.45	0.04	0.05
2	2.78	0.81	1.89	0.43	0.03	0.05
3	3.03	0.96	2.76	0.34	0.04	0.04
4	2.36	0.45	–	0.36	0.04	0.03
5	3.00	0.81	–	0.41	0.035	0.04

Hot deformation was performed by various methods. Initially, by compression at temperatures of 900 °C, 950 °C, 1050 °C, the samples were deformed on a jaw plastometer at two deformation rates 1-3 and 30 s^{-1}. The loading was performed once and twice. Deformation in one pass was 25-51 %, and in two passes reached 80 % without destruction. Later hot forging and pressing were performed in conditions close to industrial. A pneumatic power hammer with a nominal dropping weight of 50 kg and a hydraulic forging press with a nominal force of 1.6 MN were used. The deformation rate on the power hammer reached 6000 mm/s, and it reached 40 mm/s on the press. In addition, in industrial conditions to obtain pipe workpieces with an outer diameter of 82 mm, inner diameter of 30 mm, length of 180 mm we carried out deformation of Ø 90 cast iron samples using a press with a force of 16 MN at a speed of 70-350 mm/s, as well as Ø50 samples on 20-70 piercing mill in rolls with notches.

To determine the effect of the workpiece size ratio (height/diameter) on the formation of a composite (fibrous) structure during hot forging of cast rods with a diameter of 38 mm made of cast iron with 2.4 % of C; 3.2 % of V; 1.78 % of Cr. There were cut workpieces of different heights (38 mm and 76 mm of h). After preliminary annealing, the samples were upset on a pneumatic hammer with a nominal dropping weight of 50 kg ($\Sigma\varepsilon$ = 55 %), as well as with a lower deformation rate of 15-20 times on a hydraulic tensile testing machine with a nominal force of 35 MN ($\Sigma\varepsilon$ = 50 %).

2. Theoretical and experimental study of a compositional structure in white iron.

In experimental samples of cast iron alloyed with vanadium, a carbide transformation is observed in eutectic cementite as shown on Fig. 1 *a*, which occurs without diffusion exchange with austenite and has the form of $[(Fe, V)\,3C]_{KI} \rightarrow 0.06\,(V, Fe)\,C + 0.02$ austenite $+ 0.92\,[(Fe, V)\,3C]_{KII}$ (in cementite K_I the concentration of vanadium corresponds to the level before the transformation, and in cementite K_{II} the concentration of vanadium corresponds to the level after the transformation, so $K_I > K_{II}$) [8].

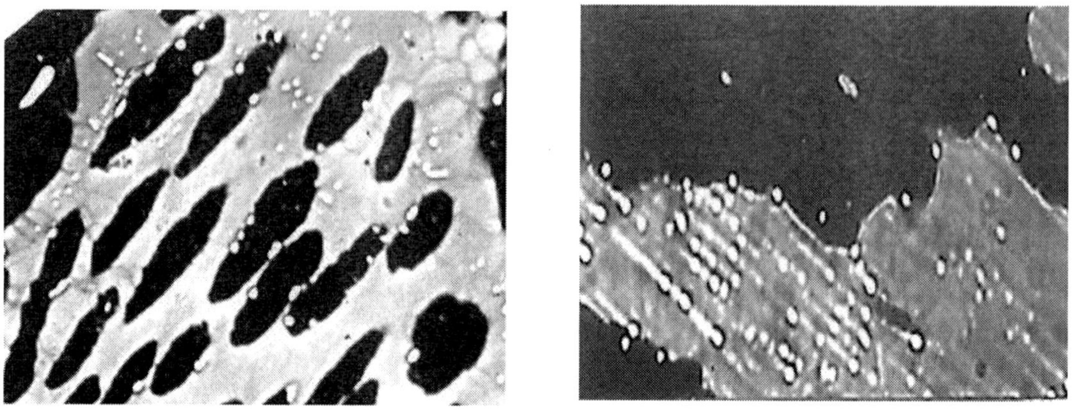

a) b)

Figure 1. Isolation of vanadium carbides in alloyed cementite:
a) multistage high-temperature annealing (1500^{x}),
b) the process of hot deformation in the sliding strips (2500^{x}) [8, 9].

The minimum concentration of vanadium in iron carbide, which causes its oversaturation and decomposition into vanadium carbides, austenite and less alloyed cementite, largely depends on factors that change the level of its stability. At isothermal holding in cementite containing less than 2-2.5 % of V, carbide transformation is not observed. Pre-processing, which stimulating an increase in the defect of iron carbide, increases the intensity of its decomposition.

A specific feature of this phase transformation is the fact that it occurs directly in the process of hot deformation. During loading in austenite near crystals of still supersaturated cementite, clusters of dislocations are created and stresses arise that exceed the level of critical values of σ shift in iron carbide. In the surface layer of cementite dislocations are formed which form sliding lines, on which the particles of the stable phase of VC carbide are released (Figure 1, b).

The presence of VC crystals directly in the cement crystal enhances the action of external forces in relation to it, as it were, contributing to the "penetration" of stresses that cause the generation of dislocations in the cementite.

This transformation is similar to intermittent autocatalytic reactions associated with the formation of dislocations on the background of the reaction. The difference is that the stresses caused by the release of VC crystals are insufficient for the formation of dislocations in the cementite when external stresses are absented. Deformation is a kind of stimulator of phase transformations, due to which the plasticity of cast iron increases by 2-3 times.

This fact was emphasized by the test results on the plastometer, namely the nature of the true yield curves as shown on Figure 2, a. In non-alloy cast irons, regardless of the eutectic level, the yield stress increases with increasing deformation degree as can be seen on Figure 2, b.

Alloying causes an increase in deformation resistance. A characteristic feature of the appending of vanadium is not only more intensive hardening even at low deformation degrees, but also a gradual cyclic unhardening as shown on Figure 2, a.

With increasing speed, the resistance to deformation of cast iron increases. However, the type of yield curve is remained regardless of the test temperature. Attention should be paid to the "pulsating" decrease in yield stress in vanadium-alloyed cast iron during compression (Figure 2, a).

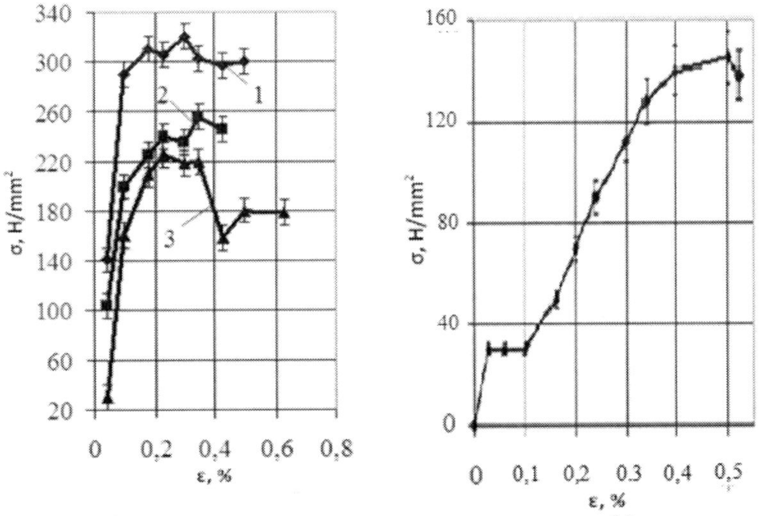

a) b)

Figure 2. True yield curves of alloys 1 a and 4 b shown on Table 1, which were annealed before testing on the plastometer at a speed of 30 s^{-1} and at temperatures of 900°C (1), 980°C (2) and 1050 C (3).

Increased plasticity and susceptibility to plastic flow is provided by the development of carbide transformation in vanadium-saturated cementite. This leads to the formation of fibers from flattened eutectic colonies after forging as shown on Figure 3.

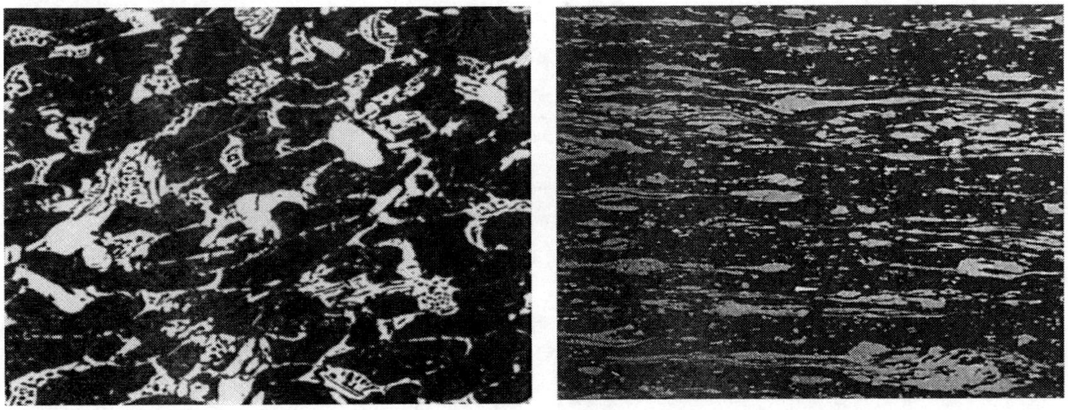

a) b)

Figure 3. Microstructure of dactylated cast iron before (*a*) and after (*b*) hot forging, 350^x.

Crushing of the eutectic mesh helps to reduce the resistance to deformation under repeated loading. Although the nature of the curve is preserved, the yield strength decreases by more than 2 times as shown on Figure 4.

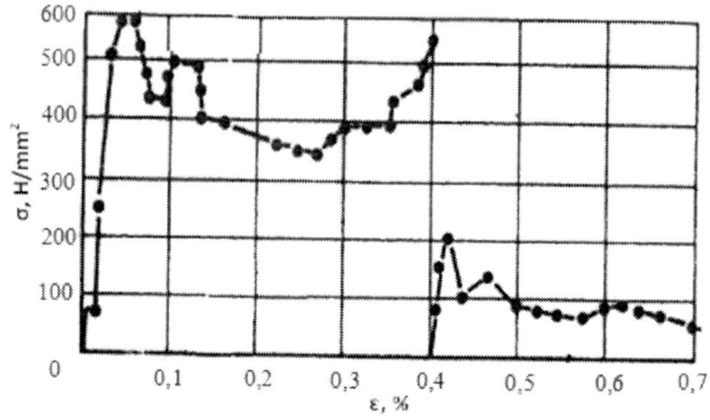

Figure 4. Flow curves at repeated loading of №3 alloy (Table 1) at a deformation rate of 30 s^{-1} and at a temperature of 900°C.

The structure of cast iron after deformation resembles stripes in steels, and we can estimate the distribution of the deformation degree on the sample height by the width b between the flattened eutectic colonies.

The deformation degree significantly affects the carbide striation of cast iron workpieces. Modelling in the licensed software product QForm [6, 10] allows us to predict the distribution of plastic deformations in cast iron samples depending on the parameters of hot upsetting and the size ratios of the initial workpieces (Figure 5).

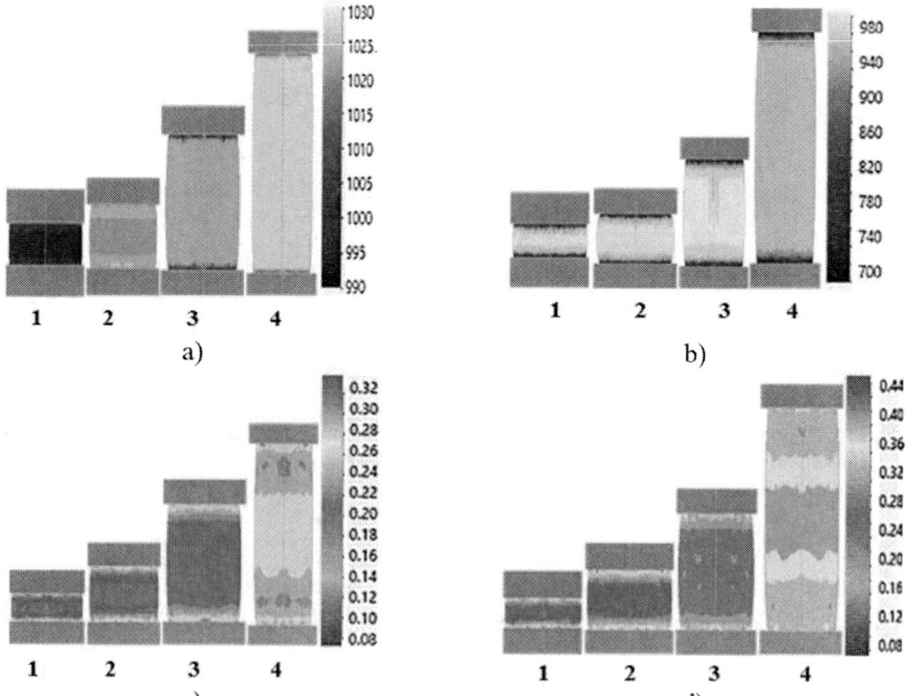

Figure 5. Temperature distribution (a, b) and plastic deformations (c, d) during upsetting on the power hammer (a, c) and on the press (b, d) at $\varepsilon = 20$ %: $1 - h_1/d_1 = 0.25$; $2 - h_2/d_2 = 0.5$; $3 - h_3/d_3 = 1.0$; $4 - h_3/d_3 = 2.0$.

As a model alloy used one containing 1.9 % of C. This alloy is heterophase and has a carbide mesh of secondary cementite in the structure, which is held at selected temperatures, it means as close as possible to the structural state of experimental cast iron. The ratio of height to diameter of the sample (h/d) was equal to 0.25, 0.5, 1.0, 2.0, and the deformation degree ε was simulated by 10 %, 20 % and 30 %.

The mathematical modelling results of the temperature distribution showed that increasing the deformation rate (compared to the press and the power hammer) at all stages of deformation is the deformation heating of the workpiece, which indicates a higher intensity of deformation on the power hammer. During the upsetting on the press, the workpiece is cooled in contact zones with the tool surface.

The results of quantitative microstructural analysis by the transverse method showed that a more homogeneous structure is formed during forging on a power hammer, and with increasing height of the sample austenite branches flatten to a greater extent and are parallel to the sample surface, it means perpendicular to the compressive stress. During forging at lower speeds on a tensile testing machine, the structure of the cast iron changes unevenly, especially in the workpiece with $h/d = 2$. In the central part of this sample, the eutectic mesh is completely crushed as can be seen on Figure 6.

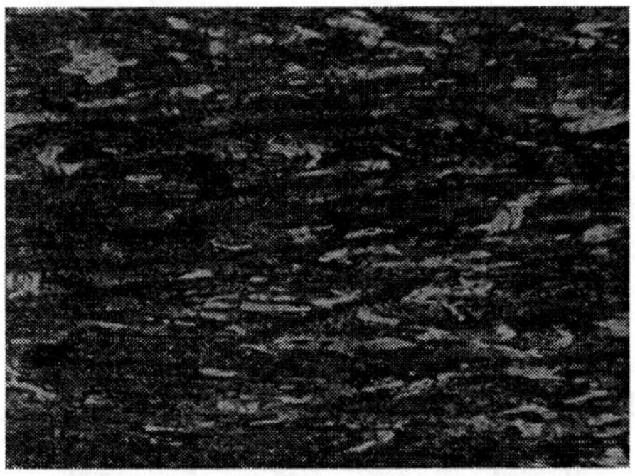

Figure 6. The microstructure of the sample central part with $h/d = 2$ after compression on the press, 100^x.

The study of the behavior of white dactylated cast irons during compression allowed us to plan and successfully conduct hot pressing in industrial conditions to obtain a pipe workpiece.

Cast iron with a diameter of 90 and 55 mm was used as the initial roll bars. The level of mechanical properties of the samples cut from the workpiece of Ø 55 mm has the following characteristics: ultimate tensile strength is 590 MPa, yield strength is 475-480 MPa, δ is 7.5-8 %, KCU is 3.0 MJ/m^2.

This allows us to recommend pressing and piercing to obtain a pipe workpiece from this alloy.

Pressing was performed at temperatures of 850 °C, 900 °C, 950 °C, 1000 °C and 1050 °C at a speed of 70 and 350 mm/s. External inspection of the pressed pipes did not reveal surface defects and traces of destruction as can be seen on Figure 7.

Figure 7. Macrostructure of extruded pipes and general view of templates.

The calculation of the average pressing pressures at different deformation modes revealed the following trends. The maximum values of the pressing pressure at the output of the workpiece in the deformation zone are 1.5-2 times higher than the average levels of the pressing pressure at the constant stage (Table 2). Increasing the pressing speed leads to an increase in the average pressing pressure, both at the initial moment of deformation and at a constant stage.

Table 2. Influence of temperature and deformation rate on the average level of pressing pressure.

Deformation temperature, °C	Average level of pressing pressure, MPa*		
	Pressing speed, mm/s		
	350	350	70
	furnace heating	induction heating	induction heating
1050	1480-1530 / 1000-1030	910-1030 / 595-640	790-850 / 505-575
1000	–	1150-1185 / 740-770	880-960 / 700-740
950	1365 / 890	1140-1185 / 710-810	910-970 / 710-740
900	–	1345-1480 / 890-970	1110-1185 / 810-865
–	–	1630-1780 / 1090-1110	1285-1335 / 913-970

* A numerator is the maximum values and a denominator is the pressure at a constant stage.

Furnace heating, in comparison with induction, leads to a significant increase in the average pressing pressure of white cast iron, especially at high temperatures. This is due to the fact that longer holding, especially above 950 °C leads to the rapid development of spheroidization and coalescence of deformed cementite.

As a result, the colonies of ledeburite become cementite monolithic (Figure 8, a).

Comparative analysis of the microstructure of pipes pressed at low speed by induction heating showed that with decreasing deformation temperature from 1050 °C to 850 °C the viscous flow of cementite and formation of carbide stripes is replaced by crushing of eutectic colonies (Figure 8, b, c). Moreover, at a temperature of 1050 °C, dynamic recrystallization of cementite is noticeable (Figure 8, c).

The nature of the release of vanadium carbides in cementite also changes, namely with increasing deformation temperature, the number of nucleation centers of VC particles decreases, and the size of carbide crystals increases. In the whole temperature range at low deformation rate the austenitic matrix retains its plasticity. Cementite crystals are divided and the solid solution fills the gaps between them. No cracks are detected. Increasing the deformation rate changes the picture of structural transformations. At high deformation temperatures of 950-1050 °C the influence of deformation heating is appeared. In all this temperature range there is a viscous flow of cementite and the formation of carbide fibers directed along the axis of deformation as shown on Figure 8, d, e, f.

At a deformation rate of 350 mm/s, even at a temperature of 1050 °C, dynamic recrystallization does not develop.

Reducing the deformation temperature to 850 °C can lead to brittle fracture of eutectic colonies. Low deformation temperatures in combination with high pressing rate cause a fast strengthening of the austenitic matrix and reduce its plasticity. The solid solution no longer has time to fill the gaps in the cementite, and microcracks are formed.

The nature of structural transformations during pressing is in good agreement with changes in the average pressing pressures of white cast iron depending on the deformation parameters. The change in the viscous flow of ledeburite colonies by the brittle fracture of hardened cementite and solid solution causes an increase in the work of deformation so the average pressing pressure increases. This phenomenon is observed both at decrease in deformation temperature, at increase in deformation rate and increase in period of high-temperature holding at heating of workpieces under deformation. In addition, the strengthening of the phase components largely depends on the diffusion phenomena.

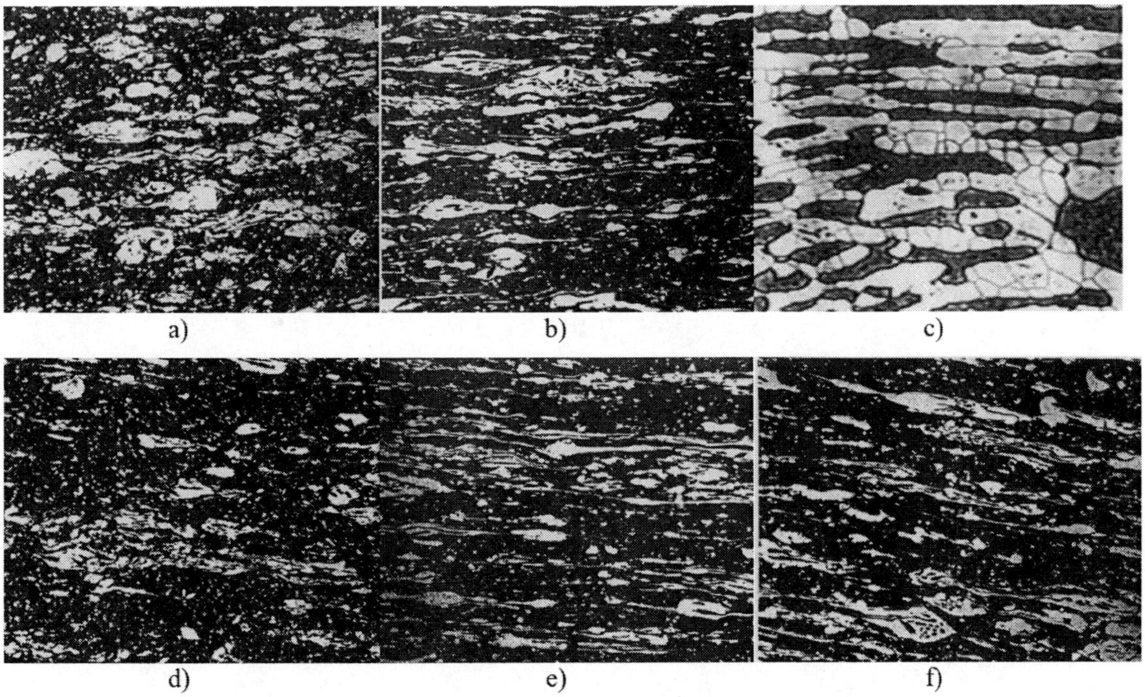

Figure 8. Microstructure of longitudinal samples of pipes at furnace and induction heating under different conditions of pressing: *a, b, d, e, f* – ×100; *c* – ×850.

The decrease in the diffusion mobility of atoms with a decrease in the deformation temperature, the formation of stratification zones or the lag of diffusion processes at high deformation rates inevitably cause the strengthening of structural components and a decrease in their plastic properties. The average pressing pressure increases.

Studies have shown that the optimal conditions for pressing white cast iron workpieces are low speed (70 mm/s), maximum deformation temperature (1050 °C) and the use of high-speed (induction) heating.

Similar results were obtained when piercing solid deformed cast iron workpieces of Ø50 mm and a length of 200 and 250 mm, which was carried out on "20-70" piercing mills in rolls with a line at a temperature of 900-1100 °C. Optimal in terms of microstructural criteria of the obtained sleeves and ensuring sufficient plasticity were piercing temperatures of 900-950 °C. The rings cut at the beginning, middle and the end of the sleeves have a dense homogeneous macrostructure without any visible defects. The formation of the structure occurs in accordance with the previously described patterns in the developed deformations of dactylated cast irons, so there is a plastic flow of cementite, the release of VC, as well as the crushing of carbide inclusions. It should be emphasized once again that an increase in the processing temperature above 950 °C, especially with such a loading scheme, causes an increase in the particle size of eutectic carbides as shown on Figure 9.

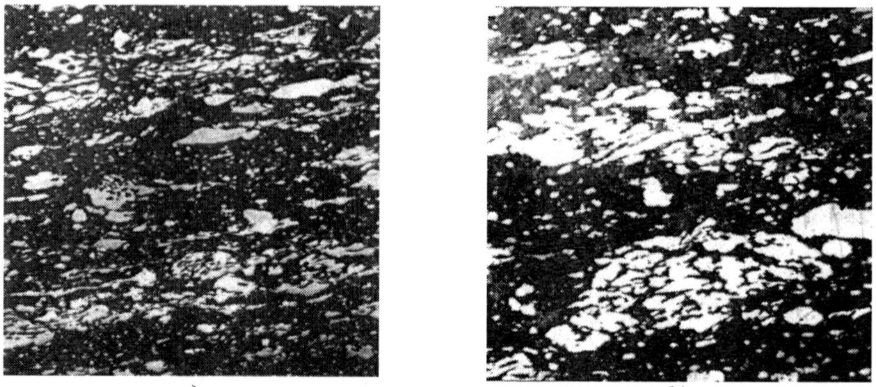

a) b)

Figure 9. The microstructure of the sleeves obtained by the piercing at different temperatures of 950 °C (*a*) and 1050 °C (*b*), ×500.

The formation of the matrix structure around the fibers of eutectic carbides in experimental alloys significantly depends on the modes after deformation annealing and further hardening, with which you can vary the properties of cast iron in a wide range.

3. Conclusions

- The paper shows the possibility of successful deformation of dactylated white cast irons by hot compression, which contributes to the formation of a composite structure in the form of carbide fibers.
- Dimensional *h*/*d* ratios and deformation rate affect the deformed structure in the workpiece volume. The most homogeneous structure is formed at a higher rate of deformation achieved by free forging on a power hammer, namely austenite branches and strips of flattened eutectic colonies are parallel to the surface of the sample, and perpendicular to the compressive stress.
- Pressing of pipes from the deformed white cast iron in the range of temperatures of 950-1050 °C with a speed of 70-350 mm/s allows us to receive high-quality pipe workpiece. The heating period of the initial workpiece under deformation should be minimal. After pressing, the pipe has a characteristic fibrous location of eutectic carbides.
- Piercing of cast iron workpiece in the temperature range of 900-950 °C provides high-quality sleeves. Eutectic carbides are extracted in the direction of deformation and divided.
- Economically alloyed ledeburite cast irons by plastic deformation and further heat treatment can be widely used as composite materials, such as mill rolls, guide rolls, etc.

References

[1] Taran-Zhovnir Y 1998 Stroenye evtektyk y sozdanye novykh splavov evtektycheskoho typa *Suchasne materialoznavstvo XXI storichchya* pp 176–197

[2] Silman G, Zhukov A and Epstein L 1972 Kristallizatsiya vanadiyevykh chugunov s invertirovannoy strukturoy karbidnoy evtektiki *Liteynyye svoystva splavov* pp 201–204

[3] Shcherbedinskiy G 2005 Chugun kak perspektivnyy material XXI stoletiya *MiTOM* **7** pp 83–93

[4] Dojka M, Kondracki M, Studnicki A et al. 2018 Crystallization Process of High Chromium Cast Iron with the Addition of Ti and Sr *Archives of Foundry Engineering* **18 (2)** pp 57–64

[5] Taran Y, Nizhnikovskaya P, Danichek O, Mironova T, Loyferman M, Belorusov S, Demchenko G and Khizhnyak D 1989 Deformiruyemyye ekonomnolegirovannyye belyye chuguny *MiTOM* pp 35–43

[6] Chukhlib V, Klemeshov E, Gubskyi S, Okun A and Biba N 2020 Theoretical and Experimental Studies of Changes in the Workpiece Shape During Narrow Die Indentation *Advances in Design, Simulation and Manufacturing III. DSMIE 2020. Lecture Notes in Mechanical*

Engineering. doi.org/10.1007/978-3-030-50794-7_35
[7] Mironova T and Ryabchiy M 2008 Upravleniye formirovaniyem struktury v belykh ledeburitnykh chugunakh na razlichnykh etapakh deformatsionnogo peredela *Metallurgicheskaya i gornorudnaya promyshlennost* **4** pp 79–81
[8] Mironova T and Proidak S 2019 Peculiarities of Alloying Effect on the Eutectic Cementite Behavior Under Hot Rolling *New Trends In Production engineering* **2** pp 289–300. doi: 10.2478/ntpe-2019-0093
[9] Myronova T, Donskaya T and Sydorova A 2012 O mekhanyzmakh vlyyanyya fazovykh perekhodov na povedenye évtektycheskykh karbydov pry deformyrovanyy *Visnyk Dnipropetrovs'koho universytetu* **20 (2)** pp 97–104
[10] Mironova T, Kutsova V, Chukhleb V et al 2015 Vliyaniye sposobov i rezhimov goryachey obrabotki davlenim na strukturnyye izmeneniya v belykh chugunakh *Metallurgicheskaya i gornorudnaya promyshlennost* **5** pp 52–57

Experimental technique of defining the elastic modulus of bulk materials in laboratory conditions

Valentyn Astakhov[1*], Dmytro Brovko[1] and Oleksandr Eremenko[1]

[1]Kryvyi Rih National University, Vitaliy Matusevych St. 11, Kryvyi Rih, Ukraine, 50036.

[*]E-mail: astvi51@gmail.com

Abstract. Defining the elastic modulus of bulk materials is an important issue in quality assurance of design and construction of motorway bases. The analysis of current regulatory framework shows that there are no applicable methods for defining the elastic modulus of loose (bulk) materials in laboratory conditions. Moreover, if new or underinvestigated materials are used for motorway bases, it is necessary either to abandon their use or define their compliance with the regulations after motorway base construction. In view of the latter, an experimental technique and a device for defining the elastic modulus of bulk materials in laboratory conditions are developed. Following this technique, the elastic modulus of 5 mm – 20 mm, 20 mm – 40 mm crushed dump blast furnace and steelmaking slags was defined. Comparison of the obtained results with the recommended design values of the elastic modulus of similar materials shows good convergence (the difference is within 8%). There is no connection between the diameter of a stamp transmitting load to a sample and the specified value of the elastic modulus. Deviations do not exceed 10% for Ø50 mm, Ø120 mm, Ø150 mm stamps.

1. Introduction

1.1. Background

Road construction is one of the most material-intensive industries. Creation of a durable and qualitative road design involves reliable base construction providing an even underlay for the road surface and transmitting the road surface load onto a sub-grade. Motorway bases are made of bulk materials – crushed stone, gravel, crushed stone and sand mixtures. To ensure high road performance, base material must have sufficient strength and rigidity (elastic modulus). Strength and reliability of a base is achieved by its layer-by-layer compaction for the material to create elastic strength through loaded road surfacing, and using the material the strength of which is not lower than regulated. A quality indicator of base compaction is its elastic modulus. Internationally, load-bearing capacity of road surfacing as a whole is evaluated in terms of base elastic modulus.

1.2. Analysis of current regulatory framework for defining the elastic modulus of bulk materials

Current regulatory requirements and recommendations provide for defining the elastic modulus of loose bulk materials (road construction materials) by static methods using a rigid stamp or dynamic methods [1] - [4]. It is decisive that all these methods and the described techniques relate to field tests of materials, namely areas with a compacted material, the elastic modulus of which is to be defined. Although this approach accurately reflects real stress-strain properties of road structure layers, it does not allow defining them for new or underinvestigated materials.

According to the regulations [4] on testing samples of loose (crushed stone, gravel, sand) materials, a stamp of 25 cm – 35 cm in diameter is to be applied.

Another regulation (see Annex B, [3]) describes a method of making samples of loose materials to define the elastic modulus in both field and laboratory conditions. The schematic diagram of sampling to define the elastic modulus in laboratory conditions is shown in figure 1.

Content from this work may be used under the terms of the Creative Commons Attribution 3.0 licence. Any further distribution of this work must maintain attribution to the author(s) and the title of the work, journal citation and DOI.
Published under licence by IOP Publishing Ltd

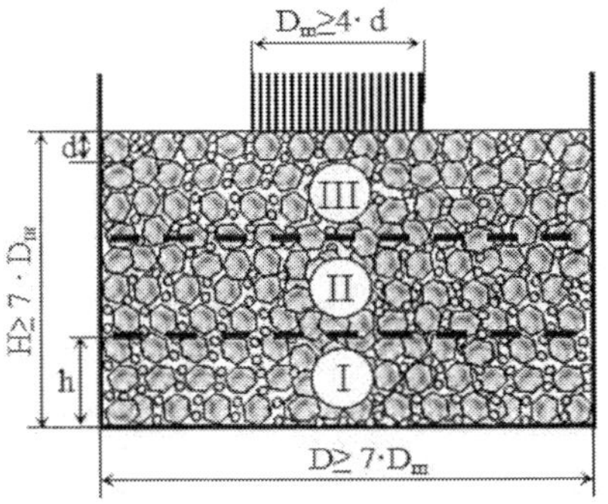

Figure 1. The dimensions chart of a mould and sequence of compacting road construction materials when defining the elastic modulus [3]: I, II, III – layers of loose materials; d – a diameter of a maximum grade; h – a thickness of a compacted layer; Dш – a stamp diameter; D – a mould diameter; H – a total compaction thickness.

According to the given scheme, for the most widespread crushed stone grade of 20 mm – 40 mm a stamp size should be $D_ш \geq 4 \cdot D = 4 \cdot 40 = 160$ mm. Respectively, a mould diameter cannot be less than $7 \cdot D_ш$, i.e. $D = 7 \cdot 160 = 1120$ mm. The thickness of a compacted layer must be no less than the width of a mould, so the height of the mould must be no less than 1120 mm. Thus, the regulations recommend making a cylindrical mould higher and wider than one meter. If we take into account the requirement contained in [4] that a stamp diameter is to be within 25 cm – 35 cm, the size of a mould is to be doubled.

Therefore, the drafters believe that a 2100 mm high and 2100 mm wide mould (designed for a 300 mm stamp diameter) is to be produced in a base of road surfacing and afterwards filled with crushed stone to accurately simulate crushed stone behaviour (20 mm – 40 mm grade) in laboratory conditions. If taken literally, the mould size exceeds the size of many bathrooms of residential buildings of the 1960s. The thickness of the material layer to be created is also unreasonable. Its estimated value is 2100 mm. In Ukraine, there is hardly any road (except for artificial embankments) where the thickness of a base layer is 2 metres. According to Annex F [5], the minimum possible thickness of a slag crushed stone base varies within 12 cm – 15 cm. For the crushed stone graded 40 mm – 60 mm, the mould dimensions will change proportionally, up to 3 m in diameter and in height. Thus, in real life there are no such roads and operating conditions of their bases to be reproduced in laboratory conditions.

In addition, to fill a cylindrical 2100 mm high and 2100 mm wide mould, 7.27 m^3 of uncompacted crushed stone or, provided the bulk density ρ = 1200 kg / m^3, crushed stone weighing 8.7 t will be required. Annex B [3] states that crushed stone should be placed in layers with subsequent compaction of each individual layer. According to the same annex, which is mandatory, the following load levels must be applied to each layer:

- 5 MPa – 15 MPa for soils and soil mixtures;
- 15 MPa – 25 MPa for sands, sandy and fine-grained mixtures;
- 20 MPa – 40 MPa for medium - and coarse-grained.

Due to compaction, the volume of the analytical sample should be increased by 30% - 40%. That is, the volume of the sample material will increase to 12 t.

Another question is what means can create pressure inside the mould of specified dimensions to compact the material layers? Let us consider the above-mentioned crushed stone 20 mm – 40 mm grade. Its compaction requires a pressure of 20 MPa as it is a medium-sized material.

$A = \pi \cdot D^2 / 4 = 3.14 \cdot 2.1^2 / 4 = 3.46$ m^2 – compaction area.

$F = \sigma \cdot A = 20 \cdot 10^6 \cdot 3.46 = 69.2 \cdot 10^6 H = 7.054 \cdot 10^6 kg = 7054$t.

Thus, the drafters make it incumbent to create a load of more than 7 kt in order to create a sample to define the elastic modulus of 20 mm – 40 mm crushed stone in laboratory conditions. How to create this load? What stamp is needed to transmit it by weight and size? How to put crushed stone into the

IOP Conf. Series: Materials Science and Engineering 1164 (2021) 012006 doi:10.1088/1757-899X/1164/1/012006

mould? How to remove it after the experiment? Where to store such an amount of crushed stone? In this case, these questions are rhetorical.

Current regulations contain [1] - [4] different information on the methods of field tests aimed at defining the elastic modulus of a road structure or base layers by static load using a rigid stamp. So, in [1] the road structure elastic modulus is defined. To do this, the design static load is applied through a rigid stamp to road structures. Then, the deflection value is specified. Afterwards, the stamp is unloaded and the reverse deformation value is specified. The latter value is the basis for calculating the elastic modulus of road structures. It should be noted that the regulation in question does not indicate the amount of load to be applied for testing. There is only a rather generalized expression – a design load. The annexes of this regulation comprise the recommended specification of stamping equipment. They state that the capacity of load generating facilities is to be 60 kN. This is the only way to find out what is meant by the term design load. These assumptions are confirmed by the information given in table B [5], where the design load imposed on a first category road is equal to 65 kN. Equivalent to a stamp size of 30 cm, this load creates a pressure of 0.9 MPa on the surface. For a fourth category road this value is 0.5 MPa.

In another regulation [2] the load value is given indirectly. It is said that the pressure gauge value should be (160 ± 5) kgf / cm^2 or 15.7 MPa. This load value created by the hydraulic cylinder through a rigid stamp is to be applied to the pavement or the surface. Next, elastic deformation and then the elastic modulus are defined by the difference of the indicators when load is applied and then released.

According to [6] when defining the elastic modulus of pavement, load is applied gradually through a rigid stamp, namely: Q1 = 0.25Qmax; Q2 = 0.5Qmax; Q3 = 0.75Qmax; Q4 = Qmax. At each load level, the elastic deformation value is determined and a chart of the elastic deflection value caused by the load is plotted. After drawing a tangent to this chart, the elastic modulus value is determined. It is suggested to define Qmax value following the instructions given in [7]. According to the latter, Qmax value = 0.5 MPa for base surfaces and Qmax = 0.2 MPa for a sub-grade. This document provides a clear answer to the question of the load levels to be applied to different layers of a road structure, but it was abolished in 2019. The alternative regulation does not contain this information.

The regulation [3] contains the same algorithm for determining the elastic modulus as the previous one [6], referring to the repealed regulation [7] and plotting. It should be noted that the determination of any characteristics via plotting leads to a significant error of the obtained values.

The greatest clarity regarding load intensity and its application method was presented in [7] (abolished in 2019). Thus, the design load on a base is 0.5 MPa, which is to be applied gradually (3 levels – 5 levels), with exposure at each level and unloading. The elastic deformation rate is defined for each level. The obtained values provide for plotting, but the value of the elastic modulus is defined analytically by calculating the elastic deformation rate obtained for the highest load level.

A review of the available regulations on defining the elastic modulus of loose road construction materials in laboratory conditions indicates a lack of a single approach to this issue. The regulations contain mutually contradictory instructions; load levels can differ significantly.

2. Research methods

As there is no real mechanism of estimating the elastic modulus of loose and bulk materials, it is advisable to start developing real methods for defining this parameter in laboratory conditions. The following method of defining the elastic modulus is the engineering one. It requires further research, comparison of research findings with the results of field tests and the formation of an appropriate calculation mechanism. The current regulatory requirements [1] - [4] to a sample creation and altered test methods enabling its conduct in laboratory conditions are taken as a basis. Apart from that, a device for making a sample and defining the elastic modulus was developed.

The schematic diagram of a stand for defining the elastic modulus of loose materials is shown in figure 2.

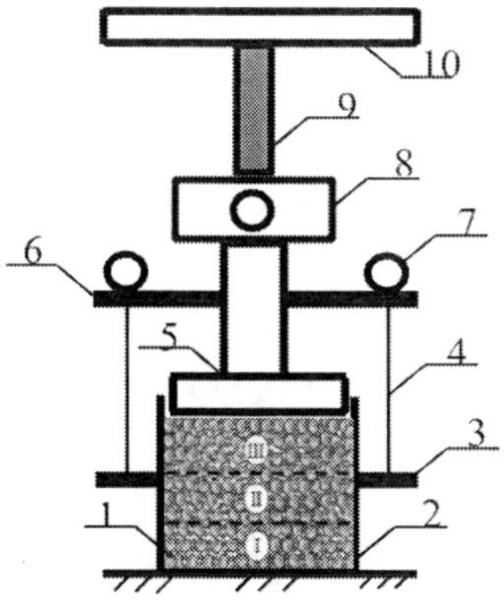

Figure 2. The schematic diagram of a stand for defining the elastic modulus of loose materials: 1 – a layered compacted sample of material; 2 – a cylindrical mould; 3 – a support for a dial test indicator extension; 4 – an extension of a dial test indicator; 5 – a stamp; 6 – a bracket for dial test indicator fixing; 7 – a dial test indicator (division value – 0.01 mm); 8 – a dynamometer; 9 – a load generating device (hydraulic jack); 10 – a fixed rest

In accordance with the requirements [8], a cylinder with a plunger and a bottom with the same dimensions as in crushing tests (Ø150 mm) was used for the tests. Additionally, clip brackets were made and put on the cylinder to support the indicator extension. A similar holder was made for the plunger to attach the indicators. During the tests, 4 dial test indicators (0.01 mm division value) were used. They were placed on four sides of the plunger in orthogonally related directions. The number and location of the indicators should ensure greater accuracy of the tests and avoid off-center application of load. Pressure was produced by a hydraulic jack, while its rate was controlled by a ДОСМ-3-1 dynamometer.

The tests were done for crushed dump blast furnace and steelmaking slags graded 5 mm – 20 mm, 20 mm – 40 mm. Size grading was carried out in accordance with the requirements [8]. Two analytical samples weighing 4 kg each were taken from each grade. Analytical samples for dry basis tests were dried to a constant weight. An overall view of the analytical samples by grades is shown in figure 3 and figure 4.

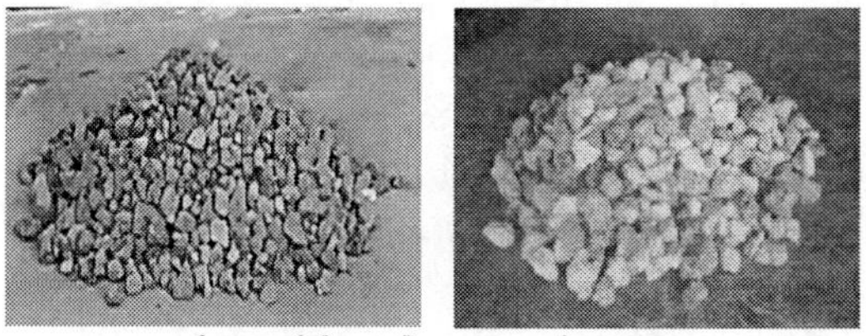

dump blast furnace slag, mm
5 – 20 20 – 40

Figure 3. The analytical sample of crushed dump blast furnace slag.

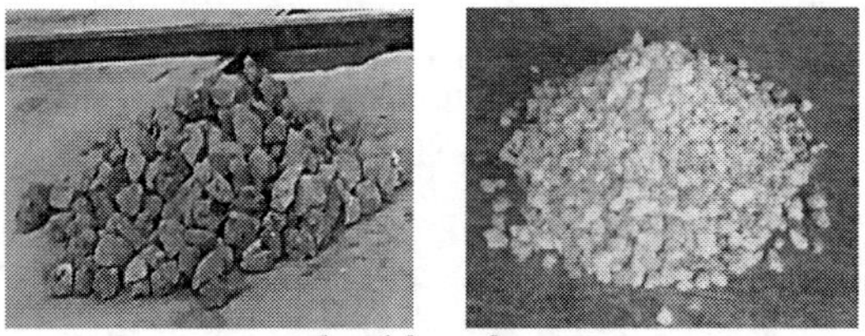

steelmaking slag, mm
5 – 20 20 – 40

Figure 4. The analytical sample of crushed steelmaking slag.

A test sample was prepared taking into account the requirements of Annex B [3]. The cylinder (inner diameter – 150 mm) was conditionally divided into three parts throughout its height and filled with crushed stone layer-by-layer. When the layer was formed, it was compacted on a hydraulic press П-125 with the load meeting the requirements of Annex B [3]. Then, the next layer was produced and the procedures above were repeated. 5 mm – 20 mm crushed stone was compacted with the load of 20 MPa, and 20 mm – 40 mm grade with the load of 30 MPa. After compaction, the outer surface of the sample was covered with a layer of screened sand (up to 3 mm thick) to level off the surface and avoid possible off-center application of load. The overall process of making the test sample is shown in figure 5.

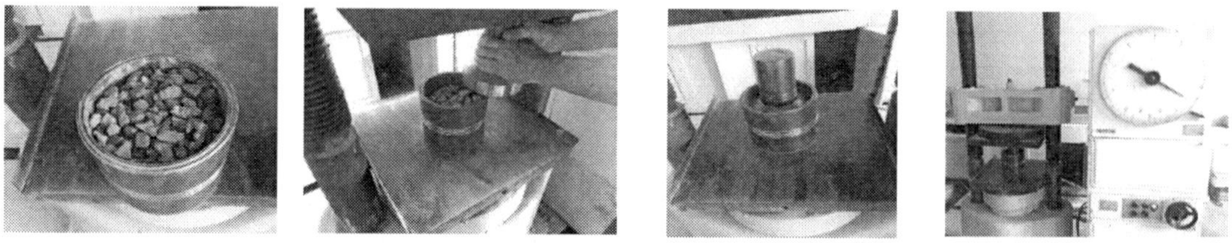

Figure 5. Making the test sample for elastic modulus defining.

The sample created in this way was tested for compaction sufficiency in accordance with the requirements [1]. To do this, the load from 0 MPa to 0.5 MPa was applied recording the sensors readings in 0.1 MPa after attenuation of deformations (no more than 0.01 mm within 20 s to 30 s) at each load value. The total modulus of deformation was determined by data processing. Having repeated the same procedures for the second time, a new value of the total modulus of deformation was defined. The mixture is considered to be sufficiently compacted if the ratios of the first and second modulus of deformation do not exceed 2 [1].

The elastic modulus was determined using the methods described in [7] and in [1]. The load level of 0.5 MPa was chosen as the design load to be imposed on the surface of the road structure base through a rigid stamp. This load level is mentioned in the regulations [3], [6] and partly in [1]. In addition, this load level is specified in regulations of neighbouring countries (Belarus, Russia).

In accordance with the requirements [7], the load was applied level-by-level, with a level value of 0.1 MPa. Load application time – 30 s, pauses after unloading – 30 s, time spent on loading and unloading – 10 – 20 s. The indicators readings were taken after maintaining the specified load level, as well as after the pause following unloading. Discrepancy of the readings was used to determine elastic deformation of the structure. Due to the fact that the regulations [3] require applying a load through a stamp with the diameter being three or more times smaller than the mould diameter, the load was applied three times through stamps of different diameters, namely: Ø50 mm, Ø120 mm, 150 mm. With a mould (cylinder) diameter of 150 mm, the first stamp (Ø50 mm) met the regulatory requirements. Two other stamp diameters were adopted to compare the test results with a stamp diameter meeting the requirements with the results to be obtained when testing with stamps of other diameters. An overall view of a stand defining the elastic modulus of slag crushed stone with stamps of different diameters is shown in figure 6.

stamp Ø50 mm stamp Ø120 mm stamp Ø150 mm

Figure 6. A stand for defining the elastic modulus of slag crushed stone

3. Results and discussion

Based on the obtained values of elastic deformation of the samples, the elastic moduli of the considered grades of slag, steelmaking crushed stone and three values of stamp diameters were calculated by formula 7.1 [4]:

$$E_л = \frac{\pi \cdot \rho \cdot D \cdot \left(1 - \mu^2\right)}{4 \cdot f}, \tag{1}$$

where μ – Poisson ratio ($\mu = 0{,}25$); ρ – surface design pressure ($\rho = 0{,}5$ MPa); D – stamp diameter, m; f – elastic deflection (elastic deformation), m.

Using this formula (1) and the research results, the elastic modulus was calculated (see Table 1)

Table 1. Values of elastic modulus of crushed dump blast furnace slag and steelmaking slag.

Crushed stone grade, mm	Observed elastic modulus, MPa			Theoretical value of elastic modulus [7], MPa
	Ø stamp – 150 mm	Ø stamp – 120 мм	Ø stamp – 50 mm	
		crushed dump blast furnace slag		
5 – 20	470	449	433	350 – 450
20 – 40	487	441	466	350 – 450
		crushed steelmaking slag		
5 – 20	398	392	357	350 – 450
20 – 40	469	436	427	350 – 450

Table 1 shows not only the results of calculations of the experimental elastic modulus of crushed stone of different grades but also the theoretical value of this parameter. Comparison of theoretical and experimental values indicates their good convergence. Experimental data either fall within the range of theoretical values or slightly (up to 8%) exceed the upper limit of values. The latter can be explained by the fact that design (theoretical) values are usually smaller than experimental ones because they assume variability and heterogeneity of studied materials by underestimating their real physical and mechanical parameters.

Comparison of the obtained experimental values of the elastic modulus defined by using stamps of different diameters indicates that the elastic modulus values are not impacted by a stamp diameter significantly. The difference in values does not exceed 10%. Given the convenience of the experiment and stability and predictability of the indicators, it is more appropriate to conduct an experiment with a larger stamp diameter.

It should be noted that this technique is experimental. It is worth mentioning that crushed stone in the middle of the cylindrical mould, provided that the stamp diameter is equal to the inner diameter of the cylindrical mould, is in a three-axis stress state, which may differ from operating conditions of a road structure. It should be borne in mind that during real operation a road structure base takes a load of vehicles through a surface layer (asphalt or cement concrete) of 150 mm – 300 mm thick. A surface will also constrain base distortion not only under a car wheel, along with the base material located on the sides and a sub-grade material.

To determine the degree of reliability of the results obtained, additional research is needed. The latter should consist in conducting field tests to determine the elastic modulus of base layers of a real road structure made of similar crushed stone. In order to compare, base layers must be compacted to the same state as the material of the laboratory sample. This can be controlled by determining the modulus of base deformation of a road structure. The last requirement is caused by the fact that crushed stone with different degrees of compaction will predictably have different elastic moduli.

If elastic modulus values defined through laboratory and field tests vary significantly (more than 15%), the calculation formula is to be amended. At present, the suggested method allows for experimental tests to determine the elastic modulus of loose and bulk materials, while regulations make such operations physically impossible.

4. Conclusions

The analysis of current regulatory requirements and guidelines for defining the elastic modulus of loose and bulk materials indicates that it is physically impossible to carry out such operations in laboratory conditions. The developed engineering technique and device for defining the elastic modulus in laboratory conditions show a good convergence with the recommended calculated values of the elastic modulus of similar materials. The performed tests of slag crushed stone allowed obtaining the elastic modulus for differently graded crushed stone. The results either fall into the range recommended by the regulations, or slightly exceed the limit value (deviation of up to 8%). It is proved that different stamp diameters do not impact test results. The deviation of the elastic modulus values for stamps of Ø50 mm, Ø120 mm, Ø150 mm do not exceed 10%.

References

[1] DSTU B V.2.3-42:2016 Automobile Roads. Determination of stress-strain behaviour of a sub-grade and pavement (Kyiv: SE UkrNDNC) 43

[2] SOU 45.2-00018112-042:2009 Automobile Roads. Determination of transport and operational parameters of road surfacing (Kyiv: State Agency of Automobile Roads of Ukraine) 47

[3] SOU 45.2-00018112-059:2010 *Road construction materials. Methods of elastic moduli determination* (Kyiv: State Agency of Automobile Roads of Ukraine) 28

[4] M 218-02070915-645:2008 *Methods of laboratory determination and establishment of design parameters of road construction materials and sub-grades* (Kyiv: State Agency of Automobile Roads of Ukraine. National Transport University) 61

[5] DBN B.2.3-4:2015 *Automobile Roads. Part I Design. Part II Construction* (Kyiv: Ministry of Regional Development, Building and Housing and Communal Services of Ukraine) 113

[6] SOU 45.2-00018112-062:2010 *Test methods on a ring stand* (Kyiv: State Agency of Automobile Roads of Ukraine) 45

[7] VBN B.2.3-218-186-2004 *Transport facilities. Non-rigid road surface. Amended* (Kyiv: State Agency of Automobile Roads of Ukraine) 79

[8] DSTU B V.2.7-71-98 GOST 8269.0-97 1999 *Crushed dense rocks and process wastes used for construction works. Methods of physical and mechanical testing* (Kyiv: State Committee of Construction, Architecture and Housing Policy) 47

IOP Conf. Series: Materials Science and Engineering 1164 (2021) 012007 doi:10.1088/1757-899X/1164/1/012007

Perpendicular to steel reinforcing bars axis displacements of rebars for determination the torsional stiffness of reinforced concrete elements with normal cracks

T Azyzov[1], D Kochkarev[2], L Tsyhanenko[3], H Tsyhanenko[3] and N Srybniak[3]

[1]Uman State Pedagogical University named after Pavlo Tychyna, st. Sadova, 2, Uman, 20300, Ukraine
[2]National University of Water Resources and Environmental Management, st. Soborna, 11, Rivne, 33028, Ukraine
[3]Sumy National Agrarian University, st. Herasyma Kondratieva, 160, Sumy, 40000, Ukraine

*E-mail: taljat999@gmail.com

Abstract. The article provides a technology for determining the displacement of the end of the reinforcing bar embedded in concrete for measuring the shifting force in the longitudinal reinforcement. Dependences of the displacement of the end of the reinforcing bar on different protective layers are shown. A formula to define the displacement of steel reinforcing bars due to the shifting force is given. The formula was obtained by the numerical computation approximation using three-dimensional finite elements. A method for determining the torsional stiffness of reinforced concrete element with normal crack is presented. In this method the pliability of the reinforcement perpendicular to its axis is also considered.

1. Research analysis and problem-setting

The redistribution of forces between separate items of coverings, bridges and other complex statically indeterminate systems in spatial behavior depends on their torsional and bending stiffness relation [2, 14, 21]. Spatial, normal and inclined cracks have a significant impact on the stiffness in reinforced concrete structures [8].

Bending stiffness of reinforced concrete elements with cracks have been researched wide enough. Most of the work related to torsional deformations involves a spatial crack. However, such techniques are not suitable for calculation of rotational movements of elements with normal cracks resulting from bending moments.

The torsional stiffness of reinforced concrete elements with normal cracks has been researched by the authors of this article [1, 2, 18, 19]. According to these researches, the torsional stiffness should be determined by cutting the longitudinal reinforcement first, then determining mutual displacement of the faces. Then the shifting force and stiffness of the element with normal cracks are determined. The forepart of the problem is approximate when solving analytically and rather time-consuming when solving by modeling three-dimensional finite elements. In this connection, in [2, 18, 19] the empirical formula for the displacement of the reinforcing bar by the force perpendicular to its axis is used to determine the shifting force. However, this formula is derived from tests of embedded details clamped in a mass concrete. It does not take into account the influence of protective layer depth of the reinforcing bar.

Based on these issues method development for the determination of the value of shifting force in longitudinal reinforcing bar for calculation the torsional stiffness of concrete elements with normal cracks becomes relevant.

In relation to the above-mentioned target, the object of this article is to develop the method for the determination of shifting force in longitudinal reinforcement according to its location in the cross-

Content from this work may be used under the terms of the Creative Commons Attribution 3.0 licence. Any further distribution of this work must maintain attribution to the author(s) and the title of the work, journal citation and DOI.
Published under licence by IOP Publishing Ltd

IOP Conf. Series: Materials Science and Engineering 1164 (2021) 012007 doi:10.1088/1757-899X/1164/1/012007

section of the concrete element and method for the determination of torsional stiffness of normally cracked concrete element.

2. Presentation of the basic material

Consider a concrete element with normal cracks (Figure 1).

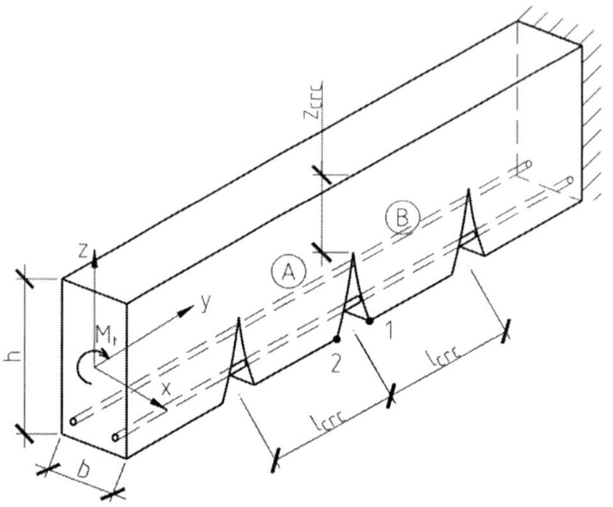

Figure 1. Model of reinforced concrete element with normal cracks subjected to torsion.

Figure 1 shows the distance between normal cracks via l_{crc} (which can be determined by any of well-known method, including the method in regulations). From block *A* to block *B* the torque is transmitted through the uncracked part of concrete with z_{crc} height and through the longitudinal reinforcing bars due to the shifting force arising in the reinforcing bars. The torsional stiffness of the element with a normal crack is calculated without any difficulty after determination of shifting force in longitudinal reinforcing bars. Scheme of the shifting forces action in longitudinal reinforcement in normal crack is in Figure 2 [1].

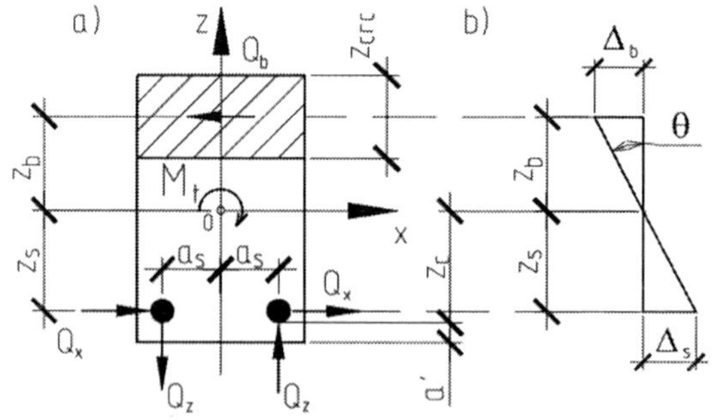

Figure 2. Scheme of forces (a) and rotation (b) in a section with a crack.

Consider separately the movements caused by local deformation. At first glance, it seems most accurate to solve this problem by modelling three-dimensional finite elements using software complexes like «Ansys», «Abacus», «Lira», etc. However, there are major obstacles in the detailed analysis. The most important of these is the correct modelling of the bond between the concrete and the anchor under the action of a perpendicular to its axis load. The reason is that when the anchor works under the transverse load, one part of concrete under the metal rod is crumpled, and the opposite part «secedes» from concrete practically without any resistance (Figure 3).

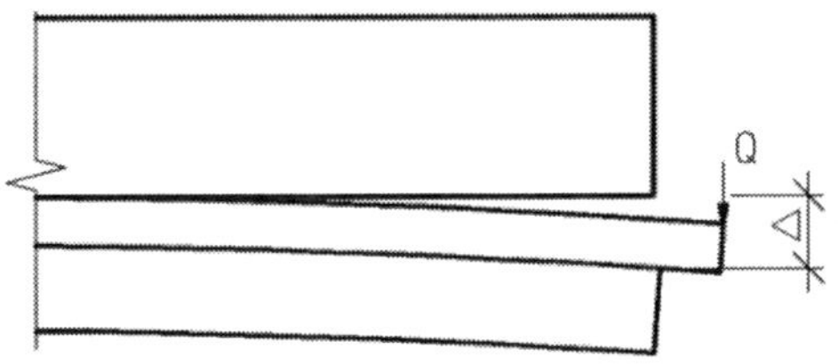

Figure 3. Deformation model of a reinforcing rod under the action of a transverse load.

Thereby, the authors of the article [1, 2] have used the empirical formula to define the movement of the anchor under the transverse load used to determine the pliability of embedded steel details [20] This formula was obtained from experimental data processing and is used to determine cross movement of a reinforcing bar subjected to a force perpendicular to its axis:

$$\Delta_{loc} = 1000\frac{Q^2}{d_s^3 E_c^3} + \frac{Q}{d_s E_c}, \tag{1}$$

where d_s and E_c – respectively reinforcing bars diameter and modulus of deformation of concrete; Q – force applied to the reinforcing bar in a direction perpendicular to its axis.

In the correlation formula (1) all the details of complex stress-strain state of the system reinforcement-concrete have been integrally taken into consideration. However, the proximity of the reinforcement to the lateral and bottom surface of the cross-section of the element has not been taken into account (see Figure 2).

From the point of view of the authors, it is correct to modeling the transverse movement of the reinforcing bar as it is shown in Figure 4.

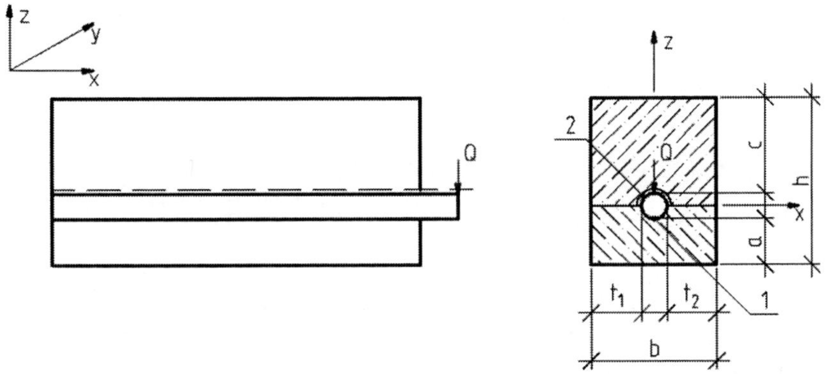

Figure 4. Model for determining the displacement of the reinforcing bar subjected to a load perpendicular to the reinforcing bar axis.

In Figure 4 the gap between the finite elements of the reinforcing bar and the finite elements of concrete is labeled «2». This is to eliminate the resistance of finite elements on the upper half of the bar diameter of the displacement of the reinforcing bar vertically under the load Q.

Vertical movement at point *1* below the reinforcing bar (along the Z axis in figure 4) depends on the thickness of the lateral protective layers t_1 and t_2, the lower protective layer a and the upper protective layer c.

It is obvious that a dependence can be obtained from a series of calculations using three-dimensional finite elements:

$$\Delta = Q \cdot f(a,t,c), \quad (2)$$

where Δ – vertical movement under the reinforcing bar.

To reduce the number of calculations it can be assumed that t_1 and t_2 are equal. This is appropriate while taking into account that basically the reinforcing bars in the cross section are disposed symmetrically (see Figure 2).

To obtain the dependency (2) you can first obtain separately dependencies of the form:

$$\Delta = f_1(a) \text{ at } t = const;\ c = const;\ \Delta = f_2(a) \text{ at } a = const;\ c = const;$$
$$\Delta = f_1(a) \text{ at } t = const;\ c = const. \quad (3)$$

Then, using these dependencies, obtain a complete dependence (2).

To obtain dependencies (3), a numerical experiment has been carried out with a modelling according to scheme in figure 4. This modeling in a well-known program «Lira» uses three-dimensional finite elements. Thus the modulus of elasticity of finite elements of concrete takes the value E_c=25 000 MPa. For any dependence (3), it is easy to switch from the basic modulus of elasticity to the modulus of elasticity of the element under considering by its multiplying by a factor equal to these elastic modules ratio.

Figure 5 shows the displacement relationship *Δ=f(a)* at C=10 mm=const, and in figure 6 at C=49 mm and fixed values t=const.

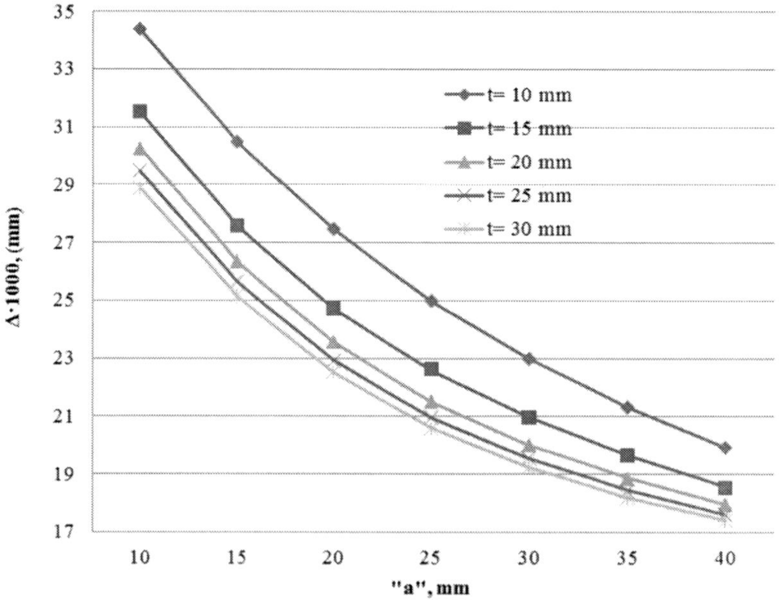

Figure 5. Displacement dependence *Δ=f(a)* at *C = 10 mm =const* and fixed values of *t=const.*

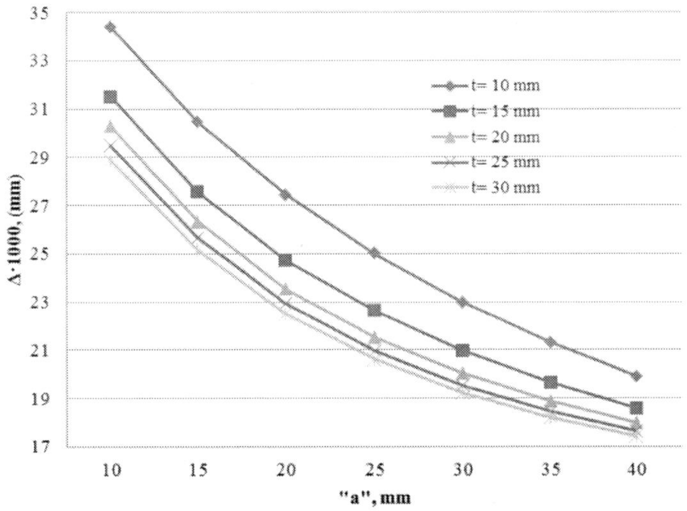

Figure 6. Displacement dependence *Δ=f(a)* at *C = 40 mm =const* and fixed values of *t=const*.

A comparison of figures 5 and 6 shows that *Δ=f(a)* is similar. However, the offset value of the end of the reinforcing bar at *C=40* mm is much bigger than the similar displacement at *C=10* mm which confirms the above-mentioned assumption about the dependence of displacement of the end of the reinforcing bar on all three sizes *a, c, t* (Figure.4).

In order to obtain the approximation dependence, a calculation has been made in the software complex Lira using the three-dimensional finite elements of the model according to Figure 4. A total of 125 analytical models have been counted. Values *t* ranged from 10 to 30 mm; values *a* from 10 to 40 mm and values *C* from 20 to 150 mm. The formula has been obtained as a result of the numerical experiment):

$$\Delta_b = (\alpha_1 C^2 + \alpha_2 C + \alpha_3) t^{\beta_1 C^2 + \beta_2 C + \beta_3} \alpha^{(\gamma_1 C^2 + \gamma_2 C + \gamma_3)t + \delta_1 C^2 + \delta_2 C + \delta_3}, \quad (4)$$

where the values of approximation coefficients are:

$$\alpha_1 = 1.34850E-02; \quad \alpha_2 = -2.61701; \quad \alpha_3 = 230.5999993$$
$$\beta_1 = -3.14300E-05; \quad \beta_2 = 0.00620; \quad \beta_3 = -0.42800$$
$$\gamma_1 = 7.94800E-07; \quad \gamma_2 = -0.00018; \quad \gamma_3 = 0.00724$$
$$\delta_1 = -3.41300E-05; \quad \delta_2 = 0.00826; \quad \delta_3 = -0.69060.$$

The maximum error of Δ_b values obtained by formula (4) compared to the values obtained by calculations using three-dimensional finite elements is 7%, thus enabling usage of formula (4) in practical calculations.

Formula (4) is obtained for base values: modulus of elasticity of concrete $E_{c,b}$=25000 MPa; modulus of elasticity of steel reinforcement E_s=200000MPa; shear force value Q_b=10 κN.

Δ value for the case is obtained on the base values using the formula:

$$\Delta = \Delta_b \frac{E_{c,b}}{E_c} \frac{Q_b}{Q}, \quad (5)$$

Knowing the displacement of the end of the reinforcing bar by the shifting force, we can determine the shear modulus of the reinforcement. This shear modulus of the reinforcement is equivalented to the shear modulus of concrete and the pliability from crumpling is taken into account. This pliability will be taken into account using the coefficient $K_{nag} < 1$.

Consider the method of determining the K_{nag} coefficient. For this purpose, consider the deformation of the cantilevered nog (reinforcing bar subjected to shear by a transverse force Q acting perpendicularly to its axis (Figure 7).

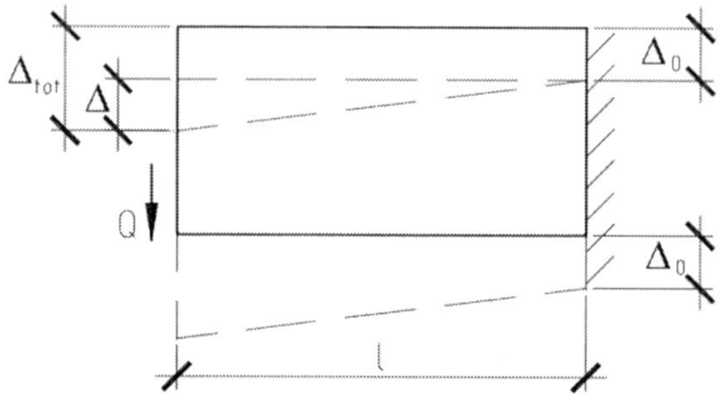

Figure 7. Scheme for determining the K_{nag} coefficient.

Δ shear displacement of an element in length l is defined by the known shear formula:

$$\Delta = \frac{Q \cdot l}{G \cdot A}, \tag{6}$$

where A – cantilever bar sectional area.

However, the reinforcing bar shifts not only under the action of shear, but also under the action of the crumpling of concrete beneath its surface at the embedment point. The displacement caused by the crumpling of concrete is marked Δ_0 in Figure 7. This displacement previously determined by the empirical formula (1), is determined by the approximation formula (4) based on the research conducted in this article.

The total displacement of the end of the bar is equal to (see Figure 7):

$$\Delta_{tot} = \Delta + \Delta_0 . \tag{7}$$

As we are considering the single width cross-section (the relative angle rotation in a normal crack is considered below) then the cantilever length l in Figure 7 should be taken equal to one.

Define the shear modulus of conditional equivalent bar in length $l=1$ as G_{ekv}. The displacement of this bar is equal to Δ_{tot} (i.e. taking into account not only the shift Δ but also the shift of support Δ_0). The displacement of the equivalent bar would then be equal to:

$$\Delta_{tot} = \frac{Q \cdot (l = 1)}{G_{ekv} \cdot A_s} . \tag{8}$$

From this it is easy to find G_{ekv}.

As we know the shear modulus of the equivalent bar, we determine the $K_{nag}<1$ coefficient:

$$K_{nag} = \frac{G_{ekv}}{G_s} . \tag{9}$$

This coefficient takes into account the displacement of the reinforcing bar due to the crumpling of concrete under its surface. As we know K_{nag}, G_{ekv}, G_s, G_c values it is easy to find the torsion center of the cross-section with a crack:

$$Z_c = \frac{Z_{crc} b(h - a^l - Z_{crc}/2) + 2A_s d_s}{Z_{crc} b + 2A_s \alpha}; \tag{10}$$

$$\alpha = \frac{G_s}{G_b} K_{nag}, \tag{11}$$

where a^I – the concrete cover (to reinforcement) (see Figure 2,a); α– the ratio of the shear modulus of the reinforcement G_s and the concrete G_b, taking into account the compliance of reinforcement in the shifting force direction.

The expression (11) differs from the generally accepted definition of the position of the center of rigidity only in the fact that the ratio of shear modulus α is multiplied by the coefficient $K_{nag}<1$. This coefficient takes into account the crumpling of concrete under the reinforcing bar when a force is applied to it, perpendicular to its axis.

When turning the cross-section about the center of rigidity its external rotation torque M_t is perceived due to the pure torsion resistance $M_{s,b}$ and to the shear resistance when the whole cross-section M_ω is turned. The moment perceived by pure torsion is determined by the formula:

$$M_{s,b} = \theta(GJ_b + 2GJ_s), \tag{12}$$

where GJ_b – torsional stiffness of concrete rectangle with sides Z_{crc} and b about the centre of mass; GJ_s – torsional stiffness of one reinforcing bar.

Moment from shear due to turning will be determined by the formula (see Figure 2):

$$M_\omega = Q_b Z_b + 2Q_x Z_x + 2Q_z a_s . \tag{13}$$

At the same time, the value of the shear force Q_b, is determined by the well-known shear formula:

$$Q_b = \Delta_b G_b A_b , \tag{14}$$

where Δ_b – the shift of a rectangle with area $A_b=Z_{crc}\cdot b$ from the Q_b. force Δb value is determined according to the scheme in Figure 2,b from the turn to θ angle:

$$\Delta_b = \theta \cdot Z_b . \tag{15}$$

Then Q_b is calculated according to the formula:

$$Q_b = \theta \cdot Z_b G_b A_b . \tag{16}$$

Forces Q_x и Q_z: are defined similarly:

$$Q_x = \theta \cdot Z_s G_s A_s ; \; Q_z = \theta \cdot a_s G_s A_s . \tag{17}$$

It should be noted that in formulas (17) the shear modulus of the reinforcement G_s must be multiplied by the described above K_{nag} coefficient.

We finally obtain the formula for the external torque by substitution values Q_b, Q_x, Q_z by (16) and (17) in the equation (13) taking into account the external moment M_t equal to the sum of moments according to (12) and (13).

$$M_t = \theta\left[GJ_b + 2\cdot GJ_s + Z_b^2 G_b A_b + 2\cdot G_s A_s (Z_s^2 + a_s^2)\right]. \tag{18}$$

All the values in square brackets in formula (18) are known. Therefore, if we know the angular turning θ it is easy to determine M_t – the part of torque perceived by the concrete part or reinforcing bar. With a known part of torque per the reinforcing bar, it is not difficult to determine the value of the shifting force Q_x and Q_z.

The problem with a different amount of longitudinal reinforcement is solved similary.

After determining the shifting forces, the mutual displacement of the faces of a normal crack is determined by formula (4), but with the known Qx and Qz values. In this case K_{nag} for determining Q_x differs from K_{nag} for determining Q_z because values c, t and a for different directions of shifting force differ. Thus, the displacements $\Delta_{loc,x}$ in the direction of the X axis and displacements $\Delta_{loc,z}$ in the direction of the Z axis will be obtained.

The mutual displacement in the crack is determined by the expression:

$$\Delta_{loc} = 2\sqrt{\Delta_{loc,x}^2 + \Delta_{loc,z}^2} \,. \tag{19}$$

The right of equation (19) is multiplied by 2 because concrete is crumpled on both sides by shifting forces.

After determining the full movement by (19) it is not difficult to determine the torsional stiffness B_{crc} of the element with normal cracks located at a distance l_{crc} from each other:

$$B_{crc} = \frac{\Delta_{bl}}{\Delta_{bl} + \Delta_{loc}} B_{t,0} \,, \tag{20}$$

where Δ_{bl} – displacement from the torsion of entire block with l_{crc} in length and overall depth of section ; $B_{t,0}$ – torsional stiffness of an element without cracks (initial torsional stiffness).

To determine the torsional stiffness using the above mentioned procedure, one has to set several values of the angular turning θ and to diagram «M_t-θ» by (18). With its help it is easy to determine external torque value M_t corresponding to the present angular turning of cross section θ.

Knowing the M_t value which corresponds to the given angle θ, with formula (18) the part of torque perceived by the shifting force Q_x *or* Q_z is determined, then the torsional stiffness of the element with a crack is determined by the above mentioned method.

The analysis of numerical calculations and comparison with the results by formula (1) allows us to make the following conclusion for proposing an approximate calculation methodology (technique). Change of displacement Δ from C parameter becomes linear after the certain value of the concrete top layer C. It is evident in the graph shown in Figure 8.

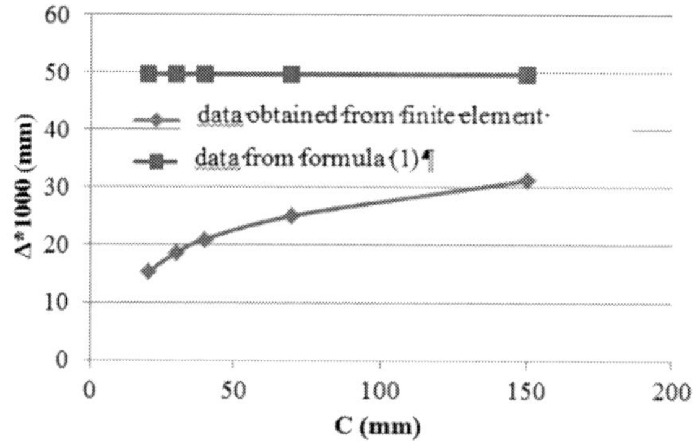

Figure 8. Dependence of reinforcing bar displacement Δ on the concrete top layer C at t=25 mm; a=25 mm: *1* – data obtained from finite element analysis (FEA); *2* – data from formula (1).

In Figure 8 the horizontal line shows the movement value determined by the formula (1) in accordance with [7]. Considering that after the C value of 70 mm the dependence in Figure 8 becomes almost linear and the also the fact that that the depth of cross section of actual beams is unlikely to be less than 100 mm, we take on a linear dependence $\Delta = f(C)$.

Mark the displacement at value $C=0$ by Δ_0, and the displacement equal to the displacement determined by (1) by Δ_∞. For practical calculations we can take approximately $\Delta_0 = \Delta_\infty/2$ taking into account the graph in Figure 8. This assumption is quite reasonable as the experimental formula (1) is derived from the test in the mass concrete (theoretically speaking at a high C value). Having the Δ values at C=70 and at C=150 and considering the linear distribution, it is not difficult to determine the value of the C value at which the Δ value is equal to the Δ_∞ value. Denote it through C. So, for the graph in Figure 8 C_∞=385 mm is obtained.

The displacement value at the given value of the concrete top layer C is determined from the obvious expression:

$$\Delta = \Delta_0 + \frac{\Delta_\infty - \Delta_0}{C_\infty} C. \tag{21}$$

Similar Δ values are easily obtained when the shifting force is directed along the upward vertical and in the horizontal direction. Thus, when the shifting force acts in the upward vertical direction, the dependence $\Delta=f(C)$ has the form shown in Figure 9.

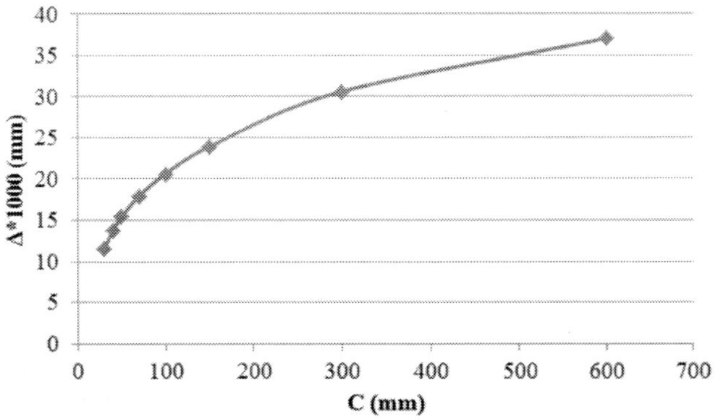

Figure 9. Dependence of reinforcing bar displacement Δ on the value of the concrete top layer C at $t=25mm$; $a=25mm$. Shifting force is directed along the upward vertical.

When the shifting force acts horizontally, the displacement Δ is essentially independent of the size of the upper protective layer and is about 0,01 mm.

Knowing the reinforcement displacement under the action of the shifting force upward vertically, downward vertically and horizontally, it is not difficult to determine K_{nag} with formulas (8) and (9) for each of these directions. Knowing the K_{nag}, it is not difficult to determine the torsional stiffness of an element with a normal crack according to the above method.

3. Conclusions and research perspectives

A new technique has been developed for determining of shifting forces in longitudinal reinforcement and torsional stiffness of reinforced concrete elements with normal cracks. An approximation formula for determining the reinforcement displacement subjected to the shifting force perpendicular to the axis of the reinforcement has been given. To determine the mutual displacement of the normal crack a flat turn of the cross-section about the torsion center has been considered. The external torque is perceived through pure torsion and shear in the longitudinal reinforcement and concrete

The parts of the external torque applied to the uncracked concrete, the horizontal and vertical components of the shifting forces in the longitudinal reinforcement are proportional to their shear and torsional stiffnesses. After determining the shifting forces in the longitudinal reinforcement, the total displacement in the crack and then the torsional stiffness of the element are determined.

Experimental testing of the proposed method is expected in the future.

References

[1] Azizov T and Kochkarev D 2020 Determination of the shifting force in the reinforcement against torsion of reinforced concrete elements with normal cracks *Sciences of Europe* **1 (58)** pp 36–40

[2] Azizov T, Jurkowska N and Kochkarev D 2019 Basis of calculation on torsion for reinforced concrete structures with normal cracks *Concrete - Innovations in Materials, Design and Structures* pp 1718–1725

[3] Azizov T N, Orlova O M and Nahaichuk O V 2019 Calculation of turning rigidity of I-section reinforced concrete elements with normal cracks *IOP Conference Series: Materials Science*

and Engineering 708

[4] Cowan H J and Armstrong S 1955 Experiments on the Strength of Reinforced and Prestressed Concrete Beams and of Concrete-Encased Steel Joints in Combined Bending and Torsion *Magazine of Concrete Research* **6** (19) pp 3–20

[5] Hsu T T C 1958 Torsion of Structural Concrete A Summary on Pure Torsion *Symposium on Tor sion* **8** pp 165–178

[6] Ilker Kalkan and Saruhan Kartal 2017 Torsional Rigidities of Reinforced Concrete Beams Subjected to Elastic Lateral Torsional Buckling. *World Academy of Science, Engineering and Technology International Journal of Civil and Environmental Engineering* **11** (7) pp 969–972

[7] Jikai Zhou, Wei Shen and Shifu Wang 2017 Experimental study on torsional behavior of FRC and ECC beams reinforced with GFRP bars *Construction and Building Materials.* **152** pp 74–81

[8] Kochkarev D, Azizov T and Galinska T 2020 Design of Effective Statically Indeterminate Reinforced Concrete Beams *Lecture Notes in Civil Engineering* **73** pp 83–93

[9] Kongjian Shen, Shui Wan, Y L Mo and Zhengwen Jiang 2018 Theoretical analysis on full torsional behavior of RC beams strengthened with FRP materials. *Composite Structures.* **183** pp 347–357

[10] Luis F A, Bernardo Catia S B Taborda and Jorge M A Andrade 2018 Generalized Softened Variable Angle Truss Model for PC Beams under Torsion *International Journal of Concrete Structures and Materials* **12**:62 https://doi.org/10.1186/s40069-018-0285-0

[11] Luis F A, Bernardo Jorge M A Andrade 2020 A unified softened truss model for RC and PC beams under torsion *Journal of Building Engineering* **32** pp 2–13

[12] Meyyada Y Alabdulhady, Lesley H Sneed and Christian Carloni 2017 Torsional behavior of RC beams strengthened with PBO-FRCM composite – An experimental study *Engineering Structures.* **136** pp 393–405

[13] Mohammed Sirage Ibrahim, Esayas Gebreyouhannes, Abdulkerim Muhdin and Abrham Gebre 2020 Effect of concrete cover on the pure torsional behavior of reinforced concrete beams. *Engineering Structures.* (**216**) Pp 2–12

[14] Gornov V N 1950 Issledovanie prochnosti i zhyostkosti sbornyh zhelezobetonnyh perekrytij iz lotkovyh nastilov *Materialy i konstrukcii v sovremennoj arhitekture* M. Strojizdat

[15] Elagin E G 1991 Raschet peremeshchenij zhelezobetonnyh sterzhnej pryamougol'nogo secheniya na stadiyah raboty s treshchinami pri sovmestnom kratkovremennom dejstvii momentov i prodol'noj sily *Stroitel'naya mekhanika i raschet sooruzhenij* **4** pp 26–31

[16] Karpenko N I 1996 *obshchie modeli mekhaniki zhelezobetona* M Strojizdat 416

[17] Kouen G Dzh 1972 *Kruchenie v obychnom i predvaritel'no napryazhennom zhelezobetone*: Per. s angl Izd-vo literatury po stroitel'stvu 104

[18] Melnyk O V 2016 *Zhorstkist ta mitsnist korobchastykh zalizobetonnykh elementiv za dii deformatsii kruchennia* Uman 116

[19] Melnyk O S 2013 *Kruchennia v zalizobetonnykh elementakh porozhnystoho trykutnoho profiliu z normalnymy trishchynamy* Uman 109

[20] *Rekomendacii po proektirovaniyu stal'nyh zakladnyh detalej dlya zhelezobetonnyh konstrukcij* 1984 [razrab. NIIZHB Gosstroya SSSR] Strojizdat 87

[21] Ulickij B E, Potapkin A A, Rudenko V I, Saharova I D and Egorushkin YU M 1967 Prostranstvennye raschyoty mostov (Moscow: Transport) 404

IOP Conf. Series: Materials Science and Engineering 1164 (2021) 012008 doi:10.1088/1757-899X/1164/1/012008

Calculation of a multi-span frame for stability

P Balduk[1,*], H Balduk[1], O Yaremenko[1] and N Yaremenko[2]

[1]Odesa State Academy of Civil Engineering and Architecture, 4, Didrihsona str., Odesa 65029, Ukraine
[2]Odesa National Maritime University

[*]E-mail: pavel9baldook@gmail.com

Abstract. The loss of stability of the first kind as applied to the longitudinal bending of the supports of a flat one-story free multispan frame under the action of vertical forces is considered. The elements of the frame have arbitrary stiffnesses constant along the length of the elements. The system of nonlinear equations of displacement method is solved. The matrix of coefficients of the system of equations is presented in block form. Formulas of generalized coefficients for each block of matrix are determined, and then one or another element of general stability equation matrix. A slight correction of generalized coefficients formulas makes it possible to extend the methodology to calculate the stability of non-free frames (overpasses). Compact notation of generalized coefficients of stability equation matrix makes it possible to considerably simplify development of algorithms and computer programs designed for solution of such tasks. The algorithm is implemented in Excel. To verify the obtained results, test calculations of stability frames for which the exact solution is known have been performed. Stability calculation of plane frame of periodic structure has been done by the suggested method. The diagram of the boundary curve has been made and the areas of frame stability and instability under the action of the given system of forces have been determined. Stability calculations of the model in the software package ANSYS, with subsequent comparison of the calculation results, have been performed.

1. Introduction
For the load-bearing capacity of engineering structures consisting of flexible compressed and compressed-curved elements, as a rule, the possibility of loss of stability of the structure as a whole or of its individual elements is a decisive factor. Multispan single-story frames are one of the most widespread of such structures. The stability of the compressed frame elements is reduced to the determination of the critical value of the longitudinal force in them and then the calculated length and flexibility of these elements. In the previously considered works, the exact way of calculating the simplest frames and approximate ways of calculating the frames of symmetric contour with certain restrictions on the applied load were used. In this paper, for the model of a free single-storey multispan frame with rigid connection at the nodes, the derivation of expressions allowing to determine precisely the critical value of an arbitrary system of nodal concentrated forces is considered. An algorithm and a computer program have been developed to determine the critical load and analyze the factors that change it.

2. Literature review
The question of determining the critical load at the loss of stability in multi-span frames has been reflected in numerous works of domestic and foreign scientists.

A separate group includes works where, as a rule, the classical displacement method is used to solve such problems [1], and the value of critical load is determined from the resulting transcendental equation, the exact solution of which is a complex mathematical problem. In this connection, approximate methods are used, in which the multi-span frame is represented by a simplified calculation scheme, in which restrictions are introduced either on the configuration or on the applied load. An approximate method for calculating free multi-span single-story frames in the case of loading

Content from this work may be used under the terms of the Creative Commons Attribution 3.0 licence. Any further distribution of this work must maintain attribution to the author(s) and the title of the work, journal citation and DOI.
Published under licence by IOP Publishing Ltd

all nodes with the same load was proposed in [2]. This method is applicable only to the frames of symmetric contour having the same linear stiffness of ledgers of the same level and is reduced to the replacement of multi-span frame by equally stable one-span frame. In [3] there was used an approximate method of designing frames of symmetrical contour proposed earlier by N.V. Kornoukhov, but with asymmetrical distribution of stiffness of the columns of the given storey and with asymmetrical loading of its nodal load. A study of the stability [3] of multi-span non-free frames made it possible to develop an approximate method based on changing the design scheme of such frames by splitting them into separate T-frames.

A large series of works are devoted to the development of engineering methods of calculation for the stability of frame structures, the application of which is possible only with the use of modern software computing complexes. These are works [8-13] and many others. To determine the critical force and the reduced length of columns in one-story frames, the method of stiffness distribution [4] has been developed, but its contribution of adjacent elements is taken into account indirectly by summing the stiffnesses of elements in the upper and lower parts of the structure frame. The work [4] forms the theoretical basis for the calculation of stability loss in BS EN 1993 [5].

To assess the stability of multi-storey frames, the method [6] offers a modification of the reduced length ratio, which allows you to take into account the load in the longitudinal bending of neighboring columns.

The EBCS 2: 1995 documents introduce the concept of a replacement frame to determine the floor longitudinal bending load.

In the paper [7], when designing for the stability of thin reinforced concrete columns in free frames, correction of the moment of inertia of the column cross section is used. However, this approach gives a significant error in the design of irregular frame frames. In [8, 9], a method of engineering design of spatial frame frames for stability with regard to physical nonlinearity is given. In the paper [10] an engineering method of calculation of frame elements of variable height was developed, which allowed to simplify the test of such rods for stability without using for this purpose modern computing complexes. The study [11] presents an approximate method in which the entire frame structure is idealized as an equivalent multilayer beam that takes into account all deformations, while using the continuum approach and the transfer matrix method to analyze the lateral stability of buildings. In [12], an improved method for simplified stability analysis of free and non-free frames, taking into account the effects of vertical interaction of the columns, is proposed. The stability matrix for a one-story block is derived, and then the system stability matrix is obtained taking into account the compatibility of the floor blocks. In [13], the issue of determining the optimum ratio of frame frame columns for stability is considered and an engineering method for determining the critical force is proposed.

3. Research aim and objectives

The purpose of this paper is to develop an approach for determining the critical value of any system of concentrated forces acting on a flat free single-story multi-span frame with rigid nodes.

4. Research results

Let's consider the calculation for stability by displacement method of a flat one-story multispan frame, the model of which is shown in Figure 1. We consider the loss of stability of the first kind as applied to the longitudinal bending of the columns. A system of vertical forces acts on the frame in the nodes. In general case, the elements of the frame have arbitrary geometrical and physical-mechanical characteristics constant along the length of the elements.

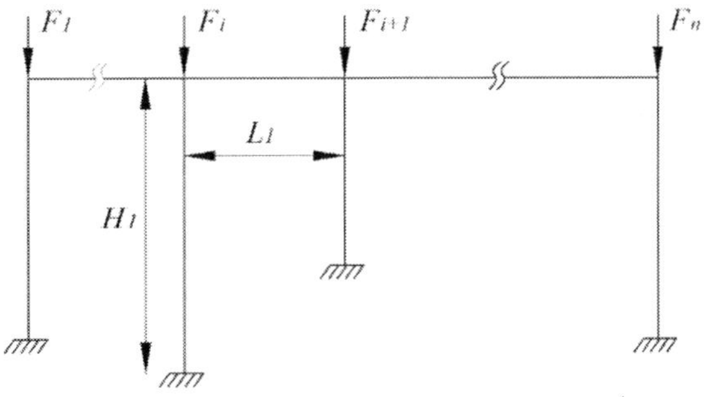

Figure 1. Flat one-story multi-span frame.

Since the main unknowns of the displacement method are the angular displacements of the rigid nodes and their independent linear displacements, this frame model has the number of unknowns of the displacement method equal to n+1, where n – number of posts in the frame. The basic system of the model displacement method is shown in Figure 2.

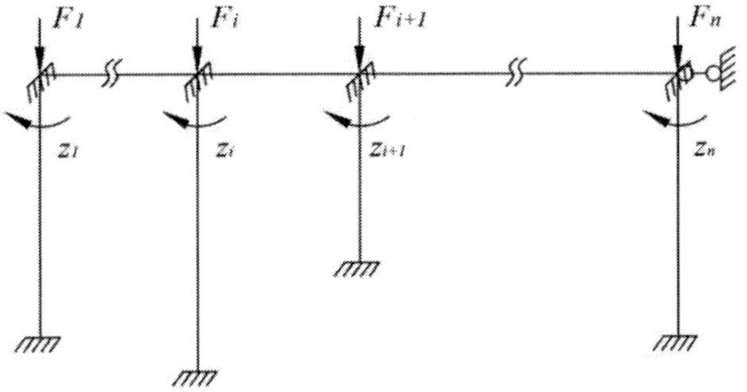

Figure 2. Basic system of the displacement method.

The system of canonical equations of the displacement method in matrix form is:

$$\begin{vmatrix} r_{1,1}(v) & r_{1,2}(v) & \dots & r_{1,n}(v) & r_{1,n+1}(v) \\ r_{2,1}(v) & r_{2,2}(v) & \dots & r_{2,n}(v) & r_{2,n+1}(v) \\ \dots & \dots & \dots & \dots & \dots \\ r_{n,1}(v) & r_{n,2}(v) & \dots & r_{n,n}(v) & r_{n,n+1}(v) \\ r_{n+1,1}(v) & r_{n+1,2}(v) & \dots & r_{n+1,n}(v) & r_{n+1,n+1}(v) \end{vmatrix} \begin{bmatrix} Z_1 \\ Z_2 \\ \dots \\ Z_n \\ Z_{n+1} \end{bmatrix} = - \begin{bmatrix} R_1(v) \\ R_2(v) \\ \dots \\ R_n(v) \\ R_{n+1}(v) \end{bmatrix} \quad (1)$$

The applied forces are related to each other by a certain ratio and change in proportion to one parameter F_0:

$$F_1 = f_1 \cdot F_0; \dots F_2 = f_2 \cdot F_o; \dots F_n = f_n \cdot F_0. \quad (2)$$

Nodal load application leads to the fact that the load coefficients $R_i(v)$ are equal zero. Then the system of equations (1) has the form:

$$\begin{vmatrix} r_{1,1}(\nu) & r_{1,2}(\nu) & \dots & r_{1,n}(\nu) & r_{1,n+1}(\nu) \\ r_{2,1}(\nu) & r_{2,2}(\nu) & \dots & r_{2,n}(\nu) & r_{2,n+1}(\nu) \\ \dots & \dots & \dots & \dots & \dots \\ r_{n,1}(\nu) & r_{n,2}(\nu) & \dots & r_{n,n}(\nu) & r_{n,n+1}(\nu) \\ r_{n+1,1}(\nu) & r_{n+1,2}(\nu) & \dots & r_{n+1,n}(\nu) & r_{n+1,n+1}(\nu) \end{vmatrix} \begin{bmatrix} Z_1 \\ Z_2 \\ \dots \\ Z_n \\ Z_{n+1} \end{bmatrix} = 0. \quad (3)$$

Determination of the critical load consists in finding the roots of the general stability equation, which we obtain by equating the determinant of the coefficient matrix (3) to zero, provided that $Z_i \neq 0$:

$$D(\nu) = \begin{vmatrix} r_{1,1}(\nu) & r_{1,2}(\nu) & \dots & r_{1,n}(\nu) & r_{1,n+1}(\nu) \\ r_{2,1}(\nu) & r_{2,2}(\nu) & \dots & r_{2,n}(\nu) & r_{2,n+1}(\nu) \\ \dots & \dots & \dots & \dots & \dots \\ r_{n,1}(\nu) & r_{n,2}(\nu) & \dots & r_{n,n}(\nu) & r_{n,n+1}(\nu) \\ r_{n+1,1}(\nu) & r_{n+1,2}(\nu) & \dots & r_{n+1,n}(\nu) & r_{n+1,n+1}(\nu) \end{vmatrix} = 0. \quad (4)$$

By determining the parameters $\nu_{i,kp}$, let's find the values of the critical forces $F_{i,kp}$

$$F_{i,kp} = \frac{\nu_{i,kp}^2 EI_{Ci}}{H_i^2}. \quad (5)$$

Let us construct the unit diagrams of the bending moments in the basic system of the displacement method (Figure 3 a, b). The diagrams are constructed taking into account the longitudinal bending from the action of the system of forces, the effect of compressive forces is taken into account using the corresponding nonlinear functions $\varphi_2(\nu_i)$, $\varphi_4(\nu_i)$, $\eta_2(\nu_i)$:

$$\varphi_2(\nu_i) = \frac{\nu_i \cdot (tg\nu_i - \nu_i)}{8tg\nu_i \cdot \left(tg\frac{\nu_i}{2} - \frac{\nu_i}{2}\right)}; \quad \varphi_4(\nu_i) = \frac{\left(\frac{\nu_i}{2}\right)^2 tg\frac{\nu_i}{2}}{3\left(tg\frac{\nu_i}{2} - \frac{\nu_i}{2}\right)}; \quad \eta_2(\nu_i) = \frac{\left(\frac{\nu_i}{2}\right)^3}{3\left(tg\frac{\nu_i}{2} - \frac{\nu_i}{2}\right)},$$

where $\nu_i = \alpha_i H_i = H_i\sqrt{F_i/EI_{Ci}}$.

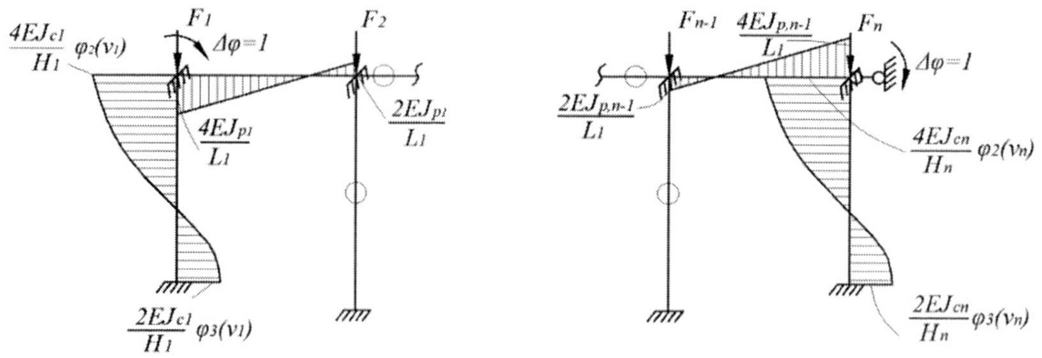

Figure 3, a. Unit bending moment diagrams.

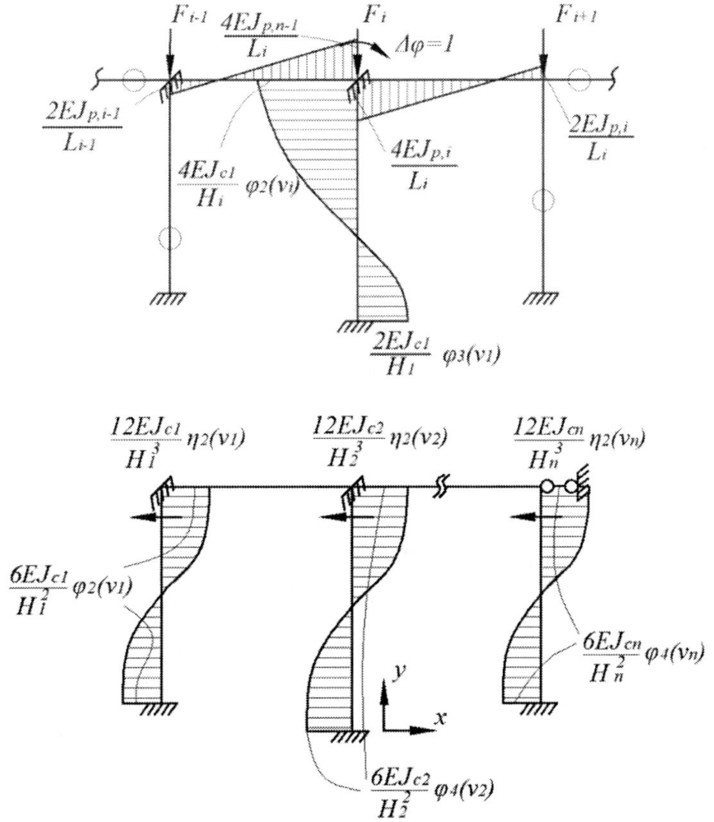

Figure 3, b. Unit bending moment diagrams.

For the considered frame model the symmetric matrix of coefficients of equation (4) with dimensions $(n+1)\times(n+1)$ can be represented in block form:

$$D(v)=\left\|\begin{array}{cccc|c} r_{1,1}(v) & r_{1,2}(v) & \dots\dots\dots & r_{1,n}(v) & r_{1,n+1}(v) \\ r_{2,1}(v) & r_{2,2}(v) & \dots\dots\dots & r_{2,n}(v) & r_{2,n+1}(v) \\ \dots & \dots & \dots\dots\dots & \dots & \dots \\ r_{n,1}(v) & r_{n,2}(v) & \dots & r_{n,n}(v) & r_{n,n+1}(v) \\ \hline r_{n+1,1}(v) & r_{n+1,2}(v) & \dots & r_{n+1,n}(v) & r_{n+1,n+1}(v) \end{array}\right\|=0, \tag{6}$$

where

block I – symmetric ribbon matrix with dimensions $n\times n$;

block II – column with size n;

block III – transposed block II;

block VI – consists of one element.

The physical meaning of elements of I and II blocks - reactive moments in superimposed rigid links, elements of III and VI blocks - reactions in superimposed linear links.

Let us introduce the following coefficients of variation of model characteristics:
h_i – in the length of the columns; s_i – on the value of the bending stiffness of the columns; k_i – by the length of the ledgers; p_i – on the value of the bending stiffness of the ledgers; f_i – by force value.

With this in mind, we can write:

$$H_i = H_0 h_i;\ EI_{Ci} = EI_{C_0} s_i;\ L_i = L_0 k_i;\ EI_{Pi} = EI_{P_0} p_i;\ F_i = F_0 f_i\,. \tag{7}$$

In accordance with the known method of static finding of the displacement method coefficients and introduced variation coefficients, let us determine the form of the record of the generalized coefficients of these blocks $(1 \le i \le n)$:

$$r_{i,i} = \frac{4EI_{Ci}}{H_i}\varphi_2(v_i) + \frac{4EI_{P(i-1)}}{L_{i-1}} + \frac{4EI_{Pi}}{L_i} = \frac{4EI_{Co}}{H_o}\cdot\frac{s_i}{h_i}\varphi_2(v_i) + \\ + \frac{4EI_{Po}}{L_o}\cdot\frac{p_{i-1}}{k_{i-1}} + + \frac{4EI_{Po}}{L_o}\cdot\frac{p_i}{k_i} = a\cdot\frac{s_i}{h_i}\varphi_2(v_i) + b\cdot\frac{p_{i-1}}{k_{i-1}} + b\cdot\frac{p_i}{k_i}. \tag{8}$$

$$r_{i,i-1} = \frac{2EI_{Pi}}{L_i} = \frac{2EI_{P1}p_i}{L_1 k_i} = \frac{b}{2}\cdot\frac{p_i}{k_i}. \tag{9}$$

$$r_{i,n+1} = -\frac{6EI_{C_i}}{H_i^2}\varphi_4(v_i) = -\frac{6EI_{C_0}}{H_0^2}\frac{s_i}{h_i^2}\varphi_4(v_i) = d\cdot\frac{s_i}{h_i^2}\varphi_4(v_i). \tag{10}$$

$$r_{n+1,n+1} = \frac{12EI_{C1}}{H_1^3}\eta_2(v_1) + \frac{12EI_{C2}}{H_2^3}\eta_2(v_2) + \ldots + \frac{12EI_{Cn}}{H_n^3}\eta_2(v_n) = \\ = \frac{12EI_{C0}}{H_0^3}\cdot\left[\frac{s_1}{h_1^3}\eta_2(v_1) + \frac{s_2}{h_2^3}\eta_2(v_2) + \ldots + \frac{s_n}{h_n^3}\eta_2(v_n)\right] = \\ = c\cdot\left[\frac{s_1}{h_1^3}\eta_2(v_1) + \frac{s_2}{h_2^3}\eta_2(v_2) + \ldots + \frac{s_n}{h_n^3}\eta_2(v_n)\right]. \tag{11}$$

In formulas (8) ÷ (11) the following symbols are used:

$$a = \frac{4EI_{C_0}}{H_0};\ b = \frac{4EI_{P_0}}{L_0};\ c = \frac{12EI_{C_0}}{H_0^3};\ d = \frac{6EI_{C_0}}{H_0^2}. \tag{12}$$

Formulas of 4 generalized coefficients (8) ÷ (11), allow obtaining the entry of any of the elements of block matrices of the general stability equation.

A slight correction of the terms in formulas (8) ÷ (11) will allow to use the proposed expressions when forming the matrix of coefficients for non-free frames (Figure 4) (as a rule, these are overpass structures) which models are a particular case of free frame model (Figure 1).

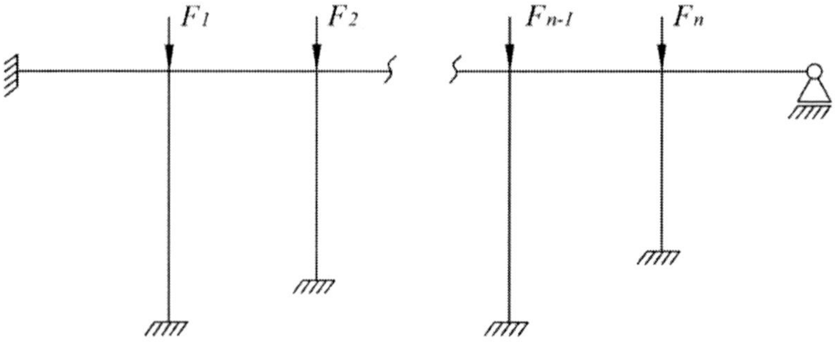

Figure 4. Particular cases of the frame calculation model.

All nonlinear parameters v_i of functions $\varphi_2(v_i)$, $\varphi_4(v_i)$, $\eta_2(v_i)$, included in the general stability equation (4) are reduced to a single argument v_0:

$$\frac{v_i}{v_0} = \frac{\alpha_i H_i}{\alpha_0 H_0} = \frac{H_i}{H_0}\sqrt{\frac{F_i}{EI_{Ci}}\frac{EI_{C0}}{F_0}} = \frac{H_0 h_i}{H_0}\sqrt{\frac{F_0 f_i}{EI_{C0} s_i}\frac{EI_{C0}}{F_0}}$$

$$v_i = v_0 h_i \sqrt{\frac{f_i}{s_i}};\ v_i = H_0 \sqrt{\frac{F_0}{EI_{C0}}}. \tag{13}$$

Then all coefficients $r_{i,k}(v)$ of the determinant (4) will be functions of only one parameter v_0:

$$r_{i,k}(v) = \Phi_{i,k}(v_0) \tag{14}$$

and equation (4) will have the following form:

$$D(v_0) = \left|\Phi_{i,k}(v_0)\right| = 0\,. \tag{15}$$

The critical state equation of the system (15) is a transcendental equation. Such an equation can be solved only by fitting or the method of successive approximations. The smallest positive value v_0 determines the critical value of any of the concentrated forces of the system. The exact solution of the equation at $n > 2$ is a complex mathematical problem, the solution of which is possible only with the use of computer technology.

It is not always reasonable to use universal computational complexes, such as ANSYS, LIRA and others, for many reasons. The obtained compact form of generalized coefficients of the general stability equation matrix allows to facilitate significantly the creation of algorithms and writing of computer programs intended for solving such problems.

For the numerical realization of the obtained expressions a program was written in Excel tables, the choice of which is due to their availability.

The resulting program makes it easy to analyze the effect of changes in geometric and physical-mechanical parameters of model elements on the values of critical forces of the system.

To verify the calculation results, a number of test frame stability calculations with exact or approximate solution has been performed [2; 8].

Stability calculation of plane frame of periodic structure has also been done. Frames of such type are frameworks of widely spread building structures (or their fragments).

The model of the considered frame has a constant height at six identical spans (Figure 5), for it the number of unknown methods of displacements is eight.

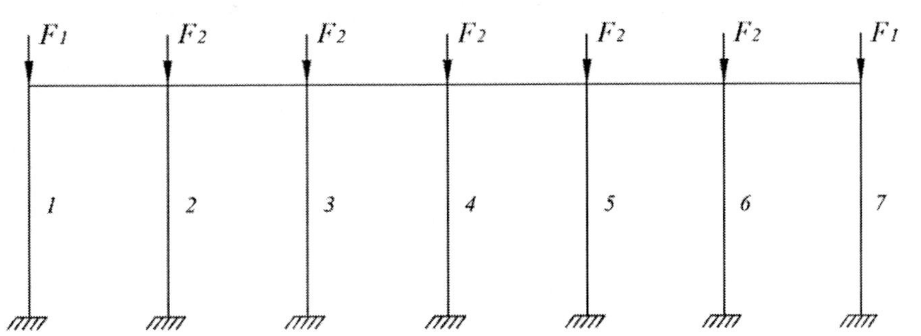

Figure 5. Frame model with six equal spans.

For this frame the following initial data are taken: all columns (6 m height) and all beams (12 m span) are made of steel with the modulus of elasticity $E = 2 \cdot 10^{11}$ *Pa*. Columns have a rectangular cross-section (*a=0,2 m*; *в=0,5 m*), the moment of inertia of the cross section of the ledger is 2.5 times greater than the moment of inertia of the cross section of the column. The following model parameters are set: $H_1 = H_0 = 6м$; $EI_{C1} = EI_{C0}$; $L_1 = L_0 = 12м$; $EI_{P1} = 2.5EI_{P0}$; $F_1 = F_0$; $F_2 = F_0 f_2$.

Calculations to determine the critical force at values of the force variation coefficient $f_2 - 0{,}5;\ 1{,}0;\ 2{,}0$. The partial values of the forces are also determined $F_{1,kp}$; $F_{2,kp}$. The results of the calculations are summarized in the table 1.

Table 1. Critical force values.

№		F_1/EI_{CO}	F_2/EI_{CO}
1	$F_{1,kp}$	0,769	0
2	$f_2 = 0{,}5$	0,363	0,181
3	$f_2 = 1$	0,233	0,233
4	$f_2 = 2$	0,135	0,271
5	$F_{2,kp}$	0	0,322

On the basis of the obtained results, the stability boundary curve plot (Figure 6) was plotted and the stability and instability regions of the frame under the action of a given system of forces were determined F_1 and F_2.

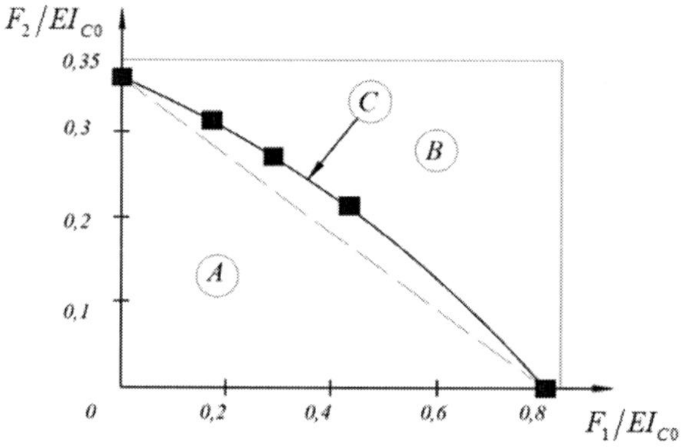

Figure 6. Boundary curve plot: A – stability region; B – instability region; C – boundary curve.

To verify the obtained results, the model was calculated for stability in the ANSYS software package [9].

The critical loads corresponding to the first two forms of stability loss were obtained (Figure 7). According to calculations, the first form of stability loss corresponds to the critical force $F_1 = 5638$ kN; the second form of loss of stability - the critical force $F_2 = 17402$ kN.

From the author's calculation the results obtained are practically the same.

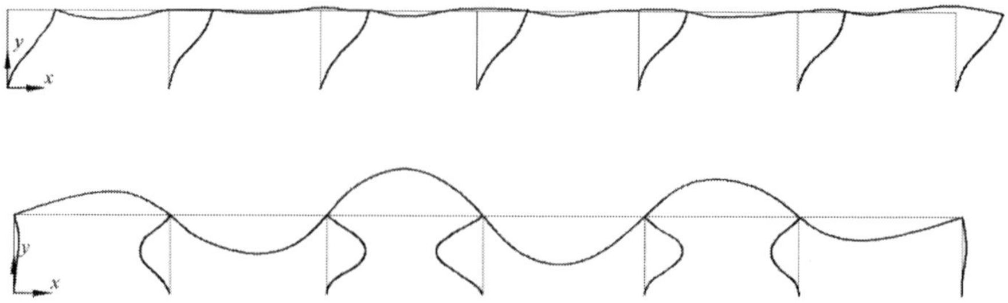

Figure 7. The first and second forms of frame stability loss

5. Conclusions

The procedure for determining the critical forces in the problem of calculating the stability of a flat multi-span frame by the displacement method has been considered. It has been found that the structure of the matrix of coefficients of the general stability equation with n unknowns for the considered frames has a block structure. A technique for obtaining four generalized coefficients which allows one to easily obtain any of the elements of the block matrices of the general stability equation has been proposed. Numerical implementation of the developed algorithm has been performed in Excel spreadsheets. Verification of the algorithm was carried out by calculations of test examples. Critical forces for a six-span model of a frame of regular structure are determined. A boundary curve is plotted and the stability zone for this model is determined. Comparison of the results of calculations of the first two critical forces using the author's method with the results of numerical analysis in the ANSYS software package showed their complete coincidence (discrepancy is 0.1 %).

References

[1] Smyrnov A, Aleksandrov A, Lashchenykov B and Shaposhnykov N 1984 *Stroitelnaia mekhanika. Dinamika i ustojchivost sooruzhenij* (M.: Stroiyzdat) 416

[2] Snytko N K 1968 *Ustojchivost sterzhnevykh sistem v upruho-plastycheskoj oblasti* (M.: Stroiyzdat) 250

[3] Darkov A V and Shaposhnykov N N 2008 *Stroitelnaia mekhanika* (S.-P.: Lan) 650

[4] Wood R H 1974 Effective lengths of columns in multi-storey buildings Parts 1, 2, 3 *Struct. Engr 52* (7, 8, 9) pp 235–244 295–302 341–346.

[5] BSI 1993-1 2005 *Design of steel structures general rules and rules for buildings* (London BSI)

[6] Webber A Orr J J Shepherd P Crothers K 2015 The effective length of columns in multi-storey frames *Engineering Structures* **102** pp 132–143.

[7] Ewnetie A and Zerayohannes G 2015 Investigation on applicability of substitute beam - Column frame for design of reinforced concrete sway Frames *Journal of EEA* **33** pp 13-25.

[8] Son M P 2010 *Eksperimentalno-teoreticheskoe issledovanie ustojchivosti prostranstvennykh ramnykh sistem i razrabotka inzhenernoj metodiki opredeleniya kriticheskoj sily s uchetom nelinejnosti* Diss. na sois. uch. step. kand. tekh. nauk 149

[9] Son M P, Kaiumov R A and Mukhamedova Y Z 2015 Issledovanie ustojchivosty prostranstvennykh ramnykh sistem *Vestnyk Kazanskoho tekhnolohicheskoho unyversiteta* **18** (17) pp 203-206 https://cyberleninka.ru/article/n/issledovanie-ustoychivosti-prostranstvennyh-ramnyh-sistem

[10] Askynazy V Iu 2017 *Prostranstvennaya ustojchivost elementov stalnykh ramnykh konstruktsij peremennoj zhestkosti* Diss. na sois. uch. step. kand. tekh. nauk 199

[11] Kanat Burak Bozdoran Duygu Ozturk 2010 An approximate method for lateral stability analysis of wall-frame buildings including shear deformations of walls Sadhana *Indian Academy of Sciences* **35** (3) pp 241–253 https://www.researchgate.net/publication/226922269_An_approximate_method_for_lateral_stability_analysis_of_wall-frame_buildings_including_shear_deformations_of_walls

[12] Quanwang L, Aming Z and Hao Z 2016 A simplified method for stability analysis of multi-story frames considering vertical interactions between stories *Journal of Structural Engineering* **19** (4) pp 599-610 ttps://journals.sagepub.com/doi/abs/10.1177/1369433216630191

[13] Repina O A 2013 Opredelenie optimalnoho sootnosheniya kolonn pri raschete prostranstvennoho ramnoho karkasa na ustojchivost *SWorld* pp 17-26 https://www.sworld.com.ua/konfer33/697

[14] Basov K A 2005 *ANSYS Spravochnyk polzovatelya* DMK Press 640

Nonlinear problems of equilibrium for axisymmetric membranes

A Bedov[1], R Vagapov[2], A Gabitov[2,*] and A Salov[2]

[1]Moscow National Research State University of Civil Engineering, 26, Yaroslavskoye Shosse, Moscow, 129337, Russia
[2]Ufa State Petroleum Technological University, 1 Kosmonavtov St., Ufa, 450064, Russia

*E-mail: gabitov.azat@mail.ru

Abstract. The problems for equilibrium of circular and annular membranes made of elastic-orthotropic and elastoplastic isotropic materials are considered in this paper in a geometrically and physically nonlinear formulation meeting the Föppl-Karman assumptions. The outer contour is considered to be stationary, while the inner one being free, or reinforced by either an elastic ring or a fixed disk. The load is assumed to be uniformly distributed over the membrane surface.

The results of calculations of circular and annular membranes with large deflections in comparison with thickness, deformations and squares of angular displacement to be comparable with each other but small in comparison with the unit are presented in the first part thereof. The resulting boundary value problems are reduced to the Cauchy problem to be numerically integrated by the shooting method under the iterative Steffensen algorithm. For circular membrane structures an analytical solution as a power series similar to the Prescott series for a circular membrane is assumed either.

The results of numerical solutions of problems are presented as dimensionless characteristic functions. Analysis of this information will enable to solve the problem of the support, being optimal in a certain sense when the membrane is close to be equal in strength.

Problems of elastoplastic equilibrium of isotropic membranes are presented in the second part thereof. The idealized Prandtl diagram is used as the main characteristic of the mechanical properties of the membrane material. The Tresca-Sant-Venant and Huber-Mises-Hencky conditions were used as criteria for the transition of a material from an elastic state to a plastic one. Solutions of physically nonlinear equilibrium problems for axisymmetric membranes are also presented as dimensionless characteristic functions.

1. Relevance

The results of calculating the problems of equilibrium of circular and annular membranes with greater deflections as compared with thickness of membrane, deformations and squares of angular displacement to be comparable with each other but small in comparison with the unit are presented herein. Quite a very large number of papers are devoted to such problems but we do not focus thereon referring the reader to the detailed review contained in [1, 2, 3]. Main attention was paid by the researchers to the solution of equilibrium problems for isotropic membranes in a geometrically nonlinear formulation meeting the well-known Föppl assumptions. The main objective of the first part hereof is to study under the same assumptions the effect of elastic orthotropy of the membrane material on the stress-strain behavior. While the second part of the paper specifies the solutions of several problems of elastoplastic equilibrium of axisymmetric isotropic membranes on the basis of the deformation theory of plasticity. When setting these solutions the calculation methods proposed in [4, 5, 6, 7] were used.

Content from this work may be used under the terms of the Creative Commons Attribution 3.0 licence. Any further distribution of this work must maintain attribution to the author(s) and the title of the work, journal citation and DOI.
Published under licence by IOP Publishing Ltd

2. Calculation of circular and annular membranes

We consider the problem of elastic equilibrium of an orthotropic membrane under a uniform load q. Let us refer the membrane to the cylindrical coordinate system rΘz (Figure 1). The dependences between membrane deformations and displacements are described by equations:

$$\varepsilon_\Theta = \frac{u}{r},\ \varepsilon_r = \frac{du}{dr} + \frac{1}{2}\left(\frac{dw}{dr}\right)^2. \tag{1}$$

Here, *r*, *u*, *w* are the radial coordinate, the corresponding displacement and deflection of the membrane related to the R_1-radius of the outer contour of the membrane. For the case of polar symmetric orthotropy, the dependences between deformations and dimensionless strains are represented as

$$\varepsilon_\Theta = \frac{n_\Theta}{\lambda} - \nu n_r,\ \varepsilon_r = -\nu n_\Theta + n_r. \tag{2}$$

Here, $n_\Theta = N_\Theta / E_r H$; $n_r = H_r / E_r H$; $\lambda = E_\Theta / E_r$; $\nu = \nu_{\Theta r} = \nu_{r\Theta}\,(E_r/E_\Theta)$ where, N_θ and N_r are strains, H is a membrane thickness, E_Θ and E_r are elastic modulus of the materials thereof in both annular and radial directions, $\nu_{\theta r}$ and $\nu_{r\Theta}$ are Poisson's ratios. The equilibrium equations for a membrane element may be presented as

$$n_\Theta = \frac{d}{dr}(n_r r),\ \frac{d}{dr}(n_r r \sin\varphi) = 2Q_r, \tag{3}$$

where φ is an angle between the z-axis and normal to the deformed surface (Figure 1); Q is dimensionless load parameter determined by $Q = qR_1/2E_rH$ Each equation from table 1 depending on geometric scheme of the membrane and kind of load case may be easily reduced to one of the equations (4) or (5). The given table is covering all kinds of problems solved herein.

The new variables are used in the table 1: $y = Bn_r r^2$, $x = 1/2r^2$. Here, B is an arbitrary constant not yet defined, introduced to give both the resolving equation and the subsequent solution of greater generality. In addition, the following notations have been introduced: $\alpha = 1/4(\lambda\text{-}1)$, $\beta = (E_rH/E_cA_c)R_1$ where A_c is the sectional area of the ring supporting the membrane, and E_C is the elastic modulus of the material thereof. The values corresponding to the inner and outer contours are marked by indices 0 and 1, respectively.

Equations (4) and (5) under the corresponding boundary conditions are integrated numerically. Therewith, the boundary value problem is reduced to the Cauchy problem with the use of the shooting method based on the iterative Steffensen algorithm [8, 9, 10].

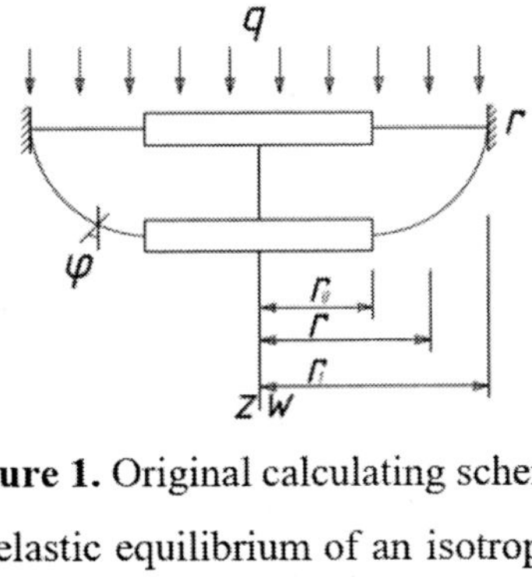

Figure 1. Original calculating scheme.

The solution to the problem of elastic equilibrium of an isotropic ($\alpha = 0$) of the uniformly loaded circular membrane under large deflections (4) may also be expressed as infinite power series proposed by D. Prescott [8]:

$$y=\sum_{n=1}^{\infty}a_n x^n \ .$$

Table 1. Kinds of problems solved.

Calculated schemes	Resolving equations	Boundary conditions
a 		$x=0:y=0$ $x=x_1:2y_1'-(1+\nu\lambda)\frac{y_1}{x_1}=0$
b 	$y''-\alpha\frac{y}{x^2}+\frac{x^2}{y^2}=0$ (4)	$x=x_0:2y_0'-(1+\nu\lambda)\frac{y_0}{x_0}=0$ $x=x_1:2y_1'-(1+\nu\lambda)\frac{y_1}{x_1}=0$
c 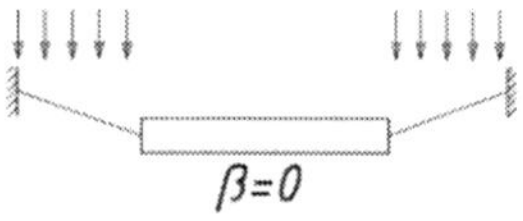		$x=x_0:2y_0'-(1+\nu\lambda)\frac{y_0}{x_0}=0$ $x=x_1:2y_1'-(1+\nu\lambda)\frac{y_1}{x_1}=0$
d 	$y''-\alpha\frac{y}{x^2}-\frac{(x-x_0)^2}{y^2}=0$ (5)	$x=x_0:2y_0'-(1+\nu\lambda)\frac{y_0}{x_0}=\beta\frac{y_0}{(2x_0)^{1/2}}$ $x=x_1:2y_1'-(1+\nu\lambda)\frac{y_1}{x_1}=0$
e 		$x=x_0:y=y_0$ $x=x_1:2y_1'-(1+\nu\lambda)\frac{y_1}{x_1}=0$

In problems of annular membranes equilibrium (5) a similar method may be applied, with the same series therewith being formally obtained, but relating to a new argument:

$$y=\sum_{n=1}^{\infty}a_n(x-x_0)^n \ .$$

An expression for the general term of a similar series is obtained:

$$a_n x^n=-\frac{x^n}{n(n-1)a_1^2}\sum_{k=n-1}^{k=2}[k(k-1)a_k\sum_{i=1,\,j=n-k+1}^{i=n-k+1,\,j=1}a_j a_i]\ . \tag{6}$$

This expression enables to get values a_n starting with $n=3$. The first coefficient of the series is found from the boundary conditions of the problem, and $a_2=-1/2a_1^2$. Reasonable good convergence (within 3%) of analytical and numerical solutions for the problems of elastic equilibrium of annular membranes is achieved when 10 terms of the series are taken into account.

Below are the results of the numerical solution of specific problems (for membranes with the inner contour radius of $r_0=0.5$) as characteristic functions:

$$f_1 = \frac{n_\Theta}{Q^{\frac{2}{3}}} = \left(\frac{\lambda}{4}\right)^{\frac{1}{3}} \left[2y' - \frac{y}{x}\right],\ f_2 = \frac{n_r}{Q^{\frac{2}{3}}} = \left(\frac{\lambda}{4}\right)^{\frac{1}{3}} \frac{y}{x};$$

$$f_4 = \frac{w}{Q^{\frac{1}{3}}} = \left(\frac{4}{\lambda}\right)^{\frac{1}{3}} \int_x^{0.5} \frac{(x - x_0)}{y} dx, \tag{7}$$

$$f_3 = \frac{u}{Q^{\frac{2}{3}}} = \left(\frac{1}{2\lambda}\right)^{\frac{2}{3}} \left[2y' - \frac{(1+\nu\lambda)y}{x}\right](2x)^{\frac{1}{2}}.$$

The analysis of solving all the problems indicated in Table 1 also enabled to solve the problem of the support, in a certain sense the optimal support, when the membrane is close to being equal in strength. Maximum radial ($\beta = 0$) and ($\beta = \infty$) tangential stresses correspond to extremes c and e, in both cases resulting in large values of stress intensity. It should also be noted that for these cases we have the largest rangeability in the stress intensity during the transition from the inner contour of the membrane to the outer one. All the above, as well as the fact that the stress functions in both cases are increasing from the outer contour to the inner one, made it possible to take the equality of radial tangential stresses on the inner contour of the membrane as the criterion of "optimality". As a result, an analytical dependence was obtained to determine the optimal supporting parameter that for the case of an isotropic annular membrane coincides with the expression obtained in [4, 7, 9] with a similar criterion of "optimality":

$$\beta = \frac{1 - \nu\lambda}{\lambda r_0}. \tag{8}$$

For membranes made of materials with strong orthotropy of mechanical properties ($\nu\lambda > 1$), the solution of problems using this criterion is possible only with additional radial load case of the ring.

Under the calculations it was assumed $\lambda = 3.152$; $\nu\lambda = 1.18$ [4]. Characteristic functions for an isotropic membrane (dotted line) made of material with Poisson's ratio $\nu = 0.37$ are also shown in Figure 2-4 to demonstrate the effect of anisotropy [6, 11].

Let us consider elastoplastic problems of equilibrium of isotropic membranes under a uniform load. As the main characteristic of the mechanical properties of the membrane material, the idealized Prandtl diagram is used, thereby enabling to apply solutions for membranes made of materials with a very small strain hardening modulus (steel, aluminum and alloys thereof) [7].

The biggest modulo constant shear stress condition (the Tresca-Sant-Venant condition) may be used as a criterion for the transition of a material from an elastic state to a plastic one, to be written as:

$$\max\left(|n_r|, |n_\Theta|, |n_r - n_\Theta|\right) = n_s = \frac{\sigma_s}{E} \tag{9}$$

or constant shear stress intensity condition (the Huber-Mises-Hencky condition), which may be written as

$$n_r^2 + n_\Theta^2 - n_r n_\Theta = n_s^2. \tag{10}$$

Here, σ_s is the yield point of the membrane material. When solving the elastoplastic problem for a uniformly loaded solid membrane, both conditions coincide and are simplified to

$$n_r = n_\Theta = n_s, \tag{11}$$

which is proved using the first of the equilibrium equations (3) and the condition of equality of stresses in the center of the membrane [5]. This plasticity condition (11) is simultaneously a solution

IOP Conf. Series: Materials Science and Engineering 1164 (2021) 012009 doi:10.1088/1757-899X/1164/1/012009

for stresses. The annular area of plastic deformation, first appearing at the center of the membrane with increasing load, is extending to the periphery [8].

Geometric equations (1) using the basic relations of the theory of small elastoplastic deformations in the plastic area are transformed into the resolving equation with respect to ε_Θ

$$\frac{d\varepsilon_\Theta}{dr}+\frac{1}{r}\left[\varepsilon_\Theta-\frac{\varepsilon_\Theta(2n_r-n_\Theta)-(1-2\nu)\left(n_r^2-n_\Theta^2\right)}{(2n_\Theta-n_r)}\right]=-\frac{1}{2r}\left(\frac{dw}{dr}\right)^2. \quad (12)$$

Comparing this equation with the solution obtained in [2], where the Tresca-Sant-Venant plasticity condition was used we come to the conclusion that assuming the smallness of deformations and thereby neglecting the effect of changes in the membrane thickness on the stress state and form thereof, the compressibility will affect the very law of thickness variation only [9, 11].

The solutions for the plastic area for deflection and radial displacement are obtained as a result of integration, taking into account expression (11) of the second of the equilibrium equations and ratio (12). These solutions are represented (although the constants of integration and parameters $\boldsymbol{x_s}$ и $\boldsymbol{y_s}$ depend on the external load) as functions similar to the above characteristic ones for convenience of conjugation with an elastic solution:

$$f_3=\left(\frac{1}{2}\right)^{\frac{2}{3}}\left[c_2-2\frac{x_s^2}{y_s^2}x\right](2x)^{\frac{1}{2}},\quad f_4=\left(\frac{1}{2}\right)^{\frac{1}{3}}\left[c_1-2\frac{x_s}{y_s}x\right]. \quad (13)$$

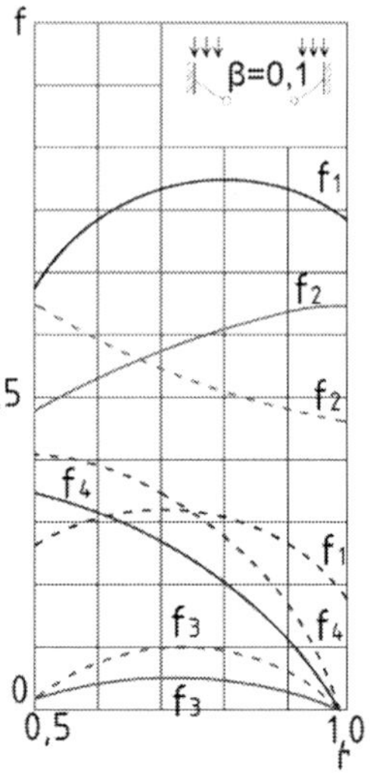

Figure 2. Characteristics of the stress-strain behavior of an annular membrane with a rigid inner contour.

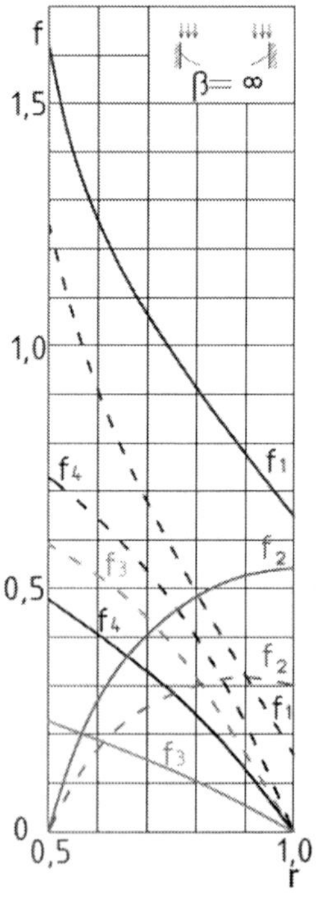

Figure 3. Characteristics of the stress-strain behavior of an annular membrane with a free inner contour.

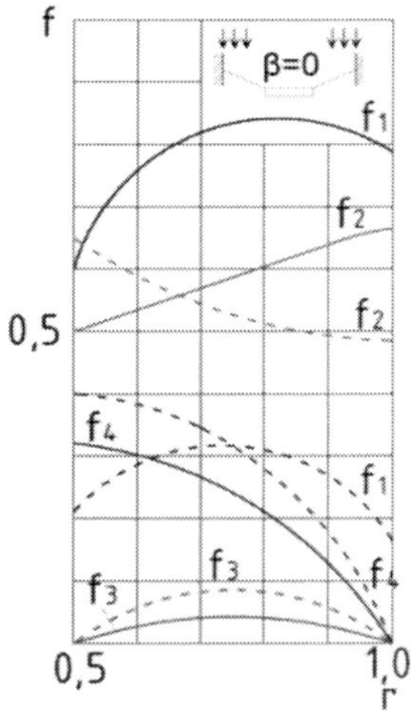

Figure 4. Characteristics of the stress-strain behavior of an annular membrane with a supported inner contour.

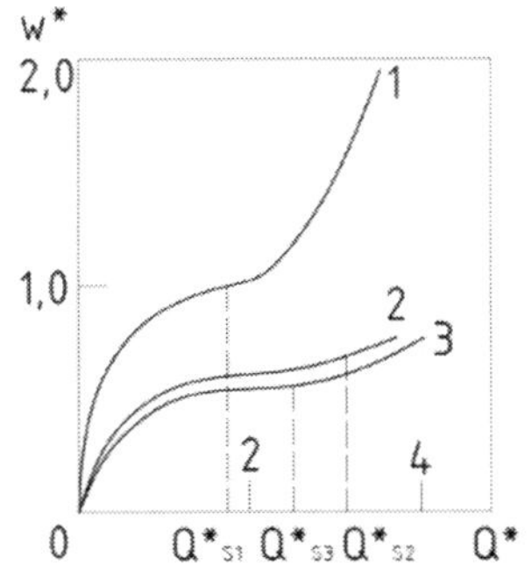

Figure 5. Elastoplastic problems solution.

The values related to the boundaries of the area are denoted here and below by s index. The constants of integration are found under the conditions for continuity of deflections and displacements at the boundary of the elastic and plastic areas [1, 12]. For this it is initially required to set up a solution in the field of elastic strains. Equation (4) is numerically integrated under the boundary conditions:

$$x = x_S : \frac{y_S'}{y_s} = \frac{\left(1 + \frac{n_\Theta^s}{n_r^s}\right)}{2x_s}, \quad x = x_1 : u_1 = 0. \tag{14}$$

Given the radius of the plastic area, with a certain load corresponding thereto, we solve the elastic problem $(Q_s = (n_s / f_2^S)^{3/2})$. Furthermore, using the conditions for continuity of components of strains and displacements, we set up an elastoplastic solution. Figure 5 (curve 1) shows the dependence of the deflection at the center of the membrane $(w_0^* = wn_s^{-1/2})$ on the load parameter $(Q_0^* = Qn_s^{-3/2})$ for a solid membrane made of material with $\nu = 0.3$.

It should also be noted that under $r_s \to 1$, i.e. in the transition to the limit equilibrium, we have a difference in the value characterizing the strains, which is explained by simultaneous fulfillment of $n_r = n_\Theta = n_s$ and $\varepsilon_\Theta = 0$ under $r_s = 1$. When solving the equilibrium problem for an annular solid uniformly loaded membrane under the A. Föppl assumptions, this difference is nonremovable [8].

3. Problems of elastoplastic equilibrium of isotropic membranes

Let us further turn to the solution of the problems of elastoplastic equilibrium of annular membranes supported by elastic rings along the inner contour. A detailed series of problems under the condition of Huber-Mises-Hencky plasticity for a noncompressible material was considered earlier [5, 8]. No restrictions were imposed on the compressibility of the membrane material when solving the problem using the Tresca-Sant-Venant plasticity condition. The fairly good convergence of the results again confirms the thesis formulated under the analysis of the equation (12).

When considering the general geometric scheme, we neglect the transverse dimensions of the ring in comparison with the membrane dimensions. The problems of a supported membrane in an elastic approach were considered in article [6, 12]. Proceeding from the analysis of these solutions, depending on the relative stiffness of the support, these problems may be said to have two solutions: the first covers situations of relatively more rigid support (from absolutely rigid, when $\beta = 0$, to "optimal"), the second one – from "optimal" support to the free boundary situation when $\beta = \infty$. The problem enabling occurrence of deformations in the support is also reduced to the problem of supporting the relatively light stiffness (both solutions are given here).

The first plastic deformations occur in the annular area near the support and propagate under the load increasing [12, 13]. The above plasticity condition for the first situation takes the form of $n_r = n_s$,

IOP Conf. Series: Materials Science and Engineering 1164 (2021) 012009 doi:10.1088/1757-899X/1164/1/012009

and for the second one $n_\Theta = n_s$. Both conditions give solutions in the field of plastic deformations for one of the strains only, the second one may be obtained from the first equation of the system (3). In the first case we get $n_\Theta = n_s$, and in the second one $n_r = (n_s r + c_3) / r$.The constant of integration is determined from the condition of conjugation of strains at the boundary of areas: $c_3 = - \gamma n_s r_s$, where $\gamma = n_r / n_\Theta$. The solutions for deflections and displacements in the area of plastic deformations are also represented as functions similar to the characteristic ones:

$$f_4 = \left(\frac{1}{2}\right)^{\frac{1}{3}} \left\{ c_4 - 2\frac{x_s}{y_s}\left[x - 2x_0 ln(2x)^{\frac{1}{2}} \right] \right\},$$
$$f_3 = (4)^{\frac{1}{3}} \left\{ c_5 - 2\frac{x_s^2}{y_s^2}\left[x - 4x_0 ln(2x)^{\frac{1}{2}} + \frac{x_0^2}{x} \right] \right\}(2x)^{\frac{1}{2}}. \quad (15)$$

Value γ, introduced as the ratio of radial and annular forces at the boundary of areas is a value that characterizes the relative stiffness of the supporting ring.

Given by r_s (in both problems) and γ (in the second case only), we solve the problem of elastic equilibrium in area $r_0 \leq r \leq 1$. Further, the constants of integration are determined from the conjugation conditions. Then the second of the expressions (15) will enable to determine the radial displacement of the inner contour of membrane $(f_3)_0$, based upon which it is possible to find the value of the relative stiffness of the ring reinforcing the membrane [6] corresponding to a given γ.

4. Results

The results of solving the first (curve 3 is corresponding to $\beta = 0.5$) and the second (curve 2 - $\beta = 1.4$) problems are given in Figure 4 for annular membrane with $r_0 = 0.5$ and ν=0.3.

References

[1] Vagapov R F 1988 *Review of work on membranes calculation theory* Deposited to RSRIIC state building of USSR № 7964 M 40

[2] Trofimov V I 1975 *Large-span spatial coatings made of aluminum sheet* (Moscow Stroyizdat) 166

[3] Engel' H 2007 *Carrier systems* (Moscow: AST) 344

[4] Grigor'ev A S, Trushina V M and Shadrin V A 1982 Equilibrium of rectangular and circular orthotropic membranes with large deflections *Modern problems of mechanics and aviation* (Moscow: Mechanical engineering) pp 106-113

[5] Ivanov M A 1973 Calculation of membrane coatings of circular outline in plan view *Suspension roofs* M **8** pp 43-70

[6] Vagapov R F and Grigor'ev A S 1987 Equilibrium of axisymmetric membranes *Static and dynamics of flexible systems* (Moscow: Science) pp 186-193

[7] Gabitov A I, Ryazanova V A, Salov A S, Gaisin A M and Timofeev A A 2020 Manufacture of construction materials by energy-saving technology through the example of the Bashkir region *IOP Conference Series: Materials Science and Engineering* **907** **(1)** 012049

[8] Grigor'ev A S 1951 Study of circular membrane operation at large deflections beyond the limit of elasticity *Inge Sat* **9** pp 99-112

[9] Bedov A I, Gabitov A I, Gaisin A M, Salov A S and Chernova A R 2018 CAD in interdisciplinary integration as a tool to increase specialist training quality in construction education *MATEC Web of Conferences* **251** 02011

[10] Dzh Ortega and V Rejnbolt 1975 *Iterative methods for solving nonlinear systems of equations with many unknowns* (Moscow: MIR) 558

[11] Grigor'ev A S, Konovalov M B, Trushina V M and Vagapov R F 1987 Problems of equilibrium of membranes and membrane structures Book: *Proceedings of the 14th All-Union Conference on the Theory of Plates and Shells* pp 255-260
[12] Vagapov R F and Grigor'ev A S 1985 Problems of equilibrium of axisymmetric membranes// *Structural Mechanics and Analysis of Constructions* **3** 66, **5** 82
[13] Stupishin L U and Kolesnikov A G 2014 Geometric Nonlinear Orthotropic Shallow Shells Investigation *Applied Mechanics and Materials* **501-504** pp 766-769

IOP Conf. Series: Materials Science and Engineering 1164 (2021) 012010 doi:10.1088/1757-899X/1164/1/012010

An analytical solution to the problem of stability loss of the rod under the combined action of concentrated force and dead weight

Yu Krutii[1*], S Bekshaev[1] and V Osadchiy[1]

[1] Odesa State Academy of Civil Engineering and Architecture, 4 Didrihson Street, 65029, Odesa, Ukraine

*Email: yurii.krutii@gmail.com

Abstract. The exact solution of the differential equation of buckling of the prismatic column under the combined action of a longitudinal concentrated end force and its own weight is constructed. Formulas are written for the state parameters of the rod, expressed in terms of the initial parameters. For six cases of ideal boundary conditions, finite analytical formulas are obtained that establish the dependence of the critical force on the weight of the rod.

1. Introduction

The rods are a widespread model in the design of structural elements in machine engineering, shipbuilding, aircraft, tool engineering and other fields of technology. In the form of a rod model can be considered not only a separate structural element, but also the structure as a whole. For example, the columns of frame structures, steel supports, antennas of various designs, industrial high-rise structures in the form of chimneys, water towers, etc. can be represented as the rods. The theory of rod calculation for stability has been studied in detail in the works [1-3] and many others.

The dead weight of short vertical rods is usually neglected, as it is considered small in comparison with the applied load. However, it is extremely necessary to take into consideration its own weight when calculating the stability of high-rise and massive framed structures. A number of publications have been devoted to solving the problem of the loss of rod stability under the action of its own weight, among which we single out the works [4, 5].

Despite the relevance and the undoubted practical significance, up to this day, the exact solution to the problem of stability loss of the column under the simultaneous action of concentrated force and dead weight has not been found. The mathematical model of such a problem will be a differential equation with variable coefficients. It is quite clear that its exact solution has the information of a qualitative nature and forms the most complete picture of the physical phenomenon of stability loss. However, the variability of the coefficients of the equation introduces significant mathematical difficulties in the procedure for constructing such a solution. It can probably explain the lack of the exact solution to this problem in the scientific literature. In this work, these difficulties were overcome. The results described below are obtained on the basis of the exact solution of the differential stability equation.

For example, figure 1 shows the general diagram of a hinged rod, and figure 2 shows the forces that arise in its element. However, all the formulas, obtained below, will be valid under any boundary conditions.

Content from this work may be used under the terms of the Creative Commons Attribution 3.0 licence. Any further distribution of this work must maintain attribution to the author(s) and the title of the work, journal citation and DOI.
Published under licence by IOP Publishing Ltd

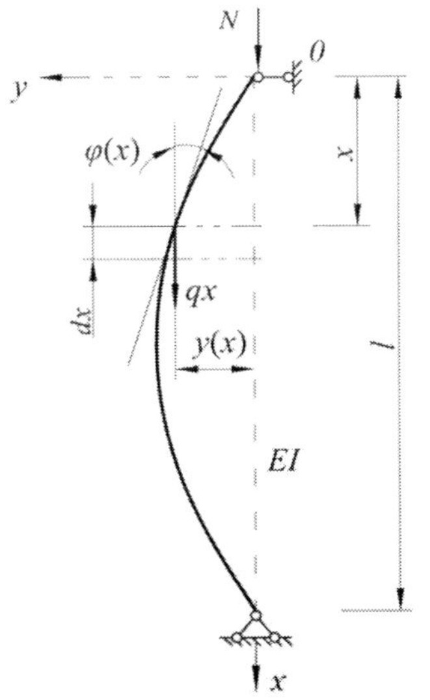

Figure 1. Design scheme of the column.

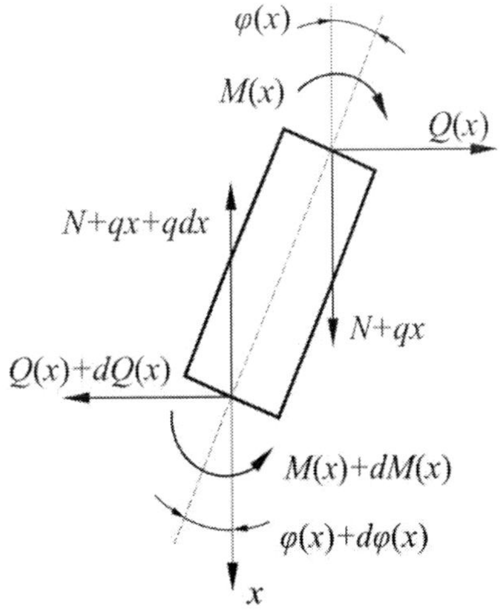

Figure 2. The internal forces acting upon the element of a column.

List of symbols:
EI – the flexural rigidity of a road;
N– constant axial load;
qx – the variable axial (compressive) force, where q–weight per unit length of beam;
$y(x)$ – the displacement of the axis point of the column with coordinate x (deflection);
$\varphi(x)$ – the angular displacement;
$M(x)$ – the bending moment;
$Q(x)$ – the transverse force.
The differential equation of rod stability in our case has the form [1-3]

$$EIy''''(x)+((N+qx)y'(x))'=0\,. \tag{1}$$

The stress-strain state of the rod is completely described by displacements $y(x)$, $\varphi(x)$ and internal forces $M(x)$, $Q(x)$. We will briefly name them as rod state parameters. As you know [1-3], they are interconnected by formulas:

$$\varphi(x)=y'(x)\,; \tag{2}$$
$$M(x)=-EI\varphi'(x)\,; \tag{3}$$
$$Q(x)=-EI\varphi''(x)-(N+qx)\varphi(x)\,. \tag{4}$$

The purpose of the work is to establish the dependence of the critical force on the dead weight of the rod in an analytical form. To achieve the goal, the following tasks are formulated and solved:
- find the exact solution to the equation (1);
- get the formulas for the state parameters of the rod;
- get the final calculation formulas for various boundary conditions.

2. Materials and methods

Earlier in the work [6], the method of direct integration of differential equations with continuous variable coefficients has been developed. Using this method, the exact solutions for a number of differential equations of statics and dynamics were also constructed there. The essence of the proposed method can also be considered, for example, according to the publications [7–11], where a number of mechanics problems of a deformable solid body is solved with its help.

In this publication, at the initial stage, using the direct integration method, it was possible to determine the form of the equation solution (1). It turned out that the solution should be sought in the form of double power series with unknown coefficients. Due to this, the problem was reduced only to the determination of these coefficients.

3. Exact solution of the differential stability equation and formulas for the rod state parameters

Integrating the equation (1) and carrying out the replacement (2), we obtain

$$EI\varphi''(x)+(N+qx)\varphi(x)=-Q(0), \tag{5}$$

or

$$l^2\varphi''(x)+\left(K+S\frac{x}{l}\right)\varphi(x)=-\frac{l^2}{EI}Q(0), \tag{6}$$

where

$$K=\frac{Nl^2}{EI}, \tag{7}$$

$$S=\frac{ql^3}{EI} \tag{8}$$

– dimensionless parameters. In addition to the equation (6), we will also consider an equivalent system of differential equations

$$\frac{d\Phi(x)}{dx}=H(x)\Phi(x)+h, \tag{9}$$

where

$$\Phi(x)=\begin{pmatrix}\varphi(x)\\ M(x)\end{pmatrix}=\begin{pmatrix}\varphi(x)\\ -EI\varphi'(x)\end{pmatrix}, \tag{10}$$

$$H(x)=\begin{pmatrix}0 & -\frac{1}{EI}\\ \frac{EI}{l^2}\left(K+S\frac{x}{l}\right) & 0\end{pmatrix},\quad h=\begin{pmatrix}0\\ Q(0)\end{pmatrix}.$$

We identify the fundamental solutions of the homogeneous equation by $Z_n(x)$ (n=1,2), corresponding to (6), and the particular solutions of the non-homogeneous equation by $Z_*(x)$ (6), for which we take the representation

$$Z_*(x)=-\frac{l^2}{EI}Q(0)Z_3(x). \tag{11}$$

Then the functions $Z_n(x)$ (n=1, 2, 3) must satisfy the equations:

$$l^2Z_n''(x)+\left(K+S\frac{x}{l}\right)Z_n(x)=0\ \ (n=1,2); \tag{12}$$

$$l^2Z_3''(x)+\left(K+S\frac{x}{l}\right)Z_3(x)=1. \tag{13}$$

We add the following boundary conditions to these equations:

$$\begin{pmatrix} Z_1(0) & Z_2(0) \\ lZ_1'(0) & lZ_2'(0) \end{pmatrix} = \begin{pmatrix} 1 & 0 \\ 0 & 1 \end{pmatrix}; \tag{14}$$

$$\begin{pmatrix} Z_3(0) \\ Z_3'(0) \end{pmatrix} = \begin{pmatrix} 0 \\ 0 \end{pmatrix}. \tag{15}$$

We will look the functions $Z_n(x)$ $(n=1, 2, 3)$ in a dimensionless format in the form of double power series

$$Z_n(x) = \left(\frac{x}{l}\right)^{n-1} \sum_{k=0}^{\infty}\sum_{j=0}^{\infty} c_{n,k,j}(-K)^k(-S)^j \left(\frac{x}{l}\right)^{2k+3j}, \tag{16}$$

or

$$Z_n(x) = \left(\frac{x}{l}\right)^{n-1} \left(c_{n,0,0} + \sum_{j=1}^{\infty} c_{n,0,j}(-S)^j \left(\frac{x}{l}\right)^{3j} + \sum_{k=1}^{\infty} c_{n,k,0}(-K)^k \left(\frac{x}{l}\right)^{2k} \right.$$
$$\left. + \sum_{k=1}^{\infty}\sum_{j=1}^{\infty} c_{n,k,j}(-K)^k(-S)^j \left(\frac{x}{l}\right)^{2k+3j} \right), \tag{17}$$

where $c_{n,k,j}$ $(n=1, 2, 3)$ $(k=0, 1, 2,\ldots)$ $(j=0, 1, 2\ldots)$ – dimensionless coefficients, that have to be determined. While we assume that the series (16) uniformly converge and the operation of their double differentiation is possible.

Due to this choice, as the analysis of formula (17) shows, the condition (15) is certainly carried out, and the condition (14) will be satisfied if we put

$$c_{1,0,0} = c_{2,0,0} = 1. \tag{18}$$

Inserting (16) into the equations (12), (13), we have

$$\sum_{k=0}^{\infty}\sum_{j=0}^{\infty} f_{n,k,j}\, c_{n,k,j}(-K)^k(-S)^j \left(\frac{x}{l}\right)^{n+2k+3j-3} - \sum_{k=0}^{\infty}\sum_{j=0}^{\infty} c_{n,k,j}(-K)^{k+1}(-S)^j \left(\frac{x}{l}\right)^{n+2k+3j-1}$$
$$- \sum_{k=0}^{\infty}\sum_{j=0}^{\infty} c_{n,k,j}(-K)^k(-S)^{j+1} \left(\frac{x}{l}\right)^{n+2k+3j} = \delta_n \quad (n=1,2,3),$$

where

$$f_{n,k,j} = (n+2k+3j-2)(n+2k+3j-1), \quad \delta_1 = \delta_2 = 0, \quad \delta_3 = 1. \tag{19}$$

In the second sum, we shift the index k by one, so, replace k with k-1. We perform the same operation in the third sum with the index j. As a result, we get

$$\sum_{k=0}^{\infty}\sum_{j=0}^{\infty} f_{n,k,j}\, c_{n,k,j}(-K)^k(-S)^j \left(\frac{x}{l}\right)^{n+2k+3j-3} - \sum_{k=1}^{\infty}\sum_{j=0}^{\infty} c_{n,k-1,j}(-K)^k(-S)^j \left(\frac{x}{l}\right)^{n+2k+3j-3}$$
$$- \sum_{k=0}^{\infty}\sum_{j=1}^{\infty} c_{n,k,j-1}(-K)^k(-S)^j \left(\frac{x}{l}\right)^{n+2k+3j-3} = \delta_n \quad (n=1,2,3).$$

Further, after the generations, we come to the necessity of fulfilling the equality

$$f_{n,0,0}c_{n,0,0}\left(\frac{x}{l}\right)^{n-3} + \sum_{k=1}^{\infty} (f_{n,k,0}c_{n,k,j} - c_{n,k-1,0})(-K)^k \left(\frac{x}{l}\right)^{n+2k-3} + \sum_{j=1}^{\infty} (f_{n,0,j}c_{n,0,j} - c_{n,0,j-1})(-S)^j \left(\frac{x}{l}\right)^{n+3j-3}$$

$$+\sum_{k=1}^{\infty}\sum_{j=1}^{\infty}(f_{n,k,j}\,c_{n,k,j}-c_{n,k-1,j}-c_{n,k,j-1})(-K)^k(-S)^j\left(\frac{x}{l}\right)^{n+2k+3j-3}=\delta_n\ \ (n=1,2,3).$$

Hence, taking into account the equalities (18), (19), we obtain the following set of recurrence formulas for determining the desired dimensionless coefficients:

$$c_{n,0,0}=\frac{1}{(n-1)!}\ (n=1,2,3); \tag{20}$$

$$c_{n,k,0}=\frac{c_{n,k-1,0}}{(n+2k-2)(n+2k-1)}\ (k=1,2,3,...); \tag{21}$$

$$c_{n,0,j}=\frac{c_{n,0,j-1}}{(n+3j-2)(n+3j-1)}\ (j=1,2,3,...); \tag{22}$$

$$c_{n,k,j}=\frac{c_{n,k-1,j}+c_{n,k,j-1}}{(n+2k+3j-2)(n+2k+3j-1)}\ (k=1,2,3,...)(j=1,2,3,...). \tag{23}$$

Thus the coefficients of the series (16), (17) can be considered known.

Note the fact that the formulas (21), (22) can also be written in explicit form:

$$c_{n,k,0}=\frac{1}{(n+2k-1)!}\ (k=1,2,3,...); \tag{24}$$

$$c_{n,0,j}=\frac{n(n+3)...(n+3j-3)}{(n+3j-1)!}\ (j=1,2,3,...). \tag{25}$$

The proof of the series (16) convergence is based on the following estimates

$$c_{n,k,j}\le\frac{1}{(2k)!}\cdot\frac{3^j\,j!}{(3j)!}\ (n=1,2,3)(k=0,1,2,...)(j=0,1,2,...), \tag{26}$$

which can be obtained by method of induction, based on the formulas (20), (24), (25), (23). Taking into consideration (26), we have

$$|Z_n(x)|\le\left(\frac{x}{l}\right)^{n-1}\sum_{k=0}^{\infty}\sum_{j=0}^{\infty}c_{n,k,j}K^kS^j\left(\frac{x}{l}\right)^{2k+3j}\le\left(\frac{x}{l}\right)^{n-1}\left(\sum_{k=0}^{\infty}\frac{1}{(2k)!}\left(\sqrt{K}\frac{x}{l}\right)^{2k}\right)\left(\sum_{j=0}^{\infty}\frac{3^j\,j!}{(3j)!}\left(\sqrt[3]{S}\frac{x}{l}\right)^{3j}\right)=$$

$$=\left(\frac{x}{l}\right)^{n-1}\cosh\sqrt{K}\frac{x}{l}\sum_{j=0}^{\infty}\frac{3^j\,j!}{(3j)!}\left(\sqrt[3]{S}\frac{x}{l}\right)^{3j}.$$

Moreover, for the radius of series convergence in powers of the variable $(\sqrt[3]{S}\,x/l)^3$, according to the D'Alembert criterion, we obtain

$$R=\lim_{j\to\infty}(3j+1)(3j+2)=\infty.$$

Thus, it is proved that the series (16) converge uniformly. The convergence of series, as a result of single and double differentiation, is proved in a similar way (16).

Therefore, we can state that two solutions $Z_n(x)$ (n=1,2) of the homogeneous equation, corresponding one (6), as well as a particular solution $Z_*(x)$ of the non-homogeneous equation (6), were found.

Each of the solutions $Z_n(x)$ (n=1,2) generates a vector according to the formula (10)

$$\Phi_n(x) = \begin{pmatrix} Z_n(x) \\ -EI\, Z_n'(x) \end{pmatrix} \; (n = 1,2)$$

–solving a homogeneous system of differential equations corresponding to (9). Then the matrix

$$\Omega(x) = \begin{pmatrix} Z_1(x) & Z_2(x) \\ -EI\, Z_1'(x) & -EI\, Z_2'(x) \end{pmatrix}, \tag{27}$$

developed with these vectors, will also satisfy a homogeneous system. Moreover, taking into account (14), we find

$$\Omega(0) = \begin{pmatrix} 1 & 0 \\ 0 & -\dfrac{EI}{l} \end{pmatrix}, \; |\Omega(0)| = -\frac{EI}{l} \neq 0 .$$

Therefore, the matrix (27) is the fundamental matrix [12] for the system (9).

Multiplying $\Omega(x)$ on the right by a constant matrix $\Omega^{-1}(0)$, we obtain a new fundamental matrix $\Lambda(x) = \Omega(x)\Omega^{-1}(0)$ for which

$$\Lambda(0) = \begin{pmatrix} 1 & 0 \\ 0 & 1 \end{pmatrix}. \tag{28}$$

As it is known [12], a fundamental matrix, possessing the characteristic (28), is determined definitely and is called a matrizant. In this case, the general solution of the homogeneous system, corresponding to (9), is written in the form $\Phi(x) = \Lambda(x)\Phi(0)$.

Forming the vector using (11)

$$\Phi_*(x) = -Q(0)\frac{l^2}{EI}\begin{pmatrix} Z_3(x) \\ -EI\, Z_3'(x) \end{pmatrix},$$

we obtain a particular solution of the non-homogeneous system (9) that satisfies the condition (15). This is easily verified by substitution.

Therefore, the general solution of the system of differential equations (9) can be written as follows

$$\Phi(x) = \Lambda(x)\Phi(0) + \Phi_*(x) .$$

Hence, for the rotation angle and bending moment, we will have:

$$\varphi(x) = \varphi(0)Z_1(x) - M(0)\frac{l}{EI}Z_2(x) - Q(0)\frac{l^2}{EI}Z_3(x) ; \tag{29}$$

$$M(x) = -\varphi(0)EI\, Z_1'(x) + M(0)lZ_2'(x) + Q(0)l^2 Z_3'(x) . \tag{30}$$

Next, integrating the equality (2), we obtain the formula for displacement

$$y(x) = y(0) + \int_0^x \varphi(x)dx ,$$

or

$$y(x) = y(0) + \varphi(0)\int_0^x Z_1(x)dx - M(0)\frac{l}{EI}\int_0^x Z_2(x)dx - Q(0)\frac{l^2}{EI}\int_0^x Z_3(x)dx . \tag{31}$$

We rewrite the formulas (29)-(31) in a different form by means of the following dimensionless functions

$$X_n(x)=\frac{1}{l}\int_0^x Z_n(x)dx,\ \tilde{X}_n(x)=lX_n'(x),\ \hat{X}_n(x)=l\tilde{X}_n'(x)\ \ (n=1,2,3)\,. \tag{32}$$

As a result, we obtain the final formulas for the rod state parameters expressed in terms of the initial parameters:

$$y(x)=y(0)+\varphi(0)lX_1(x)-M(0)\frac{l^2}{EI}X_2(x)-Q(0)\frac{l^3}{EI}X_3(x)\,; \tag{33}$$

$$\varphi(x)=\varphi(0)\tilde{X}_1(x)-M(0)\frac{l}{EI}\tilde{X}_2(x)-Q(0)\frac{l^2}{EI}\tilde{X}_3(x)\,; \tag{34}$$

$$M(x)=-\varphi(0)\frac{EI}{l}\hat{X}_1(x)+M(0)\hat{X}_2(x)+Q(0)l\hat{X}_3(x)\,; \tag{35}$$

$$Q(x)=Q(0)\,, \tag{36}$$

where

$$X_n(x)=\left(\frac{x}{l}\right)^n\sum_{k=0}^{\infty}\sum_{j=0}^{\infty}\frac{c_{n,k,j}}{n+2k+3j}(-K)^k(-S)^j\left(\frac{x}{l}\right)^{2k+3j}\ \ (n=1,2,3)\,, \tag{37}$$

$$\tilde{X}_n(x)=Z_n(x)\ \ (n=1,2,3)\,, \tag{38}$$

$$\hat{X}_n(x)=\left(\frac{x}{l}\right)^{n-2}\sum_{k=0}^{\infty}\sum_{j=0}^{\infty}(n+2k+3j-1)c_{n,k,j}(-K)^k(-S)^j\left(\frac{x}{l}\right)^{2k+3j}\ \ (n=1,2,3)\,. \tag{39}$$

We note that the formula for the transverse force (36) follows directly from the equalities (4), (5).

The presence of the formulas (33)-(39), (17), (20)-(23) allows us to investigate the rod stability in the case of the combined action of a concentrated force and dead weight under any boundary conditions.

4. Analytical presentation for longitudinal force and rod weight

Directly from the equalities (7), (8) we find:

$$N=K\frac{EI}{l^2}\,; \tag{40}$$

$$ql=S\frac{EI}{l^2}\,. \tag{41}$$

These formulas establish a direct analytical dependence of the longitudinal force N and the weight of the rod ql on other parameters of the mechanical system that is being considered.

These formulas are well known in stability theory [1-3]. There will be the formula of the form (40), studying the stability of the rod under the action of concentrated force without taking its own weight into account. There will be the formula of the form (41) when the stability of the rod loaded only with its own weight is considered. Moreover, the critical values of dimensionless parameters K and S depend only on the boundary conditions for the rod with constant stiffness.

As a matter of fact, these formulas are shown to remain correct even in the case of the combined action of a concentrated force and rod own weight. However, in our case, the critical value of parameter K, in addition to the boundary conditions, will also depend on the weight of the rod. The critical value of parameter S will not depend only on the boundary conditions, but also on the applied axial force.

5. Analytical dependence of the critical force on the weight of the rod

We establish the analytical dependence of the critical force on the weight of the rod for six ideal boundary conditions listed in table 1.

Characteristic equations, corresponding to different boundary conditions, admit a uniform notation

$$\eta_0(S)-\eta_1(S)K+\eta_2(S)K^2-\eta_3(S)K^3+...=0, \quad (42)$$

where $\eta_k(S)$ (k=0,1,2…) – dimensionless coefficients. The left side of this equation will always be an absolutely convergent series.

Based on the formulas (37)-(39), (16) we can write:

$$X_n(l)=\sum_{k=0}^{\infty}\gamma_{n,k}(S)(-K)^k; \quad (43)$$

$$\tilde{X}_n(l)=\sum_{k=0}^{\infty}\tilde{\gamma}_{n,k}(S)(-K)^k; \quad (44)$$

$$\hat{X}_n(l)=\sum_{k=0}^{\infty}\hat{\gamma}_{n,k}(S)(-K)^k, \quad (45)$$

where

$$\gamma_{n,k}(S)=\sum_{j=0}^{\infty}\frac{c_{n,k,j}}{n+2k+3j}(-S)^j, \quad (46)$$

$$\tilde{\gamma}_{n,k}(S)=\sum_{j=0}^{\infty}c_{n,k,j}(-S)^j, \quad (47)$$

$$\hat{\gamma}_{n,k}(S)=\sum_{j=0}^{\infty}(n+2k+3j-1)c_{n,k,j}(-S)^j. \quad (48)$$

Table 1. Types of rod ends fixings, corresponding boundary conditions and characteristic equations.

Case	Type of ends fixing $x=0$	Type of ends fixing $x=l$	Boundary condition	Characteristic equation
1	Pinned	Pinned	$y(0)=0;\ M(0)=0;$ $y(l)=0;\ M(l)=0.$	$X_1(l)\hat{X}_3(l)-\hat{X}_1(l)X_3(l)=0$
2	Free	Clamped	$M(0)=0;\ Q(0)=0;$ $y(l)=0;\ \varphi(l)=0.$	$\tilde{X}_1(l)=0$
3	Clamped	Clamped	$y(0)=0;\ \varphi(0)=0;$ $y(l)=0;\ \varphi(l)=0.$	$X_2(l)\tilde{X}_3(l)-\tilde{X}_2(l)X_3(l)=0$
4	Pinned	Clamped	$y(0)=0;\ M(0)=0;$ $y(l)=0;\ \varphi(l)=0.$	$X_1(l)\tilde{X}_3(l)-\tilde{X}_1(l)X_3(l)=0$
5	Sliding restraint	Clamped	$\varphi(0)=0;\ Q(0)=0;$ $y(l)=0;\ \varphi(l)=0.$	$\tilde{X}_2(l)=0$
6	Sliding restraint	Pinned	$\varphi(0)=0;\ Q(0)=0;$ $y(l)=0;\ M(l)=0.$	$\hat{X}_2(l)=0$

We suggest to consider the case of a hinge supported rod in detail:

$$y(0)=0;\ M(0)=0;\ y(l)=0;\ M(l)=0.$$

In this case, the initial parameters $y(0)$, $M(0)$ are known in advance. Realizing with the help of (33), (35) the conditions at the end $x=l$, as for $\varphi(0)$, $Q(0)$ we will have the system:

$$\begin{cases} lX_1(l)\varphi(0) - \dfrac{l^3}{EI}X_3(l)Q(0) = 0; \\ -\dfrac{EI}{l}\widehat{X}_1(l)\varphi(0) + l\widehat{X}_3(l)Q(0) = 0. \end{cases}$$

Writing down the condition for the solvability of the system, we obtain the characteristic equation

$$X_1(l)\widehat{X}_3(l) - \widehat{X}_1(l)X_3(l) = 0. \qquad (49)$$

The characteristic equations for all the boundary conditions are given in table 1.

Based on the representations (43), (45) and multiplying and subtracting the series on the left side of the equation (49), we have it in the form (42), where

$$\eta_k(S) = \sum_{i=0}^{k}(\gamma_{1,i}(S)\widehat{\gamma}_{3,k-i}(S) - \widehat{\gamma}_{1,i}(S)\gamma_{3,k-i}(S)) \ (k = 0,1,2,...).$$

Transforming the last expression taking into account (46), (48), we obtain

$$\eta_k(S) = \sum_{i=0}^{k}\sum_{j=0}^{\infty}\sum_{\nu=0}^{j}\left(\frac{2(k-i)+3(j-\nu)+2}{2i+3\nu+1} - \frac{2i+3\nu}{2(k-i)+3(j-\nu)+3}\right)c_{1,i,\nu}\,c_{3,k-i,j-\nu}(-S)^j.$$

Thus, according to the formula (8), calculating the value of the parameter S for a specified rod and finding the corresponding smallest positive root of the equation (42), we will have a pair of significance points (S, K) at which the rod will lose stability. In this case, the critical force value can be found by the formula (40).

It is easy to set the range of values for the dimensionless parameters K and S. As it is known [1-3], for a weightless rod (case $S = 0$), the critical force is solved by the formula (40), where $K = \pi^2$. It is quite clear that the calculated value of the critical force for a weighted rod (case $S \neq 0$) should be less in comparison with a weightless rod. Hence, we conclude that in our case $0 \leq K \leq \pi^2$.

It is also known [1-5] that in the absence of axial force (case $K = 0$), the critical weight of the rod is determined by the formula (41), moreover, for a hinge supported rod $S = 18.5687$. If the axial load is applied (case $K \neq 0$), the calculated value of the critical rod weight should be less compared with the case when the axial force is absent. In other words, for our case the values $0 \leq S \leq 18.5687$ are relevant.

In figure 3 there are the calculating results of the parameter K, corresponding to some given values of the parameter S from the specified interval in increments of 1,85687. Considering the parameter K as a function of a variable S, as well as having many values of this function that correspond to the values of an explanatory variable $0 \leq S \leq 18.5687$, we approximate the function K by a polynomial. In this case, the degree of the approximating polynomial is chosen from the condition that the determination coefficient is equal to 1. As a result, we obtain

$$K = 9.8700 - 0.5004S - 0.0017S^2.$$

The value 9.8700 differs from the number π^2 only by an amount 0.0004. After the replacement, we get an approximate formula

$$K = \pi^2 - 0{,}5004S - 0{,}0017S^2. \qquad (50)$$

The graph of the found dependence is shown in Figure 3.

Thus, a multiplicity of value pairs (S, K), at which the rod loses the stability, forms the line of critical values, shown in figure 3.

Substituting the value (50) into the formula (40) instead of the parameter K and taking into account (8), we finally obtain

$$N_{\mathrm{cr}} = \pi^2 \frac{EI}{l^2} - 0,5004(ql) - 0,0017(ql)^2 \frac{l^2}{EI}. \tag{51}$$

This formula establishes the dependence of the critical force N_{cr} on the dead weight of the rod ql. It is important to note that when considering a weightless rod (case $q = 0$), from the formula (51) we obtain the well-known Euler formula.

S	K
0	π^2
1.8569	8.9352
3.7137	7.9890
5.5706	7.0310
7.4275	6.0613
9.2844	5.0798
11.1412	4.0866
12.9981	3.0819
14.8550	2.0659
16.7118	1.0385
18.5687	0.0000

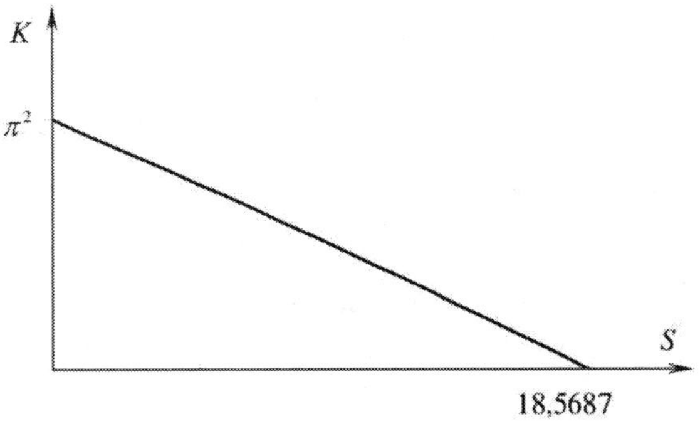

Figure 3. Calculating results and corresponding critical line for the case 1 from table 1.

Similar calculations were carried out for other cases of boundary conditions. The formulas for calculating the coefficients of the characteristic equations in the form (42) for all six cases are presented in table 2. The final formulas for the critical forces are presented in table 3.

Table 2. Formulas for calculating the coefficients of the characteristic equations.

Case	Coefficients of characteristic equations $(k = 0, 1, 2, \ldots)$
1	$\eta_k(S) = \sum_{i=0}^{k} \sum_{j=0}^{\infty} \sum_{\nu=0}^{j} \left(\frac{2(k-i)+3(j-\nu)+2}{2i+3\nu+1} - \frac{2i+3\nu}{2(k-i)+3(j-\nu)+3} \right) c_{1,i,\nu}\, c_{3,k-i,j-\nu} (-S)^j$
2	$\eta_k(S) = \sum_{j=0}^{\infty} c_{1,k,j} (-S)^j$
3	$\eta_k(S) = \sum_{i=0}^{k} \sum_{j=0}^{\infty} \sum_{\nu=0}^{j} \left(\frac{1}{2i+3\nu+2} - \frac{1}{2(k-i)+3(j-\nu)+3} \right) c_{2,i,\nu}\, c_{3,k-i,j-\nu} (-S)^j$
4	$\eta_k(S) = \sum_{i=0}^{k} \sum_{j=0}^{\infty} \sum_{\nu=0}^{j} \left(\frac{1}{2i+3\nu+1} - \frac{1}{2(k-i)+3(j-\nu)+3} \right) c_{1,i,\nu}\, c_{3,k-i,j-\nu} (-S)^j$
5	$\eta_k(S) = \sum_{j=0}^{\infty} c_{2,k,j} (-S)^j$
6	$\eta_k(S) = \sum_{j=0}^{\infty} (2k+3j+1) c_{2,k,j} (-S)^j$

Table 3. Critical Forces for various cases of boundary conditions.

Case	Formulas for Critical Force
1	$N_{cr} = \pi^2 \frac{EI}{l^2} - 0.5004(ql) - 0{,}0017(ql)^2 \frac{l^2}{EI}$
2	$N_{cr} = \frac{\pi^2}{4} \frac{EI}{l^2} - 0.2968(ql) - 0{,}0023(ql)^2 \frac{l^2}{EI}$
3	$N_{cr} = 4\pi^2 \frac{EI}{l^2} - 0.5003(ql) - 0{,}0004(ql)^2 \frac{l^2}{EI}$
4	$N_{cr} = \frac{\pi^2}{0{,}699^2} \frac{EI}{l^2} - 0.3442(ql) - 0{,}0008(ql)^2 \frac{l^2}{EI}$
5	$N_{cr} = \pi^2 \frac{EI}{l^2} - 0.5001(ql) - 0{,}0011(ql)^2 \frac{l^2}{EI}$
6	$N_{cr} = \frac{\pi^2}{4} \frac{EI}{l^2} - 0.7028(ql) - 0{,}0020(ql)^2 \frac{l^2}{EI}$

6. Results and discussion

Whenever the study of a physical phenomenon is reduced to a differential equation, the key question is constructing its exact (analytical) solution. However, the researchers often face a well-known mathematical problem, which is the lack of a universal method for integrating differential equations with the variable coefficients. Probably, this can explain the predominant use of approximate methods. This completely refers to the problem of loss of rod stability under the combined action of a concentrated force and dead weight. In this paper, these difficulties were overcome. As a result, the following results were obtained:

1. The exact solution of the differential stability equation is constructed;
2. Formulas for displacements and internal forces in the rod are obtained;
3. For six cases of the boundary conditions in an analytical form, the formulas for the critical force are obtained.

In fact, the solution of the original problem is reduced only to the implementation of the given boundary conditions and the solution of the obtained characteristic equation.

7. Conclusions

The problem of the loss of rod stability under the combined action of a concentrated force and dead weight is solved. Existence of exact formulas for the rod state parameters is a real alternative to using approximate methods for solving this class of problems. It is quite clear that based on the exact solution of the differential equation, a more established picture of the stability loss can be obtained in comparison with the approximate methods. It is the exact solution that carries the qualitative information and forms the most complete picture of the physical phenomenon that is being studied.

References

[1] Timoshenko S P and Gere J M 1961 *Theory of Elastic Stability* (New York: McGraw-Hill) 541
[2] Vol'mir A S 1967 *Ustoychivost' deformiruyemykh sistem* (Moscow: Nauka) 984
[3] Bazant Z P and Cedolin L 1991 *Stability of Structures* (New York: Oxford University Press) 1008
[4] Chai Y H and Wang C M 2006 An application of differential transformation to stability analysis of heavy columns *Int. J. Structural Stability and Dynamics* **06 (03)** pp 317–332

[5] Krutii Yu and Vandynskyi V 2019 Exact solution of buckling problem of the column loaded by self-weight *IOP Conference Series: Mat. Sci. and Eng.* **708** (**1**) 012062

[6] Krutii Yu S 2016. *Rozrobka metodu rozv'yazannya zadach stiykosti i kolyvan' deformivnykh system zi zminnymy neperervnymy parametramy* Sc. D. diss. Lutsk National Technical University.

[7] Krutii Yu S 2018 Construction of a solution of the problem of stability of a bar with arbitrary continuous parameters *J. Mat. Sci.* **231** pp 665-677

[8] Shvab'yuk V I,Krutii Yu S and Sur'yaninov M G 2016 Investigation of the Free Vibrations of Bar Elements with Variable Parameters Using the Direct Integration Method *Strength of Materials* **48(3)** pp 384-393

[9] Krutii Yu 2015 Analysis of longitudinal oscillations for systems with continuous variable parameters using force integration method *Technical journal* **9(4)** pp 420-425

[10] Krutii Yu, Surianinov M and Vandynskyi V 2018 Development of the method for calculation of cantilever construction's oscillations taking into account own weight *Eastern-European J. of Enterprise Technologies* **3(7)** pp 13-19

[11] Krutii Yu, Surianinov M and Vandynskyi V 2019 Analytic Formulas for the Cantilever Structures' Natural Frequencies with Taking into Account the Dead Weight *Materials Science Forum* **968** pp 450-459.

[12] Gantmacher F R 1959 *The Theory of Matrices* (Chelsea, New York: American Mathematical Soc.) 374

Study of the fracture surface of concretes reinforced with basalt fiber coated with titanium and zirconium dioxides. Fiber-reinforced concrete composites

V Gots[1], O Berdnyk[1,*], A Maystrenko[1], P Palchik[1] and N Amelina[1]

[1]Department Technology of building constructions and wares, Kyiv National University of Construction and Architecture, Povitroflotskyj avenue, 31, 03037, Kyiv, Ukraine

*Email: kseniareznik87@gmail.com

Abstract. It is impossible to imagine the development of modern technologies without the use of efficient and affordable fibrous materials based on minerals, glass, and other fibers. The intensive development of the production of fibers and materials based on them worldwide determines the formulation of significant scientific and technical problems of obtaining them with a given set of properties and high-quality indicators, optimization of processing, and rational use. However, the possibility of modifying the composition of the rock with other components that could significantly improve the physical and chemical properties of the fibers is not always taken into account. For the first time, a study was conducted to research the coating of basalt fibers, such as TiO_2, ZrO_2 and their effect on the physical and chemical characteristics of aluminosilicate melts made from rocks, fiber formation processes, structure, and physical and chemical properties of fibers. Studies have shown that fibers modified with oxides of TiO_2 and ZrO_2 have higher chemical resistance in aggressive environments compared with unmodified fibers obtained from melts of similar component groups.

1. Introduction

The study of the basalt fiber surface morphology and its elemental composition before and after interaction with the alkaline medium allows us to determine the details of the etching of basalt fiber in an aggressive environment and the impact on this process when coating its surface. Several different methods are used to determine the effect of an alkaline environment on glass fibers. The etching kinetics of basalt fiber with zirconium dioxide coatings has a more complex dependency. In the first two weeks from the beginning of etching, the coating performs its protective function. But in the next two weeks, the etching of the coated fiber accelerates. According to the references [1], due to the weak connection with the surface, the formed corrosion layer, after reaching a defined thickness, begins to peel off, opening access to the alkaline surface of the basalt fiber, and the process of corrosion layer formation starts from the beginning.

2. Materials and methods of the research

2.1. Study of the fracture surface of concretes reinforced with basalt fiber

Two problems hinder the widespread use of basalt fiber for reinforcing composites with aggressive matrices, primarily the cement ones. The first problem is the degradation of the properties of basalt fiber in the alkaline environment of hardening concrete due to the slow alkaline etching of basalt fiber, followed by a decrease in the fiber mechanical strength [2,3]. The second problem is to ensure the isotropic distribution of the fiber over the volume of the concrete. The solution of both problems has [4,5,6] the prospect of solving by surface modification of basalt fibers.

Research of alkali resistance of basalt fibers, including those modified with titanium and zirconium dioxides in an alkaline environment of hardening concrete was performed. The Figure 1 shows the

Content from this work may be used under the terms of the Creative Commons Attribution 3.0 licence. Any further distribution of this work must maintain attribution to the author(s) and the title of the work, journal citation and DOI.
Published under licence by IOP Publishing Ltd

research results of the concrete density change depending on the water-cement ratio for the cement brand used in the experiment.

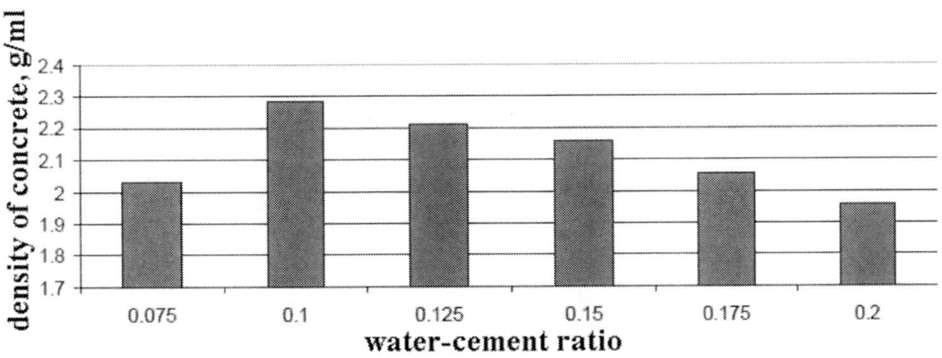

Figure 1. Effect of water-cement ratio on the density of fiber-reinforced concrete before adding basalt fibers.

At a low water-cement ratio (0.075), the amount of water was insufficient to prepare a homogeneous concrete mixture. The samples prepared from this mixture had a highly porous structure, and the removal from the mold resulted in mechanical damages to the samples. At a ratio of 0.1, the water amount was already sufficient to form strong bonds, but the mixture mobility was low. With a further increase of the water-cement ratio, there was a decrease in the concrete density, which was associated with an increase in pores filled with water. During hardening and storing water-filled pores convert into pores filled with air. Another reason may be the mobility reduction of the concrete mixture due to the introduction of the fiber. Electron microscope (EM) images of concretes reinforced with basalt fibers with and without coatings show that the fiber distribution is homogeneous enough over the volume of the concrete matrix, with individual fibers isolated from each other by a layer of the concrete matrix.

The Figures 2 a and b show the electron microscope images of the destruction of concrete reinforced with basalt fiber (uncoated) after two years of exposure. The figure shows a thick layer with an uneven surface, areas of peeling, deep depressions on the surface of the basalt fiber. These morphological features indicate the process of fiber etching in a strongly alkaline medium of hardening concrete (pH = 13).

It has previously been shown [7,8,9] that the process of etching of the basalt fiber in an alkaline medium involves the partial dissolution of the aluminosilicate composition of the fiber and the formation of a corrosion layer consisting of iron and calcium compounds. Severe changes in morphology, elemental, and phase composition, which occur to basalt fiber in an alkaline environment, lead to a loss of mechanical strength, which, in turn, significantly reduces the positive effect of its addition. Notably, the main changes in fiberglass occur in the first month of hardening, so when storing samples in the laboratory air atmosphere, the difference in morphology of the fracture surface of the samples, which lasted for three months and two years, is almost invisible.

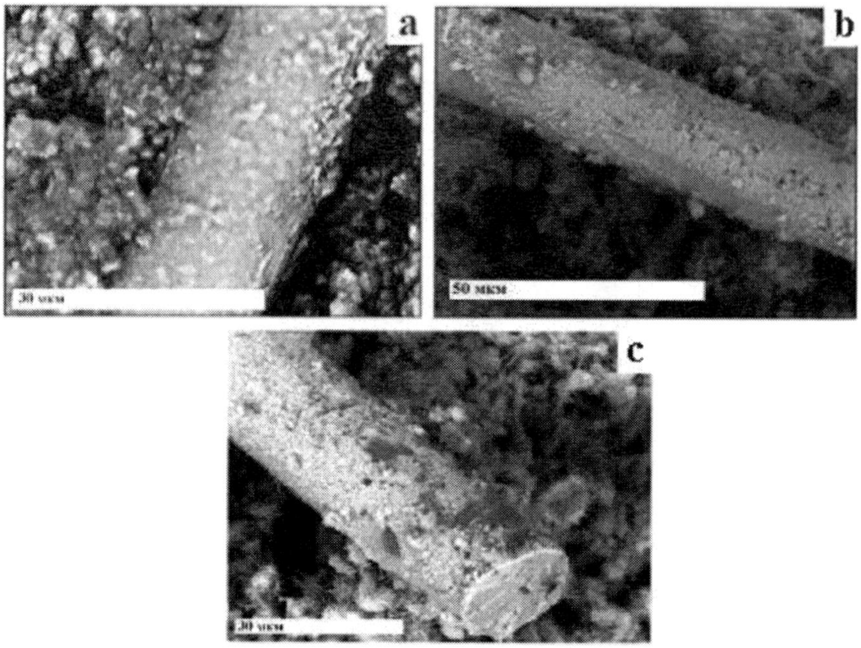

Figure 2. Electron microscope images of concrete reinforced with uncoated basalt fiber: examples of fiber degradation.

A comparative analysis of the concrete matrix morphology and the reinforcing fiber surface shows a significant amount of needle crystals in the concrete matrix. However, needle crystals are not observed on the surface of the fiber. It can be assumed that the thick surface layer consists mainly of corrosion products, but we cannot exclude hydrated components of a cement. Note that there is a strong enough connection between the basalt fiber and the concrete matrix (Figure 2, a).

2.2. *Study of the fracture surface of concretes reinforced with basalt fiber*

The Figure 3 shows the cross-section of the destruction of concrete reinforced with basalt fiber with one- and two-layer ZrO_2 coating after two years of aging. Electron microscopic examination of the morphology of the basalt fiber with ZrO_2 coating showed that there are "bridges" on its surface.

The analysis of Figure 3 a, shows that the aging of basalt fibers coated with zirconium dioxide in a concrete matrix does not lead to significant morphology changes, as compared to basalt fibers without protective coating. The coating retains its integrity, although there is peeling observed in some areas. Such observation is especially true for areas with "bridges" of a width up to 1.5-2 μm (Figure 3 c). In places of exfoliation, a juvenile unprotected surface of basalt fiber forms and can become subject to further alkaline etching.

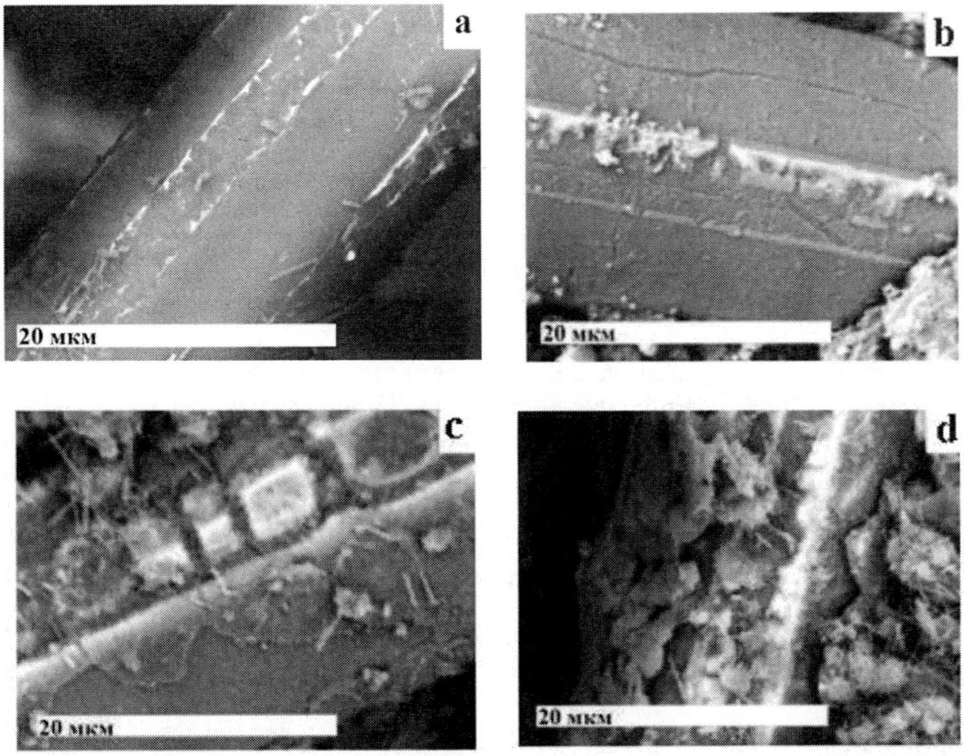

Figure 3. Electron microscope images of the fracture surface of the concrete reinforced with the zirconium dioxide ZrO_2 coated fiber.

The analysis of electron microscope images also shows that there is no strong adhesion between the basalt fiber with ZrO_2 coating (1 layer) and the concrete matrix. Notably, the zirconium dioxide coating preserves the integrity, relative inertness, and relief uniformity of hardening concrete in the alkaline environment, which prevents the adhesion of concrete matrix components to the fiber and provides poor adhesion at the boundary "coated fiber - concrete matrix". Accretion of zirconium dioxide coating layers, which leads to an even higher amount of surface defects, and increases the roughness of the fiber relief, contributes to the adhesion of the concrete matrix, the formation of large units of concrete matrix components, and needle crystals on the surface. As a result, the bonding between the surface-modified fiber and the concrete matrix becomes stronger (Figure 3, d). Notably, the electron microscope images did not reveal traces of friction of cement or coated sand particles on the fiber during the preparation of fiber concrete samples. There were smaller exfoliation areas near the stress concentrates (and, accordingly, cracks) in the coating. These areas may occur due to the weakening of the connection with the basalt base resulting from alkaline etching, and coatings weekly bound with the area surface peel off after destruction of the sample.

Cut analysis of the fracture surface of the concrete reinforced with titanium dioxide coated basalt fiber indicates the traces of deeper interaction of the alkaline medium of hardening concrete on the fiber surface than of basalt fiber coated with zirconium dioxide (Figure 3). A large number of areas with exfoliation of a corrosion layer confirm this research. Comparison of the behavior of the basalt fiber with and without TiO_2 coating in the concrete matrix allows us to conclude that TiO_2 coating can protect the fiber from the interaction of alkali to some extent.

Figure 4. Electron microscope images of the fracture surface of the concrete reinforced with titanium dioxide coated fiber.

Thus, according to electron microscopy data, the basalt fiber coated with titanium dioxide occupies an intermediate position between the uncoated fiber and the zirconium dioxide-coated fiber when compared by the corrosion resistance in the concrete matrix. These data correspond with the reference data on alkali resistance of metal oxides [10]. Analysis of electron microscopy data also shows a noticeable bond between the concrete matrix and the surface modified with titanium dioxide fiber (Figure 4 a, b).

It should be noted that the main purpose of these studies was to identify changes in the fiber after its durable stay in the concrete matrix. At this stage of research, achieving maximum strength was not a priority. To see the full picture of the research, it was decided to abandon the addition of a third component, which could improve the strength characteristics of fiber concrete, for example, SiO_2 [11]. The introduction of basalt fiber resulted in a concrete compressive strength decrease due to the growing number of voids. The compressive strength of concrete reinforced with basalt fiber with a single layer of zirconium dioxide after two years is 26 ± 4 MPa. Samples reinforced with basalt fiber with TiO_2 coating showed higher compressive strength (33 ± 8 MPa) compared with concrete samples reinforced with ZrO_2 coated fiber. At the same time, the compressive strength of non-reinforced concrete brand M200 after two years of etching should be $\sim$ 40 MPa [12]. The strength decrease of concrete reinforced with titanium dioxide coated fiber is due to the weaker interaction between the fiber and the cement matrix, like electron microscope images of the cross-sectional fiber concrete surface show (Figure 3). Assumably, sufficient adhesion between the fiber and the concrete matrix, in this case, contributes to good load transfer and, as a result, to higher strength, compared with those of concrete reinforced with ZrO_2 coated basalt fiber.

Thus, it was shown that the preliminary modification of the surface of basalt fiber by applying oxide coatings increases the stability of the fibers in the alkaline environment of the concrete matrix. Among the "uncoated fiber - titanium dioxide-coated fiber - zirconium dioxide-coated fiber" unprotected fiber is subject to the highest corrosion in the concrete matrix.

3. Summary

During the studies, techniques for alkaline etching of ZrO_2 and TiO_2 coated basalt fiber were developed. These techniques allow increasing the strength of basalt fiber in an alkaline environment. Complex physical and chemical studies of the original basalt fiber before and after etching in an alkaline environment were also performed. It was shown that during the etching of the basalt fiber in a solution of sodium hydroxide, the aluminosilicate framework dissolves, and a layer of hexagonal plates forms. Signs of fiber etching in a calcium hydroxide solution appear much later than in a sodium hydroxide solution. It is shown that a small amount of CO_2 in both alkaline solutions leads to the formation of calcium carbonate particles on the fiber surface. Application of the ZrO_2 and TiO_2

coatings slows down the process of the basalt fiber etching in both alkaline solutions and the ZrO_2 coating provides more reliable resistance to alkali than the TiO_2 coating. Also, studies have shown that the previous surface modification of basalt fibers by applying oxide coatings increases the stability of the fibers in the alkaline environment of the concrete matrix. Among the "uncoated fiber - coated fiber from titanium dioxide-coated fiber from zirconium dioxide" unprotected fiber is subject to the highest corrosion in the environment of the concrete matrix.

References

[1] Berdnyk O Yu, Lastivka O V, Maystrenko A A and Amelina N O 2020 Processes of structure formation and neoformation of basalt fiber in an alkaline environment *IOP Conf. Series: Materials Science and Engineering Innovative Technology in Architecture and Design* **907** 012036 https://iopscience.iop.org/article/10.1088/1757-899X/907/1/012036/pdf

[2] Gots V I, Lastivka O V, Berdnyk O Yu, Tomin O O and Shilyuk P S 2020 Corrosion resistance of polyester powder coatings using fillers of various chemical nature *Key Engineering Materials* **864** pp 115-121 https://doi.org/10.4028/www.scientific.net/KEM.864.115 https://www.scintific.net/KEM.864.115

[3] Bondarenko O P, Guzii S G, Zaharchenko K D and Novoselenko E D 2015 Development protective materials based glass & slag containing portland compositions *Eastern-European Journal of Enterprise Technologies* **6/11(78)** pp 41-47 doi: 10.15587/1729-4061.2015.56577

[4] Krivenko P V, Guzii S G and Bondarenko O P 2019 Alkaline Aluminosilicate Binder-Based Adhesives with Increased Fire Resistance for Structural Timber Elements *Key Engineering Materials* **808** pp 172-176 doi:10.4028/www.scientific.net/KEM.808.172

[5] Krivenko P, Petropavlovskyi O, Kovalchuk O, Rudenko I and Konstantynovskyi O 2019 Enhancement of alkali-activated slag cement concretes crack resistance for mitigation of steel reinforcement corrosion *E3S Web of Conferences* **166** 06001 DOI: 10.1051/e3sconf/202016606001

[6] Krivenko P V, Kovalchuk G Yu and Kovalhuk O Yu 2005 Heat-resistant cellular concretes based on alkaline cements *Proceedings of the International Conference on the Use of Foamed Concrete in Construction* pp 97-104

[7] Katrijn Gijbels, Pavel Krivenko, Oleksandr Kovalchuk, Anton Pasko, Sonja Schreurs, Yiannis Pontikes and Wouter Schroeyers 2020 The influence of porosity on radon emanation in alkali-activated mortars containing high volume bauxite residue *Construction and Building Materials* **230** 116982. https://doi.org/10.1016/j.conbuildmat.2019.116982

[8] Kovalchuk O, Gelevera O and Ivanychko V 2019 Studying the influence of metakaolin on self-healing processes in contact-zone structure of concretes based on the alkali- activated Portland cement *Eastern-European Journal of Enterprise Technologies* **5/6** **(101)** pp 33-40 doi:10.15587/1729-4061.2019.160959

[9] Gao S L, Mäder E and Plonka R 2007 Nanostructured coatings of glass fibers Improvement of alkali resistance and mechanical properties *Acta Materialia* **55** **(3)** pp 1043-1052

[10] Tretiakov Yu D 2007 *Neorhanycheskaia khymyia Khymyia* M Yzd-vo MHU 3-kh t T **1** 249

[11] Jantzen C M, Brown K G and Pickett J B 2010 Durable Glass for Thousands of Years *International Journal of Applied Glass Science* **1** **(1)** pp 38-62

[12] Lipatov Y V, Gutnikov S I, Manylov M S, Zhukovskaya E S and Lazoryak B I 2015 High alkali-resistant basalt fiber for reinforcing concrete *Materials & Design* **73** pp 60-66

IOP Conf. Series: Materials Science and Engineering 1164 (2021) 012012 doi:10.1088/1757-899X/1164/1/012012

Analysis of the reliability of the plates of complex shapes for the cams of automatic half-hose machines by the criterion of fatigue strength

L Berezin[1,*], M Rubanka[1], K Savchenko[2] and O Polishchuk[3]

[1]Kyiv National University of Technologies and Design, 2 Nemyrovycha-Danchenka Street, 01011 Kyiv, Ukraine
[2]G.S. Pisarenko Institute for Problems of Strength of the National Academy of Sciences of Ukraine, 2 Timiryazevs'ka str., 01014 Kyiv, Ukraine
[3]Khmelnytskyi National University, 11 Instytuts'ka str., Khmelnytskyi, 29016, Ukraine

*E-mail: berezin.lm@knutd.edu.ua

Abstract. The purpose of the work is to develop the algorithm for calculation of the probability of failure-free operation of flexible plates of complex shapes for the cams, using the criterion of fatigue strength, to make decisions on minimization of their sizes with the subsequent evaluation of the implemented measures. The studies are performed on the example of a plate for the cam, the configuration of which provides both beam and double console deformations at the same time.

The comprehensive approach is presented, which includes dynamic analysis of the needle impact on the cam surface, computer simulation of geometric parameters of the plate by the finite element method (FEM) according to the condition of strength, and classical principles of mathematical statistics and Theory of Probability. The finite-element model with a volumetric 8-node element is used, which allows, with ease of calculations, to obtain the results with a high accuracy when determining the stress-deformed state of the study object. The computer simulation is performed using Code_Aster software with free access. Also, in the calculation of reliability, the random nature of the multifactorial influence on the value of the plate fatigue limit and the loading variations of the cam are considered. When determining the equivalent forces, taking into account the cyclical operation of the machine, the forces that affect the cams during the production of one product are taken as a loading. The numbers of cycles corresponding to these loadings are calculated by the formulas, which are made as a result of the analysis of trajectories of needles motion concerning the cams when knitting different sections of the product. When calculating equivalent stresses, only stresses that influence the accumulation of fatigue failures are used, that is, the values greater than the half of the detail fatigue limit.

The proposed computational approach can be used to generate the cams with flexible work plates of different configurations, including under the equal stress condition.

1. Introduction

In the automatic half-hose machines, technological trajectories of the needles are set by the movements of their heels after the cams, accompanied by impact loadings and accumulation of fatigue damages. It is obvious that the intensification of the speed modes of the automatic half-hose machines and the enhancement of technological capabilities lead to their further growth. A promising solution that can increase the reliability of the automatic half-hose machines is to reduce the impact force in a "needle-cam" pair, primarily by increasing the flexibility of the cam.

Content from this work may be used under the terms of the Creative Commons Attribution 3.0 licence. Any further distribution of this work must maintain attribution to the author(s) and the title of the work, journal citation and DOI.
Published under licence by IOP Publishing Ltd

2. Literature review

The bibliography that concerns this way of improvement is presented in [1, 2, 3], where the design features of the cams with flexible work plates (FWPs) are mainly considered. Some works [4, 5] are devoted to the validation of FWPs by the criterion of static strength. However, given that FWPs are classified as the elements that are cyclically loaded and limited in size, we consider it appropriate to analyze them for fatigue strength in probabilistic figuration.

3. Research methodology

The purpose of the work is to propose the algorithm for calculation of the probability of failure-free operation of flexible plates of complex shapes for the cams, using the criterion of fatigue strength, to make decisions on minimization of their sizes with the subsequent evaluation of the implemented measures. The studies are performed on the example of a plate for the cam, the configuration of which provides both beam and double console deformations at the same time and the dimensions of which are obtained in [4] as the result of the computer simulation.

Classically, according to the criterion of fatigue strength, the probability of failure-free operation of the part is determined depending on the quantile [6]:

$$u_p = -\frac{\bar{n}-1}{\sqrt{\bar{n}^2 v_{-1D}^2 + v_a^2}}, \tag{1}$$

where $\bar{n} = \bar{\sigma}_{-1D}/\bar{\sigma}_a$ – the coefficient of safety on the average values of the calculated stress $\bar{\sigma}_a$ and the detail fatigue limit $\bar{\sigma}_{-1D}$; accordingly, v_{-1D} and v_a – the coefficients of their variation.

In the probabilistic aspect, in the calculations for the plates of complex shapes for the cams, σ_{-1D} and σ_a are considered as random values, for which the mean values $\bar{\sigma}_{-1D}$ and $\bar{\sigma}_a$ have been determined earlier.

The transition from the material $\bar{\sigma}_{-1}$ to the detail $\bar{\sigma}_{-1D}$ according to the medial values of fatigue is an important step in the calculation of fatigue. The traditional provisions, given in [7], are used:

$$\bar{\sigma}_{-1D} = \frac{\bar{\sigma}_{-1}}{K}, \tag{2}$$

where K – the total coefficient of multifactorial impact, which is determined by the formula:

$$K = \left(\frac{k_\sigma}{k_{d\sigma}} + \frac{1}{k_{F\sigma}} - 1\right)\frac{1}{k_V \cdot k_A}. \tag{3}$$

The coefficients k_σ, $k_{d\sigma}$, $k_{F\sigma}$, k_A and k_v, provided in (3), respectively characterize the influence of concentration of normal stresses, absolute dimensions (scale factor), quality of processed surface, the anisotropy of material and surface strengthening on $\bar{\sigma}_{-1D}$. The values of coefficients are determined according to the tables and graphs, which, in turn, are made according to the results of the experiments, or calculated in the absence thereof.

The dependence [8] is used in the calculation of k_σ:

$$k_\sigma = 1 + q_\sigma(\alpha_\sigma - 1), \tag{4}$$

where q_σ – the coefficient of metal sensitivity to the concentration of normal stresses; α_σ – theoretical coefficient of concentration.

Since for the structural bearing steel SHKH 15, from which the FWP of the cam is made, the limit strength is σ_B = 2410 MPa [9], q_σ = 0.025 is chosen, which is reasonable at σ_B > 1300 MPa.

According to (4), it is possible to take $k_\sigma = 1$. Considering the dimensions of the dangerous section of the FWP console, it is also obvious that there is no influence of the scale factor, i.e. $k_{d\sigma}=1$.

In terms of quantity, the influence of the quality of the detail surface on its fatigue limit is determined by the formulas:

$$k_{F_\sigma} = 1 - 0.22 \lg R_Z \left(\lg \frac{\sigma_B}{20} - 1 \right) \text{ at } R_Z > 1 \text{ mcm; } (a)$$

$$k_{F_\sigma} = 1 \text{ at } R_Z \le 1 \text{ mcm } (b) \tag{5}$$

where R_Z – the roughness of the surface of the plate.

For the FWP of the cam, the average height of fine irregularities is $R_Z \approx 3.2$ mcm [10]. Taking (5, a), we have:

$$k_{F_\sigma} = 1 - 0,22 \lg 3.2 \left(\lg \frac{2410}{20} - 1 \right) = 0.88 .$$

The value of the anisotropy coefficient is determined according to the table [7] at $\sigma_B > 1200$ MPa: $k_A = 0.8$.

Considering the instructions, provided in [7, 8], the following values of the coefficient k_v are recommended: 1.0 – without surface strain hardening; 1.15 – with surface strain hardening. In accordance with the technology of FWPs manufacturing, the value $k_v = 1.15$ should be selected.

Applying the calculated and accepted values of coefficients to the (3), we have $K = 1.24$. As a rule, in the calculations, we have $K = 1.5...3$. The deviation occurs, first of all, because of the small dimensions of the dangerous section of the plate compared to most real parts. The reduction K is also influenced by the technology of the FWP production. Then, according to (2), we have $\overline{\sigma}_{-1D} = 770/1.24 = 621$ MPa.

To determine the statistical characteristics of FWP loading, the range of its loadings (forces F_i and corresponding number of cycles of loading z_i) resulted from the needles actions in different speed modes is presented separately for the needle-lifting cam and the stitch cam, which brings the needles down. The numbers of modes are listed in Table 1.

The forces are calculated by the formula from [1], the reliability of which is verified experimentally:

$$F_{\max} = V_x \cdot tg\alpha \sqrt{\frac{m_{red} \cdot C_{red}}{1+K_c}} + \frac{1}{1+K_c}(F_r + 2h \cdot V_x \cdot tg\alpha \cdot m_{red}) , \tag{6}$$

where V_x – the horizontal component of the needle heel speed, which is equal to the circular speed of the points on the needle cylinder surface; α – the angle of inclination of the working surface of the cam to the horizontal line; m_{red}, C_{red} – the reduced mass and rigidity in the "needle-surface cam" pair; K_C – coefficient that takes into account the deformation of the bend of the needle rod at the moment of impact; F_r – the resistance force of the needle movement in the race, which is created specifically to eliminate the willful lowering of the needles; h – the damping coefficient.

In the calculations, the reduced mass m_{red} is considered equivalent to the mass of the needle and C_{red} is calculated as in case of the series connection of rigidity of the needle during the interaction with a non-deformable cam (which is determined in the manner similar to h, considering the frequency characteristics of the oscillograph chart [1] and the rigidity of FWP of the cam, calculated in [4]).

The results of the calculations of needle loadings on FWPs of the needle-lifting and stitch cams by the formula (6) at different speed modes are presented in Table 1. Analytical determination of stresses in the FWP of the cam as a statically indeterminate spatial construction of complex shapes with two rigid structures is voluminous and uninformative in the end result. Therefore, the computer simulation is used, which also makes it possible to eliminate most of the assumptions made in the analytical approach [11]. A finite-element model of the FWP is developed (Figure 1), where a volumetric 8-node finite element is used, which is simple for calculations and allows obtaining results with a high accuracy when determining the stress-deformed state of the study object. The simulation is performed using Code_Aster software with free access [12,13,14]. The maximum stresses at the FWPs points for all six loading modes are presented in Table 1, and sample screenshots – in Figure 2.

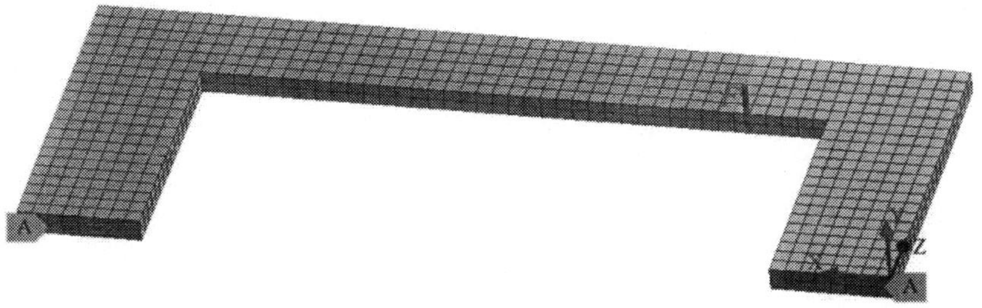

Figure 1. The model of the flexible plate for the cam according to the finite element method: F – force; A – areas of the hardcover of the plate.

a)

b)

Figure 2. The distribution of the fields of stresses σ according to Mises in the flexible plate for the cam at the following loadings: *a)* mode 1 and *b)* mode 6, Table 1.

Since the fatigue strength calculations take into account the stresses at $\sigma_i > 0.5\sigma_{-1D}$ (stresses $\sigma_i < 0.5\sigma_{-1D} = 0.5 \cdot 621 = 310.5$ MPa do not affect the accumulation of fatigue failure [8]), then in the subsequent actions the loadings in the modes 1 and 4 in Table 1 are neglected. It should be noted that since $\sigma_a < \sigma_{-1D}$, the FWP for the cam has the loading on the gigacycle area of the curve of fatigue.

Considering the cyclicity of the automatic half-hose machines work, it is advisable to take forces acting on the cams during the production of one product (sock) as the loading. The number of cycles of loading z_i is set for typical socks with jacquard weaving on the ankle and foot areas, with classic heel and toe, which are knitted during the direct and reverse rotations of the needle cylinder and with additional technological rows. The selection of needles by the cam with an incomplete closing of the third system during the knitting of spandex of the product, by the ornamented gates of three systems during the formation of the pattern on the ankle areas (the average number of included gates is 2,23 gates [15]), by the malfunction of gates on the foot area in the second and third systems, the devices for turning on and off the needles when knitting heel and toe tabs, as well as the number of needles and knitting systems, are considered. The calculation of z_i takes into account the results of the analysis of trajectories of the movement of the needles in relation to the cams when knitting different sections of the products. The formulas for determination of z_i in accordance with the modes of loading and the corresponding calculated numbers are presented in Table 2.

Table 1. Summary information about loading of the cams.

Loading modes			F_i,	σ_i,	z_i,	$p_i = \frac{z_i}{\Sigma z_i}$
i	Type of cam	V_x, m/s (areas of the product)	N	MPa	cycles	
1	Needle-lifting $\alpha = 38.0^o$	$V_x = 0.6$ m/s (heel and toe)	10.54	254.7	24824	-
2		$V_x = 1.1$ m/s (spandex)	13.14	317.53	14112	0.148
3		$V_x = 1.3$ m/s (ankle, foot and technological rows)	14.18	342.66	81480	0.852
4	Stitch $\alpha = 47.5^o$	$V_x = 0.6$ m/s (heel and toe)	11.74	283.7	15836	-
5		$V_x = 1.1$ m/s (spandex)	15.42	372.63	9366	0.274
6		$V_x = 1.3$ m/s (ankle, foot and technological rows)	16.87	407.67	24790	0.726

The real loading with an obvious regularity of alternation of different levels during the cycle of knitting of one typical product is replaced by that one, which is equivalent in the degree of accumulation of the fatigue failure, by the formula [16]:

$$\overline{F} = \sum_i p_i \cdot F_i , \tag{7}$$

where $p_i = z_i / \sum z_i$ – the relative occurrence of the loading F_i (the calculated values are given in Table 1).

Table 2. Formulas for calculating the numbers of cycles loading of the cams in accordance with the technological modes.

Cam	Loading mode		Number of cycles of loading z_i	
	i	V_x, m/s (areas of the product)	Formula for calculation	Number
Needle-lifting	1	$V_x = 0.6$ m/s (heel and toe)	$z_1 = (n_{hl} + n_{toe})(2k^{0-1305} + 4k^{0-1306}) / m$	24824
	2	$V_x = 1.1$ m/s (spandex)	$z_2 = 2n_{sp}(k_2^{0-1305} + k_2^{0-1306} + k_2^{0-1308})$	14112
	3	$V_x = 1.3$ m/s (ankle, foot and technological rows)	$z_3 = \frac{3}{r}(n_{ank} + n_{ft} + 2n_{tr})(\kappa_3^{0-1305} + \kappa_3^{0-1306} + \kappa_3^{0-1308})$	81480
Stitch	4	$V_x = 0.6$ m/s (heel and toe)	$z_4 = 2(n_{hl} + n_{toe})(\kappa_4^{0-1305} + \kappa_4^{0-1306}) / m$	15836
	5	$V_x = 1.1$ m/s (spandex)	$z_5 = 1.33n_{sp}(k_5^{0-1305} + k_5^{0-1306} + k_5^{0-1308})$	9366
	6	$V_x = 1.3$ m/s (ankle, foot and technological rows)	$z_6 = (0.91n_{ank} + 1.12n_{ft} + 2n_{tr})(\kappa_6^{0-1305} + \kappa_6^{0-1306)} + {} + (0.91n_{ank} + 0.68n_{ft} + 2n_{tr})\kappa_6^{0-1308} / r$	24790

Note. In Table 2, the following notations are used:
1. $m = 2$ – the number of simultaneously selected needles for one reverse movement of the cylinder; r – the number of cams, which make the looped rows at one rotation of the cylinder;
2. n_{hl}, n_{toe}, n_{sp}, n_{ank}, n_{ft}, n_{tr} – the number of looped rows on the following areas of the product: heel and toe, spandex, calf, foot, and technological rows;
3. κ_1^{0-1305}, κ_1^{0-1306}, κ_4^{0-1305}, κ_4^{0-1306} – the number of included needles of the positions 0-1305 and 0-1306 when knitting the sock heel and toe in the modes 1 and 4; κ_2^{0-1305}, κ_2^{0-1306}, κ_2^{0-1308}, κ_5^{0-1305}, κ_5^{0-1306}, κ_5^{0-1308} – the number of included needles of the positions 0-1305, 0-1306 and 0-1308 when knitting the spandex in the modes 2 and 5; κ_3^{0-1305}, κ_3^{0-1306}, κ_3^{0-1308}, κ_6^{0-1305}, κ_6^{0-1306}, κ_6^{0-1308} – the number of needles of the positions 0-1305, 0-1306 and 0-1308 when knitting the ankle, foot and technological rows in the modes 3 and 6 accordingly.

Applying the values, presented in Table 1, to the (7), we have $\overline{F}_{lif}$=14.03 N and $\overline{F}_{st}$=16.47 N for the needle-lifting cam and the stitch cam respectively. Using the computer simulation, the corresponding equivalent stresses $\overline{\sigma}_{lif}$ = 339.04 MPa (Figure 3, *a*) and $\overline{\sigma}_{st}$ = 398.0 MPa (Figure 3, *b*) are determined according to the median values of equivalent stresses.

The analysis of fatigue limits of materials and details indicates their variations due to the structural heterogeneity of the material of one melting, changes in mechanical characteristics in different melts, variations of dimensions, random changes in the modes of mechanical, thermal or chemical treatment, etc., which are taken into account in the general coefficient of variation. According to the rule of quadratic addition of separate coefficients of variation, the coefficient of variation of the detail fatigue limit is determined by the formula [6]:

$$v_{\sigma_{-1D}} = \sqrt{v_{\sigma_{\max}}^2 + v_{\overline{\sigma}_{-1}}^2 + v_{\alpha_\sigma}^2}, \tag{8}$$

where $v_{\sigma_{\max}}$, $v_{\overline{\sigma}_{-1}}$ and v_{α_σ} – separate coefficients of variation of the maximum stresses in the zone of concentration of average (in one melting) fatigue limits of smooth laboratory specimens, theoretical coefficient of concentration of stresses α_σ.

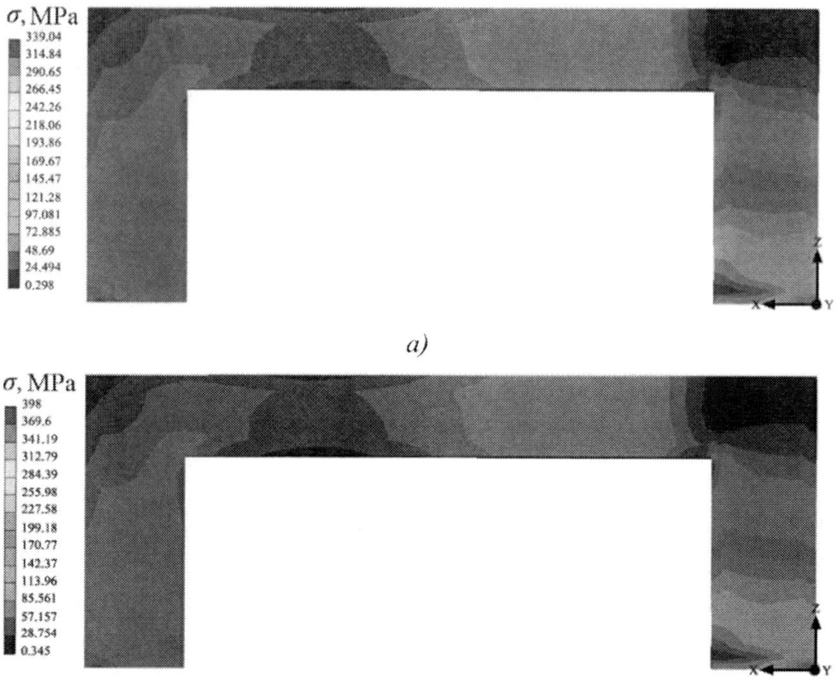

Figure 3. The distribution of the fields of stresses σ according to Mises in the flexible plate for cam at the following loadings: *a)* $\overline{F}_{lif}$=14.03 N and *b)* $\overline{F}_{st}$=16.47 N.

According to [6], in the first approximation $v_{\sigma_{\max}} \leq 0.1$ we chose $v_{\sigma_{\max}} = 0.1$. Considering the linear dependence between the fatigue limits and the limit strengths of the materials [8], it is acceptable to assume that $v_{\bar{\sigma}_{-1}} = v_{\sigma_B}$, where v_{σ_B} – the coefficient of variation of the metal strength limit considering the number of all melts. Since for the alloy steel the indicator v_{σ_B} is within the range 0.1…0.16, we used $v_{\bar{\sigma}_{-1}} = 0.13$ in the calculations.

Considering the small sizes of the dangerous section of the plate, the effect of v_{α_σ} on $v_{\sigma_{-1D}}$ is neglected. Therefore, according to (8), we finally have $v_{\sigma_{-1D}} = 0.164$.

4. Results

According to (1), standard deviations of the operating loads are $S_{lif} = 1.375$ N for the needle-lifting cam and $S_{st} = 1.416$ N for the stitch cam. Accordingly, the coefficients of variation are $v_{alif} = S_{lif} / \overline{F}_{lif} = 0.098$ and $v_{ast} = S_{st} / \overline{F}_{st} = 0.086$. The calculated values of coefficients of variation fall under the assertion provided in [6] that $v_{\bar{\sigma}_{-1}} > v_a$.

Therefore, at (1) for the needle-lifting cam at the assurance coefficient at the average stresses $n_{lif} = \overline{\sigma}_{-1D} / \overline{\sigma}_{lif} = 621/339.04 = 1.832$ and at $v_{\sigma_{-1D}} = 0.164$ and $v_a^{lif} = 0.098$, we have $u_{P_{lif}} = -4.95$, which allows determining the probability of failure-free operation $p(t) \to 1$ using the tables [6].

Accordingly, for the stitch cam $n_{st} = \overline{\sigma}_{-1D} / \overline{\sigma}_{st} = 621/398.0 = 1.560$, $v_{\sigma_{-1D}} = 0.164$, $v_a^{st} = 0.086$, $u_{P_{st}} = -2.077$ and $p(t) = 0.981$, which is sufficient for the technical systems.

If we set a goal that both needle-lifting cam and stitch cam have the same probability of failure-free operation, for example, $p(t) = 0.981$, then, concerning the needle-lifting cam, we should solve the inverse task on determination of geometric parameters of its FWP, which would provide the corresponding equivalent stress $\overline{\sigma}_{lif} = 398.0$ MPa. The adjustment is made by the width of the console beam B, where we have the dangerous section. The screenshot at such a stress at $B = 3.17$ mm is presented in Figure 4.

Figure 4. The distribution of the fields of stresses σ according to Mises in the flexible plate for cam at the loading $\overline{F}_{lif}$ =14.03 N and the width of the console beam of the plate $B = 3.17$ mm.

Using the results of similar calculations, the graphical dependence of $p(t)$ on B (Figure 5) is constructed, which is convenient to use when designing a plate for the needle-lifting cam at any given probability of failure-free operation.

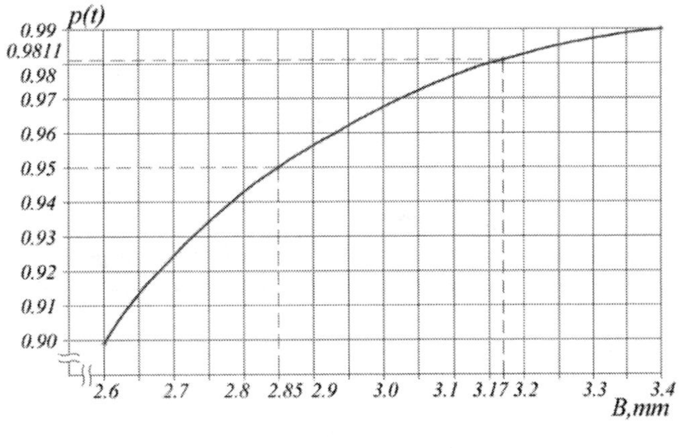

Figure 5. The dependence of probability of failure-free operation $p(t)$ on the width of the console beam of the plate *B*.

5. Conclusions

The comprehensive approach to estimation of the probability of failure-free operation of plate of complex shape for the cam of automatic half-hose machine, using the criterion of fatigue strength, is proposed, which includes dynamic analysis, computer simulation by the finite element method (FEM), and principles of mathematical statistics and Theory of Probability. On the example of the plate of the cam as the detail, the sizes of which are limited, the feasibility of using the strength analysis in probabilistic figuration is shown, which in comparison with the traditional calculations (that consider the normalized coefficients of safety) provides a given level of reliability together with minimization of sizes. To describe the stress state of the plate as a statically indeterminate spatial construction of complex shape with two rigid structures, the finite element method is used in Code_Aster software. The proposed computational approach can be used to generate the cams with flexible work plates of different configurations, including under the equal stress condition.

References

[1] Berezin L M 2013 *Evaluation of the durability and reliability of knitting mechanisms of automatic half-hose machines (*Kyiv: KNUTD) 191
[2] Pipa B F and Pleshko S A 2012 *Improvement of working bodies of mechanisms of knitting circular knitting machines* (Kyiv: KNUTD) 471
[3] Patel N S 2015 Modelling, design and analysis of cam and follower-a review *International Journal of Engineering Studies and Technical Approach Electronic Materials* **1** pp 1-7
[4] Berezin L M and Savchenko K V 2019 Modelling of a flexible grinder of a complex shape form on the criteria of deformation and strength *Herald of Khmelnytskyi national university Electronic Materials* **6** pp 24-27
[5] Berezin L M 2016 Calculation of the flexibility plate of the cam according to the criterial of rigidity and longevity *Bulletin of the Kyiv National University of Technologies and Design Electronic Materials* **3** pp 68-73
[6] Reshetov D N, Ivanov A S and Fadeev V Z 1988 *Reliability of machines* (Moscow: Vyshaia shkola) 238
[7] State Standart 25.504-82 1982 *Strenqth calculation and testinq. Methods of fatique strenqth behaviour calculation* (Moscow: Standartinform Publ) 29
[8] Kogaev V P and Drozdov Yu N 1991 *Durability and wear resistance of machine details* (Moscow: Vyshaia shkola) 319
[9] Troshchenko V T and Sosnovskiy L A 1987 *Fatigue Strength of Metals and Alloys* (Kyiv: Naukova Dumka) 808
[10] Chaban V V, Pipa B F and Prokopenko A K 2014 *The effect of lubrication and metal cladding nanotechnology on increasing the efficiency of knitting machines* (Kyiv: KNUTD) 176
[11] Suriyaninov N G and Pavlenko I V 2013 *Calculation of plates by numerical-analytical method of boundary elements* (Düsseldorf: LAB LAMBERT Academic Publishing) 178
[12] Aubry J P 2013 *Beginning with Code_aster* (Paris: Framasoft) 329
[13] Code_Aster https://www.code-aster.org/V2/spip.php?rubrique2
[14] Analyse de la tenue sismique des grands réservoirs métalliques https://code-aster.org/doc/default/fr/man_u/u2/u2.06.11.pdf
[15] Amro M and Voloshchenko V P 1981 The study of the loading mechanism of the pattern of hosiery machines OZD *University News. Light industry technology Electronic Materials* **2** pp 113-117
[16] Konovalov L V 1981 *Loadiness, fatigue, reliability of metallurgical machinery details* (Moscow : Metallurgiya) 280

Contact heat transfer of a cutting diamond wheel with a boundary layerof air

A Bespalova[1,*], **O Faizulyna**[1], **V Lebedev**[2], **O Frolenkova**[2] **andT Chumachenko**[2]

[1]Odessa State Academy of Civil Engineering and Architecture, 4, Didrichsona St., Odessa, 65029, Ukraine
[2]Odessa National Polytechnic University, 1, Shevchenko avenue, Odessa, 65044, Ukraine

[*]E-mail: bespalova.a.v.2015@gmail.com

Abstract. Cutting of natural and artificial building materials is most often carried out with diamond cutting wheels on a metal base at cutting speeds of about 50-80 m/s. The intensity of the cutting process causes a significant heat release, as a result of which the wheel temperature rises to unacceptable values. The value of these unacceptable temperatures is about 600 - 650°C.

At these temperatures, graphitization of diamond grains occurs, i.e. loss of diamond layer and loss of cutting properties. In addition, a thin diamond wheel (thickness 1 - 3 mm) is deformed, which leads to jamming and its tensile strength at these temperatures is reduced by half, which creates the risk of rupture by centrifugal forces. In this work, it is taken into account that during the rotation of the wheel, a boundary layer of air is created around it, which is stationary relative to the wheel. Consequently, contact heat transfer occurs between the wheel and the boundary layer, and then convective heat transfer occurs between the boundary layer and the surrounding air. This scheme allows you to more accurately determine the time of safe operation of the diamond wheel. Contact heat transfer between the wheel and the boundary layer is not effective enough to lower the temperature. When air with a negative temperature is introduced into the boundary layer by means of a Rank-Hillsch tube, the wheel temperature decreases by about 10%.

When a sprayed coolant (fog cooling) is introduced into the boundary layer by means of an ejector tube, the wheel temperature decreases by 25%, which ensures an increase in the time of continuous operation.

1. Introduction

Cutting of natural and artificial building materials is most often carried out with diamond cutting wheels on a metal base at cutting speeds of about 50-80 m/s. The intensity of the cutting process causes a significant heat release, as a result of which the wheel temperature rises to unacceptable values. The value of these unacceptable temperatures is about 600 - 650 °C.

At these temperatures, graphitization of diamond grains occurs, i.e. loss of diamond layer and loss of cutting properties. In addition, a thin diamond wheel (thickness 1 – 3 mm) is deformed, which leads to jamming and its tensile strength at these temperatures is reduced by half, which creates the risk of rupture by centrifugal forces.

Thus, the heating temperature of the wheel should not exceed 600 °C. Therefore, the operating time of a diamond cutting disc is the time during which it is heated during continuous operation to a temperature of 600 °C. The longer this time, the higher the resistance of the diamond blade. In the present work, mathematical modeling is performed taking into account contact heat transfer between a rotating wheel and a boundary layer

The simulation of the process of interaction of the wheel with the environment is carried out according to the results of which it is possible to determine the time of wheel performance. However, in this paper, convective heat transfer between the wheel and the surrounding air is considered at a

Content from this work may be used under the terms of the Creative Commons Attribution 3.0 licence. Any further distribution of this work must maintain attribution to the author(s) and the title of the work, journal citation and DOI.
Published under licence by IOP Publishing Ltd

time when the heat transfer process is more complex. When the wheel rotates around it, a boundary layer of air is created that is stationary relative to the wheel. Consequently, contact heat transfer occurs between the wheel and the boundary layer, and then convective heat transfer occurs between the boundary layer and the surrounding air. This scheme allows you to more accurately determine the time of safe operation of the diamond wheel.

2. References review

A significant number of research works devoted to this subject firstly consider convective heat transfer between the wheel and air, moreover, at high Reynolds numbers, which does not correspond to our case. Works [1-12] precisely consider precisely such cases, therefore, the data presented in these works cannot be used in our studies.

2.1. Research Methodology

The purpose of this work is to investigate the contact heat transfer process between a wheel and a boundary layer of air, based on which to determine the possibilities of cooling a rotating wheel by changing the thermophysical characteristics of the boundary layer.

The tasks to be solved in this article are as follows.

1. Mathematical modeling of the heat transfer process to determine the intensity of the latter.
2. Mathematical modeling of the wheel cooling process when changing the thermophysical characteristics of the boundary layer.

Calculations are carried out in accordance with the scheme presented in Figure1.

Here is a solution for a thin rotating wheel heated at the end in the contact area and cooled from the side surfaces as a result of contact heat exchange with the boundary layer. Figure 1 [13, 14].

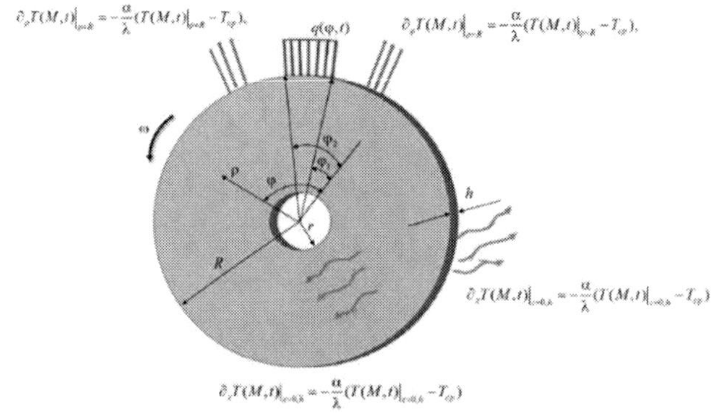

Figure 1. Design scheme for cooling of a rotating wheel.

A thick wheel with thickness h rotates in a plane XOY with angular velocityω. At the end of the circle, within the limits of the contact arc, a heat source of intensity $q(\varphi, t)$ is defined, depending on the cutting conditions. On the lateral surfaces of the wheel and outside the contact arc, heat transfer occurs at the end according to the Newton-Richmann law, and on the lateral surfaces of the wheel heat transfer is considered not with the environment, but with the boundary layer, the temperature of which can be varied over a wide range up to -50 °C.

The boundary-value heat conduction problem for a thin wheel in the presence of heat transfer through the side surfaces, taking into account the angular velocity in the polar coordinate system (ρ,φ), has the form:

$$\partial_t T = \alpha(\partial_\rho^2 + \rho^{-1}\partial_\rho + \rho^{-2}\partial_\varphi^2)T + \upsilon_\rho \partial_\rho T + \omega \partial_\varphi T - \frac{2\alpha_*}{\rho c b}(T - T_{bl}),$$
$$\partial_\rho = \frac{\partial}{\partial \rho},\quad \partial_\varphi = \frac{\partial}{\partial \varphi},\quad \partial_t = \frac{\partial}{\partial t},\quad T \equiv T(\rho,\varphi,t). \tag{1}$$

Initial condition

$$T(\rho,\varphi,t)\big|_{t=0} = T_0. \tag{2}$$

Boundary conditions

$$\lambda\partial_\rho T(M,t)\Big|_{\rho=R} + \alpha(T(M,t)\big|_{\rho=R} - T_{cp}) = 0,\ \varphi \not\subset [\varphi_1,\varphi_2],$$
$$\partial_\rho T(M,t)\Big|_{\rho=R} = -\frac{q(\varphi,t)}{\lambda},\varphi \not\subset [\varphi_1,\varphi_2], \tag{3}$$

where T_{bl} boundary layer temperature; T_0 initial wheel temperature; α_*–coefficient of convective heat transfer between rotating disc and boundary layer; c – specific heat, (Jkg·grad); ρ - substance density (kg/m^3) ρc – (J/m^3·deg); α – coefficient of heat output; λ – coefficient of thermal conductivity; b – wheel thickness; T_{cp}– ambient temperature.

Using substitution

$$T(M,t) = \Theta(M,t)\exp\left(-\frac{\omega\rho}{2a}\varphi - \frac{\omega^2}{4a}t\right).$$

The boundary problem equations (1) - (3) is modified to the form:

$$\partial_t\Theta(M,t) = \alpha\left(\partial_r^2 + r^{-1}\partial_\rho + r^{-2}\partial_\phi^2\right)\Theta(M,t) - \frac{2\alpha}{\rho c h}(\Theta(M,t) - T_{cp}),\quad M = M(r,\phi). \tag{4}$$

$$\Theta(M,t)\big|_{t=0} = \Theta_0,$$
$$\lambda\partial_\rho\Theta(M,t)\big|_{\rho=R} = -\alpha(\Theta(M,t)\big|_{\rho=R} - T_{cp}),\ \varphi \notin [\varphi_1,\varphi_2], \tag{5}$$
$$\partial_\rho\Theta(M,t)\big|_{\rho=R} = f(\varphi,t),\ f(\varphi,t) = \frac{q(\varphi,t)}{\lambda}\partial_r \exp\left(\frac{1}{2\alpha}rv + \frac{1}{4\alpha}\right)\sum_{j=1}^{2}\upsilon_j^2,\ \varphi \notin [\varphi_1,\varphi_2].$$

In order to avoid the application of the Laplace transform and the difficulties with its treatment, we proceed as follows. We divide the time interval T into M intervals of length $h = TM^{-1}$ and replace the time derivative by the difference relation

$$\partial_t\Theta(r,\varphi,t) = \frac{\Theta_j(r,\varphi) - \Theta_{j-1}(r,\varphi)}{h},\ \Theta_j(r,\varphi,jh),\ j = 1, 2, ...$$

then

$$\partial_\rho^2 + r^{-1}\partial_\rho + r^{-2}\partial_\varphi^2)\Theta_j(r,\varphi) - \mu^2\Theta_j(r,\varphi) = F_j(r,\varphi), \tag{6}$$

$$\Theta(M,t)\big|_{t=0} = \Theta_0,$$
$$\lambda\partial_\rho\Theta(M,t)\big|_{\rho=R} = -\alpha(\Theta(M,t)\big|_{\rho=R} - T_{cp}),\ \varphi \notin [\varphi_1,\varphi_2], \tag{7}$$
$$\lambda\Theta(M,t)\big|_{\rho=R} = f(\varphi,t),\ f(\varphi,t) = \frac{q(\varphi,t)}{\lambda}\partial_\rho \exp\left(\frac{1}{2\alpha}rv + \frac{t}{4\alpha}\sum_{j=1}^{2}\upsilon_j^2\right),\ \varphi \notin [\varphi_1,\varphi_2],$$

where

$$\mu^2 = \frac{2\alpha}{\rho ach} + \frac{1}{h\alpha}, \quad F_j(r,\varphi) = T_{cp} - (h\alpha)^{-1}\Theta_{j-1}(r,\varphi).$$

Let us construct a discontinuous solution [15] of the heat equation for an unbounded plane $0 \le \rho \prec \infty$, $|\varphi| < \pi$ containing a circular defect occupying the region $r = R, -\pi \le \varphi \le \pi$, upon transition through which they suffer discontinuities of continuity of the first kind, the temperature $\Theta_j(\rho,\varphi)$ and the heat flux $\partial_\rho\Theta_j(\rho,\varphi)$ with given jumps.)

$$\begin{aligned} &\Theta_j(R+0,\varphi) - \Theta_j(R-0,\varphi) = \langle \Theta(R,\varphi) \rangle, \\ &\partial_\rho\Theta_j(R+0,\varphi) - \partial_\rho\Theta_j(R-0,\varphi) = \langle \partial_\rho\Theta_j(R,\varphi) \rangle, \end{aligned} \tag{8}$$

those, a solution that satisfies the heat equation everywhere except for defect points. At these points, jumps in temperature and heat flux are set.

To construct a discontinuous solution of equation (6) with jumps equation (8), we apply the finite Fourier transform in the variable φ:

$$\Theta_j(\rho,\varphi) = \sum_{n=-\infty}^{\infty} \Theta_{j,n}(\rho)e^{in\phi}, \Theta_{j,n}(\rho) = \Phi_n[\Theta_j] \equiv \frac{1}{2\pi}\int_{-\pi}^{\pi} \Theta_j(\rho,\varphi)e^{in\phi}d\phi, j = 1,2,...$$

and, Hankel transform, according to a generalized scheme [15]:

$$\Theta^j_{n,a} = \int_0^\infty \rho\Theta_{j,n}(\rho)J_n(a\rho)d\rho = \left(\int_0^{R=0} + \int_{R+0}^\infty\right)\rho\Theta_{j,n}(\rho)J_n(a\rho)d\rho,$$

where $J_n(z)$ – Bessel function.

As a result, we obtain:

$$\Theta^j_{n,a} = \frac{R}{a^2+\mu^2}\left\{J_n(\alpha R)\langle \partial\Theta_{j,n}(R)\rangle - \langle \Theta_{j,n}(R)\rangle \partial_r J_n(\alpha R)\right\} + \frac{F^j_{n,\alpha}}{a^2+\mu^2}.$$

where $F^j_{n,\alpha} = \int_0^\infty \rho F_{j,n}(\rho)J_n(\alpha\rho)dp$, $F_{j,n}(\rho) = \Phi_n[F_j(\rho,\varphi)]$.

After reversing the Hankel transform, the required discontinuous solution of the heat equation in Fourier transforms can be written as:

$$\begin{aligned} &\Theta_{j,n}(\rho) = R\left\{\langle \partial\Theta_{j,n}(R)\rangle G_{n,\lambda}(\rho,R) - \langle \Theta_{j,n}(R)\rangle \partial_\rho G_{n,\lambda}(\rho,R)\right\} + F_{j,n}(\rho), \\ &G_{n,\lambda}(\rho,R) = \int_0^\infty \frac{\alpha J_n(\alpha\rho)J_n(\alpha R)}{\alpha^2+\tilde{\mu}^2}d\alpha, F_{j,n}(\rho) = \int_0^\infty \frac{\alpha J_n(\alpha\rho)F^j_{n,\alpha}}{\alpha^2+\tilde{\mu}^2}d\alpha. \end{aligned} \tag{9}$$

Taking into account that:

$$\left\{\langle \Theta_{j,n}(R)\rangle, \langle \partial\Theta_{j,n}(R)\rangle\right\} \equiv \frac{1}{2\pi}\int_{-\pi}^{\pi} \left\{\langle \Theta_j(R,\psi)\rangle, \langle \partial\Theta_j(R,\psi)\rangle\right\} e^{in\psi}d\psi.$$

We write the discontinuous solution equation (9) in the form:

$$\Theta_j(\rho,\varphi) = \frac{R}{2\pi}\left\{\int_{-\pi}^{\pi} \chi_j^{1,-}(\psi)G_{n,\lambda}(\rho,R)e^{in\psi}d\psi - \int_{-\pi}^{\pi} \chi_j^{2,-}(\psi)\partial_R G_{n,\lambda}(\rho,R)e^{in\psi}d\psi\right\} + F_{j,n}(\rho),$$

$$\chi_j^{1-}(\psi) = \langle \Theta_j(R,\psi)\rangle, \quad \chi_j^{2-}(\psi) = \langle \partial\Theta_j(R,\psi)\rangle.$$

Using the inverse finite Fourier transform formula, by φ, and the addition theorem [16].

$$\sum_{n=-\infty}^{\infty} J_n(\alpha r)J_n(\alpha R)e^{in\varphi} = J_0(\alpha, r_*), \quad r_*^2 = r^2 + R^2 - 2rR\cos\varphi.$$

we obtain:

$$\Theta_j(r,\varphi) = \frac{R}{2\pi}\left\{\int_{-\pi}^{\pi} \chi_j^{1,-}(\psi)K(\rho,R,\varphi-\psi)d\psi - \partial_\rho \int_{-\pi}^{\pi} \chi_j^{1,-}(\psi)K(\rho,R,\varphi-\psi)d\psi\right\} + F_j(\rho,\varphi)\,.$$

If we use equation (17) from [2], we obtain

$$K(\rho,R,\varphi) = \int_0^{\infty} \frac{\alpha J_0(\alpha,r_*)}{\alpha^2+\mu^2} d\alpha = K_0(r_*\mu)\,.$$

Then:

$$\Theta_j(r,\varphi) = \frac{R}{2\pi}\left\{\partial_R \int_{-\pi}^{\pi} \chi_1^{1,-}(\psi)K_0(r_*,\bar{\mu})d\psi - \int_{-\pi}^{\pi} \chi_2^{1,-}(\psi)K_0(r_*,\bar{\mu})d\psi\right\} + F_j(\rho,\varphi)\,. \quad (10)$$

We illustrate the technique of using discontinuous solutions to solve the boundary value problem. This technique is based on the idea that the boundary of the wheel $\rho = R$ be considered a defect. We consider the third boundary value problem:

$$\Theta_j(R-0,\varphi) + \kappa\partial_\rho\Theta_j(R-0,\varphi) = f_j(\varphi)\,. \quad (11)$$

Given that outside the domain specified in the conditions of the problem, the solution is identical to zero, i.e. a $\rho \succ R, \Theta_j(\rho,\varphi) \equiv 0$, и $(\Theta_j(R+0,\varphi) = \partial_\rho\Theta_j(R+0,\varphi) = 0)$, then on the basis of equation (11), we can write the following:

$$\chi_j^{2,-} = \langle\partial_\rho\Theta_j(R,\varphi)\rangle = -\kappa^{-1}(f_j(\varphi) - \Theta_j(R-0,\varphi), \kappa = \lambda/\alpha\,. \quad (12)$$

As a solution to the boundary value problem equations (6), (12), we will use the discontinuous solution equation (10) with the jumps obtained here:

$$\Theta_j(r,\varphi) = \frac{R}{2\pi}\left\{-\partial_R \int_{-\pi}^{\pi} \chi_j^{1,-}(\psi)K_0(r_*,\bar{\mu})d\psi + \int_{-\pi}^{\pi} \frac{f_j(\varphi) - \Theta_j(R-0,\varphi)}{\kappa} K_0(r_*,\bar{\mu})d\psi\right\} + F_j(\rho,\varphi)\,. \quad (13)$$

To get an equation to determine an unknown boundary value: $\chi_j^{1,-}(\psi) = \Theta_j(R-0,\varphi)$, make the limit transition in equation (13) $\rho \to R-0$, considering that:

$$\lim_{\rho\to R} K_0(\mu\sqrt{\rho^2+R^2-2\rho\cos(\varphi-\psi)} = K_0(\mu 2^{1/2}R\sqrt{1-\cos(\varphi-\psi)} = K_0(\mu 2R|\sin(\varphi-\psi)/2|).$$

And that $\sin(\varphi-\psi)/2 \approx (\varphi-\psi)/2$, at $\varphi \to \psi$, as well as representing the MacDonald function as:

$$K_0(z) = \ln\left(\frac{2}{z}\right) + \ln\left(\frac{2}{z}\right) + \sum_{m=0}^{\infty}\left(\frac{z}{2}\right)^m \frac{1}{\Gamma(m+1)m!} + \sum_{m=0}^{\infty}\left(\frac{z}{2}\right)^m \frac{\psi(m+1)}{(m!)^2}\,.$$

We can say that:

$$\left[\partial_\rho \int_{-\pi}^{\pi} \chi_j^{1,-}(\psi) K_0(r_* \mu) d\psi\right]_{\rho=R\pm 0} = \frac{\nu \chi_j^{1,-}(\varphi)}{2}; \ \nu = -1 - \frac{\gamma}{2},$$

where γ – Euler constant.

As a result, we obtain:

$$\Theta_j(\rho,\varphi) = \frac{R}{2\pi}\left\{\int_{-\pi}^{\pi} \frac{f_i(\psi) - \chi_j^{1,-}(\psi)}{\kappa} K_0(\mu|\sin(\varphi-\psi)/2|) d\psi\right\} + \frac{\nu\chi_j^{1,-}(\varphi)}{2} + F_j(R,\varphi).$$

This relation can be written as an integral equation:

$$\chi_j^{1,-}(\phi) + \frac{R}{2\pi}\int_{-\pi}^{\pi} \frac{f_j(\psi) - \chi_j^{1,-}(\psi)}{\kappa(\nu-2)} K_0(\mu|\sin(\phi-\psi)/2|d\psi\} + \frac{\nu\chi_j^{1,-}(\phi)}{2} = F_j(R,\phi).$$

Using the MacDonald function representation, in the form:

$$K_0(\mu|\sin(\varphi-\psi)/2) = \int_0^\infty \frac{\alpha}{\alpha^2+\mu^2} \sum_{n=-\infty}^{\infty} J_n(\alpha R) J_n(\alpha R) e^{in(\phi-\psi)} da$$

we apply the finite Fourier transform to this equation, taking into account its properties and the convolution theorem:

$$\Phi_n\left[\partial_\varphi^k f\right] = (in)^k \Phi_n[f], \Phi_n[f] = \int_{-\pi}^{\pi} f(\varsigma) h(\varphi-\varsigma) d\varsigma = 2\pi f_n h_n, \Phi_n\left[e^{im\varphi}\right] = \frac{1}{2\pi}\int_{-\pi}^{\pi} e^{i\varphi(m-n)} d\varphi = \delta_{m,n}$$

We obtain:

$$\chi_{j,n}^{1,-} = \tau_n^{-1}[F_{j,n} - f_{j,n}], \ \tau_n = 1 - \frac{R}{\kappa(\nu-2)} h_n,$$

$$h_n = \int_0^\infty \frac{a}{a^2+\mu^2} \sum_{m=-\infty}^{\infty} J_m(aR) J_m(aR) \int_{-\pi}^{\pi} e^{i(m-n)\varphi} d\varphi = \int_0^\infty \frac{\alpha[J_m(\alpha R)]^2}{a^2+\mu^2},$$

$$f_{j,n} = \int_{-\pi}^{\pi} f_j(\Psi) e^{in\varphi} d\varphi, \ F_{j,n} = \int_{-\pi}^{\pi} F_{j,n}(R,\varphi), e^{in\varphi} d\varphi.$$

Inverting the Fourier transform [16], we obtain:

$$\chi_{j,n}^{1,-} = \int_{-\pi}^{\pi} (F_j(\eta) - f_j(\eta)) \sum_{n=-\infty}^{\infty} \frac{e^{-in(\phi-\eta)}}{\tau_n} d\eta. \quad (14)$$

We introduce the unknown function $\psi(\varphi)$, then the boundary conditions equation (7) are written as follows:

$$\Theta_j(R+0,\varphi) + \kappa\partial_\rho\Theta_j(R+0,\varphi) = g_-(\varphi) + \psi_+(\varphi),$$

$$g_-(\varphi) = 0, \ |\varphi| \in (\varphi_1,\varphi_2), \ g_-(\varphi) = g(\varphi), \ |\varphi| \in (\varphi_1,\varphi_2), \quad (15)$$

$$\psi_+(\varphi) = \psi(\varphi), \ |\varphi| \in (\varphi_1,\varphi_2), \psi_+(\varphi) = 0, \ |\varphi| \in (\varphi_1,\varphi_2).$$

If we assume that the functions $g_-(\varphi)$ and $\psi_+(\varphi)$ are known, then we pass to the third main boundary-value problem, the solution of which was obtained above. Using equation (14), we can write the following:

$$\Theta_j(R-0,\varphi) = -\int_{-\pi}^{\pi}(g_-(\eta)+\psi_+(\eta))\Psi(\varphi-\eta)d\eta + \int_{-\pi}^{\pi}F_j(\eta)\Psi(\varphi-\eta)d\eta,$$

$$\Psi(\varphi-\eta) = \sum_{n=-\infty}^{\infty}\frac{e^{-in(\varphi-\eta)}}{\tau_n}d\eta.$$

Now satisfying the first boundary condition from equation (7) and taking into account equation (15) we obtain the integral equation:

$$\frac{\partial}{\partial\varphi}\int_{\varphi_1}^{\varphi_2}\Psi_j(\eta)\Psi(\varphi-\eta)d\eta = F(\varphi),$$

$$F(\varphi) = \int_{-\pi}^{\pi}F_j(\eta)\Psi(\varphi-\eta)d\eta - \frac{\partial}{\partial\varphi}\left(\int_{-\pi}^{\varphi_1}+\int_{\varphi_2}^{\pi}\right)g(\eta)\Psi(\varphi-\eta)d\eta.$$

In Figures 2–4 show plots of temperature changes depending on the polar angle φ and radius ρ and various values of the temperature of the boundary layer T_{bl}; moreover, Figure 2 show plots at an angular velocity of $\omega r = 50 m/s$, and in Figure 3 at $\omega r = 80 m/s$. The temperature of the boundary layer was varied using the Ranque-Hills tube. The temperature was determined at an operating time of 60 s.

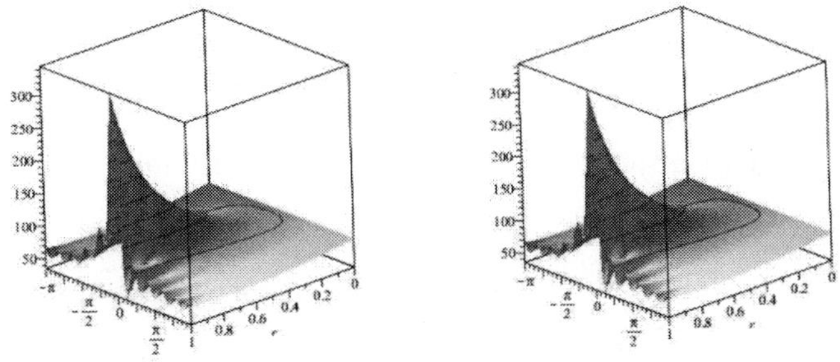

a b

Figure 2. a) The temperature of the wheel at the temperature of the boundary layer + 200 °C;
b) the temperature of the wheel at the temperature of the boundary layer –500 °C.
V_{wheel} = *50 m/s*.

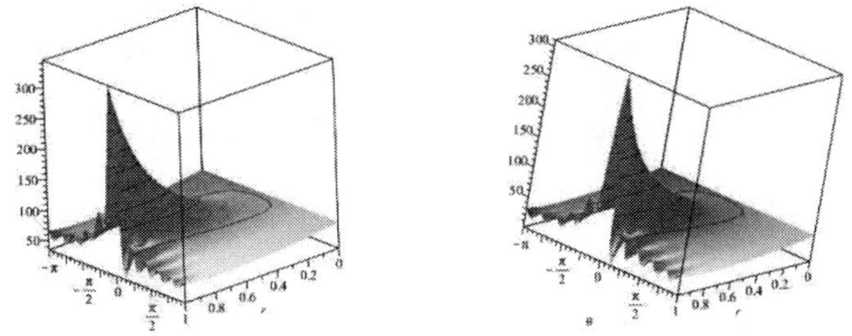

a b

Figure 3. a) The temperature of the wheel at the temperature of the boundary layer + 20° C;
b) the temperature of the wheel at the temperature of the boundary layer -50° C.
V_{wheel} = *80 m/s*.

Additional mathematical modeling of the process of introducing the boundary layer of atomized coolant (fog) showed that the temperature of the wheel decreases significantly, as shown in Figure 4.

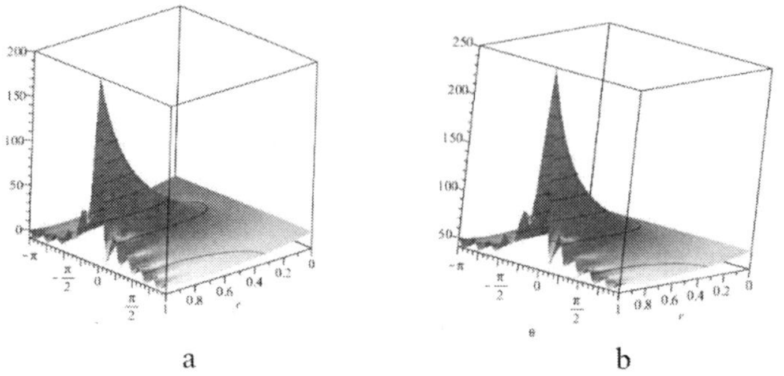

Figure 4. a) Circle temperature when sprayed coolant is introduced into the boundary layer (fog). $V_{wheel} = 50$ *m/s*; b) circle temperature when sprayed coolant (fog) is introduced into the boundary layer. $V_{wheel} = 80$ *m/s*.

3. Conclusions

Contact heat transfer between the circle and the boundary layer is not effective enough to reduce the temperature of the wheel

When air with negative temperature is introduced into the boundary layer by means of a Ranque-Hillsch tube, the wheel temperature decreases by about 10%.

A slight decrease in temperature during contact heat transfer between the cutting disc and the boundary layer is explained by a low coefficient of thermal conductivity of air. When a sprayed coolant is introduced into the boundary layer using an ejector tube (fog cooling), the wheel temperature decreases by 25%, which ensures an increase in the time of continuous operation.

4. Results

As a result of the study, it was found that to increase the time of continuous operation of the wheel, cooling of the boundary layer must shall be carried out using an ejector tube.

References

[1] Shintaro Imayama R J, Lingwood P and Alfredsson H 2014 The turbulent rotating-wheel boundary layer *Europ J of Mech B/Fluids* **48** pp 245-53
[2] Appelquist E, Schlatter P, Alfredsson P and Lingwood R 2018 Turbulence in the rotating-wheel boundary layer investigated through direct numerical simulations *Europ. J of Mech B/Fluids* **70** pp 6-18
[3] Cebeci T and Abbott D 1975 Boundary layers on a rotating wheel AIAA J **1(6)** pp 561-67
[4] Shintaro Imayama R J 2012 Experimental study of the rotating-wheel boundary-layer flow *Techn. Reports from Royal Inst. of Techn KTH Mech SE-100 44* 60
[5] Parthasarathy R N 2002 Analysis of the turbulent boundary layer on a rotating wheel *Microsystem Techn* **8(4–5)** pp 278-81
[6] Kohama Y 1984 Studies of the rotating-wheel boundary-layer flow *Acta Mech* **50** pp 193-99
[7] Imayama S 2014 Studies of the rotating-wheel boundary-layer flow *Techn. Reports from Royal Inst of Techn KTH Mech SE-100 44* Stockholm 58
[8] Shevchuk I V 2009 Convective heat and mass transfer in rotating wheel systems *Springer Verlag* 239
[9] Mehmood A and Usman M 2018 Heat transfer enhancement in rotating wheel boundary layer *Thermal Sci Year* **22(6)** pp 2467-82
[10] Ong C L and Owen J M 1991 Computation of the flow and heat transfer due to a rotating disc *Int.l J. of Heat and Fluid Flow* **12(2)** pp 106-15
[11] Harmand S, Pellé J, Poncet S and Shevchuk I 2013 Review of fluid flow and convective heat transfer within rotating wheel cavities with impinging jet *Int. J. of Thermal Sci.* **67** pp 1-30
[12] Turkyilmazoglu M 2018 Fluid flow and heat transfer over a rotating and vertically moving wheel *Phys, of Fluids* **30** pp 603-5

[13] Lebedev V, Bespalova A and Dashkoskaya O 2016 Regularities of dust formation during stone cutting for construction works *Odes'kyi Politech. Univer. Pratsi* **2(49)** pp 24-30
[14] Bespalova A, Lebedev V, Tonkonogyi V, Morozov Y and Frolenkova O 2019 Cutting stone building materials and ceramic tiles with diamond disc *Procceedings of the 2nd Int. Conf. jn Design, Simul., Manufacturing: The Innov. Exch. DSMIE-2019* pp 510-21
[15] Bespalova A, Lebedev V, Frolenkova O and Chumachenko T 2019 Cutting stone and ceramic building materials with diamond discs. *Sci. letters of acad. society of M Baludansky* **7** pp 9-17
[16] Bespalova A, Lebedev V, Morozov Y, Chumachenko T and Klymenko N 2019 Mathematical modeling of the process of the interaction of the cutting diamond wheel with the environment *Grabchenko's Intern. Conf. on Advance Manuf. Process, InterPartner-2019* pp 3-14

The influence of time on the bearing capacity of piles in water-saturated clay soils

A Novskiy[1*], I Bichev[1], V Novskiy[1], E Yeresko[1]

[1]Odessa State Academy of Civil Engineering and Architecture, 4, Didrichson str., Odessa, 65029, Ukraine

[*]E-mail: novskiva@gmail.com

Abstract. The change in the bearing capacity of piles over time is an established fact that is carefully studied by many researchers. Depending on the type of soils and their condition, there is both a decrease (in sandy) and an increase (in clay) in the bearing capacity of piles over time. The most effective increase in bearing capacity occurs in water-saturated, clayey soils with the property of thixotropy. The results of experimental studies of soils with piles in natural conditions, carried out with different periods of their rest, are of considerable value in the study of this phenomenon. During the construction of a multi-storey residential complex in Odessa, which began in 2019, it became necessary to increase the load on the pile foundation made of prismatic piles with a length of 16 m and a cross-section of 350×350 mm. According to research in 2019, the bearing capacity of the piles was 1500 kN. In February 2021, static soil tests were carried out with the existing piles from under the grillage of the building, which was installed by 90%, while the bearing capacity of the piles was 1800 kN. Thus, for the period from July 2019 to February 2021, the bearing capacity of C160.35-11 piles at the construction site in water-saturated soils in Odessa increased from 1500 kN to 1800 kN, which is 20%.

1. Introduction

The change in load-bearing capacity of piles over time is an established fact that requires comprehensive study. Depending on the type of soil and their condition, there is a decrease (in sand) and increase (in clay) of piles bearing capacity over time. The results of experimental studies of soils by piles in natural conditions, conducted with different periods of their rest are considerably valuable in the study of this phenomenon. Unfortunately, such studies are rare.

During the construction of a multi-storey residential complex with underground parking (Figure 1), at the address: 6, Fontanskaya doroga str., Odesa there was a need to increase the load on the pile foundations of prismatic piles 16 m long and 350 mm cross section. Therefore, tests of single piles from the existing strip pile foundation were performed.

2. Analysis of recent research and publications

The analysis of works [1-4] shows that in clay soils of soft and fluid-plastic consistency, as the consolidation and thixotropic hardening, there is an increase in the bearing capacity of piles over time. In July 2019, tests of prismatic piles (Figure 2) were performed before the construction of a multi-storey residential complex with underground parking at the address: 6, Fontanskaya doroga str., Odesa. The allowable design load on piles C160.35-11 was confirmed by static tests of soils by 4 piles, the rest of which ranged from 10 to 14 days. According to these researches, the bearing capacity of the piles was 1500 kN [5].

Content from this work may be used under the terms of the Creative Commons Attribution 3.0 licence. Any further distribution of this work must maintain attribution to the author(s) and the title of the work, journal citation and DOI.
Published under licence by IOP Publishing Ltd

Figure 1. General view of a multi-storey residential complex.

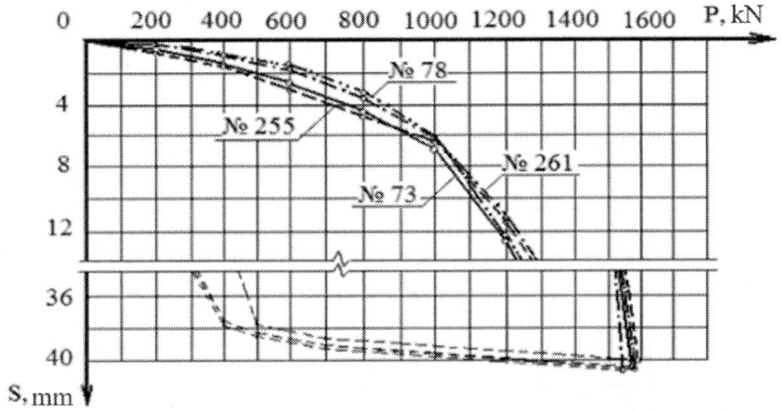

Figure 2. The results of soil tests by piles in 2019.

3. The aim of the work

Conducting static testing of soils with prismatic piles on the construction of a multi-storey residential building with underground parking, which at the time of the study was built by 90%. The foundations of the building are made of prismatic reinforced concrete piles with a length of 16 m and a cross section of 350×350 mm, which were immersed in water-saturated clay soils in July 2019. The purpose of the study is to determine the bearing capacity of piles after their rest for 19 months and inclusion in the work of an existing building and compare it with the results of research in 2019, ie to establish the effect of time on bearing capacity of piles in water-saturated clay soils.

4. Research methodology

Static tests of soils with full-scale prismatic piles are performed in accordance with the requirements of DSTU (State Standards of Ukraine) BV.2.1-27:2010 (GOST 5686-94) "Piles. Determination of bearing capacity based on the results of field tests" [6], and DBN (Ukrainian building norms) B.2.1-10-09 "Bases and foundations of buildings and structures. Amendment №1" [7].

The maximum loads during the control tests were brought to 1800 kN. Axial compressive static load on the pile was transmitted by two hydraulic jacks DGO (Jack hydraulic lightweight)-100-2. The emphasis for the jacks during the test was a pile cap, loaded with the structures of the building. The study of soils with piles was carried out according to the method of control testing. Loading was performed in steps equal to 1/10 of the maximum load, with the endurance of each stage to the conditional stabilization, which was taken as 0.1 mm per hour of observation. The first two stages were taken within 1/5 of the maximum load. Determination of pile movements during the tests was performed on two prognometers with a division value of 0.01 mm, installed on the reference system. The pressure in the system of hydraulic jacks DGO-100 during the tests was recorded by a manometer class 1.5. During the tests, the journal was kept according to the form given in the annex "Zh" of DSTU [6]. According to the results of motion measurements, the subsidence of the piles was determined, which allowed to plot the dependence of the subsidence of piles on the load in time S = f (P, t) and graphs of the dependence of subsidence on the load S = f (P).

5. The results of research

In March 2021, the builders set a task to clarify the bearing capacity of existing working piles 18 months after their immersion and construction of 90% of aboveground structures. As a result, static soil tests were performed with existing pile caps. The fragment of soil tests by a pile is presented in Figure 2. Technological piles №256, 119 in section 2 and №108, 301 in section 1 were selected for static tests. The piles are based on Engineering Geological Elements-6 (EGE-6) – brownish-brown loam, heavy, solid. The link of piles and engineering-geological section is shown in Figure 3, and physical and mechanical properties of engineering geological elements in Table. 1.

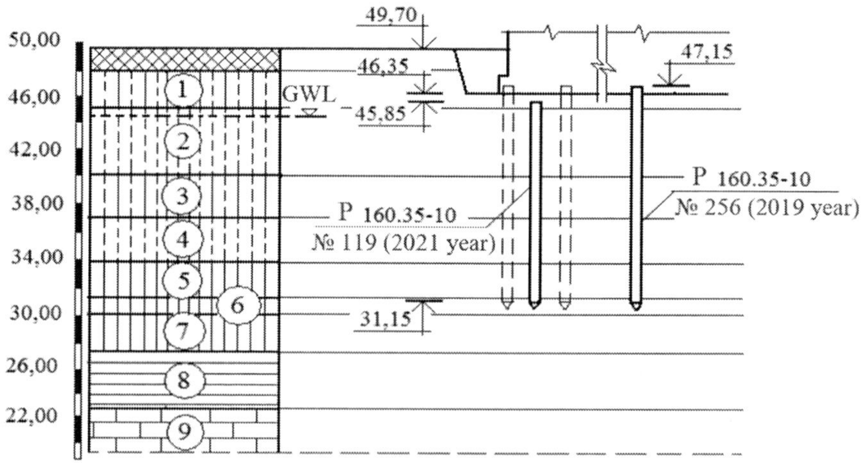

Figure 3. Link of piles and engineering geological section.

Control tests of soils by pile №279 were performed on 18.02.21, the maximum compressive load was increased to 1800 kN, while the stabilized subsidence was 7.72 mm.

Control tests of soils by pile №119 were performed on 19.02.21, the maximum compressive load was increased to 1800 kN, while the stabilized subsidence was 8.25 mm.

Control tests of soils by pile №108 were performed on 20.02.21, the maximum compressive load was increased to 1800 kN, while the stabilized subsidence was 8.73 mm.

Control tests of soils by pile №301 were performed on 22.02.21, the maximum compressive load was increased to 1800 kN, while the stabilized subsidence was 10.22 mm.

The fragment of tests of soils by a pile is given in Figure 4, and graphs of dependence between piles subsidence and vertical compressive loadings are presented in Figure 5.

Table 1. Physico-mechanical characteristics of soils.

№ EGE	Name EGE	ρ_s, g/cm³	ρ, * g/cm³	W*	W_L	W_p	φ,* deg	c,* kPa	E,* MPa
1	Medium loess loam	2.70	1.74	0.20	0.33	0.21	21	22	10
			1.86	0.28			18	16	5
2	Light loess loam	2.69	1.70	0.18	0.26	0.18	20	16	8
			1.81	0.26			17	12	4
3	Medium loess loam	2.71	1.88	0.22	0.36	0.22	21	21	9
4	Light loess loam	2.70	1.79	0.27	0.29	0.19	11	10	4
5	Heavy loam	2.72	1.82	0.21	0.38	0.23	21	23	13
6	Heavy loam	2.72	1.74	0.22	0.39	0.23	20	24	12
7	Heavy loam	2.72	1.86	0.21	0.41	0,24	21	25	11
8	Red-brown clay	2.73	1.84	0.22	0.45	0.27	19	45	16

*Above the line is the value for the state of natural humidity, below the line is the value for the water-saturated state.

Figure 4. Fragment of test of soils by a pile from under a pile cap.

The bearing capacity of prismatic piles according to the results of their tests by pressing load is determined by the formula:

$$F_d = \gamma_c \times F_u / \gamma_g ,$$

where: γ_c – coefficient of operation conditions, in case of pressing load $\gamma_c = 1.0$;

γ_g – soil reliability coefficient $\gamma_{eq} = 1.0$;

F_u – the value of the ultimate resistance according to the results of static tests.

Thus, according to the results of tests of 4 piles in 2021 value F_u is taken equal 1800 кN [8].

The bearing capacity of prismatic piles C160.35-11 is equal to:

$$F_d = 1.0 \times 1800/1.0 = 1800 kN .$$

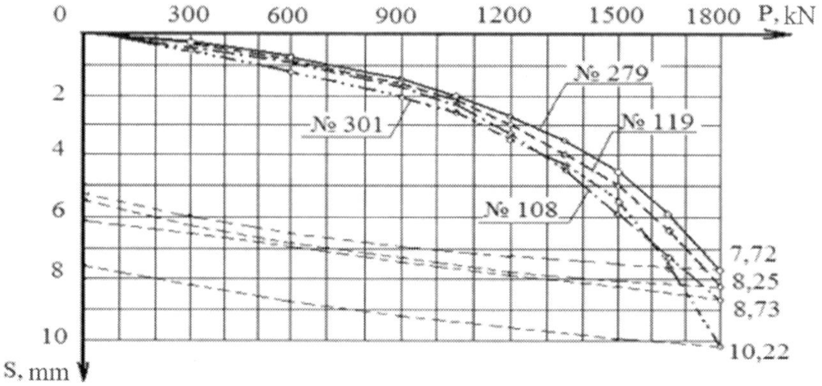

Figure 5. The results of soil tests by piles in 2021.

6. Conclusions

According to the results of static tests of water-saturated clay soils by prismatic piles C160.35-11 on the construction site of a multi-storey residential complex at the address: 6, Fontanskaya doroga str., Odesa, their bearing capacity during the tests in 2019 was 1500 kN, and during the tests of 2021 was 1800 kN.

Thus, for the period from July 2019 to February 2021, the bearing capacity of C160.35-11 piles in water-saturated clay soils increased from 1500 kN to 1800 kN, which is 20%.

References

[1] Tugaenko Yu F, Novsky A V, Tkalich A P and Vasilevskaya L A 2007 Differential approach to assessing the increase in friction forces along the lateral surface of piles in time *Galuzeve Machinery Buduvannya, Budivnistvo* **19** pp 114-120

[2] Tugaenko Yu F and Tkalich A P 2006 Results of experimental studies of the stress-strain state of the "pile – foundation soil" system *Problems of soil mechanics and foundation engineering in difficult soil conditions:* Proceedings of the international, scientific and technical conference dedicated **1** pp 142-147

[3] Tababilov R R 2019 Anticipation of the increase in the bearing capacity of piles over time *Young scientist* **21 (259)** pp 148-150

[4] Paramonov V N and Tikhomirova L K 2000 Change in the bearing capacity of driven piles over time *Urban Reconstruction and Geotechnical Construction* **1** pp 127-131

[5] Technical report on design and survey works on soil testing with piles on the site of the projected object: *"Construction of a multi-storey residential complex with underground public premises at the address of 6, Fontanskaya doroga str., Odesa"* 31

[6] DSTU BV.2.1-1-95 (GOST 5686-94) *Soils Field test methods with piles State Committee of Ukraine for Urban Planning and Architecture* 57

[7] DBN B.2.1-10-2009 Change №1 *Bases and foundations of buildings. Basic design provisions. Ministry of Regional Development of Ukraine* 55

[8] Conclusion based on the results of static tests of soils with prismatic piles at the object *"Construction of a multi-storey residential complex with underground parking, built-in public buildings and service station, with a store of related goods and administrative premises at: 6, Fontanskaya doroga str., Odesa (6 stages of construction)"* 38

IOP Conf. Series: Materials Science and Engineering 1164 (2021) 012015 doi:10.1088/1757-899X/1164/1/012015

VIBROABRASIVE TREATMENT OF ULTRA-HARD CERAMICS

V I Burlakcov[1,*], G U Burlakova[1], V A Mazur[1] and O V Manoylov[1]

[1]Priazovskyi State Technical University, 7, Universytets'ka St., Mariupol, 87555, Ukraine

*E-mail: mahaon1963@gmail.com

Abstract. The article reveals a possibility of treating ultra-hard nitride ceramics by means of vibro-abrasive treatment. It was proved that the value of residua; stresses of the second kind in ceramic samples sufficiently reduced after such treatment. Also it was confirmed experimentally that material removal increased with time. The article contains Difractograms of the surfaces. The surface quality of nitride ceramics was thoroughly analyzed. The grade of abrasive material was determined practically.

1. Introduction

Foreign and domestic experience shows that the role of tools is growing more and more at operations with mechanical treatment, which are characterized with increased thermo-power loads at high speed cutting, at treatment of parts, made of quenched, heat-resistant steels and alloys and also various composite materials.

The work has been constantly gone on, in search of creating maximally universal tool materials, that would be able to ensure high exploitation indices of metal treating tools at their different exploitation conditions. The quality of treated surfaces and intensification of production have always been of paramount importance for researchers in the domain of metal treatment. Therefore, it is necessary to seek ways of improving these factors. Application of new high-strength materials as tool cutting plates seems to be one of these ways.

A promising way is application of ultra-hard nitride ceramics as cutting plates for treating tools. This material is strong and wear-resistant enough, that is why the problem of its treatment seems to be quite urgent. It is known that strength of ultra-hard ceramics is quite comparable with that of diamonds and its treatment is very problematic. Hence some problems may arise for treating ultra-hard ceramics for manufactures of cutting plates, like improvement of the quality of treatment, raising of productivity, obtaining radii at the vertex of the cutting plate, and also presence of residual stresses.

A great number of researchers dealt with solution of problems with treatment of ultra-hard nitride ceramics. Among them such scholars like, Childs T. Maekawa K., Obikawa T., Yamane Y., Sumiya H., Harano K., Ishida Y., Sugihara T., Tanaka H., Enomoto T., Rogov V. A., Shkarupa M. I., Grishin D. K. , Halpin T., Byrne G., Barry J., Ahearne E., Yamaguchi H., Srivastava, A. K., Tan, M., Hashimoto F. and many others, who investigate ultra-hard materials, their production and application may be mentioned. They investigate treatment with diamonds of technical ceramics, strength of cutting ceramics, they study the influence of the state of blades edges of cutting tools upon efficiency of cutting tools Also, they study rapid diamond treatment of technical ceramics. So far, the researchers have failed in succeeding in finding a complex solution of the problems, which they are facing. Thus, we may conclude that the problem of treating ultra-hard ceramics remains quite urgent. One of such solutions is vibration treatment.

Content from this work may be used under the terms of the Creative Commons Attribution 3.0 licence. Any further distribution of this work must maintain attribution to the author(s) and the title of the work, journal citation and DOI.
Published under licence by IOP Publishing Ltd

2. Experimental metods

Samples with cylindrical shape were treated at experimental studies, they were made of ultra-hard nitride ceramics, on the basis of cubical boron nitride (hereinafter CBN) "composite 05 IT"(NK18–20 HPa) Ø 7,0mm in height and 3.2 mm in width.

As workspace suspension on the basis of monocorundum, with diamond cut according to FEPA F100 (acc. to SSU 12 grain size – 150–125mcm) comprising 90% of the entire volume, added with powders of various artificial diamonds (UDA, AC6, ACM) 5%, 10% and 15% in concentration of the entire abrasive volume.

Figure 1 represents the diagram of the laboratory unit. The unit was assembled on a rectangular platform (1) and had four spring supports (2). An unbalanced vibrator-motor was installed between the supports (3) an operating chamber being place don the platform (4).

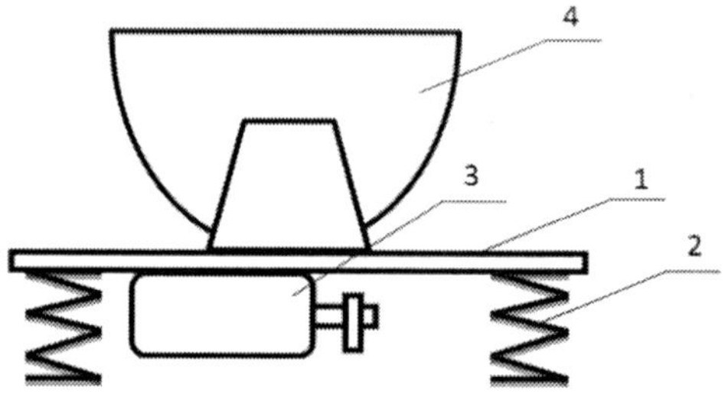

Figure 1. Laboratory vibro-unit.

The samples were treated for 10, 20 and 40 minutes. The amplitude of vibration was 5mm, the frequency was–25-30Hz. A ceramic plate in its original state was considered to be a reference sample.

In order to determine the values of alternations of internal stress due to power and thermal influence occuring at treatment of ultra-hard ceramics, ceramic samples without any treatment and after treatment with different time spans 9after 10, 20 and 40 minutes) were investigated, after that an X-ray inspection was carried out with application of DRON- 4-3 diffractometer. The results of the X-ray analysis were processed with application of MathCAD program.

It was found out that the value of residual stresses of the second type in the samples decreased from 697 MPa prior treatment to 409 MPa after 40 minutes treatment.

For determination of the influence of vibration treatment upon tools strength cutting picks were used, equipped with cutting plates in initial state and after treatment with different time spans. A blank, made of 45 (45-50 HRC). Steel was ground.

3. The results of investigations

At vibro-abrasive treatment material is removed due to the action of vibration force, because of simultaneous action of three mechanisms: shearing, tearing and micro-cutting. The amount b of removed material is but one of the principal indices of productivity of any treatment method. Vibro-abrasive treatment of ultra-hard ceramics is not an exception. The index, that determines the value of material removal and, thus, productivity as well is the time of treatment [1]. The diagram, showing the dependence of the value of material removal upon time is shown in Figure 2.

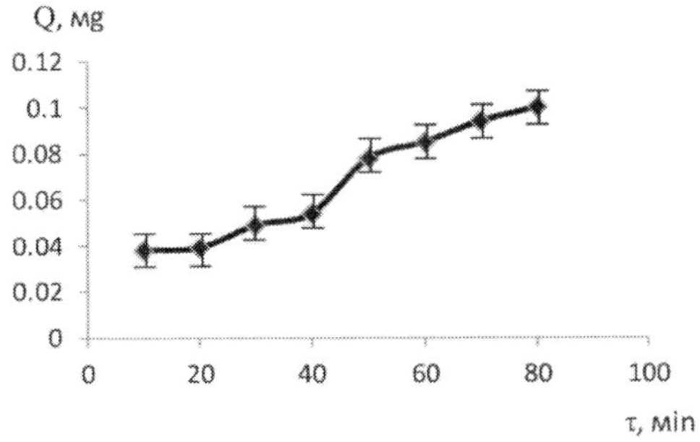

Figure 2. Dependence of the value of material removal upon time of treatment.

The experiments confirmed the correctness of the theory that the longer the treatment lasted the more material would be removed. In this case ultra-hard ceramics will not be an exception.

The surface quality, determined by roughness is another important productivity index. The surface roughness was measured in five directions, pictured in Figure3, in order to exclude random results.

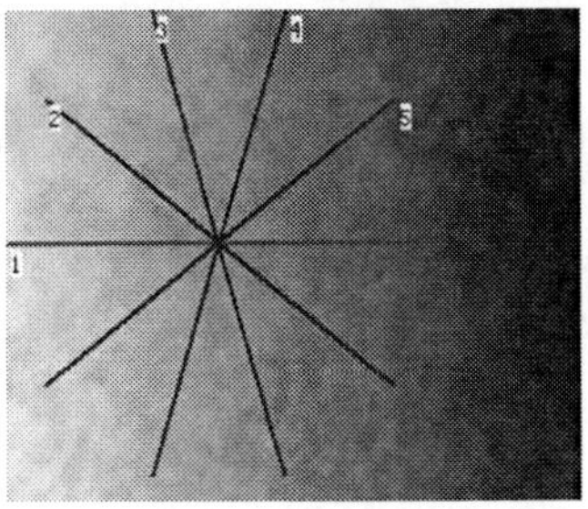

Figure 3. Orientation of directions for measuring the height of micro-unevenness on the surfaces of each specimen. (×500).

Figure 4 represents profilometry of specimens' surfaces after 10, 30, and 40 of treatmnet in one direction.

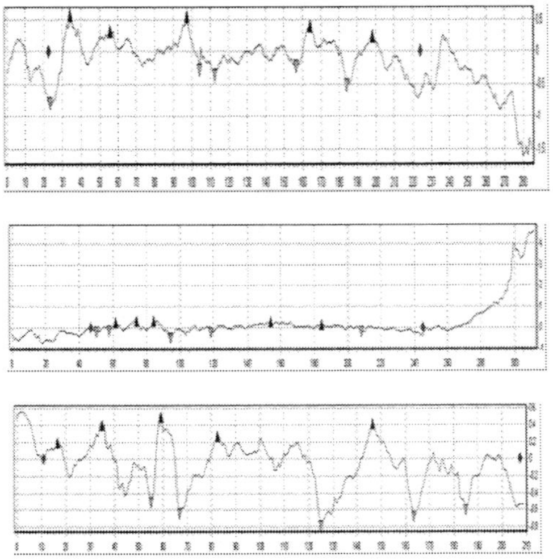

After treatment for 10 minutes, Ra 0.21

After treatment for 30 minutes, Ra 0.19

After treatment for 40 minutes, Ra 0.11

Figure 4. Profilometry diagrams of the surfaces of ultra-hard ceramics after treatment.

The investigations showed that in accordance with profilometry diagrams of surface treatment (Figure 4) after 10, 30 and 40 minutes treatment the smallest roughness turned out to be Ra (3) =0.13mcm after 30 minutes treatment.

These conclusions can be verified by images of the fragments of surfaces beyond which the treatment had been done. It can be seen that the most even surface was the one, represented in Figure 5-в that corresponds to 30 minutes treatment. Below (in Figure 6) there are fragments of the surfaces after treatment [2].

Figure 5. The state of the surface of the specimen (×500).

Analyzing the profilometry diagrams we arrive at a conclusion that the highest quality of treatment was made by means of the mixture of monocorrundum and the powder of artificial diamond UDD, 10% in concentration. The average value of Ra was = 0.11. The diagrams confirm the figures (see Figure 6), executed in 3d format. On them the differences of specimens' profile, treated with different materials are quite visible.

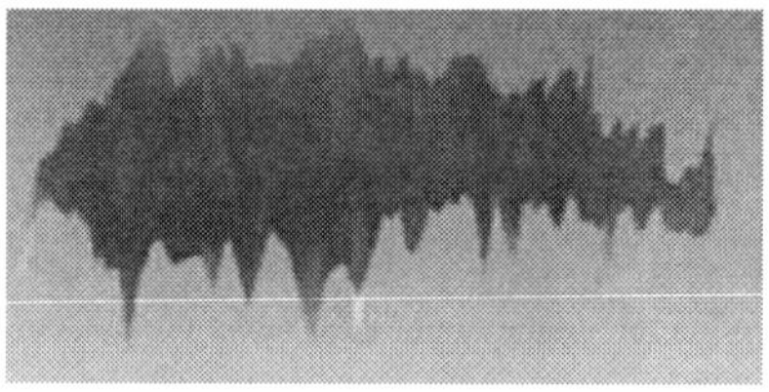

Figure 6. The surface profile of the plate after treatment in 3D mode, the time of treatment was 30 minutes, as the tool the mixture of the powder of monocorrundum and UDD artificial diamond was used (×1,000).

Application of diamond tools ensured precision and quality of treatment, bigger service life and reliability of operation of machines and instruments [3]. Beside the problem of application of diamond as an additive for treatment, it was important to determine the amount of added powder. In Figure 7 dependence of powder concentration upon the intensity of treatment is shown.

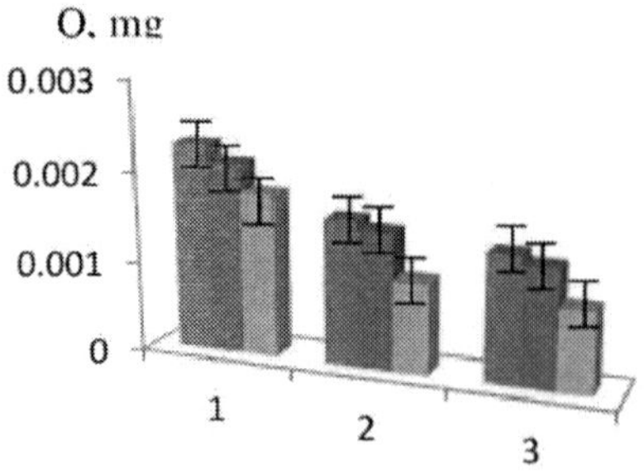

Figure 7. The influence of diamond powder concentration upon the value of material removal.

Analyzing this dependence we may arrive at a conclusion regarding the advisability of applying a certain concentration of diamond powder. At vibro-abrasive treatment of nitride ceramics with application of diamond powders in concentration 10% and 15% the value of material removal is approximately equal and it's sufficiently higher than at application of 5% amount. The same picture is observed at application of mixtures with all diamond powders used. Finally, we may conclude that as the difference in removed material with 10% and 15% is about the same it would be rational to use 10% additive, due to economical reasons. The investigation showed that the most productive treatment was vibro-abrasive treatment with application of mixtures, containing the powder of UDD artificial diamond (see Figure 8). It can be explained by the capability of diamond particles to maintain its cutting ability due to constant micro-shearing of contact sections as a result of vibration and shock loads in the treatment area.

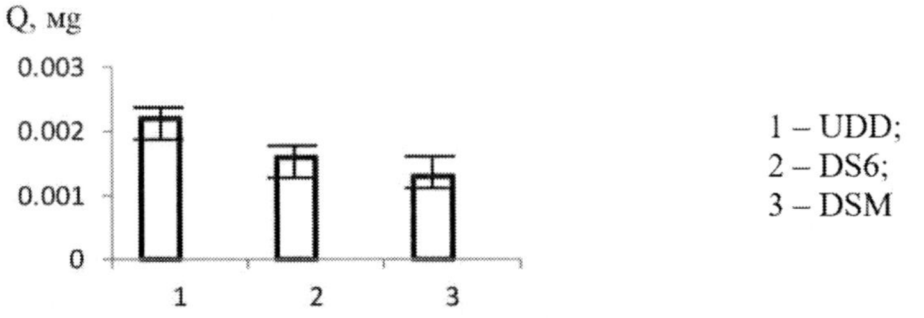

Figure 8. The influence of the diamond powder grade upon the value of material removal at the treatment of ceramics.

Application of mixtures with monocorrundum and diamond powders at different concentrations for vibro-abrasive treatment has a sufficient influence upon the value of removal from NPUM (non-regrowth plates made of polycrystalline ultra-hard materials) specimens. As can b e seen in the figure application of diamond powders in suspension alongside with monocorrundum raises the value of material removal drastically.

In our investigation some attention was paid to the presence of residual stresses in ceramic specimens. The results of experiments, aimed at removal of the concentration of residual stresses are shown in Difractograms, in which peaks, caused by alternations of the matrix by residual stresses were shown as extremes. Difractograms [4] were made at different double reflection angles for the same specimens, made of ultra-high nitride ceramics are shown in Figure 9.

Analyzing the Difractograms of the surface of cutting plates, made of ultra-hard ceramics we can conclude that the highest index, characterizing the value of residual stresses is seen on the plate that did not undergo treatment and it sufficiently reduced alongside with the growth of the treatment time. The identical picture can be observed for the analysis of difracograms, prepared at different double reflection angles. The character of difractograms testifies that by applying vibro-abrasive treatment it is possible to reduce internal stresses in the material.

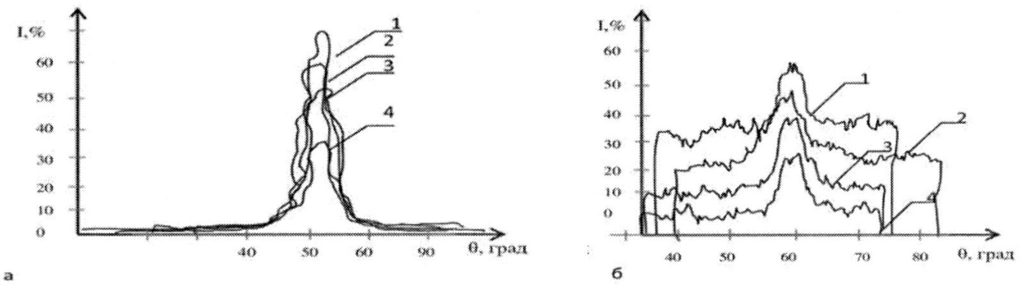

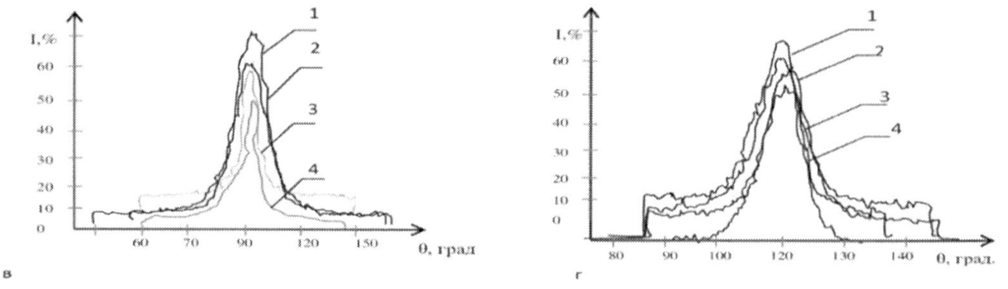

1 – with initial conditions; 2 – at 20 minutes treatment; 3 – at 30 minutes treatment; 4 – at 40 minutes treatment, with reflection angles a – 53^0: б – 63^0; в – 97^0: г – 124^0.

Figure 9. Difractograms of ceramic specimens, made of "05IT composite".

4. Discussion

Analyzing the data, obtained experimentally it is necessary to mention a solution for the problem, which we believe to be quite interesting, that lies in reduction of the concentration of internal stresses in ultra-hard ceramics. It can increase the service lives of cutting plates, made of ultra-hard ceramics, virtually, by 2-2.5 times.

Of interest also is the result of material removal, which was shown by artificial diamonds of manifested UDD grade, it can be explained only by their fine dispersity (as compared to DS6 and DSM diamond grades) and, hence, by growth in the contact patch of the abrasive material with the part. The time of treatment, determined during the investigation is also of interest.

5. Conclusions

So, it was found out that for intensification of treatment it is necessary to add an artificial diamond powder of UDD grade into the abrasive mixture, amounting to 10% of the entire volume.

The surface of the specimen, treated with the mixture of monocorrundum and UDD diamond powder in 10% concentration possesses more uniform surface in comparison with other specimens, the parameter of roughness of ceramic plates after vibration treatment can be to Ra = 0.11, it being substantially lower, as compared to other specimens, the presence of residual stresses in the cutting plates, made of ultra-hard ceramics leads to a decrease of their service lives, while abrasive treatment reduces the concentration of internal stresses within 40 minutes of treatment from 697MPa to 409MPa, the character of reduction of residual stresses being the same on difractograms while watching at different double reflection angles, it characterizing the stability of the process of relaxation, whilst further treatment of ceramic parts in time will not produce a positive result, as it means excessive consumption of working time, power and mechanical resources.

The situation in favour of decreasing the value of internal stresses in ultra-hard ceramics can be changed only by application of vibro-abrasive treatment.

References

[1] Childs T and Yamane Y 2000 Metal Machining, Theory and Applications 1st ed (London: Arnold Publishers) pp 81 - 117

[2] Sumiya H and Ishida Y 2014 Mechanical properties of nano-polycrystalline cBN synthesized by direct conversion sintering under HPHT *Diamond and Related Materials* **41** pp 14-19

[3] Sugihara T and Enomoto T 2017 Development of novel CBN cutting tool for high speed machining of Inconel 718 Focusing on Coolant Behaviors *Procedia Manufacturing* pp 436-442

[4] Rogov V A and Grishin D K 2008 Comparative analysis of mechanical treatment of superhard ceramic materials *Bulletin of RUDN Series "Engineering Research* pp 15-21

[5] Halpin T and Ahearne E 2009 The performance of polycrystalline cubic boron nitride tools in continuous, semi-interrupted, and interrupted hard machining Proc. IMechE *B: J Engineering Manufacture* **223** pp 947-953
[6] Grigoriev S N and Burton D 2012 Influence of power loads on the stress-strain state of cutting plates made of oxide ceramics *Vestnik Mashinostroeniya* **1** p 67-71
[7] Luo S Y and Tsai Y Y 1999 Wear characteristics in turning high hardness alloy steel by ceramic and CBN tools *J. Mater Process Technol* **88** pp 114-121
[8] Baron Yu M 2004 Influence of the state of the edges of the blades on the efficiency of cutting tools *Instrument and technology* pp 1-11
[9] Yamaguchi H and Hashimoto F 2014 Magnetic Abrasive Finishing of cutting tools for high-speed machining of titanium alloys *CIRP Journal of Manufacturing Science and Technology* № 7(4) pp 299-304
[10] Rakesh S and Rajiv K 2015 Continuous and Interrupted Hard Turning Using CBN-L Tools at Moderate Cutting Speeds *International Journal of Emerging Technology and Advanced Engineering* **5** pp 41-46

The influence of the process-induced deformed state on straightness of prestressing strands after manufacture

B M Ozerianskyi[1*] **and I M Chaiun**[1]

[1]Odesa Polytechnic State University, Institute of Mechanical Engineering and Transport, Shevchenko av., 1, Odessa, Ukraine.

*E-mail: Bogdan.ozer@ukr.net

Abstract. Based on the study of elastic-plastic deformation, the expressions of longitudinal force N_t, bending M_b and torque M_t moments in wire-cross sections subject to the process parameters of rope torsion are obtained: wire tensioning efforts, untwisting and preformation.

A technique for determining the technological internal force factors (TIFF) in the rope cross-sections: the longitudinal forces $N_x^T(N_{t_i})$; bending $M_y^T(N_{t_i};M_{b_i};M_{t_i})$, $M_z^T(N_{t_i};M_{b_i};M_{t_i})$ and torque $M_x^T(N_{t_i};M_{b_i};M_{t_i})$ moments. Being guided by the theorem on unloading and FEM, the external deformed state of the rope unloading from TIFF is studied. The expression of the rope unloading deformations vector $|DK| = |\varepsilon, \theta, \chi, \zeta|^T$ as a function of the $F = (N_x^T;M_x^T;M_y^T;M_z^T)$ TIFF vector using the global stiffness matrix of 4×4 rope cross-section corresponding to the elastic deformation is obtained.

Based on the bending deformations, an expression was χ and ζ obtained, the parameter h (h is the segment arrow at the chord length of 1 m), which, according to the regulatory standard, characterizes the degree of the rope straightness after manufacture. The connection between the parameters of torsion (tension, untwisting, and preformation) and the parameter h of straightness is established. It is determined that the unstraightness is directly affected only by the deviation from the uniformity of tension. The developed theory allows to determine the permissible deviation from the non-uniformity of tension with the permissible value of the parameter h of straightness. For example, a prestressing strand of a structure 1+6, having a diameter of 6.9 mm is considered to be of high quality according to the standard, if $h \le 25$мм. This value h corresponds to the deviation of the tension of $\Delta\kappa = 0,8$ one of the 6 wires of the layer. With an increase in deviation of tension, h increases almost rectilinearly to $\Delta\kappa \simeq 3$. With further increase, a loop is $\Delta\kappa$ formed – the loss of stability.

The obtained theory of the external deformed state of unloading provides further opportunity to study the internal deformed-stress state of the rope wires due to process-induced forces, which opens the prospect of even a deeper study of operational strength and durability of ropes.

1. Introduction

Prestressing strands are subject to the requirements of some strength and deformation characteristics [1, 2]. Numerical values of the characteristics vary from the state immediately after manufacture to a stable operating state.

Characteristics immediately after manufacture depend on the torsional tension (up to 10% of the breaking force of the wire [3]) and parameters of their curvature and cabling [4]:

$$\tilde{b} = \frac{\sin^2\alpha}{r}; \tag{1}$$

$$\tilde{t} = \frac{\sin\alpha(\kappa_o + \cos\alpha)}{r}, \tag{2}$$

Content from this work may be used under the terms of the Creative Commons Attribution 3.0 licence. Any further distribution of this work must maintain attribution to the author(s) and the title of the work, journal citation and DOI.
Published under licence by IOP Publishing Ltd

where α and r – torsion angle and radius;
$\pm\kappa_o$ – cabling parameter ("+" twisting, "–" untwisting), provided by the ratio of the angular velocities of the coil and the rotor.

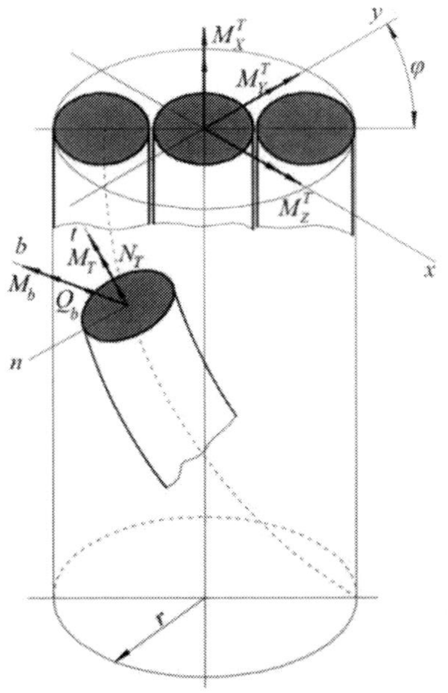

Figure 1. – Tense state.

In each wire, the internal forces of elasticity (Figure 1) are reduced to the longitudinal force N_t, bending M_b and M_t torque moments, which, being normalized to the axes of the rope cross-section, express its process-related internal force factors (TIFF). Their expressions given in [5–7] are specified and simplified in this work. TIFF provide a basis for determining the characteristics that are set by the standards [1–2].

On the basis of refined TIFF when using FEM to investigate the residual deformation after unloading from the action of TIFF and their relationship with the parameter that characterizes the straightness of the rope after manufacture.

2. Presentation of the core material

Figure 1 shows the stress state of the wires and the rope TIFF. Their expressions are obtained subject to the action of external loads

$$N_x^T = \sum_1^n N_{t_i} \cos\alpha_i \; ; \tag{3}$$

$$M_x^T = \sum_{i=1}^s (M_{b_i} \sin\alpha_i + M_{t_i} \cos\alpha_i + N_{t_i} r_i \sin\alpha_i) \; ; \tag{4}$$

$$M_y^T = \sum_{i=1}^s (M_{b_i} \cos\alpha_i - M_{t_i} \sin\alpha_i + N_{t_i} r_i \cos\alpha_i) \sin\varphi_i \; ; \tag{5}$$

$$M_z^T = \sum_{i=1}^s (-M_{b_i} \cos\alpha_i + M_{t_i} \sin\alpha_i - N_{t_i} r_i \cos\alpha_i), \tag{6}$$

where N_x^T is a longitudinal force in the cable cross-section;

M_x^T and M_y^T, M_z^T – torque and bending moment in the cable cross-section;

N_{t_i} – longitudinal force in the i-th wire caused by torsional tension;

M_{b_i} and M_{t_i} – bending and torque moments in the i-th wire cross-section, which emerge due to torsion of the wire line.

The central (core) wire takes only longitudinal force N_t, caused by torsional tension.

In the elastic state

$$M_b^y = EJ\tilde{b}^y; \tag{7}$$

$$M_t^y = GJ_0\tilde{t}^y, \tag{8}$$

where $\tilde{b}^y$ and $\tilde{t}^y$ – torsional deformations of bending and twisting in the stage of elastic deformation; $EJ = E\pi\delta^4/64$; $GJ_0 = G\pi\delta^4/32$ – cross-sectional bending and twisting rigidity of the wire with diameter δ.

Deformations $\tilde{b}^y$ and $\tilde{t}^y$ are determined by expressions similar to formulas (1), (2):

$$\tilde{b}^y = \frac{\sin^2\alpha^y}{r}; \tag{9}$$

$$\tilde{t}^y = \frac{\sin\alpha^y(\kappa_o + \cos\alpha^y)}{r}, \tag{10}$$

where α^y – angles in edge elastic state during torsion.

The edge elastic state of the wire cross-section for the specified process deformation is determined by reaching the yield parameters σ_T and ε_T in boundary points of the wire cross-section:
through the stress intensity

$$\sigma_i = \sqrt{\sigma^2 + 3\tau^2} = \sigma_T(1-\kappa); \tag{11}$$

through strain intensity

$$e_i = \sqrt{e^2 + \frac{1}{3}\gamma^2} = \varepsilon_T(1-\kappa), \tag{12}$$

where $\kappa \leq 0{,}1$ – the wires torsional tension factor;

σ_T and ε_T stress and deformation of yield limit in accordance with linearly schematized diagram of wire stretching (Figure 2, *a*). In Figure 2, *b* a diagram and formulas (11) i (12), are represented in relative terms to obtain generalized study results.

$$\bar{\sigma}_i = \sqrt{\bar{\sigma}^2 + \bar{\tau}^2} = 1-\kappa; \tag{13}$$

$$\bar{e}_i = \sqrt{\bar{e}^2 + \bar{\gamma}^2} = 1-\kappa, \tag{14}$$

where $\bar{\sigma}_i = \sigma_i/\sigma_T$; $\bar{\sigma} = \sigma/\sigma_T$; $\bar{\tau} = \tau/\tau_T$; $\bar{e}_i = e_i/\varepsilon_T$; $\bar{e} = e/\varepsilon_T$; $\bar{\gamma} = \gamma/\gamma_T$.

Parameters of torsional deformation for the edge elastic state

$$\xi_y = \frac{\gamma_{max}}{e_{max}} = \frac{\tilde{t}^y}{\tilde{b}^y}. \tag{15}$$

Similarly for the final state

$$\xi = \frac{\tilde{t}}{\tilde{b}}. \tag{16}$$

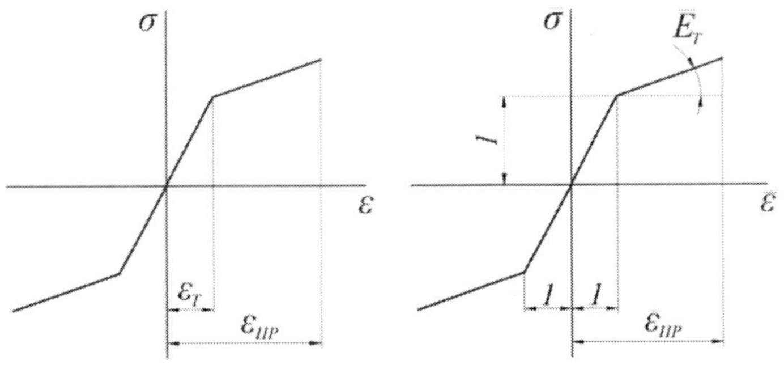

Figure 2. – Schematized diagram of the wire deformation.

Using the formulas (9), (10) we obtain

$$\xi_y = \frac{\sin\alpha^y \cos\alpha^y}{r} / \frac{\sin^2\alpha^y}{r} = ctg\alpha^y. \tag{17}$$

In the presence of untwisting-twisting

$$\xi_y = \frac{\kappa_o + \cos\alpha^y}{\sin\alpha^y}. \tag{18}$$

With respect to the deformation parameter ξ_y (17), (18), the strain intensity in the edge elastic state of the wire cross-section is expressed as follows

$$e_i = e\sqrt{1+\frac{1}{3}\xi_y^{\ 2}} = \varepsilon_T(1-\kappa). \tag{19}$$

In relative terms

$$\bar{e}_i = \bar{e}\sqrt{1+\frac{1}{3}\xi_y^{\ 2}} = 1-\kappa. \tag{20}$$

On the basis of (20) we obtain the curvature $\tilde{b}^y$, at which the wire will be in the edge elastic state due to bending with torsion during pre-stretching

$$\tilde{b}^y = \frac{2\sqrt{3}\,\varepsilon_T(1-\kappa)}{\delta\sqrt{3+\xi_y^{\ 2}}}. \tag{21}$$

Based on (21), determining of the torsion angles that correspond to the edge elastic state of the wire cross-section, is performed by successive approximations based on the equation

$$\sin^2\alpha^y = \frac{2\sqrt{3}\,r\,\varepsilon_T(1-\kappa)}{\delta\sqrt{3+\xi_y^2}}. \tag{22}$$

Table 1, on the basis of (22), shows the calculated values α^y; $\tilde{b}^y$; ξ^y; $\bar{M}_b^y$ (7); $\bar{M}_t^y$ (8) for the wires of prestressing strand of diameter $d = 6.9мм$; $\delta_1 = 2.5мм$; $\delta_2 = 2.2мм$; $\alpha = 9^\circ$; $r = 2.35мм$; $E = 2\cdot10^5\ H/мм^2$; $\sigma_T = 1550\,H/мм^2$; the torsional tension factor $\kappa = 0.1; 0$.

As can be seen from Table 1, in the absence of untwisting ($\kappa_0 = 0$), the value in the edge elastic state has torque. At $\kappa_o = -\cos\alpha$ the wires are deformed only by bending with pretension.

Table 1. Parameters of the edge elastic state

κ_0 [a]	α^y, *degr*	$\tilde{b}^y \times 10^4\, mm^{-1}$	ξ^y	$\bar{M}_b^y$	$\bar{M}_t^y$
0	1.46	2.796	39.23	0.0397	0.8991
	1.65	3.511	34.71	0.0483	0.9987
-1	7.00	63.364	-0.0611	0.8994	-0.0371
	7.38	70.405	-0.0644	0.9993	-0.0372
$-\cos\alpha$	7.00	63.37	-0.0599	0.9	0
	7.38	70.41	-0.0633	1	0

[a] For each κ_0 the upper lines at $\kappa = 0{,}1$; lower ones at $\kappa = 0$.

The contour of the boundary elastic zone of the wire cross-section is escribed by a circle

$$\left(\bar{M}_b^y\right)^2 + \left(\bar{M}_t^y\right)^2 = \left(1-\kappa\right)^2, \tag{23}$$

where bending (7) torque moments (8) of the edge elastic state are in the relative dimensionless form:

$$\bar{M}_b^y = \frac{EJ\tilde{b}^y}{\sigma_T W_b} = \frac{\sqrt{3}\left(1-\kappa\right)}{\sqrt{3+\xi_y{}^2}}; \tag{24}$$

$$\bar{M}_t^y = \frac{\sqrt{3}GJ_0\tilde{t}^y}{\sigma_T W_0} = \frac{\xi_y\left(1-\kappa\right)}{\sqrt{3+\xi_y{}^2}}. \tag{25}$$

The formation of the spiral shape of the wire ends in the stage of elastic-plastic deformation by bending with torsion during the pretension associated with the torsional tension of the wires.

It is known [8] that in the case of joint bending and torsion in the elastic-plastic stage, the hypothesis of flat sections and straight radii is unacceptable because in the process of deformation the boundary of the elastic region does not remain a circle (23). No exact solutions of such a problem in the theory of plasticity have been obtained up to date. Therefore, we will follow the path of an approximated solution.

With a completely plastic (flexible) cross-section, which is possible with very large deformation and using the model of a perfectly plastic rod ($\bar{E}_T = 0$), the limit curve of dependence that connects the bending and torque moments, without longitudinal force, i.e., for $\kappa = 0$ is close to the ellipse [8].

In this case, that is, when bending with torsion after pretension ($\kappa \neq 0$), we assume that the boundary curve with acceptable accuracy can also be taken as escribed by the ellipse of such an equation

$$\frac{\left(\tilde{\bar{M}}_b^t\right)^2}{\tilde{\bar{M}}_b^2} + \frac{\left(\tilde{\bar{M}}_t^b\right)^2}{\tilde{\bar{M}}_t^2} = \left(1-\kappa\right)^2. \tag{26}$$

The numerators of equation (26) represent the expressions of bending $\tilde{\bar{M}}_b^t$ and torque $\tilde{\bar{M}}_t^b$ moments corresponding to very large deformations $\bar{e}_{i\max}$; $\bar{e}_{\max}$; $\bar{\gamma}_{\max}$, at joint bending and torsion of ideally plastic rod. In [8], and earlier by other authors, such expressions were obtained in solving the problem with $k=0$ by integrating through complete elliptic integrals.

Denominators in (26) $\tilde{\bar{M}}_b = 16/3\pi$ and $\tilde{\bar{M}}_t = 4/3$ – the limit values of bending and torque in the case of separate bending and torsion of a perfectly plastic rod.

For any values of deformations in the interval $(1 \div \infty)$ we approximately take the relationship between bending and torque moments that are also described by the equations of ellipses, which occupy intermediate positions between the circle (23) and the ellipse (26). In this case, for the limit deformation intensities (20)

$$\bar{e}_{i\max} = \bar{e}_{\max}\sqrt{1+\frac{1}{3}\xi^2}$$

the equations of the ellipses have the form of

$$\frac{\left(\bar{M}_b^t\right)^2}{\bar{M}_b^2} + \frac{\left(\bar{M}_t^b\right)^2}{\bar{M}_t^2} = (1-\kappa)^2, \tag{27}$$

where $\bar{M}_b$ – is the bending moment under the action of bending only, i.e., when the intensity of deformation in the boundary points of the rod cross-section $\bar{e}_{i\max} = \bar{e}_{\max}$;

$\bar{M}_t$ – torque under the action of torsion only, i.e., when $\bar{e}_{i\max} = \bar{\gamma}_{\max}$.

The studies performed indicate that in the range of values inherent in the torsional deformations of the wires, the load parameters can be assumed equal to the deformation parameters, i.e. $\lambda_b = \tilde{t}/\tilde{b} = \xi$; $\lambda_t = \tilde{b}/\tilde{t} = 1/\xi$.

Then, taking into account these parameters on the basis of equation (27) we obtain the expressions of bending and torque moments in the circular cross-section of the rod during torsional bending after pretension in the following form

$$\bar{M}_b^t = \frac{\bar{M}_b \bar{M}_t \sqrt{1-\kappa}}{\sqrt{\lambda_b^2 \bar{M}_b^2 + \bar{M}_t^2}}; \tag{28}$$

$$\bar{M}_t^b = \frac{\bar{M}_b \bar{M}_t \sqrt{1-\kappa}}{\sqrt{\lambda_t^2 \bar{M}_t^2 + \bar{M}_b^2}}, \tag{29}$$

where $\bar{M}_b = M_b/\sigma_T W_b$ and $\bar{M}_t = M_t/(\sigma_T/\sqrt{3})W_0$ – expressions of bending and torque moments in the separate action.

In computations according to formulas (3–6) $\bar{M}_b$ and $\bar{M}_t$ are conveniently used in the form of functions of technological deformations of wire torsion: $\bar{M}_b = f\left(\bar{e}_{\max}; \bar{E}_T\right)$; $\bar{M}_t = f\left(\bar{\gamma}_{\max}; \bar{G}_T\right)$, where

$$\bar{e}_{\max} = \frac{\sin^2\alpha}{\varepsilon_T r}\cdot\frac{\delta}{2}.\ \bar{\gamma}_{\max} = \frac{\tilde{t}\,\delta}{2\sqrt{3}\varepsilon_T} = \frac{\sin\alpha\left(\kappa_o + \cos\alpha\right)}{2\sqrt{3}r\varepsilon_T}\delta. \tag{30}$$

In terms of physical content, the bending moment is expressed by an integral equation

$$M_b = \int_A \sigma y\, dA. \tag{31}$$

Having assumed the deformation from torsional bending as a variable integration, expression (31) takes the form

$$M_b = \frac{4r^3}{\sin^6\alpha}\int_0^{e_{\max}} \sigma e\sqrt{e_{\max}^2 - e^2}\,de\,. \qquad (32)$$

In relative terms

$$\overline{M}_b = \frac{M_b}{\sigma_T W_b} = \frac{128 r^3 \varepsilon_T^3}{\pi\delta^3 \sin^6\alpha}\int_0^{\overline{e}_{\max}} \overline{\sigma}\,\overline{e}\sqrt{\overline{e}_{\max}^2 - \overline{e}^2}\,d\overline{e}\,. \qquad (33)$$

Stress in the cross-section for relative form in the zones:
elastic $\overline{\sigma} = \overline{e}$; elastic-plastic

$$\overline{\sigma} = 1 - \overline{E}_T + \overline{e}\,\overline{E}_T\,. \qquad (34)$$

Taking into account (34)

$$\overline{M}_b = \frac{128\, r^3\varepsilon_T^3}{\pi\delta^3\sin^6\alpha}\left[\int_0^1 \overline{e}^2\sqrt{\overline{e}_{\max}^2 - \overline{e}^2}\,d\overline{e} + \int_1^{\overline{e}_{\max}} \left(1 - \overline{E}_T + \overline{e}\,\overline{E}_T\right)\overline{e}\sqrt{\overline{e}_{\max}^2 - \overline{e}^2}\,d\overline{e}\right].$$

After integration and transformations, the bending moment $\overline{M}_b$ takes the following form

$$\overline{M}_b = \frac{4}{\pi\overline{e}_{\max}^3}\left[\left(1 - \overline{E}_T\right)\left(\frac{\overline{e}_{\max}^2\sqrt{\overline{e}_{\max}^2 - 1}}{2} + \frac{\left(\overline{e}_{\max}^2 - 1\right)^{3/2}}{3} + \frac{\overline{e}_{\max}^4}{2}\arcsin\frac{1}{\overline{e}_{\max}}\right) + \frac{\pi}{4}\overline{E}_T\,\overline{e}_{\max}^4\right]. \qquad (35)$$

Torque $\overline{M}_t$ in physical terms is expressed by an integral equation

$$M_t = \int_A \tau\rho\,dA = 2\pi\int_0^{0,5\delta} \tau\rho^2\,d\rho\,, \qquad (36)$$

where the current radius ρ and its differentiation operator

$$\rho = 0,5\delta\frac{\gamma}{\gamma_{\max}}\,;\ d\rho = \frac{0.5\delta}{\gamma_{\max}}d\gamma\,. \qquad (37)$$

Having applied (37) in (36), we obtained

$$M_t = \frac{\pi\delta^3}{4\gamma_{\max}^3}\int_0^{\gamma_{\max}} \tau\gamma^2\,d\gamma\,. \qquad (38)$$

In relative terms

$$\overline{M}_t = \frac{M_T}{\tau_T W_0} = \frac{4}{\overline{\gamma}_{\max}}\int_0^{\overline{\gamma}_{\max}} \overline{\tau}\,\overline{\gamma}^2\,d\overline{\gamma}\,, \qquad (39)$$

Tangential stresses in the zones:
elastic $\overline{\tau} = \overline{\gamma}$;
elastic-plastic zone

$$\overline{\tau} = 1 - \overline{G}_T + \overline{\gamma}\overline{G}_T\,. \qquad (40)$$

Taking into account (40)

$$\bar{M}_t = \frac{4}{\bar{\gamma}_{\max}^3}\left[\int_0^1 \bar{\gamma}^3 d\bar{\gamma} + \int_1^{\bar{\gamma}_{\max}} \left(1-\bar{G}_T+\bar{\gamma}\bar{G}_T\right)\bar{\gamma}^2\, d\bar{\gamma}\right]. \tag{41}$$

After integration and transformation

$$\bar{M}_t = \frac{4}{\bar{\gamma}_{\max}^3}\left[\frac{1}{4}+\frac{\left(1-\bar{G}_T\right)\left(\bar{\gamma}_{\max}^3-1\right)}{3}+\frac{\bar{G}_T}{4}\left(\bar{\gamma}_{\max}^4-1\right)\right]. \tag{42}$$

For the formulas of TIFF (3) - (6) bending and torques moments M_{b_i} and M_{t_i} shall be represented in absolute terms:

$$M_b = \frac{\bar{M}_b \bar{M}_t \sqrt{1-\kappa}}{\sqrt{\lambda_b^2 \bar{M}_b^2 + \bar{M}_t^2}}\sigma_T \frac{\pi\delta^3}{32};\; M_t = \frac{\bar{M}_b \bar{M}_t \sqrt{1-\kappa}}{\sqrt{\lambda_t^2 \bar{M}_t^2 + \bar{M}_b^2}}\sigma_T \frac{\pi\delta^3}{16\sqrt{3}}. \tag{43}$$

The whole set of rope unloading efforts is represented by the vector

$$F = \begin{vmatrix} N_x^T \\ M_x^T \\ M_y^T \\ M_z^T \end{vmatrix}, \tag{44}$$

whose components are computed using formulas (3)–(6). Deformation of the rope under the action of the force vector (44) on the basis of the unloading theorem [9] gives grounds to consider it elastic. For its analytical description we apply FEM in a similar way to [10]:

$$|G|\cdot|DK| = F. \tag{45}$$

Here the stiffness matrix

$$|G| = \begin{vmatrix} G_{11} & G_{12} & G_{13} & G_{14} \\ G_{21} & G_{22} & G_{23} & G_{24} \\ G_{31} & G_{32} & G_{33} & G_{34} \\ G_{41} & G_{42} & G_{43} & G_{44} \end{vmatrix} = \sum_1^S \Lambda_i |K_i|\cdot|G\Pi_i||K_i|^T, \tag{46}$$

where G_{11}; G_{22} and G_{33}; G_{44} – is longitudinal, torsional and bending stiffnesses of the rope; $G_{12} = G_{21};\ldots;G_{34} = G_{43}$ – stiffness of influence.

Rope unloading deformations vector based on (45) and (46)

$$|DK| = \begin{vmatrix} \varepsilon \\ \theta \\ \chi \\ \zeta \end{vmatrix} = \frac{1}{|D|}\begin{vmatrix} A_{11} & A_{12} & A_{13} & A_{14} \\ A_{21} & A_{22} & A_{23} & A_{24} \\ A_{31} & A_{32} & A_{33} & A_{34} \\ A_{41} & A_{42} & A_{43} & A_{44} \end{vmatrix}\cdot\begin{vmatrix} N_x^T \\ M_x^T \\ M_y^T \\ M_z^T \end{vmatrix}, \tag{47}$$

Deformations of rope based on (47):

$$\varepsilon = (N_x^T A_{11} + M_x^T A_{12} + M_z^T A_{13} + M_y^T A_{14})/|D|; \tag{48}$$

$$\theta = (N_x^T A_{12} + M_x^T A_{22} + M_z^T A_{23} + M_y^T)/|D|; \tag{49}$$

IOP Conf. Series: Materials Science and Engineering 1164 (2021) 012016 doi:10.1088/1757-899X/1164/1/012016

$$\chi = (N_x^T A_{13} + M_x^T A_{23} + M_z^T A_{33} + M_y^T A_{34}) / |D| ; \qquad (50)$$

$$\zeta = (N_x^T A_{14} + M_x^T A_{24} + M_z^T A_{34} + M_y^T A_{44}) / |D| . \qquad (51)$$

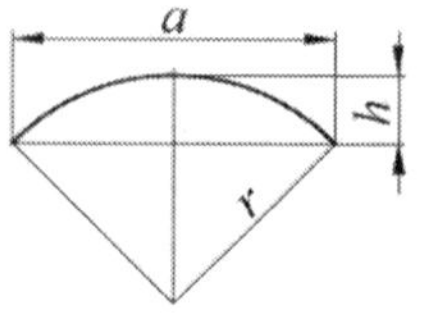

Figure 3 – Segment arrow.

The degree of the rope linearity after manufacturing after standard [1] is controlled by the segment arrow (Figure 3)

$$h = r - \sqrt{r^2 - \frac{a^2}{4}} , \qquad (52)$$

which under $a = 1$ *m* standard [1] shall not accede 25 *mm*,

where radius r is determined based on (50) and (51): $r = 1/\sqrt{\chi^2 + \zeta^2}$.

Thus, theoretically obtained deformations of the rope associated with the unloading due to TIFF. They include bending deformations χ and ζ, which determine the radius of curvature and the parameter h of the rope linearity control after manufacture. Numerical studies based on the obtained theory indicate that the most important technological argument of the ropes linearity after manufacture is the uniformity of wires tension during torsion. This confirms the urgent requirement in [2] for the level and uniformity of wire tension in the manufacture of the quality ropes.

By way of example in Figure 4 and in Table 2 for prestressing strand 1+6 with diameter 6.9 *mm* (Figure 5) shows the dependence of the parameter of straightness h, associated with the deviation of the tension of the wire №2 (Figure 5) from the accepted level determined by the coefficient $\kappa = 0{,}1$. Figure 4 shows that with uniform tension of all wires $h = 0$. When the wire No.2 tension deviates from the total tension by $\kappa = 0{,}1$ more than 20%, the arrow h exceeds GOST [1] permissible value $[h] = 25$ *mm*.

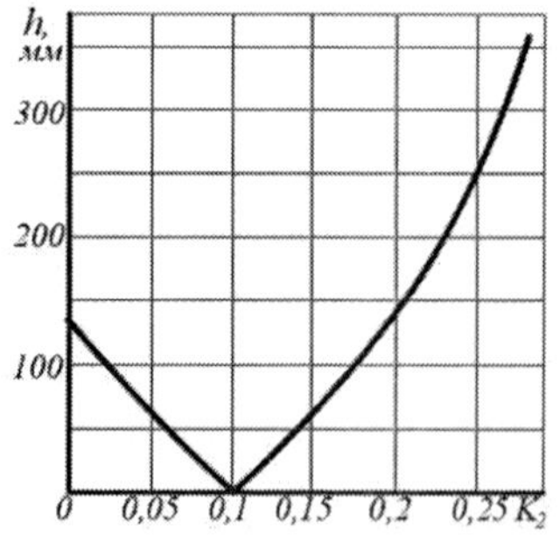

Figure 4. Dependence h on tension.

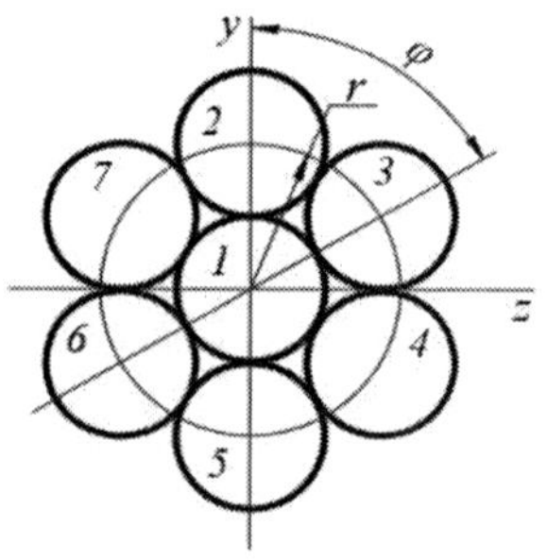

Figure 5. The rope cross-section.

Table 2. Force and deformation parameters of the rope unloading with rope diameter 6.9 mm

No.	N_x^T, H	M_x^T (Hмм)	M_y^T (Hмм)	M_z^T (Hмм)	$\varepsilon \cdot 10^3$	$\theta \cdot 10^4$ (мм$^{-1}$)	$\chi \cdot 10^4$ (мм$^{-1}$)	$\zeta \cdot 10^4$ (мм$^{-1}$)	h, мм
1	$\kappa_0 = -\cos\alpha$; $\kappa = 0.1$ for all wires; with perforation								
	4.252	1.907	~ 0	~ 0	1.640	1.988	~ 0	~ 0	0
2	$\kappa_0 = -1$; $\kappa = 0.1$ for all wires; with preformation								
	4.252	1.907	~ 0	~ 0	1.640	1.988	~ 0	~ 0	0
3	$\kappa_0 = 0$; $\kappa = 0.1$ for all wires; without preformation								
	4.252	17.486	~ 0	~ 0	5.100	10.33	~ 0	~ 0	0
4	$\kappa_0 = -1$; $\kappa = 0{,}1$ for all wires apart from the 2-nd, for which $\kappa = 0.08$; without preformation								
	4.136	951	~ 0	401	1.405	1.440	– 0.165.	1.99	25
5	$\kappa_0 = -1$; $\kappa = 0.1$ for all wires apart from the 2-nd, for which $\kappa = 0.05$; without preformation								
	3.961	856	~ 0	1.003	1.359	1.346	– 0.421	4.99	64
6	$\kappa_0 = -1$; $\kappa = 0.1$ for all wires apart from the 2-nd, for which $\kappa = 0{,}15$; without preformation								
	4.543	1.174	~ 0	-1003	1.517	1.670	0.396	-4.99	64
7	$\kappa_0 = -1$; $\kappa = 0.1$ for all wires apart from the 2,3,7-th $\kappa = 0{,}08$; without preformation								
	3.903	824	~ 0	803	1.342	1.313	-0.334	4.04	51
8	$\kappa_0 = -1$; $\kappa = 0.1$ for all wires apart from the 2-nd $\kappa = 0.05$ and 5-th $\kappa = 0.15$;without prefor-mation								
	4.252	1015	~ 0	2007	1.440	1.509	-0.819	9.97	134
9	$\kappa_0 = -1$; $\kappa = 0.1$ for all wires apart from the 2-nd, for which $\kappa = 0.3$; without preformation. Using formula (52) $a > r$: $h > r = 490$мм : a loop is formed								

By way of example in Figure 4 and in Table 2 for prestressing strand 1+6 with diameter 6.9 *mm* (Figure 5) the dependence of the linearity parameter *h*, associated with the deviation κ_2 of the wire No.2 tension (Figure 5) from the accepted level determined by the coefficient $\kappa = 0{,}1$. Figure 4 shows that with uniform tension of all wires $h = 0$. When the wire No.2 tension deviates from the total tension by $\kappa = 0{,}1$ more than 20%, the arrow *h* exceeds GOST [1] permissible value of 25 *mm*. Unstraightness occurs both in the case of undertensioning $\kappa_2 < \kappa = 0{,}1$, and as a result of excessive tensioning in relation to the set level.

As can be seen from Table 2 in the first three states of torsion, the parameter of unstraightness $h = 0$. After manufacture and unloading due to TIFF, the rope will be shortened and untwisted in proportion to the deformations ε and θ but will be straight. At the same time, untwisting produces $\kappa_0 = -\cos\alpha$ and $\kappa_0 = -1$ virtually the same positive effects. In the absence of a untwisting ($\kappa_0 = 0$), as practical experience shows, the untwisting of the rope increases sharply after the manufacture. The impact of preformation should be further investigated.

3. Conclusions and recommendations

Improved method for determining the longitudinal force N_t, bending M_b and torque M_t moments in the ropes with the general process parameters of torsion and the method for determining the integrated internal forces in the cross-sections of the rope TIFF: longitudinal force of $N_x^T(N_{t_i})$; torque $M_x^T(N_{t_i};M_{b_i};M_{t_i})$ and bending moments $M_y^T(N_{t_1};M_{b_i};M_{t_i})$ and $M_z^T(N_{t_i};M_{b_i};M_{t_i})$. Algorithm of computation M_b and M_t; initially $\bar{M}_b$ and $\bar{M}_t$ in a relative dimensionless form under the condition of their individual action directly due to torsional deformations of bending and twisting of wires; further expressions $\bar{M}_b^t$ and $\bar{M}_t^b$ under the condition of joint action during the previous tension in the form of k coefficient of the torsional tension of the wires. Numerical values $\bar{M}_b^t$ i $\bar{M}_t^b$ for $\kappa = 0$ in the interval α ($8-25^\circ$) virtually coincide with the solution [9] obtained by another way (integration through complete elliptic integrals).

Using the unloading theorem, a technique for studying the external deformed state of the rope unloading due to TIFF was developed based on FEM. The expression of the rope unloading deformations vector $|DK| = |\varepsilon, \theta, \chi, \zeta|^T$ as a function of the $F = (N_x^T; M_x^T; M_y^T; M_z^T)$ TIFF vector using the global stiffness matrix of 4×4 rope cross-section corresponding to the elastic deformation is obtained. On the basis of bending deformations χ i ζ the expression, of h parameter is obtained. In accordance with the regulatory standard, parameter h characterizes straightness of a rope after manufacturing. The connection between the parameters of torsion (tension, untwisting, and preformation) and the parameter h of straightness is established. The unstraightness is directly affected only by the deviation from the uniformity of tension. The developed theory allows to establish the admissible deviation from non-uniformity of tension with admissible straightness value of parameter h (h is an arrow of a segment for the chord length of 1 *m*). For example, a prestressing strand of 1+6 structure, *6.9 mm* in diameter according to the regulatory standard is considered to be of high quality if $h \le 25$*мм*. This value h corresponds to the tension deviation $\Delta\kappa = 0.8$ of one of the 6 wires of the layer. With increasing deviation in tension h increases almost rectilinearly to $\Delta\kappa \approx 3$. With further increase, a loop is $\Delta\kappa$ formed – the loss of stability.

The developed theory of the external deformed state of unloading provides further opportunity to study the internal deformed-stress state of the rope wires due to process-induced forces, which opens the prospect of even a deeper study of operational strength and durability of ropes.

Referenses

[1] ISO 9434 – 4:1991 *Steel for the prestressing of concrete* (Part 4: Stand, NEQ)
[2] Gost R 53772 – 2010 *Steel prestressing, seven-wire, stabilized strands* (Moscow: Standartinform)
[3] Bukshtein M A 1973 *Manufacture of steel ropes* (Moscow: Metallugizdat) 360
[4] Glushko M F 1966 *Steel fall ropes* Repr. of publication text 1966 (Odessa: Astroprint) 2013 Astroprint 336
[5] Chaiun I M and Pishnyak O A 2009 *Relation of the rope pre-tensioning elements with its residual stress state* Steel ropes (Odessa: Astroprint) **7** pp 128-135
[6] Chaiun I M, Pishnyak O A and Nepomnyashiy A V 2013 Preliminary deformed state of twisted ropes *Steel ropes* (Odessa: Astroprint) **9** pp 141-155
[7] Chaiun I M and Nepomnyashiy A V 2013 Technological deformed state of the well-logging cable elements *Works Odessa Polyt. Univ* **2** (**41**) pp 77-82
[8] Serensen S B, Kogayev St P and Shneiderovich R M 1975 *Carrying capacity and computation of machine parts strength* (Moscow: Mashinostroyenie) (Machinebuilding) 488
[9] Malinin N N 1975 *Applied Theory of Plasticity and Creeping* (Moscow: Mashinostroyenie) 399
[10] Chaiun I M 2003 *Bearing capacity of lifting ropes and belts* (Odessa: Astroprint) 233

Effect of alloying elements on structure, metastability and properties of Fe-Cr-Mn (Ni-free) corrosion-resistant austenitic-ferritic steels

Y Cheylyakh[1,*], A Cheiliakh[1] and W Kaiming[2]

[1]Pryazovskyi State Technical University, 7, Universytets'ka St., Mariupol, 87555, Ukraine
[2]Wuhan University of Science and Technology, 947, Heping Av., Qingshan District, Wuhan 430081, Hubei, P.R. China

*E-mail: aleksandr.cheylyakh@gmail.com

Abstract. Investigated was chromium influence within 15-21.4 wt.% and manganese influence within 8-11.4 wt.% upon phase transformations, formation of structure and mechanical properties of corrosion resistant nickel-free (Fe-Cr-Mn) steels of austenite-ferrite structural class. It was found out that by the variation of chromium and manganese content it was possible to modify efficiently the content of ferrite and austenite in the structure, the degree of deformation metastability of austenite and control the mechanical properties of steels. An improved complex of strength, plasticity properties and impact toughness is reached at optimal content of alloying elements (Cr, Mn) in new nickel-free steels, it can be explained by positive impact of deformation induced martensite $\gamma\rightarrow\alpha'$ transformation at testing during their formation.

1. Introduction

In the world practice, Cr-Ni steels of 18/8 (10Cr18Ni9Ti) type, austenite structural class, 08Cr22Ni6Mo2 austenitic-ferritic class, etc. are widely used as effective stainless (corrosion-resistant) steels. These steels are very expensive since they contain 6-11% Ni [1-3]. Therefore, in many countries research and development are being carried out on low-nickel and nickel-free steels capable of adequately replacing the above-mentioned Cr-Ni corrosion-resistant steels, retaining a sufficient set of mechanical properties and corrosion resistance [4-6].

With the objective of adequate replacement of expensive and scarce Fe-Cr-Ni corrosion-resistant steels, Fe-Cr-Mn (nickel-free) steels of austenitic-ferritic class have been developed for the ability to work in corrosive environments of low and medium aggressiveness including sea water [7]. However, the literary data on implementation of the deformation induced martensite transformation during the tests its role in formation their properties are scarce [7-10].

The objective of this work is to investigate the effect of chromium content and quenching parameters on the phase composition, the austenite metastability and properties of nickel-free corrosion-resistant steels of the austenitic-ferritic class.

2. Research and test methods

Variable correlations of alloying elements, mainly Mn, Cr, C, Si, were developed for smelting new alloyed steels with experimental compositions with the objective of obtaining different phase correlations of austenite and ferrite amounts in the structure of experimental steels. Steels were smelted in "IST-016" induction furnace, possessing basic chromium-magnesite refractory lining and then they were cast into shell moulds in the shape of blanks for specimens. Ferroalloys were used at smelting (low carbon ferro-chromium, ferro-silicon, ferro-molybdenum, as well as metallic manganese and copper, aluminium and

Content from this work may be used under the terms of the Creative Commons Attribution 3.0 licence. Any further distribution of this work must maintain attribution to the author(s) and the title of the work, journal citation and DOI.
Published under licence by IOP Publishing Ltd

silica-calcium for deoxidation and refining of steel. Specimens for investigation and testing of mechanical properties were made from moulded blanks by means of machining and then they underwent the specified heat treatment modes. Chromium content in the smelted steels was varied within 15 – 21.4 wt.% limits, while manganese content within 8-11.4 wt.% in numerous combinations for obtaining different phase quantitative correlations between austenite and ferrite. Also investigated was standard 12Cr18Ni9Ti (18-10) chromium-nickel steel in hot rolled state.

Investigated were specimens of developed stainless steels on Fe-Cr-Mn basis, with chromium content varying within the specified limits. The specimens underwent quenching from 1,050 °C temperature into water. Investigation of microstructure of stainless steels specimens was carried out on an optical metallographic microscope "NEOPHOT-21" (manufactured by Carl Zeiss Firm) and "MMT14", at magnifications of x100-400. The phase composition (amount of ferrite, austenite) was determined by x-ray method on difractometer "DRON-3" in Fe_{α}-radiation.

Micro-hardness test of phase and structures composition was studied on micro-hardness tester "NOVOTEST TC-MKB-1" by pressing in a pyramidal indenter with an apex angle of 136° under a load of 50 gf. Hardness was determined by the Rockwell method on the B scale (HRB) with a load of 100 kgf (GOST 9013-59). Tensile tests were carried out on 5-fold samples with a diameter of the workpiece part of 6 mm (the estimated length of the workpiece was 30 mm) on a УММ–5 testing machine with a maximum load of 50 kN, tensile rate being 2 mm·min^{-1} in accordance with GOST 1497-84 and ISO 6892-2015. Dynamic bending (impact strength) tests were performed on samples with a cross section of 10×10 mm, 55 mm length and U-shaped (R0.25 mm, depth 2 mm) on a "MK-30" impact testing machine with maximum potential energy of 300 J (GOST 9454-78).

3. Research results and discussion

The work presented the task solution of creating of new nickel-free stainless steels, based on the results of the previous investigations of such steel carried out at PSTU [7]. As the bulk of pump parts and other facilities operating in sea water are manufactured by means of casting, the objective was to create steels, obtained by smelting and cast part for subsequent exploitation in cast state following heat treatment.

The results of the previous investigations proved that it would be advisable to develop nickel-free and low-nickel steels of austenite-ferrite structural class on Fe-Cr-Mn basis. Therefore, for the development of the compositions of new steels, an alloying system Fe-Cr-Mn with a balanced content of chromium, manganese, carbon. The results of the previous investigations proved that it would be advisable to develop nickel-free and low-nickel steels of austenite-ferrite structural class on Fe-Cr-Mn basis. Therefore, for the development of the compositions of new steels, an alloying system Fe-Cr-Mn with a balanced content of chromium, manganese, carbon.

Alongside with these elements it would be advisable to alloy steels with small additions of silicon, copper, molybdenum or vanadium and nickel. The content of carbon, chromium and manganese should be strictly well-balanced in order to ensure austenite-ferrite phase composition with an opportunity to control the number of phases by selecting the above-mentioned alloying components and heat treatment modes. The carbon content should be as little as possible, not exceeding 0.12 weight % (hereinafter wt.%) to ensure high corrosion resistance, mechanical properties and weldability. Chromium and manganese content was planned to be varied within the specified parameters for adjusting the amount of austenite and ferrite, it having no effect upon the degree of deformation metastability of austenite. Silicon was introduced into steel with the objective of strengthening ferrite and austenite and also for improving heat and corrosion resistance in case new steels are applied not only as stainless but also as heat resistant.

Steels were alloyed with little additions of copper with the idea of improving plasticity, corrosion resistance and additional austenite stabilization. Molybdenum was introduced in small amounts for increasing corrosion resistance under the seawater conditions. Small nickel additions (less than 1 wt.%)

into chromium-manganese steels could be optional (recommendable) for improving corrosion resistance and impact elasticity (especially at low temperatures). It is worthwhile mentioning here that according to the requirements of GOST 5632-2014 chromium-manganese corrosion resistant steel may have residual nickel mass share up to 2.0 wt.%, that is not alloying with nickel.

Developing new steels was based on the idea of creating a metastable state of the austenite component, which in the process of testing of mechanical properties and exploitation is capable of transforming into deformation martensite, due to deformation induced martensite $\gamma \rightarrow \alpha'$ transformation at testing (hereinafter DIMTT). Such transformation ensures additional and essential improvement of mechanical properties and even the entire complex of them, which cannot be realized on the well-known analogues - stainless (corrosion-resistance) chromium-nickel steels of the austenite class (steel 18-8 or 18-10 – AI304 (10Cr18Ni9Ti)) and austenite-ferrite class (8Cr22Ni6Mo2, GOST 5632-2014).

3.1 Effect of chromium and manganese content upon the microstructure of the developed and investigated corrosion-resistant Fe-Cr-Mn steels

The microstructure of the investigated cost-saving alloyed 2-phase (austenitic-ferritic) steels with different chromium and manganese content and small additions of copper, silicon, nickel and molybdenum in some of them is summarized in Figure 1. It represents polyhedral (oval-shaped) austenite grains with closed ferrite inclusions of irregular shape. Austenite grains are clearly visible, round-shaped, the ferrite component is represented as inter-layers between austenite grains. Also observed are dispersed inclusions chromium carbides particles $(Cr,Fe)_{23}C_6$. Detachment of σ-phase on grain boundaries is not excluded when the chromium content grows.

Manganese is an austenite forming element and in our case with the content varying (8 to 11.4 wt.%) austenite is stabilized for transformation at cooling, having been introduced instead of costly and deficient nickel. As a result of heating up to 1,050 °C prior to quenching chromium carbides are mainly dissolved. Due to heating the amount of austenite is increased, the degree of stability of γ-phase growing at the same time. The increase of stability can also be explained by the growth of the degree solubility of carbides, their enrichment with carbon and chromium and, also, by possible redistribution of manganese content between ferrite and austenite at heating.

The microstructure of 10Cr15Mn10 steel (15.0 wt.% Cr) represents austenite grains of angular and elongated shape with ferrite and carbide inclusions. As a result of heat treatment (quenching at 1,050 °C) the austenite amount in the structure of steel increases (see Figure 1 a).

The microstructure of 09Cr17Mn9 steel (16.59 wt.% Cr) represents an austenite matrix with dendrite austenite grains and ferrite zones around the austenite grains. The results of the x-ray structural phase analysis showed that austenite amount was austenite (A) = 52%, while ferrite (F) = 48% (Figure 1b). In the structure of 09Cr18Mn10 steel (17.26 wt.% Cr) austenite is partially twinned, deformation twins are observed, while some austenite grains reveal the signs of ε'- and α' – martensite's. This is explained by deformation metastability, i.e. the capability of $\gamma \rightarrow \varepsilon'$ and $\gamma \rightarrow \alpha'$ to DIMTT in the thin surface layer of microsection at grinding and polishing in the process of preparation of microsections (see Figure 1c). A tendency is observed towards an increase of ferrite content in steels with increased from 17 wt.% to 20 wt.% content of chromium, for instance for 09Cr17Mn9 steel (16.59 wt.% Cr) (Figure 1b) the content of F = 47%, A = 52%, and for steel grade 09Cr20Mn9 with 20 wt.% Cr (Fig. 4) F = 62%, A = 38%.

The microstructure of 09Cr20Mn10 steel (20.7 wt.% Cr) is not homogeneous, dendrite structure can be traced in dark grains. The content of the austenite phase is reduced (A=38%) while the ferrite phase increases (F=62%) it is explained by the influence of ferrite-generating chromium. 09Cr21Mn10 steel that has the biggest chromium content 20.6 wt.% the microstructure represents an austenite matrix with ferrite areas of irregular shape (see Figure 1d). Carbide inclusions (C) are also seen. According to the data of x-ray structural phase analysis that steel contains - 69% A and 31% F.

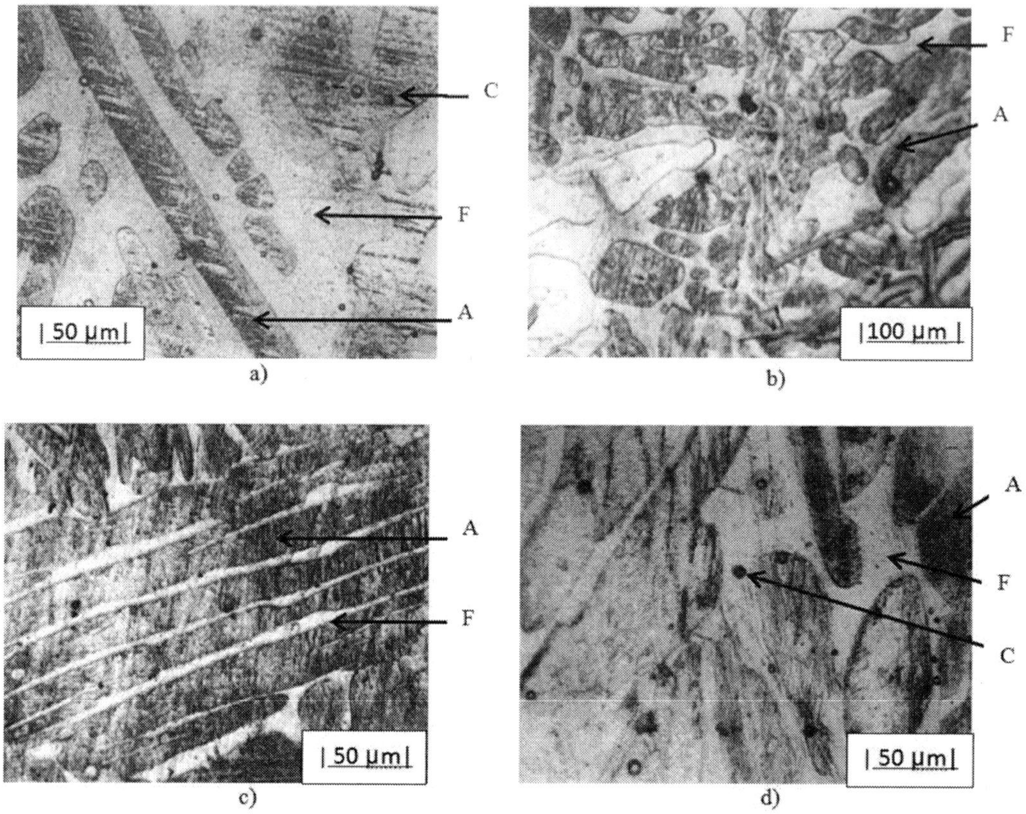

Figure 1. The microstructure of corrosion-resistant steels with different contain of chromium and manganese after quenching at temperature 1,050 °C: a) 10Cr15Mn10; b) 09Cr17Mn9; c) 09Cr18Mn10; d) 09Cr21Mn10; (A – austenite; F – ferrite; C – carbide).

Having analyzed the data mentioned above we may suppose that in 09Cr21Mn10 steel austenite microhardness is increased from 331-335 HV to 376 HV, it being explained by strengthening under the influence of chromium. However, austenite microhardness in 09Cr20Mn10 steel is just 316 HV at its smaller amount (A = 38%) as compared to other steels. It is quite obvious that chromium is distributed unevenly inside steels structure, especially between ferrite and austenite, accordingly ferrite microhardness is increased under the influence of chromium.

On the whole austenite grains possess a polyhedral (oval) shape, while in 10Cr15Mn10 steel or in 09Cr17Mn9 dendrite austenite grains are observed quite often. It is quite typical of the microstructure of cast metal. Particularly, in steels with the highest chromium content (20.7 wt.%) 09Cr21Mn10 and 09Cr20Mn10, ferrite is observed as a branched matrix, which is typical of cast metal.

The microstructure of standard 12Cr18Ni10Ti steel (taken for a comparative analysis) (Fig. 6) mainly austenite, though presence of small δ-ferrite grains can't be excluded, it being allowed by the requirements of GOST 5949-75.

The microhardness of chromium- nickel austenite - 278 HV is slightly lower than of chromium-manganese (320-386 HV), as it does not contain Mn, that strengthens austenite much more than Ni, accordingly, Cr – Mn austenite in the structure of new steels has greater microhardness - 320-386 HV.

Such difference in the value of austenite microhardness can be explained by existence of $\gamma\leftrightarrow\alpha$ transformations at heating for quenching under the influence of manganese (areas of existence of these phases can be different temperature-concentration fields) and the heating temperature 1,050 °C. These factors and heating conditions cause A→F or F→A transformations, ferrite and austenite are enriched with various alloying elements (Cr, Mn, Si, Cu, Mo) during holding during heating for quenching, it changing the correlation between ferrite and austenite [7, 11]. The structural transformations in austenite-ferrite steels are complicated enough it being determined by the presence of two-phase components with different types of lattices and, accordingly, different diffusion rate of the alloying elements, that are present in them. With an increase in the quenching temperature there happens not only correlation in the amounts of austenite and ferrite but also an alternation in the phases: ferrite is enriched with chromium, as compared to austenite. The chromium content in austenite at that can be in the area, where according to diagrams of Potak-Sagalevich, or Scheffler's diagrams martensite transformation at cooling is possible. The highest austenite stability in chromium-nickel austenite-ferrite steels is ensured at the average chromium content > 21 wt.%.

The analysis of changes in microhardness of the phase components, depending on chromium content and under its influence shows the following (see Table 1).

Table 1. Phase composition, hardness and microhardness of the structural components of the investigated steels after quenching at 1,050°C, in water.

Steel grade	Hardness, HRB	Phases content, %		Microhardness, HV	
		F	A	F	A
10Cr15Mn10	95	57	43	277	331
09Cr17Mn9	95	47	52	279	386
09Cr18Mn10	96	21	79	244	335
09Cr20Mn10	94,5	61	38	277	316
09Cr21Mn10	94	31	69	241	376
12Cr18Ni10Ti	89	0	100	-	277

At relatively smaller amount of ferrite 21-31% in the steels structure (09Cr18Mn10 and 09Cr21Mn10 steels) its microhardness has the smallest values 241-244 HV (see Table 1). As the amount of ferrite is increased up to 57-61% (10Cr15Mn10 and 09Cr20Mn10 steels) its microhardness grows up to 277 HV (Table 1). It can be explained by the solid solution strengthening of ferrite, evidently under the influence of its greater content of chromium and nickel. A similar picture is observed when austenite microhardness is changed, due to the changes in phase correlation under the influence of alloying. At the smallest austenite content (A = 38 %) in 09Cr20Mn10 steel its microhardness is 316 HV, with an increase of its content to 69-78% its microhardness grows up to 335-376 HV (see Table 1). This can be explained by a combined stabilizing action of minor additions of copper and nickel.

It can be seen in Table 1 that austenite microhardness is higher than ferrite, it can be explained by a greater opportunity of carbon solving (chiefly), as well as austenite forming alloying elements in the face-centered cubic lattice, it coordinating quite well with the data from literature. In order to analyze the data obtained graphs of dependence of mechanical properties upon chromium content (within 15-21 wt.% Cr) were compiled at different fixed and close manganese contents).

3.2 Effect of chromium and manganese content upon the mechanical properties of the developed and investigated corrosion-resistant Fe-Cr-Mn steels

Chromium content was varied within 15.0 % - 21.4 wt.% limits with the objective of altering the phase correlation between austenite and ferrite, as it known that chromium is a strong ferrite-forming element and also and also for optimization of the chemical composition for improving the mechanical properties and corrosion resistance of the developed and investigated Fe-Cr-Mn steels. Figure 2, a represents alternations in the mechanical properties under the influence of chromium within 15.0–21.4 wt.% at a constant (within a narrow interval) alloying with the smallest in the investigated range amount of manganese (8-9 wt.%).

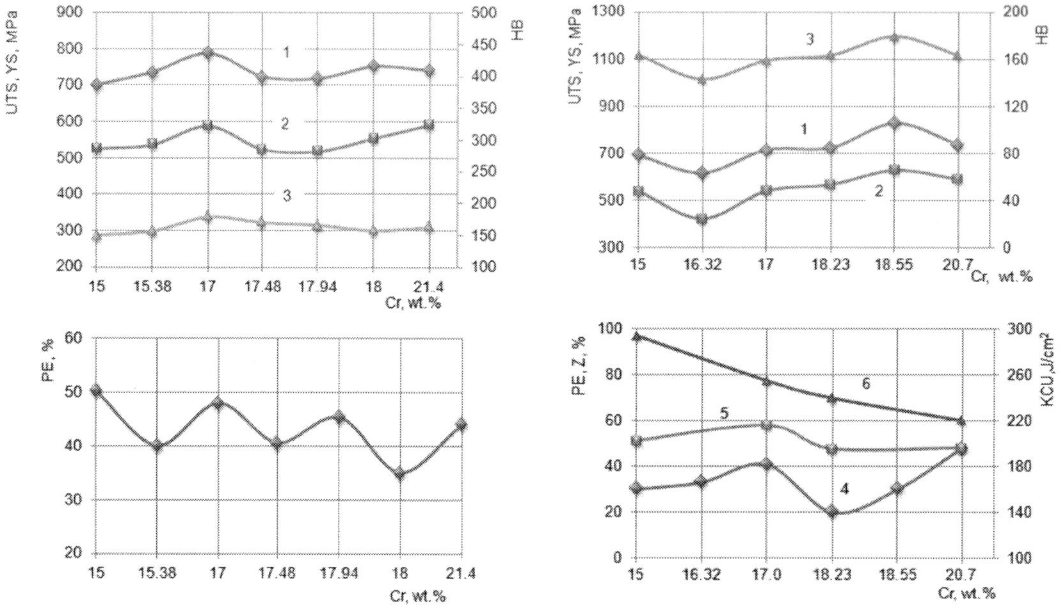

Figure 2. Effect of chromium at manganese content 8-9 wt.% (a) and 9-10 wt.% (b) upon the mechanical properties of the investigated Fe-Cr-Mn steels: 1 – ultimate tensile strength (UTS); 2 – yield strength (YS); 3 – hardness, HB; 4 - percentage elongation (PE); 5 – Z; 6 – KCU.

The strength properties are changed with the presence of maximum (UTS=789 MPa, YS=589 MPa) at chromium content in steel 17 wt.%, while the percentage elongation is altered within a narrow interval 40-51 %. Still, there is a tendency towards reduction in percentage elongation (PE) from 51 % to 37-41 % with an increase of chromium in steels. At bigger alloying of steels with manganese within 9-10 wt.% limit, with an increase of chromium content the strength properties are changed in accordance with the curved lines with maximums UTS=830 MPa; YS=620 MPa), corresponding to 18.55 wt.% Cr (Figure 2, b). Hardness changes correlate with changes in strength.

Percentage reduction of area (Z) is also changed in the shape of a curved line with (Z=50%) maximum, however. It is observed in steels at 17 wt.% Cr content (see Figure 2, b). Percentage elongation has the highest values (PE=30%) at 17 wt.% Cr content and (PE=49%) at 20.7 wt.% Cr. Impact strength (KCU) is decreased from 294 J/cm^2 (2.94 MJ/m^2) to 220 J/cm^2 (2.2 MJ/m^2) with the increase of chromium content. At the highest alloying of steels with manganese (within the investigated concentration range) 10-11 wt.% when the chromium content is raised the strength properties are changed likewise drastically, still, the maximum (UTS=790 MPa; YS=610 MPa) is shifted towards lower chromium content 16.79 wt.% (Figure 3).

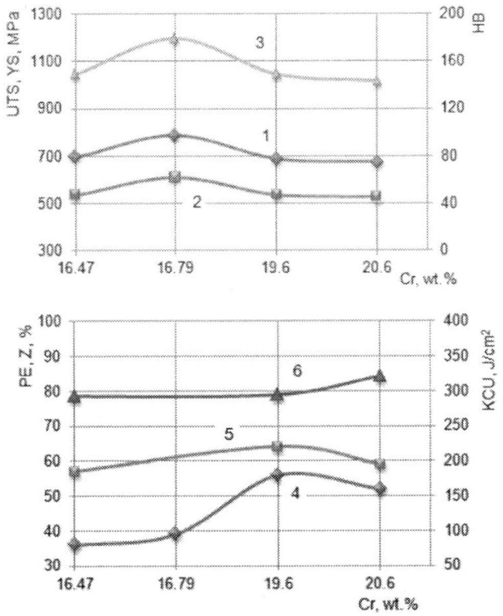

Figure 3. Chromium influence at manganese content 10-11 wt.% upon the mechanical properties of the investigated Fe-Cr-Mn steels: 1 – UTS, 2 – YS, 3 – HB, 4 – PE, 5 – Z, 6 – KCU.

At that the highest values of plastic characteristics: PE=64 % and Z=56 % correspond to chromium content 19.6 wt.%, while the impact strength KCU=320 J/cm^2 (3.2 MJ/m^2) reaches its highest value when the content of chromium is 20.6 wt.%. However, for the highest strengthen indices quite high plasticity characteristics were reached: PE=39 %; Z~60 %; and for impact strength KCU=290 J/cm^2 (2.9 MJ/m^2).

A comparison of the characteristics of mechanical properties of the developed nickel-free stainless steels on Fe-Cr-Mn basis with the well known throughout the world and widely applied chromium-nickel stainless steels of 18-8, 18-10 types, like (10-12)Cr18Ni9Ti (austenite class) and 08Cr21Ni6Mo2Ti (austenite-ferrite class) shows that the new steels possess greater strength properties (UTS, YS) at 180-240 MPa, the plasticity properties being similar or even higher (PE, Z), impact strength values being also similar or higher (KCU) (see Table 2). It is also quite worthwhile mentioning here that we used for our investigations and testing molded steels, while standard steels are manufactured by hot rolling. It is quite evident that the advantages of the new steels are bound to be higher after hot rolling or forging. For cast state of these new steels the reached advantages are quite high and are in good accordance with the requirements of the existing standards (GOST 7350-77 and AISI 304, ASTM A240).

It is important to point out here that passing of $\gamma \rightarrow \alpha'$ DIMTT plays a major part for formation of such lofty complex of mechanical properties of all developed and investigated nickel-free corrosion-resistant (stainless) steels. This transformation leads in the process of testing to the effects of additional self-strengthening and self-relaxation of stresses, thus drastically simultaneously increasing strength, plasticity and impact strength. The improved complex of properties of the designed cost-saving alloyed Fe-Cr-Mn steels can be explained primarily by that.

Table 2. Comparative characteristics mechanical properties (tensile, impact strength, ISO 6892-2015) of designed and standard corrosion-resistant steels after quenching at 1,050 (1050) °C.

Steel grade	Ultimate Tensile Strength (UTS), MPa	Yield Strength (YS), MPa	Percentage elongation (PE), %	Percentage reduction of area (Z), %	Impact strength (KCU), MJ/m^2
		Designed Fe-Cr-Mn (Ni-free) steels			
09Cr17Mn9	790	590	51	38	2.4
09Cr19Mn10	830	620	30	47	2.3
09Cr17Mn11	790	610	39	60	2.9
09Cr20Mn11	690	530	56	64	2.9
		Standard Fe-Cr-Ni steels			
10Cr8Ni9Ti (GOST7350-77)	530	230	38	55	2.5
AISI 304 (ASTM A240)	510	205	43	-	-
08Cr21Ni6Mo2Ti	588	340	20	-	0.6

4 Conclusions

1. New cost-saving alloyed Fe-Cr-Mn (nickel-free) corrosion-resistant steels of austenitic-ferritic structural class were developed that possess an improved level of mechanical properties.

2. By varying the chromium content within 15 – 21.4 wt.% the manganese content within 8 – 11.4 wt.% in various combinations it was possible to achieve different phase relations between ferrite and austenite content, the governing influence belonging to chromium, while manganese exerting influence chiefly upon the degree of austenite metastability.

3. It was found out that by regulating the content of chromium and manganese in steels it was possible to control the relationship between the amounts of austenite and ferrite, as well as the degree of metastability of γ-phase. For replacement of costly nickel in steel it is advisable to add manganese, some little additions of copper and silicon and perform micro-alloying with strong carbide-forming elements, it allowing reaching of a complex of mechanical properties that can be classified as high enough.

4. Determined were optimal compositions of cost-saving alloyed Fe-Cr-Mn (nickel-free) corrosion-resistant steels (in chromium, manganese and carbon content with regard to additional alloying with small additions of copper, silicon and micro-alloying with carbide-forming elements) that ensure an improved complex of mechanical properties after quenching.

5. The improved complex of strength, plasticity properties and impact strength in the developed corrosion-resistant (stainless) Fe-Cr-Mn steels can be explained, on the one hand, by the number of austenite and ferrite phases and, on the other hand, by the kinetics and the volume of development of deformation martensite $\gamma \rightarrow \alpha'$ transformation, occurring in the process of testing of mechanical properties.

6. Comparative testing has shown that the developed cost-saving alloyed (nickel-free) corrosion-resistant steels are superior to the existing costly and scarce chromium-nickel steels of 18-8 and 18-10 (10Cr8Ni9Ti) type of austenite class and 21-6 (08Cr21Ni6Mo2Ti) of austenitic-ferritic class, being a worthy alternative to them for practical applications.

References

[1] Svistunova T V and Shlyamnev A P 2010 Development and use of corrosion-resistant steels and alloys *Metal Science and Heat Treatment* **52 (7-8)** pp 300-308 doi: 10.1007/s11041-010-9272-z

[2] Ahmad Z 2017 *Principles of Corrosion Engineering and Corrosion Control* (Butterworth-Heinemann) 51

[3] Khatak H S and Raj B 2019 *Corrosion of Austenitic Stainless Steels* (Imprint Woodhead Publishing) 71

[4] Salamci E, Candan S and Kabakci F 2017 Effect of microstructure on corrosion behavior of dual-phase steels *Kovove Mater* **55** pp 133–139 doi: 10.4149/km 2017 2 133

[5] Shipilov S, Jones R, Olive J-M and Rebak R 2020 *Environment-Induced Cracking of Materials* (Imprint: Elsevier Science) 21

[6] Bannykh O A 2005 Economical stainless nitrogen steels *Promising substitutes of light alloys Metal Science and Heat Treatment* **47 (7-8)** pp 261-265 doi: 10.1007/s11041-005-0064-9

[7] Cheylyakh A P and Sheichenko G V 2002 Structure and physicomechanical properties of new nickel-free corrosion-resistant austenitic-ferritic steels *Russian Metallurgy* **5** pp 452-458

[8] Cheylyakh A P 2002 Kinetics of deformation-induced martensitic transformation $\gamma \rightarrow \alpha'$ and mechanical properties of steels and irons with different contents of metastable austenite *The Physics of Metal and Metallography* **94 (4)** pp 378-387

[9] Ryoo D-V, Kang N and Kang C-H 2018 Effect of Ni content on tensile properties and strain-induced martensite transformation for 304 stainless steel *Materials Science and Engineering* V **529** pp 2277-2281

[10] Fylyppov M A, Fylyppenkov A A and Plotnykov H N 2009 *Yznosostoikye staly dlia otlyvok* (Ekaterynburh: UHTU-UPY) 358

[11] Grachev S V, Mal'tseva L A, Mal'tseva T V 2000 Austenite-ferrite corrosion-resistant steel for high-strength wire **42 (11-1)** pp 419-422 doi: 10.1023/a:1010404410829

The use decorative concretes with the addition of highly dispersed chalk for volumetric and spatial elements of urban space

Svitlana Chepurna[1], Olha Popova[1], Tetiana Ryshchenko[1] and Olena Dudka[1]

[1]O M Beketov National University of Urban Economy in Kharkiv, Kulikovska descent 12, 61002 Kharkiv, Ukraine

Abstract. In recent years, the attitude to the external appearance of the urban environment and its elements has changed dramatically. Interest in new constructive solutions, style of urban space design and modern materials which improve the quality of urban space has increased. Installed that the most effective material is decorative concrete, which is used to create a variety of volumetric and spatial elements of engineering improvement, architectural and urban design. The possibility of creating decorative concrete with the addition of highly dispersed chalk and the natural pigment that is distinguished by high operational, decorative, aesthetic properties and solves environmental and economic problems has been considered in the article. The results obtained indicate the possibility of combined use of highly dispersed chalk and yellow natural pigment (ocher) to create colored concrete with high corrosion resistance, frost resistance and decorative and aesthetic properties. The relationship between the durability and corrosion resistance of the test concrete on the size and nature of the pores has been established.

1. Introduction

In recent years, changing the appearance of most cities in Ukraine that were characterized by monotony and sameness of buildings, especially residential areas, has been observed. Today, for creating an artistic expression, comfort and harmony of the urban environment, as well as an increase in the manifold and aesthetic qualities of residential areas are used various means of architectural and urban design, which are represented by stairs, ramps, fences, small architectural forms. These various means are added uniqueness, originality, originality, and scale to individual territories and as a whole the city [1, 2].

The diversity used elements of urban design are depended on the type of used building materials and products. The volumetric and spatial elements of the urban environment are exposed to various climatic factors and mechanical stress in according to various defects to connected with processes of maintenance are observed with their surface. These defects are caused by the influence of temperature fluctuations, air humidity, exposure to wind, precipitation and solar radiation, the presence of household and industrial impurities in the atmosphere (dust, soot, carbon black) [3].

Therefore, to create various elements of engineering improvement, architectural details, materials are used to are characterized by high-performance properties, especially weather resistance, wear resistance, mechanical resistance, durability, corrosion resistance, but at the same time have high aesthetic and decorative qualities and solve environmental and economic problems. Concrete and products of concrete are matched all these characteristics. Architectural forms made of concrete are practical, easy to install, are allowed to create a variety of textures and shapes of either complexity, and concrete is much cheaper than marble or granite. But products of concrete have one disadvantage of their color is gray and unattractive [4, 5, 6, 7].

Accordingly, decorated concretes are given wide use. According to in December 2019 report by Value Market Research, the global decorative concrete market will grow from $ 15.4 billion in 2020 to $ 20.5 billion by 2025, with a CAGR of 5.9% from 2020 to 2025 year [8]. Due to the fact that decorative concrete can be used to produce a variety of architectural products not only flat but also curved, with smooth conjugation of surfaces, the required color, and high-quality texture, to are allowed creating new

Content from this work may be used under the terms of the Creative Commons Attribution 3.0 licence. Any further distribution of this work must maintain attribution to the author(s) and the title of the work, journal citation and DOI.
Published under licence by IOP Publishing Ltd

architectural elements in buildings and structures, while providing high artistic and aesthetic expressiveness such elements [8].

One of the ways are given decorative concrete sufficient brightness is developing the composition of a concrete mixture on Portland cement using pigments, but obtaining bright saturated colors such concretes are impossible because of the fact that the brightness of the pigment color is muffled by the gray color of the cement. Consequently, such concretes don't better use to create objects of architecture and urban design. It will add greyness and uncomfortable areas. In order to avoid this effect and obtain decorative concrete with high properties, white cement is used in that case the cost of products is increased significantly, but at the same time, such concretes must retain decorative and operational characteristics during the time and contribute to the elimination of hazardous pollutants from the urban environment and be similar to natural materials, such as marble or granite [9, 10, 11].

The properties of decorative cements largely depend on the quality of the cement and the properties of the pigment. Therefore, pigments must be resistant to alkalis, sunlight and atmospheric influences, have been a high coloring power, but at the same time, a minimal effect is affected on the physical and mechanical properties of concretes. The quantity of mineral or synthetic pigments by decorative concrete should not exceed 15%, and natural pigments by decorative concrete are no more than 0.5% 12, 13]. An effective method of getting concrete, characterized by significant whiteness, with increased durability characteristics, is the use of carbonate additives, in particular, highly dispersed chalk.

Chalk is a lithological type of limestone that is a mixture of particles of organic calcite (small fragments of skeletons, or shells of mollusks), to have a negative surface charge due to the fact that the presence of free silicic acid, and small particles of chromogenic calcite and it is had positive surface charge. Against the background of the bulk (about 80%), chalk is represented by cocolitis and their parts (powdered calcite) [6, 14, 15, 16]. (Figure 1).

Figure 1. Electron microscopic image of the surface of natural chalk (Slavyansk field, Ukraine).

Accordingly, the cementation of rock is caused by both condensation and crystallization bonds between particles [15]. The microstructure of chalk is a fine-grained mass consist of particle size 0.01 mm. According to [15, 17, 18], the main part of chalk (95-98%) is represented by particles from 0.05 to 0.005 mm. The chalk has low solubility, does not form crystalline hydrates and does not interact with water, but easily disperses when dispersed. In this case, the chalk contains the cations that are part of most clinker minerals [19, 20].

The surface of the chalk in aqueous dispersions has an excess negative charge, as evidenced by the small negative value ζ- potential, while the equipotential chalk point corresponds pH from 5 to 7 [7, 17]. According to [15, 19, 21] is covered with free silicic acid, so the mechanism of interaction of chalk with clinker minerals and products of their hydration is significantly different from the known mechanisms of interaction of other carbonate rocks, including limestone.

Particles of highly dispersed chalk that were added into the composition of concrete, are placed in the pores and create hydrophobic areas to prevents the movement of water into the concrete. Highly dispersed chalk doesn't only contribute to the compaction and plasticization of the mixture, but also it is affected the formation of the phase composition of calcium hydrosilicates [13, 20, 22]. In this case, the main products of hydration are low-basic hydrosilicates of the CSH (I) type and calcium hydrocarbosilicates that it leads to an increase in the strength and frost resistance of such compounds [23, 24]. Accordingly, the use of highly dispersed chalk in cement compositions is led to changes in the phase composition of the cement stone due to the fact that the formation of complex compounds, including $CaCO_3$ and Ca $(OH)_2$, and it also is lea to the acceleration and complete hydration of clinker minerals.

In the process of hydration, highly dispersed chalk enters into chemical interaction with aluminum-containing clinker minerals C_3A and C_4AF, cubic calcium hydroaluminate $3CaO \times Al_2O_3 \times 6H2O$ and a complex compound calcium hydrocarboaluminate $C_3A \times CaCO_3 \times 11H_2O$ are formed, provided that chalk is represented by grains d <42 μm [15, 25, 26]. With larger grains of highly dispersed chalk is formed.a small amount of cubic C_3AH_6. In the presence of highly dispersed chalk, the character of crystallization of calcium hydrosilicates can also change, namely: instead of large fibrous crystals are formed small ones with a more developed surface, also it is led to the mixing of strength characteristics and is provided an increase in frost resistance, sulfate resistance, and water resistance [26, 27, 28,29, 30].

Based on the above, it was assumed that the particle size of highly dispersed chalk is matched the size of the capillaries of the cement stone, thus a significant part of the capillary pores volume can be filled with calcite, to is poorly soluble in water, it will increase the density and it will reduce the permeability of the cement stone.

The purpose of the study is to develop a composition of color concrete with the addition of highly dispersed chalk for volumetric and spatial elements of urban space.

2. Research methods and materials

2.1. Materials

For experimental studies Portland cement was used (CEM I 42.5 N), as a coarse aggregate, granite crushed stone of Znamenivsky Quarry (Kirovogradska region), it has fractions 2.5 ÷ 5 mm and 5 ÷ 10 mm, with a density of 2590 kg / m^3 were used, as a fine aggregate quartz sand of the Bezlyudivsky Quarry of Kharkiv region with the size of Mk = 0.9, bulk density 1370 kg/m^3 were used. As an additive chalk of the Slavyansk deposit is used, pigment is used ocher, it is clay colored with iron oxides which has various shades from light yellow, golden to dark, red and even brown. In concrete, a mixture of highly dispersed chalk is introduced as a chalk suspension [5, 6, 7].

For the study, a series of samples with different percentages of highly dispersed chalk were produced that are consider at Table 1.

Table 1. Composition of concrete samples [5, 6, 7].

Sample number	Material costs,%		W/C
	Cement	Chalk	
1	100	-	0.466
2	90	10	0.460
3	80	20	0.462
3	70	30	0.468
5	60	40	0.472

The degree whiteness of the cement is determined depending on the light reflectance as a percentage of the absolute scale and the grade must be at least: I - 80%, II - 75%, III - 68% [32].

The required quantity of pigment that it was introduced into the concrete mixture, is determined by the need color. An increase in the quantity of pigment to a certain level are led to an increase in the intensity color. The development of concrete mixtures for the manufacture of various volumetric and spatial elements for engineering improvement, architectural and urban design should be had high level operational and decorative characteristics.

2.2. Method for determining the density

The research of density was determined according to DSTU B V. 2.7-114-2002 [31] before testing the compressive strength. The tests were carried out on concrete samples - cubes 100×100×100 mm in relatively normal humid conditions at a temperature of (20 ± 2)0 C and relative humidity of at least 95%.

The density of the samples was found by the formula (1):

$$\rho_w = \frac{m}{V} \cdot 1000 \tag{1}$$

where m - mass of samples, kg; V – volume of sample, m^3.

2.3. Method for the determination of corrosion resistance

To determine the corrosion resistance, an express method was used, developed by M.I. Strelkov [32] and beams 70×70 ×280 mm were made. The basis, concentration diagram of the four-component system (Na +, Mg2 +, Cl-, SO42-) was used proposed by M.I. Strelkov [32]. According to [32], the high aggressiveness of this system is due to the formation of compounds with a high molecular weight as a result of chemical interaction between the salts of the system: NaCl, Na_2SO_4, $MgCl_2$, $MgSO_4$, and products of hydration cement.

To increase the corrosion resistance of the cement stone, finely dispersed chalk was used as a mineral additive. Samples of cement stone were made from Portland cement without additives and with the addition of chalk: 1 - control sample made of Portland cement without additives; 2, 3, 4, 5 - with the addition, respectively, of 10, 20, 30, 40% of chalk from the total mass of cement and chalk at W/C=0.45. After solidification under normal conditions for 28 days, the samples were crushed and for the following researches, fractions of 0.14-0.315 mm were, after which a quantity such particles were placed in measuring cylinders with aggressive solutions. It was a solution to various salts. For the researches, solutions were prepared to contain salts of NaCl, Na_2SO_4, $MgCl_2$, $MgSO_4$ that are shown in Table 2.

Table 2. Compositions of an aggressive solutions.

Liquid number	Compositions of the aggressive solutions
1.1	$0.4NaCl+0.2Na_2SO_4+0.27MgCl_2+0.13MgSO_4$
1.2	$0.42NaCl+0.4Na_2SO_4+0.13MgCl_2+0.,27MgSO_4$
1.3	$0.13NaCl+0.27Na_2SO_4+0.2MgCl_2+0.4MgSO_4$
1.4	$0.1NaCl+0.1Na_2SO_4+0.4MgCl_2+0.4MgSO_4$
1.5	$MgCl_2$
1.6	$0.2MgCl_2+0.8MgSO_4$

For the research, the solutions were selected according to the results of determining the corrosion resistance by the express method.

The samples of degree stability were established by the value of the stability coefficient (Cs) as was determined by the formula (2):

$$Cs = \frac{f_{strenght}^{3month}}{f_{strenght}^{28}} \tag{2}$$

where $f_{strenght}^{3month}$ – compressive strength of the sample after immersion in an aggressive solution during 3 months, MPa;

$f_{strenght}^{28}$ – compressive strength of samples after 28-day hardening before immersion in an aggressive solution, MPa.

3. Results of experimental studies

According to research results, it was found that an increase in density is observed in samples № 2, №3, modified with 10% and 20% highly dispersed chalk, while there is an increase in compressive strength for sample № 2, and for sample № 3, the strength remains the same as the control sample. The density of samples № 4 and № 5, modified with 30% and 40% highly dispersed chalk, has decreased (Figure 2).

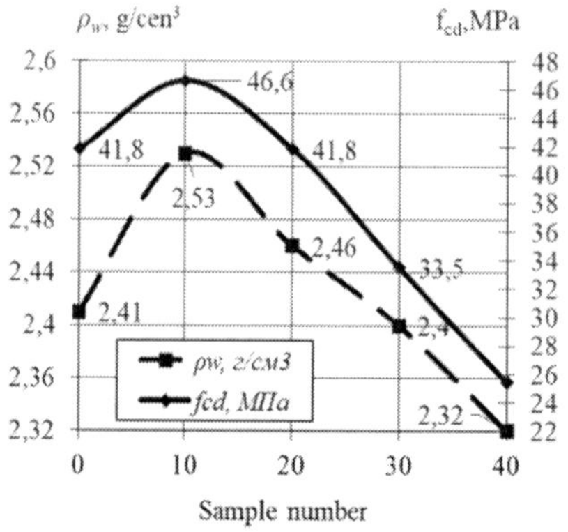

Figure 2. The dependence of compressive strength (f_{cd}) and density (ρ_w) concrete samples from chalk component.

The research has shown that the higher the percentage of the quantity of highly dispersed chalk in concrete samples, the greater the whiteness of the sample. Sample № 5 that is modified with 40% of highly dispersed chalk has the highest whiteness, according to [32], it can be classified as grade II, wherever the light reflectance is 77%. Samples № 2 and №3, modified respectively with 20% and 30% of highly dispersed chalk, have a light reflectance of 70% and 73%, respectively, that it allows classifying as III grade. When a yellow pigment in the quantity of 0.3% had added, samples № 4 and №5 had a bright yellow color, the other samples had a saturated color (Table 3).

Table 3. Factor durability of cement stone.

Sample number	**Initial compressive strength, f_{cd}, MPa**	**Index durability of cement stone, % in an aggressive liquid (liquid number)**					
		1.1	1.2	1.3	1..4	1.5	1.6
1	41.8	72	78	62	85	76	67
2	46.6	126	115	103	98	94	96
3	41.8	123	99	98	98	85	93
4	33.5	98	95	88	86	93	91
5	25.4	86	91	76	78	87	86

The research on corrosion resistance has shown that samples № 1 and №2 modified 10% and 20% of highly dispersed chalk, are the most stable in all corrosive environments (the stability coefficient Cs is outreach 85%), while cement stone without additives is not very stable in these solutions.

4. Discussion of results obtained in the study

To a large extent, the density of concrete is determined by the strength on the adhesion of the cement stone by aggregate, that depends on the shape, roughness and surface cleanliness of aggregate grains, as well as adhesion. The corrosion resistance of concrete is depended on the phase composition of the cement stone, since the solubility and reactivity of its individual phases is differed significantly. The structure of concrete is determined the speed of penetration to aggressive ions and is removal of corrosion products, that is, the course of corrosion processes.

Since all reactions are occurred on the surface of the cement grain, so crystallization of new formations, as a rule, is occurred on the surface of the grain, gradually filling the free space between the particles of cement grains (pore space), while forming a new structure of the cement stone, which it is represented by calcium hydrocarboaluminate, low-basic hydrosilicates. Considering that the particles of highly dispersed chalk are formed supplemental crystallization centers, accordingly, there is a change in the quantitative ratio of structure-forming and destructive factors during the synthesis the strength of cement stone and concrete. In this case, there is an increase in the number of growths of crystals of newgrowths, which increases the volume concentration of newgrowths and the density of the gel in the pore space between the particles of cement stone [6]. As the volume of capillary pores decreases, hydration products, such as calcium hydrocarbonaluminate, fill the voids between the cement particles, which is due to the structure of the chalk, increasing the packing density and leading to an increase in the sample density [6]. The appearance of calcium hydrocarboaluminate $C_3A \times CaCO_3 \times 12H_2O$, the formation of which is due to the interaction of C3A with CaCO3 chalk, is confirmed by X-ray studies (Figure 3).

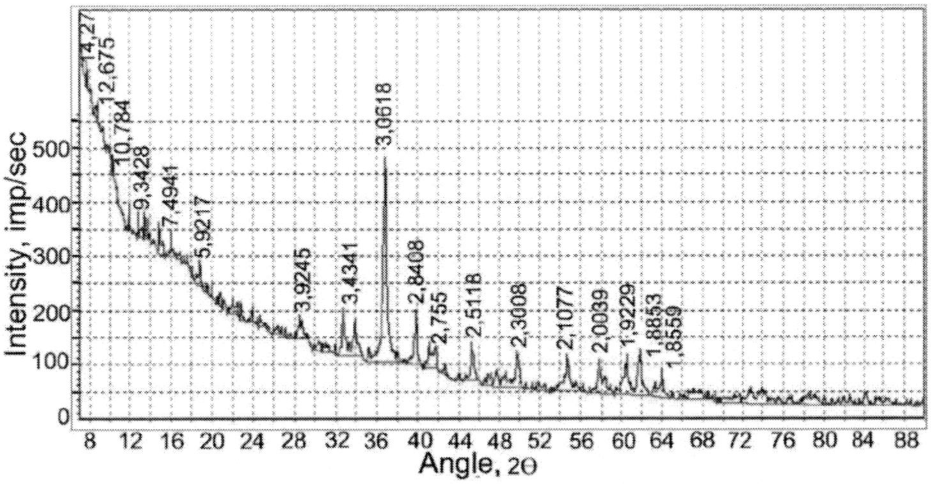

Figure 3. Radiograph of cement stone modified with 10% highly dispersed organogenic calcite (chalk).

The introduction of highly dispersed chalk, the particle size of which is in the range of from 0.54 to 2.21 μm, is helped to reduce the content of micro- and capillary pores filled with calcite, so that it is poorly soluble in water. An increase in the volume of capillary pores will increase the density of the concrete and will reduce the permeability of the cement stone. Since the volume of capillary pores usually does not exceed 40%, therefore, so that chalk is filled them, the total volume of the binder can be increased due to dispersed chalk by no more than 40%. A decrease in the number of growths is increased capillary pores, which leading to an increase in the looseness of the concrete structure and, accordingly, to a decrease in density. Therefore, in terms of capillarity, the concrete samples under research can be classified as dense.

5. Conclusions

According to the studies, carried out, it was found that highly dispersed chalk is affected growths in cement composites, which is led to a change in the phase composition of the cement stone due to the formation of complex compounds, including $CaCO_3$ and $Ca\ (OH)_2$, and also it will lead to the acceleration and complete hydration of clinker minerals. An increase in concrete density is allowed to improve its physical and mechanical characteristics, in particular, compressive strength, permeability, frost resistance, water resistance, corrosion resistance, that will lead to the durability of structures. The relationship between the density and permeability of the samples has been established, that it is allowed to predict the strength and deformation properties of decorative concrete, as well as crack resistance, frost resistance, and corrosion resistance during their operation.

The obtained results are showed that concrete samples with the addition of highly dispersed chalk have had a bright white color, that it allowed to use it as an alternative to white cement, the introduction of yellow ocher pigment does not reduce the physical, mechanical, operational and decorative characteristics of the samples. The dosage of the pigment is formed 0.1 to 0.3% by weight of the cement, depending on the desired color.

Due to the use of concretes modified highly dispersed chalk, and in combination with organic or natural pigments, it can be expanded the range of decorative ones that will be used to create and improve the urban and architectural environment.

References

[1] Dedenyova E, Demina O, Volkova A and Krivitskaya A A 2015 *Collected scientific works of Ukrainian State University of Railway Transport* **152** pp 187-192
[2] Dedeneva E B, Demina O I, Stelmakh A A and Rachkovsky A V 2014 *Priority areas of science and technology* pp 60-64
[3] Olga Bazhenova and Maxim Kotelnikov 2018 *E3S Web of Conferences* **33** pp 1-6
[4] Kochevikh M O, Blazhis G R, Gonchar O A and Vyalin D 2016 *Visnyk KhNTU INzhenerni nauky* pp 54-59
[5] Chepurna S M, Zhydkova T V and Chepurna M Ie 2018 *Zbirnyk naukovykh prats «Suchasni tekhnolohii ta metody rozrakhunkiv u budivnytstvi»* **10** pp 195-204
[6] Svitlana Chepurna, Tetiana Zhydkova and Olha Popova1 2020 *Trans Tech Publications Ltd* (Zurich, Switzerland) pp 27-35
[7] Chepurna S, Borziak O and Zubenko S 2019 *Materials Science Forum* **968** pp 82-88
[8] Decorative Concrete Market by Type (Stamped Concrete, Stained Concrete, Colored Concrete, Polished Concrete, Epoxy Concrete, Concrete Overlays), Application, End-use Sector, and Region *Global Forecast to 2025* https://www.marketsandmarkets.com/Market-Reports/decorative-concrete-market-189792263.html
[9] Luigi Cassar, Carmine Pepe, Giampietro Tognon, Gian Luca Guerrini1 and Rossano Amadelli 2003 *11th Int. Congr. on the Chemistry of Cement* (Durban)
[10] Mindess S, Young JF and Darwin D 2002 Concrete 2nd edn. Pearson (Upper Saddle River)
[11] Cassar L, Pepe C and Tognon Getal 2003 *11th International congress on the chemistry of cement* (Durban, South Africa) pp 41–11
[12] Trofimov B Ya and Kramar L Ya 2010 *Vestnik of the South Ural State Univ.* pp 36-38
[13] Marco L 1983 *Raw materials for decorative concrete* (Moscow: Stroyizdat)
[14] Arshinnikov D I and Svidersky V A 2015 *Technology audit and production reserves* **4** pp 7-11
[15] Gorkova M et al. 1962 *Priroda prochnosti i deformatsionnyie osobennosti mela i nekotoryih melopodobnyih porod.* (Moscow: Akademiya nauk USSR)
[16] . Kozlova V K, Manokha A M, Lykhosherstov A A and others 2012 *Tsement i eho primenenie* **3** pp 53-57, 60
[17] Poluektova V A, Lomachenko V A, Stolyarova Z V, Lomachenko S M and others 2014 *Basic Research: Technical Sciences* **9** pp 1205–1209
[18] Kuznetsova T V, Kudriashov Y V and Tymashev V V 1958 *Fyzycheskaia khymyia viazhushchykh veshchestv* (Hosudarstvennoe izdatelstvo literatury po stroitelstvu i arkhitekture USS) pp 5-16, 49-66
[19] Marushchak U, Sanytsky M, Mazurak T and Olevych Y 2016 *Eastern-European Journal of Enterprise Technologies* **6/6 (84)** pp 50-57
[20] Sanytsky M, Kropyvnytska T and Kotiv R 2014 *Advanced Materials Research* **923** pp 42-47
[21] Krivenko P V, Sanytsky M, Kropyvnytska T and Kotiv R 2014 *Advanced Materials Research* **897** pp 45-48
[22] Ustinova A A, Potapova E N and Konyashkina A Yu 2017 *Advances in chemistry and chemical technology* **XXXI (3)** pp 114-116

[23] Marushchak U, Sanytsky M, Sydor N and Braichenko S 2018 *MATEC Web of Conferences* **230** 03012
[24] Kononova O V and Cherepov V D 2013 *Sovremennye problemy nauki i obrazovaniya: Tekhnycheskye nauky* **1** pp 227-234
[25] Samchenko S V and Makarov E M 2013 *Tekhnyka i tekhnolohiya silikatov* **3** pp 27-29
[26] Matschei T, Lothenbach B and Glasser F P 2007 *Elsevier Magazine: Cement and Concrete Research* **7 (4)** pp 551–558
[27] Mchedlov-Petrosian O P and Olhynskyi A H 1971 *Eksperimental'noe issledovanie mineraloobrazovaniya* (Nauka)
[28] Loshak V V, Cherkasov S V and Vlasov V V 2011 *Sci. Bull. Of Voronezh State Arch. and Constr. Univ.* pp 3-4, 61-66
[29] Guziy S and Terenchuk S 2016 *Science Rise* vol. 9 **2 (26)** pp 49-54
[30] Borziak O S, Chepurna S M, Plugin A A, Zavalniy O V and Dudin O A 2019 *IOP Conference Series: Materials Science and Engineering (MSE)* **708** pp 1-5
[31] *DSTU B V. 2.7- 114 – 2002.* 2002 *Budivelni materialy. Sumishi betonni. Metody vyprobuvan (State standard of Ukraine, Kyiv)*
[32] Strelkov M Y and Zaslavskyi Y N 1987 *Uskorennaia otsenka ahressyvnosty k betonu vodnykh rastvorov s uchetom mnohokomponentnosty ykh sostava* (Lenynhrad, PromstroiNYYproekt)
[33] *HOST 965-89.* 1990 *Portland cements, white. Specifications (*Moscow: Stroyizdat)

IOP Conf. Series: Materials Science and Engineering 1164 (2021) 012019 doi:10.1088/1757-899X/1164/1/012019

Reliability assessment of cylindrical joints with tensioned roll band of the rolling mill

O N Checheneva[1], A N Pashkov[1], E E Balakhnina[1], E I Sizova[1] and V V Devyatiarova[1,*]

[1]National Research Technological University MISiS, 4,Leninsky pr., Moscow, 119991, Russia

*E-mail: vikdev@yandex.ru

Abstract: The work addresses to the problem of ensuring the operability of a composite roll made in the form of a cylindrical joint with the tension of a carbide band and a steel axis. When considering the wear processes, all the characteristics related to the surface microgeometry are combined by a dimensionless complex roughness parameter, which, along with the roughness class, includes the features of the surface treatment technology. To determine the contact pressure associated with the tension, the Lame formula for calculating thick-walled cylinders is used. When a hard-alloy bandage is placed on a steel roll with the tension between the contacting surfaces, a plastic unsaturated or saturated contact is performed, for which a method for calculating the tension values shall be proposed. Formulas are given for determining the maximum permissible torque that does not cause a change in the strength of the joint with tension under conditions of plastic saturated and unsaturated contact. In order to test the strength of the connection elements with tension, the theory of the greatest tangential stresses is applied for two dangerous points located on the inner surfaces of the covered and the covering parts: a formula is obtained for determining the maximum permissible calculated value of tension from the strength condition of the covered part. An example of fitting with the tension of a band made of hard alloy VK8 on a working roll made of improved steel 45 with a diameter of d = 100 mm is considered. The band is installed using a press with an H7/p6 fit to transmit a torque of M = 5 kNm. The minimum tension value is obtained from the condition of fixing the contact surfaces without their relative slippage, and the maximum tension value is obtained from the condition of the strength of the steel axis. As a result of the calculations, it was found that the minimum tolerance for this attachment does not provide the transmission of the specified torque since its value is less than the minimum allowable tension. Therefore, to ensure reliable operation of the connection with tension, it is recommended to use the H7/s6 fit, which provides the transmission of the specified torque without the strength of the connection elements.

1. Introduction

Recently, the reliability of technological machines and equipment has become one of the main engineering problems, which is becoming increasingly important for metallurgical production [1-4]. Insufficient reliability of technological machines and their components and devices not only leads to significant downtime of equipment but also significantly increases the cost of their operation. Increasing requirements for the quality of technological machines and equipment in order to reduce material, labor, and financial costs for maintenance and repair leads to the need for modernizing and reconstructing equipment [5-10]. At the same time, attention shall be paid to equipment for the base metal production [11-14].

2. Setting of a Problem

In the process of rolling workpieces made of refractory metals, the working rolls undergo significant thermal and mechanical pressures. Therefore, for high-temperature rolling of hard-to-deform

Content from this work may be used under the terms of the Creative Commons Attribution 3.0 licence. Any further distribution of this work must maintain attribution to the author(s) and the title of the work, journal citation and DOI.
Published under licence by IOP Publishing Ltd

materials, carbide rolls are often used. In order to save expensive carbide material, the working roll is often produced composite, consisting of a steel roll and a carbide band [15-16]

The cylindrical connection with the tension of the band with the rolling roll should ensure that the contact surfaces are fixed without their relative slippage. This occurs by assigning the appropriate tightnesses. Due to the spread of the values of the coefficient of friction and tension, when an external (excessive) load and vibration are applied, a relative displacement of the band relative to the roll may occur, which negatively affects the further ability to transmit the full load. Under the action of variable loads, especially at the moment of starting and stopping, micro-displacements and fretting corrosion occur on the mating surfaces at the ends of the connected parts [17-20].

3. Purpose and Objectives of the Study

When considering the wear processes, all the characteristics related to the surface microgeometry can be combined by a complex dimensionless parameter [17-19]

$$\Delta_r = \frac{R_{\max}}{r_{eq} \cdot \sqrt[v]{b}}. \tag{1}$$

Here r_{eq} is the reduced average radius of curvature of the vertices of the projections, equal to

$$r_{eq} = \frac{r_l \cdot r_t}{r_l + r_t},$$

where r_l и r_t are the average values of the radii of curvature of the vertices of the projections in the longitudinal and transverse directions.

It is fundamentally important here that, along with the roughness class, the surface treatment technology also plays a great role, on which the other characteristics included in the formula (1) depend. For the treated surfaces, which are characterized by an equilibrium roughness, the parameters of the reference curve of the profile take fairly stable values of $b \approx v \approx 2$.

The contact pressure p_c, MPa, is related to the tension N, m, the Lame dependence, which is derived in the course "Resistance of materials" [21]

$$p_c = \frac{N}{d \cdot C} = \frac{N}{d \cdot \left(\frac{C_1}{E_1} + \frac{C_2}{E_2}\right)}. \tag{2}$$

Here N-tension in the joint, m;
d-roll diameter, m;
C-coefficient, which is determined by the formula t

$$C = \frac{C_1}{E_1} + \frac{C_2}{E_2} = \frac{\left(\frac{d^2 + d_1^2}{d^2 - d_1^2} - \mu_1\right)}{E_1} + \frac{\left(\frac{d_2^2 + d^2}{d_2^2 - d^2} + \mu_2\right)}{E_2}, \tag{3}$$

where d_1 - the diameter of the hole of the covered part (for a solid roll $d_1 = 0$);
d_2 – outer diameter of the covering part, m;
E_1 and E_2 – elastic modulus of the material of the covered (of the roll) and covering (sleeve) surface, MPa;
μ_1 and μ_2 are the Poisson coefficients of the material of the covered (of the roll) and covering (sleeve) surface.

4. Results of the study

When the carbide band is placed on a steel roll with the tension between the contacting surfaces, as a rule, a plastic unsaturated or saturated contact occurs.

Plastic unsaturated contact occurs in the case

$$N > \frac{14,5 \cdot d \cdot \Theta^4 \cdot HB^5 \cdot C}{\Delta^2_{шер}}. \quad (4)$$

Here $\Theta = (1\text{-}\mu^2)/E$ - elastic constant for a less rigid body, MPa^{-1};
HB – hardness of a less solid body, MPa;
Δ_{abr} – a complex characteristic of roughness.

Plastic saturated contact occurs in the case when

$$N > 0,124 \cdot K_{сб} \cdot HB \cdot d \cdot C, \quad (5)$$

where K_{as} is the assembly coefficient; for thermal assembly $K_{as} = 1$, for press assembly – $K_{as} = 0.5$ [20].

The maximum permissible moment $[M]$, which does not cause a change in the strength of the joint with tension in the conditions of plastic contact, is determined by the formula

$$[M] = \left[\frac{f_m \cdot N}{d \cdot C} + K_S \cdot \sqrt{\Delta_{шер}} \cdot \sqrt[4]{\frac{N}{HB \cdot C}} \right] \cdot \frac{\pi \cdot d^2 \cdot l}{2}, \quad (6)$$

where f_m is the molecular component of the coefficient of friction, the value of which for practical calculations can be assumed to be equal to $f_m = 0.12$ [17-19];
K_S – coefficient depending on the type of contact; for unsaturated plastic contact $K_S = 0.21$, for saturated - $K_S = 0.45$ [20].

Calculations show that for the connection of the band with the roll in the formula (6), the second term can be neglected. Then, for both types of contact, the formula for the permissible moment takes the following form

$$[M] = \left(\frac{f_m \cdot N}{d \cdot C} \right) \cdot \frac{\pi \cdot d^2 \cdot l}{2}. \quad (7)$$

The minimum allowable design tension value $N_{min,p}$ is found when the permissible torque value is equated with the corresponding values of the technological resistance moments M, i.e.

$$N_{min.p} \approx \frac{2 \cdot M \cdot C}{f_m \cdot \pi \cdot d \cdot l} + \Delta N, \quad (8)$$

where ΔN is the reduction of tension during press assembly [5];
$\Delta N = 12$ µm at the roughness parameter $Ra = 2.5...1.25$ µm, $\Delta N = 8$ µm at $Ra = 1.25...0.63$ µm, $\Delta N =$ µm at $Ra = 0.63...0.32$ µm.

The maximum design tension $N_{max,\,p}$ is determined from the solution of the problem of calculating thick-walled cylinders [22] and is reduced to testing for strength at dangerous points where the greatest tensile and compressive stresses act (Figure 1):

- on the inner surface of the enclosing part

$$\sigma_1 = \sigma_{t2} = p_{max} \cdot \frac{d_2^2 + d^2}{d_2^2 - d^2}; \quad \sigma_3 = \sigma_r = -p_{max}, \quad (9)$$

- on the inner surface of the covered part

$$\sigma_1 = \sigma_{t2} = p_{\max} \cdot \frac{d_2^2 + d^2}{d_2^2 - d^2}; \quad \sigma_3 = \sigma_r = -p_{\max}. \tag{10}$$

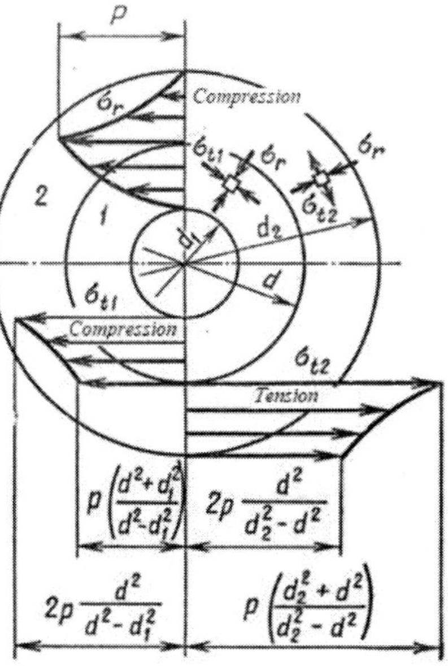

Figure 1. Stresses in parts connected with tension.

According to the theory of the greatest tangential stresses, the strength condition has the form [21]

$$\sigma_{экв} = \sigma_1 - \sigma_3 \le \sigma_т. \tag{11}$$

Given this expression from formulas (9) and (10), we shall obtain

$$p_{\max,2} \le \sigma_{т2} \cdot \frac{d_2^2 - d^2}{2 \cdot d_2^2}; \quad p_{\max,1} \le \sigma_{т1} \cdot \frac{d^2 - d_1^2}{2 \cdot d^2}. \tag{12}$$

Of the two values p_{max} , the lower value limits the amount of allowable pressure determined by the formula (2). The dangerous element, as a rule, is the covered part [21], and therefore from (12) for a solid roll ($d_1 = 0$), we get $p_{max} = \sigma_{т1}/2$. Then the maximum permissible calculated value of the tension $N_{max, p}$ according to formula (2) will be equal to

$$N_{\max.p} = p_{\max} \cdot d \cdot \left(\frac{C_1}{E_1} + \frac{C_2}{E_2} \right) = \frac{d \cdot \sigma_{т1}}{2} \cdot \left(\frac{C_1}{E_1} + \frac{C_2}{E_2} \right). \tag{13}$$

Because building connections smooths roughness, the maximum actual tension N_{Max} take more received by the formula (16), at the height of the roughness of the mating surfaces, i.e.

$$N_{\max} = N_{\max.p} + 1.2(Rz_1 + Rz_2). \tag{14}$$

Usually, the surfaces of the covered parts (rolls) are treated with a roughness of $Rz_1 = 0.4\ldots3.2$ µm, and the covering (holes) - with a roughness of $Rz_2 = 0.8\ldots6.3$ µm. Therefore,

$$N_{\max} = N_{\max.p} + (2 \ldots 10)\ \mu m. \tag{15}$$

According to the calculated values of the tension N_{max} и N_{min}, the corresponding standard fit is selected; most often, H7/p6, H7/r6, H7/s6, H7/t7, H7/z6, H7u7 landings are used.

5. Discussion of the research results

An example of fitting with the tension of a band made of hard alloy VK8 [23] on a working roll made of improved steel 45 [23]with a diameter of d = 100 mm is considered. The band is installed using a press with a H7/p6 fit to transmit a torque of M = 5 kN·m. The outer diameter of the band d_2 = 170 mm, the length of the landing surface l = 145 mm. The landing surface is treated according to the 6th class of cleanliness, for which the complex roughness characteristic of the Δ_r = 0,5. The surface hardness of the roll HB = 2500 MPa. Check whether the specified torque can be transmitted.

For the H7/p6 landing [24], we find the lower and upper limit deviations of the roll and the hole, and then the lower N_{low} = 2 μm and the upper N_{high}= 59 μm limit values of the tightness (Figure 2).

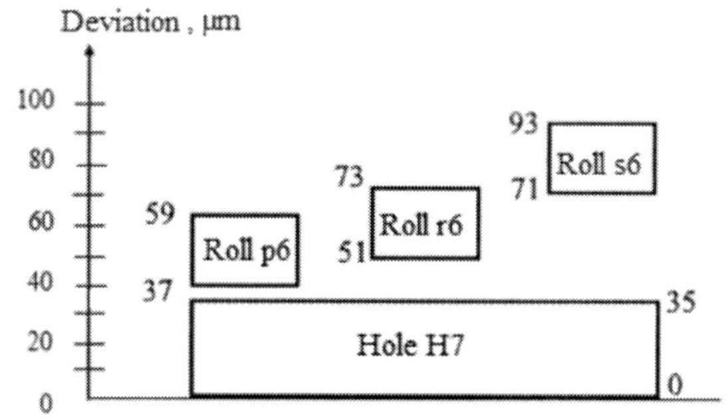

Figure 2 Roll tolerance fields and holes with a diameter of 100 mm for tight fit H7/p6, H7/r6, H7/s6.

Substituting the corresponding values of the diameters and mechanical characteristics of the materials in the formula (3), we obtained the value of the coefficient C = 0.7·10-5 MPa^{-1}.

To determine the type of contact, we use the formula (5) to determine the amount of tension that characterizes the transition condition to plastic contact, taking into account that when pressing C_{as} = 0.5.

Therefore, in connection with the tension, both unsaturated (N_{low} = 2 μm) and saturated (N_{high}= 59 μm) plastic contact can be performed. For the maximum amount of tension, according to the formula (7), we obtain the maximum permissible moment [M]

$$[M]=\left(\frac{f_m\cdot N}{d\cdot C}\right)\cdot\frac{\pi\cdot d^2\cdot l}{2}=\left(\frac{0.12\cdot 59\cdot 10^{-6}}{0.1\cdot 0.7\cdot 10^{-5}}\right)\cdot\frac{\pi\cdot 0.1^2\cdot 0.145}{2}\approx 0.026\ \text{MN}\cdot\text{m}=26\ \text{kN}\cdot\text{m}.$$

For the minimum amount of tension, using the formula (7), we obtain the minimum allowable moment [M]

$$[M]=\left(\frac{f_m\cdot N}{d\cdot C}\right)\cdot\frac{\pi\cdot d^2\cdot l}{2}=\left(\frac{0.12\cdot 2\cdot 10^{-6}}{0.1\cdot 0.7\cdot 10^{-5}}\right)\cdot\frac{\pi\cdot 0.1^2\cdot 0.145}{2}\approx 0{,}001\ \text{MN}\cdot\text{m}=1\ \text{kN}\cdot\text{m}.$$

Since $[M]_{min}$ = 1 кН·м < M= 5 кН·, the minimum tolerance for this fit does not provide the transmission of the specified torque.

During the press assembly, it is necessary to switch to a different type of fit to ensure the minimum allowable tension, which can be estimated using the formula (8)

$$N_{\min.p} \approx \frac{2 \cdot M \cdot C}{f_m \cdot \pi \cdot d \cdot l} + \Delta N = \frac{2 \cdot 5 \cdot 10^{-3} \cdot 0.7 \cdot 10^{-5}}{0.12 \cdot \pi \cdot 0.1 \cdot 0.145} + 12 \cdot 10^{-6} \approx 0.025 \text{ m} = 25\ \mu\text{m}.$$

The value of the tension reduction ΔN = 12 µm is taken for the 6th class of surface treatment purity, i.e. for Ra = 2.5...1.25 µm.

For the H7/r6 landing, we find [8] the lower N_{low} = 51 µm and the upper N_{high} = 73 µm values of the limit deviations (see Figure 2), and we determine the lower N_{low} = 16 µm and the upper N_{high}= 73 µm of the tension values, which is also not enough to transmit the specified torque.

For the H7/s6 landing, we find [24] the lower N_{low} = 71 µm and the upper N_{high} = 93 µm values of the limit deviations (see Figure 2), and we determine the lower N_{low} = 136 µm and the upper N_{high}= 93 µm of the tension values. This landing ensures the transmission of the specified torque.

Tight joints are used in various technological machines [25-29]. The proposed method can be useful in continuing the work [30-36].

6. Conclusions

When considering the wear processes, all the characteristics related to the surface microgeometry are combined by a dimensionless complex roughness parameter, which, along with the roughness class, includes the features of the surface treatment technology. To determine tensed contact pressure, the Lame formula for calculating thick-walled cylinders is used. Formulas are given for determining the maximum permissible torque that does not cause a change in the strength of the joint with tension under conditions of plastic saturated and unsaturated contact. In order to test the strength of the connection elements with tension, the theory of the greatest tangential stresses is applied for two dangerous points located on the inner surfaces of the covered and the covering parts: a formula is obtained for determining the maximum permissible calculated value of tension from the strength condition of the covered part. An example of fitting with the tension of a band made of hard alloy VK8 on a working roll made of improved steel 45 with a diameter of d = 100 mm is considered. The band is installed using a press with a H7/p6 fit to transmit a torque of M = 5 kN·m. Calculations have shown that the minimum tolerance for this landing does not provide the transmission of the specified torque. Therefore, to ensure reliable operation of the connection with tension, it is recommended to use the H7/s6 fit, which provides the transmission of the specified torque without the strength of the connection elements.

Referenses

[1] Chicheneva O N and Zarapin Y L 1995 Strain resistant of metals in electroplastic forming *Steel in Translation* **25 (3)** pp 36-38

[2] Chicheneva O N, Savchenko V S and Zarapin Y L 1995 Mathematical model of the electroplastic rolling process *Steel in Translation* **25 (5)** pp 53-55

[3] Zarapin A Yu, Shur A I and Chichenev, N A, 1999 Improvement of the unit for rolling aluminum strip clad with corrosion-resistant steel *Steel in Translation* **29 (10)** pp 69-71

[4] Durelli A J, Chichenev N A and Clark J A 1972 Developments in the optical spatial filtering of superposed crossed gratings - Spatial-filtering techniques are used to obtain individually, as separate patterns in a simple and precise manner, the whole field of displacement components and of their time and space derivatives *Experimental Mechanics* **12 (11)** pp 496-501 doi: 10.1007/BF02320745

[5] Bardovsky A D, Gerasimova A A and Basyrov I I 2019 Study of oscillating process of harp screens Lecture *Notes in Mechanical Engineering* 9783319956299 pp 133-139 doi: 10.1007/978-3-319-95630-5_14

[6] Zhil'tsov A P, Vishnevskii D A, Kozachishen V A and Bocharov A V 2018 Razrabotka algoritma i komp'yuternoi programmy dlya rascheta nadezhnosti oborudovaniya i proizvodstvennogo riska v metallurgicheskoi otrasli *Chernye metally* **11** pp 27-33

[7] Gorbatyuk S M, Zarapin A Y and Chichenev N A 2018 Retrofit of vibrating screen of Catoca Mining Company (Angola) *Mining Informational and Analytical Bulletin* **1** pp 143-149 doi: 10.25018/0236-1493-2018-1-0-143-149

[8] Albagachiev A Y, Keropyan A M, Gerasimova A A and Kobelev O A 2020 Determination of rational friction temperature in lengthwise rolling *CIS Iron and Steel Review* **19** pp 33-36 doi: 10.17580/cisisr.2020.01.07

[9] Chichenev N A 2018. Reengineering of the Slab-Centering Unit of a Roughing Mill Stand *Metallurgist* **62 (7-8)** pp 701–706

[10] Chichenev N A 2015 Import-replacing re-engineering of the drive of the rollers in the intermediate roller table of a continuous bloom caster *Metallurgist* **58 (9-10)** pp 892–895

[11] Kobelev O A and Tyurin V A 2007 Production of large plates *Steel in Translation* **37 (9)** pp 727-729 doi: 10.3103/S096709120709001X

[12] Keropyan A M, Kuziev D A and Krivenko A E 2020 Process Research of Wheel-Rail Mining Machines Traction *Lecture Notes in Mechanical Engineering* pp 703-709. doi: 10.1007/978-3-030-22063-1_75

[13] Kouziyev D, Krivenko A, Chezganova D and Valeriy B 2019 Sensing of dynamic loads in the open-cast mine combine *E3S Web of Conferences* **105** 03014 doi: 10.1051/e3sconf/201910503014

[14] Jiawen L, Timushev S, Klimenko D and Krivenko A 2019 Modeling pressure pulsation fields in a screw centrifugal pump *Proceedings of the 26th International Congress on Sound and Vibration ICSV*

[15] Naumova M G, Morozova I G and Borisov P V 2020 Study of metal surface with color image obtained with laser marking *Solid State Phenomena* **299** pp 943-948

[16] Kostetskii B I *et al* 1975 *Nadezhnost' i dolgovechnost' mashin* (Kiev: Tekhnika) 405

[17] Kragel'skii I V *et al* 1977 *Osnovy raschetov na trenie i iznos* (Moscow: Mashinostroenie) 526

[18] Ikramov U A and Levitin M A 1984 *Osnovy triboniki*: *Uchebnoe posobie dlya vuzov* (Tashkent: Ukitovchi) 183

[19] Zhirkin Yu V 2002 *Nadezhnost', ekspluatatsiya, tekhnicheskoe obsluzhivanie i remont metallurgicheskikh mashin*: *Uchebnik* (Magnitogorsk: MGTU) 330

[20] Feodos'ev V I 1999 *Soprotivlenie materialov*: *Uchebnik* (Moscow MGTU im. N.E.Baumana) 592

[21] *Detali mashin* 2002 Uchebnik Pod red. O.A. Ryakhovskogo (Moscow: MGTU im. N E Baumana) 544

[22] GOST 3882-74 (ISO 513-75) 2008 *Splavy tverdye spechennye Marki* (Moscow: Standartinform) 10s.

[23] GOST 1050-74 2008 *Stal' uglerodistaya kachestvennaya konstruktsionnaya* (Moscow: Standartinform) 32

[24] GOST 25347-82 2004 *Edinaya sistema dopuskov i posadok. Polya dopuskov i rekomenduemye posadki* (Moscow: IPK Izdatel'stvo standartov) 24

[25] Surianinov M and Shyliaiev O 2018 Calculation of plate-beam systems by method of boundary elements *International Journal of Engineering and Technology (UAE)* **7 (2)** pp 238-241 doi: 10.14419/ijet.v7i2.23.11927

[26] Krutii Y, Suriyaninov M and Vandynskyi V 2017 Exact solution of the differential equation of transverse oscillations of the rod taking into account own weight *MATEC Web of Conferences* **116** 02022 doi: 10.1051/matecconf/201711602022

[27] Kondratenko V E, Devyatiarova V V, Albul S V and Kartyshev D S 2020 Improving methodology for calculating scaffolding formwork of monolithic slabs in building constructions *IOP Conference Series: Materials Science and Engineering* **971 (5)** 052037 doi: 10.1088/1757-899X/971/5/052037

[28] Kondratenko V E, Devyatiarova V V, Albul S V and Valeeva L M 2020 Method of calculating volumetric scaffold of monolithic slab formwork *IOP Conference Series: Materials Science and Engineering* **971 (5)** 052036 doi: 10.1088/1757-899X/971/5/052036

[29] Kondratenko V E, Sedykh L V and Surkova R Y 2020 Effective design features of rotor shafts *IOP Conference Series: Materials Science and Engineering* **971 (4)** 042010 doi: 10.1088/1757-899X/971/4/042010

[30] Kobelev O A, Zinov'ev A V and Tsepin M A 2009 Effective production of large pipe blanks *Steel in Translation* **39 (6)** pp 501-505 doi: 10.3103/S0967091209060163

[31] Nazaratin V V, Kobelev O A, Efimov M V, Selyutin A A and Yavtushenko P M 2013 Analysis of technologies used to make hollow ingots and prospects for their improvement *Metallurgist* **56 (9-10)** pp 672-678 doi: 10.1007/s11015-013-9634-z

[32] Shatalov R L and Kulikov M A 2020 Influence of Outer Parts of a Strip on the Deformation and Force Parameters of Thin-Sheet Rolling *Metallurgist* **64 (7-8)** pp 687-698 doi: 10.1007/s11015-020-01045-1

[33] Titov V N, Saifullaev S D, Skripalenko M M, Ternovykh A and I Sidorov A A 2020 Using Deform-2d Software to Study Heat-Insulation Materials as Protection of Air Tuyeres Against Burnout *Metallurgist* **64 (5-6)** pp 388-395 doi: 10.1007/s11015-020-01007-7

[34] Korchunov A G, Polyakova M A, Konstantinov D V and Dabalá M 2019 Mechanical properties of prestressing strands and how they tend to change under thermo-mechanical treatment *CIS Iron and Steel Review* **18** pp 14-19 doi: 10.17580/cisisr.2019.02.03

[35] Keropyan A M, Kuziev D A and Krivenko A E 2020 Process Research of Wheel-Rail Mining Machines Traction *Lecture Notes in Mechanical Engineering* pp 703-709 doi: 10.1007/978-3-030-22063-1_75

[36] Kouziyev D, Krivenko A, Chezganova D and Valeriy B 2019 Sensing of dynamic loads in the open-cast mine combine *E3S Web of Conferences* **105** 03014 doi: 10.1051/e3sconf/201910503014

Study of metal changes in form during forging of on-off valves

V Chukhlib[1,*], O Duvanskii[1], N Biba[2], A Ashkelianets[1], S Gubskii[1] and A Okun[1]

[1]Department of Computer Modeling and Integrated Forming Technologies, National Technical University "Kharkiv Polytechnic Institute", Kyrpycheva str., 2, Kharkiv, Ukraine, 61002
[2] MICAS Simulations Ltd., 107 Oxford Road, Oxford, OX4ER, UK

[*]E-mail: profdnepro@gmail.com

Abstract. An overview of the existing technological scheme for the production of on-off valves is given. The analysis of the existing shape of the on-off valves was carried out with its classification by the output hole. Depending on the shape of the on-off valve output hole, a new technological process is proposed for obtaining on-off valve with a tapered part of the output hole using free forging. To develop the main forging technological scheme, an analysis of the part of the workpiece that is directly exposed to the technological tool was performed by determining the displaced bulks. For this, the entire bulk of the metal under the influence of the deforming tool was divided into characteristic zones, both before deformation and after its implementation. The analysis of the shape change showed the dependence of the metal flow parameters relates to the main geometric and technological parameters, and for this, the main graphic changes were built. On the basis of these studies, a new technological forging process of on-off valve forged pieces with the maximum approximate shape to the finished product can be developed.

1. Introduction

On-off valves are a type of pipeline valves designed to open, close or change the flow direction of working fluid or gas depending on the requirements of the technological process [1].

There are several technologies for the manufacture of on-off valves. Welding, when a sheet of metal is bent and a longitudinal weld is made. Sheet forming, when the workpiece is placed in the finished form of the die and a forged piece is made using the press or power hammer. Casting, when the hot metal is poured into a mold close to the shape of the finished part. And forging, when the part is covered by surplus in the simplest form of a cylinder or a parallelepiped. We will consider the technology of forging because the parts are critical. High requirements for the physical and mechanical metal properties are imposed for these parts. The parts will be most appropriate to obtain by the technology of forging [2].

This type of parts is made of corrosion-proof steels, because they often work in aggressive environment. Corrosion-proof include steels and alloys containing more than 12 % of Cr, as well as additionally alloyed with Ni, Mo, Cu, Si, Ti, Nb, N and some other elements. Their content depends on the aggressiveness of the corrosive environment and the requirements for the physical and mechanical properties of steels and alloys.

Corrosion-proof steels and alloys are used for the manufacture of technological equipment operating under various, usually highly aggressive corrosive conditions (inorganic and organic acids, their mixtures, solutions of alkalis and salts, sea and mineralized formation water, humid atmosphere, etc.) and mechanical loading (static, dynamic, cyclic or combined) [1-7].

Corrosion-proof steels do not differ much from regular carbon steels in terms of forging technology. The main difference is in the various temperature modes of heating and cooling. It is recommended to gradually increase the temperature so that the workpiece can heat up evenly and completely. This will avoid defects in the structure of the metal during forging and further cooling. In addition, corrosion-proof steel has low thermal conductivity and requires more air hardening time, in contrast to regular steel grades [4-9].

Appropriate materials and design solutions are used to ensure the durability of valves using in

Content from this work may be used under the terms of the Creative Commons Attribution 3.0 licence. Any further distribution of this work must maintain attribution to the author(s) and the title of the work, journal citation and DOI.
Published under licence by IOP Publishing Ltd

special operating conditions. Body parts of valves for corrosive and aggressive environment are usually made of AISI 321 and CF-16F or S31635, S32100 steels. At an operating pressure of up to 0.6 MPa and an operating temperature of up to 120 °C membranous (diaphragm) valves with a body covered with a protective layer of rubber, polyethylene, fluoroplastic or enamel are widely used for corrosive environment.

Both manual and mechanized drives (electric, pneumatic or electromagnetic) are used to control the valves [2].

Different types of on-off valves (taps, valves and latches) are used for corrosive and aggressive environment. The required type of valves is selected based on a comparison of technical and economic indicators of competing options for possible solutions. In addition to the general parameters (such as construction length and height, weight, cost, etc.) for corrosive and aggressive environment such parameters as air tightness, reliability, durability due to corrosion proof and others are important. Recently new designs (membranous (diaphragm) valves, ball valves, hose valves) which differ from traditional ones are widely used.

2. Technology of on-off valves production

The main problem of the existing on-off valves forging technology is the high coefficient of metal consumption, which exceeds the weight of the finished part, sometimes even several times. Because of this, most of the metal used to make forged pieces goes to waste. In addition, the removing processes of all allowances and surpluses are labor-intensive, which leads to another problem of the approximation of the forged piece shape to the workpiece shape. This will reduce the labor-intensive of the part manufacture, and most importantly, improve the quality of the finished part, as the need to remove the surpluses will be less and the metal fibers will not be destroyed during cutting. Therefore, the study of this problem can be considered essential.

The main issue of the study is to investigate the metal flow and the bulk redistribution during forging of on-off valves. First of all, we perform the operation of piercing in a cast workpiece. Piercing is a forging operation to obtain a cavity in the workpiece due to the free displacement of the metal. The operation is usually used to obtain blind and through cavities in the manufacture of forged pieces with an axial hole as shown on Figure 1 [1-11].

On-off valves are manufactured both in Ukraine and abroad by such companies as Zetkama (Poland), Ayvaz (Turkey), ODE (Italy), Genebre (Spain), Breeze (Ukraine), Interval (Belarus), MSO (Czech Republic), Dnipropress Stal (Ukraine) and others.

Analyzing the range of these bodies, it was decided to divide them into 3 types: bodies with an acute angle of taper of the hole (Figure 1, *a*), bodies with a right angle of taper of the hole (Figure 1, *b*) and bodies with an obtuse angle of taper of the hole (Figure 1, *c*).

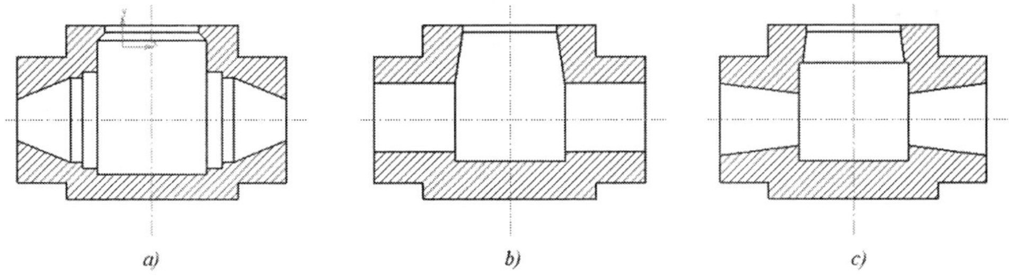

Figure 1. Types of on-off valve bodies.

In this article we will consider the option of making a first type part (Figure 1, *a*). As can be seen on Figure 1, the part has a ledge on the sides, which can be got by the operation of broaching. Also in this case, the taper of the inner hole with an acute angle can be got.

Forged pieces of these parts are made by one similar technology, when the shape of the part is forging with allowances for overall dimensions in the form of the simplest Figure (a cube, parallelepiped, etc.) as shown on Figure 2, *a*. We offer another technology that will move closer to the shape of the forged piece to the shape of the finished part, as can be seen on Figure 2, *b*. In contrast to standard technology, in which there are two technological operations of forging (upsetting and broaching), our technology will have more forging transitions (upsetting, piercing, broaching on the mandrel and broaching the ends of the workpiece without the mandrel). This technology is more complex and labor-intensive, but it can reduce the metal consumption and increase the physical and mechanical properties of the metal part.

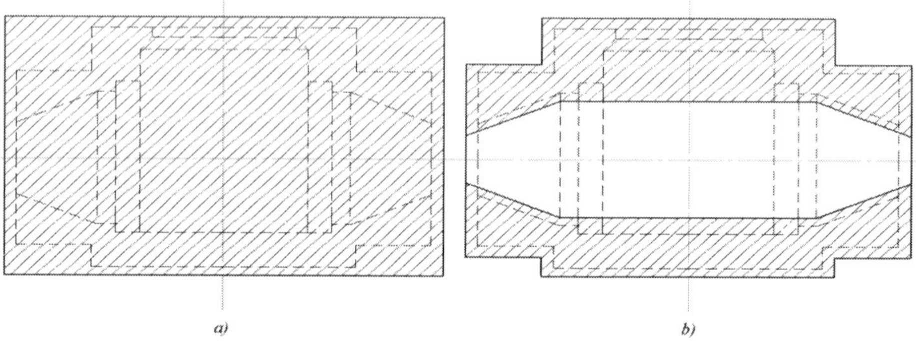

Figure 2. Example of closing the forged piece by surplus according to the existing technology (*a*) and the proposed technology (*b*).

3. Computer modelling of on-off valves forging

To provide the study, the simulation of the broaching process with a hole in the QForm software was performed [12]. QForm is professional engineering software for modelling and optimization of metal forming processes. The program is based on the finite element method and is developed by Micas Simulations.

The initial workpiece is a cylinder with a hole, as shown on Figure 3. 10 experiments were performed with different input data (the deformation degree ε, the canting angle α and the diameter of the inner hole $d_{general}$). The main dimensions are length l (1000 mm) and outer diameter $D_{general}$ (1000 mm) remained unchanged. The constant value was also the value of the hammer die moving f (100 mm). The input data of the experiments are shown in Table 1.

Table 1. Input data for forging of on-off valve bodies.

α, °	ε, %	$d_{general}$, mm	f, mm	l, mm
30	10	300	100	1000
		400		
		500		
		600		
		700		
	20	300		
		400		
		500		
		600		
		700		

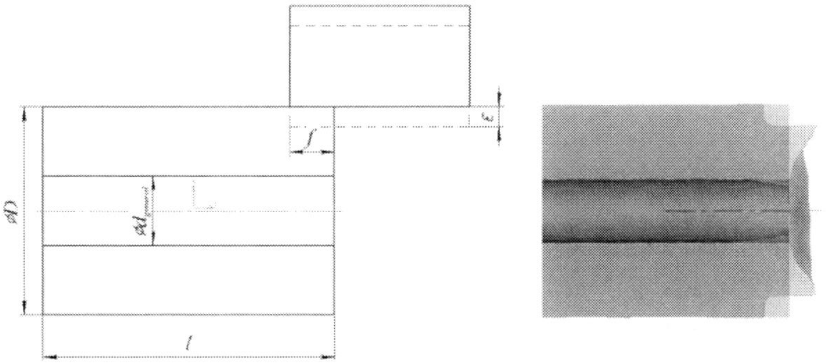

Figure 3. Forging transition of pre-pierced workpiece and workpiece after deformation.

The initial workpiece and the forging scheme of a cylindrical workpiece with a hole are shown on Figure 4.

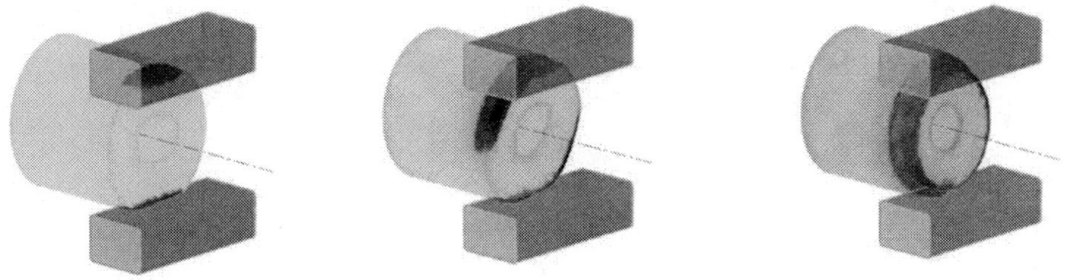

Figure 4. Some forging transitions of forged parts.

The bulk redistribution after the broaching operation of a cylindrical workpiece is shown on Figure 5. For a more detailed analysis, the forged part was divided into some amount of conditional bulks. It can be seen how the bulk of metal displaced by the impact was redistributed in 2 areas of forged piece.

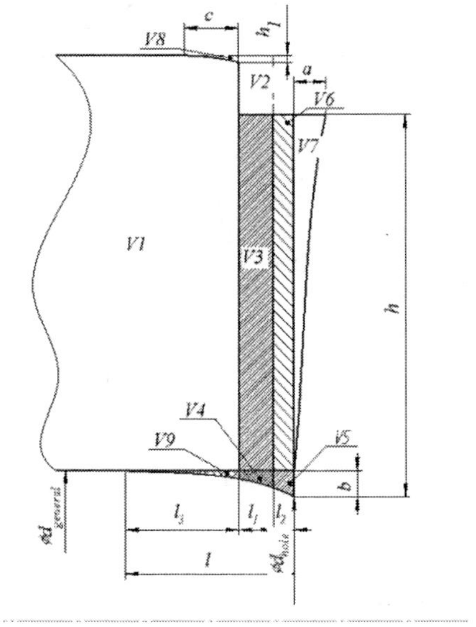

Figure 5. Distribution into displaced forging bulks.

Various values of parameters were obtained during the modelling, which show the nature of the bulk displacement as a result of broaching the ends, which at this stage are quite difficult to predict as shown in Table 2.

Table 2. The value of the workpiece dimensions after forging.

ε, %	$d_{general}$, mm	d_{hole}, mm	a, mm	b, mm	c, mm	l, mm	l_3, mm	l_1, mm	h_1, mm
	300	294	49	306	0	167	67	100	0
	400	382	49	258	0	210	110	100	0
10	500	460	42	220	0	329	229	100	0
	600	535	39	184	140	320	220	100	2
	700	606	28	147	341	509	400	100	15
	300	232	111	284	0	294	194	100	0
	400	295	103	252	0	246	146	100	0
20	500	354	91	223	0	322	222	100	0
	600	422	79	192	124	358	258	100	7
	700	490	53	157	258	382	282	100	22

During modelling of forging, it was found that the redistribution of metal bulk was following from zone *V*2 flows to zones *V*7, *V*5, *V*4 and *V*9. In the zone *V*7 from the top to the inner diameter, a shrinkage cavity is formed.

To analyze the metal flow, graphs of the dependences of some workpiece parameters from the hole diameter and the deformation degree were built. Several dependencies were derived to create graphs.

$$\frac{S}{\frac{D}{2}} = \frac{\frac{D-d_{general}}{2}}{\frac{D}{2}} = \frac{D-d_{general}}{D} = 1 - \frac{d_{general}}{D}, \tag{1}$$

$$\frac{a+l_2+l_1}{l_1} = 1 + \frac{a+l_2}{l_1}. \tag{2}$$

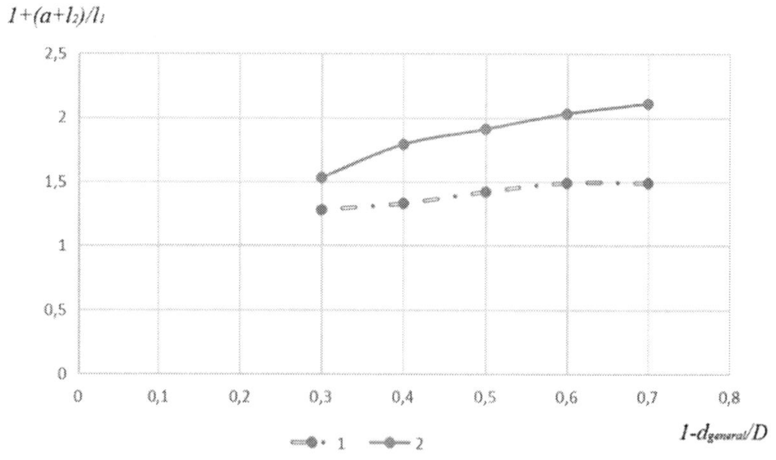

Figure 6. Graph of under die zone elongation depending on the forged piece diameter and the deformation degree: Line 1 – 10%, Line 2 – 20%.

The nature of the change in the length of the forged piece in the under die zone can be seen on

Figure 6. It should be noted that at 20 % elongation is greater than at 10 %. Also, comparing the lines on the graph, we see that regardless of the deformation degree, they have approximately the same nature of motion as the hole diameter decreases, the relative value of the elongation of the workpiece increases. However, in the value of the relative diameter on Line 1 (Figure 6) can be seen a slight change in the nature of the movement of this line, which may be a value in the range of error, and may indicate changes that can be traced in future studies.

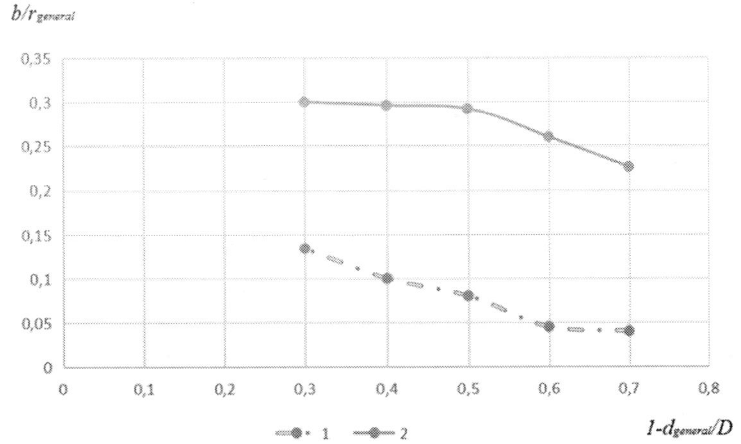

Figure 7. Relative graph of the thickening of the inner diameter (*b*) depending on the diameter of the forged piece and the deformation degree: Line 1 – 10 %, Line 2 – 20 %.

The graph, as shown on Figure 7, characterizes the change in the size of the inner diameter, namely the displacement of the bulk from the under die zone to the zone of narrowing and shrinkage cavity formation. By changing the parameters, we can note the following: with a greater deformation degree we obtain greater values of the thickness of the inner diameter of the forged piece. It can also be observed that with a larger diameter we obtain larger values of redistribution and metal flow. So with a smaller diameter we obtain smaller values of redistribution and metal flow, which build up into a certain pattern with these parameters of forging.

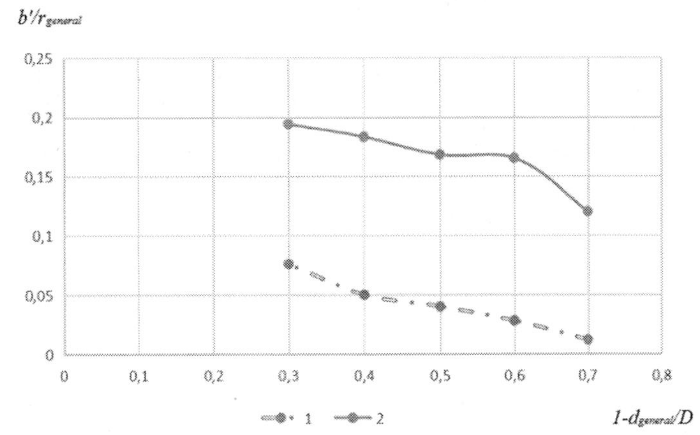

Figure 8. Relative graph of the thickening of the inner diameter (*b'*) depending on the diameter of the forged piece and the deformation degree: Line 1 – 10 %, Line 2 – 20 %.

In general, the graph, as shown on Figure 8, can be said that it is similar to the previous one, except that in this case we take not the minimum diameter, but the diameter that corresponds to the extreme point of contact of the hammer die and forged piece. The character is similar, at 20 % of deformation we receive greater values of thickening of diameter against 10 %. And also at increase in diameter of a forged piece hole the relative thickening also increases.

4. Conclusions

- An analytical review of scientific sources on the manufacture of on-off valves and their varieties was performed. A comparison of the existing technology and the proposed one was carried out, and the proposed technology was used to model the process of forging these parts.
- The analysis of the results showed that, in general, positive results were obtained, and the developed technology can save metal by approximating the forged piece shape to the shape of the finished part. However, it is not entirely clear in which ranges it will be operating, and in which not and at what parameters of forging, which means that this research needs further study.

References

[1] Seynov S 2002 *Truboprovodnaya armatura. Issledovaniye. Proizvodstvo. Remont* (Moskva: Mashinostroyeniye) 391

[2] Mustafin F et al. 2007 *Truboprovodnaya armatura* (Ufa: GUL RB UPK) 326

[3] Kamenskikh A 2004 Analiz neispravnostey truboprovodnoy armatury voznikayushchikh v protsesse ekspluatatsii *Truboprovodnaya armatura i oborudovaniye* **6 (15)** p 74

[4] Stasovs'kiy Y, Chukhlib V and Boyarkin V 2013 *Resursozberezhennya ta yekologiya v protsesakh obrobki metaliv tiskom* (Dnipropetrovs'k: Porogi) 353

[5] Chukhlib V, Klemeshov E, Gubskyi S, Okun A and Biba N 2020 Theoretical and Experimental Studies of Changes in the Workpiece Shape During Narrow Die Indentation *Advances in Design, Simulation and Manufacturing III. DSMIE 2020. Lecture Notes in Mechanical Engineering*. doi.org/10.1007/978-3-030-50794-7_35

[6] Puzyr R, Kukhar V, Maslov A and Shchipkovskyi Y 2018 The development of the method for the calculation of the shaping force in the production of vehicle wheel rims *International Journal of Engineering and Technology* **7(4.3 Special Issue 3)** pp 30-34 doi: 10.14419/ijet.v7i4.3.20128

[7] Kukhar V, Kurpe O, Klimov E, Balalayeva E and Dragobetskii V 2018 Improvement of the method for calculating the metal temperature loss on a coilbox unit at the rolling on hot strip mills *International Journal of Engineering and Technology* **7(4)** pp 35-39 doi:10.14419/ijet.v7i4.3.19548

[8] Artiukh V, Kukhar V and Balalayeva E 2018 Refinement issue of displaced volume at upsetting of cylindrical workpiece by radial dies *MATEC Web of Conferences* **224** 01036 doi:10.1051/matecconf/201822401036

[9] Chukhleb V, Klemeshov Y, Grinkevich V, Yaroshenko O, Khalezova T, Frolov A and Dyya K 2017 Promyshlennoye vnedreniye resursosberegayushchego protsesa kovki s ratsional'nym rezhimom formoizmeneniya umen'shennoy pribyl'noy chasti slitkov iz koroziionnostoykikh staley *Obrabotka materialov davleniyem* **1 (44)** pp 123-130

[10] Syzrantsev V, Syzrantseva K and Beloborodov A 2004 Sovremennyye metody rascheta i diagnostikiustalosti truboprovodnoy armatury *Armaturostroyeniye* **6 (32)** pp 62-65

[11] Milenin A, Walczyk W and Pietrzyk M 2012 Numerical Modeling of Microstructure Evolution During Forging of Crank Shafts *Steel research international* **8** pp 808-816

[12] Biba N, Maximov A, Stebunov S and Vlasov A 2012 The model for simulation of thermally, mechanically and physically coupled problems of metal forming *Proceedings of the 14th international conference on metal forming* **8** pp 1363-1366

Investigation of the surface properties of quartz-based dispersed materials

Yu Danchenko[1,*], V Andronov[2], S Vazhynskyi[2], I Khmyrov[2] and A Khmyrova[2]

[1]Kharkiv National University of Civil Engineering and Architecture, 40, Sumska str., Kharkiv 61002, Ukraine
[2]National University of Civil Defense of Ukraine, 94, Chernyshevska str., Kharkiv 61023, Ukraine
[*] E-mail: u_danchenko@ukr.net

Abstract. Researched quartz disperses materials (Ukraine): mineral - quartzite of the Ovruch deposit and quartz sand of the Novovodolazsky deposit and products of processing of quartz minerals - dinas and marshalite. The mineral composition, surface morphology, dispersion and specific surface area were experimentally studied. The structure of the hydroxyl-hydrate surface layer was studied by thermogravimetric (TG) analysis and differential thermal analysis (DTA). The number and acid-base characteristics of surfaces active centers were investigated by the methods of pK-metric and pH-metric. It is established that the mechanisms of physical and chemical adsorption of water molecules on the surface of all quartz dispersed materials are identical and do not depend on the mineral nature of impurities. It is shown that the thickness of the surface hydroxyl-hydrate layer of quartz dispersed materials is determined by the amount of chemically adsorbed water, i.e. the thickness of the hydroxyl layer and depends on the nature of the associated minerals. It is established that the acid-base surface characteristic of quartz materials significantly depends on the chemical nature of the accompanying minerals and does not depend on the thickness of the surface hydroxyl-hydrate layer.

1. Introduction and literature review

Quartz dispersed materials are widely used in various fields: to create composite materials for various purposes, as adsorbents for purification of water and other liquids, as abrasives, etc. [1-4]. Due to many properties, composite materials based on polymers and quartz fillers are effectively used in the construction and restoration of roads, bridges, hydraulic structures, residential buildings [5-9]. In some cases, the operational and physical and mechanical properties of such materials are not inferior to known materials - concrete, stone, metals, wood. The presence of quartz disperse fillers provides strength to mechanical loads [4, 7], resistance to water and other aggressive environments [8, 9], resistance to high temperatures [4, 9-11], thermal insulation, sound and vibration protection properties [5, 6, 8, 12]. The use of inorganic materials, including quartz, is due to the chemical nature and acid-base properties of the surface of the dispersed particles. This determines the efficiency and mechanisms of interaction of the surface of dispersed materials with the polymer.

2. Literature review

The high-energy surface of dispersed inorganic materials is characterized by the presence of a layer of surfaces active centers (SAC) [13-18], which, in terms of acid-base theory, can be divided into Lewis centers and Bransted centers. It is established that the surfaces of solid oxides are characterized by a wide range of acid-base SACs of alkaline (MeO^-, O^{2-}, $OH^{\delta-}$), acid (Me^{n+}, $OH^{\delta+}$), as well as neutral [15, 16]. The acidity function of SAC can vary widely. For example, the authors of the study [17] found that on the surface of quartz SiO_2 there are acid ($pK_a = 2.5$), neutral ($pK_a = 6.4$) and basic ($pK_a = 8.8$) silanol groups ≡SiOH, negatively charged oxygen atoms $=O^{2-}$ ($pK_a = -0.2$) and positively charged silicon atoms Si^{n+} ($pK_a = 14.2$), as well as siloxane groups Si—O—Si. In [19], the acid-base equilibrium on the SiO2 surface was investigated using quantum chemical modeling of PAC. Quantum chemical models of anhydrous and hydrated SACs are proposed, the structural formulas of which include the

Content from this work may be used under the terms of the Creative Commons Attribution 3.0 licence. Any further distribution of this work must maintain attribution to the author(s) and the title of the work, journal citation and DOI.
Published under licence by IOP Publishing Ltd

central element of the Si4 + crystal lattice with coordination numbers 4 and 6, oxygen atoms, OH-groups and water molecules. Anhydrous isolated centers contain one, two (geminal center) or three OH-groups and are connected to the central element. The vicinal centers contain two central elements and two OH-groups connected by a hydrogen bond. In addition to the three types of anhydrous centers, the subject of the study were hydrated centers that contain from one to five water molecules and are attached to the anhydrous center in accordance with the known mechanism of hydration of surface OH-groups [20-23]. It is shown [19] that on the surface of SiO_2, regardless of the coordination number of Si^{4+}, the smallest parameters of pK_a and, accordingly, the largest acidic properties are characterized by vicinal anhydrous centers $Si_2O_5(OH)_2$ (pK_a = 9.2) and $Si_2O_9(OH)_2$ (pK_a = 11.9). Centers with the coordination number of the Central element 6 show less acidity than centers with the coordination number 4. The most basic are characterized by isolated active centers with one OH-group SiO_3OH (pK_a = 11.9), SiO_5OH (pK_a = 13.1). As a result of calculations, it was found that the acidity of the isolated centers increases in direct proportion to the increase in the number of OH-groups. It is established that with the increase in the number of adsorbed water molecules, the acidity of all types of SAC increases. The identical character of growth of acidity of SAC at increase in thickness of both hydroxyl, and hydration layers of a surface is revealed. The calculated values of the free Gibbs energy of the processes of deprotonation of the centers are positive and vary from 52.6 to 74.7 kJ / mol.

Determining the acid-base characteristics of the active centers of the surface layer of dispersed materials is an important scientific task. For this purpose, methods are used that allow to determine the number and acid-base parameters of SAC experimentally. For example, methods of IR spectroscopy [24], gas adsorption [20, 26] or color indicators [25, 27-29]. In [15, 16, 23, 30-33] the method of potentiometric titration of suspensions is used. Experimental methods are informative, but time-consuming and do not allow to determine the chemical nature of active centers. To elucidate the chemical nature of SAC and study the protolytic equilibrium on the surface of SiO_2 [19, 21] and in other mixed and pure oxides [22], the authors propose a theoretical method of quantum chemical modeling.

The spectrum of SAC of different chemical structure forms the surface hydroxyl-hydrate layer (SHHL) of dispersed inorganic material, which is characterized by a certain thickness and acid-base properties [13-16]. SHHL of inorganic materials consists of two main layers. The lower hydroxyl dense layer is formed from water molecules chemically bound to the surface. The upper is a less dense hydrate layer, which consists of adsorbed water molecules on the surface of the lower hydroxyl layer due to hydrogen bonds. The presence of SHHL provides physicochemical interactions that occur on the surface of dispersed materials. In [15, 16] the structure and acid-base properties of SHHL of clay and oxide inorganic materials were studied. It was found that the acid-base surface properties of oxide materials based on TiO_2, Al_2O_3, CaO, Fe_2O_3 correlate with the acid-base properties and nature of oxides and do not depend on the thickness of SHHL. It has been shown that the number of chemically and physically bound water molecules does not affect the acid-base properties of the surface of oxide materials [16]. It was also found that the determining factor in the acid-base surface properties of clay fillers is the mineral composition. The thickness of the SHHL affects the acidity of the surface to a small extent. It was found that clay fillers, which include montmorillonite as the dominant mineral, are characterized by higher acidity than fillers consisting of kaolinite. In mixed clay fillers, the acid-base properties of the surface depend on the chemical nature of impurities and associated minerals [15]. Data on the surface of quartz dispersed materials in modern studies are rare, they are unsystematized and incomplete.

Thus, the purpose of this scientific work is to study the properties of SAC and SHHL of the most popular quartz dispersed materials. This will create opportunities to predict the nature and mechanism of interaction of the surface of quartz dispersed materials on the phase separation surface and will create the preconditions for their effective use in practice.

3. Materials of research

As materials for research the dispersed materials on the basis of quartz (Ukraine) are chosen: mineral – quartzite of the Ovruch deposit (QO) and quartz sand of the Novovodolazsky deposit (QN) and products of processing of quartz minerals – dinas (DS) and marshalite (MT). Quartzite of the Ovruch deposit is a gray-yellow mineral powder, the main raw material for making dinas and ferrosilicon in Ukraine and Eastern Europe. Dense, hard, fine-grained rocks, which consist mainly of β-quartz grains. Quartz sand of the Novovodolazhsky field of the PK-015-3 brand is a powdery white material. Dinas is a light gray powdery material, a product of quartzite firing with the addition of limestone and iron scale to increase the content of CaO and Fe_2O_3 in the resulting product. The firing process is carried out at temperatures of 1400–1750 °C. Marshalite - produced brands "A" and "B", a natural dusty quartz material of gray-white color, the main mineral of which is β-quartz SiO_2 with small admixtures of chalcedony, opal, carbonates and clay minerals.

4. Methods of research

The mineral composition of quartz dispersed materials was studied on an X-ray diffractometer DRON-2 (Russia). Particle size and surface morphology were determined using a scanning electron microscope brand JSM-6390LV (Japan). The specific surface was determined by the BET method (Brunauer, Emmett, Teller). Thermogravimetric analysis and differential thermal analysis using SDT Q600 and Instruments (USA) thermal analyzer were used to study SHHL.

Physic-chemical experimental methods – pK-metric and pH-metric – were used to study the acid-base properties of SAC and SHHL of quartz dispersed materials. Both methods are described in detail in studies [15, 16].

5. Study of mineral composition, size particles and surface area of quartz dispersed materials

The results of the study of mineral composition, size particles and surface area of quartz dispersed materials are presented in Table 1 and on Figures 1-4.

Table 1. Properties of quartz materials.

Quartz dispersed material	Mineral composition	Surface area (m^2/g)	Preferred particle size (μm)
QO	β-quartz SiO_2• Montmorillonite $(Al,Mg)_2(OH)_2[Si_4O_{10}]{\cdot}nH_2O^*$ Kaolinite $Al_2Si_2O_5(OH)_4^*$ Albite $Na_2O{\cdot}Al_2O_3{\cdot}6SiO_2^+$ Orthoclase $K_2O{\cdot}Al_2O_3{\cdot}6SiO_2^+$	3.03	10–20
QN	β-quartz SiO_2• Montmorillonite $(Al,Mg)_2(OH)_2[Si_4O_{10}]{\cdot}nH_2O^*$	2.91	10–15
DS	β-quartz SiO_2• Cristobalite α-SiO_2• Tridymite SiO_2•	2.16	20–30
MT	β-quartz SiO_2• Microcline $K_2O{\cdot}Al_2O_3{\cdot}6SiO_2^*$	2.75	5–10

*Remarks: • – basic mineral phase; + - side mineral phase; * - traces.*

X-ray phase analysis of QO shows the presence in addition to the main phase of β-quartz impurities of accessory minerals: potassium and sodium feldspar (orthoclase and albite). Kaolinite and montmorillonite are identified in trace amounts. Isometric particles of irregularly shaped QO have a cancerous fracture, the surface without porous. Dispersion exceeding 10–20 μm, large fraction (40–

50 μm) is present in the sample up to 10%. The fine fraction is present in small quantities, which mostly consists of particles smaller than 5 μm (Figure 1).

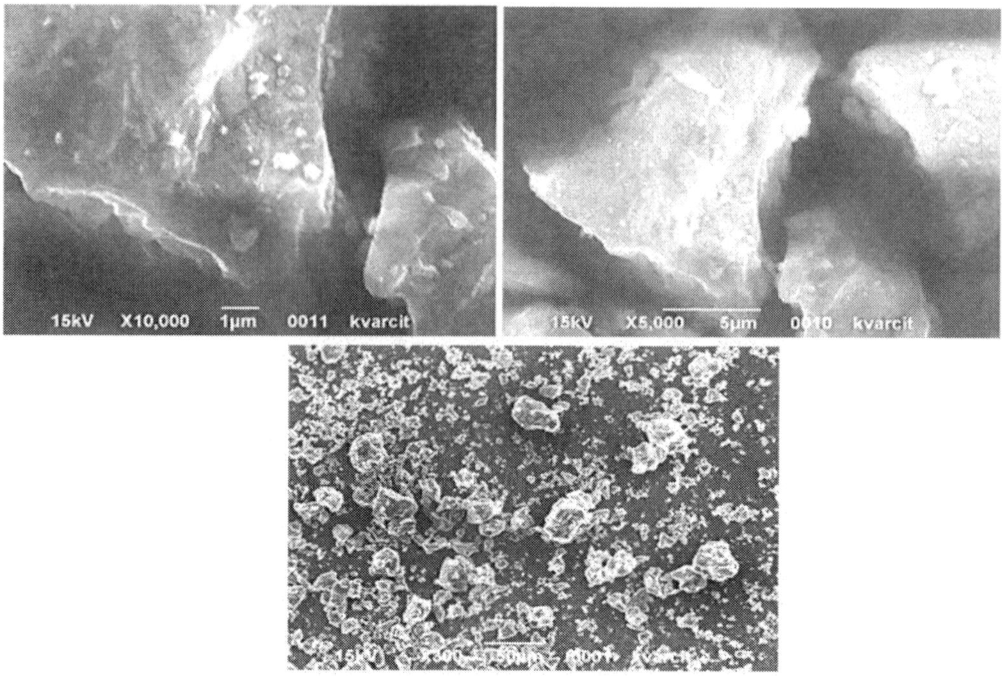

Figure 1. Microphotographs for quartzite of the Ovruch deposit (QO).

The QN diffraction pattern is represented by a set of pronounced diffraction maxima corresponding to β-quartz SiO_2 of high crystallinity. Weak peaks of montmorillonite indicate a low content of clay component in the sand sample, which is confirmed by scanning microscopy: microphotographs clearly visible fragments of quartz grains with a smooth without porous surface. Clay minerals are not observed, respectively, we can talk about sufficient purity of sand, consisting of SiO_2 crystallized in the tetragonal system (β-quartz). The preferred dispersion of the powder is 10-15 μm. A sufficient fraction (2–4 μm) is present in sufficient quantities, and large particles with a size of 25–30 μm are found (Figure 2).

According to the results of X-ray phase analysis, in addition to low-temperature tridymite, unregenerated quartz and weakly structured cristobalite are present in the DS sample. The photomicrographs show thin needle-like crystals of tridymite grown on isometric quartz grains and bound to a weakly structured cristobalite. The preferred dispersion of DS is 20–30 μm. The size of the crystals of the tridymite phase varies in the range of 0.15–0.3 μm (Figure 3).

Highly dispersed MT powder with a predominant dispersion of 5–10 μm was used for the study. The size of large particles is 50-60 microns. The thin fraction of 0.2–2 μm is in small quantities (Figure 4). According to the results of X-ray phase analysis, the main crystalline phase of the powder is β-quartz with a high degree of crystallinity. A microcline is present in the sample as trace impurities.

Figure 2. Microphotographs for quartz sand of the Novovodolazsky deposit (QN).

Figure 3. Microphotographs for dinas (DS).

Figure 4. Microphotographs for marshalite (MT).

As can be seen from the results presented in Table 1, the specific surface area and particle size of mineral quartz dispersed materials QO (3.03 m^2/g and 10–20 μm) and QN (2.91 m^2/g and 10–15 μm) are identical. Heat treatment of quartzite (firing) produces a DS material with a smaller specific surface area (2.16 m^2/g) and, due to sintering, with a larger particle size (20–30 μm). Machining of quartzite (grinding) produces an MT material with a smaller particle size (5–10 μm) and a slightly smaller specific surface area (2.75 m^2/g).

6. Study of the dynamics of mass loss of quartz dispersed materials

In Table 2 shows the dynamics of mass loss of quartz materials in the process of increasing temperature.

Table 2. Loss of mass of quartz materials in the process of temperature increase.

Quartz dispersed material	Weight loss of the materials in the temperature range (%)			
	0–150°C	150–300°C	300–600°C	600–900°C
QO	0.3	0.1	0.9	0.5
QN	0.3	0.2	0.5	0.2
DS	1.0	0.6	1.6	0.2
MT	0.3	0.2	0.6	0.3

From the results presented in table 2 it follows that the processes of weight loss when heated of all quartz materials occur identically. Thus, the most intense weight loss in all cases is observed in the temperature range of 300–600 °C. In the temperature range up to 150 °C in materials QO, QN and MT weight loss has the same value (0.3%). This indicates that the surface of these materials contains approximately the same amount of physically adsorbed water. The DS surface is characterized by three times more physically adsorbed water (weight loss 1%). Obviously, this is due to the fact that DS is a product of quartzite firing. Thus, it can be argued that the mechanisms of physical and chemical adsorption of water molecules on the surface of all quartz dispersed materials are identical and do not depend on the mineral nature of impurities. Obviously, the adsorption mechanism is affected only by the method (thermal or mechanical) processing of the material.

It should be noted that at temperatures of 0–300 °C there are processes of dehydration of the surface of the dispersed material, i.e. the loss of physically adsorbed water molecules (hydrate layer). At temperatures of 300–900 °C, the processes of surface dehydroxylation occur, i.e. the loss of chemically bound water molecules (hydroxyl layer). The total weight loss in the entire temperature range, i.e. up to 900 °C, determines the thickness of SHHL. Thus, the thickness of the hydrated surface layer of quartz dispersed materials increases in a row:

$$QO(0.4\%) < QN(0.5\%) < MT(0.5\%) < DS(1.6\%) . \quad (1)$$

The thickness of the hydroxyl layer of quartz dispersed materials increases in a row:

$$QN(0.7\%) < MT(0.9\%) < QO(1.4\%) < DS(1.8\%) . \quad (2)$$

The thickness of SHHL of quartz dispersed materials increases in a row:

$$QN\ (1.2\%) < MT(1.4\%) < QO(1.8\%) < DS(3.4\%) . \quad (3)$$

As can be seen from the obtained series (1), the thickness of the surface hydrate layer depends only on the method of processing quartz dispersed materials and, obviously, does not depend on the mineral composition. Series (2) indicates that the thickness of the surface hydroxyl layer is 2–3 times greater than the thickness of the hydrate layer for all quartz materials. The amount of chemically adsorbed water depends on the mineral nature of the impurities. The presence of clay minerals montmorillonite and kaolinite determines the greatest thickness of the hydroxyl layer QO. As can be seen from row (3), the thickness of SHHL quartz dispersed materials is determined by the amount of chemically adsorbed water, i.e. the thickness of the hydroxyl layer and, accordingly, also depends on the nature of the associated minerals.

7. Study of pK-metric and pH-metric methods for quartz dispersed materials

According to the results of the pK-metric method, the following indicators were calculated: the number of acid SACs ($\sum q_A$); the number of alkaline SAC ($\sum q_B$); the total number of SAC ($\sum q = \sum q_A + \sum q_B$); integral index ($Q = \sum qA / \sum qB$), which characterizes the surface of quartz material as neutral ($Q \approx 1$), acidic ($Q > 1$) or alkaline ($Q < 1$). According to the results of the pH-metric method, the following indicators were calculated: equilibrium value of the hydrogen index of the aqueous suspension of quartz material (pH^{eq}_{susp}); the value of the isoadsorption point of the aqueous suspension of quartz material, at which there is no mass transfer of ions from the solution to the surface of the solid ($pH_{IIP} = pH_0 + (pH^{eq}_{susp} - pH_0) \cdot (pH_0 - 6)$); an indicator indicating the width of the pK SAC range on the surface of the quartz material ($\Delta pH = pH_{IIP} - pH^{eq}_{susp}$); an indicator that characterizes the duration of the equilibrium state in the aqueous suspension of quartz material (t^{eq}). The results of pK-metric and pH-metric experimental studies of quartz dispersed materials are presented in Table 3 and Table 4.

The presented results show that the data obtained by different methods (pK-metric and pH-metric) correlate with each other. Thus, according to the results of pK-metric analysis (table 3), the alkalinity of quartz materials increases in a number:

$$QN(Q \approx 0.80) < QO(Q \approx 0.61) < DS(Q \approx 0.35) < MT(Q \approx 0.28) . \quad (4)$$

According to the results of pH-metric analysis (table 4), the alkalinity of the surface of quartz materials, in accordance with the value of pH_{IIP}, increases in a number:

$$QN(7.16) < QO(8.94) < DS(9.51) < MT(9.74) . \quad (5)$$

From the results presented in table. 1, 3 and 4 it follows that the acid-base surface characteristics of quartz materials significantly depend on the chemical nature of the accompanying minerals. The highest acidity is characterized by the surface QN (Q≈0.80; pH_{IIP}=7.16), which contains montmorillonite. This is confirmed by the known fact [15] that the presence of montmorillonite determines the acidic properties of the clay dispersed material. On the QN surface, in addition to Bronsted's neutral SACs, strongly acidic SACs with pK≈-0.5–1.5 (possibly partly Lewis character) and alkaline SACs with pK≈9–10 were detected. The weakly alkaline nature of the QO surface (Q≈0.61; pH_{IIP}=8.94) is due to the fact that in the presence of montmorillonite the composition includes impurities of kaolinite. Apparently, this is due to the fact that the surface of dispersed kaolinite is known to have an alkaline reaction [15]. The surfaces of DS (Q≈0.35; pH_{IIP}=9.51) and MT (Q≈0.28; pH_{IIP}=9.74) differ in the greatest alkalinity. This is due to the presence on the surface of a large amount of SAC with pK≈9–11.

Table 3. Results of study of quartz materials pK-metric method.

Quartz dispersed materials	$\sum q_A \cdot 10^{-12}$ (1/sm^2)	$\sum q_B \cdot 10^{-12}$ (1/sm^2)	$\sum q \cdot 10^{-12}$ (1/sm^2)	$Q = \sum q_A / \sum q_B$
QO	48.51	80.14	128.65	0.61
QN	152.23	189.52	341.75	0.80
DS	79.13	226.09	305.22	0.35
MT	45.74	163.36	209.10	0.28

Table 4. Results of study of quartz materials pH-metric method.

Quartz dispersed materials	pH_0 for distilled water	pH^{eq}_{susp}	pH_{IIP}	ΔpH	t^{eq} (s)
QO	7.45	8.48	8.94	0.46	360
QN	7.25	7.18	7.16	0.02	40
DS	7.50	8.84	9.51	0.67	120
MT	7.10	9.50	9.74	0.24	300

A comparison of all the obtained regularities (1–5) allows us to state that the acid-base surface properties of quartz materials do not depend on the SHHL thickness.

8. Conclusions

The results of the study allow us to formulate the following conclusions. For the most popular quartz dispersed materials quartzite of the Ovruch deposit (QO), quartz sand of the Novovodolazsky deposit (QN), dinas (DS) and marshalite (MT), the mineral composition, surface morphology, predominant size and specific surface area of the particles are determined. The structure of the SHHL of quartz materials was investigated using thermogravimetric (TG) and differential thermal analysis (DTA). The number and acid-base characteristics of SACs were studied by pK- and pH-metric.

It is established that the mechanisms of physical and chemical adsorption of water molecules on the surface of all quartz dispersed materials are identical and do not depend on the mineral nature of impurities. Obviously, the mechanism of adsorption of water molecules is affected only by the method (thermal or mechanical) treatment of the dispersed material. It is also found that the thickness of the surface hydrate layer depends only on the method of processing quartz dispersed materials and does not depend on the mineral composition. It is shown that the thickness of the surface hydroxyl layer is 2-3 times greater than the thickness of the hydrate layer for all quartz materials, and the amount of chemically adsorbed water depends on the mineral nature of the impurities. The presence of clay minerals montmorillonite and kaolinite determines the greatest thickness of the hydroxyl layer of quartzite. The thickness of SHHL quartz dispersed materials is determined by the amount of chemically adsorbed water, i.e. the thickness of the hydroxyl layer and, accordingly, also depends on the nature of the associated minerals.

It is established that the acid-base characteristic of the surface of quartz materials significantly depends on the chemical nature of the accompanying minerals. It has been experimentally shown that the surface QN, which contains montmorillonite, is characterized by the highest acidity. On the QN surface, in addition to neutral SACs, strong acid SACs with pK≈-0.5–1.5 and alkaline SACs with pK≈9–10 were detected. The weakly alkaline nature of the QO surface is due to the fact that in the presence of montmorillonite the composition includes impurities of kaolinite. The DS and MT surfaces differ in the highest alkalinity due to the presence of a large amount of SAC with pK≈9–11 on the surface. It was found that the acid-base surface properties of quartz dispersed materials do not depend on the thickness of SHHL.

References

[1] Melnyk L, Bessarab O, Matko S and Malyovanyy M 2015 Adsorption of heavy metals ions from liquid media by palygorskite *Chem. & Chem. Tech.* **9(4)** pp 467-470

[2] Malyovanyy M, Sakalova G, Chornomaz N and Nahurskyy O 2013 Water sorption purification from ammonium pollution *Chem. & Chem. Tech.* **7(3)** pp 355-358

[3] Danchenko Yu, Andronov V, Teslenko M, Permiakov V, Rybka E, Meleshchenko R and Kosse A 2018 Study of the free surface energy of epoxy composites using an automated measurement system. *East.-Europ. J. of Ent. Tech.* **1 12(91)** pp 9-17

[4] Pozdieiev S, Pidgoretskiy Y, Nekora O, Sidnei S and Tyshchenko O 2018 Research of explode exposure at the relief vent system structures with soft transparent material *Intern. J. of Eng. and Tech. (UAE)* **7 (4.3) 3** pp 298-302

[5] Starokadomsky D L 2008 About influence of non-modified nanosilica on physico-chemical properties of epoxy-polymer composites *Russ. J. Appl. Chem.* **12** pp 2045-2051

[6] Saienko N V, Demidov D V, Popov Y V, Bikov R A, Yunis B and Saienko L V 2019 Effect of mineral filler compounds on vapor permeability and hydroscopic properties of water-based polymer dispersions *Mater. Sci. Forum.* **968** pp 89-95

[7] Cardiano P, Ponterio R C, Sergi S, Lo Schiavo S and Piraino P 2005 Epoxy-silica hybrids as stone conservation materials *Polymer* **46(6)** pp 1857-1864

[8] Saienko N V, Demidov D V, Popov Y V, Bikov R A, Yunis B and Saienko L V 2019 Effect of mineral filler compounds on vapor permeability and hydroscopic properties of water-based polymer dispersions *Mater. Sci. Forum.* **968** pp 89-95

[9] Danchenko Y, Andronov V, Obizhenko T, Kosse A and Khmyrov I 2018 The influence of inorganic fillers on the protective properties of epoxy polymer composite materials *Intern. J. of Eng. and Tech. (UAE)* **7(4.3) 3** pp 279-283

[10] Dasari A, Yu Z Z, Cai G P and Mai Y-W 2013 Resent developments in the fire retardancy of polymeric materials *Prog. Polym. Sci.* **38** pp 1357-1387

[11] Danchenko Y M, Popov Y V and Barabash E S 2016 Influence of acid-base properties of the surface of polymineral fillers on the structure and characteristics of epoxy composites *Voprosy Khimii I Khimicheskoi Tekhnologii* **3(107)** pp 53-60

[12] Skripinets A, Danchenko Y and Kabus' A 2015 A research on technological and physicochemical laws of manufacturing vibration-absorbing products based on epoxy-urethane polymer compositions. *East.-Europ. J. of Ent. Tech.* **3(11)** pp 4-8

[13] Zapata-Massot C and Le Boley N 2007 Effect of the mineral filler on the surface properties of co-ground polymeric composites *Part. Part. Char.* **24** pp 339-344

[14] Janczuk B and Bialopiotrowicz T 1988 Components of surface free energy of some clay mineral *Clay and Clay minerals* **36(3)** pp 243-248

[15] Danchenko Yu, Andronov V, Sopov V, Khmyrov I and Khryapynskyy A 2018 Acid-basic surface properties of clay disperse fillers *MATEC Web of Conf.* **230**

[16] Danchenko Yu, Andronov V, Barabash E, Rybka E and Khmyrova A 2019 Acid-basic surface properties of dispersed fillers based of metal oxides TiO_2, Al_2O_3, CaO and Fe_2O_3 *IOP Conference Series: Materials Science and Engineering* **708** 012083

[17] Horvath J and Dolnik V 2001 Polymer wall coatings for capillary electrophoresis *Electrophoresis* **22** pp 644-655
[18] Fei-Zhou Li, Zheng-Lin Lu, Zhi-Huai Yang and Kai Qi 2015 Surface interaction energy of ceramic materials with epoxy resin *Polimery* **60 (7-8)** pp 468-471
[19] Danchenko Yu, Andronov V, Rybka E and Skliarov S 2017 Investigation into acid-basic equilibrium on the surface of oxides with various chemical nature *East.-Europ. J. of Ent. Tech.* **4 12(88)** pp 17-25
[20] Kuzema P O, Starokadomsky D L, Tkachenko O O, Tertykh V A 2020 Reinforcement of epoxy polymers with hydride-silylated fumed silica *Chem., Phys. and Tech. of Surf.* **11** pp 484-491
[21] Kravchenko A A, Demianenko Ye M, Hrebeniuk A H and Lobanov V V 2014 Kvantovokhimichne modeliuvannia protolitychnoi rivnovahy poverkhni kremnezemu *Khimiia, fizyka ta tekhnolohiia poverkhni* **5(1)** pp 16-20
[22] Hunko V M 2010 Kvantovokhimicheskii analiz adsorbtsii I mekhanizmov khimicheskikh reaktsii na poverkhnosti tverdykh tel *Khimiia, fizika ta tekhnolohiia poverkhni* **1 (1)** pp 5-18
[23] Ikonnikova K V, Ikonnikova L F, Minakova T S and Sarkisov Yu S 2011 Teoriia I praktika pH-metricheskogo opredeleniia kislotno-osnovnykh svoistv poverkhnosti tverdykh tel 85
[24] Chukin G D 2008 Khimiia poverkhnosti i stroieniie dispersnogo kremniezema 172
[25] Brei V V 2008 Korreliatsiia mezhdu siloi osnovnykh tsentrov katalizatorov I ikh aktivnostiu v testovoi reaktsii razlozheniia 2-metil-3-butin-2-ola *Teoreticheskaia i eksperimentalnaia khimiia* **44 (5)** pp 310-314
[26] Yekimova I A and Minakova T S 2013 Adsorbtsionnyie issledovaniia parov vody na oksidakh I ftoridakh shchelochnozemelnykh metallov i magniia *Vestnik TGASU* **2** pp 263-275
[27] Weissenbach K and Mack H 2005 Silane Coupling Agent. Functional fillers for Plastics *WILEY-VCH Verlag GmbH & Co. KGaA Weinheim* pp 59-83
[28] Sytnikov P A, Riabkov Yu I, Riazanov M A, Belykh A G, Vaseneva I N, Fedoseev M S and Tereshatov V V 2013 Vliianiie kislotno-osnovnykh svoistv poverkhnosti oksida aliuminiia na reaktsionnuiu sposobnost s epoksidnymi soiedineniiami *Izvestiia Komi nauchnogo tsentra UrO RAN*, **3(15)** pp 19-26
[29] Sorochkina E A, Smotraiev R V, Kalashnikov Yu V and Gruzdeva E V 2013 Kislotno-osnovnyie svoistva poverkhnosti sfericheski granulirovannykh sorbentov na osnove gidratirovannykh oksidov tsirkoniia I aluminiia *Voprosy khimii i khimicheskoi tekhnologii* **6** pp 102-104
[30] Danchenko Yu, Kachomanova M and Barabash Ye 2018 The acid-base interaction role in the processes of the filled diane epoxy resin structuring *Chem. & Chem. Tech.* **12(2)** pp 188-195
[31] Danchenko Yu, Andronov V, Barabash E, Obigenko T, Rybka E, Meleshchenko R and Romin A 2017 Research of the intramolecular interactions and structure in epoxyamine composites with dispersed oxides *East.-Europ. J. of Ent. Tech.* **6 12 (90)** pp 4-12
[32] Danchenko Yu, Andronov V, Kariev A, Lebedev V, Rybka E, Meleshchenko R and Yavors'ka D 2017 Research into surface properties of dispersive fillers based on plant raw materials, *East.-Europ. J. of Ent. Tech.* **5 12 (89)** pp 20-26
[33] Danchenko Yu, Kariev A, Andronov V, Cherkashina A, Lebedev V, Shkolnikova T, Burlutskyi O, Kosse A, Lutsenko Yu and Yavors'ka D 2020 Research of chemical nature and surface properties of plant disperse fillers *East.-Europ. J. of Ent. Tech.* **6 1 (103)** pp 32-41

Acknowledgements
The authors are grateful to Dr. Sayenko Sergey U. from Institute of Solid State Physics, Materials Science and Technologies NAS of Ukraine and Bakevich Olexandr S. from Ukrainian State University of Chemical Technology for assistance in conducting the experiment. The work is supported by the Ministry of Education and Science of Ukraine (State registration number 0114 U 004376).

Method for calculating the shape of rolls for helical rolling

S M Gorbatyuk[1,*], T V Deniskina[1], E E Balakhnina[1] and O N Chicheneva[1]

[1]National University of Science and Technology «MISIS», Moscow, Russia

*E-mail: sgor02@mail.ru

Abstract. The paper is devoted to solving the problem of determining the shape of the rolls of helical rolling mills, depending on the specified profile of the deformation zone. A universal calculation method has been proposed, thanks to which it is possible to determine the shape of the working surface of a roll for all types of helical rolling mills (with mushroom-shaped, cup-shaped, barrel-shaped and disc rolls), any relative arrangement of the rolling axis and rolls axes, and various locations of the deformation zone on the rolling axis. The proposed method is implemented as a standalone exe-application with a simple intuitive interface. The application allows you to output the calculation results into txt-files, which can then be imported into CAD systems to create 3D roll models.

1. Introduction

Recently, the development of different industry sectors [1-12] requires the solution of various technological production problems [13-27] and also increases the requirements for the efficiency and quality of production technologies [28-40]. One of the most promising and profitable methods to produce high-quality seamless pipes and rods is helical rolling.

In helical rolling, metal deformation occurs between obliquely located rolls, the axes of which are inclined relative to the rolling axis by a certain angle. Points on the surface of the billet move along a helical line, performing both rotational and translational movements. This method of metal forming is widely used for the manufacture of products in the form of bodies of revolution (balls, axes, pipes, etc.)

When considering many issues related to the helical rolling process, geometric parameters are the basis. This is explained by the fact that the shape and size of the deformation zone significantly affect the conditions of rolling, the required shape of the rolls, etc.

Helical rolling mills are generally classified according to the shape of the work rolls into four main types (figure 1): *a* - with barrel-shaped rolls, *b* - with mushroom-shaped rolls, *c* - with cup-shaped rolls, *d* - with disc rolls.

2. Calculation model of the geometric parameters of the helical rolling process

The calculation method is based on the description of the relative position of the rolling and roll axes by the values of the angle α and the distance ρ between them (figure 2).

This was done because during helical rolling the rolling axis and roll axis are skew lines, the relative position of which is uniquely determined by the values of the angle and the distance between them. Also, this approach is interesting in that α and ρ do not depend on the method of adjusting the work rolls and, therefore, are convenient for analysing processes in stands of various designs.

Content from this work may be used under the terms of the Creative Commons Attribution 3.0 licence. Any further distribution of this work must maintain attribution to the author(s) and the title of the work, journal citation and DOI.
Published under licence by IOP Publishing Ltd

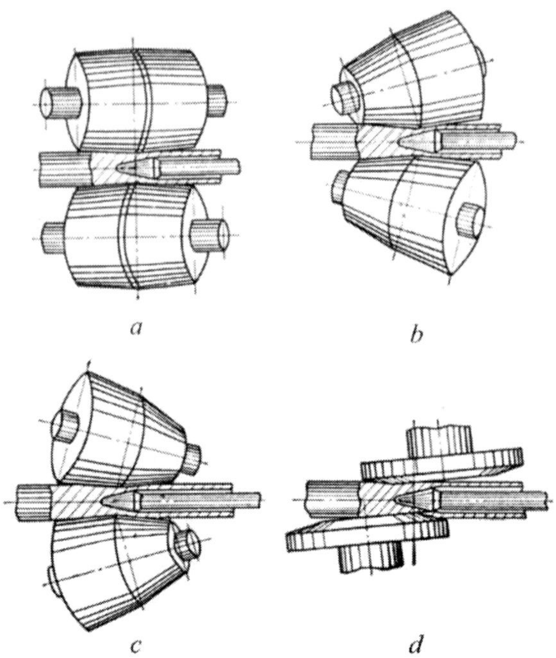

Figure 1. Types of rolls for helical rolling:
a - barrel-shaped, *b* - mushroom-shaped, *c* - cup-shaped, *d* - disc rolls.

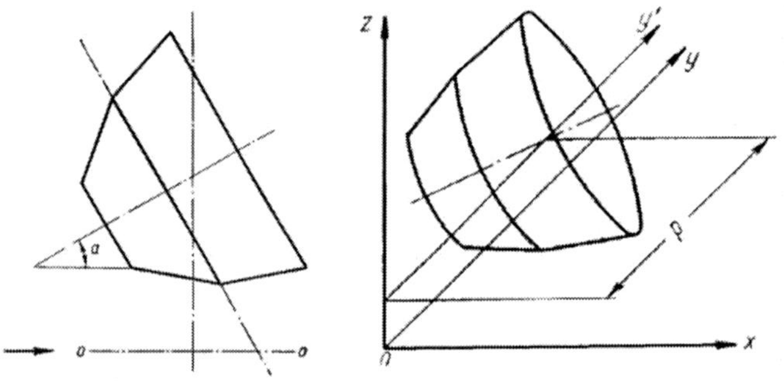

Figure 2. Parameters of the relative position of the rolling and roll axes.

Thus, if we introduce the assumption that each *i*-th section of the deformation zone is processed by the *i*-th section of the roll, then to calculate the exact values of the roll radius we need to know:
α – angle between rolling and roll axes;
ρ – distance between rolling and roll axes;
x_i and y_i – coordinates of the points of the deformation zone.

For a given section of the roll, the roll radius is gradually increased by the iterative method, and the condition is checked whether the points of the circle formed by this radius are outside the billet

$$\begin{cases} \Delta Z \le |Z_{roll}| \\ \Delta Y \le |Y_{roll}| \end{cases}$$

As soon as the condition is met, the calculation will be interrupted.

Figure 3 represents a diagram that shows that for all points lying outside the billet, either $\Delta Y > Y_{roll}$, or $\Delta Z > Z_{roll}$. Only for one point of the circle ΔY will be equal to Y_{roll}, and ΔZ will be less than Z_{roll}, thus, at this point the condition will be fulfilled and the calculation for this section will be interrupted.

The smallest possible radius at which the condition is satisfied on the entire circle will be the necessary one.

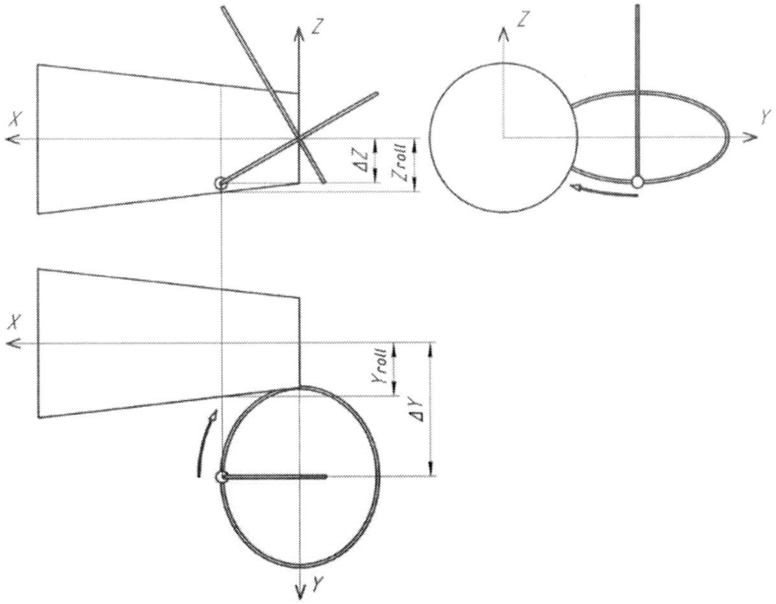

Figure 3. 2D-scheme of calculation method.

Figure 4 shows the current point on the circle with coordinates X_0, Y_0, Z_0, and the arrow shows the direction of reference. The roll radius *R* increases until one of the points of the circle touches the surface of the billet.

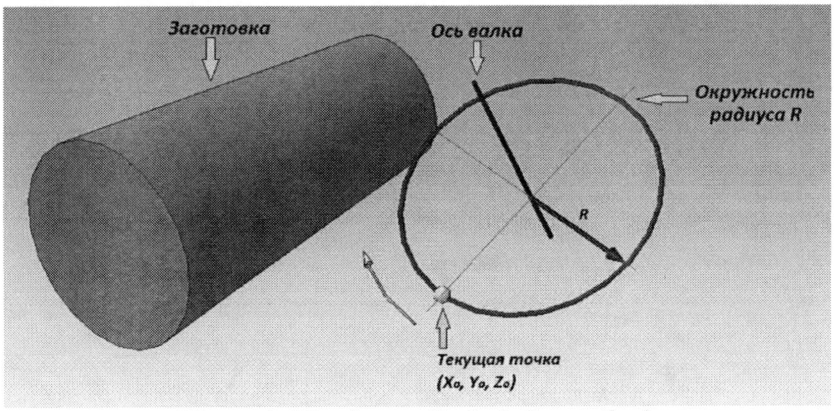

Figure 4. 3D-scheme of calculation method.

It is important to note that, *firstly*, the calculation is performed for each section of the roll, and not for the deformation zone, which makes it possible to get rid of additional transformations in the transition from the coordinates associated with the rolling axis to the coordinates associated with a roll axis. *Secondly*, a different mathematical model is used, based on simple formulas, which makes it possible to exclude errors of the roll caliber at values of $\alpha > 60°$ and the correction of the deformation zone associated with these errors [40], thus, the roll bends around the billet without penetrating anywhere ... And, *thirdly*, this method is suitable for constructing rolls of any type of mill, including even the cases of $\alpha = 0°$ and $\alpha = 90°$.

3. Application for calculating the shape of rolls

To implement the computational method on a computer, it was decided to use Microsoft Visual Studio 2010 with .NET Framework 4.0 as a programming environment, and C# as a language. All this allowed us to create an application with an intuitive interface (figure 5).

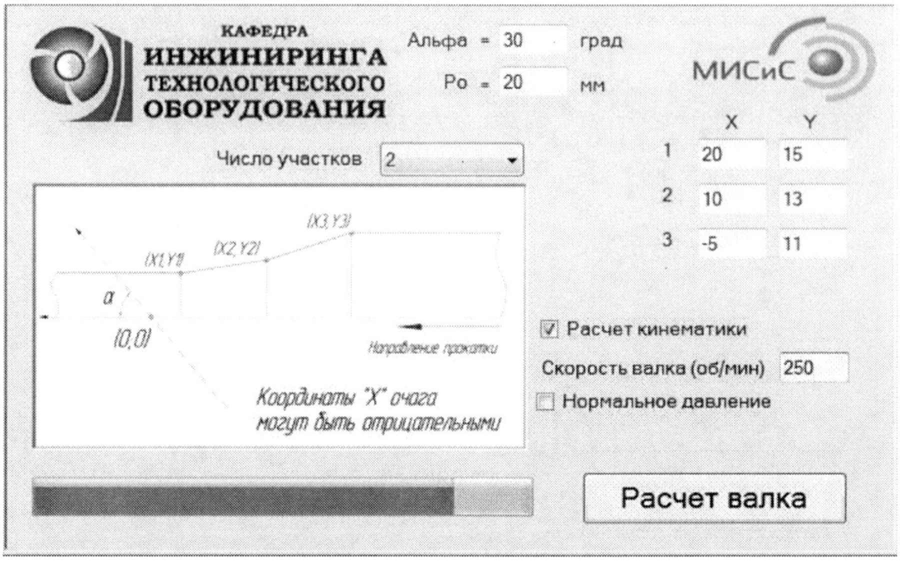

Figure 5. Application interface.

The basic procedure for the program:

1. The values of α and ρ and the geometry of the deformation zone profile are specified.
2. The necessary calculations are made: defining the boundaries of the roll and checking the data for erroneous values.
3. The *i*-th section of the roll is taken.
4. The radius of the roll in this section is calculated by the iterative method.
5. A calculation is made for each next section of the roll.
6. Data processing and output is in progress.

After entering the required data, pressing the "Roll calculation" button starts the calculation process. Progress will be reflected to the left of the button - in the progress bar. At the end of the calculation, a folder with calculation txt-files opens (figure 6). The resulting txt-files can be used by CAD systems as input data for modeling (figure 7).

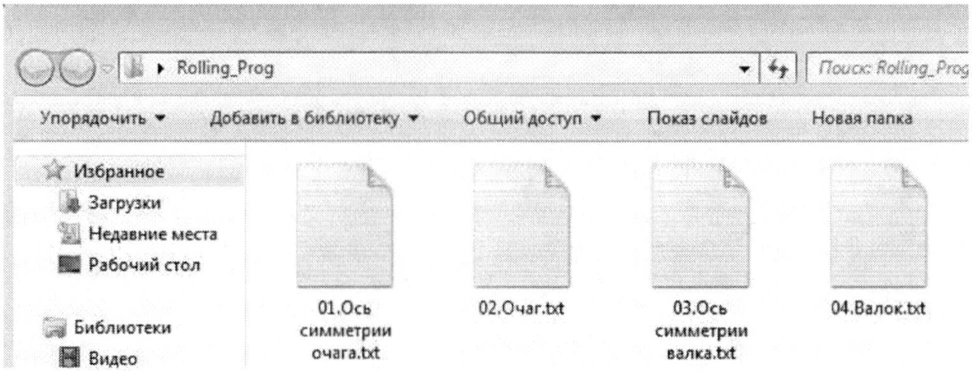

Figure 6. Output files.

Figure 7. Visualization of the rolling process of a 6 mm bar obtained using the application and CAD KOMPAS-3D V13.

References

[1] Gorbatyuk S M, Pashkov A N, Morozova I G and Chicheneva O N 2020 Technologies for applying Ni-Au coatings to heat sinks of SiC-Al metal matrix composite material *Materials Today: Proceedings* **38** pp 1889-1893 DOI:10.1016/j.matpr.2020.08.581

[2] Valiullin I I, Kobelev O A, Naumova M G, Morozova I G, Pashkov A N, Nagovitsyn V A and Goloshchapov K V 2021 Application of a Discrete Double-Row Scraper Cleaner as an Efficient Procedure of Cleaning of Belt Conveyors Aimed at Transporting Bulk Materials *Metallurgist* DOI:10.1007/s11015-021-01124-x

[3] Kerimov R I and Shakhov S I 2020 Use of Metallized Raw Materials in Electric Furnace Steelmaking *Metallurgist* **64 (1-2)** pp 128-135 DOI:10.1007/s11015-020-00974-1

[4] Sivak B A, Shakhov S I, Vdovin K N, Rogachikov Y M and Kerimov R I 2020 Development of a System for Electromagnetic Stirring of Liquid Steel in Molds of Billet and Bloom CCMS *Metallurgist* **63(9-10)** pp 910-914 DOI:10.1007/s11015-020-00909-w

[5] Shakhov S I and Vdovin K N 2019 Electromagnetic Stirring in Continuous-Casting Machines *Steel in Translation* **49(4)** pp 261-264 DOI:10.3103/S0967091219040120

[6] Shakhov S I, Vdovin K N, Kerimov R I, Rogachikov Y M and Shakhov D S 2020 Improvement of Built-In Electromagnetic Stirring in the Molds of Bloom Continuous-Casting Machines *Metallurgist* **64(5-6)** pp 410-417 DOI:10.1007/s11015-020-01010-y

[7] Jiawen L, Timushev S, Klimenko D and Krivenko A 2019 Modeling pressure pulsation fields in a screw centrifugal pump *Proceedings of the 26th International Congress on Sound and Vibration, ICSV* **2019**

[8] Keropyan A M, Kuziev D A and Krivenko A E 2020 Process Research of Wheel-Rail Mining Machines Traction *Lecture Notes in Mechanical Engineering* **2020** pp 703-709 DOI:10.1007/978-3-030-22063-1_75

[9] Kouziyev D, Krivenko A, Chezganova D and Valeriy B 2019 Sensing of dynamic loads in the open-cast mine combine *E3S Web of Conferences* **105** 03014 DOI:10.1051/e3sconf/201910503014

[10] Krivenko A E and Khanh Z K 2020 Influence of power fluid temperature in hydraulics on operating efficiency of hydraulic mining excavators *Gornyi Zhurnal* **2020(12)** pp 78-81 DOI:10.17580/gzh.2020.12.18

[11] Kobelev O A and Tyurin V A 2007 Production of large plates *Steel in Translation* **37(9)** pp 727-729 DOI:10.3103/S096709120709001X

[12] Pimenov G A, Kostyukov G A, Ryabov P S, Rogal' V D and Kobelev O A 1991 Making of large-sized solid-forged blanks *Tyazheloe Mashinostroenie* **1991(9)** pp 21-24

[13] Albagachiev A Y, Keropyan A M, Gerasimova A A and Kobelev O A 2020 Determination of rational friction temperature in lengthwise rolling *CIS Iron and Steel Review* **19** pp 33-36 DOI:10.17580/cisisr.2020.01.07

[14] Kobelev O A, Zinov'ev A V and Tsepin M A 2009 Effective production of large pipe blanks *Steel in Translation* **39(6)** pp 501-505 DOI:10.3103/S0967091209060163

[15] Kobelev O A, Sotnikov A I and Kobelev A V 1998 Analysis of the Electrocapillary Curves and Adsorption Process Models in Copper || Boron + Aluminium Oxide Melt *Zeitschrift fur Physikalische Chemie* **203(Part_1_2)** pp 143-158 DOI:10.1524/zpch.1998.203.Part_1_2.143

[16] Kobelev O A, Deev A V and Sotnikov A I 1995 Electrocapillary curve of copper in boroaluminate melt *Rasplavy* **1995(1)** pp 39-45

[17] Surianinov M and Shyliaiev O 2018 Calculation of plate-beam systems by method of boundary elements *International Journal of Engineering and Technology(UAE)* **7(2)** pp 238–241 DOI:10.14419/ijet.v7i2.23.11927

[18] Korneeva I, Neutov S and Surianinov M 2017 Experimental studies of fiber concrete creep *MATEC Web of Conferences* **116** 02021 DOI:10.1051/matecconf/201711602021

[19] Yu L, Ignatov Y, Ivannikov A, Khotchenkov E and Krasnoshtanov D 2019 Common features in the manifestation of natural and induced geodynamic events in the eastern regions of Russia and China *IOP Conference Series: Earth and Environmental Science* **324(1)** 012004 DOI:10.1088/1755-1315/324/1/012004

[20] Sukhorukova M A and Ivannikov A L 2020 Vehicle accident risk assessment in mines *Mining Informational and Analytical Bulletin* **2020(6)** pp 224-232 DOI:10.25018/0236-1493-2020-61-0-224-232

[21] Nikitin A G, Demina E I, Zhivago E Y, Dvornikov L T and Saruev L A 2018 Energy-saving method of cutting a pre-bent bar on the shear machine *IOP Conference Series: Materials Science and Engineering* **411(1)** 012053 DOI:10.1088/1757-899X/411/1/012053

[22] Gurin I A, Lavrov V V, Spirin N A and Nikitin A G 2017 Web-technologies for construction of automated information-modeling systems of technological processes in metallurgy *Izvestiya Ferrous Metallurgy* **60(7)** pp 573-579 DOI:10.17073/0368-0797-2017-7-573-579

[23] Nikitin A G and Chainikov K A 2011 Deformation of elastic cylindrical element in pneumatic shock absorber *Russian Engineering Research* **31(8)** pp 742-743 DOI:10.3103/S1068798X11080168

[24] Snitko S A and Yakovchenko A V 2017 Influence of the Axial Reduction Conditions on the Variation in the Thickness of a Wheel Rim at the Initial Stage of Rolling *Metallurgist* **61(5-6)** pp 387-393 DOI:10.1007/s11015-017-0505-x

[25] Snitko S A, Yakovchenko A V, Pilipenko V V and Ivleva N I 2020 Modeling of ring billets rolling on radial-axial ring-rolling mill *Izvestiya Ferrous Metallurgy* **63(8)** pp 665-673 DOI:10.17073/0368-0797-2020-8-665-673

[26] Eron'ko S P, Danilov V L, Kuklev A V, Tkachev M Y, Tinyakov V V and Mechik S V 2020 Experience of Design and Industrial Application of Systems for the Driven Feed of Slag-Forming Mixtures into the Crystallizers of Slab CCM *Metallurgist* **64(3-4)** pp 214-222 DOI:10.1007/s11015-020-00986-x

[27] Eron'ko S P, Danilov V L, Tkachev M Y, Tinyakov V V and Ponomareva E A 2020 Model Studies and Modernization of a Manipulator for Tapping Spout Replacement in Continuous Steel Casting *Metallurgist* **64(3-4)** pp 301-308 DOI:10.1007/s11015-020-00996-9

[28] Artiukh V, Mazur V and Pokrovskaya E 2016 Influence of Strip Bite Time in Work Rolls Gap on Dynamic Loads in Strip Rolling Stands *MATEC Web of Conferences* **86** 01030 DOI:10.1051/matecconf/20168601030

[29] Gorbatyuk S, Efremov D, Albul S and Kirillova N 2020 The effect of the deformation method of the billets on mechanical properties of molybdenum bars and wire *Key Engineering Materials* **864 KEM** pp 204–210 DOI:10.4028/www.scientific.net/KEM.864.204

[30] Sedykh L V, Albul S V, Efremov D B and Sukhorukova M A 2020 Application of additive technologies for manufacturing a wear-resistant steel pipe outlet *IOP Conference Series: Materials Science and Engineering* **971(2)** 022002 DOI:10.1088/1757-899X/971/2/022002
[31] Kondratenko V E, Devyatiarova V V, Albul S V and Valeeva L M 2020 Method of calculating volumetric scaffold of monolithic slab formwork *IOP Conference Series: Materials Science and Engineering* **971(5)** 052036 DOI:10.1088/1757-899X/971/5/052036
[32] Kondratenko V E, Devyatiarova V V, Albul S V and Kartyshev D S 2020 Improving methodology for calculating scaffolding formwork of monolithic slabs in building constructions *IOP Conference Series: Materials Science and Engineering* **971(5)** 052037 DOI:10.1088/1757-899X/971/5/052037
[33] Keropyan A, Albul S and Zarapin A 2020 Problem of Increasing Tractive Effort of Railway Locomotives in Conditions of Arctic and Continental Shelf Regions *Lecture Notes in Mechanical Engineering* pp. 651-658 DOI:10.1007/978-3-030-22063-1_69
[34] Osadchiy V A, Albul S V, Kuprienko N S and Kirillova NL 2020 Future developments of a roll forming mill design algorithm *IOP Conference Series: Materials Science and Engineering* **709(3)** 044079 DOI:10.1088/1757-899X/709/4/044079
[35] Kukhar V V 2015 Producing of elongated forgings with sharpened end by rupture with local heating of the workpiece method *Metallurgical and Mining Industry* **7(6)** pp 122-132
[36] Kukhar V V and Vasylevskyi O V 2014 Experimental research of distribution of strains and stresses in work-piece at different modes of stretch-forging with rotation in combined dies *Metallurgical and Mining Industry* **6(3)** pp 71-78
[37] Markov O E, Kukhar V V, Zlygoriev V N, Shapoval A A, Khvashchynskyi A S and Zhytnikov R U 2020 Improvement of Upsetting Process of Four-Beam Workpieces Based on Computerized and Physical Modeling *FME Transactions* **48(4)** pp 946-953 DOI:10.5937/fme2004946M
[38] Albagachiev A Y, Keropyan A M, Gerasimova A A and Kobelev O A 2020 Determination of rational friction temperature in lengthwise rolling *CIS Iron and Steel Review* **19** pp 33-36 DOI:10.17580/cisisr.2020.01.07
[39] Chichenev N A, Gorbatyuk S M, Naumova M G and Morozova I G 2020 Using the similarity theory for description of laser hardening processes *CIS Iron and Steel Review* **19** pp 44-47 DOI:10.17580/cisisr.2020.01.09
[40] Kobelev O A, Albul S V and Kirillova N L 2020 Research and development of broaching methods on mandrel of large-sized pipe forgings *IOP Conference Series: Materials Science and Engineering* **709(4)** 044104 DOI:10.1088/1757-899X/709/4/044104

Improvement of mixing technology of concrete mortar components in mixers

M N Erofeev[1], I N Kravchenko[1,2], Yu A Kuznetsov[3] and V V Goncharenko[3,*]

[1]Federal State Budgetary Science Institution «Blagonravov Institute of Mechanical Engineering of the Russian Academy of Sciences». 101990, Moscow, Maly Kharitonievsky Lane, 4
[2]Federal State Budgetary Educational Institution of Higher Education «Russian State Agrarian University – Moscow Timiryazev Agricultural Academy», 127550, Moscow, Timiryazev St., 58
[3]Federal State Budgetary Educational Institution of Higher Education «Orel State Agrarian University named after N.V. Parakhin», 302019, Orel, General Rodin St., 69

[*]E-mail: vovaniii@bk.ru

Abstract. Methods of calculations and methods of modeling the mixing of components of construction mixtures are presented. Experimental studies were carried out using a prototype mixer. Compressive strength, workability, and concrete uniformity coefficient in strength were taken as the main defined characteristics of the concrete mixture. On the basis of the conducted researches the dependences for an estimation of design and layout decisions and parameters of work of mixers and technological cycles, and also their influence on quality and structural - rheological properties of ready products have been established. The obtained data confirm the hypothesis about the possibility of improving the quality of concrete mixtures by using a gravity-forced method. It has been established, that application of gravity-forced method of mixing allows to increase concrete strength by 5-10 %, and of mixture uniformity coefficient - by 1.4-2.5 %. At the same time, the duration of concrete mixtures preparation can be reduced by 1.5-2.0 times compared with the known methods.

1. Introduction

The development of the construction industry is associated with the improvement of constructions of technological equipment, new constructional materials and methods of their production.

The main challenges faced by developers and designers include the choice of design-layout scheme of the main unit (mixer) with the determination of its geometrical and kinematic parameters, the calculation of time parameters and power consumed during mixing [1-5]. The structural scheme choice is associated with the formulation of requirements for the operation mode (cyclic), mixed materials compositions. In this case, geometrical and kinematic parameters are mainly determined by the required capacity of the finished product output.

Due to the complexity of formalizing the kinetics of the mixing process and determining the power consumption [6-9], the experimental studies that allow obtaining experimental data, on the basis of which one can establish the real dependences of the design parameters, performance and quality of concrete mortar mixtures, are of great importance.

The mixing materials process is a mechanical process, as a result of which the constituent mixtures are evenly distributed in the mixed volume [3, 10, 11]. The mixing process is opposite to the separation process, i.e. separation of the mixture components from each other, which is predetermined by the imperfect action of the mixer (the influence of gravitational, inertial, centrifugal forces), the influence of the grain size difference of the mixing materials and other reasons. The actual mixing process usually consists of an overlap of mixing and separation.

The mixing quality is characterized by the degree of approximation of the mixture to the material with some ideal order of distribution of individual components [12-14]. For example, under

Content from this work may be used under the terms of the Creative Commons Attribution 3.0 licence. Any further distribution of this work must maintain attribution to the author(s) and the title of the work, journal citation and DOI.
Published under licence by IOP Publishing Ltd

production conditions, the quality of concrete mixture preparation is checked by periodic sampling of the prepared mixture directly at its unloading from the mixer. In this case, the main indicators of the quality of concrete mixture are: workability, density, porosity, and delaminatability (uniformity).

2. Materials and methods

To solve the problem of obtaining concrete mixtures of high qualities, it is necessary to study the process of mixture formation during the interaction of the mixers working bodies with the particles of mixed materials. During the initial period of the mixer operation, the quality of mixing, estimated by the mixing degree (Fig. 1), is increased mainly due to convective mixing of components – moving groups of particles from one position to another, transfer, embedding, sliding layers, deformation and abrasion of the layer portion as a result of mechanical movement of the mixer working bodies (section I of the curve). At this time the mixing process takes place at the macro-volume level (macro-mixing).

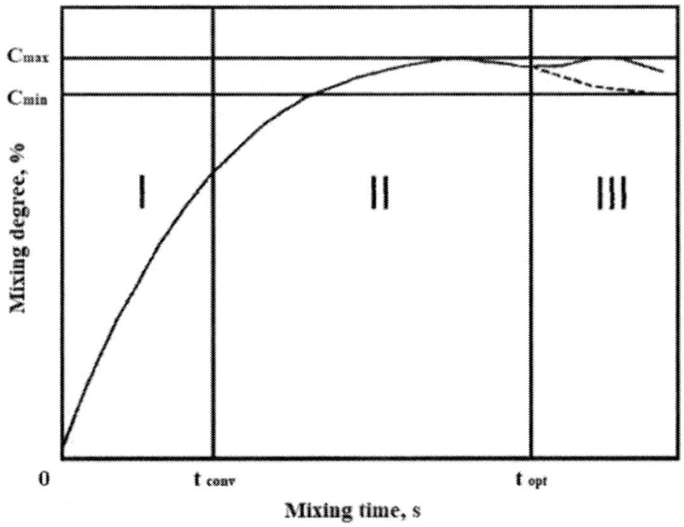

Figure 1. Concrete mixing process: I – convective mixing; II – diffusion mixing; III – balancing the mixing and separation process.

The interface between the components is not yet large, hence the proportion of diffusion mixing - gradual redistribution of particles of different components across the interface - will also be insignificant. After this, further mixing makes no sense since the mixture quality remains constant (section III of the curve). The redistribution rate of individual particles in sections II and III depends not only on the nature of material movement [15], but also on its structural and rheological properties (particle size, size of interfacial surface, adhesion forces, etc.).

The mixing ratio is determined by the structural and rheological properties of the mixture, the kinematic parameters and the design features of the mixer. The value of this coefficient averaged over the mixer volume and over a certain mixing time can be determined experimentally. Using similarity theory and dimensional analysis, then it is possible to calculate the value of the mixing ratio for a mixer with other parameters. Since a theoretical determination of the mixing ratio is not possible, it is proposed to obtain it from experimental data.

Experimental studies were carried out using a prototype mixer [16], the layout of which is shown in Figure 2. The mixer works as follows. Electric motor 7 through the reducer 8 transmits torque to the working body made in the form of a screw with blades 11, 12, 13 rigidly mounted on it. In the position (a) the construction mixtures are prepared (mixing), This process is accompanied by the auxiliary operations (dosing and loading of the components). Mixing is carried out with combined (gravity-forced) method: under the influence of an auger the mixtures components are transported to the top of the installation body, after which under the influence of gravitational forces they are dropped down. The forced mixing is carried out by the rotating blades. In the position (b) the mechanized laying of prepared mixture is carried out. Since the installation is mounted on a movable platform, it has maneuverability and smaller dimensions compared with the known ones [2, 3, 17].

Concrete compressive strength, workability, and the uniformity coefficient of concrete strength (coefficient of variation) are accepted as the main characteristics of the concrete mixture which should to be determined [18]. To analyze the mixing process, sampling was performed at various time periods during the process of the concrete mixture preparation. After a specified time, the mixer was stopped and sampling was performed into a special cast, while the workability of the mixture was determined by mobility (cone slump) and rigidity of concrete mixture in accordance with GOST 10181-2014, using for this purpose standard cone and technical viscosity analyzer. The resulting control samples were tested for strength in accordance with GOST 10180-2012 (table 1).

Table 1. Means of measurement used for samples testing.

Designation of the product characteristic (parameter) to be determined	Designation of the measuring instruments
Environmental parameters	Multimeter Testo 622
Mass measurement	Electronic balance GF-1200
Geometrical dimensions	Calipers, type SHC-I-0.1-2
Rupture load determination	Tensile machine 2054 P-5

3. Research results and analysis

On the basis of the performed studies we obtained approximating dependencies of the uniformity coefficient of the strength (coefficient of variation) from the mixing time, adequately describing a mathematical model of the mixing process in a gravity-pressure mixer (see figure 2).

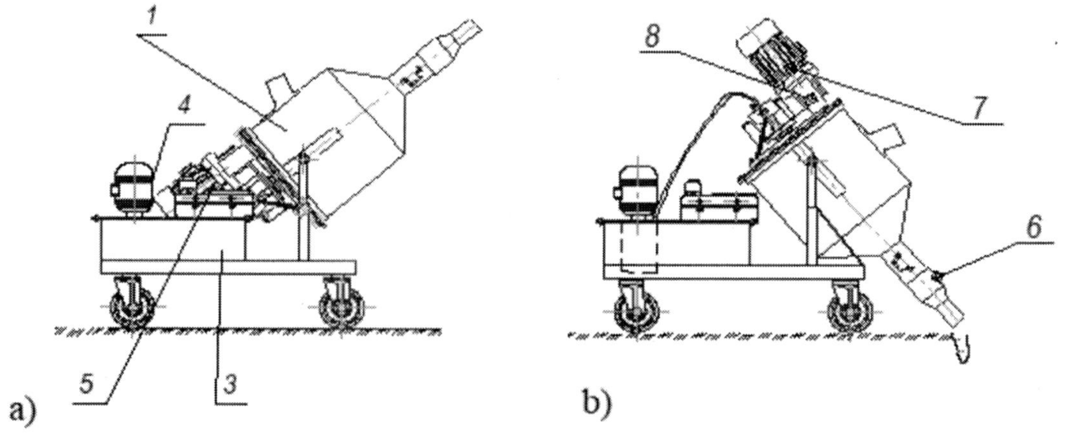

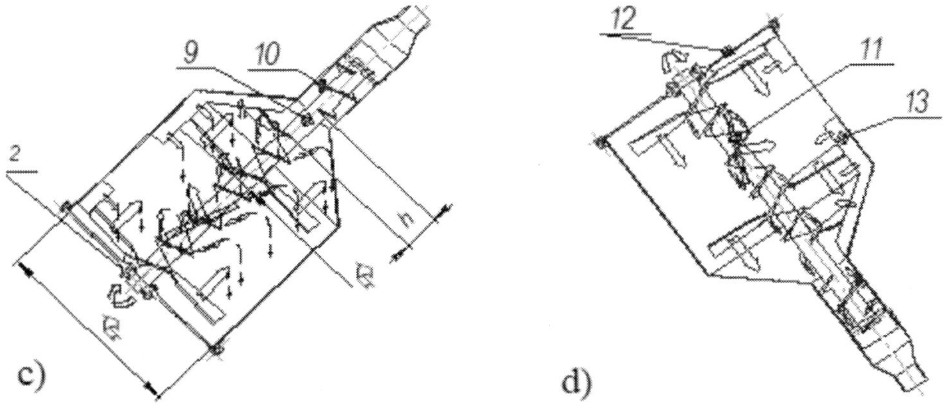

Figure 2. Structural and technological features of the gravity-forced mixer in the preparation of various compositions and samples of concrete mortar mixtures: a - during the mixtures preparation; b - during unloading into the construction (surface); c – moving direction of the mixtures components during mixing; d – during unloading; 1 – body in the form of a tipping cylindrical container with a conical bottom; 2 – support bearing of the mixer shaft; 3 – moving platform; 4 – tank for water storage; 5 – water pump for water supply; 6 – nozzle of adjustable diameter at the outlet pipe of the mixer; 7 – electric motor; 8 – reducer; 9 – shaft; 10 – auger with screw winding; 11 – mixing blade; 12 – bottom blades; 13 – blades of the cone part of the mixer drum.

The obtained data confirm the hypothesis of improving the quality of concrete mixtures prepared by gravity-forced method. For numerical estimation of quality improvement, the graphical dependences of strength growth and change in the uniformity coefficient of concrete strength were built (table 2). Experimental points, approximating curves and equations describing them, as well as the value of approximation reliability R^2 are shown on the graphs. Based on these dependencies we can conclude about a greater increase in the quality of concrete in the early mixing time, which can be explained by the manifestation of diffusion mixing already at the initial stage.

It should also be noted that the maximum increase in the concrete strength corresponds to the early curing time (figure 3), which leads to a reduction in the curing time of construction mixes, which is especially important in the conditions of a high-speed construction.

As a result of analysis of the experimental results, it is obvious that the use of gravity-forced mixing leads to the reduction in the mixing time of concrete mixtures until they reach their design strength (figure 4).

Based on the mathematical model of the mixing process for gravity-forced mixer, numerical equations of mixing process kinetics were obtained. It will allow to determine the mixing time more accurately, depending on the required coefficient of strength uniformity V_R. So, for example, for qualitative mixing of concrete mixture, i.e. achievement of variation coefficient $V_R = 10\%$, it is required not less than 100 s (figure 5).

From the obtained numerical equations of the mixing process kinetics it is clear that the uniformity coefficient of concrete strength (coefficient of variation) V_R depends mainly on the used mixer.

The considered approaches of mathematical modeling make it possible to analyze the mixing process and to control it at the design stage, and also allow solving the problems of rationalization of the working process of mixers. The experimental studies results have confirmed the correctness and adequacy of theoretical studies to improve the quality of concrete due to combined mixing in gravity-forced mixer.

Table 2. Dependencies of the coefficient of variation and the strength of the concrete mortar mixtures CMM on the mixing time.

<table>
<tr><th rowspan="2">Sample number</th><th rowspan="2">Curing time, day</th><th colspan="2">Parameters and properties of CMM</th></tr>
<tr><th>The variation coefficient change, %</th><th>Strength growth, MPa</th></tr>
<tr><td>1</td><td>28</td><td>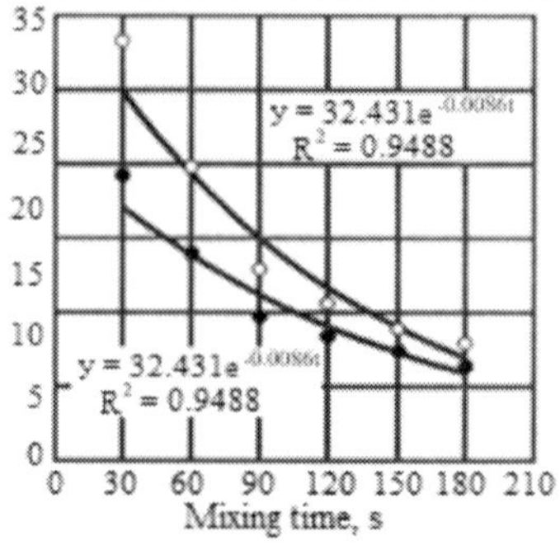
</td><td>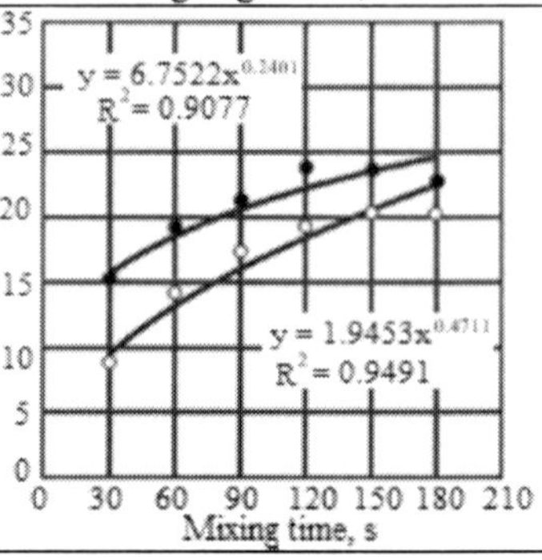
</td></tr>
<tr><td rowspan="2">2</td><td>7</td><td>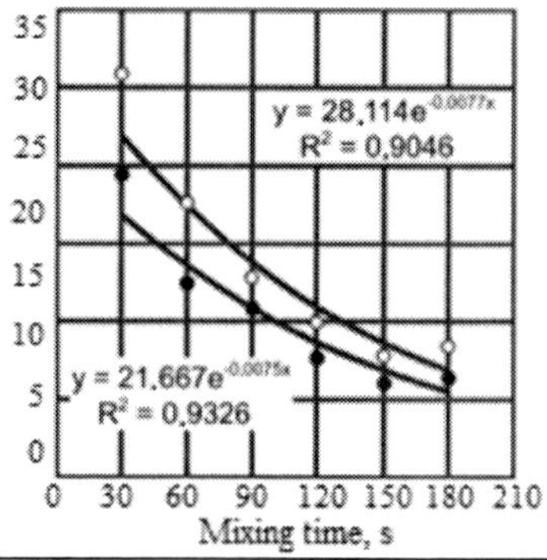
</td><td>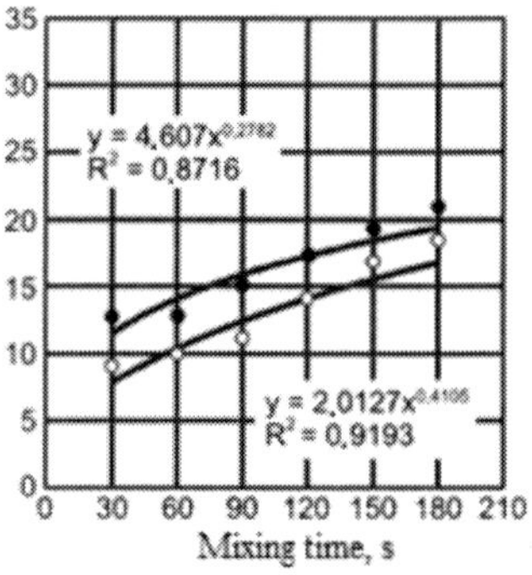
</td></tr>
<tr><td>28</td><td>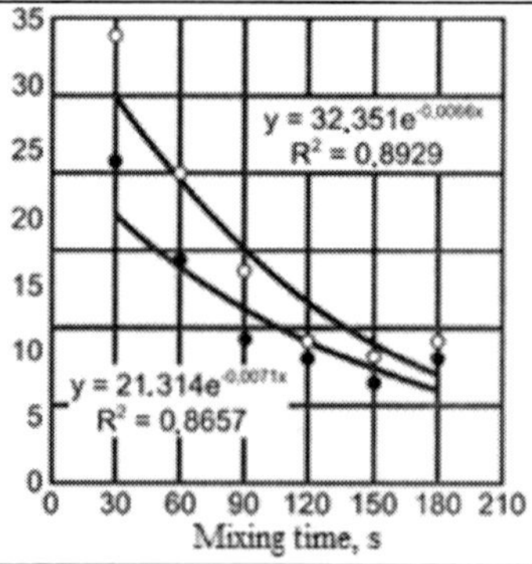
</td><td>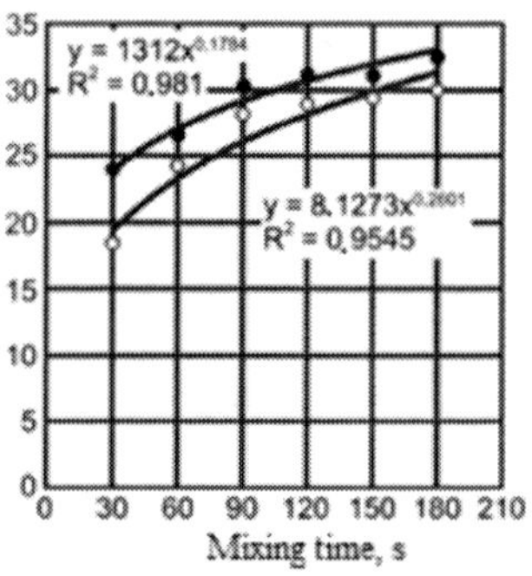
</td></tr>
<tr><td>3</td><td>7</td><td>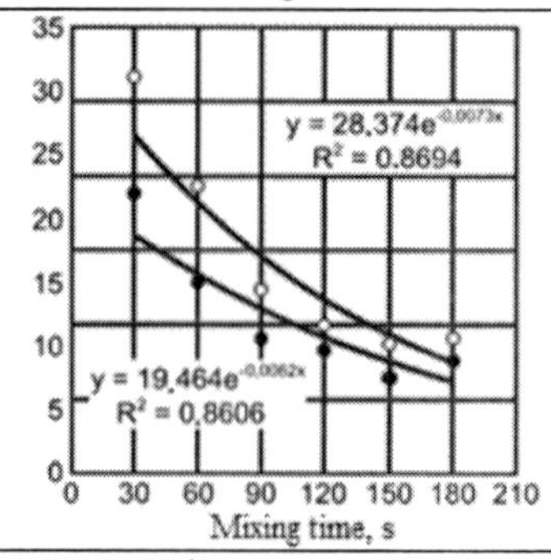
</td><td>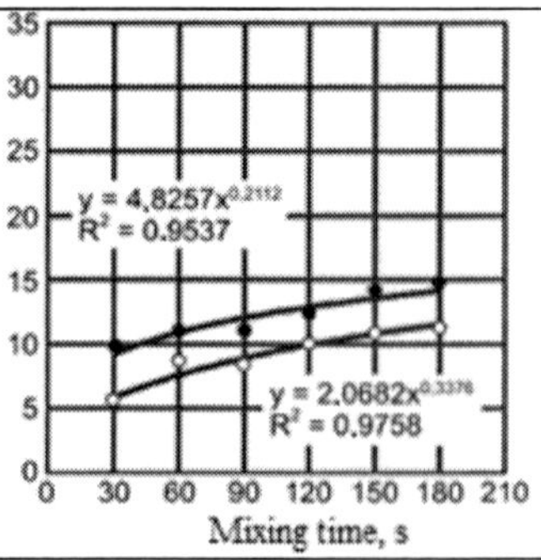
</td></tr>
<tr><td colspan="4">Note: mixture composition №1 – M200; W/C = 0.6; C:S:G = 1:2.5:3.9; №2 – M300; W/C = 0.47; C:S:G = 1:1.8:2.8; №3 – M200; W/C = 0.28; C:S:G = 1:2.6:6.4; ● – gravity-forced mixer; ○ – standard mixture</td></tr>
</table>

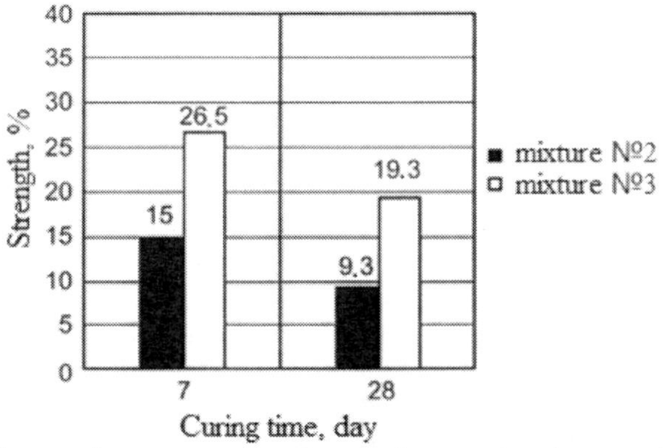

Figure 3. Dependence of concrete strength growth on curing time at mixing time 180 s.

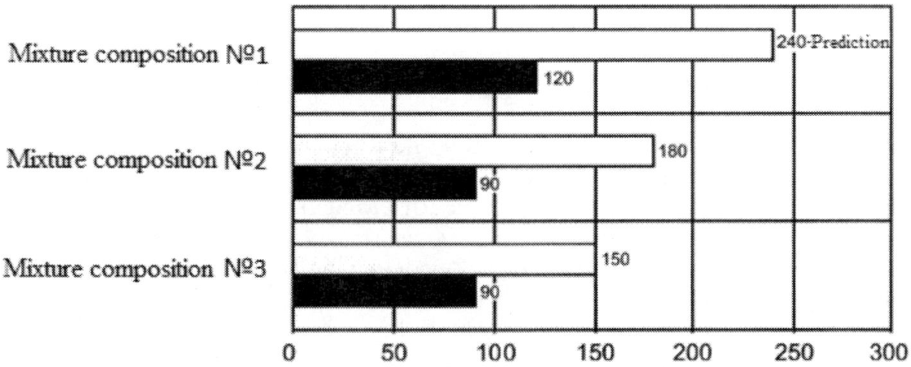

Figure 4. Mixing time until the design strength is reached.

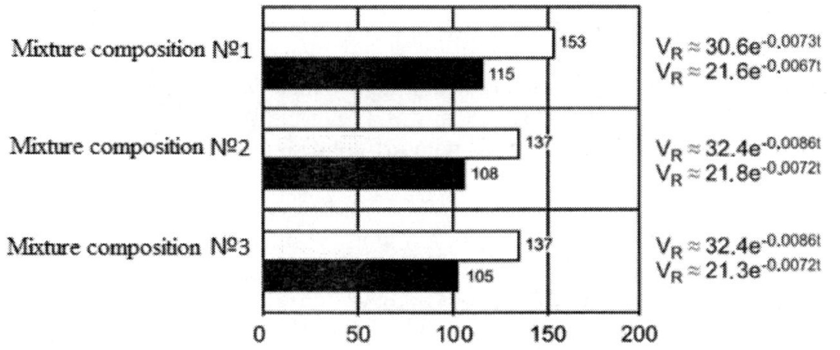

Figure 5. Mixing time of concrete mixtures until the coefficient of variation is reached $V_R = 10\%$ (equations of mixing kinetics are given on the right).

4. Conclusion

The use of gravity-forced method improves the quality of mixing, characterized by the strength and coefficient of uniformity of the hardened concrete. In this case, the strength of the fast hardening concrete increases by 5-10%, and the coefficient of the mixture uniformity – by 1.4-2.5%. The use of this mixing method leads to increased productivity, as the time of concrete mixtures preparation can be reduced by 1.5-2 times compared with serial gravity mixers.

References

[1] Bogomolov A A 2007 Modern mixing equipment for the construction mixtures production *Bulletin of Kharkov National Automobile and Highway University* **38** pp 31-35

[2] Efremov I M, Lobanov D V, Figura K N and Komarov I V 2011 Modern concrete mixing equipment in the aspect of the study of mechanical activation methods of concrete mixtures in the mixing processes intensification *Systems. Methods. Technologies* **3(11)** pp 19-27

[3] Lozovaya S Y, Sablin V S and Tkacheva O V 2013 Intensification of homogenization of construction mixtures in static mixers *Fundamental Research* **8-6** pp 1326-1331

[4] Trostin V P, Erofeev M N, Markov D V, Kravchenko I N and Bumarskov D A 2014 Interactive information system «Influence assessment of technical solutions of technological equipment on the efficiency of mobile concrete mortar complexes» (DIS «Equipment Assessment BRK 1.0») *State Registration Certificate for Computer Software Program* № 2014618238

[5] Kravchenko I N, Erofeev M N and Fedorov A O 2020 Efficiency estimation of the activators introduction in the building materials production *Collection of scientific papers of the 3rd International Scientific and Practical Conference: Resource Conservation and Ecology of Building Materials, Products and Constructions. Kursk: Southwest State University* pp 97-102

[6] Serebrennikov A A 1999 Methods of mathematical modeling of mixing processes *Problems of adaptation of machinery to rigorous conditions. Tyumen: Tyumen Industrial University* pp 214- 218

[7] Barantseva E A, Mizonov V E, Fedosov S V and Khokhlova Yu V 2008 Kinetics mathematical model of the blade mixing of bulk materials *Building Materials* **2** pp 12-13

[8] Bakin M N, Kapranova A B and Verloka I I 2014 Modern methods of mathematical description of the mixing process of bulk materials *Fundamental Researches* **5-5** pp 923-927

[9] Paramonova M S 2018 General methods of modeling the processes of the bulk materials mixing *Izvestiya Tula State University. Technical Sciences* **12** pp 582-584

[10] Zolotukhin V I, Varyash G M, Gordeev E I, Gordeeva N A and Medvedev R B 2013 Intensification of mixing processes of thixotropic materials *Izvestiya Tula State University. Technical Sciences* **5** pp 141-149

[11] Mudrov A G 2016 Methods of intensification of processes in construction technologies *Proceedings of Kazan State University of Architecture and Engineering* **2 (36)** pp 233-240

[12] Voronin V V, Adigamov K A, Petrenko S S and Sizyakin R A 2012 Criteria and methods to estimate the quality of bulk materials mixing *Engineering Journal of Don* **4-2(23)** 36

[13] Voronin V V, Adigamov K A, Cherkasov R I, Sizyakin R A and Gapon N V 2015 Mixing quality estimation of bulk materials with different size fractions *Modern problems of science and education* **2-2** p 169

[14] Verigin A N, Panferov A A, Emelyanov M V and Nezamaev N A 2015 Quality of mixing of multicomponent disperse materials *Bulletin of the Saint Petersburg State Institute of Technology (Technical University)* **31(57)** pp 75-83

[15] Emelyanova I A, Kovrevsky A P and Blazhko V V 2015 Analysis of the motion of particles of dry concrete mixtures after contact with the blade of the concrete mixer working body, working in cascade mode *Land transport-technological complexes and means. Proceedings of the International Scientific and Technical Conference. Tyumen: Tyumen State Oil and Gas University* pp 90-94

[16] Praslov V P, Trostin V P, Erofeev M N [et.al.] 2008 Patent RU 2332302, IPC B28C 5/14. Installation for preparation and laying of construction mixtures №2006126547/03 **24**

[17] Smolina I O, Smolin D O and Demin O V 2013 On the modeling process of mixing bulk materials in blade mixers *Modern High Technologies* **8-1** pp 93-94

[18] Evseev A V 2015 Mathematical model of the determinate formation of the mixture homogeneity of bulk materials *Izvestiya Tula State University. Technical Sciences* **11(1)** pp 92-100

Efficient technologies for making gypsum binders

A I Gabitov[1,*], A S Salov[1], V A Ryazanova[1], A A Timofeev[1] and V A Timofeev[1]

[1] Ufa State Petroleum Technological University, 1 Kosmonavtov St., Ufa, 450064, Russia

*E-mail: gabitov.azat@mail.ru

Abstract. A number of various proposed options and flow charts applied for the production of gypsum binders are considered herein. Large volumes of inorganic waste amount to millions of tons therefore the phosphogypsum utilization to get effective gypsum binders is an urgent issue. The production of phosphogypsum binders both in our country and abroad is indicated to go in three main directions - burnt, autoclaved and anhydrite (hard-burnt) binders.

One of main methods for phosphogypsum utilization is the production of gypsum binders characterized by a wide variety of the proposed options and flow charts being implemented. The production of phosphogypsum binders both in our country and abroad goes in three main directions, i.e. burnt, autoclaved and anhydrite (hard-burnt) binders [1-3].

The production of phosphogypsum binders is associated with certain difficulties due to chemical composition and physical and mechanical properties of the raw materials. The moisture content of the material, as well as phosphoric and sulfuric acid residues found in phosphogypsum, the soluble salts thereof and other impurities complicate the dehydration process negatively affecting the final properties of gypsum binders [3].

Analysis of the dehydration of unwashed phosphogypsum being made in a wide range of acidity thereof has shown that one of reasons for the deterioration of binding properties is unstable phase composition thereof, in particular, formation of a lot of insoluble anhydrite influenced by acidic phosphate and fluoride compounds.

The hemihydrate obtained by dehydration of "reagent grade" calcium sulfate in a solution of pure phosphoric acid is hydrating far more slowly as compared to the hemihydrate obtained by dehydration of chemically pure gypsum in distilled water. Delayed hydration and, accordingly, low strength characteristics of the binder are due to the formation of phosphate films on the crystal surface that are slowing down the hemihydrate dissolution, with the concentration of hemihydrate in the solution of the hardening mixture being reduced more than twice from 7-8 to 3 g/l [4].

The activity of impurities was also due to formation of complex slightly soluble salts, "poisoning" of active centers of the growing crystal with impurity ions, formation of impermeable films on the crystal surface, and fixation of impurities on the surface of faces during nucleation and growth of crystals [1,5].

According to P.F.Gordashevsky's studies the content of phosphoric acid in an amount of 0.1-0.3% of the mass of the binder is slightly increasing the setting time. Just like dibasic calcium phosphate has little effect on the setting time either, causing the hydration acceleration. The addition of soluble phosphates up to 0.5% even slightly increases the binder strength. Fluorine compounds have a pronounced effect on the properties of the binder producing the acidic environment of the sludge along with phosphoric and sulfuric acids. The addition of soluble fluorides and fluorosilicates NaF and Na_2SiF_6 amounting even to 0.1% is reducing the strength of the gypsum binder, while an increase in the content thereof to 0.4% is reducing almost twice the strength properties of the binder. The content of soluble fluorides also increases the normal density and decreases the compressive strength of the

Content from this work may be used under the terms of the Creative Commons Attribution 3.0 licence. Any further distribution of this work must maintain attribution to the author(s) and the title of the work, journal citation and DOI.
Published under licence by IOP Publishing Ltd

samples. The emission of fluorine gases at the heat treatment is complicating the technology, thereby resulting in the increased corrosion of the equipment due to higher acidity of the material. Organic impurities found in phosphogypsum make the resulting gypsum dark and reduce the strength thereof. Salts of sodium and potassium are likely to appear on the surface of drying products as efflorescence.

Low strength properties of the binder obtained by burning phosphogypsum are also associated with the high water demand during tempering (over 100%). According to P.F. Gordashevsky the main reason for the high water demand of the phosphogypsum binder is the large internal porosity of dehydrated gypsum and the crystalline structure thereof resulting in high content of extended needle-shaped crystals. A significant reduction in the water demand of the binder is achieved by phosphogypsum grinding (from 130 to 60%) [5,6].

There are different views on impurities affecting the hardening mechanism of phosphogypsum binder. Anyway, despite the different views the majority of scientists unanimously think that it is unlikely to obtain a binder with satisfactory properties without prior purification of phosphogypsum from impurities [7].

The maximum total amount of Na^+ and K^+ should not exceed 0.15%. The content of P_2O_5 impurity (aqueous) according to various sources should not exceed 0.1-0.5% depending on the method of phosphogypsum processing and the requirements for the binder properties.

The known methods of phosphogypsum purification from impurities to be actually used may be divided into the following types: 1 - washing of phosphogypsum with water; 2 - adding of neutralizing additives; 3 - thermal method.

Methods of the first and second groups and the combination thereof provide a high degree of purification. Additional washing of phosphogypsum with water is the most common method for extracting soluble impurities.

The most promising technologies are those where the harmful impurities of phosphogypsum are neutralized by putting them into slightly soluble or insoluble compounds.

Most of the third group methods are based on burning phosphogypsum to soluble anhydrite with further hydration thereof and repeated burning to hemihydrate. They are not widely used since the increased energy cost and scarce additives are required therefor, and also the constancy of the binder properties is not provided under the changing phosphogypsum composition.

The high moisture content and dispersion of phosphogypsum were the prerequisites for the processing thereof in an autoclave as a pulp with constant mixing. The phase transition in an autoclave is used for cleaning the binder and reducing the impurities included in the crystal lattice of the dihydrate. The main problem in the autoclave method of pulp processing is the formation of dense isometric crystals of α-hemihydrate, providing the binder with low values of normal density and high strength indicators. The hydrothermal method of dehydration of gypsum under pressure with active additives i.e. hemihydrate crystallization regulators (HCR - according to P.F. Gordashevsky) is currently more widely used [8,9]. The analysis of dependence of the crystallization character, and α-hemihydrate properties on the HCR nature and concentration, the process temperature and the acidity of the medium has been made.

A number of companies like ICI (England), Mitsubishi (Japan), "CdF Chimie" (France) [8] and others were involved in the development of the technology for producing high-quality gypsum binders. Currently, the α-hemihydrate is generally made from phosphogypsum by "Giulini Chemie GmbH" method (Germany), being applied in Germany ("Babkock BSchH", "Knapsak", etc.), France, Scotland, Ireland, Russia (Voskresensk Industrial Group "Minudobreniya") and others. The manufacturing process includes: phosphogypsum washing; making of working pulp; autoclave pulp processing; autoclaved product filtration; drying and grinding of the finished product. Carboxymethylcellulose is used as an active additive therein [10].

There are actually no fundamental differences in industrial versions of process flow for making autoclave binders from phosphogypsum. There are slight differences in the methods and equipment for washing, filtration and drying. The additives recommended for regulating the growth of hemihydrate crystals also differ [11].

VNIIstrom named after P.P. Budnikov has developed an efficient method for making a high-strength water-proof binder from phosphogypsum. Phosphogypsum is dehydrated in an autoclave with a hydraulic component (Portland cement), therewith the calcium hydroxide being released in the autoclave under hydration of the hydraulic component is neutralized by acidic impurities of phosphogypsum, i.e. the latter act as a pozzolanic additive preventing hydrosulfoaluminate destruction of the system, thereby enabling Portland cement use with no extra addition of pozzolanic additives thereto, and no prior washing of phosphogypsum [12,13].

However, the high cost of the resulting binder and the complicated technological process prevent from wide application of the autoclave method for processing phosphogypsum into a binder.

Adequate studies on the technology of processing phosphogypsum into anhydrite binders were made in the USSR more than 30 years ago. An anhydrite binder with a breaking strength up to 25 MPa (with sodium sulfate addition - 2%) was obtained after 7 days, with the phosphogypsum being burnt in 8m furnace at 600-1000°C [14].

Paper [61] indicates that S^{6+} ion may be substituted by P^{5+} ion, and a higher hydration activity and the strength of the anhydrite formed with P_2O_5 are noted. Under the results of analyzing the properties of anhydrite binders, it was concluded that the unwashed phosphogypsum may be used for making anhydrite binders, although according to other paper [15] a slight strength reduction of the binder was found with more than 0.2% of fluorine or more than 0.8% of P_2O_5.

The development of phosphogypsum processing production therewith is prevented by the need for high-temperature burning, as well as by the related disadvantages of phosphoanhydrite binders i.e. delayed setting and the need to use hardening activating additives.

The wide-scale industrial production of anhydrite binders in some countries mainly was on the basis of natural anhydrite and fluoroanhydrite i.e. hydrofluoric acid production waste, when mechanochemical activation of raw materials only is required for obtaining a finished binder [16].

The most process operations for binders making from phosphogypsum are complicated, resulting in the fact that phosphogypsum binders often cannot compete in cost with a binder obtained from natural gypsum raw materials [17, 18]. "Knauf" experts consider that phosphogypsum binders will have advantages under the following conditions:

- remote location of phosphogypsum processing enterprises from deposits of natural gypsum feedstock;
- arranging the binder production in a phosphoric acid plant thereby solving the problem of transport and discharge of wastewater, as well as agreement of the phosphogypsum processes with the phosphoric acid production;
- close location of potential consumers.

The Research and Educational Center for Innovative Technologies at the Architectural and Construction Institute of the Ufa State Petroleum Technical University is currently making active research in this area. ACI USPTU specialists make R&D support and quality control of materials and work done on the most of the construction sites in the city of Ufa and the Republic of Bashkortostan.

Figure 1. The Research and Educational Center for Innovative Technologies

Innovative laboratory-tested concrete compositions with the application of waste, by-products and end products of chemical and petrochemical industries with high performance properties modified by a number of multifunctional chemical additives and modifiers developed and patented by the University staff are being introduced. Today, the alternative binders, nano-based fillers and compositions of effective superplasticizers are being developed under significant expansion of the composition selection automation. Certificates of registration of programs enabling optimization and evaluation of the quality of the materials applied were obtained [19-20].

References

[1] Mirsayev R N, Babkov V V, Yunusova S S, Kuznetsov L K, Nedoseko I V and Gabitov A I 2004 *Phosphogypsum waste of the chemical industry in the production of walling items* (Moscow: Chemistry)

[2] Ermilova E Yu, Kamalova Z A and Rakhimov R Z 2016 The study of the composite cement effect with a complex mineral additive on the physical and mechanical and other operational properties of mortars *News of KazGASU* **1 (35)** pp 165-171

[3] Udalova E A, Gabitov A I, Shuvaeva A R, Nedoseko I V, Chernov A R and Yamilova V V 2016 Actual status and promising opportunities for application of phosphogypsum for the binder making *History and Education of Natural Science* **4** pp 55-58

[4] Ivanitsky V V, Klassen P V, Novikov A A et al. 1990 *Phosphogypsum and application thereof* (Moscow: Chemistry)

[5] Gabitov A I, Udalova E A, Salov A S, Chernova A R Pyzhjanova D V and Yamilova V V 2017 Historical aspects of production and application of hollow ceramic blocks *History of science and technology* **6** pp 58-65

[6] Ermilova E Yu, Kamalova Z A, Rakhimov R Z and Gulyaeva R I 2017 The study of the effect of complex additives based on carbonate rocks and thermally-activated polymineral clay on the composition of hydration products of composite cement stone *News of KazGASU* **1 (39)**

[7] Mirsayev R N, Akhmadulina I I, Babkov V V, Nedoseko I V, Gaitova A R and Kuzmin V V 2010 Gypsum slag compositions from industrial waste in construction technologies // Construction materials **7** pp 4-6

[8] Narkevich I P and Pechkovsky V V 1984 *Utilization and disposal of waste in the inorganic substance technology* (Moscow: Chemistry) 239

[9] Salov A S, Chernova A R and Kuzmina A Yu 2015 Problems of concrete quality at cast-in-situ construction *Current topics of technical, humanities and science Papers of the International Science and Technology Conference* pp 71-74

[10] Meshcheryakov Yu G and Fedorov S V *2007 Industrial processing of phosphogypsum* (St.-Petersburg, Stroyizdat SPb Publ) 104
[11] Rakhimov R Z 2008 Ways for reducing the cement intensity of construction products *Popular Concrete Science. St.-Petersburg - LLC "Stroybetony"* **107 (21)** pp 24-28
[12] Babkov V V, Sakhibgareev R R, Chuikin A E, Terekhov I G and Salov A S 2005 Physical and chemical aspects of cement composites hardening at late stages *Bashkir Chemical Journal* **12 (4)** pp 124-129
[13] Gaitova A R, Akhmadulina I I, Pechenkina T V, Pudovkin A N and Nedoseko I V 2014 Nanostructural aspects of hydration and hardening of gypsum and gypsum-slag compositions based on calcium sulfate dihydrate *Construction materials* **1-2** pp 46-51
[14] Ratinov V B and Stekanov D I 1984 Physical and chemical fundamentals for making high-strength cast stones *Construction materials* **11** pp 6-7
[15] Khaliullin M I, Rakhimov R Z, Sabanina Yu V, Nurieva E M, Korolev E A and Bakhtin A I 2006 The influence of aging on the physical, mechanical and structural properties of multiphase gypsum binders *News of higher educational institutions. Construction* **10** pp 25-29
[16] Khaliullin M I, Altykis M G and Rakhimov R Z 2000 Composite anhydrite water-resistant binders *Construction materials* **12** pp 34-35
[17] Anisimova S V, Korshunov A E and Emelyanov D N 2016 Properties of gypsum suspensions with water-soluble acrylic polymers *Construction materials* **7**
[18] Shelikhov N.S, Rakhimov R Z, Sagdiev R R and Stoyanov O V 2016 The influence of mineral additives on properties of the low-temperature burning hydraulic binders *Bulletin of Kazan Technological University* **5**
[19] Pudovkin A N, Gabitov A I, Salov A S and Chernova A R 2018 *Program for calculating the grain size and grain composition of sand* Certificate of state registration of the computer program No. 2018617134 / right holder FGBUU from UGNTU; claimed 24/04/2018; reg. 19.06.2018
[20] Sinitsyna E A, Pudovkin A N, Nedoseko I V and Salov A S 2020 *The program for calculating the specific surface area of the raw mixture and the average surface particle size* Certificate of state registration of the computer program No. 2020663812/ right holder FGBUU from UGNTU; claimed 26.10.2020; reg. 02.11.2020

The problem of plane bending a direct composite beam of arbitrary cross-section and the prerequisites for its approximate analytical solution

S B Koval'chuk[1,*], O V Goryk[1] and A V Antonets[1]

[1]Poltava State Agrarian Academy, 1/3 H. Skovorody Str., Poltava, 36003, Ukraine

[*]E-mail: stanislav.kovalchuk@pdaa.edu.ua

Abstract. The approach to the reduction of the spatial problem of plane bending a composite discrete-inhomogeneous beam of arbitrary cross-section to the approximate two-dimensional bending problem of the equivalent multilayer beam has been discussed here. The result is represented in the form of relations for determination the characteristics of the equivalent multilayer structure by physical and mechanical materials characteristics of the original beam's phases and the system of static, geometric and physical relations of the corresponding two-dimensional problem. The obtained equations are similar to the plane problem equations of the elasticity theory, but instead of stresses, they contain internal efforts consolidated to the main plane of the beam. The equations of the approximate two-dimensional problem were used to solve the problem of static bending a composite console of arbitrary structure with a load on the free end, taking into account the uniform change of the temperature field. The given system of equations and relations is the starting point for the construction of non-classical deformation models and solving a wide range of problems concerning the deformation of a direct composite beams.

1. Introduction

Beams of discrete-inhomogeneous, composite structure are increasingly used in various fields of mechanical engineering and construction. The research of mechanics deformation of such elements contributes to the development of methods for their design, which is a necessary condition for effective practical implementation of composites. However, the construction of theoretical models of composite beams is complicated by anisotropy of mechanical properties of their material, heterogeneity of the structure, significant susceptibility to transverse displacements, uneven thermal expansion of composite components, even with a uniform temperature field, etc.

For most traditional structural materials, the model of a linear-elastic body is acceptable when solving problems of mechanics. In [1, 2] it was experimentally established that carbon plastics, even at elevated temperatures, shows an almost linear relationship between stresses and strains before failure. Despite the significant physical nonlinearity of the reinforced concrete matrix material [3], for such composites, with some caveats, it is also permissible to assume the linear-elastic work of the components. It has become topical to develop deformation models of composite elements based on the linear theory of elasticity.

The exact solutions of spatial deformation problems of composite beams have been obtained only for simple shapes and cross-sectional structures (circle, annular multilayer structure) [4-7] and in cases of loading (tension-compression, pure bend).

A lot of works [8-35] have suggested exact solutions of plane problems of elasticity theory concerning composite beams, directly or indirectly. Such solutions preferably allow one of the types of external loads to be taken into account. There are also works devoted to bending by force and moment at the end [8-12], uniformly distributed normal load on the longitudinal faces [8, 13-21], linearly distributed load [22-25], load distributed by the law of sinusoids [26, 27]. Several works consider more complex types and combinations of loads. In works [28-31], the load is considered as the sum of

Content from this work may be used under the terms of the Creative Commons Attribution 3.0 licence. Any further distribution of this work must maintain attribution to the author(s) and the title of the work, journal citation and DOI.
Published under licence by IOP Publishing Ltd

the power series, in [32, 33, 35] it is studied as trigonometric series, and in [34] it is discussed as their combinations. Most of these works are devoted to composite homogeneous or continuously inhomogeneous beams with different types of elastic materials symmetry. Only in some works [9, 10, 12, 17, 21, 25, 35] multilayer beams are presented.

Plane solutions for composite beams are relatively simple and give precise distributed components of the stress-strain state. They can be directly used only for the calculation of beams with a rectangular cross-section. Their structure is formed by a package of flat layers perpendicular to the force plane. However, in practice, composite beams have a more complex shape and cross-sections structure.

Non-classical models of bending a composite beams, for example [36, 37, 38, 39], are also constructed as approximate plane problem solutions of the elasticity theory. In some works, while constructing such models, there is a possibility to take into account the arbitrary structure of the cross-sections. In particular, the construction of an iterative model in [38] is carried out by introducing static and kinematic hypotheses at the level of the spatial problem. Next, integrating by the width of the section, the three-dimensional equations of the elasticity theory are reduced to generalized two-dimensional ones, but without taking into account the susceptibility of beam materials to transverse compression. The combination of such an approach with methods for constructing exact solutions of plane problems of the elasticity theory for multilayer beams is quite promising. This will allow to obtain practical ratios for determining stress-strain state when bending composite beams with an arbitrary cross-sectional structure.

2. Materials, hypotheses and research methods

Consider a direct composite beam with a constant structure and cross-sectional dimensions relative to length – see figure 1. The beam consists of m discrete continuous single-connected or multi-connected phases $P_1, P_2, P_3 \ldots, P_k, \ldots, P_m$, which include the matrix, reinforcement, adhesive layers and other elements of the composite, made of appropriate different materials (figure 1, a). The phases of the beam are rigidly connected on common surfaces, there is no relative displacement and separation. The beam has a longitudinal plane of symmetry, relative to which its cross-sections are symmetrical in shape and structure.

The beam is under the action of external loads distributed on its longitudinal faces with a width of $b_{1,2}$ and the ends – see figure 1, b. Normal and tangential loads are distributed symmetrically with respect to the main plane of the beam (xOz). Longitudinal lateral cylindrical surfaces, with generators $\upsilon_{1,2}(z)$ are free from external loadings. The figure 1, b shows the loads acting only in the accepted positive directions. Despite this conditionality, we assume that the load system is balanced.

The material of the composite beam's phases can be homogeneous or continuously inhomogeneous (functionally gradient) in the plane of the cross section. It can also be isotropic or orthotropic with planes of elastic symmetry parallel to the coordinate planes of the coordinate system xyz. Physico-mechanical characteristics of the materials of the beam phases are known:

$$\left\| E_x^{[k]}, E_y^{[k]}, E_z^{[k]}, G_{xz}^{[k]}, G_{zy}^{[k]}, G_{yx}^{[k]}, \nu_{xz}^{[k]}, \nu_{zy}^{[k]}, \nu_{xy}^{[k]}, \vartheta_x^{[k]}, \vartheta_y^{[k]}, \vartheta_z^{[k]}, \rho^{[k]} \right\| = \left\| S_a^{[k]} \right\|,$$

where $S_a^{[k]} = \text{const}$ – within a homogeneous phase and $S_a^{[k]} = f(y,z)$ – in the general case of a continuous-inhomogeneous phase.

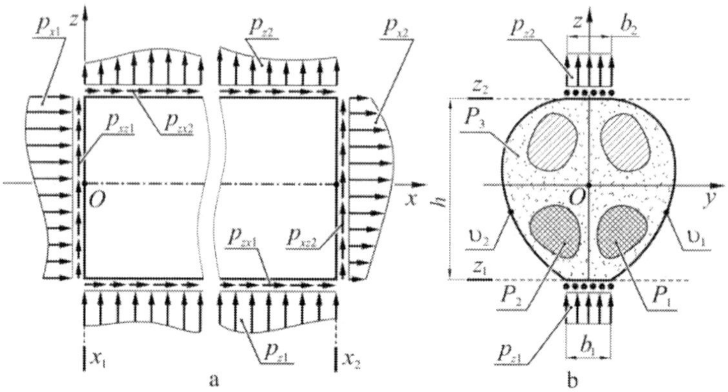

Figure 1. Scheme of a composite beam: (a) side view; (b) cross section, right view.

For the whole beam, the characteristics of the inhomogeneous material will be piecewise continuous functions $\mu_a^S = \mu_a^S(y,z)$ with discontinuities of the first kind at the boundaries of the phases, which can be conditionally represented in the following form:

$$\mu_a^S = \sum_{k=1}^{m}\left(p_k S_a^{[k]}\right), \tag{1}$$

where p_k – characteristic function of phase k :

$$p_k = p_k\left(K'(y,z)\right) = \begin{cases} 1, & K' \in P_k; \\ 0, & K' \notin P_k. \end{cases} \tag{2}$$

For the analytical description of the characteristic functions (2), the Heaviside function can be used. For example, for an arbitrary phase of a composite beam, which is limited in cross-section on the sides by continuous curves $\upsilon_{k,l-1}(z)$ and $\upsilon_{k,l}(z)$, and at the bottom and top by horizontal lines z_{k-1} and z_k, the characteristic function can be written as follows:

$$p_k = \left[H\left(y-\upsilon_{k,l-1}\right) - H\left(y-\upsilon_{k,l}\right)\right]\left[H\left(z-z_{k-1}\right) - H\left(z-z_k\right)\right]. \tag{3}$$

For a beam with an arbitrary cross-sectional structure, it is always possible to divide it in height by z_k horizontal lines into a number of generalized layers. Within these layers, the characteristic functions of the phases can be represented as (2). This gives a generalized approach to the analytical representation of the mechanical characteristics functions of an inhomogeneous beam (3).

Due to the accepted condition of absolutely rigid contact of phases, the material of the considered beam is continuous. Accordingly, the classical equilibrium equations and geometric relations of the elasticity theory are valid for such an element:

$$\frac{\partial\sigma_x}{\partial x} + \frac{\partial\tau_{zx}}{\partial z} + \frac{\partial\tau_{yx}}{\partial y} + \mu^\rho\left(F_x^V - \frac{\partial^2 u}{\partial t^2}\right) = 0, \quad \frac{\partial\tau_{xy}}{\partial x} + \frac{\partial\sigma_y}{\partial y} + \frac{\partial\tau_{zy}}{\partial z} + \mu^\rho\left(F_y^V - \frac{\partial^2 v}{\partial t^2}\right) = 0,$$
$$\frac{\partial\tau_{xz}}{\partial x} + \frac{\partial\tau_{yz}}{\partial y} + \frac{\partial\sigma_z}{\partial z} + \mu^\rho\left(F_z^V - \frac{\partial^2 w}{\partial t^2}\right) = 0, \tag{4}$$

$$\varepsilon_x = \frac{\partial u}{\partial x}, \qquad \varepsilon_z = \frac{\partial w}{\partial z}, \qquad \varepsilon_y = \frac{\partial v}{\partial y},$$
$$\gamma_{xz} = \frac{\partial w}{\partial x} + \frac{\partial u}{\partial z}, \qquad \gamma_{zy} = \frac{\partial v}{\partial z} + \frac{\partial w}{\partial y}, \qquad \gamma_{yx} = \frac{\partial v}{\partial x} + \frac{\partial u}{\partial y}, \tag{5}$$

where F_x^V, F_y^V, F_z^V – projections of bulk forces per unit mass of beam material; μ^ρ – function of density distribution of beam materials; t – time.

The physical dependences of Hooke's law also remain valid within individual phases. However, for all inhomogeneous beams, they must take into account the change in mechanical characteristics in its cross-section. This can be achieved by replacing the elastic constants in the classical equations with the distribution functions of physical and mechanical characteristics:

$$\varepsilon_x = \frac{\sigma_x}{\mu_x^E} - \frac{\mu_{yx}^v \sigma_y}{\mu_y^E} - \frac{\mu_{zx}^v \sigma_z}{\mu_z^E} + \mu_x^\vartheta \Delta T, \qquad \gamma_{xz} = \frac{1}{\mu_{xz}^G} \tau_{xz},$$
$$\varepsilon_y = \frac{\sigma_y}{\mu_y^E} - \frac{\mu_{xy}^v \sigma_x}{\mu_x^E} - \frac{\mu_{zy}^v \sigma_z}{\mu_z^E} + \mu_y^\vartheta \Delta T, \qquad \gamma_{zy} = \frac{1}{\mu_{zy}^G} \tau_{zy}, \tag{6}$$
$$\varepsilon_z = \frac{\sigma_z}{\mu_z^E} - \frac{\mu_{yz}^v \sigma_y}{\mu_y^E} - \frac{\mu_{xz}^v \sigma_x}{\mu_x^E} + \mu_z^\vartheta \Delta T, \qquad \gamma_{yx} = \frac{1}{\mu_{yx}^G} \tau_{yx},$$

where ΔT – change in temperature field inside the beam.

Given the received load (figure 1) static boundary conditions on the longitudinal faces and ends of the beam will be written as

$$\sigma_x |_{x=x_\varsigma} = (-1)^\varsigma p_{x,\varsigma}, \qquad \tau_{xz} |_{x=x_\varsigma} = (-1)^\varsigma p_{xz,\varsigma}, \qquad \tau_{xy} |_{x=x_\varsigma} = 0, \tag{7}$$

$$\tau_{zx} |_{z=z_\varsigma} = (-1)^\varsigma p_{zx,\varsigma}, \qquad \sigma_z |_{z=z_\varsigma} = (-1)^\varsigma p_{z,\varsigma}, \qquad \tau_{zy} |_{z=z_\varsigma} = 0, \ \varsigma = 1, 2. \tag{8}$$

For a symmetrical section, the side curves of its contour are distinguished only by a sign:

$$\upsilon_2(z) = -\upsilon_1(z) = \upsilon(z), \tag{9}$$

where $\upsilon(z)$ – an arbitrary piecewise continuous function that does not change the sign when $z \in (z_1, z_2)$.

Then, based on geometric considerations, we write the guide cosines of the normal to the lateral cylindrical surfaces:

$$m_y = (-1)^\varsigma \left(1 + \left(\frac{d\upsilon}{dz}\right)^2\right)^{-1/2}, \qquad n_z = -\frac{d\upsilon}{dz}\left(1 + \left(\frac{d\upsilon}{dz}\right)^2\right)^{-1/2}, \ \varsigma = 1, 2. \tag{10}$$

Given the absence of loads on the side surfaces, the static boundary conditions for them, taking into account (10), will be written as

$$\left(\tau_{yx} + (-1)^{\varsigma+1} \tau_{zx} \frac{d\upsilon}{dz}\right)\Big|_{y=(-1)^\varsigma \upsilon} = 0, \ \left(\tau_{yz} + (-1)^{\varsigma+1} \sigma_z \frac{d\upsilon}{dz}\right)\Big|_{y=(-1)^\varsigma \upsilon} = 0, \ \left(\sigma_y + (-1)^{\varsigma+1} \tau_{yz} \frac{d\upsilon}{dz}\right)\Big|_{y=(-1)^\varsigma \upsilon} = 0. \tag{11}$$

Equations (4),(5) and (6) constitute a complete system of equations of the elasticity theory spatial problem for the considered composite beam. Their analytical solution in general in accordance with the boundary conditions (7), (8), (11) encounters significant mathematical difficulties.

However, by introducing some physical assumptions about the properties of phase materials, the dimension of such a problem can be reduced, which simplifies its research and analytical solution.

3. Reduction of the spatial problem of plane bending a composite discrete-inhomogeneous beam of arbitrary cross-section to the approximate two-dimensional problem

We consider the beam narrow enough to neglect the deformation of the composite's phases along the axis Oy. This is equivalent to the assumption of absolute stiffness of the materials of all composite beam's phases in the specified direction:

$$\mu^{\nu}_{yx}, \mu^{\nu}_{xy}, \mu^{\nu}_{yz}, \mu^{\nu}_{zy}, \mu^{\vartheta}_{y} = 0, \qquad \mu^{E}_{y}, \mu^{G}_{zy}, \mu^{G}_{yx} \to \infty. \tag{12}$$

We also assume a uniform distribution of the temperature field along the axis Oy:

$$\Delta T = \Delta T(\eta, \xi, t). \tag{13}$$

Substituting (12) for the physical dependences (6), we have:

$$\begin{aligned} &\varepsilon_x = \frac{\sigma_x}{\mu^E_x} - \frac{\mu^{\nu}_{zx}\sigma_z}{\mu^E_z} + \mu^{\vartheta}_x \Delta T, \quad \varepsilon_z = \frac{\sigma_z}{\mu^E_z} - \frac{\mu^{\nu}_{xz}\sigma_x}{\mu^E_x} + \mu^{\vartheta}_z \Delta T, \quad \varepsilon_y = 0, \\ &\gamma_{xz} = \frac{1}{\mu^G_{xz}}\tau_{xz}, \qquad \gamma_{zy} = 0, \qquad \gamma_{yx} = 0. \end{aligned} \tag{14}$$

Taking into account (14), the geometric relations (5) have the form:

$$\begin{aligned} &\varepsilon_x = \frac{\partial u}{\partial x}, \qquad \varepsilon_z = \frac{\partial w}{\partial z}, \qquad 0 = \frac{\partial v}{\partial y}, \\ &\gamma_{xz} = \frac{\partial w}{\partial x} + \frac{\partial u}{\partial z}, \qquad 0 = \frac{\partial v}{\partial z} + \frac{\partial w}{\partial y}, \qquad 0 = \frac{\partial v}{\partial x} + \frac{\partial u}{\partial y}. \end{aligned} \tag{15}$$

Solving the third, fifth and sixth relations (15), taking into account the assumption of plane deformation of the beam, we obtain:

$$\begin{aligned} &v = 0, \qquad u = u(x,z,t), \qquad w = w(x,z,t), \\ &\varepsilon_x = \varepsilon_x(x,z,t), \qquad \varepsilon_z = \varepsilon_z(x,z,t), \qquad \gamma_{xz} = \gamma_{xz}(x,z,t). \end{aligned} \tag{16}$$

Solving (14) with respect to stresses and integrating the width of the section (within $y = -\upsilon(z)\ldots\upsilon(z)$), taking into account (13) and (16) we have:

$$\bar{\sigma}_x = \bar{\mu}^S_{xx}\varepsilon_x + \bar{\mu}^S_{zx}\varepsilon_z + \bar{\mu}^{\beta}_x \Delta T, \qquad \bar{\sigma}_z = \bar{\mu}^S_{zz}\varepsilon_z + \bar{\mu}^S_{xz}\varepsilon_x + \bar{\mu}^{\beta}_z \Delta T, \qquad \bar{\tau}_{xz} = \gamma_{xz}\bar{\mu}^G_{xz}, \tag{17}$$

where $\bar{\mu}^S_{xx},\ldots,\bar{\mu}^G_{xz}$ – functions of the consolidated elastic characteristics of the inhomogeneous material of the composite beam:

$$\begin{aligned} &\bar{\mu}^S_{xx} = \int_{-\upsilon}^{\upsilon}\left\{\mu^E_x\mu^E_x\left[\mu^E_x - \mu^E_z\left(\mu^{\nu}_{xz}\right)^2\right]^{-1}\right\}dy, \qquad \bar{\mu}^S_{zz} = \int_{-\upsilon}^{\upsilon}\left\{\mu^E_x\mu^E_z\left[\mu^E_x - \mu^E_z\left(\mu^{\nu}_{xz}\right)^2\right]^{-1}\right\}dy, \\ &\bar{\mu}^S_{xz} = \bar{\mu}^S_{zx} = \int_{-\upsilon}^{\upsilon}\left\{\mu^E_x\mu^E_z\mu^{\nu}_{xz}\left[\mu^E_x - \mu^E_z\left(\mu^{\nu}_{xz}\right)^2\right]^{-1}\right\}dy, \qquad \bar{\mu}^G_{xz} = \int_{-\upsilon}^{\upsilon}\mu^G_{xz}dy, \\ &\bar{\mu}^{\beta}_x = -\int_{-\upsilon}^{\upsilon}\left\{\mu^E_x\frac{\mu^E_x\mu^{\vartheta}_x + \mu^E_z\mu^{\nu}_{xz}\mu^{\vartheta}_z}{\mu^E_x - \mu^E_z\left(\mu^{\nu}_{xz}\right)^2}\right\}dy, \qquad \bar{\mu}^{\beta}_z = -\int_{-\upsilon}^{\upsilon}\left\{\mu^E_z\frac{\mu^E_x\mu^{\vartheta}_z + \mu^E_x\mu^{\nu}_{xz}\mu^{\vartheta}_x}{\mu^E_x - \mu^E_z\left(\mu^{\nu}_{xz}\right)^2}\right\}dy; \end{aligned} \tag{18}$$

$\bar{\sigma}_x,\ldots,\bar{\tau}_{zy}$ – functions of the consolidated stresses along the width of the beam's cross-section:

$$\bar{\sigma}_x = \int_{-\upsilon}^{\upsilon} \sigma_x dy,\ \bar{\sigma}_z = \int_{-\upsilon}^{\upsilon} \sigma_z dy,\ \bar{\sigma}_y = \int_{-\upsilon}^{\upsilon} \sigma_y dy,\ \bar{\tau}_{zx} = \int_{-\upsilon}^{\upsilon} \tau_{zx} dy. \tag{19}$$

In equation (18) it is taken into account that for a composite beam with isotropic and orthotropic phases: $\mu_x^E \mu_{zx}^{\nu} = \mu_z^E \mu_{xz}^{\nu}$, as a result we have $\bar{\mu}_{zx}^S = \bar{\mu}_{xz}^S$.

Solving (17) with respect to deformations and comparing with the corresponding expressions (14) we obtain the summary relations of Hooke's law for a composite beam:

$$\varepsilon_x = \frac{\bar{\sigma}_x}{\bar{\mu}_x^E} - \frac{\bar{\mu}_{xz}^{\nu}}{\bar{\mu}_x^E}\bar{\sigma}_z + \bar{\mu}_x^{\vartheta}\Delta T,\ \ \varepsilon_z = \frac{\bar{\sigma}_z}{\bar{\mu}_z^E} - \frac{\bar{\mu}_{xz}^{\nu}}{\bar{\mu}_x^E}\bar{\sigma}_x + \bar{\mu}_z^{\vartheta}\Delta T,\ \ \gamma_{xz} = \frac{1}{\bar{\mu}_{xz}^G}\bar{\tau}_{xz}. \tag{20}$$

and ratios for consolidated mechanical characteristics:

$$\begin{aligned}
&\bar{\mu}_x^E = \left(\bar{\mu}_{zz}^S\right)^{-1}\left(\bar{\mu}_{xx}^S\bar{\mu}_{zz}^S - \bar{\mu}_{xz}^S\bar{\mu}_{zx}^S\right),\quad \bar{\mu}_{xz}^{\nu} = \left(\bar{\mu}_{zz}^S\right)^{-1}\bar{\mu}_{zx}^S,\quad \bar{\mu}_x^{\vartheta} = \left(\bar{\mu}_{xx}^S\bar{\mu}_{zz}^S - \bar{\mu}_{xz}^S\bar{\mu}_{zx}^S\right)^{-1}\left(\bar{\mu}_{zx}^S\bar{\mu}_z^{\beta} - \bar{\mu}_{zz}^S\bar{\mu}_x^{\beta}\right),\\
&\bar{\mu}_z^E = \left(\bar{\mu}_{xx}^S\right)^{-1}\left(\bar{\mu}_{xx}^S\bar{\mu}_{zz}^S - \bar{\mu}_{xz}^S\bar{\mu}_{zx}^S\right),\qquad\qquad \bar{\mu}_z^{\vartheta} = \left(\bar{\mu}_{xx}^S\bar{\mu}_{zz}^S - \bar{\mu}_{xz}^S\bar{\mu}_{zx}^S\right)^{-1}\left(\bar{\mu}_{xz}^S\bar{\mu}_x^{\beta} - \bar{\mu}_{xx}^S\bar{\mu}_z^{\beta}\right).
\end{aligned} \tag{21}$$

We equate the right-hand sides of the corresponding relations (14), (20) and solve the obtained equations with respect to stresses $\sigma_x,\sigma_z,\tau_{xz}$. As a result, we obtain the dependences for the transition from the consolidated stresses $\bar{\sigma}_x,\bar{\sigma}_z,\bar{\tau}_{xz}$ to distributed across the width of the cross-section:

$$\begin{aligned}
\sigma_x &= \frac{\mu_x^E}{\mu_x^E - \mu_z^E\left(\mu_{xz}^{\nu}\right)^2}\frac{1}{\bar{\mu}_x^E\bar{\mu}_z^E}\left[\begin{aligned}&\left(\mu_x^E - \mu_z^E\mu_{xz}^{\nu}\bar{\mu}_{xz}^{\nu}\right)\bar{\mu}_z^E\bar{\sigma}_x + \left(\mu_z^E\mu_{xz}^{\nu}\bar{\mu}_x^E - \mu_x^E\bar{\mu}_{xz}^{\nu}\bar{\mu}_z^E\right)\bar{\sigma}_z + \\ &+\left[\mu_z^E\mu_{xz}^{\nu}\left(\bar{\mu}_z^{\vartheta} - \mu_z^{\vartheta}\right) + \mu_x^E\left(\bar{\mu}_x^{\vartheta} - \mu_x^{\vartheta}\right)\right]\bar{\mu}_x^E\bar{\mu}_z^E\Delta T\end{aligned}\right],\\
\sigma_z &= \frac{\mu_x^E\mu_z^E}{\mu_x^E - \mu_z^E\left(\mu_{xz}^{\nu}\right)^2}\frac{1}{\bar{\mu}_x^E\bar{\mu}_z^E}\left[\begin{aligned}&\left(\mu_{xz}^{\nu} - \bar{\mu}_{xz}^{\nu}\right)\bar{\mu}_z^E\bar{\sigma}_x + \left(\bar{\mu}_x^E - \mu_{xz}^{\nu}\bar{\mu}_z^E\bar{\mu}_{xz}^{\nu}\right)\bar{\sigma}_z + \\ &+\left[\mu_{xz}^{\nu}\left(\bar{\mu}_x^{\vartheta} - \mu_x^{\vartheta}\right) + \left(\bar{\mu}_z^{\vartheta} - \mu_z^{\vartheta}\right)\right]\bar{\mu}_x^E\bar{\mu}_z^E\Delta T\end{aligned}\right],\\
\tau_{xz} &= \frac{\mu_{xz}^G}{\bar{\mu}_{xz}^G}\bar{\tau}_{xz}.
\end{aligned} \tag{22}$$

Integrate the equilibrium equation (4) along the width of the beam's cross section (within $y = -\upsilon(z)\ldots\upsilon(z)$). Taking into account the rule of the definite integral differentiation respect to the parameter and equations (9), (16), (19), we obtain:

$$\begin{aligned}
&\frac{\partial\bar{\sigma}_x}{\partial x} + \frac{\partial\bar{\tau}_{zx}}{\partial z} + \bar{F}_x^V - \bar{\mu}^{\rho}\frac{\partial^2 u}{\partial t^2} = -\left(\tau_{yx} - \frac{d\upsilon}{dz}\tau_{zx}\right)\Big|_{y=\upsilon} + \left(\tau_{yx} + \frac{d\upsilon}{dz}\tau_{zx}\right)\Big|_{y=-\upsilon},\\
&\frac{\partial\bar{\tau}_{xy}}{\partial x} + \frac{\partial\bar{\tau}_{zy}}{\partial z} + \bar{F}_y^V - \bar{\mu}^{\rho}\frac{\partial^2 v}{\partial t^2} = -\left(\sigma_y - \frac{d\upsilon}{dz}\tau_{zy}\right)\Big|_{y=\upsilon} + \left(\sigma_y + \frac{d\upsilon}{dz}\tau_{zy}\right)\Big|_{y=-\upsilon},\\
&\frac{\partial\bar{\tau}_{zx}}{\partial x} + \frac{\partial\bar{\sigma}_z}{\partial z} + \bar{F}_z^V - \bar{\mu}^{\rho}\frac{\partial^2 w}{\partial t^2} = -\left(\tau_{yz} - \frac{d\upsilon}{dz}\sigma_z\right)\Big|_{y=\upsilon} + \left(\tau_{yz} + \frac{d\upsilon}{dz}\sigma_z\right)\Big|_{y=-\upsilon},
\end{aligned} \tag{23}$$

where $\bar{\tau}_{xy},\bar{\tau}_{zy}$ – functions of the consolidated stresses along the width of the beam's csection (similarly (22)); $\bar{F}_x^V$, $\bar{F}_y^V$, $\bar{F}_z^V$, $\bar{\mu}^{\rho}$ – summary components of bulk forces and density distribution function:

$$\bar{F}_x^V = \int_{-\upsilon}^{\upsilon} \left(\mu^\rho F_x^V\right) dy,\ \bar{F}_y^V = \int_{-\upsilon}^{\upsilon} \left(\mu^\rho F_y^V\right) dy,\ \bar{F}_z^V = \int_{-\upsilon}^{\upsilon} \left(\mu^\rho F_z^V\right) dy,\ \bar{\mu}^\rho = \int_{-\upsilon}^{\upsilon} \mu^\rho dy$$

The right-hand sides of equations (23) are similar to the corresponding boundary conditions (11), taking into account which we have:

$$\frac{\partial \bar{\sigma}_x}{\partial x} + \frac{\partial \bar{\tau}_{zx}}{\partial z} + \bar{F}_x^V - \bar{\mu}^\rho \frac{\partial^2 u}{\partial t^2} = 0, \quad \frac{\partial \bar{\tau}_{xy}}{\partial x} + \frac{\partial \bar{\tau}_{zy}}{\partial z} + \bar{F}_y^V - \bar{\mu}^\rho \frac{\partial^2 v}{\partial t^2} = 0,$$
$$\frac{\partial \bar{\tau}_{zx}}{\partial x} + \frac{\partial \bar{\sigma}_z}{\partial z} + \bar{F}_z^V - \bar{\mu}^\rho \frac{\partial^2 w}{\partial t^2} = 0. \tag{24}$$

In the case where the load on the lateral cylindrical surfaces $p_{x,\varsigma}, p_{y,\varsigma}, p_{z,\varsigma} \neq 0$, but symmetrical: $p_{x,\varsigma} = p_x^\wedge$, $p_{z,\varsigma} = p_z^\wedge$, $p_{y,\varsigma} = (-1)^\varsigma p_y^\wedge$, in the right parts of the first and third equations (23) will be the sums of projections of the corresponding external loads on the plane of the beam symmetry (xOz),and the second equation will be zero, as in the absence of loads:

$$\frac{\partial \bar{\sigma}_x}{\partial x} + \frac{\partial \bar{\tau}_{zx}}{\partial z} + \bar{F}_x^V - \bar{\mu}^\rho \frac{\partial^2 u}{\partial t^2} = -2p_x^\wedge \sqrt{1+\left(\frac{d\upsilon}{dz}\right)^2}, \quad \frac{\partial \bar{\tau}_{xy}}{\partial x} + \frac{\partial \bar{\tau}_{zy}}{\partial z} + \bar{F}_y^V - \bar{\mu}^\rho \frac{\partial^2 v}{\partial t^2} = 0,$$
$$\frac{\partial \bar{\tau}_{zx}}{\partial x} + \frac{\partial \bar{\sigma}_z}{\partial z} + \bar{F}_z^V - \bar{\mu}^\rho \frac{\partial^2 w}{\partial t^2} = -2p_z^\wedge \sqrt{1+\left(\frac{d\upsilon}{dz}\right)^2}.$$

That is, the loads on the lateral cylindrical surfaces will directly enter the system of consolidated static equations and will affect the stress distribution along the cross-section height.

Integrating the static conditions (7) and (8) over the width of the section, we write:

$$\bar{\sigma}_x\big|_{x=x_\varsigma} = (-1)^\varsigma q_{x,\varsigma}, \qquad \bar{\tau}_{xz}\big|_{x=x_\varsigma} = (-1)^\varsigma q_{xz,\varsigma}, \qquad \bar{\tau}_{xy}\big|_{x=x_\varsigma} = 0, \tag{25}$$

$$\bar{\tau}_{zx}\big|_{z=z_\varsigma} = (-1)^\varsigma q_{zx,\varsigma}, \qquad \bar{\sigma}_z\big|_{z=z_\varsigma} = (-1)^\varsigma q_{z,\varsigma}, \qquad \bar{\tau}_{zy}\big|_{z=z_\varsigma} = 0,\ \varsigma = 1,2. \tag{26}$$

The second equation (24) and the boundary conditions (25) and (26) will be satisfied if

$$\bar{\tau}_{xy} = 0, \qquad \bar{\tau}_{zy} = 0, \tag{27}$$

however, this does not mean that the stresses τ_{xy}, τ_{zy} are zero.

Thus, for a consolidated two-dimensional problem based on (15), (16), (20), (24), (27) we can write the following system of equations:

$$\frac{\partial \bar{\sigma}_x}{\partial x} + \frac{\partial \bar{\tau}_{zx}}{\partial z} + \bar{F}_x^V - \bar{\mu}^\rho \frac{\partial^2 u}{\partial t^2} = 0, \quad \frac{\partial \bar{\tau}_{zx}}{\partial x} + \frac{\partial \bar{\sigma}_z}{\partial z} + \bar{F}_z^V - \bar{\mu}^\rho \frac{\partial^2 w}{\partial t^2} = 0, \tag{28}$$

$$\varepsilon_x = \frac{\partial u}{\partial x}, \qquad \varepsilon_z = \frac{\partial w}{\partial z}, \qquad \gamma_{xz} = \frac{\partial w}{\partial x} + \frac{\partial u}{\partial z}, \tag{29}$$

$$\varepsilon_x = \frac{\bar{\sigma}_x}{\bar{\mu}_x^E} - \frac{\bar{\mu}_{xz}^V}{\bar{\mu}_x^E}\bar{\sigma}_z + \bar{\mu}_x^\vartheta \Delta T,\ \ \varepsilon_z = \frac{\bar{\sigma}_z}{\bar{\mu}_z^E} - \frac{\bar{\mu}_{xz}^V}{\bar{\mu}_x^E}\bar{\sigma}_x + \bar{\mu}_z^\vartheta \Delta T,\ \ \gamma_{xz} = \frac{1}{\bar{\mu}_{xz}^G}\bar{\tau}_{xz}. \tag{30}$$

The system of eight equations (28-30) contains 8 unknown functions: $\bar{\sigma}_x$, $\bar{\sigma}_z$, $\bar{\tau}_{zx}$ – consolidated internal efforts (dimension N/m); ε_x, ε_z, γ_{xz} – averaged linear and angular deformations; u, w –

averaged displacements. Efforts $\bar{\sigma}_x$, $\bar{\sigma}_z$, $\bar{\tau}_{zx}$, in addition to the specified system of equations, must meet the summary boundary conditions (25), (26).

In an arbitrary section of the beam, the integral conditions of equilibrium of the beam part, cut off by the cross-section with the coordinate x, must be fulfilled:

$$\int_{z_1}^{z_2} \bar{\sigma}_x dz = -\sum_{\varsigma=1}^{2} \int_0^x q_{zx,\varsigma}|_{x=\theta} \, d\theta + N_1, \qquad \int_{z_1}^{z_2} \bar{\tau}_{xz} dz = -\sum_{\varsigma=1}^{2} \int_0^{\eta} q_{z,\varsigma}|_{\eta=\theta} \, d\theta + Q_1,$$
$$\int_{z_1}^{z_2} (\bar{\sigma}_x z) dz = -\sum_{\varsigma=1}^{2} \int_0^x \left[(x-\theta) q_{z,\varsigma}|_{x=\theta} + z_\varsigma q_{zx,\varsigma}|_{x=\theta} \right] d\theta + Q_1 x + M_1, \tag{31}$$

where N_1, Q_1, M_1 – components of internal force factors from loads in the initial section:

$$N_1 = -\int_{z_1}^{z_2} q_{x1} dz, \; Q_1 = -\int_{z_1}^{z_2} q_{xz1} dz, \; M_1 = -\int_{z_1}^{z_2} (z q_{x1}) dz. \tag{32}$$

The above approach actually reduces the problem of bending a composite beam with an inhomogeneous cross-section of arbitrary structure to the problem of bending some equivalent rectangular multilayer section of unit width. In the case where the contours of the homogeneous phases of the composite beam consist of vertical and horizontal lines, the generalized layers of such an equivalent beam will have constant physical and mechanical characteristics throughout the thickness. Otherwise, the generalized layers of the equivalent beam will be constantly inhomogeneous or functionally gradient in thickness.

Let the structure of an equivalent multilayer beam consist of m generalized layers under conditions $z = z_{bd1}, z_{bd2}, z_{bd3}, \ldots, z_{bd,k}, \ldots, z_{bd,m}$. Then, at the boundaries of the layers, based on the conditions of absolutely rigid connection of the original beam's phases, which provide equality of full effort and displacement at each point of the common boundary, the following conditions must be met:

$$\bar{\tau}_{xz}^{[k]}|_{z=z_{bd,k}} = \bar{\tau}_{xz}^{[k+1]}|_{z=z_{bd,k}}, \qquad \bar{\sigma}_z^{[k]}|_{z=z_{bd,k}} = \bar{\sigma}_z^{[k+1]}|_{z=z_{bd,k}}, \tag{33}$$

$$w^{[k]}|_{z=z_{bd,k}} = w^{[k+1]}|_{z=z_{bd,k}}, \qquad u^{[k]}|_{z=z_{bd,k}} = u^{[k+1]}|_{z=z_{bd,k}}, \; k = \overline{1, m-1}. \tag{34}$$

The system of equations (28-30) is similar in form to the equations of the plane problem of the elasticity theory, but instead of stresses in static and physical dependences it contains their equivalent. Using this similarity, the solution of this system can be obtained by analogy with the known solutions of the plane problem of the elasticity theory for multilayer beams.

4. Solution of the problem of static bending a composite console of arbitrary structure with a load on the free end taking into account the uniform change of the temperature field

As an example of solving the obtained system of equations, consider the solution of the problem of the composite console's static bending (figure 2) with an arbitrary load on the initial free end ($q_{z,\varsigma}, q_{zx,\varsigma} = 0$), a uniform change in temperature field ($\Delta T = \text{const}$) and in the absence of bulk forces.

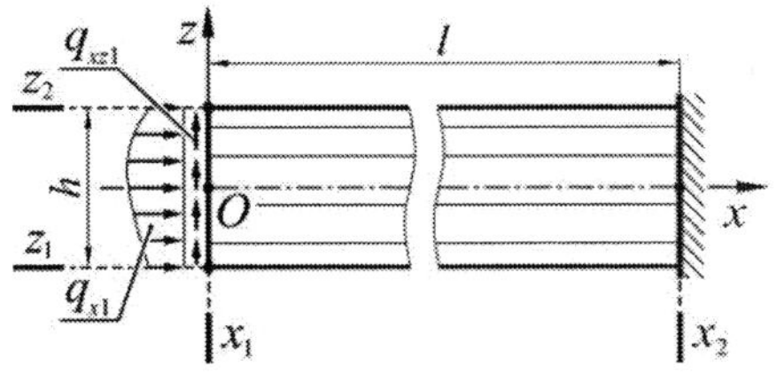

Figure 2. Scheme of a composite beam.

Similarly to [12], where a similar problem for a multilayer beam is considered, but without taking into account thermal deformations, we solve the solution of the problem with respect to the function of tangential stresses $\bar{\tau}_{xz}$. According to the second integral condition (31):

$$\bar{\tau}_{xz} = \bar{\tau}_{xz}(z). \tag{35}$$

According to the construction of the solution in [21] by successive solution of static, physical and geometric equations, we obtain

$$\bar{\sigma}_x = \bar{\mu}_x^E \left(\frac{N_1 + N^T}{\bar{B}_0} - \frac{Q_1}{\bar{B}_2} zx - \frac{M_1 + M^T}{\bar{B}_2} z - \bar{\mu}_x^\vartheta \Delta T \right), \ \sigma_z = 0, \ \bar{\tau}_{xz} = \frac{Q_1}{\bar{B}_2} \int_{z_1}^{z} \left(\bar{\mu}_x^E z \right) dz, \tag{36}$$

$$\begin{gathered} \varepsilon_x = -\frac{Q_1}{\bar{B}_2} zx - \frac{M_1 + M^T}{\bar{B}_2} z + \frac{N_1 + N^T}{\bar{B}_0}, \\ \varepsilon_z = \frac{Q_1}{\bar{B}_2} \bar{\mu}_{xz}^\nu zx + \frac{M_1 + M^T}{\bar{B}_2} \bar{\mu}_{xz}^\nu z - \frac{N_1 + N^T}{\bar{B}_0} + \left(\bar{\mu}_{xz}^\nu \bar{\mu}_x^\vartheta + \bar{\mu}_z^\vartheta \right) \Delta T, \quad \gamma_{xz} = \frac{Q_1}{\bar{B}_2} \frac{1}{\bar{\mu}_{xz}^G} \int_{z_1}^{z} \left(\bar{\mu}_x^E z \right) dz, \end{gathered} \tag{37}$$

$$\begin{aligned} u = & -\frac{Q_1}{\bar{B}_2} \left[\frac{x^2 z}{2} - \int_{z_1}^{z} \left(\frac{1}{\bar{\mu}_{xz}^G} \int_{z_1}^{z} \left(\bar{\mu}_x^E z \right) dz - \int_{z_1}^{z} \left(\bar{\mu}_{xz}^\nu z \right) dz \right) dz + \bar{D}_2 \frac{z - z_1}{h} \right] + \\ & + \frac{N_1 + N^T}{\bar{B}_0} x - \frac{M_1 + M^T}{\bar{B}_2} zx + u|_{x=0,z=z_2} \frac{z - z_1}{h} + u|_{\eta=0,z=z_1} \frac{z_2 - z}{h}, \end{aligned} \tag{38}$$

$$\begin{aligned} w = & \frac{Q_1}{\bar{B}_2} \left[\frac{x^3}{6} + x \left(\int_{z_1}^{z} \left(\bar{\mu}_{xz}^\nu z \right) dz + \frac{\bar{D}_2}{h} \right) \right] + \frac{M_1 + M^T}{\bar{B}_2} \left[\frac{x^2}{2} + \int_{z_1}^{z} \left(\mu_{xz}^\nu z \right) dz \right] - \\ & - \frac{N_1 + N^T}{\bar{B}_0} \int_{z_1}^{z} \bar{\mu}_{xz}^\nu dz + \Delta T \int_{z_1}^{z} \left(\bar{\mu}_{xz}^\nu \bar{\mu}_x^\vartheta + \bar{\mu}_z^\vartheta \right) dz + u|_{x=0,z=z_1} \frac{x}{h} - u|_{x=0,z=z_2} \frac{x}{h} + w|_{z=z_1,x=0}, \end{aligned} \tag{39}$$

where N^T, M^T – internal force and moment from thermal stresses

$$N^T = \Delta T \int_{z_1}^{z_2} \left(\bar{\mu}_x^E \bar{\mu}_x^\vartheta \right) dz, \quad M^T = \Delta T \int_{z_1}^{z_2} \left(\bar{\mu}_x^E \bar{\mu}_x^\vartheta z \right) dz; \tag{40}$$

$\bar{B}_0$, $\bar{B}_2$, $\bar{D}_2$ – integral characteristics of rigidity of a beam cross-section

$$\bar{B}_0 = \int_{z_1}^{z_2} \bar{\mu}_x^E dz, \ \bar{B}_2 = \int_{z_1}^{z_2} \int_{z_1}^{z} \left(\bar{\mu}_x^E z \right) dz dz, \ \bar{D}_2 = \int_{z_1}^{z_2} \left(\frac{1}{\bar{\mu}_{xz}^G} \int_{z_1}^{z} \left(\bar{\mu}_x^E z \right) dz - \int_{z_1}^{z} \left(\bar{\mu}_{xz}^\nu z \right) dz \right) dz; \tag{41}$$

$u|_{x=0,z=z_2}$, $u|_{\eta=0,z=z_1}$, $w|_{z=z_1,x=0}$ – unknown movements of the extreme points the initial end of the beam.

Solution (36-39) is obtained under the condition that the distance z_O' from the lower longitudinal surface to the axis of the beam (Ox) is determined by the ratio

$$z_O' = \int_0^h \left(\bar{\mu}_x'^E z \right) dz \bigg/ \int_0^h \bar{\mu}_x'^E dz, \tag{42}$$

where $\bar{\mu}_x'^E$ – distribution function of the longitudinal elastic modulus for the case when the beginning of the coordinate system is on the bottom line of the section.

It should be noted that for a narrow multilayer beam considered in [12], the solution (36-39) will be accurate, although it gives values consolidated by the width of the cross-section values of internal efforts instead of stresses. For a beam of arbitrary structure, these relations constitute an approximate solution of the problem, and do not give all the components of the stress state. However, having the distribution of the consolidated efforts (36), according to (22) we can find an approximate distribution of stresses in the output beam. Using these relations according to the spatial equations of equilibrium (4), we can obtain an approximate distribution of stresses $\tau_{\eta y}, \tau_{\xi y}, \sigma_y$ that will correspond to the physical assumption (12).

Rigid fixing of the end face of a beam with a coordinate $x = x_2$, it is possible to model conditions:

$$u\,|_{x=l,z=z_1} = 0, \;\; u\,|_{x=l,z=z_2} = 0, \;\; w\,|_{z=z_1,x=l} = 0. \tag{43}$$

Applying (38) and (39) under conditions (43) and solving the obtained equations with respect to the unknown displacements of the initial end, we obtain

$$u\,|_{\eta=0,z=z_1} = \frac{Q_1}{\bar{B}_2}\frac{l^2 z_1}{2} - \frac{N_1 + N^T}{\bar{B}_0} l + \frac{M_1 + M^T}{\bar{B}_2} l z_1, \;\; u\,|_{x=0,z=z_2} = \frac{Q_1}{\bar{B}_2}\frac{l^2 z_2}{2} - \frac{N_1 + N^T}{\bar{B}_0} l + \frac{M_1 + M^T}{\bar{B}_2} l z_2,$$
$$w\,|_{z=z_1,x=0} = \frac{Q_1}{\bar{B}_2}\left(\frac{l^3}{3} - \frac{l\bar{D}_2}{h}\right) + \frac{M_1 + M^T}{\bar{B}_2}\frac{l^2}{2}. \tag{44}$$

Substitution of expressions (44) to relations (38) and (39) gives the completed functions of distribution longitudinal and cross movements of the composite console.

The last expression (44) gives the transverse movement of the lower fiber in the initial cross-section of the console, which is close in value to its deflection arrow:

$$f \approx w\,|_{z=z_1,x=0} = \left(\frac{Q_1 l^3}{3\bar{B}_2} + \frac{M_1 l^2}{2\bar{B}_2}\right) - \frac{Q_1 l \bar{D}_2}{h\bar{B}_2} + \frac{M^T l^2}{2\bar{B}_2}.$$

Analyzing this expression, we can clearly see the components of the pure bending of the console, corresponding to the plane sections hypothesis, the component of the transverse shear deformation and compression, as well as the component of thermal stresses due to uneven expansion the phases of composite.

5. Conclusion

The paper proposes some approach to the reduction of the plane bending spatial problem of composite discrete-inhomogeneous beam of arbitrary cross-section to the approximate two-dimensional problem. This approach is based on the assumption of absolute stiffness of the phases along the width of the beam.This assumption allows us to move from the original spatial problem to the equivalent plane problem of bending a multilayer beam of unit width with continuously inhomogeneous layers. The research resulted in the necessary ratios for determining the physical and mechanical characteristics of the equivalent beam's layers according to the structure and material properties of the original beam's phases, and a system of static, geometric and physical relationships for the approximate two-dimensional problem. An example realization of the obtained equations and relations is given. Namely, the completed solution of the problem of static bending a composite console of arbitrary structure with a load on a free end surface taking into account a uniform change of temperature field is presented.

The results of solving the problem of bending the composite console can be directly used to predict the strength and rigidity of such elements. The given system of equations and relations is the starting point for the construction of non-classical deformation models and solving a wide range of problems concerning the deformation of a direct composite beams.

References

[1] Kucher N K, Zarazovskii M N and Danil'chuk E L 2013 Deformation and strength of laminated carbon-fiber-reinforced plastics under a static thermomechanical loading *Mechanics of Composite Materials* **48**(6) pp 669-80
[2] Danil'chuk E L, Kucher N K, Kushnarev A P, Potapov A M, Rudnitskii N P, Samusenko A A and Filatov V E 2015 Deformation and Strength of Unidirectional Carbon-Fiber-Reinforced Plastics at Elevated Temperatures *Strength of Materials* **47**(4) pp 573-8
[3] Neutov S, Sydorchuk M and Surianinov M 2019 Experimental Studies of Reinforced Concrete and Fiber-Reinforced Concrete Beams with Short-Term and Long-Term Loads *Mater. Sci. Forum* **968** pp 227-33
[4] Lekhnitsky S G 1977 *The theory of elasticity of an anisotropic body* (Moscow: Science) 416
[5] Soos E 1963 Sur le problème de Saint-Venant dans le cas des barres hétérogènes avec anisotropie cylindrique *Bull. Math. Soc. Sci. Math. Phys. de la R.P.R.* **55**(7) pp 61-75
[6] Muskhelishvili N I 1966 *Some basic problems of the mathematical theory of elasticity* (Moscow: Science) 708
[7] Koval'chuk S B, Gorik A V, Pavlikov A N and Antonets A V 2019 Solution to the Task of Elastic Axial Compression-Tension of the Composite Multilayered Cylindrical Beam *Strength Mater.* **51**(2) pp 240-51 doi: 10.1007/s11223-019-00070-z
[8] Lekhnitskii S G 1968 *Anisotropic Plate* (New York: Gordon and Breach) 534
[9] Gerstner R W 1968 Stresses in a Composite Cantilever *Journal of Composite Materials* **2**(4) pp 498-501
[10] Cheng S, Wei X and Jiang T 1989 Stress distribution and deformation of adhesive-bonded laminated composite beams *ASCE J. Eng. Mech* **115** pp 1150-62
[11] Zhao L, Chen W Q and Lü C F 2012 New assessment on the Saint-Venant solutions for functionally graded beams *Mechanics Research Communications* **43** pp 1-6
[12] Goryk O V and Kovalchuk S B 2018 Elasticity theory solution of the problem on plane bending of a narrow layered cantilever beam by loads at its end *Mechanics of Composite Materials* **54**(2) pp 179-190 doi:10.1007/s11029-018-9730-z
[13] Jiang A M and Ding H J 2005 The analytical solutions for orthotropic cantilever beams (I): Subjected to surface forces *Journal of Zhejiang University* A **6**(2) pp 126-31
[14] Ding H J, Huang D J and Wang H M 2006 Analytical solution for fixed-fixed anisotropic beam subjected to uniform load *Appl. Math. Mech.* **27**(10) pp 1305-10
[15] Huang D J, Ding H J and Chen W Q 2007 Analytical solution for functionally graded anisotropic cantilever beam under thermal and uniformly distributed load *Journal of Zhejiang University* A **8**(9) pp 1351-55
[16] Zhong Z and Yu T 2007 Analytical solution of a cantilever functionally graded beam *Composites Science and Technology* **67**(3-4) pp 481-88
[17] Wang M and Liu Y 2010 Analytical solution for bi-material beam with graded intermediate layer *Composite Structures* **92** pp 2358-68
[18] Daneshmehr A, Momeni S and Akhloumadi M 2012 Exact elasticity solution for the density functionally gradient beam by using airy stress function *Applied Mechanics and Materials* **110** pp 4669-76
[19] Yang Q, Zheng B, Zhang K and Li J 2014 Elastic solutions of a functionally graded cantilever beam with different modulus in tension and compression under bending loads *Applied Mathematical Modelling* **38**(4) pp 1403-16
[20] Benguediab S, Tounsi A, Abdelaziz H and Meziane M 2017 Elasticity solution for a cantilever beam with exponentially varying properties *Journal of Applied Mechanics and Technical Physics* **58**(2) pp 354-61
[21] Goryk A V and Koval'chuk S B 2018 Solution of a Transverse Plane Bending Problem of a Laminated Cantilever Beam Under the Action of a Normal Uniform Load *Strength of Materials* **50**(3) pp 406-18

[22] Esendemir U, Usal M R and Usal M 2006 The effects of shear on the deflection of simply supported composite beam loaded linearly *J. Reinf. Plast. Compos* **25** pp 835-46
[23] Huang D J, Ding H J and Chen W Q 2007 Analytical solution for functionally graded anisotropic cantilever beam subjected to linearly distributed load *Applied Mathematics and Mechanics* **28**(7) pp 855-60
[24] Daouadji T H, Henni A H, Tounsi A and El Abbes A B 2013 Elasticity Solution of a Cantilever Functionally Graded Beam *Applied Composite Materials* **20**(1) pp 1-15
[25] Koval'chuk S B and Goryk A V 2019 Solution of the problem on an elastic bending of a multilayer narrow beam by normal linear loads on longitudinal faces *International Applied Mechanics* **55**(2)
[26] Pagano N J 1969 Exact solutions for composite laminates in cylindrical bending *J. Compos. Mater.* **3** pp 398-411
[27] Sankar B V 2001 An elasticity solution for functionally graded beams *J. Composites Science and Technology* **61**(5) pp 689-96
[28] Silverman I K 1964 Orthotropic beams under polynomial loads *ASCE Journal of the Engineering Mechanics Division* **90** pp 293-319
[29] Hashin Z 1967 Plane anisotropic beams *J. Appl. Mech.* **34**(2) pp 257-62
[30] Jiang A M and Ding H J 2005 The analytical solutions for orthotropic cantilever beams (II): solutions for density functionally graded beams *Journal of Zhejiang University* A **6**(3) pp 155-8
[31] Ding H J, Huang D J and Chen W Q 2007 Elasticity solutions for plane anisotropic functionally graded beams *International Journal of Solids and Structures* **44**(1) pp 176-196
[32] Huang D J, Ding H J and Chen W Q 2009 Analytical solution and semi-analytical solution for anisotropic functionally graded beam subject to arbitrary loading // *Science in China Series G: Physics, Mechanics and Astronomy* **52**(8) pp 1244-56
[33] Nie G J, Zhong Z and Chen S 2013 Analytical solution for a functionally graded beam with arbitrary graded material properties *Composites* B **44** pp 274-82
[34] Zhang L, Gao P and Li D 2012 New Methodology to Obtain Exact Solutions of Orthotropic Plane Beam Subjected to Arbitrary Loads *Journal of Engineering Mechanics* **138**(11) pp 1348-56
[35] Koval'chuk S B 2020 Exact Solution of the Problem on Elastic Bending of the Segment of a Narrow Multilayer Beam by an Arbitrary Normal Load *Mech. Compos. Mater.* **56**(1) pp 55-74
[36] Thurnherr C, Groh R M J, Ermanni P and Weaver P M 2016 Higher-order beam model for stress predictions in curved beams made from anisotropic materials *International Journal of Solids and Structures* **97-98** pp 16–28 doi: 10.1016/j.ijsolstr.2016.08.004
[37] Sayyad A S and Ghugal Y M 2017 Bending, buckling and free vibration of laminated composite and sandwich beams: A critical review of literature *Composite Structures* **171** pp 486-504
[38] Piskunov V G, Goryk A V and Cherednikov V N 2000 Modeling of transverse shears of piecewise homogeneous composite bars using an iterative process with account of tangential loads. 1. Construction of a model *Mechanics of Composite Materials* **36**(4) pp 287-96
[39] Shvabyuk V I, Rotko S V and Uzhegova O A 2014 Bending of a Composite Beam with a Longitudinal Section *Strenght of Materials* **46**(4) pp 558-66

Influence of technological factors on the properties of building materials on the basis of expanded alunitized high-silica mountain rocks

V I Gots[1], P P Palchik[1,*], S P Palchik[1] and O Y Berdnik[1]

[1]Kyiv National University of Construction and Architecture, 31, Povitroflotsky Avenue, 03037, Kyiv, Ukraine

*E-mail: PPPalchik@ukr.net

Abstract. The influence of technological factors in the production of building materials on the basis of expanded alunitized high-silica rocks on their physical-mechanical and thermal-technical characteristics is determined. It is determined that the production of wall materials with high physical and mechanical and thermophysical characteristics, as well as the ability to obtain products with certain predetermined geometric dimensions, are determined by a set of initial indicators, both properties of raw materials and processes, as well as a certain sequence of individual elemental operations. It is shown that the production of wall materials with high physical and mechanical and thermal characteristics can be achieved by reproducing in the volume of the material of a porous system of closed type with homogeneous, evenly distributed throughout the material pores. The influence of the nature of pores and their textural characteristics of alunitized highly siliceous rocks on the performance properties of building materials has been studied. It is shown that the extension of the service life of these materials without reducing their basic physical and mechanical, thermal and thermal-technical indicators directly depends on the porosity coefficient, the values of their textural values and the nature of the pores. The influence of technological factors on the process of pore formation of alunitized highly siliceous rocks, their textural characteristics and stability of parameters during operation under the influence of aggressive environmental factors are considered. It is determined that the increase in the operating temperature of materials created in compliance with these principles is determined by the morphology and structural characteristics of the source materials by their chemical and mineralogical composition, as well as the influence of process parameters.

1. Introduction

Modern requirements of documents of state structures in the field of construction show a steady tendency to increase the requirements for thermal protection of buildings and structures in conditions of intensive exposure to aggressive environmental factors. The necessity of maintaining the declared properties is also emphasized: physical-mechanical, thermophysical, decorative -finishing, acoustic and other characteristics of materials during all term of operation under the condition of complex influence of external aggressive factors. These trends necessitated research, both scientific and chemical - technological in nature to create new types of materials for use in the external structures of buildings and structures. These materials should have low average density and at the same time increased physical and mechanical characteristics, and also to provide invariability of the properties during all term of operation at simultaneous influence of aggressive factors of external environment (corrosion, thermal, mechanical, atmospheric factors, etc.). One of such materials is construction products based on swollen alunitized high-silica rocks. The starting materials for their production are non-scarce, quite widely and evenly distributed in the world rocks - alunite ores, alunitized rocks and their structural analogues (jarosite and others). In addition to high technical performance should take into account their lower cost, compared with other types of structural thermal insulation materials, the

Content from this work may be used under the terms of the Creative Commons Attribution 3.0 licence. Any further distribution of this work must maintain attribution to the author(s) and the title of the work, journal citation and DOI.
Published under licence by IOP Publishing Ltd

environmental friendliness of both the materials themselves and the process of their production. The main active ingredient, which determines the technological,

Alunite is a natural mineral that has the general formula Am (XO_4)pZg at m;p=2;1. Chemical formula of alunite - $KSO_4 \cdot Al_2(SO_4)_3 \cdot 2Al_2O_3 \cdot 6H_2O$. Alunite belongs to the group of basic sulfates. All its structural analogues are characterized by the phenomenon of isomorphic substitution by anions and cations, as well as the formation of solid solutions. But the most widespread in nature are structural analogues of alunite - jarosite (sodium form of alunite and mixed K-Na form of alunite (1,2). The hardness of alunite is 4.5…5, density 2.6…2.82 (pure varieties), alunitized rocks - 2.39…2.89 t/m^3. At room temperature, alunite is almost insoluble in water, low-concentration acids, ammonia, but easily soluble in alkalis, and hydrofluoric acid. At 1000^{0}C it acquires limited solubility in hydrochloric and nitric acids, at 2000^{0}C it is completely soluble in concentrated hydrochloric acid (1).

Alunite is a fairly widespread mineral in the world, which forms strong and extended layers of alunite and alunite-kaolin rocks. Alunite deposits and their structural analogues are located in Ukraine, USA, China, Italy and many other countries. The content of alunite deposits is usually in the range of 10…70%. Typical impurities of alunite ores are: quartz, kaolin, dikit, pyrophyllite, opal, chalcedony, iron oxides and others. When heated, alunite consistently undergoes a number of structural transformations that make significant changes in its physical-mechanical, chemical and textural characteristics. According to the results of X-ray structural and differential thermal analyzes (Figure 1…3) the first irreversible structural transformations in the crystal lattice of alunite begin at temperatures of 350^{0}C and up to 475…480^{0}C the disordered state of matter remains. The dehydration region is in the range of 490… 600^{0}C. Increasing the temperature to 750^{0}C leads to the formation of $KSO_4 \cdot Al_2(SO_4)_3$. Further increase in temperature leads to the formation of potassium sulfate and alumina. The results of X-ray diffraction analysis also confirm the process of destruction of the crystal lattice and the formation of the amorphous phase, which is confirmed by the increase in the value of the continuous background.

2. Problem status overview

The aim of the research was to determine the parameters and sequence of elemental operations in the technological process of manufacturing porous building materials with predetermined geometric dimensions, as well as to determine the influence of technological parameters on the formation of pore structure of alunitized high-silica rocks. A study of the impact on the process of pore formation of the chemical and mineralogical composition of raw materials and their relationship with the parameters of the technological process. The formation of the cellular structure of building materials on the basis of alunitized high-silica rocks is due to the process of chemical interaction of alunite with the accompanying silica components of the source materials under thermal loading. Changing the chemical composition of alunite as the main mineral of this process.

Table 1 Chemical composition of potassium alunite under thermal loading.

№	Burning temperature, ^{0}C	Content,%			
		K_2O	Al_2O_3	SO_3	H_2O
1	95	10.16	28.01	29.14	26.08
2	200	11.80	28.64	45.41	2.19
3	400	10.60	36.98	38.81	13.53
4	800	17.80	63.90	16.96	0.00
5	1000	17.80	64.12	16.91	0.00
6	1200	18.12	64.90	16.04	0.00
7	1400	18.40	63.68	15.45	0.00

This process is the main in the multi-stage process of obtaining the cellular structure of the studied materials with predetermined textural characteristics. The formation of aluminosilicate melt in the material, under thermal load on the material samples and its properties (viscosity, surface tension forces and other characteristics) allow to regulate the formation of the pore system and, most importantly, to ensure the formation of its predetermined textural characteristics. This makes it possible to use in the production of cellular building materials rocks with a low content of directly alunite, which provides an expansion of the raw material base due to substandard ores and leads to a sharp reduction in production costs. Obtaining building materials with specified properties is possible by choosing the optimal ratio, in terms of the problem, alunite and siliceous component of raw materials at which the maximum intensity of gas evolution will occur at the time of formation of aluminosilicate melt with design characteristics. It should also be noted that the technological parameters of this main process are significantly influenced by minor impurities that are present in the raw materials, namely - clay minerals, impurities of carbonates and iron oxides.

3. Results

As a result of temperature loading (heating of initial materials, depending on the set purpose, to temperature 1000... 1400^{0}C) fine-porous material with vitreous walls of which pores are within 5...8 mm, average density 1500...1700 kg/m^3 turns out. The presence in alunitized rocks of foreign impurities in the form of silica inclusions, clay, carbonate materials and inclusions of iron, titanium and other chemical elements makes significant changes in the main process of pore formation. In this case, it is possible to obtain fine-grained material with a low average density in the range of 500...700 kg/m^3. This allows the use of low-quality alunite ores for the production of fine-grained and cellular building materials in the form of backfill insulation. In this case, the use of expanded alunitized rocks has undeniable advantages over traditional expanded clay with increased strength, fire resistance, resistance to aggressive environments and durability. Both highly enriched samples of alunite ores and natural samples of alunite-containing rocks of various deposits located both on the territory of Ukraine and other countries were used for research. The results of chemical and mineralogical analyzes of the studied samples are given in table 2.3.

Table 2 The content of basic oxides of alunite ores and alunitized rocks.

№	Material name	SiO_2	AL_2O_3	SO_3	Fe_2O_3
1	Alunitized rock №1	86.31	4.92	3.76	0.98
2	Alunitized rock №2	68.16	11.08	11.34	2.50
3	Substandard alunite ore	80.2	18.26	7.8	0.6
4	Alunite ore	31.93	28.58	28.5	9,0,7

Table 3 Mineral composition of alunite ores and alunitized rocks.

№ of sample	Name of the material	Alunite,%	Quartz,%	Oxidation,%	Carbo-Nati,%	Clay materials,%
1	Alunitized breed №1	8.7	76.1	0.9	0.7	9.6
2	Alunitized breed №2	28.9	54.8	1.4	2.9	11.2
3	Substandard alunite	36.1	42.6	3.6	2.4	12.7
4	Alunite ore	47.4	32.4	6.2	5.1	2.4

The thermal load according to the set mode occurred in the temperature range 1250...1500^{0}C. As a result of determination of physical and mechanical characteristics of the samples received after firing the optimum ratio of the basic minerals of initial materials at which the most significant effect is reached is defined. The interval of concentrations of components of alunitized rocks outside which leads to change of dynamics of gas formation and viscosity of aluminosilicate melt and, as a result or decrease in coefficient of swelling of material, or even leads to process of shrinkage is established. It was found that samples with an alunite content within 16... 38% provide a stable cellular structure at a temperature load above 1280^{0}C. In the process of thermal loading SO_3 is removed, the average density decreases. The process of formation of the cellular structure is also influenced by the presence of other impurities. The results of determining the physical and mechanical characteristics of the swollen samples after thermal loading are given in table 4.

Table 4 Physical-mechanical properties of swollen samples of alunite - containing high-silica rocks.

№	Name of the material	Burning temperature, ^0C	Average density, kg / m^3	Water absorption, %	Clamping strength, MPa
1	Alunitized breed №1	1290	840	15.8	14.1
2	Alunitized breed №2	1340	560	16.2	11.6
3	Substandard alunite	1290	510	17.9	10.8
4	Alunite ore	1360	410	19.4	8.3
5	Expanded clay (control sample)	1290	640	27.4	4.6

Heat resistance of materials based on swollen alunitized rocks is high, which is determined by the significant temperature of its formation. On the other hand, high temperature is a significant negative factor in the technological process, which leads to a decrease in technical and economic indicators of production and the deterioration of the competitiveness of the material. Therefore, the identification of ways to reduce the temperature of the thermal load is a priority way to improve the technology of obtaining materials of cellular structure on the basis of alunitized high-silica rocks.

As can be seen from the above indicators, the main factor that determines the formation of the cellular structure is an aluminosilicate melt capable of holding gaseous products of thermal destruction of alunite. Thus, the process of regulating the parameters of the technological process is possible by influencing the temperature of formation of aluminosilicate melt, its viscosity, physico - chemical properties, as well as the process of thermal destruction of alunite. The decrease in the temperature of thermal destruction of alunite can be associated with a decrease in the temperature of aluminosilicate melt as a result of chemical interaction of alunite with silica on the one hand and with a decrease in the decomposition temperature of its sulfate-containing components on the other hand. These processes are significantly influenced by impurities that are part of alunitized rocks and, primarily the influence of various forms of silica, as well as carbonate components. Influence of silica content. A significant influence on the course of physico - chemical reactions on the formation of the cellular structure of the material is the presence of silica in the raw material. Different forms of silica have different effects on the textural characteristics of the pore system and on the parameters of the technological process, its total duration and cost-effectiveness of the obtained building materials. In the presence of silica in amorphous or submicrocrystalline form, the process of thermal destruction of alunite in all allotropic formations is significantly accelerated in comparison with its crystalline modifications, such as quartz. The shape of the silica also affects the temperature of formation of aluminosilicate solution during thermal loading.

Accordingly, move to the zone of lower temperatures and all subsequent processes that determine the formation of the specified textural characteristics of the cellular structure of the material. A characteristic feature of such transformations is the intensification of the formation of closed-type pores. As the temperature increases, this process accelerates to maximum values at temperatures of 1280…1320 ^0C.

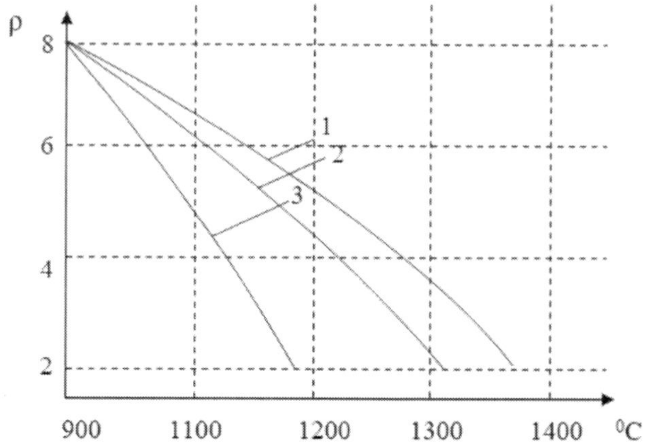

Figure 1. Influence of silica forms on the temperature of aluminosilicate melt formation: 1- Alunitized rocks (crystalline silica); 2- Alunitized rocks (submicrocrystalline silica); 3- Alunite ores (amorphous silica).

3.1 Influence of carbonate inclusions.

It is known that under the influence of oxides and salts of alkali and alkaline earth metals intensify the processes of sintering, crystallization, as well as the formation of a porous structure. According to a number of studies (5,6), the effect of lowering the temperature of the formation of the liquid phase is determined in the presence of carbonate impurities in the raw materials. According to research, the introduction of calcium oxides in the charge in the amount of 8...19% significantly reduces the temperature of formation of the liquid phase, as well as shifts towards low temperatures the formation of closed-type pores and promotes their uniform distribution throughout the material. Therefore, the addition of carbonate components to the source materials in obtaining a cellular structure is technologically feasible and cost-effective.

4. Conclusions

According to the obtained results of determination of physical and mechanical characteristics of cellular structure of building materials on the basis of alunitized high-silica rocks and their comparison with similar characteristics of other porous aggregates (expanded clay gravel) it is possible to draw conclusions about certain advantages of the first. This gives grounds for determining the scope of use of the obtained materials as light porous building materials in the construction of buildings and structures.

References

[1] Iler R K 1956 *The Colloidal chemistry of silica and silicates* (Moscow: Gosstroyizdat) 324
[2] Mitsyuk G M 1971 *Behavior of Silica in Hydrothermal Conditions* Scientific Opinion
[3] Mitsyuk G M 1972 *Geochemistry*
[4] Kolesnikov E A 1974 Swelling of low-melting clays *Glass and ceramics* **5** pp 28- 30
[5] Labutin G V 1965 Alunites ed. *Metallurgy*
[6] Priev E R 1975 *Physical-chemical processes of formation of cellular structures in alunite - silica compositions and crucible clays* Abstract Cand. Diss.
[7] Kashkay M A 1970 *Alunites, their genesis and the use* V.1...2 "Nedra"

Calculation of coast-protecting structure from wave influence

A Grishin

Odessa National Maritime University, Odessa, Ukraine

E-mail: a619grin@gmail.com

Abstract. A coupled system consisting of a shore protection structure, a soil base and an aquatic environment is considered. The influence of the aquatic environment on the oscillations of the structure was taken into account. The wave processes arising in the system from the action of surfing and breaking waves were investigated. The developed methodology and software package allow the calculation of various structures interacting with a deformable soil environment from the action of static and dynamic loads, taking into account the real properties of their materials, such as elasticity, viscosity and plasticity. The theory of plastic flow was used. Discretization of the computational area was performed by the finite element method. Studies were carried out on the basis of the obtained schemes of plastic zones, diagrams of stresses and oscillations at various points of the structure.

1. Introduction

When calculating shore protection structures in the normative document [1], it is recommended that dynamic wave actions be replaced by a static load, which is to some extent justified for standing waves. But in the case of breaking and breaking waves, under the influence of which the structures in question are usually located, it is necessary to use dynamic calculation methods. The coastal protection structure together with the soil mass forms a connected system, which is an inertial elastoplastic model that takes into account the wave processes arising in it. The theory of plastic flow with hardening is used, based on the Mises maximum principle [2]. For a flexible or rigid structure, the Geniev condition [3] is applied as a function of loading, and the Coulomb-Mohr condition [4] for soil.

2. Formulation of the problem

The technique for solving elastoplastic problems under dynamic loading in the above formulation is described in [5], therefore, here we present only the basic equations.

The equation of motion of the system at a time is found from the ratio of the principle of virtual work, which has the following form

$$\int_{\Omega}[\delta\varepsilon]^T \sigma d\Omega + \int_{\Omega}[\delta u]^T(\rho\ddot{u} + c\dot{u} - Q)d\Omega - \int_{S_q}[\delta u]^T q ds = 0. \quad (1)$$

It is believed that deformations of the system occur at small elongations, shifts, and rotation angles. Therefore, the relationship between the increments of displacements and deformations is determined by linear Cauchy relations

$$d\varepsilon_{ks} = \frac{1}{2}\left(du_{k,s} + du_{s,k}\right). \quad (2)$$

In this case, the postulate of summing the increments of elastic and plastic deformation also holds.

$$d\varepsilon_{ks} = d\varepsilon_{ks}^{(e)} + d\varepsilon_{ks}^{(p)}. \quad (3)$$

The components of the tensor of increments of elastic strain are related to the components of the tensor of increments of stress Hooke's law

Content from this work may be used under the terms of the Creative Commons Attribution 3.0 licence. Any further distribution of this work must maintain attribution to the author(s) and the title of the work, journal citation and DOI.
Published under licence by IOP Publishing Ltd

$$d\varepsilon_{ks}^{(e)} = C_{ksmn}^{(e)} d\sigma_{mn} . \tag{4}$$

The increments of plastic deformations are defined as

$$d\varepsilon_{ks}^{(p)} = d\lambda f_{\sigma_{ks}} , \; d\lambda = const > 0 . \tag{5}$$

The Genius condition in invariant form can be written as follows

$$f = 3\left(\sigma_c - \sigma_p\right)\sigma_0 + 3\sigma_i^2 - \sigma_p\sigma_c = 0 . \tag{6}$$

The Coulomb-Moore condition is written as

$$f = \left(\sigma_0 - \frac{\sigma_i}{\sqrt{3}} \sin\psi\right)\sin\varphi + \sigma_i \cos\psi - c\cos\varphi = 0 . \tag{7}$$

where, σ_0, σ_i, ψ – are invariants of the stress tensor; σ_p, σ_c – plasticity limits under tension and compression; φ, c – angle of internal friction and adhesion.

Derivatives of the loading function are given in [5]. Transforming equations (3) ÷ (7), we find the equations of state:

$$\begin{aligned} d\sigma_{ks} &= D_{ksmn}^{(e)} d\varepsilon_{mn}, \; at \;\; f = 0, \; d'f \le 0 \;\; or \;\; f < 0; \\ d\sigma_{ks} &= D_{ksmn}^{(ep)} d\varepsilon_{mn}, \; at \;\; f = 0, \; d'f > 0. \end{aligned} \tag{8}$$

The above initial equations have a complex form and can only be solved by numerical methods. This work uses direct step-by-step procedures. For their implementation, it is necessary to perform two main stages: discretization of the initial equations both in time and in the area occupied by the system; building an iterative process to determine a solution with a predetermined accuracy.

For the first discretization, the given time interval [0, T] on which the solution is determined is divided into N time intervals of length Δt and the source equations are considered at each discrete time t_n instant. For the second discretization, the region of the system is divided by computers according to a special program into finite ones, and along infinite boundary of the soil massif into infinite isoparametric elements. If the array is limited to a finite region, then the waves reflected from its boundary can distort the results of the solution. After discretization at the moment of time t_n the following matrix equation is obtained

$$M\ddot{\delta}_n + C\dot{\delta}_n + K(\delta)\delta_n = Q_n . \tag{9}$$

Here: M – matrix of distributed masses of the system; C – damping matrix; $K(\delta)$ – non-linear stiffness matrix; Q_n – vector of nodal loads; $\delta_n, \dot{\delta}_n, \ddot{\delta}_n$ – respectively, nodal displacements, speeds and accelerations.

An iterative process is constructed to determine the solution, which is implemented using the implicit modified Newmark method. It is unconditionally stable, which makes it possible to increase the length of the time step *Δt*. The modification of the method is caused by the nonlinearity of elastoplastic problems and is described in detail in [5].

3. The main material and results

Based on the proposed methodology, a software package was developed in the Delphi system that allows numerically implementing various dynamic tasks. Usually, before a dynamic load is applied to a system, a static load acts on it, from which plastic zones form in some areas. Therefore, it is not possible, as is done in elastic calculation, to separately perform static and dynamic calculations and summarize the results, since the principle of superposition is not applicable here. It is necessary to perform a joint calculation of the system from the action of both static and dynamic loads. The software package allows such a calculation.

Consider the solution to the problem, the design scheme of which is shown in Fig. 1. The shore protection wall of the corner type is under the action of breaking waves.

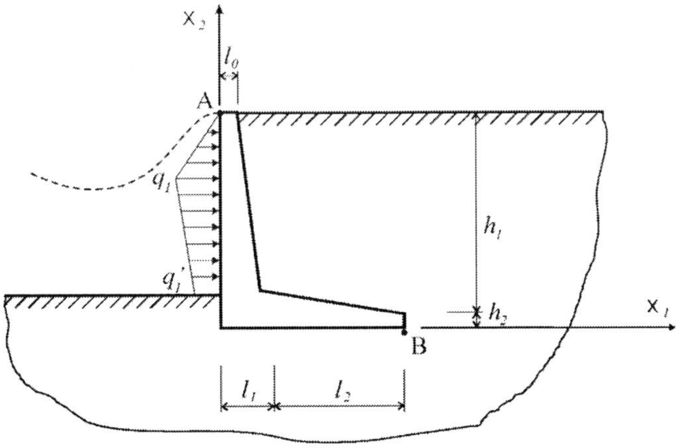

Figure 1. The design scheme of the system.

The following geometric dimensions of the structure were used in the calculation: $h_1 = 9.5$ m; $h_2 = 0.5$ m; $l_1 = 1.0$ m; $l_2 = 4.0$ m. Wave load: $q_1 = 0.075$ MPa; $q_1' = 0.017$ MPa. Characteristics of concrete walls: $E_c = 300000$ MPa; $\mu = 0.2$; $\sigma_p = 1.8$ MPa; $\sigma_c = 18$ MPa. Soil properties: $E_0 = 40$ MPa; $\mu = 0.3$; $c = 0.01$ MPa; $\varphi = 26°$. The length of the time step $\Delta t = 0.001$ sec., The duration of the load 0.1 sec. The solution is determined on the time interval $T = 3$ sec.

First, a static calculation is performed on the effect of the dead weight of the wall and backfill. From this load point A, shown in Fig. 1, receives displacements $u_1 = -1.242$ cm, $u_2 = -4.456$ cm. Plastic zones form in the soil mass. Then, taking into account the stress-strain state determined from the action of its own weight, a dynamic calculation was performed from the action of the wave load. In Fig. 2 shows the plastic zones that formed in the soil mass from both loads at time t = 0.48 sec. On the display screen during the oscillation of the system, it is possible to observe the formation of new plastic zones and the closure of those that arose earlier, and the plastic regions that formed after the static calculation go into the elastic stage. Consequently, the system is under complex loading, therefore, deformation plasticity theories cannot be applied to such problems.

In Fig. 3 shows a plot of the time variation of stress σ_{22} at point B shown in Fig. 1. At this point, from the action of a static load, a voltage of –0.072 MPa arises. On the time interval $T = [0, 3]$ sec. loading occurs, which lasts 0.1 seconds. and free vibrations taking up the rest of the time. During loading, compressive stresses σ_{22} vary to –0.078 MPa. Then, at the stage of free oscillations, unstable voltage fluctuations arise for 0.45 seconds, which then stabilize and decay.

In Fig. 4 and Fig. 5 shows the diagrams of the horizontal and vertical oscillations of point A. After the action of the static load, the horizontal movements occur in a sharp jump in the direction of the wave load and at $t = 0.16$ sec. reach a value of 0.979 cm. Then this point moves in the opposite direction and then it oscillates around a position equal to –1.5 cm. It should be noted that such oscillations do not occur near a static equilibrium position of –1.242 cm, but this position has shifted, caused by the formation of plastic deformations. The same picture is observed for vertical vibrations. In the initial period of free oscillations, point A moves up to $u_2 = -4.1$ cm, then its oscillations occur near a position equal to $u_2 = -4.54$ cm, which is not equal to the static $u_2 = -4.456$ cm.

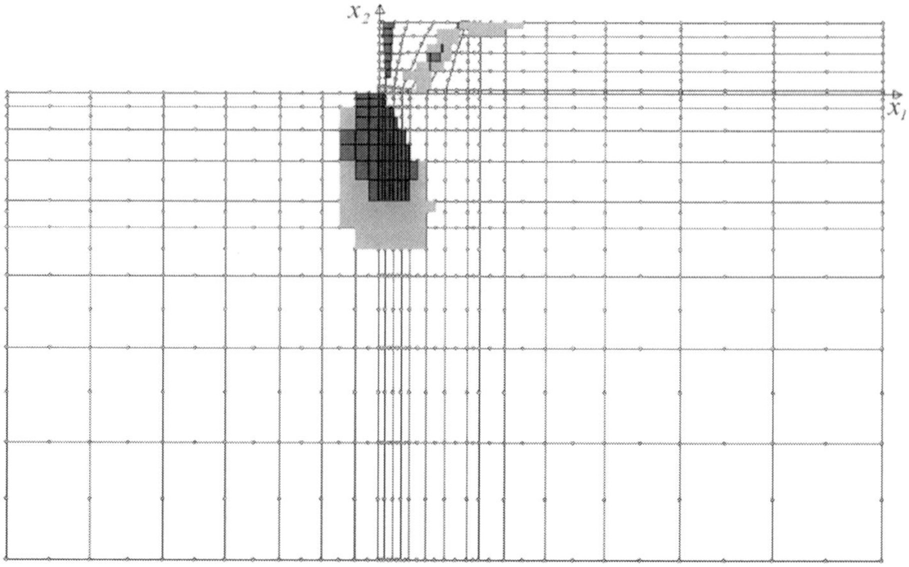

Figure 2. The formation of plastic zones in the system at time $t = 0.48$ sec.

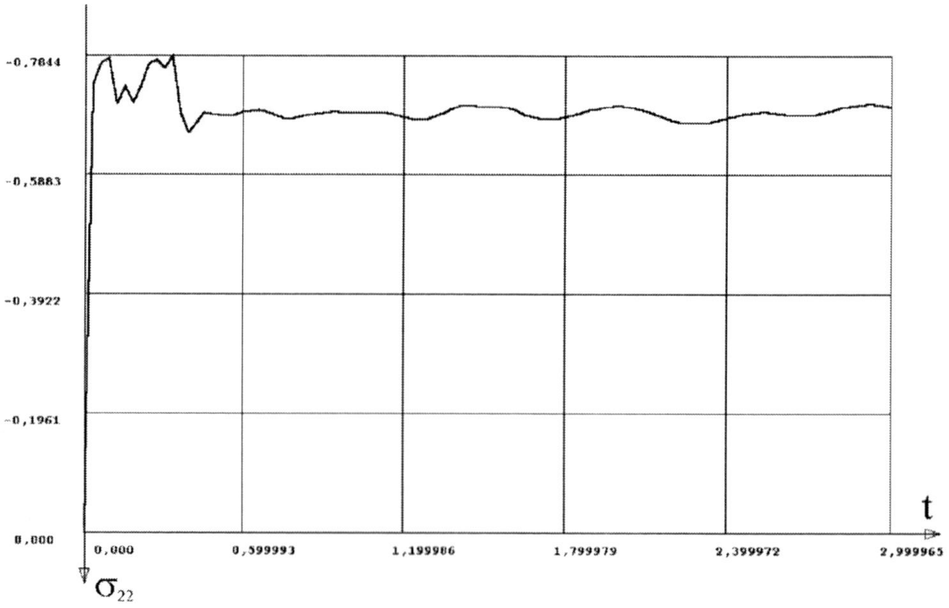

Figure 3. Diagrams of stress σ_{22} at point B (at 10 MPa).

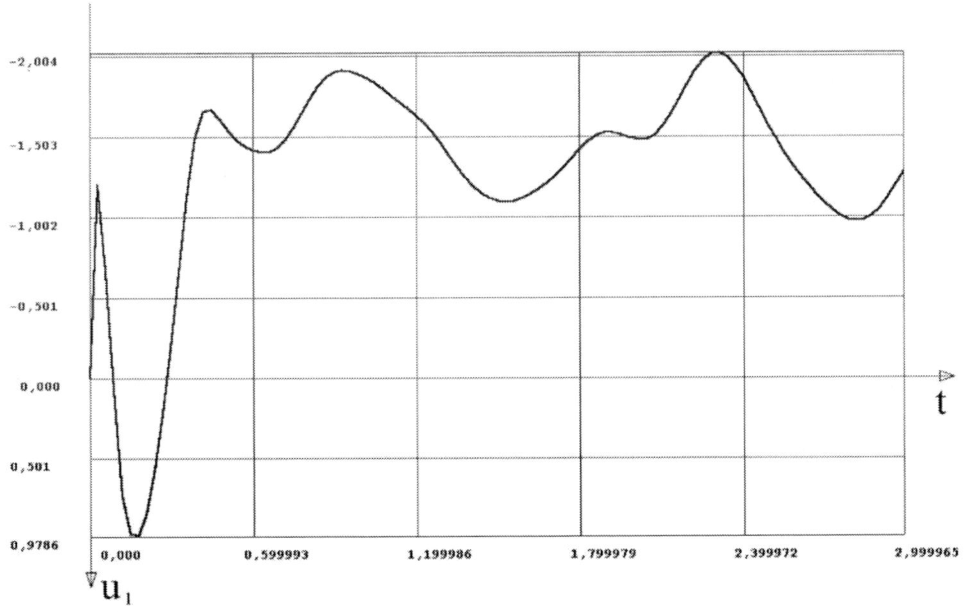

Figure 4. Diagrams of horizontal oscillations of point *A* (in cm).

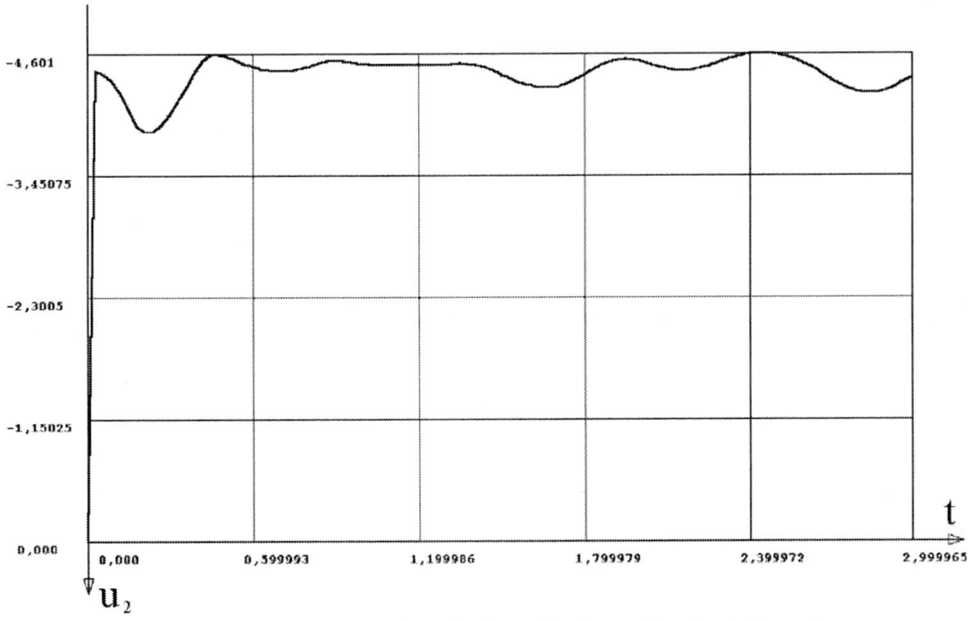

Figure 5. Diagrams of vertical oscillations of point *A* (in cm).

4. Findings

In conclusion, it should be noted that the proposed methodology and software package make it possible to calculate shore protection structures from the combined action of static and dynamic loads, taking into account the occurrence of plastic zones both in the structure and in the soil mass. When solving dynamic problems, it is necessary to take into account the deformations and stresses obtained from static loads, since they affect the oscillatory process that occurs when the system is dynamically loaded. In addition, the wall and soil environment are interconnected. The change in time of the properties of one element of the system affects the stress-strain state of another. Therefore, only their joint calculation will allow to obtain the correct results.

References

[1] SNiP 2.06.04-82*1986 *Loads and impacts on hydraulic structures (wave, ice and from ships)* (Moscow: TsITP Gosstroy SSSR) 40
[2] Ivlev D and Bykovtsev G 1971 *The theory of hardening plastic body* (Moscow: Nauka) 231
[3] Geniev G, Kissyuk V and Tyupin G 1974 *The theory of plasticity of concrete and reinforced concrete* (Moscow: Stroyizdat) 316
[4] Fadeev A 1987 *The finite element method in geomechanics* (Moscow: Nedra) 221
[5] Grishin A and Fedorova E 2002 *Nonlinear dynamic problems of calculating port hydraulic structures* (Odessa) 126

Features of production and application of glued wooden beams in cottage construction

S G Guzii

PhD, academic, senior researcher Department of fire prevention in settlements National University of Civil Defence of Ukraine, 94 Chernishevska str., Kharkiv 61023, Ukraine

Email: guziy@nuczu.edu.ua

Abstract. This paper discusses the technological features of the production and use of laminated veneer lumber in cottage construction. The main properties of aluminosilicate glue for gluing pine lamellas have been determined. Changes in both the dynamic and plastic viscosity of aluminosilicate glue are given depending on the spindle rotation speed and shear rate. Equations describing these dependencies are given. These dependencies influence the choice of the method for applying the aluminosilicate glue to the substrate surface. It has been shown that a change in dynamic viscosity from 1662 to 1268 cP in the operating speed range from 20 to 30 RPM and shear rate from 4.2 to 6.3 1/s provides a small shear force from 69.9 to 79.88 dynes/cm^2, which makes it easy to apply the glue with a brush or a special roller.
It is noted that aluminosilicate glue with a density of 1.54 g/cm^3 has a low surface tension of 54 mN/m, which affects the contact angle and the tendency to spread on a pine base. It has been found that an increase in the spreading coefficient results in more complete wetting. It is shown that with an increase in the curing time of the glue joints, the shear strength along the fibers and the elastic modulus on the 28th day increase by 1.63/1.1 times as compared with the seven-day age. An increase in strength and elastic modulus is associated with the passage of crystallization processes in an aluminosilicate gel with the subsequent formation of zeolite-like formations. It is noted that the increase in the value of the breaking load of the samples at the age of 28 days is 1.63 times higher than after 7 days of curing, for determining the values of the shear modulus and elasticity does not have such a decisive importance and this decrease is 5.1%.
Key words: aluminosilicate glue, glued laminated timber, pine, physical, mechanical and deformation properties, colloidal-chemical properties, rheokinetic properties.

1. Introduction

The growing global demand for wood products used in construction is primarily due to the fact that wood is a renewable and environmentally friendly material. No other area of building technology has offered so many innovative solutions in the last 20 years as in the processing and use of wood.
Industrial construction from wood confidently strides across the planet: buildings and structures using modern materials and structures made of wood have come to cities, where multi-story residential areas, public, social, sports and administrative buildings are being erected. The wooden residential complex 79&Park in Stockholm, Sweden entered the top 20 best buildings in the world in 2018. The world record for the height of wood-based objects is updated almost every year. After the official opening on March 1, 2019, the 84.5 m high wooden skyscraper in Norway, Mjes Tower, which includes residential apartments, a hotel, offices and a swimming pool, will become the world's tallest timber building, surpassing the university dormitory by more than 30 m. Vancouver, Canada (built in 2017). Based on large-scale scientific research and testing in the United States, at the end of 2018, at the state level, it was allowed to build wooden buildings up to 18 floors. With the use of modern glued structures, airports, sports arenas, shopping and office centers, cultural objects, hundreds of grandiose projects are being built all over the world now at the design or construction stage [1]. The same trend is typical for Ukraine because wooden houses are still the most popular in private construction. The main reasons prompting many to move to a cottage from a "concrete box" - the desire for maximum comfort and care for their own health. In our case, the study of the main characteristics of mineral adhesives for gluing beams, as

Content from this work may be used under the terms of the Creative Commons Attribution 3.0 licence. Any further distribution of this work must maintain attribution to the author(s) and the title of the work, journal citation and DOI.
Published under licence by IOP Publishing Ltd

well as the characteristics of glued beams themselves for the purpose of their use in the designs of ergonomic modular low-rise houses and cottages is an urgent scientific and practical task.

2. Literary review and problem statement.

In modern projects of wooden houses, in addition to solid wood, glued beams are widely used [2, 3]. For gluing of similar products and designs on their basis of designs use, mainly wood of a pine of a radial saw and organic glues (Akzo Nobel, the Netherlands; Klebchemie, Kiilto, Finland and others).
To date, there are four types of glue for the production of glued beams:

1. Polyurethane adhesive. It has excellent manufacturability, hardens quickly.
2. Melamine glue. It is widely used in Europe, recently it has become widespread in Ukraine due to its good adhesion characteristics. After polymerization it becomes transparent. It has a wide operating range of weather conditions - relative humidity up to 80%.
3. EPI glue (EPI system - polymer isocyanate emulsion). Meets both domestic and some foreign standards for the production of wall beams. EPI glue does not contain formaldehyde and belongs to the safest group of adhesive compositions. Is universal and transparent.
4. Resorcinol adhesives. They differ in that after polymerization they acquire a dark shade. Became widespread in the United States and Japan in the gluing of floor beams. Used only outdoors.

Sawing of slats is carried out by a special technology that takes into account the angle of the annual rings to reduce humidity and stress in the wood. After drying, the pre-treated slats are glued and pressed under high load and at a temperature of about 200 degrees. In the process of making glued beams, the glue under pressure is pushed deep into the pores of the wood, which provides a strong grip. Then, after the polymerization of the glue, the beam is processed and cut to the desired size.
In addition to high physical and mechanical properties [4, 5], the above adhesives are characterized by high flammability. The basic requirement for fire safety in construction is to limit the spread of fire, smoke and maintain the load-bearing capacity of the structure for a certain period of time, and an important way to ensure these principles is to manufacture fire-resistant building structures and study their fire resistance, including fire resistance. structures. Of the 7106 buildings completely destroyed in 2020, 2263 (31.9% of the total number of destroyed buildings and structures) had a V degree of fire resistance, 1531 (21.6%) had a III degree of fire resistance, 841 (11.8%) - IIIa degree of fire resistance , 431 (6.1%) - IV degree of fire resistance, 246 (3.5%) - IIIb degree of fire resistance, 108 (1.5%) - IVa degree of fire resistance, 86 (1.2%) - II degree of fire resistance and 12 (0.2%) - I degree of fire resistance. That is, the largest losses occur in buildings of III degree of fire resistance, to which there are quite high requirements for building structures [6].
An alternative to organic-based adhesives are aluminosilicate adhesives, which, in comparison with the above, are characterized by non-flammability [7, 8].
The main purpose of the study is to study the rheological, colloidal-chemical, physical-mechanical and deformation properties of aluminosilicate for obtaining products in the form of glued beams for their use in ergonomic modular low-rise cottages.

3. Materials and methods of research

For gluing pine lamellas and determining the strength, deformation, rheological and colloidal-chemical properties, we used aluminosilicate glue type KB3 manufactured by Geofip LLC (Kropyvnytskyi, Ukraine).
Given this original feature of aluminosilicate adhesives, they were used for gluing beams from prepared pine slats measuring 270x50x3000 mm, taking into account the requirements for gluing, namely:
1. Humidity of wood - 10-16%.
2. After grouting - clear geometry. Take into account the uneven shrinkage of wood depending on the structure of the rings.
3. After grouting and before applying the glue it is necessary to measure the humidity of the slats for sorting. Humidity between the slats should not exceed 2%.

4. No more than two hours should elapse after grouting before gluing the slats. After two hours gluing is forbidden.
5. To type for gluing of a lamella by fibers in opposite directions. To compensate for torsion, relieve internal stress and deformation of the rafter layers.
6. Conditions in the shop - humidifiers for humidity and the temperature corresponding to the TU of the technological cycle of gluing. In accordance with these conditions, the ratio of glue to hardener is selected.
7. Parameters on gluing: - assembly time - pressing time.
8. A pressure of 10 kg/cm^2 must be maintained throughout the pressing time. The quality of gluing should be stronger than the body of wood.

To eliminate internal stresses arising in the glued joints, "cold" gluing was carried out. On pre-prepared wood samples (surface roughness Rz 60, wood moisture 12%) aluminosilicate adhesive was applied with a brush. The glued samples were compressed with clamps at a load of 0.6 MPa for 4 hours. After the release of the crimping pressure, further curing of the glue line occurred at ambient temperature of 20 ± 2°C and a relative humidity of $65 \pm 5\%$ (Fig. 1). The determination of the strength of the samples after 7 and 28 days of hardening was carried out on a universal testing machine with a servo drive UMT-10 (China) in the range of breaking forces from 0 to 8 kN at a loading rate of 0.5 mm/s with fixing of stresses in the diagram like F-Δl (Fig. 2) [9].

a) b)

Figure 1. Brush application of KB3 aluminosilicate glue on pine lamellas (a) and fixation of glued laminated timber in Waima (b)

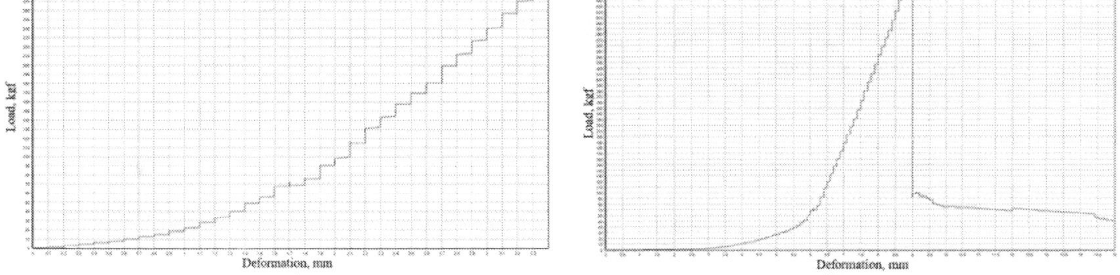

a) b)

Figure 2. Diagrams fracture after 7 (a) and 28 (b) days of hardening adhesive pine compounds

To analyze the effect of aluminosilicate glue on colloidal-chemical, rheokinetic properties, the following parameters were chosen [7]: work of adhesion, cohesion and wetting forces, spreading and wetting coefficients, dynamic and plastic viscosity, surface tension and angular wetting of pine wood substrates.

The numerical values of the parameters were obtained on devices and according to the methods used in the study of varnishes and paints and varnishes (LKM). Dynamic viscosity was determined with a Brookfield LVDV2T viscometer using an LV-3C spindle. Surface tension, contact angle, work of the forces of adhesion, cohesion and wetting, as well as the values of the spreading and wetting coefficients were calculated using the formulas given in the literature [10]. The error in measuring the values of surface tension and contact angle did not exceed 10%. The physical-mechanical and deformation properties of the glue joints were determined on the 7th and 28th days of the glue hardening according to [11, 12].

4. Results of researches

In fig. 3 shows the architectural solutions of ergonomic modular low-rise buildings, which use curved and rectangular glued beams in the form of beams and posts. In this paper, a variant of using glued laminated timber on aluminosilicate glue in the form of racks in a security house is considered (Fig. 1, g, h).

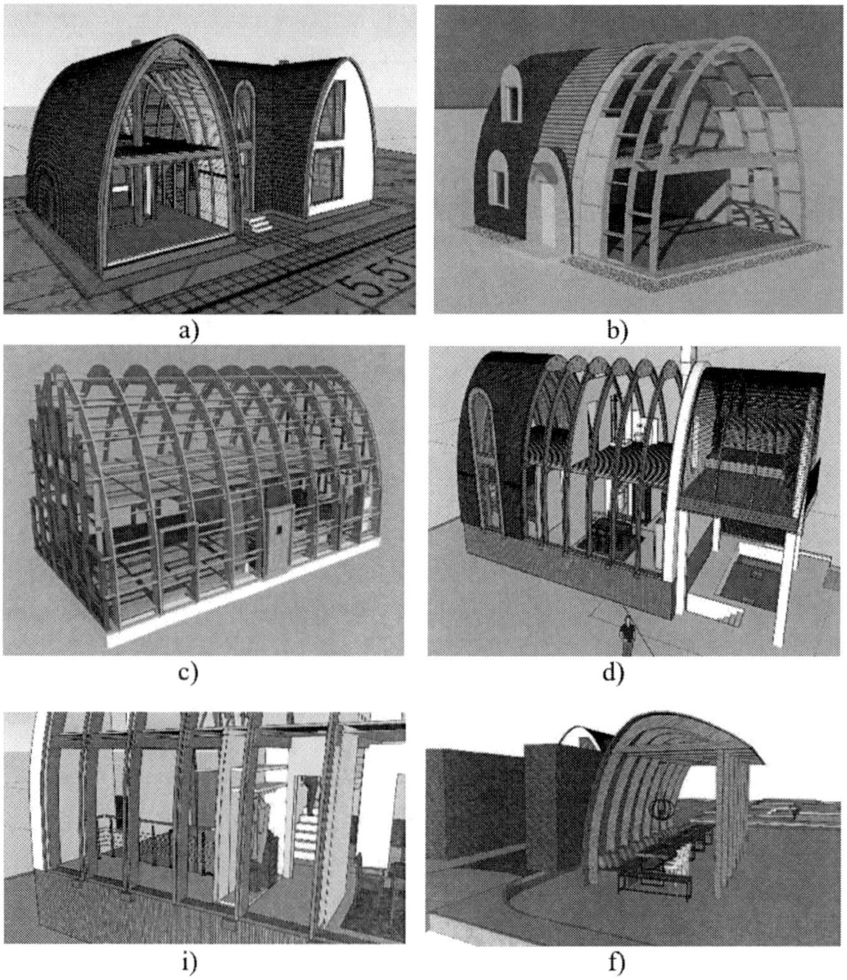

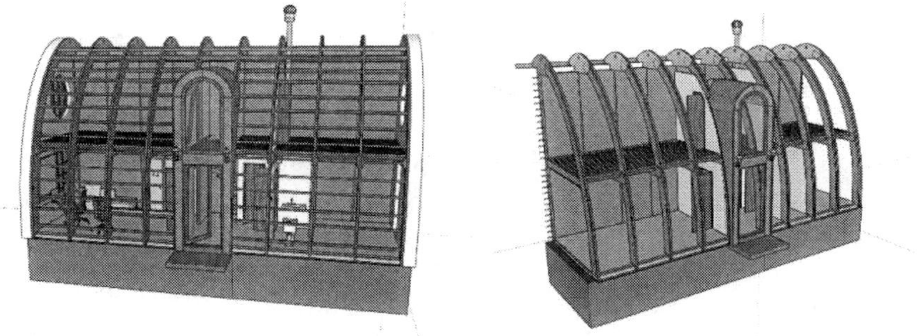

g) h)

Figure 3. Architectural solutions ergonomic modular low-rise buildings: and - a general view; b, c, d, i - fragments of the framework; f - arbor; g, h - guard house

The amount of adhesive applied per unit surface area depends, first of all, on its viscosity, and secondly, on the required thickness of the glue line, the temperature of the base and the environment, as well as the quality of the base surface preparation. Aluminosilicate adhesive is characterized by a density of 1.54 g/cm^3, has a small surface tension of 54 mN/m, pH = 11.77 and EMF = -250 mV.

Changing the dynamic viscosity from 1662 to 1268 cP in the operating speed range from 20 to 30 RPM (Fig. 4) and shear rates from 4.2 to 6.3 1 / s (Fig. 5) provides a slight shear force from 69.9 to 79.88 dyne/cm^2, which allows you to easily, using a brush or a special roller to apply glue to the pine slats.

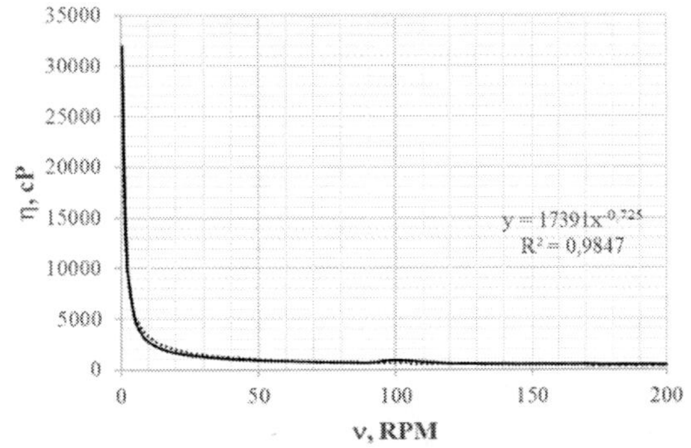

Figure 4. Kinetics of changes in the dynamic viscosity of aluminosilicate glue from speed and torque of the spindle

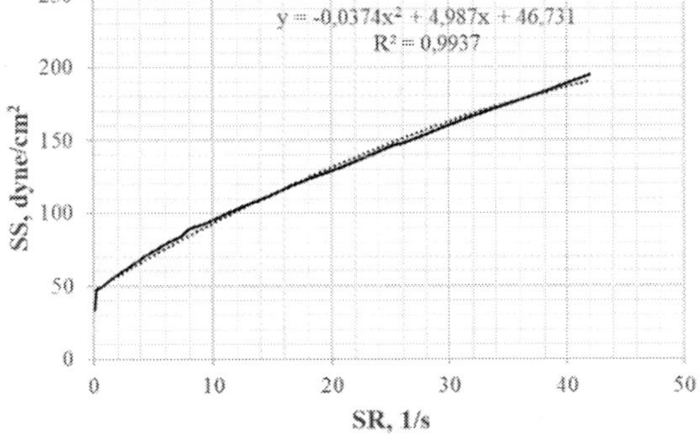

Figure 5. Kinetics of changes in the shift force of aluminosilicate glue on the shear rate

The Figure 6 shows the wetting angles of aluminosilicate glue on a pine substrate.

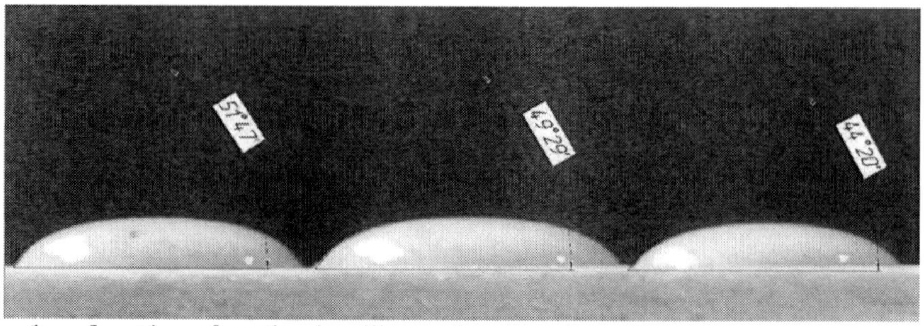

Figure 6. Angles of wetting of an aluminosilicate glue on a pine substrate

As seen in Fig. 6, the aluminosilicate glue has a fairly low wetting angle and is prone to slight spreading on a pine base.
In fig. 7 gives data on the change of the adhesive strength along fibers of pine compound.

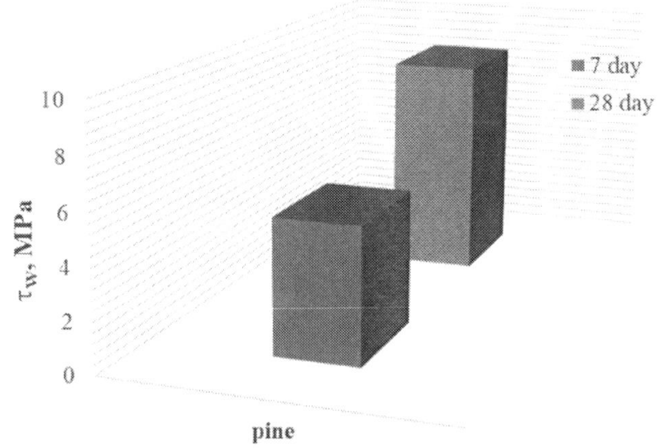

Figure 7. Change in the density of the protective composition depending on the concentration of the Fe-containing filler

As seen from Fig. 7, on the 7th day of curing, the shearing strength along the fibers is 5.31 MPa, and on the 28th day after curing, it is 8.66 MPa. The increase in strength is due to the passage of crystallization processes in an aluminosilicate gel with the subsequent formation of zeolite-like formations.

5. The discussion of the results.

It is known that more completely the substrate is wetted by an adhesive, the greater loss of energy occurs (from the wetting condition): $S=s_c-s_a-s_a\text{-}c\geq 0$, where s_c– an adhesive; s_a– substrate; s_a-c – section adhesive-substrate.
The higher the spreading coefficient "s", the more complete the wetting is. And for this it is necessary that the free surface energy at the adhesive-substrate (s_a-c) interface be minimal, which is possible only if the molecular nature of the adhesive and the substrate is close. Considering that the molecular nature of the mineral adhesive and the substrate is far from close, many factors must be taken into account for complete wetting and spreading of the substrate, namely: surface tension, contact angle, work of forces of adhesion, cohesion, wetting, etc. According to the criteria, this assumption can be displayed as follows: $\Theta av<75°$, $s\rightarrow max$ and $f\rightarrow min$.
The chemical and colloidal properties of the protective composition are shown in table 1.

Table 1
Chemical and colloidal properties of the protective composition

Wood	Θav	cosΘav	Wa, mN/m	Wk, mN/m	Ww, mN/m	s	f, mN/m
pine	48°32′	0.66497	89.91	108	35.91	0.8325	-18.09

Note: Θav – the average value of the wetting angle of the substrate with the adhesive; Wa, Wk, Ww – work of adhesion forces, cohesion and wetting, mN/m; s, f – coefficients of wetting and spreading of the adhesive over a substrate surface, mN/m.

The significance of the obtained results of the work is quite obvious and, first of all, is aimed at practical originality, which will help us to solve the problem of stabilizing the rheokinetic properties aluminosilicate glue, namely: the values of dynamic and plastic viscosity in the range of operating speeds when applying a protective composition to substrates surface; ensure the formation of colloidal-chemical structures with minimum wetting angles and maximum values of the spreading coefficient.

The deformation characteristics of adhesive joints are shown in Table 2.

Table 2
The values of the shear modulus G and the modulus of elasticity E of the aluminosilicate glue, determined from experimental data

Substrate material	Sample age, day	Aluminosilicate glue					
		Experimental and calculated values				Geometric parameters of samples	
		F, [kgf]	Δx, [cm]	G, [MPa]	E, [MPa]	A, [cm^2]	l, [cm]
pine	7	220	0.16	76	228	5.48	3.03
	28	400	0.28	79.9	239.7	5.42	3.03

From the data of Figure 1 and table 2, it follows that with an increase in the curing age of the aluminosilicate glue, there is a more tendency to an increase in the absolute values of the deformation value from 0.16 to 0.28 mm, starting from which the section of the F-Δl diagram has an almost linear section. The value of the load, upon reaching which the diagram becomes linear, is F = 10 kgf. Obviously, this is the load, under the action of which the samples compensate for all the initial errors and imperfections. At the same time, the upper limit of deformation of the samples decreases (respectively 8.1 mm; 7.0 mm (Fig. 1)), which may indicate an increase in the stiffness of the adhesive joint. A significant increase in the value of the breaking load of the samples at the age of 28 days by 1.63 times higher than after 7 days of curing for determining the values of the shear modulus and elasticity does not have such a decisive importance and this decrease is 5.1%.

6. Conclusions

- The carried-out studies make it possible to determine the conditions and methods of using aluminosilicate glue for the manufacture of glued beams in the construction of ergonomic modular low-rise houses and cottages.
- Methods for applying aluminosilicate glue to pine lamellae from changes in dynamic and plastic viscosity depending on shear rates and shear rates have been determined.
- The angles of wetting of aluminosilicate glue applied to a base made of wood were determined, which made it possible to assess the possibilities of resistance of the material on surfaces depending on the magnitude of surface tension.
- All the main chemical-colloidal properties of aluminosilicate glue have been obtained, which will ensure the durability of the glue seam in real operating conditions.

- The strength and deformation properties of glued joints of pine lamellas have been determined, which make it possible to predict the operation of glued beams in the designs of ergonomic modular low-rise houses and cottages.
- Further work will be aimed at obtaining the characteristics of the cured glue: modulus of elasticity along the X, Y and Z axes; Pausson's coefficients in XY, YZ and XZ; tensile and compressive strength in the X and Y axes; yield point. These data will make it possible to assess the stress-strain state of laminated veneer lumber under the influence of both static and dynamic loads.

Acknowledgements

The author would like to thank the architects Sergii and Pavlo Bezverkhih for the sketches of wooden modular structures provided (Certificate of copyright registration № 630092 for the work "Collection of architectural works ergonomic modular low-rise buildings of architects Bezverkhih").

References

[1] Accessible electronic resource: https://lesprominform.ru/jarticales.html&d=5235
[2] Wooden structures and details. Ed. V.M. Khrulev. Stroyizdat. Moscow (1983) 288 p.
[3] D. Fendel, G. Werener. Wood (Chemistry, Ultrastructure, Reaction). Walter de Gruyter Berlin. New York. 1984.
[4] Technology and equipment for wood protection. Ed. A.I. Raseva. GOU VPO MGUL Moscow (2010) 171 p.
[5] O. Ulker. Wood Adhesives and Bonding Theory. Adhesives – Applications and Properties (2016).
[6] Otrosh Y., Kovalov A., Semkiv O., Rudeshko I., Diven V. (2018). Methodology remaining lifetime determination of the building structures. *MATEC Web of Conferences*. **Vol. 230** No 02023.
[7] S. Guzii, O. Guzii, V. Lashchivskiy (2020). Investigation of the Rheokinetic Properties and Penetration Depth of Aluminosilicate Adhesive in Pine Wood. *Solid State Phenomena*. **Vol. 321** PP. 97-103.
[8] Guzii S., Krivenko P., Bondarenko O., Kopylova T. (2019). Study on physico-mechanical properties of the modified alkaline aluminosilicate adhesive-bonded timber elements. *J. SSP*. **Vol. 296** PP. 112-117. doi:10.4028/www.scientific.net/SSP.296.112.
[9] GOST 16483.9-73. Wood. Methods for determination of modulus of elasticity in static bending
[10] Glazacheva E.N., Uspenskaya M.V. (2015). Colloidal chemistry. Methodical instructions for laboratory work. *St. Petersburg: ITMO University*, 62 p.
[11] GOST 16483.5-73. Wood. Methods for determination of ultimate shearing strength parallel to grain.
[12] GOST 26454-85. Adhesives. Method for determination of elasticity modulus of adhesive in adhesive bond.

Application of parallel computing in finite element analysis of constructions

S I Homenyk[1] and V Yu Kozub[1,*]

[1]Zaporizhzhia national university, Zaporizhzhia, 69600, Ukraine

[*]E-mail: v.y.kozub@gmail.com

Abstract To optimize finite elements computational processes of the "MIRELA +" computing complex, parallel programming algorithms have been developed for creating stiffness matrices of finite elements and calculating the stress-strain. The parallel data processing is implemented using OpenMP library. On the basis of computational methods of experiments, the use of parallel computing algorithms at the beginning of solving problems with traditional finite element algorithms for large-dimensional grids has been established.

1. Introduction

One of the most common methods of calculating structures is the finite element method. Depending on the type of problem in finite-element modeling, the calculated grids reach large sizes, in particular, for composite multilayer structures, structures with stress concentrators, composite structures. In this case, the time of calculation of the stiffness matrix is a significant part of the total time to solve the problem.

The need to solve complex three-dimensional problems with minimal costs leads to the need to optimize computational processes. This is especially true of the application of the method of superelements, in which the construction of solving equations for individual superelements can be performed in parallel, which significantly reduces the calculation time [1].

The use of several computing units to solve a computational problem simultaneously is a parallel computation. Provided that the computational problem can be divided into separate, autonomous from each other, parts that can be processed simultaneously, thereby ensuring the solution of problems in less time with several computing units than with a single computing unit

Usage of several computing units to solve a computational problem simultaneously is defined as parallel computation. Provided that the computational problem can be divided into separate, autonomous from each other, parts that can be processed simultaneously, thereby ensuring the solution of problems in less time with several computing units than with a single computing unit

Computing units can be a single computer with multiple processors/cores, such as laptops with dual-core or quad-core processors or a supercomputer created from a large number of such computers (or nodes) connected by a network that creates a cluster. Each node consists of several cores that are connected in a socket. For example, the supercomputer Stampede2 is a cluster of 4200 nodes, each of which contains 68 cores (a total of 285,600 cores).

2. Advantages and disadvantages

Parallel programs have the following features:

- management of work of set of processes is carried out;
- data exchange between processes is organized;
- determinism of behavior is lost due to asynchrony of data access;
- non-local and dynamic errors predominate;
- there is a possibility of deadlocks;
- there are problems with the scalability of the program and balancing the load of computer nodes.

When using multiprocessor systems, it is expected that the speed of calculations will increase n times while using n processors instead of one. In fact, the acceleration is much less than expected. The decrease

Content from this work may be used under the terms of the Creative Commons Attribution 3.0 licence. Any further distribution of this work must maintain attribution to the author(s) and the title of the work, journal citation and DOI.
Published under licence by IOP Publishing Ltd

in productivity growth in parallel computing is due to two factors: the first is related to the properties of the program algorithm (these properties were studied in 1967 by supercomputer developer Gene Amdal) [2], and the second - to the technical properties of the computer system.

The essence of the problem is that each parallel program contains some part of non-parallel code. On systems with shared memory, this is the part of the code that is executed only by the main thread, and on systems with distributed memory, it is the part of the code that is executed by all processors. Assume that the ratio of the execution time of the non-parallel part of the program to the total execution time is $f\,(0 < f < 1)$. If the execution time on one processor is equal to t_s, then the execution time of the parallel part is – $(1\text{-}f)t_s$, and the execution time of the serial part is - ft_s. Ideally, when using n processors, the execution time of the parallel part will be $(1\text{-}ft_s)/n$, and the execution time of the serial part will remain ft_s.

Thus, the acceleration of program execution on n processors will be:

$$S(n)=\frac{t_s}{ft_s+(1-f)t_s\,/\,n}=\frac{n}{1+(n-1)f}. \qquad (1)$$

Expression (1) is a formalized definition of Amdahl's law. At n, which goes to infinity, the acceleration will be $1/f$. It follows from this law that if part of the hardware in the program is 5 it is impossible to get more than 20 times the acceleration program, no matter how many processors are used.

When partitioning a large task into smaller subtasks that need to run in parallel on different processors, there is often a situation where there's a need to wait for some processors to complete some execution stage. In this case, the total execution time will be determined by the slowest subtask. Therefore, when developing a parallel program, it is extremely important to distribute computing as evenly as possible between processors. Achieving a load-balanced state is a non-trivial task

In shared-memory systems, where information is exchanged between processors using variables stored in shared memory, process synchronization is especially important. If several processes simultaneously try to modify the same variable, then it's required a special mechanism for synchronizing the execution of processes to ensure the determinism of program execution.

The efficiency of parallel programs on systems with distributed memory significantly depends on the communication environment. The communication environment is characterized by two parameters: bandwidth, which determines the number of bytes transmitted per unit time, and latency, which determines the time spent preparing for the transmission of the message. For example, for a Gigabit Ethernet network bandwidth in MPI programs is about 70 Mbps, and the latency is about 40 μsec [3]. For a large number of tasks in which data exchange is infrequent and small in size, these parameters are quite satisfactory, however, in cases where the program transmits many small messages, such parameters become unacceptable, and the scalability of the program is extremely low. In any case, communication operations are much slower than accessing local memory, so the most effective will be those parallel programs in which exchanges are minimized.

3. Approaches

Approaches to parallelization in FEM can be divided into two categories - approaches with distributed memory and shared memory, which shown in Figure 1 and Figure 2 respectively. When working with the first category, a high-performance model of parallel programming Message Passing Interface (MPI) is usually used. MPI is an easily scalable means of parallel programming, but on the condition of workload distribution done, which usually leads to large-scale changes in the program [4,5]. An MPI program is a collection of independent processes, each of which runs its own program (not necessarily the same), written in C/C++ or Fortran. The processes of an MPI program interact with each other by calling communication procedures. Despite significant advances in the development of programming technology using the message transfer mechanism, the complexity of programming using this technology is still too high in the context of getting quick results.

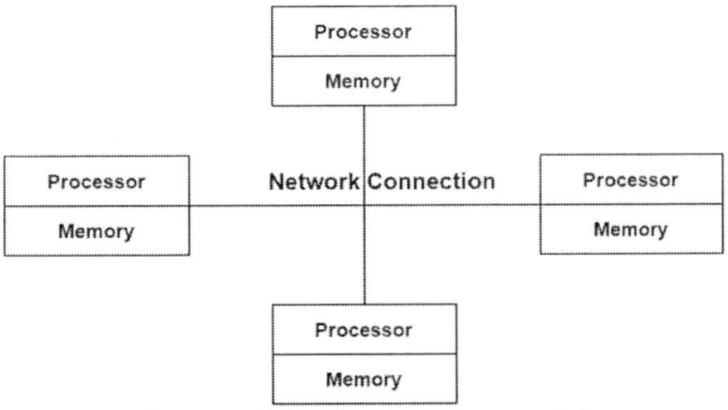

Figure 1. Distributed memory model.

Parallelization in the second category is mostly done using special compiler directives. The OpenMP standard [5, 6] was developed to provide basic concurrency constructs in Fortran and C / C ++ programs. The use of OpenMP gives limited control over threads compared to the more fundamental Pthreads standard [7]. However, OpenMP provides greater development speed due to ease of use.

The ability to quickly create effective parallel programs is a strong argument when choosing programming tools. The technology was originally designed so that the user could work with a single text for parallel and sequential programs. The usual compiler on the serial machine simply "does not notice" the OpenMP directives as they are located in comments (except for environment variables and special functions). An additional feature of OpenMP is the possibility of progressive, "incremental" parallelization of the program. Based on sequential code, it is possible to step by step add directives describing parallel constructions. With this approach, there is no need to immediately write a complete parallel program - its development is consistent. This simplifies both programming process and program debugging.

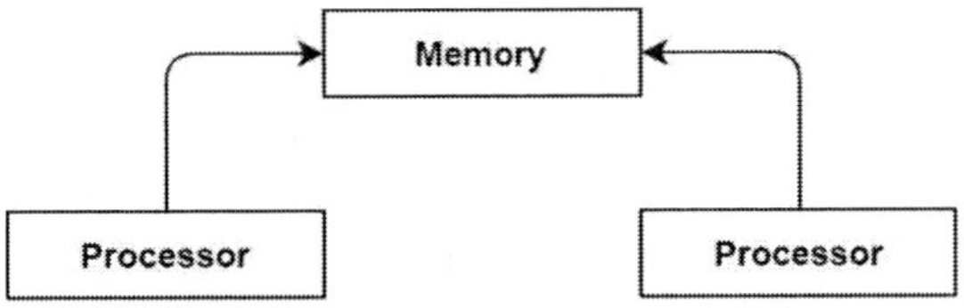

Figure 2. Shared memory model.

It is worth noting that programs on MPI can be executed on systems with shared memory no less efficiently than on OpenMP. From this point of view, MPI technology is more versatile. However, programming in OpenMP, firstly, is much more convenient than using MPI technology, and secondly, OpenMP technology is much more economical to spend RAM on systems with shared memory. For tasks that require a large amount of RAM, it often turns out that it is impossible to run so many processes on a computer system to load all the computing cores.

4. Parallelization in "MIRELA+"

The "MIRELA +" computing system is designed to solve problems in the deformed solid mechanics. To eliminate the effect of "false shift" in the complex used a moment scheme of finite elements, the essence of which is to approximate fields of displacement and deformation in the form of decomposition in a series of power functions of a special kind, that allows a fairly simple relationship between decomposition components and avoid unnecessary decomposition components of deformations

$$u_i = \sum_{p,q,r=0}^{m,n,l} w_i^{(pqr)} \psi^{(pqr)} = \sum_{p,q,r=0}^{m,n,l} w_i^{(pqr)} \frac{(x_1)^p}{p!} \frac{(x_2)^q}{q!} \frac{(x_3)^r}{r!},$$

$$\varepsilon_{ij} = \sum_{s,t,g} e_{ij}^{(stg)} \psi^{(stg)} ,$$

$$\varepsilon_{ij} = \mathbf{F}_{ij}^{s'} w_{s'} = \mathbf{F}_{ij}^{s'} \mathbf{A}\mathbf{u}_{s'} , \tag{2}$$

where **A** is the transformation matrix that establishes the relationship between functions of the form of a finite element and the power basis functions.

Similarly for the function of changing the volume by decomposition by power functions allows considering the low compressibility of materials (such as elastomers) [8]

$$\theta_i = \sum_{\alpha,\beta,\gamma=0}^{m-1,n-1,l-1} \xi^{(\alpha\beta\gamma)} \psi^{(\alpha\beta\gamma)} ,$$

$$\xi^{(\alpha\beta\gamma)} = e_{ii}^{(\alpha\beta\gamma)} g^{ii}$$

$$\theta = \mathbf{F}_{\theta}^{s'} w_{s'} = \mathbf{F}_{ij}^{s'} \mathbf{A}\mathbf{u}_{s'} \tag{3}$$

For problems of structure statics, the variational equation of equilibrium of a linearly elastic body has the form

$$\delta \iiint_v \left(\mu g^{kj} g^{li} \varepsilon_{kl} \varepsilon_{ij} + \frac{\lambda}{2} \theta^2 \right) dv + \iiint_v \rho P^i \delta u_i dv + \iint_s F^i \delta u_i ds , \tag{4}$$

where μ, λ, ρ - mechanical characteristics of the material, P^i, F^i- body and surface forces respectively.

Given the approximation methods used the variation of the deformation energy has the form

$$\delta W = \iiint_v \delta \{e_{ij}\}^T \{\psi\} \mu g^{ik} g^{jl} \{e_{kl}\}^T \{\psi\} dv + \iiint_v \{\xi\}^T \{\psi\} \lambda \delta \{\xi_{kl}\}^T \{\psi\} dv$$

$$\mathbf{H}^{ijkl} = \int_{-1}^{1} \int_{-1}^{1} \int_{-1}^{1} 2\mu g^{ik} g^{jl} \{\psi_{ij}\} \{\psi_{kl}\}^T \sqrt{g} dx_1 dx_2 dx_3$$

$$\mathbf{H}^{\theta} = \int_{-1}^{1} \int_{-1}^{1} \int_{-1}^{1} \lambda \{\psi_{\theta}\} \{\psi_{\theta}\}^T \sqrt{g} dx_1 dx_2 dx_3$$

$$\delta W = \delta \mathbf{u}_{s'}^T \mathbf{K}^{s't'} \mathbf{u}_{t'} + \delta \mathbf{u}_{s'}^T \mathbf{K}_{\theta}^{s't'} \mathbf{u}_{t'} , \tag{5}$$

where $\mathbf{K}^{s't'} = \mathbf{A}^T \mathbf{F}^T \mathbf{H} \mathbf{F} \mathbf{A}$, $\mathbf{K}_{\theta}^{s't'} = \mathbf{A}^T \mathbf{F}_{\theta}^T \mathbf{H}^{\theta} \mathbf{F}_{\theta} \mathbf{A}$ - stiffness matrices.

The procedure for forming the stiffness matrix of a finite element consists of the following steps:

- calculation of displacement coefficients and coefficients of matrix A;
- numerical integration according to the Gaussian scheme.

The following procedures are performed for integration points:

- calculation of transition matrices from global coordinate system to the local coordinate system and the inverse transformation matrix at integration points;
- calculation of partial derivatives for the formation of a matrix F, which connects the displacement and deformation in a finite element.

The implementation of parallel processing in the study was performed using the OpenMP library. Joint processing of large data sets requires autonomy of operations on them for each cycle of formation of the stiffness matrix and calculation of the parameters of the stress-strain state.

Parallelism is achieved by dividing the iterations of the computational cycle into blocks, which are evenly and parallel distributed over the workflows. In OpenMP, the DO pre-loop directive is used to implement such an execution scheme [9]. The SCHEDULE (STATIC) option specifies the block-cyclic distribution of iterations in blocks of size determined by dividing the number of iterations by the number of threads. The next lines show the example of parallelized loop:

```
!$OMP PARALLEL PRIVATE(S, SN)
!$OMP DO SCHEDULE(STATIC) ORDERED
```

```
    DO 1 I=1,N
    <Parallel loop code>
!$OMP ORDERED
    WRITE (*) RESULT
!$OMP END ORDERED
    1 CONTINUE
!$OMP END DO
!$OMP END PARALLEL
```

The PRIVATE directive allows defining variables (arrays) that must be local to each thread. While enter the parallel region, a separate instance is created for each thread that has nothing to do with the original variable outside the parallel region. By default, the compiler defines variables that were initialized to the parallel section as shared (SHARED) for all threads.

Before the end of the loop, the procedure of outputting the results of the calculation of the finite element is performed, which must be performed in the same order as in the sequential version of the loop. The sequential access behavior is provided by the ORDERED directive.

To support parallelism the program uses a fork-join model which shown in Figure 3. It means execution forks parallel at designated points in the program to "join" (merge) at the next point and resume sequential execution. Sequential region of the OpenMP program is executed only by the main thread (master, №0). When entering a parallel region (!$OMP PARALLEL), the master thread generates additional threads (Fork operation is performed). After generation, each thread receives its own unique number. All generated threads execute the same code corresponding to the parallel region. When exiting (!$OMP END PARALLEL) from the parallel region, the main thread waits for the completion of other threads, and further execution of the program continues only master (Join operation is performed)

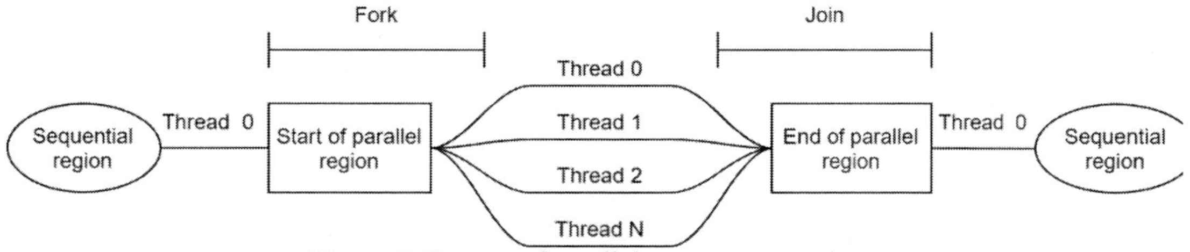

Figure 3. Process of parallel program execution.

As an example, for verification and testing, the calculation of a square plate clamped along the contour under the action of the bending load is presented. The computing complex "MIRELA+" with an application of the moment scheme of finite elements is used for calculation. Testing was performed on a device equipped with an Intel i7-9750H processor (6 cores and 12 threads) and 16 GB of RAM. Figure 4 shows speed comparison of execution for meshes with different count of finite elements.

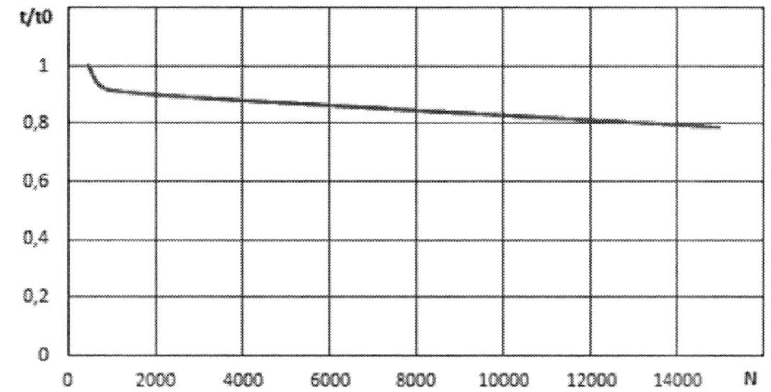

Figure 4. Speed comparison of execution for meshes with different count of finite elements.

5. Conclusion

With the advent of multi-core processors, shared memory systems have become the most popular type of multiprocessor system, and OpenMP, in turn, provides one of the most convenient software development interfaces for such systems. The OpenMP version of the program can also be compiled like a sequential program using the compiler parameter "non-parallel" (OpenMP directives will be ignored by the compiler), which provides portability of code to different platforms. With the growing popularity of SMP computers, the importance of having efficient and cross-platform parallel codes is growing.

The paper analyzes the existing technical means of parallelization, strengths and weaknesses of the implementation of parallel solutions, using the OpenMP directives developed a parallel version of the calculation procedure to perform target calculations based on the "MIRELA+" computing system. Performance tests in the presented example showed satisfactory acceleration using parallel sections.

References

[1] Jarzebski P, Wisniewski K and Taylor R 2015 On parallelization of the loop over elements in FEAP *Comput. Mech.* **56** pp 77–86
[2] Amdahl G M 1967 Validity of the single-processor approach to achieving large scale computing capabilities *In AFIPS Conference Proceedings* **30** pp 483-485
[3] Datsyuk V N, Datsyuk O V, Bukatov A A and Vinogradova S A 2017 HPC Programming Guide http://rsuib.cc.rsu.ru/tutor/high_performance_computing/index.html [Online]
[4] Akhmetova D, Iakymchuk R, Ekeberg O and Laure E 2017 Performance Study of Multithreaded MPI and OpenMP Tasking in a Large Scientific Code *Performance Study of Multithreaded MPI and OpenMP Tasking in a Large Scientific Code* pp 756-765
[5] Pantalé O 2005 Parallelization of an object-oriented FEM dynamics code: influence of the strategies on the Speedup *Adv. Eng. Softw.* **36** pp 361–373
[6] OpenMP Architecture Review Board. Openmp 3.0 api specifications. https://www.openmp.org/wp-content/uploads/OpenMP-API-Specification-5-1.pdf 2021 [Online]
[7] Jia R and Sunde´n B 2004 Parallelization of a multi-blocked CFD code via three strategies for fluid flow and heat transfer analysis *Comput Fluids* **33** pp 57–80
[8] Kirichevskiy V V, Doknyak B M, Kozub Yu G etc. 2005 Finite element method in obtained complex “MIRELA+” (Kyiv: Naukova dumka) 402
[9] Kopysov S, Kuzmin I, Nedozhogin, N, Novikov A, Rychkov V, Sagdeeva Yu and Tonkov L 2014 Parallel implementation of finite element algorithms on graphics accelerators in the software complex FEStudio *Computer research and modeling* **6** pp 79-97

Research of the process of water absorption by thermally modified wood

Y Tsapko[1], O Bondarenko[1], O Horbachova[2,*] and S Mazurchuk[2]

[1]Kyiv National University of Construction and Architecture, Povitroflotskyj ave., 31, Kyiv, 03037, Ukraine
[2]National University of Life and Environmental Sciences of Ukraine, Heroiv Oborony str., 15, Kiyv, 03041, Ukraine,

*Email: gorbachova.sasha@ukr.net

Abstract. The analysis of the process of thermal modification of wood, which was obtained by a controlled process of pyrolysis of wood heating (> 180 °C), which causes some chemical changes in the structures of cell wall components (lignin, cellulose and hemicellulose), to increase its durability. It is proved that in the process of thermal modification there is a decomposition of hemicelluloses and amorphous part of cellulose, reduction of water absorption, and also the quantity of substances which are the environment for development of fungi decreases. In addition lignin and psevdolihnin formed by the process of polymerization and redistribution in cell volume and cell walls provide higher density, hardness, increase the hydrophobicity (water repelling), thereby reducing their ability to absorb moisture and swell. Polymerised lignin fills the inside of the cell – formed closed porous structure with low ability to bind water. It was established that the most effective option to reduce such substances are temperature and exposure time. The results of water absorption, dependence on which the calculated diffusion coefficient in the water absorption. Thermal modification of wood reduces water absorption by more than 10 times within 6 hours, which allows it to be used on objects with high humidity.

1. Introduction

It is possible to increase the level of operation of facilities where wooden building structures are used by modifying it, the essence of which is to give the wood the ability to resist moisture, the spread of biodamage, which contributes to the destruction of wood and accelerate the destruction process. Given the above, when modifying wood there are difficulties with the technological modes of its processing, namely, as time, temperature. This is primarily due to the fact that the modification process may not be achieved, and the use of thermodified wood leads to its destruction [1].

Knowledge of physicochemical properties of thermo-modified wood, indicators of its quality, mechanism of action on temperature materials, allows to make a choice taking into account economic indicators, duration and safety of application, ecological aspects, etc. [2, 3]. And the use of protection allows you to maintain their functions during operation for a specified period of time [4-7].

Therefore, development of technological modes of modification of wood, water absorption investigation, the impact of the structure of wood, this process is insoluble component ensuring the stability of building structures of wood and determine the need for resistance to fracture.

2. Analysis of recent research and publications

In recent years, in the direction of research on thermal modification of wood, there are works aimed at studying the equilibrium moisture during modification, dimensional stability, durability and mechanical properties of modified wood [8]. Mass loss, wettability, wood color, and chemical transformations have since been extensively studied, while work is needed to focus on quality control, modeling, and the study of the causes of improved wood properties.

Content from this work may be used under the terms of the Creative Commons Attribution 3.0 licence. Any further distribution of this work must maintain attribution to the author(s) and the title of the work, journal citation and DOI.
Published under licence by IOP Publishing Ltd

Thermomodified wood is one of the best materials for lining [9]. Due to this humidity measurement conducted thermally modified heat and low moisture content of thermally modified wood in comparison with the standard spruce. But no, that was the extent and modes of modification of wood.

Increased use of thermally modified wood [10] led to the need for reliable quality control, including control of products deviations within certain limits, allowing you to control third party in the event of certification and regulation of the complaints and demands of consumers. However, it is not specified what methods are needed to characterize the change in quality in terms of improved target properties of modified wood during industrial production.

A study of changes in swelling and surface roughness of wood alder and elm normal after heat treatment at two different temperatures and duration [11]. It was found that the parameters of swelling and surface roughness differed significantly for two temperatures and two durations of heat treatment. The values of swelling and surface roughness decreased with increasing processing temperature and processing time, but did not specify the technological parameters.

The characteristics of wood due to weather conditions change over time, especially the color, which reflects chemical changes [12]. The change of color and reflectivity of wooden surfaces due to artificial weathering, obtained with the help of a solar box chamber, which simulates external conditions and subsequent leaching of water, was carried out. As the weathering time increases, the untreated surfaces of the wood samples darken, while the modified samples lighten to have a similar color or, in any case, to reduce the chromatic difference that was at the beginning of the weathering tests.

During the heat treatment is many chemical reactions that lead to changes in the components of the primary cell walls of wood and darkening material. Among other changes, thermodified wood is more resistant to fungal decay and more stable at times than untreated, making it suitable for use indoors and outdoors as cladding, flooring, floors, garden furniture and window frames [13].

Laboratory tests have shown a positive effect of thermal modification on the durability, dimensional stability and thermal conductivity of wood [14]. The monitoring results showed that the elements and windows of thermally modified spruce have a significantly lower moisture content in the wood compared to windows of unmodified, and that the wax also has a positive effect on humidity.

Natural aging is usually a relatively slow process because the artificial aging plays a role important in assessing the results by reducing the time compared to the natural weathering conditions. The approach is to protect the surface with various types of commercial products such as water-based solvents with a high solids content, powder coatings and substance-free products [15].

One approach is wood modification, a set of processes that give the treated material more ability to cope with damage caused by the external environment by increasing the duration of treatment. The process is also performed to enhance the physical, mechanical or aesthetic properties of wood and derived products with the advantage that they are not as harmful to users and the environment as natural wood [16].

Thus, from the literature it is established that thermal modification of wood is able to withstand destruction. All this gives grounds to say that it is advisable to conduct a study to determine the parameters that provide resistance to water absorption by thermodified wood. Therefore, research in this area is the unresolved part of ensuring the sustainability of building designs, which resulted in the need for research.

3. The purpose of this work

The purpose of this work is to study the diffusion of moisture in wood and establish effectiveness against water absorption during thermal modification.

4. Materials and methods of research

Samples of untreated and thermodified hornbeam wood with a size of 25×20×20 mm were used to establish the diffusion of moisture by wood after thermal modification. Heat treatment of wood was performed at a temperature of 200 °C for 1 to 6 hours (Figure 1).

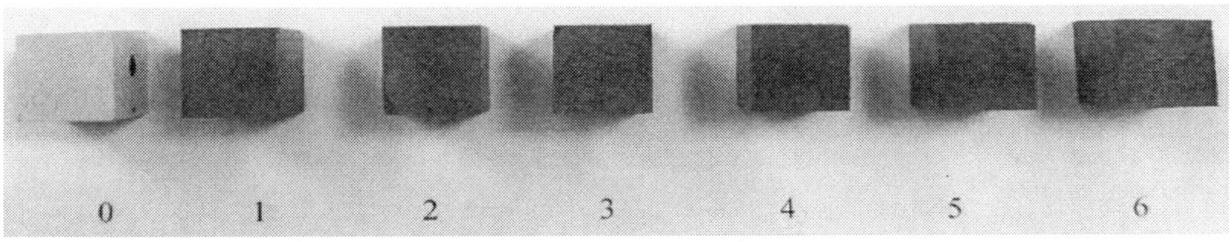

Figure 1. Model samples of hornbeam wood for testing: 0 - untreated; heat-treated at a temperature of 200 °C for: 1 – 1 hour; 2 – 2 hours; 3 – 3 hours; 4 – for 4 hours; 5 – for 5 hours; 6 – for 6 hours.

To obtain the values of moisture diffusion in wood, special equipment was developed and manufactured (Figure 2).

The test sample was fixed in a special cuvette so that it was in a humid environment above the water (Figure 2).

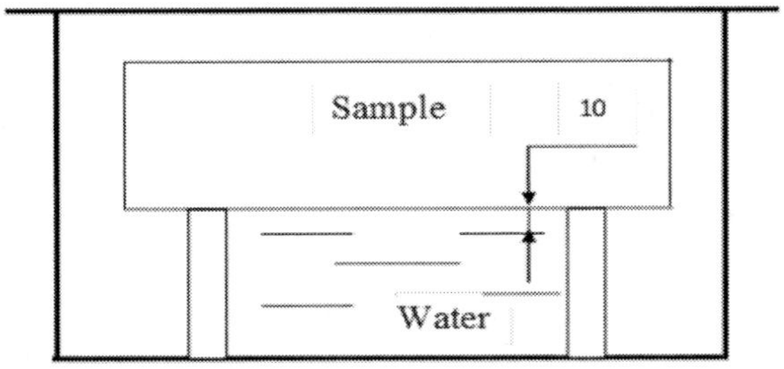

Figure 2. Device for testing the absorption of moisture by wood.

After a certain period of time, the sample was weighed on scales and the amount of water absorbed was determined.

5. Modeling of moisture diffusion parameters by wood

Since the effect of changing the moisture concentration in the sample is in the range from C_{max} to C_f (Figure 3), the value of the diffusion coefficient in this zone can also be considered insignificant, which gives reason to assume the moisture diffusion coefficient constant throughout the process of moving the front from the body surface to its center.

The advance of the phase transformation front in most processes, as a rule, is very slow, so the distribution of moisture across the spent zone in the first approximation is entrusted to the corresponding stationary distribution at the instantaneous position of the front. For a flat body, such a stationary profile is linear.

Consider the water absorption of a flat body with a thickness of $2h$, on the surface of which a constant moisture concentration C_{max} is maintained, and on the moving front of the phase transformation with a coordinate $Z(\tau)$, a constant concentration C_f value is set throughout the process.

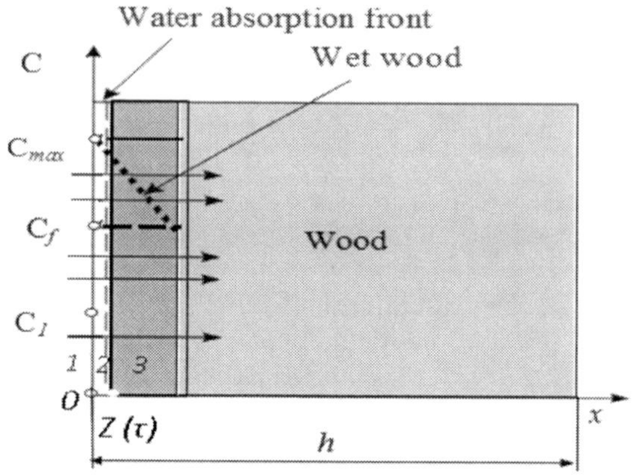

Figure 3. Scheme of promotion of the moisture front in the capillary-porous material (wood): 1 – external environment with a humidity of 100%, $x<0$; 2 – area of the wood sample with humidity, $0<x\leq Z(\tau)$ (Z – coordinate of the water absorption front, m); 3 – material of the wood sample, $Z\leq x\leq h$ (h – half the thickness of the sample, m).

The process of diffusion of moisture into the wood in the stationary approximation can be described by the equation of the form:

$$\frac{\partial^2 C}{\partial x^2} = 0, \tag{1}$$

with boundary conditions: at $x=0$, $C=C_{max}$; at $x=Z(\tau)$, $C=C_f$ gives the following distribution of concentration on the surface of the sample:

$$C = C_{\max} - \frac{C_{\max} - C_f}{Z} \cdot x. \tag{2}$$

The amount of moisture entering the wood can be described by:

$$C_1 = C_{\max} \cdot \frac{Z}{h}, \tag{3}$$

where $C_1 = C_{\max} - m(\tau) \cdot C_{\max}$.

Taking into account (3) the amount of moisture remaining in the wood is described by the following relationship:

$$m(\tau) = 1 - \frac{Z}{h}. \tag{4}$$

The derivative of the dependence (2) gives the rate of change of mass in the coordinate:

$$\frac{\partial C}{\partial x} = \frac{C_{\max} - C_f}{Z}, \tag{5}$$

and the amount of moisture absorbed by the wood during time $d\tau$ and responsible movement front $C_{max}dZ$ value that corresponds to the intensity of moisture absorption equation:

$$C_{\max} \cdot \frac{dZ}{d\tau} = D \cdot \frac{C_f - C_1}{Z}. \tag{6}$$

The integration of which under the initial condition $Z|_{\tau=0} = 0$ gives the coordinate for the current coordinate of the front:

$$Z = \sqrt{2D \cdot \frac{C_f - C_1}{C_{max}} \cdot \tau}\ . \quad (7)$$

The proportion of moisture in the wood corresponds to:

$$1 + m(\tau) = \sqrt{2\frac{D}{h^2} \cdot \frac{C_f - C_1}{C_{max}} \cdot \tau}\ . \quad (8)$$

Given the quasi-stationary process of moisture diffusion, the concentration of the substance at the front of the phase transformation can be considered as follows:

$$C_f = C_1 + \frac{C_{max}}{C_1} \cdot (C_{max} - C_1)\ . \quad (9)$$

Given the above, the proportion of moisture in the wood can be calculated by the equation:

$$m(\tau) = 1 + \sqrt{2\frac{D}{h^2} \cdot \frac{C_{max}}{C_1} \cdot \tau}\ . \quad (10)$$

Relation (10) can be used to experimentally determine the diffusion coefficient if the relative humidity in a material, such as wood, is measured during the experiment.

6. Research results

In the Table 1 shows the results of water absorption by wood.

Table 1. The results of experimental determination of water absorption by wood.

№	Exposure time, days									
sample	0	1	2	5	6	7	8	12	21	26
0	7,41	8,01	8,35	8,74	8,72	8,75	8,73	8,75	8,76	8,76
1	7,55	8,03	8,31	8,52	8,59	8,54	8,57	8,63	8,70	8,72
2	7,34	7,81	7,99	8,31	8,35	8,35	8,38	8,33	8,38	8,38
3	7,38	7,87	8,07	8,29	8,32	8,33	8,30	8,44	8,45	8,44
4	7,11	7,52	7,72	7,99	8	8	8,00	8,07	8,11	8,12
5	7,16	7,56	7,74	7,87	7,89	7,9	7,93	7,99	7,99	8,00
6	7,01	7,39	7,6	7,76	7,8	7,81	7,79	7,86	7,86	7,86

According to the obtained experimental data, the share of moisture in the wood that was thermodified was calculated, which are shown in Figure 2.

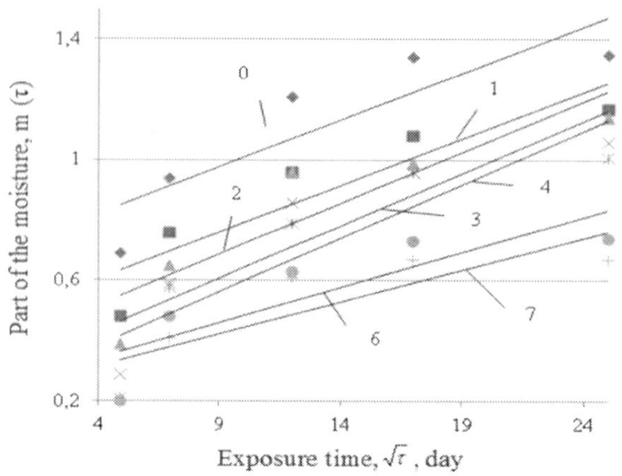

Figure 4. Change of water absorption of thermodified wood: 0 - untreated; heat-treated at a temperature of 200 °C for: 1 – 1 hour; 2 – 2 hours; 3 – 3 hours; 4 – for 4 hours; 5 – for 5 hours; 6 – for 6 hours.

After measuring the tangent of the angle of inclination of the obtained line, the diffusion coefficient (D) is calculated for known values C_1, C_{max} and h as a function of:

$$D = \frac{h^2}{2} \cdot \frac{C_{max}}{C_1} \cdot tg^2\alpha , \quad (11)$$

which is for wood depending on the time of thermal modification, Table 2.

Table 2. The value of the diffusion coefficient in water absorption of thermodified wood.

Indicator	Time of thermal modification of wood, hours						
	0	1	2	3	4	5	6
Moisture diffusion coefficient, $м^2$/s	0,024916	0,00045	0,00044	0,00013	0,00011	0,00004	0,000006

As can be seen from the table, when the wood is thermally modified for 6 hours, the water absorption is reduced by more than 10 times, which allows it to be used on objects with high humidity.

7. Conclusions

The obtained dependences allow, on the results of experimental researches, to carry out calculations of coefficient of diffusion of moisture in wood and to establish norms of application of a building design. The results of the research will also allow to purposefully solve further problems related to the creation of new means and methods of protection of organic materials and conditions of their operation at various sites.

References

[1] Lo Monaco A, Pelosi C, Agresti G, Picchio R and Rubino G 2020 Influence of thermal treatment on selected properties of chestnut wood and full range of its visual features *Drewno* **63 (205)** pp 5-24

[2] Gots V I, Berdnyk O Y, Rogozina N O and Maystrenko A A 2019 Production of modified basalt fibre for heat-insulating products manufacturing *IOP Conference Series: Materials Science and Engineering (MSE)* **708** 012082 doi:10.1088/1757-899X/708/1/012083

[3] Berdnyk O Yu, Lastivka O V, Maystrenko A A and Amelina N O 2020 Processes of structure formation and neoformation of basalt fiber in an alkaline environment *IOP Conf. Series: Materials Science and Engineering. Innovative Technology in Architecture and Design (ITAD)* **907** 012036 doi:10.1088/1757-899X/907/1/012036
[4] Tsapko Yu, Tsapko A and Bondarenko O 2020 Determination of the laws of thermal resistance of wood in application of fire-retardant fabric coatings *Eastern-European Journal of Enterprise Technologies* **2/10 (104)** pp 13-18 doi: 10.15587/1729-4061.2020.200467
[5] Tsapko Yu, Tsapko A, Bondarenko O 2020 Research of conditions of removal of fire protection from building construction *Key Engineering Materials* **864** pp 141-148 doi:10.4028/www.scientific.net/KEM.864.141
[6] Tsapko Yu, Horbachova O, Tsapko A, Mazurchuk S, Zavialov D and Buiskykh N 2021 Establishing regularities in the propagation of phase transformation front during timber thermal modification *Eastern-European Journal of Enterprise Technologies* **1/10 (109)** pp. 30-36 doi: 10.15587/1729-4061.2021. 225310
[7] Tsapko Yu, Rogovskii I, Titova L, Bilko T, Tsapko A, Bondarenko O and Mazurchuk S 2020 Establishing regularities in the insulating capacity of a foaming agent for localizing flammable liquids *Eastern-European Journal of Enterprise Technologies* **5/10 (107)** pp 51-57 doi: 10.15587/1729-4061.2020.215130
[8] Esteves B and Pereira H 2009 Wood modification by heat treatment *Bioresources* **4 (1)** pp 370-404
[9] Humar M, Lesar B and Kržišnik D 2020 Moisture performance of façade elements made of thermally modified norway spruce wood *Forests* **11 (3)** 348 doi: 10.3390/f11030348
[10] Humar M, Repič R, Kržišnik D and Lesar B 2020 Quality control of thermally modified timber using dynamic vapor sorption (dvs) analysis *Forests* **11 (6)** p 666 doi: 10.3390/f11060666
[11] Aytin S and Korkut P 2016 Effect of thermal treatment on the swelling and surface roughness of common alder and wych elm wood *Journal of Forestry Research* **27 (1)** pp 225-229 doi: 10.1007/s11676-015-0136-7
[12] Pelosi G, Agresti L, Lanteri R, Picchio E, Gennari E and Lo Monaco A 2020 Artificial weathering effect on surface of heat-treated wood of ayous *The 1st International Electronic Conference on Forests (IECF)*
[13] Ugovšek B, Šubic G, Humar M, Lesar B, Thaler N, Brischke C, Jones D and Lozano J I 2016 Performance of windows and façade elements made of thermally modified norway spruce (picea abies) in different climatic conditions *In Proceedings of the WCTE 2016-World Conference on Timber Engineerin* doi: 10.1007/s11998-016-9871-8
[14] Ugovšek B and *et al* 2019 Short-term performance of wooden windows and facade elements made of thermally modified and non-modified Norway spruce in different natural environments *Wood Material Science and Engineering* **14** pp 42-47
[15] Bonifazi G, Serranti S, Capobianco G, Agresti G, Calienno L, Picchio R, Lo Monaco A, Santamaria U and Pelosi C 2017 Hyperspectral imaging as a technique for investigating the effect of consolidating materials on wood *Journal of Electronic Imaging* **26 (1)** 011003
[16] Jones D, Sandberg D, Goli G and Todaro L 2019 Wood Modification in Europe: a state-of-the-art about processes, products and applications p 123

Acknowledgements
Authors express their gratitude to the Ministry of Education and Science of Ukraine for financial support of the research, that was performed in the framework of budget funding No. 0121U001007, as well as for the development of the theme of research according to the program of scientific cooperation COST Action FP 1407 "Understanding the modification of wood through an integrated scientific and environmental approach" of the European Union's framework program HORIZON 2020.

IOP Conf. Series: Materials Science and Engineering 1164 (2021) 012031 doi:10.1088/1757-899X/1164/1/012031

Determination of the strength of the support sections of concrete beams with BFRP under the action of shear forces

I Karpiuk[1,*], Ye Klymenko[1], V Karpiuk[1], A Posternak[1], O Maistrenko[1] and A Tselikova[1]

[1]Odessa State Academy of Civil Engineering and Architecture, 4 Didrihson Street, 65029, Odesa, Ukraine

*E-mail: irina.carpyuk@gmail.com

Abstract. In a few foreign codes (ACI, Eurocode 2, JSCE, CSA, CNR), the principles for calculating flexible concrete elements with (BFRP) are preserved the same as for calculating concrete structures with steel reinforcement. Experimental studies have not confirmed these principles, which indicates the relevance of research. Two calculation models are proposed for determining the strength near the support sections of concrete beams with BFRP. The first model considers the calculation of beams with average (*a/d=2*) and large (*a/d=3*) shear spans along the main inclined crack, taking into account the experimentally established length of its projection l_{crf1}, the coefficients φ_{c2}, φ_{c3} and φ_{c4} the magnitude of tensile stresses in the transverse fittings. The second computational model provides for the calculation of these beams with small shear spans (a/d≤1) along a compressed inclined strip using the experimentally established coefficient k_{f5}. The use of the presented design models provides good convergence (coefficient of variation ν=8%) of the calculated and experimental values of strength near the support sections of concrete beams with BFRP.

1. Introduction

Recently, concrete structures with non-metallic composite reinforcement (FRP) are increasingly used in the construction of bridge spans, in hydraulic and transport construction, in the arrangement of treatment facilities, chemical and food industry facilities, as well as special-purpose buildings. The advantages of using basalt-plastic reinforcement (BFRP) are the low cost of the main raw material (basalt fibers), due to the large reserves in the world of basalt, as well as their unique physicochemical and mechanical properties.

High strength, resistance to chemical and physical corrosion, dielectric and dimagnetic properties, low weight and low thermal conductivity of FRP contribute to its increasing use instead of steel reinforcement. However, the widespread use of FRP for reinforcing concrete structures is complicated by the insufficient number of studies carried out on this topic, as well as experimentally confirmed author's and normative methods for calculating their bearing capacity. And if there are relatively many publications devoted to the calculation of the strength of normal sections of elements with FRP, then the question of determining the bearing capacity of their inclined sections is still open. The existing theory of calculating structures reinforced with non-metallic composite reinforcement is based not on experimentally substantiated design schemes, but on various kinds of analogies. Therefore, research in this area is important and relevant.

2. Analysis of recent publications

In the Odessa State Academy of Civil Engineering and Architecture, complex experimental and theoretical studies of the bearing capacity of concrete beam elements with BFRP are carried out [1, 2]. Comparison [3] of the obtained experimental data [1, 2] of the bearing capacity of inclined sections of the indicated structures with the results of calculations according to the recommendations of the European EUROCODE-2 [4], Japanese JSCE [5], American ACI [6], Canadian CSA [7], Italian CNR [8] design codes, showed significant discrepancies between them (variation coefficients υ fluctuate

Content from this work may be used under the terms of the Creative Commons Attribution 3.0 licence. Any further distribution of this work must maintain attribution to the author(s) and the title of the work, journal citation and DOI.
Published under licence by IOP Publishing Ltd

within 59-85%). These norms, as a rule, underestimate their real bearing capacity by several times. The poor convergence of experimental data with the results of calculations is due to the fact that these norms are based not on real physical schemes of the operation of such structures, but on the classical (more than a century old) model of the truss analogy W. Ritter - E. Mörch and its various modifications, arched analogy or a combination of truss and arched analogies. A remarkably good convergence (υ = 17.5%) of experimental data and calculation results was shown by A.A. Gvozdev, A.S. Zalesova, A.F. Ilyin et al. As amended by SNiP [9] for beams reinforced with steel reinforcement [10]. Obviously, this method is advisable to adapt and apply to the design of concrete beams with FRP.

The aim of this work is to improve the existing design models based on the method of limiting forces, bearing capacity of inclined sections of concrete beam structures with FRP.

Research methods. Before the start of laboratory experiments, predictions were made for assessing the bearing capacity of prototypes-beams by modeling their stress-strain state in the PC "Lira - SAPR". Experimental samples - the beams were reinforced with basalt-plastic reinforcement BFRP and manufactured in accordance with the planning theory according to the three-factor, three-level Box-Behnken B3 plan. The beams were tested according to a four-point scheme under a static effect of a load, which was applied in steps of up to $0{,}1F_{ult}$. The load holding at each stage was 15 minutes with taking readings from measuring instruments, measuring the width of the opening of normal and oblique cracks. The experimental data of the bearing capacity of the beams were compared with the results of calculations according to the recommendations of the existing national design standards and author's methods. For further improvement, the method of limiting efforts was chosen, which turned out to be the closest to the physical picture of the work of concrete beams with BFRP.

3. Main material and results

The performed comparisons [3, 9, 11] of the experimental and calculated values of the bearing capacity of the inclined sections of the experimental concrete beams with steel reinforcement showed their unsatisfactory, in general, convergence. The coefficients of variation υ vary from 17.5% (the method of limiting efforts by A. A. Gvozdev, A. S. Zalesov, A. F. Ilyin as amended by SNiP 2.03.01-84*) to 65.4% (Eurocode - 2). These discrepancies are presented [3] also in graphical form, in the form of the influence of the most significant design factors on the bearing capacity of inclined sections of reinforced concrete beams. Thus, the method of limiting efforts, laid down in the previously existing SNiP [9] and based on the experimentally provided physical picture of the operation of reinforced concrete beam structures, is in better agreement with the available experimental data (ASD series) [10, 12] than the European standards EN [4] , which are based on a modified truss analogy.

Experimental data [1, 2, 3] on the study of the bearing capacity of concrete beams, reinforced with BFRP, also do not agree with the results of calculations according to foreign design standards. The coefficients of variation υ range from 59.0% (CNR) to 85.2% (EN). Especially inadequately they reflect the influence of the relative shear span *a/d*. Thus, a comparison of the experimental and calculated values of the bearing capacity of concrete beams with BFRP showed that a modified truss analogy or a modified theory of compression fields cannot serve as adequate design schemes for determining the strength of their inclined sections.

4. Features of the calculation of the bearing capacity of inclined sections of concrete beams with basalt-plastic reinforcement

Taking into account the recommendations of A. A. Gvozdev, A. S. Zalesova, A. F. Ilyin as amended by SNiP [9], calculation of concrete beams with large (a/d=2) shear spans, reinforced in sufficient quantity ($\rho_{fw} \geq 0{,}0025$, $\rho_{lf} \geq 0{,}0100$) with basalt-plastic reinforcement BFRP (Figure 1), on the combined action of shear force and bending moment to ensure their bearing capacity along an inclined crack is recommended to be carried out along the most dangerous inclined section from the condition:

$$V_f \leq V_{fc} + V_{fw} + V_{f,inc} \,. \tag{1}$$

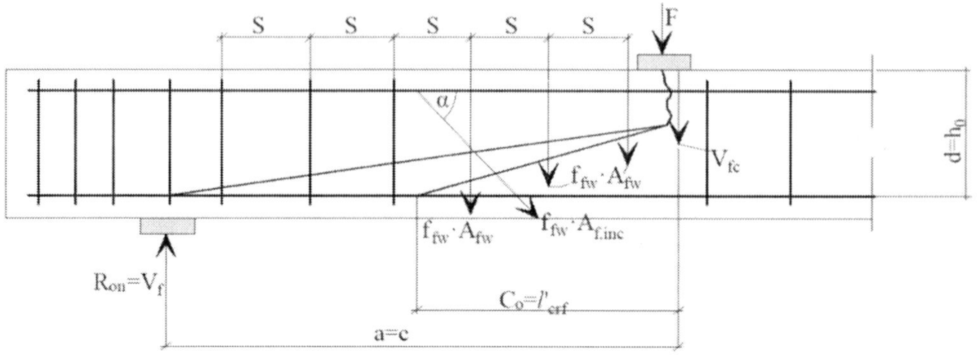

Figure 1. Scheme, near the support section of a concrete beam, reinforced with BFRP, when determining the bearing capacity of its inclined section.

Where $V_{fc}, V_{fw}\, i\, V_{f,inc}$ - are the components of the shear force, which are perceived, respectively, by concrete, transverse basalt-plastic reinforcement and bends.

The transverse force V_f in condition (1) is determined from the external load located on one side of the inclined section under consideration.

The component of the shear force perceived by the concrete in the compressed zone is proposed to be determined using the improved formula of A.S. Zalesova and A.F. Ilyina:

$$V_{fc} = \frac{\varphi_{c2}(1+\varphi_f+\varphi_n) f_{ctd} \cdot b_w \cdot d^2}{a}, \tag{2}$$

where a = c is the length of the projection of the most dangerous inclined section on the longitudinal axis of the element (span of the cut);

φ_{c2} - coefficient that comprehensively takes into account the influence of the shear span, the concrete class and the amount of transverse reinforcement BFRP. This coefficient, in contrast to its prototype φ_{b2}=2 in SNiP [9] for heavy concrete, takes into account the influence of design factors on the value of V_{fc}. in a differentiated manner. The φ_{c2} coefficient was obtained experimentally. Having experimentally established adequate values of the projection length of a dangerous inclined crack $l'_{crf} = c_0$ (experimental-statistical dependence (3), as well as confirmed by direct measurements of deformation ε_{fw} and modeling of SSS stress σ_{fw} in transverse reinforcement bars, reliable values of the components V_{fw}. were determined.

Hence the experimental significance $V_{fc} = V_f^{exp} - V_{fw}$. At the same time $V_{f,inc} = 0$, since there were no bends in the beams.

The length of the projection of a dangerous inclined crack can be represented by the following experimental-statistical dependence:

$$\hat{Y}(l_{crf1}) = 176 + 25X_1 + 9X_2 - 16X_3 - 9X_1^2 + 8X_2^2 - 9X_3^2 - 9X_1X_3, мм, \upsilon = 5.4\%. \tag{3}$$

The experimental-statistical dependence of the variable coefficient φ_{c2} has the form

$$\hat{Y}(\varphi_{c2}) = 1{,}38 - 0{,}25X_1 - 0{,}12X_2 - 0{,}22X_1X_3, \upsilon = 5.2\%, \tag{4}$$

which, after replacing the coded variables with natural values of design factors, turns into the following form:

$$\varphi_{c2} = 1.38 - 0,25\left(\frac{a}{h_0} - 2\right) - 0.12\left(\frac{C - 35M\Pi a}{15M\Pi a}\right) - 0.22\left(\frac{a}{h_0} - 2\right)\left(\frac{\rho_{fw} - 0.0072}{0.0043}\right). \tag{5}$$

Taking into account the admissible extrapolation of the obtained results, dependence (5) is valid for the following values of design factors:

- the relative span of the cut (factor X_1), a/d=1.0-3,3;
- concrete class (X_2) in MPa from C12 / 15 to C45 / 55;
- transverse reinforcement coefficient (X_3) ρ_{fw} =0.0018-0.0126.

For practical use, the length of the projection of a dangerous inclined crack in concrete beams with FRP, including BFRP, under static load action, is recommended to be determined by the experimental-statistical dependence (3), which, after replacing the coded variables with their natural values, takes the form:

$$l'_{crf1} = [176 + 25(a/h_0 - 2) + 9\left(\frac{C - 35M\Pi a}{15M\Pi a}\right) - 16\left(\frac{\rho_{fw} - 0.0072}{0.0043}\right) - 9(a/h_0 - 2)^2 + \\ + 8\left(\frac{C - 35M\Pi a}{15M\Pi a}\right)^2 - 9\left(\frac{\rho_{fw} - 0.0072}{0.0043}\right)^2 - 9(a/h_0 - 2)\left(\frac{\rho_{fw} - 0.0072}{0.0043}\right)]\text{мм}. \tag{6}$$

Obviously, dependence (6) is also valid when the design factors change within the same limits as in expression (5).

The forces perceived by transverse reinforcement are traditionally determined by the formula (7):

$$V_{fw} = q_{fw} \cdot l'_{crf1}, \tag{7}$$

where q_{fw} is the intensity of the transverse reinforcement or the force in the transverse rods (clamps) per unit length of the beam, determined by the formula (8):

$$q_{fw} = \sigma_{fw} \cdot \frac{A_{fw}}{S}, \tag{8}$$

where σ_{fw} - stresses in the rods of transverse non-metallic composite reinforcement, which are crossed by a dangerous inclined crack.

Based on the results of in-situ measurements of deformations, as well as by modeling the stress-strain state, it was found that the stresses in the transverse bars BFRP, in beam structures with large shear spans ($a/d = 3$), are $0,10f_{fk}$, and in beams with average shear spans ($a/d = 2$) $\sigma_{fw} = 0.15 f_{fk}$. In this case, the relative deformations εfw in them before the destruction of the experimental specimens - beams were, respectively, 0.0018 and 0.0028. These data are in good agreement with the fib recommendations [14, 15, 16] to limit the relative deformations of transverse nonmetallic reinforcement to 0.20-0.25%.

In experimental beams with small shear spans ($a/d \leq 1$), direct measurements and modeling of stress-strain state showed that in vertical transverse bars BFRP compressive stresses reached, on average, - 363MПa ≈ $-0.45f_{fk}$, and relative deformations $\varepsilon_{fw} \approx -0.0084$. It is obvious that the probability of destruction near the support sections of such elements from a cut of concrete along an inclined crack ($max\tau_{cxz} \leq f_{ck}/2$) or along an inclined compressed strip ($max\sigma_{cz} = (1.0\text{-}1.5) f_{ck}$) is the same. However, given that the height of the compressed concrete zone above the top of a dangerous inclined crack under a concentrated force applied at the end of the shear span is artificially reduced (by 25% or more) due to shear deformations in concrete, the probability of its destruction according to the design scheme (Figure 1), nevertheless, remains with concrete shear and without taking into account the work of transverse reinforcement ($\sigma_{fw} = 0$).

Let us check such a possibility of destruction near the support sections of beams with small shear spans a little lower.

Shear force perceived by the bends:

$$V_{f,inc} = q_{fw,inc} \cdot l_{crf1} \cdot \quad (9)$$

When determining $q_{fw,inc}$ it is necessary to take into account the angles of inclination α and it is allowed to take stresses in them $\sigma_f = 0.25 f_{ck}$ with $\varepsilon_{f,inc} = 0.0045$ according to the fib recommendations [14, 15, 16] and under the condition of reliable anchors (bends) on the supports and in the spans of structures.

In beam structures with large and medium shear spans, in which non-metallic composite transverse reinforcement is installed according to calculation, the following condition must be met:

$$q_{fw} \geq \frac{\varphi_{c3}(1+\varphi_f+\varphi_n)\cdot f_{ctd}\cdot b}{2}, \quad (10)$$

as well as the corresponding design requirements.

Calculation of concrete beams reinforced only with longitudinal non-metallic composite reinforcement (without transverse rods and clamps) for the combined action of shear forces and bending moments in order to ensure their bearing capacity along an inclined crack is also recommended to be performed along the most dangerous section from the condition:

$$V_f = \frac{\varphi_{c4}(1+\varphi_n)f_{ctd}b\cdot d^2}{\alpha^{l}}, \quad (11)$$

in which, the right-hand side should not exceed the value and should not be less

$$\varphi_{c3}(1+\varphi_n)f_{ctd}\cdot b\cdot d\,. \quad (12)$$

It is recommended to determine the φ_{c3u} coefficients taking into account their decrease in comparison with the φ_{b3} coefficients from SNiP [9] according to the formulas:

$$\varphi_{c3} = \frac{\varphi_{c2}}{\varphi_{b2}}\varphi_{b3}, \quad \varphi_{c4} = \frac{\varphi_{c2}}{\varphi_{b2}}\varphi_{b4}, \quad (13)$$

Moreover, for heavy concrete φ_{b3}= 0,6 and φ_{b4}= 1.5.

Comparison of experimental data and calculated values of bearing capacity of inclined sections of concrete beams with BFRP performed in [3] according to the method of limiting forces improved for FRP showed, in general, their satisfactory convergence (υ = 11%). The largest discrepancies between the experimental and calculated values of the breaking shear force were observed in basalt concrete beams with small shear spans ($a \leq d$). To improve the convergence of the experimental and predicted values of the bearing capacity near the support sections of concrete beams with BFRP with small shear spans ($a \leq d$), we calculate it, according to the recommendations of T.I. Baranova [17], according to the scheme of short cantilevers, transformed in our case to the form (Figure 2). The validity of this approach is confirmed by the results of field experiments using distribution plates of width l_{sup} under concentrated forces and on supports, as a result of which an artificial reduction in the cut span occurred. This conclusion is also confirmed by the results of modeling the stress-strain state of concrete and transverse reinforcement, which in all beams with such shear spans experienced compression defloration.

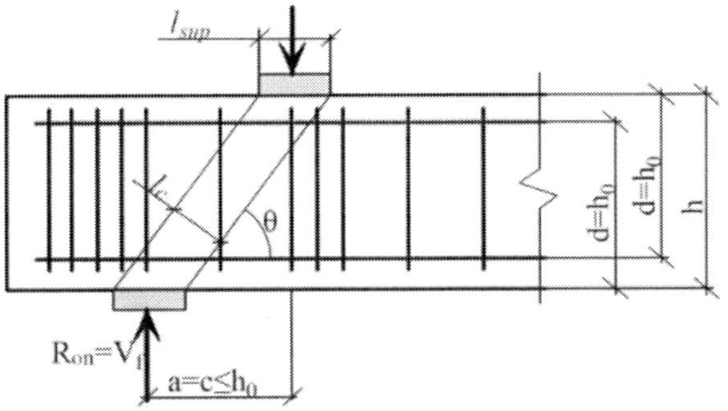

Figure 2. Diagram of the bearing capacity near the support section of a concrete beam, reinforced with BFRP, with a small shear span ($a \le d$) along an inclined compressed strip between the concentrated force and the support.

The bearing capacity of the inclined compressed strip of the considered beam is determined from the condition:

$$V_f \approx F \le k_f \cdot \varphi_{w2} \cdot f_{cd} \cdot b \cdot l_c \cdot sin\theta \le 3.5 f_{ctd} \cdot b \cdot h_0,$$
$$k_f \cdot \varphi_{w2} \cdot f_{cd} \cdot b \cdot l_c \cdot sin\theta \ge \frac{\varphi_{c4}(1+\varphi_n) f_{ctd} \cdot b \cdot d^2}{a}, \quad (14)$$

where θ is the angle of inclination of the compressed strip on the support section of the beam to its longitudinal axis;

l_c - the width of the calculated compressed strip. which is determined by the formula:

$$l_c = l_{sup} \cdot sin\theta, \quad (15)$$

where l_{sup} is the length of the load transfer area along the beam span;

φ_{W2} - коэффициент, учитывающий влияние поперечной арматуры или хомутов, расположенных по высоте балки (при их наличии), и определяется по формуле:

$$\varphi_{w2} = 1 + 5\alpha \cdot \mu_{w1}, \quad (16)$$

where $\alpha = E_f/E_c$; $\mu_{w1} = A_{fw}/(b \cdot s_w)$,

where A_{fw} - cross-sectional area of the clamps (if any) in one horizontal plane;

s_w - the distance between the specified clamps (if any) along the height of the beam along the normal to them. In this case, only horizontal clamps or those inclined at an angle of not more than 45° to the horizontal are taken into account in compliance with the relevant design requirements.

Instead of the coefficient k_f, when calculating short cantilevers according to the method [17], its constant value equal to 0.8 is used. For concrete beams with a small shear span and short cantilevers reinforced with BFRP, inversely recalculated, the experimental values of the coefficient k_f were found by the formula

$$k_f = \frac{V_{uf}}{\varphi_{w2} \cdot f_{ck} \cdot b \cdot l_c \cdot sin\theta}, \quad (17)$$

subject to conditions (14).

The processing of the values of the coefficient k_f obtained in this way using the least squares method made it possible to obtain an adequate experimental-statistical dependence for the indicated concrete beams or short cantilevers:

$$\hat{Y}(k_f) = 1.67 - 0.25X_2 - 0.21X_3 - 0.08X_2X_3,\ \upsilon = 2.5\%, \tag{18}$$

replacing the coded variables with natural values of design factors, within the framework of the carried out full-scale experiment, made it possible to obtain an empirical expression for the coefficient k_f, which is fair taking into account the permissible extrapolation outside the limits of the concrete class change from C12/15 to C45/55 and the coefficient of transverse reinforcement, ρ_{fw} from 0.0018 to 0.0126:

$$k_f = 1.67 - 0.25\left(\frac{C - 35М\Pi a}{15М\Pi a}\right) - 0.21\left(\frac{\rho_{fw} - 0.0072}{0{,}0043}\right) - 0.08\left(\frac{C - 35М\Pi a}{15М\Pi a}\right)\left(\frac{\rho_{fw} - 0.0072}{0.0043}\right). \tag{19}$$

The geometric interpretation of the dependence of the coefficient k_f on the ratio of design factors is shown in Figure 3.

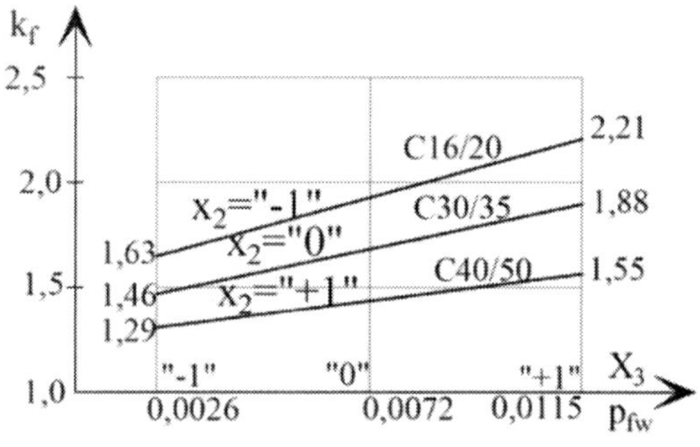

Figure 3. Influence of the class of concrete and the amount of transverse reinforcement on the value of the variable coefficient in condition (14).

Thus, the use of experimentally established coefficients for beams reinforced with BFRP, with large (*a/d=3*), medium (*a/d=2*), as well as kf for elements with small (*a/d≤1*) shear spans, allowed to bring together the experimental and calculated values of their bearing capacity near the reference sections, to a satisfactory convergence (υ=7,9%).

The design of concrete T-beams and I-beams reinforced with BFRP, with the combined action of a shear force and a bending moment, is performed to ensure the strength of the inclined strips between the inclined cracks in their walls in the vicinity of the support sections according to the formula:

$$\text{where } V_f \approx F \le 0{,}3\varphi_{w1} \cdot \varphi_{c1} \cdot f_{cd} \cdot b_w \cdot d,, \tag{20}$$

where - coefficient that takes into account the effect of transverse bars or stirrups normal to the longitudinal axis on the bearing capacity near the support sections of these beams. It is recommended to determine it by formula (16) and take equal to no more than 1.1;

b_w - wall thickness of T-section or I-section.

The coefficient φ_{c1} is recommended to be determined by the method of A. S. Zalesova and A. F. Ilyin, stated in [9], with the adjusted value of the coefficient β^*:

$$\varphi_{c1} = 1 - \beta^* \cdot f_{cd}, \tag{21}$$

where $\beta^* = (\varphi_{c2}/\varphi_{b2})\beta$,

φ_{b2} and β - experimental values of the coefficients from [9]. For heavy concrete, their values are, respectively, 2 i 0.01.

5. Conclusions:

1. Comparison of the experimental values of the bearing capacity of inclined sections of concrete beams reinforced with BFRP with the results of calculations performed in accordance with the recommendations of the existing national design codes of foreign countries [3] showed unsatisfactory convergence: the coefficient of variation υ using Eurocode-2 [5] was 85, 2%, Japanese JSCE [5] 77.9%, American ACI [6] 81.7%, Canadian CSA [7] 61.1%, Italian CNR [8] 59%. Basically, these norms significantly underestimate the real bearing capacity near the supporting sections of the experimental samples - beams. Based on the foregoing, the basic principles on which the considered national standards for the design of concrete structures with FRP are based cannot be adopted for calculating the bearing capacity of the considered sections of beams reinforced with BFRP, without making appropriate changes and adjustments to them.

2. In order to search for the closest to reality physical models of the operation of experimental reinforced concrete beams (series 1, ACD), which are analogs of concrete beams reinforced with BFRP, a comparative analysis of the data (their actual bearing capacity with its calculated values) was made, which was determined by the method boundary efforts [9], as well as using a modified truss analogy [4, 13]. It was found that the method of boundary forces [9] provides a significantly better convergence (υ = 17.5%) of the experimental and calculated values of the bearing capacity of concrete beams with steel reinforcement in comparison with the modified truss analogy [4, 13] (υ=65.4%). This method is better than European and others, as studies have shown [10, 11], design standards reflect the physical picture of the work of the investigated beams under load.

3. Considering the above, the bearing capacity of inclined sections of concrete beams reinforced with BFRPF is recommended to be determined by the method of boundary forces. At the same time, the calculation of beams with large $(a/d=3)$ and medium $(a/d=2)$ shear spans should be performed along an inclined crack using variable coefficients $\varphi_{c2}, \varphi_{c3}, \varphi_{c4}$, taking into account the real projection length of a dangerous inclined crack l'_{crf1} and a significant reduction in tensile stresses σ_{fw} to $(0.10-0.15)f_{fk}$ in transverse reinforcement. The bearing capacity near the support sections with small $(a \le d)$ shear spans must be determined as for short cantilevers along an inclined compressed strip between the concentrated force and the support using a variable coefficient k_f. Compliance with these requirements allows for satisfactory convergence of the calculated and experimental values of the bearing capacity of inclined sections and near the support sections of concrete beams reinforced with BFRP, to carry out their reliable and, at the same time, more economical design.

References

[1] Karpiuk V, Tselikova A, Khudobych A, Karpiuk I and Kostyuk A 2020 Study of strength, deformability property and crack resistance of beams with BFRP *Eastern-European journal of enterprise technologies* **4/7(106)** pp 42-53

[2] Rusu I, Khudobych A,. Tselikova A, Karpiuk V, Karpiuk I and Zavoloka M 2020 Features of stress-strain behaviour of basalt fibre reinforced concrete beam structures *Journal of Engineering Science* (Kishinev. Moldova) **XXVII (2)**

[3] Karpiuk I, Karpiuk V, Klimenko E, Tselikova A and Khudobych A 2020 Comparative analysis of research and calculated values of bearing capacity of reinforced concrete and basalt concrete beams according to recommendations of national design standards *Bulletin of odessa state academy of civil engineering and architecture collection of scientific works* **80** pp 43-49

[4] EN 1992-1-1:2004, Eurocode 2 - *Design of Concrete Structures Part 1 General rules and rules* (CEN)

[5] JSCE, 1997 *Recommendation for Design and Construction of Concrete Structures Using Continuous Fiber Reinforcing Materials* Tokyo (Japan: Japan Society of Civil Engineers)

[6] ACI 440.1R-03, 2003 *Guide for the Design and Construction of Concrete Reinforced with FRP Bars* (American Concrete Institute)
[7] CAN/CSA-S6-00, 2019 *Canadian High Bridge Design Code* (Canadian Standards Association)
[8] CNR-DT 203/2006 *Guide for the Design and Construction of Concrete Structures Reinforced with Fiber-Reinforced Polymer Bars* (Rome, Italy)
[9] SNiP 2.03.01 84* *Betonnie i jelezobetonnie konstrukcii* Gosstroi SSSR (M.: CITP Gosstroya SSSR)
[10] Karpyuk V M, Somina Yu A, Kostyuk A I and Maistrenko O F 2018 *Osoblivosti naprujeno deformovanogo stanu irozrahunku zilizobetonnih konstrukcii za dii ciklichnogo navantajennya visokih rivniv* Monografiya. (Odesa: ODABA) 234
[11] Karpiuk V, Karpiuk I, Tselikova A and Khudobych A 2020 Calculating model of the bearing ability of the substructed areas of baltic basalt concrete structures *Science and education a new dimension Natural and Technical Science* Hungary **VIII (29)** pp 16-22
[12] Karpiuk V, Orešković M, Somina Yu and Kostiuk A 2019 *Basis of Force and Deformation-Force Resistance of Reinforced Concrete at the Complex Stress-Strain State* Monograph (Varaždin: Croatia) 99
[13] DSTU B V.2.6-145-2010. *Konstrukcii budinkiv i sporud. Zahist betonnih i zalizobetonnih konstrukcii vid korozii. Zagalni tehnichni vimogi* (K.: Minrehionbud Ukrayiny)
[14] Fib bul. 55, ModelCode 2010. *First complete draft* (Switzerland: fib)
[15] Fib bul.40, 2007 *FRP reinforcement in RC structures. Technical report TG9.3* (Switzrland: fib)
[16] FIB Bulletin 40 2007: *FRP reinforcement in RC structures. -International Fed-eration for Structural Concrete*
[17] Baranova T I, Zalesov A S 2003 *Karkasnosterjnevie raschetnie modeli i injenernie metodi rascheta jelezobetonnih konstrukcii* (Moscow: ASV) 240

Some features of oscillations and stability of reinforced cylindrical shells under action of movable inertial loading

D Klyushnyk[1,*], A Demianenko[1] and A Zelensky[2]

[1] Dniprovsk State Agrarian and Economic University, 25, Efremova str., Dnipro 49027, Ukraine
[2] Prydneprovsk State Academy of Civil Engineering and Architecture, 24a, Chernichevskogo str., Dnipro 49600, Ukraine

[*]E-mail: kdv_ua@i.ua

Abstract. This paper describes some features of the mathematical models for the elastic shell with movable load and for the elastic elements of changeable length. In these systems two forms of eigenoscillations - the eigen component and the accompanying one, displaced in phase to the right angle correspond to every frequency of the system. The accompanying component is caused by the mobile inertia load or by the changeable length and they are non-trivial when those factors are present. As well as for objects with time-varying length, these problems lie in outside of the scope of the classical problems of mathematical physics due to that the eigenfrequencies and eigenforms will be presented as time-nondependent functions. This non-classical section of the mathematical physics is waiting for its development, new researches and generalizations. The main attention of this work is paid to the elastic systems with the mobile inertia load. The problem of oscillations and stability of thin cylindrical envelope with regular stringer set is being considered under assumption that they are regularly placed with a small step from one to another. Also, we assume that there is unsymmetrical stream of mobile inertia load. There is a wide variety of problems in area of jet engine, cooling systems, flying, and floating vehicle framework's design, which boil down to such computational schemes. The linear theory of structural-orthotropical envelopes with equilibrium equations and geometrical Donnel-Mushtary relations are being used to obtain of the discussed problem formulation. We assume that ribs of the shell are considered as Kirchhoff-Klebsh bars. The accurate solution of the problem is obtained by using of two-waves representation for motion and it has been applied for research on influence of geometry, elastic and structural parameters on the oscillation and stability properties of the shell.

1. Introduction and literature review

174 years have gone since the day of the first formulation of the problem of acting of movable loading on the elastic structure and building, after decay of Chester bridge in England in May 1847. During this time a lot of problems of dynamical impact of the movable loadings are different by their nature, behavior and influence on the elastic structures, systems, and buildings have been considered, solved and verified.

State-of-the-art of technology, increased road traffic intensity, intensification of flaw process require more accurate mechanical and mathematical models, that could reflect the essence of the phenomenon more comprehensively and precisely, so there is necessity of improvements and modifications for traditional modeling concepts or new concepts and methods of researches.

In the well-known review [10], dedicated to 100th anniversary of the problem formulation, the famous scientists in mechanical engineering Ya.G. Panovko wrote: "The problem of dynamic acting of movable load, 100th anniversary of which we have celebrated in 1947, by today did not lost their up-to-date status, the life still set new tasks and caused by this following motion of the theory to forward".

In agile XX-XXI centuries significant increasing of masses and velocities of motion formulates a wide variety of new problems, and requires their solutions by developing new approaches in the

Content from this work may be used under the terms of the Creative Commons Attribution 3.0 licence. Any further distribution of this work must maintain attribution to the author(s) and the title of the work, journal citation and DOI.
Published under licence by IOP Publishing Ltd

mechanical and mathematical modeling, renewing and improving of old methods to make allowable more comprehensive representation for quantitative and qualitative features of the kinematical and dynamical properties of the system motion.

Nowadays the keen interest to this problem arose due to the intensively usage of information technologies, that allows to research the mathematical model and to analyze their results more deeply and comprehensive. The traditional representation of mechanical systems under movable inertia loads has been changed significantly. The simple examples of those systems are bridges with moving vehicles, different kind of pipelines, thin-walled structures with additional strengthening bars, plates or envelopes loaded by moving liquid or gas.

As the problem of this class, we could treat the dynamical problems of the variable-length and time-dependent length objects, the dynamic of the objects under longitudinal motion like threads, wires, profiled rod in the rolling process, strips and chain saws, belts, cables of mining lifts and some others.

In dependency of the analytical model of inertia properties of the elastic structure and acting movable loading we could use the four variants of statement of the problem of influence of the movable loading on the elastic structures and buildings [7,11]. The most complex for applying in the practice is a fourth variant, that considers both inertia forces of the structure and inertia forces of the movable loading. Research on the qualitative and quantitative properties of the motion of such objects could be reduced to analysis of the following mathematical model

$$L\left(x,t,\frac{\partial}{\partial t},\frac{\partial}{\partial x}\right)\cdot w = L_1\left(\frac{\partial}{\partial t},\frac{\partial}{\partial x}\right)\cdot q(x,t)\,, \tag{1}$$

where

$$q(x,t) = -\frac{q_0+q_1}{g}\cdot\frac{\partial^2 w}{\partial t^2} - 2\cdot\frac{q_1 v}{g}\cdot\frac{\partial^2 w}{\partial t\partial x} - \frac{q_1 v^2}{g}\cdot\frac{\partial^2 w}{\partial x^2} \tag{2}$$

with the respective boundary and initial conditions under assumption of constant motion velocity.

2. Mechanical, mathematical models and features

The main features of the mathematical models of this problem, at first, is the presence of the inertia operator q(x,t) in some of its form in the differential equation. It is distinguishing detail, that the loading is dependent from the load intensity $q_1(x)$, the velocity v of the loading stream, the elastic displacement w(x,t).

Moreover, it is clearly, there is a strong dependency between loading intensity and the acceleration w_{tt} (x,t), velocity of angular deformation w_{tx}(x,t) and variation of curvature of the elastic line of the object w_{xx}(x,t), that is in such systems the acting force is following after the system behavior and changing its value and direction during the deformation.

Thereby, acting force applied to the elastic object caused by movable mass is not predefined and depends on the state of the system. It is the second feature of the dynamic problem of the elastic system in the inertia force field of movable loadings. Third significant features of these problems is that the mathematical model should contain the mixed derivative of odd-order by time in one of its form that represents Coriolis acceleration of the moving inertial loading and does not allow to separate space and time variables by using the Fourier schema in the field of real functions.

An aerodynamic and hydrodynamic action of a liquid or a gas applied to the elastic object could be reduced to the same kind of the inertia operator. The velocity of the liquid stream in pipelines of the aircrafts is ranged between 50-80 m/s, and 200-250 m/s for gases and aircraft failures due to the instability of their pipelines attain 60% from the total number of failures [6,7].

The movable loading could be distributed uniformly or by some law, that could be discrete or continuously distributed with discrete inclusions with the constants or time-varying velocity.

It is known a number of applied mathematical researches, where successive approximation structure has been built.

At the beginning the rough approximation should be built, then the mechanical model and the respective mathematical model and choose the technology of analysis of the mathematical model to refine rough approximation and to get more accurate solution at the followed steps.

The quality of rough approximation has a critical sense to get more accurate solution quickly. The advantage of the rough approximate models and solutions is simplicity, transparency and evident.

2.1. Mechanical, mathematical models and research methods

As it is known, Fourier method of mathematical physics allows to get solutions of some class of partial differential equations in the explicit form [4,5]. Only in relatively simple cases it is possible to build up the solutions of partial differential equations as a sum of partial solutions in the form of product of the separated functions. To those equations belong the equations of oscillations of the string, a beam and some others. The direct applying of this method to the dynamic problem of elastic systems under movable inertia loading is not possible in general cases.

That is why some authors tried to use this method by its modifying and generalizing. One of the first publications was H. Steuding [13], where the lateral oscillations of the beam under movable distributed and concentrated loadings have been considered. The second one G.W. Housner [12] proved that the general solution of the partial differential equation of elastic oscillations under movable inertia loading could be obtained as a linear combination of partial solutions, those contain symmetric and antisymmetric forced forms shifted by 90 degrees in their phase. Moreover, antisymmetric forced forms occurred due to the mixed derivative of odd order by time and Coriolis' inertia forces caused by movable loading and related through them to symmetrical forced forms. The symmetrical forced forms under non-movable loading are matched to the eigenforms of the loaded system. Two above works began the method of two-wave representation of the elastic system oscillations under movable inertia loading and its physics interpretation had been provided by O. A. Goroshko [4-6].

Using the method of two-wave representation of the oscillations for research of those system, that allows in some cases to obtain analytical solutions, the general solution of the differential equations could be found as a sum of two infinity series the first series is a classical part of the solution and the second one is the part of the solution caused by presence of odd-order by time mixed derivative and inertia of movable loading, that could not be discovered by using of traditional direct methods of mathematical physics. The forms of the first group are called as eigenforms and the forms of the second group known as accompanying oscillation forms of the elastic system. Accompanying oscillations could be non-trivial if the elastic system is loaded by movable inertia loading. The modes of the first group called eigenmodes, and the modes from the second one are called as accompanying modes of the elastic system oscillations. Accompanying modes are induced and non-trivial when the movable inertia loading is present.

Today more penetrating and thorough research on the dynamic problem of elastic system under movable inertia loading by the method of two-wave representation is supported by modern information technologies, that was never used before, especially in the days of H. Steuding, G. W. Housner, Ya.G. Panovko and others.

2.2. A simple case of analogue between mathematical models of the dynamic problem of elastic objects under movable loading and the static problem

As for analogues of mathematical models, it is easy to see, that the problem of lateral oscillation of the beam under uniformly distributed inertia loading in the critical mode could be reduced to the problem of solving of the differential equation [7]

$$E \cdot J_{\min} \cdot w^{IV}(x) = -m \cdot V_{cr}^2 \cdot w^{II}(x), \tag{3}$$

with respective boundary and initial conditions, and, as it is known, could be reduced to the solving of the following differential equation

$$E \cdot J_{\min} \cdot w^{IV}(x) = -F_{cr} \cdot w^{II}(x) . \tag{4}$$

Analysis of these equations shows that the mathematical models of these problems are identical, that is to say some mathematical analogy exists and by using of this analogy we will get the approximate values of the critical speed of motion of the loading when the pinned beam will have buckling failure

$$F_{cr} = \frac{\pi^2 \cdot E \cdot J_{\min}}{(\nu \cdot l)} = m \cdot V_{cr}^2 \text{ or } V_{cr} = \frac{\pi}{\nu l} \cdot \sqrt{\frac{E \cdot J_{\min}}{m}} . \tag{5}$$

In the formulas from (3) – (5):

E – Young's modulus of longitudinal elasticity of the beam material;

w (x) - deflection of arbitrary cross-section of beam;

m - mass of unit of beam length;

J_{min} - the axial moment of inertia of the cross-section;

F_{cr} is the critical value of the compression force by Euler;

V_{cr} - the critical speed of the movable loading;

ν - coefficient of the effective length of the beam, which depends on the conditions of fixation of cross-sections.

3. Differential equation of casing radial oscillations and its solution

The problem of eigenoscillations and stability, strengthened in longitudinal direction by the regular stringer set of cylindrical shell with radius R, thickness h and length L, is considered. The stream of mass with constant intensity q1 moves along stringers with a constant velocity V. The stringers having the same geometry, placed with a small step between them and unsymmetrically to middle surface of the shell. Long measures theory of the shell with equations of balance and geometrical correlations in the shape of Donnel-Mushtary are used during compiling the equations of moving. The stringer set could be modelled by using of Kirchhoff – Klebsh approach. In this case the stiffness of stringers should be connected to the elasticity properties of the shell. If we will neglect tangential components of inertial forces, then strength factors which reduce to the middle surface of strengthened casing, will presented in the following form [2,9]

$$N_1 = B(\varepsilon_1 + \mu\varepsilon_2) + \left(\frac{EF}{l}\right)\varepsilon_1 - \left(\frac{ES}{l}\right)\kappa_1,$$

$$N_2 = B(\varepsilon_2 + \mu\varepsilon_1),\ N_{12} = 0{,}5(1-\mu)B\varepsilon_{12}, \tag{6}$$

$$M_1 = D(\kappa_1 + \mu\kappa_2) + \left(\frac{EJ}{l}\right)\kappa_1 - \left(\frac{ES}{l}\right)\varepsilon_1,$$

$$M_2 = D(\kappa_2 + \mu\kappa_1),\ M_{12} = \left[(1-\mu)D + \frac{GI_p}{2l}\right]\kappa_{12} . \tag{7}$$

In this wording our task reduces to decision the equations system of designed casing relatively the function of crook bend W and the function of effort Φ [9]

$$L_1(W) = \frac{1}{R} \cdot \frac{\partial^2 \Phi}{\partial x^2} - \frac{a_3 S}{hl}\left(\frac{\partial^4 \Phi}{\partial x^2 \partial y^2} - \mu\frac{\partial^4 \Phi}{\partial x^4}\right) + \frac{Z}{h}, \tag{8}$$

$$L_2(\Phi) = -\frac{1}{R} \cdot \frac{\partial^2 w}{\partial x^2} - \frac{a_3 S}{hl} \cdot \left(\mu\frac{\partial^4 w}{\partial x^4} - \frac{\partial^4 w}{\partial x^2 \partial y^2}\right), \tag{9}$$

where L_1 and L_2 have following form

$$L_1=\frac{D}{h}\left[b_1\cdot\frac{\partial^4}{\partial x^4}+2\cdot b_2\cdot\frac{\partial^4}{\partial x^2\partial y^2}+b_3\cdot\frac{\partial^4}{\partial y^4}\right],$$
$$L_2=\frac{1}{E}\left[a_1\cdot\frac{\partial^4}{\partial x^4}+2\cdot a_2\cdot\frac{\partial^4}{\partial x^2\partial y^2}+a_3\cdot\frac{\partial^4}{\partial y^4}\right], \quad (10)$$

and

$$a_1=\left[1+(1-\mu^2)\cdot F/hl\right]\cdot a_3;\ a_2=1+\mu(1-a_3);\ a_3=1/(1+F/hl);$$

$$b_1=1+12\left(1-\mu^2\right)\cdot\left[J/h^3l-\left(S/h^2l\right)^2\cdot a^3\right];\ b_2=1+3(1-\mu)I_p/h^3l;\ b_3=1.$$

By applying operators to the equation (8), using, that inertial forces of casing, ribs and mass streams, and distributing Z along the shell, equation of the motion for functions W and Φ could be written as below

$$L_2L_1(W)=\frac{1}{R}\cdot\frac{\partial^2}{\partial x^2}L_2(\Phi)-\frac{a_3S}{hl}\left[\frac{\partial^4}{\partial x^2\partial y^2}-\mu\cdot\frac{\partial^4}{\partial x^4}\right]\cdot L_2(\Phi)-\frac{1}{h}L_2L_3(W), \quad (11)$$

where operator L_3 has got a shape

$$L_3(W)=\frac{q}{g}\cdot\frac{\partial^2W}{\partial t^2}+\frac{q_1}{g}\left(\frac{\partial^2W}{\partial t^2}+2\cdot V\cdot\frac{\partial^2W}{\partial t\partial x}+V^2\cdot\frac{\partial^2W}{\partial x^2}\right), \quad (12)$$

and $L_2(\Phi)$ will be used as a right part of equation (9), q,q_1 - the weight of shell and mass streams, F, J, S - the area, moment of inertia and statical moment of stringer's cross-section area relatively middle surface of panelling, I_p - moment of inertia rib's cross-section for torsion, l - the rib's step.

4. Results of solutions

Partial solution of equation (11) will have following form

$$W(x,y,t)=\varphi(x,y)\cdot\cos\omega t+\psi(x,y)\cdot\sin\omega t. \quad (13)$$

After the substitution expression (13) into equation (11) and after introducing of complex function of real variables

$$\Phi_1(x,y)=\varphi(x,y)+i\psi(x,y), \quad (14)$$

will come to the partial differential equations for function $\Phi_1(x,y)$ that does not have derivatives by time. Satisfying to the arrangement of recurring by the circular coordinate, function $\Phi_1(x,y)$ could be presented in the following form

$$\Phi_1(x,y)=\sum_m F_m(x)\cos\left(\frac{m\cdot y}{R}\right), \quad (15)$$

and after substitution (15) into equations for function $\Phi_1(x,y)$ will come to the usual differential equation of eighth order for function $F_m(x)$. The general solution under assumption that the roots of characteristic equation are simple, Fm(x) could be written as below [3,8]

$$F_m(x)=\sum_{i=1}^{8}C_i\cdot e^{k_ix}, \quad (16)$$

where C_i – arbitrary constants, k_i- the roots of characteristic equation eighth degree with complex coefficients which have a shape [3,4]

$$C_0k^8 + C_1k^7 + C_2k^6 + iC_3k^5 + C_4k^4 + iC_5k^3 + C_6k^2 + iC_7k + C_8 = 0 . \qquad (17)$$

Using standard programs to define the roots of equation, for long or middle length shells the roots of equation (17) could be divide on two categories, that satisfy following conditions

$$|k_1| \ge |k_2| \ge |k_3| \ge |k_4| >> |k_5| \ge |k_6| \ge |k_7| \ge |k_8| \qquad (18)$$

At the table there are significances of characteristic equation's (17) roots for casing with means L=50 cm, R=15 cm, h=0.05 cm, d_1=1 cm, d_2=0.98 cm, the ribs quantity in the form of tubes with external and inside diameters d_1 and d_2, n=15.

Table 1. Roots of the characteristic equation (17) of eighth degree with complex coefficients

V=83 m/s ω_{15}=863 1/s	k_1	k_2	k_3	k_4	k_5	k_6	k_7	k_8
Re k	-24.50	24.50	-24.60	24.60	-3.64	3.64	0	0
Im k	-19.34	-19.34	19.35	19.35	1.01	1.01	-4.15	1.72

With considerations (18) according to the method Lobachevskogo-Greffe, looking for all roots of equation (17) reduce to the decision of two equations of fourth order

$$C_0 \cdot k^4 + C_1 \cdot k^3 + C_2 \cdot k^2 + i \cdot C_3 \cdot k + C_4 = 0, \qquad (19)$$

$$C_4 \cdot k^4 + i \cdot C_5 \cdot k^3 + C_6 \cdot k^2 + i \cdot C_7 \cdot k + C_8 = 0, \qquad (20)$$

After evaluation of roots of equations (19) and (20), we may set for elementary approach of roots (17), which in future we may define more precisely by one of iteration methods with any required accuracy.

After determining of the characteristic equation's (17) roots and satisfying the function (16) to the boundary conditions of pinned shell ends for Fm(x) in the form

$$F(x) = F_{II}(x) = F_{IV}(x) = F_{VI}(x) = 0 \quad \text{for } x=0.1,$$

will have the equation

$$\Delta = \begin{vmatrix} 1 & 1 & 1 & 1 & 1 & 1 & 1 & 1 \\ k_1^2 & k_2^2 & k_3^2 & k_4^2 & k_5^2 & k_6^2 & k_7^2 & k_8^2 \\ k_1^4 & k_2^4 & k_3^4 & k_4^4 & k_5^4 & k_6^4 & k_7^4 & k_8^4 \\ k_1^6 & k_2^6 & k_3^6 & k_4^6 & k_5^6 & k_6^6 & k_7^6 & k_8^6 \\ e^{k_1} & e^{k_2} & e^{k_3} & e^{k_4} & e^{k5} & e^{k_6} & e^{k_7} & e^{k_8} \\ k_1^2e^{k_1} & k_2^2e^{k_2} & k_3^2e^{k_3} & k_4^2e^{k_4} & k_5^2e^{k_5} & k_6^2e^{k_6} & k_7^2e^{k_7} & k_8^2e^{k_8} \\ k_1^4e^{k_1} & k_2^4e^{k_2} & k_3^4e^{k_3} & k_4^4e^{k_4} & k_5^4e^{k_5} & k_6^4e^{k_6} & k_7^4e^{k_7} & k_8^4e^{k_8} \\ k_1^6e^{k_1} & k_2^6e^{k_2} & k_3^6e^{k_3} & k_4^6e^{k_4} & k_5^6e^{k_5} & k_6^6e^{k_6} & k_7^6e^{k_7} & k_8^6e^{k_8} \end{vmatrix} = 0, \qquad (21)$$

from equality to "0" (21), which determine the frequencies $\omega_{nm}$of shell oscillations.

Finally, the general solution could be written in the following form

$$W(x,y,t) = \sum_{nm} a_{nm} \left(\mathrm{Re}[\Phi_1(x,y)] \cdot \cos(\omega_{nm}t + \alpha_{nm}) + \mathrm{Im}[\Phi_1(x,y)] \cdot \sin(\omega_{nm}t + \alpha_{nm}) \right). \qquad (22)$$

The analysis of the shell oscillations and stability, conducted by means of given method for casings with means L = 50 cm, L/R = 3,4,5,6, L/h = 2500, n = 8,10,15, 20, where n is number of stringers, for tubes with inside diameter d_1=0,98 cm and external diameter d_2=1 cm as stringers, which have shown that frequencies and critical speeds have got lesser meanings with maximal number of district waves mn. Besides, if a rib's number is increase, frequencies of oscillations slightly increase, but at that time critical speeds slow down.

5. Discussion of results obtained in the study

The obtained exact solution of the problem shows that the vibrations of the reinforced cylindrical shell in the field of the moving inertia loads could be presented as a superposition of two groups of oscillations. The oscillations of those groups will incorporate oscillations with the frequencies ω_{nm}, but oscillations of second group φ_{nm} (x, y), known as accompanying oscillations, are shifted in phase by a right angle relatively to the respective oscillations from the first group ψ_{nm} (x, y), which is known as natural or eigenoscillations.

$$\varphi_{nm}(x,y) = \mathrm{Re}\left[\Phi_1(x,y)\right],\ \psi_{nm}(x,y) = \mathrm{Im}\left[\Phi_1(x,y)\right].$$

Note, that the accompanying vibrations are caused by a moving inertial load and are non-trivial only if the inertial loading is present. The natural forms of natural oscillations in the absence of a moving inertial loads pass into the eigenoscillations modes of the shell.

6. Conclusions

An analysis of the dependence of the main lowest oscillation frequency on the flow velocity showed that the Coriolis forces of inertia of the moving load significantly affect the fundamental vibration frequency of the shell. Already at average flow velocities, the value of the fundamental oscillation frequency, determined considering of the Coriolis inertia forces, is 20–25% lower than the value obtained without taking them into account.

An analysis of the forms of the intrinsic and accompanying vibrations of the shell shows their significant dependence on the velocity of the flows, the ratio of the moving and stationary masses of the system, and, at average speeds of movement of the load, the main form of intrinsic vibrations passes from the nodal to the nodal, the node of which moves with increasing speed. The role of the accompanying oscillations increases significantly with increasing speed and the ratio of the moving and stationary masses of the system.

If this is neglected, then a fairly distorted picture of the shell motion can be obtained not only from the quantitative, but also from the qualitative side.

References

[1] Bolotin V V 1961 *Nonconservative problems of elastic stability* (Moscow: Fizmatgiz) 339
[2] Volmir A C 1967 The stability of deformable systems *Moscow: Nauka* pp 328-338
[3] Zelensky A G 2019 Mathematical Theory of Transversally Isotropic Shells of Arbitrary Thickness at Static Load *Materials Science Forum Trans Tech Publications Ltd* **968** pp 496–510
[4] Goroshko O A, Demianenko A G and Chigov G G 1975 Exact solution some problem of strengthened plates and casings dynamics with mobile inertial loading *The 10-th Int. Conference* Tbilisy **2** pp 98-109
[5] Goroshko O A and Savin G N 1971 Introduction to mechanics of deformable one-dimension bodies of varying length *Naukova dumka* Kyiv pp 32-40.
[6] Goroshko O A, Demianenko A G and Kiba S P 1991 Two-wave processes in the mechanical systems *Lybid* Kyiv pp 83-94
[7] Demianenko A G 2019 Dynamics of Elastic Objects under Movable Inertial Loading *Materials Science Forum Trans Tech Publications Ltd* **968** pp 427-436.
[8] Karnovsky I A and Lebed E 2016 Theory of Vibration Protection *Springer* pp 191-194.
[9] Manevich A I 1970 To the stability of eccentric strengthened cylindrical casing *The 7-th. Int. Conference* Moscow pp 403-407

[10] Panovko Ya 1947 G Historical esse of development of the theory of the dynamical effect of the movable loading *Leningrad Works of KMAIA* pp 8-38
[11] Panovko Ya G and Gubanova I I 1987 Stability and oscillation of elastic systems *Moscow Nauka* pp 277-294
[12] Housner G W 1952 Bending Vibrations of a Pipe Line Containing Flowing Fluid *Journal of Applied Mechanics Trans ASME* **19 (2)** pp 205-209
[13] Steuding H 1934 Die Schwingung von Trager bei bewegten Lasten *Jng.Archiv* pp 275-305

IOP Conf. Series: Materials Science and Engineering 1164 (2021) 012033 doi:10.1088/1757-899X/1164/1/012033

Effect of critical speed in machining of the main shaft of cone crushers on accuracy of treated surfaces

V E Kondratenko[1], V V Devyatiarova[1,*], L V Sedykh[1] and F U Madgoziev[1]

[1]National Research Technological University MISiS, 4, Leninsky prospect, Moscow, 119991, Russia

[*]E-mail: vikdev@yandex.ru

Abstract: One of the industry development factors is the advancement of design and technological solutions of production applications. Mining industry makes heavy use of cone crushers, with the main shaft being one of the main components under serious loads, thus requiring repair or replacement. The paper suggests the method to determine the critical speed of the main shaft of cone crushers in machining and the effect, which the type of fixing on metal-cutting machines has on the accuracy of its surfaces.

1. Introduction

Ore preparation in mining industries predetermines considerably the final technological and economic indicators of an enterprise. Cone crushers ensure the most efficient crushing of hard and abrasive ores, which include copper and iron ores, and gold-containing quartzites [1-3].

Cone crushers are widely used in mining enterprises for fine, medium and primary crushing (Figure 1).

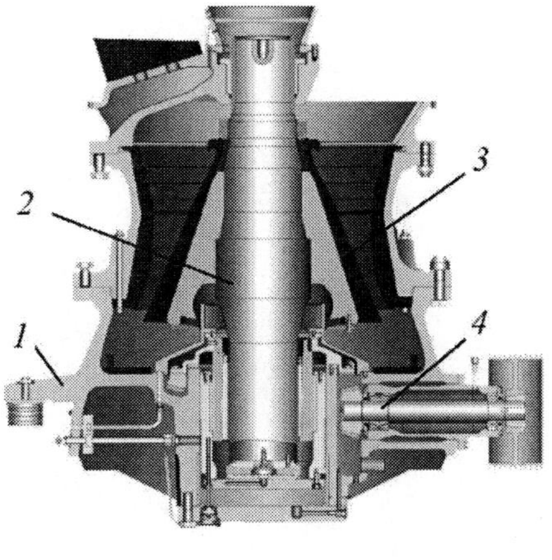

Figure 1. Cone crusher diagram: *1* – base; *2* – main shaft; *3* – mantle;*4* – drive shaft.

Main shaft 1 of any type of the cone crusher is a large-sized turned part with a solid loading, located in the center of the shaft. Two main processing methods are used to produce such parts with a given accuracy – turning and circular grinding [4-9].

Object of the paper – is to offer the method to determine the critical speed of the main shaft, designed for cone crushers, during its cutting in turning and circular machines.

2. Main Part

The specificity of the process turning or circular-grinding operation is that the main shaft, based in the chuck jaws and the center of the tailstock of the machine (Figure 2) [10, 11] performs not only rotational movement during machining. The shaft axis bends in rotary mode and performs a precessional motion. When the angular velocity of rotation reaches a certain value, called the critical

Content from this work may be used under the terms of the Creative Commons Attribution 3.0 licence. Any further distribution of this work must maintain attribution to the author(s) and the title of the work, journal citation and DOI.
Published under licence by IOP Publishing Ltd

speed, shaft deflections become quite substantial. Hence, it is recommended that the operating speed fall within the range of (stiff shaft) (flexible shaft). Not counting the gyroscopic effect, the critical speed of the shaft is equal to the natural frequency of transverse (bending) vibrations of the system p [12-18].

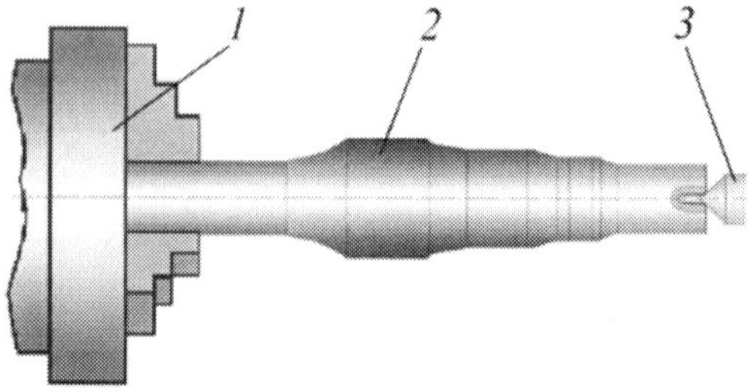

Figure 2. View of the main shaft mounting on the turning machine: 1 – chuck; 2 – main shaft; 3 – movable tailstock center.

Let us review the design model of the main shaft with the lumped mass M, which is in the center of span (Figure 3).

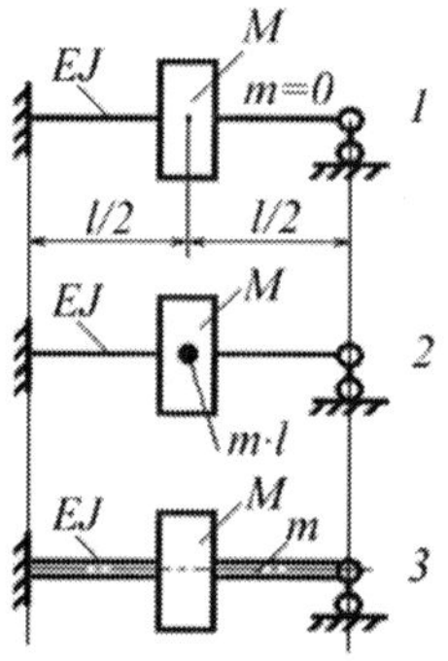

Figure 3. Main shaft design model: *1* – simple; *2* – complex; *3* – precise.

To define the natural frequency of system vibration, let us consider several design model options:
1. Simple, with no account of the shaft mass.
2. Complex, considering the shaft mass as lumped in the center (*m* – mass of the length unit).
3. Precise, accounting for shaft mass, distributed along its length.
Design models 1 and 2 are the one-degree-of-freedom systems. [19-27]. Their natural frequencies and critical speeds consequently equal to, s^{-1}:

$$p_1 = \omega_{c1} = \sqrt{\frac{C_f}{M}}, \quad (1)$$

$$p_2 = \omega_{c2} = \sqrt{\frac{C_f}{M + m \cdot l}}, \quad (2)$$

where:
M – lumped mass, kg;
m – mass of the shaft length unit, kg/m;
l – shaft length, m;
C_f – shaft flexural rigidity. The shaft inverse compliance, which is equal to the displacement of the shaft centre by the unit force, is determined by Mohr integral, calculated by Vereshchagin method [2]

In the following, we make the bending-moment diagram from the external force F (Figure 3, 1), and the bending-moment diagram from the unit force, in the main statically determinate system of force [28-36] method (Figure 3, 2).

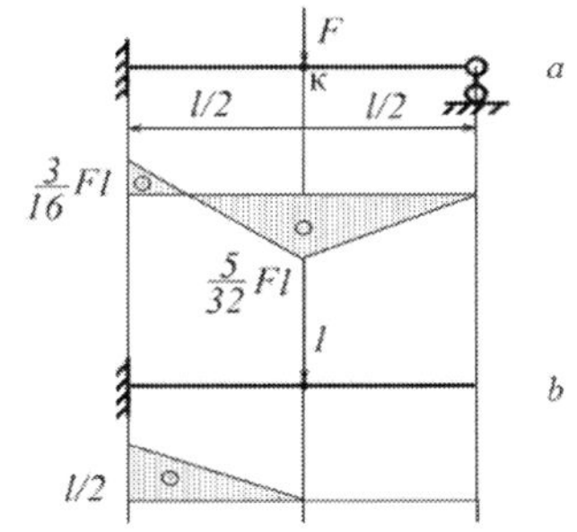

Figure 4. Bending-moment diagrams: a – from external force F; b – from unit force $\overline{F} = 1$.

Vertical movement of point K (see Figure 4) equals to Mohr integral, calculated by Vereshchagin method, m:

$$y_{\kappa} = \int_{l}^{0} \frac{M_F \cdot \overline{M}}{E \cdot J} dz = \frac{1}{E \cdot J}\left[2 \cdot \frac{3}{16} F \cdot l \cdot \frac{l}{2} - \frac{5}{32} F \cdot l \cdot \frac{l}{2}\right] \cdot \frac{l}{12} = \frac{7}{768} \cdot \frac{F \cdot l^3}{E \cdot J}.$$

External force is determined from the formula, N:

$$F = C_f \cdot y_{\kappa}.$$

The resulting shaft inverse compliance is, N/m:

$$C_f = \frac{768 \cdot E \cdot J}{7 \cdot l^3}, \quad (3)$$

where:

E – Young's modulus of steel $2 \cdot 10^5$ MPa $= 2 \cdot 10^{11}$ N/m^2

$J = \pi D^4/64$ – Shaft axial moment of inertia, m^4;

Substituting expression (3) in formulas (1) and (2), we get the value of shaft critical speeds for the first (see Figure 3, 1) and the second design models (see Figure 3, 2) respectively:

$$\omega_{c1} = \sqrt{\frac{768 \cdot E \cdot J}{7 \cdot l^3 \cdot M}}, \quad (4)$$

$$\omega_{c1} = \sqrt{\frac{768 \cdot E \cdot J}{7 \cdot l^3 \cdot (M + ml)}}. \quad (5)$$

Pursuant to the first design model (see Figure 3, 1), which disregards the shaft mass, formula (1) disregards the shaft mass respectively. The second design model and formula (5) account for inertia characteristics of the shaft mass, as a lumped mass. Evidently, the value of the natural frequency of the third design model (see Figure 3, 3), which considers the inertia characteristics of the mass, distributed along the shaft length, should be halfway of the interval between and, that is:

$$\omega_{c3} \approx \frac{\omega_{c1} + \omega_{c2}}{2}. \quad (6)$$

Validity of this statement is illustrated with the example of calculating the natural frequencies of transverse vibrations or critical speeds of the shaft.

Using the data proposed in Table 1, we calculate the values of critical speeds using the approach formulated in the article [14].
Moment of inertia and area of section of the hollow shaft are calculated from the formulas:

$$J = \frac{\pi D^4}{64}\left(1-\left(\frac{d}{D}\right)^4\right), \quad (7)$$

$$A = \frac{\pi D^2}{64}\left(1-\left(\frac{d}{D}\right)^2\right). \quad (8)$$

Table 1. Parameter points for critical speeds calculation.

Shaft type	l, m	M, kg	D, m	d, m	J, m^4	A, m^2	m, kg/m
Solid shaft	2	5	0,02	–	$0{,}75 \cdot 10^{-8}$	$3 \cdot 10^{-4}$	2,4
Hollow shaft	2	5	0,02	0,016	$0{,}44 \cdot 10^{-8}$	$1{,}13 \cdot 10^{-4}$	0,864

Values of solid ω_c and hollow ω_c shafts, calculated from the formulas (4) and (5) are specified in Table 2.

Table 2. Critical speeds values for design models (see Figure 3).

Shaft type	ω_{c1}, s^{-1}	ω_{c2}, s^{-1}	ω_{c3}, s^{-1}
Solid shaft	$\omega_{c1} = 255$	$\omega_{c2} = 178$	$\omega_{c3} \approx 216$
Hollow shaft	$\omega_{c1} = 192$	$\omega_{c2} = 166$	$\omega_{c3} \approx 179$

Results of calculation using the suggested method, evidenced validity of the statement, set out in the formula (6).

3. Conclusions

The proposed method of calculation facilitates largely the complexity of determining the critical speed of the main shaft of the cone crusher.

The following result was obtained upon comparative analysis of calculations made in the present paper and the article [14]: the average critical speed for machining the main shaft of cone crushers, fixed in the chuck and the center of the tailstock, exceeds six times that of the part with the same dimensions, fixed in the centers. Based on these data, we conclude that in order to reach a specified surface accuracy while increasing the processing performance, it is more efficient to fix the parts, which design type is that of the main shaft of the cone crushers, in the chuck and the center of the tailstock.

4. Results of the study

When the carbide band is placed on a steel roll with the tension between the contacting surfaces, as a rule, a plastic unsaturated or saturated contact occurs.

References

[1] Bardovsky A D, Valeeva L M and Basyrov I I 2020 A plant with a rotary jet grinder to produce small fractions of mineral raw material *IOP Conf. Series: Materials Science and Engineering* **971** 052004 doi:10.1088/1757-899X/971/5/052004
[2] Aleksandrov A V,. Potapov V G and Derzhavin B P 2007 *Resistance of materials: a textbook for universities* (Moscow: Higher School) 560
[3] Bibikov P Y, Bardovskiy A D and Keropyan A M 2019 Investigation of press classification process of weak rocks *Materials Today: Proceedings* **19** pp 2552-2554 doi: 10.1016/j.matpr.2019.08.207

[4] Biderman V L 2017 *Theory of mechanical vibrations*: *Textbook for universities* (URSS Physico-mathematical heritage: Physics (Mechanics)) 416

[5] Birger I A 1993 *et al. Calculation of the strength of machine parts*: Handbook Birger I A Shorr B F Iosilevich G B - 4th ed reprint and dop (M.: Mashinostroenie) 640

[6] Chichenev N A, Gorbatyuk S M, Naumova M G and Morozova I G 2020 Using the similarity theory for description of laser hardening processes *CIS Iron and Steel Review* **19** pp 44-47 doi: 10.17580/cisisr

[7] Dragobetskii V V, Shapoval A A, Mospan D V, Trotsko O V and Lotous V V 2015 *Excavator bucket teeth strengthening using a plastic explosive deformation*. Metallurgical and Mining Industry **7 (4)** pp 363-368

[8] Evart T Ye, and Puchkov V P, Skhirtladze A G and Pris N M 2017 Modeling the choice of a rational method for manufacturing machine-building parts blanks *Bulletin of MSTU Stankin* **2(41)** pp 89-92

[9] Gerasimova A A, Keropyan A M and Girya A M 2018 Study of the Wheel–Rail System of Open-Pit Locomotives in Traction Mode *Journal of Machinery Manufacture and Reliability* **47(1)** pp 35-38

[10] Jiawen L, Timushev S, Klimenko D and Krivenko A 2019 *Modeling pressure pulsation fields in a screw centrifugal pump* Proceedings of the 26th International Congress on Sound and Vibration (ICSV)

[11] Kerimov R I and Shakhov S I 2020 Use of Metallized Raw Materials in Electric Furnace Steelmaking *Metallurgist* **64 (1-2)** pp 128-135 doi: 10.1007/s11015-020-00974-1

[12] Keropyan A M, Kuziev D A and Krivenko A E 2020 Process Research of Wheel-Rail Mining Machines Traction *Lecture Notes in Mechanical Engineering* pp 703-709 doi: 10.1007/978-3-030-22063-1_75.

[13] Kobelev O A and Tyurin V A 2007 Production of large plates *Steel in Translation* **37 (9)** pp 727-729 doi: 10.3103/S096709120709001X

[14] Kondratenko V E, Devyatiarova V V, Albul S V and Kartyshev D S 2020 Improving methodology for calculating scaffolding formwork of monolithic slabs in building constructions *IOP Conference Series: Materials Science and Engineering* **971 (5)** 052037 doi: 10.1088/1757-899X/971/5/052037

[15] Kondratenko V V, Sedykh L V, Mirzakarimov A and Aleksakhin A 2020 Static analysis and strength calculation of drive shaft of large-scale cone crusher *E3S Web of Conferences* **193** 01038 doi: 10.1051/e3sconf/202019301038

[16] Kondratenko V E, Sedykh L V and Surkova R Y 2020 Effective design features of rotor shafts IOP Conference Series: Materials Science and Engineering **971 (4)** 042010 doi: 10.1088/1757-899X/971/4/042010

[17] Kouziyev D, Krivenko A, Chezganova D and Valeriy B 2019 Sensing of dynamic loads in the open-cast mine combine *E3S Web of Conferences* **105** 03014 doi: 10.1051/e3sconf/201910503014

[18] Krivenko A E and Khanh Z K 2020 Influence of power fluid temperature in hydraulics on operating efficiency of hydraulic mining excavators *Gornyi Zhurnal* **12** pp 78-81 doi: 10.17580/gzh.2020.12.18

[19] Leontiev N. N., Sobolev D. N., and Amosov A. A. 1996. *Fundamentals of structural mechanics of rod systems* (M.: Publishing House of the Association of Construction Universities) 541

[20] Krutii Y, Kovrov A, Otrosh Y and Surianinov M 2020 Analysis of forced longitudinal vibrations of columns taking into account internal resistance in resonance zones *Materials Science Forum **1006** MSF* pp 79-86

[21] Krutii Y, Suriyaninov M and Vandynskyi V 2018 Analytic formulas for the natural frequencies of hinged structures with taking into account the dead weight *MATEC Web of Conferences* **230** 02016 doi: 10.1051/matecconf/201823002016

[22] Mazhirin E A, Chichenev N A and Zadorozhnyi V D 2009 Extending the life of disk cutters for thick strip *Steel in Translation* **39 (1)** pp 84-85 doi: 10.3103/S0967091209010215

[23] Naumova M G, Basyrov I I and Aliev Kh B 2018 Reengineering of the ore preparation production process in the context of "almalyk MMC" *JSC MATEC Web of Conferences* **224** 01030 doi:10.1051/matecconf/201822401030
[24] Naumova M G, Morozova I G and Borisov P V 2019 Investigating the features of color laser marking process of galvanic chrome plating in order to create a controlled color image formation at given marking *Materials Today Proceedings* **19** pp 2405-2408 doi: 10.1016/j.matpr.2019.08.044
[25] Osadchiy V A, Albul S V, Kuprienko N S and Kirillova N L 2020 Future developments of a roll forming mill design algorithm *IOP Conference Series: Materials Science and Engineering* **709 (4)** 044079 doi: 10.1088/1757-899X/709/4/044079
[26] Pimenov G A, Kostyukov G A, Ryabov P S, Rogal' V D and Kobelev O A 1991 Making of large-sized solid-forged blanks *Tyazheloe Mashinostroenie* **9** pp 21-24
[27] Sedykh L V, Albul S V, Efremov D B and Sukhorukova M A 2020 Application of additive technologies for manufacturing a wear-resistant steel pipe outlet *IOP Conference Series: Materials Science and Engineering* **971 (2)** 022002 doi: 10.1088/1757-899X/971/2/022002
[28] Shakhov S I and Vdovin K N 2019 *Electromagnetic Stirring in Continuous-Casting Machines* Steel in Translation **49 (4)** pp 261-264 doi: 10.3103/S096709121904012
[29] Sivak B A, Shakhov S I, Vdovin K N, Rogachikov Y M and Kerimov R I 2020 Development of a System for Electromagnetic Stirring of Liquid Steel in Molds of Billet and Bloom CCMS *Metallurgist* **63 (9-10)** pp 910-914 doi: 10.1007/s11015-020-00909-w
[30] Surianinov M, Lazarieva D and Kurhan I 2020 Stability of orthotropic plates *E3S Web of Conferences* **166** 06004 doi: 10.1051/e3sconf/202016606004
[31] Turkes E, Orak S and Neseli S 2011 Linear analysis of chatter vibration and stability for orthogonal cutting in turning *Journal of Refractory Metals and Hard Materials* pp 163-169
[32] Turman E 2006 Konusnyye drobilki i effektivnost' ikh primeneniya v gornodobyvayushchey promyshlennosti *Mining industry* **2 (66)** pp 50-57
[33] Zagirnyak M, Zagirnyak V, Moloshtan D, Drahobetskyi V and Shapoval A 2019 A search for technologies implementing a high fighting efficiency of the multilayered elements of military equipment *Eastern-European Journal of Enterprise Technologies* **6 (1-102)** pp 33-40 doi: 10.15587/1729-4061.2019.183269
[34] Zarapin A Yu, Shur A I and Chichenev N A 1999 Improvement of the unit for rolling aluminum strip clad with corrosion-resistant steel *Steel in Translation* **29 (10)** pp 69-71.
[35] Zhetesova G S, Yurchenko V V, Nikonova T YU, Zharkevich O M and Modelina Ye D 2020 Osobennosti bazirovaniya pri upravlenii tochnost'yu sistemy stanok-prisposobleniye-instrument-detal' *Mechanical engineering technology* **1** pp 34-42
[36] Surianinov M, Lazarieva D and Kurhan I 2020 Stability of orthotropic plates *E3S Web of Conferences* **166** 06004 doi: 10.1051/e3sconf/202016606004

Study of Stress-Strain Behavior of the Laminated Plate Damaged by Delamination

A Kondratiev[1,*], L Smovziuk[2], M Shevtsova[2], T Nabokina[4] and A Tsaritsynskyi[2]

[1]O M Beketov National University of Urban Economy, 17 Marshal Bazhanov st., Kharkiv 61002, Ukraine

[2]National Aerospace University "Kharkiv Aviation Institute", 17 Chkalova st., Kharkiv 61070, Ukraine

[*]E-mail: andrii.kondratiev@kname.edu.ua

Abstract. The process of manufacturing of the structures of polymeric composite materials is accompanied by various technological defects. Delamination defects have the greatest impact on the bearing capacity of the composite structures. The method for assessment of the stress-strain behavior of a panel structure made of polymeric composite materials, which is damaged by delamination, has been developed. This method allows determining the degree of stress concentration in the area of step change in thickness and identifying the most dangerous point where the structural failure may start. In contrast to the classical theory of laminated plates, the proposed model gives an opportunity to simplify three-dimensional problem by setting the displacement field on the interfaces of layers and their linear interpolation over the plate thickness, taking into account the transverse shear deformation. Stress-strain behavior of the plate with delamination was determined according to the computational model of the plate with a cut for the case when the layers in the defect area cannot take up the applied load. According to the results obtained, distribution of stresses over the plate thickness changes significantly in the area of step transition to the delamination and near it. At the distance, it approaches the stress distribution in a plate of the uniform thickness. It allowed confirming the previous conclusions that maximum rise in stresses may occur in the weakened layer near the delamination, initiating the failure of the structure. Reliability of the developed mathematical model is confirmed by comparison with the results of experimental studies. The resulting maximum error does not exceed 7% and demonstrates good convergence of results. With the use of the developed method, it is possible to solve the important practical tasks of studying the loss of performance of the panel structures made of polymeric composite materials with delaminations and determining the optimal methods for repair of such defects.

1. Introduction

High physical and mechanical characteristics, as well as a number of special properties inherent in polymeric composite materials, are the main reasons for their wider application in all branches of technology [1, 2].

Transition to composites leads to combining the technological processes for creation of the material and finished product, independent of each other for the traditional materials, into one process operation [3]. This specific property of the composite materials explains the fact that properties of the materials and structures made of the same components under the same conditions may differ significantly [4].

In the course of manufacturing of components of the polymeric composite materials, there are complex physico-chemical and thermo-physical processes running in the half-finished products [3]. These processes are accompanied by the mechanical phenomena, which are the causes of occurrence of the various initial technological defects in the unloaded condition [5]. Defects such as delaminations have the greatest impact on the bearing capacity of composite structures [6, 7]. These defects are caused by the layered structure of the polymeric composite material itself. This structure represents an alternation of layers of the reinforcing material with high strength and stiffness

Content from this work may be used under the terms of the Creative Commons Attribution 3.0 licence. Any further distribution of this work must maintain attribution to the author(s) and the title of the work, journal citation and DOI.
Published under licence by IOP Publishing Ltd

characteristics with matrix interlayers featuring significantly lower strength and stiffness [3]. Low values of the transverse tensile strength of the composite material and resistance to interlayer shear also lead to frequent occurrence of initial delamination defects [8]. This process is promoted by the difference in the thermo-physical properties of the composite material's components and low adhesive strength of the binder materials [9]. Most of these defects appear at the stage of the technological process [6]. At the structure operation stage, defect formation process may continue under the impact of heating, local loads, shocks and vibrations [10, 11].

Therefore, strength of the structures made of composite materials in the process of operation significantly depends on the presence of delamination defects in the product.

2. Literature Review

At present time, studies of the effect of delaminations on the performance of composite structures are carried out in several main directions [5–7]:

– assessment of the influence of size, configuration and location of local delaminations over the thickness and area of typical structural elements made of polymeric composite materials on the stress-strain behavior and critical load;

– assessment of the influence of the above factors on the dynamic characteristics of the structure;

– investigation of the conditions corresponding to transition of the defect boundaries into unstable state and initiation of the delamination development.

Exact mathematical problem setting with regard to analysis of discontinuity defects such as delaminations in the composite elements is rather complex and has no correct solution within the existing analytical models [12]. Usage of the multiscale model in order to take into account the impact of delamination of the layers on the strength of composite structures is proposed in the paper [13]. The paper [14] deals with obtaining of the analytical dependences, which allows assessing the quality of technological processes for the formation of half-finished products and goods of polymeric composite materials by the level of defects of the considered class, in contrast to other models. However, disadvantage of this work is that the analytical dependences obtained give an opportunity to assess only the level of decrease in the composite physical and mechanical properties in the presence of local discontinuities in it. When modeling the delamination in [15, 16], it is assumed that interlaminar stresses in the laminated composite are similar to stresses in the adhesive joint. Nevertheless, there are no reliable mathematical models of discontinuity defects of the composite material constructed in this way. The paper [17] shows the essential role of stability calculations for delamination defects. The method of stability calculation for thin-walled structural elements with the delamination defects under compression is proposed in [18]. The papers [6, 19], using the example of a compressed multilayer plate with through delamination between layers, give the solutions of the problems of stability loss and the process of damage. The studies have shown that for the adequate solving of the class of problems under consideration it is reasonable to use the numerical models based on the finite-element method. The paper [20] proposes the finite-element mathematical model, which allows to analyze the delamination behavior from the moment of buckling to the loss of bearing capacity. Another direction for the analysis of composite elements' discontinuity defects in the form of delamination is proposed in [21]. In this paper, well-known mathematical models of the multilayer medium are reduced to the system of differential equations with variable coefficients, with the use of various numerical and analytical methods. The result is the geometrically nonlinear system of equations, but the lack of the true solution of this system is a disadvantage of this work. In all papers cited above, the development of so-called "admissible" delamination is studied. Such delamination is understood as a defect the shape and dimensions of which allow the product to be operated under the compressive force for a certain time. During this time, the defect develops, increasing its size up to the failure. The paper [7] assumes the beginning of local delamination under the regulated compressive load as the exhaustion of the bearing capacity of a composite structural element. Under these conditions, tolerances for the input parameters are established. This paper deals with the numerical example for the composite element with the specified structure and shape of the delamination. Comparison with the results of finite-element modeling showed the satisfactory agreement of results. Expediency of using the numerical implementation of the problem is confirmed by the comparison with the results of experimental strain-gage measurements [19]. However, the main disadvantage of this approach is that the computational model is created for the strictly defined structure and cannot be arbitrarily extended to similar elements [2, 7]. Presence of the local and edge effects caused by delamination imposes certain conditions for the solvability of boundary problems and determines the instability of calculations in the process of numerical implementation of the problem. In recent times, the mathematical apparatus of spline functions is widely used for solving such problems [22]. The advantage of this approach lies in high stability of spline approximations with regard to various kinds of local perturbances, fast convergence, simplicity and ease of implementation of the algorithms. The most adequate assessment of the bearing capacity of composite structures with the considered technological defects is possible with the use of experimental studies [23, 24]. It should be recognized that in order to assess the strength and stiffness of structural elements made of polymeric composite materials it is necessary to subject them to direct mechanical tests [25, 26]. Among the experimental methods for the deformation measurement, the method of tensometry is widely spread [25]. However, the scope of its application is limited to areas of the structure with the regular stress-strain behavior, since the strain gage measures the average value of deformation within the base only. With rare exceptions, the methods of nondestructive testing of structural elements made of polymeric composite materials do not allow reliably predicting the moment of failure [27]. In this case, only the weakest point of the structure can be identified [26].

In view of this, the urgent task today is to develop a methodology for assessment of the stress-strain behavior of panel structure of polymeric composite materials, which is damaged by delamination.

3. Materials and Methods

The problem of determining the stress-strain behavior of the damaged panel structure is solved in the linear setting with the use of methods of the elasticity theory and the basic dependencies of the mechanics of composites. The layerwise theory is chosen for the calculation of stress-strain behavior. Stress-strain behavior of the plate with delamination is determined according to the computational model of the plate with a cut for the case when the layers in the defect area cannot take up the applied load. Reliability of the developed mathematical model is confirmed by comparison with the results of experimental studies. Results of experimental studies presented in the paper were obtained in the laboratory conditions using standard equipment, instruments and work tools. In the process of experimental studies, the method of double exposure holographic interferometry with the registration of holograms in counter-propagating beams was used.

4. Results

Let's consider a flat rectangular plate with the specified delamination consisting of k orthotropic layers. For each layer, physical and mechanical characteristics of the material and the fiber orientation angle relative to the axes of the common coordinate system (x, y, z) are known. We assume that the coordinate plane (x, y) coincides with the bottom surface of the plate, and z-axis is directed upwards. The planes of elastic symmetry of the materials of each layer are mutually perpendicular, and one of the planes of elastic symmetry at each point of each layer is parallel to the outer parallel planes of the plate. Position of the coordinate system and geometric parameters of the plate are shown in Figure 1.

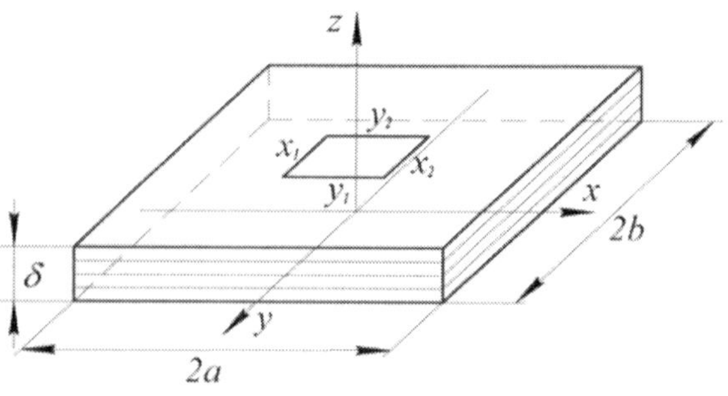

Figure 1. Geometric parameters of the plate and delamination.

The delamination, which in practice has an arbitrary shape, will be approximated by the rectangle with an area equal to the area of the real delamination and considered flat in the unloaded condition. The lines x_1, x_2, y_1, y_2 in Figure 1 correspond to the boundary where the plate thickness changes stepwise by the value of the delamination depth $\bar{\delta}$.

We consider the pressure with the intensity p uniformly distributed over the surface as an effective load.

The layerwise theory, which allows simplifying three-dimensional problem by setting the displacement field on the interfaces of layers and their linear interpolation over the plate thickness [28, 29], is chosen for the calculation of stress-strain behavior.

The following assumptions were made:

– the normal displacement ω_i in each layer of the plate does not depend on z coordinate, therefore: $\varepsilon_z=0$ and $\omega_i=\omega(x, y)$;

– normal stresses σ_z, acting in sections parallel to the median plane are negligible compared to the main stresses and they are not taken into account in the calculations;

– deformations ε_x, ε_y, γ_{xy}, γ_{xz}, γ_{yz} are approximated by linear functions within the layer;

– all layers of the plate are orthotropic in the axes (x, y).

In the process of calculation of the plate, conditions for its edges' fixing correspond to the boundary conditions of free bearing and suppose the absence of deflection and displacement along the edge and free movement across the edge. The fixing conditions as a set of static and geometric boundary conditions are written as follows:

at $x=\pm a$: $\omega=0$, $v=0$, $\theta_y=0$, $N_x=0$, $M_x=0$;

at $x=\pm b$: $\omega=0$, $u=0$, $\theta_x=0$, $N_y=0$, $M_y=0$.

The advantage of such boundary conditions consists in fairly accurate modeling of the common panel fixation patterns in real structures of various applications [3, 30**Ошибка! Закладка не определена.**].

Similar to the method proposed in [28, 29], we define the displacements u_i, v_i, ω_i of the certain point on the i-th surface of the plate in the directions x, y, z as a function of variables x, y. In this case, owing to acceptance of the hypothesis of inextensible normals the displacement ω_i is the same for all layers of the plate. The displacements of points of the i-th surface are specified so as to satisfy the adopted boundary conditions at the edges of the plate. The i-th surface of the plate is understood as the interface of the "i – 1"-th and "i"-th layers.

In the event if the plate contains the layers with the same geometric dimensions and properties of the material, calculations can be reduced by grouping the same layers into one "mathematical" layer. As a result, the number of "mathematical" layers can be significantly less than the number of actual layers of the material.

Within the i-th layer the displacements u_x^i, u_y^i are approximated by linear functions. The displacement u_z^i remains constant for all layers of the plate and equals ω.

The displacement u_x^i is defined as follows:

$$u_x^i = u_i + \left(u_{i+1} - u_i\right) \cdot \left(z - z_i\right) / \delta_i \,. \tag{1}$$

We define the displacements u_y^i in a similar way:

$$u_y^i = v_i + \left(v_{i+1} - v_i\right) \cdot \left(z - z_i\right) / \delta_i \,. \tag{2}$$

The stress-strain behavior is determined in the framework of the linear theory. We assume that arising deformations are small and they are associated with displacements by the linear geometric relationships. Then, the deformations in the i-th layer are determined by the geometric dependencies:

$$\begin{gathered} \varepsilon_x^i = \frac{\partial u_x^i}{\partial x};\ \varepsilon_y^i = \frac{\partial u_y^i}{\partial y}; \\ \gamma_{xy}^i = \frac{\partial u_x^i}{\partial y} + \frac{\partial u_y^i}{\partial x};\ \gamma_{xz}^i = \frac{\partial u_x^i}{\partial z} + \frac{\partial u_z^i}{\partial x};\ \gamma_{yz}^i = \frac{\partial u_y^i}{\partial z} + \frac{\partial u_z^i}{\partial y}. \end{gathered} \tag{3}$$

Since each layer is orthotropic one in the (x, y) axes and, therefore, has three planes of elastic symmetry, the solution to equations of the generalized Hooke's law with regard to stresses $\sigma_x^i, \sigma_y^i, \tau_{xy}^i, \tau_{xz}^i, \tau_{yz}^i$ appears as follows:

$$\begin{aligned} &\sigma_x^i = b_{11}^i \cdot \varepsilon_x^i + b_{12}^i \cdot \varepsilon_y^i; && \tau_{xz}^i = b_{44}^i \cdot \gamma_{xz}^i; \\ &\sigma_y^i = b_{21}^i \cdot \varepsilon_x^i + b_{22}^i \cdot \varepsilon_y^i; && \tau_{yz}^i = b_{55}^i \cdot \gamma_{yz}^i. \\ &\tau_{xy}^i = b_{13}^i \cdot \varepsilon_x^i + b_{23}^i \cdot \varepsilon_y^i; && \end{aligned} \tag{4}$$

Stiffness matrix coefficients included in the equations (4) are calculated by the known formulas [30].

To solve the problem, we define the total deformation energy of the elastic system as the difference between the potential energy and the external force energy:

$$A = U + V\,, \tag{5}$$

where U – potential deformation energy; V – external force potential.

Potential deformation energy of the i-th layer is defined as follows:

$$U_i = \frac{1}{2} \cdot \iint\limits_S \int\limits_{z_i}^{z_{i+1}} \left(\sigma_x^i \cdot \varepsilon_x^i + \sigma_y^i \cdot \varepsilon_y^i + \tau_{xy}^i \cdot \gamma_{xy}^i + \tau_{xz}^i \cdot \gamma_{xz}^i + \tau_{yz}^i \cdot \gamma_{yz}^i\right) \cdot dx \cdot dy \cdot dz\,, \tag{6}$$

where Z_i и Z_{i+1} – coordinates of the surfaces bounding the i-th layer.

Then, potential deformation energy of the plate as a whole is equal to the sum of the potential deformation energies of all layers:

$$U = \sum_{i=1}^{k} U_i\,. \tag{7}$$

For the final determination of total energy of the elastic system, it is necessary to define the work of external forces applied to the plate. In case if the transverse uniformly distributed load with intensity p acts on the plate, the potential of external forces is determined by the formula:

$$V_p = -\iint\limits_S p \cdot \omega_k \cdot dx \cdot dy\,. \tag{8}$$

For the determination of the plate's stress-strain behavior under action of the applied load, we use the Ritz-Timoshenko method, which allows obtaining an approximate solution in the displacements based on the Lagrange variational principle.

In accordance with the chosen method, we define the displacements u_i, v_i, ω of the i-th surface of the plate in the form below:

$$\begin{aligned} u_i(x,y) &= \left(y^2-b^2\right)\cdot\left(u_{1,i}+u_{2,i}x+u_{3,i}y+u_{4,i}x^2+u_{5,i}xy+\ldots+u_{st,i}y^{st}\right); \\ v_i(x,y) &= \left(x^2-a^2\right)\cdot\left(v_{1,i}+v_{2,i}x+v_{3,i}y+v_{4,i}x^2+v_{5,i}xy+\ldots+v_{st,i}y^{st}\right); \\ \omega(x,y) &= \left(x^2-a^2\right)\cdot\left(y^2-b^2\right)\cdot\left(\omega_1+\omega_2x+\omega_3y+\omega_4x^2+\omega_5xy+\ldots+\omega_{st}y^{st}\right), \end{aligned} \tag{9}$$

where st – degree of the polynomial which determines the plate displacement; $u_{i,j}$, $v_{i,j}$, ω_i – unknown coefficients to be determined.

We obtain the variational equation by equating to zero the first variation of total energy of the system.

Consequently, the resolving system of equations will be as follows:

$$\frac{\partial A}{\partial u_{i,j}}=0,\ \frac{\partial A}{\partial v_{i,j}}=0,\ \frac{\partial A}{\partial \omega_i}=0 \quad (i=1\ldots n)\ . \tag{10}$$

Since the total energy is the quadratic function of displacements, the equations (10) represent a complete system of the linear algebraic equations for the coefficients being determined. After finding of the coefficients $u_{i,j}$, $v_{i,j}$, ω_i as a result of this system's solving, we can determine, with the use of formulas (9), the displacements, and then deformations and stresses, that is, we obtain a complete solution to the problem.

To determine the degree of the polynomial which provides the required error in calculating the stress-strain behavior of the plate, we have made preliminary calculations of the maximum deflection ω, maximum normal stress σ_x and maximum shear stress τ_{xz} of the square orthotropic plate. The results demonstrate the good convergence of the values of deflection and stresses. With the holding of the term of series st=6, the error in determining the deflection is 0.042%, normal stresses 0.28%, and transverse shear stresses 0.1%.

When passing from the computational model of the whole plate to the pattern of the plate with delamination, we use the approach proposed in [29], and introduce the auxiliary function F, which ensures the defect modeling:

$$F=1-H(x-x_1)\cdot H(x-x_2)\cdot H(y-y_1)\cdot H(y-y_2)\,, \tag{11}$$

where x_1, x_2, y_1, y_2 – coordinates of the left and right boundaries along x and y axes of the step change in the plate thickness in the i-th layer (Figure 1); H – the Heaviside function.

Using the function F, we define the stiffness matrix coefficients of layers containing the defect, $\overline{b_{ij}}$ as:

$$\overline{b_{ij}}=H\cdot b_{ij}\,, \tag{12}$$

where b_{ij}, – stiffness matrix coefficients for the defect-free layer.

This setting of stiffness properties of the package allows modeling a delamination with any geometric and physico-mechanical parameters.

For the study of stress-strain behavior of the multilayer plate with delamination, we choose the square orthotropic plate with geometric dimensions of 100×100×2.8 mm. The plate is made of AeroGlass–163 glass cloth and "Epicur" hot cured epoxy binder. The reinforcement structure is [0°,90°]. Side of delamination is $\bar{a}$=20 mm in size, with the depth $\bar{\delta}$ of 0.7 mm. Physical and mechanical characteristics of the glass cloth material are as follows: $E_1=E_2=60$ GPa; $G_{12}=6$ GPa; $G_{13}=G_{23}=4$ GPa; $\nu_{12}=0.28$; $\delta_0=0.35$ mm [29].

For clarity, Figure 2 shows the plate section in the area of delamination, which characterizes its position.

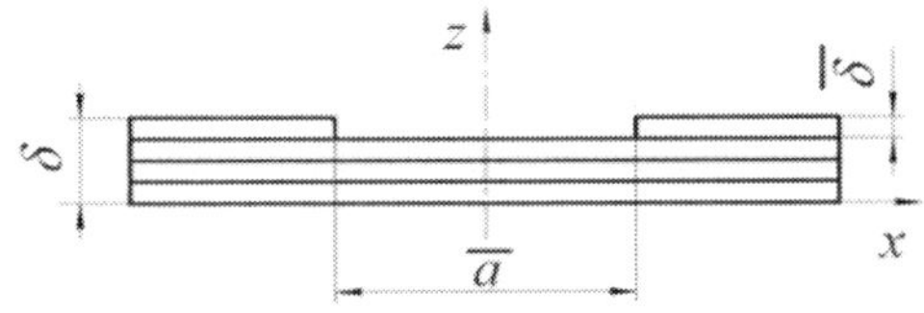

Figure 2. Location of delamination in the plate.

First of all, let's consider the pattern of change of stress distribution over the plate thickness during transition from the defect-free structure to the defective one with a delamination. Graphs of distribution of normal stresses σ_x and transverse shear stresses τ_{xz} in the plate section at $y=0$ are shown in Figure 3. Stress distribution in the defect-free plate is shown by dashed line, while solid line is used for its indication in the delaminated plate. As shown by Figure 3, distribution of stresses over the plate thickness, changing significantly in the delamination area and near it, at the distance approaches the stress distribution in the defect-free plate.

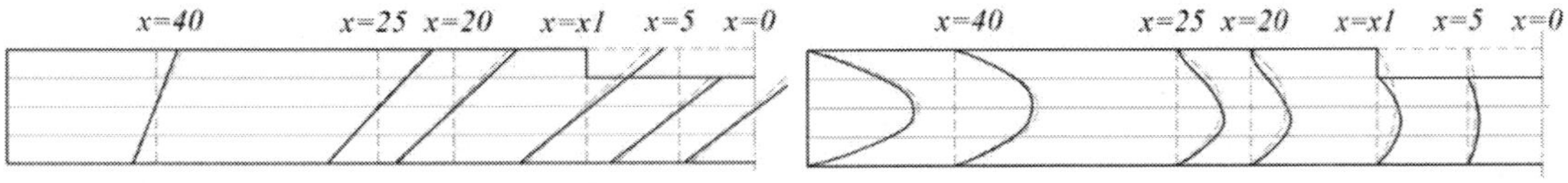

Figure 3. Distribution of normal stresses and transverse shear stresses over the plate thickness.

For assessment of the degree of change in the maximum normal stresses, we compare them for the defect-free plate with stresses in the damaged plate with the delamination of the relative area S_{delam} 5 and 15%. Distribution patterns are shown in Figure 4.

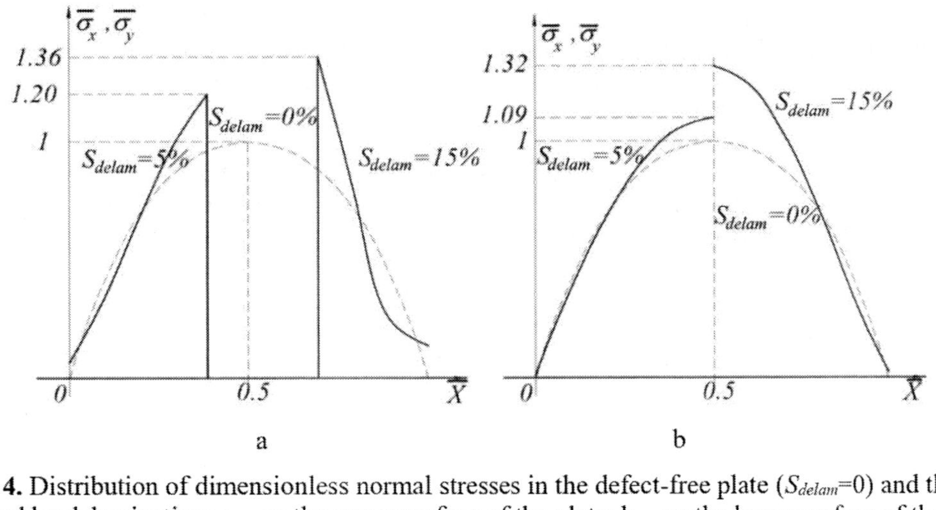

Figure 4. Distribution of dimensionless normal stresses in the defect-free plate ($S_{delam}=0$) and the plate damaged by delamination: a – on the upper surface of the plate; b – on the lower surface of the plate.

For the damaged plate with delamination, it is necessary to take into account the increase in shear stresses and transverse shear stresses because of their effect on strength of the structure. Their distribution patterns are shown in Figure 5.

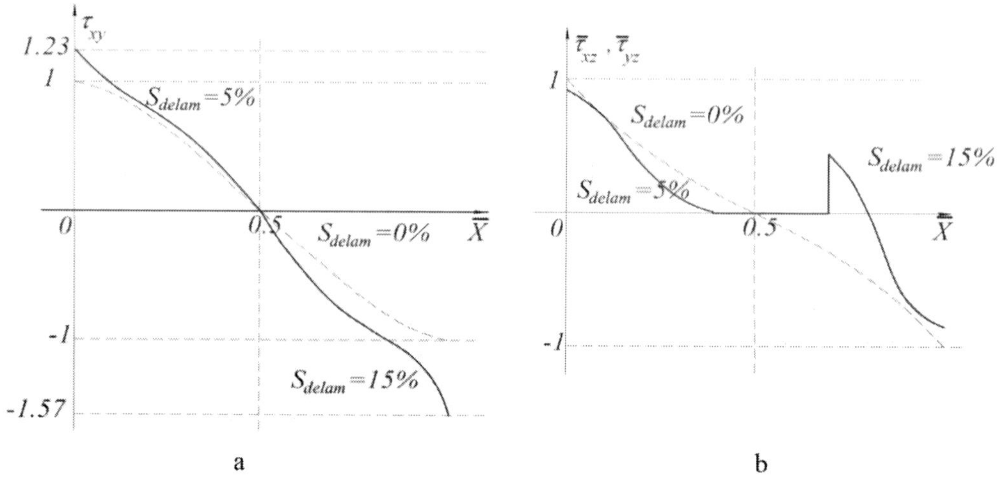

Figure 5. Distribution of shear stresses in defect-free plate (S_{delam}=0) and plate damaged by delamination: a – $\bar{\tau}_{xy}$ on the upper surface of the plate; b – $\bar{\tau}_{xz}$ in the layer with delamination.

Figures 5 and 6, as before, show the stress distribution in the plate by the dashed line. Vertical dashed lines in the figure indicate the boundary of the change in thickness of the delamination. The patterns of changes in stress distributions are presented in Figures 4 and 5 in the form of graphical dependencies of distribution of the dimensionless normal and shear stresses in the plate under study, determined as follows:

$$\bar{\sigma} = \frac{\sigma}{\sigma_0^{max}}; \quad \bar{\tau} = \frac{\tau}{\tau_0^{max}}, \tag{13}$$

where σ, τ – stresses acting in the plate; σ_0^{max}, τ_0^{max} – maximum stresses in the defect-free plate.

Experimental studies were carried out to verify the results of theoretical calculations using the proposed mathematical model. Since the stress field in the damaged plates with delamination is not the uniform one, it is impossible to use the tensometry research method. At the same time, a number of optical methods allow obtaining the displacement field in the structure under study and compare it with the analytical results [19, 25].

In the process of experimental studies, the method of double exposure holographic interferometry with the registration of holograms in counter-propagating beams was used. Comparison of the analytical results with the values obtained by the holographic method was carried out for two types of samples:

– defect-free samples;

– samples with a cut (delamination) (Figure 2).

For each type of samples, holographic interferograms were obtained and graphs of normal displacements along the central section of the plate were constructed. The results are shown in Figure 6.

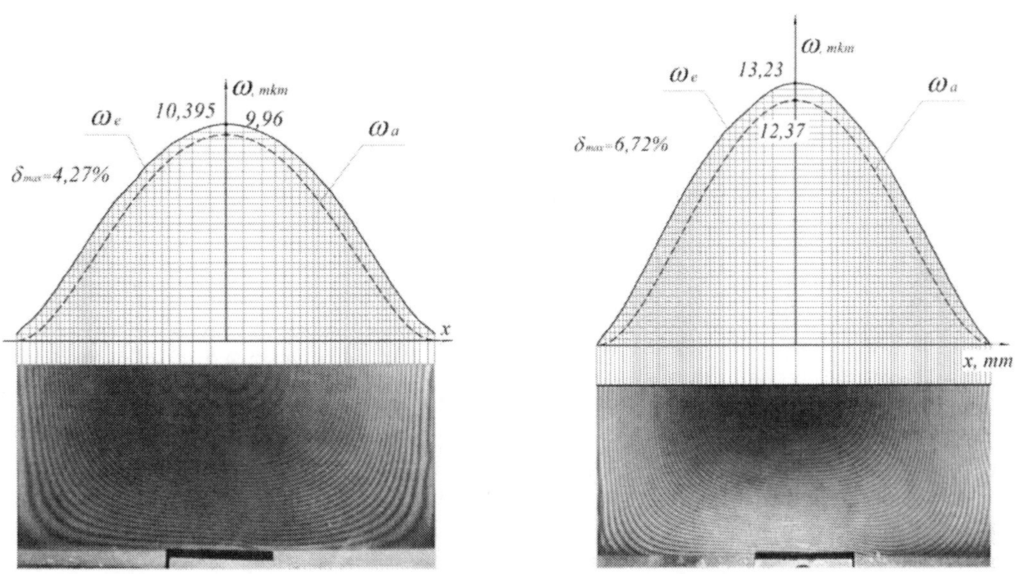

a б

Figure 6. Deflection of the defect-free plate and plate damaged by delamination under action of the uniformly distributed pressure *p*: ω_e – experimental distribution; ω_a – theoretical (analytical) distribution; δ_{max} – maximum relative error.

Therefore, the results of experimental studies of the deformed state of the defect-free plate and plate damaged by delamination showed the degree of correspondence of the displacement fields obtained in practice to the predicted theoretical results.

5. Discussion

In contrast to the classical theory of laminated plates [30], the proposed model gives an opportunity to simplify three-dimensional problem by setting the displacement field on the interfaces of layers and their linear interpolation over the plate thickness, taking into account the transverse shear deformations. Consideration of these factors allows proceeding to the consideration of flexible plates widely used in practice [2, 3].

The developed method allows modeling a plate actually with any geometric parameters in the plan and thickness, as well as combined physico-mechanical characteristics of the composite by setting the relevant stiffness matrix coefficients. In contrast to existing works, it gives an opportunity to determine:

– degree of stress concentration in the area of step change in thickness

– the most dangerous point where the structural failure may start.

According to the obtained results, distribution of stresses over the plate thickness changes significantly in the area of step transition to the delamination and near it. At the distance, it approaches the stress distribution in the plate of the uniform thickness. It allowed confirming the conclusions of [6, 7, 19] that the maximum rise in stresses may occur in the weakened layer near the delamination, initiating the failure of the structure. In this case, the pattern of the change in stresses on the lower surface of the plate, without a sharp change in thickness, is smoother than on the upper one. These features are to be taken into account in the designing of composite structures of various applications [2, 8].

The proposed mathematical model is characterized by the certain field of applicability, in which the accuracy of results corresponds to the required one. Based on the comparison of results with the literature data [29, 30], the error of the theoretical determination of the deflection does not exceed 3%, and that of stresses - 8%. This corresponds to sufficiently high accuracy of the description of a multilayer plate stress-strain behavior. In order to improve the accuracy of calculations, it is enough to increase the number of "mathematical" layers. However, this will lead to the significant increase in the number of variables and the time of calculation. It should also be taken into account that the proposed modeling of a plate with delamination is possible only for the case when the layers in the defect area cannot take up the applied load. This can happen if [5]:

– a delamination occurs because of non-uniform impregnation of the reinforcing material and is characterized by the lack of binder in the defect area, which provides redistribution of the external loads between individual fibers;

– delamination resulting from the violation of the joint work of layers lies in the compressed zone when the plate is bent under action of the transverse load and loses its stability, since the stresses in the defect area exceed the certain critical level.

The experimental method chosen for the study, compared to existing works [19, 23, 24], allowed obtaining the results in the form of a continuous displacement field. It gives an opportunity to better understand the behavior of the structure in the area of irregularity.

The displacement fields obtained experimentally correspond to the predicted theoretical results. The resulting maximum error does not exceed 7% and demonstrates the good convergence of results.

It allows concluding that the proposed method can be applied for the assessment of stress-strain behavior of a panel structure made of polymeric composite materials, which is damaged by delamination.

6. Conclusions

The method for assessment of the stress-strain behavior of a panel structure made of polymeric composite materials, which is damaged by delamination, has been developed. This method allows determining the degree of stress concentration in the area of step change in thickness and identifying the most dangerous point where the structural failure may start. With the use of the developed method, it is possible to solve the important practical tasks of studying the loss of performance of the panel structures made of polymeric composite materials with delaminations and determining the optimal methods for repair of such defects.

The obtained results can be further used for the calculation of delaminations in the sandwich panels. However, in this case the results of calculations using the proposed method are to be assessed separately, in particular, if the filler is the discrete one.

References

[1] Fomin O, Logvinenko O, Burlutsky O and Rybin A 2018 Scientific Substantiation of Thermal Leveling for Deformations in the Car Structure *International Journal of Engineering & Technology* **7** **4.3** pp 125-29 http://doi.org/10.14419/ijet.v7i4.3.19721

[2] Kondratiev A 2019 Improving the Mass Efficiency of a Composite Launch Vehicle Head Fairing with a Sandwich Structure *Eastern-European Journal of Enterprise Technologies* **6** **7(102)** pp 6–18 https://doi.org/10.15587/1729-4061.2019.184551

[3] Kondratiev A, Gaidachuk V, Nabokina T and Tsaritsynskyi A 2020 New Possibilities in Creating of Effective Composite Size-stable Honeycomb Structures Designed for Space Purposes *Advances in Intelligent Systems and Computing* **113** pp 45-59 https://doi.org/10.1007/978-3-030-37618-5_5

[4] Bychkov A S and Kondratiev A V 2019 Criterion-based Assessment of Performance Improvement for Aircraft Structural Parts with Thermal Spray Coatings *Journal of Superhard Materials* **41** **1** pp 53-59 http://doi.org/10.3103/S1063457619010088

[5] Heslehurst Rikard Benton 2014 *Defects and Damage in Composite Materials and Structures* (CRC Press Boca Raton FL) 177

[6] Bokhoeva L A, Rogov V E, Chermoshentseva A S and Lobanov D V 2016 Stability and Process of Destruction of Compressed Plate of Layered Composite Materials with Defects *IOP*

Conference Series: Materials Science and Engineering **142** 012077 https://doi.org/10.1088/1757-899X/142/1/012077

[7] Kondratiev A, Kovalenko V, Tsaritsynskyi A and Nabokina T 2021 Modeling of Failure in Laminated Plates by Delamination Buckling In: Tonkonogyi V. et al. (eds) Advanced Manufacturing Processes II. InterPartner 2020. *Lecture Notes in Mechanical Engineering.* pp. 149–158 https://doi.org/10.1007/978-3-030-68014-5_15

[8] Moure M M, Sanchez-Saez S and Barbero E 2014 Analysis of Damage Localization in Composite Laminates Using a Discrete Damage Model *Composites. Part B* **66** pp 224-232 https://doi.org/10.1016/j.compositesb.2014.05.015

[9] Mustafa L M, Ismailov M B and Sanin A F 2020 Study on the Effect of Plasticizers and Thermoplastics on the Strength and Toughness of Epoxy Resins *Naukovyi Visnyk Natsionalnoho Hirnychoho Universytetu* **4** pp 63-68 https://doi.org/10.33271/nvngu/2020-4/063

[10] Ugrimov S, Smetankina N, Kravchenko O and Yareshchenko V 2021 Analysis of Laminated Composites Subjected to Impact *Lecture Notes in Networks and Systems* **188** pp 234-246 https://doi.org/10.1007/978-3-030-66717-7_19

[11] Fomin O, Lovska A, Kulbovskyi I, Holub H, KozarchukI and Kharuta V 2019 Determining the Dynamic Loading on a Semi-wagon when Fixing it with a Viscous Coupling to a Ferry Deck *Eastern-European Journal of Enterprise Technologies* **2 7(98)** pp 6-12 http://doi.org/10.15587/1729-4061.2019.160456

[12] Lemanski S L, Wang J, Sutcliffe M P F, Potter K D and Wisnom M R 2013 Modelling Failure of Composite Specimens with Defects under Compression Loading *Composites Part A* **48** pp 26-36 http://dx.doi.org/10.1016/j.compositesa.2012.12.007

[13] Dimitrienko Yu I, Dimitrienko I D and Sborschikov S V 2015 Multiscale Hierarchical Modeling of Fiber Reinforced Composites by Asymptotic Homogenization Method *Applied Mathematical Sciences* **9 (145)** 7211-7220 http://dx.doi.org/10.12988/ams.2015.510641

[14] Kondratiev A, Gaidachuk V, Nabokina T and Kovalenko V 2019 Determination of the Influence of Deflections in the Thickness of a Composite Material on its Physical and Mechanical Properties with a Local Damage to its Wholeness *Eastern-European Journal of Enterprise Technologies* **4 1(100)** pp 6-13 http://doi.org/10.15587/1729-4061.2019.174025

[15] Goodsell J, Pagano N J, Kravchenko O and Pipes B 2013 Interlaminar Stresses in Composite Laminates Subjected to Anticlastic Bending Deformation *Journal of Applied Mechanics* **80 (4)** pp 1020-1026 https://doi.org/10.1115/1.4007969

[16] Barakhov K, Dvoretska D and Poliakov O 2021 One-Dimensional Axisymmetric Model of the Stress State of the Adhesive Joint *Lecture Notes in Networks and Systems* **188** pp 310–319 https://doi.org/10.1007/978-3-030-66717-7_26

[17] Zhou R, Guan Z, Jia Y, Wang K and Zhang C 2015 Effects of Delamination on Compressional Properties of Composite Laminate *Beijing Hangkong Hangtian Daxue Xuebao/Journal of Beijing University of Aeronautics and Astronautics* **41 (2)** pp 311-317 https://doi.org/10.13700/j.bh.1001-5965.2014.0116

[18] Muthurajan K G, Sankaranarayanasamy K, Tiwari S B and Nageswara B Rao 2006 Post-buckling of a Thin Film Strip Delamination in a Composite Laminate *Trends in Applied Sciences Research* **1** pp 48-60 https://doi.org/10.3923/tasr.2006.48.60

[19] Chermoshentseva A, Pokrovskiy A and Bokhoeva L 2016 The Behavior of Delaminations in Composite Materials – Experimental Results *IOP Conference Series: Materials Science and Engineering* **116** 012005 https://doi.org/10.1088/1757-899X/116/1/012005

[20] Timonin A M 2013 Finite-layer Method: a Unified Approach to a Numerical Analysis of Interlaminar Stresses, Large Deflections and Delamination Stability of Composites. Part 1. Linear Behavior *Mechanics of Composite Materials* **49(3)** pp 231-244 https://doi.org/10.1007/s11029-013-9339-1

[21] Grigorenko Y M and Vasilenko A T 2002 Some Approaches to the Solution of Problems on

Thin Shells with Variable Geometrical and Mechanical Parameters *International Applied Mechanics* **38** pp 1309-1341 https://doi.org/10.1023/A:1022693101758

[22] Grigorenko Y M and Kryukov N N 1995 Solution of Problems of the Theory of Plates and Shells with Spline Functions (survey) *International Applied Mechanics* **31** pp 413-434 https://doi.org/10.1007/BF00846794

[23] Pernicea M F, Nelson V De Carvalhob, Ratcliffe J G and Halletta Stephen R 2015 Experimental Study on Delamination Migration in Composite Laminates *Composites Part A* **73** 20-34 https://doi.org/10.1016/j.compositesa.2015.02.018

[24] Foraboschi P 2013 Layered Plate with Discontinuous Connection. Exact Mathematical Model *Composites. Part B* **47** 365-378 https://doi.org/10.1016/j.compositesb.2012.11.004

[25] Shah V 2007 *Handbook of Plastics Testing and Failure Analysis* (Consultek Brea, California) 648

[26] Kondratiev A and Slivinsky M 2018 Method for Determining the Thickness of a Binder Layer at its Non-uniform Mass Transfer Inside the Channel of a Honeycomb Filler Made from Polymeric Paper *Eastern-European Journal of Enterprise Technologies* **6 5 (96)** pp 42-48 https://doi.org/10.15587/1729-4061.2018.150387

[27] Cárdenas D, Elizalde H, Marzocca P, Abdi F, Minnetyan L and Probst O 2013 Progressive Failure Analysis of Thin-walled Composite Structures *Composite structures* **95** pp 53-62 https://doi.org/10.1016/j.compstruct.2012.05.022

[28] Kassegne S K and Reddy JN 1999 Local Behavior of Discretely Stiffened Composite Plates and Cylindrical Shells *Mechanics of Composite Materials and Structures* **5** pp 432-438.

[29] Smovziuk L and Shevtsova M 2016 *Prompt Repair of Damaged Aircraft Skin Panels* (National Aerospace University "Kharkiv Aviation Institute" Publ., Kharkov) 135

[30] Ugrimov S V and Shupikov A N 2015 Layered Orthotropic Plates. Generalized Theory. *Composite Structures* **129 1** pp 224-235 https://doi.org/10.1016/j.compstruct.2015.04.004

Laboratory tests of the steel fiber concrete road slab model

M Surianinov[1,*], S Neutov[1], I Korneieva[1], D Kirichenko[1]

[1]Odesa State Academy of Civil Engineering and Architecture, 4, Didrihsona str., Odesa 65029, Ukraine
[*] E-mail: sng@ogasa.org.ua

Abstract. The results of experimental studies of bearing capacity and cracking resistance of road slab models made of steel-fiber concrete according to the span loading scheme are presented. The bearing capacity of the models has been fixed at 27.96 kN, which is 2 times higher than the load corresponding to the beginning of cracking (14.8 kN). At the moment of bearing capacity loss, the height of compressed zone was 1 cm, which is one third of its original size. The relative strains before the first crack appeared were 0.1175×10-4 and 0.235×10-4 in the compressed and stretched zones, respectively, and similar limiting relative strains reached values of 3.715×10-4 and 9.235×10-4. The maximum crack opening width was substantially less than in the same tests of the slabs without dispersion reinforcement and was 0.3 mm, which is explained by the more ductile properties of steel fiber concrete, which prevent crack opening. The final deflection in the middle of the plate model span was 10.4 mm, which is 5.5 times greater than the deflection before cracking began (1.9 mm).

1. Introduction

Road slabs are a good alternative to conventional asphalt pavement. They are used for the construction of temporary and permanent roads, airfields, warehouses and industrial sites, as well as where you want to get a durable surface that can withstand the weight of heavy machinery in the shortest possible time. Road slabs are also used in private construction, for example, for arrangement of yard area. A wide range of slabs allows you to choose the product with the most suitable performance characteristics.

The first road slabs appeared in the early twentieth century. It was an experiment in which it was wanted to compare the strength of conventional asphalt and concrete slabs. The experiment lasted for several decades, as it was necessary to compare the durability of the pavement. It turned out that asphalt needed to be replaced after 10 years, and every 3-4 years it had to be repaired. The concrete slabs lasted about 40 years without serious deformation [1]. All concrete slabs are characterized by high frost resistance, strength, resistance to rainfall and mechanical damage, high durability. The maximum load depends on what kind of concrete was used in the production. Slabs are produced in different sizes and shapes, different thicknesses, with different types of reinforcement and using different grades of concrete.

The undeniable advantage of concrete road slabs is the possibility of their multiple use. If the temporary roadway is no longer in use, its fragments are dismantled and transported to another place for subsequent laying of a new route. Used products are only visually different from the new fragments. The characteristics of durability remain unchanged, and the cost of recycled raw materials is 30%-40% lower than the price of new products. This makes it possible to reduce the cost of new roads without compromising their quality and durability.

2. Recent researches analysis

Theoretical and experimental studies of road slabs are reflected in the publications of many researchers. Thus, in [1] the calculation of road surface slabs for operational loads by the method of B. N. Zhemochkin was made. For temporary road slabs, 16 variants of loading are considered; for permanent road slabs and prestressed - 12 variants of loading. The slabs are calculated as structures on an elastic basis. The conducted research shows that during the operation, reinforced concrete slabs of

Content from this work may be used under the terms of the Creative Commons Attribution 3.0 licence. Any further distribution of this work must maintain attribution to the author(s) and the title of the work, journal citation and DOI.
Published under licence by IOP Publishing Ltd

temporary and permanent roads are subjected not only to bending moments and transverse forces in two orthogonal directions, but also to torsional moments. In this case it is impossible to avoid the joint action of torsional and bending moments. In this case it is impossible to avoid the joint action of torsional and bending moments. In this regard, it is recommended that the load-bearing capacity of the slabs for a given reinforcement and concrete class be checked for the strength of the normal and inclined sections as well as for the combined effect of torsional and bending moments.

The peculiarity of cement concrete pavements of highways is the presence of transverse and longitudinal joints, necessary for the possibility of deformation of the pavement without its destruction during operation. These joints serve as a potential fracture zone not only for the pavement, but also for all the layers of the pavement, and also affect the comfort and convenience of passenger traffic. The paper [2] considers the causes and physical processes that caused the destruction of transverse and expansion joints of compression. The technological causes and characteristic errors that are allowed during construction works, leading to the premature destruction of joints depending on the operating conditions, are analyzed in detail. On the basis of the material presented and the experimental data obtained in the course of full-scale inspections of cement concrete pavements of highways the mathematical model for predicting the destruction of expansion joints in compression in time has been proposed.

The study of road slabs is considered in detail in the work of Belarusian scientists [3] on the calculation of the cross-section of serial reinforced concrete slabs for road pavements. The determination of bearing capacity of road slabs of several types according to the strength of normal sections in the short and long direction, as well as the bearing capacity under the joint action of torsional and bending moments is studied. The calculation procedure of reinforced concrete road surfacing slabs according to the strength of normal sections and on the joint action of torsional and bending moments which has been developed on the basis of the experimental and theoretical researches has been proposed.

In the study of crack resistance in the standards of different countries different formulas are used to calculate the distance between cracks and the width of cracks occurring in bending elements [4]. Most of these formulas are based on the analysis of test results of beams or one-sided slabs. However, the behavior of reinforced concrete slabs with double-sided reinforcement has significant differences, and, therefore, the methods developed for beams cannot be applied directly to such slabs. A new theoretical equation for calculating the crack spacing for road slabs, the validity of which is confirmed by the results of experiments, is proposed in the paper.

Interesting results of the work, the purpose of which was to study the effect of steel fiber on the structural behavior of the slabs (the possibility of redistribution of forces, the nature of cracks and bearing capacity), are presented in the paper [5]. The experiments were carried out on octagonal slabs, which were freely supported on four edges and were loaded with a point load in the center.

The study of the behavior of reinforced concrete slabs exposed to fire is devoted to [6]. Let's also note the publications [7-11], where the calculations and modeling of the work of road pavement made of reinforced concrete were performed.

3. The purpose of this work is a laboratory study of models of steel-fiber concrete road slabs to develop recommendations for the manufacture of full-size natural structures of combined reinforcement slabs using steel fiber.

4. Materials and methods

The design of the road slab model (Figure 1) was developed, the requirements of the regulatory documents operating in Ukraine were taken into account in the design [12-14]. The geometric dimensions of the model are determined by the technological features of manufacturing and the conditions of correctness of the comparison of the results. The slab models are reinforced with a spatial framework consisting of two meshes (Figure 2).

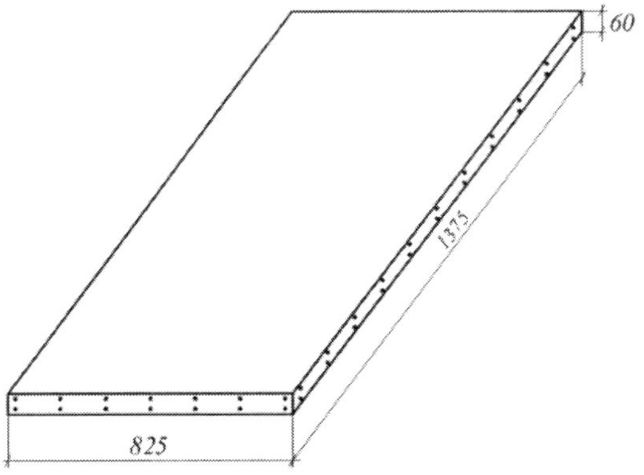

Figure 1. Road slab model.

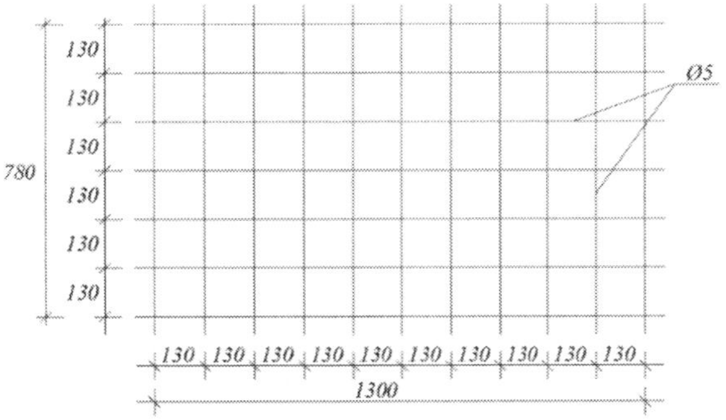

Figure 2. Road slab model reinforcement.

The slab models were tested on a specially assembled stand (Figure 3). The force device of the stand consists of four racks connected in pairs by beams, on which the road slab is supported. The load is produced by means of a 100 kN jack, the lower plane of which presses on two-level cross-beam system and through it on the slab. The design and dimensions of the cross-beam system are referenced to the road slab loading schemes given in the regulatory literature [15, 16]. The loading was performed by applying two concentrated vertical strip loads along the width of the slab (Figure 4). The upper rod of the jack rests on an I-beam. The forces applied to the plate are controlled with a 500 kN sample dynamometer.

Figure 3. Test stand.

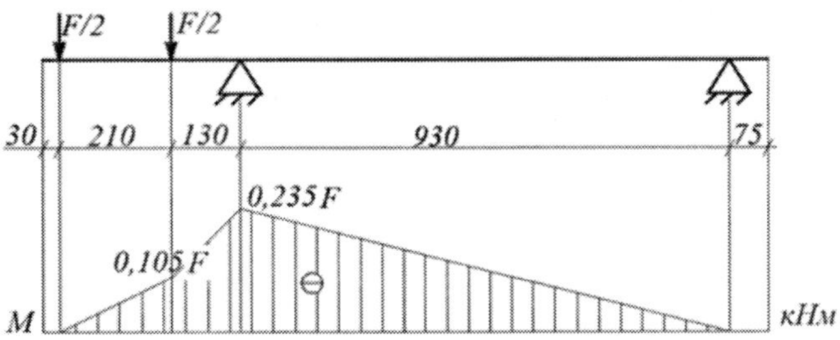

Figure 4. Test scheme with bending moment diagram.

Taking into account the adopted loading scheme, two rows of indicators were installed on the top surface of the slab (Figure 5). The first group of four indicators (2, 3, 4 and 5) are mounted on the cantilever loading section of the model, and the other group (7, 8, 9 and 10) in the upper span of the slab. Indicators 1, 6 are mounted on the side faces in the support area of the cantilever section. A pair of indicators 11, 12 recorded the deformations of the compressed zone in the span section of the slab.

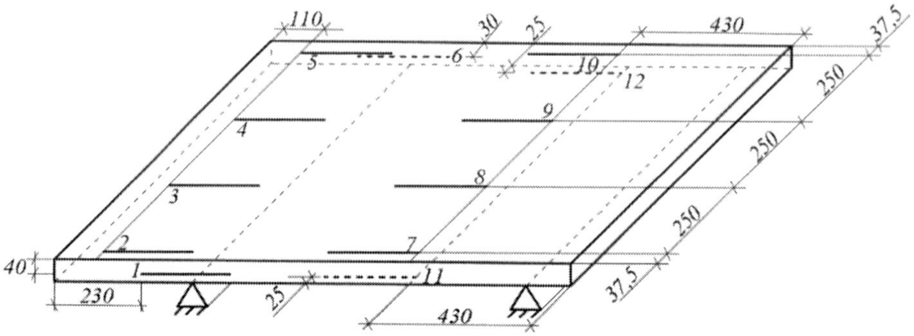

Figure 5. Indicator layout.

During the tests, the load, longitudinal deformations, deflections, and cracking process were recorded.

5. Research results

A series of road slabs modeli of dimensions 1375x825x60 made of concrete C20/25 and reinforcement Bp1 with the addition of steel fibers 1% of the volume of the product. The composition of the mixture was determined by the authors in previous studies [17, 18].

The load was applied in steps of 0.05 of the destructive load in accordance with the generally accepted methodology. Each loading step ended with a five-minute exposure time, at the beginning and the end of which readings were taken on the measuring instruments.

Figure 6 shows the values of relative deformations of individual fibers of concrete. As can be seen from the graphs, indicators 1-10 work in tension, and 11 and 12 in compression.

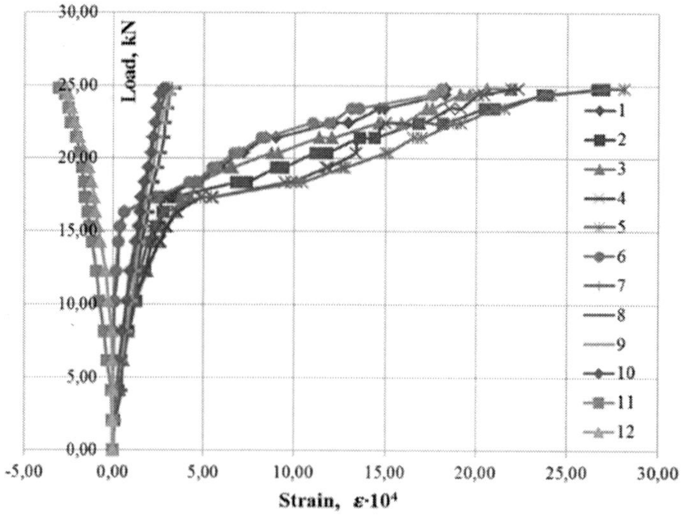

Figure 6. Relative deformation according to indicator readings.

The nature of the change in the deformations shown in Fig. 6, can be divided into two clearly defined stages. At the first, initial stage, all indicators without exception operate linearly. The values of relative deformations in the compressed and stretched zones, in the loaded cantilever and unloaded span of the slab differ very insignificantly. Such unity lasts until the load of 14.27 kN, corresponding to the moment of crack formation. After the formation of cracks (at the second stage), all the indicators were clearly divided into two groups. One group consisted of indicators 7-12, located in the span of the slab. They practically did not react to the cracking process, continuing to work linearly up to the breaking load of 24.78 kN. The relative deformations of this group of indicators at this stage increased 2.1 times, from 1.2×10-4 to 2.6×10-4.

The second group included indicators located on the cantilevered loaded section of the slab. With the beginning of cracking, the strain growth rate (deformation pattern) changed sharply, and the curves on the graph deviated significantly from the linear law of change. The relative deformations of the cantilever section at the second stage increased twelve times, from 2.1×10-4 to 25×10-4.

Figure 7 shows for comparison the relative deformations obtained with strain gauges. The results show that the strains measured by mechanical and electrical instruments are identical.

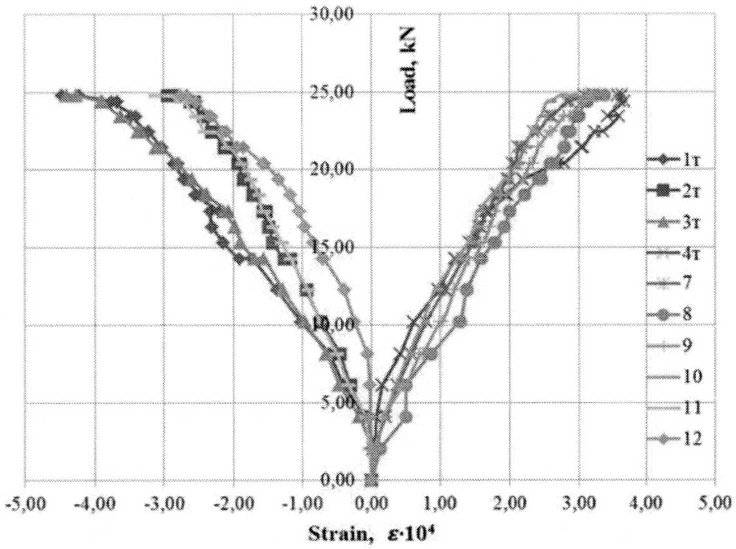

Figure 7. Relative strain according to the readings of indicators 7-12 and strain gauges 1t-4t.

Figure 8 shows the averaged deflections obtained using Maximov deflectometers in the cantilever and span sections of steel-fiber-concrete road slab models. The results show that up to the moment of crack formation, the deflections of the span and cantilever sections change linearly and practically by the same value equal to 5 mm. After the formation of cracks, the deflections in the span section of the plates continue to change linearly. By the time the load-bearing capacity of the slabs is lost, these deflections reach the value of 13 mm. At the cantilever section, the deflections growth rate is significantly higher. By the time of failure, the deflections of the cantilever loaded section increased 4.5 times in comparison with the first stage, reaching 25 mm.

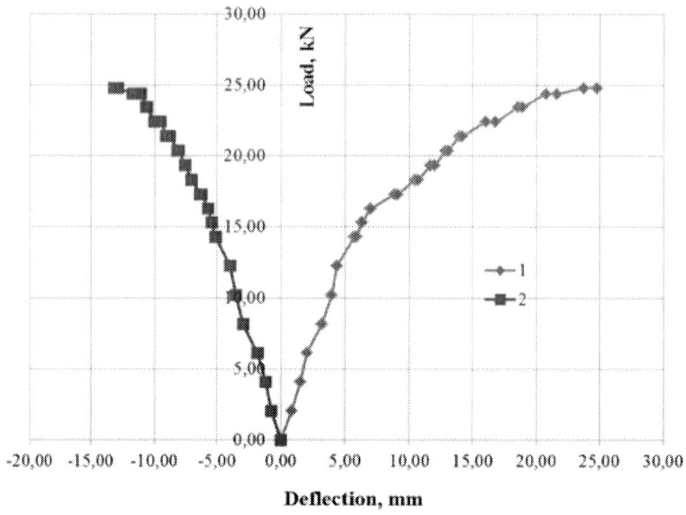

Figure 8. Deflections in the span and on the console.

6. Conclusions

Thus, laboratory tests were carried out on a series of model road slabs made of steel fiber concrete with 1% reinforcement. The tests were carried out according to one of the two loading schemes recommended by the regulatory documents acting in Ukraine - with the application of load to the cantilever part of the slab.

It was found that the deformations of concrete in the span unloaded section of the slabs almost at all stages of loading change according to the linear law, and by the time of failure reach the value of 3.2×10-4.

The deformation of concrete in the cantilever section of the slab after cracking increases from 2.1×10-4 to 25×10-4, i.e. by a factor of 12. The beginning of the cracking process in the models of road slabs under cantilever loading occurs at the load level of 0.58 of the destructive load. The maximum deflections of the cantilever loaded section (25 mm) are almost 2 times greater than those in the middle of the slab span (13 mm).

The developed methodology and test results of road slab models will be used to test full-size natural structures of combined reinforcement slabs using steel fibers.

References

[1] Semeniuk S D 2017 Raschet plyt pokrytyia avtomobylnykh doroh na ekspluatatsyonnye nahruzky sposobom B N Zhemochkyna *Ynzhenerno-heotekhnycheskye yzыskanyia, proektyrovanye y stroytelstvo osnovanyi, fundamentov y podzemnыkh sooruzhenyi: sbornyk trudov Vserossyiskoi nauchno-tekhnycheskoi konferentsyy* pp 139-145

[2] Fotyady A A, Mordvyn S S, Hnezdylova S A and Seleznev Yu V 2020 O protsessakh razrushenyia deformatsyonnykh shvov tsementobetonnykh pokrytyi avtomobylnykh doroh *Promyshlennoe y hrazhdanskoe stroytelstvo* **12** pp 17-22 DOI:10.33622/0869-7019.2020.12.17-22

[3] Semeniuk S D and Belyi N V 2008 Raschet sechenyi zhelezobetonnыkh plyt seryy B.3.503.1–1 dlia pokrytyi avtomobylnykh doroh *Vestnyk Belorussko–Rossyiskoho unyversyteta* **3(20)** pp 166 – 176

[4] Rizk Emad and Marzouk Hesham 2008 A new formula to calculate crack spacing for plates and two-way concrete slabs *Canadian Society for Civil Engineering* **2** pp 1350-1360

IOP Conf. Series: Materials Science and Engineering 1164 (2021) 012035 doi:10.1088/1757-899X/1164/1/012035

[5] Fall D, Shu J, Rempling R, Lundgren K and Zandi K 2014 Two-way slabs: Experimental investigation of load redistributions in steel fibre reinforced concrete *Engineering Structures* **80** pp 61–74
[6] Said M A, Hazem M F E and Alaa G R 2013 Behavior of one-way reinforced concrete slabs subjected to fire *Alexandria Engineering Journal* **52(4)** pp 749-761 https://doi.org/10.1016/j.aej.2013.09.004
[7] Pushkarenko H H and Cherniakevych V Y 2002 Modelyrovanye dorozhnoi plytы y ee raboty v pokrytyy *Maryisk. hos. tekhn. un-t* **504**
[8] Makarov E V, Popov A N and Podolsky V P 2018 Theoretical Calculation of Repaired Aerodrome Pavement Deflected Mode Basis *International Journal of Applied Engineering Research* **13(4)** pp 2000-2005
[9] Pal A and Miklos K 2003 Modelling of reinforced soil *Periodica polytechnica* **47(2)** pp 169-174
[10] Van Dam T, Taylor P and Fick G 2010 *Sustainable concrete pavements: a manual of practice Institute for Transportation USA* lowa State University research park 167
[11] Dô Ba Chuong 2007 *Thiét ké duong ô tô tâp 1* Nhà xuât bân giâo duc 159
[12] DBN V.2.6-98:2009 2011 *Betonni ta zalizobetonni konstruktsii. Osnovni polozhennia* [chynnyi vid 2011-07-01] Minrehionbud Ukrainy 71
[13] DSTU B.V.2.6-2:2009 2010 *Vyroby betonni i zalizobetonni. Zahalni tekhnichni umovy* [chynnyi vid 2010-06-30] Minrehionbud Ukrainy 34
[14] DSTU B.V.2.6-122:2010 2011 *Plyty zalizobetonni z nenapruzhenoiu armaturoiu dlia pokryttia miskykh dorih* [chynnyi vid 2011-07-01] Minrehionbud Ukrainy 23
[15] DSTU B.V.2.7-214:2009 2010 *Betony. Metody vyznachennia mitsnosti za kontrolnymy zrazkamy* [chynnyi vid 2009-12-22] Minrehionbud Ukrainy 43
[16] DSTU B.V.2.6-7-95 1997 *Vyroby budivelni betonni ta zalizobetonni zbirni. Metody vyprobuvan navantazhuvanniam. Pravyla otsinky mitsnosti, zhorstkosti ta trishchynostiikosti* [chynnyi vid 1995-16-11] Derzhavnyi komitet Ukrainy u spravakh mistobuduvannia i arkhitektury 34
[17] Neutov S F, Boiadzhy A A and Korneeva I B 2018 Opredelenye osnovnykh fyzyko-mekhanycheskykh kharakterystyk stalefybrobetonnoi smesy optymalnoho sostava *World science* (Warsaw Poland) **5 (33)** pp 26-30
[18] Surianinov M, Neutov S and Korneieva I 2020 Comparative analysis of strength and deformation of reinforced concrete and steel fiber concrete slabs *E3S Web of Conferences* **166** № 06003

Study of deformation properties of alkali activated concretes using active aggregates

O Kovalchuk[1,*] and V Zozulynets[1]

[1]Scientific Research Institute for Binders and Materials, Kyiv National University of Construction and Architecture, Kyiv, 03037, Ukraine

[*]Email: kovalchuk.oyu@gmail.com

Abstract. Results of study of deformation properties of fine-grain concrete are shown using slag alkali activated cement and active aggregate. It had been shown that expansion deformations of specimens, supplying process of alkaline corrosion of aggregate in concrete, directly combined with component composition and conditions of hardening of material. It was shown that using alkaline component in the state of dry salt shrinkage deformations are varying in the shorter ranges comparing to concrete with the soluble glass. Introduction of active mineral admixture leads to the decreasing of deformations comparing to the compositions without admixture. Different conditions of hardening of the specimens are also influence well on the development of shrinkage deformations. It is shown that drying of specimens with active process of alkaline corrosion of concrete makes it possible to stop development of expansion deformations in concrete. Hydrophobization of the dried specimens make it possible to store linear characteristics of specimens. Hydrophobization of without drying leads to the intensification of structure formation processes and higher rates of development of deformations. Traditional method of protection of concrete constructions (covering by painting materials) is not able to prevent, but also possible to activate development of destructive processes of alkaline corrosion of concrete.

1. Introduction

The problem of use aggregates containing active grains is a target of studies of different scientists all over the world for a various years [1-5]. Today this problem rises sharper because of replacements of gas fuel in the cement production by coal, resulting in rising of specific content of alkalis in the cement. In general, cements are meeting requirement of standards by alkalis content [6-8], but increasing of their specific content is dangerous because of possibility of alkaline corrosion of aggregates in concrete [9-11].

Modern production is oriented on constant search of cheaper materials and reducing of materials costs [12-14]. Some time it can be realized by replacement of aggregated by other, cheaper materials. But quality assurance during such replacement is providing not every time in the satisfied level – only a few characteristics are verifying, like crashing, frost resistance, compressive strength, fraction of aggregates, and presence of needle-like grains and so on. However, such cheap aggregated also can content active grains that limit their application. Combining of two factors – higher alkalis content in the cement and presence of active grains in the aggregate composition leads to the risk of development of destructive processes in the hardened concrete body – alkaline corrosion of concrete [15].

It is well-known possibility to prevent development of alkaline corrosion of concrete by modifying of concrete mix by active mineral admixtures [16-18], which are rich on active silica that makes it possible to change destructive gain of the process on constructive and also shift this processes in time. But such modification have to be done on the stage of concrete mix design and there are have to be information about presence of active grains in the aggregate composition that is impossible because of underlined higher. Moreover, introduction of such active mineral admixtures as a partial cement replacement could provide some strength drop [19-22].

Content from this work may be used under the terms of the Creative Commons Attribution 3.0 licence. Any further distribution of this work must maintain attribution to the author(s) and the title of the work, journal citation and DOI.
Published under licence by IOP Publishing Ltd

From the point of view of exploitation of constructions under influence of alkaline corrosion of concrete, today is only one solution proposed – construction has to be destroyed and rebuilt. Such approach is very financial- and source demanding and needs a lot of time and logistic sources (for example, in the USA the Hoover Dum had been totally reconstructed).

From the rationality of repairing work point of view, it is interesting to study possibility to stop the destructive process in the concrete and provide possibility to repair existing constructions with alkaline corrosion processes. The previous works shows possibility to stop strength development, making it possible to predict possibility to regulate structure formation processes in the destined direct [23-26]. However, the most actual question is possibility to stop expansion deformations in material.

The aim of the study is determination of regulations of influence of alkali activated cement mix design (type and characteristics of alkaline component and alumina silicate component) and hardening and storing conditions on deformation properties of fine grain concretes on their basis using active aggregate.

2. Materials and test methods

For investigation of possibility to influence on internal corrosion of alkali activated concrete gain the basalt rock was chosen as an active aggregate. The aggregate, taking into account results of previous studies [27-29] was represented by the fraction 0-2.5 mm. Chemical composition of basalt rock is given in Table 1.

Table 1. Chemical composition of basalt rock.

	Oxides content, % by mass												Mass loses	Σ, %
	SiO_2	Al_2O_3	Fe_2O_3	FeO	TiO_2	MnO	CaO	MgO	P_2O_5	K_2O	Na_2O	SO_3		
Basalt	50.42	14.0	6.14	8.37	2.66	0.243	8.04	5.56	0.316	0.71	2.27	0.07	0.77	99.57

As main calcium - alumina silicate component of alkali activated cement was chosen granulated blast furnace slag (Kamenske, Ukraine), ground to the specific surface 450±20 m^2/kg by Blaine. Chemical composition of the slag is given in Table 2.

Table 2. Chemical composition of granulated blast furnace slag.

	Oxides content, % by mass						Mo
	SiO_2	Al_2O_3	CaO	MgO	MnO	TiO_2	
GGBS	37.90	6.85	44.6	5.21	0.106	0.35	-

As a hydrophobization agent was use admixture on the basis of silicon hydro siloxane liquid 136-157M, produced by «ANTALCOM» ltd.

As an active mineral admixture was used metakaolin from Glukhovetsky factory, ground to the specific surface 1000 m^2/kg by Blaine.

As alkaline components were used sodium carbonate (soda ash) in the state of dry salt and soluble glass with silicate modulus M_c=1,0, obtained by correlation of modulus of initial product – soluble glass with M_c=2,96. Characteristics of initial high-modulus soluble glass are given in Table 3.

Table 3. Chemical composition and properties of initial soluble glass.

Initial density of soluble glass, kg/m^3	Silicate Modulus	Oxides content, % by mass in liquid state			
		SiO_2	R_2O_3	Na_2O	CaO
1400	2.96	28.5	0.19	9.37	0.15

As a reference material and also as an alumina silicate component of alkali activated cements was taken OPC Grade 42.5 manufactured by PAT "Volyncement". Chemical-mineralogical composition of cement is given in Table 4.

Table 4. Chemical-mineralogical composition of OPC grade 52.5.

Oxides Content, % by mass								Minerals content, % by mass			
SiO_2	Al_2O_3	CaO	MgO	Fe_2O_3	SO_3	R_2O	Mass loses	C_3S	β-C_2S	C_3A	C_4AF
21.82	5.30	65.91	1.11	4.86	0.99	0.22	0.2	61.2	17.7	5.9	15.1

Specimens were prepared from cement : sand mortar in the ratio 1:2.25 according to DSTU B V.2.7-185 demands using mixer «Hobart". Tests of specimens were done according to DSTU B V.2.7-181 recommendations. Deformations were tested on the specimens-beams 2.5×2.5×25.4 cm using mentioned ratio. On the tool with indicator type IP-04 with discretion rate 0.01 mm.

In accordance to the test methodic of influence hardening conditions on properties of alkali activated concrete specimens were storied in normal conditions (20±2°C and relative humidity 95±5%) to the moment of expansion appearance. After that specimens were divided onto four groups. One group was dried to the constant mass and then returned to the curing chamber to be storied in normal conditions. Second group also was dried to the constant mass and covered by hydrophobic agent and also returned to the chamber. Third group was not dried, but cover by hydrophobic agent and then returned to the camber. Control specimens keep studying in normal conditions.

3. Test results and discussions

Tests of shrinkage/expansion deformations were done using cement systems, chosen from the previous study results.

Systems under study:

1. Alkaline OPC (OPC + soluble glass)
2. Alkaline OPC (OPC + soluble glass)+ 10% metakaolin (MK)
3. Slag alkali activated cement (LCEM-1) with extra alkaline component
4. Slag alkali activated cement (LCEM-1) with extra alkaline component +10% MK
5. OPC Grade 42.5
6. OPC Grade 42.5 + 10% MK

Alkaline component in the system LCEM-I was represented by dry salt of sodium carbonate. Test results are given in Table 5 and on Figure 1. Analysis of obtain results shows that all systems under study are characterized by shrinkage deformations in all diapason of study. It can be noted that the highest shrinkage belongs to the systems using soluble glass (alkaline OPC), and the lowest – slag alkali activated cement LCEM-I. This could be explained by presence in the systems with soluble glass higher content of gel phases, which, as it well-known, are present significant less in the systems with dry alkaline component.

Table 5. Shrinkage/expansion deformations of fine-grain concretes using basalt rock aggregates.

№	Composition	Shrinkage/expansion deformations, mm/m, at the age, days				
		2	7	28	90	180
1	Alkaline OPC (OPC + soluble glass)	–0.23	–0.45	–0.63	–0.55	–0.51
2	Alkaline OPC (OPC + soluble glass) + 10% metakaolin (MK)	–0.30	–0.49	–0.61	–0.49	–0.36
3	Alkali activated cement (LCEM-1) with extra alkaline component	–0.21	–0.33	–0.44	–0.43	–0.40
4	Slag alkali activated cement (LCEM-1) with extra alkaline component + 10% MK	–0.15	–0.26	–0.31	–0.33	–0.32
5	OPC Grade 42.5	–0.25	–0.41	–0.50	–0.44	–0.40
6	OPC Grade 42.5 + 10% MK	–0.19	–0.27	–0.41	– 0.30	–0.23

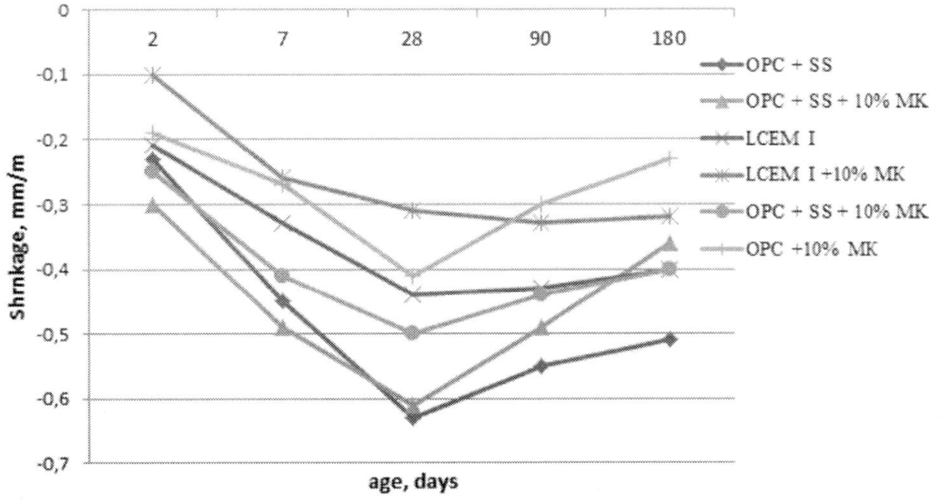

Figure 1. Shrinkage/expansion deformations of fine-grain concretes using basalt rock aggregates.

It has to be mentioned that all systems are characterized by reducing of shrinkage in time, showing continuation of structure formation processes and development of ASR reaction in the concrete body.

The best properties has the system on slag alkali activated cement (LCEM-I) themselves, and also with the metakaolin admixture. They were characterized by slow development of shrinkage deformations and some changes of process because of processes of alkaline corrosion of aggregate gain and development of expansion deformations

Tests of influence of hardening conditions and storing on deformation properties of fine grain concretes were done on the bases of slag alkali activated cement system (LCEM-I) with and without metakaolin admixture. Test results are given in Table 6 and on Figure 2.

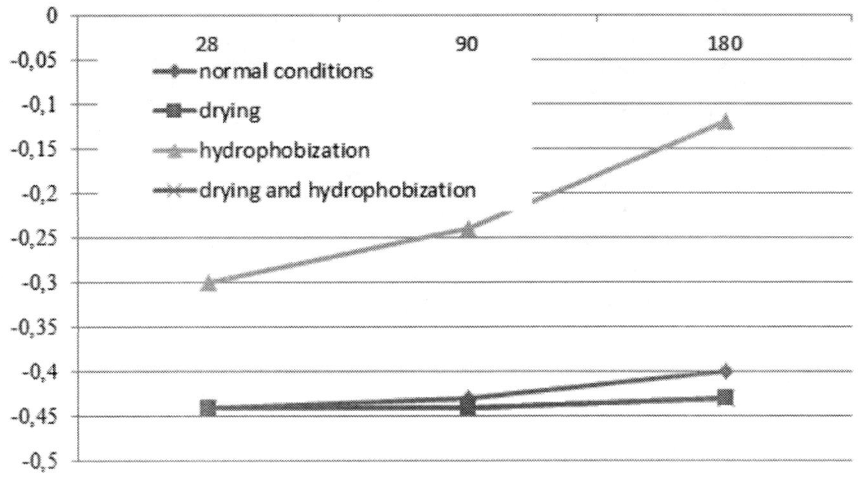

a)

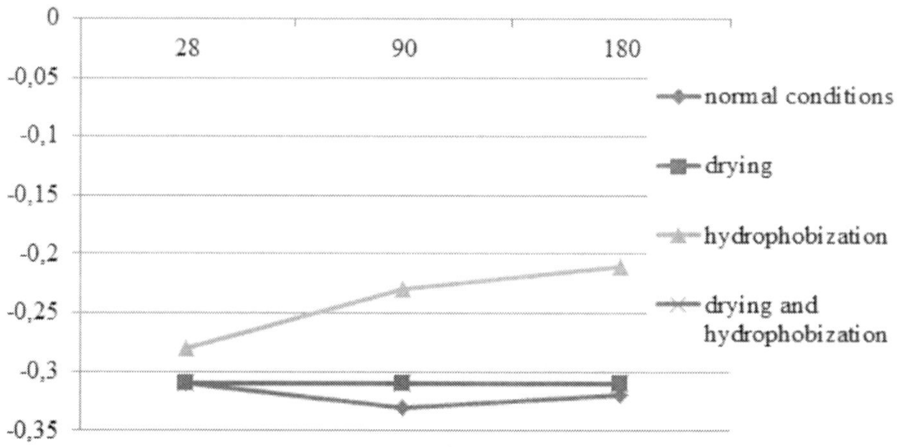

b)

Figure 2. Shrinkage/expansion deformations of fine-grain concretes depending from hardening conditions a) on the LCEM-I basis; b) on the LCEM-I basis and metakaolin admixture.

Analysis of the obtain results showed that hardening conditions are influence well on deformations variations. Thus, it was shown that in case of drying of specimens to the constant mass expansion of the specimens is stopping. This could tell about stop of structure formation processes inside the material because of vanishing of liquid phase, that is necessary condition for the reaction. At the same time, it has to be mentioned that difference between dried specimens and specimens, hydrophobizated after drying, from the point of view of deformations is not significant. This also can prove that hygroscopic characteristics of the concrete are not high enough to replace the water content necessary to continue reaction after being dried. Hydrophobization in that case can serve as a guaranty of storing of low moisture content and reducing of risk of destructive reaction in the future.

Table 6. Shrinkage/expansion deformations of fine-grain concretes depending from hardening conditions.

Composition	Hardening conditions	Shrinkage/expansion deformations, mm/m, at the age, days		
		28	90	180
Slag alkali activated cement (LCEM-1)	normal conditions	–0.44	–0.43	–0.40
	drying	-0.44	-0.44	-0.43
	hydrophobization	-0.3	-0.24	-0.12
	drying and hydrophobization	-0.44	-0.44	-0.43
Slag alkali activated cement (LCEM-1) +10% MK	normal conditions	–0.31	–0.33	–0.32
	drying	-0.31	-0.31	-0.31
	hydrophobization	-0.28	-0.23	-0.21
	drying and hydrophobization	-0.31	-0.31	-0.31

Hydrophobization of fine grain concrete without drying leads to intensification of specimen's expansion. Combining this information with the strength properties, obtain as a result of previous studies, the conclusion can be done about intensification of structure formation processes via self-steaming of the specimens. That fact could be dangerous from the point of view of traditional methods of concrete structures protection, namely – covering of construction by the paint cover.

4. Conclusions

Thus, it was shown that from the point of view of shrinkage/expansion deformations the best systems are alkali activated cement LCEM-I as it is and with metakaolin admixture (-0.44 and -0.31 mm/m correspondently at the age of 28 days of normal hardening), which, beside this, also showed minimal shrinkage drop at the later terms of hardening, witnessing about decreasing of development of ASR reaction in concretes using such systems. Drying of concrete specimens makes it possible to stop expansion deformation in concrete, and following hydrophobization makes it possible to provide stability of linear dimensions, opening possibility to prolong exploitation terms of constructions under influence of alkaline corrosion of concrete.

References

[1] Shtark I 2010 *Alkaline corrosion of concrete* ed P V Krivenko (Kyiv)
[2] Stark J, Freyburg E, Seyfarth K, Giebson C and Erfurt D 2010 70 years of ASR with no end in sight? (Part 2) *ZKG International* **63(5)** pp 55-70
[3] Malek R I A and Roy D M 1983 Effect of Slag Cements and Aggregate Type on Alkali-Aggregate Reaction and its Mechanism *Proc. 6th Int. Conf. Alkalis in Concrete, Research and Practice (Denmark)* pp 223-230
[4] Tang M-S and Han S F 1991 Effect of $Ca(OH)_2$ on Alkali-silica Reaction *J. of Chinese Silicate Society* pp 160-166
[5] Malek R I A and Roy D M 1984 Alkali-Aggregate Reaction Resistance of Granulated Blast Furnace Slag Cement *Slag Cements Workshop P. State University, Mater. Res. Laboratory, Dec-March*
[6] Pushkarova E and Gonchar O 2005 Peculiarities of synthesizing artificial zeolite on a fly ash basis and their usage for modification of special destined concrete *Proc. Int. Conf. on Achieving Sustainability in Construction* pp 53-60
[7] Bilchenko A V, Kislov O H, Synkovska O V and Ihnatenko A V 2019 Durability of reinforced concrete constructions is the dasis of briges life cycle *Naukovyi Visnyk Budivnytstva (Kharkiv)* **93(3)** pp 140-144
[8] Kawamura M and Kodera T 2005 Effects of externally supplied lithium on the suppression of ASR expansion in mortars *J. Cement and concrete research* **35(3)** pp 494–498
[9] Pushkareva K K, Gonchar O A and Kaverin K O 2019 The role of the crystallo-chemical factor in the evaluation and improvement of the nanomodification efficiency of mortar and concrete *IOP Conference Series: Materials Science and Engineering* **708(1)** 012102
[10] Berdnyk O Y, Lastivka O V, Maystrenko A A and Amelina N O 2020 Processes of structure formation and neoformation of basalt fiber in an alkaline environment *IOP Conference Series: Materials Science and Engineering* **907(1)** 012036
[11] Hunger K J 2007 The contribution of quartz and the role of aluminium for understanding the AAR with greywacke *J. Cement & Concrete Research* **37** pp 1193–1205
[12] Runova R, Gots V, Rudenko I, Konstantynovskyi O and Lastivka O 2018 The efficiency of plasticizing surfactants in alkali-activated cement mortars and concretes *MATEC Web of Conferences* **230** 03016 DOI: 10.1051/matecconf/201823003016
[13] Rudenko I I et al. 2018 Efficiency of redispersible polymer powders in mortars for anchoring application based on alkali activated Portland cements *Key Engineering Materials* **761** pp 27-30 DOI: 10.4028/www.scientific.net/KEM.761.27
[14] Borziak O, Chepurna S, Zidkova T, Zhyhlo A and Ismagilov A 2018 Use of a highly dispersed chalk additive for the production of concrete for transport structures *MATEC Web of Conference* **230** 03003 DOI: 10.1051/matecconf/201823003003

[15] Ramachandran V S 2003 Alkali-Aggregate Expansion Inhibiting Admixtures *J. Cement & Concrete Composites* **20** pp 149–161
[16] Ramlochan T, Thomas M and Gruber K 2000 The Effect of Metalkaolin on Alkali-Silica Reaction in Concrete *J. Cement & Concrete Research* **30** pp 339-344
[17] Kryvenko P, Cao H, Petropavlovskyi O, Weng L and Kovalchuk O 2016 Applicability of alkaliactivated cement for immobilization of lowlevel radioactive waste in ion-exchange resins *Eastern European Journal of Enterprise Technologies* **1(6)** pp 40-45
[18] Krivenko P V, Gelevera A G, Petropavlovsky O N and Kavalerova E S 2005 Role of metakaolin additive on structure formation in the interfacial transition zone "Cement – Alkali-susceptible aggregate" *Proc. Int. Symp. Non-Traditional Cement and Concrete II (Brno)* pp 83-95
[19] Winnefeld F et al. 2020 RILEM TC 247-DTA round robin test: sulfate resistance, alkali-silica reaction and freeze–thaw resistance of alkali-activated concretes *Mater Struct* ***53*** 140
[20] Angulo-Ramirez D, Gutierrez R and Medeiros M 2018 Alkali-activated Portland blast furnace slag cement mortars: performance to alkali-aggregate reaction *Construction and building materials* **179** pp 49-56
[21] Shi Z, Shi C, Zhao R and Wan S 2015 Comparison of alkali–silica reactions in alkali-activated slag and Portland cement mortars *J. Materials and Structures* **48(3)** pp 743-751
[22] Nicolas R and Provis J 2015 The interfacial transition zone in alkali-activated slag mortars *Frontiers in materials* **2** 70
[23] Krivenko P, Gelevera O, Kovalchuk O, Bumanis G and Korjakins A 2019 Alkali-aggregate reaction in alkali-activated cement concretes *IOP Conference Series: Materials Science and Engineering* **660(1)** 012002
[24] Lu D, Mei L, Xu Z, Tang M and Fournier B 2006 Alteration of alkali reactive aggregates autoclaved in different alkali solutions and application to alkali–aggregate reaction in concrete:(I) Alteration of alkali reactive aggregates in alkali solutions *J. Cement and concrete research* **36(6)** pp 1176-1190
[25] Feng X, Thomas M D A, Bremner T W, Balcom B J and Folliard K J 2005 Studies on lithium salts to mitigate ASR-induced expansion in new concrete: a critical review *J. Cement and Concrete Research* **35(9)** pp 1789-1796
[26] Kropyvnytska T, Semeniv R and Ivashchyshyn H 2017 Increase of brick masonry durability for external walls of buildings and structures *MATEC Web of Conferences* **116** 01007
[27] Krivenko P, Petropavlovsky O, Kovalchuk O 2018 A comparative study on the influence of metakaolin and kaolin additives on properties and structure of the alkali-activated slag cement and concrete *EEJET* **1/6(91)** 33-39 DOI: 10.15587/1729-4061.2018.119624
[28] Krivenko P V et al. 1997 Durability of Alkaline Portland Cement Concretes made with Alkali-Reactive Aggregates *Proc. ACI Int Conf on Durability of Concrete (Sydney)*
[29] Krivenko P, Petropavlovskyi O, Kovalchuk O, .Rudenko I and Konstantynovskyi O 2019 Enhancement of alkali-activated slag cement concretes crack resistance for mitigation of steel reinforcement corrosion *E3S Web of Conferences* **166** 06001 DOI: 10.1051/e3sconf/202016606001

Acknowledgement
Present results were obtained within the project «Development of technological methods of prevention and stop of alkaline corrosion of concrete using active aggregates» ordering by the Ministry of Education and Science of Ukraine (registration number 0119U002580).

An analytical solution to the problem of thermoelastic bending of a multilayer beam with different temperature of longitudinal faces

S B Koval'chuk[1,*] and O V Goryk[1]

[1]Poltava State Agrarian Academy, 1/3 H. Skovorody str., Poltava, 36003, Ukraine

*E-mail: stanislav.kovalchuk@pdaa.edu.ua

Abstract. An exact analytical solution of the quasistatic problem of thermoelasticity is presented for a section of a narrow multilayer beam with different temperatures of the longitudinal lower and upper faces and a heat flow of arbitrary height across the sections through the lateral faces. The solution was obtained for the entire package of layers by sequentially solving the heat equation for an inhomogeneous beam, taking into account the ideal thermal contact of the layers and the system of equations of the plane problem of the theory of elasticity under the assumption of a rigid connection of the layers. To take into account the inhomogeneity of the beam, a continuum approach is used, in which the multilayer material is considered continuous with variable physical and mechanical characteristics. The resulting relations take into account the orthotropy of the physico-mechanical properties of the materials of the layers and their compliance with the transverse shear and compression strains. An example of implementation of a solution for a five-layer beam with combined rigid and articulated fastening of the ends is given.

1. Introduction

Bars and beams are one of the most common elements of engineering structures. For composite beams, a multilayer structure with a pronounced difference in the mechanical characteristics of the layers is typical. With proper design, this provides a significant gain in the specific strength and stiffness of the composite beam in comparison with a homogeneous analogue. However, the combination of dissimilar materials in the structural element has certain disadvantages, one of which is the occurrence of temperature deformations and stresses. In contrast to homogeneous in multilayer boards even in a uniform temperature field, the occurrence of temperature stresses and displacements is inevitable. This can significantly reduce the reserve of the bearing capacity of the element for the perception of the useful external load. Therefore, the development of theories of bending of composite bars in the direction of taking into account temperature deformations is an important direction in the mechanics of structurally inhomogeneous elements.

Along with high-tech glass and carbon fiber, the reinforced concrete widely used in construction also belongs to composites. Despite the nonlinearity of the elastic characteristics of polymer matrices, reinforced plastics exhibit an almost linear relationship between stresses and strains up to failure [1]. As experimental studies show, for example [2, 3], when calculating reinforced concrete beams, it is important to take into account the nonlinearity of the mechanical characteristics of the concrete matrix. However, with certain reservations, the phase materials of such composites can be considered as linearly elastic materials.

To date, a significant number of scientific papers have been published on the problems of elastic deformation of composite beams with various methods of solution, types of loading and fastening, as well as features of the properties and structure of materials. If we consider the analytical direction of solution methods, then refined bending models are quite developed for multilayer beams, for example [4-7]. The main approaches to the construction of bending models of various orders can be found in the review [8].

No less important for the analytical direction, is to obtain accurate solutions of the equations of elasticity theory for the problems of deformation of composite bars, which give a more accurate and

Content from this work may be used under the terms of the Creative Commons Attribution 3.0 licence. Any further distribution of this work must maintain attribution to the author(s) and the title of the work, journal citation and DOI.
Published under licence by IOP Publishing Ltd

complete picture of the distribution of stress-strain state (SSS) characteristics. Most of these solutions were obtained for a separate simple type of load: load at the end [9-13], uniform [9, 14-21], linear [22-25] or sinusoidal [26, 27] normal load on the longitudinal faces. In a number of works, more complex laws of changing the intensity of the load in the form of the sum of a power [28, 29] or trigonometric [30-32] series, as well as their combination [33], are considered.

Among the mentioned works, the majority is devoted to beams made of functionally gradient materials (FGM) and a relatively small number to multilayer beams [10, 11, 13, 20, 21, 25, 32]. The temperature component of stresses and displacements was considered only in [15, 35] devoted to FGM beams. If we do not take into account the classical theory of temperature bending of a bimetallic strip [36], which in some cases corresponds to the theory of elasticity, then for multilayer beams exact analytical solutions are practically absent.

In this work, an exact analytical solution is proposed for the quasistatic problem of thermoelasticity of a multilayer beam in an important case for practice of different temperatures of longitudinal surfaces and symmetric heat transfer through side faces. In this case, a complete system of equations of thermoelasticity is applied, including the heat equation.

2. Problem formulation

Let us consider a section of a multi-layer beam with a length l consisting of m heterogeneous layers of the same width b (figure 1). The layers are rigidly interconnected and have perfect thermal contact. The cross section of the beam is constant along the length of the section and has the shape of a narrow rectangle (figure 1b), the dimensions of which meet the condition: $b << h << l$.

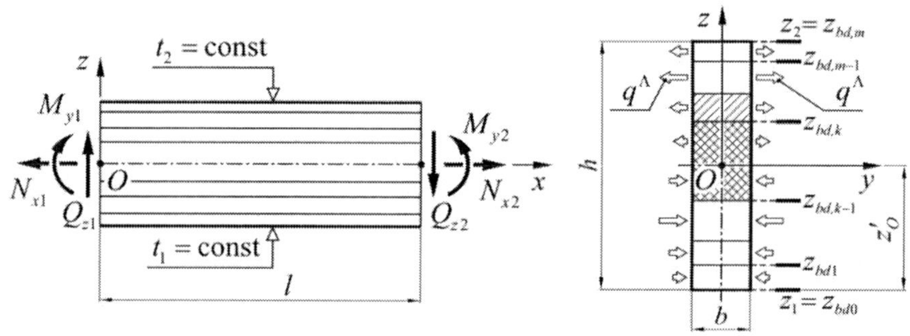

Figure 1. Scheme of a section of a bar and its cross section.

The longitudinal edges of the beam are free from external loads (figure 1a), and the forces in the extreme sections are reduced to the resultant $N_{x,\varsigma}, Q_{z,\varsigma}, M_{y,\varsigma}$ $(\varsigma = 1, 2)$ and constitute a balanced system of forces.

We consider that during the manufacture of the beam, all its layers had the same temperature $T_0 = \text{const}$, and during operation on the surface of the lower ($z = z_1$) and upper ($z = z_2$) longitudinal faces, a uniform temperature $t_\varsigma = \text{const}$ is maintained. On the lateral longitudinal surfaces there is a symmetric heat exchange with the external environment. The heat flow density through these surfaces is constant along the length of the section and varies arbitrarily along the section height: $q^\wedge = q^\wedge(z)$.

The bar layers are homogeneous orthotropic or functionally gradient (FG) in thickness. The orthotropy planes of the mechanical and thermophysical characteristics are parallel to the coordinate planes. For an arbitrary k th layer, independent physical and mechanical characteristics are specified:

$$\left\| E_x^{[k]}, E_z^{[k]}, G_{xz}^{[k]}, \nu_{xz}^{[k]}, \vartheta_x^{[k]}, \vartheta_z^{[k]}, \lambda_{0x}^{[k]}, \lambda_{0z}^{[k]} \right\| = \left\| S_a^{[k]} \right\|, \tag{1}$$

where $E_x^{[k]}, E_z^{[k]}$ – longitudinal and transverse elastic moduli of k th layer; $G_{xz}^{[k]}$ – shear modulus; $\nu_{xz}^{[k]}$ – Poisson ratio; $\vartheta_x^{[k]}, \vartheta_z^{[k]}$ – linear thermal expansion coefficients; $\lambda_{0x}^{[k]}, \lambda_{0z}^{[k]}$ – thermal conductivity at initial temperature T_0.

Within a homogeneous layer, the physico-mechanical characteristics are constant $S_a^{[k]} = \text{const}$, and for the FG layer, the law of their change along the section height is specified $S_a^{[k]} = S_a^{[k]}(z)$.

For the entire section, an arbitrary characteristic, similar to [13, 20, 25, 31], can be represented as a piecewise continuous function $\mu_a^S(z)$:

$$\mu_a^S = \sum_{k=1}^{m} \left\{ S_a^{[k]} \left[H\left(z - z_{bd,k-1}\right) - H\left(z - z_{bd,k}\right) \right] \right\} , \tag{2}$$

where $H(z)$ – Heaviside function.

We consider that the axis Ox passes through the center of stiffness of each section of the bar, the coordinate z'_O (figure 1b) of which, relative to the lower edge of the section of the bar, is determined by the ratio

$$z'_O = \int_0^h \left(\mu_x'^E z\right) dz \Big/ \int_0^h \mu_x'^E dz , \tag{3}$$

where $\mu_x'^E$ – distribution function of the longitudinal elastic modulus for the case when the origin of the coordinate system is on the lower face of the section.

Let us determine the distribution of the temperature field and the field of stresses, deformations, and displacements associated with it within the considered section of the multilayer bar under the assumption that its thermal state is steady-state and upon transition to it the materials of all layers are deformed elastically.

3. The solution of heat conduction problem

When solving this problem, the priority is to obtain the distribution of the temperature field inside the beam, which gives the necessary initial data for determining its SSS.

In the case of a steady state of heat and in the absence of internal heat sources, the heat equation for a composite straight-line bar [34], after a corresponding replacement of the coordinate notation ($\eta \to x, \xi \to z$) and taking into account equation (2), will take the form

$$\mu_{0x}^{\lambda} \frac{\partial^2 T}{\partial x^2} + \frac{\partial}{\partial z}\left(\mu_{0z}^{\lambda} \frac{\partial T}{\partial z} \right) + \mu_{0y}^{\lambda} \frac{\partial^2 T}{\partial y^2} = 0 , \tag{4}$$

where $T = T(x, y, z)$ – temperature field distribution function.

Integrating equation (4) over the section width and dividing the resulting expression by b, we have

$$\mu_{0x}^{\lambda} \frac{\partial^2 \bar{T}}{\partial x^2} + \frac{\partial}{\partial z}\left(\mu_{0z}^{\lambda} \frac{\partial \bar{T}}{\partial z} \right) + \mu_{0y}^{\lambda} \frac{1}{b}\left(\frac{\partial T}{\partial y} \Big|_{y=y_1}^{y=y_2} \right) = 0 , \tag{5}$$

where y_1, y_2 – the coordinates of the left and right longitudinal side faces of the bar; $\bar{T}$ – temperature averaged over the width of the section

$$\bar{T} = \frac{1}{b} \int_{y_1}^{y_2} T dy .$$

In the case of symmetric boundary conditions of the second kind on the lateral faces [34]:

$$\left(\mu_{0y}^{\lambda} \frac{\partial T}{\partial y} \right) \Big|_{y=y_\varsigma} = (-1)^{\varsigma+1} q^{\Lambda}, \ \varsigma = 1,2 , \tag{6}$$

where q^{Λ} – heat flow density through the side faces of the bar.

In view of equation (6), equation (5) takes the following form

$$\mu_{0x}^{\lambda} \frac{\partial^2 \overline{T}}{\partial x^2} + \frac{\partial}{\partial z}\left(\mu_{0z}^{\lambda} \frac{\partial \overline{T}}{\partial z} \right) = \frac{2q^{\Lambda}}{b} . \tag{7}$$

Thus, we will seek the desired temperature distribution as a solution to equation (7), which corresponds to boundary conditions of the first kind on the longitudinal faces:

$$T|_{z=z_1} = t_1 = \text{const}, \qquad T|_{z=z_2} = t_2 = \text{const} . \tag{8}$$

The condition of symmetry of heat transfer conditions on the lateral longitudinal surfaces together with conditions equation (8) let us suggest that the solution to equation (7) will have the form $T = \overline{T} = \overline{T}(z)$. Then the partial differential equation (7) is transformed to an ordinary inhomogeneous differential equation

$$\frac{d}{dz}\left(\mu_{0z}^{\lambda} \frac{dT}{dz} \right) = \frac{2q^{\Lambda}}{b} . \tag{9}$$

By double integration equation (9) we obtain

$$T = \frac{2}{b}\int_{z_1}^{z}\left(\frac{1}{\mu_{0z}^{\lambda}} \int_{z_1}^{z} q^{\Lambda} dz \right) dz + C_0 \int_{z_1}^{z} \frac{1}{\mu_{0z}^{\lambda}} dz + C_1 , \tag{10}$$

where C_0, C_1 – unknown constants.

Substituting solution equation (10) into equation (8) and solving the obtained equations, we have

$$C_1 = t_1, \qquad C_0 = \frac{\Delta t - Q^{\Lambda}}{B_0^{\lambda}} , \tag{11}$$

where $B_0^{\lambda}, Q^{\Lambda}, \Delta t$ – constant:

$$B_0^{\lambda} = \int_{z_1}^{z_2} \frac{1}{\mu_{0z}^{\lambda}} dz, \qquad Q^{\Lambda} = \int_{z_1}^{z_2}\left(\frac{1}{\mu_{0z}^{\lambda}} \int_{z_1}^{z} q^{\Lambda} dz \right) dz, \qquad \Delta t = t_2 - t_1 . \tag{12}$$

In view of equation (11), solution equation (10) takes on following form

$$T = \frac{2}{b}\int_{z_1}^{z}\left(\frac{1}{\mu_{0z}^{\lambda}} \int_{z_1}^{z} q^{\Lambda} dz \right) dz + \frac{1}{B_0^{\lambda}}\left(\Delta t - \frac{2Q^{\Lambda}}{b} \right) \int_{z_1}^{z} \frac{1}{\mu_{0z}^{\lambda}} dz + t_1 . \tag{13}$$

It should be noted that based on the general solution equation (10), one can also obtain particular solutions in the case of a given heat flow density (conditions of the second kind) or a given law of heat exchange with the environment (conditions of the third kind) on longitudinal faces [34].

Relation equation (13) gives the temperature field distribution inside the considered section of the multilayer beam and allows us to go on to determine the associated SSS. The obtained solution satisfies the heat equation and boundary conditions on the longitudinal surfaces and is accurate if the temperature distribution in the extreme sections of the section corresponds to equation (13).

4. The solution of the elasticity problem

The accepted initial conditions for determining the SSS of the considered section of the beam allow us to apply the equations of the plane problem of the linear theory of elasticity. The Navier equations and the Cauchy relations for the considered multilayer beam will be similar to the classical ones. However, instead of elastic constants, Hooke's law must contain functions equation (2), which allow one to take into account the inhomogeneity of the:

$$\varepsilon_x = \frac{1}{\mu_x^E}\left(\sigma_x - \mu_{xz}^{\nu}\sigma_z\right) + \mu_x^{9}\left(T - T_0\right), \qquad \varepsilon_z = \frac{\sigma_z}{\mu_z^E} - \frac{\mu_{xz}^{\nu}\sigma_x}{\mu_x^E} + \mu_z^{9}\left(T - T_0\right), \qquad \gamma_{xz} = \frac{1}{\mu_{xz}^G}\tau_{xz} \quad (14)$$

where the dependence $\mu_x^E \mu_{zx}^{\nu} = \mu_z^E \mu_{xz}^{\nu}$ is taken into account for orthotropic material.

On the longitudinal edges of the beam, the solution must meet uniform static conditions

$$\sigma_z \left.\right|_{z=z_\varsigma} = 0, \qquad \tau_{xz} \left.\right|_{z=z_\varsigma} = 0, \ \varsigma = 1, 2 \quad (15)$$

In the leftmost section of the section, we require the fulfillment of integral conditions [13]:

$$\int_{z_1}^{z_2} \left(\sigma_x \left.\right|_{x=0}\right) dz = \frac{N_{x1}}{b}, \qquad \int_{z_1}^{z_2} \left(z\sigma_x \left.\right|_{x=0}\right) dz = -\frac{M_{y1}}{b}, \qquad \int_{z_1}^{z_2} \left(\tau_{xz} \left.\right|_{x=0}\right) dz = -\frac{Q_{z1}}{b} \quad (16)$$

Having solved the Navier equations with respect to normal stresses, taking into account equation (15), we obtain the following relations

$$\sigma_x = -\int_0^x \frac{\partial \tau_{xz}}{\partial z} dx + \sigma_x \left.\right|_{x=0}, \qquad \sigma_z = -\int_{z_1}^{z} \frac{\partial \tau_{xz}}{\partial x} dz \quad (17)$$

In the absence of load on the longitudinal faces of the bar, the transverse force: $Q_z = Q_{z1} = \text{const}$, which, with equation (16), allows us to make an assumption

$$\tau_{xz} = \tau_{xz}(z) \quad (18)$$

In view of equation (18), solutions of equation (17) take the form

$$\sigma_x = -\frac{d\tau_{xz}}{dz} x + \sigma_x \left.\right|_{x=0}, \qquad \sigma_z = 0 \quad (19)$$

Substituting equation (19) into equation (14) we obtain the following relations for strains

$$\varepsilon_x = -\frac{1}{\mu_x^E}\left(\frac{d\tau_{xz}}{dz}x - \sigma_x \left.\right|_{x=0}\right) + \mu_x^{9}\left(T - T_0\right), \ \varepsilon_z = \frac{\mu_{xz}^{\nu}}{\mu_x^E}\left(\frac{d\tau_{xz}}{dz}x - \sigma_x \left.\right|_{x=0}\right) + \mu_z^{9}\left(T - T_0\right), \ \gamma_{xz} = \frac{\tau_{xz}}{\mu_{xz}^G} \quad (20)$$

Having solved the Cauchy relations for linear strains with respect to longitudinal u and transverse w displacements, taking into account equation (20), we obtain

$$u = -\frac{1}{\mu_x^E}\left(\frac{d\tau_{xz}}{dz}\frac{x^2}{2} - \sigma_x \left.\right|_{x=0} x\right) + \mu_x^{9}\left(T - T_0\right)x + u\left.\right|_{x=0},$$

$$w = x\int_{z_1}^{z}\left(\frac{\mu_{xz}^{\nu}}{\mu_x^E}\frac{d\tau_{xz}}{dz}\right)dz - \int_{z_1}^{z}\left(\frac{\mu_{xz}^{\nu}}{\mu_x^E}\sigma_x \left.\right|_{x=0}\right)dz + \int_{z_1}^{z}\left\{\mu_z^{9}\left(T - T_0\right)\right\}dz + w\left.\right|_{z=z_1}. \quad (21)$$

Substituting equation (21) and the third dependence equation (20) into the Cauchy relation for angular strains, we obtain the following integral-differential expression

$$\frac{dw\left.\right|_{z=z_1}}{dx} - \left[\frac{d}{dz}\left(\frac{1}{\mu_x^E}\frac{d\tau_{xz}}{dz}\right)\right]\frac{x^2}{2} + \left[\frac{d}{dz}\left(\frac{\sigma_x \left.\right|_{x=0}}{\mu_x^E}\right) + \frac{d}{dz}\left\{\mu_x^{9}\left(T - T_0\right)\right\}\right]x +$$

$$+\left[\frac{du\left.\right|_{x=0}}{dz} - \frac{\tau_{xz}}{\mu_{xz}^G} + \int_{z_1}^{z}\left(\frac{\mu_{xz}^{\nu}}{\mu_x^E}\frac{d\tau_{xz}}{dz}\right)dz\right] = 0. \quad (22)$$

Identity equation (22) will be valid for all points of the beam section only if the expressions in square brackets are equal to some constant:

$$\frac{d}{dz}\left(\frac{1}{\mu_x^E}\frac{d\tau_{xz}}{dz}\right)=K_0,\ \frac{d}{dz}\left(\frac{\sigma_x|_{x=0}}{\mu_x^E}\right)+\frac{d}{dz}\left\{\mu_x^9\left(T-T_0\right)\right\}=K_1,\ \frac{du|_{x=0}}{dz}-\frac{\tau_{xz}}{\mu_{xz}^G}+\int_{z_1}^{z}\left(\frac{\mu_{xz}^v}{\mu_x^E}\frac{d\tau_{xz}}{dz}\right)dz=K_2 \quad (23)$$

In view of equation (23), expression equation (22) is transformed into a linear differential equation:

$$\frac{dw|_{z=z_1}}{dx}=K_0\frac{x^2}{2}-K_1x-K_2 \quad (24)$$

Equations (23), (24) allow us to determine the desired function τ_{xz}, as well as unknown integration functions $\sigma_x|_{x=0}, u|_{x=0}, w|_{z=z_1}$.

When solving the first equation (23), taking into account the boundary conditions equation (15) and the third condition equation (16), we obtained

$$\tau_{xz}=-\frac{Q_{z1}}{bB_2}\int_{z_1}^{z}\left(\mu_x^E z\right)dz, \qquad K_0=-\frac{Q_{z1}}{bB_2} \quad (25)$$

It is taken into accounts that in the adopted coordinate system [13]:

$$B_2=\int_{z_1}^{z_2}\int_{z_1}^{z}\left(\mu_x^E z\right)dzdz, \qquad \int_{z_1}^{z_2}\left(\mu_x^E z\right)dz=0 \quad (26)$$

Relation equation (25) is similar to the solution for shear stresses when bending a multilayer cantilever with a free end load. Thus, temperature strains do not directly affect the distribution of shear stresses and strains; however, they can affect their magnitude through a change in the shear force in statically indeterminate elements.

The general solution of the second equation (23)

$$\sigma_x|_{x=0}=-\mu_x^E\mu_x^9\left(T-T_0\right)+K_1\mu_x^E z+C_0\mu_x^E \quad (27)$$

Substituting equation (27) into conditions equation (16) and solving the obtained equations with respect to unknowns K_1 and C_0 taking into account equation (26), we have

$$C_0=\frac{N_{x1}+N^T}{bB_0}, \qquad K_1=\frac{M_{y1}-M^T}{bB_2} \quad (28)$$

where N^T, M^T, B_0 – constant:

$$N^T=b\int_{z_1}^{z_2}\left[\mu_x^E\mu_x^9\left(T-T_0\right)\right]dz, \qquad M^T=b\int_{z_1}^{z_2}\left[\mu_x^E\mu_x^9 z\left(T-T_0\right)\right]dz, \qquad B_0=\int_{z_1}^{z_2}\mu_x^E dz \quad (29)$$

The quantities N^T and M^T have the dimension of force and moment, respectively. This allows us to consider them as the temperature components of the longitudinal force and bending moment inside the section of the beam.

Substituting equation (28) into equation (27) we obtain

$$\sigma_x|_{x=0}=\mu_x^E\left[\frac{N_{x1}+N^T}{bB_0}+\frac{M_{y1}-M^T}{bB_2}z-\mu_x^9\left(T-T_0\right)\right] \quad (30)$$

As we see in equation (30), the temperature component of normal stresses can be clearly distinguished, which at $T-T_0=0$ is equal to zero, which leads (30) to the relation obtained in [13].

The solution of the third equation (23) taking into account equation (25)

$$u|_{x=0}=-\frac{Q_{z1}}{bB_2}\int_{z_1}^{z}\left[\frac{1}{\mu_{xz}^G}\int_{z_1}^{z}\left(\mu_x^E z\right)dz-\int_{z_1}^{z}\left(\mu_{xz}^v z\right)dz\right]dz+K_2\left(z-z_1\right)+u|_{x=0,z=z_1} \quad (31)$$

Express the constant K_2 through the displacement of the upper fiber in the initial section. Accepting in equation (31) $z = z_2$ and solving relatively to K_2, we obtain

$$K_2 = \frac{Q_{z1}}{b}\frac{D_2}{hB_2} + \frac{1}{h}\left(u\,|_{x=0,z=z_2} - u\,|_{x=0,z=z_1}\right), \quad (32)$$

where D_2 – constant:

$$D_2 = \int_{z_1}^{z_2}\left[\frac{1}{\mu_{xz}^G}\int_{z_1}^{z}\left(\mu_x^E z\right)dz - \int_{z_1}^{z}\left(\mu_{xz}^{\nu} z\right)dz\right]dz. \quad (33)$$

In view of equation (32), relation (31) takes the following form

$$u\,|_{x=0} = -\frac{Q_{z1}}{bB_2}\left[\int_{z_1}^{z}\left(\frac{1}{\mu_{xz}^G}\int_{z_1}^{z}\left(\mu_x^E z\right)dz - \int_{z_1}^{z}\left(\mu_{xz}^{\nu} z\right)dz\right)dz - \frac{D_2}{h}\left(z - z_1\right)\right] + u\,|_{x=0,\,z=z_2}\frac{z - z_1}{h} + u\,|_{x=0,\,z=z_1}\frac{z_2 - z}{h}. \quad (34)$$

The solution of equation (24) after substituting (25), (28) and (32) is obtained in this form:

$$w\,|_{z=z_1} = -\frac{Q_{z1}}{bB_2}\left(\frac{x^3}{6} + \frac{D_2}{h}x\right) - \frac{M_{y1} - M^T}{bB_2}\frac{x^2}{2} - u\,|_{x=0,\,z=z_2}\frac{x}{h} + u\,|_{x=0,\,z=z_1}\frac{x}{h} + w\,|_{z=z_1,\,x=0}. \quad (35)$$

Solutions equations (25), (30), (34) and (35) together with (19), (20) and (21) allow us to write the final relations for the desired functions of stresses, deformations, and displacements.

Taking into account equations (19), (25) and (30) for stresses, we have

$$\sigma_x = \mu_x^E\left[\frac{Q_{z1}}{bB_2}zx + \frac{N_{x1} + N^T}{bB_0} + \frac{M_{y1} - M^T}{bB_2}z - \mu_x^{\vartheta}\left(T - T_0\right)\right], \quad \sigma_z = 0, \quad \tau_{xz} = -\frac{Q_{z1}}{bB_2}\int_{z_1}^{z}\left(\mu_x^E z\right)dz. \quad (36)$$

Strain equation (20) after substitution (25) and (30):

$$\varepsilon_x = \frac{Q_{z1}}{bB_2}zx + \frac{M_{y1} - M^T}{bB_2}z + \frac{N_{x1} + N^T}{bB_0},$$

$$\varepsilon_z = -\mu_{xz}^{\nu}\left(\frac{Q_{z1}}{bB_2}zx + \frac{M_{y1} - M^T}{bB_2}z + \frac{N_{x1} + N^T}{bB_0}\right) + \left(\mu_{xz}^{\nu}\mu_x^{\vartheta} + \mu_z^{\vartheta}\right)\left(T - T_0\right),$$

$$\gamma_{xz} = -\frac{Q_{z1}}{bB_2}\frac{1}{\mu_{xz}^G}\int_{z_1}^{z}\left(\mu_x^E z\right)dz. \quad (37)$$

Displacements equation (21) taking into account (34), (35):

$$u = \frac{Q_{z1}}{bB_2}\left[\frac{x^2 z}{2} - \int_{z_1}^{z}\left(\frac{1}{\mu_{xz}^G}\int_{z_1}^{z}\left(\mu_x^E z\right)dz - \int_{z_1}^{z}\left(\mu_{xz}^{\nu} z\right)dz\right)dz + \frac{D_2}{h}\left(z - z_1\right)\right] + $$
$$+ \frac{N_{x1} + N^T}{bB_0}x + \frac{M_{y1} - M^T}{bB_2}xz + u\,|_{x=0,\,z=z_2}\frac{z - z_1}{h} + u\,|_{x=0,\,z=z_1}\frac{z_2 - z}{h},$$

$$w = -\frac{Q_{z1}}{bB_2}\left[\frac{x^3}{6} + x\left(\int_{z_1}^{z}\left(\mu_{xz}^{\nu} z\right)dz + \frac{D_2}{h}\right)\right] - \frac{M_{y1} - M^T}{bB_2}\left(\frac{x^2}{2} + \int_{z_1}^{z}\left(\mu_{xz}^{\nu} z\right)dz\right) - $$
$$-\frac{N_{x1} + N^T}{bB_0}\int_{z_1}^{z}\mu_{xz}^{\nu}dz + \int_{z_1}^{z}\left[\left(\mu_{xz}^{\nu}\mu_x^{\vartheta} + \mu_z^{\vartheta}\right)\left(T - T_0\right)\right]dz + u\,|_{x=0,\,z=z_1}\frac{x}{h} - u\,|_{x=0,\,z=z_2}\frac{x}{h} + w\,|_{z=z_1,\,x=0}. \quad (38)$$

The conditions of absolutely rigid connection of the layers imply the equality of the total stress and displacement at each point on the common surface of arbitrary two layers, which requires the following static and kinematic conditions:

$$\tau_{xz}^{[k]}\big|_{z=z_{bd,k}} = \tau_{xz}^{[k+1]}\big|_{z=z_{bd,k}}, \qquad \sigma_z^{[k]}\big|_{z=z_{bd,k}} = \sigma_z^{[k+1]}\big|_{z=z_{bd,k}}, \quad k=\overline{1,m-1}, \tag{39}$$

$$w^{[k]}\big|_{z=z_{bd,k}} = w^{[k+1]}\big|_{z=z_{bd,k}}, \qquad u^{[k]}\big|_{z=z_{bd,k}} = u^{[k+1]}\big|_{z=z_{bd,k}}. \tag{40}$$

In the solutions for τ_{xz}, u and w, the functions of the elastic characteristics μ_a^S, which have a discontinuity at the boundaries of the layers, are found only in integrand expressions. This ensures the continuity of the respective SSS components and the fulfillment of conditions equations (39) and (40).

Thus, relations equations (36), (37) and (38) satisfy all the equations of the plane problem of the theory of elasticity, boundary conditions on longitudinal surfaces, and conditions of absolutely rigid connection of layers. The resulting solution is accurate provided that in the extreme sections of the section the distribution of internal forces or external load corresponds to the obtained distribution of stresses.

The displacement solution contains three unknown displacements of the extreme points of the initial section of the beam section: $u|_{x=0,z=z_1}, u|_{x=0,z=z_2}, w|_{z=z_1,x=0}$ (initial kinematic parameters) and three forces in the initial section: N_{x1}, Q_{z1}, M_{y1} (initial static parameters). For a bar, which consists of several sections, these constants make it possible to coordinate the deformations of the considered section with its neighboring parts. In the case of a bar consisting of one section using these constants, you can simulate various types of fastening of the extreme sections.

In the general case, six unknowns $u|_{x=0,z=z_1}, u|_{x=0,z=z_2}, w|_{z=z_1,x=0}$ and N_{x1}, Q_{z1}, M_{y1} allow you to set independent movements of four arbitrary points of the extreme sections (two points in the left and right sections). In [21], a method for modeling various fastenings of the ends of a multilayer beam was proposed, which can be applied in the case under consideration.

Based on the principle of superposition, the obtained solution can be superimposed on the solutions [20, 25, 32], which allows one to obtain the SSS ratios for the case of multilayer beam bending by an arbitrary load on the longitudinal faces, taking into account the temperature conditions considered here.

5. Example of realization of the relations obtained

Let us determine the temperature field distribution and the resulting SSS for a five-layer beam with rigid fastening of the left end and pivotally fixed fastening of the right end (figure 2). The materials of the layers and their physical and mechanical characteristics:

- CFRP (P_1): $E_x^{[1]} = 142.80\,GPa$, $E_z^{[1]} = 9.13\,GPa$, $G_{xz}^{[1]} = 5.49\,GPa$, $\nu_{xz}^{[1]} = 0.320$, $\vartheta_x^{[1]} = 0.0\,K^{-1}$, $\vartheta_z^{[1]} = 27.7\cdot10^{-6}\,K^{-1}$, $\lambda_{0z}^{[1]} = 0.385\,W(m\cdot K)^{-1}$;
- GRP (P_2, P_4): $E_x^{[2,4]} = 36.8\,GPa$, $E_z^{[2,4]} = 11.00\,GPa$, $G_{xz}^{[2,4]} = 4.50\,GPa$, $\nu_{xz}^{[2,4]} = 0.351$, $\vartheta_x^{[2,4]} = 7.8\cdot10^{-6}\,K^{-1}$, $\vartheta_z^{[2,4]} = 49.9\cdot10^{-6}\,K^{-1}$, $\lambda_{0z}^{[2,4]} = 0.460\,W(m\cdot K)^{-1}$;
- wood (P_3): $E_x^{[3]} = 12.80\,GPa$, $E_z^{[3]} = 0.625\,GPa$, $G_{xz}^{[3]} = 0.617\,GPa$, $\nu_{xz}^{[3]} = 0.360$, $\vartheta_x^{[3]} = 5.4\cdot10^{-6}\,K^{-1}$, $\vartheta_z^{[3]} = 34.0\cdot10^{-6}\,K^{-1}$, $\lambda_{0z}^{[3]} = 0.107\,W(m\cdot K)^{-1}$;
- aluminum alloy (P_5): $E_x^{[5]} = E_z^{[5]} = 72.00\,GPa$, $G_{xz}^{[5]} = 26.9\,GPa$, $\nu_{xz}^{[5]} = 0.338$, $\vartheta_x^{[5]} = \vartheta_z^{[5]} = 22.9\cdot10^{-6}\,K^{-1}$, $\lambda_{0z}^{[5]} = 169.0\,W(m\cdot K)^{-1}$.

The initial temperature of the beam $T_0 = 293K$. On the longitudinal surfaces of the beam constant temperature: $t_1 = 285K, t_2 = 273K$. Heat is exchanged through the lateral faces with the external environment, while the heat flow density is variable along the section height:

$$q^{\Lambda} = 12760.43z^2 + 66.57z - 11.94 \quad (41)$$

The position of the center of stiffness of the section (figure 2b) is determined according to equation (3): $z'_O = 0.026078m$.

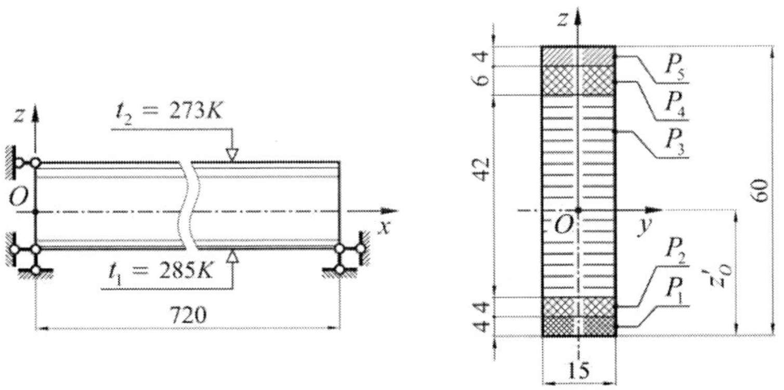

Figure 2. The scheme of kinematic restrictions and the cross-section of the beam (dimensions in mm).

The functions of physical and mechanical characteristics are formed according to equation (2):

$$\mu_a^S = S_a^{[1]}\left(H(z+0.02608)-H(z+0.02208)\right)+S_a^{[2]}\left(H(z+0.02208)-H(z+0.01808)\right)+ \\ +S_a^{[3]}\left(H(z+0.01808)-H(z-0.02392)\right)+S_a^{[4]}\left(H(z-0.02392)-H(z-0.02992)\right)+ \\ +S_a^{[5]}\left(H(z-0.02992)-H(z-0.03392)\right), \quad (42)$$

Using equations (41), (42) according to equation (12), the constants necessary for determining the distribution of the temperature field are calculated:

$$B_0^{\lambda} = 0.42467\, m^2K/W, \qquad Q^{\Lambda} = -0.11762\, K\cdot m, \qquad \Delta t = -12K \quad (43)$$

The desired temperature T distribution function was obtained according to equation (13) using equations (41), (42), (43) and given initial data. Graphs of the distribution of temperature T and heat flow density q_z over the section height along with the distributions $1/\mu_{0z}^{\lambda}$ and q^{Λ} are shown in figure 3.

The constants necessary for determining the SSS of the bar are determined according to equations (26), (29) and (33) using equation (42) and the obtained function T:

$$B_0 = 1764.80\cdot10^6\, N/m, \qquad B_2 = -9296.26\cdot10^2\, N\cdot m, \qquad D_2 = -1244.11\cdot10^{-6}\, m^3, \\ N^T = -3130.29N, \qquad M^T = -77.3489\, N\cdot m. \quad (44)$$

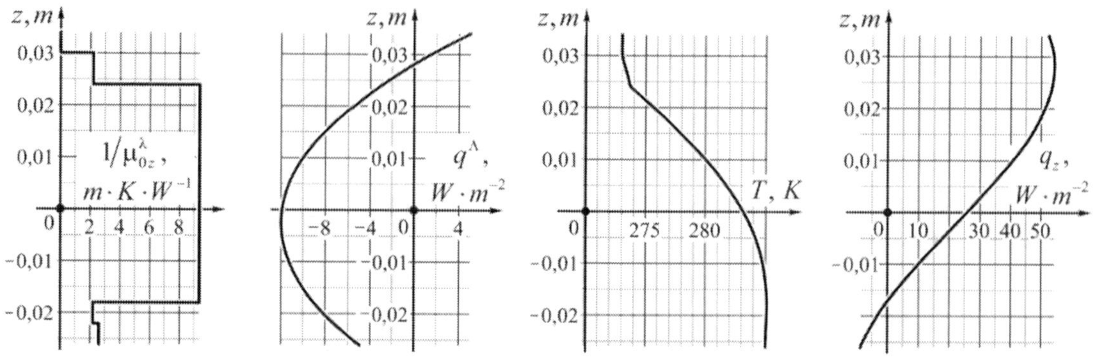

Figure 3. Initial data and results of solving the heat conduction problem

The problem under consideration is statically indeterminate; therefore, the kinematic ($u|_{x=0,z=z_1}, u|_{x=0,z=z_2}, w|_{z=z_1,x=0}$) and static ($N_{x1}, Q_{z1}, M_{y1}$) initial parameters are unknown. To determine them, we used the conditions of the static equilibrium of the beam and the kinematic conditions corresponding to the accepted types of fastenings.

Similarly to [21], rigid and simple fastening of the ends of the beam were modeled according to the diagram in figure 2a:

$$u|_{x=0,z=z_1} = 0, \qquad u|_{x=0,z=z_2} = 0, \qquad w|_{x=0,z=z_1} = 0, \qquad u|_{x=l,z=z_1} = 0, \qquad w|_{x=l,z=z_1} = 0 \; . \tag{45}$$

The missing equations are obtained from the conditions of static equilibrium

$$-N_{x1} + X_s = 0, \qquad Q_{z1} + Z_s = 0, \qquad -M_{y1} + Z_s l + X_s(-z_1) = 0 \; . \tag{46}$$

According to equation (45), the kinematic parameters are equal to zero, and the static ones are obtained by a joint solution of equations (45) and (46) with allowance for equation (38):

$$N_{x1} = 1307.59 N, \qquad Q_{z1} = 207.31 N, \qquad M_{y1} = -115.16 N \cdot m \; . \tag{47}$$

Substituting the initial data, the corresponding functions equation (42), the constants equations (44), (45), (47) and the temperature distribution T in equations (36) and (38), we obtain the desired distribution functions of stresses and displacements.

Normal σ_x and shear τ_{xz} stress distributions for individual sections and fibers are shown in figure 4.

An abrupt change in the longitudinal elastic modulus at the boundaries of the layers (figure 4a) leads to discontinuities of the function σ_x. According to figure 4b, c, d the distribution of σ_x on the height of various sections is disproportionate, however, for each longitudinal fiber, in particular the outermost fibers (figure 4e), the normal stresses vary according to their own linear law. The shear stresses are constant along the length of the beam and vary according to a parabolic law along the thickness of each layer (figure 4f).

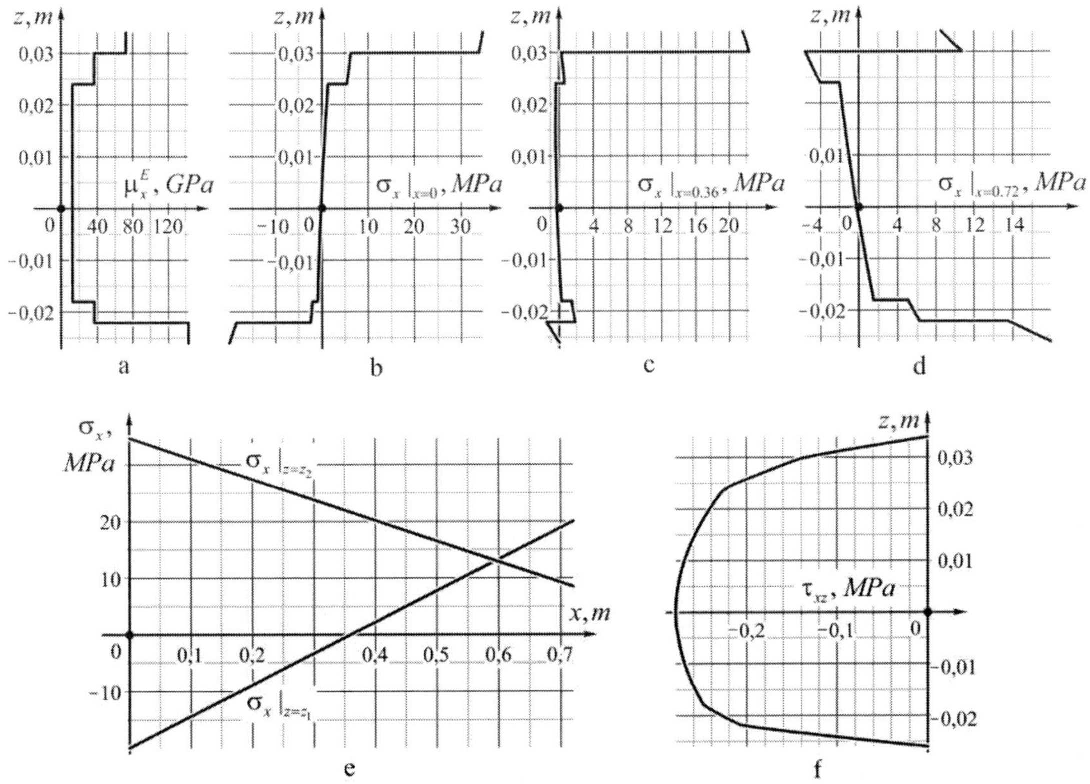

Figure 4. Graphs of normal and shear stresses.

Graphs for stresses σ_x show that even with a relatively small temperature change in the layers of the composite beam, significant tensile and compressive longitudinal forces can occur.

It is important to note the ambiguous effect of such stresses on the margin of strength of the beam. For example, under the action of a useful uniform load down, in the middle part of the beam, they will increase strength due to thermal tensile forces in the upper layer, but near the supports they will decrease on the contrary. Thus, for given temperature conditions of work of a multilayer beam by a certain alternation of layers, it is possible to increase its strength due to the effect of temperature redistribution of stresses between layers and sections.

The distributions of the longitudinal u and transverse w displacements for the extreme fibers and middle sections are shown in figure 5.

Graphs figure 5a, c shows the correspondence of the obtained solution to kinematic conditions equation (45).

Due to the presence of a transverse force in the graph of the distribution of longitudinal displacements u (figure 5b), one can observe a curvature of the cross sections, which is absent with other methods of securing the beam. The total curvature of the cross section is slightly enhanced by the displacement component of the displacement variable (figure 5d), which in this case is caused by the Poisson effect and thermal expansion of the layers.

The graphs of the transverse displacements w of the lower and upper fibers (figure 5c) show a slight decrease in the stiffness reserve of the beam for the case of a downward payload. This, obviously, can be avoided by changing the order of alternating layers.

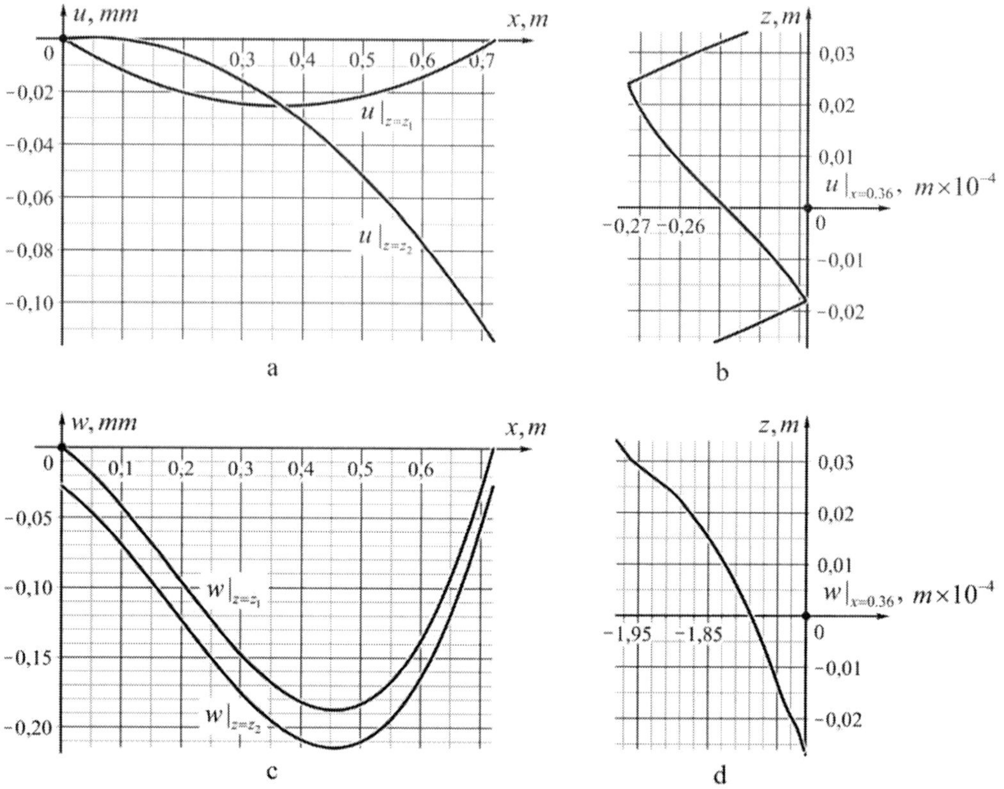

Figure 5. Graphs of longitudinal and transverse displacements.

6. Summary

Thus, an exact analytical solution to the problem of plane thermoelastic bending of a multilayer beam section has been constructed, which includes a solution to the heat conduction problem equation (13) and a solution to the theory of elasticity equations (36)-(38). The obtained solution allows us to determine the temperature field distribution inside the multilayer beam and the associated SSS with taking into account the different temperatures of its longitudinal surfaces and an arbitrary law of change in the heat flux density through the side faces. The solution is built for a beam with an arbitrary number of homogeneous or continuously inhomogeneous layers, it takes into account the orthotropy of the physical and mechanical characteristics and the flexibility of the transverse shear and compression of their materials.

The results of solving the test problem of thermoelastic bending of a five-layer bar with rigid and hinged fastening of the ends showed that a relatively small temperature change for such an element can cause significant stresses (up to 17% of the permissible) and significant deflections (14.8% of the permissible). Moreover, depending on the type of payload, the consideration of temperature stresses and displacements can lead to both a decrease and an increase in the strength and rigidity of the beam.

The obtained general solution can be used to refine the assessment of the strength and stiffness of multilayer beams taking into account temperature deformations, as well as to develop applied methods for calculating such elements.

References

[1] Kucher N K, Zarazovskii M N and Danil'chuk E L 2013 Deformation and strength of laminated carbon-fiber-reinforced plastics under a static thermomechanical loading *Mech. Compos. Mater.* **48**(6) pp 669–680

[2] Neutov S, Sydorchuk M and Surianinov M 2019 Experimental Studies of Reinforced Concrete and Fiber-Reinforced Concrete Beams with Short-Term and Long-Term Loads *Mater. Sci. Forum.* **968** pp 227–233
[3] Pavlikov A, Kosior-Kazberuk M and Harkava O 2018 Experimental testing results of reinforced concrete beams under biaxial bending *Int. J. Eng. & Technol.* **7**(3.2) pp 299–305
[4] Zhen W and Wanji C 2016 A global higher-order zig-zag model in terms of the HW variational theorem for multilayered composite beams *Compos. Struct.* **158** pp 128–136
[5] Zhao D, Wu Z and Ren X 2019 New Sinusoidal Higher-Order Theory Including the Zig-Zag Function for Multilayered Composite Beams *J. Aerosp. Eng.* **32**(3)
[6] Shvabyuk V I, Rotko S V and Uzhegova O A 2014 Bending of a Composite Beam with a Longitudinal Section *Strength Mater.* **46**(4) pp 558–566
[7] Piskunov V G, Goryk A V and Cherednikov V N 2000 Modeling of transverse shears of piecewise homogeneous composite bars using an iterative process with account of tangential loads. 1. Construction of a model *Mech. Compos. Mater.* **36**(4) pp 287–296
[8] Sayyad A S and Ghugal Y M 2017 Bending, buckling and free vibration of laminated composite and sandwich beams: A critical review of literature *Compos. Struct.* **171** pp 486–504
[9] Lekhnitskii S G 1968 *Anisotropic Plate* (New York: Gordon and Breach)
[10] Gerstner R W 1968 Stresses in a Composite Cantilever *J. Compos. Mater.* **2**(4) pp 498–501
[11] Cheng S, Wei X and Jiang T 1989 Stress distribution and deformation of adhesive-bonded laminated composite beams *ASCE J. Eng. Mech.* **115** pp 1150–1162
[12] Zhao L, Chen W Q and Lu C F 2012 New assessment on the Saint-Venant solutions for functionally graded beams *Mech. Res. Commun.* **43** pp 1–6
[13] Goryk A V and Kovalchuk S B 2018 Elasticity theory solution of the problem on plane bending of a narrow layered cantilever bar by loads at its end *Mech. Compos. Mater.* **54**(2) pp 179–190
[14] Ding H J, Huang D J and Wang H M 2006 Analytical solution for fixed-fixed anisotropic beam subjected to uniform load *Appl. Math. Mech.* **27**(10) pp 1305–1310
[15] Huang D J, Ding H J and Chen W Q 2007 Analytical solution for functionally graded anisotropic cantilever beam under thermal and uniformly distributed load *J. Zhejiang Univ.-Sci. A.* **8**(9) pp 1351–1355
[16] Zhong Z and Yu T 2007 Analytical solution of a cantilever functionally graded beam *Compos. Sci. Technol.* **67**(3-4) pp 481–488
[17] Daneshmehr A, Momeni S and Akhloumadi M R 2012 Exact elasticity solution for the density functionally gradient beam by using airy stress function *Appl. Mech. Mater.* **110-116** pp 4669–4676
[18] Yang Q, Zheng B L, Zhang K and Li J 2014 Elastic solutions of a functionally graded cantilever beam with different modulus in tension and compression under bending loads *Appl. Math. Model.* **38**(4) pp 1403–1416
[19] Benguediab S, Tounsi A, Abdelaziz H H and Meziane M A A 2017 Elasticity solution for a cantilever beam with exponentially varying properties *J. Appl. Mech. Tech. Phys.* **58**(2) pp 354–361
[20] Goryk A V and Koval'chuk S B 2018 Solution of a Transverse Plane Bending Problem of a Laminated Cantilever Beam Under the Action of a Normal Uniform Load *Strength Mater.* **50**(3) pp 406–418
[21] Koval'chuk S and Goryk A 2019 Exact Solution of the Problem of Elastic Bending of a Multilayer Beam under the Action of a Normal Uniform Load *Mater. Sci. Forum.* **968** pp 475–485
[22] Esendemir U, Usal M R and Usal M 2006 The effects of shear on the deflection of simply supported composite beam loaded linearly *J. Reinf. Plast. Compos.* **25** pp 835–846
[23] Huang D J, Ding H J and Chen W Q 2007 Analytical solution for functionally graded anisotropic cantilever beam subjected to linearly distributed load *Appl. Math. Mech.* **28**(7) pp 855–860
[24] Daouadji T H, Henni A H, Tounsi A and El Abbes A B 2013 Elasticity Solution of a Cantilever Functionally Graded Beam *Appl. Compos. Mater.* **20**(1) pp 1–15

[25] Gorik A V and Koval'chuk S B 2020 Solving the Problem of Elastic Bending of a Layered Cantilever Under a Normal Load Linearly Distributed over Longitudinal Faces *Int. Appl. Mech.* **56**(1) pp 65–80
[26] Pagano N J 1969 Exact Solutions for Composite Laminates in Cylindrical Bending *J. Compos. Mater.* **3** pp 398–411
[27] Sankar B V 2001 An elasticity solution for functionally graded beams *Compos. Sci. Technol.* **61**(5) pp 689–696
[28] Jiang A M and Ding H J 2005 The analytical solutions for orthotropic cantilever beams (II): solutions for density functionally graded beams *J. Zhejiang Univ.-Sci. A.* **6**(3) pp 155–158
[29] Ding H J, Huang D J and Chen W Q 2007 Elasticity solutions for plane anisotropic functionally graded beams *Int. J. Solids Struct.* **44**(1) pp 176–196
[30] Huang D J, Ding H J and Chen W Q 2009 Analytical solution and semi-analytical solution for anisotropic functionally graded beam subject to arbitrary loading *Sci. China Ser. G-Phys. Mech. Astron.* **52**(8) pp 1244–1256
[31] Nie G J, Zhong Z and Chen S 2013 Analytical solution for a functionally graded beam with arbitrary graded material properties *Compos. B. Eng.* **44** pp 274–282
[32] Koval'chuk S B 2020 Exact Solution of the Problem on Elastic Bending of the Segment of a Narrow Multilayer Beam by an Arbitrary Normal Load *Mech. Compos. Mater.* **56**(1) pp 55–74
[33] Zhang L, Gao P and Li D 2012 New Methodology to Obtain Exact Solutions of Orthotropic Plane Beam Subjected to Arbitrary Loads *J. Eng. Mech.* **138**(11) pp 1348–1356
[34] Koval'chuk S B 2019 Zadacha termopruzhnosti dlia kompozytnoho brusa iz ploskoiu vissiu dovilnoi formy u pryrodnii systemi koordynat *Mizhvuz. zb. Naukovi notatky* **68** pp 30–40
[35] Sankar B V and Tzeng J T 2002 Thermal stresses in functionally graded beams *AIAA J.* **40** pp 1228–1232
[36] Timoshenko S P 1953 *The collected papers.* (New York-London-Toronto: McGraw-Hill)

Stress-strain state of elastomeric vibroseismic insulators

Yu G Kozub[1,*], V I Dyrda[2] and G A Kozub[1]

[1]Luhansk Taras Shevchenko National University, Starobilsk, 92703, Ukraine
[2]Institute of Geotechnical Mechanics by N. Polyakov of National Academy of Sciences of Ukraine, Dnipro, 49005, Ukraine

*E-mail: kosub.yg@gmail.com

Abstract. A method for solving problems of viscoelastic deformation of structures made of weakly-compressible elastomers, based on the application of the hereditary Boltzmann-Volterra theory, is considered. To consider the weak compressibility and eliminate the "false shift" effect, the moment scheme of the finite element is used.

1. Introduction

Every year over a hundred earthquakes occur on the planet, the victims of which are people who die under the rubble of buildings. In addition, the development of technological equipment and transport systems lead to an increase in the vibration impact on buildings and structures. Therefore, the creation and improvement of various structures that reduce the risk of destruction during vibration and seismic impacts [1,2], is extremely important. One of the most effective devices are vibration and seismic insulators based on rubber-metal structures (Figure 1). The use of such vibration-seismic protection systems allows: to ensure protection of objects during earthquakes and man-made impacts, while the estimated cost of construction is reduced by 3–6%, the material consumption of buildings and structures is reduced by 5–10%; the scope of application of standard structures is expanding by building up areas with increased seismicity, the permitted building height buildings is also increasing [3].

Vibroseismic insulators protect machines and buildings against seismic impacts not only in horizontal and vertical planes, but also from torsion. It is believed that it is torsion in combination with unfavorable factors, in particular with the vertical component of shocks, that is the main cause of catastrophic destruction during earthquakes. In addition, the use of rubber-metal layered vibration insulators allows protecting buildings and people in them from the effects of the subway, road and rail transport [4].

Figure. 1. General view of rubber-metal high-damping structures of rubber seismic blocks on natural rubber.

The main damper in such structures is an elastomeric element, the dynamic properties of which depend on the initial stresses arising under static loading.

2. Result of research

Due to the specificity of the mechanical properties of rubber, the determination of the parameters of the stress-strain state of rubber vibration insulators is a difficult task.

The finite element method (FEM) [5-8] is a universal numerical method for calculating rubber vibration seismic insulators, which allows one to considering the asymmetry of loads and fixings. The use of finite element models makes it possible to obtain a complete picture of the stress-strain state. At the same time, the traditional FEM does not allow considering such a property of rubber as the weak

Content from this work may be used under the terms of the Creative Commons Attribution 3.0 licence. Any further distribution of this work must maintain attribution to the author(s) and the title of the work, journal citation and DOI.
Published under licence by IOP Publishing Ltd

compressibility of the material; therefore, we will use a specially developed moment finite element scheme for weakly compressible materials [7]. The peculiarity of the moment scheme lies in the triple approximation of displacement vector components, components of deformation tensor and volume change function. When expanding in a Maclaurin series the functions of displacements, functions of deformations and functions of volume change in the vicinity of the origin of the local coordinate system of an element, only the associated components are retained in the expansion of these functions. In addition, the use of MSFE avoids the "false shift" effect

In the study of compressible and weakly compressible elastomers, a special place is occupied by the integral law of the state, which connects small increases in stresses and strains based on the ordinary Hooke's law in metric of deformed volume. It is widely used in solving static and quasi-static problems in the field of large deformations for nonlinear-elastic and thermo-viscoelastic bodies that require integration by parameter [7]. The increment in stresses and strains is related by the ratio

$$d\sigma^{ij} = 2\mu\left(G^{mi}G^{nj}d\varepsilon_{mn} - \frac{1}{3}G^{ij}d\theta\right) + BG^{ij}d\theta \quad (1)$$

For a linear material, the volume compression function is determined by the formula:

$$\theta = \varepsilon_{ii} = I_1(\hat{\varepsilon}), \quad (2)$$

where $I_1(\hat{\varepsilon}) = I_1$ is the first invariant of the Cauchy-Green's strain tensor.

Applying a triple approximation of displacements, deformations and volume change functions using power functions of a special form

$$\psi^{(pqr)} = \frac{(x_1)^p}{p!}\frac{(x_2)^q}{q!}\frac{(x_3)^r}{r!}, \quad (3)$$

and establishing a connection between the corresponding components of the expansions, one can write down the equilibrium equations

$$[A]^T\left[F_{ij}^s\right]^T\left[C^{ijkl}\right]\left[F_{kl}^t\right][A]\{u_t\} + [A]^T\left[F_\theta^s\right]^T\left[C^\theta\right]\left[F_\theta^t\right][A]\{u_t\} = \{R_s\},$$

$$\left[G^{st} + G_\theta^{st}\right]\{u_t\} = \{R_s\}, \quad (4)$$

where A is a matrix of correspondence between functions of the form of a finite element and basis power functions;

F_{kl}^t, F_θ^s are transformation matrices of the expansion of displacements and expansions of deformations and the function of changing the volume, respectively;

$G^{st} + G_\theta^{st}$ are the stiffness matrix of a weakly compressible solid.

To considering the nonlinear behavior of the material, the volume deformation function can be calculated by the formula

$$\theta = \frac{(dV - dv)}{dv} = \sqrt{\frac{G}{g}} - 1 = \sqrt{I_3\left(\hat{G}^x\right)} - 1. \quad (5)$$

Then for nonlinear weakly compressible material the modified Hooke's law (1) has the form:

$$\sigma^{ij} = \int_0^{\varepsilon_{kl}}\left(G^{mi}G^{nj} - \frac{1}{3}G^{ij}G^{mn}\right)d\varepsilon_{mn} - \int_0^{\hat{G}^\times} B\left(\sqrt{I_3} - 1\right)dG^{ij}. \quad (6)$$

The strain tensor can be represented as the sum of linear and nonlinear terms

$$\varepsilon_{mn} = \varepsilon_{mn}^{L} + \varepsilon_{mn}^{N}, \tag{7}$$

where $\varepsilon_{mn}^{L} = \frac{1}{2}\left(C_n^j u_{j,m} + C_m^i u_{i,n}\right)$ - linear terms of strain tensor,

$\varepsilon_{mn}^{N} = \frac{1}{2} u_{j,m} u_n^j$ - nonlinear terms of strain tensor, u_i - displacement vector components,

C_n^j - transformation tensor components.

The variation of the strain energy in this case has the form

$$\delta W = \iiint\limits_{v} \left(\sigma_L^{ij} + \sigma_N^{ij}\right)\delta\left(\varepsilon_{ij}^{L} + \varepsilon_{ij}^{N}\right)dv, \tag{8}$$

In this case, the solution can be obtained using the modified Newton-Kantorovich method.

To describe the viscoelastic properties of rubber, the hereditary Boltzmann-Volterra theory and the stress tensor are used in the form of an integral equation using the instantaneous and equilibrium elastic characteristics of the material.

$$\sigma^{ij}(t) = E^{ijkl}\left(\varepsilon_{kl}(t) - \int_0^t K(t-\tau)\varepsilon_{kl}(\tau)d\tau\right). \tag{9}$$

The Rabotnov's fractional-exponential nucleus is used as a Volterra integral equation nucleus

$$K(t-\tau) = \lambda(t-\tau)^{\alpha}\sum_{n=0}^{\infty}\frac{(-\beta-\lambda)^n (t-\tau)^{n(1+\alpha)}}{\Gamma\left[(n+1)(1+\alpha)\right]}, \tag{10}$$

where $-1<\alpha<0$; $\beta>0$; $\lambda>0$ - material constants.

Most often, structures made of highly elastic materials work in conditions of large deformations, when linear dependences no longer work. When calculating thin-layer elastomeric elements of structures operating in conditions of limited deformation, it is necessary to teach the weak stylistics of the elastomer. To describe the behavior of highly elastic materials, considering compressibility and nonlinearity, special laws of state are required, which considering the properties of material at large deformations.

One of the distinctive properties of elastomers is that when deformed in a highly elastic state, the balance between forces and displacements is established over a period of time. The high viscosity, the pronounced relaxation nature of the stresses, the geometrically and physically nonlinear nature of the deformation require the application of the mathematical apparatus of the nonlinear three-dimensional theory of viscoelasticity. The relationship between the components of stress and strain tensors for nonlinear viscoelastic weakly compressible material can be taken as Hooke's law, replacing the elastic constants - the compression modulus and the shear modulus - by Volterra integral operators:

$$\tilde{\mu}\varphi = \mu\left(\varphi(t) - \int_{-\infty}^{t} K_{\mu}(t-\tau)\varphi(\tau)d\tau\right),$$
$$\tilde{B}\varphi = B\left(\varphi(t) - \int_{-\infty}^{t} K_{b}(t-\tau)\varphi(\tau)d\tau\right). \tag{11}$$

In this way, all the terms due to nonlinear and viscoelastic properties can be selected.

To solve the problem of viscoelastic deformation, we use the spatiotemporal approximation of displacements. Moving at any point of the element is approximated as:

$$u=\sum_{i=1}^{M} u_i(t)N_i(x_k), \quad (k=1,2,3) \qquad (12)$$

where u_m is time-dependent approximation of the displacement of the i -th node;
$N_i(x_k)$ is the basic functions of spatial variables.

The time domain can also be divided into linear elements, then enter the basic functions:

$$u(t)=\sum_{m=1}^{\infty} u_m N(t), \qquad (13)$$

where u_m is displacement in the m-th node, ie at time t_m.

The solving equations of viscoelastic deformation can be represented as

$$\int_{t_n}^{t_{n+1}} N_n\left\{G\left[u(t)-\int_0^{t_n} R(t-\tau)u(\tau)d\tau\right]\right\}dt=\int_{t_n}^{t_{n+1}} N_n P dt \qquad (14)$$

Integration is performed only on the n-th period of time. Therefore, we can use the approximation of displacements (13) for $u(t)$.

$$K\int_{t_n}^{t_{n+1}} N_n\left[u_n(1-t)+u_{n+1}t-\int_0^{t_n} R(t-\tau)\left(u_n\left(1-\frac{\tau-t_n}{\Delta t_n}\right)+u_{n+1}\left(\tau-\frac{tLn}{\Delta t_n}\right)\right)d\tau\right]dt=\int_{t_n}^{t_{n+1}} N_n P dt.$$

Then, assuming that the matrix G is constant on the element, equation (14) after integration takes the form:

$$Gu_{n+1}\left[\frac{1}{3}-\frac{R_2}{\Delta t}+R_1\frac{t_n}{\Delta t}\right]=-Gu_n\left[\frac{2}{3}-R_1\left(1+\frac{t_n}{\Delta t}\right)+\frac{R_2}{\Delta t}\right]+P, \qquad (15)$$

where

$$R_1=\int_0^{t_n} K(t-\tau)d\tau$$

$$R_2=\int_0^{t_n} K(t-\tau)\tau d\tau \qquad (16)$$

Thus, a system of equations of hereditary viscoelasticity is obtained

On the basis of the considered approach, the characteristics of the stress-strain state of a seismic support with a diameter of $d=500$ mm and a height of a rubber layer of $h=50$ mm are determined. Rubber grade 2959, elastic modulus $G=0.7$ MPa. Poisson's ratio ν=0.499. Parameters of the Rabotnov's nucleus α =-0,60, β=1.72, λ=0.38. Loading P= 30 kN.

Figures 2-5 show the results of solving the problem of the stress-strain state of the structure.

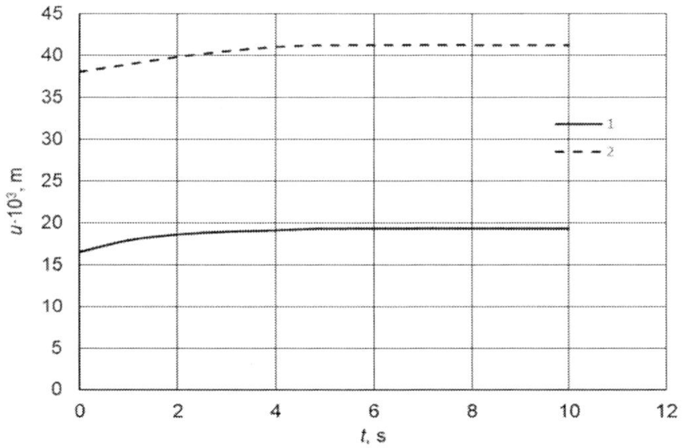

Figure. 2. Dependence of axial (1) and radial (2) displacements on time.

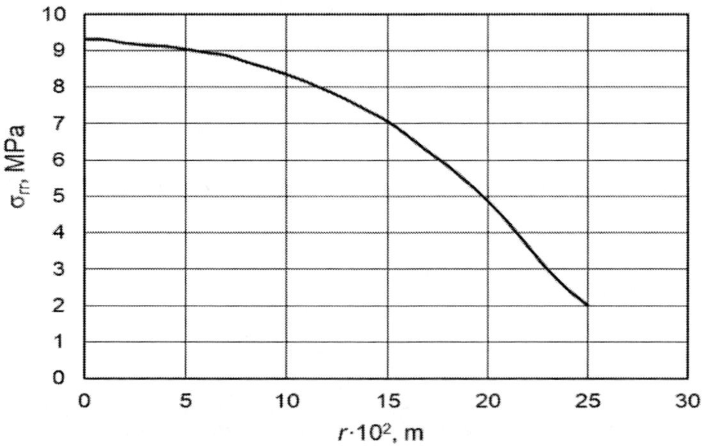

Figure 3. Distribution of radial stresses.

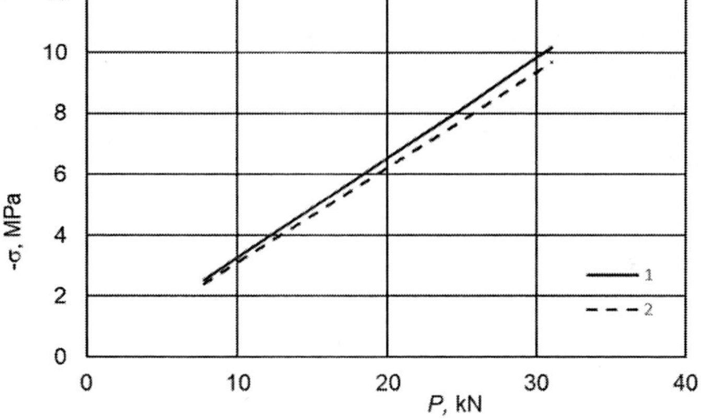

Figure 4. Dependence of axial (1) and (2) radial stresses.

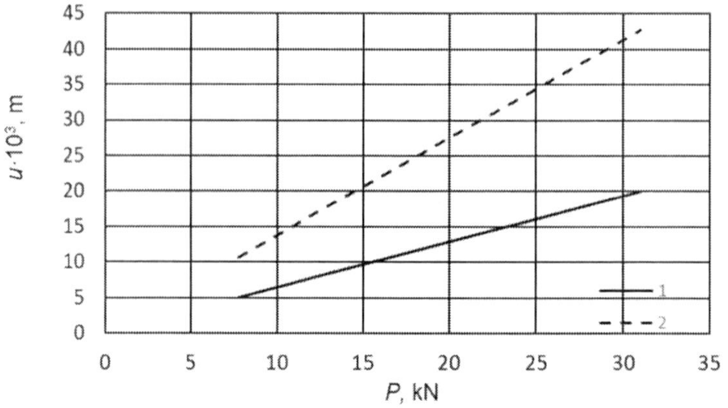

Figure 5. Dependence of axial (1) and radial displacements (2) on the load.

3. Conclusion

The damping elements of the considered vibration seismic insulators operate under conditions of preliminary stresses caused by the initial and installation loads, which affect the dissipative properties of such structures. To calculate the initial stress state of an elastomeric element, its viscoelastic properties and weak compressibility of the material are taken into account. The analysis of the distribution of compressive stresses makes it possible to evaluate the stiffness properties of the vibration insulator.

Dependences of axial and radial displacements on time and on load have been determined; dependences of axial and radial stresses; the distribution of radial stresses is constructed. The analysis of the distribution of compressive stresses makes it possible to evaluate the stiffness properties of the vibration isolator.

References

[1] Bulat A F, Dyrda V I, Nemchinov Yu I, Lisitsa N I, Lisitsa N N and Tymko N V 2010 Vibroseismic protection of machines and structures using rubber blocks *Geotechnical Mechanics* **85** pp 128-132

[2] Dyrda V I, Lisitsa N I, Novikova A V, Grebenyuk S N, Kozub Yu G and Bova A A 2012 Calculation of the stress-strain state of rubber-metal seismic supports *Methods for solving applied problems in the deformed solid mechanics* **13** pp 152-158

[3] Kozub Y, Dyrda V and Lisitsa N 2013 Substantiation of parameters and calculation of vibration insulator *TEKA* **13** pp 107-114

[4] Dyrda V I, Lisitsa N I, Marienko N G and etc. 2009 Justification and selection of parameters of rubber-metal seismic supports *Geotechnical Mechanics* **84** pp 17-23

[5] Kirichevskiy V V 2002 Finite element method in mechanics of elastomers (Kyiv: Naukova dumka) 605

[6] Comité Européen de Normalisation “Eurocode 8 Design of Structures for Earthquake Resistance Part 1 General Rules, Seismic Actions and Rules for Buildings”EN 1998-1 2004 (CEN, Brussels)

[7] Kirichevsky V V, Dokhnyak B M, Kozub Yu G, Gomenyuk S I, Kirichevsky R V and Grebenyuk S N 2005 Finite element method in the computing complex "MIRELA +" (Kyiv: Naukova Dumka) 402

[8] Malkov V M 1998 Mechanics of multilayer elastomeric structures. Monograph. (SPb: Publishing house of SPbU) 320

Microcrack appearance and coefficient of tension intensity ceramsite concrete on multicomponent binding

S A Kravchenko[1], O O Posternak[1,*], A I Kostyuk[1] and I A Stolevich[2]

[1]Odessa State Academy of Civil Engineering and Architecture Didrihsona str., 4, Odessa, Ukraine, 65029

*E-mail: alex.bk@ukr.net

Abstract. The results of experimental research of microcrack appearance and coefficient of tension intensity of experimental patterns from ceramsite concrete on multicomponent binding and quartz sand are given. Optimal compositions of construction and heat insulation ceramsite concrete on multicomponent binding are presented. Methodology of experimental researches, parameters of experimental patterns with description of technology of their making, results of basic patterns tests and their analysis, and also volume, aims and results of auxiliary standards tests are described. In the process of tests the longitudinal and transversal deformations of patterns, value of crack appearance loading, size of the destroying loading were measured.The done experimental researches of limits of microcrack appearance and coefficient of tension intensity of patterns from ceramsite concrete on multicomponent binding allowed to educe the characteristic features of their tensely-deformed state, to define the high and lower bound of crack appearance, deformability and size of the destroying loading of experimental patterns.

1. Introduction

Over the past decade, a study of the basic physical and mechanical properties of lightweight concrete on various aggregates and binders has been carried out [1-4], however, the parameters of the fracture mechanics of these concretes have remained unexplored.

The destruction of concrete has its own characteristics associated with the parameters of the fracture mechanics of individual components of the concrete structure (aggregate, matrix, contact zone between the aggregate and the matrix).

It is known that the phenomenon of destruction of concrete and reinforced concrete is a complex multistage process of the appearance, growth and development of cracks. Hence, it becomes necessary to study concrete and reinforced concrete from the standpoint of fracture mechanics, which in the broad sense of this concept includes that part of the science of the strength of materials and structures associated with the study of the bearing capacity of a body, taking into account the initial distribution of cracks, as well as with the study of various patterns of their development.

The use of fracture mechanics requires the determination of an additional concrete characteristic, which is characterized by crack propagation on the support. There are several such characteristics in fracture mechanics, which take into account the influence of irreversible deformations at the end of the crack in different ways. These parameters include the complexity of their experimental determination and structural analysis. To simplify the application of fracture mechanics to concrete, based on previous studies, a "stress intensity factor" and a corresponding crack propagation criterion are commonly used.

However, the stress intensity factor has not yet been sufficiently studied for concretes with multicomponent binders and research values are necessary.

2. Analysis of recent research and problem statement

The emergence of main cracks, as is known, involves the processes of microfracturing, which are observed in the areas of the so-called pre-fracture. The description of the processes of deformation and cracking of concrete at this stage became possible due to the use of methods of fracture mechanics and

Content from this work may be used under the terms of the Creative Commons Attribution 3.0 licence. Any further distribution of this work must maintain attribution to the author(s) and the title of the work, journal citation and DOI.
Published under licence by IOP Publishing Ltd

the theory of cracks. It can be noted that thanks to the research of Y. V. Zaitsev, P. I. Vasiliev, V. G. Orekhov, A. P. Pak, E. N. Peresypkin, L. P. Trapeznikov and other scientists, carried out back in the 70s of the 80s of the 20th century, the priority of domestic science in the field of concrete fracture mechanics is generally recognized today.

When determining the parameter of the stress intensity factor, standards and recommendations were developed [5-16], which recommend the use of many variations of concrete specimens with artificial cuts, produced in different ways. An artificial cut cannot be tied in its final natural fracture area. This drawback of an artificial cut occurs when loading concrete samples: as can be seen from the experiments carried out at the end of the cut, its continuation always develops in the form of a stable natural crack. The increase in the section is quite significant and the calculation of the stress intensity factor along the length of the section without taking into account the increment due to the increase in the crack leads to an underestimated value of the stress intensity factor.

In [17], the authors argue that in an inhomogeneous body, cracks can propagate not only in one or another fold material, but also in the zone of their contact. If the strength of the contact zone is high enough (higher than the strength of each of the materials), then on the contact surface the crack will not go along this surface, but will propagate deep into one or both materials according to the laws of crack development in homogeneous bodies. If the strength of the contact zone is insufficient, then the crack will be directed along the contact surface. In addition, in an inhomogeneous material, depending on the ratio of the properties of its components and the characteristics of the contact zone of these components, cracks can develop in different areas of the material.

To determine the stress intensity factor in laboratory conditions, you can use the method of S. N. Leonovich, O.V. Popov, K.A. Piradov [18], the essence of the method consists in determining the maximum loads that destroy control samples during compression and calculating the critical values of stress intensity. The challenge for the authors was to find a way to determine the stress intensity factor in already existing structures.

Identifying the previous unresolved parts of the common problem:

1. The use of methods for determining the stress intensity factor in heavy concrete based on cement binder is very numerous, but still insufficiently standardized for lightweight concrete based on multicomponent binders.

2. Based on the analysis of the results of studies of cracks from the point of view of fracture mechanics, the expediency of using the data in heavy concrete on a cement binder is obvious, for which special requirements are imposed. There are not enough publications on the use of lightweight concrete in scientific periodicals.

3. Goals and objectives of the study

The aim of the work was to conduct experimental studies of the influence of the factors of the composition of expanded clay concrete on a multicomponent binder on microcracking and the stress intensity factor. To achieve this goal, the following tasks were solved:

- to determine the lower and upper boundaries of the micro-thickness setting in the structural and heat-insulating expanded clay concrete on a multicomponent binder;
- to reveal the appearance of crack reproduction along the notch to determine the stress intensity factor in structural and heat-insulating expanded clay concrete on a multicomponent binder.

4. Materials and methods for studying expanded clay concrete on a multicomponent binder.

The research methods were the use of previous experience in certain micro-gaps and stress intensity factors, which included the determination of the upper and lower boundaries of cracking in accordance with the current standards [16].

The Odessa State Academy of Civil Engineering and Architecture has developed a technology for low-cement concretes. The technology is based on the use of a four-component binder, which, in addition to portland cement and ground quicklime, also contains an active mineral additive (fly ash) and gypsum [4].

IOP Conf. Series: Materials Science and Engineering 1164 (2021) 012039 doi:10.1088/1757-899X/1164/1/012039

The materials used in the studies had the following characteristics:

- expanded clay gravel 5 ... 10 mm, not fractionated by the Kulindorovsk industrial concern "Into-Stroy", grades by bulk density M 600, conditional strength in the cylinder equal to 2.8 ... 3.0 MPa;
- quartz sand of the Kremenchuzhzky quarry;
- cement M 400 of the Krivorizhzka plant - DSTU B V.2.7-112-2002;
- fly ash of the Chelyabinsk CHPP - GOST 25818-91;
- quicklime lime of the Kulindorovsky plant, the content of active calcium oxide CaO-75%;
- building gypsum - DSTU B V.2.7-104-2000;
- superplasticizer S-3 - TU-2481-001-51831493-00.

Based on the dependences obtained in earlier studies [1-3], as well as on the basis of the requirements [19] for lightweight concretes, the optimal compositions of expanded clay concrete on multicomponent binder and quartz sand were assigned, which are shown in Table 1.

Table 1. Optimal compositions of expanded clay concrete on a multicomponent binder.

Design strength, MPa	Aggregate-structural factor, r	Material expense for 1 m^3 of concrete							
		Cement, kg/m^3	Lime, kg/m^3	Fly ash, kg/m^3	Ceramsite, kg/m^3	Sand, kg/m^3	Gypsum, kg/m^3	S-3, %	Water, l
10	0.5	100	110	100	495	500	25	0.3	205
12,5	0.3	110	150	105	460	210	25	0.3	225
15	0.3	150	125	145	475	215	25	0.3	235
20	0.3	200	105	190	455	205	25	0.3	220
25	0.3	220	150	200	460	200	25	0.3	230

Investigations to determine the boundaries of microthickness in expanded clay concrete on a multicomponent binder were reduced to constructing a diagram of the state of concrete, expressed by the curve of changing the transit time of ultrasonic vibrations in concrete. Determination of the lower R_T^0 and upper R_T^v boundaries of the area of microthickness of structural concretes was carried out by the ultrasonic method in accordance with the methodology [20] using the UK-10PMS device in parallel with the determination of their strength and deformation characteristics. The measurement of the transit time of ultrasonic pulses τ_σ through the prototype with increasing load was carried out in the direction perpendicular and diagonal with respect to the direction of axial compression. In the first case, the ultrasonic probes were installed on opposite faces of the sample strictly opposite each other, and the instrument readings were taken at three positions of the probes: above, in the middle, and at the bottom of the sample. In the second case, the probes were also mounted on opposite faces of the sample, one on top and the other at the bottom of the sample. Thus, a large area of uniaxial stress state of the concrete prism was covered.

One of the most important elements of improving the quality of prefabricated load-bearing wall structures is the optimization of the composition of concrete, aimed at increasing its crack resistance. To experimentally determine the critical stress intensity factor, two types of samples were used: notched beam samples (Figure 1) and notched cubes (Figure 2) in accordance with DSTU B 2.7-227: 2009 [16].

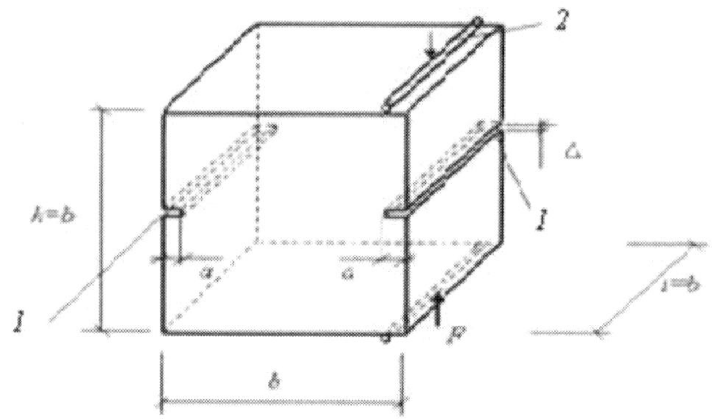

Figure 1. Determination of K_{Ic} on eccentrically compressed cubes with side slots. 1 – initial notch; 2 – metal bar.

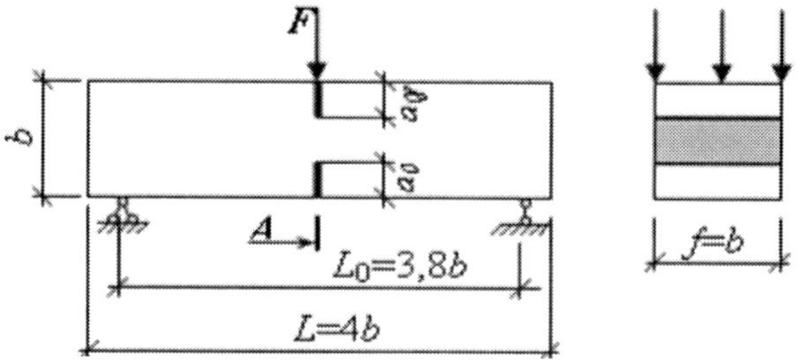

Figure 2. Three-point test scheme for notched beam specimens.

The study of the stress intensity factor in expanded clay concrete on a multicomponent binder was carried out on samples of cubes 10 × 10 × 10 cm with notches and beams 10 × 10 × 40 cm with notches, which were tested according to the three-point bending scheme, respectively, according to DSTU B.2.7-227: 2009 [16]. The samples were tested on a PG-50 hydraulic press, loading them with a low load until the appearance of the crack reproduction along the notch.

5. The results of the study of expanded clay concrete on a multicomponent binder

5.1. Microcracking

Figure 3 shows the results of studies of the dependence of the transit time of ultrasonic pulses on the stress levels of axial compression. This figure shows the processes of compaction of the material (this is evidenced by a decrease in the pulse transit time). The limit R_T^0 correspond to the shortest time of ultrasound passage through the sample. Further, decompaction occurs, the development of micro-ruining (above the limit R_T^0) and their next intensive development. The intersection of the curve with the ordinate axis R_T^v corresponds to the upper boundary of the microfracture setting and reflects significant changes in the stress and strain state of the material under load. The limit R_T^v stands out when microfracture begins to prevail over compaction processes.

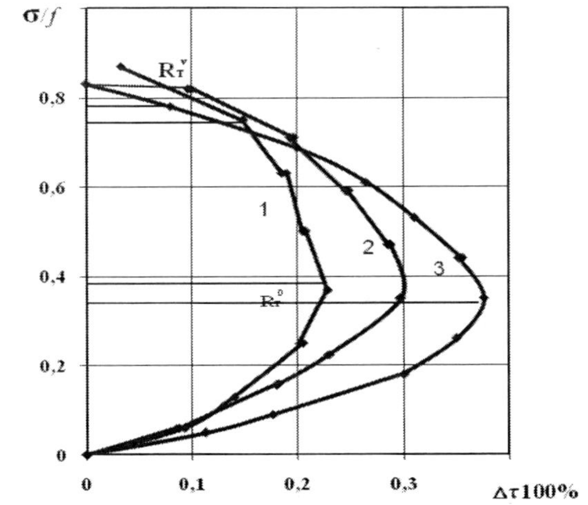

Figure 3. The dependence of the increments of the passage time of the ultrasonic pulse on the levels of axial compression stresses: 1 – $f = 6MPa$; 2 – $f = 8MPa$; 3 – $f = 10MPa$.

The relative measurement of the time of passage of ultrasonic pulses through the concrete was determined by the formula:

$$\Delta\tau = \frac{(\tau_\sigma - \tau_0)}{\tau_0}, \tag{1}$$

where τ_0 - is the transit time of ultrasonic pulses through concrete under tension σ;

τ_0 - the same, before the start of the load application.

The stress levels corresponding to the lower and upper boundaries of microcracking of concrete on a multicomponent binder are higher than those for heavy concrete. With an increase in the strength (class) of the concretes under study, an increase in the indicated parametric levels is observed, similar to ordinary expanded clay concrete.

Thus, an increase in the limits of microcracking of expanded clay concrete on a multicomponent binder in comparison with conventional expanded clay concrete can be explained, firstly, by the structural features of this concrete due to the presence of a porous aggregate and filler, and secondly, by the increased adhesion strength of the multicomponent binder with the aggregate.

5.2. Stress intensity factor

Fracture of all samples took place along the crack formed from the notch. The nature of the destruction was the same as for ordinary lightweight concrete - the surface of the destruction passed through both the mortar and the aggregate grains, that is, the destruction of the "matrix-filler" type took place, when the crack cuts the matrix and the filler [18]. The failure of the cube specimen was the same as for the bar specimen.

The processing of the experimental results was carried out according to the formulas:

- for a beam sample:

$$k_{IC} = \frac{3F_{cr}L}{2bh_{nt}^2}\ell^{\frac{1}{2}}\left(\frac{1-\ell}{h}\right)^2\left[1{,}93 - 3{,}07\left(\frac{\ell}{h}\right) + 14{,}53\left(\frac{\ell}{h}\right)^2 - 25{,}11\left(\frac{\ell}{h}\right)^3 + 25{,}8\left(\frac{\ell}{h}\right)^4\right]. \tag{2}$$

- for a cube sample:

$$k_{IC} = \frac{F_{cr}}{bh^{1/2}}\left[18,3\left(\frac{\ell}{b}\right)^{\frac{1}{2}} - 430\left(\frac{\ell}{b}\right)^{\frac{3}{2}} + 3445\left(\frac{\ell}{b}\right)^{\frac{5}{2}} - 11076\left(\frac{\ell}{b}\right)^{\frac{7}{2}} + 12967\left(\frac{\ell}{b}\right)^{\frac{9}{2}}\right], \tag{3}$$

where: L - span of the beam; ℓ – the length of the cut (in the conducted experiments it was 0.01-0.03 m); b - the width of the sample; h -cross-sectional height of the specimen (excluding the notch length, $h_{nt} = (h - \ell)$ for a bar specimen).

The main data (averaged over 6 samples of each composition) on the geometric dimensions of the samples, the value of the breaking (critical) load and the obtained values k_{IC}, together with the value of the tensile strength f_{ctd} obtained by breaking the rest of the sample, are shown in Table 2.

Table 2. Test results of beams and cubes with a notch

Design strength, MPa	F_{cr}, MN	L, m	h, m	b, m	k_{IC} MN/m$^{3/2}$	f_{ctd}, MPa
1	2	3	4	5	6	7
			Beam specimens			
10	0.0016				0.275	0.83
12.5	0.0015				0.268	0.9
15	0.0013	0.4	0.102	0.101	0.257	0.87
			Cube specimens			
10	0.0012				0.134	-
12.5	0.0028	0.101	0.102	0.101	0.181	-
15	0.0025				0.168	-

As can be seen in Figure 4, for a beam specimen, the change in the aggregate-structural factor in the concrete mixture affects the crack resistance insignificantly. For cube specimens, a change in the aggregate-structural factor in a concrete mixture affects crack resistance more noticeably than in the case of testing beams. The very dependence of crack resistance on the content of the aggregate-structural factor has a complex "wavy" character. At the same time, it can be stated that an increase in the aggregate-structural factor (including fine aggregate) leads to a decrease in crack resistance, and its decrease - to an increase in crack resistance.

Analysis of the results also shows that there is a close correlation between the tensile strength f_{ctd} and the critical stress intensity factor k_{IC}. The results obtained - the relationship between these two quantities - is shown in Figure 5. As can be seen, all experimental points are satisfactorily approximated by the dependence:

$$k_{IC} = \alpha f_{ctd}, \tag{4}$$

where $\alpha = 0.265\text{м}^{1/2}$ is the value characterizes the angle of inclination of the approximating straight line.

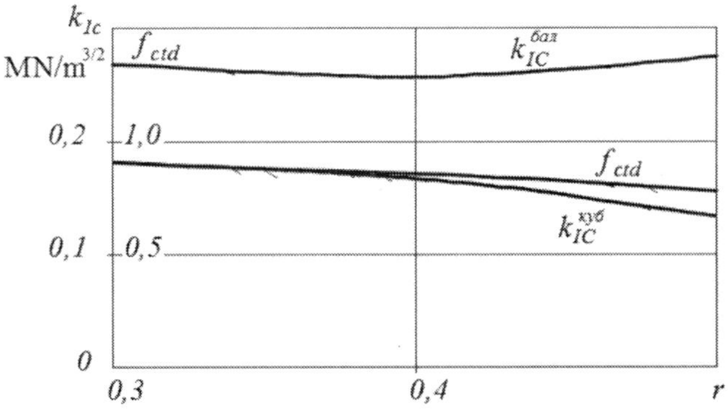

Figure 4. Graph of changes in the characteristics of concrete from the aggregate-structural factor.

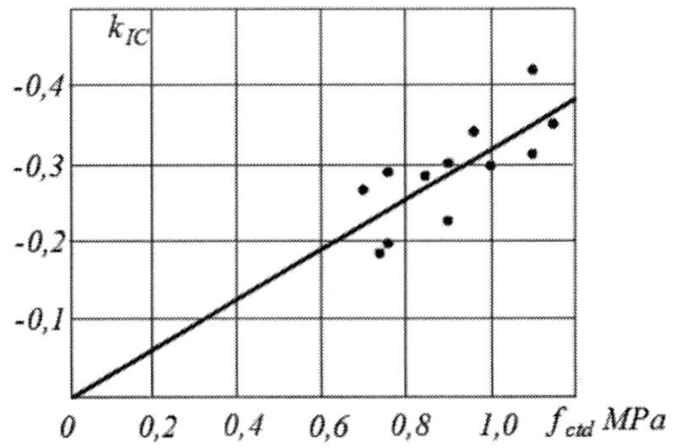

Figure 5. Dependence k_{IC} on tensile strength of concrete.

6. Conclusions

1. The increased limits of micro-thickness setting allow us to recommend the developed expanded clay concrete based on a multicomponent binder for structures designed for the combined effect of force factors and unfavorable environmental influences.

2. The area of stress state, located within the limits of micro-blasting, is of fundamental importance for assessing many important processes of deformation and strength of concrete. It was revealed that the most effective in terms of strength and crack resistance are concrete warehouses with the highest content of filler and aggregate, which should be recommended for the manufacture of structures.

3. The warehouses of expanded clay concrete on a multicomponent binder given in the article can be recommended for the manufacture of structures, taking into account the observance of the technology of their manufacture.

4. The results of studies of micro-thinning are presented, which are the initial link in determining the main properties of expanded clay concrete on a multicomponent binder and require further study of the effect of micro-thinning and stress intensity factor on the operation of structures made of this concrete.

References:

[1] Stolevich A S, Makarov S V, Stolevich I A, Madi K M and Kravchenko S A 2006 Structural lightweight concrete *Bulletin of ODABA* **21** pp 246-255

[2] Kamal Madi, Kravchenko S A and Stolevich A S 2005 Research of expanded clay concrete on a multicomponent binder *Bulletin of ODABA* **17** pp 101-105

[3] Dorofeev V S, Stolevich I A, Zinchenko S V, Kravchenko S A and Stolevich A S 2009 Influence of composition factors on the physical and mechanical properties of expanded clay

IOP Conf. Series: Materials Science and Engineering 1164 (2021) 012039 doi:10.1088/1757-899X/1164/1/012039

concrete on a multicomponent binder *Bulletin of ODABA* **34** pp 92-102

[4] Kravchenko S A, Stolevich I A, Kostyuk A I, Stolevich A S 2011 Properties of expanded clay concrete on a multicomponent binder *Science and technology collection Resource economy materials, construction, building and construction* **21** pp 32-37

[5] ACI committee 446, Fracture mechanics of concrete: concepts, models and determination of material properties 1990 *Concrete International* **12 (12)** pp 67-70 http://www.civil.northwestern.edu/people/bazant/PDFs/Papers/P90.pdf.

[6] Siregar A P N, Rafiq M I and Mulheron M 2019 Experimental investigation of the effects of aggregate size distribution on the fracture behaviour of high strength concrete *Constr. Build. Mater* **150** pp 252–259 https://doi.org/10.1051/matecconf/201925804011

[7] Alam M R, Azad MA and Kadir M A 2010 Fracture toughness of plain concrete specimens made with industry-burnt brick aggregates *Journal of Civil Engineering (IEB)* **38(1)** pp 81-94 http://mail.jce-ieb.org/doc_file/3801007.pdf

[8] Leonovich S N 1994 Fracture mechanics of the lightweight concrete *Proc. of 2-nd International Scientific Conference Durability and service life of bridge structures* pp 167-170

[9] Vonk R A A 1993 Micromechanical investigation of softening of concrete loaded in compression *Heron.* **38 №3** 94

[10] Chan H C, Cheung Y K and Huang Y P 1992 Crack Analysis of Reinforced Concrete Tension Members *Journal of Structural Engineering, ASCE.* **118 No. 8** pp 2118-2132 DOI: 10.1061/(asce)0733-9445(1992)118:8(2118)

[11] Neville A M 1971 Hardened Concrete: Physical and Mechanical Aspects *ACI Monograph* **6** 260

[12] Laffan M J 2012 Testing the toughness of polymer matrix composites *Failure Mechanisms in Polymer Matrix Composites* DOI:10.1533/9780857095329.1.110

[13] Shukla A and Dally JW 2010 Experimental Solid Mechanics. *Knoxville TN: College House Enterprises*

[14] Anderson T L 2005 Fracture Mechanics Fundamentals and Applications *3 rd Ed., Boca Raton, FL: Taylor & Francis*

[15] USAF Damage Tolerant Design Handbook: Guidelines for the Analysis and Design of Damage Tolerant Aircraft Structures 2002.

[16] DSTU B V.2.7-227: 2009 2010 Education materials. *Betony. Methods for the value of the characteristics of trischinostiykost (viscosity ruinuvannya) with static navantazhenny* (Kyiv: Ministry of Regional Development of Ukraine) 23

[17] Ashrabov A A and Zaitsev Y V 1983 About the conditions for the occurrence of structural cracks in the process of destruction of lightweight concrete *Improving the efficiency and quality of concrete and reinforced concrete. Abstracts. report IX All-Union. conf. for concrete and reinforced concrete* pp 207-215

[18] Piradov K A 1998 *Theoretical and experimental foundations of the mechanics of destruction of concrete and reinforced concrete* (Tbilisi) 355

[19] Recommendations for the design of compositions and production technology of structural lightweight concrete for load-bearing structures of large-panel buildings 1982 *TsNIIEP* 56

[20] Methodical recommendations for determining the mechanical characteristics of concrete under short-term and long-term loading 1984 *NIIZhB* 48

Influence of the chemical composition of blast-furnace slag on the whiteness of decorative slag-alkaline cements

P V Krivenko[1,*], A G Gelevera[1], O Yu Kovalchuk[1] and N V Rogozina[1]

[1]Kiev National University of Construction and Architecture, Research Institute of Binders and Materials named after V D Glukhovsky, 03037 Povitroflotskyi Avenue, 31, Kyiv, Ukraine

*Email: pavlo.kryvenko@gmail.com

Abstract. The construction industry is demanding more and more quality decorative cements. The demand for them and the requirements for their performance are constantly growing. But since decorative cements are based on white Portland cement, their production is associated with the shortcomings of the production of all clinker cements – low environmental friendliness, high energy consumption and high prices. They are not always able to provide decorative ecological and comfortable coatings with increased performance. In addition, many countries do not produce it and have to import it.

An effective alternative to decorative clinker cements can be decorative slag-alkaline cement obtained from industrial waste. It also provides a number of special properties – a wide range of colors, color fastness, high strength, high adhesion, durability and many others.

But the problem associated with the use of slag-alkaline cements as decorative cements with high whiteness (≥ 70%) is the instability of the chemical composition of the slag and, first of all, the different presence of iron oxides in it.

The paper shows the regularities of the influence of the chemical composition of blast-furnace slags on the whiteness of an artificial slag-alkaline stone. A new method for determining the whiteness of hardened materials is proposed. Possibilities of obtaining decorative alkali-activated cements with a wide range of whiteness – from 70 to 97% are shown.

1. Introduction

Industrial development and safety policies require new building materials that are economical and environmentally friendly. In particular, this applies to decorative cements used to enhance the architectural expressiveness of buildings and to carry out finishing work. The demand for them and the requirements for their performance are constantly growing [1]. Decorative cements are usually made on the basis of white cements [2]. However, such cements have all the disadvantages of Portland cement, including high energy demand, high cost and low environmental friendliness in their production. White Portland cements need to be imported. They, also, are not always able to provide decorative ecological and comfortable coatings with increased performance [3-5]. Decorative slag-alkaline cement produced from industrial waste can be an effective alternative. It also provides a number of special properties – a wide range and color stability, high strength, high adhesion, durability, etc. [6-12].

But the problem associated with the use of slag-alkaline cements as decorative cements with high whiteness (≥ 70%) is the unstable chemical composition of the slag. In particular, this concerns the presence of iron oxides in it, the content of which can range from 0,13% to 2,95%. In addition, in the initial periods of hardening, slag-alkali compositions can acquire a blue-green color. In the literature [13], this is associated with the formation of compounds of the $FeSO4 \cdot 7H2O$ type and this was the reason for concerns about the possibility of obtaining white decorative cements on their basis.

In previous studies in this direction [14-16], insufficient attention was paid to the influence of individual additives and their interaction on the decorative and physical-mechanical properties of slag-alkali cements.

Content from this work may be used under the terms of the Creative Commons Attribution 3.0 licence. Any further distribution of this work must maintain attribution to the author(s) and the title of the work, journal citation and DOI.
Published under licence by IOP Publishing Ltd

2. Purpose and objectives of research

Based on the analysis of literary sources and theoretical prerequisites, the aim of the work is to study the possibility of effective control of the decorativeness of slag-alkaline cements (slag alkali-activated cements), regardless of the composition of the slags. Research data are aimed at choosing such additives that should have maximum bleaching properties and would enhance the stabilization of bleaching processes during hydration and hardening of slag in the presence of alkalis. The research data are aimed at determining both the individual and the simultaneous effect of several additives on whiteness and physical and mechanical characteristics, as well as determining the optimal content of these additives in slag-alkali decorative compositions.

3. Materials and research methods

Three types of granulated blast furnace slags were used in the study: Ukrainian metallurgical plants (slag M and slag K) and one – from the Russian Federation (slag T). All slags were ground in a ceramic mill to obtain a specific surface area of 4500 cm^2/g (according to Blaine). All investigated slags had a glass content in the range from 50 to 60%. The chemical composition of the studied slags is shown in Table 1.

Table 1. Chemical composition of slag.

Slag type	Main oxides content, % by mass											
	CaO	MgO	Al_2O_3	SiO_2	MnO	SO_3	FeO	$Na_2O + K_2O$	TiO_2	Σ	M_b	M_a
Slag T	43.5	6.3	7.0	43	0.054	–	0.42	–	0.26	100.53	1.00	0.16
Slag M	47.0	4.6	6.8	38.9	0.55	1.8	0.35	–	–	100.00	1.13	0.17
Slag K	48.5	4.6	4.6	38.2	0.81	1.1	1.65	0.79	–	100.25	1.24	0.12

A low-hygroscopic alkaline compound натрия sodium metasilicate pentahydrate – $Na_2O{\cdot}SiO_2{\cdot}5H_2O$ was used as an alkaline component.

Titanium dioxide (TiO_2), KH 84 class kaolin used in the paper industry with a whiteness of 84%, and calcium carbonate ($CaCO_3$) in powder form with a whiteness of 90% were used as bleaching additives.

Determination of the whiteness of decorative slag alkali-activated cements was carried out by scanning the surface of the cement samples with high resolution and then measuring the whiteness by comparison with a reference sample with a whiteness of 98% in PhotoShop Soft.

Technological and physical-mechanical properties of slag alkali-activated cements were determined in accordance with the current Ukrainian standards.

4. Research results and discussion

4.1. *Mechanical characteristics of slag alkaline-activated cements*

At the first stage of the research, the compressive strength of the slag-alkaline cement was determined depending on the amount of the alkaline component. Slag (M) was assigned as the base aluminosilicate component of slag-alkaline cement for comparison with other slags based on the minimum content of iron oxides in it.

The compressive strength of slag-alkaline cement is shown in Figure 1.

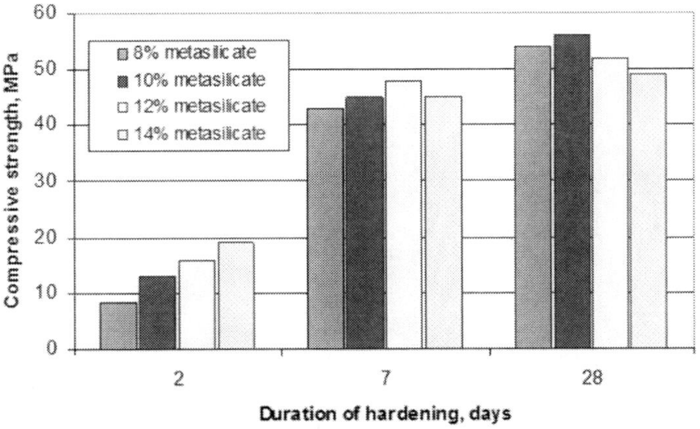

Figure 1. Compressive strength of slag-alkaline cement depending on the content of the alkaline component. Slag type – M.

The analysis of the given results showed that all compositions meet the requirements for cement of strength classes 42.5...52.5. High values were achieved with an alkaline activator content of 8 and 10% by weight – they turned out to be almost the same at the age of 28 days and both cement compositions met the requirements for cements of strength class 52.5. The results of the study show that the compressive strength after 28 days decreases with an increase in the content of sodium metasilicate by more than 10%.

Considering the higher early strength (after 2 days) of the cement containing 10 wt% sodium metasilicate, it was chosen for further research.

4.2. *Whiteness of slag alkali-activated cements*

The next stage of the study was to test the brightness of alkali-activated slag cements made from three slags, and to investigate possible ways to increase it by adding bleaching additives such as TiO_2, $CaCO_3$ and kaolin clay with a high degree of disease. In the studies, the following limits were adopted for varying the amount of bleaching additives in cement: TiO_2 from 2.5 to 7.5% by weight, kaolin clay from 5 to 15% by weight and $CaCO_3$ from 15 to 25% by weight.

The results of the study of the degree of whiteness of the investigated slag-alkaline cements are shown in Figure 2.

Analysis of the above results allowed us to reveal that slag alkali-activated cements based on slag (T) and slag (M) adhered to similar patterns, and the best bleaching effect for these two cement compositions was achieved with the addition of TiO_2. Slag (K) showed the lowest whiteness characteristic of all cement compositions. The best whitening result was obtained with calcium carbonate, but it was close to the system without any bleaching agents. All other additives showed a decrease in cement performance by whiteness.

The obtained data on the degree of whiteness for a part of the investigated slag alkali-activated cements indicate that they meet the requirements for decorative materials, that is, they have a whiteness degree of 70% and higher.

However, the addition of large amounts of TiO_2 is not economically efficient. On the other hand, it is necessary to use some fillers in order to reduce possible shrinkage deformations of slag materials based on slag alkali-activated cements. Moreover, such fillers should not reduce the degree of whiteness.

For this, a set of experiments was carried out aimed at identifying the combined effect of various bleaching agents. Based on the results of previous studies, TiO_2, kaolin clay and $CaCO_3$ were selected as such additives. The experiments were carried out according to a two-factor mathematical design. According to research results, TiO_2 was recognized as the best whitening agent.

The ranges of variation for all additives have been set according to previous results in terms of maintaining the required strength grades of 42.5 and 52.5. The maximum content of $CaCO_3$ was taken to be 25%, for kaolin clay – 15%. The TiO_2 content in the experiments was taken in the range of 2.5...7.5%.

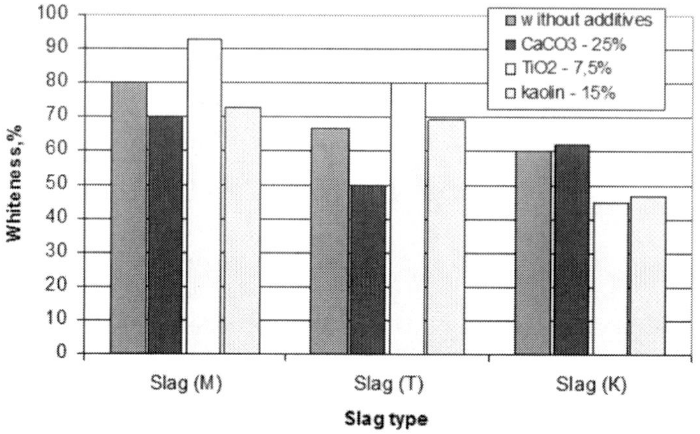

Figure 2. The degree of whiteness of slag alkali-activated cement, depending on the type and amount of bleaching additives. The amount of sodium metasilicate – 10%.

The results of studies of the degree of whiteness of the investigated slag alkali-activated cements with TiO_2 and calcium carbonate powder are shown in Figure 3, and with TiO_2 and kaolin – in Figure 4.

The analysis of the results obtained allowed us to show that in the case of the combined use of TiO_2 and $CaCO_3$, a degree of whiteness can be achieved, which varies within 85...97%, which meets the standard requirements for high-quality decorative materials.

The $CaCO_3$ content in slag-alkaline cement made with slag (M) was the main one, and the titanium dioxide content was minor (corrective). And for slag-alkaline cement based on slag (T), the interaction of the aforementioned additives was important.

In the case of slag (K), when the TiO_2 content was 2.5 to 5.0 wt.%, the calcium carbonate content was more important. After increasing the TiO_2 content to 5.0...7.5 wt.%, The combination of the above additives was the most effective. The combination of kaolin and white TiO_2 pigment also made it possible to obtain a high degree of whiteness of the studied alkali-activated slag cements. It can be concluded that the whiteness of the slag alkali-activated test cements increased rapidly with increasing titanium dioxide content, while the addition of kaolin clay allowed the whiteness to be changed only within narrow limits.

The experiments carried out allow us to conclude that slag-alkaline cement made from slag (K) has an insufficient degree of whiteness compared to slag-alkaline cements made from slag (T) and slag (M). This can be explained by the different chemical composition of the feedstock, and first of all, by the presence of iron oxides contained in the slag: 0.35% – in the slag (M), 0.42% in the slag (T) and 1.65% in the slag (K). Thus, it can be assumed that the content of iron oxide in the raw material is the main reason that explains why the whiteness tended to decrease. The small difference between the whiteness degree of slag-alkali cements made from slag (T) and slag (M) can be explained by the relatively small difference in the presence of iron oxide in these slags (Table 1).

IOP Conf. Series: Materials Science and Engineering 1164 (2021) 012040 doi:10.1088/1757-899X/1164/1/012040

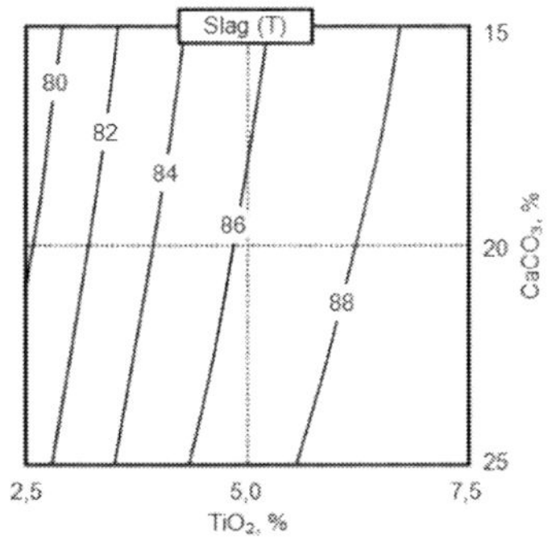

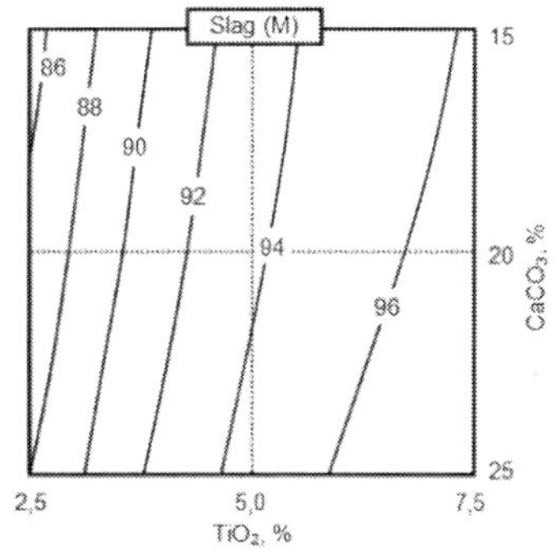

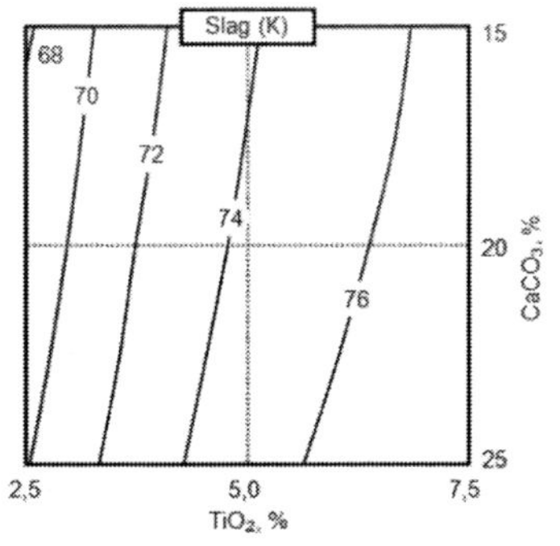

Figure 3. Isolines of whiteness of slag-alkaline cements depending on the type of slag and the amount of TiO_2 and $CaCO_3$ additive.

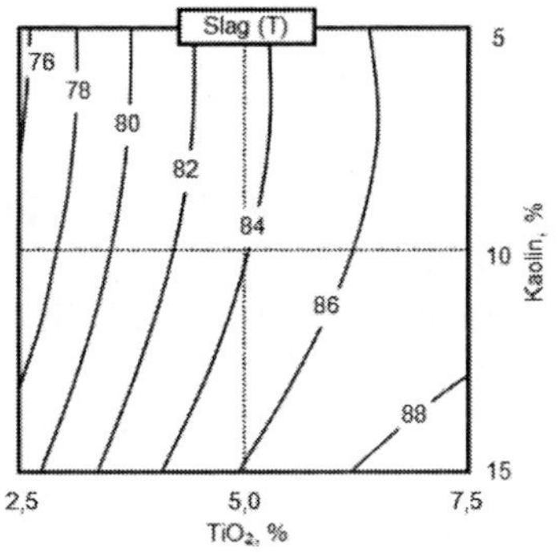

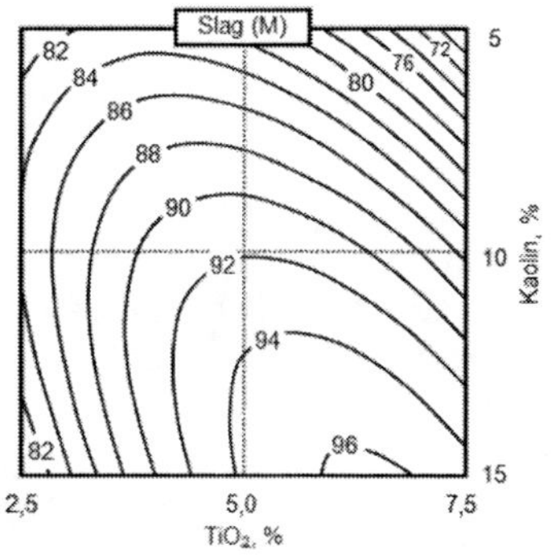

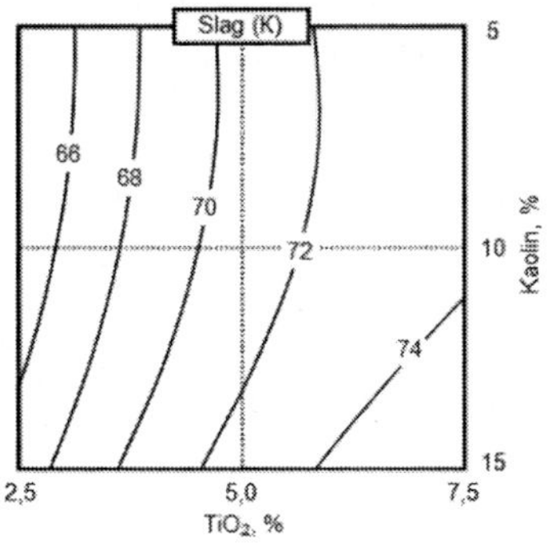

Figure 4. Isolines of whiteness of slag-alkali cements depending on the type of slag and the amount of kaolin and TiO_2 addition.

4.3. Influence of iron oxides on the whiteness of slag-alkaline decorative cements

The presence of iron oxides in slag-alkaline decorative cements can impair their whiteness for two reasons:

- due to the formation of compounds of the type $FeSO_4{\cdot}7H_2O$ [13] and the acquisition of a blue-green color;
- due to the fact that iron oxides have a color from brown to dark brown, they thereby reduce the whiteness of the cement.

But if in the first case this color appears in the early stages of hardening and gradually disappears in air due to carbonization over the surface [17], then in the second case, the decrease in whiteness depends on the amount of iron oxides and can only be corrected with appropriate bleaching additives and/or the choice of slag low in iron oxides.

To reveal the role that *FeO* plays in relation to the whiteness of slag-alkali cements, the following experiment was performed – *FeO* powder was added to slag (M) to simulate slag compositions (T) and (K). The proportions of the optimal cement composition did not change.

The results are shown in Figure 5.

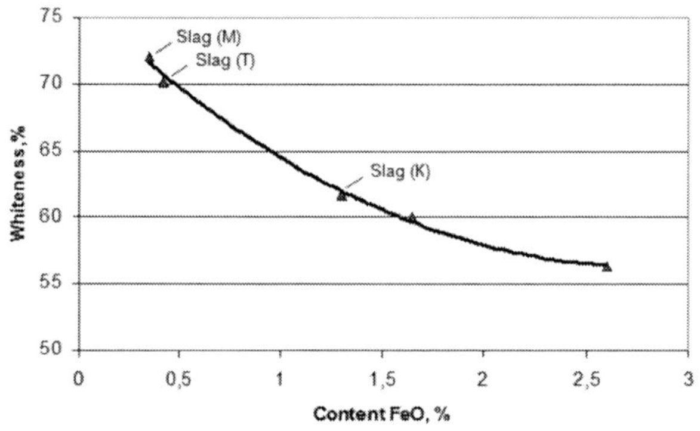

Figure 5. Influence of FeO on the whiteness of alkali-activated slag cements. No whitening additives.

The results of this study allow us to conclude that with an increase in the *FeO* content in the slag, a decrease in the whiteness of the cement is observed.

Thus, the assumption about the role played by the composition of the slag and, in particular, the content of *FeO* in it for the degree of whiteness, was correct.

5. Conclusions

The research results lead to the following conclusions:

- The content of iron oxide in the raw material plays a decisive influence on the whiteness of alkali-activated slag cements. The easiest way to improve whiteness is to reduce the amount of iron oxides in alkali-activated slag cement. This can be realized by choosing slags with a low iron oxide content, grinding the slag in mills with ceramic lining and ceramic grinding bodies, and by introducing bleaching additives.
- The main corrective effect on the whiteness of alkali-activated slag cements is produced by TiO_2. Its amount can vary depending on the required brightness of the final product and the content of iron oxides in the blast furnace slag. Some TiO_2 can be replaced with white mineral additives (calcium carbonate and kaolin clay) without significant loss of whiteness. The use of TiO_2 additives alone and in combination with $CaCO_3$ and kaolin additives made it possible to obtain decorative slag-alkaline cements with a whiteness of 70...97%.
- The obtained decorative slag-alkali cements have a strength class of 42.5...52.5.

– A new method for determining the whiteness of decorative cements has been proposed by scanning the surface of cement samples, followed by measuring the degree of whiteness by comparing it with a reference sample with a whiteness of 98% in PhotoShop Soft.

References

[1] Semenov V M 2010 Paintwork materials for the protection of concrete and cement surfaces *Paint and varnish industry* **11** pp 23-27
[2] White cement [Electronic resource] *Portal of Budpostach* Access mode: https://pp-budpostach.com.ua/a118411-belyj-tsement.html.
[3] Berezina L A and Abdulberov Z A 2009 Application of kaolin by IMERYS Minerals in the manufacture of paints and varnishes *Paintwork materials and their use* **1-2** pp 61-65
[4] Levashova Yu S, Kosenko N O and Lebedeva O S 2019 Investigation of the parameters of the microclimate of large crowds *Scientific Bulletin of Construction* **93(3)** pp 217-221
[5] Petrenko G S 2010 Modern pigments and fillers for the paint and varnish industry *Paint and varnish industry* **9** pp 30-36
[6] Krivenko P V 1994 Alkaline Cements: *materials First Intern. Conf. Alkaline Cements and Concretes* pp 111-129
[7] Krivenko P V, Petropavlovsky O N, Gelevera A G, Vozniuk G V and Pushkar V I 2009 Commercially-produced alkaline cements and their efficiency *Scientific-Technical Collection Resource-Saving Materials, Structures, Buildings and Structures* **18** pp 64-71
[8] Krivenko P 2014 Alkali-Activated Materials – 55 Years of Experience *Proc. of the International Conference on Alkali-Activated Cement and Concretes* pp 1-5
[9] Shi C, Krivenko P V, Della Roy 2012 *Alkaline activated cements and concretes* (in Chinese, Authorized translation from English):Monograph Taylor & Francis 326
[10] Fernández-Jiménez A, Garcia-Lodeiro I, Maltseva O and Palomo A 2019 Hydration mechanisms of hybrid cements as a function of the way of addition of chemicals. *Journal of the American Ceramic Society* **102(1)** pp 427-436
[11] Krivenko P, Petropavlovsky O, Kovalchuk O, Pasko A and Lapovska S 2018 Design of the composition of alkali activated Portland cement using mineral additives of technogenic origin *Eastern-European Journal of Enterprise Technologies* **4/6 (94)** pp 6-15
[12] Krivenko P V, Petropavlovsky O N, Gots V I and Rostovskaya G S 2009 Alkali activation of Composite Cement *Ibausil. Internationale Baustofftagung* **1** pp 445–456
[13] Chaouche M, Gao X X, Cyr M, Cotte M and Frouin L 2017 On the origin of the blue/green color of blast-furnace slag-based materials: Sulfur K-edge XANES investigation. *J Am Ceram Soc.* **100** pp 1707-1716 doi:10.1111/ jace.14670
[14] Krivenko P, Petropavlvskyy O, Puskar V and Ostrovska L 2011 Decorative alkaline cements *IV Intern. Symp: Non-Traditional cement & Concrete* (Brno) pp 257- 265
[15] Kryvenko P, Sanytsky M, Kropyvnytska T and Kotiv R 2014 Decorative multi-component Alkali Activated Cements for restoration and finishing works *Advanced Materials Research* **897**Trans Tech Publications, Switzerland pp 45-48 doi:10.4028/www.scientific.net/AMR.897.45
[16] Krivenko P V and Kovalchuk A Yu 2019 Management of the decorative properties of alkali cements *Scientific Bulletin of Construction* **2(95)** pp 280-285
[17] Kovalchuk O Yu 2014 Research of carbonization of alkaline concretes *Collection Building materials, products and sanitary ware* **52** pp 19-24

Acknowledgements
Authors also express their gratitude to the Ministry of Education and Science of Ukraine for financial support of this research that is carried out within the budgetary financing of topic with registration No 1020U001010 and implementation period 2021…2022.

Stability of compressed rods when their stiffness changes according to the law of the fourth power

Yu Krutii[1,*], M Surianinov[1], O Shyliaiev[1]

[1]Odesa State Academy of Civil Engineering and Architecture, 4, Didrihsona str., Odesa 65029, Ukraine

[*] E-mail: yurii.krutii@gmail.com

Abstract. Based on the previously obtained representations for the state parameters of a rod with arbitrary continuous bending stiffness, the stability problem for a family of rods whose stiffness varies according to the law of the fourth power is solved. The spectrum of critical forces is determined and formulas for curved forms of equilibrium are derived. We propose to consider the stability coefficient as a function of the variable α, and having a set of values of this function corresponding to the values of the independent variable $0 < \alpha \le 1$, approximate the function K by a polynomial. The result is the expression $K_2 = 0.0001\alpha + 80.7626\alpha^2 + 0.0002\alpha^3$. The power of the polynomial was chosen from the condition that the coefficient of determination should not be less than 0.9999. The directions for further research are presented: introduction of the results of this work into the practice of calculations for stability of real objects; application of general formulas for state parameters to the study of stability of rods with other laws of change in transverse stiffness encountered in practice.

1. Introduction

In many structures, variable stiffness rods are used to reduce material consumption. These are power line supports, wind turbine poles, antenna supports, crane arms, drilling towers, trestle supports, chimneys, structures in aircraft construction, bridge construction, machine building, etc. The problem of determining critical loads and forms of stability loss for such structures has always been one of the most important ones in mechanics. In recent years, this problem has become even more topical due to the use of new materials for lightweight structures, for which the problem of stability loss is of major importance. Nevertheless, most works on this subject are based on the use of approximate methods. Analytical solutions are limited, which is explained by the difficulties encountered in integrating differential equations with variable coefficients, to which the mathematical side of the problem is reduced.

2. Literature review

Fundamental studies of the problem of the stability of compressed rods have been performed by many famous scientists, among whom we should mention the works of L. Euler, F. S. Yasinsky, A. N. Dinnik, S. P. Timoshenko, A. S. Volmir, A. R. Rzhanitsyn, and others.

The attempt to find the optimal form of a rod compressed by a longitudinal force belongs to J. L. Lagrange [1]. To find the maximum compressive critical force of a rod of circular cross-section at its smallest volume, he introduced a value called efficiency. This problem has not yet been solved in general form [2, 3].

The stability of solid rods of variable cross-section with a power law of stiffness change has been studied most completely by A. N. Dinnik [4], and numerical results of calculations for such rods are given in work [5]. A review of the most frequent laws of stiffness change for different variants of rod fastening under short- and long-term load action is made in [6, 7]. As for the development of

Content from this work may be used under the terms of the Creative Commons Attribution 3.0 licence. Any further distribution of this work must maintain attribution to the author(s) and the title of the work, journal citation and DOI.
Published under licence by IOP Publishing Ltd

calculation methods, it should be noted the works [8 - 14], in which many interesting results were obtained, which served as a basis for further research.

In the works of recent years, the use of numerical methods prevails, and, more often, it is the method of finite elements with implementation in popular software packages. Thus, in [15], a finite-element analysis of a beam with a variable cross-section in the COSMOS/M program is performed. Stability and dynamics of a thin-walled composite rod are considered in [16]; the problem is solved by the Galerkin method. The same issues with respect to metal boring mandrels are investigated in [17] on the basis of a modified Adomian decomposition method with subsequent verification in ANSYS.

Stability studies based on exact solutions of the differential equation of the curved rod axis are extremely rare in the literature [18].

3. The purpose of this work is to determine the spectrum of critical forces and formulas for curved forms of equilibrium of an elastic homogeneous centrally compressed rod, rigidly clamped at the ends, when the moment of inertia of the cross-section changes according to the law of the fourth degree.

4. Materials and Methods

The methods of structural mechanics and the theory of stability of compressed rods are used. The mathematical side of the problem is provided by applying the theory of differential equations and the theory of series.

5. Results and Discussion

At Euler loss of stability, the critical force is determined from the differential equation of the bent axis of the rod [10]

$$(E(x)I(x)y''(x))'' + Ny''(x) = 0, \tag{1}$$

where $E(x)I(x)$ – переменная изгибная жесткость стержня в точке x;

$E(x)$ – modulus of elasticity of the rod material;

$I(x)$ – moment of inertia of the cross section;

N – constant compressive load;

$y(x)$ – the unknown function representing the transverse displacement (deflection) of the rod section at the point x (figure 1).

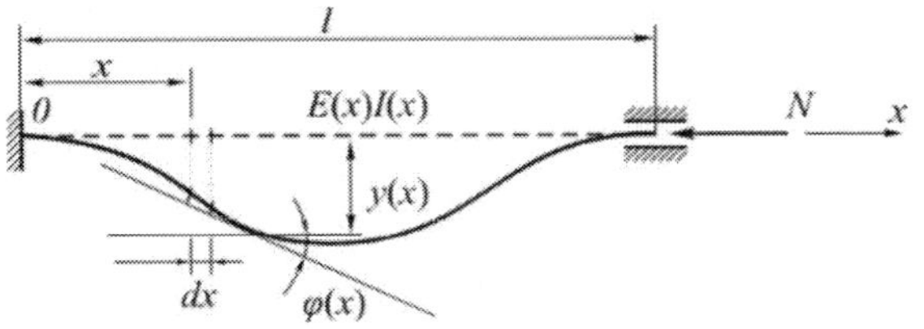

Figure 1. Calculation scheme.

In [19, 20] we obtained the exact solution of equation (1) in the case of arbitrary continuous transverse stiffness. Here we will assume that the moment of inertia changes according to the law of the fourth power

$$I(x) = I_0\left(1-(1-\alpha)\frac{x}{l}\right)^4,$$

where I_0 – moment of inertia of the cross section of the rod at the point $x = 0$;

α – a constant that satisfies the condition $0 < \alpha \le 1$.

The stress-strain state of the rod is fully characterized by the parameters: displacement $y(x)$, angle of rotation $\varphi(x)$, bending moment $M(x)$ and shear force $Q(x)$.

For the parameters we will use the representations from [19-21]. Before writing out these representations, note that due to the homogeneity of the rod $E(x)=E=const$. Also we will introduce a notation of $u=1-(1-\alpha)x/l$. Then, as applied to this case, the above formulas in dimensionless format take the form:

$$y(x)=y(0)+\varphi(0)lX_2(x)-M(0)\frac{1}{N}(1-X_1(x))-Q(0)\frac{l}{N}\left(\frac{x}{l}-X_2(x)\right); \tag{2}$$

$$\varphi(x)=\varphi(0)\tilde{X}_2(x)+M(0)\frac{1}{Nl}\tilde{X}_1(x)-Q(0)\frac{1}{N}(1-\tilde{X}_2(x)); \tag{3}$$

$$M(x)=\varphi(0)NlX_2(x)+M(0)X_1(x)+Q(0)lX_2(x); \tag{4}$$

$$Q(x)=Q(0); \tag{5}$$

$$X_n(x)=\alpha_{n,0}(x)-K\alpha_{n,1}(x)+K^2\alpha_{n,2}(x)-K^3\alpha_{n,3}(x)+...(n=1,2); \tag{6}$$

$$\alpha_{n,0}(x)=\left(\frac{x}{l}\right)^{n-1},\quad \alpha_{n,k}(x)=\frac{1}{l^2}\int_0^x\int_0^x\frac{1}{u^4}\alpha_{n,k-1}(x)dxdx\ \ (k=1,2,3...); \tag{7}$$

$$N=K\frac{EI_0}{l^2}. \tag{8}$$

Here $X_n(x),\tilde{X}_n(x)=lX_n'(x)$ $(n=1,2)$ – dimensionless functions and K – dimensionless oscillation factor to be determined.

Using a specially derived formula

$$\int_0^x\int_0^x\frac{1}{u^3}\left(\frac{x}{u}\right)^i dxdx=\frac{1}{(i+1)(i+2)}u\left(\frac{x}{u}\right)^{i+2}\ \ (i=1,2,3...), \tag{9}$$

integrals (7) can be calculated explicitly. The proof of this formula is beyond the scope of this paper. Let us point out only one of the possible variants of such a proof. By differentiating both parts of formula (9) twice, we arrive at an identity. This means that the left part of the formula is equal to the right part up to an arbitrary polynomial of the first degree. However, considering the fact that both these parts together with their first derivatives turn to zero when $x=0$, we conclude that the above polynomial is an identical zero.

First, by a direct twofold integration we obtain

$$\alpha_{1,1}(x)=\frac{1}{l^2}\int_0^x\int_0^x\frac{1}{u^4}\alpha_{1,0}(x)dxdx=u\left(\frac{1}{2!}\left(\frac{x}{lu}\right)^2+\frac{1-\alpha}{3!}\left(\frac{x}{lu}\right)^3\right).$$

Then, using formula (9), we successively find:

$$\alpha_{1,k}(x)=\frac{1}{l^2}\int_0^x\int_0^x\frac{1}{u^4}\alpha_{1,k-1}(x)dxdx=u\left(\frac{1}{(2k)!}\left(\frac{x}{lu}\right)^{2k}+\frac{1-\alpha}{(2k+1)!}\left(\frac{x}{lu}\right)^{2k+1}\right)\ \ (k=2,3,4,...);$$

$$\alpha_{1,k}(x)=\frac{1}{l^2}\int_0^x\int_0^x\frac{1}{u^4}\alpha_{2,k-1}(x)dxdx=\frac{1}{(2k+1)!}u\left(\frac{x}{lu}\right)^{2k+1}\ \ (k=1,2,3,...).$$

According to formula (6) we will now have:

$$X_1(x)=u\sum_{k=0}^{\infty}(-1)^kK^k\left(\frac{1}{(2k)!}\left(\frac{x}{lu}\right)^{2k}+\frac{1-\alpha}{(2k+1)!}\left(\frac{x}{lu}\right)^{2k+1}\right)=u\left(\cos\sqrt{K}\frac{x}{lu}+\frac{1-\alpha}{\sqrt{K}}\sin\sqrt{K}\frac{x}{lu}\right); \tag{10}$$

$$X_2(x) = u\sum_{k=0}^{\infty}\frac{(-1)^k}{(2k+1)!}K^k\left(\frac{x}{lu}\right)^{2k+1} = \frac{u}{\sqrt{K}}\sin\sqrt{K}\frac{x}{lu}. \tag{11}$$

Using the known functions $X_n(x)$ $(n = 1,2)$, we find:

$$\tilde{X}_1(x) = (1-\alpha)^2\frac{x}{lu}\cos\sqrt{K}\frac{x}{lu} - \left(\frac{(1-\alpha)^2}{\sqrt{K}} + \frac{\sqrt{K}}{u}\right)\sin\sqrt{K}\frac{x}{lu}; \tag{12}$$

$$\tilde{X}_2(x,K) = \frac{1}{u}\cos\sqrt{K}\frac{x}{lu} - \frac{1-\alpha}{\sqrt{K}}\sin\sqrt{K}\frac{x}{lu}. \tag{13}$$

To the case of both fixed ends corresponds the boundary conditions: $y(0)=0;\ \varphi(0)=0;\ y(l)=0;\ \varphi(l)=0$. Realization of these conditions with formulas (2), (3) leads to the system of equations

$$\begin{cases}(1-X_1(l))M(0) + l(1-X_2(l))Q(0) = 0;\\ \dfrac{\tilde{X}_1(l)}{l}M(0) - (1-\tilde{X}_2(l))Q(0) = 0.\end{cases} \tag{14}$$

For nontrivial coincidence of the system, its definition should be equal to zero. Hence, given equations (10)-(13), we arrive at the characteristic equation

$$2 - 2\cos\frac{\sqrt{K}}{\alpha} - \frac{\sqrt{K}}{\alpha}\sin\frac{\sqrt{K}}{\alpha} = 0.$$

After elementary trigonometric transformations, this equation is equivalent to the set of two equations:

$$\sin\frac{\sqrt{K}}{2\alpha} = 0; \tag{15}$$

$$\sin\frac{\sqrt{K}}{2\alpha} - \frac{\sqrt{K}}{2\alpha}\cos\frac{\sqrt{K}}{2\alpha} = 0. \tag{16}$$

We obtain the set of stability coefficients by combining the roots of each of these equations into a single set and arranging them in ascending order. As the calculations show, in the specified ordered set the roots of equation (15) are located at odd numbers, and the roots of equation (16) are located at even numbers.

Since equation (15) is elementary, the stability coefficients with odd numbers are determined exactly: $K_1 = (2\alpha\pi)^2$, $K_3 = (4\alpha\pi)^2, \ldots$. Consequently, by formula (8) the critical forces with odd indices are also determined exactly

$$N_1 = (2\alpha\pi)^2\frac{EI_0}{l^2},\ N_3 = (4\alpha\pi)^2\frac{EI_0}{l^2}, \ldots .$$

The transcendental equation (16) is used to find stability coefficients with even numbers $K_2, K_4, \ldots$. Finding the roots of such equation is not difficult. For this purpose there are many different software possibilities. Let us restrict ourselves here to the root of K_2. Its values corresponding to each value K_2 in increments of 0.1 are given in Table 1.

Table 1. Values of stability coefficients.

α	0,1	0,2	0,3	0,4	0,5	0,6	0,7	0,8	0,9	1,0
K_2	0.8076	3.2305	7.2687	12.922	20.191	29.075	39.574	51.688	65.418	80.763

It is quite clear that for any other values α it is possible to find stability coefficients. In order to do this without solving the characteristic equation, the idea arises to express the found table dependence between the parameters K and α analytically as well.

Considering the stability coefficient K as a function of the variable α, and having a set of values of this function corresponding to the values of the independent variable $0 < \alpha \leq 1$, we approximate the function K by a polynomial. As a result, we obtain $K_2 = 0.0001\alpha + 80.762\alpha^2 + 0.0002\alpha^3$. The polynomial degree was chosen from the condition that the coefficient of determination should not be less than 0.9999. Therefore,

$$N_2 = (0.0001\alpha + 80.7626\alpha^2 + 0.0002\alpha^3)\frac{EI_0}{l^2}.$$

It remains to write out formulas for the curved forms of equilibrium of the rod. For this purpose let us transform the formula for displacements (2) to the form

$$y(x) = -\frac{M(0)}{N}\left[1 - u\left(\cos\sqrt{K}\frac{x}{lu} + \frac{1-\alpha}{\sqrt{K}}\sin\sqrt{K}\frac{x}{lu}\right) - \eta\left(\frac{x}{l} - \frac{u}{\sqrt{K}}\sin\sqrt{K}\frac{x}{lu}\right)\right], \tag{17}$$

where $\eta = -Q(0)l/M(0)$ dimensionless parameter to be determined.

From the equations of the system (14) we find

$$\eta = \frac{1 - X_1(l)}{1 - X_2(l)} = -\frac{\tilde{X}_1(l)}{1 - \tilde{X}_2(l)} = \frac{1 - \alpha\left(\cos\frac{\sqrt{K}}{\alpha} + \frac{1-\alpha}{\sqrt{K}}\sin\frac{\sqrt{K}}{\alpha}\right)}{1 - \frac{\alpha}{\sqrt{K}}\sin\frac{\sqrt{K}}{\alpha}}.$$

After transformations we arrive at the formula

$$\eta = \frac{1 + tg^2\frac{\sqrt{K}}{2\alpha} - \alpha\left(1 - tg^2\frac{\sqrt{K}}{2\alpha} + \frac{2(1-\alpha)}{\sqrt{K}}tg\frac{\sqrt{K}}{2\alpha}\right)}{1 + tg^2\frac{\sqrt{K}}{2\alpha} - \frac{2\alpha}{\sqrt{K}}tg\frac{\sqrt{K}}{2\alpha}}. \tag{18}$$

If K – the root of equation (15), then $tg(\sqrt{K}/2\alpha) = 0$, and if K – the root of equation (16), then $tg(\sqrt{K}/2\alpha) = \sqrt{K}/2\alpha$. Substituting the found tangent values into formula (18), we find that for stability coefficients with odd indices $\eta = 1 - \alpha$, and for stability coefficients with even indices $\eta = 1 + \alpha$.

For the curved forms of equilibrium we take the representation

$$y_j(x) = C_j Y_j\left(\frac{x}{l}\right) \; (j = 1, 2, 3, \ldots),$$

where C_j – constant size multiplier; $Y_j\left(\frac{x}{l}\right)$ – dimensionless function defining the law of the curved form of equilibrium.

Then, on the basis of formula (17) we obtain:

$$C_j = -\frac{M_j(0)l^2}{EI_0} \; (j = 1, 2, 3, \ldots);$$

$$Y_j\left(\frac{x}{l}\right)=\begin{cases}\dfrac{u}{((j+1)\alpha\pi)^2}\left(1-\cos(j+1)\alpha\pi\dfrac{x}{lu}\right)(j=1,3,...)\\ \dfrac{1}{K_j}\left(1-(1+\alpha)\dfrac{x}{l}-u\left(\cos\sqrt{K_j}\dfrac{x}{lu}-\dfrac{2\alpha}{\sqrt{K_j}}\sin\sqrt{K_j}\dfrac{x}{lu}\right)\right)(j=2,4,...).\end{cases}$$

If necessary, on the basis of representations (3) - (5) it is easy to write formulas for other parameters of the rod state corresponding to the considered case.

6. Conclusions

Based on the previously obtained representations for the state parameters of a rod with arbitrary continuous bending stiffness, the stability problem for a family of rods whose stiffness varies according to the law of the fourth degree is solved. The spectrum of critical forces is determined and formulas for curved forms of equilibrium are derived.

We propose to consider the stability coefficient as a function of the variable α, and having a set of values of this function corresponding to the values of the independent variable $0<\alpha\le 1$, approximate the function K by a polynomial. As a result, we obtained the expression $K_2=0.0001\alpha+80.7626\alpha^2+0.0002\alpha^3$. The degree of the polynomial was chosen from the condition that the coefficient of determination should not be less than 0.9999.

The following directions of activity seem to be promising: introduction of the results of this work into the practice of calculations for stability of real objects; application of general formulas for state parameters to the study of stability of rods with other laws of change in transverse stiffness encountered in practice.

References

[1] Clough R W and Penzien J 1975 *Dynamics of structures* McGraw-Hill 634
[2] Seiranian A P 2003 Zadacha Lahranzha o nayvyhodneishem ochertanyy kolonny *Uspekhy mekhanyky* **2** pp 45 – 96
[3] Kanno Y 2001 Necessary and sufficent conditions for global optimality of eigenvalue optimization problems *Structural and Multidiscriplinary Optimization* **22** pp 248-252
[4] Dynnyk A N 1955 *Prodolnyi yzghy. Kruchenye* Yzdatelstvo akademyy nauk SSSR 392
[5] Senytskyi Y E and Yshutyn A S 2015 Obshchaia ustoichyvost szhatыkh sostavnыkh sterzhnei peremennoho sechenyia *Vestnyk Samarskoho hosudarstvennoho tekhnycheskoho unyversyteta. Seryia Fyzyko-matematycheskye nauky* **19(2)** pp 341-357
[6] Lytvynov S V, Klymenko E S and Kulynych Y Y 2011 Raschet na ustoichyvost sterzhnei yz ЭDT–10 pry razlychnykh varyantakh zakreplenyia *Ynzhenernyi Vestnyk Dona elektronnyi zhurnal* **2** URL: http://www.ivdon.ru/magazine/latest/n2y2011/415
[7] Kulynych Y Y 2012 *Ustoichyvost prodolno-szhatыkh sterzhnei peremennoi zhestkosty pry polzuchesty*: dys. kand. tekhn. nauk. (Rostov-na-Donu) 161
[8] Prykhodko V E 2011 Yssledovanye prodolno-szhatыkh sterzhnei peremennoi zhestkosty Voprosy proektyrovanyia y proyzvodstva konstruktsyi letatelnыkh apparatov: sb. nauch. tr. *Nats. aerokosm. un-ta ym N E Zhukovskoho KhAY* **2(66)** pp 35 – 41
[9] Rzhanytsyn A R 1991 *Stroytelnaia mekhanyka* Vysshaia shkola 439
[10] Smyrnov A F, Aleksandrov A V, Lashchenykov B Ia and Shaposhnykov N N 1984 *Stroytelnaia mekhanyka Dynamyka y ustoichyvost sooruzhenyi* Stroiyzdat 416
[11] Bazant Z and Cedolin L 1991 *Stability of Structures* New York Oxford University Press
[12] Brush D O and Almroth B O 1975 *Buckling of Bars, Plates and Shells* New York Mc Graw-Hill
[13] Seide P 1956 Axisym metrical buckling of circular cones under axial compression *Jornal of Applied Mechanics* **23** pp 625-628
[14] Timoshenko S P and Gere J M 1961 *Theory of Elastic Stability* New York: McGraw-Hill

[15] Šapalas V, Samofalov M and Šaraškinas V 2005 Fem stability analysis of tapered beam columns *Journal of Civil Engineering and Management* **11(3)** pp 211–216 doi:10.1080/13923730.2005.9636352
[16] Ma J and Ren Y 2018 Free vibration and chatter stability of a rotating thin-walled composite bar *Advances in Mechanical Engineering* **10(9)** 168781401879826 doi:10.1177/1687814018798265
[17] Zhang C, Ren Y, Ji S and Zhang J 2020 Analysis of the Vibration and Chatter Stability of a Tapered Composite Boring Bar *Shock and Vibration* pp 1–11 doi:10.1155/2020/4190806
[18] Wang C M, Wang C Y and Reddy J N 2005 *Exact solutions for buck ling of structural members* Boca Raton, Florida, USA: CRC Press LLC (CRC series in computational mechanic san dap plied analysis) 286
[19] Krutyi Yu S 2010 Zadacha Eilera v sluchae nepreryvnoi poperechnoi zhestkosty *Stroytelnaia mekhanyka y raschet sooruzhenyi* **6** pp 22-29
[20] Krutyi Yu S 2011 Zadacha Eilera v sluchae neprerыvnoi poperechnoi zhestkosty (prodolzhenye) *Stroytelnaia mekhanyka y raschet sooruzhenyi* **2** pp 27-33
[21] Krutii Y, Kovrov A, Otrosh Y and Surianinov M 2020 Analysis of forced longitudinal vibrations of columns taking into account internal resistance in resonance zones *Materials Science Forum 1006 MSF* pp 79-86

Lightweight concrete for frame-monolithic construction

L M Ksonshkevych[1,*], I V Barabash[1], K O Streltsov[1] and O M Krantovska[1]

[1]Odesa State Academy of Civil Engineering and Architecture, 4, Didrihsona st., Odesa, 65029, Ukraine

[*]E-mail: ksonshkevych@ogasa.org.ua

Abstract. The proposed article is devoted to the study of the properties (strength, impact strength, sound absorption) of lightweight concrete used for frame-monolithic construction. Characterized by the well-known advantages (reduced load on the foundation, reduced consumption of reinforcement, as well as improved heat and sound isolation), lightweight concrete requires additional technological methods to achieve the necessary mechanical characteristics - compressive strength and impact strength. An alternative to the traditional preparation of concrete mixtures is proposed a technology using a high-speed mixer-activator in the presence of ground chamot, superplasticizer C-3 with the introduction of hydrophobic expanded clay gravel up to 75% of the amount of granite crushed stone. The use of mechanical activation leads to an increase in impact strength by 44% compared to the control, and makes it possible to obtain concretes, the sound isolation level of which provides the maximum efficiency of impact noise isolation and is equal to 45.7 - 42.8 dB in three frequencies, and the sound absorption level is 39. 8 - 49.7 dB.

1. Introduction

One of the promising directions in frame-monolithic construction is the use of lightweight concrete with an average density of 1850 ... 2100 kg / m^3, in which part of the dense aggregate is replaced by porous one, for example, expanded clay gravel. The use of lightweight concrete of classes C20/25...C25/30 is advisable in almost all building structures. By decreasing the average concrete density by 20-25%, lightweight concretes will reduce the load on the foundations, decrease the consumption of reinforcement, reduce transport costs, and also provide better sound isolation of intermediate floors. At the same time, a decrease in the average density of concrete causes a decrease in its strength. It is possible to accelerate the process of structure formation of concrete, as well as to increase its compressive strength, using the mechanic activation of the binder.

The amplification of the quality of concrete, more intensive growth of strength, can be achieved through the purposeful change of the structure of the cement stone, as by activating the cement grains [1-4] and fillers [5, 6] under conditions of intense hydrodynamic influence on them, and due to the modification of their surface area by surfactants [7, 8].

Works by Y. M. Bazhenov [9], R. F. Runova [10] etc. found that the use of active mineral additives provides a real opportunity to obtain concrete with preset physical and mechanical characteristics, with significant savings Portland cement.

2. The purpose of the research

The goal is to improve the physical and mechanical characteristics of concrete by mechanical activation of Portland cement and partial replacement of crushed granite with expanded clay gravel. The objective of the research is to find out the effect of partial replacement of crushed granite with expanded clay gravel and the introduction of ground chamot into mechanically activated Portland cement on the compressive strength, sound absorption and impact strength of concrete.

Content from this work may be used under the terms of the Creative Commons Attribution 3.0 licence. Any further distribution of this work must maintain attribution to the author(s) and the title of the work, journal citation and DOI.
Published under licence by IOP Publishing Ltd

3. Research method

The impact strength of concrete (T) was evaluated by testing cube specimens with an edge of 7.07 cm on a laboratory pile driver. The sound isolation of concrete was determined by testing specimens measuring 20.5×10×5 cm on a PI-6 measuring device and a GZ-112 low-frequency signal generator. by measuring the number of dB passing through the sample. To clarify the role of mechanical activation of Portland cement on the structural and mechanical characteristics of concrete, a high-speed mixer with an adjustable number of revolutions of the working body of the mixing chamber was used.

4. Research results

The study of the mechanical properties of concrete was carried out by testing two similar series of samples: the first - using mechanical activation of cement in a high-speed mixer and the other, control - using traditional technology. To plasticize the concrete mix, the C-3 thinner was used. The hydrophobization of expanded clay granules was carried out by dipping them into an organosilicon liquid GKZH - 94 and subsequent drying at t = 90 ÷ 100 ° C. Ground chamot was used as a mineral additive to Portland cement. Quartz sand with M_{cr} = 2.5 was used as a fine aggregate. As a coarse aggregate in the experiments, we used:

a) crushed granite of 5-20 mm fractions (the bulk density of crushed stone was 1375 kg / m^3);
b) expanded clay gravel of fraction 5-20 mm (bulk density of gravel - 540 kg / m^3).

The activation time of the cement-containing slurry in the experiment was 60 sec.

The experimental design and the levels of variation of the recipe factors are shown in table 1:

Table 1. Experiment design and levels of variation.

№	Values of natural variables					Reaction	
	mixed			technological		Control	Mechanoactivation
	S, m^2/kg	S, m^2/kg	S, m^2/kg	Clay gravel, %	Chamot, %	T, J	T, J
1	250	-	-	25	10	42.37	52.4
2	-	350	-	25	10	43.15	53.8
3	-	-	450	25	10	40.95	51.1
4	250	350	-	25	20	44.76	55.05
5	-	350	-	25	30	46.25	57.3
6	250	-	450	25	30	45.8	55.2
7	250	-	-	50	30	53.35	60.5
8	-	-	450	50	30	51.2	60.1
9	250	-	450	50	20	51.1	59.8
10	250	350	450	50	10	50.75	59.2
11	250	-	-	75	10	59.6	66.4
12	-	350	-	75	10	60.25	67.5
13	-	-	450	75	10	59.15	66.2
14	-	350	450	75	20	61.4	68.4
15	250	350	-	75	10	62.1	68.95

As a result of research, it was found that the impact strength is influenced by the granulometric composition of the ground chamot. The maximum impact resistance with the same content of ground chamot and expanded clay gravel is observed at the maximum content of ground chamot with specific surfaces of 250 m^2 / kg and 350 m^2 / kg.

Mechanical activation of the binder sharply increases the impact strength of concrete at any content of expanded clay gravel and ground chamot. Thus, in concrete with 50% expanded clay gravel, activation of a binder containing 20% of ground chamot makes it possible to increase the impact strength of concrete from 50.8 to 59.4 J. At the same time, the tendency for an increase in the impact

IOP Conf. Series: Materials Science and Engineering 1164 (2021) 012042 doi:10.1088/1757-899X/1164/1/012042

strength of concrete with an increase in the content of expanded clay gravel and ground chamot remains, Figure 1.

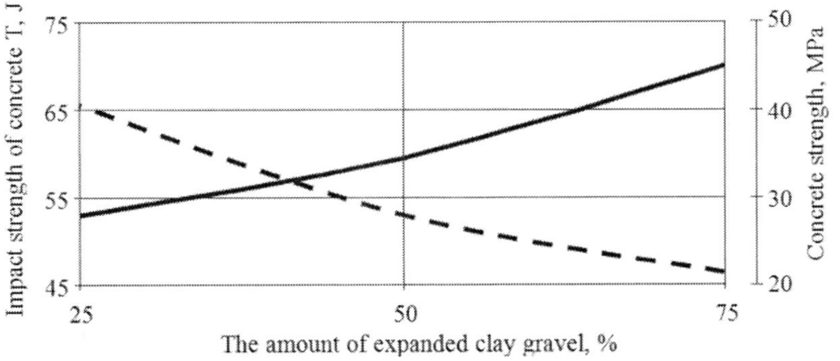

Figure 1. Influence of the amount of expanded clay gravel and ground chamot on the impact strength and compressive strength of concrete
——— impact strength; – – – – compressive strength.

In Figure 2 shows the dependence of the change in impact strength depending on the strength of concrete in compression.

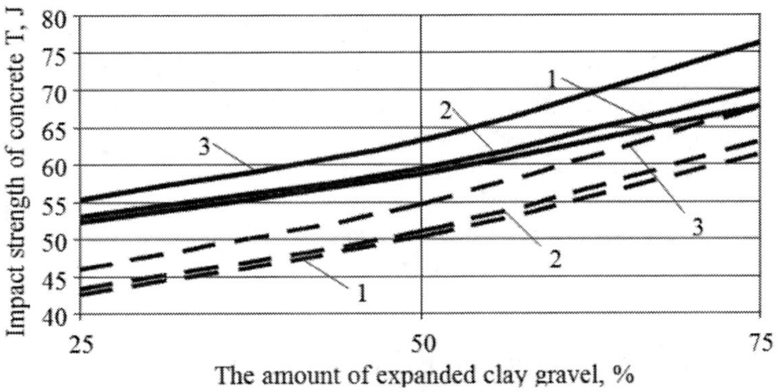

Figure 2. Influence of the amount of ground chamot and expanded clay gravel on the impact strength of concrete: 1, 2,3 - content of ground chamot, respectively 10%, 20%, 30%
——— mechanically activated astringent; – – – – traditional technology.

As mentioned above, with an increase in the amount of expanded clay gravel, there is an increase in the impact strength and a decrease in the strength of concrete in compression, and the intensity of the decrease in the strength in compression is higher than the intensity of the increase in the impact strength when the content in concrete is up to 50%.

The most important property of concrete, along with strength, is its sound isolation.

It is worth noting that with the introduction of 20% expanded clay gravel, the level of sound isolation sharply decreases by an average of 5 dB, another jump can be observed with the introduction of 60% expanded clay gravel. The sound isolation level is reduced by an average of 10 dB. The lower the level of sound isolation of concrete, the better it will soundproof the room from noises and sounds of a different nature (it absorbs more dB), Figure 3a.

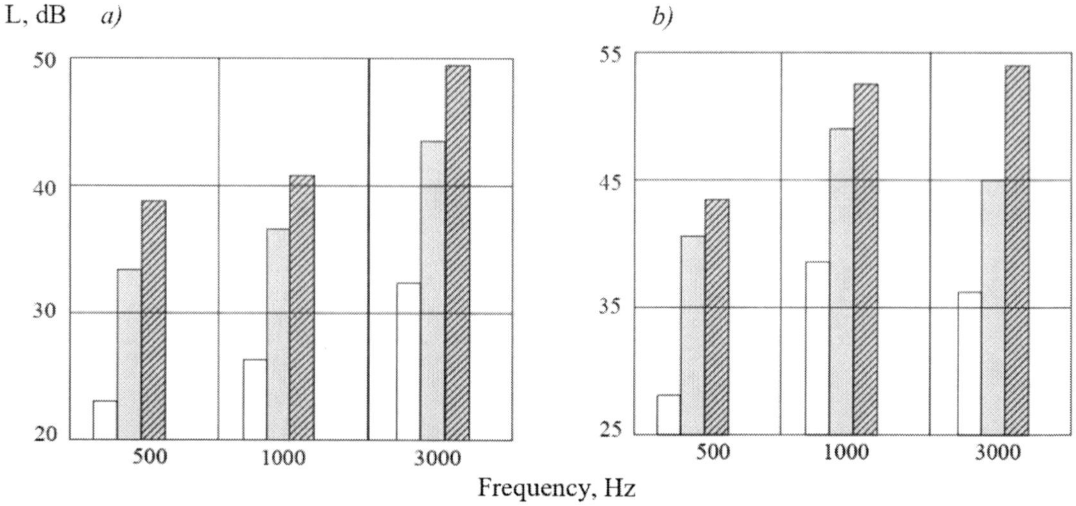

Figure 3. Influence of the content of expanded clay gravel in the aggregate mixture "granite crushed stone - expanded clay gravel" on the sound isolation of concrete:
a) mechanically activated binder; b) on mechanically activated binder
- 100 % expanded clay gravel;
– 60 % expanded clay gravel;
– 0 % expanded clay gravel.

It is known that for maximum impact sound isolation efficiency, the sound absorption of the indoor material is 26-39 dB. It was experimentally established that the sound absorption of concrete on an un mechanically activated binder varied within the following limits: frequency 500 Hz - from 23.2 to 38.2 dB, i.e. by 39%; frequency 1000 Hz - from 26 to 41.3 dB, i.e. by 37%; frequency 3000 Hz - from 32.1 to 49.5 dB, i.e. by 35%.

The given figure 3 indicates that with the introduction of a porous aggregate in the amount of 60% of expanded clay gravel, the sound absorption of the material increased by an average of 10 dB.

5. Conclusions

It has been established that the introduction of hydrophobic expanded clay gravel into the composition of heavy concrete makes it possible to obtain lightweight concrete with an average density from 2000 kg / m^3 to 1850 kg / m^3 and strength from 25 MPa to 30 MPa, respectively. Mechanical activation of the binder allows one to obtain concretes of a similar average density with compressive strength from 34 MPa to 45 MPa. Impact strength is 44% higher than that of heavy concrete.

The conducted studies allowed us to conclude that the introduction of hydrophobic expanded clay gravel (in the amount of 60% of the basic composition of concrete) instead of granite crushed stone (in the amount of 60% of the basic composition of concrete) makes it possible to obtain concretes, the level of sound isolation of which ensures the maximum efficiency of isolation of impact noise and is equal to 45.7- 42.8 dB in three frequencies. The sound absorption level is 39.8 - 49.7 dB.

References

[1] Barabash I V 2002 Mechanochemical Activation of Mineral Weavers *Training Manual* (Odessa: Astroprint) 100

[2] Fedorkin S I 1997 *Mechanoactivation of secondary raw materials in the production of building materials* (Simferopol: Tavria) 180

[3] Ksonshkevych L M 2013 *High-strength concrete on the mechanically activated binder*: diss. Cand tech Sciences: 05.23.05. (Odessa) 145

[4] Ksonshkevych LM, Krantovska O M, Petrov M M, Synii S V and Uhl A 2018 Investigation of

the structure of cement stone, obtaining and optimization of high-strength concrete on mechanically activated binder *Reliability and Durability of Railway Transport Engineering Structures and Buildings: MATEC Web Conference*: https://doi.org/10.1051/matecconf/201823003010

[5] Andriichuk O, Babich V, Yasyuk I and Uzhehov S 2017 *MATEC Web of Conferences* **116** 02001 pp 1-9

[6] Ksonshkevych L M, Barabash I V, Krantovska O M, Synii S V and Sunak P 2019 Disperse reinforced concrete with polycarboxylate additive on a mechanically activated binder *Materials Science and Engineering: IOP Conference Serie*: https://iopscience.iop.org/article/10.1088/ 1757-899X/708/1/ 012092

[7] Ramakrishnan V, Coyle W and Pande S 1980 Indian Concrete J **54:1** pp 23-26

[8] *Constitutive modeling o / high strength/ high performance concrete* FIB Bulletin 42 Sprint-Digital-Druck 2008 (Stuttgart) 125

[9] Bazhenov Y M 2011 *Concrete technology* (Moscow: Publishing ACB) 528

[10] Runova R F, Rudenko I I, Troyan V V, Tovstonis V V, Shcherbyna S P and Pashina L D 2008 *Building materials, products and sanitary equipment* **29** pp 91-97

Basics of automation in the formation of matrices of the numerical-analytical method of the boundary elements in calculation of spatial frame structures

A M Kushnir[1,*], I M Kuchmenko[1] and O A Chub[1]

[1]Odessa State Academy of Civil Engineering and Architecture, 4, Didrikhsona, Odessa 65029, Ukraine

*E-mail: amkushnir84@gmail.com

Abstract. The basic principles of the methodology for automating the formation of matrices of the numerical-analytical method of boundary elements (BEM) for the calculation of spatial frame structures are presented.

1. Research relevance

Extensive use of spatial frame structures necessitates calculations of their strength and stability in order to make a reasonable choice of parameters on which the magnitudes of stresses and deformations depend. Existing calculation methods are associated with a complex mathematical apparatus, as well as with a number of difficulties and conventions for modeling the individual elements of the bar structures.

The practice of engineering design requires reliable and alternative methods that allow determining the values of calculation forces and displacements that reflect the real work of structures.

To formulate the basic principles of the automation of the formation of matrices of the NA MBE in the calculation of spatial elastic frame structures.

2. Main part

Let us review the principle of the automation of the formation of matrices of the NA MBE using the example of a spatial frame structure shown in Figure1.

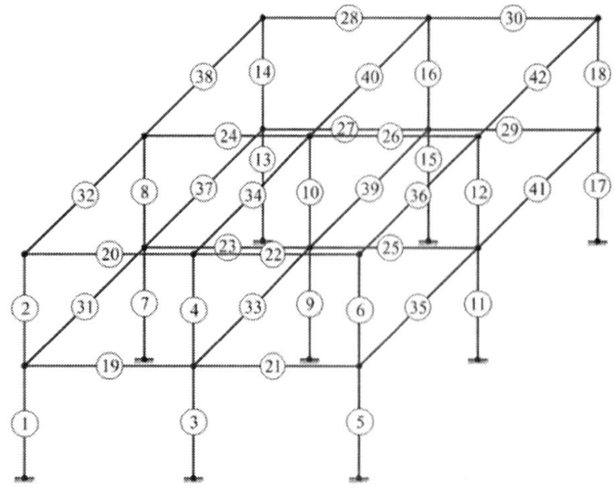

Figure 1. Numbering of elements of a spatial frame structure.

The number of elements in spatial frame structures can be determined by the formula:

Content from this work may be used under the terms of the Creative Commons Attribution 3.0 licence. Any further distribution of this work must maintain attribution to the author(s) and the title of the work, journal citation and DOI.
Published under licence by IOP Publishing Ltd

$$K = (N+1)(M+1)P + (M+1)NP + (N+1)MP, \quad (1)$$

where

K – number of elements;
N – number of steps;
M – number of stairs flights;
P – number of floors;
(N+1)(M+1)P – number of racks;
(M+1)NP – number of longitudinal crossbars;
(N+1)MP – number of transverse crossbars.

The number of elements of the spatial frame structure shown in Fig. 1, which has two floors, two flights and two steps, according to formula (1) was 42.

As a result of the research carried out for spatial frame structures, general patterns of the formation of the vector for unknown boundary parameters X* and the matrix of coefficients A* were determined.

The system of equations describing the stress-strain state of an individual element of the spatial frame structure shown in Fig. 1, in matrix form (2) looks as follows:

$$\boldsymbol{Y}(l_i) = \boldsymbol{A}(l_i)\boldsymbol{X}(0) + \boldsymbol{B}(l_i). \quad (2)$$

The vectors $\boldsymbol{Y}(l)$, $\boldsymbol{X}(0)$, $\boldsymbol{B}(l)$ included in it, before the transformations look as follows:

$$Y(l) = \begin{Vmatrix} Y_{01} \\ Y_{02} \\ Y_{03} \\ Y_{04} \\ Y_{05} \\ Y_{06} \\ Y_{07} \\ Y_{08} \end{Vmatrix} \quad X(0) = \begin{Vmatrix} X_{01} \\ X_{02} \\ X_{03} \\ X_{04} \\ X_{05} \\ X_{06} \\ X_{07} \\ X_{08} \end{Vmatrix} \quad B(l_i) = \begin{Vmatrix} B_{01} \\ B_{02} \\ B_{03} \\ B_{04} \\ B_{05} \\ B_{06} \\ B_{07} \\ B_{08} \end{Vmatrix}, \quad (3)$$

where

$$Y_{0i} = \begin{Vmatrix} EI_z v_y(l_i) \\ EI_z \varphi_z(l_i) \\ M_z(l_i) \\ Q_y(l_i) \\ EI_y v_z(l_i) \\ EI_y \varphi_y(l_i) \\ M_y(l_i) \\ Q_z(l_i) \\ EAu(l_i) \\ N(l_i) \\ GI_k \theta(l_i) \\ M_x(l_i) \end{Vmatrix} \quad X_{0i} = \begin{Vmatrix} EI_z v_y(0) \\ EI_z \varphi_z(0) \\ M_z(0) \\ Q_y(0) \\ EI_y v_z(0) \\ EI_y \varphi_y(0) \\ M_y(0) \\ Q_z(0) \\ EAu(0) \\ N(0) \\ GI_k \theta(0) \\ M_x(0) \end{Vmatrix} \quad B_{0i} = \begin{Vmatrix} B_1(l_i) \\ B_2(l_i) \\ B_3(l_i) \\ B_4(l_i) \\ B_5(l_i) \\ B_6(l_i) \\ B_7(l_i) \\ B_8(l_i) \\ B_9(l_i) \\ B_{10}(l_i) \\ B_{11}(l_i) \\ B_{12}(l_i) \end{Vmatrix}. \quad (4)$$

The system of equations describing the stress-strain state of an individual element of the spatial frame structure shown in Fig. 1, in matrix form (2) as a result of a chain of equivalent transformations looks as follows:

$$\boldsymbol{A}^* \boldsymbol{X}^* = -\boldsymbol{B}. \quad (5)$$

As a result of studying the topology of the vector of unknown boundary parameters of the resolving equation of the NA MBE, for spatial frame structures having $\boldsymbol{P}$ floors, $\boldsymbol{M}$ flights, $\boldsymbol{N}$ steps and consisting of $\boldsymbol{K}$ elements, all patterns of the location of the unknown parameters were identified.

As a result of studying the topology of the coefficient matrix A_0^* of the resolving equation of the NA MBE, for spatial frame structures having P floors, M flights, N steps and consisting of K elements, below the methodology for its formation is suggested.

Transformed matrix is proposed to be created by summing the matrix A_0^* with the matrix Z^*.

$$A^* = A_0^* + Z^*, \tag{6}$$

where A_0^* - a matrix of coefficients, consisting of compensating elements that appear in connection with the participation of dependent final parameters in the equations of equilibrium and compatibility of displacements;

Z^* - a matrix of coefficients, consisting of compensating elements that appear in connection with the participation of independent finite parameters in the equations of equilibrium and compatibility of displacements.

Decomposition of the coefficient matrix A^* into two components is adopted in connection with the possibility of the matrix blocks Z^* coinciding with the matrix blocks A_0^*, lying on the main diagonal.

The general order of the matrix for spatial frame structures can be determined by the formula:

$$P_m = \left[(N+1)(M+1)P + (M+1)NP + (N+1)MP\right] \times 12. \tag{7}$$

According to formula (7), the order of the matrix for a spatial frame structure with two floors, two flights and two steps was 504x504.

The coefficients matrix A_0^* for a spatial frame structure, formed as a result of a chain of equivalent transformations and automated in the MatLab computer mathematics system, is shown in Figure 2.

$A_0^* =$ 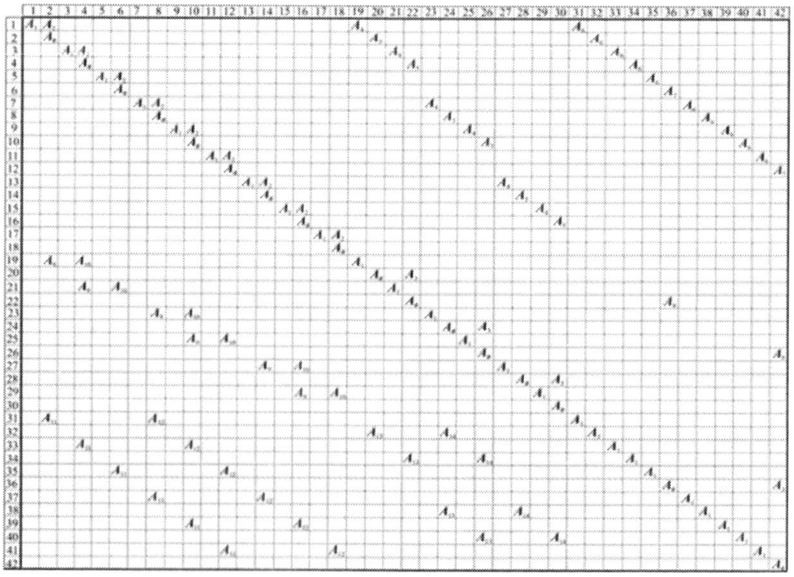

Figure 2. The coefficients matrix A_0^* for a spatial frame structure, formed in the computer mathematics system MatLab.

The coefficients matrix Z^* for a spatial frame structure, formed as a result of a chain of equivalent transformations and automated in the MatLab computer mathematics system, is shown in Figure 3.

$Z^* =$

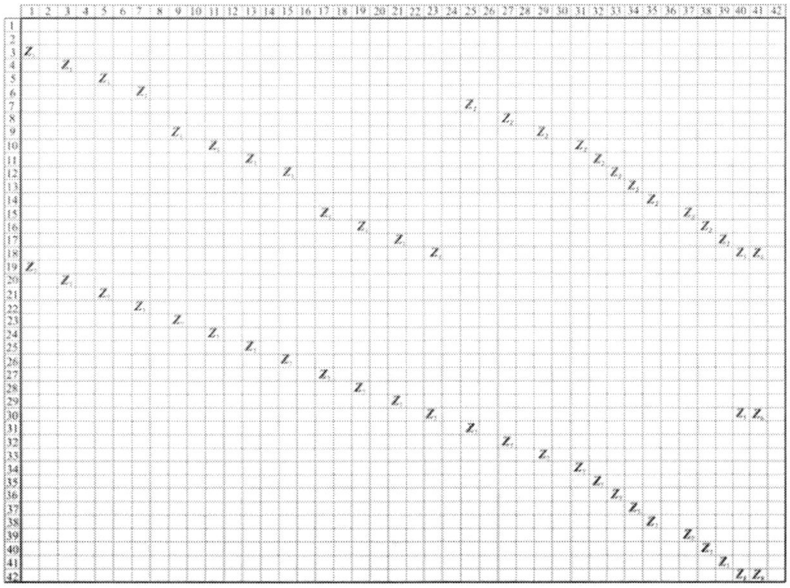

Figure 3. The coefficients matrix Z^* for a spatial frame structure, formed in the computer mathematics system MatLab computer mathematics system.

The structure of the A^* matrix formed in the MatLab computer mathematics system is shown in Figure 4.

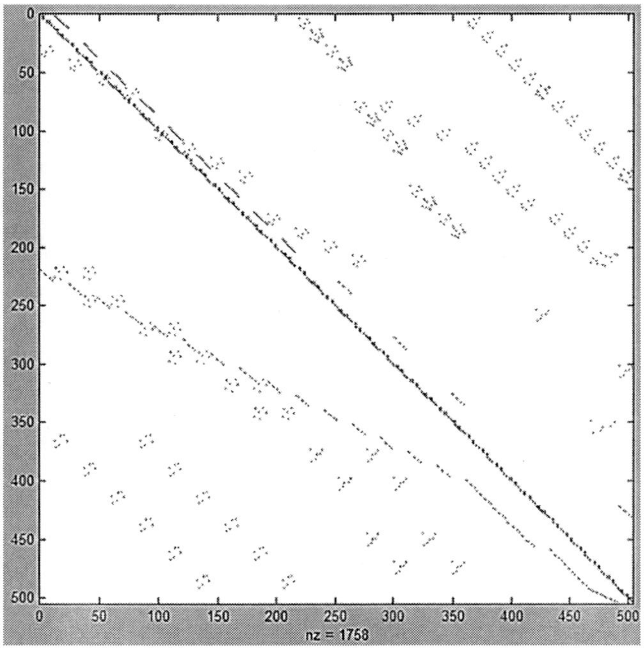

Figure 4. The structure of the A^* matrix formed in the system of computer mathematics MatLab.

IOP Conf. Series: Materials Science and Engineering 1164 (2021) 012043 doi:10.1088/1757-899X/1164/1/012043

3. Conclusions

As a result of the research and analysis of its results, the general regularities of the formation of the vector of unknown parameters X^* and the coefficients matrix A^* for the NA MBE were revealed when calculating spatial frame structures.

The order of formation of matrices included in the resolving equation of the NA MBE for the simplest spatial frame structure is formulated. A technique is proposed for automating the formation of matrices of the NA MBE.

An algorithm has been compiled and a program has been implemented that allows calculating spatial frame structures in the elastic stage of work, in the MatLab computer mathematics system.

References

[1] Dashchenko A F , Kolomiets L V, Orobei V F and Suryaninov N G 2010 *Numerical-analytical method of boundary elements* **1 2** p 451

[2] Kovrov A V, Kushnir A M and Kovtunenko A V 2010 On the application of the numerical-analytical method of boundary elements for the calculation of spatial frame structures *Odessa: OSAoEaA/Bulletin of the Odessa State Academy of Education and Architecture* **40** pp 117-123

[3] Orobei V F and Kovrov AV 2004 Solving problems of statics, dynamics and stability of bar structure. *Application of the method of boundary elements* p 122

IOP Conf. Series: Materials Science and Engineering 1164 (2021) 012044 doi:10.1088/1757-899X/1164/1/012044

Influence of modifying additives on properties of the powder coatings

V Gots[1], O Lastivka[1,*], O Berdnyk and O Tomin[1]

[1]Department of technology of building constructions and wares, Kyiv National University of Construction and Architecture, Povitroflotsky ave. 31, Kyiv, Ukraine, 03680

*E-mail: oles.lastivka@gmail.com

Abstract. The study deals influence of modifying additives on properties of the powder coating. The authors Using rheological and degassing modifying additives with different nature of the main active substance. It was found that the additives based on the acrylate polymer adsorbed on the silicon dioxide in the form of Byk-3900P and on the polyoxyethylene derivative of the castor oil in the form of Luvotix R400 decreasing the surface tension in the coating film, which, in its turn, contributes to the increase in the wettability of base during the melting of the powder coating, decrease the 'orange peel' effect during the cross-linking, reduction in pinholes on the obtained surface and improvement of mechanical characteristics of the coating. At the same time, rheological additives based on the bentonite in the form of Luvogel 4B and on the hydrophilic silicon dioxide in the form of Cab-o-sil M5 cause the increase in the surface tension in the coating film, which, in its turn, leads to the worsening of flow, appearance and mechanical characteristics of the powder coating.

1. Introduction

Powder coating is one of the most perspective materials to create a technical and economic, environmentally friendly decorative and protective coating for a wide range of building products and constructions. In the first instance, powder coatings play an esthetic role in improving the appearance of the product while ensuring high mechanical and chemical stability of the coating [1, 2].

Rapid rates of development in production of powder coatings are, in comparison to liquid paint-and-lacquer materials, evidence of their importance, high effectiveness and prospectivity [3, 4]. The increase in popularity of powder coatings can be explained by their environmental adequacy and attractiveness from the perspective of the environmental protection as well as high effectiveness related to the possibility of obtaining high quality protective and decorative coatings during the one-layer application. At the same time, it is obvious that during modern powder coating-and-lacquer materials have not exhausted all possibilities of improving pharmaceutical compositions and expanding the assortment as well as modernizing their production technology the relatively short period of their development [5].

The composition of the thermosetting powder coating contains five key components: polymer resin, hardener, pigments, fillers and functional additives [6]. In general, the polymer resin and hardener play a key role in ensuring necessary mechanical characteristics and lifespan of the powder coating. In this case, the role of functional (modifying) additives is extremely important to obtain characteristics that are often fundamentally required to meet predetermined technical specifications of the products and needs of end users.

Additives play an important role in forming properties of powder coatings and coatings based on them as they have become their integral part for several main reasons: because of control of rheological properties, surface defects as well as light and temperature stabilization of coatings. In Richard's opinion [7], a 'perfect' additive for powder coatings and coatings based on them must meet the following requirements: it must be solid and in the form of the fine powder with the crystalline melting temperature Tg > 50 °C; it must not chemically react with the adhesive and hardener; it must be 100 % reactive and effective at the low level of addition.

Thus, the modern approach to the production of powder coatings provides for the improvement of

Content from this work may be used under the terms of the Creative Commons Attribution 3.0 licence. Any further distribution of this work must maintain attribution to the author(s) and the title of the work, journal citation and DOI.
Published under licence by IOP Publishing Ltd

their properties using complex modifiers that help in obtaining high physical and mechanical as well as performance properties of the powder coating-and-lacquer material.

The purpose of this paper is to study the influence of modifying additives on the formation of properties of the powder coating.

2. Research materials and methods.

Rationale for selection of raw materials for obtaining powder coatings and coatings based on them.

The carboxylated polyester resin Crylcoat 2618-3 by Alnex has been used as a film-forming component. Its characteristics are listed in the Table 1. It is also necessary to use a structure-forming hardener together with the carboxylated polyester resin. In this article we have used Primid XL-552 by EMS-Griltech. As a filler the silicate in the form of metakaolin from Hlukhivtsi has been used that allows obtaining the powder coating with high physical and mechanical as well as performance characteristics according to the analysis [8, 9].

Table 1. Description of the film-forming component.

Type of the resin	Indices					
	Appearanc e	Glance 200/600, %	Brookfield viscosity 200°C, mPa.s	Glass transition temperature, °C	Acidity (mg KOH/g)	Hardening temperature and time
Crylcoat 2618-3	Non-transparent granules	67	4000-5200	60	30-35	180C 10min

As modifying additives the additives of two various groups have been used: 1 – rheological additives; 2 – degassing additives.

The film formation and flow of thermosetting powder coatings are regulated by two key parameters, namely, surface tension and melt viscosity [10]. The surface tension is a key driver (during the set hardening regime) that causes the flow of the powder coating. The only one resistance to the flow behaviour is the viscosity index of the melted coating within the conditions of its hardness. So, the additives controlling rheological properties of the powder coating can be more precisely classified as 'surface tension modifiers'.

As surface tension modifiers of the first group the additives of the following various nature have been used: a) – acrylate polymer adsorbed on the silicon dioxide in the form of **Byk-3900P** by Byk Germany; b) – polyoxyethylene derivative of the castor oil in the form of **Luvotix R400** by Lehmann&Voss&Co Germany; c) – modified bentonite in the form of **Luvogel 4B** by Lehmann&Voss&Co Germany; d) – hydrophilic silicon dioxide in the form of **Cab-o-sil M5** by Cabot.

To avoid the microcracks in the powder coatings, the release of air caught by them is required during the flow in course of their hardening. The reason is that during the cross-linking the viscosity of the powder coating increases drastically, which leads to the decrease in the release of volatiles caught by the coating film and affects the formation of physical and mechanical as well as performance properties of the material. Therefore, to obtain the high quality of the coating film it is appropriate to use degassing additives in the composition of coatings.

As degassing modifiers the additives of the following various nature have been used: a) – modified hydroxyl ketone in the form of ***Benzoin*** by Tech-Power Huangshan ltd; b) – modified polyethylene in the form of ***Ceraflour 962*** by Byk Germany; c) – micronized Ethylene-Bis-Stearamide in the form of ***Ceratan MA7019*** by Munzing; d) – oxygen-containing high molecular silicon compound in the form of ***Powderadd 9423*** by Lubrizol.

Research methods. Properties of decorative and protective powder coatings using various groups of modifying additives have been studied in the following sequence:

1. The powder coating of colour RAL 9016 has been applied on steel plates Ст3 (150x60 mm in size) using the white pigment titanium dioxide K-2190 by Kronos in its composition. The powder coating has been applied by electrostatic method according to ISO 1514:2016 using the spray gun Start 50.

2. The powder coating has hardened on sample plates in the polymerization oven at 180°C within 10 min.

3. The flow has been studied in accordance with ISO 8130-11:2019 [11] that establishes the method for determining rheological properties of the melted coating made of the thermal hardening powder on the plane inclined at a given angle to the horizontal. The principle of method is as follows: The powder coating was compressed into the tablet at a given pressure and placed on the inclined heated surface exposed to melting. Then the baking is observed during a given period of time and the drainage rate over the inclined surface is measured.

4. The impact resistance of the coating to the reverse impact has been determined in accordance with DSTU ISO 6272-2:2015 [12]. The purpose of this control is to check powder coatings for fast deformation and to evaluate their resistance to cracking and / or peeling out of the surface being coatinged when they are exposed to deformation due to the dropped object of a given weight with the spherical hammer head.

5. The bend strength of the coating has been determined in accordance with DSTU ISO 1519:2015 [13]. The purpose of this test is to evaluate the resistance of powder coating to cracking and (or) peeling out of the substrate (lining) by bending studied plates of the coating about the cylinder rod using the testing device equipped with a set of ball joints, each of which is linked to the cylinder rod with the relevant diameter.

3. Research results.

The composition of the powder coating has been prepared by mixing together dosed raw materials in wheel-type mixers and homogenizing the mixture (at 80...120 °C) in thermostatic screw mixers (extruders), pressing the pasty mixture out the extruder, cooling it and breaking to pieces with special mills. As a control composition the powder coating with the following ratio of raw materials has been selected: Crylcoat 2618-3 – 60 %; Primid XL-552 – 3.2 %; metakaolin – 20 %; TiO_2 – 16.8 %. Modifying additives have been added over 100 % of the mass of components in the control composition of the powder coating. Compositions of studied powder coatings based on the control composition using modifying additives as well as research results are shown in the Table 2, Table 3.

Table 2. Influence of rheological additives on properties of the powder coating.

Composition №	Rheological additives				Visual assessment*	Flow, sm	Impact resistence sm/kg	Bend strength
	BYK 3900P	Luvotix R400	Luvogel 4B	CAB-O-SIL M5				
1	0.5	-	-	-	+	3.5	60	8
2	1.0	-	-	-	+++	4.0	80	8
3	-	0.5	-	-	-	3.2	50	8
4	-	1.0	-	-	+	3.3	60	8
5	-	-	0.5	-	-	2.7	15	10
6	-	-	1.0	-	-	2	5	12
7	-	-	-	0.5	-	2.9	20	10
8	-	-	-	1.0	-	2.2	10	12
9	1.0	0.5	-	-	+++	4.5	90	8
10	1.0	-	0.5	-	+++	3.0	30	8
11	1.0	-	-	0.5	++	3.1	40	8
12	1.0	1.0	-	-	+++	5.0	80	8
13	1.0	-	1.0	-	+	2.2	15	8
14	1.0	-	-	1.0	+	2.4	20	8

* Note: - -negative, + - satisfactory, ++ - good, +++ - excellent.

IOP Conf. Series: Materials Science and Engineering 1164 (2021) 012044 doi:10.1088/1757-899X/1164/1/012044

Table 3. Influence of degassing additives on properties of the powder coating.

Composition №	Degassing additives				Visual assessment*	Flow, sm	Impact resistence sm/kg	Bend strength
	Benzoin	Ceretan MA7019	Ceraflour 962	Powderadd 9423				
1	0.3	-	-	-	++	3.5	60	8
2	0.6	-	-	-	+	4.0	80	8
3	-	0.5	-	-	-	3.3	30	8
4	-	1	-	-	+	3.4	40	8
5	-	-	0.5	-	-	2.7	15	10
6	-	-	1	-	-	2	5	12
7	-	-	-	0.5	-	2.9	20	10
8	-	-	-	1	-	2.2	10	12
9	0.3	0.5	-	-	+++	4.5	80	8
10	0.3	-	0.5	-	+++	3.0	30	8
11	0.3	-	-	0.5	++	3.1	40	8

* Note: - -negative, + - satisfactory, ++ - good, +++ - excellent.

According to the results it has been established that **rheological additives** (Table 2) have a different influence on the formation of properties of the powder coating. For example, according to the visual evaluation criterion, the most effective thing is to use the additive based on the acrylate polymer adsorbed on the silicon dioxide in the form of ***Byk-3900P***. Adding such additive in the quantity of 1 % contributes to the increase in the flow index from 3.0 to 4 cm, obtaining the smooth powder coating and prevents the formation of noticeable visual defects (Figure 1). At the same time, the powder coating without the rheological additive is characterized by small pin holes and the creation of the 'orange peel' effect (Figure 1, b). This effect has been caused by problems resulting from local differences in the surface tension (wettability of base) of the melted powder coating film, which forces the melted film to move to the direction of the upper surface tension and leads to the formation of the 'orange peel' and pin holes. However, adding the mentioned additive has a positive influence on properties of the interfacial area, by that smoothening local differences in the surface tension of base and contributes to obtaining the high quality coating without any noticeable visual defects.

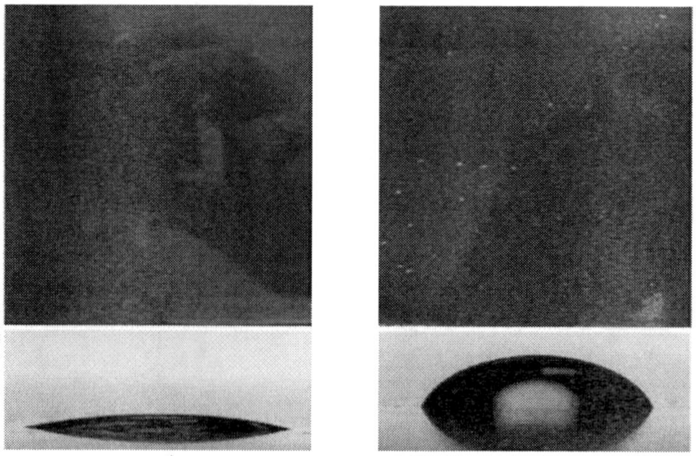

a) b)

Figure 1. Influence of rheological additive on appearance and wetting of powder coating: a) with additive Byk-3900P; b) without additive.

It should be also noted that adding the additive in the form of Byk-3900P causes the increase in the impact resistance of the coating to the reverse impact from 50 cm/kg (control composition) to 80 cm/kg (Figure 2) and bend strength (8 mm).

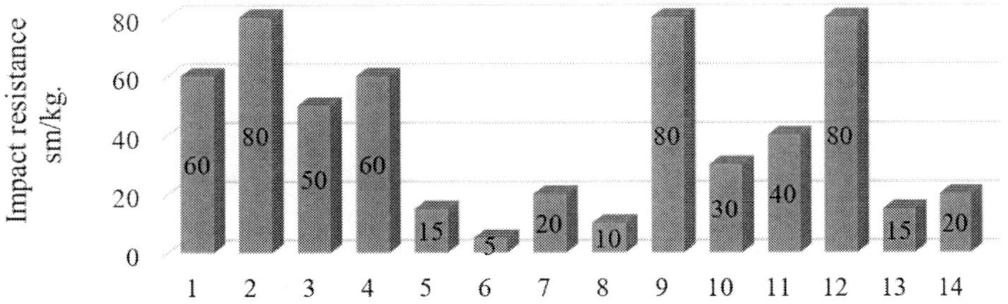

Figure 2. Impact resistance of the powder coating.

Using the additive based on the polyoxyethylene derivative of the castor oil in the form of ***Luvotix R400*** causes the increase in the flow index from 3.0 cm to 3.2…3.3 cm but it does not ensure the high smoothness of the coating (Figure 3) where the 'peel orange' is seen. Adding the additive also contributes to the increase in its impact resistance to the reverse impact from 50 cm/kg (control composition) to 60 cm/kg (Figure 2) and bend strength (8 mm).

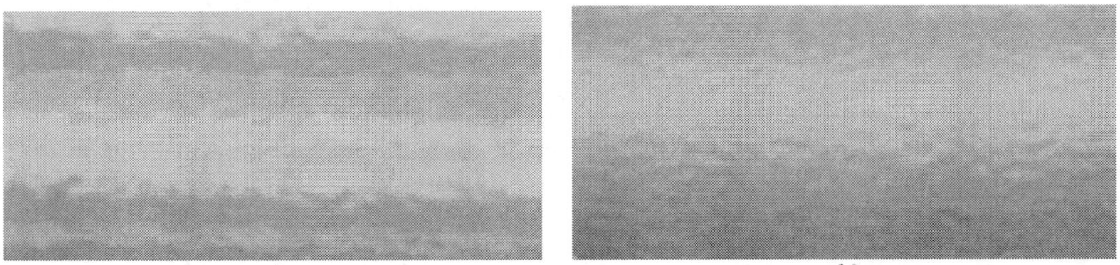

a) b)

Figure 3. Visual appearance of the powder coating using Luvotix R400,%: a) 1, 0; b) 0.5.

Using the additive based on the modified bentonite in the form of ***Luvogel 4B*** leads to the decrease in the flow index from 3.0 to 2.0 cm and worsening of smoothness of the coating with the formation of so called 'texture coating' (Figure 4). In this case, we can see the sharp decrease in the mechanical resistance: decrease in the impact resistance to the back impact from 50 to 5 cm/kg and flexural strength to 12 mm. The reason is that additives in the form of bentonite represent highly dispersible modified layered silicate with the high binding ability, adsorption and catalytic activity, which, accordingly, decreases the flow and wettability of powder systems and worsens the smoothness of the coating and their physical and mechanical characteristics.

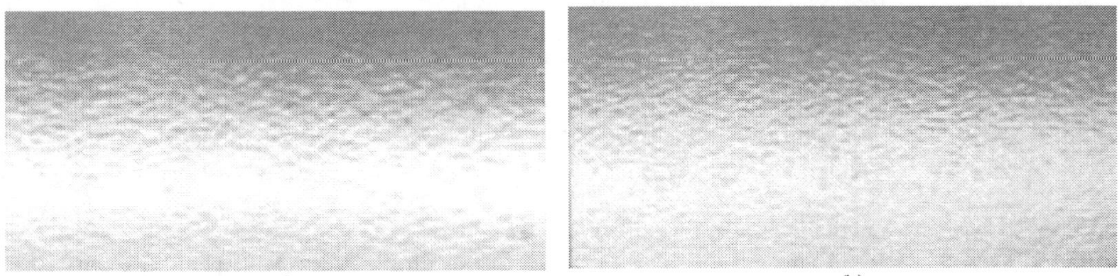

a) b)

Figure 4. Visual appearance of the powder coating using Luvogel 4B,%: a) 0.5; b) 1.0.

Adding the additive based on the hydrophilic silicon dioxide in the form of ***Cab-o-sil M5*** also leads to the decrease in the flow index from 3.0 to 2.2 cm and worsening of smoothness of the coating with the formation of so called 'texture coating' (Figure 5). In this case, we can see the decrease of the impact resistance to the back impact from 50 to 10 cm/kg and flexural strength to 12 mm.

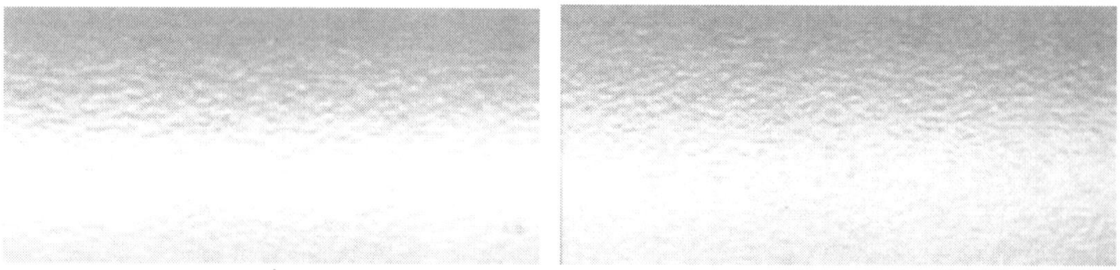

a) b)

Figure 5. Visual appearance of the powder coating using Cab-o-sil M5,%: a) 0.5; b) 1.0.

The most effective thing is to use a combination of rheological additives in the form of Byk-3900P and Luvotix R400 (compositions No. 9 and No. 12 according to the Table 2) that cause the considerable decrease in the surface tension, which, in its turn, contributes to the increase in the flow from 3.0 to 5.0 cm and improvement of quality of the powder coating in comparison to the control composition. The reason is that since there are no solvents in the powder coatings, their melts have a quite high surface tension, which can affect the quality of the polymer coating. Additives that decrease the surface tension of the powder coating improve not only the flow but also the appearance of the obtained coating while preventing the defect formation.

Thus, the additives based on the acrylate polymer adsorbed on the silicon dioxide in the form of Byk-3900P and on the polyoxyethylene derivative of the castor oil in the form of Luvotix R400 used as flow enhancers allow decreasing the surface tension in the coating film, which, in its turn, contributes to the increase in the wettability of base during the melting of the powder coating, decrease in the 'orange peel' effect during the cross-linking, reduction in pin holes on the obtained surface and improvement of mechanical properties.

Degassing additives also have a different influence on the formation of properties (Table 2) of the powder coating depending on the nature of the main active substance. For example, according to the visual evaluation criterion, the most effective thing is to use the additive based on the modified hydroxy ketone in the form of ***Benzoin.*** Adding such additive in the quantity of 0.3 % contributes to the increase in the flow index from 3.0 cm to 3.5 cм (Table 3), decrease in the formation of visual defects (pores, voids) and obtaining the defect-free, smooth powder coating (Figure 6 , b). This additive acts as a ′solid solvent′ to keep the coating film ′open′ for a quite long time so that air caught by the film can escape from it during the hardening. In this case, the impact resistance to the back impact increases from 50 cm/kg (control composition) to 60 cm/kg and flexural strength (8 mm). However, adding ***Benzoin*** in the quantity of 0.6 % is ineffective because this causes the considerable yellowing of the coating (Figure 6, c). The reason can be that during the hardening of the powder coating the modified hydroxy ketone can go into benzyl groups and their derivatives, which affects the change in the coating colour. Therefore, the desired content of ***Benzoin*** in powder coatings is 0.3 %.

Using the additives based on the modified polyethylene in the form of ***Ceraflour 962,*** micronized Ethylene-Bis-Stearamide in the form of ***Ceratan MA7019*** and oxygen-containing high molecular silicon compound in the form of ***Powderadd 9423*** do not cause the considerable decrease in the formation of visual defect (pores, voids) and obtaining the defects on the smooth powder coating (Figure 7). In this case, we can see the decrease in the impact resistance to the back impact from 50 to 5…30 cm/kg and bend strength to 10…12 mm (Table 3).

It should be also noted that the most effective way of obtaining the defect-free, smooth powder coating is using a combination of degassing additives in the form of **Benzoin** in the quantity of 0.3 % and **Ceretan MA 7019** in the quantity of 0.5 %. Such combination contributes to the acceleration of release of air caught by the film during the hardening of the coating, which ensures high degassing properties in the system where no pores and voids are formed and there are no pinholes defects. It does not cause the yellowing of the obtained coating. In this case, we can see the increase in the impact resistance to the back impact from 50 to 80 cm/kg and bend strength to 8 mm.

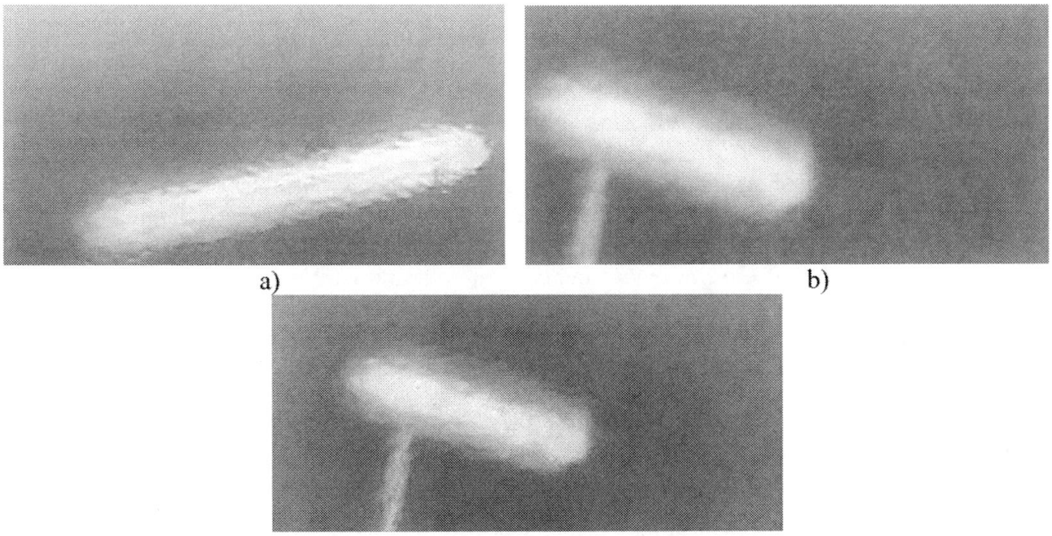

a) b) c)

Figure 6. Visual appearance of the powder coating using Benzoin:
a) control composition; b) Benzoin - 0.3 %; c) Benzoin - 0.6 %.

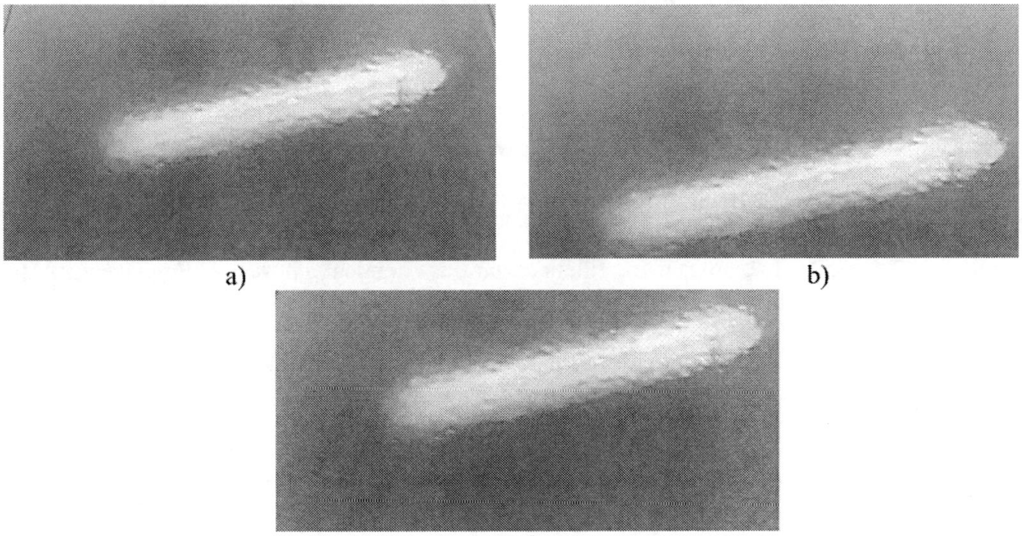

a) b) c)

Figure 7. Visual appearance of the powder coating using degassing additives:
a) **Ceraflour 962** – 0.5 %; b) **Ceratan MA7019** - 0.5 %; c) **Powderadd 9423** - 0.5 %.

4. Conclusions

Using rheological modifying additives has a different influence on properties of the powder coating depending on the nature of the main active substance. Additives based on the acrylate polymer adsorbed on the silicon dioxide in the form of Byk-3900P and on the polyoxyethylene derivative of the castor oil in the form of Luvotix R400 allow decreasing the surface tension in the coating film, which, in its turn, contributes to the increase in the wettability of base during the melting of the powder coating, decrease the 'orange peel' effect during the cross-linking, reduction in pinholes on the obtained surface and improvement of mechanical characteristics of the coating. At the same time, rheological additives based on the bentonite in the form of Luvogel 4B and on the hydrophilic silicon dioxide in the form of Cab-o-sil M5 cause the increase in the surface tension in the coating film, which, in its turn, leads to the worsening of flow, appearance and mechanical characteristics of the powder coating.

The most effective way of obtaining the defect-free powder coating is using a combination of degassing additives based on the modified hydroxy ketone in the form of **Benzoin** in the quantity of 0.3 % and micronized Ethylene-Bis-Stearamide in the form of **Ceretan MA 7019** in the quantity of 0.5 %. Such combination contributes to the acceleration of release of air caught by the film during the hardening of the coating, which ensures high degassing properties in the system where no pores and voids are formed and there are no pinholes defects. It does not cause the yellowing of the obtained coating. At the same time, adding the mentioned degassing additives as well as additives based on the modified polyethylene in the form of ***Ceraflour 962*** and oxygen-containing high molecular silicon compound in the form of ***Powderadd 9423*** separately does not cause the considerable decrease in the formation of noticeable visual defects (pores, voids) and obtaining the defect-free smooth powder coating.

References

[1] Richart D S 1990 Powder Coating – Past, Present and Future: A Review of the State of the Art. *Powder Coating* pp 16-24

[2] Gots V I, Lastivka O V, Tomin O O and Tymoshenko S A 2020 Fillers for modification of polyester powder coating. *Materials Science and Engineering. Innovative Technology in Architecture and Design* **6**

[3] Richart D S 1990 Powder Coating – Past, Present and Future: A Review of the State of the Art *Powder Coating* pp 16-24

[4] Liberto N 2012 Powder Coating: The Complete Finisher's Handbook 4th Edition. *Powder Coating Institute* 466

[5] Maty J 2004 90 Years with PCI: A Retrospective Paint & Coatings Industry Magazine 20:11 (November) pp 28-29

[6] Muller B and Poth U 2006 Coatings Compendia Coatings Formulation

[7] Richart D 1999 Industr. Paint & Powder (Powder R&D Section) Fall

[8] Gots V, Lastivka O, Tomin O and Kovalchuk O 2019 Influence of Film-Forming Components on the Corrosion Resistance of Powder Coating *Materials Science Forum* **968** pp 143-152

[9] Gots V I, Lastivka O V, Berdnyk O Yu, Tomin O O and Shilyuk P S Corrosion resistance of polyester powder coatings using fillers of various chemical nature *Key Engineering Materials* **864** pp 115-121

[10] Wagemakers' Lakfabrieken, NL Pat. 138009 1973

[11] EN ISO 8130-11-2019 *Coating powders Part 11: Inclined-plane flow test* 12

[12] DSTU ISO 6272-2:2015 *Paints and varnishes — Rapid-deformation (impact resistance) tests Part 2: Falling-weight test, large-area indenter* 12

[13] DSTU ISO 1519:2015 *Paints and varnishes — Bend test (cylindrical mandrel)* 9

IOP Conf. Series: Materials Science and Engineering 1164 (2021) 012045 doi:10.1088/1757-899X/1164/1/012045

Influence of boundary forces on the strength of a thin-walled cylindrical tank

L Dziuba[1], K Lishchynska[2,*], O Chmyr[1], V Tomenko[3]

[1]Lviv State University of Life Safety, 35 Kleparivska St., Lviv, Ukraine, 79007
[2]Hetman Petro Sahaidachnyi National Army Academy, 32 Heroes of Maidan St., Lviv, Ukraine, 79012
[3]Cherkasy Institute of Fire Safety named after the Heroes of Chernobyl of the National University of Civil Defense of Ukraine, 8 Onoprienko St., Cherkassy, Ukraine, 18034

[*] E-mail: k_lishch@meta.ua.

Abstract. The strength of a thin-walled cylindrical large-capacity tank was evaluated. The walls of the tank and the bottom are simultaneously exposed to hydrostatic pressure of the liquid and additionally gas pressure-loaded. The stress at the bottom of the tank was determined taking into account the boundary forces under the combined action of hydraulic and gas pressures. The dependences for determining the boundary forces were obtained by solving the equation of deformations compatibility of a thin-walled cylindrical shell and a round plate resting on a solid surface. The influence of gas pressure increase on the magnitude of stresses was investigated.

1. Introduction and literature review

Large-capacity cylindrical tanks are widely used in the industry for storage of liquids: petroleum, oil, chemicals, etc. Vertically placed steel cylindrical tanks are welded thin-walled structures with a flat round bottom. The walls and bottom of the liquid-filled tanks are exposed to hydrostatic pressure. External action of wind load on the tank wall can lead to buckling collapse if the tank is partially filled with liquid. In case of a shortcoming of valve devices or their failure, an additional gas pressure appears in tanks. The presence of microcracks and other material defects on the inner surfaces of the tanks contributes to the concentration of stresses and reduces the strength of the structure. Therefore, the action of various factors can lead to damaging the tanks filled with liquid. Leakage of liquid from the tank is potentially dangerous to the environment due to pollution, fire or explosion hazard [1]. This causes significant financial losses and the emergence of dangerous situations for humans and the environment.

2. Problem formulation

Today, vertical cylindrical tanks are the most convenient and rather cost-effective device for storing liquids: petroleum, oil, chemicals, etc. Vertical cylindrical tanks with low pressure are designed under the assumption that the excess pressure inside the tank is almost equal to atmospheric pressure ($p = 93.3\ kPa$) [1]. Therefore, the tank wall will only be exposed to the hydrostatic pressure of the liquid. However, in case of deficiency of valve devices or their failure the gas pressure occurs in the tanks, the value of which exceeds the value of atmospheric pressure. Thus, failing to take into account the action of additional pressure on the cylindrical tank walls adversely affects its operational reliability.

In [2] it is noted that 7% of emergencies that occur during the use of tanks are the result of structural defects, and 50% are caused by defects during installation and manufacture of tank structures. Vertical steel tank designs involve a weak weld, connecting the stationary roof to the tank walls. This weld protects the tank from the destruction of its walls when the internal pressure is increased due to the destruction of the roof. In addition, the work [2] emphasized that the world

Content from this work may be used under the terms of the Creative Commons Attribution 3.0 licence. Any further distribution of this work must maintain attribution to the author(s) and the title of the work, journal citation and DOI.
Published under licence by IOP Publishing Ltd

practice showed low efficiency of such protection against tank destruction. There were often cases when the roof of the tank did not come off, but the bottom of the tank teared off and the whole cylindrical structure rose into the air.

According to [3], uneven subsidence of the tank structure over the area and along the bottom perimeter cause additional deformations in the elements of the structure, especially in the lower junction of the wall with the edge of the bottom, which causes additional stresses. The combination of working stresses with stresses from uneven subsidence of the bottom can lead to the destruction of the wall-bottom joint.

Therefore, given the relevance of the issue of the tanks' strength in general and to avoid the destruction of the junction of the wall with it, this junction must be calculated taking into account the edge effect.

3. Analysis of publications

The strength of cylindrical, conical and spherical tanks and the influence of various factors on their strength were considered in many works, since today the problems of strength remain relevant. The study [4] covers a number of important issues related to the strength of tanks, taking into account the design features and plastic deformation of the material. The influence of weld joint defects on the strength of tanks was studied using the finite element method in the publication [5]. Stresses in cylindrical vessels under pressure are considered in the publication [6]. In this study, both the theoretical research of the main stresses in the wall of the cylindrical tank and the experimental study of these stresses were conducted. In general, the issue of the strength of tanks with different shapes, the design models of which are cylindrical, spherical and conical shells, is covered in the works [7-10]. The influence on the strength of the tank walls by the existing stress concentrators was investigated in study [11] using the finite elements method. According to [3], the most common tank accidents are brittle fractures. As an example, this paper shows the origin of a crack at the tank's wall-bottom joint, made of low-alloy steel of 09G2S-15 mark. A significant number of published works testify to the relevance of assessing the strength of tanks for various purposes, taking into account their stress.

4. Aim of Paper

The purpose of the work is to study the combined effect of hydraulic and gas pressure on the strength of a steel cylindrical thin-walled tank in the places of walls-bottom joint, taking into account the boundary. The object of the study is the change of stresses in the wall of a cylindrical steel tank, arising from the combined action of hydraulic and gas pressure, taking into account the boundary forces. The subject of the study is a large-capacity cylindrical steel tank, the bottom of which rests on a well-compacted sand cushion.

5. Materials and methods

Materials of structures of large-capacity cylindrical tanks are low-alloy steels for engineering structures of the C255, C275, C285 brands, in the form of sheet or broadband rolled metal [1]. C255 steel analogues in the European Union are S275JR, S275J0 and S275J2 steels. The mechanical characteristics of C255 steel with a sheet thickness of 10-20 *mm* are as follows: yield strength is 245 *MPa*, temporary resistance limit is 370 *MPa*.

The design model of a large-capacity tank is a thin-walled cylindrical shell with a medium radius R and wall thickness δ_1, the bottom of which is in the form of a round plate δ_2 thick resting on a compacted sand cushion (Figure 1). The tank is under the action of hydrostatic pressure q of the liquid with a density ρ that fills the tank to a height H and is additionally loaded with gas pressure p. In the case of such a load on the cylindrical shell, the dangerous cross-sections in which the greatest stresses occur are located near the joints between the wall and the bottom.

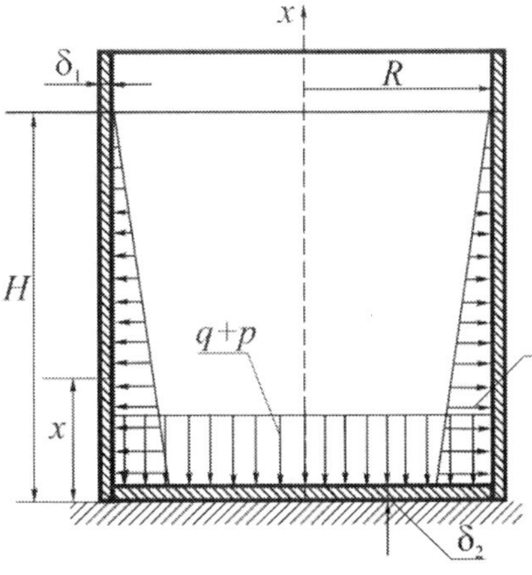

Figure 1. The design model of the cylindrical tank.

The internal forces (Figure 2) in the wall of the cylindrical tank near the wall-bottom joint in accordance with [12] are the circular force $T(x)$, meridian $M(x)$ and circular $K(x)$ bending moments:

$$T(x) = -2kRP_0e^{-kx}\cos kx + 2k^2RM_0e^{-kx}(\cos kx - \sin kx) + \left[\rho(H-x)+p\right]R,$$
$$M(x) = -\frac{1}{k}P_0e^{-kx}\sin kx + M_0e^{-kx}\cdot(\cos kx + \sin kx), \qquad (1)$$
$$K(x) = \mu M^{P_0}(x) + \mu M^{M_0}(x),$$

where k – attenuation coefficient, ρ – the density of the liquid the tank is filled with, P_0 – boundary force, M_0 – edge moment.

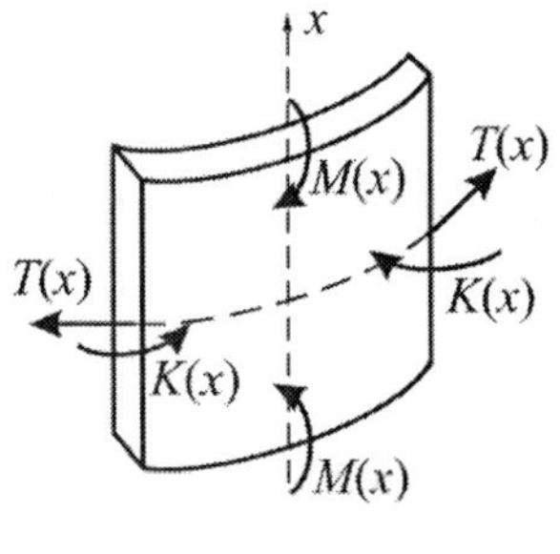

Figure 2. Internal forces in the wall of the cylindrical tank.

According to [12], the attenuation coefficient is calculated by the formula

$$k = \frac{\sqrt[4]{3\left(1-\mu^2\right)}}{\sqrt{R\delta_1}},$$

where μ – Poisson's ratio of the tank material.

The directions of the boundary forces in the wall of the cylindrical tank and its bottom are shown in Figure 3.

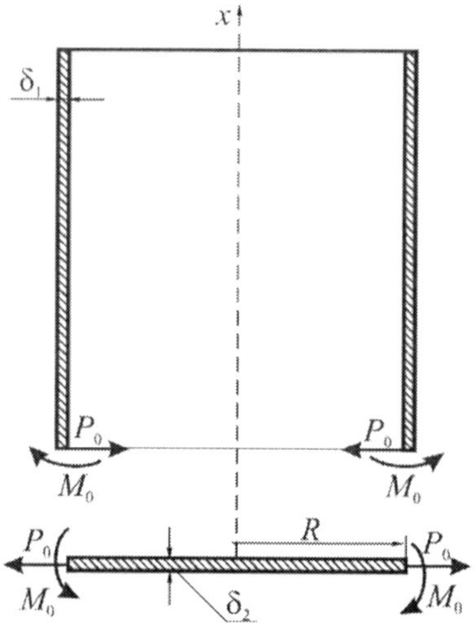

Figure 3. Boundary forces in the wall of the cylindrical tank and on the round bottom.

Research results To determine the boundary forces, let us write the equation of deformations compatibility of the tank wall with the bottom:

$$\begin{cases} \Delta_1 = \Delta_2; \\ \vartheta_1 = \vartheta_2, \end{cases} \tag{2}$$

where Δ_1, ϑ_1 – radial displacement and rotation angle of the lower edge of the wall of the cylindrical tank; Δ_2, ϑ_2 – radial displacement and rotation angle of the bottom.

Since we consider the walls of a cylindrical tank to be loaded with hydraulic q and gas pressure p (Figure 1), the expressions for radial displacements and rotation angles, taking into account the boundary force P_0 and the edge moment M_0 of the lower edge of the tank's wall and the bottom, are presented as follows

$$\Delta_1 = \Delta_1^q + \Delta_1^p + \Delta_1^{P_0} + \Delta_1^{M_0}; \quad \vartheta_1 = \vartheta_1^q + \vartheta_1^p + \vartheta_1^{P_0} + \vartheta_1^{M_0}; \tag{3}$$

$$\Delta_2 = \Delta_2^q + \Delta_2^p + \Delta_2^{P_0} + \Delta_2^{M_0}; \quad \vartheta_2 = \vartheta_2^q + \vartheta_2^p + \vartheta_2^{P_0} + \vartheta_2^{M_0}, \tag{4}$$

where Δ^q, Δ^p, Δ^{P_0}, Δ^{M_0} – radial displacements caused, respectively, by hydrostatic pressure q, gas pressure p, boundary force P_0 and edge moment M_0. Angles of rotation ϑ are denoted by similar indices.

The values of radial Δ and angular deformations ϑ included in expressions (3) for a cylindrical tank with a medium radius R, wall thickness δ_1 made of a material with modulus of elasticity E and Poisson's ratio μ, filled with a liquid with density ρ up to height H and additionally loaded with gas pressure p are as follows [12]:

$$\begin{aligned} &\Delta_1^p = -\frac{2-\mu}{2\delta_1 E} pR^2; \; \Delta_1^q = -\frac{\rho R^2}{\delta_1 E} H; \; \Delta_1^{P_0} = -\frac{2kR^2}{\delta_1 E} P_0; \; \Delta_1^{M_0} = -\frac{2k^2 R^2}{\delta_1 E} M_0; \\ &\vartheta_1^p = 0; \qquad \vartheta_1^q = -\frac{\rho R^2}{\delta_1 E}; \; \vartheta_1^{P_0} = \frac{2k^2 R^2}{\delta_1 E} P_0; \; \vartheta_1^{M_0} = -\frac{4k^3 R^2}{\delta_1 E} M_0. \end{aligned} \tag{5}$$

The bottom of the tank, which lies on a solid base, is a round plate with a thickness δ_2 and radius R (Figure 3). The ratio of the plate's radius to its thickness – R/δ_2 is a large value. According to [12], the radial displacements and rotation angles of the round plate are the following:

$$\Delta_2^q = \Delta_2^p = \Delta_2^{M_0} = 0;\ \ \vartheta_2^q = \vartheta_2^p = \vartheta_2^{P_0} = \vartheta_2^{M_0} = 0;\ \Delta_2^{P_0} = -\frac{1-\mu}{E}\cdot\frac{P_0 R}{\delta_2}. \tag{6}$$

In view of (3) – (6) in (2) we have:

$$\begin{cases} \Delta_1^q = \Delta_1^p = \Delta_1^{P_0} = \Delta_2^{M_0} = \Delta_2^{P_0}; \\ \vartheta_1^q = \vartheta_1^{P_0} = \vartheta_1^{M_0} = 0, \end{cases}$$

or

$$\begin{cases} -\dfrac{\rho R^2}{\delta_1 E}H - \dfrac{2-\mu}{2\delta_1 E}pR^2 + \dfrac{2kR^2}{\delta_1 E}P_0 - \dfrac{2k^2R^2}{\delta_1 E}M_0 = -\dfrac{1-\mu}{E\delta_2}P_0 R; \\ -\dfrac{\rho R^2}{\delta_1 E} + \dfrac{2k^2R^2}{\delta_1 E}P_0 - \dfrac{4k^3R^2}{\delta_1 E}M_0 = 0. \end{cases} \tag{7}$$

After the conversion, let us write the system of equations (7) as follows

$$\begin{cases} \left(2k + \dfrac{1-\mu}{R}\cdot\dfrac{\delta_1}{\delta_2}\right)P_0 - 2k^2 M_0 = \left(\rho H + \dfrac{2-\mu}{2}p\right); \\ 2k^2\cdot P_0 - 4k^3\cdot M_0 = \rho. \end{cases} \tag{8}$$

From the system of equations (8) let us determine the unknown boundary force P_0 and edge moment M_0:

$$M_0 = \frac{\rho\left(2k + \dfrac{1-\mu}{R}\cdot\dfrac{\delta_1}{\delta_2}\right) - 2k^2\left(\rho H + \dfrac{2-\mu}{2}p\right)}{4k^4 - \left(2k + \dfrac{1-\mu}{R}\cdot\dfrac{\delta_1}{\delta_2}\right)4k^3};\quad P_0 = \frac{\rho H + \dfrac{2-\mu}{2}p + 2k^2}{2k + \dfrac{1-\mu}{R}\cdot\dfrac{\delta_1}{\delta_2}}. \tag{9}$$

After taking into account expressions (9) for M_0 and P_0 in (1) we will obtain expressions for determining the internal forces: circular force $T(x)$, meridian $M(x)$ and circular $K(x)$ bending moments:

$$\begin{aligned} T(x) = &-2kR\frac{\rho H + \dfrac{2-\mu}{2}p + 2k^2}{2k + \dfrac{1-\mu}{R}\cdot\dfrac{\delta_1}{\delta_2}}e^{-kx}\cos kx + \\ &+2k^2R\frac{\rho\left(2k + \dfrac{1-\mu}{R}\cdot\dfrac{\delta_1}{\delta_2}\right) - 2k^2\left(\rho H + \dfrac{2-\mu}{2}p\right)}{4k^4 - \left(2k + \dfrac{1-\mu}{R}\cdot\dfrac{\delta_1}{\delta_2}\right)4k^3}e^{-kx}(\cos kx - \sin kx) + [\rho(H-x)+p]R; \end{aligned} \tag{10}$$

$$M(x) = -\frac{\rho H + \frac{2-\mu}{2}p + 2k^2}{k\left(2k + \frac{1-\mu}{R}\cdot\frac{\delta_1}{\delta_2}\right)} e^{-kx}\sin kx +$$

$$+\frac{\rho\left(2k + \frac{1-\mu}{R}\cdot\frac{\delta_1}{\delta_2}\right) - 2k^2\left(\rho H + \frac{2-\mu}{2}p\right)}{4k^4 - \left(2k + \frac{1-\mu}{R}\cdot\frac{\delta_1}{\delta_2}\right)4k^3} e^{-kx}(\cos kx + \sin kx);$$

$$K(x) = \mu \cdot M^{P_0}(x) + \mu \cdot M^{M_0}(x) = \mu \cdot M_0 .$$

Taking into account (10) for internal forces, we will calculate normal stresses σ_M and σ_T in the wall of the cylindrical tank, arising from the action of boundary forces, by the formula:

$$\sigma_M = \frac{6\cdot|M(X)|}{\delta_1^2}; \quad \sigma_T = \frac{T(x)}{\delta_1} + \frac{6\cdot|M(X)|}{\delta_1^2}. \tag{11}$$

According to [4, 12], the forces and stresses associated with the edge effect at a distance $x = \pi / k$ constitute only a few percent of their values at the edge of the shell, so it is not reasonable to take them into account at a further distance.

The influence of boundary forces and additional gas pressure on the magnitude of stresses will be considered based on the example of a steel $(\mu = 0.3)$ thin-walled tank with dimensions $R = 7.5\ m,\ H = 9\ m,\ \delta_1 = \delta_1 = 10^{-2}\ m$ filled with liquid (oil) with a density of $\rho = 760\ kg / m^3$. The curves of changes in stresses σ_M and σ_T obtained by formulas (11) taking into account (10) are shown in Figure 4. Curves 1 in Figure 4 show the change of stresses in the tank wall under additional exposure to gas pressure $p = 93.3\ kPa$, curves 2 – pressure $p = 186.6\ kPa$. Under the combined effect of gas and hydraulic pressure the stress σ_M acquires the largest value in the wall of the tank near the joint with the bottom: when $x = 0$, then $\sigma_M = 126\ MPa$ with gas pressure $p = 93.3\ kPa$, (Figure 4, a, curve 1). Double increase of the additional gas pressure (Figure 4, a, curve 2) also causes a double increase in stress σ_M from the boundary forces.

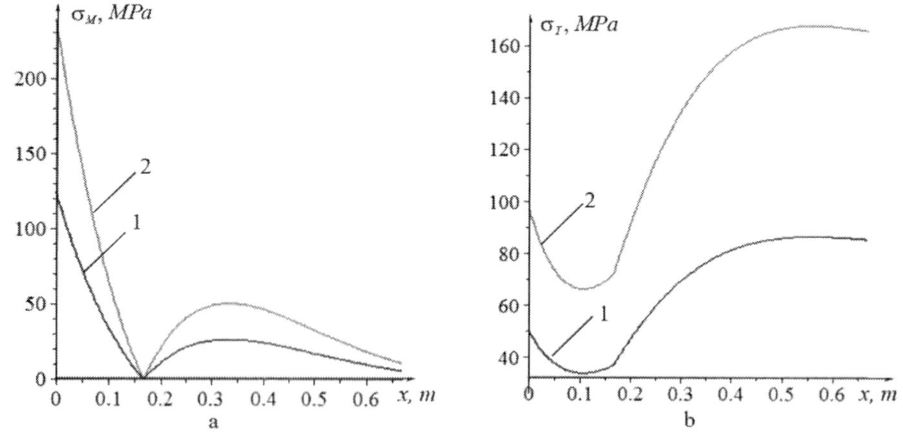

Figure 4. Graphs of changes in stresses from boundary forces in the wall of a cylindrical tank.

The stress σ_T in the tank wall near the joint with the bottom, when $x=0$, is equal to $\sigma_T = 50$ *MPa* at additional gas pressure $p = 93.3$ *kPa*, (curve 1). When the pressure doubles, they are increased almost twice as well. (Figure 4, b, curve 2). These stresses reach the greatest value at a distance of $x = 0.5$ *m* from the flat bottom of the tank. In a cylindrical tank with a spherical bottom [12], the largest circular stresses occur at a distance $x = 1.447\sqrt{R \cdot \delta_1}$. With the specified sizes of the tank in case of a spherical bottom the value x constitutes 0.396 *m*.

According to the momentless theory of shell strength, provided that the cylindrical tank is exposed only to hydrostatic pressure of the liquid, the values of the membrane stresses in the wall are much smaller. Membrane stresses are equal to: meridian – $\sigma_m = 38$ *MPa*, circular $-\sigma_\theta = 76$ *MPa*. The meridian membrane stress at a given tank load is 3.6 times less than the stress $\sigma_M = 126$ *MPa* calculated taking into account the boundary forces at the tank wall-bottom joint.

Conclusions

The strength of a thin-walled cylindrical tank was evaluated under the combined action of the hydrostatic pressure of the liquid and the additional gas pressure. To determine the stresses in the dangerous intersection near the flat bottom of the tank, the presence of boundary forces is taken into account. The dependences for determining the boundary forces were obtained by solving the equation of deformations compatibility of a thin-walled cylindrical shell and a round plate resting on a solid surface.

It was found that the meridian stresses in the tank wall at the junction with the round bottom plate are 3.6 times greater than the values of such stresses, determined by the momentless theory of shells. Therefore, taking into account additional gas pressure and boundary forces to calculate the strength of a cylindrical tank with a flat bottom increases the values of stresses and changes the correlations between them. Such factors of impact on the strength must be taken into account both during the design and during the use of the tanks.

References

[1] VBN B.2.2-58.2-94 1994 *Industry Construction Standards of Ukraine* Vertical Steel Tanks for Storage of Petroleum and Oil Products with Saturated Vapor Pressure Not Exceeding 93.3 kPa State Committee of Ukraine for Oil and Gas (Kyiv)

[2] Chernetskyi V V 2015 *Influence of Thermal Factors of Fire on Integrity of Vertical Steel Tanks with Oil Products* Thesis of Cand. Sc. (Engineering)

[3] Kondrashova O G and Nazarova M N 2004 Cause-and-Effect Analysis of Accidents in Vertical Steel Tanks *Oil and Gas Business* http: //www.ogbus.ru

[4] Annaratone D 2007 *Pressure Vessel Design* 1st ed (Berlin: Springer)

[5] Pircher M and Bridge R Q 2001 Buckling of Thin-Walled Silos and Tanks under Axial Load Some New Aspects *Journal of Structural Engineering* **127 (10)** pp 1129-1136

[6] Ibrahim A, Ryu Y and Saidpour M 2015 Stress Analysis of Thin-Walled Pressure Vessels Modern Mechanical Engineering *SciRes* **5**

[7] Gere J 2004 *Mechanics of Materials* 6th ed (Belmont: Thomson Learning Inc)

[8] Moss D and Basic M 2012 *Pressure Vessel Design Manual* Elsevier (Oxford: Butterworth-Heinemann)

[9] Razzaghi M S and Karimi F 2012 Buckling of Stiffened Thin Walled Cylindrical Tanks with Shell Imperfections due to Global Shear and Seismic Loads *Proceedings of the 15th World Conference on Earthquake Engineering, Curran Associates Inc* pp 25868-25874

[10] Tekleab B 2009 *Buckling of Thin-Walled Cylindrical & Conical Metal Tanks: Analysis* Test Evaluation & Design TU Graz

[11] Paraschiva B(G), Stefanescu M-F and Ghencea A 2016 Study of Stresses and Stress Concentrations in Pressure Vessels *Journal of Business Economics and Information Technology* **3 (1)** pp 20-44

[12] Beliaev V M and Mironov V M 2016 *Design and Calculation of the Industry Equipment Elements*. P. 1. Thin-Walled Vessels and Chemical Engineering Machinery: textbook 3rd ed Publishing House of Tomsk Polytechnic University

IOP Conf. Series: Materials Science and Engineering 1164 (2021) 012046 doi:10.1088/1757-899X/1164/1/012046

Investigation of the Stability Conditions of an Automated Control System for a Centrifugal CVT of Automotive Special Equipment

V. Lytovchenko[1,*], M. Pidhornyy[1], M. Bondarenko[1], I. Fedyuk[2] and O. Shapovalov[3]

[1]Cherkasy State Technological University (ChSTU), 460, Shevchenko blvd., Cherkasy 18006, Ukraine
[2]National University of Civil Defense of Ukraine, 94, Chernyshevska str., Kharkiv 61023, Ukraine
[3]National Academy of National Guard of Ukraine, 3, Square of the Defenders of Ukraine, Kharkiv, 61023, Ukraine

[*]E-mail: akronimail@gmail.com

Abstract. Modern continuously variable mechanical transmissions (CVT) with V-belt variators change the gear ratio due to tangential forces. In most cases, both the input and the initial links are geometrically similar wedge disks that move in a linear mode along their axis of rotation, while the process of controlling the transformation of the gear ratio is realized individually for each of the proposed solutions. The authors suggest employing direct action forces, the centrifugal force of the CVT working elements and the variator for transformation, instead of the traditionally used tangential forces. The use of the methods specified above allows reducing power loss by means of mechanical self-regulation of the continuously variable transmission.

1. Introduction and literature review

Over the past decades, automated hydraulic and hydrostatic, robotic gear shifting systems (GSS) and variators have been widely used in vehicles. Modern GSSs have temporary losses in power flow from the engine to the drive wheels. This disadvantage significantly affects the issue of vehicle transmission automation. Automated hydraulic GSSs have low efficiency, super-complex hydraulic systems, expensive workers and control bodies. Therefore, the demand for systems using hydraulics is decreasing and demand for mechanical systems is increasing.

Modern automatic V-belt variators of vehicles are structurally equipped with regulators. These are centrifugal regulators for changing the gear ratio [1, 2]. But the kinematic scheme of variators is based on an irrational transformation of centrifugal force. The use of continuously variable transmissions would be advisable for automatically operating systems. Ensuring the continuity of transmission control and the possibility of its application without stopping the vehicle makes it easy to automate control. That is why the use of self-regulated transmissions is currently one of the most promising areas in the development of new transmission types.

2. Literature review

Among the existing SV there are non-automatic, semi-automatic and automatic [4]. The disadvantage of analogs is that they are automatic or semi-automatic only in the manner of controlling the transmission and changing the gear ratio, but not in terms of adaptability to replaceable factors of influence on the transmission and not in terms of control stability, which makes it impossible for the existence of a feedback of engine transmission control. This contradicts the main purpose of the vehicle's automatic control system - to provide control of both the engine and transmission with one

Content from this work may be used under the terms of the Creative Commons Attribution 3.0 licence. Any further distribution of this work must maintain attribution to the author(s) and the title of the work, journal citation and DOI.
Published under licence by IOP Publishing Ltd

control element (pedal, lever, etc.) [5]. This condition allows the engine to do at a constant optimal mode, and the transmission to transmit the necessary power to the drive wheels, depending on the resistance of movement, which allows you to obtain high traction, speed and fuel and economic characteristics of the vehicle

In his work O. I. Dubinets considers the conditions of static equilibrium of the pulleys of the driven and driving links of V-belt variators with the use of regulators [6]. The centrifugal regulator combines two important control properties: to create an increased axial force and to adjust the axial force depending on the frequency of rotation of the pulley.

In the work of N. M. Filkin and S. A. Shvetsova, another image of the imposition of the axial force of the centrifugal regulator of the driven link at idle operation is shown [7].

An atypical solution of the CVT design without additional regulators considered in the work of I.P. Zgonnik [8]. Here is a variant of the technical solution of the V-belt variator, the geometric shape of the pulley of one or both links, this is a single-lane hyperboloid of rotation. The idea here is to basically rotate the active axle with a variable diameter wheel (main wheel) and use the rotation frequency value that is transformed in the active axle to rotate the centrifugal governor, which in turn changes the gear ratio. During this cycle, the levers of the centrifugal governor move outward or inward relative to the spring, depending on the speed of rotation.

3. Materials of research

The object of the study was a new CVT developed at the Department of Automobiles and Technologies for Their Operation, Cherkasy State Technological University, Ukraine. For its mechanical properties, it is similar to existing V-belt or chain drives. The essential difference lies in the way the gear ratio is changed. It consists in the use of centrifugal force, both of the working effort and of the manager. The control method refers to adaptive mechanisms. It can be used on small classes of cars and on motor vehicles. The simplified circuit and structure of the variator simplifies the assembly process and reduces the cost.

4. Methods of research

The synthesis of mechanisms is a crucial stage in the creation of a future variator. Synthesis is a complex problem with different solutions. Therefore, additional analysis is required to select the most appropriate option. The ambiguity of decisions during the synthesis occurs due to the fact that: firstly, at the stage of developing the technical specifications for the creation of a new mechanism (variator), as usual, it is impossible to correctly and unambiguously formulate the requirements that apply to it; secondly, the same conditions can be reproduced both by several mechanisms different in structure, and by one mechanism, which have different sizes of variator links.

Traditionally, the synthesis of mechanisms [2, 9] is carried out in the following two stages: the structure of the future mechanism is determined (structural synthesis). Designing such a structural diagram of the mechanism, on which the base, moving links, types of kinematic pairs and their mutual arrangement are indicated; for the given kinematic or dynamic properties of the mechanism, the sizes of its links are determined - parametric synthesis.

5. Study of the kinematic scheme and the principle of operation of the physical model

The authors have developed a CVT, which satisfies these conditions (Figure 1, a, b) [1, 2].

IOP Conf. Series: Materials Science and Engineering 1164 (2021) 012046 doi:10.1088/1757-899X/1164/1/012046

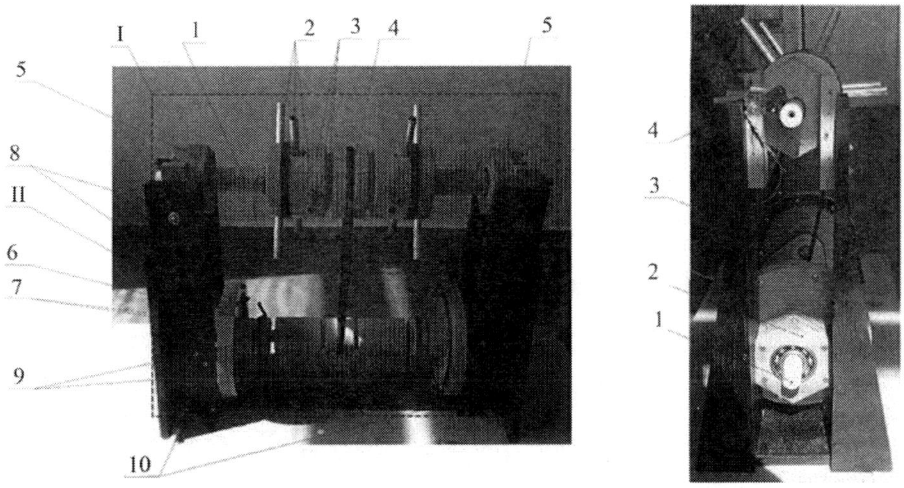

a. b.

Figure 1. Research facility as a physical model of a centrifugal variator: front view (a); left side view (b): I – multiplication link; II – reduction link; 1 – input shaft; 2 – guides; 3 – sliding sectors of the driving pulley; 4 – belt; 5 – resistance of the input shaft; 6 – shaft-gear; 7 – drove; 8 – cogwheel; 9 – exit shaft; 10 – racks.

At its core, the installation is a physical model of a centrifugal variator and is intended to demonstrate the operation of the device, and not for practical implementation. However, the current model allows experimental research. Structurally, the multiplication link is made up of the input shaft 1, guides 2, sectors of the drive pulley 3, belt 4 and supports to the shaft 5. The guides are rigidly connected from the feathers by the guilty shaft and wrap with it. The drive pulley sectors can move freely along the guides, but their extreme position is determined by the limiters. The wedge pass covers the sectors of the drive pulley, the tension on the belt also limits their movement by a certain measure. The transmission of torque and frequency of rotation occurs in the following order. The engine wraps shaft 1, the torque is transmitted to the multiplication link of the transmission. The input shaft 1 wraps around the guides 2, which are rigidly fixed on it. They, in turn, wrap the sectors of the driving pulley 3. The torque from the driving pulley is transmitted to pass 4, which connects the animation and reduction links.

The reduction link is made up of a driven pulley 5, a gear shaft 6, a carrier 7, a gear wheel 8, a secondary shaft 9, a belt and supports 10. The driven pulley is rigidly fixed to the shaft - gear. Shaft is a gear fixed to the carrier, which can turn around the axis of the secondary shaft. The gear wheel is in constant engagement with the pinion shaft and increases the torque that is transmitted from the multiplier link by a fixed amount.

A special class of automated control systems (ACS) is formed by systems to which the system proposed by the authors belongs and are capable of automatically adapting to changes in external conditions and properties of the controlled object, while ensuring the necessary quality of control by changing the structure and parameters of the controlled device [2]. They are called adaptive. As part of an adaptive (which are adapted by systems) ACS should not contain additional control devices. The device must perform the algorithm of functioning independently [9]. The algorithm for the functioning of an adaptive ACS, as usual, suggests maximizing the quality indicator, which characterizes either the properties of the control process in the ACS as a whole (speed, accuracy, and so on), or the properties of the processes that occur in the control object (productivity, achieving the highest efficiency, minimizing costs etc.). Therefore, adaptive ACS are also optimal.

The physical model described above became the basis of the prototype, on which tests were carried out in the mode of start from a place and straight motion at a distance of 200 g. The test brought the performance of a new continuously variable mechanical transmission based on a centrifugal variator (Fig. 1, a). According to the force distribution scheme (Fig. 1, b), we determine the gear ratio of the new V-belt variator.

At the beginning of the design of the scheme for the distribution of work and control efforts, it is necessary to describe the balance equation of the efforts of the input and output links [9]. Working forces include the centrifugal force of the sliding pulley, centrifugal force of the flex member (belt), belt tension, wedging force between the guide and the sliding pulley, load on the original shaft, and the weight of the driven pulley. The control forces include: the centrifugal force of the flexible element (belt), the load on the original shaft.

Figure 2. General view of the centrifugal link of the V-belt variator of the Honda Tact AF-24 scooter: 1 – engine; 2 – guide disks of the sliding pulley segments; 3 – segments of a sliding pulley; 4 – belt.

Let us indicate the equation that describes each effort:

Centrifugal force of one sliding pulley:

$$F_{CFS} = m \cdot \omega^2 \cdot R, \tag{1}$$

where m is the pulley weight, g; ω will be angular velocity of the input shaft in Hz; R is the radius of the sliding pulley's position in mm.

Centrifugal belt force:

$$F_{CFB} = m_1 \cdot \omega^2 \cdot R_1 \cdot \sin\beta, \tag{2}$$

where m_1 is the mass of the part of the belt limited by angle β, g; ω is angular velocity of the input shaft, Hz; R_1 is belt radius, mm; β is the belt's looping angle of the pulley, deg.

Centripetal force of one sliding pulley:

$$F_{CPF} = \frac{m \cdot v^2}{R}, \tag{3}$$

where m is the pulley weight, g; v is the sliding pulley's linear speed, mm/min; R is the position radius of the sliding pulley, mm.

Belt tension force:

$$F_{BTF} = 500 \cdot \frac{(2,5 - C_a) \cdot P_{DP} \cdot C_p}{C_a \cdot v} + m_a \cdot v_1^2, \quad (4)$$

where ma is the belt's length weight, g; C_a is the looping angle coefficient; C_p is the dynamic load and single-shift operating mode coefficient; P_{DP} is the rated drive power, N; v_1 is the belt's linear speed, mm/min.

The wedging force of one sliding centrifugal pulley link:

$$F_{WF} = n \cdot F_N \cdot \frac{1}{3} \cdot \kappa, \quad (5)$$

where n is summary points of contact between the guide rail and the sliding pulley; F_N is the normal force acting on the pulley, N; κ is the gearing coefficient.

Output shaft load. It is defined as the moment of resistance on the output shaft:

$$P = F_F \cdot R_{OS}, \quad (6)$$

where R_{OS} is the output pulley's radius, mm; F_F is the friction force between the shaft and the load element:

$$F_F = F_N \cdot f. \quad (7)$$

where F_N is the normal force acting on the pulley, N; f is the coefficient of friction (here accepted as f = 0,6). By substituting the values in formula 6, we will get the following equation:

$$P = F_N \cdot f \cdot R_{OS}. \quad (8)$$

Taking into account the direction of the force vectors (Fig. 1, b) and carrying out the transformation of the indicated equations (1-8), we obtain the equation of the centrifugal force of one sliding pulley:

$$F_{CFS} = \frac{\omega_{out} \cdot m \cdot \omega \cdot R_{OS}}{R}. \quad (9)$$

The total gear ratio of the variator is defined as the difference between the input power and the original power. Input power is the product of the engine shaft torque with the shaft rotation frequency. Power output is the product of the scooter's drive wheel torque times the rpm. Knowing that the torque is the product of torque to frequency, we get the equation to determine the overall gear ratio:

$$i_T = \frac{z \cdot m}{P}, \quad (10)$$

where z is the number of sliding segments of the variator centrifugal link, pcs; m - the mass of the sliding pulley, g; P - output shaft load, N.

IOP Conf. Series: Materials Science and Engineering 1164 (2021) 012046 doi:10.1088/1757-899X/1164/1/012046

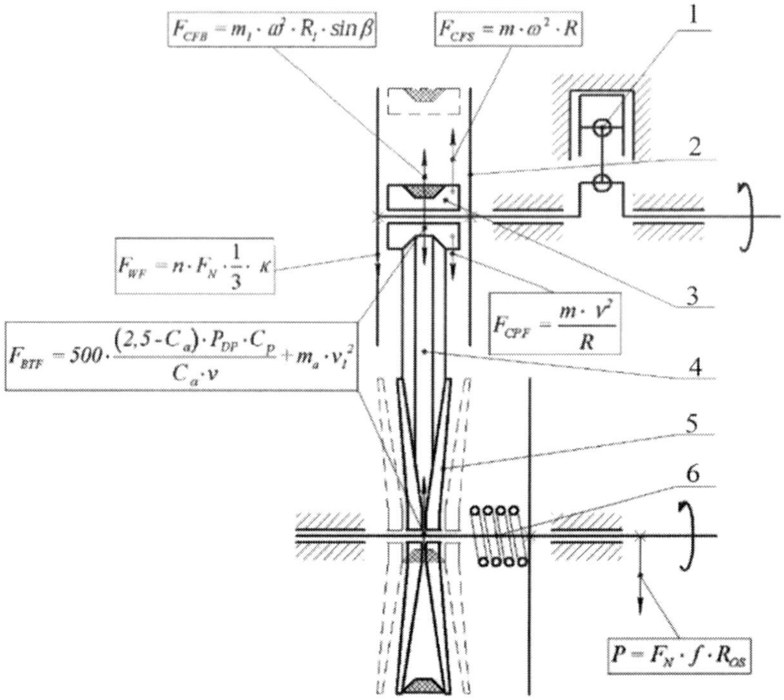

Figure. 3. Diagram of the distribution of the working and control forces of the variator with the centrifugal link of the Honda Tact AF-24 scooter: 1 – engine; 2 – guide disks of the sliding pulley segments; 3 – segments of a sliding pulley; 4 – pass; 5 – sliding disks of the output link of the variator; 6 – variator centrifugal clutch.

6. Study of the dynamics of mass loss of quartz dispersed materials

The authors carried out research on adaptive ACS. Checking the operation of the physical model of the CVT. The results of these studies are presented in thus check-list (stages of the experiment):

1) Idle start without changing the engine speed;

2) Idle start at minimum engine speed;

3) No-load start with a change in the engine speed;

4) Continuous work with load;

5) Constant load start;

6) Replaceable load start;

7) Stop mode of the mechanism to loads (braking of the driving wheel);

8) The mode of stopping the mechanism with the engine off without load.

Note that the adaptation can only be checked during continuous operation with load. All other modes simulate the operation of CVT on a vehicle and test the results of theoretical studies. For the load of the system, which is being investigated, we take three levels. The first level is no load, the second is at partial load, and the third is at the maximum load allowed by the drive engine.

Durability of a linear system with a characteristic equation:

$$a_3 \cdot p^3 + a_2 \cdot p^2 + a_1 \cdot p + a_0 = 0 \quad (11)$$

the necessary fulfillment of two conditions:

1) All coefficients of the characteristic equation must be positive;

2) The product of the average coefficients must be greater than the product of the extreme coefficients

$$a_1 \cdot a_2 > a_0 \cdot a_3 . \quad (12)$$

The coefficients of the equation include only the values of the parameters of the system, therefore, the resistance of the latter is determined only by the parameters and does not depend on its mill. The tables of the results of the experiment (table 1) have the following structure. In the left column there are written circulation frequencies, which are given to the input shaft and are recorded by a frequency meter. The top line shows the values of the load on the original shaft. In the place of the section of the corresponding row and column, the recorded measurement results for each of the three replicas. Based on the measurement results, we calculate the arithmetic mean value for each mill of the system and accept it as valid for further calculations. The above table is compiled for each objective function. Further, using the table in which the already recorded research results, we build a Microsoft Office Excel graph (Fig. 4) of the required dependencies.

For the dependence of the output shaft rad/s on the rad/s of the input shaft, it is necessary to carry out an approximation to the polynomial form and examine it for movement resistance using Vishnegradskiy's criteria.

Table 1. Measurement of the ratio of the frequency of revolutions of the output and input shafts.

Time [s]	The frequency of rotation of the crankshaft of the internal combustion engine [rad/s]	The frequency of rotation of the driving wheel [rad/s]	The frequency of rotation of the crankshaft of the internal combustion engine [rpm.] (at a load of 40 [N·m])	The frequency of rotation of the driving wheel [rpm.] (at a load of 40 [N·m])
1	2	3	4	5
1	5.429	0	531.783	54.4
2	47.67	1.1333	530.816	55.533
3	89.906	2.2667	529.463	56.667
4	132.13	3.4	527.818	57.8
5	174.337	4.5333	525.882	58.933
6	216.521	5.6667	523.657	60.067
7	258.677	6.8	521.145	63.267
8	300.798	7.9333	518.349	64.4
9	342.881	9.0667	515.274	65.533
10	384.918	10.2	511.926	66.667
11	426.904	11.333	508.313	67.8
12	468.835	12.467	504.442	68.933
13	514.402	13.6	500.324	70.067
14	583.452	14.733	494.572	71.2
15	653.327	15.867	491.59	72.333
16	681.569	17	493.266	73.467

1	2	3	4	5
17	686.546	18.133	494.932	74.6
18	682.627	19.267	496.59	75.733
19	675.545	20.4	498.238	76.867
20	666.989	21.533	499.877	78
21	658.357	23.5	503.028	79.133
22	650.42	25.05	506.415	80.267
23	641.807	26.6	509.834	81.4
24	634.095	28.15	513.291	82.533
25	626.24	29.7	516.787	83.667
26	619.663	31.25	520.327	84.8
27	612.223	32.8	523.914	85.933
28	606.549	34.35	527.55	87.067
29	599.717	35.9	531.238	88.2
30	593.206	37.45	534.98	89.333
31	588.971	39	538.776	90.467
32	583.598	40.55	546.535	91.6
33	577.02	42.1	550.498	92.733
34	570.598	43.65	554.514	93.867
35	566.558	45.2	558.582	95
36	563.219	46.75	562.698	96.133
37	559.134	48.3	566.859	97.267
38	554.304	49.85	571.06	98.4
39	548.737	51.4	575.295	98.592
40	542.457	52.95	579.558	98.717
41	536.297	53.267	601.044	99.042

Based on the measurement results, we plot the graphs of the dependences nout = f(nin) [rad/s]. These graphs are shown in Figure 4-7.

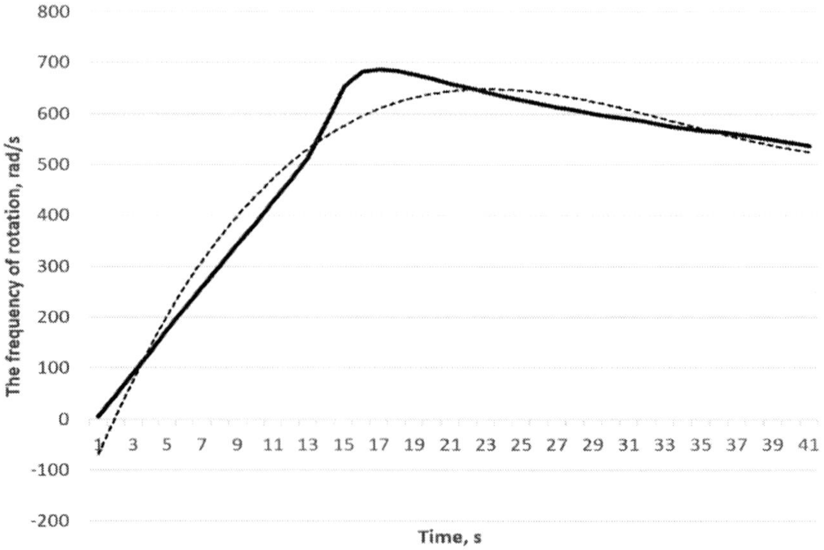

Figure. 4. Graph of changes in the frequency of rotation of the engine shaft. Approximation of dependence plots.

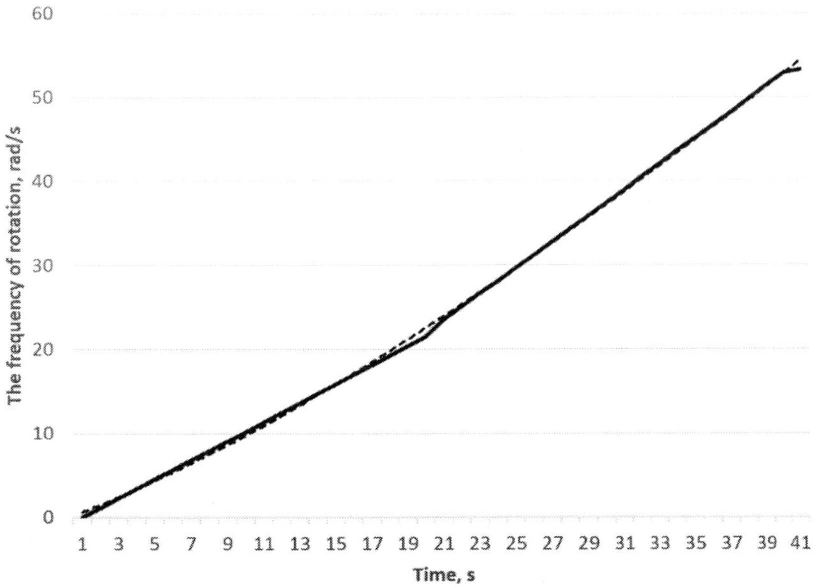

Figure. 5. Graph of changes in the frequency of rotation of the drive wheel. Approximation of dependence plots.

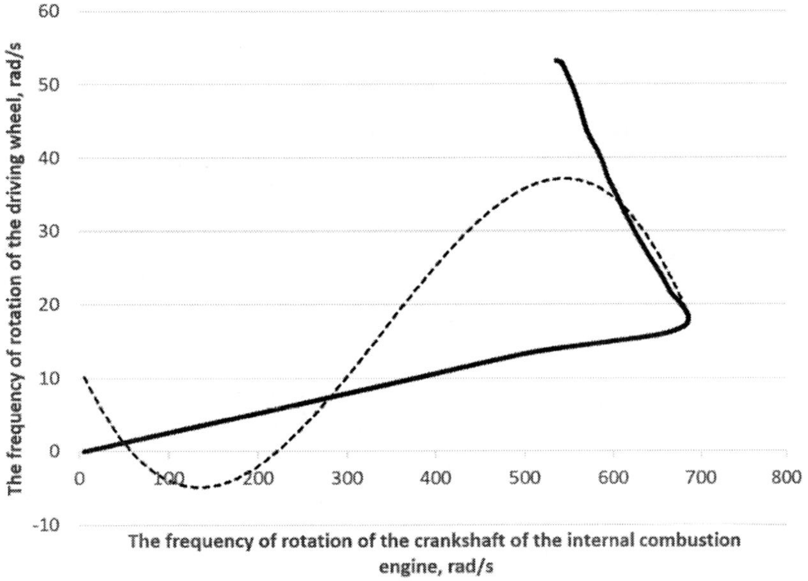

Figure. 6. Graph of the dependence of the frequency of rotation of the engine shaft to the frequency of rotation of the driving wheel. Approximation of dependence plots.

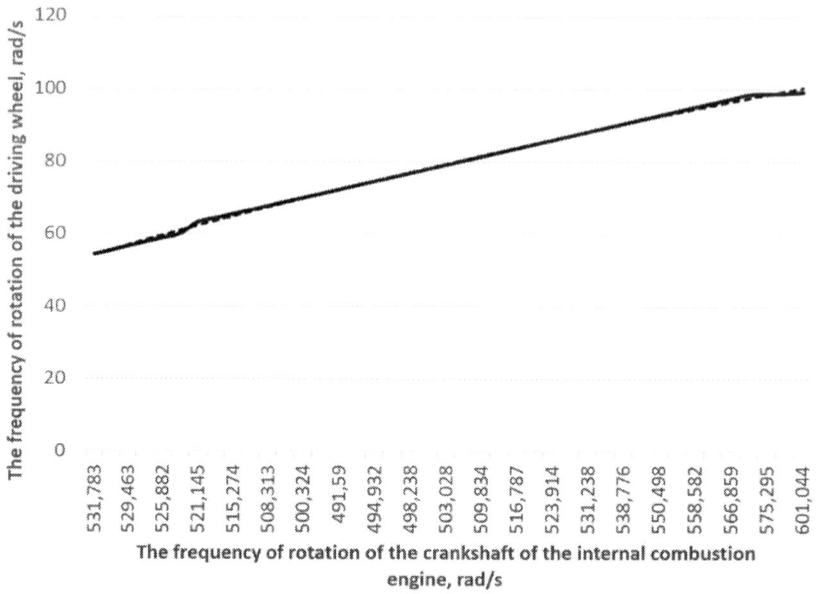

Figure. 7. Graph of the dependence of the frequency of rotation of the engine shaft to the frequency of rotation of the driving wheel (at a load of 40 [N.m]). Approximation of the graphs of dependence.

7. Study the durability of the variator

The obtained characteristic equations of the CVT control system must be checked for durability.

To study the stability of the system, we choose the criterion of Vishnegradsky's resistance as such, which allows us to conduct a reliable study and is quite easy to use.

Vishnegradskiy's resistance criteria are as follows:

1) All coefficients of the polynomial characteristic equation of the system must be positive. As can be seen from the equations shown in fig. 6-8, this condition is fulfilled;

2) The product of the extreme coefficients of the polynomial equation must be less than the product of its average coefficients:

$$\begin{aligned}
&y=0.0271x^3-2.7477x^2+83.574x-150.27;\\
&\quad 0.0271\cdot150.27>2.7477\cdot83.574;\\
&y=0.0313x^3+0.016x^2+14.652x+3.3057;\\
&\quad 0.0313\cdot3.3057>0.016\cdot14.652;\\
&y=0.0002x^3+0.019x^2+0.8214x+0.2004;\\
&\quad 0.0002\cdot0.2004>0.019\cdot0.8214;\\
&y=0.00006x^3+0.0056x^2+1.3801x+52.879;\\
&\quad 0.00006\cdot52.879>0.0056\cdot1.3801.
\end{aligned} \tag{13}$$

According to the graph of the change in the speed of the engine shaft, the first criterion of Vishnegradsky's resistance is not met (Fig. 4). This is due to the unstable dynamics of changes in the acceleration of the internal combustion engine. But on the graph of the change in the speed of the driving wheel and the graphs of the dependences of the speed of the engine shaft to the speed of the driving wheel, the criterion of resistance is fulfilled. As can be seen from equations 3, both conditions of the Vishnegradskiy resistance criterion are satisfied. Therefore, it is safe to say that this CVT control system is stable and adaptive.

8. Conclusions

The CVT proposed by the authors has proved to be stable and adaptive. The results of experimental studies confirm the operational stability of the control system for the centrifugal variator based on a CVT. Within this system, the working link performs the function of a control system. The operating parameters of the system are at the same time its control parameters. The investigation has determined the conditions for the resistance of the ACS according to Vishnegradskiy's criteria. According to the operating principle, the proposed system is stable due to the affinity of mechanical operation and control. Therefore, the system is able to perform its functions (dynamic transformation of the engine power to the driving wheels), even under a significant perturbation influence. Hence, it has the ability to adapt to external factors, namely, a change in the frequency of the engine shaft revolution and the load on the driving wheels.

References

[1] O. V. Spiridonov 2008 *Methods of increasing the efficiency of the operation of automatic V-belt variators in the transmission of motorized vehicles* Abstract dis. kand. tech. sciences - Specialty 05.02.02-05 - Kovrov 26

[2] V V Lytovchenko and M V Pidhornyy 2019 Investigation of step-by-step mechanical transmission on adaptation *Materials of the **VII-th** International Scientific and Practical Internet Conference Problems and Prospects for the Development of Automobile Transport* pp. 96-100

[3] Rvsmaster.ru 2021 Sequentronic and Speedshift robotic gearbox for Mercedes-Benz. [online] Available at: <https://rvsmaster.ru/articles-magazine/robotizirovannaja-korobka-peredach-sequentronic-i-speedshift-dlja-mercedes-benz/>. (in Russian).

[4] A I Leonov and A F Dubrovsky 1984 *Mechanical stepless non-friction transmissions of continuous action* (Moscow: Mechanical engineering) 192

[5] M Podrigalo, O Koriak and V Verbytskyi 2020 Evaluation of the stability of the automotive engine under conditions of constant angular speed of the crankshaft *Collection of scientific works of the National Academy of the National Guard of Ukraine* **1 35** pp 57-68

[6] O I Dubinets 2004 *Scientific bases of mechanics of V-belt variators as systems with control connections* Abstract dis. doctor tech. sciences 05.02.02 Mechanical Engineering Odessa 31

[7] N M Filkin and S A Shvetsov 2008 Calculation of the V-belt variator control system (regulators) *Successes of modern natural science* **4** pp 140-142

[8] P D Balakin and I P Zgonnik 2009 Synthesis of adaptive autovariator for mechanical drives of technological and transport machines *Dynamics of systems, mechanisms and machines, Omsk Scientific Bulletin* **1** pp 8-11

[9] V V Lytovchenko and M V Pidhornyy 2018 Structural synthesis of the centrifugal variator belt tension synchronizer *Proceedings of the international scientific-practical internet conference "Problems and prospects of road transport development* **6** pp 117-124

Scientifically grounded approach to the prescription of additives for obtaining complex-modified concretes

A I Adylkhodzhaev[1], I M Makhamataliev[1], V M Tsoy[1], V I Kondrashchenko[2]

[1]Department of Construction of Buildings and Structures of the Tashkent Transport University, 100067, Tashkent, Adilkhodjaev st., 1, Uzbekistan
[2]Department of Construction Materials and Technologies of Russian University of Transport, 127994, Moscow, Obrazcova st, 15, Russia

Annotation. The results of theoretical and experimental studies on the development of a scientifically grounded method of prescribing plasticizing chemical additives and mineral fillers in the selection of the compositions of complex-modified concretes (CMB) at the design stage are presented. The classification of plasticizing additives according to the degree of decrease in the surface tension of water upon their introduction and the activity of mineral additives according to the indicator of the reduced hydration activity, allowing to obtain highly economical MPC with the required properties.

1. Introduction

At present, concrete has finally established itself as the main building material in world construction practice. This was facilitated by such its main advantages as a wide range of properties, high manufacturability, availability of a raw material base, architectural attractiveness, versatility of application, environmental friendliness and profitability [1].

At the same time, over the past decades, a revolutionary nature of the development of concrete science has been observed, which manifests itself in the emergence of new types of concrete with unique properties, called new generation concretes (BNP).

The new generation concretes include high performance concrete (HPC), a distinctive feature of which is their achievement of maximum performance, strength, density and durability [2]. According to [3], HPC is an engineering material in which one or more of its specific properties are improved through the informed selection of components, design of the composition and maintenance of hardening concrete.

A striking example of new generation composites are modern concretes, realized at the Scientific Research Institute of Concrete and Reinforced Concrete (NIIZhB, Russia) [4]. These materials brilliantly realize the potential functional properties of the components of concrete mixes and concretes. The extremely high capabilities of concrete and reinforced concrete are realized by the authors in the technologies reactive powder concrete (RPC), as well as macrodefect free concrete (MDC) [4].

In France and the USA, the technology of ultra-functional concretes (UHPC) is widely used in construction practice, which are distinguished by high compressive strength (200 MPa) and bending strength (50 MPa) [5, 6].

Self-compacting concrete (SCC) also belongs to the category of BNP. In such concretes, the main technological problem associated with the minimization of material, energy and labor resources while achieving the specified properties of concrete is radically solved [6, 7].

In the domestic practice of concrete science, new generation concretes also include multicomponent high-quality concretes (MVB), developed by a team of authors under the leadership of professor A.I. Adylkhodzhaev [8]. These concretes contain TPP fly ash, JK-08 super plasticizing additive and are distinguished by high strength and performance properties.

As you know, the selection of compositions of traditional concrete without additives is not a big deal, the methods of their calculation are classic and are given in textbooks and teaching aids on

Content from this work may be used under the terms of the Creative Commons Attribution 3.0 licence. Any further distribution of this work must maintain attribution to the author(s) and the title of the work, journal citation and DOI.
Published under licence by IOP Publishing Ltd

building materials science [9]. A distinctive feature of modified concretes (MB) is the presence in their composition of a plasticizing chemical additive or a finely dispersed mineral filler; the methodology for the purpose of compositions with their use is also reflected in educational and scientific publications [10, 11,16-17].

As for the complex-modified concretes (CMB), it should be noted that despite the large number of scientific works devoted to this type of concretes, the issue of developing a method for selecting their composition with the appointment of a rational type of mineral fillers in combination with plasticizing chemical additives practically remains unexplored.

The authors of the article for the first time made an attempt to develop a scientifically grounded methodology for the selection of mineral fillers and plasticizing chemical additives for MPC.

2. Research results

According to the proposed method, plasticizing chemical additives are assessed according to the criterion representing the degree of decrease in the surface tension of water upon their introduction.

This approach is based on the ability of the plasticizing additive to reduce the surface tension of water at a certain temperature of the solution. In our opinion, the technological parameters of preparation, the amount of the added filler and the properties of the activated filled binder and concrete significantly depend on the chemical composition, structure and plasticizing ability of the added additive.

For experimental studies, the following types of local and most accessible plasticizing chemical additives and the corresponding dosage limits in percent of the mass of cement (on dry matter) were adopted: technical lignosulfonate (LST) - (0.10-0.25)%, sodium vat liquid -carboxymethyl cellulose (KNZH) - (0.3-0.9)%, waste water caprolactam (SVK) - (0.3-0.9)%, superplasticizer C-3 - (0.3-0.9)% [12, 18-19].

At the first stage of research, the effect of chemical additives on the degree of decrease in the surface tension of water was studied according to the method [11]. As a result of the experiments, isotherms of the surface tension of aqueous solutions of chemical additives were obtained depending on their dosage (Fig. 1), which are a family of falling curves with a pronounced zone of transition from vertical to inclined and horizontal sections.

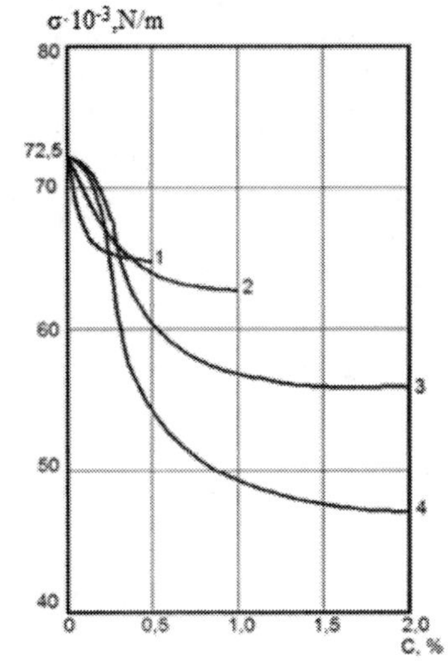

Figure 1. Isotherms (at t = 200C) of the surface tension σ of chemical additives depending on their concentration C in an aqueous solution
1, 2, 3, 4 - chemical additives LST, KZhN, SVK, S-3.

This kind of curves in Figure 1 is explained by the fact that, in accordance with the Gibbs equation [13-15], with an increase in the concentration of chemical additives, their content at the interface increases, leading to a decrease in the surface tension of aqueous solutions. Moreover, at low concentrations of chemical additives, the effect of reducing the surface tension of aqueous solutions is more significant than at higher dosages. The characteristic region on the isotherms, related to the transition from the vertical to the horizontal section of the decrease in the value of σ, indicates the maximum saturation of the adsorption layer with molecules of chemical additives, which, as was found, corresponds to the region of their rational concentrations.

Thus, the isotherms of changes in surface tension from the concentration of aqueous solutions of chemical additives can be used as the basis for their characteristics in terms of the degree of plasticizing ability and, consequently, according to their surface-active properties - the lower the surface tension isotherm is, the more pronounced the plasticizing ability of the additive. Moreover, the range of rational dosages of chemical additives can be characterized by the relative index of surface tension, which is the ratio of the surface tension at the maximum saturation of the adsorption layer with the molecules of the chemical additive to the surface tension of the solution without the additive. In particular, for the studied additives, the indicator is the following values: LST - 0.90-0.92; CNZh - 0.88-0.90; SVK - 0.78-0.80; C-3 - 0.67-0.70.

The analysis of the obtained values of the relative index of surface tension allows us to propose a classification of chemical additives according to the effect of plasticization, taking into account their surface-active properties, given in table 1.

Table 1. Classification of plasticizing chemical additives by surface-active properties.

Relative indicators of surface tension, rel. units	Characteristics of the additive by the effect of plasticization
$\bar{\sigma} > 0.95$	Weak plasticizer
$0.85 < \bar{\sigma} \leq 0.95$	Medium plasticizer
$0.75 < \bar{\sigma} \leq 0.85$	Strong plasticizer
$\bar{\sigma} \leq 0.75$	Superplasticizer

With regard to the studied chemical additives, the relative surface tension index allows us to arrange the additives studied by us in the following decreasing series according to the effect of plasticization: C-3> SVK> CNZH ≥ LST, which characterizes a decrease in their plasticizing ability.

The expediency of such a classification of plasticizing chemical additives lies in the fact that their effectiveness is assessed by a specific numerical value of the indicator - a relative indicator of surface tension, an analogue of which can also be used in the scientifically grounded choice of dispersed mineral fillers for cement concretes.

The proposed method for the selection of mineral additives for MPC uses the classification of fillers for cement concretes according to their hydration activity [14-16, 20-21], which makes it possible to most accurately assess the contribution of the surface activity of mineral fillers to the course of the processes occurring in the hydrated system.

As a basis for determining the indicator of the hydration activity of mineral fillers, experimentally obtained graphical dependences of the distribution of adsorption centers q on their surface are used depending on the acidity constant pKa, shown in Figure 2 for basalt filler and TPP fly ash.

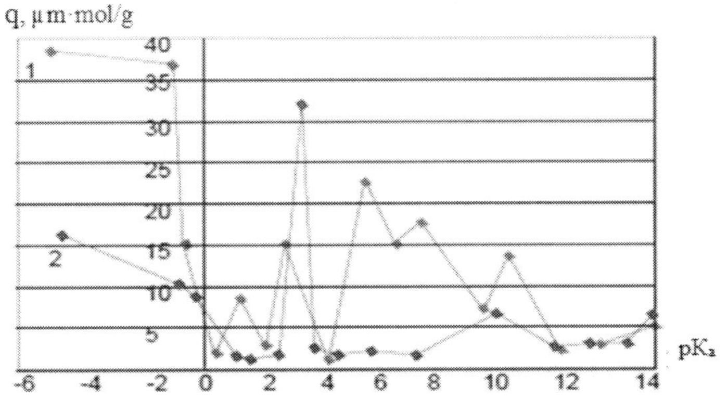

Figure 2. Distribution of centers of adsorption q on the surface basalt filler (1) and TPP fly ash (2) depending on from the acidity constant pKa.

Table 2. Content of adsorption centers on the surface mineral fillers.

No. p/p	Mineral filler	Number of Pi·10³ centers, mg-eq/g, in the ranges of pKa values				Total number of centers, mEq/g
		-4…0	0…7	7…13	≥ 13	
		P_{ol}	P_{kb}	P_{ob}	P_{kl}	P
1	Quartz sand	8.04	9.11	8.75	1.88	27.78
2	Sand dune	4.12	7.08	9.95	1.07	22.22
3	Gliezh	13.22	16.47	10.08	2.87	42.64
4	Basalt	23.41	22.15	11.16	1.96	58.68
5	OEP	41.18	5.48	9.34	1.14	57.14
6	OMP	6.61	23.88	16.37	4.32	51.18
7	Fly ash	43.14	27.61	11.77	5.32	87.84
8	CP	102.08	24.88	12.62	2.14	141.72

Notes. 1. Abbreviations: OEP, wastes from electric smelting production; OMP - wastes from copper-smelting production; CP is a zeolite-containing rock. 2. The total number of centers is $P = P_{kb} + P_{kl} + P_{ol} + P_{ob}$.

To assess the surface properties of mineral fillers, it seems advisable to use a generalized criterion - an indicator of the reduced hydration activity P_{pga}, calculated by the formula

$$P_{pga} = P_{kb} + P_{kl} + 0.33P_{ol} - 0.1P_{ob}, \quad (1)$$

where P_{kb}, P_{kl}, P_{ol}, P_{ob} – the number of adsorption centers, respectively, in the areas $0 \le pK_a < 7$; $pK_a \ge 13.0$; $-4 < pK_a < 0$; $7.0 \le pK_a < 13$ в 10^{-3} mg-ekv/g.

This criterion, which characterizes the acid-base properties of the surface of mineral fillers, makes it possible to scientifically classify mineral fillers according to the degree of their effect on cement systems.

In general, the following classification of mineral fillers is proposed according to the P_{pga} criterion - an indicator of the reduced hydration activity characterizing their potential efficiency in cement systems, assessed by a reduction in cement consumption (Table 3).

Table 3. Classification of mineral fillers by indicator the reduced hydration activity of P_{pga} .

Type of mineral filler	Criterion values of P_{pga}	Reducing cement consumption, %
Weakly active	$0 \le P_{pga} < 10$	≤ 10
Medium active	$10 \le P_{pga} < 25$	10-20
Highly active	$25 \le P_{pga} < 50$	20-30
Super active	$P_{pga} \ge 50$	30-50

For the mineral fillers used in the studies, the indicators of the reduced hydration activity of P_{pga} are given in table. four.

Table 4. P_{pga} Criterion Values for Mineral Fillers.

№ p/p	Mineral filler	PKa values, rel. units.				Transformed data		P_{pga}, rel. units.
		-4...0	0...7	7...13	≥ 13			
		P_{ol}	P_{kb}	P_{ob}	P_{kl}	$0.33P_{ob}$	$0.1 P_{ol}$	
1	Sand	8.04	9.11	8.75	1.88	2.65	0.87	12.77
2	Sand dune	4.12	7.08	9.95	1.07	1.36	0.99	8.52
3	Gliezh	13.22	16.47	10.08	2.87	4.36	1.01	22.39
4	Basalt	23.41	22.15	11.16	1.96	7.72	1.12	30.71
5	OEP	41.18	5.48	9.34	1.14	13.59	0.93	19.28
6	OMP	6.61	23.88	16.37	4.32	2.18	1.64	28.74
7	Fly ash	43.14	27.61	11.77	5.32	14.23	1.18	46.68
8	CP	102.08	24.88	12.62	2.14	33.68	1.26	59.44

Note: OEP - wastes from electric smelting production; OMP - wastes from copper-smelting production; CP - zeolite-containing rock

Comparative analysis given in table. 4 mineral fillers according to the Ppga criterion makes it possible to rank their effectiveness in cement systems and to characterize them according to the degree of activity: sand dune - weakly active; quartz sand, glezh, OEP - medium active; basalt, OMP, fly ash of Angren TPP - highly active and zeolite-containing rock - super active.

The developed classification of mineral fillers according to the proposed criterion for assessing the acid-base properties of the surface of mineral fillers P pga showed a high convergence of the data obtained with the results of previous studies from the standpoint of assessing their effectiveness in the design of various types of cement concretes and mortars, which made it possible to put it in the basis of the developed methodology scientifically a reasonable choice of these additives to obtain rational compositions of MPC [16].

3. Conclusions

The proposed classification of plasticizing additives according to the relative index of surface tension and mineral fillers according to their surface-active properties made it possible to rank such modifiers, respectively, according to the plasticizing effect and activity, which serves as the basis for a scientifically grounded choice of modifiers when obtaining, in particular, highly economical MPCs with the required properties.

References

[1] Usherov-Marshak A V 2009 Modern concrete and its technologies *Sat. "Concrete and Reinforced Concrete"* (SPb, Ed. "Slavutich") pp 20-24
[2] Aitchin P-C and Neville A 1993 High-Performance Concrete Demystified *Coner. Intern.* **15** no.1 pp 21-26
[3] Edvard G and Nawy P 2001 Fundaments of High Performanse Concrete *Sec. ed. Willy* 302p
[4] Batrakov V G 1998 Modified concrete. Theory and practice (M) 768 p
[5] Walraven J 2006 Concrete for a new centure *Proc. of the ist fib congress* pp 11-22
[6] Bazhenov Yu M 2005 Concrete technology of the XXI century Academic readings of the RAASN *New scientific directions of building materials science Part 1* (Belgorod) pp 9-20
[7] Hillemeiez B, Buchenau G and Herr R 2006 Spezialbeton, Betonkalander *Ernst Sbhn* pp 534-549
[8] Adylkhodzhaev A I, Makhamataliev I M, Tsoi V M and other Innovative materials and technologies in construction *Monograph under the General editorship of Adylkhodzhaeva A.I.* (T.: "Fan va technology") 2016 pp 292 p
[9] Rybiev I A 2004 Building materials science *Textbook. Manual for universities* (M: Higher school) 701
[10] Bazhenov Yu M 2011 *Concrete technology* (M.: Publishing house of ASV) 524
[11] Dvorkin L V 1991 *Cement concretes with mineral fillers* (K.: Budivelnik) 136
[12] Batrakov *V G* 1998 *Modified concrete. Theory and practice. 2nd ed. Rev. and add.* (M.) 768
[13] Haskova T N 2003 Colloidal chemistry: Surface phenomena and dispersed systems Textbook T N Khaskova, P M Kruglyakov (Penza, PGASA) 152
[14] Adylkhodzhaev A I, Makhamataliev I M, Kadyrov I A and Ruzmetov F Sh 2019 To the Question of the Influence of the Intensity of Active Centers on the Surface of Mineral Fillers on the Properties of Fine-Grained Concrete *International Journal of Innovative Technology and Exploring Engineering (IJITEE)* **8** Issue-9S2 ISSN: 2277-3075 (Online) pp 219-222
[15] Adylkhodzhaev A I, Makhamataliev I M, Tsov V M, Turgaev J F and Ruzmetov F Sh 2019 Assesment of Reinforcement Corrosion in High-Filled Ash-Containing Concrete *International Journal of Innovative Technology and Exploring Engineering (IJITEE)* **8** Issue-12 ISSN: 2278-3075 (Online) pp 4464-4466
[16] Adylkhodzhaev A I, Makhamataliev I M, Tsov V M, Turgaev J F, Umarov K S 2019 Theoretical bases of optimization of concrete microstructure with application of the imporved of mathematical planning of experiments *International Journal of Advanced Science and Technology (IJAST)* **8** Issue-9S2 ISSN 2207-6360 (Online) pp 210-219
[17] Anvar A, Shaumarov S, Elena S, Ulugbek S 2019 New method for diagnostic of heat engineering and mechanical properties of cellular concrete *International Journal of Engineering and Advanced Technology* **9** (1) pp 6885–6887
[18] Zabihullaevich S U, Ishratovich K S 2019 Some aspects of the photo-optical method of estimation composition of light *concrete International Journal of Engineering and Advanced Technology* **8**(5) pp 1924–1927
[19] Adilkhodjaev A, Shaumarov S, Kadir U 2020 New structure assessment method cell concrete *International Journal of Advanced Science and Technology* **29**(5) pp 1889–1893
[20] Adilkhodjaev A, Tsoy V, Khodzhaev S, Shaumarov S, Umarov K 2020 Research of the influence of silicon-organic hydrophobizer on the basic properties of cement stone and mortar *International Journal of Advanced Science and Technology* **29**(5) pp 1918–1921
[21] Adilkhodjaev A, Makhamataliev I, Tsoy V, Shaumarov S, Ruzmetov F 2020 Features of Forming the structure of cement concrete on second crushed stone from concrete scrap *International Journal of Advanced Science and Technology* **29**(5) pp 1901–1906

Modelling of wind impacts on silos and silo parks

A Makhinko[1] and N Makhinko[2,*]

[1]ETUAL LLC, 1, Bortnytska street, Petropavlivske, Boryspil district, Kiev region, 08316, Ukraine
[2]National Aviation University Ave Liubomyra Huzara 1, Kyiv, 03058, Ukraine

[*]Email: pasargada1985@gmail.com

Abstract. This paper devoted computer modelling of wind impacts on the silos for grain storage in order to obtain their aerodynamic characteristics. This modelling was made in the system of computer analysis Ansys Workbench for various wind conditions and different layouts of silos: separately located silo and silo parks by varying their sizes and mutual arrangement at various angles of the wind pressure. Within the numerical experiment it was obtained not only a huge array of data about the aerodynamic resistance of the silos, but also developed a special technique of the mathematical modelling of the wind effects on the selected class of constructions. This technique is a single software module for the platform Ansys Workbench, which automatically forms an optimal calculation area, makes an estimation of the wind loads on the silo and exports obtained loads in the software package SCAD for calculating the strength and deformability. To achieve this goal it was made the calculation of the same tasks on meshes of different dimensions, set the dimensions of the mesh in different places of the calculation area and installed the minimum required time of computer simulations.

1. Introduction

Wind impacts on empty cylindrical grain storage silos are one of the main types of loading, which often leads to their destruction. To determine the magnitude and nature of the wind load distribution on the surface of the silo, standard methods are used [1-5]. In addition, many theoretical and experimental studies have been conducted on this issue [6-10]. However, the possibility of using these materials in the design of modern silos and silo parks raises a number of questions. This is primarily because the basic information about the wind flow of cylindrical structures was obtained for models with a Reynolds number much smaller than the corresponding value characteristic of silos. Secondly, due to the corrugated walls and the presence of a large number of ribbed stiffeners, the silos have a rather rough structure, which in turn provokes the appearance of complex wall effects that increase the load on the silo body. The third reason is the question of the absence or insufficiency in the design rules of information regarding the consideration of wind loads on cylindrical structures, which are partially or completely in the wind shadow of other objects. This is a typical situation for silo parks being part of hopper complex.

The modern development of computer technology and software implementation of the control-volume method makes it possible to perform computer modelling of wind impacts on structures, avoiding these disadvantages. In particular, it is possible to implement flow modes with high Reynolds numbers that cannot be reproduced in wind tunnels [11-13].

2. Computational aerodynamics verification tests

2.1. Choosing a computational mesh

In order to be able to verify the calculation procedure, a computer modelling of a two-dimensional flow along cylinder (2D notation) with a diameter of $d_0 = 100$ mm with Reynolds number Re = 3900 was carried out. These output parameters are quite commonly used for test activities [6-10], that meet

Content from this work may be used under the terms of the Creative Commons Attribution 3.0 licence. Any further distribution of this work must maintain attribution to the author(s) and the title of the work, journal citation and DOI.
Published under licence by IOP Publishing Ltd

the selected initial criteria. At Re = 3900, the airflow is characterized by a laminar separation zone, in which the transition to turbulence occurs in the shear range, which accordingly causes the appearance of large-scale vortices and complex flows in the separation zone. Unfortunately, these experiments set values of different parameters and in different zones, which caused extremely large variance of data. The representative characteristics were the means sharing of the aerodynamic coefficients along the contour and the drag coefficient.

Given that the ANSYS CFX software package does not provide for classical two-dimensional aerodynamic calculations, a pseudo-two-dimensional statement is applied when a single cell of size is generated at the height of the cylinder. The quality of the mesh (Figure 1) was taken in six variants (from A to F), and the number of points per circle was constant and equal to 180.

a) b)

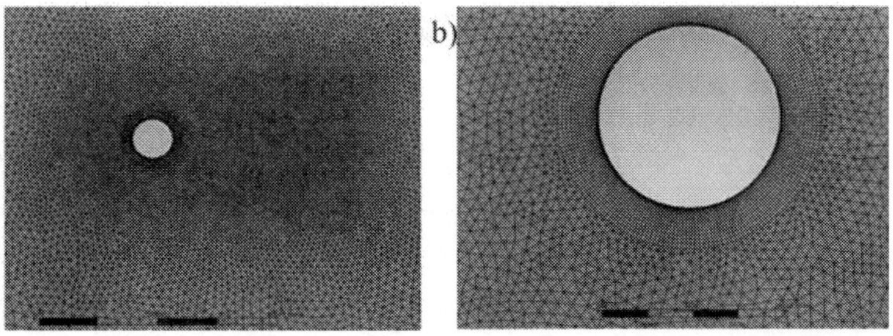

Figure 1. Mark F of the computational mesh of domain (a) and its enlarged fragment (b) around the cylinder.

Table 1 shows the modelling results for different variants of the calculated meshes, and Figure 1 shows a visual stratification of the cells of the computational domain and on a larger scale near the cylindrical body for meshes D and F.

The most consistent with the experimental values showed the variant of the E mesh, which was adopted as a reference for further research. Visualization of the instantaneous wind flow velocity in the flow of a single cylinder within the test task are shown in Figure 2a. In addition, Figure 2b presents a graphical comparison of the pressure coefficients for a single cylinder in the 2D formulation at Re = 3900 for the considered mesh options.

2.2. Two-dimensional flow along round cylinder

The turbulent flow around the cylinder in the 2D formulation was considered in two variants - an isolated body in free airflow and a cylinder near the obstacle, which was selected as the infinitely long rough screen [14]. The airflow model was adopted with constant physical properties. Four high-speed wind modes were selected to fit the cylinder's supercritical flow mode. Each mode was characterized by its own Reynolds number. The diameter of the silo was assumed to be 3500 mm, and the distance from its center to the screen plane was $G = 0.7D$.

The computational mesh for solving the set tasks was taken for option E. A general picture of the turbulent flow around an isolated cylinder is shown in Figure 3.

On the basis of the constructed mesh, the problem of turbulent flow around both isolated cylinder and cylinder near the obstacle, was solved in two formulations: stationary and unsteady. When solving the problem in the stationary formulation, the installed wind flow was considered, and the results of the calculation focused the aerodynamic characteristics of the cylindrical body.

Table 1. Parameters of the computational mesh for 2D flow along cylinder.

Name/Mesh	A	B	C	D	E	F
d_0 / Δ_{min} [a]	20	20	20	20	100	20
d_0 / Δ_{max} [b]	2	2	2	2	2	2
d_0 / y_0 [c]	50	100	200	500	1000	1000
γ_0	1.05	1.10	1.10	1.10	1.10	1.25
n_y [d]	20	20	20	20	20	20
d_0 / y_n [e]	19.8	16.36	32.70	81.75	163.50	14.41
n_{nodes} [f]	40560	33632	76836	324352	1141240	31008
$n_{elements}$ [g]	36627	29702	72895	318725	1130214	27080
Orthog, Angle	70.7 OK	70.3 OK	69.9 OK	61.6 OK	56.9 OK	69.5 OK
Exp, Factor	4 OK	4 OK	4 OK	5 OK	6 OK	4 OK
Aspect Ratio	81 OK	74 OK	78 OK	77 OK	80 OK	89 OK
C_D [h]	0.488	0.666	0.773	0.820	0.828	0.846

[a, b] Minimum and maximum mesh size.
[c] Size of the first wall cell of the boundary layer.
[d] number of wall cells.
[e] the size of the last wall cell and the mesh size of the internal volume of the computational domain.
[f] number of mesh points.
[g] the number of cells (elements) of the mesh.
[h] drag coefficient.

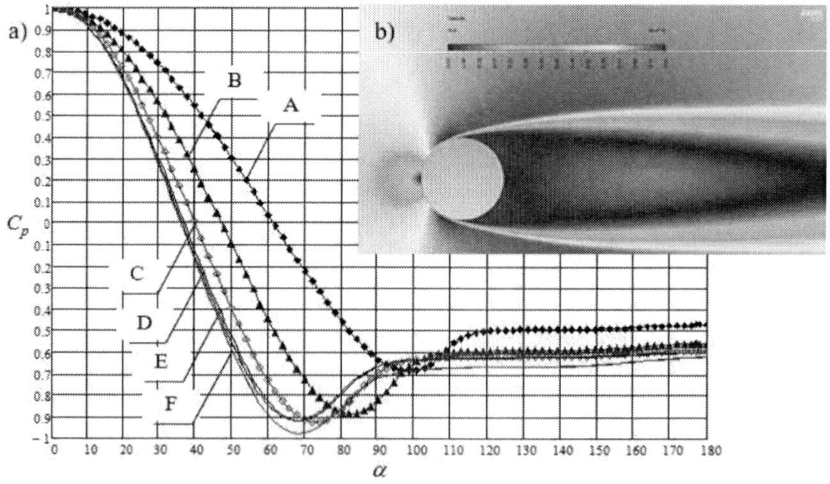

Figure 2. Visualization of the instantaneous wind flow velocity in the flow of a single cylinder in the framework of test task (a) and the pressure coefficient (b) for a single cylinder in the 2D formulation at Re = 3900.

Unsteady calculation made it possible to track the change in aerodynamic characteristics over time, and to determine a number of frequency parameters of the turbulent flow around the cylinder, such as the frequency of failure of vortices from its surface. Considering the long duration of unsteady calculation and the attraction of considerable computer resources, it was possible to switch to it only after adjusting models for tasks in a stationary settin.

Figure 3 shows a characteristic feature of the trail is the presence of transverse (Karman street) and sufficiently intense longitudinal vortices. The obtained results are in full agreement with the experimental data known from the literature and the results of calculations of the cylinder flow by the methods of computational aerodynamics [7, 8, 9, 15].

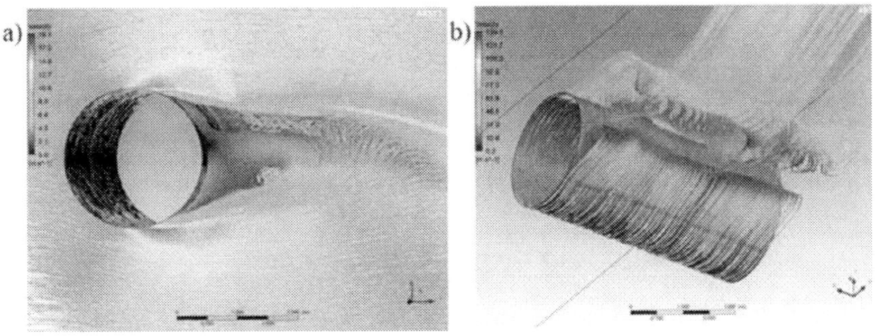

Figure 3. The general instantaneous picture of the flow around an isolated cylinder at $Re = 2{,}33 \cdot 10^6$ (a) a cylinder near an infinitely long screen at $Re = 1{,}17 \cdot 10^7$ or 50 m/s (b).

3. Numerical experiment of vortex systems modelling

3.1. In the plane flow (2D modelling)

The real design decisions of the master plans of grain terminals or hopper enterprises, involve a simple arrangement of silos in the form of small groups with parallel, perpendicular placement or some angular deviation. Preferably, silo parks include silos of the same diameter and height, which greatly simplifies their study.

With airflow, the silos in the local group interact with each other and form some velocity spectrum inside and around the formed arrangement. The nature of the silos placement is reflected in the pressure distribution on their surface and, of course, differs from those that are set for single, non-windproof structures and declared in the design standards.

Thus, the study focuses on the study of the kinematic spectrum and the evaluation of the interference interaction of group models of two, three and four silos with different geometric dimensions and distance. In the first formulation, two-dimensional (plane) numerical modelling of wind flow is performed, which does not depend on the coordinate of the normal to the plane under consideration. The cylindrical bodies of the same diameter were considered as storage capacities. Height does not matter in this experiment. The airflow direction was fixed: the cylinders were placed in a line perpendicular to the flow, with variable variations in the location of the objects within the group. The position of each body of the flow was expressed parametrically along two axes by the ratio of the characteristic distance between the centers of the cross sections to a given diameter.

Variables in these experiments were the accepted body diameters and four variants of the flow Reynolds number, which respectively determined the speed of its movement at the inlet. The main quantitative characteristics of the conducted numerical experiments were to obtain the distributions of the aerodynamic coefficient for each of the cylindrical bodies in the group and to establish the dependence of its values upon changing the size and initial parameters that determine the wind speed. In the first stage, the pairwise arrangement of cylindrical bodies was considered - parallel and perpendicular to the flow and at diagonal placement. In the following, we analyzed a group of three objects with a similar arrangement and taking into account the separated position of one of the cylinders. The third stage was to study the flow conditions of the formation of the leeward domain for a four-cylinder system. The analysis of the received results made it possible to obtain a visual representation of the spatial behavior of the airflow in the vicinity of cylindrical bodies with the help of isofields of the wind flow (Figure 4).

It is possible to observe the characteristic direction of the flow, the formation of vortex systems in the conditions of limited space, the point of flow separation on the surface of the cylindrical body, the formation of the vortex shedding, as well as qualitative changes in the distribution of isofields in different variants of the Reynolds number values. The color spectrum of the lines characterizes the velocity field gradient - from minimum values to maximum. The illustration of the complete cycle of the flow phenomenon of groups of cylindrical bodies is very important for the possibility of aerodynamic optimization and reduction of wind loads for the silo park in real design tasks.

The graphs of the change of the aerodynamic coefficient along the perimeter of cylindrical bodies at different Reynolds numbers showed a considerable divergence of values in quantitative equivalent (Figure 5).

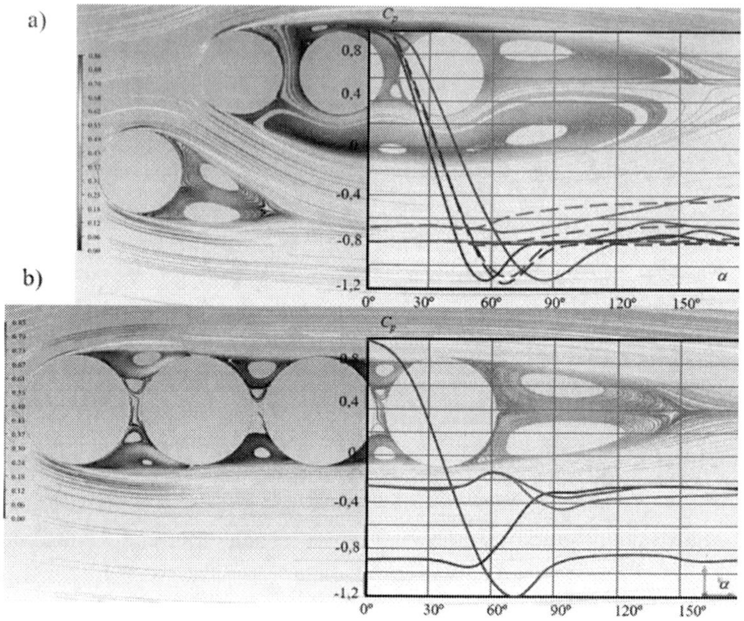

Figure 4. Visualization of the airflow around a system of two cylinders (a) and of three and four cylinders (b).

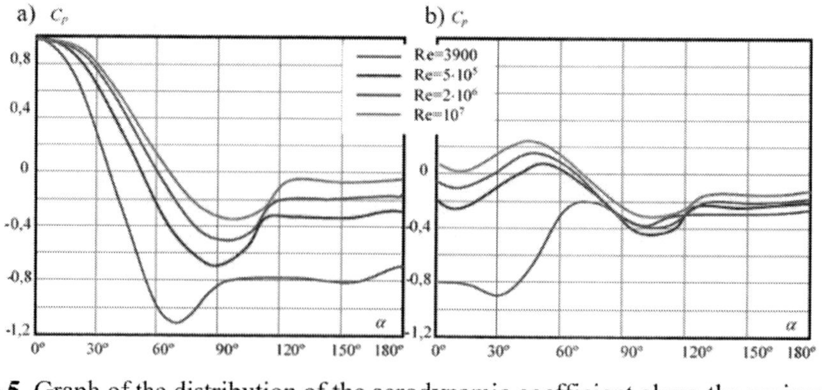

Figure 5. Graph of the distribution of the aerodynamic coefficient along the perimeter I (a) and II (b) of the cylinders with their sequential location and distance between the centers of body (b).

The results of a large-scale computer modeling of wind impacts on cylindrical storage tanks allow us to formulate a special aerodynamic atlas, which shows graphs of the distribution of the wind pressure coefficient over the surface of the silos for the vast majority of real silo park layouts.

3.2. 3D modelling of spatial flow around cylindrical models

Certainly, the numerous variants of silo park configurations considered cannot fully describe the actual construction in the vicinity of the storage tank (grain dryers, warehouse buildings, etc.). In addition, the process line can combine capacities of different configuration, diameter and height. In this case, it is not possible to limit the problem of numerical modeling of wind impacts only in 2D formulation. Therefore, 3D modelling of spatial flows in the vicinity of groups of cylindrical bodies (Figure 6) was performed.

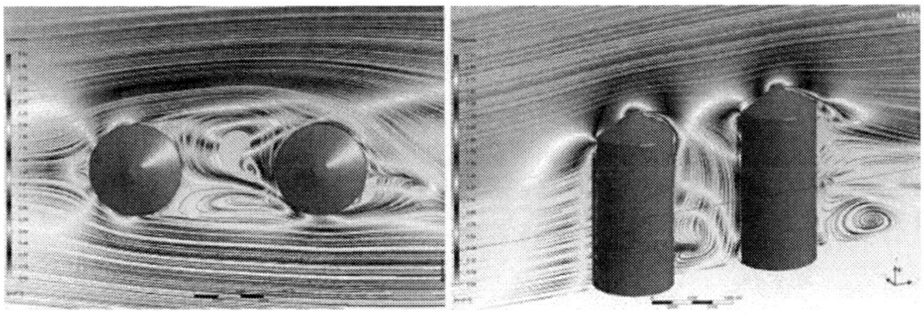

Figure 6. Velocity isofields in spatial flow modeling for a group of twin-tandem cylinders.

In the future, followed by determination of the dependences of the distributions of aerodynamic coefficients, which showed that a large number of calculations and the involvement of a large number of computational devices were necessary for the exact solution of this problem. The comparisons of the results obtained prove the rationality of using a two-dimensional approach to determine the basic aerodynamic characteristics.

4. Conclusions

In the framework of the numerical experiment, not only a huge array of data on the aerodynamic drag of the silos was obtained, but also a special method for mathematical modeling of wind impacts on the selected class of structures was developed. The method is the only software module for the Ansys Workbench platform that automatically generates the optimum computational domain, performs an estimation of wind loads on the silo, and exports the resulting loads to the strength analysis programs to perform strength and deformation calculations.

References

[1] DBN V.1.2-2:2006 2007 *Navantazhennia i vplyvy. Normy proektuvannia* (Kyiv: Minrehion Ukrainy) 75
[2] EN 1991-1-4:2005 2010 *Actions on structures Part 1-4: General actions - Wind actions* Eurocode 1 (Brussels: CEN) 149
[3] Makhinko N 2019 Stochastic calculaton of the wall of steel vertical silos *Proceedings of the National Aviation University* **1**(78) pp 54-9
[4] Maleki S and Mehretehran A M 2019 3D wind buckling analysis of steel silos with stepped walls *Thin-Walled Structures* **142** pp 236-61
[5] Iwicki P, Rejowski K and Tejchman J Stability of cylindrical steel silos composed of corrugated sheets and columns based on FE analyses versus Eurocode 3 approach *Eng. Fail. Anal.* **57** pp 444-69
[6] Belostotskii A, Akimov P and Afanaseva I 2017 *Computational aerodynamics for problems of construction* (Moskva: ASV) 720
[7] Makhinko A V 2012 *Probabilistic design of communication towers* (Poltava: PoltNTU) 410
[8] Norberg C 1978 *Effects of Reynolds number and a low-intensity freestream turbulence on the flow around a circular cylinder* (Göteborg: Chalmers University of Technology 87/2) 54
[9] Kravchenko A G and Moin P 2000 Numerical studies of flow over a circular cylinder at

ReD=3900 *Phys. Fluids* **12**(3) pp 68-73
[10] Lapenko O, Makhinko A and Makhinko N 2019 Stochastic description of loads on the steel storage capacities *Materials Science Forum* **968**(2) pp 528-33
[11] Hillewaere J, Dooms D, Quekelberghe B, Degroote J, Vierendeels J, Roeck G, Lombaerta D and Degrandea G 2012 Unsteady Reynolds averaged Navier–Stokes simulation of the post-critical flow around a closely spaced group of silos *J. Fluids. Struct.* **30** pp 51-72
[12] Hillewaere J, Degroote J, Lombaert G, Vierendeels J and Degrande G 2015 Wind-structure interaction simulations of ovalling vibrations in silo groups *J. Fluids. Struct.* **59** pp 328-50
[13] Makhinko N 2018 Stress-strain state of the storage silos under the action of the asymmetric load *Matec Web of Conf.* **230** 02018
[14] Makhinko A and Makhinko N 2015 Numerical simulation of flow around circular cylinder at high reynolds numbers *Resource-saving materials, structures, buildings and structures* **30** pp 371-83
[15] Isaev S A, Zhdanov V L and Niemann, H J 2002 Numerical study of the bleeding effect on the aerodynamiccharacteristics of a circular cylinder *J. Wind Eng. Ind. Aerodyn.* **90(11)** pp 1217-26

Improvement of methods for assessing the reliability of axle boxes for freight cars

I Martynov[1], A Trufanova[1], V Shovkun[1,*], A Babenko[1] and N Kladko[1]

[1]Ukrainian State University of Railway Transport, 7, Feuerbach sq., Kharkiv, 61050, Ukraine

*E-mail: vadimshovkun62@gmail.com

Abstract. The box bearing unit of the freight car is a complex mechanical system, which is influenced by various factors. Under normal operation, the parameters characterizing the functional state of the components of the axle box should be within the prescribed limits throughout the standard service life. Mathematically, this corresponds to the placement of the axle box elements in the space of admissible states. The railroad model consists of a set of submodules: track and elastic ballast layer. The model allows simulate the microgeometry of the track, the characteristics of the inequalities, and the elasticity of the base under the track. Subsequently, the main parameters characterizing random processes were determined: the magnitude of the mathematical expectation, the variance, as well as the minimum and maximum values of the effort. The results indicate that random processes are distributed according to normal law. According to the results of the research, for each of the random loading processes of the axle boxes, appropriate correlation functions were constructed and the space of "admissible states" for different operating modes was con-structed.

1. Introduction

Railway transport is the main transport artery of Ukraine. It is the railway transport that performs overwhelming majority of cargo and passenger transportation both for Ukrainian and in the interstate connection.

Traffic safety is a priority for railways. Its provision depends on the well-coordinated work of all structural railway transport sub-divisions, but one of the key factors is the reliable operation of wagons. Failures of wagon structural elements may cause delays in the delivery of goods to consumers due to uncoupling of wagons along the route, as well as significant additional costs for restoring the railway infrastructure operability.

One of the most critical structural elements of a freight wagon is an axle box unit with roller bearings. As the long-term experience of rolling stock operating shows, during the period from 2009 to 2019 axle box units caused 2339 cases of wagons uncoupling along the route due to excessive heating. At the same time, up to 1000 cases of axle box unit failures were additionally detected annually by means of remote-control devices for wheelsets and by wagon inspectors basing on external signs.

Bearings constitute the main structural elements of the axle box unit. Ensuring bearing durability under dynamic radial and axial loading is challenging. When calculating the strength and reliability of structural elements of axle box units, simplified schemes are often used, but they don't take into account the number of operating factors.

A freight wagon axle box unit is a complex mechanical system. It consists of an axle box housing (or adapter), in which the outer and inner bearing rings, rolling elements, a separator and sealing devices are located, preventing the dirt and moisture ingress into the unit inner cavity.

Random nature of the impact on the axle box unit is determined by random values of the load parameters, random distribution of loads in time and at different points of the system, random combination of loads and many other factors (overload, different track conditions, climatic conditions, level of running gears wear and tear, repair and technical service quality). *Literature review*

The fundamental principles of calculating the strength and reliability of bearings of all types are set out in [1]. It should be noted that, until recently, there was no unified methodological approach to both

Content from this work may be used under the terms of the Creative Commons Attribution 3.0 licence. Any further distribution of this work must maintain attribution to the author(s) and the title of the work, journal citation and DOI.
Published under licence by IOP Publishing Ltd

the choice of the bearing type and the methods of calculating the latter for strength and reliability in relation to railway axle box bearings. Only the study [2] of O.M. Savchuk, Professor of Ukrainian State University of Railway Transport named after Academician V. Lazaryan, proposed a unified method for calculating bearings with cylindrical rollers installed in resilient housings of bearing supports.

The issue of selection and standardization of reliability indicators of carriage cylindrical axle box bearings, taking into account the peculiarities of their operation, was considered in the article [3].

There is also a noteworthy work [4], in which an attempt to determine the theoretical and actual longevity of bearings was made. Research is devoted to the analysis of axle box bearing damage [5, 6, 7, 8].

Research of I. E. Martynov, Professor of Ukrainian State University of Railway Transport [9, 10, 11], is devoted to the substantiation of introduction of double-row tapered cassette bearings on the railways of Ukraine. He also determined the percentage of double row tapered bearings operation resource. But in his research, average loads were used to calculate the bearings operation life.

Among the many known methods, devoted to the determination of certain indicators of the technical systems reliability, it is necessary to outline the so-called theory of "emissions", the foundations of which were created by the American researcher S. Rice [12] and his followers [13, 14]. With regard to mechanical systems, this theory was developed further in the works of Professor V.V. Bolotin [15, 16] and his students [17]. However, the abovementioned works do not take into account the combined action of vertical and horizontal forces acting on the axle box unit. Therefore, the improvement of methods for calculating the axle box unit durability is an urgent task.

2. Development of a mathematical model of axle box unit operation

The axle box unit of a freight wagon is a complex mechanical system. Its parameters change during operation. These changes are associated with material ageing, accumulation of fatigue and corrosion damage, frictional wear and tear, strength characteristics deterioration, etc.

External loads q_i affect the axle box unit when the wheel pair passes the irregularities of the superstructure of the track. They are different in their origin and are random in nature. That is, the loads take random values $q_1, q_2, \ldots, q_i$ from a certain range of possible external loads Q. The change in these loads over time is a random process $v(t)$. Stochastic process of the axle box unit operation leads to a change in its elements condition. As a result, the axle box unit takes on the corresponding condition u_i from the range of possible conditions U. The range U is chosen in such a way that with its help, within the framework of the selected design scheme, the technical condition of the axle box unit is fully described.

Operator L connects each instantiation of elements from the volume of external loads with the instantiation of elements in the volume of possible conditions U

$$Lu = Q, \tag{1}$$

The random nature of the axle box unit behavior is due to the spread of its own properties as well as the acting loads. During normal operation, the parameters characterizing the axle box elements functional condition must be within the established limits during the entire standard service life. Mathematically, this corresponds to finding elements in the admissible volume Ω of the range of possible conditions U. Overrange of the random process $v(t)$ of the axle box unit operation outside permissible volume Ω leads to its failure.

At the initial moment of time (the moment when the wagon starts moving), the random process of the axle box unit operation $v(t)$ with a probability equal to 1 will be in the admissible volume Ω of the range of possible conditions U, i.e. $P(0) = 1$. Since axle box units are highly reliable products, runouts from this volume in the time interval [0, t] are highly doubtful, i.e.

$$P(t) = P\{v(\tau) \in \Omega;\ \tau \in [0,t]\}. \tag{2}$$

As a result of the load action there may be contact bearing resistance in the volume between the rollers and the raceways during axle box bearings operation. They lead to subsequent destruction of the bearing.

Then the permissible volume Ω will be set by limiting the maximum contact bearing resistance $\sigma(t)$ that arise between the rolling elements in the bearing and depend on the bearing steel ultimate strenght

$$\sigma(t) < \sigma_{CT}, \tag{3}$$

where σ_{CT} is ultimate compressive strength of bearing steel.

The maximum contact bearing resistance $\sigma(t)$ will depend on the magnitude of the vertical and horizontal loads affecting the axle box bearing together. Then the range of possible conditions U will be two-dimensional, and, according to the theory of random processes, the main statistical characteristics of a random process of combined action of radial $P_{rad}$and axial loads P_{axial}take the following form

$$\overline{P} = \overline{P}_{rad} + \overline{P}_{axial}, \tag{4}$$

$$D_P = D(P_{rad}) + D(P_{axial}) + 2k_r\sqrt{D(P_{rad}) \cdot D(P_{axial})}, \tag{5}$$

$$R_{(P)} = R(P_{rad}) + R(P_{axial}) + 2R[(P_{rad})(P_{axial})], \tag{6}$$

where $\overline{P}, \overline{P}_{rad}, \overline{P}_{axial}$ are mathematical expectations of random processes of total, radial and axial loading $PP_{rad}P_{axial}$ respectively ;

$D(P), D(P_{rad}), D(P_{axial})$ are variances of corresponding loading processes;

$R_{(P)}, R(P_{rad}), (P_{axial})$ are correlation functions of corresponding loading processes;

k_r is correlation coefficient of the corresponding loading processes;

$R[(P_{rad})(P_{axial})]$ is cross-correlation function of the processes that $P_{rad}P_{axial}$.

$$\overline{N} = \frac{1}{2\pi}\sqrt{-\frac{\ddot{R}_{(P)\tau=0}}{R_{(P)\tau=0}}}. \tag{7}$$

The mathematical expectation of random process runouts is determined by the formula. The distribution of the load vector will be a surface in the form of a hill (Figure 1).

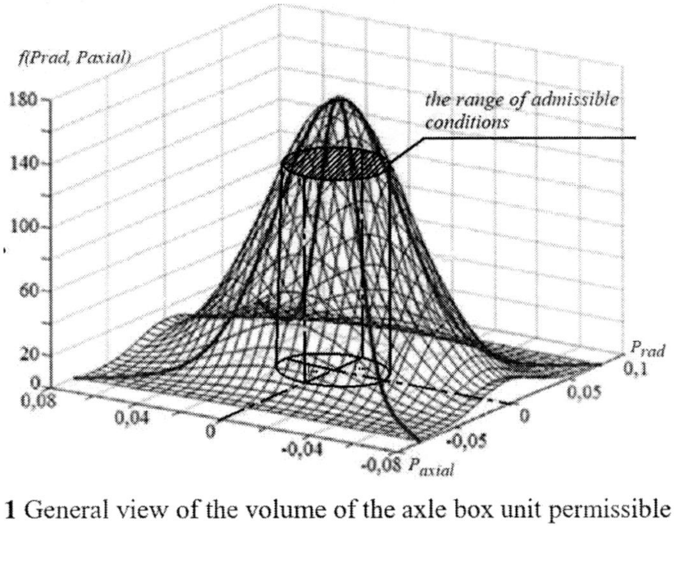

Figure. 1 General view of the volume of the axle box unit permissible conditions.

If you cut this surface with a plane parallel to the plane $P_{rad}oP_{axial}$ (horizontal plane), you will get an ellipse in the section. The equation of the ellipse projection on the plane will have the form $P_{rad}oP_{axial}$

$$\frac{\left(P_{rad}-P_{axial}\right)^2}{D_{P_{rad}}}-\frac{2\left(P_{rad}-\bar{P}_{\mathrm{rad}}\right)\left(P_{axial}-\bar{P}_{\mathrm{axial}}\right)}{\sqrt{D_{P_{rad}}\cdot D_{P_{axial}}}}+\frac{\left(P_{axial}-\bar{P}_{\mathrm{axial}}\right)^2}{D_{P_{axial}}}=const\,. \tag{8}$$

To determine the failure rate for two combinations of loads, it is advisable to represent the random values as a random vector on the plane. It is advisable to represent vector components $P_{rad}P_{axial}$ as random normal statistically dependent processes. To bring the equation to the canonical form, it is necessary to move the origin of coordinates to the point with coordinates $(\bar{P}_{rad}\bar{P}_{axial})$ and rotate the coordinate axes by angle φ. The form of the normal law on the plane will have the form, where $D\left(P'_{rad}\right),D\left(P'_{axial}\right)$ are the main standard deviations.

$$f\left(P'_{rad},P'_{axial}\right)=\frac{1}{2\pi\sqrt{D\left(P'_{rad}\right)\cdot D\left(P'_{axial}\right)}}\cdot exp-\left[\frac{(P'_{rad})^2}{D\left(P_{rad}\right)}-\frac{(P'_{axial})^2}{D\left(P_{axial}\right)}\right]. \tag{9}$$

To determine the density of the probability distribution, we will use the statement that for a stationary random process the correlation with its first derivative is equal to zero. Then

$$f\left(P'_{rad},P'_{axial},\dot{P}'_{rad},\dot{P}'_{axial}\right)=\frac{1}{4\pi^2\sqrt{DP'_{rad}\cdot DP'_{axial}\cdot D\dot{P}'_{rad}\cdot D\dot{P}'_{axial}}}\times$$
$$\times exp\left[-\frac{1}{2}\left(\frac{P'^{\,2}_{rad}}{DP'_{rad}}+\frac{P'^{\,2}_{axial}}{DP'_{axial}}+\frac{\dot{P}'^{\,2}_{rad}}{D\dot{P}'_{rad}}+\frac{\dot{P}'^{\,2}_{axial}}{D\dot{P}'_{axial}}\right)\right]. \tag{10}$$

The number of runouts of the total load change random process is the runout of combined action change vector of vertical and horizontal loads outside the volume, basing on ellipse. Dimensions are determined by the limits of the range of admissible conditions.

3. Formulation of the range of admissible conditions

To establish the limits of the range of admissible conditions, the maximum contact bearing resistance arising on the rollers under the action of radial P_{rad} and axial P_{axial} loads was obtained.

It was determined that the maximum contact bearing resistance in the contact zone of the rollers with the raceways occur at the point of transition from the moving line to the end of the roller and, at a load of 230 kN/axle is equal to 1198 MPa.

As a result, we have obtained the limiting values of radial P_{rad} and axial P_{axial} loads, which lead to contact bearing resistance in the bearing, causing its destruction. The value of this load is the radius r of the limit of the range of admissible conditions.

The random process of the loads change r affecting the axle box bearing during operation must be within the obtained range with a radius. Exceeding the obtained ultimate load in operation will result in bearing failure (Figure 2).

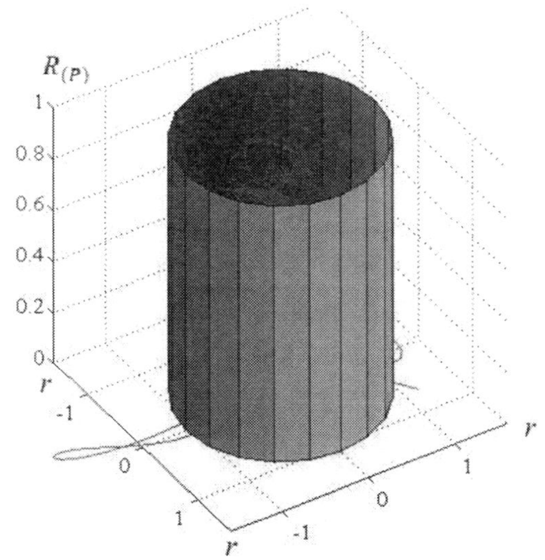

Figure. 2 Runout of the radial and axial loads action change vector beyond the range of permissible conditions.

The mathematical expectation of runout N of the total random process of changing the loads represents the runout of radial and axial loads action change vector beyond the cylinder. Having calculated the number of possible runouts beyond the permissible condition with a radius r, we will determine the axle box bearings reliability indicators for various wagon movement modes.

In the worst case, the lower bound for the probability of no-failure operation for a doubled cylindrical bearing is 0.9921.

4. Conclusions

- Method for determining the reliability indicators of freight wagon axle box bearing units has been presented, a feature of which is the probabilistic nature of the loads, and the failure is interpreted as the runout of combined bearing loading random process outside the "range of permissible conditions";
- Limits of the range of permissible conditions have been established for various loading options for freight wagon axle box unit;
- Indicators of probability of failure-free operation for freight wagon axle box bearings have been determined.

References

[1] Tedric A Harris and Michael N 2006 Essential concepts of bearing technology. *CRC press*

[2] Savchuk O 1980 Teoreticheskoye issledovaniye nagruzhennosti rolikov v pidshipnikakh busovykh uzlov podvizhnogo sostava. *Problemy mekhaniki zheleznodorozhnogo transporta: tezisy dokladov mezhdunarodnoy vsesoyuznoy konferentsii* **127** pp 11-13

[3] Pokrovskiy B K 1979 postanovke voprosa ob otsenke nadezhnosti podshipnikov kacheniya buks vagonov *Vsesoyuznyy zaochnyy institut inzhenerov transporta* 4(**97**) pp 41-49

[4] Tsyurenko V Petrov V 1982 Nadezhnost rolikovykh podshipnikov v buksakh vagonov *Transport* **96** pp 49-54

[5] Herminio Maio Graca Fernandes 2017 *Analysis of failures of rolling stock railways rolling bearings* (Porto) 121

[6] N Symonds Ilaria Corni RJK Wood Adam Wasenczuk and David Vincent 2015 Observing early stage rail axle bearing damage *Engineering Failure Analysis* **56** pp 216-232

[7] Ferreira J L A Balthazar J C Araujo A P N 2003 An investigation of rail bearing reliability under real conditions of use *Eng. Fail. Anal.* **10** pp 745-758

[8] Gerdun V Sedmak T Šinkovec V Kovše I Cene B 2007 Failures of bearings and axles in railway freight wagons *Eng. Fail. Anal.* **14** pp 884-894
[9] Martynov I 2007 Viznachennya dovgovichnosti konichnikh pidshipnikiv dlya rukhomogo skladu. Ukraïnska derzhavna akademiya zaliznichnogo transportu. *Zb. nauk. prats.* **86** pp 56-61
[10] Martynov I 2005 Viznachennya pokaznikiv nadiynosti vagonnikh buks za rezultatami viprobuvan Ukraïnska derzhavna akademiya zaliznichnogo transportu *Zb. nauk. prats.* **68** pp 191-198
[11] Martnov I 2007 Doslidzhennya napruzhenogo stanu konichnikh rolikopidshipnikiv Ukraïnska derzhavna akademiya zaliznichnogo transportu *Zb. nauk. prats.* **81** pp 83-86
[12] *Rice S O 1948* Mathematical analysis of randomnoise *BellSystemTech.* **23** pp 282-332
[13] Tikhonov V I and Khimenko V I 1987 *Vybrosy trayektoriy sluchaynykh protsessov* (Moscow: Nauka) 307
[14] Aza¨is J M and Wschebor M 2009 *Level sets and extrema of random processes and fields* (N. Y.: Wiley) 393
[15] Bolotin V 1971 *Primeneniye metodov teorii veroyatnostey i teorii nadezhnosti v raschetakh sooruzheniy* (Moscow: Stroyizdat) 255
[16] Bolotin V 1961 Statisticheskiye metody v stroitelnoy mekhanike. (Moscow: Stroyizdat) 160
[17] Fedorov D Bondarovich B 1981 *Nadezhnost rabochego oborudovaniya zemleroynykh mashin* (Moscow: Mashinostroyeniye) 279

IOP Conf. Series: Materials Science and Engineering 1164 (2021) 012050 doi:10.1088/1757-899X/1164/1/012050

Justification of conditions of application of thermo modified veneer for protection of wooden products from moisture

Y Tsapko[1], O Horbachova[2], O Bondarenko[1] and S Mazurchuk[2]

[1]Kyiv National University of Construction and Architecture, 31, Povitroflotskyj ave., Kyiv, 03037, Ukraine
[2]National University of Life and Environmental Sciences of Ukraine, 15, Heroiv Oborony str., Kiyv, 03041, Ukraine

Email: mazurchuk.s.m@ukr.net

Abstract. The analysis of the process of thermal modification of wood was carried out and it was found that the treatment causes chemical changes in the structures of wood components (lignin, cellulose and hemicellulose). It is proved that in the process of thermal modification there is a decomposition of hemicelluloses and amorphous part of cellulose, reduced water absorption, and also reduces the amount of substances that are the environment for the development of fungi, which leads to increased durability. In addition lignin and psevdolihnin formed by the process of polymerization and pererozpodilynyu in cell volume and cell walls provide higher density, hardness, increase the hydrophobicity (water repelling), thereby reducing their ability to absorb moisture and swell. It has been experimentally established that in comparison with unmodified veneer, which gained 22.41 % of moisture in two weeks, veneer modified at 250 and 280 °C gained 2.5 times less moisture - on average 8.6 %. It is established that with the increase of the modification temperature to 280 °C, the water absorption decreases more than twice. The results of the research will also allow to purposefully solve further problems on the creation of new methods of thermal modification of wood and the conditions of its operation at various sites.

1. Introduction

In construction is more intense search for new highly effective wood protection from destruction. But protection today must not only ensure the normalized stability of wood, but also to preserve its operational parameters, to solve environmental safety and durability.

To increase the level of wood exploitation, it is possible by modifying it, the essence of which is to give the wood the ability to resist moisture, bio-damage, which contributes to the destruction of wood and accelerate the process of destruction. But when modifying wood, there are difficulties with the technological modes of its processing, namely, such as time, temperature. This is due to the fact that the modification process may not be achieved, and the use of such wood leads to its destruction [1].

Knowledge of physicochemical properties of thermodified wood, its quality indicators, the mechanism of action on temperature materials, allows to make a choice taking into account economic indicators, duration and safety of application, environmental aspects, etc. [2-5].

Therefore, the development of technological measures for wood modification, study of water absorption and impact on this process is an unresolved component of ensuring the stability of wood and determine the need for research.

2. Analysis of recent research and publications

Thermodified wood is becoming one of the best materials for cladding building structures and products [6]. In this regard, the humidity of thermally modified facades was measured and the low moisture content of thermally modified wood was revealed in comparison with the reference spruce. But it is not specified what was the degree and modes of wood modification. Mass loss, wettability, wood color and chemical transformations have since been widely studied [7], while work is needed to focus on quality control, modeling and study of the causes of improved properties of modified wood.

Extensive use of thermally modified wood has led to the need for quality control, including control of product deviations, which allows it to be carried out by a third party in the case of certification and

Content from this work may be used under the terms of the Creative Commons Attribution 3.0 licence. Any further distribution of this work must maintain attribution to the author(s) and the title of the work, journal citation and DOI.
Published under licence by IOP Publishing Ltd

regulation of complaints and consumer requirements [8]. However, it is not specified what methods are needed to characterize the change in quality in terms of improved target properties of modified wood during industrial production.

In [9] studied the swelling and surface roughness after heat treatment of wood. Established that the parameters of swelling and surface roughness vary considerably for the two two-temperature heat treatment durations. The value of swelling and surface roughness decreased with increasing treatment temperature and treatment time, but not listed under any technical indicators.

The assessment of changes in color and reflectivity wood surfaces through artificial weathering [10] obtained from the solar box camera that mimics the conditions and further leaching water. As the weathering time increases, the untreated surfaces of the wood samples darken, while the modified samples lighten to have a similar color or, in any case, to reduce the chromatic difference that was at the beginning of the weathering tests.

During heat treatment, many chemical reactions occur, leading to changes in the components of the primary cell wall of the wood and darkening of the material. Among other changes, thermomodified wood becomes more resistant to fungal decay and more stable than raw size, making it suitable for use indoors and outdoors as cladding, decking, flooring, garden furniture and window frames [11]

Laboratory tests have shown a positive effect of thermal modification on the durability, dimensional stability and thermal conductivity of wood [12]. The monitoring results showed that the elements and windows of thermally modified spruce have a significantly lower moisture content in the wood compared to windows of unmodified, and that the wax also has a positive effect on humidity.

Natural aging is usually a relatively slow process because the artificial aging plays a role important in assessing the results by reducing the time compared to the natural weathering conditions. The approach is to protect the surface with various types of commercial products such as water-based solvents with a high solids content, powder coatings and substance-free products [13].

One approach is the modification of wood – a set of processes that provide the treated material over the ability to cope with the damage caused by the environment, by increasing the duration of treatment. The process is also performed to enhance the physical, mechanical or aesthetic properties of wood and derived products with the advantage that they are not as harmful to users and the environment as natural wood [14-16].

Thus, from the literature it is established that thermal modification of wood can give it the opportunity to resist destruction. All this gives reason to believe that it is appropriate to conduct a study on the determination of the parameters that ensure its stability Therefore, research in this area is the unresolved part of ensuring the sustainability of building designs, which resulted in the need for research.

3. The purpose of this work

The purpose of this work is to study the process of water absorption by thermodified oak veneer and substantiate the effectiveness of counteracting water absorption during thermal modification.

4. Materials and methods of research

To establish the efficiency of operation of the facades of furniture products at different temperature and humidity conditions, we used samples of oak veneer, which had previously undergone a heat treatment process. Thermal modification was performed in a convective chamber, for the reliability of the result, the veneer was fixed between the plates of porcelain. In general, the material was kept at temperatures of 160, 220, 250 °C for 10, 20 and 30 minutes and 280 °C for 10 minutes (Figure 1-4).

Samples of thermodified oak veneer with a tangential cross-sectional surface 0.7 mm thick and 155x15 mm in size were used to study moisture absorption. The amount of water bending was determined on samples measuring 155x150 mm, 0.7 mm thick, which have undergone preliminary thermal modification according to the above parameters.

a b c

Figure 1. Samples of oak veneer after thermal modification at a temperature of 160 °C for: a – 600 s; b – 1200 s; c – 1800 s.

a b c

Figure 2. Samples of oak veneer after thermal modification at a temperature of 220 °C for: a – 600 s; b – 1200 s; c – 1800 s.

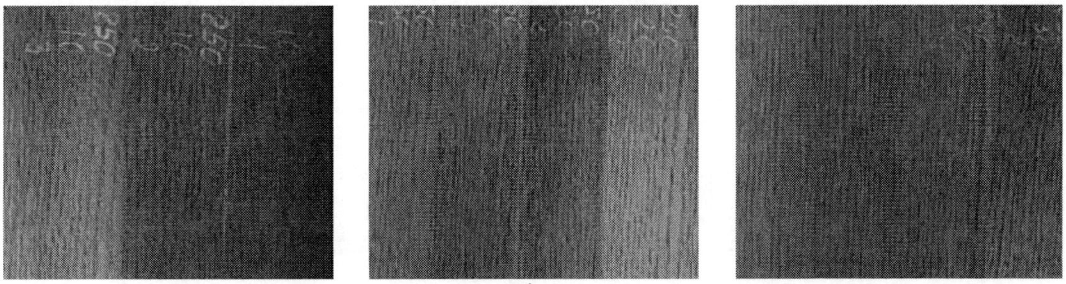

a b c

Figure 3. Samples of oak veneer after thermal modification at a temperature of 250 °C for: a – 600 s; b – 1200 s; c – 1800 s.

Figure 4. Samples of oak veneer after thermal modification at a temperature of 280 °C for 600 s.

Moisture absorption of wood – the ability of dry wood to absorb water vapor. Moisture absorption is defined as the change in mass during holding of a material in a humid environment for a certain time. The essence of the method is that strips of thermodified oak veneer were placed in a desiccator (Figure 5) with a saturated solution of sodium, which was impregnated with filter paper to increase the humidity.

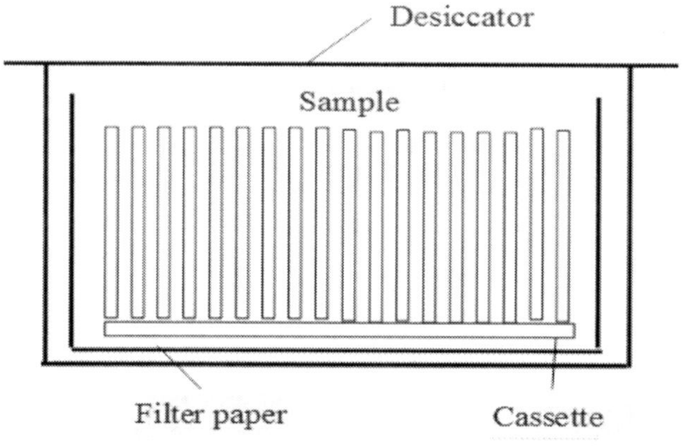

Figure 5. Desiccator for veneer moisture absorption tests.

The study carried out periodic weighing of veneer at the appointed time, namely after 2, 3, 6, 9, 13 days after placement in the desiccator.

The experiment was completed when the difference in mass of the last weighing's did not exceed 0.01 g.

Water absorption – the ability of wood to absorb water in direct contact with it (Figure 6).

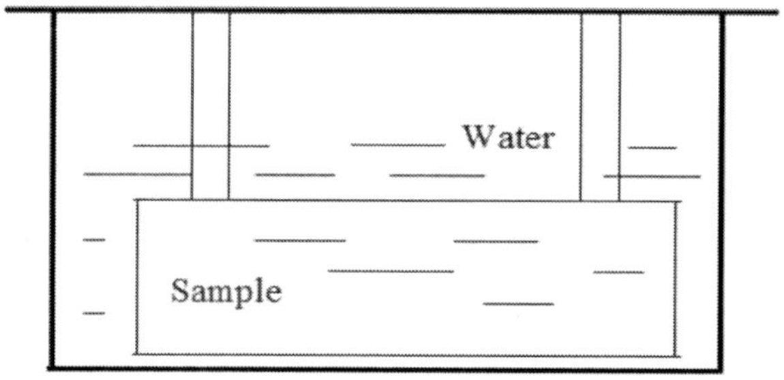

Figure 6. Device for testing the absorption of water by veneer.

The research method provides for the installation of veneer in the cassette, its immersion and exposure of the material in water with periodic weighing. The first recording was carried weight gain after 2 hours, then at 1, 2, 3, 6, 9 days. Weighing was stopped when the difference in readings was not less than 0.05 g.

5. Research results

In Figure 7 shows the process of moisture absorption of veneer samples.

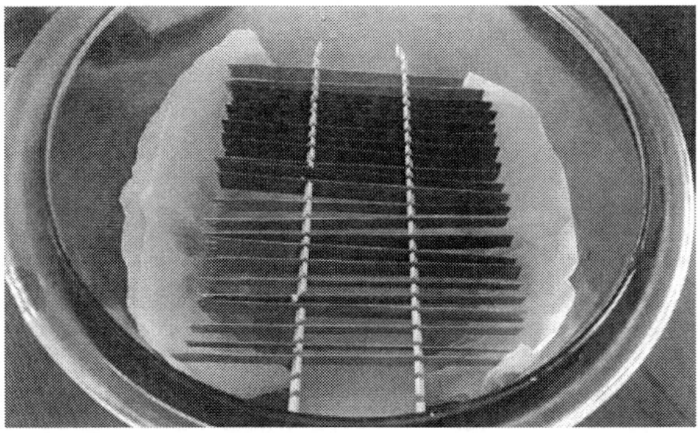

Figure 7. Keeping the test samples in a wet environment.

The results of studies of moisture absorption showed that on the 13th day of aging in a humid environment, the mass of all samples did not increase – Table 1.

It was determined that the samples of oak veneer gained the most moisture from the environment in the first two days. The weight of unmodified veneer increased by 0.06 g, similar results showed veneer modified at lower temperatures. Thermodified veneer at a temperature of 250 °C for 20, 30 min and 280 °C in the first two days gained 2 times less – 0.03 g. Over the next 11 days, the control samples gained another 0.09 g, slightly less samples modified at temperature 160 °C, and modified at 220 °C and above not more than 0.03 g. Compared with unmodified veneer, which in two weeks gained 22.41 % moisture, veneer modified at 250 and 280 °C gained 2.5 times less moisture – an average of 8.6 %.

Table 1. The results of studies of moisture absorption of veneer.

Thermomodification mode temperature / duration	Mass of samples after keeping in a wet environment, g						Absorbed moisture,%
	Exposure in a wet environment, days						
	0	2	3	6	9	13	
Control	0.67	0.73	0.76	0.80	0.82	0.82	22.41
160°C /10 minutes	0.64	0.71	0.72	0.73	0.76	0.76	18.12
160°C /20 minutes	0.60	0.64	0.66	0.67	0.70	0.70	16.74
160°C /30 minutes	0.62	0.68	0.70	0.71	0.72	0.72	15.85
220°C /10 minutes	0.65	0.70	0.71	0.71	0.73	0.73	11.61
220°C /20 minutes	0.60	0.66	0.67	0.67	0.67	0.67	11.78
220°C /30 minutes	0.63	0.66	0.67	0.68	0.69	0.69	9.58
250°C /10 minutes	0.63	0.68	0.69	0.69	0.70	0.70	10.66
250°C /20 minutes	0.60	0.63	0.64	0.65	0.65	0.65	8.87
250°C /30 minutes	0.66	0.69	0.70	0.70	0.72	0.72	8.37
280°C /10 minutes	0.56	0.59	0.60	0.60	0.61	0.61	8.59

In Figure 8 shows the process of water absorption by veneer.

The effect of high temperature also had a positive effect on reducing the water absorption of veneer in Table 2.

Figure 8. Exposure of samples in water.

Table 2. Change in the mass of samples of thermodified veneer in water

Thermomodification mode temperature / duration	Mass of samples after keeping in a wet environment, g							Absorbed moisture,%
	Exposure in a wet environment, days							
	0	0,1	1	2	3	6	9	
Control	6.21	10.32	11.07	11.93	12.01	12.13	12.30	98.00
160 °C /10 minutes	6.41	10.55	11.17	11.86	11.93	12.38	12.44	94.33
160 °C /20 minutes	6.25	10.34	10.98	11.53	11.62	11.99	12.09	93.33
160 °C /30 minutes	6.27	10.32	10.89	11.55	11.63	11.86	11.99	91.33
220 °C /10 minutes	6.07	9.32	10.18	11.09	11.19	11.48	11.59	91.33
220 °C /20 minutes	6.14	9.56	10.43	11.29	11.22	11.58	11.70	90.67
220 °C /30 minutes	6.14	9.34	10.28	11.23	11.22	11.58	11.66	89.67
250 °C /10 minutes	5.85	7.62	8.52	10.17	10.34	10.90	11.14	90.33
250 °C /20 minutes	5.97	8.14	9.16	10.70	10.80	11.14	11.27	89.00
250 °C /30 minutes	5.87	7.61	8.48	10.12	10.34	10.81	10.93	86.00
280 °C /10 minutes	5.64	6.82	7.47	9.07	9.28	10.27	10.44	85.00

According to the results of research, it is seen that in the first two hours, all samples took a significant amount of water. Thus, unmodified veneer and thermodified at temperatures of 160 and 220 °C absorbed 50…66 % of its initial weight, and modified at 250 and 280 °C – 20…36 %. The nature of moisture accumulation in the following days during the continuation of the experiment is interesting. Control samples of veneer and modified at lower temperatures with each subsequent weighing showed smaller values. But the samples modified at temperatures of 250 and 280 °C for the second day gained almost as much water as in the first two hours, then the difference decreased.

Control samples of veneer for the whole experiment gained 98 % of the initial mass in Figure 9.

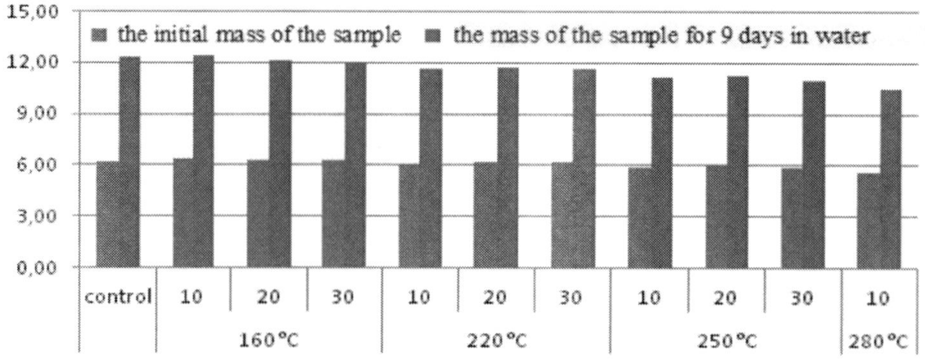

Figure 9. The results of water absorption of veneer of various degrees of modification.

The best results showed samples of veneer thermodified at a temperature of 280 °C, the weight of which increased by 85%.

6. Conclusions

Thus, as a result of experimental researches the positive influence of thermal modification of a veneer on water absorption is established. It is established that with the increase of the modification temperature to 280 °C, the water absorption decreases more than twice. The results of the research will also allow to purposefully solve further problems on the creation of new methods of thermal modification of wood and the conditions of its operation at various sites.

References

[1] Lo Monaco A, Pelosi C, Agresti G, Picchio R and Rubino G 2020 Influence of thermal treatment on selected properties of chestnut wood and full range of its visual features *Drewno* **63 (205)** pp 5-24

[2] Tsapko Yu, Tsapko A and Bondarenko O 2020 Modeling of thermal conductivity of reed products *IOP Conf. Series: Materials Science and Engineering. Innovative Technology in Architecture and Design* **907** 012057 doi:10.1088/1757-899X/907/1/012057

[3] Tsapko Yu, Tsapko A and Bondarenko O 2020 Research of conditions of removal of fire protection from building construction *Key Engineering Materials* **864** pp 141-148 doi:10.4028/www.scientific.net/KEM.864.141

[4] Tsapko Yu, Rogovskii I, Titova L, Bilko T, Tsapko A, Bondarenko O and Mazurchuk S 2020 Establishing regularities in the insulating capacity of a foaming agent for localizing flammable liquids *Eastern-European Journal of Enterprise Technologies* **5/10 (107)** pp 51-57 doi: 10.15587/1729-4061.2020.215130

[5] Tsapko Yu, Horbachova O, Tsapko A, Mazurchuk S, Zavialov D and Buiskykh N 2021 Establishing regularities in the propagation of phase transformation front during timber thermal modification *Eastern-European Journal of Enterprise Technologies* **1/10 (109)** pp 30–36 doi: 10.15587/1729-4061.2021. 225310

[6] Humar M, Lesar B and Kržišnik D 2020 Moisture performance of façade elements made of thermally modified norway spruce wood Forests **11 (3)** p 348 doi: 10.3390/f11030348

[7] Esteves B and Pereira H 2009 Wood modification by heat treatment *Bioresources* **4 (1)** pp 370-404

[8] Humar M, Repič R, Kržišnik D and Lesar B 2020 Quality control of thermally modified timber using dynamic vapor sorption (dvs) analysis *Forests* **11 (6)** p 666 doi: 10.3390/f11060666

[9] Aytin S and Korkut P 2016 Effect of thermal treatment on the swelling and surface roughness of common alder and wych elm wood *Journal of Forestry Research* **27 (1)** pp 225-229 doi: 10.1007/s11676-015-0136-7

[10] Pelosi G, Agresti L, Lanteri R, Picchio E, Gennari E and Lo Monaco A 2020 Artificial weathering effect on surface of heat-treated wood of ayous *The 1st International Electronic Conference on Forests (IECF)*

[11] Ugovšek B, Šubic G, Humar M, Lesar B, Thaler N, Brischke C, Jones D and Lozano J I 2016 Performance of windows and façade elements made of thermally modified norway spruce (picea abies) in different climatic conditions *In Proceedings of the WCTE 2016-World Conference on Timber Engineerin* doi: 10.1007/s11998-016-9871-8

[12] Ugovšek B, Šubic G, Starman J, Rep G, Humar M, Lesar B, Thaler N, Brischke C, Meyer-Veltrup L, Jones D, Häggström U and Lozano J I 2019 Short-term performance of wooden windows and facade elements made of thermally modified and non-modified Norway spruce in different natural environments *Wood Material Science and Engineering* **14** pp 42-47

[13] Bonifazi G, Serranti S, Capobianco G, Agresti G, Calienno L, Picchio R, Lo Monaco A, Santamaria U and Pelosi C 2017 Hyperspectral imaging as a technique for investigating the effect of consolidating materials on wood *Journal of Electronic Imaging* **26 (1)** 011003

[14] Jones D, Sandberg D, Goli G and Todaro L 2019 *Wood Modification in Europe: a state-of-the-*

art about processes, products and applications 123
[15] Råberg U, Edlund M-L, Terziev Nasko and Johan Land C 2005 Testing and evaluation of natural durability of wood in above ground conditions in Europe – an overview *Journal of Wood Science* **51** pp 429v440
[16] Chong J, Yang Z, Juqing C and Lu G 2019 *The antibacterial properties and safety of a nanoparticle-coated parquet floor* Coatings **9 (6)** 403

Acknowledgements
Authors express their gratitude to the Ministry of Education and Science of Ukraine for financial support of the research, that was performed in the framework of budget funding No. 0121U001007, as well as for the development of the theme of research according to the program of scientific cooperation COST Action FP 1407 "Understanding the modification of wood through an integrated scientific and environmental approach" of the European Union's framework program HORIZON 2020.

Optimization of the calculated scheme

I Medved[1,*], M Surianinov[2], Y Otrosh[3] and O Pirohov[3]

[1]Volodymyr Dahl East Ukrainian National University, 59-a Central Avenue, m. Sevurodonetsk, 93400, Ukraine
[2]Odessa State Academy of Civil Engineering and Architecture, 4, Didrihsona str., Odessa 65029, Ukraine
[3]National University of Civil Defense of Ukraine, 94, Chernyshevska str., Kharkiv 61023, Ukraine

*E-mail: iw.medwed@yandex.ua

Abstract. Any calculation is preceded by the stage of creating a design scheme of a real design, which would most accurately respond to real operating conditions. The more optimally drawn up the calculation scheme, the less time consuming there will be stages of calculation and designing the corresponding design. There are practically no results of the use of such an approach in the calculations of building structures. This article an attempt is made to show the possibility and feasibility of using "search for solutions" (add-in for Microsoft Excel) at the stage of selecting the optimal parameters of the design scheme.

1. Introduction and literature review

Most often, the add-on "solution search" is used in solving optimization problems of the economy (simplex method, transport task, etc. [1, 2, 3]. There are practically no results of the use of such an approach in the calculations of building structures. The paper [4] considers a class of problems for covering an area with a set of geometric objects of a given figure. To formalize the problem, the concept of configuration space of geometric objects is used, the generalized variables of which are metric and placement parameters. The authors consider the method of solving the problem of optimizing the allocation of limited resources of the project as a problem of location of rectangular objects, where the placed objects have variable metric characteristics, which are subject to functional dependencies [5].

2. Literature review

Building structures are very diverse in its intended purpose and use. The reliability and safety of their work depends on many factors: the geometric sizes used by materials, external loads and their combinations, etc. All these parameters determine the internal efforts, stresses and deformations that arise in the structures that determine their strength, rigidity and stability [6, 7].

In order to ensure the strength, rigidity and stability of buildings and their structural elements, appropriate calculations are performed [9].

In the area of the theory of calculation of building structures there is a permanent refinement to the actual work of these structures, i.e. Create such calculation schemes that most accurately meet the actual operating conditions [10]. In solving these tasks, a huge role belongs to the introduction of a computer [9].

The study of operations implies mathematical modeling of economic processes [7, 8]. The integrity of the decisions made [1].

When solving a specific task, the use of methods for studying operations involves the construction of mathematical models for the decision-making tasks in difficult situations or under uncertainty. For a quantitative evaluation of the study, a mathematical model of the operation is required.

The operation model is a fairly accurate description of the operation with the help of a mathematical apparatus (various kinds of functions, equations, systems of equations and inequalities, etc.).

Content from this work may be used under the terms of the Creative Commons Attribution 3.0 licence. Any further distribution of this work must maintain attribution to the author(s) and the title of the work, journal citation and DOI.
Published under licence by IOP Publishing Ltd

The operation model is an analytical dependence of the target function from dependent (controlled) variables, which in certain limits we can choose at our discretion and choose the range of their change.

The preparation of the operation model requires a deep understanding of the essence of the described phenomenon and knowledge of the mathematical apparatus. The efficiency of the operation is quantitatively expressed as a numerical value of the target function.

Among the models of research of operations should be noted primarily a large class of optimization models. Such tasks occur when trying to optimize the planning and management of complex systems.

Any calculation is preceded by the stage of creating a design scheme of a real design, which would most accurately respond to real operating conditions. The more optimally drawn up the calculation scheme, the less time consuming there will be stages of calculation and designing the corresponding design. If the effectiveness criterion represents a linear function, and the variables in the system of restrictions are also linear, then such a task is a linear programming task. Of the listed methods of mathematical programming, linear programming is the most common and developed [1,2,3].

"Solution Search" is an add-in for Microsoft Excel, which can be used in the tasks of calculating building structures. With it, it is possible to find the optimal value (maximum or minimum) of the formula contained in one cell, called the target, taking into account restrictions on variable values in other cells.

Simply put, using the "Solution Search" superstructure, you can determine the maximum or minimum value of one cell, changing other cells.

3. Materials of research

Analysis of the latest achievements and publications. Most often, the add-on "solution search" is used in solving optimization problems of the economy (simplex method, transport task, etc. [1, 2, 3]. There are practically no results of the use of such an approach in the calculations of building structures.

Consider the simplest building structure. To illustrate the idea of the proposed approach, a simple statically defined farm on two supports is intentionally selected (Figure 1). This is done in order to make the idea of the proposed approach due to the complexity and cumbersome.

In Kurs, construction mechanics known formula to determine the numerical value of the generalized internal power factor F along the corresponding line of influence [11]:

$$F = \sum q_i w_i + \sum P_i y_i + \sum M_i \tan \alpha_i \,. \tag{1}$$

Using the add-on "Decision Search", we define the optimal geometric dimensions of the farm at a given external load. As an optimization parameter, we select a minimum of force in individual farm rods [11]. For all selected rods, the target feature will look:

$$F = P/2 \times y_2 + P \times y_3 + \ldots + P \times y_7 + P/2 \times y_8 , \tag{2}$$

where: $y_2, y_3, \ldots, y_8$ - ordinates of the corresponding line of influence.

Now you can clarify the task. We find such values a, b, c, d at which the target function (2) will take the minimum value at a given external load P = 100kN.

Obviously, the real operating conditions of the design are imposed on variable values of certain limitations.

Let those restrictions be the following:

d = 2.2 m-throttle farm on supports;

2 (a + b + c)= 18 m - farm persolem,

In addition, we will require the dimensions of a, b, c were a multiple 1.5 m.

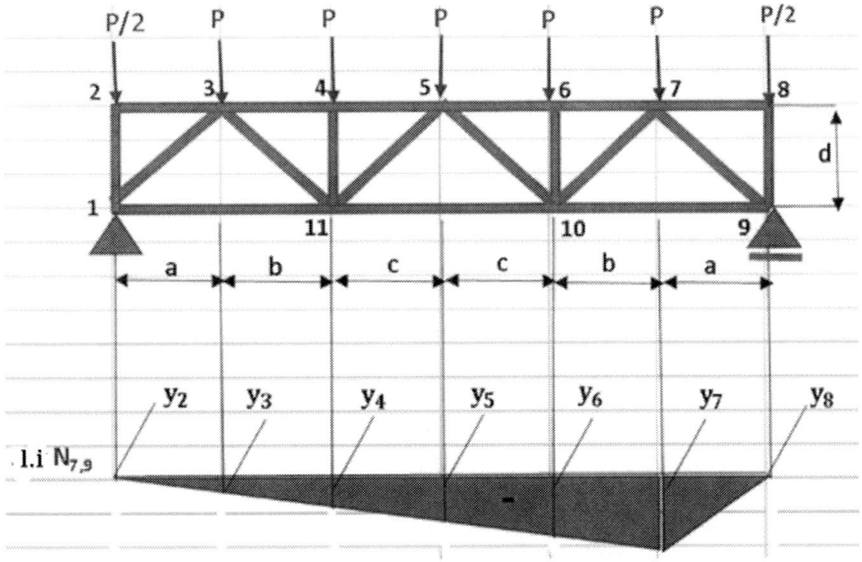

Figure 1. Calculated scheme and the line of influence of the longitudinal force.

$$y_2 = 0,$$

$$y_3 = \frac{a \times \sqrt{a^2 + d^2}}{2 \times (a+b+c) \times d},$$

$$y_4 = \frac{(a+b) \times \sqrt{a^2 + d^2}}{2 \times (a+b+c) \times d},$$

$$y_5 = \frac{(a+b+c) \times \sqrt{a^2 + d^2}}{2 \times (a+b+c) \times d},$$

$$y_6 = \frac{(a+b+2c) \times \sqrt{a^2 + d^2}}{2 \times (a+b+c) \times d},$$

$$y_7 = \frac{(a+2b+2c) \times \sqrt{a^2 + d^2}}{2 \times (a+b+c) \times d},$$

$$y_8 = 0.$$

The selected restriction system is entered into the solution search parameters table (Figure 2).

IOP Conf. Series: Materials Science and Engineering 1164 (2021) 012051 doi:10.1088/1757-899X/1164/1/012051

Figure 2. Table of solutions search parameters.

The calculation results are presented in Figure 3.

	a	b	c	d	i	j	k
	1.5	3	4.5	2.2	1	2	3
y	0	0.10086	0.30258	0.60516	0.90774	1.109461	0
P	50	100	100	100	100	100	50

302.5802	Target function
18	2(a+b+c)
1.5	1.5*i
3	1.5*j
4.5	1.5*k

Figure 3. Numerical values of controlled variables and target function.

The value of the target function corresponds to the minimum value of the longitudinal force in the rod 7-9 ($N_{7\text{-}9}$ (min) = - 302.6kN) at a = 1.5m, b = 3m, C = 4.5m.

Similar results were obtained when searching for a minimum of effort in rods 10-11 and 5-6 (Figure 4, 5).

	a	b	c	d	i	j	k
	1.5	3	4.5	2.2	1	2	3
y	0	0.340909	1.022727	2.045455	1.022727	0.340909	0
P	50	100	100	100	100	100	50

477.2727	Target function
18	2(a+b+c)
1.5	1.5*i
3	1.5*j
4.5	1.5*k

Figure 4. Numerical values of controlled variables and target function. For N $_{10\text{-}11}$ (min)

	a	b	c	d	i	j	k
	1.5	3	4.5	2.2	1	2	3
y	0	0.170455	0.511364	1.022727	1.534091	0.5113636	0
P	50	100	100	100	100	100	50

375	Target function
18	2(a+b+c)
1.5	1.5*i
3	1.5*j
4.5	1.5*k

Figure 5. Numerical values of controlled variables and target function.

For N $_{5\text{-}6}$ (min)

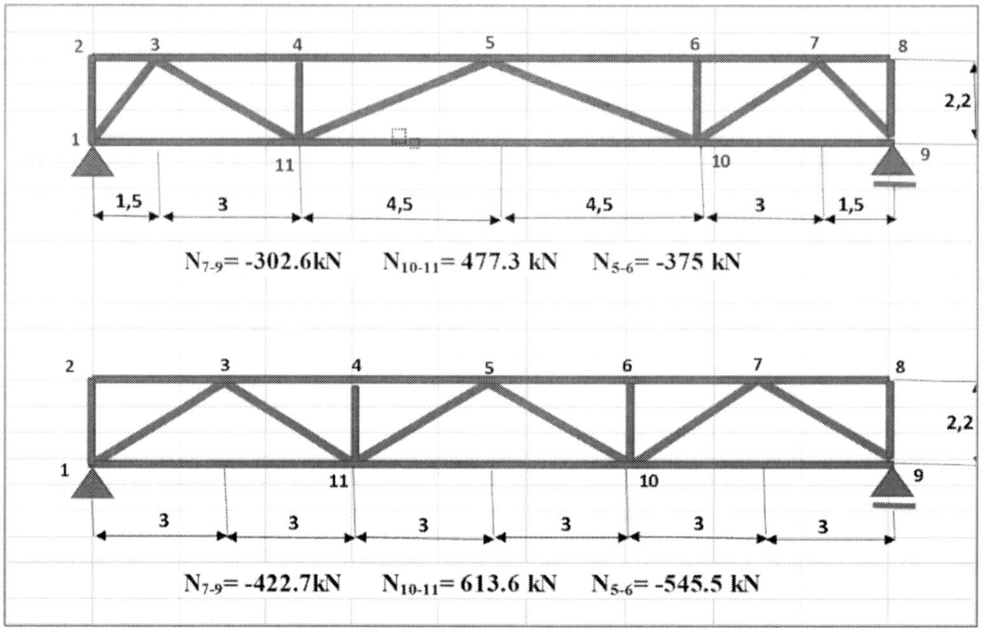

Figure 6. Comparative analysis of the calculated schemes.

4. Conclusions. This article shows the possibility and feasibility of using "search for solutions" at the stage of selecting the optimal parameters of the design scheme. After the calculated scheme is determined, more powerful generally accepted means of calculating building structures can be attracted for its complete calculation [9].

References

[1] Kremer M Sh 1999 *Research of operations in economics* Textbook settlement (M.: Marketing) 270

[2] Karmanov V G 1986 *Mathematical programming* (M.: Nauka)

[3] Romanyuk T P, Tereshchenko T O, Prisenko G V and Gorodkova I M 1996 *Mathematical program* Navch. posibnik (K.: IZMH) 1996

[4] Yakovlev S, Kartashov O, Komyak V, Shekhovtsov S, Sobol O and Yakovleva I 2019 Modeling and simulation of coverage problem in geometric design systems *15th International Conference on the Experience of Designing and Application of CAD Systems CADSM* **8779303** pp 20–23

[5] Chub I, Novozhylov M and Murin M 2013 Optimization problem of allocating limited project resources with separable constraints *Cybernetics and Systems Analysis* **49(4)** pp 632–642

[6] Makhinko A and Makhinko N 2021 Resource-saving steel hopper for loading of railway cars *IOP Conf. Series: Materials Science and Engineering* **1021** 012021

[7] Makhinko A and Makhinko N 2019 To the calculation of the optimal level of reliability by using economic indicators *International Conference Current Issues of Civil and Environmental Engineering* **47**

[8] Kondratiev A and Gaidachuk V 2019 Weight-based optimization of sandwich shelled composite structures with a honeycomb filler *Eastern-European Journal of Enterprise Technologies* **1(97)** pp 24 – 33

[9] Bashynska O, Otrosh Y, Holodnov O, Tomashevskyi A and Venzhego G 2020 Methodology for Calculating the Technical State of a Reinforced-Concrete Fragment in a Building Influenced by High Temperature *In Materials Science Forum. Trans Tech Publications Ltd* **1006** pp 166-172

[10] Kovalov A, Otrosh Y, Rybka E, Kovalevska T, Togobytska V and Rolin I 2020 Treatment of Determination Method for Strength Characteristics of Reinforcing Steel by Using Thread Cutting Method after Temperature Influence *In Materials Science Forum. Trans Tech Publications Ltd.* **1006** pp 179-184
[11] Darkov A V and Shaposhnikov N N 1986 *Structural mechanics* (M.: Higher. shk) 608

IOP Conf. Series: Materials Science and Engineering 1164 (2021) 012052 doi:10.1088/1757-899X/1164/1/012052

Investigation of impulse load attenuation in closed-cell foam in the framework of couple stress elasticity

O Mikulich, V Shvabyuk

Applied Mathematics and Mechanics Department, Lutsk National Technical University, 75 Lvivska Street, Lutsk, Ukraine

shyprao@gmail.com

Abstract. The article analyzes the processes of attenuation of impulse loads in foam materials with closed-cell. In the framework of the couple stress elasticity, the analysis of the distribution of dynamic radial stresses in foam media with tunnel cavities of constant cross-section under the action of an impulse load of different duration applied to the cavity boundary in the radial direction is performed. An analytical-numerical approach is used for calculating of the dynamic stresses, which is based on the application of the Fourier transform over time and the modified boundary integral equation method. Numerical calculations of the dynamic stresses were performed using the method of mechanical quadrature and collocation method. The modified algorithm for discrete inverse Fourier transform was used for the calculation of the originals of the dynamic radial stresses. Based on the analysis of numerical calculations, the influence of the impulse duration on the attenuation rate of the pulse load in foam media is established.

1. Introduction

Foam materials are now becoming increasingly popular in construction, as they have good heat-retaining, noise and vibration-absorbing properties. Polyurethane-based polymer foams are characterized by the lightness of the obtained foam material, the absence of the effect of temperature and humidity, water resistance.

Therefore, for selecting the optimal parameters of such foam materials for ensuring vibration and noise protection of structural elements, it is necessary to have information on the rate of attenuation of non-stationary influences in the respective foam materials.

The use of the model of the classical theory of elasticity, which is suitable for the study of elastic deformations in structural materials: steel, concrete, aluminum, gives quite large errors for the case of the study of mechanical effects in porous, granular, fibrous, etc. materials, which is confirmed by the results of experiments. This is due to the need of accounting for the structural heterogeneity of their structure, and therefore it is necessary to use refined models of the continuous mechanics, which accounts for the discrete structure of these materials [1, 2].

However, this consideration leads to significant complications of the equations describing the relationship between stresses and strains in such media, and, at the same time, to a significant complication of the methods of constructing solutions of the corresponding classes of problems. Since within the classical mechanics, there are a large number of problems that need to be solved, much less attention is paid to refined models and, together with them, to the development of analytical and

Content from this work may be used under the terms of the Creative Commons Attribution 3.0 licence. Any further distribution of this work must maintain attribution to the author(s) and the title of the work, journal citation and DOI.
Published under licence by IOP Publishing Ltd

numerical methods for constructing solutions of the corresponding problems. But the relevance and importance of the described problems necessitate the search for approaches to their solution in the framework of refined models of continuous mechanics.

2. Statements of the problem

To study the processes of attenuation of the impulse load in foam materials, we investigate the distributions of dynamic radial stresses in foam media under the action of according load applied to the boundary of the tunnel cavity. Based on the analysis of the values of the maximum normalized dynamic stresses at different internal points of the medium, we analyze the influence of the impulse duration on the rate of its attenuation.

Consider the foam medium with the tunnel cavity of constant cross-section (Fig. 1, a). To the boundary of the cavity the non-stationary load in the normal direction is applied (Fig. 1, b).

Denote by L the boundary of a cross-section of the cavity. The centre of gravity is placed at the origin of a Cartesian coordinate system Ox_1x_2.

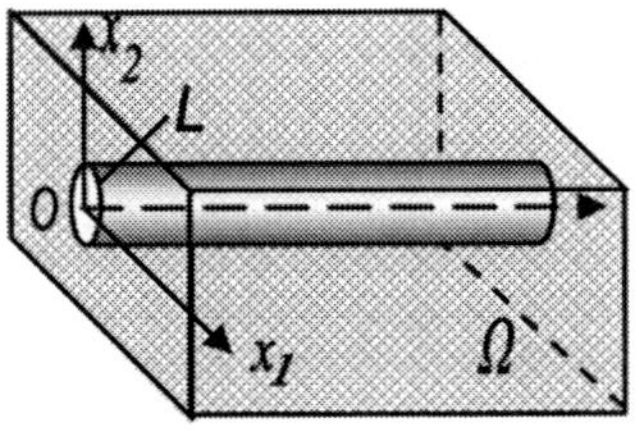

Figure 1, a. 3D full geometry of the problem.

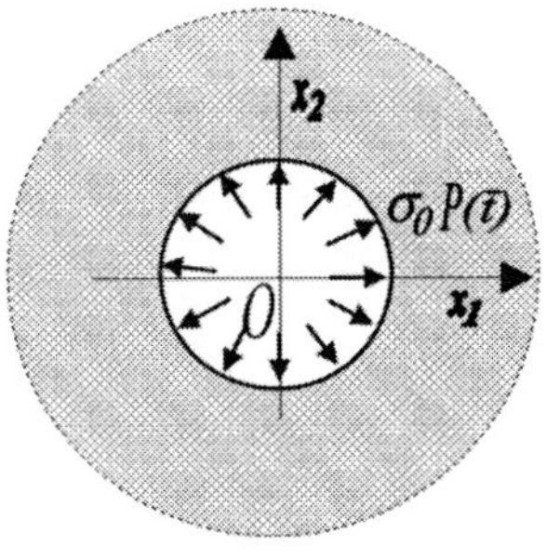

Figure 1, b. 2D plane geometry of the problem.

The problem consists in the determination and comparing of the radial dynamic stresses in the different internal points of foam medium with tunnel cavity under the action of impulse load different duration, which is applied in normal to the boundary of cavity direction.

For accounting for the influence of the microstructure of the foam material on the stress distribution in the structurally inhomogeneous material, we use the apparatus of couple stress elasticity theory [3], which makes it possible the accounting for the influence of shear-rotational deformations of microparticles with some simplification. In numerical calculations, the values of the elastic characteristics of the microstructure of the material are chosen according to the works of R.S. Lakes [4,5] obtained in Eringen's notation [1]. For the case of micropolar stress elasticity in Novatsky's notation [2], these elastic characteristics are given in S. Hassanpour [6].

The boundary conditions of the problem are written:

$$\sigma_n\big|_L = \sigma_0 P(t),\ \ \tau_{sn}\big|_L = 0; \tag{1}$$

where σ_0 is a constant characterized the value of the maximum intensity of applied load, $P(t)$ is known a function describes the change of load intensity over time. In the case of the action of impulse load the change of the load intensity over time is chosen in the form [7]:

$$P(t) = p_* \bar{t}^{\,-n_*} e^{-\alpha_* \bar{t}},\ \bar{t} > 0,\ n_* > 0, \tag{2}$$

where α_*, n_*, p_* are the constants, $\bar{t} = \dfrac{t \cdot c_l}{a}$ is the time parameter, $c_l = \sqrt{\dfrac{(\lambda + 2\mu)}{\rho}}$ is the speed of the expansion wave, a is character size (for the case of tunnel cavity of circular cross-section $a = R$, where R is the radius of cavity), λ and μ are Lame parameters.

3. Solution of the problem

To ensure the accuracy of numerical calculations under the action of non-stationary load for the determination of dynamic radial stresses, we use the analytical-numerical approach [8], which is based on the use of Fourier transform for time variable and modified boundary integral equation method. The use of such an approach makes it possible to obtain analytical dependences for the determination of dynamic stresses in integral form. Numerical calculation occurs only at the last stage of determining the dynamic stresses.

3.1. Analytical approach for solving of the problem

For the solving of the non-stationary problem in the framework of couple stress elasticity we use the equation of the motion in displacement form [8, 9]:

$$(\lambda + 2\mu)\, grad\, div\, \vec{u} + rot\, rot\left(\frac{B}{4}\Delta\vec{u} - \mu\vec{u}\right) = \rho\frac{\partial^2 \vec{u}}{\partial t^2} \tag{3}$$

where $\vec{u} = \vec{u}(x,t) = \{u_j(x,t)\}, j = 1,2$ is a vector of displacements, B is the constant of microstructure of a material, Δ is Laplace operator, λ is Lame parameter, μ is shear modulus.
Applying the Fourier transform [10]:

$$\tilde{f}(x,\omega) = \int_{-\infty}^{\infty} f(x,t)e^{-i\omega t}dt, \tag{4}$$

to the motion equation of couple stress elasticity (3) we obtained the equations, which are equivalent to the equations of time-harmonic motion with cyclic frequency ω [8]. Further solving of the problem we realized for Fourier transforms of displacements and stresses, which we denote with symbol '~'.
In the framework of using of modified boundary integral equation method [8] we wrote the representation of displacements for the second exterior problem as [8]:

$$\tilde{u}_i = \int_L p_j U^*_{ji}(x, x^0)ds + \int Q_j U^*_{ji}(x, x^0)d\Omega \tag{5}$$

where $p_j, j = 1, 2$ are unknown potential functions. The fundamental functions U^*_{ji} for couple stress elasticity were built regarding Sommerfield radiation condition in form:

$$U^*_{jk} = U^{*(CTE)}_{jk} + U^{*(CSE)}_{jk}, \tag{6}$$

where elements with index *(CTE)* correspond to the case of classical theory of elasticity, and elements with index *(CSE)* correspond to the case of couple stress elasticity [8].
For the determination of the unknown functions $p_j, j = 1, 2$ we satisfy the Fourier transforms of the boundary condition (1) and apply Plemelj-Sokhotski formulas for the limits, when internal point tends to the boundary. We obtain the system of integral equations in form:

$$\begin{aligned} &\frac{Re(q)}{2} + PV\int_L (f_1(x,x^0)qd\zeta + f_2(x,x^0)\bar{q}d\bar{\zeta}) = \sigma_0\tilde{P}(\omega); \\ &\vartheta\frac{Im(q)}{2} + PV\int_L (g_1(x,x^0)qd\zeta + g_2(x,x^0)\bar{q}d\bar{\zeta}) = 0; \end{aligned} \tag{7}$$

where $pds=-iqd\zeta$, $p=p_1+ip_2$ is the unknown complex function, $\zeta=x_1^0+ix_2^0$, ϑ is the constants: $\vartheta=1-\frac{B\rho}{4E^2}(1+\nu)^2\omega^2$ for the case of plane deformation. Here the integrals are understood in the sense of Cauchy principal value.
The representation for determination of Fourier transforms of radial stresses in the internal points of medium were built by substituting the dependencies (5) to the stress formulas [2] in form:

$$\tilde{\sigma}_r = PV\int_L (h_1(x,x^0)qd\zeta + h_2(x,x^0)\bar{q}d\bar{\zeta}); \tag{8}$$

where $h_j(x,x^0)$, $j=1,2$ are known functions [8].

3.2. Numerical approach for stress calculations
The system of integral equations (7) is solved numerically using the methods of mechanical quadrature and collocation method. After discretization of the boundary of the cross-section of the tunnel cavity, the system of integral equations is reduced to a system of linear algebraic equations. A refined Gaussian method was used to solve the resulting system of equations.
After determining the unknown functions from the system of equations (7), the Fourier images of dynamic radial stresses are calculated numerically by formula (8).
For calculation of originals of the dynamic radial stresses the modified discrete Fourier transform is used:

$$\sigma_r(t_k)=\frac{2}{T}\sum_{n=0}^{K-1}\left(Re\,\tilde{\sigma}_r(\omega_n)exp\left(2\pi i\frac{nk}{K}\right)-\tilde{\sigma}_r(\omega_n)\right) \tag{9}$$

where K is the number of elements of selection, $\omega_n=\frac{2\pi n}{t}$ are frequencies, $t_k=\frac{k\cdot t}{K}$, $k=\overline{0,K-1}$ is time moments.

4. Analysis of numerical calculation results
Let's study the distribution of dynamic radial stress in internal points of foam media under the action of the impulse load applied to the boundary of tunnel cavity (Fig. 2).

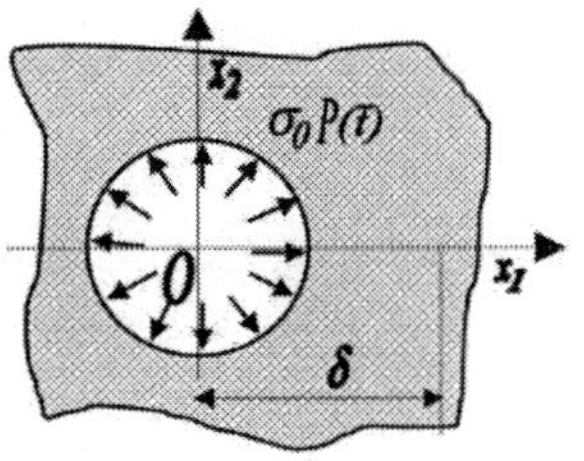

Figure 2. Cross-section of tunnel cavity.

Numerical calculations are performed for the case of WF300 and WF110 foam media. Numerical calculations are performed for the case when the radius of the tunnel cavity is $R=3\ell$, where $\ell=\sqrt{\frac{\gamma+\varepsilon}{4\mu}}$, l is scale effect in couple stress elasticity.
Micropolar properties of these materials are present in [6] based on the works of R.S. Lakes [4, 5]. The values of elastic classical and micropolar properties of these foams are chosen according to [6]. For the foam WF300 the values of the physics characteristics of were chosen: the density of foam was

ρ=380 kg/m^3, Young's modulus was E=655.5 MPa, Poisson's ratio ν=0.13, Cosserat Twist coefficients ε=502.2 N, γ=185.6 N. For the foam WF110 the values of the physics characteristics of were chosen: the density of foam was ρ=110 kg/m^3, Young's modulus was E=215 MPa, Poisson's ratio ν=0.43, Cosserat Twist coefficients ε=16.47 N, γ=20.28 N.

The distributions of normalized dynamic radial stresses are performed for seven internal points of foam medium for different distance from the boundary of the cavity and different duration of impulse. For numerical calculations values of constants of the impulse load in (2) are chosen as: $\alpha_* = 5,\ p_* = 47,\ n_* = 2$ for the case of $\bar{t}_* = 2$, and $\alpha_* = 2.5,\ p_* = 11.7,\ n_* = 2$ for the case of $\bar{t}_* = 4$, and $\alpha_* = 1.25,\ p_* = 2.89,\ n_* = 2$ for the case of $\bar{t}_* = 8$, and $\alpha_* = 0.9375,\ p_* = 1.64,\ n_* = 2$ for the case of $\bar{t}_* = 12$. Here $\bar{t}_*$ is dimensionless parameter, which according to the impulse duration.

The distribution of normalized dynamic radial stress in WF300 foam for the case of different durations of impulse load is present in Fig. 3. Here the curve 1 according to the case of $\delta = R$ (on the boundary of the cavity), curve 2-7 according to the case of $\delta = \{2.5;\ 3;\ 5;\ 7;\ 10;13\}R$ (in the internal points of the medium).

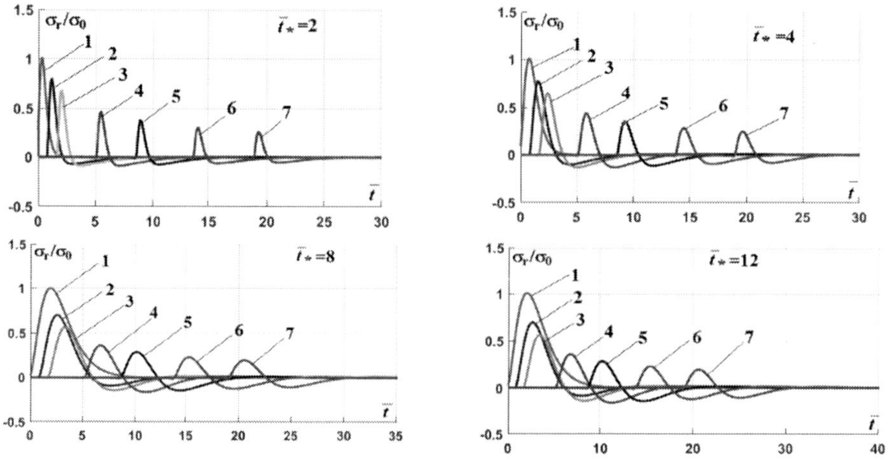

Figure 3. The distribution of normalized dynamic radial stress in WF300 foam for the case of different durations of impulse load.

The values of the maximum normalized stresses at different points of the medium for WF300 foam are presented in Table 1.

Table 1. Maximum values of normalized radial stresses σ_r / σ_0 for the case of WF300.

Distance \ Impulse duration	$\bar{t}_* = 2$	$\bar{t}_* = 4$	$\bar{t}_* = 8$	$\bar{t}_* = 12$
$\delta = R$	1	1	1	1
$\delta = 2.5R$	0.8017	0.7734	0.6989	0.6978
$\delta = 3R$	0.6806	0.6500	0.5640	0.5611
$\delta = 5R$	0.4719	0.4375	0.3620	0.3581
$\delta = 7R$	0.3797	0.3541	0.2874	0.2843
$\delta = 10R$	0.3053	0.2866	0.2270	0.2282
$\delta = 13R$	0.2626	0.2469	0.1989	0.1959

Similar results were obtained for the case of foam WF 110. The results are shown in Table 2.

Table 2. Maximum values of normalized radial stresses σ_r / σ_0 for the case of WF110.

Impulse duration / Distance	$\bar{t}_* = 2$	$\bar{t}_* = 4$	$\bar{t}_* = 8$	$\bar{t}_* = 12$
$\delta = R$	1	1	1	1
$\delta = 1.5R$	0.4519	0.4503	0.0442	0.4489
$\delta = 2R$	0.2542	0.2533	0.2426	0.2415
$\delta = 4R$	0.06354	0.0633	0.06321	0.06312
$\delta = 6R$	0.02823	0.02814	0.02809	0.02801
$\delta = 9R$	0.01254	0.01251	0.0125	0.01247
$\delta = 12R$	0.00705	0.007036	0.007023	0.007013

For analyzing the attenuation processes of the impulse load in foam depending on impulse duration, the corresponding curves are constructed, which are shown in Fig. 4.

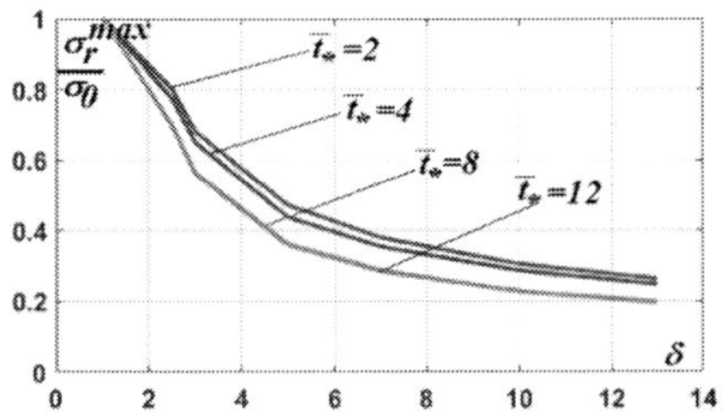

Figure 4, a. Attenuation of impulse load in WF300 closed-cell foam.

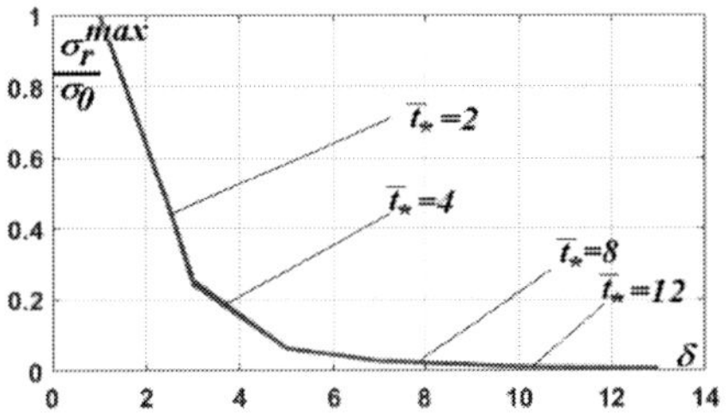

Figure 4, b. Attenuation of impulse load in WF110 closed-cell foam.

5. Conclusion

Analysis of the numerical results shown in Fig. 3 confirms the congruence of the obtained results with the basic principles of wave mechanics - until the impulse does not reach the appropriate internal points the radial stresses are equal to zero.

Analysis of numerical data of Tables 2 and 3 and curves in Fig. 4 confirm the vibration-absorbing properties of foam materials. For medium stiffness foam (WF110) there is no dependence of the impulse duration on its attenuation. Numerical calculations show that a layer of material with a thickness of 30ℓ is sufficient for almost complete attenuation of the impulse.

For high stiffness foam (WF300) there is a little dependence of the impulse damping on its duration. For an impulse load with a long pulse duration ($\bar{t}_* \geq 8$), the damping processes are almost the same for these types of foams. For impulse loads with a short impulse duration, the attenuation rate is 7-10% lower than in the case of a pulse load with a long impulse duration. Numerical calculation shows that in a layer of material with a thickness of 30ℓ the impulse attenuates on 70-80%.

These numerical results are explained by the fact that in foam materials of medium stiffness the shear-rotational deformations of the microparticles of the medium are higher than in foam of high stiffness.

References

[1] Eringen A C 1999 *Microcontinuum Field Theory. I. Foundations and Solids* (Springer, New York)

[2] Nowacki W 1986 *Theory of Asymmetric Elasticity* (Pergamon-Press, Oxford) p 126

[3] Hadjesfandiari A R and Dargush G F 2011 Couple stress theory for solids *International Journal of Solids and Structures* vol 48(18) pp 2496-2510

[4] Anderson W B and Lakes R S 1994 Size effects due to Cosserat elasticity and surface damage in closed-cell polymethacrylimide foam *Journal of Materials Science* vol 29(24) pp 6413–6419
[5] Anderson W B, Lakes R S and Smith M C 1995 Holographic evaluation of warp in the torsion of a bar of cellular solid *Cellular Polymers* vol 14(1) pp 1–13
[6] Hassanpour S, Heppler G 2015 Micropolar elasticity theory: A survey of linear isotropic equations, representative notations, and experimental investigations *Mathematics and Mechanics of Solids* vol 22(2) pp 224-242
[7] Sulym H, Mikulich O and Shvabyuk V 2020 Modelling of impulse load influence on stress state of foam materials with positive and negative Poisson's ratio *Acta Mechanica et Automatica* vol 14 no 2(52) pp 79-83
[8] Mikulich O, Shvabyuk V, Pasternak Ia and Andriichuk O 2018 Modification of boundary integral equation method for investigation of dynamic stresses for couple stress elasticity *Mechanics Research Communications* vol 91 pp 107-111
[9] Savin G N and Shulga N A 1967 Dynamic plane problem of the moment theory of elasticity *Applied mechanics* vol 3(6) pp 216-221
[10] Ramamohan K, Kim D and Hwang J 2010 *Fast Fourier Transform: Algorithms and Applications* (Springer, New York)

Nonbearing Wall infill Under Seismic Impacts In Terms Of Non-Linear Static Analysis

Murashko Oleksiy [1], Kubiiovich Mikola[2], Bezushko Denys[3], Arsiriy Andrii[4]
[1] Department of Reinforced concrete structures and transport structures, Odessa State Academy of Civil Engineering and Architecture, Odessa, Ukraine
[2] Private company "Kapitel-M", Odessa, Ukraine
[3] Department of Civil Engineering and Architecture, Odessa National Maritime University, Odessa, Ukraine
[4] Department of Metal, wooden and plastic structures, Odessa State Academy of Civil Engineering and Architecture, Odessa, Ukraine

E-mail: alexeymurashko@gmail.com

Abstract. The presence of wall infill during seismic impacts is one of the key factors that significantly affect the seismic stability of buildings. Also, irregularity of configuration in plan and height can often lead to more serious damage comparing to regular buildings, due to torsional oscillations and stress concentration on individual elements. The purpose of this article is to study the behavior of various irregularities in the shape of buildings under seismic impacts in terms of non-linear static analysis, taking into account the non-bearing wall infill. For this purpose, five typical geometric shapes of the building plan used in Ukraine and abroad (rectangular, + shape, L-shaped, T-shaped and 4-shaped) were adopted.
The calculations were performed using the Pushover Analysis (nonlinear static analysis) method in CAD Lira-SAPR 2019. The main parameter that forms the basis for seismic stability analysis is the story drift under seismic impacts. Obtained results indicate the need to take into account the effect of wall infill, as this can lead to an increase in the seismic stability of the building and stress concentration on individual elements, which can lead to their collapse.

1. Introduction

Ensuring the reliability and safety of construction in seismic areas is a complex engineering challenge. Modern design norms [1-2] require consideration of the behavior of structural material beyond elasticity. Within the existing systems of assessment of actual seismic stability, used in seismic regions of different states, normative documents regulate the consideration of such factor as "influence of irregularity" [3]. There are two types of structural irregularity; vertical irregularity (also called height irregularity) and irregularity in plan (also called plan asymmetry). Vertical irregularity usually refers to the uneven distribution of mass over the height of a multi-story structure or geometric displacements that change story plan between adjacent stories. Plan irregularity is usually considered as irregular distribution of rigidity or masses in a plan, which results in a torsional reaction of the structure when it is exposed to seismic excitation. Buildings with plane irregularity quite often receive serious damage during earthquake, since response of the structure is not only transverse but also torsional. According to American FEMA-154 [4], the degree of irregularity in the plan is up to 92% (in maximum area for highly seismic zones). According to Swiss system [5] the irregularity is equal up to 100% (both in the zone of the minimum and maximum area), and according to New Zealand system [6-7], this parameter also varies within fairly wide limits.

1.1 Analysis of recent studies

International building code experience is very contradictory when considering the influence of non-bearing wall infill, the consequences of earthquakes are considered as decisive argument. According to the results of their analysis, the influence of non-bearing wall infill on the seismic stability of frame buildings was identified and systematized [8] with taking into consideration the nonlinear behavior of reinforced concrete and requirements the second group of limit states [9].

1.2 Purpose of the work

The purpose of the work is to study the behavior of various irregularities of the configuration of buildings under seismic impacts, taking into account the non-bearing wall infill. This work focuses on buildings with flat slab frame.

2. Main part

This numerical study is the final third stage of the study. The first stage of the study consisted of full-scale testing of multi-story buildings with non-bearing wall infill. The second stage is laboratory studies of wall infill fragments on horizontal

Content from this work may be used under the terms of the Creative Commons Attribution 3.0 licence. Any further distribution of this work must maintain attribution to the author(s) and the title of the work, journal citation and DOI.
Published under licence by IOP Publishing Ltd

shear effects.

The first stage. Number of full-scale and experimental tests was conducted in order to obtain reliable information about behavior of structures with irregular shape in plane considering non-bearing wall infill under earthquakes and vibrations. The tests were performed in three stages: June 15, 2015, August 5, 2015, and August 31, 2015 in the course of in-situ concreting, construction of nonbearing wall infill and obtaining of access to the facility, taking into account the possibility of using a crane to strike the building with wrecking ball attached to hook (tests were conducted with the participation of Murashko A.V., Mikhailova A.A., Mikhailova N.A.). Oscillations were excited also due to strikes by a tamper (Peri formwork beam) and recording of oscillations from wind effects (Fig. 1).

Strikes with wrecking ball stroke central loading core of a building at the level of cover and on the reinforced concrete parapet wall at the level of the second floor. Tampers were also used to strike central loading core at different altitudes (Fig. 2).

Obtained results proved that the method for excitation of oscillation affects only oscillation amplitude while values of oscillation periods remain unchanged. The obtained value of the oscillation periods is 1.63 sec. was compared with the numerical model -1.6556 sec. The second translational and third torsional mode of their self-oscillations could not be excited.

Figure 1. Object of field studies, a) - Photo fixation of building

Figure 2 - Tamper strikes in the elevator section

Obtained results proved that the method for excitation of oscillation affects only oscillation amplitude while values of oscillation periods remain unchanged. The obtained value of the oscillation periods is 1.63 sec. was compared with the numerical model -1.6556 sec. The second translational and third torsional mode of their self-oscillations could not be excited.

The second stage. Experimental tests were carried out in the laboratory of the Department of Reinforced Concrete and Transport Structures of the Odessa State Academy of Civil Engineering and Architecture. Four identical fragments with dimensions 1200x1200x100 mm were prepared (Fig.3.). Fragments of the masonry were tested on a special stand under a monotonous incremental loading (0.4 kN and then to the destructive load - Psection = 4 kN. The tests include the following steps:

- Horizontal concentrated static load is applied in steps of 10% of the destructive load at the level of the upper part of the frame/
- Horizontal low-cycle focused alternating load (fig.4), increased by 10% of the destructive load at the level of upper part of frame (1st stage 10% left, 2nd stage 10% right, 3rd stage 20% left, 4th stage 20% right).

Upon testing of fragments, the displacement of the upper area of the frame in the plane of horizontal seismic loads and the relative displacements along the diagonals of panel were measured.

Figure 3. Experimental samples

Analyzing the result of the experimental tests, we see that the fragments perceive horizontal forces up to 4 tons under deformations up to 13 mm.

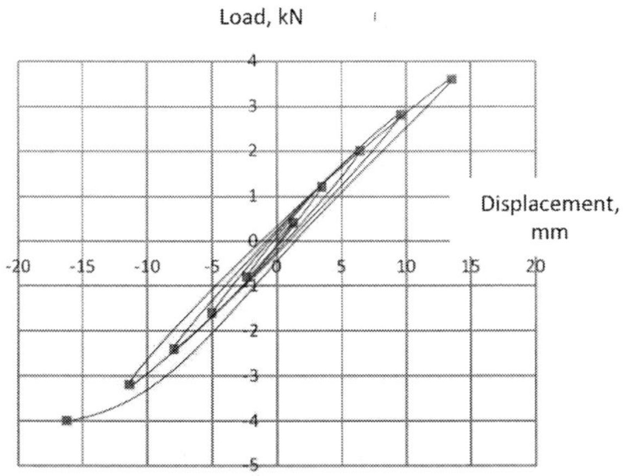

Figure 4. Diagram of load/deformation at horizontal low-cycle concentrated alternating load

The third stage is numerical simulation based on conventional idea of building behavior in an earthquake: all wall infill were modeled by loads considering non-bearing wall infill in explicit form.

For calculations, buildings with dimensions 36x36 m2 are used, bays are multiple of 6 meters. The structural scheme of buildings: girdless framework structure from reinforced concrete. Story height is 3m. Building has 4 stories. Columns are square with the size in the section 400x400 mm. Working reinforcement has diameter 20 class A500C. Floor slabs are made of monolithic reinforced concrete, reinforcement has diameter of class A500C (Fig. 5). When modeling and calculating 5 models in LIRA-SAPR, 2 types of calculations were performed: Linear spectral and Non-linear (Pushover)

Pushover Analysis allows identification of errors that can be made in the design process. At this stage, the calculation results can be divided into 2 stages. In the first stage, the percentage of the seven-point seismic impact that results in maximum story drift (collapse prevention level) was analyzed, and in the second stage, the percentage of the seven-point seismic impact that result in dimensional instability of the system (progressive collapse) was analyzed.

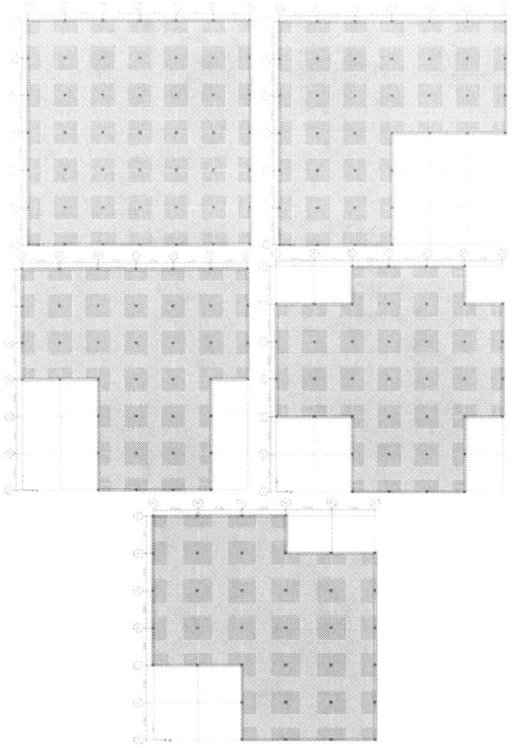

Figure 5. Schemes of irregularities in the shape of buildings in plan: F-0: square shape; F-3: L shape; F4: T shape; F-5: (+) shape; F-6: "four digit" shape

Results. Numerical modeling for buildings of various shapes disregarding non-bearing wall infill provided similar results: 50% of the seven-seismic impact. The simulation results for the same shapes in the plan, considering non-bearing wall infill, showed results that significantly change depending on the shape in the plan (the results are shown in Fig. 6 and Table 1).

Shapes F4 and F-5 r showed greatest increase in seismic stability, and shapes F-6 showed the smallest increase in seismic stability.

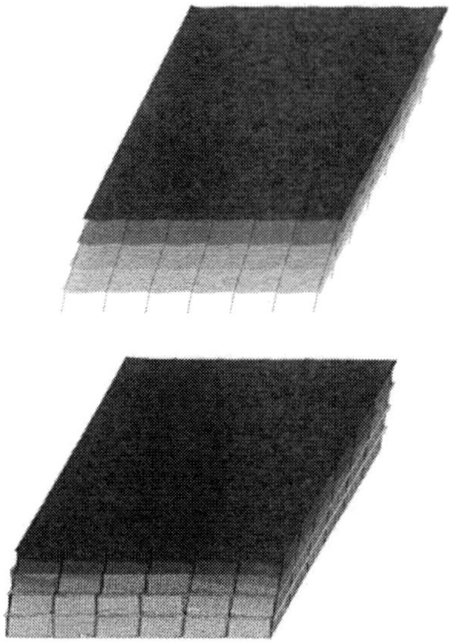

Figure 6. View of the design model of the building during destruction, with non-bearing wall infill (top) and without it (down).

Table 1

Shape	Part of the seismic impact perceived by the building at 7 points, [%]		
	Without non-bearing wall infill	With non-bearing wall infill	Exceeding of maximum drift Complete destruction
F-0	50	61	91
F-3	50	57	57
F-4	50	64	94
F-5	50	66	70
F-6	50	55	56

Acknowledgements

1. The results of previous field and laboratory experiments have been confirmed by the numerical modeling described in this study
2. Non-bearing wall infill has a significant influence on drift value: the increase in area of maximum drift was within the range from 10% to 32%.
3. Consideration of non-bearing wall infill results in redistribution of seismic impacts on columns in the central part
4. Upon designing buildings and structures, it is necessary to take into account the effect of non-bearing wall infill by performing duplicate calculations.

References

[1] DBN B.1.1-12: 2014. Construction in seismic regions of Ukraine. [Valid from 2014-10-01]. K .: Ministry of Regional Development, Building and Housing and Communal Services, 2014. VI, 84, p.110 (Building norms of Ukraine).

[2] DBN B 1.1-12: 2006. Construction in seismic regions of Ukraine. [Valid from 02.01.2007]. - K.: Ministry of Construction of Ukraine, 2006. - p.80.

[3] O.V. Murashko, O.S. Gubanov, V.S. Dorofeev, I. Benradi, Seismic resistance of buildings with jointless ossature without girders and joints at different height configurations. Science and construction. - 2018. - No. 2. - p. 25-31.

[4] [4] Rapid Visual Screening of Buildings for Potential Seismic Hazards A Handbook FEMA 154, Edition 3 / January 2015. – 388 p.

[5] Vérification de la sécurité parasismique des bâtiments existants Concept et directives pour l'étape 1 Richtlinien des BWG. – Directives de l'OFEG – Dirretive dell'UFAEG Berne, 2005, Deuxième édition.

[6] Assessment and Improvement of the Structural Performance of Buildings in Earthquakes Prioritisation Initial Evaluation Detailed Assessment Improvement Measures Recommendations of a NZSEE Study Group on Earthquake Risk Buildings June 2006 Including Corregendum No. 1.

[7] NZS 1170.5-2004. Earthquake actions - New Zealand.

[8] O. Murasko, I. Benradi, M. Abdelhadi. Approval of the developed visual assessment of seismic resistance, taking into account the irregular wall infill. Bulletin of Odessa state academy of civil engineering and architecture. – Odessa. – 2020. – Issue 78. –pp. 34–40.

[9] Iakovenko I., Kolchunov Vl. (2017). The development of fracture mechanics hypotheses applicable to the calculation of reinforced concrete structures for the second group of limit states. Journal of Applied Engineering Science, vol. 15(2017)3, article 455, pp. 366–375.

Portland cement easy treatability as the basis of concrete structure stable quality

A P Nikolaev[1,*], E V Kondrachenko[1] and V I Kondrachenko[2]

[1]O. M. Beketov National University of Urban Economy, 17, Marshala Bazhanova St., Kharkov, 61002, Ukraine
[2]Russian University of Transport, Obraztsova St., House. 9, Building 9, Moscow, GSP-4, 127994, Russia

*Email: apnikolaev1946@gmail.com

Abstract An opportunity is discussed to improve operational parameter stability of concrete and reinforced concrete articles on Portland cement basis in heavy-duty environment. Demonstration of binding properties of cements with identical normative parameters was shown to be dissimilar under amended operational conditions. Such behavior may be supposedly connected to insufficiently stable relation between alkalinity and dispersity of consumed Portland cement.

It was found feasible to arrange preventive control of incoming Portland cement with due consideration of its application conditions. The results of such control were shown to be efficient at substantiated correction of applied Portland cement activity.

The purpose of this research has been expansion of methodological basis of mineral binder quality control for stabilization of consumer properties of concrete and reinforced concrete structures produced on their basis.

1. Introduction

Efficient maintaining of concrete and reinforced concrete structure operation parameter stability depends substantially on stable properties of applied Portland cement (PC) [1,2]. It was previously found that physical-mechanical strength of tested cement beam samples was in direct proportion to their content of new hydrate formations [3]. Such characteristics as modulus of elasticity, creep, fracture strength, freeze resistance, durability, etc. are known to be connected not only to concentration of these new formations, but also to their formation rate on the basis of cement rock [4,5]. It follows that in order to ensure necessary stability level of concrete and reinforced concrete structure operational properties, corresponding dynamics of their hardening must be maintained at least at initial maturing stages. Nevertheless, substantially different conditions of PC control and application preclude direct control of this dynamics without additional analysis of consumed binder properties.

PC of identical activity are known to often have different relations between their clinker content and dispersion which are in inverse relationship [6]. Such cements cannot by definition have identical water interaction dynamics at any conditions, including at the same normative activity values. The value of this relation is not subdued to regular control, it is usually determined by on-site needs, economic feasibility and equipment capabilities of a particular facility [7]. Besides, Portland cement clinker (PCC) with its mineral properties being formed at all stages of its manufacture does not belong to materials with easily predicted behavior under various conditions of interaction with water.

The described multi-factored character of PC consumer properties formation requires selection of direct control methods of its setting and hardening dynamics, so that their confidence level was sufficient to correct this dynamics with known means. One of the main problems in arrangement of such control is considered to be insufficient methodological basis for PC binding activity testing under conditions different from standard. Another input control problem is too long procedure of PC physical-mechanical activity determination [8]. Due to clumsy cement quality determination time is unreasonably lost and material resources overspent, trying to ensure necessary product properties stability. This is especially evident in application of PC from different vendors, which substantiates

Content from this work may be used under the terms of the Creative Commons Attribution 3.0 licence. Any further distribution of this work must maintain attribution to the author(s) and the title of the work, journal citation and DOI.
Published under licence by IOP Publishing Ltd

feasibility of search for additional alternative analysis methods of their properties after different hardening conditions.

2. Equations and materials

Behavior of identical class cements under various hardening conditions was performed on PC 500 from vendors Kryvyi Rih-Cement, Podolsk-Cement and Lafarge-Cement. Additional materials were quartz flour (below 50 μm), LST plastifier and C-3 superplastifier PC setting and hardening dynamics was controlled by penetration methods using a plastometer and a durometer [9,10]. In order to expedite physical-mechanical properties evaluation of cements in mixtures of different composition and different preparedness, tested samples were thermos matured at 45 °C for 9 hours. The samples were sealed to ensure 100 % humidity under hardening.

3. Main part

Table 1 specifies the results of air filtration time through tested cement samples (0.012 kg) in PSH unit measurement cell and the values of their specific surface as calculated from these values. Additionally normal cement paste density was determined on the basis of tested PC.

Table 1. Dispersion properties and water consumption of tested PC.

№	PC vendor	Air filtration time, s	$S_{sp.}$, m^2/kg	Normal density, %
1	Kryvyi Rih-Cement 1	133	300	26
2	Kryvyi Rih-Cement 2	125	290	25
3	Podolsk-Cement	203	370	25
4	Lafarge-Cement	432	540	31

As seen from Table 1, both PC batches from Kryvyi Rih-Cement were actually identical as regards their dispersion characteristics and water consumption. Podolsk-Cement sample was more finely ground, but this fineness did not affect its water consumption, as distinct from more finely ground PC from Lafarge-Cement whose normal density turned to be much higher than in three other samples

Dynamics of setting and initial hardening of cement paste based on the above PCs were analyzed under the following preparation and maturing conditions:

1. W/C = 0.30; t = 24 °C; mixing time – 5minutes.
2. As in 1, but W/C = 0.33.
3. As in 1, but W/C = 0.26 and C-3 -0.6 % and LST – 0.1 % added.
4. As in 1, but one third of PC was substituted for quartz flour.
5. As in 1, but every 30 minutes within 2.5 hours period after mixing cement paste was kneaded for 5 minutes each time.

The samples were poured to prepared molds and penetration depth in these was measured for conical indentor. Measurements were from time to time corrected depending on developing dynamics of cement paste setting. The value of maximum shear stress was calculated from indentor penetration depth according to applied plastometer parameters.

Figure 1 shows the results of obtained cement rock setting dynamics for all tested samples.

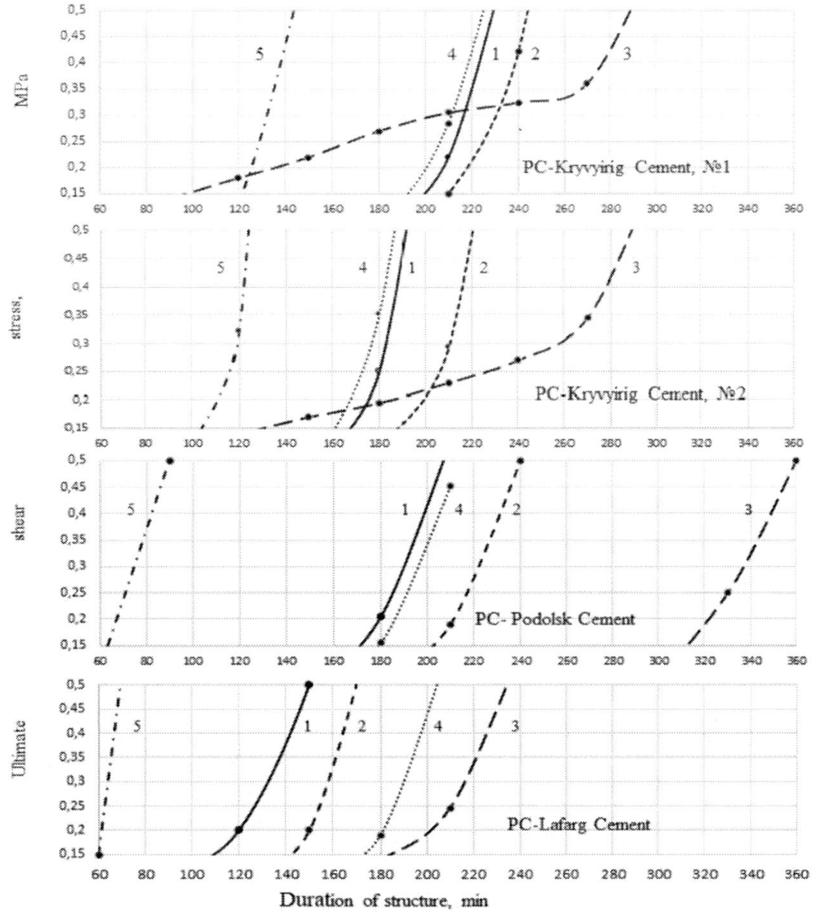

Figure 1 – Effect of hardening conditions for cement paste on the basis of PC 500 samples from various batches and vendors on their setting dynamics:
1 – W/C = 0.30; 2 – W/C = 0.33; 3 – 0.7 % complex plastifier added at W/C = 0.26; 4 – one third of PC substituted for quartz flour at W/C = 0.30; 5 – additional regular kneading within 2.5 hours at W/C = 0.31.

For netter clearness all plots are shown in one drawing on one scale. In order to visualize PC setting commencement and finish time vertical axis of maximum shear stress value begins at 0.15 and ends at 0.5 MPa. In this case any curve crossing time with plot bottom grid (0.15 MPa) corresponds to setting commencement time of tested sample, whereas crossing with plot top grid (0.5 MPa) corresponds to its setting end time.

From locations of curves in Figure 1 we can see that most active setting under conditions close to norm (Curves 1) proved to be finely ground PC from Lafarge- Cement. With W/C increased by 10% cement paste plastic condition retention time turned to be roughly identical for all PC samples (Curves 2). Effect of plastifier additives on cement behavior was not so unambiguous. Both batches of Kryvyi Rih-Cement PC responded identically to plastifiers, as distinct from Podolsk-Cement and Lafarge which demonstrated such lower demand of this additive (Curves 3), especially Sample 3 (Table 1). Substitution of one third of cement for quartz flour also showed individual response of initial structure formation dynamics for all PC. If plastic condition retention time reduced insignificantly for Kryvyi Rih-Cement PC, Podolsk-Cement PC, on the contrary, slightly increased this time, as distinct from Lafarge PC whose setting time actually doubled after introduction of quartz flour (Curves 4). Additional kneading of PC after mixing also caused different responses of cement pastes.

After determination of setting dynamics cement paste samples were sealed and stored for subsequent analysis their initial hardening dynamics (up to 3 days) under 100 % humidity. One day after preparation the samples were extracted from molds and after hardness measurement put into water. At 2nd and 3rd day of hardening we measured hardness of samples after their stay in water. Hardness value after each test was determined as an average of 8 parallel measurements. In order to compare comprehensive dynamics of sample hardening for total tested period the value of their average hardness was determined from values obtained for Days 1, 2 and 3. Besides, samples prepared by Case 1 for all four cements were also hardening under thermos maturing (45 °C within 9 hours).

The results of physical-mechanical tests for all samples are specified in Table 2.

Table 2. Physical-mechanical properties of tested PC.

№№	PC vendor	Average hardness for 1-3 days of maturing, MPa					*Thermos maturing, MPa
		W/C =0.30	W/C=0.33	Plastifier	Quartz flour	Additional kneading	
1	Kryvyi Rih-Cement 1	193	161	250	138	246	160
2	Kryvyi Rih-Cement 2	232	190	282	155	258	145
3	Podolsk-Cement	221	181	263	133	251	155
4	Lafarge-Cement	213	170	265	135	245	170

*) Note: hardness of samples was determined after 9 hours of retention at thermostat

Having analyzed PC hardening dynamics (Table 2), we may note that all cements behaved differently within three days hardening under normal conditions and else. In this finely ground cements did not actually show any advantage over others. Also PC from the same vendor demonstrated insufficient stability (Kryvyi Rih-Cement samples 1 and 2). Authors suppose some additional causes for such different activity of tested PC, which may be connected, for instance, with insufficient stability of their feed clinker.

The obtained results of cement paste structure formation under setting and initial hardening showed certain unique behavior features when PC application conditions are amended which served as information basis for purposeful correction of their activity indications. Taking Kryvyi Rih-Cement Sample 1 hardening dynamics as a basis, authors selected in relation to its value the structure formation rate for other PC by corresponding W/C increase while mixing. At the same time for stabilization of hardening dynamics inherent in Kryvyi Rih-Cement Sample 2 PC, hydration rate of other cements was, on the contrary, initiated by introduction of respective amount of calcium chloride ($CaCl_2$). Results of this correction are specified in Table 3.

Table 3. Results of PC physical-mechanical tests after their correction.

№	PC vendor	W/C, relative units	Setting time, min		Average hardness at Days 1-3, MPa	Hardness under in thermos maturing, MPa		
			commence ment	finish		$CaCl_2$, %	no additive	with additive
1	Kryvyi Rih-Cement 1	0.30	200	230	193	0.3	160	169
2	Kryvyi Rih-Cement 2	0.31	205	235	200	0.6	145	172
3	Podolsk-Cement	0.32	195	225	200	0.2	155	175
4	Lafarge-Cement	0.35	200	234	188	0.0	170	170

The results as specified in Table 3 prove that after a dedicated correction of binding activity all PCs demonstrated a closer setting and hardening dynamics at initial maturing period than their previous characteristics. Such behavior gives ground to expectation for improved stability of consumer properties of articles and structures on the basis of similar corrected binders.

4. Conclusion

The results of our work obviously proved that arrangement of due preliminary control of PC activity under its application conditions constitutes a necessary stage in guaranteed ensuring of operational properties stability in production of concrete and reinforced concrete articles. Presence of such control enables a more reasonable consumption of building materials.

References

[1] Khilazhev D R and Gorbunov S P 2015 A study in effect of active mineral additives and surfactants on peculiarities of fine concrete structure formation In: *Construction and environment: theory, practice, innovation* Chelyabinsk: PIRS) pp 204-206

[2] Nesvetaev G V and Khalezin S V 2017 A model for estimation of composition factors effect on creep of concrete *Naukovedenie Internet Magazine* **9** (**1**) p 4

[3] Volzhenskyi A V and Burov Yu S et al 1979 *Mineral binders* (Moscow: Stroyizdat Publishers) 476

[4] Batrakov V G 1998 *Modified concretes Theory and practice* (Moscow: Tekhnoproekt Publishers) 560

[5] Bazhenov Yu M and Pashchenko A A 2011 *Technology of concrete* (Moscow: LSV Publishers) 528

[6] Pashchenko A A and Myasnikova E A et al 1991 *Theory of cement* (Kyiv: Budivelnyk Publishers) 168

[7] Klassen V K 2012 *Cement production technology and optimization: a short course of lectures* (Belgorod: Belgorod State Technical University Press) 308

[8] Bolotskikh O N 2015 *European methods of cement physical-mechanical testing* 2nd ed/ (Kjarkiv UNISOFT Publishers) 87

[9] Nikolaev A P and Kondrashchenko O V 2019 Necessity of internal control of Portland cement binding properties *Visnyk Odeskoyi Derzhavnoyi Arkhytekturno-Budivelnoyi Akademiyi* **78** pp 108-117

[10] Nikolaev A P and Kondrashchenko O V 2018 Control of cement grout plastic properties *International Scientific Conference on Far East Con, ISCFEC* (Vladivostok: Russian Federation) **945** pp 70-75

IOP Conf. Series: Materials Science and Engineering 1164 (2021) 012055 doi:10.1088/1757-899X/1164/1/012055

The results of complex tests of piles during the installation of piled-raft foundations of the grain terminal in difficult soil conditions near the sea coast

O V Novsky[1], M V Marchenko[1], I I Mosicheva[1*] and V O Novsky[1]

[1]Odessa State Academy of Civil Engineering and Architecture, Odessa, 4, Didrihsona str., Odessa, Ukraine

*E-mail: imosicheva@gmail.com

Abstract. The engineering and geological conditions of the construction site for the designed terminal for processing and transshipment of grain products and its brief structural characteristics were presented; during arranging the zero cycle of the structure for the operative assessment of the accepted parameters of piles and confirmation of their calculated load, a series of dynamic and static tests of prismatic piles were carried out; requirements, normative methods, technology and devices for testing piles were given; calculations and analysis of their results and preliminary assessment of the accepted parameters of the piles were performed; the possibility of operational forecasting and adjustment of the calculated load and parameters of the piles accepted in the project in difficult geological conditions near the edge of the sea coast was shown.

1. Introduction

Due to the growing volume of imports of grain products, there is a need to build in limited areas of seaports, or near them, appropriate complexes of loading and unloading terminals for storage, processing and transshipment of such goods. As a rule, engineering-geological coasts and hydrogeological conditions of the sites allocated for these purposes have a number of negative features, including seismic.

2. Engineering and geological construction of the base

The following structure of engineering-geological layers was recorded at the construction site for the warehouse №1 of the terminal for reception, storage and shipment of grain cargoes and products of their processing in the city of Chernomorsk as a result of surveys and sampling during drilling of eight wells (Figure 1):

layer 1 – bulk layer, represented by sand, clay, greenish-gray and brown loams of hard and soft plastic consistency, with the inclusion of some construction waste;

layer 2 – gray sand, medium density;

layer 3 – greenish-gray loam, silted, plastic consistency;

layer 4 – light loam, brownish-gray, with the inclusion of sludge fluid-plastic consistency;

layer 5 – heavy loam, greenish-gray, with the inclusion of limestone land waste, from semi-hard to stiff consistency;

layer 6 – greenish-gray clay, from semi-solid to solid consistency;

layer 7 – greenish-gray clay of hard or semi-hard consistency with the inclusion of limestone land waste.

In the general assessment of engineering and geological conditions of the selected site, it should be noted that it is located near the sea coast, so soil formation has very significant variability both in depth and in plan, and construction in such conditions can generally be classified as "extremely unsuitable". In addition, the construction site has seismic category of more than seven points on the Richter scale.

Content from this work may be used under the terms of the Creative Commons Attribution 3.0 licence. Any further distribution of this work must maintain attribution to the author(s) and the title of the work, journal citation and DOI.
Published under licence by IOP Publishing Ltd

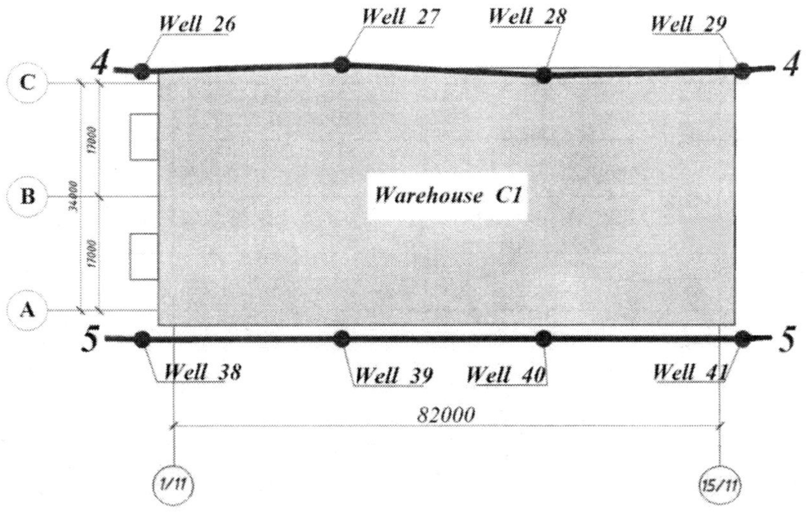

Figure 1. Plan with mapped exploration wells and geological section lines.

3. Design solution of the building foundations

The production and storage facilities are designed as double structures on pile-raft foundations by the project organization "Elevator Technologies of the Future". The total number of piles for the warehouse №1 – 794, of which: piles brands C170.35 – 180 pcs., C180.35 – 12 pcs., C190.35 – 404 pcs. and C200.35 – 198 pcs. (designation: C – prismatic pile; 170/180/190/200 – pile length, dm; 35 – pile cross-sectional size, cm) Piles are accepted as components, the lower part is 12 m long, the upper parts are of variable length.

The piles were immersed with a tubular diesel hammer, the height of the hammer lift was 1.70 m, the total weight of the hammer was 100 kN, the impact part was 50 kN, the rated impact energy was 128 kJ, and the hammer head was made of wood. Connection of piles during process of immersion was made by a welded butt seam on perimeter of final metal bonds with strengthening by figured overlays. The distance between the axes of the extreme piles on the letter axes is 82.00 m, on the digital axes – 34.00 m. The main distance between the axes of the piles located in the pile field on the square grid is 1.70 m.

The value of the estimated allowable load as per project was assumed to be equal to 950 kN. The bearing layer for piles is layer 7 – greenish-gray clay. Engineering-geological section with height reference of piles is shown in Figure 2. Due to special engineering-geological conditions, for prompt adjustment of pile parameters, the designer issued a task to perform dynamic tests of operational piles in the process of pile field formation works.

4. Methods and results of dynamic tests of soils by piles

Dynamic tests of soils by prismatic piles were performed in accordance with the requirements of regulatory documents [1, 2]. According to the technical task, ten piles were tested, from them – six C190.35 and four C200.35. The "recovery" of the piles after the dive was at least fifteen days. The number of blows at the first drive of each pile is accepted equal to three, at the second – five. The value of "failure" was determined by the ratio of the total amount of immersion (subsidence) of the pile to the number of blows. Detailed results of dynamic tests indicating the numbers of operation piles in the pile field, test dates, time of "recovery", number of blows, the value of the total immersion of the pile for the cycle of blow and the estimated "failure" for piles C190.35 and C200.35 are shown in Table 1.

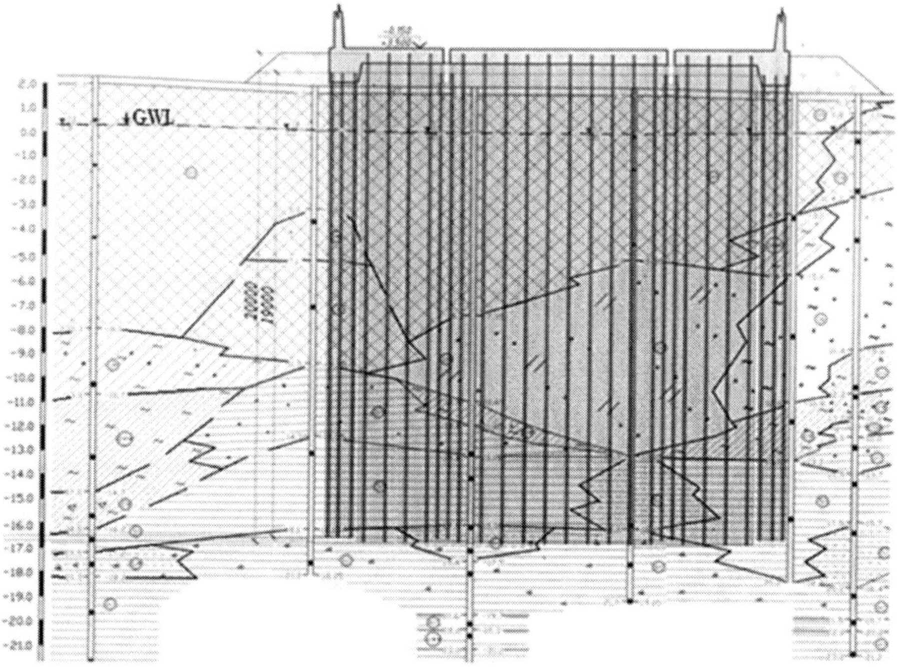

Figure 2. Vertical positioning of piles on the engineering-geological section.

The normative algorithm for determining the maximum design load on the pile based on the results of dynamic tests involves the following step-by-step actions:

a) in accordance with the requirements, if the actual "failure" in dynamic tests was $s_a > 2$ mm (see Tab. 1, for piles № 46, 180 i 508), the value of the boundary resistance was determined by the formula:

$$F_u = \frac{\eta AM}{2}\left[\sqrt{1+\frac{4E_d}{\eta As_a}\cdot\frac{m_1+\varepsilon^2(m_2+m_3)}{m_1+m_2+m_3}}-1\right], \quad (1)$$

where η – tabulated coefficient depending on the pile material, kN/m^2;

A – area limited by the outer contour of a continuous or hollow cross section of the pile trunk, m^2;

M – coefficient taken when blowing the pile with an impact hammer;

$E_d = G(H-h)$ – the estimated impact energy of the hammer, kJ, where:

G – weight of the impact part of the hammer, kN;

H – the height of the fall of the hammer, m;

h – the height of the first bounce of the impact part for the tubular diesel hammer, m;

ε – impact recovery factor when driving reinforced concrete piles with impact hammers using a wooden hammer head;

m_1 – hammer weight, kN;

m_2 – weight of head and headgear, kN;

m_3 – weight of drive head, kN.

Table 1. Results of pile dynamic load test.

Pile №	Pile immersion date	Recovery time, days	Number of blows	Total subsidence per cycle, mm	The average failure of one blow, mm
failure to drive piles 19 m long					
№67	27.08.18	7	3	5,0	1,7
			5	6,0	1,2
№84	27.08.18	8	3	3,0	1,0
			5	4,0	0,8
№46	05.09.18	15	3	6,0	2,0
			5	**12,0**	**2,4**
№255	05.09.18	15	3	3,0	1,0
			5	5,0	1,0
№326	05.09.18	15	3	3,0	1,0
			5	4,0	0,8
№389	05.09.18	15	3	5,0	1,7
			5	7,0	1,4
failure to drive piles 20 m long					
№180	27.08.18	7	3	6,0	2,0
			5	**12,0**	**2,4**
№222	27.08.18	8	3	3,0	1,0
			5	5,0	1,0
№492	05.09.18	15	3	6,0	2,0
			5	9,0	1,8
№508	05.09.18	15	3	6,0	2,0
			5	**12,0**	**2,4**

б) bearing capacity of piles according to the results of dynamic tests was determined by the formula:

$$F_d = \frac{\gamma_c F_u}{\gamma_g}, \tag{2}$$

where γ_c – coefficient of operational conditions;

γ_g – coefficient of soil reliability;

F_u – the value of the boundary resistance of the pile according to the test results.

c) piles allowable design load was determined by the formula:

$$N = \frac{F_d}{\gamma_k}, \tag{3}$$

γ_k – coefficient of reliability, which depends on the method or technique of determining the bearing capacity of the pile.

d) the maximum design load taking into account seismic influences for the city of Odessa, adjusted according to the formula:

$$N_{eq} = N \cdot \gamma_{eq}, \tag{4}$$

where γ_{eq} – seismic coefficient.

Thus, taking into account the calculations of the relevant initial values, parameters and accepted coefficients, performed at different stages, the final maximum design load according to the results of dynamic tests was 980 kN, which fully confirmed the design decisions of the choice of bearing layer, prismatic piles size (length and cross section), and also substantiated the accepted value of the piles maximum design load.

5. Methods and results of static tests of soils by piles

Tests of soils by prismatic piles were performed in full compliance with the requirements of regulatory documents [1, 3 and 4]. The "recovery" of the piles after the dive was at least fifteen days. The following equipment, devices and limit values were used during the benchmark static tests of operation piles:

a) axial compressive static load on the piles was transmitted by two jacks equipped with a manometer, using an autonomous manual pumping station;

b) the buffer for the jacks were two metal beams, located crosswise, the ends of which were attached to the reinforcement of the anchor piles (Figure 3);

c) movement (subsidence) of pile heads was measured by two deflection meter of the model with a division value of 0.01 mm;

d) the first criterion was the maximum load on the pile, accepted according to the requirements of the technical task, 1500 - 1600 kN, the second - the maximum subsidence equal to 40 mm.

The method and technology of piles static tests contained the following sequential actions and criterias:

a) the load on the pile was transmitted in stages – the initial stages were 1/5, and the subsequent stages were taken equal to 1/10 of the specified in the test program;

b) readings from the devices were taken: zero – before loading the pile, the first – immediately after the application of the degree of load, the next four – every 30 minutes after, and then every hour of observations until conditional stabilization (deformation damping), which is normalized;

c) the following recommended requirements were observed: discrepancies in the readings of the devices should not exceed 50% – for subsidence less than 1 mm, 30% – for subsidence from 1 to 5 mm and 20% – for subsidence over 5 mm;

Figure 3. Photo of soil test by pile.

d) conventional stabilization for each load level was taken as 0.01 mm in the last 60 minutes of observation;

e) in the process of piles test, a working log was kept, and based on the results of fixing the corresponding values, graphs of subsidence dependence on each load stage in time $S = f(P, t)$ and the resulting pile subsidence by load $S = f(P)$ were built.

On piles №№ 45, 66 and 85 the ultimate load was brought to the value of 1500 kN, while their stabilized subsidence was 18.87 mm, 17.33 mm and 9.23 mm. On piles №№ 254, 325 and 390, the ultimate load was increased to 1650 kN, and their stabilized subsidence was, respectively, 15.85 mm, 16.26 mm and 12.38 mm. The test results are shown in table 2.

Table 2. The main results of soil static compression load tests by piles.

Pile	Sedimentations of piles (mm) at values of load levels (kN)								
№	300	600	750	900	1050	1200	1350	1500	1650
№ 45	0.96	2.16	3.25	4.83	6.75	9.62	13.18	18.87	–
№ 66	0.82	1.80	2.95	4.50	6.26	8.45	11.94	17.33	–
№ 85	0.55	1.57	2.18	3.00	4.23	5.88	7.56	9.23	–
№ 254	0.53	1.35	–	2.40	3.22	4.65	6.40	8.70	15.85
№ 325	0.85	1.68	–	3.17	4.15	5.51	7.57	10.35	16.26
№ 390	0.48	1.10	–	2.22	3.16	4.25	5.75	7.73	12.38

The bearing capacity of prismatic piles according to the results of their tests by static compressive load was determined by the formula [4]:

$$F_d = \frac{\gamma_c F_{u,n}}{\gamma_g}, \tag{5}$$

where γ_c – coefficient of operational conditions;

γ_g – coefficient of soil reliability;

$F_{u,n}$ – the lowest value of the ultimate resistance according to the test results.

Permissible design load on the pile was determined by the formula:

$$N = \frac{F_d}{\gamma_k}, \tag{6}$$

where γ_k – coefficient of reliability.

The maximum design load on the pile, taking into account the seismic effects was determined by the formula:

$$N_{eq} = N \cdot \gamma_{eq}, \tag{7}$$

where γ_{eq} – seismic coefficient.

Thus, based on the results of static tests, taking into account regulatory requirements and calculations according to the above calculation algorithm, as well as taking into account the relevant coefficients, the maximum allowable load on the prismatic piles of the production and storage terminal for receiving, storing and transshipment grain products is 1060 kN, exceeding the design, equal to 950kN.

6. Conclusions

1. Dynamic tests of piles in difficult engineering-geological conditions are an auxiliary element of operative adjustment of the accepted design decisions in the course of production works on the arrangement of pile-raft foundations in especially difficult engineering-geological conditions.

2. At the same time, it should be noted that the necessary reliable justification of the reliability of the design decision is testing of piles by static loads.

3. Static testing of piles in particularly complex engineering and geological conditions is a mandatory regulatory element that confirms the validity of the design decisions in the construction of the zero cycle of the production and storage terminal.

4. To correlate the results of full-scale tests of piles and the real interaction of the system "base-foundation-structure", which is expressed by subsidence, there is such a stage as scientific and technical support of the construction process, including geodetic monitoring.

5. The replenished base of results of static tests in similar difficult engineering-geological conditions should serve as a basis of adjustment of the accepted standard settlement coefficients, and also effective optimization of design decisions of pile foundations.

References

[1] DSTU B.V.2.1-1-95 (GOST 5686-94) 1997 *Soils Field test methods with piles* Kyiv: State Committee of Ukraine for Urban Planning and Architecture MNTKS 57 (State Standards of Ukraine)

[2] DSTU B.V.2.1-27: 2010 2011 *Piles Determination of bearing capacity based on the results of field tests* Kyiv: Ministry of Regional Development of Ukraine 11 (State Standards of Ukraine)

[3] DBN V.2.1-10-2009 2011 *Change №1 Bases and foundations of buildings. Basic design provisions* Kyiv Ministry of Regional Development of Ukraine 55 (State Building Codes of Ukraine)

[4] DBN V.2.1-10-2018 2018 *Bases and foundations of buildings and structures. Substantive provisions* Kyiv Ministry of Regional Development of Ukraine 36 (State Building Codes of Ukraine)

IOP Conf. Series: Materials Science and Engineering 1164 (2021) 012056 doi:10.1088/1757-899X/1164/1/012056

Development of analytical calculation method for axisymmetric oscillations of circular and annular plates on variable Winkler elastic foundation

Yu Krutii[1,*], M Surianinov[1], V Osadchiy[1], V Kolomiichuk[1]

[1]Odesa State Academy of Civil Engineering and Architecture, 4, Didrihsona str., Odesa 65029, Ukraine

[*] E-mail: yurii.krutii@gmail.com

Abstract. An analytical method for calculating axisymmetric vibrations of circular and annular plates on a variable elastic Winkler base is presented. The proposed calculation method is based on the exact solution of the oscillation equation with the subsequent development of a method for its numerical implementation. To construct an exact solution, the direct integration method is used, which was previously developed by one of the authors for differential equations with continuous variable coefficients. The proposed method does not require the discretization of the considered structure and it is a real alternative to the application of approximate methods during solving this types of problems. Based on the exact solution of the differential equation, this method will allow to obtain better picture of oscillations in comparison with approximate methods. In fact, the exact solution has qualitative information and forms the most complete picture of the physical phenomenon that is researched.

1. Introduction

Circular and annular plates are widely used model in the calculation of structural elements of construction objects, machine building, shipbuilding, aircraft building, instrument-making and other fields of technology. The theory of calculating such plates is discussed in detail in the works [1-3] and many others.

One of the most actual problems is the calculation of circular and annular plates lying on a solid elastic foundation. In particular, the basement of circular structures, which rest on the ground, are calculated exactly as a circular plate on an elastic foundation. Winkler model is often used among the models of ground foundations. In this case, in integral form there is a number of modifications of this model, which allow to take into account the heterogeneous properties of the foundation, both in plan and depth. The most common modification is the model of the variable modulus of subgrade reaction. Such a model, for example, is used in calculating the stress-strain state of the structure foundations lying on the loess soils, in which subsidence is characteristic. In this case, the modulus of subgrade reaction is a variable value, depending on the coordinates in which the draft of the foundation is determined.

Calculations of the plates lying on an elastic foundation with a variable modulus of subgrade reaction are extremely rare in the scientific periodicals. For rectangular plates, these calculations are known only for static bending problems. In the article [4] the calculation of such plates is carried out by the method of finite elements, and in [5] – by Galerkin method. In our opinion, the work [6] deserves special attention, in which free oscillations of the circular plate, lying on a variable elastic foundation, are investigated. The modulus of subgrade reaction changes according to the power law in the direction of radius.

This article is devoted to the development of an analytical method for calculating on the axisymmetric oscillations of circular and annular plates with constant cylindrical stiffness D lying on the solid variable Winkler elastic foundation with a variable modulus of subgrade reaction. In figure 1 the annular plate is shown. Here are a and b – radiuses of the outer and inner contour circles of the

Content from this work may be used under the terms of the Creative Commons Attribution 3.0 licence. Any further distribution of this work must maintain attribution to the author(s) and the title of the work, journal citation and DOI.
Published under licence by IOP Publishing Ltd

plate, r – radial coordinate $(0 \le r \le a)$. In particular, with the value $b = 0$ we obtain a solid circular plate (figure 2).

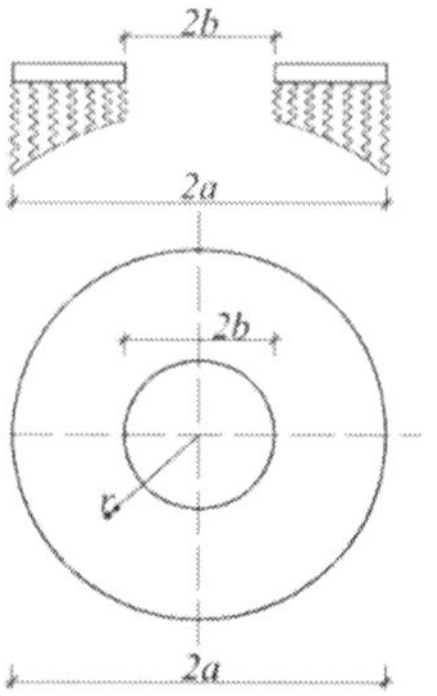

Figure 1. Annular plate on variable elastic foundation.

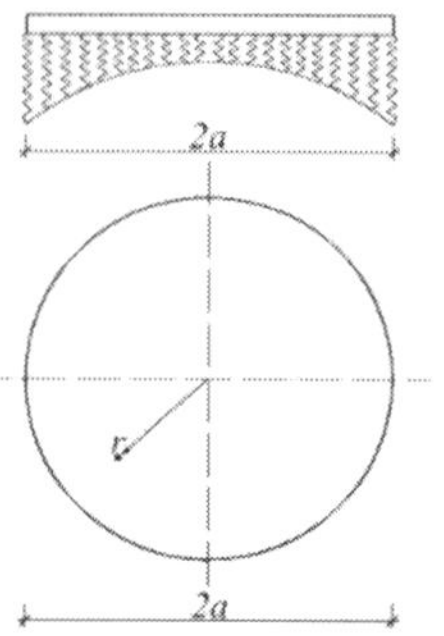

Figure 2. Solid plate on variable elastic foundation.

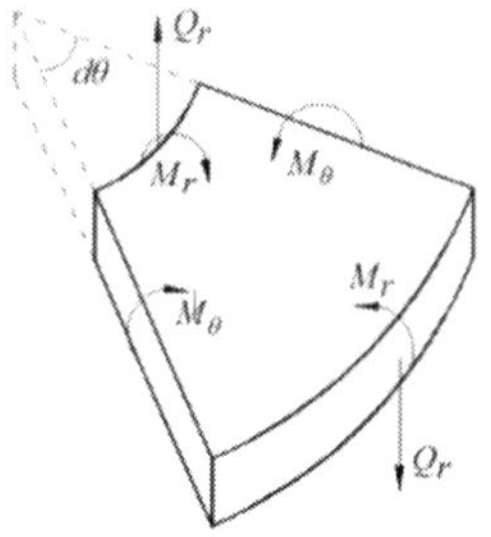

Figure 3. Bending moments and shear force in plate.

Free axisymmetric oscillations of the plates occur when the elastic $R(r,t)$ and the conditions for fixing the edges are not dependent on the polar angle θ. In such oscillations, only three dynamic internal forces act in the plate, namely, radial and circular bending moments $M_r(r,t)$ и $M_\theta(r,t)$, as well as radial shear force $Q_r(r,t)$ (figure 3). The torque moment $M_{r\theta}(r,t)$ and the circular shear force $Q_\theta(r,t)$ are zero because of the axial symmetry of the oscillations.

According to Winkler hypothesis, the reaction of the foundation $R(r,t)$ to the plate and the dynamic deflection of the plate $W(r,t)$ are associated with equality

$$R(r,t) = -k(r)W(r,t),$$

where $k(r)$ – the continuous variable modulus of subgrade reaction. Relatively $k(r)$ we take $k(r) = k_0 A(r)$, where k_0 – the value of the modulus of subgrade reaction at some characteristic point of the plate; $A(r)$ – a dimensionless continuous function that expresses the law of changing the modulus of subgrade reaction from the radial coordinate.

The differential equation of oscillations in our case has the form [7]

$$D\frac{1}{r}\frac{\partial}{\partial r}\left\{r\frac{\partial}{\partial r}\left[\frac{1}{r}\frac{\partial}{\partial r}\left(r\frac{\partial W}{\partial r}\right)\right]\right\} + k_0 A(r)W + \rho h\frac{\partial^2 W}{\partial t^2} = 0. \qquad (*)$$

The purpose of the work is to develop an analytical method of calculation on free oscillations of circular and annular plates under the conditions described above. To achieve the goal, the following tasks are formulated and solved:

- to find the exact solution of the equation (*);
- to obtain the formulas for dynamic internal force movements in the plate;
- to obtain an analytical representation for the frequency of oscillations;
- to specify analytical method of numerical realization of exact predicted solutions.

2. Materials and methods

The offered method of calculation is based on the exact solution of the oscillation equation (*), followed by the development of the method for its numerical implementation. To construct an exact solution, this publication uses direct integration method, developed in the case of differential equations with continuous variable coefficients in the work [8]. The principle of the direct integration method can also be realized, for example, according to the publications [9-13], where it solves some problems of the mechanics of the deformed solid body. It is characteristic that differential equations with variable coefficients are in the role of mathematical models of these problems.

3. The exact solution of the oscillation equation and the formulas for internal forces

When the deflection function $W(r,t)$ is found from the equation (*), the dynamic angle of rotation, as well as the dynamic forces in the plate, are determined with the known formulas [7]:

$$\varphi(r,t) = \frac{\partial W}{\partial r}; \tag{1}$$

$$M_r(r,t) = -D\left(\frac{\partial^2 W}{\partial r^2} + \frac{\mu}{r}\frac{\partial W}{\partial r}\right); \tag{2}$$

$$M_\theta(r,t) = -D\left(\mu\frac{\partial^2 W}{\partial r^2} + \frac{1}{r}\frac{\partial W}{\partial r}\right); \tag{3}$$

$$Q_r(r,t) = -D\left(\frac{\partial^3 W}{\partial r^3} + \frac{1}{r}\frac{\partial^2 W}{\partial r^2} - \frac{1}{r^2}\frac{\partial W}{\partial r}\right). \tag{4}$$

The solution (1) is searched in the form

$$W(r,t) = w(r)T(t), \tag{5}$$

where $w(r)$ is the amplitude function of deflection, depending only on r coordinate, $T(t)$ – time function. Using (5) in the equation (*) and dividing the variables there, we obtain two ordinary differential equations:

$$\ddot{T}(t) + \omega^2 T(t) = 0; \tag{6}$$

$$D\frac{1}{r}\frac{d}{\partial r}\left\{r\frac{d}{\partial r}\left[\frac{1}{r}\frac{d}{\partial r}\left(r\frac{dw}{\partial r}\right)\right]\right\} + (k_0 A(r) - \rho h\omega^2)w = 0. \tag{7}$$

It is easy to write out the solution of equation (6)

$$T(t) = T(0)\cos\omega t + \frac{\dot{T}(0)}{\omega}\sin\omega t,$$

where $T(0), \dot{T}(0)$ are the parameters of the initial conditions of movement, ω – unknown plate oscillation frequency. The main mode of oscillations is defined as the solution of the equation (7), which we rewrite in the form to shorten the record

$$D\Delta\Delta w + (k_0 A(r) - \rho h\omega^2)w = 0, \tag{8}$$

where $\Delta = d^2/dr^2 + 1/r\, d/dr$ – Laplace operator.

We denote with $X_n(r), Y_n(r)\ (n=1,2)$ the fundamental solutions of the equation (8) and for solutions $Y_n(r)$ we accept representations

$$Y_n(r) = X_n(r)\ln\frac{r}{a} + Z_n(r)\ (n=1,2), \tag{9}$$

where $Z_n(r)$ – unknown functions. Substituting (9) into the equation (8), after transformations we get

$$\left[D\Delta\Delta X_n(r) + (k_0 A(r) - \rho h\omega^2)X_n(r)\right]\ln\frac{r}{a} + D\Delta\Delta Z_n(r) +$$

$$+(k_0 A(r) - \rho h\omega^2) Z_n(r) + 4D \frac{1}{r}\frac{d^3 X_n(r)}{dr^3} = 0. \qquad (10)$$

Since $X_n(r)$ $(n=1,2)$ are fundamental solutions of the equation (8), the left part in logarithm (10) is obliged to be identically equal to zero. Therefore, instead of the equality (10), we can write:

$$D\Delta\Delta X_n(r) + (k_0 A(r) - \rho h\omega^2) X_n(r) = 0 \ \ (n=1,2); \qquad (11)$$

$$D\Delta\Delta Z_n(r) + (k_0 A(r) - \rho h\omega^2) Z_n(r) = -4D \frac{1}{r}\frac{d^3 X_n(r)}{dr^3} \ \ (n=1,2). \qquad (12)$$

The solutions of the equation (11) will be found in the form

$$X_n(r) = \sum_{k=0}^{\infty}\sum_{m=0}^{k} (-K_0)^{k-m} K^{2m} \alpha_{n,m,k-m}(r), \qquad (13)$$

where $\alpha_{n,m,k-m}(r)$ $(n=1,2)$ are unknown functions, which we consider to be continuous with derivatives from the first to the fourth order; $K_0 = a^4 k_0 / D$ is a known dimensionless parameter; K is an unknown dimensionless parameter, which is connected with the oscillation frequency by the formula

$$K^2 = a^4 \rho h\omega^2 / D. \qquad (14)$$

So we assume that the series (13), as well as similar series, made up from the first four derivative functions $\alpha_{n,m,k-m}(r)$ $(n=1,2)$, converge uniformly. In this case, a series differential operation will be possible.

Substituting (13) into the equation (11), we have

$$\sum_{k=0}^{\infty}\sum_{m=0}^{k} (-K_0)^{k-m} K^{2m} \Delta\Delta\alpha_{n,m,k-m}(r) - \frac{1}{a^4} A(r) \sum_{k=0}^{\infty}\sum_{m=0}^{k} (-K_0)^{k-m+1} K^{2m} \alpha_{n,m,k-m}(r) -$$

$$-\frac{1}{a^4}\sum_{k=0}^{\infty}\sum_{m=0}^{k} (-K_0)^{k-m} K^{2(m+1)} \alpha_{n,m,k-m}(r) = 0.$$

In the second and third sums, we shift the index k by one, that is, we replace k by $k-1$. The same operation is carried out in the third sum with index m. As a result, we get

$$\sum_{k=0}^{\infty}\sum_{m=0}^{k} (-K_0)^{k-m} K^{2m} \Delta\Delta\alpha_{n,m,k-m}(r) - \frac{1}{a^4} A(r) \sum_{k=1}^{\infty}\sum_{m=0}^{k-1} (-K_0)^{k-m} K^{2m} \alpha_{n,m,k-m-1}(r) -$$

$$-\frac{1}{a^4}\sum_{k=1}^{\infty}\sum_{m=1}^{k} (-K_0)^{k-m} K^{2m} \alpha_{n,m-1,k-m}(r) = 0.$$

Then, after the transformation, it is necessary to implement the equality

$$\Delta\Delta\alpha_{n,0,0}(r) + \sum_{k=1}^{\infty} K^{2k}\left(\Delta\Delta\alpha_{n,k,0}(r) - \frac{1}{a^4}\alpha_{n,k-1,0}(r)\right) + \sum_{k=1}^{\infty}(-K_0)^k\left(\Delta\Delta\alpha_{n,0,k}(r) - \frac{1}{a^4}A(r)\alpha_{n,0,k-1}(r)\right) +$$

$$+\sum_{k=2}^{\infty}\sum_{m=1}^{k-1}(-K_0)^{k-m}K^{2m}\left(\Delta\Delta\alpha_{n,m,k-m}(r) - \frac{1}{a^4}\alpha_{n,m-1,k-m}(r) - \frac{1}{a^4}A(r)\alpha_{n,m,k-m-1}(r)\right) = 0.$$

To satisfy it, we equate to zero all the variable coefficients at the powers of $(-K_0)^{k-m}K^{2m}$ $(k=0,1,2,...)(m=0,1,2,...,k)$. In this case, in the corresponding equations, we move from the operational form of record to the usual one:

$$\frac{1}{r}\frac{d}{dr}\left\{r\frac{d}{dr}\left[\frac{1}{r}\frac{d}{dr}\left(r\frac{d\alpha_{n,0,0}(r)}{dr}\right)\right]\right\}=0\ (n=1,2); \tag{15}$$

$$\frac{1}{r}\frac{d}{dr}\left\{r\frac{d}{dr}\left[\frac{1}{r}\frac{d}{dr}\left(r\frac{d\alpha_{n,k,0}(r)}{dr}\right)\right]\right\}=\frac{1}{a^4}\alpha_{n,k-1,0}(r)\ \ (k=1,2,3,...); \tag{16}$$

$$\frac{1}{r}\frac{d}{dr}\left\{r\frac{d}{dr}\left[\frac{1}{r}\frac{d}{dr}\left(r\frac{d\alpha_{n,0,k}(r)}{dr}\right)\right]\right\}=\frac{1}{a^4}A(r)\alpha_{n,0,k-1}(r)\ \ (k=1,2,3,...); \tag{17}$$

$$\frac{1}{r}\frac{d}{dr}\left\{r\frac{d}{dr}\left[\frac{1}{r}\frac{d}{dr}\left(r\frac{d\alpha_{n,m,k-m}(r)}{dr}\right)\right]\right\}=\frac{1}{a^4}\alpha_{n,m-1,k-m}(r)+\frac{1}{a^4}A(r)\alpha_{n,m,k-m-1}(r) \tag{18}$$

$$(k=2,3,4,...)\ (m=1,2,3,...,k-1).$$

As $\alpha_{n,0,0}(r)(n=1,2)$, we will select the following functions

$$\alpha_{n,0,0}(r)=\left(\frac{r}{a}\right)^{2n-2}\ \ (n=1,2). \tag{19}$$

It is easy to verify that each of them satisfies the equation (15). Further, integrating the equations (16)-(18) and assuming integration constants are zero, we obtain:

$$\alpha_{n,k,0}(r)=\frac{1}{a^4}\int_0^r\frac{1}{r}\int_0^r r\int_0^r\frac{1}{r}\int_0^r r\alpha_{n,k-1,0}(r)\,drdrdrdr\ \ (k=1,2,3,...); \tag{20}$$

$$\alpha_{n,0,k}(r)=\frac{1}{a^4}\int_0^r\frac{1}{r}\int_0^r r\int_0^r\frac{1}{r}\int_0^r rA(r)\alpha_{n,0,k-1}(r)\,drdrdrdr\ \ (k=1,2,3,...); \tag{21}$$

$$\alpha_{n,m,k-m}(r)=\frac{1}{a^4}\int_0^r\frac{1}{r}\int_0^r r\int_0^r\frac{1}{r}\int_0^r r(\alpha_{n,m-1,k-m}(r)+A(r)\alpha_{n,m,k-m-1}(r))\,drdrdrdr \tag{22}$$

$$(k=2,3,4,...)\ (m=1,2,3,...,k-1).$$

Formulas (20)-(22) are recurrent. Functions $\alpha_{n,k,0}(r)$, $\alpha_{n,0,k}(r)$, $\alpha_{n,m,k-m}(r)$, which we call the generating ones, are sequentially defined with the help of these formulas according to the known initial function $\alpha_{n,0,0}(r)$. For such functions, the equality (11) is satisfied identically by the formation.

As we can see, this equation is similar to the equation (11) intrinsically. The difference is only that the equation (12) contains a non-zero right part. Therefore, acting like the previous one, we can obtain the solutions of the equation (12). For brevity sake, we will write only the final formulas, which determine the solutions that are being searched for:

$$Z_n(r)=\sum_{k=0}^{\infty}\sum_{m=0}^{k}(-K_0)^{k-m}K^{2m}\beta_{n,m,k-m}(r); \tag{23}$$

$$\beta_{n,0,0}(r)=\alpha_{n,0,0}(r)=\left(\frac{r}{a}\right)^{2n-2}\ \ (n=1,2); \tag{24}$$

$$\beta_{n,k,0}(r)=\frac{1}{a^4}\int_0^r\frac{1}{r}\int_0^r r\int_0^r\frac{1}{r}\int_0^r r\left(\beta_{n,k-1,0}(r)-\frac{4a^4}{r}\frac{d^3\alpha_{n,k,0}(r)}{dr^3}\right)drdrdrdr\ (k=1,2,3,...);\qquad(25)$$

$$\beta_{n,0,k}(r)=\frac{1}{a^4}\int_0^r\frac{1}{r}\int_0^r r\int_0^r\frac{1}{r}\int_0^r r\left(A(r)\beta_{n,0,k-1}(r)-\frac{4a^4}{r}\frac{d^3\alpha_{n,0,k}(r)}{dr^3}\right)drdrdrdr\ (k=1,2,3,...);\qquad(26)$$

$$\beta_{n,m,k-m}(r)=\frac{1}{a^4}\int_0^r\frac{1}{r}\int_0^r r\int_0^r\frac{1}{r}\int_0^r r\left(\beta_{n,m-1,k-m}(r)+A(r)\beta_{n,m,k-m-1}(r)-\frac{4a^4}{r}\frac{d^3\alpha_{n,m,k-m}(r)}{dr^3}\right)drdrdrdr;\qquad(27)$$

$$(k=2,3,4,...)\ (m=1,2,3,...,k-1).$$

Series (13), (23) converge uniformly. Here are the demonstration of convergence, for example, for series (13).

Firstly, note that the generating functions $\alpha_{n,k,0}(r)$ $(k=1,2,3,...)$ can be found by formulas (19), (20) in an explicit form:

$$\alpha_{n,k,0}(r)=c_{n,k,0,0}\left(\frac{r}{a}\right)^{2n+4k-2}\quad(k=1,2,3,...);\qquad(28)$$

$$c_{n,k,0,0}=\frac{1}{(2^{2k}(n+2k-1)!)^2}.\qquad(29)$$

For other generating functions, based on the properties of certain integrals, we obtain estimates:

$$\alpha_{n,0,k}(r)\le\frac{\gamma}{a^4}\int_0^r\frac{1}{r}\int_0^r r\int_0^r\frac{1}{r}\int_0^r r\alpha_{n,0,k-1}(r)\,drdrdrdr\ (k=1,2,3,...);\qquad(30)$$

$$\alpha_{n,m,k-m}(r)\le\frac{1}{a^4}\int_0^r\frac{1}{r}\int_0^r r\int_0^r\frac{1}{r}\int_0^r r(\alpha_{n,m-1,k-m}(r)+\gamma\alpha_{n,m,k-m-1}(r))drdrdrdr\qquad(31)$$

$$(k=2,3,4,...)\ (m=1,2,3,...,k-1).$$

where $\gamma=\max\limits_{0\le r\le a}A(r)$. Performing sequential operations, prescribed by recurrent formulas (30), (31) for the specified values of indices k, m and taking into account the equality (28), (29), concerning the generating functions (20)-(22), we have general result

$$\alpha_{n,m,k-m}(r)\le\frac{\gamma^{k-m}C_k^m}{(2^{2k}(n+2k-1)!)^2}\left(\frac{r}{a}\right)^{2n+4k-2}\quad(k=1,2,3,...)\ (m=1,2,3,...,k),\qquad(32)$$

where C_k^m is the number of combinations from k to m.

Taking into account (32), for the series (13) we have

$$|X_n(r)|\le\left(\frac{r}{a}\right)^{2n-2}\sum_{k=0}^{\infty}\frac{1}{(2^{2k}(n+2k-1)!)^2}\left(\frac{r}{a}\right)^{4k}\sum_{m=0}^{k}C_k^m(\gamma K_0)^{k-m}K^{2m}=$$

$$=\left(\frac{r}{a}\right)^{2n-2}\sum_{k=0}^{\infty}\frac{(K^2+\gamma K_0)^k}{(2^{2k}(n+2k-1)!)^2}\left(\frac{r}{a}\right)^{4k}.$$

Using the D'Alembert principle to calculate the convergence radius of the last power series, we get

$$R=\frac{2^4}{K^2+\gamma K_0}\lim_{k\to\infty}((n+2k)(n+2k+1))^2=\infty.$$

Thus it is proved that the series (13) converge uniformly. Similarly, it is possible to prove uniform convergence for the series (23), as well as for similar series, where the functions of variable coefficients will be derived from the initial and generating functions $\alpha_{n,k}(r),\ \beta_{n,k}(r)\ (k=0,1,2,...)$ from the first to the fourth order.

Now let's prove that $X_n(r),\ Y_n(r)\ (n=1,2)$ is linearly independent. Let us first assume that the identity has been fulfilled

$$C_1X_1(r)+C_2X_2(r)+C_3Y_1(r)+C_4Y_2(r)=0\,, \tag{33}$$

and among the constants C_1,C_2,C_3,C_4 are not all zero. Taking into account (9) and separately equating to zero the expression with logarithm, we get

$$C_3X_1(r)+C_4X_2(r)=\sum_{k=0}^{\infty}\sum_{m=0}^{k}(-K_0)^{k-m}K^{2m}\left[C_3\alpha_{1,m,k-m}(r)+C_4\alpha_{2,m,k-m}(r)\right]=0\,. \tag{34}$$

This identity can only be performed under the following conditions:

$$C_3\alpha_{1,m,k-m}(r)+C_4\alpha_{2,m,k-m}(r)=0\ (k=0,1,2,...)(m=0,1,2,...,k)\,.$$

In particular, when $k=0,\ m=0$, should be performed: $C_3+C_4(r/a)^2=0$. Hence

$$C_3=C_4=0\,. \tag{35}$$

Then from the formula (34) we conclude that the solutions $X_n(r)(n=1,2)$ are linearly independent. Considering this fact, from the identity (33) taking into account (35) we get: $C_1=C_2=C_3=C_4=0$. Consequently, the solutions $X_n(r),\ Y_n(r)\ (n=1,2)$ are also linearly independent.

Thus, we can state that the formulas (9), (13), (19)-(27) define four fundamental solutions $X_n(r),\ Y_n(r)\ (n=1,2)$ of the equation (8). In addition, based on the analysis of the listed formulas and the results of the work [8], we conclude that $X_n(r),\ Z_n(r)\ (n=1,2)$ functions are dimensionless. Consequently, the functions $Y_n(r)\ (n=1,2)$ will also be dimensionless.

As a result, we will have for the general solution of the equation (8)

$$w(r)=C_1X_1(r)+C_2X_2(r)+C_3Y_1(r)+C_4Y_2(r), \tag{36}$$

where C_1,C_2,C_3,C_4 are arbitrary constants having a bend dimension. In order to highlight a dimensionless multiplier in the right part of the formula (36), let's put $C_n=a\lambda_n\ (n=1,2,3,4)$, where λ_n is arbitrary dimensionless constants. Then the formula for amplitude bends (36) can be written as:

$$w(r)=a\,w_0(r)\,; \tag{37}$$

$$w_0(r)=\lambda_1X_1(r)+\lambda_2X_2(r)+\lambda_3Y_1(r)+\lambda_4Y_2(r)\,, \tag{38}$$

where $w_0(r)$ is dimensionless function.

In a similar form, we also present formulas for the first three derivatives of the $w(r)$ function:

$$\frac{dw}{dr}=\tilde{w}_0(r)\,, \tag{39}$$

$$\tilde{w}_0(r)=\lambda_1\tilde{X}_1(r)+\lambda_2\tilde{X}_2(r)+\lambda_3\tilde{Y}_1(r)+\lambda_4\tilde{Y}_2(r); \tag{40}$$

$$\frac{d^2w}{dr^2}=\frac{1}{a}\hat{w}_0(r)\,, \tag{41}$$

$$\breve{w}_0(r) = \lambda_1 \breve{X}_1(r) + \lambda_2 \breve{X}_2(r) + \lambda_3 \breve{Y}_1(r) + \lambda_4 \breve{Y}_2(r); \tag{42}$$

$$\frac{d^3 w}{dr^3} = \frac{1}{a^2} \hat{w}_0(r), \tag{43}$$

$$\hat{w}_0(r) = \lambda_1 \hat{X}_1(r) + \lambda_2 \hat{X}_2(r) + \lambda_3 \hat{Y}_1(r) + \lambda_4 \hat{Y}_2(r), \tag{44}$$

where

$$\tilde{X}_n(r) = a\frac{dX_n(r)}{dr},\ \breve{X}_n(r) = a^2 \frac{d^2 X_n(r)}{dr^2},\ \hat{X}_n(r) = a^3 \frac{d^3 X_n(r)}{dr^3}\quad (n = 1,2), \tag{45}$$

$$\tilde{Y}_n(r) = a\frac{dY_n(r)}{dr},\ \breve{Y}_n(r) = a^2 \frac{d^2 Y_n(r)}{dr^2},\ \hat{Y}_n(r) = a^3 \frac{d^3 Y_n(r)}{dr^3}\quad (n = 1,2). \tag{46}$$

Unlike the derivative functions $X_n(r), Y_n(r)$, the functions (45), (46) are dimensionless [8]. Then we will call them dimensionless derivatives. However, taking into account (9), for dimensionless derivatives (46) we get:

$$\tilde{Y}_n(r) = \tilde{X}_n(r)\ln\frac{r}{a} + \frac{a}{r}X_n(r) + \tilde{Z}_n(r);$$

$$\breve{Y}_n(r) = \breve{X}_n(r)\ln\frac{r}{a} + \frac{2a}{r}\tilde{X}_n(r) - \left(\frac{a}{r}\right)^2 X_n(r) + \breve{Z}_n(r)$$

$$\hat{Y}_n(r) = \hat{X}_n(r)\ln\frac{r}{a} + \frac{3a}{r}\breve{X}_n(r) - 3\left(\frac{a}{r}\right)^2 \tilde{X}_n(r) + 2\left(\frac{a}{r}\right)^3 X_n(r) + \hat{Z}_n(r),$$

where

$$\tilde{Z}_n(r) = a\frac{dZ_n(r)}{dr},\ \breve{Z}_n(r) = a^2 \frac{d^2 Z_n(r)}{dr^2},\ \hat{Z}_n(r) = a^3 \frac{d^3 Z_n(r)}{dr^3}$$

are dimensionless derivative functions $Z_n(r)$.

Thus, the amplitude function of bends and its first three derivatives are expressed with the dimensionless functions $w_0(r), \tilde{w}_0(r), \breve{w}_0(r), \hat{w}_0(r)$. In this case, the formulas for the dynamic movements (5), (1) and the dynamic forces (2)-(4) taking into account (37), (39), (41), (43) will appear as:

$$W(r,t) = a\, w_0(r) T(t); \tag{47}$$

$$\varphi(r,t) = \tilde{w}_0(r) T(t); \tag{48}$$

$$M_r(r,t) = -\frac{D}{a}\left(\breve{w}_0(r) + \mu \frac{a}{r}\tilde{w}_0(r) \right) T(t); \tag{49}$$

$$M_\theta(r,t) = -\frac{D}{a}\left(\mu \breve{w}_0(r) + \frac{a}{r}\tilde{w}_0(r) \right) T(t); \tag{50}$$

$$Q_r(r,t) = -\frac{D}{a^2}\left(\hat{w}_0(r) + \frac{a}{r}\breve{w}_0(r) - \left(\frac{a}{r}\right)^2 \tilde{w}_0(r) \right) T(t). \tag{51}$$

As a result, we can state that the dynamic movements of $W(r,t), \varphi(r,t)$ and internal forces $M_r(r,t), M_\theta(r,t), Q_r(r,t)$ are expressed with dimensionless fundamental functions of the $X_n(r), Y_n(r)\ (n = 1,2)$ equation (8) and their dimensionless derivatives. This makes it possible to operate only dimensionless values when calculating plates for oscillations.

4. Analytical representation for oscillation frequency

Directly from the equality (14) we find

$$\omega = \frac{K}{a^2}\sqrt{\frac{D}{\rho h}}\,. \tag{52}$$

The formula (52) establishes the analytical dependence of the oscillation frequency on other parameters of the mechanical system, which is considered. In fact, the frequency definition is reduced to finding a dimensionless parameter K, which we will call the coefficient of oscillations. Since the fundamental functions $X_n(r), Y_n(r)\ (n=1,2)$ depend exactly on the coefficient of oscillations, the frequency equations will be used to find it, which will be obtained after the implementation of the desired boundary conditions.

5. Analytical method of numerical realization of exact solutions

Formally, the exact formulas (37)-(44), (47)-(51) are obtained, which are necessary for the research of free oscillations of circular and annular plates under different boundary conditions. However, the implementation of these formulas in practice is connected with necessity of a repeated integral evaluation (20)-(22), (25)-(27), which define the generating functions. These evaluations can be difficult to do. Since the variable modulus of subgrade reaction is often set by polynomials [6], we suggest to specify the analytical method for evaluating the aforementioned integrals for this particular case. In fact, this means the representation of generating functions by polynomials.

As it is clear from the formulas (28), (29), the generating functions $\alpha_{n,k,0}(r)\ (k=1,2,3,...)$ are already represented by polynomials. When considering other generating functions, we assume that the dimensionless function a is defined by a polynomial

$$A(r) = A_0 + A_1\left(\frac{r}{a}\right) + A_2\left(\frac{r}{a}\right)^2 + ... + A_s\left(\frac{r}{a}\right)^s. \tag{53}$$

In this case, it is directly from formula (21) taking into account (19), (53), that the generating functions $\alpha_{n,0,k}(r)\ (k=1,2,3,...)$ a will also be polynomials. The smallest and largest degrees of these polynomials will be respectively equal to $2n+4k-2$ and $ks+2n+4k-2$. Consequently, the functions $\alpha_{n,0,k}(r)$ can be represented in the form

$$\alpha_{n,0,k}(r) = \left(\frac{r}{a}\right)^{2n+4k-2}\sum_{j=0}^{ks} c_{n,0,k,j}\left(\frac{r}{a}\right)^j \quad (k=1,2,3,...), \tag{54}$$

where $c_{n,0,k,j}$ – coefficients, which have to be determined. Then

$$\alpha_{n,0,k-1}(r) = \left(\frac{r}{a}\right)^{2n+4k-6}\sum_{j=0}^{(k-1)s} c_{n,0,k-1,j}\left(\frac{r}{a}\right)^j. \tag{55}$$

In this case, for setting the polynomials (53) and (55) we will have

$$A(r)\alpha_{n,0,k-1}(r) = \left(\frac{r}{a}\right)^{2n+4k-6}\sum_{j=0}^{ks} e_{n,0,k-1,j}\left(\frac{r}{a}\right)^j, \tag{56}$$

where

$$e_{n,0,k-1,j} = \sum_{i=0}^{j} A_{j-i}\, c_{n,0,k-1,i}\,, \tag{57}$$

moreover

$$A_{j-i}=0,\ \text{if}\ j-i>s;\ c_{n,0,k-1,i}=0,\ \text{if}\ i>(k-1)s. \tag{58}$$

Put its value (56) into the formula (21) instead of the $A(r)\alpha_{n,0,k-1}(r)$ its value and integrate it. As a result, we get

$$\alpha_{n,0,k}(r)=\left(\frac{r}{a}\right)^{2n+4k-2}\sum_{j=0}^{ks}\frac{e_{n,0,k-1,j}}{p_{n,k,j}^{2}}\left(\frac{r}{a}\right)^{j}, \tag{59}$$

where $p_{n,k,j}=(2n+4k+j-4)(2n+4k+j-2)$. Comparing the equality (54), (59) and taking into account (57), we will have the following recurrence formula for required coefficients

$$c_{n,0,k,j}=\frac{\sum_{i=0}^{j}A_{j-i}\,c_{n,0,k-1,i}}{p_{n,k,j}^{2}}\quad(k=1,2,3,...)(j=0,1,...,ks). \tag{60}$$

At the initial stage of calculations, when $k=1$, the formula (60) requires the initial values $c_{n,0,0,i}\ (i=0,1,2,...,j)$. Putting $k=1$ into the formula (55) and comparing the result with the equality (19), we find

$$c_{n,0,0,0}=1. \tag{61}$$

The remaining initial values, as it is clear from the formula (58), when $k=1$, will be zero

$$c_{n,0,0,i}=0\ (i=1,2,3,...,j). \tag{62}$$

Consequently, the coefficients of the polynomials (54) are fully determined by the formulas (60)-(62), taking into account the conditions (58).

Analyzing the formula (22) taking into account (28), (29), (53), (54), we conclude that the generating functions $\alpha_{n,m,k-m}(r)$ are also polynomials, their smallest and largest degrees will respectively be equal to $2n+4k-2$ and $(k-m)s+2n+4k-2$. So we can write down

$$\alpha_{n,m,k-m}(r)=\left(\frac{r}{a}\right)^{2n+4k-2}\sum_{j=0}^{(k-m)s}c_{n,m,k-m,j}\left(\frac{r}{a}\right)^{j}\ (k=2,3,4,...)\ (m=1,2,3,...,k-1), \tag{63}$$

where $c_{n,m,k-m,j}$ – required coefficients. Moving in the formula (63) the indices k and m by one, we get

$$\alpha_{n,m-1,k-m}(r)=\left(\frac{r}{a}\right)^{2n+4k-6}\sum_{j=0}^{(k-m)s}c_{n,m-1,k-m,j}\left(\frac{r}{a}\right)^{j}, \tag{64}$$

and moving by one only index k, we have

$$\alpha_{n,m,k-m-1}(r)=\left(\frac{r}{a}\right)^{2n+4k-6}\sum_{j=0}^{(k-m-1)s}c_{n,m,k-m-1,j}\left(\frac{r}{a}\right)^{j}. \tag{65}$$

Moreover, for setting the polynomials (53) and (65) we find

$$A(r)\alpha_{n,m,k-m-1}(r)=\left(\frac{r}{a}\right)^{2n+4k-6}\sum_{j=0}^{(k-m)s}e_{n,m,k-m-1,j}\left(\frac{r}{a}\right)^{j}, \tag{66}$$

where

$$e_{n,m,k-m-1,j}=\sum_{i=0}^{j}A_{j-i}\,c_{n,m,k-m-1,i}\,, \tag{67}$$

moreover

$$A_{j-i}=0,\ \text{if}\ j-i>s;\ \ c_{n,m,k-m-1,i}=0,\ \text{if}\ i>(k-m-1)s\,. \tag{68}$$

In the formula (22), replace $\alpha_{n,m-1,k-m}(r)$ and $A(r)\alpha_{n,m,k-m-1}(r)$ with their values (64) and (66), and then integrate. As a result, we get

$$\alpha_{n,m,k-m}(r)=\left(\frac{r}{a}\right)^{2n+4k-2}\sum_{j=0}^{(k-m)s}\frac{c_{n,m-1,k-m,j}+e_{n,m,k-m-1,j}}{p_{n,k,j}^{2}}\left(\frac{r}{a}\right)^{j} \tag{69}$$
$$(k=2,3,4,...)\,(m=1,2,3,...,k-1)\,.$$

Comparing the equality (63) and (69) with each other, and also taking into account (67), we come to the next recurrence formula for required coefficients

$$c_{n,m,k-m,j}=\frac{c_{n,m-1,k-m,j}+\sum_{i=0}^{j}A_{j-i}\,c_{n,m,k-m-1,i}}{p_{n,k,j}^{2}} \tag{70}$$
$$(k=2,3,4,...)(m=1,2,3,...,k-1)(j=0,1,2,...,(k-m)s)\,.$$

At the first stage of calculations according to the formula (70), when $k=2$, $m=1$, the initial values $c_{n,0,1,j}\,(j=0,1,2,...,s)$ and $c_{n,1,0,i}\,(i=0,1,2,...,j)$. will be used. The first of them are found out according to the formula (60). For the second ones, when $i=0$, we find by formula (29)

$$c_{n,1,0,0}=\frac{1}{(2^{2}(n+1)!)^{2}}\,. \tag{71}$$

The remaining initial values will be zero

$$c_{n,1,0,i}=0\ (i=1,2,3,...,j)\,, \tag{72}$$

that follows directly from the condition (68) at the values $k=2$, $m=1$.

Thus, using formulas (70)-(72), taking into account conditions (68), the coefficients of the polynomials (63) are completely determined.

In a similar way, we can obtain the representations in the form of polynomials for the generating functions (25)-(27). So we will write the final formulas:

$$\beta_{n,k,0}(r)=d_{n,k,0,0}\left(\frac{r}{a}\right)^{2n+4k-2}\quad(k=1,2,3,...)\,;$$

$$d_{n,0,0,0}=1,\ \ d_{n,k,0,0}=\frac{d_{n,k-1,0,0}}{p_{n,k,0}^{2}}-4(2n+4k-3)\frac{c_{n,k,0,0}}{p_{n,k,0}}\quad(k=1,2,3,...)\,;$$

$$\beta_{n,0,k}(r)=\left(\frac{r}{a}\right)^{2n+4k-2}\sum_{j=0}^{ks}d_{n,0,k,j}\left(\frac{r}{a}\right)^{j}\quad(k=1,2,3,...)\,;$$

$$d_{n,0,0,0}=1,\ \ d_{n,0,k,j}=\frac{\sum_{i=0}^{j}A_{j-i}\,d_{n,0,k-1,i}}{p_{n,k,j}^{2}}-4(2n+4k+j-3)\frac{c_{n,0,k,j}}{p_{n,k,j}}\quad(k=1,2,3,...)(j=0,1,...,ks),$$

where

$$A_{j-i}=0,\text{ if } j-i>s;\ d_{n,0,k-1,i}=0,\text{ if } i>(k-1)s;$$

$$\beta_{n,m,k-m}(r)=\left(\frac{r}{a}\right)^{2n+4k-2}\sum_{j=0}^{(k-m)s} d_{n,m,k-m,j}\left(\frac{r}{a}\right)^{j}\ (k=2,3,4,\ldots)\ (m=1,2,3,\ldots,k-1);$$

$$d_{n,1,0,0}=\frac{1}{(2^2(n+1)!)^2}-\frac{4(2n+1)}{(2^2(n+1)!)^3};$$

$$d_{n,m,k-m,j}=\frac{d_{n,m-1,k-m,j}+\sum_{i=0}^{j}A_{j-i}\,d_{n,m,k-m-1,i}}{p_{n,k,j}^{2}}-4(2n+4k+j-3)\frac{c_{n,m,k-m,j}}{p_{n,k,j}}$$

$$(k=2,3,4,\ldots)\ (m=1,2,3,\ldots,k-1)\ (j=0,1,2,\ldots,(k-m)s),$$

where

$$A_{j-i}=0,\text{ if } j-i>s;\ d_{n,m,k-m-1,i}=0,\text{ if } i>(k-m-1)s.$$

As a result, all the generating functions are represented by the polynomials. Thus, in the same way, the analytical method for the numerical implementation of exact solutions is proposed.
The format of this publication does not allow to consider numerical examples here. The authors plan to devote the following article to this question.

6. Results and discussion

Whenever the research of any physical phenomenon is reduced to a differential equation, the key question is constructing its exact (analytical) solution. However, the researchers often deal with known mathematical problem, there is no universal method for integrating differential equations with variable coefficients. This may explain the predominant use of approximate methods. This includes the problem of free axisymmetric oscillations of circular and annular plates on a variable elastic foundation. In this work, for the case where the foundation is described by Winkler model with continuous variable modulus of subgrade reaction, these difficulties were overcome. As a result, there are the following results:

1. The exact solution of the differential equation of oscillations has been set;
2. Formulas for dynamic movements and internal forces in the plate have been obtained;
3. The analytical representation for the oscillation frequency has been obtained;
4. The analytical method of numerical realization of predicted exact solutions has been proposed.

As a matter of fact, the solution of the initial problem is only reduced to the realization of desired boundary conditions and finding appropriate dimensionless oscillation coefficients from the obtained frequency equation.

7. Conclusions

A new analytical method of calculation for free axisymmetric oscillations of circular and annular plates on a continuous variable elastic foundation has been developed. The proposed method does not require the discretization of the considered structure and it is a real alternative to the application of approximate methods during solving this types of problems. Based on the exact solution of the differential equation, this method will allow to obtain better picture of oscillations in comparison with approximate methods. In fact, the exact solution has qualitative information and forms the most complete picture of the physical phenomenon that is researched.

References

[1] Timoshenko S and Woinowsky-Krieger S 1959 *Theory of Plates and Shells* McGraw-Hill

[2] Ponomarev S, Biderman V, Likharev K, Makushin V, Malinin N and Feodos'yev V *1956 Raschety na prochnost' v mashinostroyenii* Mashgiz
[3] Vaynberg D, Vaynberg Ye 1970 *Raschet plastin* Budível'nik
[4] Witt M 1974 Rozwiązanie płyty spoczywajacej na podłożu sprężystym o zmiennym współczynniku podatnosci metoda elementow skonczonych *Pr. nauk. Inst. inz. Lad. Pwr* **13** pp 143-149
[5] Mofid M and Noroozi M 2009 A plate on Winkler foundation with variable coefficient *Transaction A: Civil Engineering* **16** pp 249-255
[6] Doronin A and Soboleva V 2014 Sobstvennyye kolebaniya krugloy plastinki, lezhashchey na peremennom uprugom osnovanii tipa Vinklera *Vestnik Nizhegorodskogo universiteta im. Lobachevskogo* **4** pp 254-258
[7] Filippov A 1970 *Kolebaniya deformiruemyh system* Mashinostroenie
[8] Krutii Y 2016 *Rozrobka metodu rozv'yazannya zadach stiykosti i kolyvan' deformivnykh system zi zminnymy neperervnymy parametramy* Sc. D. diss., Lutsk National Technical University
[9] Krutii Y 2018 Construction of a solution of the problem of stability of a bar with arbitrary continuous parameters *J. Math. Sci* **231** pp 665-677
[10] Shvab'yuk V, Krutii Y and Sur'yaninov M 2016 Investigation of the free vibrations of bar elements with variable parameters using the direct integration method *Strength of Materials* **48** pp 384-393
[11] Krutii Y 2015 Analysis of longitudinal oscillations for systems with continuous variable parameters using force integration method *Technical journal* **9** pp 420-425
[12] Krutii Y, Surianinov M and Vandynskyi V 2018 Development of the method for calculation of cantilever construction's oscillations taking into account own weight *Eastern-European Journal of Enterprise Technologies* **93** pp 13-19
[13] Krutii Y, Surianinov M, Vandynskyi V 2019 Analytic formulas for the cantilever structures' natural frequencies with taking into account the own weight *Materials Science Forum* **968** pp 450–459

Evaluation of electromagnetic radiation shielding characteristics of facing building vaterials

O V Panova[1], G Iu Krasnianskyi[1] and I O Aznaurian[1,*]

[1]Kyiv National University of Construction and Architecture, 31, Povitroflotskyj ave., Kyiv, 03037, Ukraine

[*]E-mail: aznaurian.io@knuba.edu.ua

Abstract. The article presents the results of theoretical studies of the protective properties of composite facing materials based on a dielectric matrix with an electrically conductive non-magnetic filler in a wide frequency range of incident electromagnetic radiation. Expressions are got for the values of the transmission, reflection and absorption, and the electromagnetic radiation shielding efficiency calculated on their basis. The dependences of the permittivity and electrical conductivity of the composite on the volume fraction of the electrically conductive additive required for calculations were got based on the hypothesis of similarity, considering the nonzero conductivity of the dielectric matrix. Satisfactory agreement between the calculation results and the measured shielding characteristics of specimens of metal silicate materials based on calcium hydro silicates and copper powder was established. The results presented show the adequacy of the proposed calculation method and indicate that it can be used for preliminary estimates of the shielding characteristics when designing electromagnetic radiation shields based on composite facing materials.

1. Introduction

In modern conditions, with a significant increase in the electromagnetic man-made load on human health, protection from electromagnetic radiation (EMR) is becoming an increasingly important task. It is necessary to consider a relatively new factor in the deterioration of the electromagnetic environment – an increase in the levels of EMR of ultrahigh frequencies, the sources of which are wireless communications and other radio-technical devices. The World Health Organization has extended the ALARA principle (As low as reasonably achievable) to electromagnetic influences. This minimization of technogenic impact is not always possible. In any case, it should be determined by the maximum permissible levels of intensity of electromagnetic radiation, which are contained in the main regulatory acts [1,2].

Besides the effect on human health, EMR penetrating the premises can both disable and affect the quality of operation of electronic equipment. With the development of information technologies using microwave radiation, serious problems have arisen associated with electromagnetic interference.

The most effective way to protect against the harmful effects of EMR is shielding. From this point of view, metal screens are the most effective. Their disadvantage is that, because of the high electrical conductivity of metals, they strongly reflect EMR, which is not always acceptable. EMR absorbent shields are usually designed based on heterogeneous materials, the properties of which can be varied over a wide range by varying the composition and manufacturing technology.

Common composite materials intended for EMR shielding are materials based on dielectrics with metal additives because of their high electrical conductivity and good mechanical properties [3-6]. At the same time, their use is limited by their high weight. This disadvantage is partially eliminated by using carbon black, graphite, charcoal as a conductive component [7-9].

EMR shields with magnetic metal additives, which have high absorption coefficients in the microwave range, are also used [3]. Of particular interest are metal-polymer materials containing finely dispersed iron ore concentrate [10-13].

Content from this work may be used under the terms of the Creative Commons Attribution 3.0 licence. Any further distribution of this work must maintain attribution to the author(s) and the title of the work, journal citation and DOI.
Published under licence by IOP Publishing Ltd

Recently, EMR shields based on polymeric materials with carbon nanofillers such as graphene, carbon nanotubes, and graphite nanoplates have become the object of many studies [3,14-17]. These materials are distinguished by high electrical and thermal conductivity, lightness, and high corrosion resistance.

Almost all building materials have the property of EMR shielding. At the same time, the effectiveness of such shielding is insufficient. To normalize the indicators following the recognized international standards for electromagnetic safety, specially designed building materials should be used as protective screens in building constructions. In particular, we have previously shown the fundamental possibility of using facing materials based on metal silicate composites as effective EMR shields [5,6].

When designing EMR shields with specified characteristics, a preliminary assessment of the protective properties of the material is required. Such calculations are usually carried out based on previously measured values of electrical conductivity and dielectric permittivity of the material at various concentrations of the conductive additive. [3,14,15].

This work aims to theoretically study the protective properties of composite facing materials based on a dielectric matrix with an electrically conductive non-magnetic filler in a wide range of radiation frequencies, including the development of a mathematical model that would allow not only to correctly evaluate the values of the shielding characteristics but also to assign the optimal composition of the material providing such characteristics.

2. Results and discussion

The primary task in the study of the shielding properties of composite materials is to get theoretical dependencies that give an adequate description of the processes of EMR scattering in the composite's body.

The problem of finding the reflection R and the transmission T of a plane electromagnetic wave incident on a plane-parallel layer of matter, which is between a vacuum and an arbitrary medium, has a rigorous solution [18]. In this case, however, the resulting expressions cannot be used for practical calculations because of their complexity. Using the corresponding formulas becomes possible under certain conditions when they are significantly simplified. So, if the complex permittivity $\hat{\varepsilon}$ of the layer of the composite material containing nonmagnetic electrically conductive additives is very high, and the third medium is a vacuum, then at normal incidence of the wave on the layer under the condition $1/|\hat{\varepsilon}| << (\omega d)/c << 1/|\hat{\varepsilon}|^{1/2}$ where ω is the radiation frequency; d is the layer thickness; c is the speed of light in vacuum, expressions for R и T take the form:

$$R = 1 - \frac{4c\varepsilon''}{\omega d|\hat{\varepsilon}|^2}, \quad T = \frac{4c^2}{\omega^2 d^2 |\hat{\varepsilon}|^2}. \tag{1}$$

If $(\omega d)/c << 1/|\hat{\varepsilon}|$, then:

$$R = \frac{\omega^2 d^2}{4c^2}|\hat{\varepsilon}|^2, \quad T = 1 - \frac{\omega d \varepsilon''}{c}. \tag{2}$$

In these expressions ε'' is the imaginary part of the complex permittivity

$$\hat{\varepsilon} = \varepsilon_d + i\frac{\sigma}{\omega\varepsilon_0}, \tag{3}$$

where σ is the electrical conductivity of material; ε_d is the permittivity of a dielectric matrix; ε_0 is the vacuum permittivity (we use the SI system of units).

It should be noted that these conditions are met in a limited number of cases. So, for the example of an EMR shield considered below based on a metal silicate material 0.5 mm thick, relations (1) can be used for $\nu < 10^7$ Hz. If, as in our case, $\nu \sim 10^{10}$ Hz, (1) are applicable for EMR shields with a thickness $d < 10$ μm. Such thin EMR shields are currently based on polymer compositions containing carbon nanotubes [19]. In most cases, the thickness of the shielding material is 1 - 5 mm, and if the incident radiation belongs to the microwave range, the above expressions are not valid.

It is shown below that to estimate the transmission and reflection at normal incidence of EMR on the material, approximate relations can be applied, which are got using the expression for the amplitude of a wave propagating in a medium with absorption, which gives the solution of Maxwell's equations for this case: $E_0 \exp(-\alpha d)$, where $\alpha = (2\omega\kappa)/c$ is the absorption coefficient, κ is the imaginary part of $\sqrt{\hat{\varepsilon}} = n + i\kappa$.

Then, considering multiple reflections from the inner surfaces of the layer under the assumption that the first and third medium is a vacuum (air), we get:

$$R = R_{12}\left(1 + (1 - R_{12})^2 \exp(-2\alpha d)M\right), \quad T = (1 - R_{12})^2 \exp(-\alpha d)M \,. \tag{4}$$

In expressions (4) R_{12} is the reflectivity from a semi-infinite medium with a permittivity equal to the permittivity of a material:

$$R_{12} = \frac{(n-1)^2 + \kappa^2}{(n+1)^2 + \kappa^2}, \tag{5}$$

n и κ are determined by the relations:

$$n = \left(\frac{\varepsilon' + (\varepsilon'^2 + \varepsilon''^2)^{1/2}}{2}\right)^{1/2}, \quad \kappa = \left(\frac{-\varepsilon' + \left(\varepsilon'^2 + \varepsilon''^2\right)^{1/2}}{2}\right)^{1/2}, \tag{6}$$

where ε' is the real part of the complex permittivity.

Factor

$$M = \left(1 - R_{12}^2 \exp(-2\alpha d)\right)^{-1} \tag{7}$$

determines the contribution of multiple reflections. It should be noted that taking M into account significantly affects the values of R and T under the condition: $d < 3\delta$, where $\delta = 1/\alpha$ is the layer thickness at which the wave intensity decreases by a factor of *e*. This condition is satisfied at small κ, which determines the absorption of radiation by the material, at low frequencies of the incident radiation, or for very thin EMR shields. Otherwise, multiple reflections can be ignored.

Having determined the values of reflection and transmission, it is possible to estimate the EMR absorption from the condition: $A = (1 - R - T)$.

Thus, knowing the values of the real and imaginary parts of the dielectric permittivity of samples with different contents of the electrically conductive component, it is possible to calculate the dependences of *R*, *T*, and *A* on the volume content of the conducting additive and the frequency of the incident electromagnetic wave.

Expressions for the dependences of the electrical conductivity and permittivity of the composite on the volume fraction of the conducting component θ were got by us earlier [5,20] for a quasi-stationary electromagnetic field based on the similarity hypothesis [21,22] (analogy with the theory of phase transitions) considering the nonzero conductivity of the dielectric matrix.

At the contents of the conducting component corresponding to the subthreshold region $\tau = (\theta - \theta_c) < 0$ (θ_c is the percolation threshold), the dependencies $\varepsilon''(\omega,\theta)$ and $\varepsilon''(\omega,\theta)$ have the form:

$$\varepsilon''(\omega,\theta) = \frac{\sigma_m}{\omega\varepsilon_0}\left(B_0 h(-\tau)^{-q} + B_1\left(h^2 - \left(\frac{\omega\varepsilon_0\varepsilon_d}{\sigma_m}\right)^2 (-\tau)^{-p}\right)\right), \tag{8}$$

$$\varepsilon'(\omega,\theta) = B_0\varepsilon_d(-\tau)^{-q}, \quad (9)$$

where $h = \sigma_d/\sigma_m$ and σ_d, σ_m are the electrical conductivity of a dielectric matrix and a conductive additive; q и p are the critical indices of the percolation theory [21,23,24]; B_0, B_1 are the constants.

When the content of the conductive component is above the threshold ($\tau > 0$):

$$\varepsilon''(\omega,\theta) = A_0 \frac{\sigma_m}{\omega\varepsilon_0}\tau^t, \quad (10)$$

$$\varepsilon'(\omega,\theta) = A_1\varepsilon_d\tau^{-q},, \quad (11)$$

where t is the critical index; A_0, A_1 are the constants.

Expressions (8) - (11) are valid for $|\tau| >> \Delta$. Here

$$\Delta = \left(h^2 + \left(\frac{\omega\varepsilon_0\varepsilon}{\sigma_m}\right)^2\right)^{\frac{s}{2t}} \quad (12)$$

is the interval near the percolation threshold, where there is a transition from dependences (8), (9) to dependences (10), (11); s is another critical index. Besides, (10) is performed for such values of θ, for which $\sigma(\theta) << \sigma_m$, and (9), (11) – for $\varepsilon(\theta) << \varepsilon(\theta_c)$. Note that for θ that are far enough from θ_c relation (11) does not describe the behavior of $\varepsilon'(\theta)$. With such contents of a conducting additive, free electrons lead to the appearance of an imaginary part of ε_m and $\varepsilon(\theta)$ can become negative.

In the range of concentrations of the conducting component near θ_c, equal to Δ, the values of ε'' for samples of finite dimensions are not determined, and ε' tend to the finite limit, the value of which can be estimated from:

$$\varepsilon'(\omega,\theta_c) = a_0\varepsilon_d\left(\frac{\sigma_m}{\omega\varepsilon_0\varepsilon_d}\right)^{1-s}, \quad (13)$$

where a_0 is the constant.

From (9), (11), and (13) it follows that when approaching from both sides to the percolation threshold, a sharp increase in the real part of the permittivity should be observed. Qualitatively, this can be interpreted based on the following considerations. Each pair of the nearest conducting clusters, which are separated near the percolation threshold by a thin dielectric layer, can be considered as a capacitor. Then, when approaching the percolation threshold, the effective surface of such a capacitor increases sharply, because of which the effective capacity of the system should sharply increase.

In practical calculations, the percolation threshold θ_c is first determined based on the measured dependence $\sigma(\theta)$ at any value of ω, including direct current (θ_c does not depend on ω). Then it is taken into account, that with a uniform distribution of the particles of the conducting component over their sizes and volume of the material, the process of electric transfer in such systems is satisfactorily described within the framework of the site percolation model. In this case, the values of 1.6 and 1.0 are used for the values of t and q, respectively [21,23]. The critical indices p and s are associated with the t and q by expressions:

$$q = \frac{t(1-s)}{s} \text{ and } p = t\left(\frac{2}{s} - 1\right), \quad (14)$$

which follow from the Kramers – Kronig relations. Constants B_0, B_1, A_0, A_1 are estimated from the conditions: $\sigma(\theta) = \sigma_d$, $\varepsilon(\theta) = \varepsilon_d$ at $\theta = 0$, $\sigma(\theta) = \sigma_m$ at $\theta = 1$, $A_1 = B_0$.

Having calculated the values of R, T, and A according to the proposed method, we can estimate the total $SE_t = 10\log(1/T)$, the reflection $SE_R = 10\log(1/(1-R))$, and absorption $SE_A = SE_t - SE_R$ EMR shielding efficiency of the material:

$$SE_t = 20\log\left(\frac{(n+1)^2+\kappa^2}{4n}\right)+10\alpha d\log e-10\log M\,, \tag{15}$$

$$SE_R = 10\log\left(\frac{(n+1)^2+\kappa^2}{4n}\right)-10\log\left(1-\frac{4n\left((n-1)^2+\kappa^2\right)\exp(-2\alpha d)M}{\left((n+1)^2+\kappa^2\right)^2}\right). \tag{16}$$

To illustrate the application of the described calculation method, the previously got results of experimental studies of the electrophysical properties of metal silicate materials were used [5,20].

Metal silicate materials comprising amorphous silicate substances and metal powders [25] have several technological and functional advantages over other building materials designed to protect against EMR. The most important is the ability to control such properties of these materials as electrical conductivity, thermal conductivity, elastic modulus, etc. because of the targeted effect on the microstructure of the composite. High physical-mechanical and construction-technical characteristics make it possible to use metal silicate materials for manufacturing facing (including external) products. By changing the type of silicate component, the durability of the products can also be changed.

The feature of the electrophysical properties of metal silicate materials is that when the volume fraction of the metal component approaches a certain value (percolation threshold), the conductivity of the system increases abruptly (by 6 - 7 orders of magnitude), and the permittivity increases to values of $\sim 10^3$ [5,20]. The presence of such electrophysical characteristics suggests that with optimization of the composition, materials based on metal powders and amorphized silicates can exhibit pronounced shielding properties.

The experimental technique – preparation of specimens, measurement of their permittivity and conductivity, EMR transmission, and reflection is described in [6,26]. Amorphized dispersed calcium hydrosilicates were used as a dielectric matrix in the preparation of specimens. Copper powder with a particle size of no more than 60 μm was used as an electrically conductive additive. Special measures were taken to ensure a uniform distribution of granular copper throughout the volume of the material. The dimensions of the specimens - tiles were 125×66×5 mm. Transmission and reflection measurements were carried out at EMR frequencies in the range of 19 - 26 GHz.

It should be noted that the quasi-stationarity condition ($v << c/d$) for the specimens understudy in the showed frequency range is not strictly fulfilled. These frequencies are the limit, at which it is still possible to estimate the quantities σ and ε from relations (8) - (11).

Figure 1 shows the dependences of transmission, reflection, and absorption, and Figure 2 shows the shielding efficiency of metal silicate material specimens, calculated by the method described above at the contents of the metal additive corresponding to the subthreshold region ($\theta < \theta_c, \theta_c = 0.162$ according to the measurement results), and the frequency of the incident radiation equal to 20 GHz.

As can be seen from Figure 1, where the measured values of T, R, and A are also given for comparison, there is a satisfactory agreement between the theoretical and experimental values, which shows the adequacy of the proposed calculation method. It should be noted that the measurement error of these quantities near the percolation threshold increases because of the spread of the experimental values caused by the significant influence of fluctuations of the local parameters of the composite on the conductivity and dielectric permittivity.

Analysis of the data in Figure 1 shows that the EMR absorption by the material in this range of metal contents becomes noticeable only near the percolation threshold at $\theta > 0.15$. This is because the electrical losses that cause absorption are determined by the imaginary part of the permittivity ε''. In this case, the first term of (8) is responsible for the Joule losses associated with the conduction, the second describes the dissipation of radiation energy in the material because of the polarization of isolated conducting clusters and prevails at the radiation frequency for which the calculations and measurements were performed.

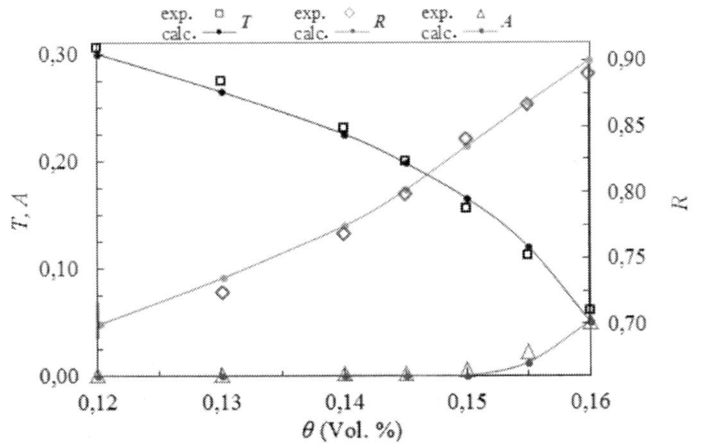

Figure 1. Dependences of the transmission T, reflection R, and absorption A on the content of the metal additive θ in the metal silicate material at $\theta < \theta_c$.

Low absorption in the subthreshold region of metal contents also leads to an increase in radiation intensity losses because of multiple reflections from the inner surfaces of the sample, which causes a significant increase in the values of R at $\theta < \theta_c$.

It should be concluded that the shielding efficiency in the subthreshold metal content range (Figure 2) is low and does not correspond to commercially acceptable values of $SEt > 20$ dB.

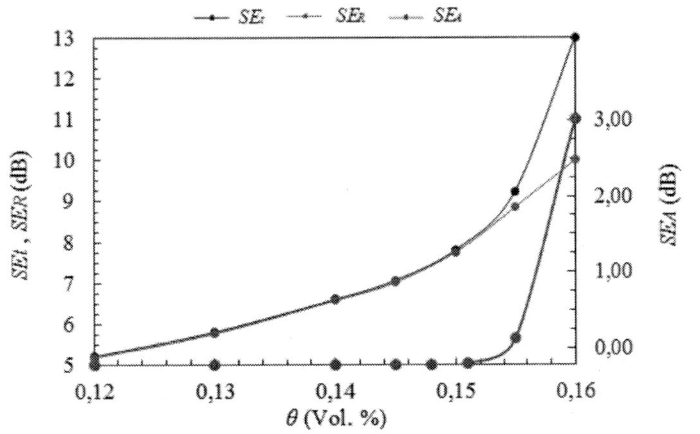

Figure 2. Dependences of the total SE_t, reflection SE_R and the absorption SE_A shielding efficiency on the content of the metal additive θ in the metal silicate material at $\theta < \theta_c$.

At $\theta > \theta_c$ (Figure 3, Figure 4), the electrical losses are determined by the conductivity, which increases with increasing θ. In this case, despite a sharp decrease of ε' with departure from the percolation threshold, the reflection value R decreases slowly because of the simultaneous increase in the electrical conductivity of the composite (Figure 3). At the same time, because of an increase in absorption, the transmission value T also decreases. As a result, the contribution of absorption to the total shielding efficiency SE_t becomes dominant (Figure 4). At $\theta > \theta_c$ practically in the entire investigated range of the metal additive content, $SE_t > 20$ dB and reaches 50 dB at $\theta = 0.2$.

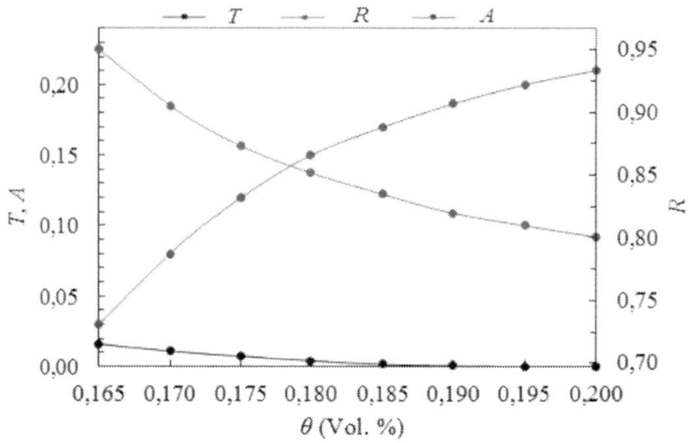

Figure 3. Dependences of the transmission T, reflection R, and absorption A on the content of the metal additive θ in the metal silicate material at $\theta > \theta_c$.

As follows from (8), with increasing frequency of the incident radiation, ε'' should decrease. Besides, at large θ and high frequencies, when the process of electro transfer in the material occurs according to the metallic type, ε'' can also decrease because of the skin effect, which leads to a decrease in electrical conductivity. As a result, as the frequency increases, the shielding efficiency should decrease.

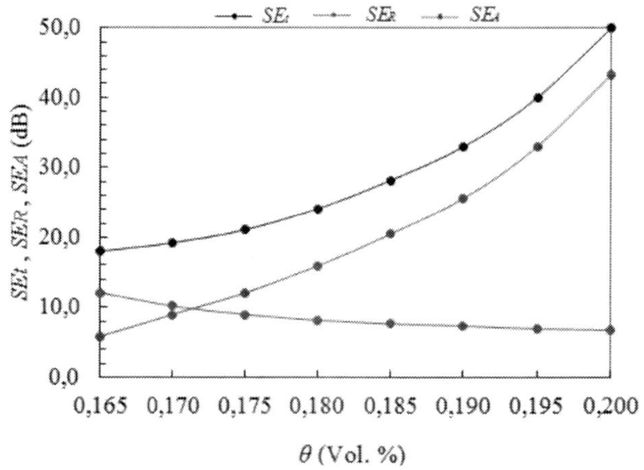

Figure 4. Dependences of the total SE_t, reflection SE_R and the absorption SE_A shielding efficiency on the content of the metal additive θ in the metal silicate material at $\theta > \theta_c$.

Comparing Figures 1, 2 with Figures 3, 4, we can conclude that there is no smooth transition of the corresponding dependences at $\theta < \theta_c$ to the dependences at $\theta > \theta_c$. This can be explained if we consider that the singularity of $\sigma(\theta)$ and $\varepsilon(\theta)$ following from (8) - (11) near the percolation threshold extends to the interval θ equal to Δ, which is determined by (12). And, since in this interval the values of ε'' are not determined, the calculation of the values of R, T, and A in it becomes incorrect. As for the experimental values, the accuracy of their determination near the percolation threshold is significantly influenced by the fluctuation of the local parameters of the composite.

In conclusion, we note that when carrying out calculations according to the proposed method, it is necessary to consider that the assumptions about the normal fall of EMR on the material and the contact of both sides of the material with air, used in obtaining the working formulas, do not correspond to the real conditions in which the shielding facing material is used. The back surface of the shield is in contact with a wall material, the impedance of which is closer to the impedance of the EMR shield and, therefore, the real contribution to the shielding efficiency by the reflection will be lower than the calculated one.

3. Conclusions

Theoretical studies of the protective properties of composite facing materials based on a dielectric matrix with an electrically conductive non-magnetic filler in a wide frequency range of incident electromagnetic radiation have been carried out. A calculation technique is proposed that allows one to estimate the values of reflection, transmission, absorption, and the EMR shielding efficiency of the material. Satisfactory agreement of the calculation results with the measured shielding characteristics of specimens of metal silicate materials based on calcium hydrosilicates and copper powder has been established. The results presented show the adequacy of the proposed calculation method and indicate that it can be used for preliminary estimates of the shielding characteristics in the design of EMR shields based on composite facing materials.

References

[1] Guidelines for limiting exposure to electromagnetic fields (100 kHz to 300 GHz) 2020 *Health Phys.* **118 (5)** 483-524

[2] Directive 2013/35/EU of the European Parliament and of the Council of 26 June 2013 on the minimum health and safety requirements regarding the exposure of workers to the risks arising from physical agents (electromagnetic fields) 2013 *Official Journal of the European Union* **56** L179

[3] Kruzelak J, Kvasnicakova A, Hlozekova K and Hudec I 2021 Progress in polymers and polymer composites used as efficient materials for EMI shielding *Nanoscale Adv.* **3** pp 123-172

[4] Chung D D L 2000 Materials for electromagnetic interference shielding *J. Mater. Eng. Perform.* **9** pp 350-354

[5] Krasnianskyi G, Maksunov S and Velichko T 1989 *Electrophysical properties of contact hardening metallosilicate materials Proc. 3-rd All-Union Scientific and Practical Conf. Slag- alkali cements, concretes and constructions (Kiev)* **1** (Kiev: KISI) pp 209-10

[6] Klapchenko V, Krasnyansky G, Glyva V and Aznaurian I 2010 Protective properties of electromagnetic screens based on metal silicate materials in the microwave range *Hygiene of populated places* **56** pp 219-226

[7] Chuhg Deborah D L 2016 *Carbon Composites: Composites with Carbon Fibers, Nanofibers, and Nanotubes* (New York: Butterworth-Heinemann) p 706

[8] Boiprav O V, Ayad H and Lynkou L M 2019 Electromagnetic Radiation Shielding Properties of Copper Containing Activated Carbon *Doklady BGUIR* **1** pp 51-55

[9] Ayad H A, Belousova E S and Pulko T A 2016 Influence of Composite Coatings Based on Powdered Charcoal on Electromagnetic Radiation Shielding *Doklady BGUIR* **3** pp 88-94

[10] Glyva V A at al 2020 The composite facing material for electromagnetic felds shielding *IOP Conf. Ser.: Mater. Sci. Eng.* **907** 012043

[11] Glyva V, Barabash O, Kasatkina N, Katsman M, Levchenko L, Tykhenko O, Nikolaiev K, Panova O, Khalmuradov B and Khodakovskyy O 2020 Studying the shielding of an electromagnetic fieldby a textile material containing ferromagnetic nanostructures *Eastern-European Journal of Enterprise Technologies* **1 10 (103)** pp 26-31

[12] Glyva V, Kasatkina N, Nazarenko V, Burdeina N, Karaieva N, Levchenko L, Panova O, Tykhenko O, Khalmuradov B and Khodakovskyy O 2020 Development and study of protective properties of the composite materials for shielding the electromagnetic fields of a wide frequency range *Eastern-European Journal of Enterprise Technologies* **2 12 (104)** pp 40-47

[13] Glyva V, Lyashok J, Matvieieva I, Frolov V, Levchenko L, Tykhenko O, Panova O, Khodakovskyy O, Khalmuradov B and Nikolaiev K 2018 Development and investigation of

protective properties of the electromagnetic and soundproofing screen *Eastern-European Journal of Enterprise Technologies* **6 5 (96)** pp 54-61

[14] Vovchenko L et al 2020 Polyethylene Composites with Segregated Carbon Nanotubes Network: Low Frequency Plasmons and High Electromagnetic Interference Shielding Efficiency *Materials* **13** 1118

[15] González M, Pozuelo J and Baselga J 2018 Electromagnetic Shielding Materials in GHz Range. *Chem Rec.* **18** pp 1-11

[16] Abbas N and Kim H T 2016 Multi-walled carbon nanotube/polyethersulfone nanocomposites for enhanced electrical conductivity, dielectric properties and efficient electromagnetic interference shielding at low thickness *Macromol. Res.* **24** pp 1084-1090

[17] Song W L at al 2014 Flexible graphene/polymer composite films in sandwich structures for effective electromagnetic interference shielding *Carbon* **66** pp 67-76

[18] Landau L D and Lifshitz E M 1984 *Electrodynamics of Continuous Media* (Oxford: Pergamon Press) 460

[19] Zi Ping Wu at al 2015 Electromagnetic interference shielding effectiveness of composite carbon nanotube macro-film at a high frequency range of 40 GHz to 60 GHz *AIP Advances* **5** 067130

[20] Klapchenko V, Krasnyansky G and Aznaurian I 2002 *Electrophysical Investigations of Building materials* (Kiev: VIPOL) 84

[21] Efros A L and Shklovskii B I 1976 Critical behaviors of conductivity and dielectric constant near the metal-nonmetal transition *Phys. Status solidi* B **76 2** pp 475-485

[22] Balagurov B Ya 1985 On the theory of dispersion of the conductivity of two-component media *JETP* **88** No 5 1664-75

[23] Kirkpatrick S 1973 Percolation and Conduction *Rev. Mod. Phys* **45** (**4**) pp 574-587

[24] Stauffer D and Aharony A 1992 *Introduction to Percolation Theory* (London: Taylor & Francis) 192

[25] Glukhovsky V D, Runova R F and Maksunov S E 1991 *Contact hardening binders and composite materials* (Kiev: Vyshcha Shkola) 242

[26] Krasnianskyi G, Aznaurian I and Kucherova G 2013 Methods of electrophysical research of concrete in the early stages of hardening *Mistobuduvannya ta terytorialne planuvannya* **50** pp 310-315

MODELING OF TRANSFORMATION OF QUALITY INDICATORS IN THE PROCESS OF MECHANICAL PROCESSING OF PRODUCTS WITH HIGH PHYSICAL AND MECHANICAL PROPERTIES

Olena Parshyna[1], Marharyta Parshyna[2], Yurii Parshyn[3], Anna Gurenko[4], Iurii Savchenko[5], Alexander Shapoval[6]

1 Department of Cybersecurity and Information Technology, University of Customs and Finance, 2/4, Vladimir Vernadsky Street, Dnipro, 49000, Ukraine
2 Department of International Economic Relations, University of Customs and Finance, 2/4, Vladimir Vernadsky Street, Dnipro, 49000, Ukraine
3 Department of Financial and Strategic Investigations, Dnipro State University of Internal Affairs, 26, Gagarina, Dnipro 49005, Ukraine
4 Department of Analytical Economics and Management, Dnipro State University of Internal Affairs, 26, Gagarina, Dnipro 49005, Ukraine
5 Department of Cybersecurity and Information Technology, University of Customs and Finance, 2/4, Vladimir Vernadsky Street, Dnipro, 49000, Ukraine
6 Department of Manufacturing Engineering, Kremenchuk Mykhailo Ostrohradskyi National University, 20, Pershotravneva Street, Kremenchuk, 39600, Ukraine

E-mail: parschina@ukr.net

Abstract. Complex technologies are technologies which consist of a certain set of stages and technological processes. Using the theoretical basis of the systemic and synergetic approaches, the technological process of manufacturing of such parts has been represented as a process of qualitative and quantitative changes of the object. Taking into account the phenomena of technological heredity, it has been assumed that physical and mechanical properties, in particular, tensile strength, proportionality limit, relative narrowing and resilience, as well as characteristics of the quality of the treated surfaces and processing accuracy can be transformed during the complex technological process. To describe the transformation processes of quality indicators, a set of experimental studies of the roughness of the treated surface during machining of parts with high physical and mechanical properties has been carried out and mathematical models have been obtained. To determine the conditions of the integrated technology, it has been proposed to use a complex of dimensionless characteristics - relative changes of the coefficients of the mathematical model. The obtained mathematical models of the processes of mechanical treating of steel 38HN3MFA make it possible to determine the processing conditions that ensure the formation of a given set of quality indicators.

1. Introduction

To ensure a high level of operational reliability of special-purpose parts, complex technologies are being developed, consisting of a certain set of stages and technological processes [1-8]. Using the

Content from this work may be used under the terms of the Creative Commons Attribution 3.0 licence. Any further distribution of this work must maintain attribution to the author(s) and the title of the work, journal citation and DOI.
Published under licence by IOP Publishing Ltd

theoretical basis of the systemic and synergetic approaches [9-19], the technological process of manufacturing such parts can be represented as a process of qualitative and quantitative changes of the object. At the same time, the issues of studying the reliability of machining processes while ensuring the required level of stability of quality indicators [20-29] and the competitiveness of finished products [30-42] remain relevant.

2. The main part of the article

Considering the technological process of manufacturing of a product in complex, taking into account the phenomena of technological heredity, it can be assumed that the physical and mechanical properties, in particular, tensile strength, proportionality limit, relative narrowing and resilience, as well as characteristics of the quality of the treated surfaces and processing accuracy can be transformed, i.e. ... inherit from one stage of a complex technological process to another.

Let us represent the general structure of a complex technology as a complex multidimensional system (Fig. 1), in which to the input of the structure the characteristics of the initial material (charge) for smelting the workpiece (M_1, M_2, ... M_n) come in, and the output provides the corresponding set of physical and mechanical properties (S_1, S_2, ... S_n) and quality indicators (K_1, K_2, ... K_n).

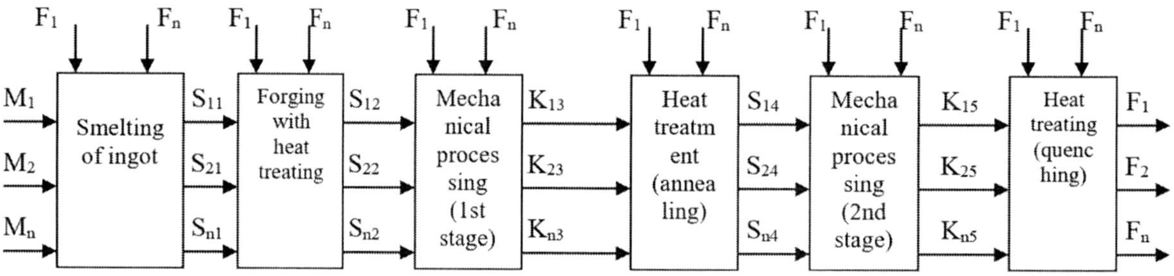

Figure 1. The general structure of a complex technology

The transformation of quality indicators is carried out as a result of the action of a set of technological factors (F_1, F_2, ... F_n), characteristic of each stage of the complex technological process. As a result, the quality of finished products is characterized by a set of quality indicators (R_1, R_2, ... R_n). This approach makes it possible to represent the integrated technology as a complex technological system, consisting of several structural cells - technological stages and connections between them.

The stages of the integrated technology are the technological processes of ingot smelting, forging, machining and heat treating. Let us consider each stage of the integrated technology as a set of elementary technological processes. It should be noted that the resulting quality indicators of the finished product are being formed step by step. Each stage of the integrated technology determines more-less the physical and mechanical properties and quality indicators of the part. Methods of obtaining workpieces and subsequent types and methods of heat treating play an important role in ensuring the specified performance characteristics. The quality indicators of the processed surfaces and the accuracy indicators are finally being formed at the stage of machining. However, to ensure the entire range of indicators of the finished product, it is necessary to consider the stages of complex technology in conjunction with the phenomena of technological heredity, i.e. transformation of quality indicators. In this regard, in order to design the optimal structure of an integrated technology, it is necessary to investigate the quantitative dependencies that describe the connections of the previous technological processes with subsequent technological operations.

According to [43-44], a mathematical model of an elementary technological process of machining can be represented in the form of a graph, the vertices of which correspond to a certain state of surface quality, and a set of arcs – to technological transitions or operations. Let us apply to this model the manifestation of technological heredity in the form of an expression proposed by the authors of [45]:

$$x_i = a \cdot x_{i-1}^{b}, \qquad (1)$$

where X_{i-1}, X_i – quality indicators for (i-1) and i-operations of the machining technological process;
a, b – coefficients of technological heredity, which show the level of transformation of quality indicators.
To describe the transformation processes of quality indicators, a set of experimental studies of the roughness of the treated surface during machining of parts with high physical and mechanical properties has been carried out and expressions have been obtained in the form:

$$Ra_i = a \cdot Ra_{i-1}^{b}, \qquad (2)$$

where Ra_i is the surface roughness at the i-th operation, μm (microns);
Ra_{i-1} – surface roughness at the preliminary operation, μm (microns).
Coefficients a and b have been obtained from regression equations. The mathematical processing of experimental researches has been carried out using the EXCEL program. Table 1 shows the results of the researches of processing of steel 38HN3MFA, obtained by the method of electroslag remelting.
The results of the research have showed that the transformation coefficients of quality indicators depend on the technological conditions of the machining processes of steel 38HN3MFA. Differences in the processes have been established, which are conditioned on not only as the result of the use of various brands of instrumental material. It should be noted that the method of the preliminary stage of complex technology also has a significant impact.
It has been experimentally identified that the surface roughness is to some extent influenced by the method of obtaining the initial metal and the type of heat treating. The physical meaning of these phenomena is explained by the fact that the ability of the metal to provide the specified physical and mechanical properties and quality indicators is initially laid at the stage of the ingot smelting, while providing a reserve for increasing the operational reliability of especially critical products [46].

Table 1

Coefficients of transformation of quality indicators during machining of steel 38HN3MFA

Tool brand	Coefficients of transformation	
	a	b
BK10XOM	2,4825	0,003
BK6OM	2,1579	0,2815
T15K6	1,2156	0,2087

With a certain degree of assumptions, one can assert that using formula (2), it is possible to characterize the quantitative and qualitative relationships of the transformation processes of quality indicators, i.e. technological heredity.
In particular, the quantitative relationships of technological heredity, depending on the choice of the method of the preliminary stage of the complex technology, are determined in formula (2) by the coefficient (b), and the main conditions of machining at a particular stage of the complex technology are determined by the coefficient (a). The obtained coefficients make it possible to quantify the degree of influence of the phenomena of technological heredity for a specific production stage of an integrated technology.
Mathematical models of the investigated processes of mechanical treating of steel 38HN3MFA have been obtained in the form:

$$Ra = a \cdot v^2 + b \cdot v + c, \qquad (3)$$

where a, b, c – coefficients of the mathematical model; v – cutting speed, m / min; Ra – roughness of the processed surface, µm (microns).
In fig. 2 the regression equations are presented – the dependence of the roughness of the processed surface of the workpieces obtained by various technological methods, in particular, electroslag remelting and double remelting with microalloying, taking into account the subsequent heat treating.
The calculation of the coefficients of the mathematical model (3) has been carried out using the least squares method through regression analysis of the EXCEL program. After determining the coefficients and building models, their significance and adequacy of the whole model have been carried out. In table 2 the mathematical models of the investigated processes of machining of workpieces from steel 38HN3MFA taking into account preliminary heat treating are being presented [47].

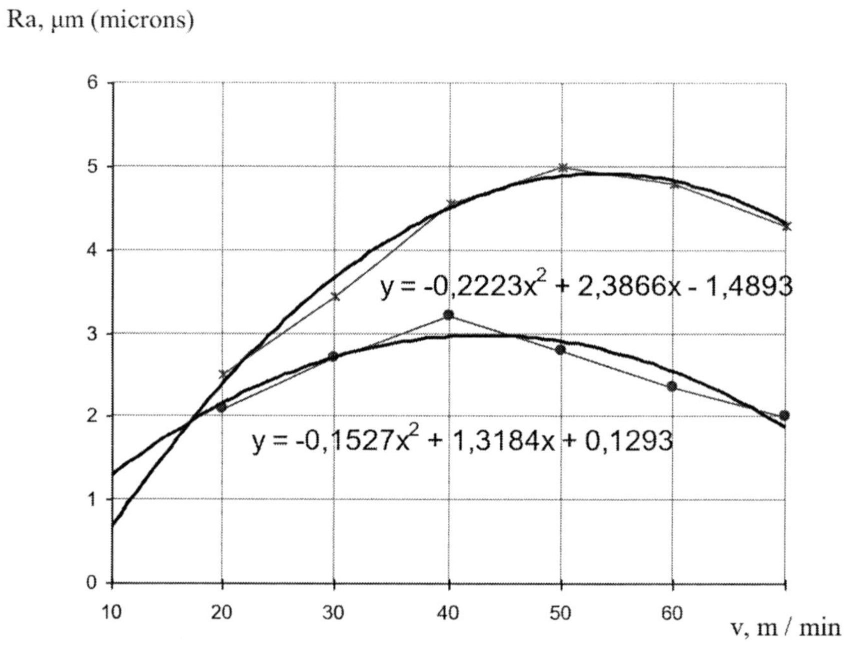

Figure 2. Regression dependences of the roughness of the machined surface on the cutting speed during machining of workpieces made of steel 38HN3MFA with VK10HOM cutting inserts
(1 – after annealing, 2 – after normalization with tempering)

The obtained partial dependences of the roughness of the processed surface on the elements of the machining modes make it possible to determine the optimal processing modes, to assess the degree of influence of technological factors on the quality indicators.
Experimental studies have shown that the processes of mechanical treating of 38HN3MFA steel with various tool materials differ significantly in terms of quality of the treated surfaces and the intensity of wear of cutting tools. To a certain extent, this is due to the peculiarities of obtaining the original workpiece, heat treating, as well as physicochemical processes that arise during mechanical processing. For machining processes, it is important to determine such technological conditions under which the specified quality indicators are being provided for a certain operation. When developing and implementing an integrated technology, it is important to determine such a sequence of operations and such processing conditions under which the degree of transfer of processing errors from the previous operation should be minimal, i.e. the phenomenon of technological heredity should be manifested to a minimum [48-49].

To determine such conditions, it has been proposed to use a complex of dimensionless characteristics - the relative changes of the coefficients of the mathematical model (3):

$$\Delta a = \frac{a_{ann}}{a_{norm}}, \quad \Delta b = \frac{b_{ann}}{b_{norm}}, \quad \Delta c = \frac{c_{ann}}{c_{norm}} \qquad (4)$$

where a_{ann}, b_{ann}, c_{ann} are the constant coefficients of the mathematical model describing the process of mechanical treating of steel 38HN3MFA after annealing;
a_{norm}, b_{norm}, c_{norm} – constant coefficients of the mathematical model describing the process of mechanical treating of steel 38HN3MFA after normalizing with stress relieving.

Table 2

Mathematical models of machining processes of the steel 38XH3MFA

Tool material grade	Heat treating of a workpiece	Mathematical model $Ra = a \cdot v^2 + b \cdot v + c$
VK6OM	annealing	$Ra = -0{,}3179 \cdot v^2 + 3{,}0321 \cdot v - 0{,}5796$
VK6OM	normalizing with stress relieving	$Ra = -0{,}1667 \cdot v^2 + 1{,}4583 \cdot v + 0{,}6429$
T15K6	annealing	$Ra = -0{,}1 \cdot v^2 + 1{,}121 \cdot v + 0{,}0071$
T15K6	normalizing with stress relieving	$Ra = -0{,}097 \cdot v^2 + 0{,}8351 \cdot v + 0{,}2143$

The analysis of the obtained dimensionless characteristics (Table 3) makes it possible to assess the degree of change of qualitative indicators from one operation to another.
Determination of the optimal conditions for mechanical treating is based on identifying a stable technological process in terms of minimizing the degree of change of qualitative indicators. Thus, the most optimal grade of instrumental material is the one for which the relative changes of the coefficients are minimal.

Table 3

Relative changes of the coefficients of the mathematical model of the machining process of steel 38HN3MFA

Tool brand	Relative changes of the coefficients		
	a	b	c
BK10XOM	1,4558	1,8102	11,518
BK6OM	1,907	2,079	0,9015
T15K6	1,0309	1,342	0,033

Similar researches have been carried out for analyzing of the wear of cutting tools of various brands used in production. Regression dependences have been obtained, which made it possible to reveal the influence of technological factors on the wear of cutting tools, taking into account the phenomena of transformation of qualitative indicators.

3. Conclusions

In conclusion, it should be noted that the obtained mathematical models of the processes of mechanical treating of steel 38HN3MFA make it possible to determine the processing conditions that ensure the formation of a given set of quality indicators. The conducted researches create the basis for the automation of the design of complex technologies for the manufacture of critical products.

4. References

[1] Parshina E.A Synergetical approach to the management of product's competitiveness. *Scientific Reports on Recourse Issues 2010*. Freiberg: Technische Universität Bergakademie Freiberg, Germany, Volume 1, 2010. Pp. 398–404.

[2] Parschina E. The Investegation of Stability of Technological Systeme of Cutting Processing. *Proceedings of the International Scientific Conference MECHANIKA-98*, June 1998. – Rzeszow: Rzeszow University of Technology, Poland, 1998. Pp.205–210.

[3] Parshyna O., Parshyn Yu. (2020). Analytical platform to provide competitiveness of ore-mining machinery manufacturing. *Mining of Mineral Deposits*, 14(3), Pp. 61–70. https://doi.org/10.33271/mining14.03.061

[4] Savchenko Iu, Gurenko A and Naumenko O 2016 The cutting-edge industrial technology of mining tool manufacturing Mining of Mineral Deposits 10 4 105–110

[5] Bast J, Gorbatyuk S M, and Kryukov I Yu 2011 Horizontal hcc-12000 unit for the continuous casting of semifinished products Metallurgist 55(1-2) 116-118

[6] Surianinov M and Krutii Y 2018 To the solution of the problem of bending of a cylindrical shell by the boundary elements method MATEC Web of Conferences 230

[7] Krutii Y, Suriyaninov M and Vandynskyi V 2018 Analytic formulas for the natural frequencies of hinged structures with taking into account the dead weight MATEC Web of Conferences 230

[8] Kondratenko V, Sedykh L, Mirzakarimov A and Aleksakhin A 2020 Static analysis and strength calculation of drive shaft of large-scale cone crusher E3S Web of Conferences 193

[9] Malinov L, Malysheva I, Klimov E, Kukhar V and Balalayeva E 2019 Effect of Particular Combinations of Quenching Tempering and Carburization on Abrasive Wear of Low-Carbon Manganese Steels with Metastable Austenite. Materials Science Forum 945 574–578

[10] Khrebtova O, Zachepa N, Zachepa I, Mykhalchenko G and ProkopenkoV 2020 Formation of Starting Torque of Double-Fed Induction Motor Proceedings of the 25th IEEE International Conference on Problems of Automated Electric Drive. Theory and Practice 9240786

[11] Khrebtova O 2020 Forming the induction motor torque when starting Technical Electrodynamics (5) 40-44

[12] Bogoboyashchii V, Vlasov A and Izhnin I 2001 Mechanism for conversion of the conductivity type in arsenic-doped p-CdxHg1-xTe subject to ionic etching Russian Physics Journal 44(1) 61-70

[13] Markov O, Kukhar V, Zlygoriev V, Shapoval A, Khvashchynskyi A and Zhytnikov R. 2020 Improvement of Upsetting Process of Four-Beam Workpieces Based on Computerized and Physical Modeling FME Transactions 48(4) 946–953

[14] Markov O E, Gerasimenko O V, Shapoval A A, Abdulov O R and Zhytnikov R U 2019 Computerized simulation of shortened ingots with a controlled crystallization for manufacturing of high-quality forgings Int J of Advanced Manufacturing Technology 103(5–8) 3057–3065

[15] Shapoval A, Savchenko I and Markov O 2021 Determination coefficient of stress concentration using a conformed display on a circle of a single radius Solid State Phenomena 316 928-935

[16] Zagirnyak M, Mamchur and Kalinov A 2010 Elimination of the influence of supply mains low-quality parameters on the results of induction motor diagnostics The XIX International Conference on Electrical Machines - ICEM 2010

[17] Zagirnyak M, Maliakova M and Kalinov A 2015 Analysis of electric circuits with semiconductor converters with the use of a small parameter method in frequency domain", COMPEL – The international journal for computation and mathematics in electrical and electronic engineering 34 3 808-823

[18] Lutsenko I 2015 Identification of target system operations Development of global efficiency criterion of target operations Eastern-European Journal of Enterprise Technologies 2 2 35–40
[19] Lutsenko I, Vihrova E, Fomovskaya E and Serdiuk O 2016 Development of the method for testing of efficiency criterion of models of simple target operations Eastern-European Journal of Enterprise Technologies 2 4 42–50
[20] Gorbatyuk S, Shapoval A, Mos'pan D and Dragobetskii V 2016 Production of periodic bars by vibrational drawing Steel in Translation 46 7 474-478
[21] Kurpe O, Kukhar V, Puzyr R, Burko V, Balalayeva E and Klimov E 2020 Electric Motors Power Modes at Synchronization of Roughing Rolling Stands of Hot Strip Mill. CONFERENCE 2020, PAEP 510
[22] Kukhar V, Grushko A and Vishtak I 2018 Shape indexes for dieless forming of the elongated forgings with sharpened end by tensile drawing with rupture Solid State Phenomena 284 408–415
[23] Grushko A, Kukhar V and Slobodyanyuk Y 2017 Phenomenological model of low-carbon steels hardening during multistage drawing Solid State Phenomena 265 114–123
[24] Haikova T, Puzyr R, Savelov D, Dragobetsky V, Argat R and Sivak R 2020 The Research of the Morphology and Mechanical Characteristics of Electric Bimetallic Contacts CONFERENCE 2020, PAEP 579
[25] Shapoval A, Drahobetskyi V, Savchenko I, Gurenko A and Markov O 2020 Profitability of production of stainless steel + zirconium metals combination adapters Key Engineering Materials 864 285-291
[26] Zarapin A, Shur A and Chichenev N 1999 Improvement of the unit for rolling aluminum strip clad with corrosion-resistant steel Steel in Translation 29(10) 69-71
[27] Artiukh V, Mazur V ahd Adamtsevich A 2017 Priority influence of horizontal forces at rolling on operation of main sheet rolling equipment MATEC Web of Conferences 106
[28] Savchenko I, Shapoval and Gurenko A 2020 Modeling dynamic parameters of hard alloys during shock wave regeneration IOP Conference Series: Materials Science and Engineerin. 969(1)
[29] Zagirnyak M, Kalinov A, Melnykov V and Kochurov I 2015 Correction of the operating modes of an induction motor with asymmetrical stator windings at vector control International Conference on Electrical Drives and Power Electronics 259
[30] Hrudkina N, Aliieva L, Markov O, Marchenko I, Shapoval A, Abhari P and Kordenko M 2020 Predicting the shape formation of hollow parts with a flange in the process of combined radial-reverse extrusion Eastern-European Journal of Enterprise Technologies 4 1 (106) 55-62.
[31] Shvab'yuk V, Krutii Y and Sur'yaninov M 2016 Investigation of the Free Vibrations of Bar Elements with Variable Parameters Using the Direct Integration Method Strength of Materials 48(3) 384-393
[32] Markov O, Gerasimenko O, Aliieva L and Shapoval A 2019 Development of the metal rheology model of high-temperature deformation for modeling by finite element method EUREKA, Physics and Engineering (2) 52-60
[33] Puzyr R, Shchetynin V, Arhat R, Sira Yu, Muravlov V and Kravchenko S 2021 Numerical modeling of pipe parts of agricultural machinery expansion by stepped punches IOP Conference Series: Materials Science and Engineering 1018
[34] Shapoval A, Kantemyrova R, Markov O, Chernysh A, Vakulenko R and Savchenko I 2020 Technology of production of refractory composites for plasma technologies Proceedings of the 25th IEEE Int. Conf. on Problems of Automated Electric Drive Theory and Practice 9240830
[35] Khrebtova O, Shapoval A, Mos'pan D, Dragobetsky V, Gorbatyuk S and Markov O 2021 Automatic temperature control system for electrocontact annealing of steel welding wire. Metallurgist 65 (3-4)

[36] Zagirnyak M, Maliakova M and Kalinov A 2016 Analysis of operation of power components compensation systems at harmonic distortions of mains supply voltage Joint International Conference – ACEMP 2015: Aegean Conference on Electrical Machines and Power Electronics, OPTIM 2015: Optimization of Electrical and Electronic Equipment and ELECTROMOTION 2015: International Symposium on Advanced Electromechanical Motion Systems 355

[37] Haikova T, Puzyr R and Levchenko R 2020 Experimental Studies on the Stress-Strain State under Drawing Aluminum–Copper Bimetal Parts Rectangular in Plan Russian Journal of Non-Ferrous Metals 61 4 404–412

[38] Zagirnyak M, Zagirnyak V, Moloshtan D, Drahobetskyi V and Shapoval A 2019 A search for technologies implementing a high fighting efficiency of the multilayered elements of military equipment Eastern-European Journal of Enterprise Technologies 6/1 (102) 33–40

[39] Dragobetskii V, Savchenko I, Pavlenko O, Parschina E, Gurenko A and Markov O 2020 Comparative Assessment of Multilayer Waveguide Manufacturing Technologies Proceedings of the 25th IEEE International Conference on Problems of Automated Electric Drive. Theory and Practice 9240865

[40] Bogoboyashchii V and Inzhin I 2000 Mechanism for conversion of the type of conductivity inp-Hg1-xCdxTe crystals upon bombardment by low-energy ions Russian Physics Journal 43 8 627-636

[41] Belas E, Bogoboyashchyy V, Grill R, Izhnin I, Vlasov A and Yudenkov V 2003 Time relaxation of point defects in p- and n-(HgCd)Te after ion milling Journal of Electronic Materials 32 7

[42] Zagirnyak M, Prus V and Nikitina A 2006 Grounds for efficiency and prospect of the use of instantaneous power components in electric systems diagnostics Przeglad Elektrotechniczny 82 12 123-125

[43] Prus V, Nikitina A, Zagirnyak M and Miljavec D 2011 Research of rnergy processes in circuits containing iron in saturation condition Przeglad Elektrotechniczny 87 3 149-152

[44] Zagirnyak M, Mamchur D and Kalinov A 2012 Comparison of induction motor diagnostic methods based on spectra analysis of current and instantaneous power signals Przeglad Elektrotechniczny 88 12 221-224

[45] Zagirnyak M, Mamchur D and Kalinov A 2012 Induction motor diagnostic system based on spectra analysis of current and instantaneous power signals IEEE SOUTHEASTCON 2014 1

[46] LutsenkoI, Fomovskaya E, Oksanych I, Koval S and Serdiuk O 2017 Development of a verification method of estimated indicators for their use as an optimization criterion Eastern-European Journal of Enterprise Technologies 2 486 17-23

[47] Salenko Y, Puzyr R, Shevchenko O, Kulynych V and Pedun O 2020 Numerical Simulation of Local Plastic Deformations of a Cylindrical Workpiece of a Steel Wheel Rim Simulation and Manufacturing III. DSMIE 2020 1 442 – 451

[48] Use of Balanced Scorecard for Enterprise Competitiveness Assessment / S.O. Faizova, M.I. Ivanova, O.L Faizova, V.L. Smiesova, O.A. Parshyna, O.O. Zavhorodnia. *Journal of Advanced Research in Law and Economics,* [S.l.], v. 11, n. 2, Pp. 349–361, mar. 2020. ISSN 2068-696X. Available at: <https://journals.aserspublishing.eu/jarle/article/view/5103>. Date accessed: 03 july 2020. doi: https://doi.org/10.14505/jarle.v11.2(48).08.

[49] Enterprise Competitiveness Management by Introducing Virtual Reality Technologies / Hanna Andrushchenko, Natalia Shandova, Zoriana Hbur, Nadiya Yavorska, Olena Parshyna / *Academy of Strategic Management Journal* / Print ISSN: 1544-1458; Online ISSN: 1939-6104 / 2019, Vol: 18 Issue: 1. (https://www.abacademies.org/articles/enterprise-competitiveness-management-by-introducing-virtual-reality-technologies-8715.html)

Influence of the processes taking place in the contact zone "masonry - plaster coating" on the destruction of the wall structure

V Paruta[1], I Grynyova[1,2], O Gnyp[1] and L Lavrenyuk[1]

[1]Odessa State Academy of Civil Engineering and Architecture, 4, Didrishson str., Odessa, Ukraine
[2] Address all correspondence to this author. E-mail: irene.grinyova@gmail.com

Abstract. To create the optimal composition of the plaster solution with the preassigned properties, it is necessary to know the processes taking place in the "masonry-coating" system, including in the contact zone between it and the masonry, the mechanism of its destruction. The purpose of the presented work was to describe the processes of occurrence and growth of cracks in the plaster coating and its contact zone with the masonry, in the process of its application, hardening and operation of the wall structure. In recent years, significant progress has been made in analyzing the reliability of plaster coatings, their behavior during operation, however, there are problems in predicting the evolution of damage in terms of initiation and growth of cracks, which ultimately leads to cracking and delamination of the plaster coating from masonry. The behavior of the plaster coating largely depends on its structure and the structure of the contact zone. The state of the "masonry-coating" system was described and this made it possible to obtain data on the process of its destruction, the appearance and spread of cracks. The system was destroyed according to two main schemes: formation and development of cracks in the coating and in the contact zone. Thus, these researches make it possible to reduce the number of defects in the contact zone with the masonry and increase the service life of the wall structure.

1. Introduction and literature review

The use of plaster coatings is due to the need to ensure a longer service life of the wall structure. To create a plaster mortar with desired properties, it is necessary to know the processes taking place in the "masonry-coating" system, including in the contact zone between it and the masonry, the mechanism of its destruction.

In recent years, there has been significant progress in the analysis of the reliability of plaster coatings [1,2], their behavior during operation, but nevertheless, there are problems in predicting the evolution of damage in terms of initiation and growth of cracks, and this leads to cracking and delamination of the plaster coating from the masonry [3].

The behavior of the plaster coating largely depends on its microstructure [4,5] and the structure of the contact zone. Plaster mortars are hierarchical composites consisting of fine aggregates and inorganic compounds - brittle minerals, cement hydration products, the structure of which and the contact zone with masonry is formed as a result of processes occurring during application, hardening, operational influence and the environment [6].

Investigation of the state of the "masonry-coating" system allows obtaining data on the process of its destruction, the appearance and propagation of cracks. The article discusses a model of the stress-

Content from this work may be used under the terms of the Creative Commons Attribution 3.0 licence. Any further distribution of this work must maintain attribution to the author(s) and the title of the work, journal citation and DOI.
Published under licence by IOP Publishing Ltd

strain state of the system. It destruction according to two main schemes: formation, development of cracks in the coating and its contact zone with the masonry.

After applying the mortar mixture to the masonry, cracks form on the surface of the plaster coating, the cause of which is the shrinkage of the cement stone due to its hydration, the suction of moisture by the masonry, its evaporation under the influence of the sun and wind (figure 1, b, c).

The concept and view of the problem disclosed by the authors in this article is a unique and independent scientific direction in previously published works [7-11], the processes occurring in a plaster coating during its application and hardening, the destruction mechanism of the system "masonry - plaster coating" are considered. It was revealed that the system is destroyed according to two main schemes: the formation and development of cracks in the plaster coating and its contact zone with the masonry. The reason for this is the stress caused by: shrinkage of the material due to moisture loss and hardening of cement, structural heterogeneity, deformation of the coating, masonry and their difference, etc.

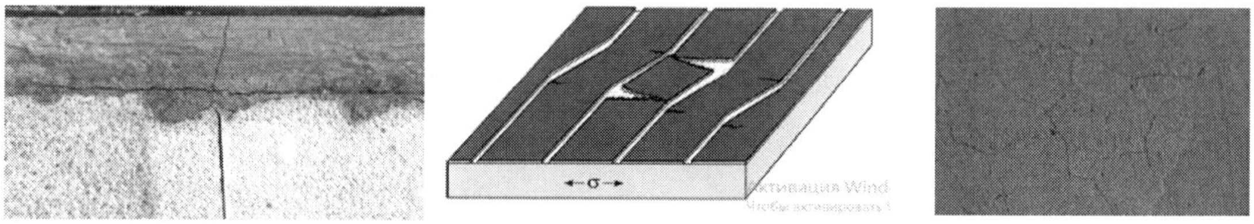

a b c

Figure 1. Cracking in the plaster coating (b, c) and its contact zone with the masonry (a)

2. The aim of the work

The aim of the work is to describe the processes of occurrence and growth of cracks in the plaster coating and its contact zone with the masonry, in the process of its application, hardening and operation of the wall structure.

3. Materials and research methods

The rate of occurrence of new cracks is determined by the stiffness of the coating. In hard plastering coatings, breaking stress is achieved as a result of less deformation. Therefore, at the initial stage of cracking, the density of cracks grows most rapidly for coatings that are characterized by the maximum modulus of elasticity.

Due to the fact that the masonry restrains the shrinkage deformations of the plaster coating, tensile stresses (σ) appear in it, due to which the cracks formed during shrinkage increase.

Their value is found from the equations [12]:

$$\sigma = \Delta\varepsilon \cdot E/1 - \mu\ ; \sigma_y = \Delta\varepsilon \cdot \upsilon \cdot E/1 - \mu \quad (1)$$

where: σ_y – tensile stresses taking into account elastoplastic deformations due to the presence of a gel; E and μ – modulus of elasticity and Poisson's ratio of plaster coating; $\Delta\varepsilon$ – difference in deformations of plaster and masonry; $\upsilon <1$ – coefficient of elastoplastic deformation of the solution under tension.

To calculate stresses (σ) and simplify, we assume that the shrinkage of the aerated concrete masonry has stopped, i.e. $\Delta\varepsilon = \varepsilon_{pl}$ (ε_{pl} – shrinkage deformations of plaster). Then for plaster mortar M25, with modulus of elasticity E =4·10³MPa [13], shrinkage deformation $\Delta\varepsilon = \varepsilon_{pl}$ =0,3...0,8 mm/m 30...80-10^{-5}; Poisson's ratio μ =0,3; coefficient of elastoplastic deformation of the solution under tension υ = 0,5 [14], with minimal shrinkage:

$$\sigma=30\cdot10^{-5}\cdot0{,}5\cdot4\cdot10^{3}/1-0{,}3=0{,}85\ \text{MPa}>0{,}3\ldots0{,}4\ \text{MPa} \quad (2)$$

- at maximum shrinkage:

$$\sigma=80\cdot10^{-5}\cdot0{,}5\cdot4\cdot10^{3}/1-0{,}3=2{,}28\ \text{MPa}>0{,}3\ldots0{,}4\ \text{MPa} \quad (3)$$

They are significantly higher than the calculated tensile resistance for the calculation of the formation of cracks equal to 0.3 MPa and the destruction stress equal to 0,4 MPa [13].

For plaster mortar M50, with modulus of elasticity E =6·10^3 MPa [13], at shrinkage deformation $\Delta\varepsilon = \varepsilon_{pl}$ =0,3...0,8 mm/m or 30...80-10^{-5}; at Poisson's ratio μ =0,3; coefficient of elastoplastic deformation of the mortar under tension v=0,5 [14], with minimal shrinkage, the stresses will be:

$$\sigma = \Delta\varepsilon \cdot \upsilon \cdot E/1 - \mu = 30 \cdot 10^{-5} \cdot 0{,}5 \cdot 6 \cdot 10^3/1 - 0{,}3 = 1{,}28\ \text{MPa} > 0{,}35 \ldots 0{,}5 \quad (4)$$

- at maximum shrinkage:

$$\sigma = \Delta\varepsilon \cdot \upsilon \cdot E/1 - \mu = 80 \cdot 10^{-5} \cdot 0{,}5 \cdot 6 \cdot 10^3/1 - 0{,}3 = 3{,}42\ \text{MPa} > 0{,}35 \ldots 0{,}5 \quad (5)$$

These stresses are also significantly higher than the calculated tensile resistance for the calculation of the formation of cracks equal to 0.35 MPa and the destruction stress equal to 0.5 MPa [13].

In the contact zone "masonry-coating", due to the suction of moisture by the masonry from the mortar mixture and its shrinkage, horizontal cracks are also formed (figure 1.a) [15, 16].

Cracks formed on the surface of the plaster coating "grow" both along the plane and into its depth, initially vertically, to the intersection with the plane of the "masonry-coating" contact zone, and then horizontally, along the contact zone. The direction of growth of these cracks is determined not only by the geometry of stresses, but also depends on the presence of previously formed shrinkage cracks in this zone.

The reasons for the destruction of the plaster coating and the contact zone are also the deformation of the wall structure due to the effect of temperature (figure 2a, b), static and dynamic loads (figure 2c) and stresses caused by them (figure 3). The plaster coating under these conditions works in shear and bending.

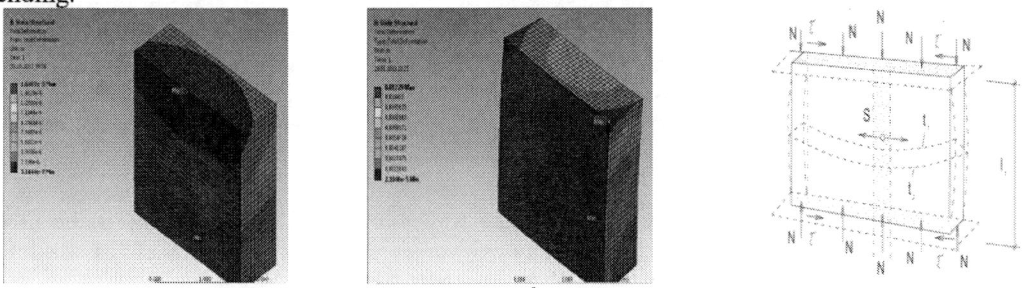

a b c

Figure 2. Deformations in a wall structure at positive (a) and negative temperatures (b), diagram of static and dynamic loads (c)

a b

Figure 3. Influence of variable factors on stresses in the plaster coating at negative (a) and positive (b) temperatures, t / m^2

At the same time, a gradual change in its properties is observed. The dynamic type of loading is characterized by the magnitude of the load, the amplitude and frequency of the load application. Even if the value of the applied load is significantly lower than the breaking stress of the contact zone, there is still a gradual decrease in its strength and the value of adhesion of the coating to the masonry. This loss of strength is called fatigue. Under the cyclic loads occur degradation processes and causes an irreversible change in the properties of the material of the contact zone.

In bearing wall structures made of autoclaved aerated concrete, deformations of the masonry are added to these deformations, due to the constant load, which reach 1.2–2 mm / m [17]. As a result, in the contact zone, cracks of normal separation and flat shear are developing (figure 4). At the initial stages, the crack propagates according to the mechanism of normal separation, then longitudinal shear begins to prevail (figure 4).

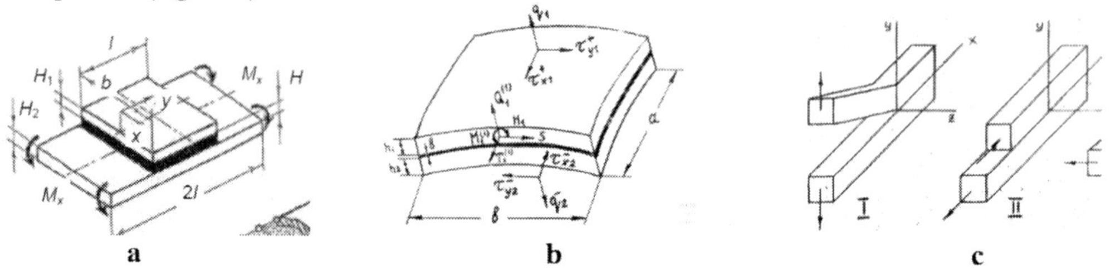

Figure 4. Contact zone of the "masonry-coating" system (a), the character of the stresses in it (b), types of cracks in it (c): I. Normal separation; II. Longitudinal shear;

Destructive stress at uniform separation (det), MPa, is determined by the formula:

$$\sigma_{det} = P/F \tag{6}$$

where: P – destructive load, H; F – the area of the contact area of coverage with masonry, m^2.

4. The results of research

The materials forming the system i.e. masonry and coating differ from each other in their properties. The difference in their coefficients of thermal expansion, moisture deformation, elastic characteristics, causes the occurrence of longitudinal shear stresses along the interface, which also leads to the development of cracks in the contact zone.

The magnitude of these deformations is determined using the equations with the initial data [18, 19, 20]:

$$\Delta L = L_0 \cdot \alpha_t \cdot \Delta t; \qquad \Delta t = t_2 - t_1 \tag{7}$$

where: ΔL – relative extension or reduction of the wall structure; L_0 – its length at the time of erection; α_t – coefficients of thermal expansion: autoclaved aerated concrete α_t =8·10^{-6}/°C, lime-sand mortar (1:4) α_t =9·10^{-6}/°C; cement-sand (1:4) α_t =10,4·10^{-6}/°C, lime-cement α_t =6·10^{-6}/°C; Δt – temperature change of the wall structure; t_1 – temperature of environment at the time of erection of the wall structure; t_2 – maximum or minimum temperature to which wall structure is subjected during the summer and winter periods.

The relative deformation of the masonry compression at -20^{0}C, made at +30°C, will be 0,4 mm/m, and total deformation:, with a length, for example, 8m:

$$\Delta L = L_0 \cdot \alpha_t \cdot \Delta t = 8 \cdot 0{,}000008 \cdot 50 = 0{,}0032\text{m} = 3{,}2\text{mm};\ \Delta t = 50°\text{C} \tag{8}$$

Relative compression strain of cement-sand plaster coating (1:4) will be 0,55 mm/m, and total deformation:

$$\Delta L = L_0 \cdot \alpha_t \cdot \Delta t = 8 \cdot 0{,}0000104 \cdot 50 = 0{,}0044\text{m} = 4{,}44\text{ mm};\ \Delta t = 50°\text{C} \tag{9}$$

Relative compression strain of lime-cement plaster coating will be 0,3 mm/m, and total deformation:

$$\Delta L = L_0 \cdot \alpha_t \cdot \Delta t = 8 \cdot 0{,}000006 \cdot 50 = 0{,}0024\text{m} = 2{,}4\text{mm};\ \Delta t = 50°\text{C} \tag{10}$$

By the action of heating to a temperature +80°C, the relative deformation of the expansion of the masonry will be 0,4 mm/m, and total deformation:

$$\Delta L = L_0 \cdot \alpha_t \cdot \Delta t = 8 \cdot 0{,}000008 \cdot 50 = 0{,}0032\text{m} = 3{,}2\text{mm};\ \Delta t = 50°\text{C} \tag{11}$$

Relative deformation of the expansion of cement-sand plaster coating (1:4) will be 0,55 mm/m, and total deformation:

$$\Delta L = L_0 \cdot \alpha_t \cdot \Delta t = 8 \cdot 0{,}0000104 \cdot 50 = 0{,}0044\text{m} = 4{,}44\text{ mm};\ \Delta t = 50°\text{C} \tag{12}$$

Relative deformation of the expansion of lime-cement plaster coating will be 0,3 mm/m, and total deformation:

IOP Conf. Series: Materials Science and Engineering 1164 (2021) 012059 doi:10.1088/1757-899X/1164/1/012059

$$\Delta L = L_0 \cdot \alpha_t \cdot \Delta t = 8 \cdot 0{,}000006 \cdot 50 = 0{,}0024\text{m}=2{,}4\text{mm};\ \Delta t = 50°\text{C} \qquad (13)$$

The difference in the deformations of the masonry and the coating causes stresses in the plaster coating and the contact zone, the magnitude of which can be determined by transforming the equation [20].

$$\Delta L : L_0 = \sigma : E \qquad (14)$$

where: ΔL – relative extension or reduction of the wall structure; L_0– length of the wall structure at the time of erection; σ – stress in H/mm^2; E – modulus of elasticity in H/mm^2.

The difference in the values of deformations, elasticity moduli of masonry and plaster coating are the cause of shear deformations and stresses (τ), in the contact zone between them (figure 5), which predetermine the development of a crack in the contact zone. The shear stress in the contact zone is equal to:

$$\tau = [\Delta T_1 \alpha_1 - \Delta T_2 \alpha_2] / \left[{}^1/_{E_1} + {}^1/_{E_2}\right] \qquad (15)$$

where: τ – shear stress due to thermal deformations, kgf/cm^2; ΔT_1, ΔT_2 – temperature difference at the time of installation and operation of the plaster coating and masonry, °C; α_1, α_2 – coefficient of thermal expansion of masonry and plaster coating; E_1, E_2 – moduli of elasticity of masonry and plaster coating, kgf/cm^2;

The contact zone works for shear along the axis of the interface and for tension-compression, perpendicular to this axis.

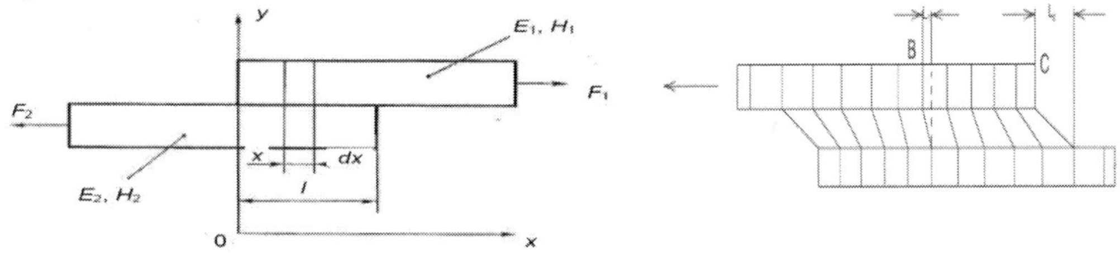

a b

Figure 5. Difference between deformations of masonry and coating (a) and shear deformation in the contact zone "masonry-plaster coating" (b)

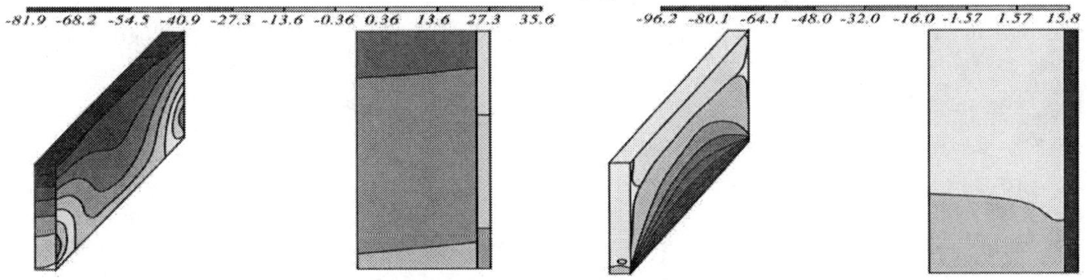

Figure 6. Deformations (a) and shear stresses in the contact zone (b)

Destructive shear stress (sd), MPa is determined according to the formula:

$$\sigma_{sd} = P/F \qquad (16)$$

where: P – destructive load, H; F – the area of the contact area of coverage with masonry, m^2

The contact zone transfers normal and tangential stresses through shear deformation and tension /compression in the z direction. The resulting stresses in the contact layer are not the same in different places of the contact area due to the presence of cracks and hollows in it.

Shear stresses are imposed on tangential stress (due to the mass of the plaster coating (figure 7):

$$\tau_k = \rho \cdot \left(-\frac{1{,}5 \cdot \delta^3}{h^2} - 4 \cdot \delta\right) \qquad (17)$$

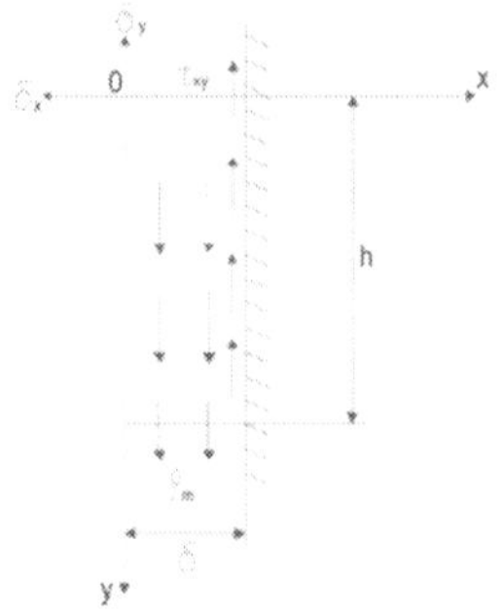

Figure 7. Shear stresses in the contact zone "masonry-plaster coating"

where: δ – plaster coating thickness, m; ρ – average density of plaster mortar, kg/m^3; τ_k – stresses in the contact zone; h – the height of the plaster coating, m;
The described actions cause tensile and compression loads (figure 8a, b), shear loads and splitting loads (figure 8.c, d) in the contact zone.

a,b c,d

Figure 8. Types of loads on the contact zone of the system "masonry-plaster coating"

Deformations of the plaster coating and the "coating-masonry" contact zone are the sum of three components: elastic deformation, reversible in phase with stress, completely irreversible - residual, highly elastic, reversible, but not in phase with stress. The competition of the above described fracture mechanisms in the coating - masonry system with an increase in the degree of deformation leads either to the formation of secondary cracks in the coating of normal separation or to the formation of shear cracks at the interface.

Due to the fact that the strength of the connection between the coating and the masonry at shear τ_k and uniform separation σ_k depends on the destructive load and the area of their contact, then with a decrease in the contact area, the crack growth and delamination of the coating from the masonry becomes easier

$$\tau_k = {}^{F_p}/_{A}; \qquad \sigma_k = {}^{F_p}/_{A} \tag{18}$$

where: F_p – destructive load, A – contact area.

The growth of the main crack occurs due to merging with pore-like cracks, which originate in the zone in front of the apex. This leads to the propagation of a crack along the coating-masonry interface, a decrease in the area of their contacts, as a result of which there is a widespread delamination of the coating (figure 9).

Figure 9. Destruction and peeling of the plaster coating from the masonry

The development of the main crack occurs until the plaster coating is detached from the base and the denudation of the masonry (figure 9).

5. Conclusion

The choice of the components of the mixture, the selection of its composition of the mixture must be made, taking into account the processes of destruction occurring in the system "masonry-plaster coating" during the operation of the wall structure. The article discusses a model of the stress-strain state of the system. Its destruction occurs as a result of the growth of many cracks formed in the zone where the local stress exceeds the local strength of the material. It collapses according to two main schemes: formation, development of cracks in the coating and its contact zone with the masonry. The reasons for the appearance and development of cracks are shrinkage of the plaster coating during hardening, deformation of the wall structure due to the effect of temperature, static and dynamic loads and stresses caused by them. Therefore, for the selection of the components of the mortar mixture, their ratio, it is necessary to take into account the factors. It is necessary to increase the water-retaining capacity of the mortar mixture and reduce the stresses in the coating and the contact zone during operation. The proposed actions will make it possible to reduce the number of shrinkage cracks at the stage of coating hardening, prevent the formation and development of cracks in it and the contact zone during the operation of the wall structure, slow down the rate of its destruction, and increase the time of its operation.

References

[1] Feng T, Xu L, Shi X, Han J and Zhang P 2021 Investigation and Preparation of the Plastering Mortar for Autoclaved Aerated Blocks Walls. *Crystals.* **11(2)** 175. https://doi.org/10.3390/cryst11020175

[2] Netinger Grubeša I, Marković B, Vračević M, Tunkiewicz M, Szenti I and Kukovecz Á 2019 Pore Structure as a Response to the Freeze/Thaw Resistance of Mortars. *Materials.*; **12(19)** 3196. https://doi.org/10.3390/ma12193196

[3] Bruce M, Marc J and Benoit B 2020 Assessing the Shrinkage Cracking Potential of Concrete Using Ring Specimens with Different Boundary Conditions *Advances in Materials Science and Engineering* vol. 2020 pp 1-13 https://doi.org/10.1155/2020/4842369

[4] Dong Y, Yang X and Gao Q 2018 Research on Effect of Pore Characteristics on Performance of Aerated Backfilling Concrete Based on Imaging Method[J]. *Materials Reports* **32(18)** pp 3128-3134 DOI:10.11896 / j.issn.1005-023X.2018.18.006

[5] Brachaczek W 2018 Influence of selected admixtures on the microstructure of renovation plaster mortars. *Architecture, Civil Engineering, Environment* **11(3)** pp 79-85. https://doi.org/10.21307/ACEE-2018-040

[6] Ni T, Yang Y, Wu D and Jiang C 2018 Influences of Environmental Conditions on the Cracking Tendency of Dry-Mixed Plastering Mortar *Advances in Materials Science and Engineering*, vol. 2018 pp 1-9. https://doi.org/10.1155/2018/9160801

[7] Paruta V A, Gnyp O P, Lavrenyuk L I, Rusev A G, Tsonda D A and Vandinsky V Yu 2014 Designing the composition of the stucco mortar for walls of aerated concrete stones, taking into account the mechanics of the destruction of the "masonry-plaster coating" system *Scientific Bulletin of Civil Engineering* **2 (76)** pp 102-110

[8] Paruta V A and Brynzin E V 2014 The formation of the structure of crack-resistant polymer-cement plaster mortar for masonry from autoclaved aerated concrete *Building materials and products* no. 3. pp 12-16

[9] Kos Z, Paruta V and all 2019 Designing plaster mortar for a high-rise building *Tehnički glasnik* **13 (4)** pp.331-336. https://doi.org/10.31803/tg-20181206210501

[10] Paruta V A 2014 Fracture mechanics of system "aerated concrete masonry - plaster covering" *Magazine of Civil Engineering* **47(3)** pp 48-55 DOI: 10.5862/MCE.47.5

[11] Paruta V, Saevskij A, Brynzin E, Gusak D and Kraeva A 2014 Theoretical bases designing of plaster solutions for wall protection of buildings and structures made of aerated concrete *Tehnički glasnik* **8 (4)** pp 346-349 https://hrcak.srce.hr/131560

[12] Khalilov R K 2007 *Issledovaniye sovmestnoy raboty stroitel'nykh materialov v sostave sovremennykh mnogosloynykh teploeffektivnykh naruzhnykh sten zdaniy* dis. *kand. tekhn. nauk : 05.23.05* (Ufa) p 221

[13] Chernyshov Ye M., D'yachenko Ye I and Neupokoyev Yu A 2000 Osobennosti stroyeniya, zakonomernosti deformirovaniya i razrusheniya vysokoporizovannykh neorganicheskikh kompozitov. Materialy shestykh akademicheskikh chteniy. *Sovremennyye problemy stroitel'nogo materialovedeniya: sb. nauchn.* pp. 572-580

[14] Chernyshov Ye M 1977 *Rukovodstvo po tekhnologii izgotovleniya yacheistogo betona plotnost'yu 250–300 kg/m*^{3} (Moscow: Stroyizdat) p 24

[15] Bobryshev A A 2003 *Otdelochnyye kleyevyye rastvory na osnove sukhikh smesey s ispol'zovaniyem kompleksnykh poroshkovykh dobavok dis. kand. tekhn. nauk:05.23.05* (Penza) p 204

[16] Frolov Yu G 1988 *Kurs kolloidnoy khimii. Poverkhnostnyye yavleniya i dispersnyye sistemy* (Moskva: Khimiya) p 464

[17] Pinsker V A, Vylegzhanin V P and Grinfel'd G I 2008 Prochnost' i deformativnost' sten iz gazobetona nizkoy plotnosti. *Yacheistyye betony v sovremennom stroitel'stve* 5 pp 6-9

[18] Voldrzhikh F 1978 *Deformatsionnyye shvy v konstruktsiyakh nazemnykh zdaniy* (Moskva: Stroyizdat) p 224

[19] Galkin S L, Sazhnev N P and Sokolovskiy L 2006 *Primeneniye yacheistobetonnykh izdeliy Teoriya i praktika.* (Minsk: NP OOO "Strinko") p 448

[20] Blazi V 2005 *Spravochnik proyektirovshchika. Stroitel'naya fizika.* (Moscow:Tekhnosfera) p 536

Special aspects of exterior design of elements of stationary and mobile observation constructions

Yu Petrovska[1,*]

[1]Department of Design and Fundamentals of Architecture and Design, Lviv Polytechnic National University, Ukraine

*E-mail: yuliana.r.petrovska@lpnu.ua

Abstract. The article systematizes scientific, informational, regulatory and reference literature on the research topic. Construction schemes of stationary, mobile and transportable Ferris wheels are considered, and their common and distinctive features and peculiarities are outlined. Designs of the world's highest Ferris wheels are illustrated, the history of their creation, types, design and functional features are described. Special aspects of construction of amusement towers as objects of the short-term panoramic observation of the area are specified. The functional purpose of metal observation constructions used in the entertainment industry is discussed. The development of the tourist industry of cities with active sightseeing attractions is analyzed. Conclusions of the scientific research are drawn and prospects for the development of observation constructions as amusement facilities and landmarks of the urban environment are described.

1. Introduction

The Ferris wheel is one of the undisputedly popular attractions in any amusement park. This amusement invaded all entertainment areas for the first time more than a hundred years ago, and nowadays it is difficult to imagine any large amusement park without it. The Ferris wheel always attracted the attention of visitors and tourists of large cities, where these tourist attractions became iconic. The Ferris wheel is very impressive, because the feeling of flying above the ground is additionally accompanied by the extreme sensations of rising high. In a short period of time, upon a rapid change of the height, visitors can observe panoramas of the area.

Amusement towers also perform the sightseeing function. These towers are entertainment facilities that have a secondary function of a short-term panoramic observation of the area. They are divided into observation towers and free-fall towers, and their height is usually over 40 meters. These towers to some extent act as substitutes for Ferris wheels. They are less metal-intensive and more compact. Most of them are focused on emotions and impressions of visitors from falling from the height in order to experience the adrenaline rush and extreme thrills. Racks of such high-rise spatial constructions fixed on a rigid base or basic platform work as a console.

In addition to the classic Ferris wheel, the similar functions are performed by modern panoramic observation facilities, including outdoor elevators with glass platforms, open platforms, terraces and balconies, amusement and observation complexes, escalators, mechanically movable constructions. These objects are very interesting in terms of design, and their value reaches tens of millions of dollars in view of the technical and engineering complexity of their manufacture with the use of concrete, steel and glass.

Today, metal observation constructions, historical monumental complexes, supports of telecommunication towers, meteorological constructions, supporting constructions for factory exhaust pipes, frames of monuments perform additional functions of observation of the territory. They are additionally included into the entertainment and tourism industry of the city where they are located.

Observation towers and Ferris wheels are one of the main highlights of amusement parks, so that is why their architectural expressiveness and design are important. Along with the safety of objects,

Content from this work may be used under the terms of the Creative Commons Attribution 3.0 licence. Any further distribution of this work must maintain attribution to the author(s) and the title of the work, journal citation and DOI.
Published under licence by IOP Publishing Ltd

manufacturers of such constructions are paying more and more attention to the aesthetics and creativity of their products.

2. Analysis of recent publications on the subject

Over the last ten years, only few publications in scientific journals have been devoted to the analysis of the construction and design of Ferris wheels. Most of the publications are contained in periodicals, which have more informative nature. Articles in popular science journals cover individual objects, the history of their creation and operation. By the example of Singapore Flyer, researcher Erica X. Y. Yap, accentuates that Ferris wheels should be considered as popular urban forms for tourists (Yap, 2012). Researchers M. Lohvyn and I. Lytovchenko touch upon the issue of change of the urbanization level that is closely related to the development of tourism. They note that cities concentrate a large number of historical and cultural objects and educational objects, including Ferris wheels that attract tourists from other countries (Lohvyn & Lytovchenko, 2017). Vira D. Kovtun considers the Ferris wheel as a component of the amusement park environment, identifies the basic concepts and technical characteristics and gives examples of their practical implementation (Kovtun, 2010). S. Bilyk and M. But developed an algorithm for finding of a rational location of supporting struts for cantilever steel Ferris wheel (Bilyk & But, 2018).

Researchers Jason Krolicki, Brandon Sullivan and Michael Willford described the story of the Vegas High Roller, which in 2014 became the highest Ferris wheel in the world. It is a giant Ferris wheel that is well integrated into the surrounding environment and that became a world-level tourist attraction. The researchers carried out a detailed analysis of potential threats for the construction, as well as developed recommendations for ensuring the smooth operation and safety of transportation of passengers. They made calculations for the design of the Ferris wheel in seismic situations, under conditions of the strong wind load and extreme temperatures (Krolicki, Sullivan & Willford, 2013). A group of researchers – Andrew Allsop, Pat Dallard and Brendon McNiven – drew attention to the design of the Ferris wheel in Singapore, which became the highest one in the world in 2007-2014 (Allsop, Dallard & McNiven, 2018). Authors of Vegas High Roller: Giant Observation Wheel Design, Jason Krolicki, Brandon Sullivan and Michael Willford, describe the design, composition and safety of the High Roller in Las Vegas, its power and passenger capacity (Krolicki, Sullivan & Willford, 2013). For this object, they made additional calculations of the comfort of conditions of stay of visitors inside cabins under different wind loads (Almufti, Willford, Ferguson, Allsop and Krolicki, 2013).

The monograph by O.Popova, I.Kondratenko, A.Rashchepkina highlights the main design aspects of modern amusement towers. The authors analyze constructive solutions for their main parts – platforms, columns, and nodes with account for the technological equipment and current engineering developments on this topic (Popov, Kondratenko & Rashchepkin, 2009). Authors of the scientific research Multi-Level Viewing Points in the Urban Structure of a City V. Kuzmych and Yu. Petrovska formed a compositional observation scheme using the example of a Ferris wheel, where a viewer observes a change in the level of height above the ground within a short period of time (Kuzmych & Petrovska, 2020).

3. Research Methods

The article systematizes scientific, informational, regulatory and reference materials on the research topic. Construction schemes of stationary and mobile foundationless Ferris wheels are analyzed. Based on the structural analysis, the differences between mobile, transportable and stationary Ferris wheels are highlighted. Designs of the world's highest Ferris wheels, history of their creation, types, design, functional features and significance for development of the tourist sphere of cities are illustrated. By the method of generalization, conclusions of the scientific research are drawn and prospects for the development of observation facilities as entertainment attractions and urban landmarks are outlined.

4. The Ferris wheel as an object of tourist interest

Urbanization plays a special part in terms of influencing on and stimulating various types of tourism activities. Tourist cities are a new form of urbanization that results from the development of functions related to the provision of tourist and recreational services and are a product of industrialization of the society. Large metropolitan cities with the significant historical and cultural potential, in particular unique points of interest, additionally contribute to the growth of the international tourism (Lohvyn & Lytovchenko, 2017). The Ferris wheel is the most popular attraction in any city, both for residents and for tourists, and entertains them with observation effects. It is a mechanized attraction where visitors move with help of different types of energy. Every city in the world wishes to have an iconic landmark that would distinguish it from others by an interesting design or construction. That is why this attraction is often the most significant and the most popular object in a city. These attractions are often located in the central part of the city or in nearby areas for the purpose of providing a better overview of its historical and architectural part.

These objects are becoming the most visited by weekend tourists. The central part of historic cities is the most interesting and attractive from the point of view of people flows, therefore the location of observation facilities in the historical part improves the attendance of cities. If the central part of a city is included in the list of UNESCO World Heritage Sites, then, in accordance with UNESCO requirements, it is prohibited to build any stationary objects or buildings over 20-30 meters high in the historical parts of such cities.

5. Ferris wheels as urban landmarks

The world's largest Ferris wheels include the Ain Dubai (United Arab Emirates, 250 meters), High Roller (Las Vegas, Nevada, 167.6 meters), Singapore Flyer (165 meters), Star of Nanchang (Nanchang, Jiangxi, 160 meters) and London Eye (London, 135 meters). All these objects are unique, have their own history, design, construction, and each of them exceeded the height of its predecessor. The last of them, Ain Dubai, reached 250 meters in height in five years of construction. Its opening is scheduled for 2021 and it will be the largest Ferris wheel in the world. In 2017, the Bailang River Bridge Ferris Wheel in China became the largest centerless Ferris wheel in the world that does not rotate axially, but reaches a height of 145 meters exceeding the world-famous London Eye by 10 meters (Figure 1) (Trejos, 2014) (Nair, 2019) (Martin, 2017).

a) Ain Dubai b) Eye of Bohai c) London Eye d) High Roller

Figure 1. Ratio of the height of the largest Ferris wheels in the world (Author's scheme).

The London Eye, one of the symbols of London, is located on the banks of the River Thames and was built in 1999. It is a giant cantilever Ferris wheel resting on two supports that are attached to the foundation on the bank of the river and inclined at the angle of 65 degrees from the horizontal line. The whole structure is held by support cables attached to the foundation on the opposite side of the river. The Ferris wheel is 120 meters in diameter and is connected with its centre by 64 cables that work like a bicycle wheel. The wheel is equipped with 32 cabins for passengers. One rotation cycle takes about 30 minutes. In 2006, the decorative digital LED lighting system was installed to make the wheel more attractive in the urban environment (Lewis, 2018).

In 2006, a new Ferris wheel – Star of Nanchang – was opened in the Chinese city of Nanchang, the capital of Jiangxi Province, which surpassed its predecessor, the London Eye, by 160 meters (Ming, Zhou & Song, 2018). This giant Ferris wheel was considered to be the largest one in the world until 2008, but the Singapore Flyer at Downtown Core, Singapore set a new world record of 165 meters. The Singapore Flyer has 28 cabins with dimensions of 4 by 7 meters, mounted on a rotating wheel structure, providing passengers with a continuous view of the panorama. They are equipped with the air conditioning system and protection from direct sunlight. Initially, the wheel rotated counterclockwise, but later on its movement direction was changed. The Singapore Flyer includes the station building located at the base of the wheel, which serves as a place of passenger boarding, as well as provides retail services – in particular, the station has shops, restaurants and a wellness center inside (Allsop, Dallard & McNiven, 2018).

6. Structural elements of stationary Ferris wheels

The Ferris wheel belongs to entertainment attractions that create a festive mood for all visitors. Its design provides for the spatial movement in a circle in one plane (gradual ascent, staying at the calculated height for a certain time and gradual descent in an arc). Every time, authors of such constructions aim to improve the wheel by increasing its functionality, safety, entertainment ability and comfort. The structural elements of the wheel and its supports are improved; they are getting equipped with stations and boarding platforms, and observation cabins are getting additionally equipped with devices for air conditioning, rest and relaxation. In their calculations, engineers take into account the components of wind and snow load, seismicity of the location, icing load, various types of mechanical loads and numerous additional factors ensuring the safe operation of the construction (Arkhipov & Arkhipova, 2008).

The bearing frame of the Ferris wheel contains a base in the form of a rigid metal structure with at least one mounted support and a rotating part of the wheel, which in its turn consists of an outer ring, spoke ropes, suspensions, drums, motor drive, support part and frame. Spoke ropes are flexible ropes that provide for the spatial rigidity of the outer ring and are equipped with fastening units. Suspensions that fix cabins for passengers are installed under the bridge. They are attached to the outer ring and have a counterweight system in each cabin. In the automatic mode, they maintain the horizontality of the cabin within the limits of the permissible deviation (not more than 3-5 degrees). The cabin is made in the form of a spatial rod-like frame of tubes in the shape of ellipsoids. Its exterior part is glazed with the toughened glass and equipped with at least two doors, comfort appliances (air conditioning, food heaters, additional lighting, etc.) (Arkhipov & Arkhipova, 2008).

The mobile attraction is equipped with devices for transportation and multiple-time quick assembly (disassembly). The attraction is transportable and installed on a specially prepared site or foundation, the construction of which allows disassembly, transportation and assembly at other place of operation. The stationary attraction is designed for one-time installation at the site of operation (Law of Ukraine, 2006).

Attractions of the rotating-and-observation type must endure maximum loads (weights of own structures, moving parts, attraction with visitors, wind and snow, as well as dynamic, vibrating, shock, emergency, temperature, seismic and other loads that occur in the process of installation and operation. Attractions must retain the specified properties in case of a repeated assembly, disassembly and transportation. Constructions of rotating attractions must be suitable for static loads for out-of-operation conditions and in the process of operation with full and partial loading. The calculations for one-sided loading are made for cases when seats are occupied by 1/4 or 3/4 of the perimeter of the attraction. Rotating attractions must be stable under conditions of one-sided loading of seats by 1/6 of the perimeter of the attraction. Driven or non-driven rotating devices must be adjusted so that they cannot exceed the maximum permissible speed (Law of Ukraine, 2006).

7. Structural elements of mobile Ferris wheels

The Ferris wheel can be transportable, mobile or stationary, have a height from 5 meters to 200 meters, have different types of cabins and seats, various configurations and complexity levels of the model, various sizes and equipment. The transportable attraction up to 55 meters high does not require any foundation. A mobile amusement park is one or more mobile or transportable attractions that are located at the place of installation for no more than a half of the season of operation in the present area (Law of Ukraine, 2006).

It is an interesting fact that the largest mobile Ferris wheel in Europe, Mirador Princess, reaches a height of 70 meters and weighs 600 tons. 42 cabins for transportation of passengers are mounted on it. This Ferris wheel was installed in 2015 in the port of Malaga in Spain. First of all, the foundation was mounted, the supporting structures were assembled and later on the wheel was fixed. Simultaneously raising all the supports to the calculated height, the engineers fixed the wheel at the target level and mounted cabins for passengers. To strengthen the foundation, bored piles were installed under the foundation of the construction. The wheel operated in 2015-2019, and was dismantled in view of termination of the lease agreement (Chaplow, 2019).

There are two approaches to the installation of support platforms of amusement towers with a shaft serving as a guide surface for sliding equipment. The amusement tower (free-fall tower) carries out advancing movement of boarding places by vertical guides. Most of manufacturers use platforms in the form of a rigid flat truss that lies freely on thoroughly leveled ground or rests on it via supports with the possibility to adjust the height. The platforms of modern amusement towers are made according to the following design scheme: the shaft is arranged on a powerful frame structure connected to the platform of six or eight beams joined in the center by a rigid core to increase the resistance to overturning (Popov, Kondratenko & Rashchepkin, 2009).

Amusement towers with no foundation still remain interesting and profitable elements of the entertainment industry. These structures must be as light, rigid and reliable as possible. At the same time, manufacturers of such attractions must reduce the cost, the main component of which is the material consumption (Archipov, 2008). The construction of such buildings consists of a shaft and a support platform. The elevator car with passengers moves along the shaft of the attraction tower, so in contrast to the classic telecommunication towers these constructions must have a high accuracy of guiding elements, which role is often performed by the main supporting racks (Popov, Kondratenko & Rashchepkin, 2009). In the process of mounting, the attraction must be leveled, have a stable position, generate uniform pressure on the ground and be firmly fixed. The site's ground must reliably withstand the calculated load of the attraction specified in the installation instruction (Law of Ukraine, 2006).

In January 2021, the world's largest amusement tower with a height of 140 meters was opened in Bollywood Amusement Park in Dubai. SkyFlyer is a giant swing that set a new world record. 421 ton of steel were used to manufacture the Skyflyer's structure. The console has a mounted rotating structure, which consists of 12 beams with 12 sets of double seats that rotate around the tower, rise and spin at a height to the accompaniment of light and music (Cairns, 2021).

8. Transformation of the shape of Ferris wheels

The classic Ferris wheel has always been the object of imitation in terms of its form, function, orientation, and content. Half a century ago, designers and engineers began to work actively in order to change, add and transform the shape of this sightseeing object, thus creating new forms of amusement facilities, which were also in demand in the tourism industry.

Two Ferris wheels, made to order by Waagner-Biro's, were installed in 1976 in amusement parks in America. The Giant Wheel is a double Ferris wheel that operated in 1973-2004 in the family theme park Hershey Park located in Hershey, Pennsylvania. Its total structure weighed over 135 tons and was fixed in a concrete slab. The structure included a shaft where a transverse beam was attached with rotating wheels at both ends that had twelve spokes. The cantilever-type bracket was supported in the middle by a high vertical support. Round cabins, which could accommodate eight passengers, were attached to each of 12 spokes. The wheels rotated clockwise (Michelson, 2016). A triple version of the Ferris wheels under the name Sky Whirl was developed in 1976. Each wheel rotated around one of three ends of the support brackets. The plane of rotation was not horizontal, so in the process of rotation of the support frame each wheel ascended and descended. When descending, one wheel occupied a horizontal position at the ground level, while other wheels remained raised and continued to rotate in an almost vertical plane at a considerable height. The attraction operated until 2000, and then it was closed and dismantled (Michelson, 2018).

One of the most unusual Ferris wheels in the world is located in Japan. This wheel is one of the main attractions of Tokyo Dome City entertainment complex. The Big O is a huge "empty" wheel, without a core and spokes that is mounted only on the side supports. Extreme and more impressive sensations of visitors are additionally provided by the roller coaster passing through the center of the Ferris wheel with its cabins racing along at a speed of 150 kilometers per hour (Big O, 2016).

Modern constructions also try to imitate the shapes and functions of classic Ferris wheels. SkyView Globen is an interesting project that aims to simulate an object of observation. The equipment is mounted on a large spherical structure of the dome of the complex for holding concerts and sports competitions. Outside the dome, completely transparent glass balls move on special rails and lift passengers to a height of 130 meters, allowing to explore the surrounding urban landscapes of Stockholm (Andersson Sören, 2020). Another Ferris wheel is mounted on a skyscraper in the Georgian city of Batumi. At the top of the building, the construction is built into the facade at the level of the 30th floor (100 meters above the ground level), and has eight capsules for observation of the area (Skrypnyk, 2018). The world's highest horizontal Ferris wheel is located on the Canton Tower in Guangzhou, China. Since the tower affords beautiful views of the big modern city, the authors took advantage of this potential and installed an observation deck at its height. Transparent Bubble Tram cabins go at an angle in a circle along the edge of the roof at the height of 460 meters above the ground. This unusual attraction consists of 16 round passenger cabins with large panoramic windows (Bubble Tram, 2020).

9. Conclusions

The Ferris wheel is an indisputable example of the classic sightseeing attraction being an attribute of any amusement park or an element of the urban environment. It is a large, vertically mounted round rotating metal structure with passenger cabins attached to its edge. The classic Ferris wheel is already over 100 years old, but today its design remains an example of stability, imitation and transformation of form. Many followers imitate its main function – a panoramic view of the surrounding area. Designers and engineers searched for new engineering forms of expression of this object, and created new attractive forms as a result of development and transformation of the form, which are interesting for and popular among tourists. The main function of the Ferris wheel is getting a bird's-eye view of the city, historical and cultural monuments, and unique landscapes. If it is not possible to mount such object in the city center, it immediately gives an impetus to new ideas and design of new forms. New amusement and observation facilities are installed on facades of high-rise buildings, including outdoor elevators with glass platforms, open platforms, terraces, escalators, mechanically moving objects made of glass, concrete and metal. Old TV towers, historical monumental complexes, and engineering structures also began to perform the observation function, getting additional opportunities for operation and reconstruction. Additional panoramic windows and balconies are arranged at the highest points of high-rise buildings, thus inducing tourists to visit these buildings.

For over a hundred years, the world's major cities have been competing for the largest and most original Ferris wheel, installing them in the most interesting places, making them higher than before, giving them different shapes and designs, changing the shape of passenger cabins and providing with different functions for comfortable panoramic view. Its form and functions are still popular today, and the Ferris wheel continues to remain an object of great significance in the urban environment.

References

[1] Yap Erica X Y 2012 Wheels of Fame and Fortune: The Travels of the Singapore Flyer *Sage Journals* Urban Studies https://doi.org/10.1177/0042098012452462

[2] Lohvyn M and Lytovchenko I 2017 Urbanizatsiyni protsesy u sviti yak stymul dlya staloho rozvytku sfery turyzmu (In Ukrainian) *Scientific Bulletin of Poltava University of Economics and Trade* **3 (81)**

[3] Kovtun Vira 2010 Side-shows as important segment in the amusement park system *Bulletin of the National Academy of Management of Culture and Arts* **4** pp 65-69

[4] Bilyk S and But M 2018 Poshuk ratsionalnoho rozmishchennya stiyok konsolnoho kolesa ohlyadu za sylovym kryteriyem (In Ukrainian) *Urban and territorial planning* pp 77-84 http://repositary.knuba.edu.ua/bitstream/handle/987654321/6028/201867-78-85.pdf?sequence=1

[5] Krolicki J, Sullivan B and Willford M 2013 Vegas High Roller: Giant Observation Wheel Design *Structures Congress 2013 American Society of Civil Engineers* https://doi.org/10.1061/9780784412848.118

[6] Allsop Andrew, Dallard Pat and McNiven Brendon 2018 The Singapore Flyer and design of giant observation wheels Singapore *Structural engineering international* pp 12-16 doi: 10.2749/101686609787398326 https://www.tandfonline.com/doi/abs/10.2749/101686609787398326?journalCode=tsei20

[7] Almufti I Willford M Ferguson M Allsop A and Krolicki J 2013 Wind Comfort Studies for the Vegas High Roller Observation Wheel *Structures Congress 2013 American Society of Civil Engineers* https://doi.org/10.1061/9780784412848.116

[8] Popov V Kondratenko I and Rashchepkin A 2009 *Bezfundamentni bashty-atraktsiony z vysokotochnym stovburom* (Vinnytsia National Technical University) 250

[9] Kuzmych V and Petrovska Yu 2020a Multi-Level Viewing Points in the Urban Structure of a City 2020 *IOP Conference Series: Materials Science and Engineering* **907** 012012 doi:10.1088/1757-899X/907/1/012012

[10] Lewis Robert London Eye 2018 *Encyclopedia Britannica* Geography & Travel Tourist Attractions https://www.britannica.com/place/London-Eye

[11] Arkhipov O & Arkhipova T 2008 Mitsnist ta nadiynist parametriv konstruktsiyi bashty atraktsionu *Collection of scientific works of Dnipro State Technical University* Technical Sciences **29** pp 159-161

[12] Ming Ma Zhou Li and Song Tao 2018 Researching and Applying of the New Structure Style for Ferris Wheels *Proceedings of IASS Annual Symposia IASS Boston Symposium: Lightweight structures* pp 1-7

[13] Law of Ukraine "On labor protection" 2006 *Order of the Ministry Of Ukraine On Emergencies And Protection Of The Population From Consequences Chernobyl Disaster Registered in the Ministry of Justice of Ukraine N 405/12279* (In eUkrainian)

[14] About the statement of Rules of a structure and safe operation of attraction quipment https://zakon.rada.gov.ua/laws/show/z0405-06#Text

[15] Chaplow Chris 2019 Malaga City – Big Wheel *About Andalucia.com* https://www.andalucia.com/cities/malaga/big-wheel-noria.htm

[16] Cairns Rebecca 2021 World's tallest swing ride now open in Dubai *CNN travel Dubai now* https://edition.cnn.com/travel/article/bollywood-skyflyer-dubai-tallest-swing-ride-hnk-spc-intl/index.html?fbclid=IwAR1NePiRVg5OebjOx8lI77Vlxs3-zKbpSof4_4czEDMcnwrUQY3-Ks1Zssw

[17] Michelson Harry 2016 More on Giant Wheel *The Amusement Parkives* https://amusementparkives.com/2016/04/16/giant-wheel-follow-up/

[18] Michelson Harry, 2018 Double and Triple Wheels *The Amusement Parkives* https://amusementparkives.com/2018/02/08/waagner-biro-double-and-triple-wheels/
[19] Big O 2016 *Tokyo Dome City Atractions* https://www.tokyo-dome.co.jp/en/tourists/attractions/information.html
[20] Andersson Sören 2020 SkyView *Stockholm Live* https://www.stockholmlive.com/en/your-visit/skyview
[21] Skrypnyk Viktor 2018 Koleso ohlyadu na khmarochrsi vstanovyly nashi maystry (In Ukrainian) Holos Ukrayiny http://www.golos.com.ua/article/309813
[22] Trejos Nancy 2014 World's tallest Ferris wheel opens in Vegas *USA TODAY*.
[23] Nair Manoj 2019 World's tallest Ferris wheel Ain Dubai to open in 2021 *UAE Gulf News* https://gulfnews.com/uae/worlds-tallest-ferris-wheel-ain-dubai-to-open-in-2020-1.1555479653856
[24] Martin James Gabriel 2017 The world's largest spokeless Ferris wheel is opening in China *Lonely Planet Writer* https://www.lonelyplanet.com/articles/bailang-river-bridge-ferris-wheel-china
[25] Bubble Tram 2020 *Canton Tower* http://www.cantontower.com/en/sightseeing/ferriswheel/

Investigation of the Conditions for the Use of Material with the Shape Memory Effect in the Sprinklers

O Petukhova[1,3], S Gornostal[1], S Sherbak[1]

[1]National University of Civil Defence of Ukraine, 94, Chernyshevska str., Kharkiv, 61023, Ukraine

[3] Address all correspondence to this author. E-mail: voda1970@gmail.com

Abstract. The expediency of using nitinol, which is a material with the shape memory effect (SME), as a sensitive element (SE) of a sprinkler has been substantiated. The influence of various parameters on the SE dimensions is analyzed and it is shown that the response time of an element made of a material with the shape memory effect depends on the dimensions of the fixing device (FD) (the diameter of the glass bulb and its thickness). Calculations have shown that the use of the material minimizes the response time of the sprinkler. It is shown that the response time depends on the voltage (or current) in the electrical network, on the size of the fixing device, the cross-sectional radius of the actuator and the critical temperature at which the actuator is triggered.

1. Introduction and literature review

One of the achievements of materials science, which have found application in fire engineering, is the discovery of the shape memory effects. The interest of scientists in studying the nature and mechanisms of SMF is explained by the desire to deepen the fundamental understanding of the inelastic behavior of solids. [1]. The practical application of research is due to the possibility of creating devices and elements with new functionality. [2-4]. Among alloys with SMF, the most studied is nitinol. It is a compound of nickel (55%) and titanium (45%). The melting point of the alloy is 1240-1310 ^{0}C, the density is 6,45 g/sm^3. The positive qualities of nitinol include: resistance to corrosion; strength; the ability to memorize and restore the form; absorption of vibration energy; low level of deformation; high elasticity; the ability to adjust the temperature to change the form. These characteristics make the material almost ideal for use as a SE or an actuator for a sprinkler. The negative qualities that limit the introduction of the material into widespread use include the high cost and insufficient manufacturability of production. A promising area of application of new alloys in fire extinguishing systems is the creation of a hybrid fixing device, which will combine the functions of locking, sensitive and actuating elements. These are exactly the qualities that are necessary to minimize the response time of the sprinkler and the introduction of the extinguishing agent into the fire.

A sprinkler is a sophisticated device designed to detect and extinguish a fire. Obviously, the sooner a fire is detected, the less damage will be incurred. The principle of operation of sprinklers is based on the ability to respond to the thermal factor of a fire. Triggering occurs when the temperature-sensitive element heats up to a certain temperature. The lower this temperature, the faster the sprinkler will open. However, if the temperature is too low, there is a high probability of false opening. This can happen, for example, under the influence of energy-intensive equipment located in the room.

The choice of the optimal opening temperature of the sprinkler is an important task. It is known that an increase in the difference between the nominal response temperature and the maximum ambient

Content from this work may be used under the terms of the Creative Commons Attribution 3.0 licence. Any further distribution of this work must maintain attribution to the author(s) and the title of the work, journal citation and DOI.
Published under licence by IOP Publishing Ltd

temperature helps to reduce the likelihood of spontaneous opening of the device. This difference must be at least 20^{0}C. But the greater this difference, the longer the time of detection of ignition. Thus, minimizing the response time of a sprinkler under specific operating conditions remains an urgent problem. One solution is to use material with shape memory effect (nitinol) to create a sprinkler focused on fast response.

Analysis of literature sources [5-8] showed that among the priority areas for reducing the response time of sprinklers, two have gained the most popularity: the choice of material and the form of its components or the choice of the installation site and the number of sprinklers. The greatest effect can be provided by the implementation of the first approach or their combination. The way to select the material and form of the sprinkler components is ensured by minimizing the weight and size characteristics and using electrical or chemical action to open the sprinkler outlet. The paper proposes the use of nitinol (material with the shape memory effect) for a functional device that combines the functions of a sensitive and an actuator, which will reduce the functional redundancy of the structure, while leaving the minimum weight and size characteristics, which is important when using this material. The aim of the work is to analyze the feasibility of using nitinol as a sensitive element of a sprinkler, to determine the parameters that have the greatest influence on the size of such a sensitive element, to build a model of the response time of a sprinkler with elements made of material with the shape memory effect.

2. Construction of the model and its analysis

The subject of research is a sprinkler sensing element made of material with the shape memory effect. The research methodology is based on the formulas proposed in [9-10], methods and assumptions proposed in [11-12]. The preparation and conduct of the experiment was carried out in accordance with the theory of experiment planning [13]. The numerical implementation of the formulas is performed in the program Maple. The verification of the adequacy of the results obtained was carried out according to the Fisher criterion [13].

In accordance with the principle of construction of a sprinkler, its opening occurs when the sensitive element is destroyed. As a rule, SE is destroyed when the temperature reaches a threshold value. The distribution of temperatures in the SE of a sprinkler with a change in the ambient temperature is described by the equation of thermal conductivity:

$$\rho c \frac{\partial t}{\partial \tau} = \mathrm{div}(\lambda\, \mathrm{grad} t), \tag{1}$$

where ρ - is the density of the SE material, kg/m^3; c - specific heat of the material SE, J/(kg×K); t (x,y,z,τ) - temperature of points (x,y,z) SE, K; τ - moment of time, c; λ (x,y,z,τ) - coefficient of thermal conductivity of the SE material at the point (x,y,z), W/(m×K), with the initial condition (for τ=0):

$$t(x,y,z,\tau) = t_0, \tag{2}$$

where t_0 - initial value of SE temperature, K.

Boundary condition of the third kind (expresses the continuity of the heat flux at the boundary)

$$\lambda \left.\frac{\partial t}{\partial n}\right|_b = \alpha (t_a - t)|_b, \tag{3}$$

where t_a (x,y,z,τ) - ambient temperature near the surface of the SE, K;

α - the heat transfer coefficient, which for forced convection depends on the similarity criteria of Reynolds (Re) and Prandtl (Pr) and lies within the limits 50 - 250 W/(m^2×K) [9]. Let us average expression (1) over the SE volume V. If ρ and c do not depend on coordinates, then the left-hand side of (1) takes the form

$$\rho c \frac{d\bar{t}}{d\tau}, \tag{4}$$

where $\bar{t}$ - the average over the volume V SE value of temperature, K, which is determined by the ratio

$$\bar{t}(\tau) \equiv \frac{1}{V} \int_V dV t(x, y, z, \tau). \tag{5}$$

The right-hand side of the averaged expression has the form of an integral over the SE volume

$$\frac{1}{V} \int_V dV \operatorname{div}(\lambda \operatorname{grad} t(x, y, z, \tau)) \cdot \tag{6}$$

Transforming (6) according to the Ostrogradsky-Gauss theorem and using the boundary condition (3), we write (6) in the form of an integral over the surface of the SE

$$\frac{1}{V} \oint_S dS_n \alpha (t_{BH} - t(x, y, z, \tau)) \cdot \tag{7}$$

The temperature value on the SE surface can be associated with its average value

$$t(x, y, z, \tau) \approx \bar{t}(\tau) + (\bar{r} - \bar{\bar{r}}) \cdot \operatorname{grad} t, \tag{8}$$

where $\bar{r}$ и $\bar{\bar{r}}$ - respectively, the radius vectors of the point (x, y, z) of the surface and the nearest point of the SE, in which the temperature is equal to its mean value $t(\bar{x}, \bar{y}, \bar{z}, \tau) = \bar{t}(\tau)$.

Let us estimate the second term on the right-hand side of (8). The value $|\bar{r} - \bar{\bar{r}}|$ is less than the cross-sectional diameter of the SE d. The grad t value is estimated using the boundary condition (3). Thus $|(\bar{r} - \bar{\bar{r}}) \operatorname{grad} t| < \left| d \frac{\alpha}{\lambda} (t_a - \bar{t}) \right| = Bi |t_a - \bar{t}|$, where Bi - Biot's dimensionless criterion, for this problem corresponds to the following value $Bi = \frac{\alpha}{\lambda} d << 1$.

The latter estimate shows that, with an accuracy of the order of Bi in expression (8) and integral (7), one can replace the temperature on the SE surface with its average volume value.
As a result, expression (1) takes the form

$$\rho c \frac{d\bar{t}(\tau)}{d\tau} = \frac{1}{V} \oint_S dS_n \alpha(x, y, z) [t_{BH}(x, y, z, \tau) - \bar{t}(\tau)] \cdot \tag{9}$$

Let us introduce the notation:

$$\overline{\alpha t_a}(\tau) = \frac{1}{S} \oint_S dS_n \alpha(x, y, z) t_a(x, y, z, \tau); \tag{10}$$

$$\bar{\alpha} \equiv \frac{1}{S} \oint_S dS_n \alpha(x, y, z), \tag{11}$$

where S - surface area of SE, m^2; $\overline{\alpha t_a}(\tau)$ and $\bar{\alpha}$ - mean the values αt and α averaged over the surface of SE, respectively.

Then expression (9) taking into account (10) and (11) takes the form

$$\rho\frac{d\bar{t}(\tau)}{d\tau}=\frac{S\bar{\alpha}}{V}\left[\frac{\overline{\alpha t_a(\tau)}}{\bar{\alpha}}-\bar{t}(\tau)\right]. \tag{12}$$

Introducing a value inverse to the time constant of the SE β, s^{-1}

$$\beta\equiv\frac{\bar{\alpha}S}{V\rho c}, \tag{13}$$

and effective ambient temperature t_a^{ef}

$$t_a^{ef}=\frac{\overline{\alpha t_a}}{\bar{\alpha}}=\frac{\oint_S dS_n\alpha t_a}{\oint_S dS_n\alpha}, \tag{14}$$

obtain (12) in the form

$$\frac{d\bar{t}}{d\tau}=\beta\left[t_a^{ef}-\bar{t}(\tau)\right]. \tag{15}$$

Neglecting the dependence of β and t_a^{ef} on $\bar{t}$, solve (15) and obtain

$$\bar{t}(\tau)=t_0e^{-\beta\tau}+\int_0^{\tau}\beta t_a^{ef}(\tau')e^{-\beta(\tau-\tau')}d\tau'. \tag{16}$$

To simplify the notation, further accept $\bar{t}(\tau)\equiv t(\tau)$ and $t_a^{ef}\equiv t_a$.

Analyzing the change in the temperature of the SE in time (16), taking into account (13), we can conclude that the opening time of the sprinkler is in direct proportion to the size of the SE. The response time of a standard sprinkler is a function of the similarity criteria Nu and Re, when calculating which the value of the characteristic size of the SE δ is determined based on the fact that the width of the fusible lock is $b=0{,}022m$, and its length $l=0{,}036m$.

If a material with SME is used as a SE, and the SE form is taken in the form of a ring placed on a glass cylinder that plays the role of a FD, then the SE dimensions also depend on the wall thickness of the FD, the value of which, in turn, depends on the water pressure in front of the sprinkler.

The dimensions of FD depend on the water pressure in front of the spray arm, and the following condition must be met:

$$\sigma_c\geq\frac{PS_p}{S_{FD}}, \tag{17}$$

where σ_c - stress compressive of material FD, at which it deforms, which will entail partial or complete opening of the sprinkler outlet, MPa; P - water pressure in the pipeline before the sprinkler, Pa; S_p - cross-sectional area of the pipeline in front of the sprinkler, m^2; S_{FD} - contact area affected by water pressure, m^2, which for a fixing device in the form of a glass cylinder is determined:

$$S_{FD}=\pi R_o^2-\pi r_i^2, \tag{18}$$

where R_o – outer radius FD, m; r_i – inner radius FD, m.

Solving together (17) and (18), determine the dependence of the dimensions of the FD on the water pressure in the pipeline

$$R_o^2 - r_i^2 \geq \frac{PS_p}{\sigma_c}. \tag{19}$$

Using the relationship between the radiuses of the FD and the thickness of its walls $h_{FD} = R_o - r_i$, where h_{FD} - wall thickness of FD; as well as the concept of the mean radius of the FD R_{FD} ($R_{FD} = R_o - 0{,}5h_{FD}$);a graphical representation of the dependency solution was obtained (17)–(19) (fig. 1). Analyzing the calculation results (fig. 1), we can conclude that for different values of the radius of the FD and the water pressure in front of the sprinkler, the required minimum dimensions of the wall thickness of the FD are much less than those made in the production. Consequently, the effect of water pressure in front of the sprinkler on the dimensions of the walls of FD can be neglected. The value of the minimum cross-sectional radius of the ring depends on how much force is required for the ring when changing its form in order for the glass cylinder to distroyed and the sprinkler to opened.
The condition for the contact of a ring made of material with SME and FD (glass cylinder) can be written in the following form [14]

$$R_o - \omega_c = R_{in}^r + \delta_r, \tag{20}$$

where ω_c - displacement FD, m; R_{in}^r - inner radius of a ring made of material with SME, m; δ_r - displacement of a ring made of material with SME, m.

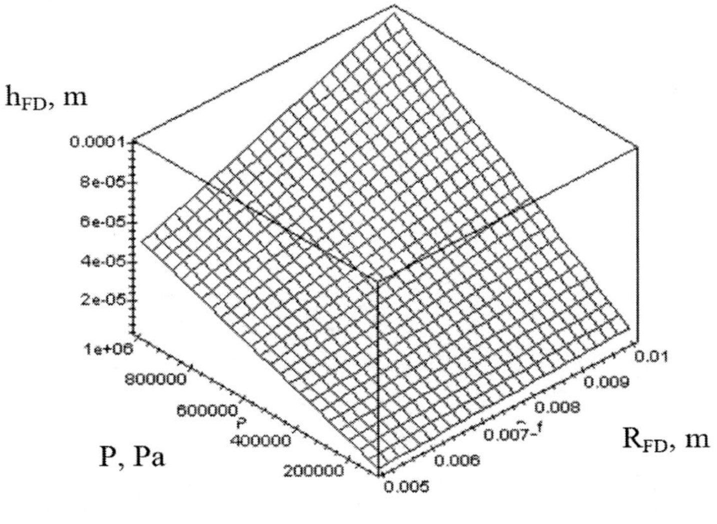

Figure 1. Determination of the minimum wall thickness (h_{FD}) depending on the pressure (P) in front of the sprinkler and the mean radius of the FD (R_{FD})

$$\omega_c = \frac{qR_{FD}^2\beta}{2Eh_{FD}}, \tag{21}$$

where q - strength of contact, N/m; R_{FD} – mean radius FD, m;
E - modulus of elasticity of the material of FD, N/m^2;
β - indicator related to size and material of FD

$$\beta = \sqrt[4]{\frac{3(1-\nu^2)}{R_{FD}^2 h_{FD}^2}},\ \text{m}^{-1},$$

where ν - Poisson's ratio of material of FD.

$$\delta_r = \frac{qR_{in}^r R_r}{S_r E_r}, \tag{22}$$

where R_{in}^r - inner radius of a ring made of material with SME, m; R_r - mean radius of a ring made of material with SME, m, $R_r = R_{in}^r + r$; r - cross-section radius of a ring made of material with SME, m; S_r - cross-sectional area of a ring made of material with SME, m^2; E_r - modulus of elasticity of a ring made of material with SME, N/m^2.

Knowing that the outer radius of the FD R_{FD} should be equal to the inner radius of the ring made of material with SME R_{in}^r, can write down the condition for the destruction of the FD

$$\frac{3q}{2\beta h_{FD}^2} = \sigma_C, \tag{23}$$

while maintaining the strength of the ring made of material with SME, which is determined

$$\frac{qR_r}{S_r} < \sigma_r, \tag{24}$$

where σ_r - permissible stress compressive of the ring made of material with SME, MPa.

It is known that the shape memory effect can be accompanied by a significant work of the sample, which determines the use of this material in sprinklers. For TiNi, work A can reach 208 kJ/m^3, while the deformation of the sample is 7%. Consequently, the correction factor must be taken within 20%. Then, we can assume that the ring is capable of creating a strength q, which depends on r as follows

$$q = \frac{F}{2\pi(R_{FD} + 0.5h_{FD})}, \tag{25}$$

where F – force that must be created of the ring made of material with SME to obtain strength of contact q, N:

$$F = \frac{AV}{l} = \frac{2A\pi^2(\frac{h_{FD}}{2} + R_{FD} + r)r^2}{l}, \tag{26}$$

where $l = 0{,}07\ (h_{FD}/2 + R_{FD} + r)$ - resizing the ring when it fires, m.

Solving expressions (23), (25) and (26) together, we can determine the value of the characteristic minimum size of a SE made of material with SME (radius of the cross-section of the ring). In fig. 2 shows a graphical representation of the dependence of the obtained value on the average radius of the FD R_{FD} and the wall thickness of the FD h_{FD}.

As seen from fig. 1 and fig. 2, even at high values of water pressure in front of the sprinkler, a small thickness wall of the FD is required. Based on the existing assortment of glass products used in sprinklers, as well as to ensure reliability requirements, when determining the minimum cross-sectional radius of the ring, it is advisable to take the wall thickness of the FD device within (0.0001÷0.005) m.

Thus, when constructing a sprinkler focused on quick response, the dimensions of the SE responsible for opening should be kept to a minimum. If the SE performs the functions of a fixing and locking device, then the possibility of reducing its size is limited, which follows from the performed analysis. This means that the functional tasks of the SE should only be the detection of ignition, and the functions of the actuating device should be performed by a separate element, which at the same time can create sufficient force to open the outlet in a minimum time.

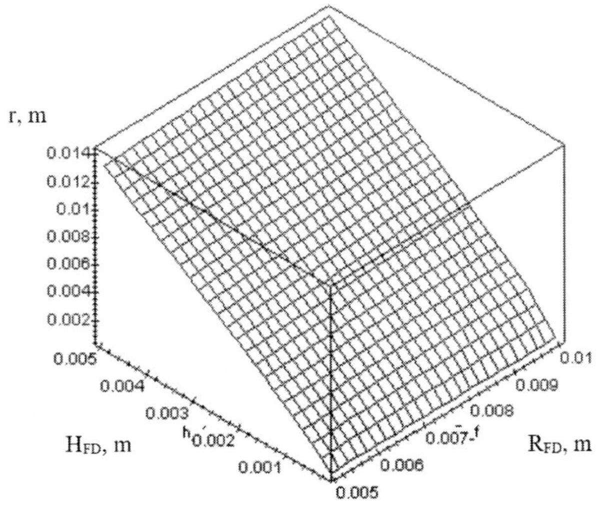

Figure 2. Determination of the minimum radius (r) of the cross-section of a ring with an SME (SE) depending on the wall thickness of the FD (h_{FD}) and the average radius of the FD (R_{FD})

To determine the response time of a sprinkler with a sensitive element made of a material with SME, an experimental study was carried out depending on three factors: the response temperature, the time constant, and the rate of temperature change. In this case, we take an additional value, which for a two-level experiment is equal to α=1.682, the total number of experiments is N=20, the number of observations in the center of the plan is n_0=6 [13]. Information on the levels of changes of the parameters is given in the table 1.

Table 1. Factor level changes

Variation interval and factor level	The rate of change in the temperature of the medium A_t (K/s)			Time constant SE T (s)	Response threshold SE tcr (K)
Zero level x_i=0	0.075	0.3	0.85	80	343
Variation interval δ_i	0.015	0.12	0.2	40	10
Lower level x_i=-1	0.06	0.18	0.65	40	333
Upper level x_i=+1	0.09	0.42	1.05	120	353
Additional points: x_i=-1,682	0.05	0.1	0.5	12.7	326
x_i=+1,682	0.1	0.5	1.2	147.3	360
Code designation	x_1			x_2	x_3

Processing the results of the experiment made it possible to obtain the regression equations:
- for the rate of change in the temperature of the medium A_t = (0.05÷0.1), K/s:

$$y_1 = 752.2 - 141.2x_1 + 39.9x_2 + 136.7x_3 + 27.1x_1^2 - 27.4x_1x_3; \tag{27}$$

- for the rate of change in the temperature of the medium A_t = (0.1÷0.5), K/s:

$$y_2 = 242.1 - 98.6x_1 + 34x_2 + 38.8x_3 + 43.9x_1^2; \tag{28}$$

- for the rate of change in the temperature of the medium A_t = (0.5÷1.2), K/s:

$$y_3 = 122.2 - 20.3x_1 + 24.3x_2 + 15.7x_3. \tag{29}$$

The verification of the adequacy of the obtained models (27)-(29) was carried out according to the Fisher criterion [13]. The general view of dependences (39)-(41) for factors at different levels is shown in fig. 3.

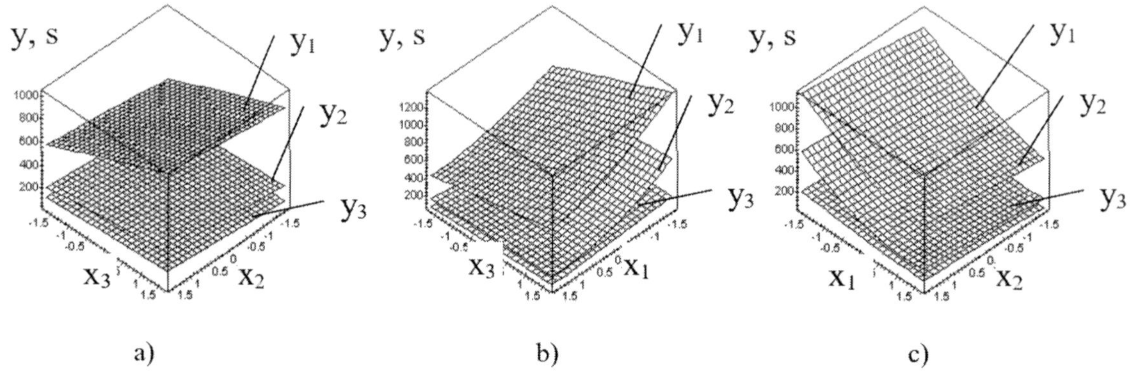

Figure 3. Dependence of the sprinkler response time with the maximum SE: a) on the threshold temperature of its operation and the SE time constant; b) on the rate of change of the ambient temperature and the threshold value of the response temperature; c) on the time constant of the SE and the rate of change in the ambient temperature.

As can be seen from Fig. 3, at the response threshold (326-360) K, the time constant (12-150) s and the rate of change in the temperature of the medium (0.05-1.2) K/s, the response time of the sprinkler having the maximum thermal SE, is in the range (60÷1200) s. For modern sprinklers, which have a glass flask as a shut-off device, the response time is in the range (200-270) s, while the sprinkler response time was determined under the condition that sprinkler was placed in a thermostat at an ambient temperature of 30±2 K is higher than the temperature of destruction of the thermal lock i.e., the time required to reach a certain temperature of the medium is not included in the given data. If we determine the response time of a modern sprinkler, which is under the same heat transfer conditions as the sprinkler under consideration, it turns out that the response time of a sprinkler with a thermal maximum SE, made from a material with SME, is less than the response time of a similar sprinkler by more than 3 times, or is in the same time interval.

3. Conclusions

The article examines the possibility of using a material with the shape memory effect to create sensitive elements of sprinklers, which will minimize their response time. The influence of various parameters on the SE dimensions is analyzed and it is shown that the response time of the SME element depends on the dimensions of the fixing device (the diameter of the glass bulb and the thickness of its walls). Based on the existing range of glass products used in sprinklers, as well as to ensure reliability requirements, when determining the minimum cross-sectional radius of the ring, it is advisable to take the FD wall thickness within (0.0001÷0.005) m. A model of the response time of an actuator made of a material with SME is obtained. It is shown that the response time depends on the voltage (or current) in the electrical network, on the dimensions of the control unit, the radius of the cross-section of the actuator and the critical temperature at which the actuator is actuated. Since an SME material is capable of «memory» at certain temperatures, its threshold should be set in the same way as for a sprinkler with thermal lock.

References

[1] Akimov O.V., Sundus Mohammed Nuri. 2015 Splavyi s effektom pamyati formyi. Istoriya poyavleniya i razvitiya, fizika protsessa ih unikalnyih svoystv, *Visnyk Natsionalnoho tekhnichnoho universytetu «Kharkivskyi politekhnichnyi instytut»* **14 (1123)** pp 42-49.

[2] Jani J. M, Leary M., Subic A., Gibson M.A. 2014. A Review of Shape Memory Alloy Research, Applications and Opportunities. *Materials & Design* **56** pp 1078–1113.

[3] Polischuk V.A., Nikolaev A. L. 2016 Opredelenie parametrov pruzhinnyih elementov s effektom pamyati formyi v protsesse termomehanicheskih vozdeystviy. *Zbirnik naukovih prats Natsionalnogo universitetu korablebuduvannya,* **1** pp 26-29.

[4] Shyi-Kaan Wu, Yi-Ching Chang. 2019 Thermal Cycling Effect on Transformation Temperatures of Different Transformation Sequences in TiNi-Based Shape Memory Alloys, *Materials (Basel)* **12 (2512)** pp 2-14.

[5] Wancho Thomas F. 2020 Sprinkler and sprinkler socket for fire extinguishing system. (Rosyski patent RU 2726760 C1)

[6] Meshman L.M., Byilinkin V.A., Didyaev A.G. 201) Sposob registratsii vremeni srabatyivaniya sprinklernogo orositelya (variantyi) i ustroystvo dlya ego realizatsii (Rosyski patent 2661858 C1).

[7] Lipik A.E., Syichev S.V., Babikov I.A. 2017 Zapornoe ustroystvo sprinklernogo orositelya s prinuditelnyim puskom. *Nedelya nauki SPbPU* pp 125.

[8] Dorosh I.V., Churakova A.A., Yamaletdinova K.Sh., Martyinova O.G. 2020 Vyibor materiala termochuvstvitelnogo elementa protivopozharnogo oborudovaniya metodom graficheskogo ekspress-analiza na primere sprinklera. *Sovremennyie naukoemkie tehnologii* **4(1)** pp 13-18.

[9] Budnik A.F. 2008 Teplomasoperenos u protsesah s matersalah dizaynu matersalsv: *Navchalniy possbnik, Sumi: Vid-vo SumDU* p 158.

[10] Troschenko V.T. 2005 Prochnost materialov i konstruktsiy. *Kiev: Akademperiodika* p 1086.

[11] Pospelov B., Meleshchenko R., Krainiukov O., Karpets K., Petukhova O., Bezuhla Y., et. al. 2020 A method for preventing the emergency resulting from fires in the premises through operative control over a gas medium *Eastern-European Journal of Enterprise Technologies* **1 (10 (103))** pp 6-13.

[12] Pospelov B., Meleshchenko R., Asotskyi V., Petukhova O., Gornostal S., Harbuz S. 2019 Development of a self-adjusting method for calculating recurrent diagrams in a space with a scalar product. *EUREKA: Physics and Engineering* **5** pp 19-28.

[13] Sidnyaev N.I., Vilisova N.T. 2011 Vvedenie v teoriyu planirovaniya eksperimenta: *ucheb. Posobie, M.:Izd-vo MGTU im N.E. Baumana* p 463.

[14] Wang D., Zhang Z., Zhang J.A., Zhou Y.M., Wang Y., Ding X.D., Wang Y.Z., Ren X.B. 2010 Strain glass in Fe-doped Ti-Ni *Acta Mate.* **58** pp 6206-6215.

IOP Conf. Series: Materials Science and Engineering 1164 (2021) 012062 doi:10.1088/1757-899X/1164/1/012062

Features of the structure formation processes and synthesis of strength for Portland cement compositions modified with nanocarbonate additives

K. Pushkarova[1], D. Hadaichuk[1], O. Honchar[1,*], L. Kushnierova[1], K. Kaverin, [1] and D. Ionov [2]

[1]Kyiv National University of Construction and Architecture, 31, Povitroflotsky Ave., 03680, Kyiv, Ukraine
[2]State enterprise "Ukrainian research, project planning and design institute of building materials and products "NIISMI"04080,68, Konstyantinovskv Street, Kyiv, Ukraine

*E-mail: gonchar.oa@knuba.edu.ua

Abstract. The work shows the results of studying the features of the structure formation process when nano-carbonate additives of different physical states are added to Portland cement and the establishment of the appropriate technology for their introduction, which provides the best conditions for the synthesis of artificial stone. It was found that the introduction of nanocarbonate additives in the form of dispersion provides better kinetics of the cement stone strength gain, and the effect of the introduction of dispersion exceeds the effect of the introduction of powder on 7th day by (43 ... 48)%, on 28th day - (27… 30)% and 365th day by (25… 29)%. The using of the dispersion is more economically efficient, while preventing the aggregation of particles provides a directed synthesis of crystal-chemically similar compounds. The nanocarbonate additive acts as a substrate for the directional formation of hydrosilicate phases, providing microreinforcement and the formation of a more homogeneous and improved structure of artificial stone at both nano- and micro- levels.

1. Introduction

According to the US Geological Survey for 2020, global cement production has reached about 5 billion tons and is expected to grow from year to year. Moreover, large increases in cement production are expected in some developing countries, such as China and India [1]. Given the above, it becomes clear why in the production of cement and concrete are increasingly using various mineral additives, including carbonates.

It is confirmed that the use of calcium carbonate has a positive effect on the hydration processes of Portland cement, its strength at an early age, microstructure and durability [2]. Thus, many studies have been conducted to clarify the mechanism of influence of calcium carbonate additives on the properties of cement paste, mortar and concrete [3], [4].

In 1938, Bessey et al. [2] with the addition of calcium carbonate in the process of hydration of cement for the first time established the formation of calcium hydrocarboaluminate, similar results were obtained in later studies [5].

It is now accepted that the density and density of the cement matrix can be increased by the introduction of calcium carbonate when it acts as a filler, ie its particle size is comparable to the particle size of cement, or even smaller, and the role of chemical reagent if involved in hydration process and this process can be accelerated due to the nucleation effect [2,6].

The number of scientific publications on the use of finely ground limestone and other carbonate additives in concrete has increased almost 6 times over the past 20 years and this topic attracts scientists from many countries. Based on studies of the effect of calcium carbonate on the properties of cement composites, many standards have been adopted in different countries, which are listed in Table 1.

Content from this work may be used under the terms of the Creative Commons Attribution 3.0 licence. Any further distribution of this work must maintain attribution to the author(s) and the title of the work, journal citation and DOI.
Published under licence by IOP Publishing Ltd

Table 1. Standards of different countries governing the use of finely ground limestone additives in cement production.

Year	Country (region)	Standard	Composition of cement with additives
1991	Brazil	NBR 11578/91	Portland cement LS containing 6-10% limestone filler
1999	Mexico	NMX C -414-0/99	Portland cement LS containing 6-35% limestone filler
2000	Europe	EN 197-1	Determined 4 types of Portland Cement LS, containing 6-20% limestone (types II / A-L and II / A-LL) and 21-35% limestone (types II / B-L and II / B-LL) respectively
2007	China	GB 175-2007	LS can be used as an activated mix in the cement industry
2007	Argentina	IRAM 50000/07	Portland cement LS may contain 25% shell rock
2008	Canada	CSA A3001-08	Portland cement containing <15% limestone filler LS
2010	Canada	CSA A3001-10	Portland cement LS is defined as GUL
2012	USA	ASTM C 595	Portland cement containing <15% limestone filler LS
2008	Ukraine	DSTU B EN 197-1:2008	Cement. Composition, technical specification and conformity criteria for conventional cements
2010	Ukraine	DSTU B B.2.7-46-2010	Cements for general construction purposes. Specifications

Finely ground limestone (or calcium microcarbonate) has a particle size (1 μm … 1 mm), and can be in the form of powder or limestone dust, widely used in cement production as a mineral additive. Although calcium carbonate has no pozzolanic activity and cannot react with alkaline substances such as $Ca(OH)_2$, the introduction of calcium carbonate into the cement can have both physical and chemical effects on the hydration process, as well as a positive effect on rheology of fresh concrete mix and physical and mechanical properties of hardened cement stone.

K. Vance et al. [7] investigated the effect of particle size of finely ground limestone during cement hydration using three types of powders of different degrees of dispersion. The finer limestone powder (average particle size 0.7 and 3 μm) significantly accelerates the hydration process of calcium silicate and increases the hydration peak (see Figure 1 [7]), as the finer limestone powder has a larger specific area and surface energy, which provides additional nucleation zones for the formation and development of calcium hydrosilicates (CSH) [7], [8]. This effect is known as the nucleation effect. In addition, the second peak on the heat curve associated with the hydration of calcium aluminate shows a significant increase in the introduction of limestone powder with a grain size of 0.7 μm, which means the formation of new hydration products such as hemicarboaluminate ($C_3A{\cdot}0.5CaCO_3{\cdot}12H_2O$) and monocarboaluminate ($C_3A{\cdot}CaCO_3{\cdot}11\,H_2O$).

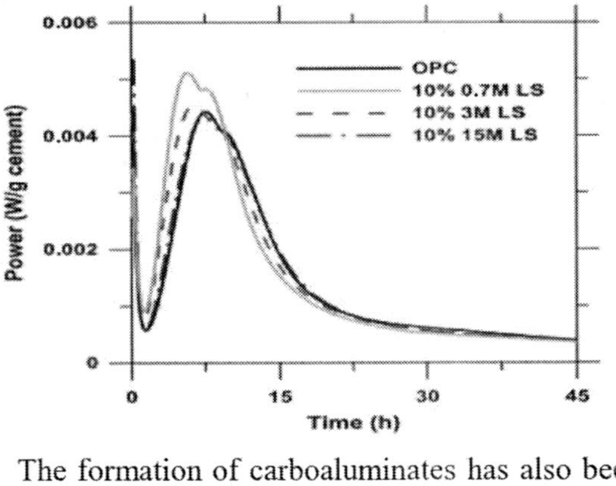

Figure 1. Influence of fine particle size of finely ground limestone (0.7… 15 μmc) on heat release during cement hydration [7].

The formation of carboaluminates has also been confirmed by many other researchers [9]. At the same time, hemicarboaluminate is not thermostable and is recrystallized to monocarboaluminate. The formation of carboaluminate depends on many factors, including the kinetics of formation of hemi- and monocarboaluminate, and the dissolution of calcium carbonate is lower with increasing pH, due to the fact that the amount of reactive calcium carbonate involved in the formation of carboaluminate is much less than the content of limestone powder [9]. The authors of [7], [10] showed that at a constant content of limestone powder, the compressive strength at an early age (up to 7 days) increases with decreasing particle size of the carbonate additive. In the later stages of hardening, the introduction of finely ground limestone can sometimes adversely affect the strength of cement compositions [11], which may be due to the fact that the physical effect of filling the pores will outweigh the chemical effect.

It is also proven that with increasing content of finely ground limestone (meaning the amount exceeding the optimal), the compressive strength and flexural strength decrease [7], [12]. On the other hand, the introduction of more carbonate additives increases the water consumption of the concrete mixture, which causes a higher effective water-cement ratio and reduces the strength.

Given the above data, it becomes clear the scientific interest in the use as carbonate additives of nanoparticles with a grain size of less than 100 nm [13]. Compared to microcarbonate, calcium nanocarbonate has a smaller particle size and a higher specific area, and thus should have a greater impact on hydration processes, as well as on the rheology of concrete mixes and on the physical and mechanical properties and durability of cement composites.

The effect of calcium nanocarbonate on the hydration process of cement depends on its content, particle size and crystal structure [14].

T. Sato et al. [15] studied the effect of calcium nanocarbonate content and particle size on cement hydration. Nanocarbonate (50… 120 nm) is very effective in accelerating the hydration of cement, especially during the induction of tricalcium silicate (C_3S) due to the manifestation of the nucleation effect. However, calcium nanocarbonate can also react with C_3S to form C-S-H gel and Ca $(OH)_2$, and this can cause more heat to be released during hydration and at an earlier time.

The crystal structure of nanocalcium carbonate can also affect the hydration process of cement. The influence of nanocarbonate additives of calcite and aragonite composition on the properties of concrete capable of self-compaction was studied [15] and the significantly higher efficiency of calcite additive addition than aragonite was shown. Comparing the effect of calcite and aragonite nanocarbonate on the change in flexural and compressive strength, it can be noted that calcite is more effective because it significantly accelerates the formation of C-S-H phases.

Physico-mechanical properties of cement composites containing nanocarbonate additives mainly depend on their content, namely when the additive content increases to a certain (optimal) amount, the strength indicators first increase, and then, when using excessive amounts of nanocarbonate decrease [16].

The reasons are that, on the one hand, nanocarbonate can accelerate the hydration process and react with C_3S and C_3A to form C-S-H and carboaluminates, and this effect is more significant with a certain amount of additive. With increasing amount of additive, the dilution effect may occur and the strength of the composite decreases.

Agglomeration of nanocarbonate particles is also possible, which can lead to loss of both physical and chemical activity [18]. In addition, a denser matrix formed by the addition of nanocarbonate cannot provide the synthesis of a sufficient number of hydration products of the hydrosilicate composition [16].

2. Materials and Methods

The aim of this research is to study the features of the structure formation process when adding nanocarbonate additives of different physical state to Portland cement and to establish the appropriate technology of their introduction to provide the best conditions for the synthesis of artificial stone.

The Portland cement CEM I 42.5 R (according to DSTU B B.2.7-46: 2010) of CRH company (PJSC "Podilsky Cement") with a specific surface area of 390 m^2 / kg according to Blaine was used in the research.

In order to form the structure of artificial stone and improve its physical and mechanical characteristics, a superplasticizer based on polycarboxylates (MS PowerFlow 3100) with a molecular weight of PEG of about 3000 units was used.

As nanocarbonate additives used technical products of the Norwegian company "Nordcalk" in the form of dry powder "Enrich C" and dispersion "Enrich C 50", the main characteristics of which are presented in table 2. The additives were introduced into the cement in terms of dry matter.

Table 2. General characteristics of nanocarbonate additives by "Nordkalk" company.

Technical characteristics	Units of measure	«Enrich C»	«Enrich C50»
Dry matter content	%	> 99	50
Average density	g / cm^3	0,75	1,45
pH of the dispersion		7…9	7…9
Appearance (at T = 250C)		White powder	White disperti
Size of the particles d50	µm	25	
Particle size distribution			
d50%	nm	130	130
d90%	nm	300	300

As a microcarbonate additive, finely ground limestone of Humentsi quarry of Khmelnytsky region was used, the chemical and mineralogical composition of which is given in Table 3.

Table 3. Chemical and mineralogical properties of finely ground limestone (microcarbonate additive).

The content of oxides, mas. %						
SiO_2	Al_2O_3	Fe_2O_3	MgO	CaO	L.O.I.	SiO_2+ R_2O_3
1,59	0,49	0,44	0,51	53,89	42,94	2,16
Mineral content, mas.%						
$CaCO_3$ - 96,16; $MgCO_3$ - `1,26; clays (SiO_2+ R_2O_3) – 2,27						

Given the influence of polycarboxylate superplasticizers and nanocarbonate additives on the rheological properties of cement paste and on the processes of structure formation of cement compositions, it was rational to first study the physico-mechanical properties of cement paste samples to eliminate the effect of fine aggregate on strength and reduce error.

Studies of the strength of cement paste were performed using cube samples 2×2×2 cm, which were tested after curing under standard conditions on a hydraulic press. The kinetics of compressive strength of the investigated compositions on the basis of cement pastes are presented in Table 4.

Table 4. Evaluation of the effectiveness of the use of micro- and nanocarbonate additives for the modification of cement compositions.

Binder composition	Strength gain of paste samples, MPa, after curing for, days			
	3	7	28	365
Portland Cement+water	43	61	74	79
Portland Cement+water + 1% MCPF 3100	64	82	94	96
Portland Cement+water + 1% MCPF 3100 + microcarbonate additive				
10%	79	85	90	92
12,5%	85	89	95	97
15%	75	76	78	79
Portland cement+water + 1% MC PF 3100 + nanocarbonate additive				
2,5% in a form of powder	54	64	94	100
3,5% in a form of powder	59	69	96	106
4,5% in a form of powder	55	63	94	99
2,5% in a form of dispersion	71	95	124	125
3,5% in a form of dispersion	77	99	129	133
4,5% in a form of dispersion	74	96	127	128

3. Results and Discussion

The analysis of the given data allows to note that in case of use of finely ground limestone as carbonate additive to Portland cement, efficiency of its action (in comparison with cement paste without limestone addition) is shown mostly at early terms of hardening and makes (17… 23)% on 3 days and (3.6… 8.5)% at 7 days. On the 28th day, the effect of the additive does not exceed 1%, or almost leveled off. The data obtained are confirmed by the results of research by other scientists [17], [18].

At the same time, using a nanocarbonate additive in the form of a dispersion, the strength gain of the artificial stone is (10… 20)%, (15… 20)%, (31… 37)% and (30… 38)% by 3,7,28 and 365 days respectively, so there is a relatively stable increase in strength over time, which indicates the direct crystallization of low-basic hydrosilicates, and it is likely that the introduction of nanocarbonate additive (relatively small amount compared with the addition of finely ground limestone) specifically contributes to the nucleation structure not only at the micro-, but also at the nanoscale.

The process of hydration of C_3S in the presence of nano-$CaCO_3$ is accelerated due to the reduction of the induction period [14]. This is due to the fact that the additional surface created by nano-$CaCO_3$ particles promotes the effect of nucleation and increasing of sites of hydration products of Portland cement, including calcium hydrosilicates.

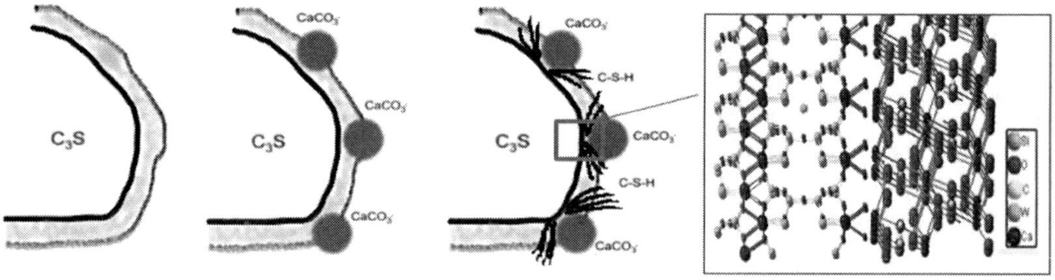

a) b) c) d)

Figure 2. Scheme of the mechanism of C_3S hydration in the induction period in the presence of a nanocarbonate additive.

Usually the induction period during C_3S hydration occurs due to the formation of a protective layer (Figure 2a), which significantly inhibits the hydration process [2]. The presence of nano-$CaCO_3$ particles near the C_3S surface weakens the protective layer (Figure 2b) and in places of weakening the permeability increases and access to the surface of the C_3S particle solution with a high concentration of Ca^{2+}, which leads to accelerated growth of C-S-H (Figure 2c). Nano-$CaCO_3$ particles, due to their crystal chemical similarity to hydrosilicate compounds (Fig. 3), act as "substrates" for C-S-H crystallization, providing additional nucleation centres (Figure 2d).

Comparison of the results of physical and mechanical tests of samples modified with nanocarbonate additives in the form of powder and dispersion (Table 4), confirms the effectiveness of the introduction of nanoadditives specifically in the form of dispersion together with a plasticizing additive, which ensures uniform distribution throughout the volume of cement paste particles [19].

The effectiveness of the nanocarbonate additive compared to control mixtures is especially noticeable on the 28-th day and later curing, due to the formation of a stable composition of products of hydration, dominated by gillebrandite, low-basic calcium hydrosilicates similar to tobermorite 11.3 A and hydrocarboaluminates.

In order to confirm the phase composition of products of hydration, electron scanning microscopy of Portland cement stone samples was performed after 28 and 365 days of hardening. The results of the study are shown in Figures 4-7.

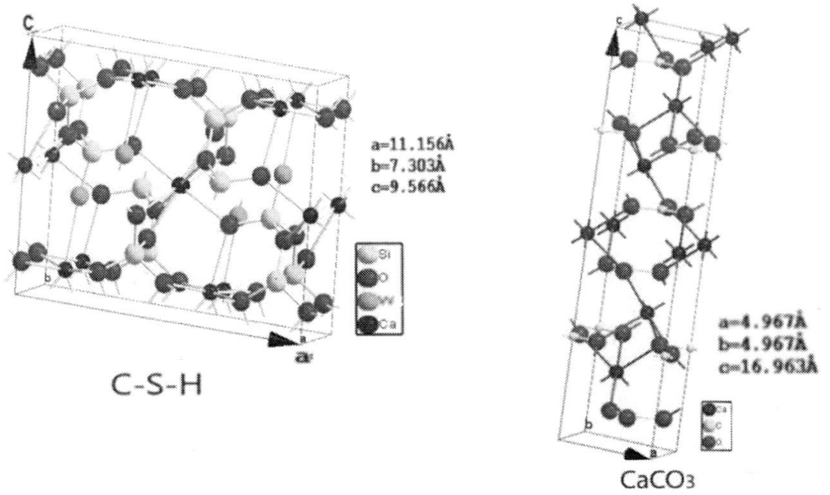

Figure 3. Parameters of crystal lattices C-S-H and $CaCO_3$.

Figure 4. - Microstructure of the spall surface of Portland cement stone modified with polycarboxylate superplasticizers after 28 days of curing: magnification: (a) × 1200 (b) × 3000.

Figure 5. Microstructure of artificial stone based on Portland cement modified with nanocarbonate dispersion in the presence of polycarboxylate plasticizer after 28 days of curing: magnification x2500 (a); x250 (b).

IOP Conf. Series: Materials Science and Engineering 1164 (2021) 012062 doi:10.1088/1757-899X/1164/1/012062

Comparison of the results of electron scanning microscopy allows us to note the directional crystallization of hydrosilicate phases in the presence of nanocarbonate additives, associated with the effect of nucleation, which is clearly manifested on the 28-th day and then recorded at later curing age.

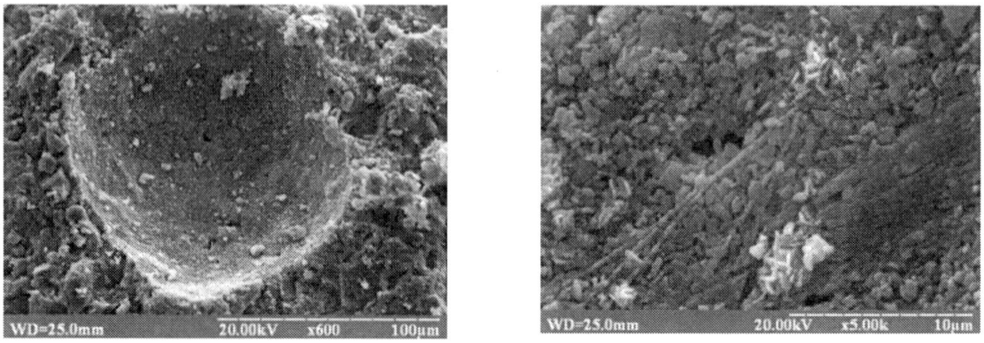

Figure 6. Photos of the spall surface of Portland cement stone modified with polycarboxylate superplasticizers after 365 days of curing: magnification (a) × 600; (b) × 5000

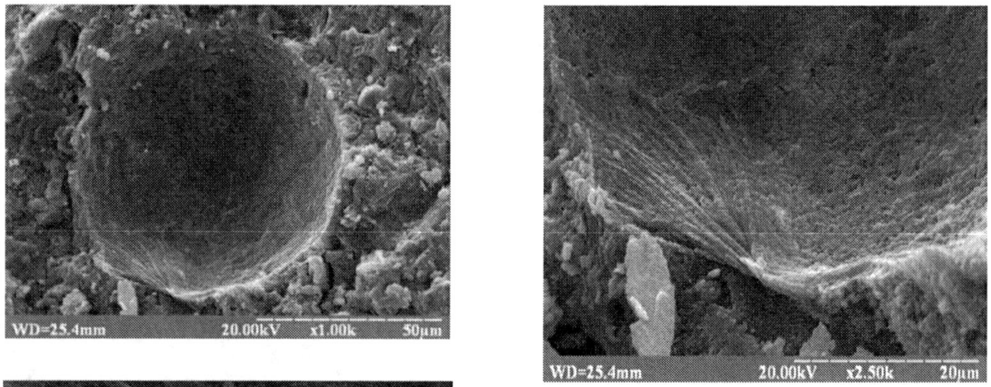

Figure 7. Microstructure of artificial stone based on Portland cement modified with nanocarbonate dispersion in the presence of polycarboxylate plasticizer after 365 days of hardening: magnification: x 1000 (a); x2500 (b) x5000 (c).

Calcium nanocarbonate acts as a substrate for the directed growth of calcium hydrosilicates, which is confirmed by theoretical calculations by comparing the parameters of the crystal lattices of minerals, and the results of electron microscopic analysis, which allows to record changes in the morphology of products of hydration. According to the Ruyet-Friedel rule, the criterion for the possibility of crystal fusion can be the value $\Delta = (a_1 - a_2) / a1$, where a_1, a_2 are the parameters of mineral lattices in the area of their fusion, and the difference between the parameters of fused crystals should not exceed 15% [20] - [22].

Calcite, which is the basis of the nanocarbonate additive, may act a substrate for the crystallization of ettringite and low-basic calcium hydrosilicates. In the $CaCO_3$ crystal lattice (Fig. 8a) there is a layer consisting of CaO atoms, which is similar to the Ca-O layer in the structure of gillebrandite (Fig. 8b), tobermorite (Fig. 8c) and ettringite (Fig. 8 d).

The length of Ca - O bonds is 2.35 A for calcite, (2.28 - 2.41) A for gillebrandite, (2.4 - 2.52) A for tobermorite and (2.43 - 2.61) A for ettringites, the value of Δ is close to zero.

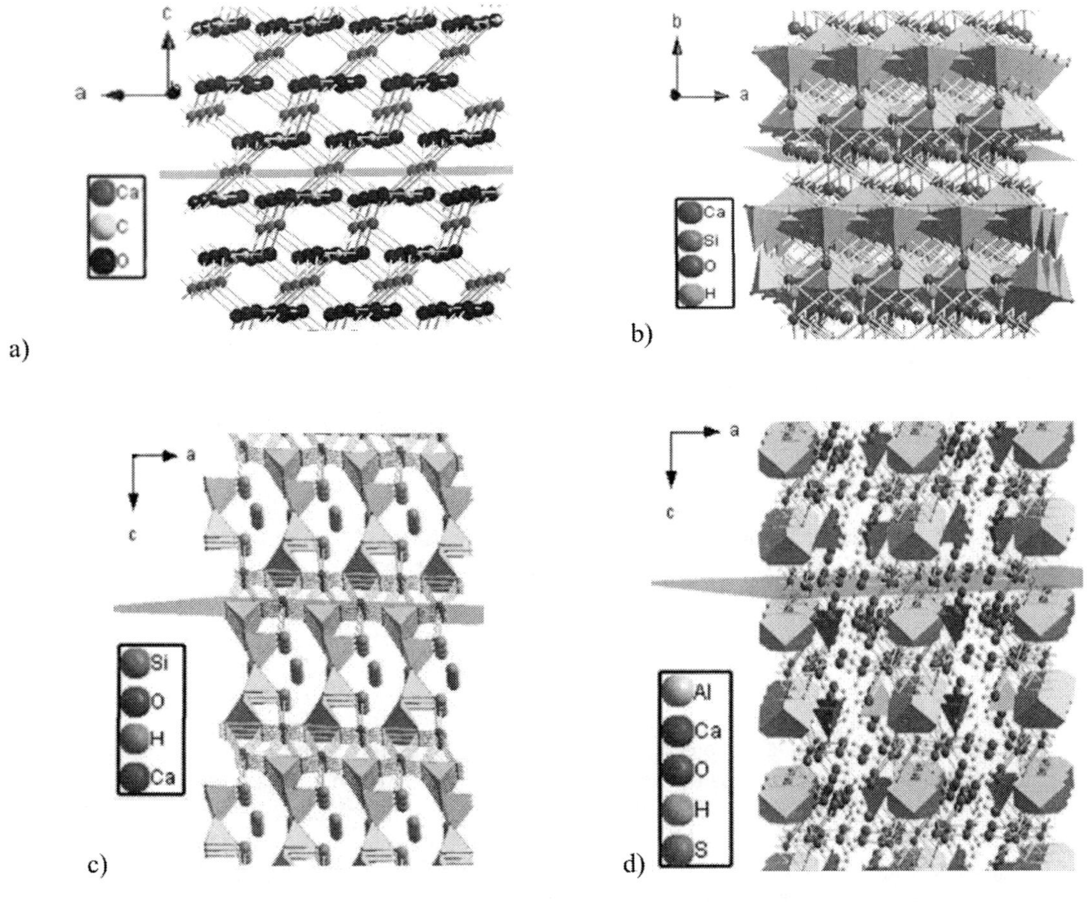

Figure 8. Diagram of the crystal structure of minerals, where the selected crystallographic area indicating the layer Ca - O: calcite (a), gillebrandite (b), tobermorite (c), ettringite (d).

Directional crystallization is possible on crystallographic planes passing through the Ca-O layer. Below are diagrams of possible fusion of calcite and ettringite (Fig. 9a) on the crystallographic planes (001) and (001), calcite and gillebrandite (Fig. 9b) on the planes (036) and (010), calcite and tobermorite (Fig. 9 c).) on the planes (036) and (001).

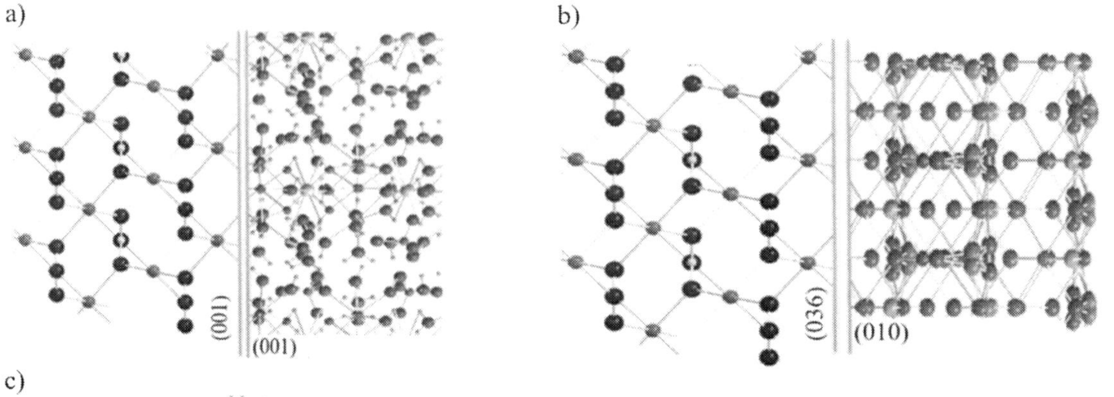

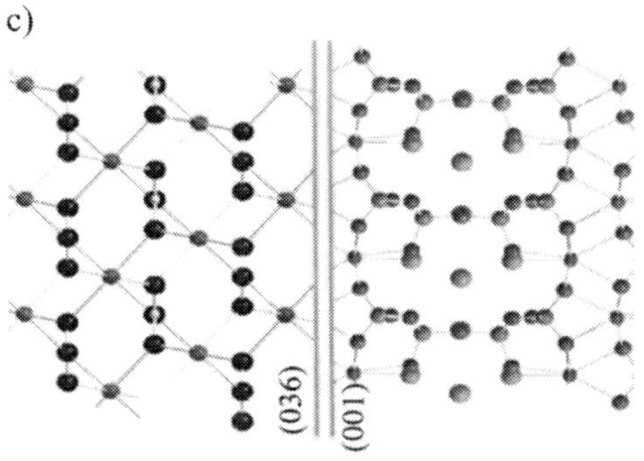

Figure 9. Conjugation schemes of calcite fusion planes with ettringite (a), gillebrandite (b) and tobermorite (c)

In particular, the crystallochemical similarity of the hydration products of cement compositions both after 28 and after 365 days can explain the stability of the increase in strength over time, and the manifestation of the nucleation effect in the presence of nanocarbonate additives increases crystal nuclei and directional crystallization of hydrosilicates (tobermorite and hillebrandite), especially in late stages of hardening of artificial stone.

4. Conclusions

The introduction of a nanocarbonate additive provides a directed synthesis of the strength of an artificial stone and the effectiveness of the nanoadditive in comparison with a microcarbonate additive is 3,7,28 and 365 days (10 ... 20)%, (15..20)%, (31 ... 37)% and (30 ... 38)%, respectively.

The introduction of nanocarbonate additives in the form of dispersion provides better kinetics of the cement stone strength gain, and the effect of the introduction of dispersion exceeds the effect of the introduction of powder on 7th day by (43 ... 48)%, on 28th day - (27… 30)% and 365th day by (25… 29)%.

The introduction of the dispersion is more economically efficient, while preventing the aggregation of particles provides a directed synthesis of crystal-chemically similar compounds. The nanocarbonate additive acts as a substrate for the directional formation of hydrosilicate phases, providing microreinforcement and the formation of a more homogeneous and improved structure of artificial stone at both nano- and micro- levels.

References

[1] Schneider M, Romer M, Tschudin M andBolio H 2011 Sustainable cement production - Present and future *Cem. Concr. Res.* **41** pp 642-650

[2] Wang D, Shi C, Farzadnia N, Shi Z, Jia H and Ou Z 2018 A review on use of limestone powder in cement-based materials: Mechanism, hydration and microstructures *Constr. Build. Mater.* **181** pp 659–672

[3] Dweck J, Buchler P, Coelho A and Cartledge F 2000 Hydration of a Portland cement blended with calcium carbonate *Thermochim Acta* **346** pp 105-113
[4] Ye G, Liu X, De Schutter G, Poppe A and Taerwe L 2007 Influence of limestone powder used as filler in SCC on hydration and microstructure of cement pastes *Cem. Concr. Compos.* **29** pp 94-102
[5] Bonavetti V, Rahhal V and Irassar E 2011 Studies on the carboaluminate formation in limestone filler-blended cements *Cem. Concr. Res.* **31** pp 853-859
[6] Soroka I and Stern N 1976 Calcareous fillers and the compressive strength of portland cement *Cem. Conctrete Res.* **6** pp 367-376
[7] Fernandes F, Manari S, Aguayo M, Santos K, Oey T, Wei Z and Falzone G 2014 On the feasibility of using phase change materials (PCMs) to mitigate thermal cracking in cementitious materials *Cement and Concrete Composites* **51** pp 14-26
[8] Kirk V, Matthew A, Tandre O, Gaurav S and Narayanan N 2013 Hydration and strength development in ternary portland cement blends containing limestone and fly ash or metakaolin *Cem. Concr. Compos.* **39** pp 93-103
[9] [Thongsanitgarn P, Wongkeo W, Chaipanich A and Poon C 2014 Heat of hydration of Portland high-calcium fly ash cement incorporating limestone powder Effect of limestone particle size *Constr. Build. Mater.* **66** pp 410-417
[10] Zajac M, Rossberg A, Le Saout G and Lothenbach B 2014 Influence of limestone and anhydrite on the hydration of Portland cements *Cem. Concr. Compos.* **46** pp 99-108
[11] Lertwattanaruk P, Sua-iam G and Makul N 2018 Effects of calcium carbonate powder on the fresh and hardened properties of self-consolidating concrete incorporating untreated rice husk ash *J. Clean. Prod.* **172** pp 3265-3278
[12] Kenai S, Soboyejo W and Soboyejo A 2004 Some engineering properties of limestone concrete *Mater. Manuf. Process.* **5** pp 949-961
[13] Turk K and Nehdi M 2018 Coupled effects of limestone powder and high-volume fly ash on mechanical properties of ECC *Constr. Build. Mater.* **164** pp 185-192
[14] Silvestre J, Silvestre N and de Brito J 2016 Review on concrete nanotechnology *Eur. J. Environ. Civ. Eng.* **20** pp 455-485
[15] Sato T and Beaudoin J 2011 J Effect of nano-$CaCO_3$ on hydration of cement containing supplementary cementitious materials *Adv. Cem. Res.* **23** pp 33-43
[16] Yes¸ilmen S, Al-Najjar Y, Balav M, S¸ahmaran M, Yıldırım G and Lachemi M 2015 Nano-modification to improve the ductility of cementitious composites *Cem. Concr. Res.* **76** pp 170-179
[17] Wu Z, Shi C, Khayat K and Wan S 2016 Effects of different nanomaterials on hardening and performance of ultra-high strength concrete (UHSC) *Cem. Concr. Compos.* **70** pp 24-34
[18] Kozlova V, Manyukha A and other 2012 Vliyanie karbonatsoderzhashhikh dobavok na svojstva kompoziczionny`kh czementov/ *Czement i ego primenenie* **3** pp53-57
[19] Kropivniczka T 2013 Vpliv karbonatnikh dobavok na vlastivosti` portlandczementu kompoziczi`jnogo *Visnik Naczionalnogo universitetu Lvivska politekhnika Teoriya i` praktika budivnicztva* **755** pp 214-220
[20] Bondarenko O, Guzij S, Zakharchenko K and Novoselenko Ye 2015 Razrabotka zashhitny`kh materialov na osnove steklo- i shlaksoderzhashhikh portland czementny`kh kompoziczij *Eastern-European Journal of Enterprise Technologies* **6/11 (78)** pp 41-47
[21] Pushkareva E, Sukhanevich M and Marczikh A 2013 Tekhnologicheskie aspekty vvedeniya uglerodny`kh pri modificzirovanii portlandczementny`kh kompoziczij//Proizvodstvo e`nergo- i resursosberegayushhikh stroitel`ny`kh materialov i izdelij *Sb. trudov I`I` nauchno-prakt. sem., Tashkent,TASI* t **1** pp 102-108
[22] Pushkareva K, Gonchar O and Kaverin K 2019 The role of the crystallo-chemical factor in the evaluation and improvement of the nanomodification efficiency of mortar and concrete *IOP Conf. Series: Materials Science and Engineering TRANSBUD-2019* **708** 012102 11

Development of tools (laboratory facilities) for researching the effect of physical properties of landfill soils on slope stability

N Rashkevich[1,*], V Strelec[1], S Shcherbak[1], S Yeremenko[2]

[1]National University of Civil Defence of Ukraine, 94, Chernishevska str., Kharkiv 61023, Ukraine
[2]Institute of Public Administration and Research in Civil Protection,18, Rybalska str., Kyiv 01011, Ukraine

[*]Email: nine291085@gmail.com

Abstract. The paper presents a developed laboratory unit and a technique for conducting experimental studies of the effect of physical properties of landfill soils, such as moisture, density, temperature, on slope stability in order to verify the reliability of a mathematical model and a methodology for preventing emergencies associated with displacement developed on its basis. Conducted field studies of humidity, density, temperature of landfill soils of typical solid waste disposal facilities. The analysis of the conditions for obtaining the maximum amount of methane in the composition of biogas has been carried out. The basic requirements for the laboratory unit have been determined and the area of the factor space for conducting experimental laboratory studies has been established.

1. Introduction and literature review

At landfills of municipal solid waste (MSW) or landfills, there are numerous cases of hazardous events, emergency situations associated with fires and landslides [1–3]. The consequences of the emergency danger include a significant area of their distribution and a significant number of dead, injured, persons with violation of the living conditions. The basis of the process of limiting the spread of the consequences of emergencies is work aimed at changing the physical properties of landfill soils [4], taking into account the provision of conditions for their safe management [3], and the impact of hazardous factors outside the object of their occurrence on the environment and the population (for example, on the state of the atmospheric air, as the main source of transport of hazardous substances that can harm the population in the area of possible damage [5, 6]).

Humidity, density, temperature of landfill soils affect both the dynamic behavior of the MSW slopes, and the composition of the generated biogas – an alternative source of energy. Facilities of utilization biogas may be incur by hazardous emergency factors – lose their technical condition [7, 8]. The desire to obtain the maximum amount of methane in biogas at waste disposal facilities is an additional factor in studying the influence of the physical properties of MSW soils on the stability of slopes, in the context of solving civil security problems.

First of all, let's note that according to the modern interpretation, emergency is a conditional level of the course of an emergency, which is achieved behind one or more dominant signs, in terms of the level of threat and/or the possibilities of counteraction. In work [9], which additionally considers liquidation energy-intensive technological equipment (operating on biogas obtained from a disposal facility) as a zone of propagation of hazard consequences, it is indicated that solving problems of assessing moisture, density, temperature of landfill soil makes it possible to determine the conditions for the absence of victims and casualties among civilians and specialists of units of the State Emergency Service of Ukraine.

The solution to the problem of assessing the moisture content of landfill soil is an analytical relationship that describes the relationship of moisture in accordance with the variation of the input and output fluid flows. The solution to the problem of assessing the density describes the relationship

Content from this work may be used under the terms of the Creative Commons Attribution 3.0 licence. Any further distribution of this work must maintain attribution to the author(s) and the title of the work, journal citation and DOI.
Published under licence by IOP Publishing Ltd

of density in accordance with variations in the load force, porosity, height of waste storage. The solution to the problem of assessing the temperature - describes the relationship of temperature in accordance with the variation in the amount of heat supplied or generated and removed from the landfill massif.

Today, in view of the development of a mathematical model and the methodology developed on its basis for the prevention of emergency situations of a cascade type of propagation associated with the displacement of landfill soil at potentially hazardous facilities, there is a need to verify their reliability. Thus, the unresolved part of the problem is the development of a simple and visual laboratory base for laboratory studies of the influence of the physical properties of landfill soils on the stability of slopes in order to solve civil security problems.

The mechanism of destruction and the mode of instability of MSW facilities are studied based on the results of field research, tests of physical models, and simulation.

With the help of ground engineering seismic exploration by the method of refracted waves, the physical and mechanical properties of the MSW facilities in natural bedding were established [10]. Field studies [11] have shown that landfills consist mainly of soil-bearing material (51 %) and plastic (31 %) with an average moisture content of 43 % at the surface and 53 % at a depth of about 11 meters.

The authors of [12], using the method of multichannel analysis of surface waves, obtained data on the speed of wave propagation in the waste mass, which were later used for extended interpretation of the sounding results - identifying solid waste as a soil-like material, obtaining information about the structure of the waste column, mechanical characteristics. The waste massif can be divided into several layers according to the age of the burial [13]. Each of the waste layers has an individual composition and shear strength characteristics. According to the results of the triaxial test [13], the MSW shear strength increases with the years. Slope stability decreases linearly with increasing external load [14].

The simulated situations [15] show the overall stability of the MSW facility without fire was 1.13; with fire, when the tensile strength has been lost, the overall stability is 1.00. Laboratory research program [16] using direct landslide (DS), triaxial and simple landslide tests, taking into account the influence of waste composition, orientation of fibrous particles, limited stress, stress rate of change, stress-strain compatibility and specific gravity showed that static strength MSW shear DS is best characterized by a cohesion of 15 kPa and a friction angle of 36° at a normal stress of 1 atm. with a decrease in the angle of friction by 5 V for each logarithmic cycle of normal stress growth.

A series of model tests of centrifuges [17, 18] demonstrated the development of failure processes in the stability of a waste array under conditions of an increased moisture level. The ratio of the critical water level and the height of the solid waste array were about 0,8 [18].

Scientists [19] investigated the shear strength by inverse analysis of failures on slopes: for normal stresses less than 200 kPa, the specific adhesion is 6 kPa, the angle of internal friction is 35°; for normal stresses over 200 kPa, the specific adhesion is 30 kPa, the angle of internal friction is 30°.

The authors of [20], based on the analysis of existing trays and their structures, developed a laboratory unit and research methodology to identify a qualitative picture of the formation of cracks on model materials and cover deposits of landslide-prone slopes moistened with atmospheric precipitation of varying intensity. The slope model was erected in a transparent tray; to change the slope angle, vertical posts with holes and metal fixing rods of the corresponding height (slope angle 25^0) were used. Work direction to study the effect of moisture, changes in the initial density, temperature were not taken into account.

Most of the works on determining the mechanism of destruction and the mode of instability of MSW facilities are based on the results of field research and simulation using computer programming, and the laboratory base used by the authors has insufficient visibility of the mutual influence of the physical properties of landfill soils, such as moisture, density, temperature on the stability of slopes. and the possibility of conducting research on the hazardous impact of landslide experimental blocks on process equipment. Modern trends in the reconstruction of MSW facilities include the placement of liquidation energy-intensive technological equipment on the territory of potentially hazardous facilities, and are sources of the emergence and spread of man-made emergencies.

The aim of the paper is to develop a laboratory unit and a methodology for conducting experimental studies of the effect of physical properties of landfill soils on slope stability in the context of solving civil security problems, taking into account the results of field studies of humidity, density, temperature of landfill soils of typical solid domestic waste disposal facilities, analysis of conditions for obtaining the maximum amount of methane as part of biogas.

2. Construction of the model and its analysis

To achieve this aim, theoretical and experimental research methods were comprehensively used: analysis and synthesis, generalization, probability theory, decision theory, field and laboratory methods of observation and measurement.

The determination of the values of the physical properties of the landfill soil and sampling were carried out by the envelope method at five typical MSW landfills at a depth of 2.5 m at different times of the year. Moisture was measured using an MG-44 soil moisture meter, density – a LAN-M soil density meter, temperature – an electronic thermometer. Mechanical characteristics of strength (angle of internal friction, specific adhesion) were measured by the single-plane cut method in accordance with DSTU B V.2.1-4-96 (GOST 12248-96) "Bases and foundations of buildings and structures. Soils. Methods for laboratory determination of strength and deformability characteristics" [21].

The developed laboratory unit (Fig. 1) includes a rectangular experimental box with rotary (1) and fixed (2) parts, rotary (3) and locking mechanisms (4), tabletop heating plates (5), sprayer (6), system tangential load (7).

Figure 1. Laboratory unit for studying the influence of physical properties of landfill soils on the stability of slopes.

The base of the experimental box (1, 2) is made of sheet iron with a thickness of 3 mm with dimensions of 800x600x600 mm of swivel and 500x600x300 mm of fixed parts. At the bottom of the rotary part, ribs are provided for rigid fixation of the base of the sliding surface. The side walls on three sides are made of 2 mm thick sheet iron, and the fourth, a viewing window, is made of 4 mm thick glass edged with silicone slats.

The rotaty mechanism (3) consists of loops, parts of the experimental box are fastened together, the tilt angle indicator is a protractor. The tilt angle is reproduced using a hydraulic bottle jack 5 t 80-030Miol 216-413 mm, which is installed on a shield with dimensions of 40x120x400 mm. The locking mechanism (4) is made in the form of a pedestal with an anchor made of a steel corner 50x50mm and a plate with dimensions of 50x500 mm and a thickness of 5 mm.

For heating the experimental material, there are electric tabletop heating plates (5) with mechanical rotary stepped power control, a 19 mm burner, 1.5 kW, 96x260x269 mm in size – 2 pcs., hiwch are installed on a metal tray under the rotary part of the experimental box. The pallet is made of iron sheet and has a recess for the jack rod.

At the unit using a pneumatic sprayer (6) with a pressure gauge and a graduated tank of 8 liters, the effect of moisture on the stability of slopes on the shift of landfill soils is experimentally investigated. In force, the sprayer is driven by creating an overpressure with a foot pump. Excess liquid flows down the box into a measuring container (not shown in the figure).

The tangential load system (7) consists of 2 blocks (rollers) interconnected by a steel cable with a diameter of 4 mm, a basket for cargo, a dial gauge GOST 577-68, as far as the landslide block is concerned, graduation 0.01 mm, stroke arrows 10 mm.

The auxiliary equipment includes: mineral wool, metal clothespins to limit heat loss during heating of the experimental material, an electric tabletop, a 10-liter measuring metal tank for heating liquid, a universal household thermometer TP-3-M1-2 with a "float chamomile" for measuring the temperature of the liquid; trowel, shovel, Wimpex electronic scales up to 50 kg, III class of accuracy, metal mold for the experimental sliding block, hand-held metal mortar, 500 mm iron graduated probe for working with experimental material.

Fig. 2 shows the construction of a metal mold and a hand mortar that are used to form the landslide blocks and the base of the sliding surface.

a) b)

Figure 2. Design of auxiliary equipment for the formation of research samples: where a) a manual metal mortar, b) a metal mold.

Research using a laboratory unit (Figure 1) is based on the assumption that the transition of the landslide experimental block to the dynamic state will be considered the onset of the emergency of the object level of propagation.

Accordingly, the experimental research methodology includes the following stages:

1) Establishment of the initial and boundary conditions for experimental studies.

1.1) Selection of the experimental area and coding of natural values of the levels of factors in dimensionless quantities. For factors with a continuous range, the encoding is determined by the formula:

$$x_i = \frac{\tilde{x}_i - \tilde{x}_{i0}}{I_i}, \tag{1}$$

where x_i – coded value of the i-th factor; $\tilde{x}_i$ – natural value of the i-th factor; $\tilde{x}_{i0}$ – natural value of the i-th factor of the main (zero) level; I_i – interval of variation of the natural value of the i-th factor.

After coding, the levels of factors take on the following values: "1" – the lower level; "+1" – top level; "0" – zero level. The center of the interval in which the experiment is supposed to be carried out is selected as the zero level.

1.2) The choice of measuring instruments or methods of laboratory determination of physical and mechanical properties for compliance with the experimental area of the factor space. Temperature is measured using a soil thermometer or thermal rod, moisture – soil moisture meter or by drying to constant weight, density – soil density meter or by the cutting ring method, strength (angle of internal friction, specific adhesion) – by the single-plane cut method.

Moisture percentage is calculated by the formula [22]:

$$w = \frac{100 \cdot (m_1 - m_0)}{m_0 - m}, \tag{2}$$

where m – mass of an empty cup with a lid, (g); m_1 – mass of wet soil with a cup with a lid, (g); m_0 – the mass of the dried soil with a glass with a lid, (g).

Density ρ (g/cm^3)is calculated by the formula [22]:

$$\rho = \frac{(m_1 - m_0 - m_2)}{V}, \tag{3}$$

where m_1 – mass of soil with a ring and plates, (g); m_0 – mass of the ring, m_2 – mass of the plates, (g); $V = \pi \cdot r^2 \cdot h$ – internal volume of the ring, (cm^3) ((r – radius, (cm); h – height, (cm)).

The angle of internal friction φ and specific adhesion C are calculated by the formulas [21]:

$$tg\phi = \frac{n \cdot \sum \tau_i \cdot \sigma_i - \sum \tau_i \cdot \sum \sigma_i}{n \cdot \sum (\sigma_i)^2 - \left(\sum \sigma_i\right)^2}, \tag{4}$$

$$C = \frac{\sum \tau_i \cdot \sum \sigma_i^2 - \sum \sigma_i \cdot \sum \tau_i \cdot \sigma_i}{n \cdot \sum (\sigma_i)^2 - \left(\sum \sigma_i\right)^2}, \tag{5}$$

where τ_i – experimental values of the structural shear resistance at different values σ_i, (MPa); $n \geq 3$ – the number of tests ($n \geq 3$).

According to the values of the tangential and normal loads measured during the test, n the tangential and normal stresses τ and σ, (MPa) are calculated, according to the formulas:

$$\tau = \frac{X}{A}, \tag{6}$$

$$\sigma = \frac{N}{A}, \tag{7}$$

where X and N – respectively, tangential and normal forces to the shear plane, (kN) A – shear area, (cm^2).

According to the values of shear deformations measured during the tests Δl, which correspond to different stresses τ, a graph of dependence $\Delta l = f(\tau)$ is plotted (Figure 3).

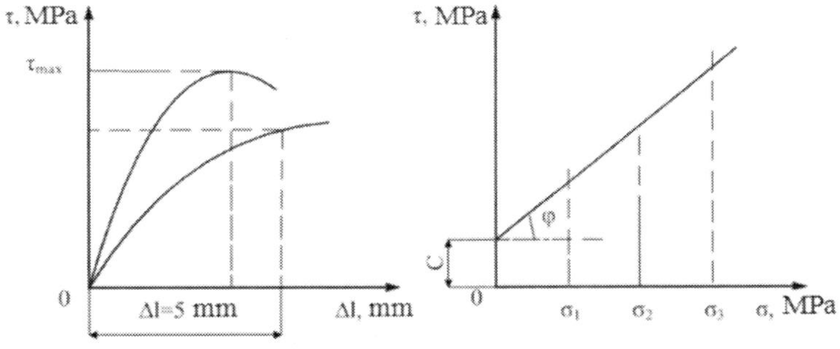

Figure 3. Graph of dependence of shear deformations Δl on stress τ : $\Delta l = f(\tau)$.

For the resistance of the soil to shear, the maximum value τ obtained from the graph $\Delta l = f(\tau)$ or diagram of the shear on a segment Δl is taken, which does not exceed 5 mm. If τ grows monotonously, then the shear resistance of the soil should be taken τ at $\Delta l = 5$ mm.

1.3) Preparation of the laboratory unit and experimental blocks.

The unit is placed on an open, flat, solid surface. The jack is installed on the type board, and its rod – to a special recess in the heating system pallet. Electrical appliances are supplied with a 220 V electric power supply. The sprayer is filled with water.

1.3.1) Selection and transportation of experimental material. The material is taken from the map of the MSW facility (depth of at least 3 m and 15 years old) and, if necessary, sorted from coarse fractions larger than 15x15 mm, and transported in plastic bags to the unit.

1.3.2) Formation of a horizontal sliding surface and experimental landslide blocks with a given density. The specified density value is achieved due to the fixed size of the molds V_{m_i} for the formation of the landslide block and sliding surface with a predetermined mass $m_{l.b.}$ and $m_{s.s.}$, accordingly:

$$m_{l.b(s.s.)} = V_{m_i} \cdot \rho_i \tag{8}$$

The experimental material is weighed to a certain mass, poured into metal molds installed on the base of the experimental box. With a manual metal mortar, the shapes are reproduced layer by layer according to the dimensions and the given density.

2) Conducting a series of experiments in accordance with the plan (matrix) of experimental studies (Table 1).

Determination of the angle of internal friction, specific adhesion and shear angle of experimental landslide blocks under conditions of different humidity, density and temperature.

2.1) Formation of research samples of a group of experiments with an initial density ρ_{in1}, kg/m^3 in accordance with the plan of experimental research.

2.1.1) Determination of temperature T_{in}, ^{0}C and humidity w_{in}, % of experimental material.

2.1.2) Determination of the angle of internal friction, the specific cohesion of the experimental material due to the cyclic shear load, which creates shear and normal stresses.

Table 1 Matrix for planning experimental studies of the influence of the physical properties of landfill soils on the stability of slopes.

№ of test	Research procedure			Planning matrix		
				x_{ρ}	x_T	$x_{w(\alpha)}$
1	1	2	3	−1	−1	−1
2	4	5	6	−1	−1	+1
3	7	8	9	−1	0	0
4	10	11	12	−1	+1	−1
5	13	14	15	−1	+1	+1
6	16	17	18	0	−1	−1
7	19	20	21	0	−1	0
8	22	23	24	0	0	−1
9	25	26	27	0	0	+1
10	28	29	30	0	+1	0
11	31	32	33	+1	−1	−1
12	34	35	36	+1	−1	+1
13	37	38	39	+1	0	0
14	40	41	42	+1	+1	0
15	43	44	45	+1	+1	+1

2.1.3) Changing the angle of inclination of the sliding surface using a jack until the moment of displacement of the landslide block. Record the value of the shift angle on the protractor of the rotary mechanism of the laboratory unit

2.2) Formation of research samples of a group of experiments with an initial density ρ_{in1}, kg/m^3 in accordance with the plan of experimental research.

2.2.1) Raising the moisture content of the landslide block and the base of the sliding surface to the specified value w_1 by uniform spraying with heated water to the specified temperature at $T_{in} = const$.

2.2.2) Determination of the angle of internal friction, the specific cohesion of the experimental material due to the cyclic shear load, which creates shear and normal stresses.

2.2.3) Changing the angle of inclination of the sliding surface using a jack until the moment of displacement of the landslide block. Record the value of the shift angle on the protractor of the rotary mechanism of the laboratory unit

2.2.4) Cyclic repetition of actions from paragraphs 2.2 inclusive for different values of humidity w_i and $T_{in} = const$ in accordance with the plan of experimental studies.

2.3) Formation of research samples of a group of experiments with an initial density ρ_{in1}, kg/m^3 in accordance with the plan of experimental research.

2.3.1) Raising the temperature of the landslide block and the base of the sliding surface to T_i with the help of the tabletop plates of the heating system, adjusting the power and heating time, provided $w_{in} = const$ (maintaining humidity). To limit heat loss with mineral wool, wrap the lower part of the unit and fix it with metal clothespins. When the set temperature T_i is reached, turn off and remove the plates from the pallet.

2.3.2) Determination of the angle of internal friction, the specific cohesion of the experimental material due to the cyclic shear load, which creates shear and normal stresses.

2.3.3) Changing the angle of inclination of the sliding surface using a jack until the moment of displacement of the landslide block. Record the value of the shift angle on the protractor of the rotary mechanism of the laboratory unit

2.3.4) Cyclic repetition of actions from paragraphs 2.2 to 2.3 inclusive for different temperature values T_i and $w_{in} = const$ in accordance with the plan of experimental studies.

2.4) Formation of research samples of a group of experiments with an initial density ρ_{im}, kg/m^3 and cyclic repetition of actions from paragraphs 2.1 to 2.3 inclusive in accordance with the plan of experimental research.

Determination of humidity, temperature and density based on the displacement of experimental blocks, taking into account a gradual increase in humidity.

2.5) Formation of research samples of a group of experiments with an initial density ρ_{im}, kg/m^3 in accordance with the plan of experimental research.

2.5.1) Determination of temperature T_m, ^{0}C.

2.5.2) Changing the inclination angle of the sliding surface α using a jack until the moment of displacement of the landslide block in accordance with the plan of experimental studies.

2.5.3) An increase in the moisture content of the landslide block and the base of the sliding surface to the fact of displacement.

2.5.4) Determination of humidity w, temperature T, density ρ of the experimental material on the fact of displacement at $T_{liq} = 25$ ^{0}C and $T_{liq} = T_{heat}$.

2.5.5) Cyclic repetition of actions from paragraphs 2.5 inclusive for different values of the shift angle α.

2.6) Formation of research samples of a group of experiments with an initial density ρ_{im1}, kg/m^3 in accordance with the plan of experimental research.

2.6.1) Raising the temperature of the landslide block and the base of the sliding surface to T_i in accordance with the plan of experimental studies using tabletop plates of the heating system, adjusting the power and heating time.

2.6.2) Cyclic repetition of actions from paragraphs 2.5 inclusive for different values of the inclination angle of the sliding surface α in accordance with the plan of experimental studies.

2.7) Formation of research samples of a group of experiments with an initial density ρ_{im1}, kg/m^3 and cyclic repetition of actions from paragraphs 2.5 to 2.6 inclusive in accordance with the plan of experimental research.

In order to determine the experimental area of the factor space of laboratory studies of the influence of the physical properties of landfill soils on the stability of slopes, field studies of moisture w, density ρ, temperature T of landfill soils (Fig. 4), as well as the angle of slopes of the body of landfill soils α (Fig. 5) of typical MSW sites were carried out at different times of the year from closed maps. The research was conducted in 2019 (it is October) to 2020 (it is August). Including the values of mechanical properties, the angle of internal friction φ, specific adhesion, were determined according to the corresponding physical state.

Figure4. Investigation of the physical state of landfill soils at the MSW landfill.

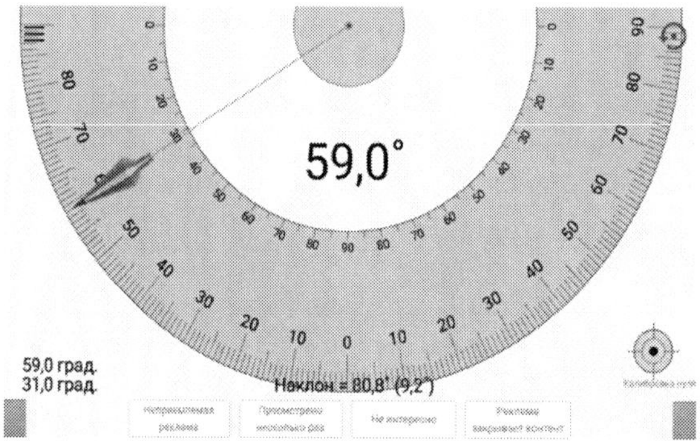

Figure 5. Mobile phone interface – investigation of the slope angle of the landfill soil body at the MSW.

The research results are reduced to the arithmetic mean of their values $\bar{x}$:

$$\bar{x} = \frac{1}{n}\sum_{i=1}^{n} x_i \ , \tag{9}$$

where x_i – indicator i of the parameter of landfill soils; n – the number of tests.

Analysis of the results shows that the dynamics of changes in humidity and temperature of landfill soils depends on the season (Fig. 6): in spring, humidity increases in accordance with an increase in atmospheric precipitation, which was typical for the year of research, and the temperature decreases, and in summer, humidity decreases with an increase in temperature. environment and landfill grounds. The lower the density of landfill soils (II object), the greater the ability to penetrate and saturate with moisture. The higher the density (III object), the less moisture is absorbed, and the more stable the temperature.

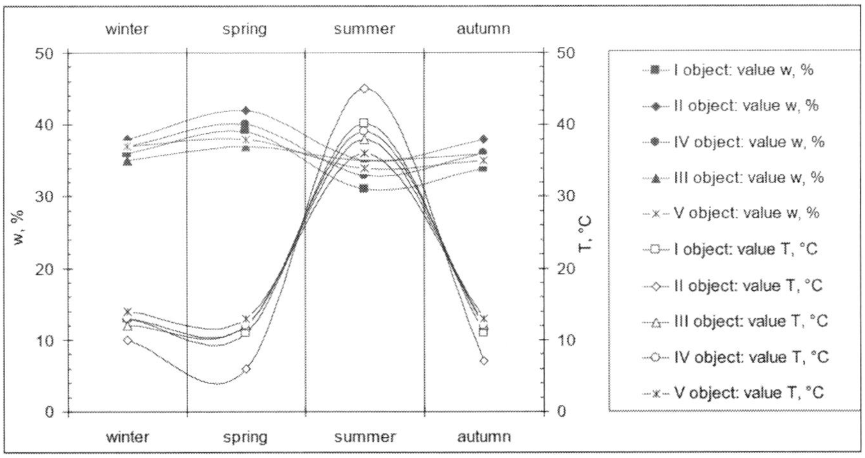

Figure 6. Dynamics of changes in humidity (upper group of curves) and temperature (lower group of curves) landfill soils in accordance with the season from 2019 to 2020.

Figures 7 and 8 show the dynamics of changes in the mechanical properties of landfill soils in accordance with the season. It has been determined that the values of the angle of internal friction and specific adhesion vary depending on the physical parameters: the higher the density (III object), the more stable the values of mechanical properties with the change of the season. Excessive moisture and high temperatures lead to instability. The object with the lowest density (II object) is characterized by more dynamic processes of changes in the angle of internal friction and specific adhesion.

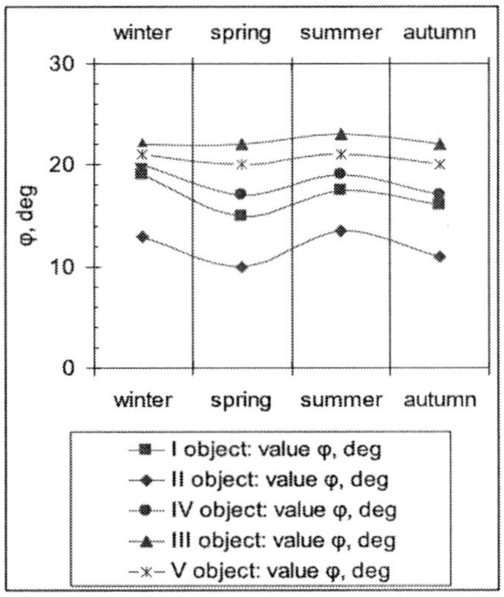

Figure7. Dynamics of changes in the angle of internal friction of landfill soils in accordance with the season from 2019 to 2020.

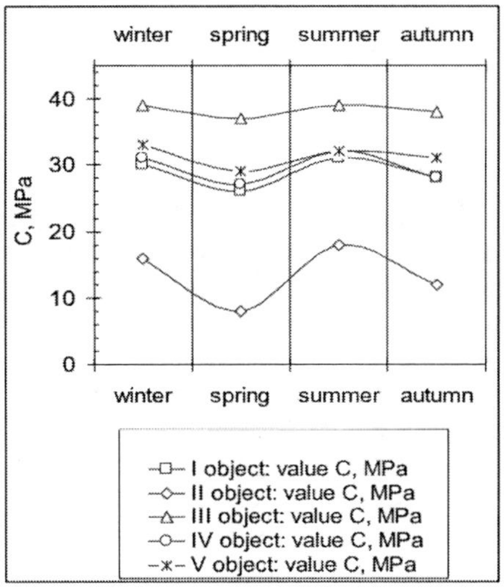

Figure8. Dynamics of changes in the specific cohesion of landfill soils in accordance with the season from 2019 to 2020.

The results of field studies of determining the angle of the slopes of the body of landfill soils at typical solid waste disposal facilities are in the range of 60^0.

The qualitative and quantitative composition of biogas is individual for each MSW facility. The process of decomposition of the organic component of waste occurs unevenly and with different intensity. On one part of the landfill, aerobic decomposition with the release of carbon dioxide may prevail, on the other, intense release of methane, and there may also be dead zones.

Biodegradation takes place under the influence of a large number of microorganisms. The main place is occupied by bacteria, which ensure the beginning of the decomposition process of the organic component and a rapid rise in temperature. First, a group of mesophilic bacteria develops, and after warming up the waste environment, a group of thermophilic bacteria begins to actively develop, which are able to break down more stable organic compounds. According to scientific research [23–25], it is customary to distinguish 5 phases of waste decomposition: phase 1 – aerobic decomposition; phase 2 – anaerobic decomposition without methane release (fermentation);phase 3 – anaerobic decomposition with variable release of methane (mixed fermentation); phase 4 – anaerobic decomposition with constant release of methane (methane fermentation);phase 5 –attenuation of anaerobic processes.

Aerobic decomposition occurs in the upper layers of the solid waste massif at a depth of 50–80 cm and, as a rule, is rather short, since its duration is limited by the amount of oxygen. This stage is characterized by the formation of carbon dioxide, water, nitrates, nitrites, nitrogen, organic residues and a large amount of heat. As the waste is compacted and isolated by soil, the aerobic phase of microbiological decomposition tends to become anaerobic – aerobic microorganisms become anaerobic. This is caused by an insufficient supply of oxygen to the waste mass to meet the conditions of the aerobic process.

Anaerobic decomposition proceeds more slowly and is accompanied by an order of magnitude less heat release. In the phase of hydrolysis under the action of bacteria, the decomposition of easily stored waste occurs, the hydrolysis of cellulose-containing wastes. In the acetogenic (acidic) phase – further decomposition of cellulose with the formation of low molecular weight acids, alcohols. The environment becomes very sour. Acids together with moisture release nutrients for methane-forming microorganisms. Then comes the methanogenic phase in which acids decompose with significant methane formation. Anaerobic microorganisms receive the energy necessary for life as a result of the decomposition of organic matter. High density slows down microbiological life.

The maximum amount of methane in the composition of biogas under the conditions of a MSW landfill can be obtained by observing temperatures within 30–40 °C, humidity within 60–80 %. Humidity, temperature is an indispensable factor for the life of methane-forming microorganisms.

As the main requirements for a laboratory unit, the possibility of conducting a series of experiments based on the use of experimental landslide blocks of different moisture, density, temperature and angle of inclination of the base of the sliding surface, as well as determining the angle of internal friction, specific adhesion is determined. Taking into account the significant variability of the problem, only general, fundamentally significant operating factors of influence and their values are reproduced, fully ensure adequate reproduction of the existing conditions for the functioning of the MSW facility.
The design features of the developed laboratory unit for studying the influence of physical properties of landfill soils (moisture, density, temperature) on the stability of slopes. In the developed laboratory unit the change of moisture of sliding experimental blocks is provided by means of a pneumatic sprayer, change of temperature is provided by means of electric tabletop heating plates, change of an angle of inclination is provided by means of a hydraulic bottle jack.

Based on the results of a series of field studies and taking into account the optimal range of the formation of the maximum amount of methane in the biogas composition, it was determined that the experimental area of the factor space of studies of the influence of the physical properties of landfill soils on the stability of slopes corresponds to: the value of moisture in the range from 30 % to 80 %, density – from 600 kg/m^3 to 1300 kg/m^3, temperature from 25 ^{0}C to 45 ^{0}C, slope angle of landfill soil body within 60^0.

Table 2 shows the natural and coded values of the factors selected for a series of laboratory experiments to study the influence of the physical properties of landfill soils on the stability of slopes. The lower level of the values of the factors corresponds to the initial physical state of the experimental blocks.

Table 2. Natural and coded values of factors selected for a series of laboratory research experiments.

Factordesignation	ρ, (kg/m^3)	T, (^{0}C)	w, (%)	x_ρ	x_T	x_w
Main level	950	35	55	0	0	0
Upper level	1300	45	80	+1	+1	+1
Lower level	600	25	30	–1	–1	–1

The inclination angle of the landslide experimental blocks is of scientific and practical interest: the upper level corresponds to 60^0, the main one – 45^0, the lower one – 30^0.

The choice of the experimental area of the factor space is associated with a thorough analysis of the a priori information. The limiting values of density close to the real position of compaction of MSW at disposal facilities, humidity are determined by the results of field studies and their optimal range for the formation of the maximum amount of methane in the biogas composition, temperature – by environmental conditions and the optimal temperature range for the formation of biogas.

3. Conclusions

A laboratory unit has been developed that allows experimental studies of the influence of the physical properties of landfill soils, such as humidity, density, temperature, on the stability of slopes in order to solve civil security problems – to check the reliability of the mathematical model and the methodology for preventing emergencies associated with displacement developed on its basis. The main elements of the unit are a rectangular experimental box with rotary and fixed parts, rotary and locking mechanisms, table heating plates, a sprayer, and a tangential load system.

A method for conducting experimental studies using the developed laboratory unit has been developed, which includes:

– establishment of the initial and boundary conditions of experimental research and coding of natural values of the levels of factors in dimensionless quantities; preparation of the laboratory facility and experimental blocks)

– carrying out a series of experiments (determination of the angle of internal friction, specific cohesion and shear angle of experimental landslide blocks under conditions of different humidity, density and temperature, determination of humidity, temperature and density based on the fact of displacement of the experimental blocks, taking into account a gradual increase in humidity). The results obtained are processed using the classical statistical method – Student's t-test.

The experimental area of the factor space of studies of the influence of the physical properties of landfill soils on the stability of slopes corresponds to the value of moisture in the range from 30 % to 80 %, density – from 600 kg/m^3 to 1300 kg/m^3, temperature from 25°C to 45°C, the angle of body slopes landfill soils within 60^0.

References

[1] Lavigne F, Wassmer P, Gomez C et al. 2014 The 21 February 2005, catastrophic waste avalanche at Leuwigajah dumpsite, Bandung, Indonesia Geoenvironmental Disasters **1** 10

[2] Korrespondent.net. Vsi novyny «svalka» na saiti Korrespondent.net. Accessible electronic resource: http://korrespondent.net/tag/3441

[3] Rashkevich N V 2020 Development of the control algorithm of the methodology of emergency prevention on landfill with liquidation energy-intensive technological equipment *Municipal economy of cities* **156** pp 188-194

[4] Rashkevich N, Cherepnev I and Kovalev I 2019 Method of fire detecting at a municipal solid waste landfill *Engineering of nature management* **3(13)** pp 102-109

[5] Rashkevich N., Goncharenko I., Anishenko L. et al. 2018 Anishenko et al. Biogas from the municspal solid waste polygon *Scientific Journal «ScienceRise»* **9 (50)** pp 39-42

[6] Pospelov B., Kovrehin V., Rybka E. et al. 2020 Development of a method for detecting dangerous states of polluted atmospheric air based on the current recurrence of the combined risk *Eastern-European Journal of Enterprise Technologies* **1/9 (107)** pp 49-56

[7] Otrosh Y, Semkiv O, Rybka E and Kovalov A 2019 About need of calculations for the steel framework building in temperature influences conditions *In Materials Science and Engineering. IOP Publishing* **708** 012065

[8] Otrosh Y, Rybka Y, Danilin O and Zhuravskyi M 2019 Assessment of the technical state and the possibility of its control for the further safeoperation of building structures of mining facilities *EDP Sciences In E3S Webof Conferences* **123** 01012

[9] Rashkevich N 2020 Formation of mathematical apparatu for the methodology of emergency prevention on solid waste landfill with technological equipment *Municipal economy of cities* **154** pp 100-107

[10] Romanov V V, Ivanov A A and Kaurkin M D 2018 Inzhenernaya seysmorazvedka pri izuchenii poligonov TBO *Nauchnaya konferentsiya v ramkakh IX Mezhdunarodnogo foruma «Ekologiya» materialy godichnoy sessii Nauchnogo soveta RAN po problemam geoekologii, inzhenernoy geologii i gidrogeologii* pp 142-146

[11] Abreu A E, Gandolfo O C and Vilar O M 2016 Characterizing a Brazilian sanitary landfill using geophysical seismic techniques *Waste Manag.* **53** pp 116-127

[12] Ofrikhter V G and Bezgodov M A 2015 Ofrikhter Otsenka geotekhnicheskikh kharakteristik TBO polevymi metodami *Akad. vestnik UralNIPproyekt RAASN* **3** pp 58-65

[13] Zhan T L T., Chen Y M and Ling W A 2008 Shear strength characterization of municipal solid waste at the Suzhou landfill, China *Engineering Geology* **9** pp 97-111

[14] Chen W, Song Y, Z Wu, Zeng J and Li C 2018 Stability of double-step muck slope under different overload conditions *European Journal of Environmental and Civil Engineering*

[15] Bray J D, Zekkos D, Kavazanjian E J et al. 2009 Shear strength of municipal solid waste *ASCE J Geotechnical Geoenvironmental Eng.* **135(6)** pp 709-722

[16] Koelsch F, Fricke K, Mahler C and Damanhuri E 2005 Stability of landfills–The Bandung dumpsite desaster. CISA (Hrsg.): *Proceedings of the 10th International Landfill Symposium, Cagliari (Italy)*

[17] Yun-min Chen, Jun-chao Li, Chun-bao Yang et al. 2017 Centrifuge modeling of municipal solid waste landfill failures induced by rising water levels *Canadian Geotechnical Journal* **54(12)** pp 1739-1751

[18] Zhu B, Wang L and Chen Y M 2018 Centrifugal model tests on static and seismic stability of landfills with high water level Proceedings of the 9th International Conference on Physical Modelling in Geotechnics. Physical Modelling in Geotechnics. United Kingdom **2**

[19] Huvaj-Sarihan N and Stark T D 2008 Back-Analyses of Landfill Slope Failures. 6-th International Conference on Case Histories in Geotechnical Engineering **12**

[20] Dzhamanbayev M Dzh, Omuraliyev S B O 2011 Vliyaniye atmosfernykh osadkov razlichnoy intensivnosti na smeshcheniye suglinistykh gruntov opolznevykh sklonov *Izvestiya Kyrgyzskogo gosudarstvennogo tekhnicheskogo universiteta im. I. Razzakova* **23** pp 46-51

[21] DSTU B V.2.1-17:2009. Osnovy ta pidvalyny budynkiv i sporud. GRUNTY. Metody laboratornoho vyznachennya fizychnykh vlastyvostey. Kyyiv Minrehionbud Ukrayiny

[22] DSTU B V.2.1-4-96 (HOST 12248-96) Osnovy ta pidvalyny budynkiv i sporud. Grunty. Metody laboratornoho vyznachennya kharakterystyk mitsnosti i deformovanosti. Derzhavnyy komitet Ukrayiny u spravakh mistobuduvannya i arkhitektury Kyyiv

[23] Min'ko O I and Lifshits A B 1992 Ekologicheskiye i geokhimicheskiye kharakteristiki svalok tverdykh bytovykh otkhodov. Ekologicheskaya khimiya **2** pp 37-47

[24] Vaysman Ya I, Korotayev V N and Petrov Yu V 2001 *Poligony deponirovaniya tverdykh bytovykh otkhodov* 150

[25] Abramov N F, Sannikov E S, Rusakov K B i dr 2004 *Metodika rascheta kolichestvennykh kharakteristik vybrosov zagryaznyayushchikh veshchestv v atmosferu ot poligonov tverdykh bytovykh i promyshlennykh otkhodov* 28

IOP Conf. Series: Materials Science and Engineering 1164 (2021) 012064 doi:10.1088/1757-899X/1164/1/012064

Theoretical and experimental studies of energetics of the dolomite decarbonization process

V I Vinnichenko[1], A N Riazanov[2], R Z Rakhimov[3] A A Riazanov[2] and O V Vinnichenko[4]

[1] Cool Clean Researches & Technologies, 37 Rue Sainte Catherine, Le Cannet, 06110, France
[2] Ufa State Petroleum Technological University, 1 Kosmonavtov st., Ufa, 450064, Republic of Bashkortostan, Russia
[3] Kazan State University of Architecture and Engineering, 1 Zelenaya st., Kazan, 420043, Tatarstan, Russia
[4] Kharkiv National University of Civil Engineering and Architecture, 40 Sumska st., Kharkiv, 61002, Ukraine

*E-mail: vvinnichenko@ukr.net

Abstract. One of the known advantages of magnesia binders is a significant reduction in energy consumption for the production of such binders compared to the production of lime and Portland cement. The theoretical costs of thermal energy for firing dolomite, which amount to 220-230 kcal/kg (for Portland cement – 420-430 kcal/kg), are almost halved under equal conditions of raw material moisture and furnace efficiency. This study examines the screening of dolomite waste and coal waste materials as raw materials for the production of dolomite cement. Coal preparation wastes in this case are both the initial raw material and, in part, can be used as fuel to ignite the raw charge. Theoretical and experimental studies of the influence of the organic component of coal waste on the process of decarbonization of dolomite have been carried out. The results of thermodynamic studies show that the participation of carbon and its gasification products in the reactions of dissociation of magnesium carbonate makes it possible to reduce the temperature of the beginning of the decomposition reaction of $MgCO_3$ and $CaCO_3$. The energy feasibility of obtaining dolomite binders has been theoretically substantiated.

Introduction

Dolomite rocks are one of the most common varieties of mineral commodities slightly developed by the construction industry. They may be used for the production of various types of binders and construction materials thereunder [1-2]. But nowadays such materials are actually not produced by the domestic industry, although a large number of construction materials and products based on magnesium carbonate (magnesites) and double calcium and magnesium carbonates (dolomites) are supplied from abroad. Various types of finishing and heat-insulating materials are made with the application of such binders [2-4]. One of the magnesia binder advantages is much less energy costs for the production thereof as compared with the production of lime and Portland cement [5]. Less gaseous pollutants are released under the dolomite firing [6-7].

The principal advantages of magnesia binders are: high mechanical strength with fast growth thereof in the initial period of hardening, better ultimate bending strength values as compared with

Content from this work may be used under the terms of the Creative Commons Attribution 3.0 licence. Any further distribution of this work must maintain attribution to the author(s) and the title of the work, journal citation and DOI.
Published under licence by IOP Publishing Ltd

other binders, dense structure of hardened magnesia stone under low true and average density, low heat conductivity, high bond strength with aggregates when making magnesia concretes and mortars, as well as high corrosion resistance [8-12]. The hardening products of some magnesia binders are known to have an extremely high resistance to seawater, alkali soil water, saline, and alkali solutions, much higher than the resistance of the hardening products of special Portland cement types [12]. The above properties indicate the advisability of developing technologies for making such binders and materials thereunder [9].

On the other hand, much coal-based waste has been accumulated. The problem of industrial waste disposal and recycling is still extremely pressing. The ecological danger of coal waste piles is widely covered [13-15]. They are both the initial commodities and at the same time they may be partially used as fuel for the construction materials industry under the integrated application of waste. In addition, waste disposal enables to solve the environmental problems, improve the lands by removing the waste piles therefrom, and reduce harmful emissions.

Theoretical and experimental energy costing for the dolomite binder firing is the aim of the paper.

1. Theoretical

1.1. Theoretical costs of heat energy

Theoretical costs of heat energy for dolomite clinker formation are the costs of heat for dissociation of $MgCO_3$, a certain amount of $CaCO_3$, and impurities dehydration [7] will be 'equation 1':

$$q_{kl} = G^C_{CaCO_3} \cdot (\Delta H_t)_1 + G^C_{MgCO_3} \cdot (\Delta H_t)_2 + G^C_{AS_2H_2} \cdot (\Delta H_t)_3 \qquad (1)$$

where $G^C_{CaCO_3}$, $G^C_{MgCO_3}$, $G^C_{AS_2H_2}$ are the content of calcium carbonate, magnesium carbonate and aluminosilicates respectively in dolomite per 1kg of clinker, $(\Delta H_t)_1$, $(\Delta H_t)_2$, $(\Delta H_t)_3$ are the enthalpy of the reactions of decarbonization of calcium carbonate, magnesium carbonate, aluminosilicate impurity.

Theoretical costs of heat energy for dolomite firing are (220-230) kcal/kg of clinker. For comparison, theoretical heat consumption for firing Portland cement clinker is (420-430) kcal/kg of clinker, on average. Actual fuel consumption for firing depends on the commodity moisture and efficiency of the heat generating unit. Fuel consumption for firing dolomite clinker is reduced by 47% as compared to Portland cement clinker, i.e. almost twice under equal conditions of the initial moisture content of the commodity and the furnace efficiency.

1.2. Thermodynamic studies

Procedure [17] was applied therefor. Initial data were taken from [17-19]. Thermodynamic analysis was made to theoretically determine the effect of the carbon involvement and gasification products thereof in the mixture on the magnesium carbonate dissociation. The following reactions are considered:

- 1. $MgCO_3 = MgO + CO_2$
- 2. $MgCO_3 + C = MgO + 2CO$
- 3. $MgCO_3 + 2OH^- = MgO + CO + H_2O + O_2$
- 4. $MgCO_3 + CO + 0,5O_2 = MgO + 2CO_2$

The results obtained are indicated in 'figure 1'.

Thermodynamic analysis of the waste organic component affecting dissociation reactions of magnesium carbonate has revealed that reaction 1 is thermodynamically possible at temperatures above 600 K, and MgO presence is thermodynamically possible in reactions 2,3,4 at lower temperatures, with the organic component containing in the waste.

The thermodynamic studies results indicate that the involvement of carbon and the products of gasification thereof in the dissociation reactions of magnesium carbonate enable to reduce the temperature of starting the decomposition reaction of $MgCO_3$.

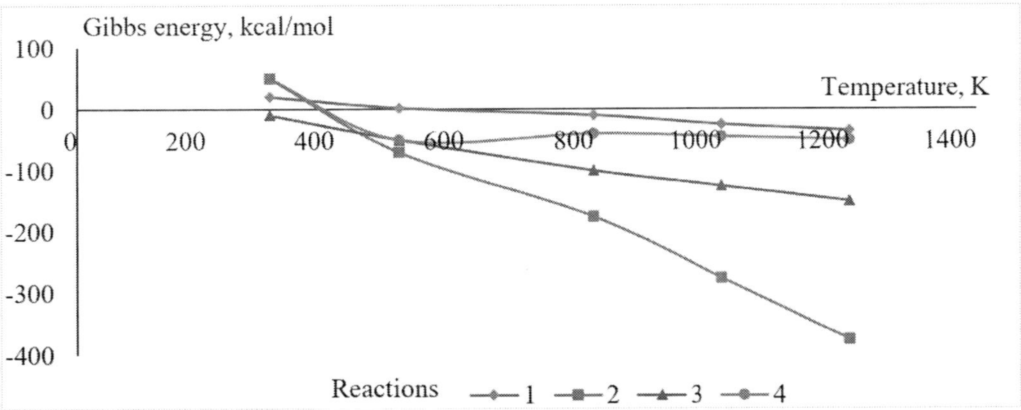

Figure 1. Action of temperature on Gibbs energy

2. Experimental

The possibility and advisability of joint application of waste coal and dolomite waste to obtain binders is first determined by the nature of the behavior under heating of the organic component of the waste coal. Gravity concentration waste of the Belorechensk concentrating mill and Krasnoyarsk dolomite were used for making the studies.

The process of thermal oxidative breakdown 'Figure 2' characterized by the release of gaseous volatile products is within the temperature range of (100-366.8) °C.

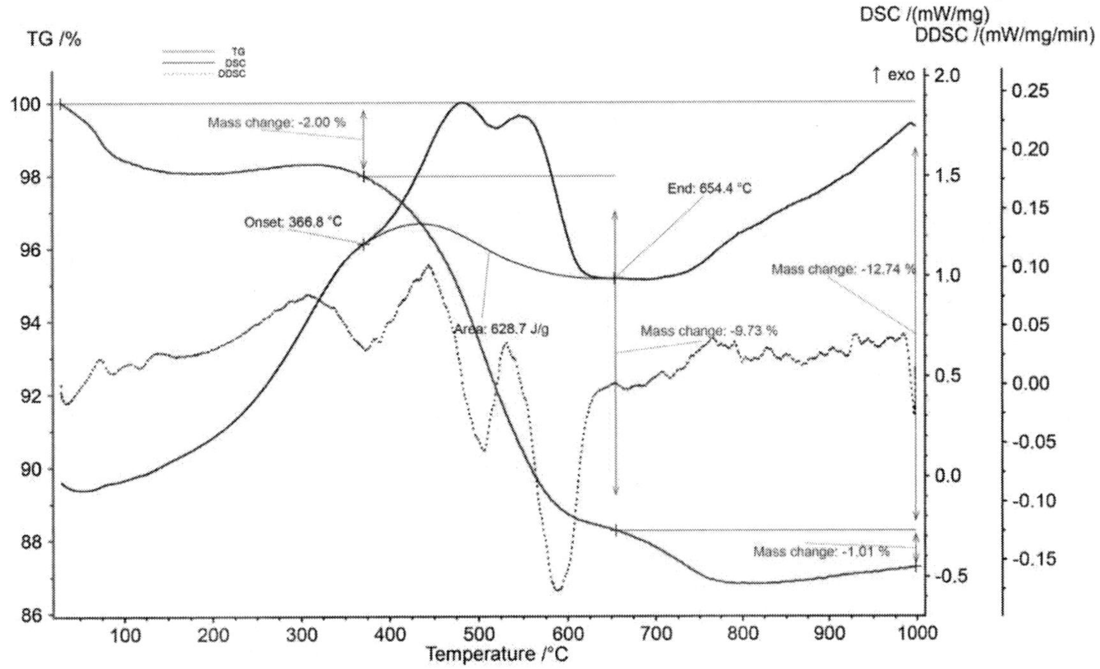

Figure 2. Waste coal heating pattern.

Under further heating the processes of thermal breakdown of the organic mass of waste is intensified accompanied by the release of volatile substances and resins, occurring together with the dehydration of aqueous oxides of aluminosilicates. The character of the effect i.e. two maxima on the

DSC curve indicates that the composition of the organic component is inhomogeneous. On the DTA curve, intense exothermic effects with a maximum at 480 °C and 550 °C correspond to this process. The process ends at 654.4 °C, therewith a 9.73 % weight loss is observed due to volatile organic compound release. The process of thermal breakdown of the organic component within 654.4-980 °C temperature range is known as semicoking with the formation of a liquid phase – the plastic state of coal – with the subsequent transformation thereof into semicoke.

Thermal transformations of dolomite upon heating are shown in 'figure 3'.

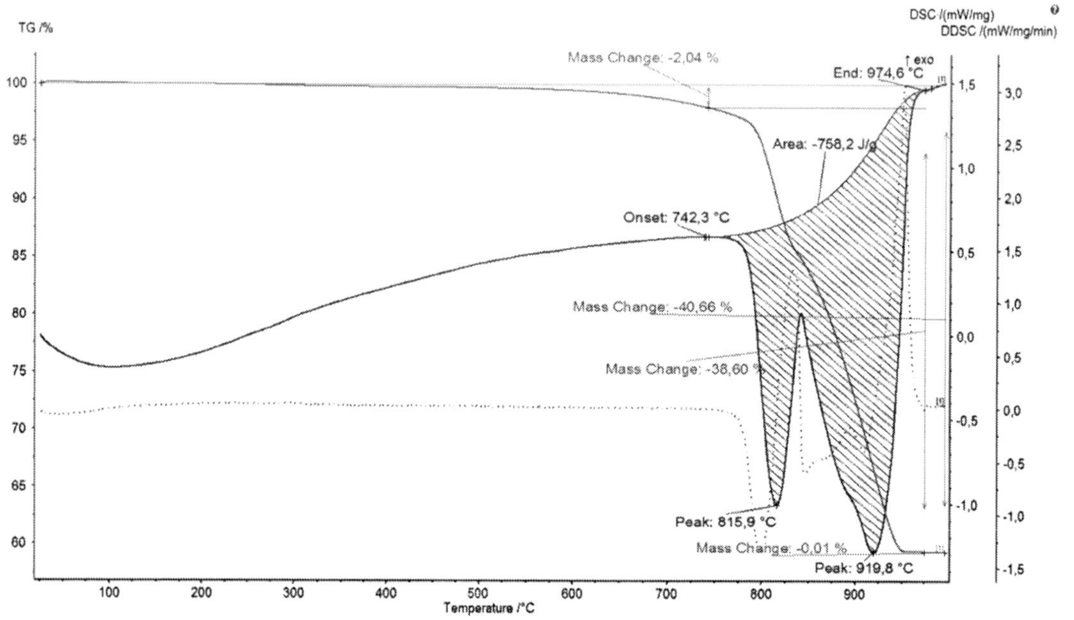

Figure 3. Dolomite heat pattern.

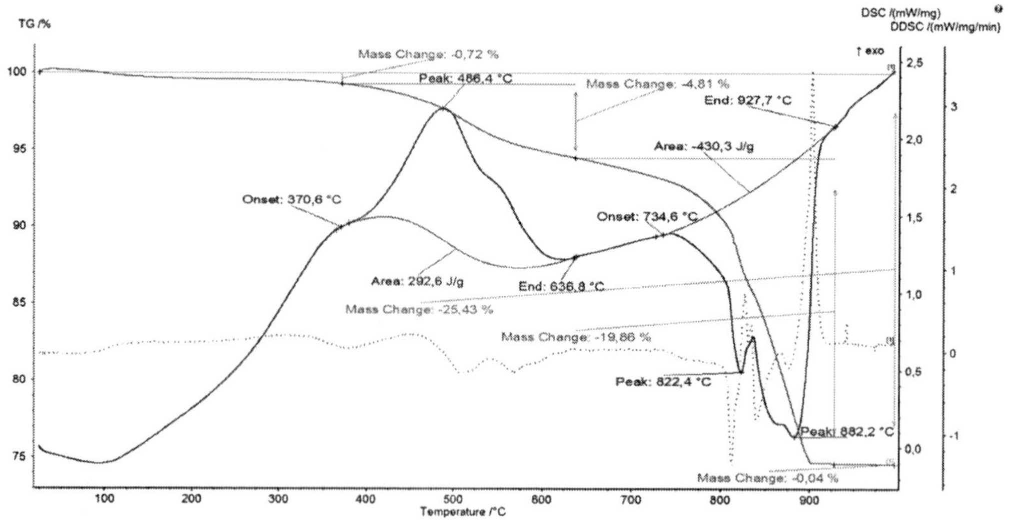

Figure 4. Heat pattern of dolomite:waste coal mixture as 1:1.

The deflections of the DTA curve after heating above 750 °C indicate the stepwise process of dolomite decarbonization. The intense endothermic effect at 815.9 °C corresponds to the magnesium carbonate dissociation, while the endothermic effect with a maximum at 919.8 °C corresponds to the calcium carbonate dissociation. Energy consumption for endothermic decarbonization reactions of magnesium and calcium carbonates is 758.2 J/g (Area – 758.2 J/g).

Waste coal adding to the dolomite, the character of thermal effects is significantly changing, although the features of thermochemical transformations of the waste organic component remain. The exothermic effects seen on the DTA curve 'Figure 4', 'Figure 5', 'Figure 6' at 486.4 °C, 489.0 °C and 485.8 °C are due to thermal oxidative breakdown of the organic component of the waste coal.

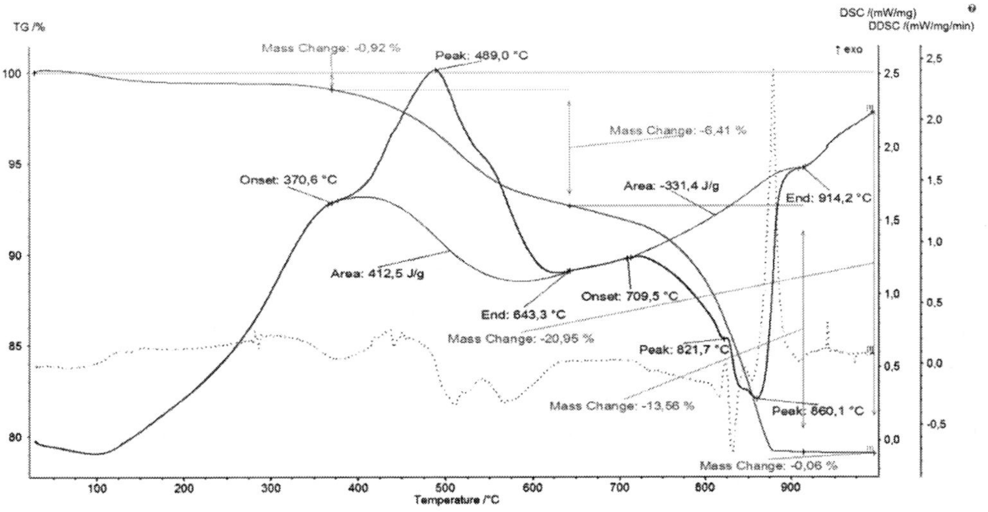

Figure 5. Heat pattern of dolomite:waste coal mixture as 1:2.

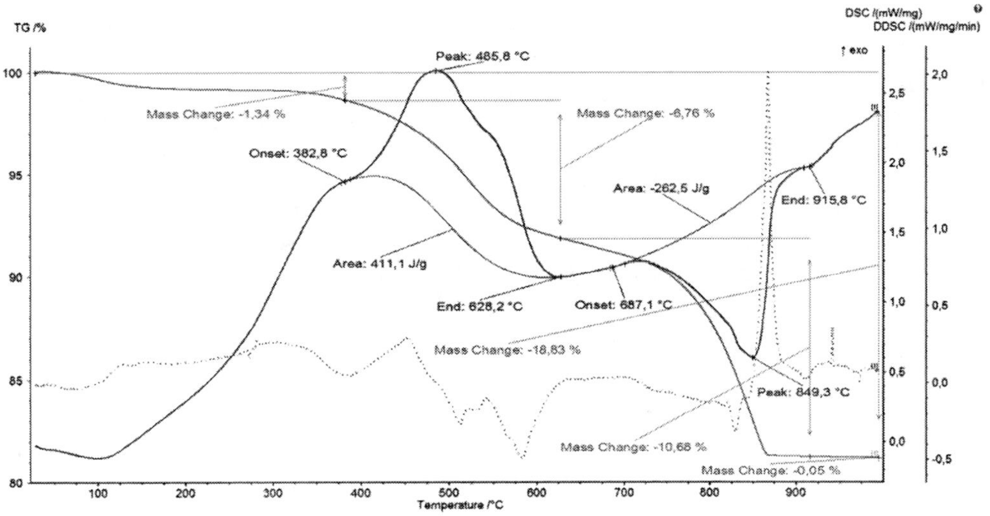

Figure 6. Heat pattern of dolomite:waste coal mixture as 1:3.

According to the heat patterns analysis the decarbonization process is intensifying. Comparison of the heating curves of dolomite 'Figure 3' and a mixture of dolomite and waste coal 'Figure 4', 'Figure 5', 'Figure 6' indicated a temperature decrease at the start of carbonate dissociation. While in the dolomite/mixture series as 1:1, 1:2, and 1:3, the temperature at the start of dissociation is: (742.3-734.6-709.5-687.1) °C.

The endothermic process of the decarbonization effect 'Figure 4', 'Figure 5', 'Figure 6' of $MgCO_3$ and $CaCO_3$ with the waste is seen to undergo significant changes. First, the curve has two distinct peaks 'Figure 4': one is the decarbonization of $MgCO_3$ at 822.4 °C and the second one is the decarbonization of $CaCO_3$ at 882.2 °C. The waste coal in the mixture increasing to 1:3, there is one exothermic peak 'figure 6' at 849.3 °C.

Combustion of the waste organic component at the required temperature and oxygen results in exothermic transformations and is accompanied by a change in the internal heat content of the system. This is confirmed by a significant reduction of area occupied by curves on the heat patterns that are recording endothermic reactions from the point of starting the heat absorption to the point of end thereof.

A very interesting dependence is seen when analyzing the heat required for the decarbonization reaction.

The total heat energy used for decarbonization of calcium and magnesium carbonates is:

- dolomite – 758.2 J/g 'Figure 3';
- dolomite:waste mixture as 1:1 – 430.3 J/g 'Figure 4';
- dolomite:waste mixture as 1:2 – 331.4 J/g 'Figure 5';
- dolomite:waste mixture as 1:3 – 262.5 J/g 'Figure 6.

Exothermic reactions in dolomite and waste mixture provide for the requirement in endothermic processes in heat energy.

3. Results

The energetic advisability of obtaining dolomite binders has been theoretically substantiated.

The thermodynamic studies results have shown that carbon and gasification products thereof involved in the dissociation reactions of magnesium carbonate enables intensification of the carbonization process.

Comparative studies of the behavior under heating of waste coal, dolomite and waste/dolomite mixture have been carried out. Decrease in the demand of energy of endothermic processes of dolomite decarbonization is found to be due to exothermic reactions of waste coal and dolomite mixture.

References

[1] Sorel S 1866 Improved composition to be used as a Cement and as a Plastic Material for Molding Various Articles *United States Patent Office. Patent 53/092, Paris, France.*

[2] Kozlova V K, Sweet T F, Dushevina A M, Chelyshev A S, and Pimenov A T 2004 Integrated use of Taenzinskoye deposit dolomites *Construction materials* No **1**. 29-31

[3] Kozlova V K, Dushevina A M, and Chelyshev A S 2003 Construction materials based on Taenzinskoye dolomite *Reliability and durability of construction materials and structures: Proceedings of the III International Scientific and Technical Conference.* Volgograd: VolgGASA, P.3, pp 108-110

[4] De Wolff P M and Walter-Levy M L 1953 Hydratations prozesse und Erhartungs eigenschaften in Systemen MgO-$MgCl_2$ [text] *Zement-Kalk-Gips* II. No **4**. pp 125-137

[5] Kasai J, Ichiba M and Nakanara M 1956 Mechanism of the Hydration of Magnesia Cement [Text] *J. of Chem. Soc. of Japan* Vol. 63 No **7**. pp 1182-1184

[6] Vinnichenko V and Ryazanov A 2018 Construction binders and environmental indicators of their production *7th International Scientific Conference "Reliability and Durability of Railway Transport Engineering Structures and Buildings"* Published online Vol 230 (Transbud-2018).

[7] Vinnichenko V and Ryazanov A 2018 Energy Efficiency of Binder Application in Concrete *International Journal of Engineering Technology* Vol 7, No **4.3** pp 335-338

[8] Vinnichenko V and Ryazanov A 2017 Ecological indices of manufacture of Portland cement clinker and production of the dolomite clinker *MATEC Web of Conferences 116, 01020*
[9] Biryuleva D K, Shelikhov N S and Rakhimov R Z 1997 Mechanism of dolomite cement hardening *Abstracts of 3 academic lectures "Current problems of construction materials science"* Saransk p 119
[10] Biryuleva D K, Shelikhov N S and Rakhimov R Z 1999 Dolomite cement modified by silicate additives *Abstracts of the anniversary International scientific and practical conference "Construction 99"* Rostov-on-Don p 49
[11] Komlev V G, Ivanov A E and Kurkina I D 1992 Development of dolomite-based magnesia from Melekhovskoye deposit *Waste disposal in the production of construction materials.* Abstracts: Penza
[12] Komlev V G and Ivanov A E 1993 Dolomite-based magnesia binder from the Melekhovskoye deposit *Resource-saving technologies of construction materials, products and structures.* Abstracts: Belgorod pp 53-54
[13] Mazuranic C, Biliuski H and Matcovic B 1982 Magnesium oxychloride cement obtained from partially calcined dolomite [Text] *J. Amer. Ceram. Soc* Vol 65 No **10** pp 523-526
[14] The land of our anxiety 2005 *According to the report on the environmental conditions in the Donetsk region in 2004. Ed. by S. Tretyakov* Donetsk OOO Tsepi "EPICENTER LTD" pp 110-120
[15] Galetsky L S Naumenko U Z and Pilipchuk A D 2002 Technogenic deposits: a new unconventional source of mineral resources in Ukraine *Environmental ecology and habitation* No. **5-6** pp 77-81
[16] Petrova L O 2002 Environmental impact of coal mining and coal processing waste *Geol. J.* Kyiv No **2** pp 81-87
[17] Babushkin V I, Matveev G M and Mchedlov-Petrosyan O P 1965 *Thermodynamics of silicates* Moscow: Gosstroyizdat p 352
[18] Glushko V P 1970 *Thermal constants of substances. A reference-guide in ten editions* Iss IV (C, Si, Ge, Sn, Pb) Part I: Tables of approved values M.: VINITI p 510
[19] Landiya N A 1962 *Calculation of high-temperature heat capacities of solid inorganic substances by standard entropy* Tbilisi p 224

IOP Conf. Series: Materials Science and Engineering 1164 (2021) 012065 doi:10.1088/1757-899X/1164/1/012065

Calculation of reinforcement to concrete adhesion in reinforced concrete elements

V M Romashko[1,*] and O V Romashko-Maistruk[1]

[1]National University of Water and Environmental Engineering, Soborna str. 11, 33028 Rivne, Ukraine

[*]Email: romashkovasyl@gmail.com

Abstract. A model and method for calculating of reinforcement to concrete adhesion in reinforced concrete elements have been developed. The basic parameters by means of which it is possible to control interaction of reinforcement with concrete at any stage of a reinforced concrete element deformation are defined. These parameters are associated with the processes of formation and development of normal cracks. A criterion is proposed by which it is possible to determine the adhesion forces of a reinforcing rod with concrete along its conditional cylindrical surface and to estimate, thus, the degree of violation of their interaction. It is recommended to calculate the specified efforts in the simplest way: by means of average stresses of reinforcement with concrete adhesion on sites between adjacent cracks. It is statistically substantiated that the function of these stresses is nonlinear with respect to normal stresses in the reinforcement. An analytical dependence is proposed to calculate the coefficient taking into account the profile of the reinforcement (its adhesion index to concrete).

1. Introduction

The general theory of reinforcement to concrete adhesion, which would meet the basic requirements of design engineers, has not yet been built. And the current situation is due to the results of numerous studies [1-4], which show that the adhesion of reinforcement to concrete depends on a very large number of different factors. It is extremely difficult, and sometimes even impossible, to take into account their joint influence in modeling the interaction of reinforcement with concrete. Therefore, it is not surprising that the methods and algorithms for calculating reinforced concrete elements and structures regulated in domestic and foreign regulations [5, 6], almost completely ignore the interaction of reinforcement with concrete.

In general, all currently known studies of reinforcement to concrete adhesion can be divided into the following four areas:

- identification of the main parameters that most affect the strength of adhesion [2];
- study of general patterns of mutual displacements of concrete and reinforcement [1, 4];
- creation and development of a general technical theory of adhesion [4];
- modeling of reinforcement to concrete adhesion with numerical methods [7].

However, the ambiguity of the results obtained in these works and the lack of a universal theory of reinforcement to concrete adhesion encourage further research on its construction. And a special role in solving this problem should be given to a generalized model of reinforcement to concrete adhesion.

2. Analysis of recent research and publications

The results of research known today, aimed at modeling the interaction of reinforcement with concrete, not only differ significantly from each other, but also often contradict each other. This situation is quite natural and understandable, because it is due to the results of experimental studies, which were different: the shape and size of the samples for different types and classes of concrete and reinforcement; reinforcement schemes and conditions of concrete compaction, and therefore the nature of the contact layer formation; technological features of concrete care; schemes and conditions of prototypes testing; hypotheses and preconditions for theoretical research and the like.

Content from this work may be used under the terms of the Creative Commons Attribution 3.0 licence. Any further distribution of this work must maintain attribution to the author(s) and the title of the work, journal citation and DOI.
Published under licence by IOP Publishing Ltd

It is indisputable that the classification of reinforcement to concrete adhesion models would allow not only to systematize the research itself, but also to organize their results in a certain way. These models could be divided into many features, the main of which should still be considered:

- scale and structure of the model;
- representations of concrete and reinforcement in the system of their contact interaction;
- test schemes and the nature of the concrete destruction at the contact boundary "concrete-reinforcement";
- mathematical solutions of reinforcement with concrete interaction schemes.

The last sign on which all models of reinforcement with concrete adhesion should be conditionally divided into analytical and numerical deserves special attention. In general, analytical models are presented in the form of mathematical equations (formulas) that characterize the functional dependence of research results on certain initial factors or parameters. By the nature of the mathematical dependences of the adhesion model can be divided into: linear (elastic), nonlinear (elastic-plastic), differentiated and variational.

Elastic models have found the widest application in researches of reinforcement with concrete adhesion. Now they represent mainly power, polynomial, exponential, fractional, logarithmic [4], trigonometric dependences, and most often spline functions [2]. But the results of experimental studies show that the diagram of the stresses of reinforcement to concrete adhesion along the area of their contact is constantly changing (Figure 1), and therefore cannot be described by a single function.

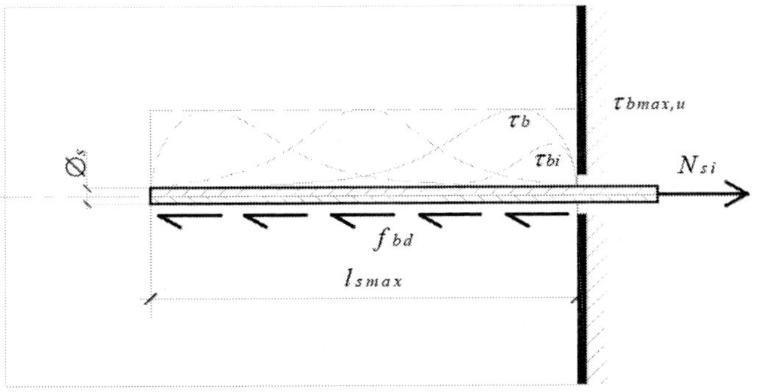

Figure 1. Diagram of the change in adhesion stresses when pulling out a reinforcing bar from concrete.

Differentiated models, unlike others, take into account the variability of the diagrams of reinforcement to concrete adhesion along the area of their contact in the sections of a reinforced concrete element [3, 8]. However, it is extremely difficult to assess the change in the stress-strain state of the contact interaction of reinforcement with concrete using such diagrams in practical calculations, even by numerical methods.

In variational models [9], concrete around the reinforcement is divided into three characteristic zones, which makes it possible to fairly accurately simulate its stress-strain state based on the solution of the symmetric-axial volumetric problem of a deformable solid mechanics. However, given the complexity of determining the mechanical characteristics of concrete in each zone, these models have not found practical application even in calculations using the finite element method.

In numerical models, studies are always carried out using specific numerical characteristics - finite elements, depending on the selected shape of the relief (profile) of the reinforcement. The contact layer of concrete is modeled, starting from the ideal adhesion scheme to the scheme of elastic-plastic damaged material [7]. The solution of the problems of reinforcement to concrete adhesion according to the given schemes is so complicated that it can be done only with the help of special computer programs. Among the latter, the most widely used software systems are ABAQUS, ANSYS, DIANA,

NASTRAN and others. All of them are endowed with rather broad functional capabilities, but for the overwhelming majority of them, the description of SSS in the zone of reinforcement to concrete adhesion remains rather complicated, which noticeably limits their widespread use in practical calculations when solving contact problems in reinforced concrete.

3. The purpose and objectives of research

These studies are aimed at developing a generalized model of reinforcement to concrete adhesion, which would be valid for any stage of reinforced concrete elements deformation and would satisfy the basic requirements of design engineers. It is appropriate to base such a model on the function of a universal parameter - the forces of the reinforcement to concrete adhesion along its conditional cylindrical surface. In this case, the main task of these studies is to obtain the specified function, taking into account the nonlinear relationship of the average bond stresses from the normal stresses in the reinforcement.

These studies are based on mathematical modeling of the deformation processes of reinforced concrete elements and structures in general, as well as analytical and numerical methods of mathematical analysis of reinforcement to concrete adhesion parameters (forces, stresses) along its conditional cylindrical surface in particular.

4. Results and discussion

It should be noted right away that the generalized model of reinforcement to concrete adhesion is based on the basic relationships of solid mechanics (SDM), expressed by the well-known system of static, geometric and physical relationships (equations) [10, 11]. The defining elements of the developed model will be considered:

- the main modeling parameter, with the help of which it would be possible to reliably estimate the reinforcement to concrete adhesion;
- the main criterion by which it would be possible to fix violations of the reinforcement to concrete adhesion;
- the most important parameters by which it would be possible to monitor the reinforcement to concrete adhesion at any stage of a reinforced concrete element deformation.

The main parameter by which it is possible to fairly easily evaluate the adhesion of reinforcement to concrete is the adhesion force N_{bd}. According to the numerous results of experimental studies, it can be argued that they cannot exceed the forces in tensile concrete N_{ct}. That is, violation of the condition

$$N_{ct} = N_{bd}, \tag{1}$$

responds not only to the appearance of the first crack, but also to all others without exception.

The appearance of the first crack does not directly depend on the adhesion of the reinforcement to the concrete. Therefore, its occurrence in a reinforced concrete element can be determined by the ultimate deformations of tensile concrete ε_{ctu}, which are fixed by the extreme criterion of its bearing strength $dM_{ct} / d(1/r) = 0$ ($dN_{ct} / d\varepsilon_{ct} = 0$). Compliance with the current norms [5, 6], they can even be taken equal $\varepsilon_{ctu} = 2f_{ctk} / E_{c0}$. The application of the criterion $dM_w / d\varepsilon = 0$ for determining the moment of the cracks appearance [12] is somewhat incorrect, since only the tensioned concrete loses its bearing strength, and not the element as a whole. In addition, the expression $dM / d\varepsilon = 0$ is not endowed with some kind of physical content, so it cannot serve as a criterion at all. The Fermat criterion $dM / d(1/r) = 0$ is endowed with a similar content, since they characterize the rigidity (EA - for centrally loaded concrete and reinforced concrete elements, EI - for bending and eccentrically loaded elements), at which these elements lose their bearing strength.

When any new crack appears, the ultimate bond force $N_{bd,cr}$ in the section between the cracks according to (1) will remain unchanged and equal to the ultimate force in tensile concrete $N_{ct,cr}$ (Figure 2). Then, for an element stretched in the center, we can write the equality

$$\tau_{bm1} \cdot s_{r1} = \tau_{bm2} \cdot s_{r2}, \tag{2}$$

where s_{r1} and s_{r2} are the distances between the cracks of the first and second levels, respectively; τ_{bm1} and τ_{bm2} - average adhesion stresses in areas between cracks of the same levels, respectively.

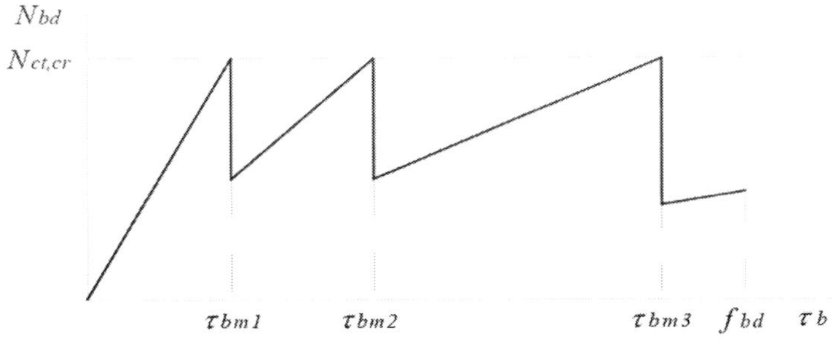

Figure 2. Changes in the adhesion forces of reinforcement to concrete with multilevel formation of normal cracks in a reinforced concrete element.

The adhesion stress of the reinforcement to the stretched concrete in the area between the cracks of the first level can be calculated by a fairly simple expression [13, 14]

$$\tau_{bmi} = \eta_1 \cdot \eta_2 \cdot f_{ctk} \cdot (\sigma_{si} / \sigma_{s,\max})^{1-1/\eta_1}, \tag{3}$$

where σ_{si} is the greatest stresses in the reinforcement at the site of its active adhesion to the stretched concrete at a certain stage of the reinforced concrete element deformation; $\sigma_{s,\max}$ - the maximum possible stresses in the stretched reinforcement on a site of its active adhesion with the stretched concrete, cannot exceed limit values ($\sigma_{s,\max} \le f_{yk}$); η_2 - the coefficient taking into account the influence of the reinforcement diameter can be taken according to current regulations [5, 6]; η_1 is the coefficient characterizing the profile of the reinforcing bar according to the adhesion index f_R (Figure 3), can be taken by the expression [15]

$$\eta_1 = 1 + 35 \cdot f_R - (12 \cdot f_R)^2, \tag{4}$$

which at the confidence interval in 2σ (σ - standard deviation) takes the form [16]

$$\eta_1 = 1 + 29 \cdot f_R - 120 \cdot f_R^2. \tag{5}$$

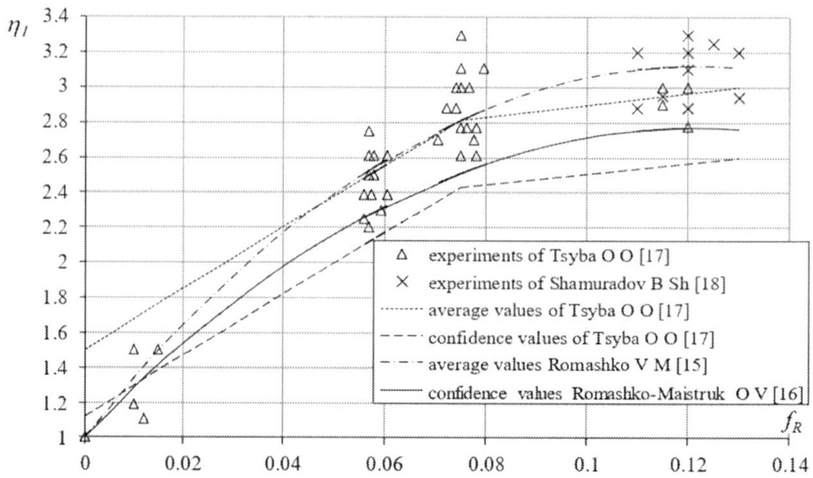

Figure 3. Relationship of the coefficient η_1 with the index of adhesion of reinforcement to concrete f_R.

Analysis of the experimental studies results by various authors [19-22] showed that the dependence of the average adhesion stresses on the stresses in the reinforcement itself is closer to the nonlinear function (3).

This is evidenced by the results of a corresponding comparison of experimental and theoretical values of the average adhesion stresses (Table 1), calculated including by the linear function (6) [12]

$$\tau_{m,i} = f_{ctm} \cdot ((\eta_1 \cdot \eta_2 - 0.4) \cdot \sigma_{si} / f_{yd} + 0.4). \tag{6}$$

Table 1. Comparison of theoretical and experimental values of reinforcement to concrete adhesion average stresses.

Authors research	Rebar profile and diameter, mm	Deviation from experimental data by formula					
		(3)			(6)		
		Δ	σ	ν ,%	Δ	σ	ν ,%
Colner V [19]	Periodic 20	1.00	1.73	1.73	1.08	5.27	4.88
Adrouche K [20]	Periodic 16	0.995	2.00	2.01	1.01	2.32	2.30
Rashedul K [21]	Periodic 20	0.994	2.29	2.30	1.051	4.48	4.26
Samoshkin A [22]	Periodic 16	0.998	2.28	2.29	1.023	2.76	2.70

Δ - mean deviation.
σ - standard deviation.
ν - coefficient of variation.

Similarly, the average adhesion stresses are determined in the areas between cracks of all subsequent levels. As the results of experimental studies show, their number for the operational stage of deformation can be limited to two levels [23].

It should be noted that equality (2) is in the form

$$\tau_{bm2,1} \cdot s_{r2,1} = \tau_{bm2,2} \cdot s_{r2,2} \tag{7}$$

remains valid for bending elements (Figure 4).

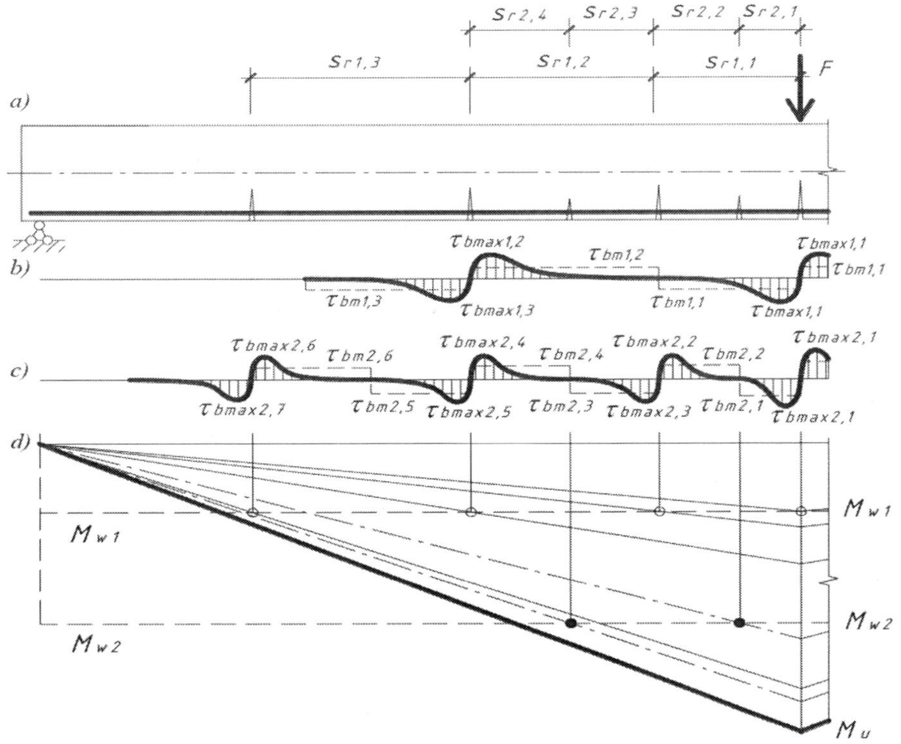

Figure 4. a) scheme of level formation of cracks, b) diagram of adhesion stresses of the first level, c) diagram of adhesion stresses of the second level, d) diagrams of moments in the bending element.

5. Conclusions

According to the obtained results, it is recommended to model the reinforcement with concrete interaction with the help of adhesion forces, as their limit values remain unchanged in the process of the reinforced concrete element deformation. Adhesion forces are suggested to be calculated in the simplest way - by means of average adhesion stresses. The function of the average adhesion stresses is nonlinear with respect to the stresses in the valve itself. It is expedient to regulate the crack formation processes in reinforced concrete elements by the parameters of reinforcement adhesion with concrete [14, 24], which will contribute to the construction of an adhesion universal theory within the generalized deformation-force model of reinforced concrete elements and structures resistance to force.

References

[1] Rehm G 1957 The fundamental law of bond *Proc. of the symposium on bond and crack formation in reinforced concrete* (Stockholm: RILEM) **2** pp 491-498

[2] Eligehausen R, Popov E P and Bertero V V 1983 *Local bond stress-slip relationships of deformed bars under generalized excitations: Report No UCB/EERC-83/23* (Berkeley: Earthquake Engineering Research Center of California University)

[3] Shima H, Chou L L and Okamura H 1987 Micro and macro models for bond in reinforced concrete *Journal of the Engineering Faculty of Tokyo University* **XXXIX(2)** pp 133-194

[4] Kholmyanskiy M M 1961 Calculation of centrally reinforced prismatic elements for adhesion *Collection of works of NIIZhB* **4** pp 122-153

[5] DSTU B B.2.6-156:2010 Designs of buildings and structures - Concrete and reinforced concrete

structures made of heavy concrete

[6] EN 1992-1-1:2004 Design of Concrete Structures Part 1-1 General rules and rules for buildings

[7] Benin A B, Semyonov A C, Semenov S G and Melnikov B E 2013 Mathematical modeling of the process of destruction of adhesion of reinforcement to concrete - Part 1: Models taking into account the discontinuity of the connection. *Engineering and construction journal* **5(40)** pp 86-99

[8] Mirza S A and Houde J 1979 Study of bond-slip relationships in reinforced concrete *ACI Journal* **76(1)** pp 19-46

[9] Nazarenko P P 2012 *Contact interaction of reinforcement and concrete under short-term loading* (Samara: SamGUPS)

[10] Romashko V and Romashko O 2017 The construction features of the deformation and force model of concrete and reinforced concrete resistance *MATEC Web of Conf* **116** 02028

[11] Romashko V and Romashko O 2019 Fundamentals of the general theory of resistance of reinforced concrete elements and structures to power influences *Materials Science Forum* **968** pp 534-540

[12] Kochkarev D V 2018 *Nonlinear resistance of reinforced concrete elements and structures to force influences* Doctoral thesis, Poltava National Technical University

[13] Romashko O and Romashko V 2018 Evaluation of bond between reinforcement and concrete *MATEC Web of Conf* **230** 02027

[14] Romashko O V and Romashko V M 2019 Model of multilevel formation of normal cracks in reinforced concrete elements and structures *IOP Conf. Ser.: Materials Science and Engineering* **708** 012069

[15] Romashko V M 2016 *Deformation-force model of resistance of concrete and reinforced concrete* (Rivne: O Zen) 424

[16] Romashko-Maistruk O V 2021 *Resistance of reinforced concrete elements to the multilevel formation of normal cracks* PhD thesis, National University of Water and Environmental Engineering

[17] Tsyba O O 2011 *Crack resistance and deformability of tensile concrete with unstressed and prestressing bar reinforcement having different relative area of crushing of transverse ribs* PhD thesis, Research Institute of Concrete and Reinforced Concrete

[18] Shamuradov B Sh 1987 *Opening width of normal cracks in reinforced concrete elements* PhD thesis, Kiev Civil Engineering Institute

[19] Colner V M 1965 Coupling of reinforcement with concrete and the strength of the embedding of rod reinforcement with a periodic profile *Concrete and reinforced concrete* **11** pp 25-27

[20] Adrouche K 1987 Influence of the constitutive parameters for steel-concrete association on bond strength under slow cyclic loading *Materials and Structures* **20** pp 315-320

[21] Rashedul Kabir Md 2014 Bond stress behavior between concrete and steel rebar Critical investigation of pull-out test via Finite Element Modeling *International Journal of Civil and Structural Engineering* **5(1)** pp 80-90

[22] Samoshkin A S and Tikhomirov V M 2017 Research of nonlinear deformation of reinforced concrete by experimental and computational methods *Izv. universities: building* **5** pp 17-27

[23] Romashko V and Romashko O 2018 Calculation of the crack resistance of reinforced concrete elements with allowance for the levels of normal crack formation *MATEC Web of Conf* **230** 02028

[24] Romashko V M 2021 General model and the mechanics of concrete elements and structures deformation *IOP Conf. Ser.: Materials Science and Engineering* **1021** 012026

doi:10.1088/1757-899X/1164/1/012066

Restriction of Cl^- and SO_4^{2-} Ions Transport in Alkali Activated Slag Cement Concrete in Seawater

P Kryvenko[1], I Rudenko[1], O Konstantynovskyi[1,*] and O Boiko[1]

[1]Kyiv National University of Construction and Architecture, 31, Povitroflotskyi av., Kyiv 03037, Ukraine

[*]E-mail: alexandrkp@gmail.com

Abstract. The relevance of alkali activated slag cement (AASC) concretes for structures operated in seawater is due to their enhanced resistance in aggressive environments. The application of high consistency fresh concretes is accompanied by negative changes in their structure with higher penetration of aggressive environments. Thus, the methods to prevent corrosion of steel reinforcement in plasticized AASC concrete are actual for investigations. It is shown, that modification of plasticized AASC concrete (consistency class S4) by the complex «alumina cement - portland cement - clinoptilolite - trisodium phosphate ($Na_3PO_4 \cdot 12H_2O$)» restrict the transport of aggressive Cl^- and SO_4^{2-} ions. The results of DTA, XRD, electron microscopy, microzond analysis show that mentioned complex limits transport of the mentioned aggressive ions due to their binding by AFm phases in hydration products, exchange with OH^- ions in the structure of clinoptilolite, formation of hydrated products of apatite group $Ca_5(PO_4)_3(OH, Cl)$. This was confirmed by qualitative reaction on Cl^- and SO_4^{2-} ions in concrete structure, as well as by assessing of surface and mass loss of steel bars embedded in AASC concrete after 9 months in seawater. It was ensured the advanced crystallization with densification of microstructure, which increases corrosion resistance of artificial stone.

1. Introduction

Current trends in construction engineering predetermine application of binder materials alternative to portland cement. It's well-known, that production of 1 ton of portland cement requires 1.5 tons of raw materials and is accompanied by 900 kg of CO_2 emission [1]. World cement production causes 6…8 % of total CO_2 emission and consumes of 12…15 % of total industrial energy [2]. Thus, production of low-emission cements, obtained by partial replacement of OPC clinker with additives, is demanded [3-6].

Alkali-activated slag cement (further, AASC) is the most perspective one in view of effective consumption of raw recourses and energy as well as responsible attitude to environment. The ecological benefits of AASC are caused not only by reduction of CO_2 emission due to application of by-products as well as waste products [7, 8, 9], but also by possibility to utilize manufacturing waters in safety building materials [10]. AASC materials are characterized by advanced strength [11], heat resistance [12], corrosion resistance [13, 14] and freeze-thaw resistance [15] in comparison with analogues based on traditional clinker cements.

The benefits of AASC cause the effectiveness of concrete constructions which are exploited in under the influence of seawater (foundations, berths, piers, coast-protecting structures, dams etc.). Durability of AASC concrete in such conditions was confirmed [16, 17].

Peculiarities of AASC hydration products cause increased protective properties of AASC concretes to steel reinforcement in comparison with portland cement. It's well-known, that the aggressive ions can be chemically adsorbed by hydrosilicates C–S–H and hydroaluminosilicate C–A–S–H gel phases at the early stages of AASC hydration [18, 19]. High content of gel-like low-calcium hydrosilicates causes advanced initial binding if compare with portland cement [20, 21]. Moreover, occlusion of Cl^- and SO_4^{2-} ions by alkaline aluminosilicates, which are analogues of natural zeolites, takes place [7]. Mentioned minerals form at late stages of structure formation [7, 22] and can be characterized both as cationites and anionites [23].

Content from this work may be used under the terms of the Creative Commons Attribution 3.0 licence. Any further distribution of this work must maintain attribution to the author(s) and the title of the work, journal citation and DOI.
Published under licence by IOP Publishing Ltd

Advanced protection of AASC concrete to steel reinforcement is also caused by presence of alkaline component which acts as cathode corrosion inhibitor [24, 25].

Durability of AASC concretes, including reinforced concretes, which are obtained from harsh (low consistency) fresh mixes, was already proved by long-term exploiting experience [7, 8, 9]. However, the modern requirements to high consistency fresh concretes are governed by practice. This way, the increasing of porosity and, consequently, permeability can be caused by changes in hardened concrete structure. This causes the increasing transport of aggressive ions in AASC concrete and risk of disturbance of reinforcement passive state.

Thus, corrosion of steel reinforcement under action of chlorides and sulfates due to their penetration in concrete structure is the main problem for structures, which are exploited in seawater. There are two main processes during corrosion action on steel reinforcement: carbonation and pitting corrosion, caused by chlorine-ions [26]. The passive state of steel reinforcement is provided at molar ratio of $Cl^-/OH^- \leq 0.6$ in pore solution [27, 28]. In one's turn, sulfate-ions don't cause immediate depassivation of steel reinforcement, but determine formation of hydrogen sulfide (H_2S) and catalyze oxidation (carbonation) of hydrate phases.

Formation of *AFm* phases (Al_2O_3-Fe_2O_3-mono) was proposed to enhance protective properties of AASC concrete, which is obtained from high consistency fresh mix, to steel reinforcement. It's well-known, that these phases can chemically bind anions (Cl^-, SO_4^{2-}, CO_3^{2-}, OH^- etc.) [30]. The effectiveness of mentioned mean was confirmed by work [31]. The *AFm* phases form at early hydration and are characterized by greater stability comparing to *AFt* phases (ettringite) while increasing alkalinity of hydration medium [32].

Thus, application of calcium aluminate cement and portland cement as additives is reasonable for formation of highly-calcium hydroaluminates $3CaO{\cdot}Al_2O_3{\cdot}10H_2O$ [33], which ensures binding of Cl^- and SO_4^{2-} ions in *AFm* phases, known as Friedel's salt $3CaO{\cdot}Al_2O_3{\cdot}CaCl_2(SO_4){\cdot}10H_2O$ [30, 34] and Kuzel's salt $3CaO{\cdot}Al_2O_3{\cdot}0{,}5CaCl_2{\cdot}0{,}5SO_4{\cdot}10H_2O$ [35] etc.

Application of aluminosilicate ionite such as clinoptilolite (zeolite) in AASC concrete can restrict transport of aggressive ions. Admixture of clinoptilolite enhances occlusion function of alkaline hydroaluminosilicates (analogues of natural zeolites), formed during AASC hydration [23].

Furthermore, application of trisodium phosphate $Na_3PO_4{\cdot}12H_2O$ (further, TSP) was proposed to enhance the inhibitory action of AASC alkaline component. TSP is the most widespread mixed corrosion inhibitor [36, 37]. Inhibitory effect of TSP is caused by dense protective film presented by iron phosphates $FePO_4$ and $FePO_4{\cdot}2H_2O$ on steel surface [38], increasing polarization resistance of steel [39] as well as due to formation of chemical stable products like hydroxyapatite $Ca_{10}(PO_4)_6(OH)_2$ [40]. Formation of crystal hydroxyapatite causes densification of structure due to filling of micropores microcracks in AASC concrete to decelerate diffusion of aggressive Cl^- and SO_4^{2-} ions [38]. At that, hydroxyapatite can bind Cl^- ions in chlorapatite $Ca_5(PO_4)_3Cl$ to decrease their content in porous solution, also while their transport from aggressive medium [41]. Effectiveness of TSP in complex with surfactants (sodium lignosulphonate, sodium gluconate) as well as calcium-containing additive (OPC clinker) for prevention of steel corrosion was shown [42]. Complex of additives (further, CA) based on salts of strong acids ensures multy-effects: reduction of water content, acceleration of crystallization, alterations in morphology of hydrated phases [43, 44].

It can be predicted advanced protective properties of AASC concrete, which is obtained from high consistency mixes and exploited in seawater, to steel reinforcement, due to application of CA which ensures chemical adsorption of Cl^- and SO_4^{2-} ions by gel-like hydration products (C-S-H and C–A–S–H), chemical binding (*AFm* phases abd hydroxyapatite) as well as occlusion (zeolite-containing components and zeolite-like alkaline hydroaluminosilicates). Proposed CA acts as corrosion inhibitor as well.

Thus, the aim of this research was to investigate the possibility of restriction Cl^- and SO_4^{2-} ions transport in plasticized AASC concrete, which is exploited in seawater, due to modification by CA including portland cement, calcium aluminate cement, TSP and zeolite.

2. Raw materials and testing techniques

Ground-granulated blast furnace slag (further, GBFS) (CaO – 47.30 %; SiO_2 – 39.00 %; Al_2O_3 – 5.90 %; Fe_2O_3 – 0.30 %; MgO – 5.82 %; SO_3 – 1.50 %; TiO_2 – 0.31 %), basicity modulus= 1.11, content of glass phase= 84.0 %, specific surface= 450 m^2/kg (by Blaine), was used as aluminosilicate component of AASC.

Polyorganohydridosiloxane (liquid 136-41) was used for milling intensification of GBFS and stabilization of AASC properties. Content of admixture was 0.1 by mass of GBFS.

Alkaline components of AASC were presented by:

- soda ash (Na_2CO_3), dry state;
- five-water sodium metasilicate ($Na_2SiO_3 \cdot 5H_2O$), dry state.

Two reference compositions of AASC were used:

- based on soda ash (GBFS – 93.50 %, soda ash – 6.50 % (3.80 % by Na_2O));
- based on sodium metasilicate (GBFS – 88.50 %, sodium metasilicate – 11.50 % (3.36 % by Na_2O)).

The AASC was also modified by CA, which components were presented by:

- portland cement CEM I 42,5 R (PJSC Ivano-Frankivskcement, Ukraine);
- calcium aluminate cement ISTRA 40 (HeidelbergCement, Germany);
- TSP ($Na_3PO_4 \cdot 12H_2O$) according to CAS № 7601-54-9
- natural zeolite (clinoptilolite) powder (by mass, %: SiO_2 – 72.5, Al_2O_3 – 13.1, Fe_2O_3 – 0.9, TiO_2 – 0.2, CaO – 2.1, MgO – 1.07, P_2O_5 – 0.003, K_2O+Na_2O – 5.03), fr. 0 - 0.1 mm, content of clinoptilolite ≤ 93.0 %, porosity 54.0 % (JSC Zeolite-Bio, Ukraine).

Content of CA was 15.00 % by mass of AASC. Contents of CA components, %: (portland cement + calcium aluminate cement) – 5.00, clinoptilolite – 5.00, TSP – 5.00. The ratio between portland cement and calcium aluminate cement was like 2.17:1.00 to form $3CaO \cdot Al_2O_3 \cdot 10H_2O$.

Surfactant was presented by sodium lignosulphonate according to CAS 8061-51-6 (pH ≥ 8.5), dosage 0.8 by mass of AASC.

The standard quartz sand according to EN 196-1 was used in AASC fine aggregate concretes (ratio AASC to sand = 1:3).

Aqueous solution of salts, which simulates 99.90 % of ones in seawater, was used in investigations. Composition of aqueous solution, % by mass of mixture: $NaCl$ – 78.70, $MgCl_2$ – 9.80, $MgSO_4$ – 5.76, $CaSO_4$ – 3.75, KCl – 1.73, $CaCO_3$ – 0.29. Total concentration of the salts was 35 g/l.

Normal consistency AASC pastes were mixed with seawater for simulation of chemical binding of Cl^- i SO_4^{2-} ions by AASC hydrate phases, as it takes place in contact zone of concrete surface with aggressive medium. Cement pastes were prepared in Hobart mixer. Consistency of AASC pastes was determined according to the national standard of Ukraine DSTU B V.2.7-185:2009. Monitoring was carried out by X-ray diffraction (XRD), differential-thermal analysis (DTA) and electronic microscope with microanalyzer.

Fresh concretes were prepared in mixer «Raimondi Iperbet» (Italy). Consistency (workability) was determined by cone slump according to the national standard of Ukraine DSTU B V.2.7-114:2002.

Performances of AASC concrete were determined on specimens 40x40x160 mm. Strength of AASC concrete was determined according to the national standard of Ukraine DSTU B V.2.7-187:2009. Transport properties of AASC concrete, regarding penetration of Cl^- i SO_4^{2-} ions, were evaluate after 28 d of hardening in normal conditions (t=20±2 °C, $R.H.$=95±5 %) and further 270 d of storing in seawater.

Penetration depth (transport) of aggressive ions was determined by qualitative reactions due to spraying of aqueous solution of corresponding salts on surface of AASC concrete specimens right away after their breaking in half. Penetration depth of chlorine-ions was determined by white sediment of silver chloride, which was formed as a result of spraying of $AgNO_3$ solution (concentration 0.1 N). Penetration depth of sulfate-ions was determined by white sediment of lead sulfate, which was formed as a result of spraying of $Pb(CH_3COO)_2$ solution (concentration 0.5 N).

Corrosion resistance coefficient (K_{cr}) of AASC concrete was determined as ratio of tensile strength in bending of specimens after storing in seawater to tensile strength in bending of specimens after storing in fresh water. According to DSTU B V.2.7-288:2011 concretes can be classified as unresistant

(K_{cr} < 0.3), low-resistant (0.3 < K_{cr} < 0.5), resistant (0.5 < K_{cr} < 0.8) and high-resistant (K_{cr} > 0.8) ones in corrosion mediums.

The state of steel reinforcement in plasticized AASC concrete was estimated according to following method. The basic steel rebars, length 120 mm and diameter from 4.1 mm to 4.3 mm, were embedded in specimens 40x40x160 mm of AASC concrete. These rebars were degreased by acetone and weighted with accuracy of ±0,001 g before embedding. After hardening of specimens in normal conditions (t=20±2 °C, $R.H.$=95±5 %) the basic rebars were reached from AASC concrete and etched during (25±5) min in 10 % solution of hydrochloric acid with adding of urotropine (1 % by acid mass) to remove rests of cement stone and products of corrosion. The reference rebars, which weren't embedded in concrete, were weighted and etched simultaneously with basic rebars. After etching the basic and reference rebars were cleansed by distilled water and were immersed in saturated solution of sodium nitrate for 5 min. Then rebars were wiped by filter paper, dried up and weighted. Mean mass loss of the basic as well as the reference rebars were calculated as ratio of mean differences in their mass before and after etching to surface area. Mass loss was calculated as a difference between mean loss of the basic and the reference rebars.

3. Results and discussions

Restriction of Cl^- and SO_4^{2-} ions from seawater in plasticized AASC concrete was considered in view of advancing both chemical binding and occlusion while application of modifiers.

3.1. Chemical binding of Cl^- and SO_4^{2-} ions

Effectiveness of CA components in AASC's for chemical binding of Cl^- and SO_4^{2-} ions was investigated.

AASC based on soda-ash

According to XRD the low-calcium hydrosilicates such as CSH(B) (d= 0.307; 0.280; 0.183 nm) and gyrolite $2CaO{\cdot}3SiO_2{\cdot}2H_2O$ (d= 0.33; 0.268; 0.180 nm) were identified in hydration products of the reference AASC mixed with seawater after 270 d of hydration (Figure 1, curve 1). Calcite (d= 0.307; 0.191; 0.160; 0.152 nm) can be fixed as well. Specified phases are typical for AASC [45].

Clorine- and sulfate-binding zeolite-like minerals, which are similar to nosean $Na_8(Al_6Si_6O_{24})(SO_4){\cdot}H_2O$, sodalite $Na_4(Si_3Al_3)O_{12}Cl$ and cancrinite $(Na,Ca)_8(Al_6Si_6O_{24})(CO_3,SO_4)_2{\cdot}2H_2O$ etc., formed due to occlusion of aggressive ions by zeolite-like alkaline aluminosilicates, can be also predicted [7, 25]. However, these hydrates were not identified because of their submicrocrystalline state in specified period.

DTA confirms formation of calcium hydrosilicates CSH(B) by endothermic effect at t= 175 (dehydration) and exothermic effect at t= 865 °C (recrystallization into wollastonite) (Figure 2, curve 1). The endothermic effects at t= 175 °C and 700 °C (stepped dehydration) and exothermic effect at t= 865 °C (recrystallization into wollastonite) are typical for gyrolite $2CaO{\cdot}3SiO_2{\cdot}2H_2O$. The endothermic effect at t= 890 °C confirms presence of $CaCO_3$.

Chloride-sulfate AFm phases, known as Kuzel's salt $3CaO{\cdot}Al_2O_3{\cdot}0{,}5CaCl_2{\cdot}0{,}5SO_4{\cdot}10H_2O$ (d= 0.83; 0.42; 0.23 nm) [30], were also identified in hydration products of AASC, modified by admixtures of portland cement and calcium aluminate cement, mixed with seawater (Figure 1, curve 2). Specified phases were formed due to chemical binding of Cl^- and SO_4^{2-} ions by calcium hydroaluminate $3CaO{\cdot}Al_2O_3{\cdot}10H_2O$, formed by co-acting of portland cement and calcium aluminate cement in hydration process.

The presence of Kuzel's salt in hydration products of modified AASC, mixed with seawater, was confirmed by endothermic effects at t= 330 °C (dehydration) and 480 °C (departure of chloride) as well as by exothermic effect at t= 1000 °C (decomposition of sulfate) (Figure 2, curve 2). Relocation of effects to higher temperatures is evidence of CSH(B) and gyrolite with advanced crystallization.

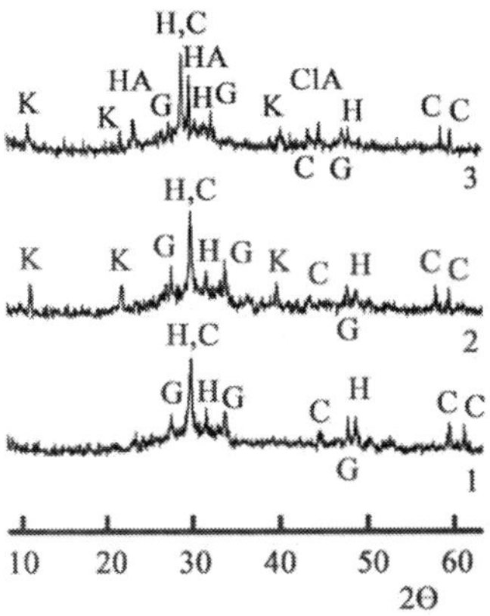

Figure 1. XRD of alkali-activated slag cement, based on soda-ash and mixed with seawater, after 270 d of hydration: 1 – the reference; 2 – co-modified by portland cement and calcium aluminate cement; 3 – co-modified by portland cement, calcium aluminate cement and TSP. Legend: H – calcium hydrosilicates CSH(B), G – gyrolite, C – calcite, K – Kuzel`s salt, ClA – chlorapatite.

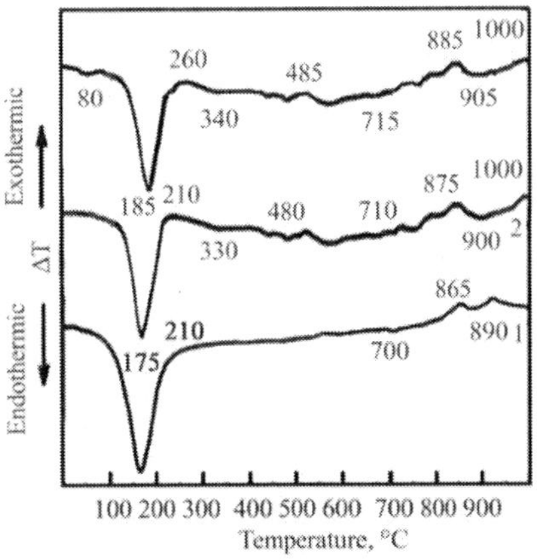

Figure 2. DTA of alkali-activated slag cement, based on soda-ash and mixed with seawater, after 270 d of hydration: 1 – the reference; 2 – co-modified by portland cement and calcium aluminate cement; 3 – co-modified by portland cement, calcium aluminate cement and TSP.

Chlorapatite $Ca_5(PO_4)_3Cl$ (d= 0.339; 0.285; 0.196 nm) was identified in hydration products of AASC mixed with seawater besides of Kuzel's salt while additional application of TSP (Figure 1, curve 3). Specified mineral was formed due to chemical binding of Cl^- ions by hydroxyapatite $Ca_{10}(PO_4)_6(OH)_2$,.

The presence of chlorapatite in hydration products of AASC co-modified by portland cement, high aluminate cement and TSP, while mixing with seawater, was confirmed by endothermic effects at t= 80 °C (dehydration) as well as by exothermic effect at t= 360 °C (Figure 2, curve 3). Advanced crystallization of specified phases was confirmed by relocation of effects to higher temperatures.

According to electron microscopy, co-modification of AASC by mentioned additives ensures formation of hexagonal thin plates of Kuzel's salt $3CaO{\cdot}Al_2O_3{\cdot}0,5CaCl_2{\cdot}0,5SO_4{\cdot}10H_2O$ (content in probe, %: CaO – 32,72, Al_2O_3 – 21,51, Cl – 10,27, SO_3 – 9,56) and long prismatic formations of chlorapatite (content in probe, %: CaO – 57.82, Al_2O_3 – 5.34, P_2O_5 – 47.82, Na_2O – 0.64) (Figure 3).

AASC based on sodium metasilicate

XRD testifies to formation of low-calcium hydrosilicates such as CSH(B) and gyrolite in hydration products of the reference AASC mixed with seawater after 270 d of hydration (Figure 4, curve 1). The presence of zeolite-like minerals (similar to nosean, sodalite, concrinite etc. by composition), which can bind Cl^- and SO_4^{2-} ions, may be assumed. The presence of mentioned hydrates was confirmed by DTA (Figure 5, curve 1).

According to XRD (Figure 4, curve 2) and DTA (Figure 5, curve 2) low-calcium hydrosilicates and Kuzel's salt were fixed in hydration products of AASC, co-modified by portland cement and high aluminate cement, while mixing with seawater. Displacement of mentioned effects to higher temperatures ensures formation of CSH(B) and gyrolite with advanced crystallization.

Figure 3. SEM images of alkali-activated slag cement, based on soda-ash mixed and with seawater, after 270 d of hydration.

The presence of chlorapatite in hydration products of AASC, co-modified by TSP and other mentioned additives (portland cement, high aluminate cement), while mixing with seawater, after 270 d of hydration was confirmed by XRD (Figure 4, curve 3) and DTA (Figure 5, curve 3). Microprobe analysis confirmed formation of hexagonal thin plates of Kuzel's salt (content in probe, %: CaO – 32,72, Al_2O_3 – 21,51, Cl – 10,27, SO_3 – 9,56) and long prismatic formations of chlorapatite (content in probe, %: CaO – 57.82, Al_2O_3 – 5.34, P_2O_5 – 47.82, Na_2O – 0.64) (Figure 6).

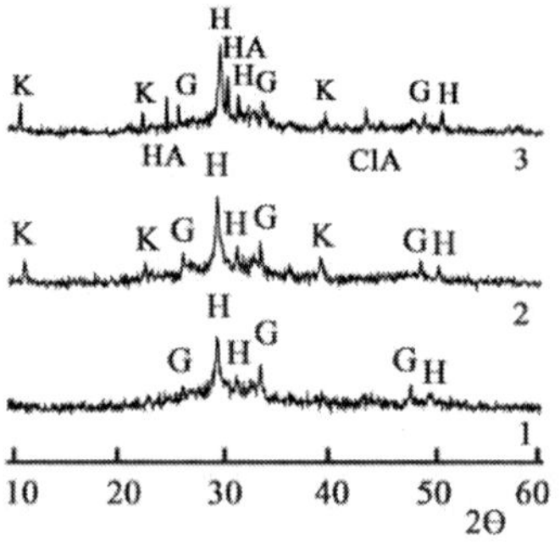

Figure 4. XRD of alkali-activated slag cement, based on sodium metasilicate and mixed with seawater, after 270 d of hydration: 1 – the reference; 2 – co-modified by portland cement and calcium aluminate cement; 3 – co-modified by portland cement, calcium aluminate cement and TSP. Legend: H – calcium hydrosilicates CSH(B), G – gyrolite, K – Kuzel`s salt, ClA – chlorapatite.

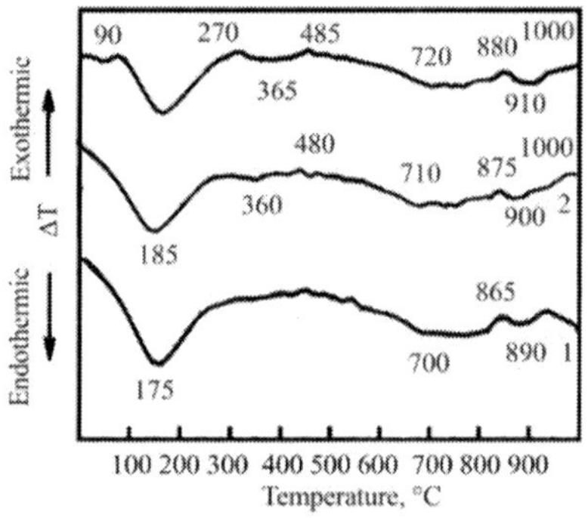

Figure 5. DTA of alkali-activated slag cement, based on sodium metasilicate and mixed with seawater, after 270 d of hydration: 1 – the reference; 2 – co-modified by portland cement and calcium aluminate cement; 3 – co-modified by portland cement, calcium aluminate cement and TSP.

Thus, co-modification of AASC by poratland cement, high aluminate cement and TSP ensures binding of Cl^- and SO_4^{2-} ions by calcium hydroaluminate $3CaO{\cdot}Al_2O_3{\cdot}10H_2O$ and hydroxyapatite $Ca_{10}(PO_4)_6(OH)_2$ in Kuzel's salt and chlorapatite agreeably. Specified structure formation of modified AASC ensures restriction of aggressive ions transport due to chemical binding in contact zone of plasticized AASC concrete with seawater.

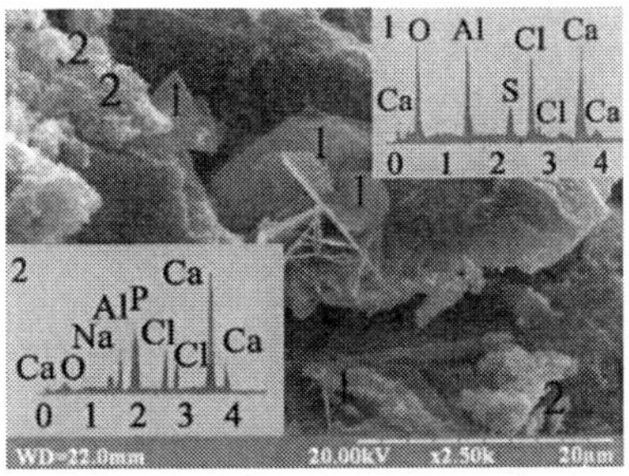

Figure 6. SEM images of alkali-activated slag cement, based on sodium metasilicate and mixed with seawater, after 270 d of hydration.

3.2 Occlusion of Cl^- and SO_4^{2-} ions

Occlusive additive of zeolite was applicated for additional restriction of aggressive ions transport in plasticized AASC concrete and enhancement of its protective properties to steel reinforcement.

Effectiveness of CA including portland cement, calcium aluminate cement, TSP and clinoptilolite, was tested while consistency class S4 (slump 160 - 210 mm) for AASC fresh concrete was provided. While mixing with seawater W/C ratios were 0.46 and 0.41 for cases of soda-ash and sodium metasilicate as alkaline components. Compressive strength of AASC concrete based on soda-ash and metasilicate corresponds to values of 40.3 and 48.7 MPa after 28 d of hardening and of 38.2 and 47.5 MPa after 270 d storing in seawater agreeably. Thus, corrosion resistance coefficient of plasticized AASC concrete based on soda-ash and sodium metasilicate corresponds to values of 0.95 and 0.97 agreeably. Dense macrostructure of AASC concrete prevents penetration of Cl^- and SO_4^{2-} ions even after 270 d in seawater.

The absence of Cl^- and SO_4^{2-} ions transport in plasticized AASC concrete was confirmed by qualitative reactions by solutions of silver nitrate ($AgNO_3$) and lead acetate ($Pb(CH_3COO)_2$) agreeably. The absence of white sediment on the surfaces of specimens confirmed this fact (Figure 7).

a) b)

Figure 7. Sections of half specimens of plasticized AASC concrete, based on sodium metasilicate and stored 270 d in seawater, after spraying of silver nitrate (a) and lead acetate (b).

3.3 The state of steel reinforcement

The effect of CA on protective properties of plasticized AASC concrete was evaluated by mass loss tests of steel rebars. Mass losses of steel rebars, reached from plasticized AASC concrete (the reference as well as modified by CA) after 270 d of storing in seawater, were fixed (Table 1).

Generally, mass loss of steel rebars was in compliance with mandatory requirements (no more than 10 g/m^2) according to DSTU B V.2.6-181 that ensures protective properties of plasticized AASC concrete to steel reinforcement. In case of AASC concrete, based on soda ash, application of CA caused 3.7 times lesser mass loss than for the reference one. For AASC concrete, based on sodium metasilicate, application of CA resulted in 3.4 times smaller mass loss. The obtained results correlate with foregoing ones, concerning enhanced activity of AASC based on sodium metasilicate, if compare with soda-ash, that was caused by enhanced density of AASC concrete structure with advanced protective properties to steel reinforcement.

Table 1. The state of steel rebars embedded in plasticized AASC concrete.

Type of AASC concrete	Loss mass, g/m^2
reference AASC concrete based on soda-ash	4.95
reference AASC concrete based on sodium metasilicate	4.32
AASC concrete based on soda-ash and modified by the complex of additives	1.34
AASC concrete based on sodium metasilicate and modified by the complex of additives	1.27

4. Conclusions

1. The enhancement of steel reinforcement protection in AASC concrete, which is obtained from high consistency mix and exploited in seawater, is possible due to restriction of Cl^- and SO_4^{2} ions transport in structure while application of complex of additives including portland cement, calcium aluminate cement, trisodium phosphate and zeolite.
2. Restriction of aggressive ions transport from seawater into plasticized AASC concrete can be provided due to their chemical adsorption by gel-like phases, chemical binding in Kuzel's salt $3CaO{\cdot}Al_2O_3{\cdot}0{,}5CaCl_2{\cdot}0{,}5SO_4{\cdot}10H_2O$ and chlorapatite $Ca_5(PO_4)_3Cl$ as well as their occluding in structure of zeolite-containing admixture and hydrates presented by alkaline hydrosilicates. Advanced crystallization with densification of artificial stone restricts mentioned ions transport from seawater.
3. The 3.4…3.7 times lesser mass loss of steel rebars embedded in AASC concretes in comparison with non-modified ones, depending on anion of alkaline component, confirms advanced protective properties of plasticized AASC concrete, modified by proposed complex of additives after influence of seawater during 9 months.

References

[1] Ali N, Jeong G J 2019 Recent progress in green cement technology utilizing low-carbon emission fuels and raw materials: a review *Sustainability* **11** 537 doi: 10.3390/su11020537
[2] Shen W, Liu Y, Yan B, Wang J, He P, Zhou C, Huo X, Zhang W, Hu G and Ding Q 2017 Cement industry of China: Driving force, environment impact and sustainable development. Renewable and Sustainable Energy Reviews **75** 618–628. doi: 10.1016/j.rser.2016.11.033
[3] Kropyvnytska T, Rucinska T, Ivashchyshyn H and Kotiv R 2020 Development of eco-efficient composite cements with high early strength Lecture Notes in Civil Engineering **47** pp 211-218. doi: 10.1007/978-3-030-27011-7_27
[4] Ivashchyshyn H, Sanytsky M, Kropyvnytska T and Rusyn B 2019 Study of low-emission multicomponent cements with a high content of supplementary cementitious materials *Eastern-European Journal of Enterprise Technologies* **4/6 (100)** pp 39-47 doi: 10.15587/1729-4061.2019.175472
[5] Borziak O S, Plugin A A, Chepurna S M, Zavalniy O V and Dudin O A 2019 IOP Conference Series: Materials Science and Engineering **708(1)** 012080 doi: 10.1088/1757-899X/708/1/012080
[6] Abyzov V A, Pushkarova K K, Kochevykh M O, Honchar O A and Bazeliuk N L 2020 Innovative building materials in creation an architectural environment *IOP Conf. Series: Materials Science and Engineering* **907** 012035 doi: 10.1088/1757-899X/907/1/012035
[7] Krivenko P 2017 Why Alkaline Activation – 60 Years of the Theory and Practice of Alkali-Activated Materials *Journal of Ceramic Science and Technology* **8** pp 323-334 doi: 10.4416/JCST2017-00042
[8] Provis J 2014 Geopolymers and other alkali activated materials: why, how, and what? *Mater Struct* **47** pp 11-25. doi: 10.1617/s11527-013-0211-5
[9] Awoyera P and Adesina A 2019 A critical review on application of alkali activated slag as a sustainable composite binder *Case Studies in Construction Materials* **11** e00268

doi: 10.1016/j.cscm.2019.e00268
[10] Kochetov G, Prikhna T, Kovalchuk O and Samchenko D 2018 Research of the treatment of depleted nickel-plating electrolytes by the ferritization method Eastern-European Journal of Enterprise Technologies 3(6-93) pp 52–60 doi: doi.org/10.15587/1729-4061.2018.133797
[11] Pavel K, Oleg P, Hryhorii V and Serhi L 2017 The Development of Alkali-activated Cement Mixtures for Fast Rehabilitation and Strengthening of Concrete Structures *Procedia Engineering* **195** pp 142-146 doi:1016/j.proeng.2017.04.536
[12] Krivenko P V, Kovalchuk G Y and Kovalchuk O Y 2005 Heat-resistant cellular concretes based on alkaline cements *Proceedings of the International Conference on the Use of Foamed Concrete in Construction*
[13] Kovalchuk O, Grabovchuk V and Govdun Ya 2018 Alkali activated cements mix design for concretes application in high corrosive conditions *MATEC Web of conferences* **230** 03007 doi: 10.1051/matecconf/201823003007
[14] Shi C Corrosion resistance of alkali-activated slag cement 2003 *Advances in Cement Research* **15(2)** pp 77-81 doi: 10.1680/adcr.2003.15.2.7
[15] Cyr M and Pouhet R 2015 The frost resistance of alkali-activated cement-based binders *Handbook of Alkali-Activated Cements, Mortars and Concretes* pp 293-318 doi: 10.1533/9781782422884.3.293
[16] Krivenko P, Petropavlovskyi O, Kovalchuk O, Cao H L and Weng L Q Efficiency of the Alkali-Activated Cement Concretes for Sea Construction *Materials Science Forum* **968** pp 3-12 doi: 10.4028/www.scientific.net/MSF.968.3
[17] Mengasini L, Mavroulidou M and Gunn M J Alkali-activated concrete mixes with ground granulated blast furnace slag and paper sludge ash in seawater environments *Sustainable Chemistry and Pharmacy* **20** 100380. doi: 10.1016/j.scp.2021.100380
[18] Ke X, Bernal S A and Provis J L 2017 Chloride binding capacity of synthetic C-(A)-S-H type gels in alkali-activated slag simulated pore solutions *1st International Conference on Construction Materials for Sustainable Future* 1-7
[19] Yuan Q, Shi C, De Schutter G, Audenaert K and Deng D 2009 Chloride Binding of Cement-Based Materials Subjected to External Chloride Environment – A Review *Constr. Build. Mater.* **23** **(1)** pp1-13 doi: 10.1016/j.conbuildmat.2008.02.004
[20] Krivenko P, Petropavlovskyi O, Rudenko I, Konstantynovskyi O and Kovalchuk A 2020 Complex multifunctional additive for anchoring grout based on alkali-activated portland cement *IOP Conference Series: Materials Science and Engineering (MSE)* **907** 012055 doi: 10.1088/1757-899X/907/1/012055
[21] Krivenko P, Gots V, Petropavlovskyi O, Rudenko I, Konstantynovskyi O and Kovalchuk A 2019 Development of solutions concerning regulation of proper deformations in alkali-activated cements, Eastern-European journal of Enterprise Technologies **5 (6-101)** pp 24–32 doi: 10.15587/1729-4061.2019.181150
[22] Krivenko P, Gelevera O, Kovalchuk O, Bumanis G and Korjakins A 2019 Alkali-aggregate reaction in alkali-activated cement concretes *IOP Conf. Ser.: Mater. Sci. Eng.* **660** 012002 doi: 10.1088/1757-899X/660/1/012002
[23] Akimkhan A 2012 Structural and Ion-Exchange Properties of Natural Zeolite *Ion Exchange Technologies*. doi: 10.5772/51682
[24] Raja P, Ismail M, Ghoreishiamiri S, Mirza J, Ismail M, Kakooei S, Rahim A 2016 Reviews on corrosion inhibitors: a short view *Chem. Eng. Commun* **203** pp 1145-1156 doi: 10.1080/00986445.2016.1172485
[25] Asaad M A, Hussein M N, Alkadhimi A M and Saleh A T 2021 Commercial and Green Corrosion Inhibitors for Reinforced Concrete Structures: A Review *International Journal of Current Science Research and Review* **4(02)** pp 89–99 doi: 10.47191/ijcsrr/v4-i2-04
[26] Criado M 2015 The corrosion behaviour of reinforced steel embedded in alkali-activated mortar *Handbook of Alkali-Activated Cements. Mortars and Concretes* pp 333-372 doi: 10.1533/9781782422884.3.333
[27] Mundra S, Bernal S A, Criado M et al. 2017 Steel corrosion in reinforced alkali-activated

materials *RILEM Tech. Lett.* **2**. pp 33-39 doi: 10.21809/rilemtechlett.2017.3
[28] Khan M S H and Kayali O 2018 Chloride binding ability and the onset corrosion threat on alkali-activated GGBFS and binary blend pastes *Eur. J. Environ. Civ. En.* **22** **(8)** pp 1023-1039 doi: 10.1080/19648189.2016.1230522
[29] Wang Z, Zhang L, Tang X, Cui Z, Xue J and Lu M 2017 Investigation of the deterioration of passive films in H_2S-containing solutions *International Journal of Minerals, Metallurgy and Materials* **24** (8) pp 943-953 doi: 10.1007/s12613-017-1482-6
[30] Baquerizo L G, Matschei T, Scrivener K L, Saeidpour M and Wadsö L 2015 Hydration states of AFm cement phases. Cement and Concrete Research **73** pp 143-157 doi: 10.1016/j.cemconres.2015.02.011
[31] Krivenko P, Rudenko I and Konstantynovskyi O 2020 Design of slag cement, activated by Na(K) salts of strong acids, for concrete reinforced with steel fittings. *Eastern-European Journal of Enterprise Technologies* **6** **(6-108)** pp 26-40 doi: 10.15587/1729-4061.2020.217002
[32] Clark B A and Brown P W 2000 The formation of calcium sulfoaluminate hydrate compounds, Part II. Cement and Concrete Research **30** pp 233–240 doi: 10.1016/S0008-8846(99)00234-3
[33] Nosovskyi Yu L 2004 Mortars based on composite binder for high-flowable floors PhD (Eng) thesis
[34] Plugin A and Runova R 2018 Bonding calcium chloride and calcium nitrate into stable hydration portland cement products: Stability conditions of calcium hydrochloraluminates and calcium hydronitroaluminates International Journal of Engineering Research in Africa **36** pp 69-73 doi: 10.4028/www.scientific.net/JERA.36.69
[35] Mesbah A, François M, Caudit-Coumes C et al. 2011 Crystal structure of Kuzel's salt $3CaO{\cdot}Al_2O_3{\cdot}1/2CaSO_4{\cdot}1/2CaCl_2{\cdot}11H_2O$ determined by synchrotron powder diffraction Cement and Concrete Research **41** pp 504-509 doi: 10.1016/j.cemconres.2011.01.015
[36] Valcarce M and Vazquez M 2010 Phosphate ions used as green inhibitor against copper corrosion in tap water *Corros Sci* **52** pp 1413-1420 doi: 10.1016/j.corsci.2009.12.015
[37] Shi J and Sun W 2014 Effects of phosphate on the chloride-induced corrosion behavior of reinforcing steel in mortars *Cem. Concr. Compos* **45** 166 doi: 10.1016/j.cemconcomp.2013.10.002
[38] Bastidas D, Criado M, Fajardo S, La Iglesia A and Bastidas J 2015 Corrosion inhibition mechanism of phosphates for early-age reinforced mortar in the presence of chlorides *Cem. Concr. Compos* **61** pp 1–6 doi: 10.1016/j.cemconcomp.2015.04.009
[39] Zhang J, Lu X, Zhang J, Zhang L, Zhu C, Zhang Y and Wu T 2019 Corrosion-Inhibition Effect of Different Phosphate Compounds for Carbon Steel in Chloride-Contaminated Mortars *Int. J. Electrochem. Sci* **14** 8601-8610 doi: 10.20964/2019.09.29
[40] Kalina L, Bílek V, Novotný R, Mončeková M, Másilko J and Koplík J 2016 Effect of Na_3PO_4 on the Hydration Process of Alkali-Activated Blast Furnace Slag *Materials 2016* **9(5)** 395 doi: 10.3390/ma9050395
[41] Bastidas D M, La Iglesia V M, Criado M, Fajardo S, La Iglesia A and Bastidas J M 2010 A prediction study of hydroxyapatite entrapment ability in concrete *Constr. Build Mater* **24** pp 2646-2649 doi: 10.1016/j.conbuildmat.2010.04.060
[42] Krivenko P, Petropavlovskyi O, Kovalchuk O, Rudenko I and Konstantynovskyi 2020 O Enhancement of alkali-activated slag cement concretes crack resistance for mitigation of steel reinforcement corrosion *E3S Web of Conferences* **166** 06001 doi: 10.1051/e3sconf/202016606001
[43] Krivenko P V, Rudenko I I, Petropavlovskyi O M, Konstantynovskyi O P and Kovalchuk A V 2019 Alkali-activated portland cement with adjustable proper deformations for anchoring application *IOP Conference Series: Materials Science and Engineering* **708(1)** 012090 doi: 10.1088/1757-899X/708/1/012090
[44] Krivenko P, Rudenko I, Petropavlovskyi O, Konstantynovskyi O and Kovalchuk A 2019 Alkali-activated portland cement with adjustable proper deformations for anchoring application *IOP*

Conference Series: Materials Science and Engineering **708(1)** 012090 doi: 10.1088/1757-899X/708/1/012090
[45] Pushkar V I 2010 *Alkaline slag cements and concretes based on them PhD* (Eng) thesis

Acknowledgments
Authors would like to acknowledge the contribution of the COST Action CA15202 «Self-Healing concrete: the path to sustainable construction» of the European Union's framework program HORIZON2020, http://www.cost.eu/COST_Actions/ca/CA15202. The authors also express their gratitude to the Ministry of Education and Science of Ukraine for financial support of this research that is carried out within the budgetary financing of topic with registration No 1020U001010 and implementation period 2020 - 2022.

Study of a dual frequency vibration press for the formation of concrete elements

A Gordienko[1], N Emelianenko[1], L Saienko[1,*], BasheerYounis[2] and O Dobrokhodova[1]

[1]Kharkiv National University of Civil Engineering and Architecture, Sumska Street 40, 61000, Kharkiv, Ukraine
[2]Kharkiv National University of Civil Engineering and Architecture, Sumska Street 40, 61000, Kharkiv, Ukraine

*E-mail: saenko_l@ukr.net

Abstract. The article presents the results of a study of a planetary drive dual frequency vibration press for the formation of concrete elements. Mathematical modeling of the movement of the working bodies (matrix and punch) has been carried out, analytical dependences have been obtained to determine the parameters of a vibration press. The results of the study indicate the ability to control the nature of the movement of working bodies, which expands the technological capabilities. The obtained dependences make it possible to improve the methodology for calculating the parameters of a planetary drive dual frequency vibration press.

1. Introduction

For current construction needs, the manufacture of such concrete materials as paving slabs, bricks, curly paving elements, blocks, etc., is relevant.

Technological sets of equipment for the formation of concrete elements by vibration pressing are widely represented on the world market by such companies as Hess, Masareport, Rosacometta, Zenit, Besser, etc. [1-7].

Sets of equipment for the manufacture of concrete products by vibrocompression occupy relatively small areas. For example, the Spanish line "Kompakta-2000" occupies a span of 18×60 m in the workshop and produces tiles of 600 ... 900 thousand m^2/year (or 5 ... 8 million concrete blocks), which is equivalent to a brick plant with a capacity of 35 ... 50 million bricks per year.

Vibration equipment of technological complexes for the production of building materials has an important place. These machines include vibration presses, the improvement of which is reflected in the quality of the formation of concrete products.

Presses have been developed and used for the manufacture of paving slabs, floor slabs, figured paving elements and other small products. The process of molding products on vibration presses is carried out due to the complex action of vibration and static pressure on the entire volume of the compacted mixture.

For the production of paving slabs, rigid concrete mixtures are used, because otherwise there is adhesion on the surface of the lower and upper voids of the punches.

The rigidity of the concrete mix leads to a decrease in density and frost resistance, if the intensity of the vibration action is not increased, which can be achieved, for example, by introducing dual frequency drives in the design of presses.

The duration of the vibration process affects performance. However, excessively prolonged vibration may, instead of a positive effect, lead to a deficiency through the delamination of the concrete mixture.

The main directions of improvement of equipment for molding concrete products by vibration pressing are:

- development of new designs and calculation methods for the formation of units;
- introduction of regulated modes of vibration and vibration pressing;

Content from this work may be used under the terms of the Creative Commons Attribution 3.0 licence. Any further distribution of this work must maintain attribution to the author(s) and the title of the work, journal citation and DOI.
Published under licence by IOP Publishing Ltd

- the use of computer and laser technologies during the design and manufacture of equipment.

The works [1-7] are devoted to the analysis of achievements in the field of design, manufacture and supply to the market of technological sets of equipment for the manufacture of concrete elements by vibration pressing. In works [7-9] data on the ways of improvement and research of vibration presses are given.

Options for improving the designs of vibration press units, including working bodies and drives, are given in the sources [4, 7, 9].

The articles [8, 9] consider the issues of expanding the technological capabilities of vibration presses (creation of machines with combined and regulated modes of operation, development and research of presses with dual frequency drives).

The analysis of studies and publications shows that to ensure high strength, wear resistance and frost resistance of concrete products made by vibration pressing, it is necessary to use rational molding modes based on the results of analytical and experimental studies. These research results are the basis for the creation of methods for calculating the parameters of vibration press systems, taking into account the properties of the processed materials [10-11].

The aim is to study the dynamics and develop a calculation methodology for a dual frequency vibration press for the formation of concrete elements.

The purpose of the research include the establishment of dependencies for determining the parameters of the movement of the working bodies of a planetary drive dual frequency vibration press based on the creation of a mathematical model of the vibration system of the press, which takes into account the static pressure on the formed concrete mixture and its rheological characteristics.

2. Materials and methods

A vibration press was adopted as an object of research, which includes a matrix, a punch, a dual frequency planetary vibration drive of the matrix (Figure 1) and a hydraulic drive for creating static pressure on the concrete mixture [7].

The design diagram of the two-mass system "vibrating table - medium - punch" is shown in Figure 2.

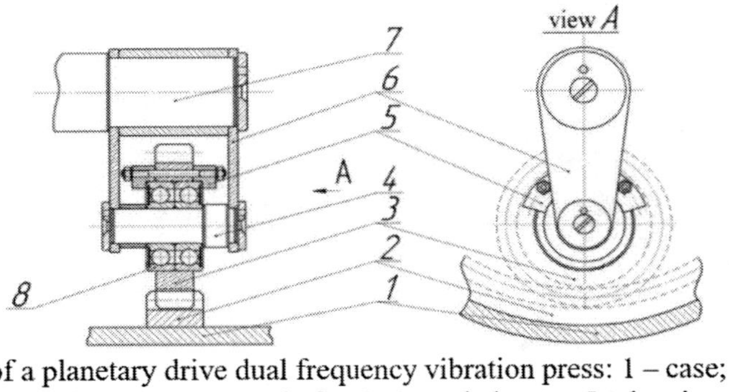

Figure 1. Scheme of a planetary drive dual frequency vibration press: 1 – case; 2, 3 – gear pair of internal gearing; 4, 7 – shafts; 5, 6 – unbalances; 8 – bearing.

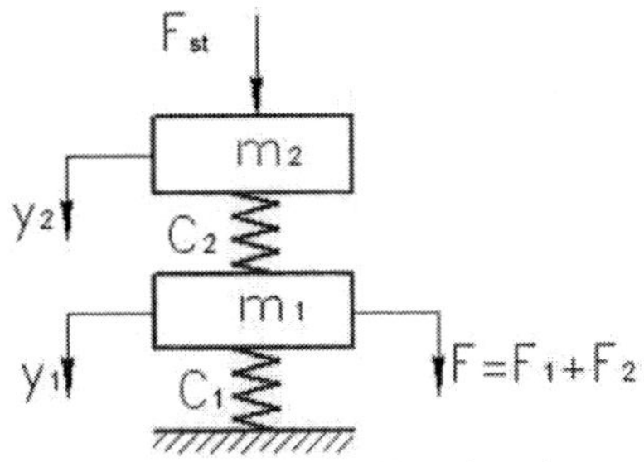

Figure 2. Physical two-mass model of the vibrational system of a vibration press.

IOP Conf. Series: Materials Science and Engineering 1164 (2021) 012067 doi:10.1088/1757-899X/1164/1/012067

Mathematical model of the "vibration press-concrete mix" system.

The differential equations of motion of a two-mass vibration press system, without considering the forces of energy dissipation, are written in the form:

$$\begin{cases} \ddot{y}_1+\omega_{01}^2 y_1+\omega_{012}^2(y_1-y_2)=\dfrac{F_1}{m_1}\sin\omega t+\dfrac{F_2}{m_1}\left\{\cos\left[(u-1)\omega t\right]-\cos\left[(u+1)\omega t\right]\right\} \\ \ddot{y}_2+\omega_{02}^2(y_2-y_1)=0 \end{cases}, \tag{1}$$

where $m_1=m_m+m_c+m_v$; m_m, m_c, m_v – respectively, the mass of the matrix, concrete mix and matrix vibrator; $F_1=m_0\cdot r_0\cdot\omega^2$; $F_2=\dfrac{1}{2}\cdot m_D\cdot r_D\cdot(u\omega)^2$ – exciting forces of the main and additional imbalances of the vibrator; ratio; m_2 – mass of vibrating parts of the block of punches; C_1 – stiffness coefficient of elastic matrix supports; $C_2=\dfrac{(P_a+P_{st})^{1+1/k}\cdot S_k}{P_a^{1/k}h_c\Pi_c}$ – concrete stiffness coefficient; P_a, P_{st} – respectively atmospheric pressure and static pressure on the concrete mix; k – adiabatic exponent; h_c – concrete bed height; S_k – contact area of punches with concrete mix; ω – angular velocity of rotation of the main unbalance; t – time.

$$\omega_{01}^2=\frac{C_1}{m_1};\ \omega_{012}^2=\frac{C_2}{m_1};\ \omega_{02}^2=\frac{C_2}{m_2}.$$

Particular solutions of the system of equations (1) were obtained in the form (2):

$$\begin{aligned} y_1&=a_1\sin\omega t+b_1\cos\left[(u-1)\omega t\right]+d_1\cos\left[(u+1)\omega t\right]; \\ y_2&=a_2\sin\omega t+b_2\cos\left[(u-1)\omega t\right]+d_2\cos\left[(u+1)\omega t\right]. \end{aligned}, \tag{2}$$

The amplitudes of the composite vibrations are found in the form:

$$\begin{aligned} a_1&=\frac{F_1\left(C_2-m_2\omega^2\right)}{\left(C_1+C_2-m_1\omega^2\right)\cdot\left(C_2-m_2\omega^2\right)-C_2^2};\quad b_1=\frac{F_2\left(C_2-m_2\omega^2(u-1)^2\right)}{T}; \\ a_2&=\frac{F_1\cdot C_2}{\left(C_1+C_2-m_1\omega^2\right)\cdot\left(C_2-m_2\omega^2\right)-C_2^2};\quad b_2=\frac{F_2\cdot C_2}{T}; \end{aligned} \tag{3}$$

$$T=(C_1+C_2-m_1\omega^2(u-1)^2)\cdot(C_2-m_2\omega^2(u-1)^2)-C_2^2;\ d_1=\frac{F_2(C_2-m_2\omega^2(u+1)^2)}{K};$$

$$K=(C_1+C_2-m_1\omega^2(u+1)^2)\cdot(C_2-m_2\omega^2(u+1)^2)-C_2^2;\ d_2=\frac{F_2\cdot C_2}{K}.$$

3. Discussion

Relative displacements of the die and the punch $\Delta y=y_1-y_2$, m. for the following values of the vibration system parameters are shown in Figure 3:

m_1 = 300 kg; F_1 = 6800 H; P_a = 0,1 MPa; S_k = 0,15 m ^{2}; ω = 157 sec^{-1};
m_2 = 100 kg; F_2 = 2300 H; P_{cm} = 0,07 MPa; u = 2.

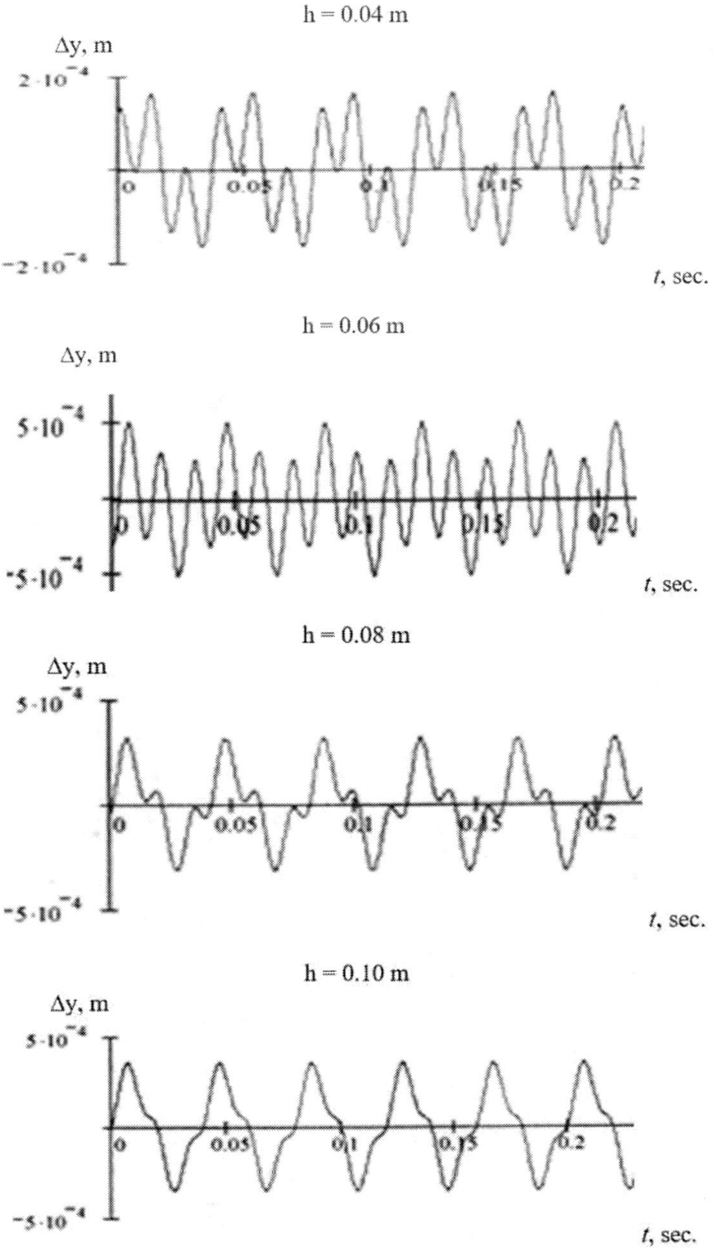

Figure 3. Graphs of the relative displacements of the matrix and punch for different values of the thickness of the layer (h, m) of the concrete mixture.

The dependences obtained on the basis of mathematical modeling for evaluating the parameters of the movement of the working bodies of the vibration press, which allow us to evaluate the effect of the thickness of the column of concrete mixture and the ratio of the frequencies of the pathogens on the nature of the relative vibrations of the matrix and punch.

The analysis of the graphs (Figure 3) shows that due to the dual frequency drive, it is possible to change the nature and range of the relative oscillations of the working bodies, expands the technological capabilities of the vibration press for the manufacture of concrete products. Based on the results of the factorial experiment carried out on the laboratory bench, a regression equation was obtained, which describes the dependence of the strength of concrete samples in compression on the following factors: the second (high) vibration frequency, static pressure and molding time.

Revealed rational ranges of parameters of the process of vibration pressing of concrete samples (for example, at a fundamental vibration frequency of 50 Hz, it is recommended to assign a high frequency value of 73 ... 75 Hz; static pressure 0.05 ... 0.06 MPa; molding time 7 ... 8 sec.).

Methodology for calculating the parameters of a planetary drive dual frequency vibration press:

1. Calculations of the productivity of a vibration press as a cyclically operating machine, m^2/h:

$$\Pi \frac{3600}{T_c} \cdot S_c ,$$

where T_c – time of one cycle of vibration press operation, sec.; S_c – the number of products that are manufactured in one cycle, m^2.

2. Frequency range selection. For molding products from fine concrete mixtures with a height of 40 ... 200 mm, a frequency range of 100 ... 50 Hz is recommended, respectively.

3. The choice of static pressure and its strength. Rational value of static pressure on concrete mix:

$$P_{st} = 0.05...0.1 \text{ MPa}.$$

Required static load force: $F_{st} = P_{st} \cdot S_k$.

4. Hydraulic pressing force of punches, H:

$$F_{hyd} = P_{st} \cdot S_k - (m_{pun} + m_{trav}) \cdot g .$$

5. The required limiting oscillation amplitude of the elements, or half-span, for dual frequency oscillations according to models (1) – (3):

$$A_{bound} \leq h_c \cdot \Pi_c \cdot \left(1 - \frac{P_a}{P_a + P_{cm}}\right).$$

6. The required total vibration force (taken from the ratio):

$$F_{vibr} \approx \frac{F_{st}}{(0.5...0.55)} .$$

7. Required maximum total static moment of mass of unbalanced vibrators:

$$S_{gr \sum \square} = m_1 - A_{gr} .$$

The vibrators are selected according to the found static moment of the unbalance mass.

8. Required force on the hydraulic cylinder rod:

$$F_{hyd} = F_{hyd} - m_2 \cdot g .$$

9. Required piston diameter at hydraulic line pressure p_{hyd}:

$$d \geq \sqrt{\frac{4 \cdot F_{hyd}}{\pi \cdot p_{hyd}}} .$$

10. Estimated power consumed by the concrete mix from the vibroblock:

$$N_1 = 12 \cdot S_k \cdot \omega_2 \cdot A_{prel} .$$

11. Dimensions of rubber elastic supports for the matrix. Total stiffness of n supports:

$$C_{M\sum\square} = m_1 \cdot \omega_1^2 \cdot \left(\frac{1}{30} \ldots \frac{1}{60}\right).$$

Stiffness coefficient of one elastic support:

$$C_{M1} = \frac{C_{M\sum\square}}{n_M}.$$

For cylindrical rubber supports with a radius R and a height h, the dimensions are related by a ratio that takes into account the properties of the material:

$$h = \frac{3 \cdot \pi \cdot G_g \cdot R^2}{C_{M1}}.$$

For rubber grade 2959 with modulus of elasticity $G_g=1.6 \cdot 10^6$ Pa = 1.6 MPa graphical dependence of the stiffness coefficient of a rubber support on its height was constructed for the given other parameters (Figure 4).

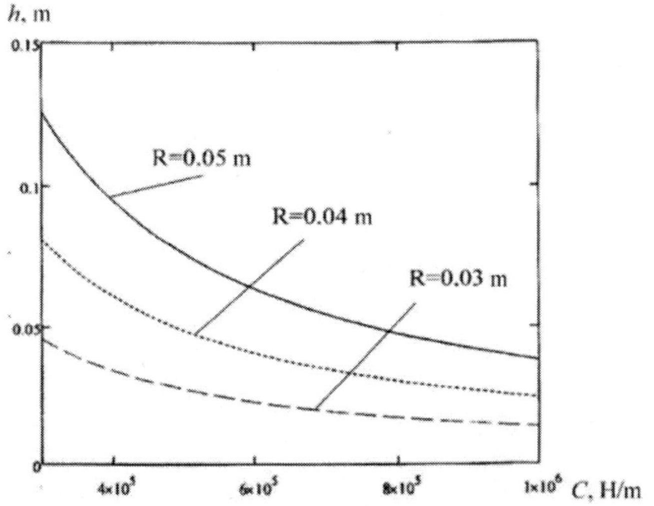

Figure 4. Dependence for determining the height of a cylindrical rubber support from its radius and stiffness coefficient.

The rational dimensions of the elastic elements are selected taking into account the shear modulus of a particular brand of rubber.

4. Conclusions

1. It was proposed, as a press drive, to use a planetary vibration drive, which makes it possible to increase the vibration intensity due to the generation of dual frequency vibrations and expands the technological capabilities.

2. The dependences were obtained on the basis of mathematical modeling, allowing to evaluate the effect of the thickness of the concrete mixture column and the ratio of the frequencies of vibration exciters on the nature of the relative vibrations of the matrix and punch of the vibration press.

3. A refined method for calculating the parameters of a vibration press with a dual frequency drive has been compiled.

References

[1] Gusev B V and Zazimko V G 191 *Vibration concrete technology* (Kiev: Builder) 160
[2] Utkin V L 2004 *New technologies of the construction industry* (Moscow: JSC Russian Publishing House) 116

[3] Nazarenko I I, Nesterenko M M, Orysenko O V and Zygun A Yu 2018 *International Journal of Engineering & Technology* **7** (4.8) pp 284-288 https://doi.org/10.14419/ijet.v7i4.8.27257
[4] Zagreba V P and Dudar I N 2009 *Formation of concrete and reinforced concrete products by the method of pulsating pressing of concrete mixtures: monograph* (Vinnytsia) 104 http://ir.lib.vntu.edu.ua/handle/123456789/9490
[5] Dvorkin L I and Zhitkovsky V V 2007 *Building materials and products* **4** pp 11-14
[6] Dvorkin L I, Zhitkovsky V V and Kaganov V O 2006 *Concrete based on super hard mixtures* (Rivne: Publishing house of RDTSNTEI) 179
[7] Lavendel E E 1981 *Vibrations in Engineering: Handbook* (Moscow: Mechanical engineering) 509
[8] Iskovich-Lototsky R D 2006 *Fundamentals of the theory of calculation and development of processes and equipment for vibro-impact pressing: monograph* (Vinnytsia) 338
[9] Nazarenko I I 2007 *Vibration machines and processes of the construction industry* (Kiev: KNUBA) 207
[10] Emelianenko N. G and Saienko L V 2016 *Scientific Bulletin of Construction* **3** (85) pp 265-267 http://nbuv.gov.ua/UJRN/Nvb_2016_3_60
[11] Emelianenko N. G and Saienko L V 2008 Bulletin of the National Technical University "Kharkiv Polytechnic Institute" **13** pp 38-41
[12] Chelomey V N 1981 *Vibrations in Engineering: A Handbook* (Moscow: Mechanical engineering 509
[13] Serdyuk L I 2003 *International periodical collection of scientific papers: Processing of dispersed materials and media* (Odessa: NPO VOTUM) **13** pp 33-39
[14] Emelianenko N G and Saienko L V 2017 *Scientific Bulletin of Construction* **4** (90) pp 285-289 http://nbuv.gov.ua/UJRN/Nvb_2017_90_4_54

Biomechanics of self-loosening of the screw that connects the intraosseous part of the implant and the abutment with the orthopedic structure attached to it

Ye Semenov[1] and M Surianinov[2,*]

[1]State Establishment «The Institute of Stomatology and Maxillo-Facial Surgery National, Academy of Medical Science of Ukraine», Rishelevskaya str., 11, Odessa, 65026, Ukraine
[2]Odesa State Academy of Civil Engineering and Architecture, 4, Didrihsona str., Odesa 65029, Ukraine

[*] E-mail: sng@ogasa.org.ua

Abstract. The results of the study of the stress-strain state in the elements of the system "intraosseous part of the implant - screw - abutment with an orthopedic structure fixed on it" under different given conditions of its functioning are presented. We have determined the stress-strain state in the crown neck area, which replaces the upper jaw central incisor, in the presence of the natural wear (0.5 mm) of the teeth and anthropoid teeth next to the implants and the same nine load variants as in the initial non-prosthetic model, i.e. a total of 54 calculation experiments have been carried out. It has been established that the occurrence of such secondary mechanical complication of dental implantation as loosening of the screw connection between the intraosseous part of the implant and the prosthetic structure fixed on the abutment influences the type of teeth clamping (bite), the angle of the preferential transfer of chewing load, the propensity and the wear ability of the teeth. The most unfavourable, in terms of the value of the torque that unwinds the fixation screw, is the direct occlusion of the teeth and the angle of the preferential load transfer of 1450 (M_{pr} = 1.16 Nm), or 3.87% of the twisting torque value. At the same conclusions of the set parameters and the value of erosion of its teeth 0.5 mm the value of the torque, which untwists, increases by 5 times and corresponds to 19.3% of the initial torque (30 Nm) and will inevitably lead to its untwisting.

1. Introduction

One of the serious problems encountered by orthopedic surgeons during the prosthetics of patients with two-stage cylindrical implants supporting fixed prosthetic structures is self-loosening of the screw that connects the intraosseous part of the implant and the abutment with the prosthetic structure fixed to it. Clinical studies have shown that this complication is diagnosed in 13% of patients. The average service life of an orthopedic structure before the onset of the above complication was 8.1±2.4 years (at the time of delivery, the connecting screw was tightened with a force of 30 Nm, which is recommended by most implant manufacturers).

Based on the examination of 53 patients with this type of complication, we found that hyperocclusion foci were detected in 100% of cases in the area of problematic prostheses, which was a result of changes in the maxillofacial apparatus some time after their placement. This was supported by the fact that there were no hyperocclusion foci in the area of the problematic constructs at the time of delivery of the orthopedic device. In 79.2 % of the patients with the studied types of complications, they were single crowns or bridges within the same functionally oriented group [13, 14].

Thus, on the basis of the above, we can conclude that studying the biomechanics of functioning of the system intraosseous part of the implant - screw, connecting it to the superstructure, in changing conditions of its functioning on the basis of mathematical modeling is a topical task of modern dentistry. At the same time, it should be noted that in terms of mathematical analysis it is a complex applied problem, since the constituent materials of the model (implant, connecting screw, cancellous

Content from this work may be used under the terms of the Creative Commons Attribution 3.0 licence. Any further distribution of this work must maintain attribution to the author(s) and the title of the work, journal citation and DOI.
Published under licence by IOP Publishing Ltd

bone, cortical bone, dentin) have different physical and mechanical properties, and the model functioning conditions are characterized by significant variability.

2. Recent researches analysis

Recently, the use of two-stage cylindrical screw implants as a support for fixed orthopedic structures has become increasingly widespread [1, 2]. However, the number of prosthetic complications during the long-term functioning of fixed orthopedic constructions based on this type of implants is also increasing [3-5]. One of the most serious complications is the loosening of the screw that connects the intraosseous part of the implant and abutment with the fixed prosthetic structure fixed to it [8]. According to our data, the frequency of this complication is 6.3% of the total number of implants placed [3]. It is associated with changes in the dentoalveolar system that occur over time, and lead to tooth erasure or formation of traumatic occlusion nodules with changes in periodontal tissues, and, as a consequence, an increase in the load on the screw that connects the intraosseous part of the implant and abutment with fixed prosthetic structure fixed to it [6, 7].

3. The purpose of this work is to study the stress-strain state in each element of the system "intraosseous part of the implant - screw - abutment with an orthopedic structure fixed on it" under different given conditions of its functioning.

4. Materials and methods

Modern biomechanical studies are based on detailed computer modeling of objects, which allows performing comprehensive computational experiments in high-performance software packages. The vast majority of these complexes are based on the use of a numerical calculation method - the finite element method (FEM) [9, 10]. This approach makes it possible to build models of any geometric complexity with high accuracy and to determine the stress-strain state of an object at all its points, which is ideal for studying the biomechanics of both the entire dentoalveolar system and its individual parts. In our case it is a fragment of the frontal area of the upper jaw with a two-stage cylindrical screw implant embedded in it, with a fixed orthopedic structure fixed on it, replacing the central incisor. All constructions and calculations were performed using the ANSYS software package [11].

5. Research results

In the process of work to achieve the goal, the following tasks were solved:

- a three-dimensional finite-element model of an upper jaw segment with four frontal teeth, including dentin volume, cortical and cancellous bone with subsequent replacement of the central incisor with an artificial crown, supported by an intraosseous two-stage cylindrical implant (Figure 1) was built;
- the stress-strain state in the upper jaw central incisor neck area at different tooth closing angles (1250, 1350, 1450), force application point corresponding to direct, normognathic, deep teeth overlapping, the given value of 0.5 mm tooth wearability (here the value of the central incisor wearability, simulating an artificial crown supported by a dental implant, is taken as equal to zero);
- the character of tangential stress τxy (one of the components of the total stress in the area of the tooth neck under given conditions), creating a torque that leads to the loosening of the screw that connects the intraosseous part of the implant and the superstructure, was studied.

The method of constructing a finite-element model, which was subjected to a multidimensional computer analysis, was described by us earlier [7, 8].

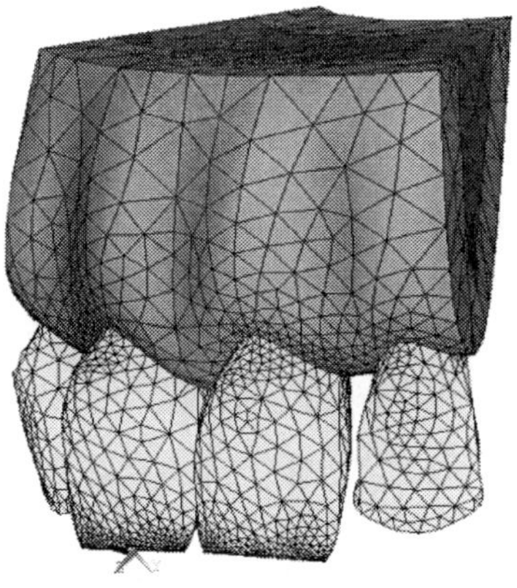

Figure 1. Fragment of a model of the dentoalveolar system.

During the study of the built model the stress-strain state in the upper jaw central incisor neck area was studied, as the value of these stresses affects the value of the torque of the screw that connects the implant to the abutment. It should be noted that at the stage of preliminary studies it was established that the value of tension in the area of the tooth neck of the central incisor of the upper jaw and the value of tension in the area of the neck of the artificial crown fixed on the abutment of the two-stage cylindrical implant replacing the central incisor of the upper jaw are comparable.

We considered three variants of teeth occlusion, which corresponded to normognathic, deep and straight bite at different angles of load transfer (125o, 135o, 145o). Also in our previous studies we have studied the stress-strain state in the maxillary central incisor tooth neck at angles of load transfer (125o, 135o, 145o) with a given value of natural tooth and antagonist teeth erosion, equal to 0.5 mm; and the value of erasability of the metal-ceramic crown, fixed on the central incisor of the upper jaw, we neglected, since it is subject to much less erosion than natural teeth [7].

The model in question was exported to SolidWorks [12] for ease of simulation of erosion. As an example, Figure 2 shows the stress fields at direct occlusion at angles 1350 and 1450 and the erasability of adjacent teeth and antagonist teeth of 0.5 mm.

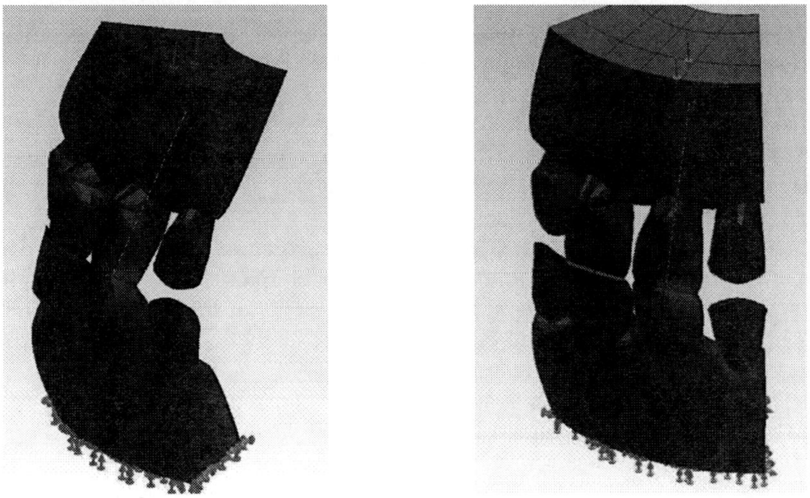

Figure 2. Stress fields at 135o and 145o straight occlusion and 0.5 mm abrasion of adjacent and antagonist teeth.

The results obtained served as the basis for the next stage of the computational experiment. The frontal incisor of the maxilla was removed from the model and replaced with a two-stage screw implant of the LIKo system with an artificial crown fixed to it. From the possible modifications of the geometric sizes of the implants (Table 1), six were chosen for the study: 1.101, 1.102, 1.201, 1.202, 1.301, 1.302.

Table 1. Geometric parameters of the implants.

№	Length, mm	Diameter, mm	Modifications
1.101	8		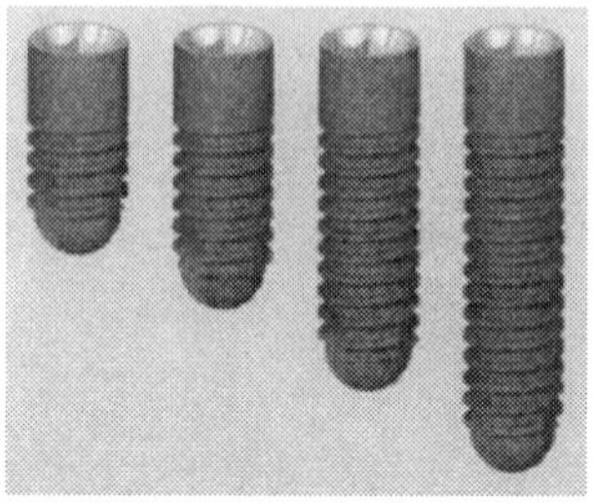
1.102	10	3.5	
1.103	13		
1.104	16		
1.201	8		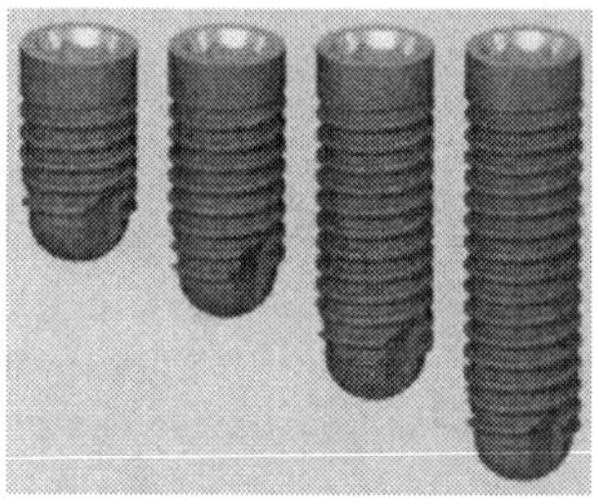
1.202	10	4.0	
1.203	13		
1.204	16		
1.301	8		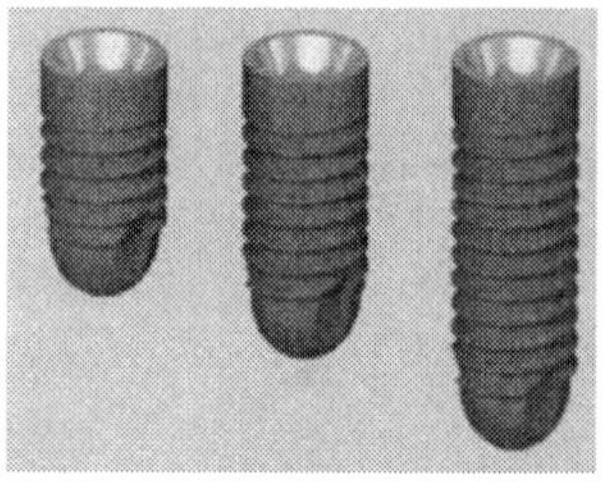
1.302	10	5,0	
1.303	13		

Thus, the experiments varied the implant diameter (three sizes) with two lengths of each implant.

6. Research results and discussion

In the course of the work we determined the stress-strain state in the neck area of the crown replacing the central incisor of the upper jaw in the presence of natural abrasion (0.5 mm) of the adjacent teeth and antagonist teeth for each of the six implant sizes under study and the same nine load variants as for the initial non-prosthetic model, that is, a total of 54 calculation experiments were performed.

Table 2 shows the results of calculations of the model fragment with implant No. 1.202. As it should be expected, the stresses and strains in the components of the maxillary system, excluding, of course, the implantation zones, in all 54 experiments practically do not differ from each other, i.e. do not depend on the geometrical parameters of the implants.

As for the stresses in the implants themselves, of course, they depend significantly on their geometry, but in all cases they are much lower than the allowable stresses, i.e. there is a significant safety margin.

Table 2. Calculation results of a model fragment with an implant № 1.202

Clamping type	Angle, deg.	Stresses, MPa				Deformation
		σ	τ_{xy}	τ_{xz}	τ_{yz}	
Direct	125	102.88	12.9	2.9	38.6	$9.247 \cdot 10^{-3}$
	135	206.6	7.5	10.5	37.0	$2.098 \cdot 10^{-2}$
	145	320.0	36.9	76.7	68.0	$5.276 \cdot 10^{-2}$
Normognathic	125	160.3	20.1	4.5	59.8	$1.015 \cdot 10^{-1}$
	135	154.0	5.6	7.9	26.8	$8.383 \cdot 10^{-2}$
	145	140.2	15.8	32.9	28.9	$6.365 \cdot 10^{-2}$
Deep overlap	125	122.38	15.3	3.45	41.92	$6.27 \cdot 10^{-2}$
	135	119.45	4.34	6.1	21.4	$4.872 \cdot 10^{-2}$
	145	117.12	13.5	28.1	24.9	$3.351 \cdot 10^{-2}$

Thus, for implant No. 1.101 (minimal geometric dimensions) the highest stresses in the series of experiments were $\sigma_{\max} = 202.4MPa$, which is 1.6 times lower than the allowable stresses.

Calculations have also shown that the presence of natural tooth erosion leads not only to an increase in normal stresses, which are mainly responsible for the strength of the structure, but also to an increase in tangential stresses in all three coordinate planes. Of greatest interest are the tangential stresses τ_{xy} in the plane perpendicular to the fixation screw, which generate the torque that ultimately leads to its unscrewing (Table 2). The tangential stresses τ_{xz} and τ_{yz}, occur in the other two coordinate planes and therefore have nothing to do with unwinding. For the maximum tangential stresses $\tau_{xy}^{\max} = 36.9MPa$ given in Table 2, the corresponding torque (diameter of the fixing screw $d_0 = 2.0mm$) was $M_{кр} = 5.79N \cdot sm$, which is about 20 % of the initial torque and inevitably leads to self-loosening of the fixing screw.

Assuming that the torque is directly proportional to the tangential stresses τ_{xy}, it is possible, using Table 2, to estimate the torque for each of the 3 types of clamping and each load transfer angle. Obviously, in these cases the torque will be less, but it will be, and so self-torquing of the screw will inevitably occur.

7. Conclusions

The dentoalveolar system in which an orthopedic structure supported by dental implants is built is not static, the changing conditions of its functioning (wearability of its teeth) contribute to the occurrence of such a secondary mechanical complication of dental implantation as a loosening of the screw connection between the intraosseous part of the implant and the superstructure. Based on these findings, we recommend the fabrication of fixed orthopedic structures with dental implants with pre-planned access to the fixation screw.

According to the results of the conducted numerical experiment with the help of the ANSYS finite-element software and the constructed three-dimensional computer model, it was established that the occurrence of such secondary mechanical complication of dental implantation as loosening of the screw connection between the intrabone implant part and the prosthetic structure fixed on the abutment is affected by the type of teeth occlusion (bite), the angle of preferential transfer of chewing load, propensity and erasibility of their teeth. The most unfavourable, in terms of the value of the torque that unscrews the fixation screw, is the direct occlusion of the teeth and the angle of the preferential load transfer of 1450 (Mpr = 1.16 Nm), or 3.87 % of the twisting torque value. With the same conclusions of the set parameters and the value of abrasion of its teeth 0.5 mm the value of torque, which untwists, increases by 5 times and corresponds to 19.3 % of the initial torque (30 Nm) and will inevitably lead to its untwisting.

References

[1] Matveeva A Y, Havriushyn S S and Borysov A H 2000 *Stomatolohycheskaia ymplantolohyia* Moskva Stomatolohyia XXI veka 63
[2] Zaboltskyi Ya V 2006 *Implantatsiia v neznimnomu protezuvanni* (Lviv: HalDent) 156
[3] Labunets V A, Sennykov O N and Semenov E Y 2009 Analyz oslozhnenyi voznykaiushchykh pry protezyrovanyy nesemnymy konstruktsyiamy zubnykh protezov, fyksyrovannykh na dvukhэtapnykh osteoyntehryrovannыkh vyntovykh ymplantatakh, ykh ustranenye y profylaktyka *Visnyk stomatolohii* **3** pp 82-84
[4] Kallus T and Bessing C 1994 loose gold screw frequently occur in full-arch prostheses supported by osseointegrated implants after 5 years *Int S. Oral maxillofac Implants* pp 169-178
[5] Bert M 1994 *Complications et Echecs en implantologie Causes Traitement Prevention* Paris editions Cd P
[6] Labunets V A, Semenov E Y and Surianynov N H 2011 Yssledovanye napriazhenno-deformyrovannoho sostoianyia v oblasty sheiky zuba tsentralnoho reztsa verkhnei cheliusty pry raznykh varyantakh smykanyia zubnykh riadov *Visnyk stomatolohy* **4** pp 70-73
[7] Semenov E Y, Labunets V A and Surianynov N H 2012 Yssledovanye napriazhenno-deformyrovannoho sostoianyia v oblasty sheiky zuba tsentralnoho reztsa verkhnei cheliusty pry raznykh klynycheskykh sostoianyiakh *Visnyk stomatolohii* **2** pp 100-104
[8] Shyrokov Yu E, Yvanov S Yu, Lomakyn M V and Chernychkyn A S 2007 Yssledovanye statystycheskoi prochnosty vyntovoho soedynenyia vnutrykostnykh y vnekostnykh chastei ymplantatov LYKO *Stomatolohyia* **1** pp 54-59
[9] Mytchell E and Ueit R 1981 *Metod konechnykh elementov dlia uravnenyi s chastnymy proyzvodnymy* (M.: Myr) 216
[10] Hallaher R 1984 *Metod konechnykh elementov. Osnovy* (M.: Myr) 428
[11] Lazarieva D V, Soroka M M and Shyliaiev O S 2020 *Pryiomy roboty z PK ANSYS pry rozviazanni zadach mekhaniky* Pid redaktsiieiu M H Surianinova: monohrafyia (Odesa: ODABA) 432
[12] Aliamovskyi A A 2015 *Ynzhenernye raschety v SolidWorks Simulatio*n Yzd 2-e (Moskva: DMK Press) 464
[13] Semenov E Y and Sennykov O N 2017 Chastota vtorychnykh oslozhnenyi dentalnykh ymplantatsyi, ykh ustranenye y profylaktyka *Dentalnaia ymplantolohyia y khyrurhyia* **4(29)** pp 36-39
[14] Semenov E Y 2017 Rezultaty yzuchenyia stomatolohycheskoho statusa patsyentov s vtorychnymy oslozhnenyiamy dentalnoi ymplantatsyy *Visnyk stomatolohii* **4(101)** pp 47-50

IOP Conf. Series: Materials Science and Engineering 1164 (2021) 012069 doi:10.1088/1757-899X/1164/1/012069

Engaging man-made waste in the closed cycle of mining production

N Gorbatyuk[1,*], A Adigamov[2], S Shakhov[3] and O Kobelev[2]

[1]V.I. Vernadsky Crimean Federal University, Prospekt Vernadskogo, 4, 295007, Simferopol, Russia
[2]National University of Science and Technology «MISiS», Leninsky av., 4, 119991, Moscow, Russia
[3]VMM Holding Company «VNIIMETMASH», Ryazanskiy prospect, 8A, 109428, Moscow, Russia

*E-mail: gorbatyuk54@bk.ru

Abstract. The problems faced by the mining and construction industries are analyzed in the article. A unified concept is formulated to eliminate the problems that arise in the region of mineral extraction, together with solving the problems of expanding the raw material base of the construction sector. The idea of rational use of man-made waste from local mining enterprises and industrial waste for the construction and backfill mixes preparation is presented. The aspects of their application as an aggregate in mixtures are considered on the example of man-made waste (halite enrichment waste). It is noted that the use of man-made waste in construction and backfill mixes must be carried out after deep additional processing. Deep processing allows to maximize the use of useful components of man-made waste and eliminate the negative consequences of harmful ones. It is established that the best way of additional processing is the activation of man-made waste in disintegrators. It is proved that activation increases the strength of structures or cemented mass by 25-30%, reduces the consumption of binder by 40-50%, while maintaining the same strength characteristics. The use of man-made waste in construction and backfill mixtures expands the raw material base for the construction and mining sectors, allows to implement the concept of sustainable development of the region, and reduces the impact on the environment.

1. Referat

In this paper, studies on the use of man-made waste in construction and backfill mixtures are carried out. The utilization of man-made waste reduces the impact of geotechnology on the environment. Sustainable mining development and gentle technologies are the main concept of the research. This concept implies ensuring the safe conduct of work, environmental safety of subsurface development, reducing losses in the process of underground mining, increasing the stability of underground structures, increasing the service life of a mining enterprise, and solving the problems of rational use of natural resources. The results of the research allowed us to propose technical and technological solutions for rational sustainable economic development, taking into account all forms of the impact of geotechnology on the environment, and suggested ways to improve the technology of extracting minerals from the subsurface.

It is necessary to take into account that in the industrial massifs of mining and processing industries, a huge amount of waste has accumulated, which has useful components in its composition. Waste can be not only raw materials for the production of additional products, but also are man-made geo-resources. High-quality preparation of man-made waste for reuse in order to maximize the use of their positive qualities; extraction of useful components; exclusion of subsequent environmental impact is a top priority.

Content from this work may be used under the terms of the Creative Commons Attribution 3.0 licence. Any further distribution of this work must maintain attribution to the author(s) and the title of the work, journal citation and DOI.
Published under licence by IOP Publishing Ltd

The use of man-made waste as an aggregate in construction or backfill material significantly expands the raw material base. The involvement of man-made waste allows them to be disposed of, which minimizes the risks of violating the ecological balance.

2. Introduction

The constant increase in the resources consumed determines the high rate of development of geotechnology and the intensification of mineral extraction, which is associated with certain risks, both industrial and environmental [1].

The problem of storing industrial waste from mining production: rocks from sinking and overburden operations and the processing sector: waste from enrichment and metallurgical processing increases every year [2]. Large areas of agricultural land are being seized for man-made massifs [3]. Industrial waste, with a visible relatively stable state during long-term storage after contact with water or other waste, becomes mobile and is carried out through the filtration systems of the earth's crust outside of industrial massifs, pollutes the environment and enters the human habitat [4]. In addition, small fractions of dump rocks, beaches of tailings dumps and slag heaps are sources of dust formation [5]. Industrial waste, including construction waste, has no less impact on the environment.

If solid man-made waste can be considered to be in a conditionally equilibrium state, then liquid, suspended matter (pulp) and gaseous waste begin to negatively affect the environment immediately after entering it.

The search for local cheap raw materials for the construction industry, which is designed to replace traditional building materials, will allow to meet their acute shortage [6, 7]. In turn, the effective use of man-made waste from mining and industrial production as raw materials for the manufacture of building materials, will create intermediate products in cyclic production and minimize the impact on the environment [8] and ensure the safety of underground workings or ground buildings and structures [9].

Waste from mining, processing and industrial production [10], as well as the secondary product obtained as a result of dismantling works in civil and industrial construction [11], may be suitable for use. The main conditions for their use are accessibility and neutral impact on the environment and human impact [12].

The use of man-made waste of all types of production directly in the extraction of minerals or in the creation of products for the construction industry will eliminate the costs of construction and operation of man-made storage facilities.

Often, it is proposed to preserve the pristine nature and mineral diversity of the Earth by transferring mining production to cosmic bodies [13]. But the lack of technical possibilities for implementation [14] and unresolved legal issues [15] postpone the implementation of this idea for the distant future.

3. Problems, objects and methods of research

The use of man-made waste from mining and construction is an important issue worldwide [16-18]. Despite the importance of the issue of re-use of man-made waste, it has a fairly local size and significance. Understanding the problems also depends on the experience of the people involved in the implementation and implementation of technologies for the reuse of man-made waste. Until now, the idea of using man-made waste has been realized by using it to prepare a backfill composite for the extraction of minerals by geotechnology with the laying of the developed space [19, 20]. The possibility of using large volumes of man-made waste can be realized in large infrastructure construction projects: the construction of roads and railways, flood protection systems-dams, etc. [21]. In addition, man-made waste can be used to prepare building composites for building foundations, monolithic structures, concrete blocks, and bricks. To implement the task of using man-made waste in civil construction, it is necessary to solve a number of the following tasks:

1. define the concept of implementing the ideas of non-waste (low-waste) production, excluding the formation of waste and assuming their use as an intermediate product in a closed cycle of the main or auxiliary production;
2. ensure the environmental cleanliness of the waste used;
3. develop and implement technologies that ensure the preparation of waste for use.

If the solution to the first problem lies in the legal and contractual sphere and consists in understanding "how flexible the customer is with regard to the use of waste" and "how much the company, which is responsible for man-made massifs, wants to dispose of them", then the implementation of the next two is in the scientific plane, and the possibility of solving them will be considered in this study.

As it was noted earlier, industrial waste, being a powerful source of disturbance of the ecological balance, can serve as raw materials for the production of construction and laying materials [22]. However, it should be remembered that the chemical components contained in man-made waste can negatively affect both the technical characteristics of the mixture and the environment [23]. To this it is necessary to add that a huge amount of waste has accumulated in the industrial massifs of mining and processing industries, which have useful components in their composition, which can be not only raw materials for the production of additional products, but also are industrial geo-resources [24]. Deep processing of man-made waste will increase the completeness of the extraction of the useful component [25]. In the context of increasing consumer demand for extracted raw materials and building materials, increasing the requirements for the completeness and complexity of the use of the mineral resource base and involving waste in the creation of materials for the construction industry is a paradigm for the further development of science [26].

Thus, the primary task is the high-quality preparation of man-made waste for use in order to maximize the use of their positive qualities, extract useful components and eliminate the subsequent impact on the environment [27, 28].

4. Results and discussions

The purpose of this study is to develop a technological and theoretical basis for the rational use of man-made waste from mining and processing production in the creation of construction and laying materials, which will improve the environmental situation of the region, develop a highly effective advanced technology for deep processing and disposal of man-made waste.

In previous studies, the rationality of the use of mechanical activation for processing tailings and off-balance ores was justified. Activation treatment allows for a 10-20 % increase in the extraction of a useful component from man-made waste. Based on the economic and technical analysis of various technologies, it was found that the most effective is the mechanochemical technology of deep processing of man-made waste from mining and processing production. This technology combines the methods of mechanical and chemical activation treatment in disintegration plants.

In addition to useful components, man-made waste contains components that can subsequently have a negative impact on composites or products made from them (softening, excessive swelling or increase in volume, caking, etc.), as well as, due to instability, disrupt the ecological balance. It is necessary to take into account a number of economic, environmental and other restrictions when using waste from mining and industrial production in the laying composite.

Therefore, deep processing of industrial waste by the method of mechanochemical activation in order to extract useful components for additional profit and neutralize components that have a negative impact on the manufactured products and the environment, is seen as a fairly promising direction of development in the field of repeated (low-waste) technologies in the production of construction and laying materials.

When creating construction and laying composites, they are currently engaged in the development of fundamentally new materials. The creation of innovative building or laying material is made by structural changes in its internal structure due to the formation of new or secondary crystal structures. This increases the strength characteristics of solidified structures or solid wood. At this stage, the most affordable and cheap method to influence the properties of the composite is the activation treatment of its components or the finished mixture as a whole [29].

Mechanical activation should be understood as one of the methods of processing the composite components separately or as a finished mixture, as a result of which the properties of the components are used most fully [30, 31].

One of the promising methods of activation is the mechanical activation treatment of the mixture components in disintegrators (Figure 1).

Due to the uneven redistribution of velocities in the working body of the disintegrator, large internal stresses arise (Figure 2), which allow us to exclude the further formation of hydrate bonds on the surface of the grains of the aggregate of the mixture [32-34].

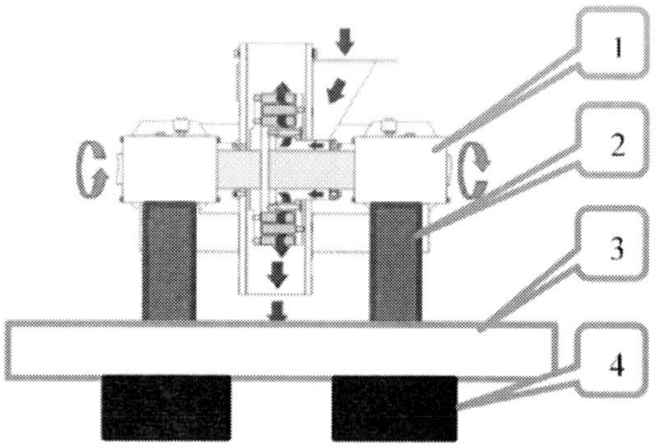

Figure 1. Disintegrator
1-disintegrator; 2-vibration exciters; 3-base; 4-vibration-isolating supports.

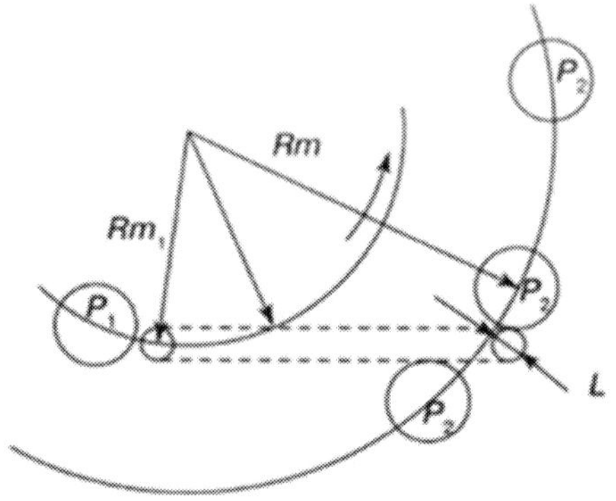

Figure 2. Motion of the material in the working body
Rm, Rm_1 – radii of the circle of pins; L – diameter of the material particle; P_1 P_2 – centers of neighboring pins of the circle.

Industrial waste is subjected to activation treatment in disintegrators before it is mixed with other components of the construction or laying composite [35, 36]. Activation processing of the ingredients leads to an increase in strength characteristics and an improvement in the uniformity of structures and artificial massifs [37]. This makes it possible to use man-made production waste as part of the construction and laying composite to create a new material [38, 39].

Consider the use of activation treatment of halite waste used as an aggregate in construction or backfill composites (Table 1, Figure 3). Since the salts are soluble in water, a saturated salt solution was used as a recluse.

The construction or laying composite obtained based on activated industrial waste is less susceptible to stratification, since it is more difficult to "give" water, is more mobile and homogeneous in composition (Figure 4). The use of industrial waste that has undergone activation treatment in disintegrators, in construction and laying composites allows to increase the strength of structures or monolith mass by 25 – 30% or reduce the consumption of binder by 40 – 50%, while maintaining their previous strength characteristics.

Table 1. Kinetics of strength gain of mixtures based on halite waste with and without activation with different content of components.

№	content of components, mass. %			ratio of saturated solution to solid	strength of samples under uniaxial compression, MPa			
					duration of hardening, days			
	waste	activated waste	cement		7	28	60	90
1	-	99	1	0.125	0.125	1.5	2.1	2.3

1a	99	-	1	0.125	0.1	1.2	1.7	1.8
2	-	98	2	0.125	0.19	2.0	2.75	3.1
2a	98	-	2	0.125	0.15	1.6	2.2	2.4
3	-	100	-	0.15	-	0.13	0.3	0.35
3a	100	-	-	0.15	-	0.1	0.25	0.25

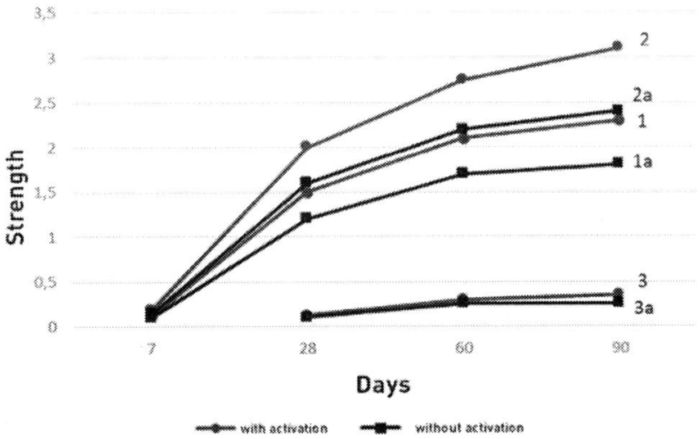

Figure 3. The kinetics of the strength set of the backfill with different content of the binder component and the aggregate (enrichment waste) astringent/waste: 1 – 1/99; 2 – 2/98; 3 – 0/100.

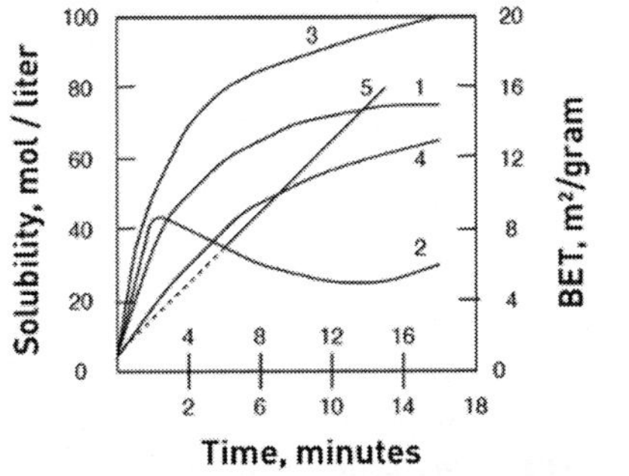

Figure 4. Effect of the duration of activation of industrial waste on 1-solubility; 2-specific surface area of non-aggregated powders; 3-specific surface area of disaggregated powders; 4-amorphization; 5 – comparison of the growth of specific surface area and strength.

5. Conclusions

The use of industrial waste (halite enrichment waste) in the construction or backfill composite becomes possible after their activation treatment in the disintegrator.

The disintegrative effect on man-made waste in combination with chemical treatment makes it possible to additionally extract useful components from man-made waste, which improves the financial stability of the enterprise. Mechanochemical technology significantly increases the mineral resource base of the mining and processing enterprise and expands the environmentally safe and affordable basis for the use of man-made waste as components in construction and building materials

The use of industrial waste (halite enrichment waste) in construction and laying materials, after activation treatment by means of a disintegrator, increases the strength characteristics of the construction monolith and the laying mass by 25 – 30% or reduces the consumption of cement by 40 – 50% while maintaining the strength characteristics of the laying array.

The use of man-made waste as an aggregate in construction or laying material significantly expands the raw material base of the construction sector, allows them to be disposed of, which minimizes the risks of environmental balance violations.

The idea of non-waste (low-waste) production allows not only to expand the raw material base for each of the industrial sectors, but also to implement the concept of sustainable development of the region. This approach, in addition to improving the economic performance of the mining and processing enterprise and the construction sector, will now lead to a high multiplier social and environmental effect.

References

[1] Khayrutdinov A et al 2020 Stress-strain behavior control in rock mass using different-stregth backfill *Mining Informational and Analytical Bulletin* **10** pp 42-55 doi:10.25018/0236-1493-2020-10-0-42-55

[2] Gorbatyuk S et al 2020 Technologies for applying Ni-Au coatings to heat sinks of SiC-Al metal matrix composite material *Materials Today: Proceedings* **38** pp 1889-1893 doi:10.1016/j.matpr.2020.08.581

[3] Kongar-Syuryun Ch et al 2021 Geotechnology using composite materials from man-made waste is a paradigm of sustainable development *Materials Today: Proceedings* **38** pp 2078-2082 doi:10.1016/j.matpr.2020.10.145

[4] Valiullin I et al 2021 Application of a discrete double-row scraper cleaner as an efficient procedure of cleaning of belt conveyors aimed at transporting bulk materials *Metallurgist* **64** pp 1340-1346 doi:10.1007/s11015-021-01124-x

[5] Rybak J et al 2021 Utilization of mineral waste: a method for expanding the mineral resource base of a mining and smelting company *Metallurgist* **64** pp 851- 861 doi:10.1007/s11015-021-01065-5

[6] Kerimov R and Shakhov S 2020 Use of metallized raw materials in electric furnace steelmaking *Metallurgist* **64(1-2)** pp 128-135 doi:10.1007/s11015-020-00974-1

[7] Niemiec D et al 2017 Selected black-coal mine waste dumps in the ostrava karviná region: an analysis of their potential use *IOP Conf. Series: Earth Environ. Sci.* **95** 042061 doi:10.1088/1755-1315/95/4/042061

[8] Rybak J et al 2021 Geomechanical substantiation of parameters of technology for mining salt deposits with a backfill *Mining Science* **28** pp 19-32 doi:10.37190/msc212802

[9] Wyjadłowski M et al 2021 Investigation of the diaphragm wall isolation of vibrations' transferred through the subsoil *IOP Conf. Series: Mater. Sci. Eng.* **1015** 012059 doi:10.1088/1757-899X/1015/1/012059

[10] Khayrutdinov M et al 2020 Cementless backfill mixtures based on water-soluble manmade waste *Bulletin of the Tomsk Polytechnic University, Geo Assets Engineering* **331(11)** pp 30–36 doi:10.18799/24131830/2020/11/2883

[11] Kawalec J et al 2017 Application of crushed concrete in geotechnical engineering - selected issues *IOP Conf. Series: Earth Environ. Sci.* **95** 022057 doi:10.1088/1755-1315/95/2/022057

[12] Zástěrová P et al 2016 Analysis the purposes of land use planning on the hard coal tailing dumps *IOP Conf. Series: Earth Environ. Sci.* **44** 022034 doi:10.1088/1755-1315/44/2/022034

[13] Khayrutdinov M et al 2020 Planetary technology. prerequisites for the formation of a new scientific discipline *Gornaya Promyshlennost* **3** pp 113–120 doi:10.30686/1609-91922020-3-113-120

[14] Gugunskiy D et al 2020 Legal models for activities on the exploration and utilization of space resources: towards the «Space-2030» Agenda *Advances in Intelligent Systems and Computing* **1100** pp 657-664 doi:10.1007/978-3-030-39319-9_73

[15] Khayrutdinov A 2020 Current issues of mining activities on celestial bodies: International law aspects *Advances in the Astronautical Sciences* **170** pp 895–902

[16] Kawalec J and Warchał T 2015 Dynamic replacement columns with aggregate transition zone stabilized by geosynthetics for embankment foundation over weak deposits *Geotechnical Engineering for Infrastructure and Development - Proceedings of the XVI European*

Conference on Soil Mechanics and Geotechnical Engineering pp 1511-1516. doi:10.1680/ecsmge.60678.vol3.221

[17] Kongar-Syuryun Ch et al 2021 Research of rheological characteristics of the mixture as a way to create a new backfill material with specified characteristics *Materials Today: Proceedings* **38** pp 2052-2054 doi:10.1016/j.matpr.2020.10.139

[18] Kongar-Syuryun Ch et al 2020 Industrial waste in concrete mixtures for construction of underground structures and minerals extraction *IOP Conf. Series: Mater. Sci. Eng.* **869** 032004 doi:10.1088/1757-899X/869/3/032004

[19] Kongar-Syuryun Ch et al 2021 Models and algorithms of choice of development technology of deposits when selecting the composition of the backfilling mixture *IOP Conf. Series: Earth Environ. Sci.* **684** 012008 doi:10.1088/1755-1315/684/1/012008

[20] Khayrutdinov M et al 2021 Improving safety when extracting water-soluble ores by optimizing the parameters of the backfill mass *Bezopasnost' Truda v Promyshlennosti* **1** pp 53-59 doi:10.24000/0409-2961-2021-01-53-59

[21] Wyjadłowski M 2007 Methodology of dynamic monitoring of structures in the vicinity of hydrotechnical works - selected case studies *Studia Geotechnica et Mechanica* **39 (4)** pp 121-129 doi:10.1515/sgem-2017-0042

[22] Sivak B et al 2020 Development of a system for electromagnetic stirring of liquid steel in molds of billet and bloom CCMS *Metallurgist* **63(9-10)** pp 910-914 doi:10.1007/s11015-020-00909-w

[23] Shakhov S and Vdovin K 2019 Electromagnetic Stirring in Continuous-Casting Machines *Steel in Translation* **49 (4)** pp 261-264 doi:10.3103/S0967091219040120/

[24] Herbut A et al 2019 The surface wave attenuation as the effect of vibratory compaction of building embankments *IOP Conf. Series: Earth Environ. Sci.* **362** 012131 doi:10.1088/1755-1315/362/1/012131

[25] Jiawen L et al 2019 Modeling pressure pulsation fields in a screw centrifugal pump *Proceedings of the 26th International Congress on Sound and Vibration ICSV26* **697**

[26] Keropyan A *et al* 2020 Process research of wheel-rail mining machines *Traction lecture notes in mechanical engineering* pp 703-709 doi:10.1007/978-3-030-22063-1_75

[27] Khairutdinov A et al 2021 The concept of geotechnology with a backfill is the path of integrated development of the subsoil *IOP Conf. Series: Earth Environ. Sci.* **684** 012007 doi:10.1088/1755-1315/684/1/012007

[28] Eronko S et al 2020 Experience of design and industrial application of systems for the driven feed of slag-forming mixtures into the Crystallizers of Slab CCM *Metallurgist* **64(3-4)** pp 214-222 doi:10.1007/s11015-020-00986-x

[29] Kouziyev D et al 2019 Sensing of dynamic loads in the open-cast mine combine *E3S Web of Conf.* **105** 03014 doi:10.1051/e3sconf/201910503014

[30] Khayrutdinov A et al 2020 Improvement of the backfilling characteristics by activation of halite waste for non-waste geotechnology *IOP Conf. Series: Mater. Sci. Eng.* **867** 012018 doi:10.1088/1757-899X/867/1/012018

[31] Kongar-Syuryun Ch et al 2021 Effect of activating treatment of halite flotation waste in backfill mixture preparation *Mining Informational and Analytical Bulletin* **1** pp 43–57 doi:10.25018/0236-1493-2021-1-0-43-57

[32] Krivenko A and Khanh Z 2020 Influence of power fluid temperature in hydraulics on operating efficiency of hydraulic mining excavators *Gornyi Zhurnal* **12** pp 78-81 doi:10.17580/gzh.2020.12.18

[33] Kobelev O and Tyurin V 2007 Production of large plates Steel in Translation **37 (9)** pp 727-729 doi:10.3103/S096709120709001X

[34] Kulikova E and Ivannikov A 2020 The terms of soils removal from the defects of the underground structures' lining *J. Phys. Conf. Series* **1425** 012062 doi.org/10.1088/1742-6596/1425/1/012062

[35] Albagachiev A et al 2020 Determination of rational friction temperature in lengthwise rolling *CIS Iron and Steel Review* **19** pp 33-36 doi:10.17580/cisisr.2020.01.07
[36] Kobelev O et al 2009 Effective production of large pipe blanks *Steel in Translation* **39 (6)** pp 501-505 doi:10.3103/S0967091209060163
[37] Pimenov G et al 1991 Making of large-sized solid-forged blanks *Tyazheloe Mashinostroenie* **9** pp 21-24
[38] Kobelev O et al 1998 Analysis of the electrocapillary curves and adsorption process models in copper ǁ Boron + Aluminium oxide melt *Zeitschrift fur Physikalische Chemie* **203** pp 143-158 doi:10.1524/zpch.1998.203.Part_1_2.143
[39] Kobelev O et al 1995 Electrocapillary curve of copper in boroaluminate melt *Rasplavy* **1** pp 39-45

MECHANICAL LOADING SYSTEMS SAFETY PROCESSES MODELING

Iurii Savchenko[1], Alexander Shapoval[2], Viktoriya Kozechko[3], Volodymyr Voskoboynik[4], Oksana Khrebtova[2], Sergii Shlyk[2]

1 Department of Cybersecurity and Information Technology, University of Customs and Finance, 2/4, Vladimir Vernadsky Street, Dnipro, 49000, Ukraine
2 Department of Manufacturing Engineering, Kremenchuk Mykhailo Ostrohradskyi National University, 20, Pershotravneva Street, Kremenchuk, 39600, Ukraine
3 Department Of Technologies of Mechanical Engineering and Materials Science, Dnipro University of Technology, Avenue Dmytra Yavornytskoho, 19, Dnipro, 49005, Ukraine
4 Department of Information Protection, Zaporizhzhya Polytechnic National University, 64, str. Zhukovsky, Zaporozhye, 69063, Ukraine

E-mail: alexshap.as@gmail.com

Abstract. Mechanical means which are directly related to the information support path (locators, observation stations, accompaniment, detection, localization, etc.) require special attention within the framework of the technical channels of receiving information. Their accurate and stable performance is of the utmost importance. Loss of the mechanical properties occurs during operation, that is material wear. To improve the quality of mechanisms and to ensure the stability of the radio-electronic means, the authors of the paper consider the possible mode of friction geoactivators in the friction unit parts of machines. The studies of the rubbing surfaces chemical composition after being processed by the geoactivator demonstrate that the components of the geoactivator diffuse into the surface layers of the material and form glass-crystalline layers, which are the solutions of the geoactivator components in the phase components of the surfaces steels and cast irons. The studies of mechanical properties have shown that the hardness and wear resistance of the surface layer increases and the roughness of contacting surfaces gets reduced.

1. Introduction

Mechanical means which are directly related to the information support path (locators, observation stations, accompaniment, detection, localization, etc.) require special attention within the framework of the technical channels of receiving information. Their accurate and stable performance is of the utmost importance. Loss of the mechanical properties occurs during operation, that is material wear. An essential obstacle to the increase of machines' and mechanisms' durability is the wear of their nodes during friction. The loss due to wear and the importance of the problem of increasing wear resistance and durability of machines can be demonstrated as follows: 80-90% of the machines failures are due to the nodes wear and the loss of funds in mechanical engineering reaches up to 5% of the national income.

To prevent the machines and mechanisms parts wear, a method of regenerating rubbing nodes, using friction geoactivators, was developed. This method is completely different from the current ones due to

Content from this work may be used under the terms of the Creative Commons Attribution 3.0 licence. Any further distribution of this work must maintain attribution to the author(s) and the title of the work, journal citation and DOI.
Published under licence by IOP Publishing Ltd

the fact that it not only reduced the friction coefficient but also allows to partially restore the functional condition of the worn surface on-the-job.

Geoactivator is a powder material, made on the basis of the domestic natural material – serpentinite, which contains a lot of chemical elements (32), with the highest of Mg (≈ 23%), Si (18 - 21%), and their oxides SiO2 (33,4 – 44,5%), MgO (25,6 – 38,0%).

Silicates play a key role in the processes which occur in the friction units. Silicates are sulfuric chemical compounds containing silica SiO2, which can create silician acid complexes of various forms (chains, hexagons, ribbons, etc.).

The properties of silicates are defined by their composition, the type of forces among the ions, the structure of crystalline lattices, and their defectiveness. Silicate hardness ranges from 1 (talc) to 6 ... 7 (forsterite) units on the Moos scale.

2. The main part of the article

The basis of silicates with small cations is a tetrahedral orthogroup [SiO4]4 with Si4 in the middle and 4O2 at the top of the tetrahedron with the ribs length of approximately 2,6 Å. Tetrahedrons in silicates can interconnect by common oxygen vertices to silica acid complexes. Every two neighboring tetrahedrons have a common vertex. At the decrease of the O/Si atomic ratio in the silicates, the number of tetrahedrons increases.

The geoactivators comprise of different forms of the magnesium silicates with a common formula of nMgO·mSiO2 or with the substitution Mg2+ ↔ Fe2+ (3Mg2+ ↔ 2Fe3+), (Mg, Fe)2[SiO4] (to be more precise, considering constitutional or crystallization water n(Mg, Fe)O·mSiO2·kH2O).

These silicates can form solid solutions of limited and unlimited solubility, as the result of substitutions of the Mg2+ lattice cations by other cations engaged in similar packaging spaces from large ions O2-. It is also possible to replace the Mg2+ cations with cations of greater valence (Al3+, Fe3+) with the occurrence of cationic vacancies. Vacancies, distortions in lattices due to differences between the ion radii of mutually replaced cations, the location of the portion of the latter in the interstices of the lattice, dislocations are the defects of silicate lattices which can change the diffusion rate, accelerate chemical reactions, increase sintering and other physicochemical characteristics.

Natural silicates which are the components of the geoactivators contain all possible phases in one way or another. Of these, forsterite and enstatite at about 50% concentration.

Forsterite has a rhombic crystal lattice with the following parameters a = 4,770 Å, b = 10,260 Å, c = 5,990 Å. Forsterite is stable within the entire temperature range with the hardness of 7 units on the Moos scale. The most important quality of the forsterite is the ability to form solid solutions.

Enstatite has a structure consisting of infinite chains of various types with a general formula of Mg2[Si2O6]. Enstatite has two polymorphic forms - enstatite and clinoestatite. The structure of main construction materials – steels and cast irons, has carbon in the form of cementite (chemical compound Fe – Fe3C) practically at any concentration, according to the Fe – C (Fe – Fe3C) state diagram. When the carbon is less than or equal to 0.8%, cementite is part of perlite, when carbon is more than 0.8% - cementite is also observed in free form.

Pic. 1 demonstrates cementite lattice. Cementite has rhombic lattice with the following parameters: a = 5,077 Å, b = 6,776 Å, c = 4,515 Å. The comparison of the cementite and forsterite lattices (table 1) showed almost complete parameters coincidence over two axes.

Geoactivator particles that lubricated in the friction zone are exposed to contact pressure, which in the micro-block of the contact spot may reach up to 1000 MPa. As a result, the destruction of geoactivator crystals and the formation of active radicals occurs. The similarity of cementite and forsterite latices allows silicate compounds to be formed on the steel surface at no significant distortion of their latices, which is very important for the diffusion processes.

At the same time, geoactivator crystals having higher hardness than the material of the part, micro polish the friction surfaces and remove oxide films, preparing the contacting surfaces to be processed by the geoactivators. At the same time, the roughness of the surface decreases along with its activation.

Under the contact loads (at the material flow limit level), the substitution of Mg crystals cations in the geoactivator crystals to the Fe cations to form solid solutions takes place. As a result of these processes, a glass-crystalline layer is formed at the friction surface, firmly merged to the surface of the part, and consisting of various compounds (from endless layers and chains to the amorphous phase), interconnected and capable of further phase transformations.

This process continues until the entire surface of the metal is saturated with a geoactivator. The process ends with the formation of the stable crystal structure of the surface layer and adjacent to the surface layers of the metal. The completion of the surface saturation process with the geoactivator is accompanied by a sharp decrease of the friction coefficient and the temperature of the friction node.

Thus, the processes occurring in the friction zone during processed by the geoactivator can be divided into four stages:

- surface activation;
- absorption of the geoactivators by the top layers of the metal;
- diffusion of the geoactivator from the top layer into the metal;
- regeneration of the friction surfaces and roughness decrease.

The studies of the chemical composition of the friction surfaces made of 45 and 45 XH steel after processing with a geoactivator for 15, 20, and 40 hours of uninterrupted operation have been conducted. Metallographic oblique grinds were made for the study.

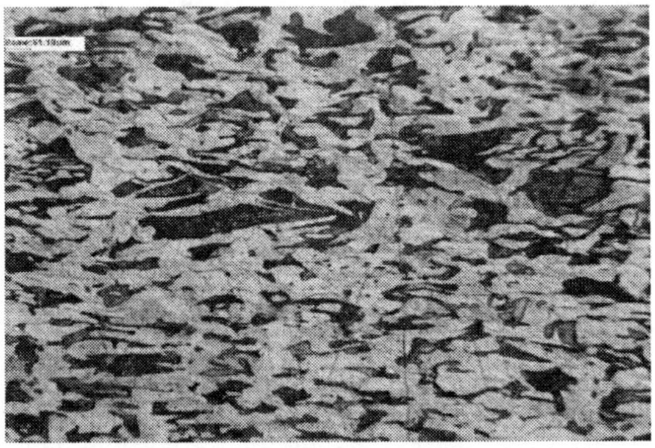

Picture 1 – Microstructure of the surface layers, x 300

Table 1 – Cementite and forsterite latices parameters

	The parameter value, Å		
Forsterite	a = 4,770	b = 10,260	c = 5,990
Cementite	c = 4,515	2 a = 10,154	b = 6,726
Parameters deviation, A	0,255	0,106	0,736
Parameters deviation, %	5	1	11

The study of the friction surfaces chemical composition processing by the geoactivator showed that the components of the geoactivator diffuse into the surface layers of the material and form glass-crystal layers, which are the solutions of the geoactivator components in the phase components of steels and cast irons. The penetration depth is 0,1…0,2 mm.

The mechanical properties of contacting surfaces are investigated: microhardness, roughness, wear resistance.

Microhardness was measured with help of the PMT-3 tool according to GOST 9450-76 at the load of 50 and 100 kg/mm2. The microhardness study at different depths of the metal surfaces treated with geoactivators allowed us to establish a certain dependency. The hardness of the surface layer revealed on the micro slice in the form of a light strip, differs from the initial hardness of the processed material and, as a rule, significantly exceeds it (Pic. 1).

The microhardness of the metal equals the initial microhardness of the metal at the depth of 0,2 mm and above. This can be seen by the microhardness and layer depth dependency curves, presented on the Pic. 2. Consideration of these curves shows that the maximum increase in hardness is observed at the depth (up to 0.04 mm), where the lower part of the modified layer ends and the subsurface layer begins.

The mechanical properties studies demonstrated that the hardness of the surface layer gets increased by 1,5…3,5 times, wear resistance gets increases 4…5 times, roughness o the contacting surfaces decreases 2,5…10 times (table 2).

Thus, the effect of geomodifiers leads to the occurrence of a unique tribological effect.

3. Conclusions

The results of the conducted studies demonstrated that adding geomodifiers based on the natural material – serpentine leads to the increase of reliability and durability of the treated friction nodes by 2…3 times, the hardness of the surface layer gets increased by 1,5…3,5 times, wear resistance gets increases 4…5 times, roughness of the contacting surfaces decreases 2,5…10 times.

The obtained results allowed us to establish the nature of the wear and disclose the mechanism of the wear-resistant layer formation with applied friction geoactivators, a change in the mechanical properties of materials, the geometric characteristics of the friction surfaces, and the operation modes of mechanisms on the wear and restoration of parts with applied geoactivators.

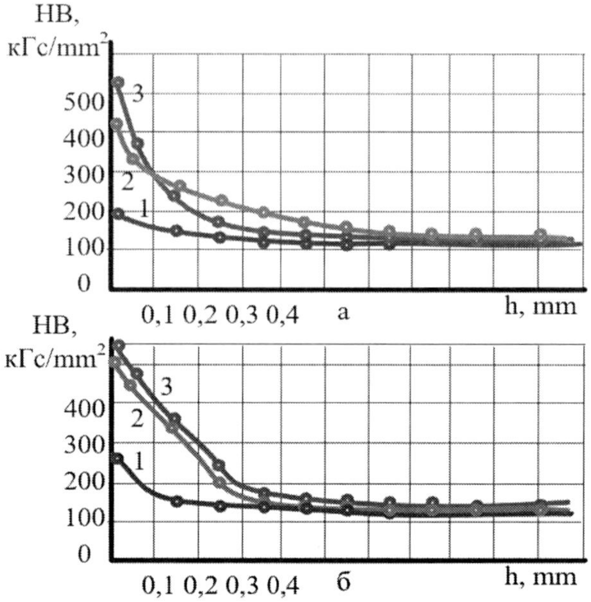

Picture 2 – Relationship of the microhardness parameter against the depth of the layer:
a – steel 45; б – steel 40XH; 1 – tests without geoactivator; 2 – tests with geoactivator №1; 3 – tests with geoactivator №2.

Table 2 – Mechanical properties of steels

Material	Mechanical properties		
	Hardness HB, kg/mm²	Roughness R_a, mcm	Wear resistance
Steel 45, initial state	179	2,5	$2{,}1\cdot10^{-11}$
Steel 45, processed by the geoactivator	550	0,8	$0{,}6\cdot10^{-11}$
Steel 40XH, initial state	200	1,25	$1{,}5\cdot10^{-11}$
Steel 40XH, processed by the geoactivator	585	0,2	$0{,}4\cdot10^{-11}$

The approach proposed by the authors will improve the quality of the radio-electronic equipment mechanical products manufacturing, will increase its durability, reduce the level of lateral deviations and ensure the reliability of the information transmission channels.

4. References

[1] Savchenko Iu, Gurenko A and Naumenko O 2016 The cutting-edge industrial technology of mining tool manufacturing *Mining of Mineral Deposits* **10 4** 105–110
[2] Bast J, Gorbatyuk S M, and Kryukov I Yu 2011 Horizontal hcc-12000 unit for the continuous casting of semifinished products *Metallurgist* **55(1-2)** 116-118
[3] Surianinov M and Krutii Y 2018 To the solution of the problem of bending of a cylindrical shell by the boundary elements method *MATEC Web of Conferences* **230**

[4] Krutii Y, Suriyaninov M and Vandynskyi V 2018 Analytic formulas for the natural frequencies of hinged structures with taking into account the dead weight *MATEC Web of Conferences* **230**

[5] Markov O E, Gerasimenko O V, Shapoval A A, Abdulov O R and Zhytnikov R U 2019 Computerized simulation of shortened ingots with a controlled crystallization for manufacturing of high-quality forgings *Int J of Advanced Manufacturing Technology* **103(5–8)** 3057–3065

[6] Shapoval A, Savchenko I and Markov O 2021 Determination coefficient of stress concentration using a conformed display on a circle of a single radius *Solid State Phenomena* **316** 928-935

[7] Zagirnyak M, Mamchur and Kalinov A 2010 Elimination of the influence of supply mains low-quality parameters on the results of induction motor diagnostics *The XIX International Conference on Electrical Machines - ICEM 2010*

[8] Zagirnyak M, Maliakova M and Kalinov A 2015 Analysis of electric circuits with semiconductor converters with the use of a small parameter method in frequency domain", *COMPEL – The international journal for computation and mathematics in electrical and electronic engineering* **34 3** 808-823

[9] Lutsenko I 2015 Identification of target system operations Development of global efficiency criterion of target operations *Eastern-European Journal of Enterprise Technologies* **2 2** 35–40

[10] Lutsenko I, Vihrova E, Fomovskaya E and Serdiuk O 2016 Development of the method for testing of efficiency criterion of models of simple target operations Eastern-European Journal of Enterprise Technologies 2 4 42–50

[11] Gorbatyuk S, Shapoval A, Mos'pan D and Dragobetskii V 2016 Production of periodic bars by vibrational drawing *Steel in Translation* **46 7** 474-478

[12] Kurpe O, Kukhar V, Puzyr R, Burko V, Balalayeva E and Klimov E 2020 Electric Motors Power Modes at Synchronization of Roughing Rolling Stands of Hot Strip Mill. *CONFERENCE 2020, PAEP* **510**

[13] Kukhar V, Grushko A and Vishtak I 2018 Shape indexes for dieless forming of the elongated forgings with sharpened end by tensile drawing with rupture *Solid State Phenomena* **284** 408–415

[14] Grushko A, Kukhar V and Slobodyanyuk Y 2017 Phenomenological model of low-carbon steels hardening during multistage drawing *Solid State Phenomena* **265** 114–123

[15] Haikova T, Puzyr R, Savelov D, Dragobetsky V, Argat R and Sivak R 2020 The Research of the Morphology and Mechanical Characteristics of Electric Bimetallic Contacts *CONFERENCE 2020, PAEP* **579**

[16] Shapoval A, Drahobetskyi V, Savchenko I, Gurenko A and Markov O 2020 Profitability of production of stainless steel + zirconium metals combination adapters *Key Engineering Materials* **864** 285-291

[17] Zarapin A, Shur A and Chichenev N 1999 Improvement of the unit for rolling aluminum strip clad with corrosion-resistant steel *Steel in Translation* **29(10)** 69-71

[18] Artiukh V, Mazur V ahd Adamtsevich A 2017 Priority influence of horizontal forces at rolling on operation of main sheet rolling equipment *MATEC Web of Conferences* **106**

[19] Savchenko I, Shapoval and Gurenko A 2020 Modeling dynamic parameters of hard alloys during shock wave regeneration *IOP Conference Series: Materials Science and Engineerin.* **969(1)**

[20] Zagirnyak M, Kalinov A, Melnykov V and Kochurov I 2015 Correction of the operating modes of an induction motor with asymmetrical stator windings at vector control *International Conference on Electrical Drives and Power Electronics* **259**

[21] Hrudkina N, Aliieva L, Markov O, Marchenko I, Shapoval A, Abhari P and Kordenko M 2020 Predicting the shape formation of hollow parts with a flange in the process of combined radial-reverse extrusion *Eastern-European Journal of Enterprise Technologies* **4 1 (106)** 55-62.

[22] Shvab'yuk V, Krutii Y and Sur'yaninov M 2016 Investigation of the Free Vibrations of Bar Elements with Variable Parameters Using the Direct Integration Method *Strength of Materials* **48(3)** 384-393

[23] Markov O, Gerasimenko O, Aliieva L and Shapoval A 2019 Development of the metal rheology model of high-temperature deformation for modeling by finite element method *EUREKA, Physics and Engineering* **(2)** 52-60
[24] Puzyr R, Shchetynin V, Arhat R, Sira Yu, Muravlov V and Kravchenko S 2021 Numerical modeling of pipe parts of agricultural machinery expansion by stepped punches *IOP Conference Series: Materials Science and Engineering* **1018**
[25] Shapoval A, Kantemyrova R, Markov O, Chernysh A, Vakulenko R and Savchenko I 2020 Technology of production of refractory composites for plasma technologies *Proceedings of the 25th IEEE Int. Conf. on Problems of Automated Electric Drive Theory and Practice* **9240830**
[26] Kondratenko V, Sedykh L, Mirzakarimov A and Aleksakhin A 2020 Static analysis and strength calculation of drive shaft of large-scale cone crusher *E3S Web of Conferences* **193**
[27] Malinov L, Malysheva I, Klimov E, Kukhar V and Balalayeva E 2019 Effect of Particular Combinations of Quenching *Tempering and Carburization on Abrasive Wear of Low-Carbon Manganese Steels with Metastable Austenite. Materials Science Forum* **945** 574–578
[28] Khrebtova O, Zachepa N, Zachepa I, Mykhalchenko G and ProkopenkoV 2020 Formation of Starting Torque of Double-Fed Induction Motor *Proceedings of the 25th IEEE International Conference on Problems of Automated Electric Drive. Theory and Practice* **9240786**
[29] Khrebtova O 2020 Forming the induction motor torque when starting *Technical Electrodynamics* **(5)** 40-44
[30] Bogoboyashchii V, Vlasov A and Izhnin I 2001 Mechanism for conversion of the conductivity type in arsenic-doped p-CdxHg1-xTe subject to ionic etching *Russian Physics Journal* **44(1)** 61-70
[31] Markov O, Kukhar V, Zlygoriev V, Shapoval A, Khvashchynskyi A and Zhytnikov R. 2020 Improvement of Upsetting Process of Four-Beam Workpieces Based on Computerized and Physical Modeling *FME Transactions* **48(4)** 946–953
[32] Khrebtova O, Shapoval A, Mos'pan D, Dragobetsky V, Gorbatyuk S and Markov O 2021 Automatic temperature control system for electrocontact annealing of steel welding wire. *Metallurgist* **65 (3-4)**
[33] Zagirnyak M, Maliakova M and Kalinov A 2016 Analysis of operation of power components compensation systems at harmonic distortions of mains supply voltage *Joint International Conference – ACEMP 2015: Aegean Conference on Electrical Machines and Power Electronics, OPTIM 2015: Optimization of Electrical and Electronic Equipment and ELECTROMOTION 2015: International Symposium on Advanced Electromechanical Motion Systems* **355**
[34] Bogoboyashchii V and Inzhin I 2000 Mechanism for conversion of the type of conductivity inp-Hg1-xCdxTe crystals upon bombardment by low-energy ions *Russian Physics Journal* **43 8** 627-636
[35] Belas E, Bogoboyashchyy V, Grill R, Izhnin I, Vlasov A and Yudenkov V 2003 Time relaxation of point defects in p- and n-(HgCd)Te after ion milling *Journal of Electronic Materials* **32 7**
[36] Zagirnyak M, Prus V and Nikitina A 2006 Grounds for efficiency and prospect of the use of instantaneous power components in electric systems diagnostics *Przeglad Elektrotechniczny* **82 12** 123-125
[37] Prus V, Nikitina A, Zagirnyak M and Miljavec D 2011 Research of rnergy processes in circuits containing iron in saturation condition *Przeglad Elektrotechniczny* **87 3** 149-152
[38] Zagirnyak M, Mamchur D and Kalinov A 2012 Comparison of induction motor diagnostic methods based on spectra analysis of current and instantaneous power signals *Przeglad Elektrotechniczny* **88 12** 221-224
[39] Zagirnyak M, Mamchur D and Kalinov A 2012 Induction motor diagnostic system based on spectra analysis of current and instantaneous power signals *IEEE SOUTHEASTCON 2014* **1**
[40] LutsenkoI, Fomovskaya E, Oksanych I, Koval S and Serdiuk O 2017 Development of a verification method of estimated indicators for their use as an optimization criterion *Eastern-European Journal of Enterprise Technologies* **2 486** 17-23

[41] Salenko Y, Puzyr R, Shevchenko O, Kulynych V and Pedun O 2020 Numerical Simulation of Local Plastic Deformations of a Cylindrical Workpiece of a Steel Wheel Rim *Simulation and Manufacturing III. DSMIE 2020* **1** 442 – 451

[42] Haikova T, Puzyr R and Levchenko R 2020 Experimental Studies on the Stress-Strain State under Drawing Aluminum–Copper Bimetal Parts Rectangular in Plan *Russian Journal of Non-Ferrous Metals* **61** **4** 404–412

[43] Zagirnyak M, Zagirnyak V, Moloshtan D, Drahobetskyi V and Shapoval A 2019 A search for technologies implementing a high fighting efficiency of the multilayered elements of military equipment *Eastern-European Journal of Enterprise Technologies* **6/1** **(102)** 33–40

[44] Dragobetskii V, Savchenko I, Pavlenko O, Parschina E, Gurenko A and Markov O 2020 Comparative Assessment of Multilayer Waveguide Manufacturing Technologies *Proceedings of the 25th IEEE International Conference on Problems of Automated Electric Drive. Theory and Practice* **9240865**

IOP Conf. Series: Materials Science and Engineering 1164 (2021) 012071 doi:10.1088/1757-899X/1164/1/012071

Ways to increase the efficiency of detection of unmanned aerial vehicles by thermal imaging devices

Yu M Shmelov[1], O V Brusakova[2], R P Yakovlev[1] and M V Petchenko[2]

[1] Kremenchuk College of Aviation of Kharkiv National University of Internal Affairs, 39600, 17/6, vul. Peremogy Kremenchuk, Poltava region, Ukraine, 39600, 17/6, Peremogy str., Kremenchuk, Poltava reg., Ukraine

[2] Kharkiv National University of Internal Affairs, 61080, 27, Leo Landau ave. Kharkiv, Ukraine

Abstract. The possibilities to increase the efficiency of detection of air objects due to the optimized choice of a functional area of an infrared range of electromagnetic waves are mathematically estimated. It is shown that there is a need to use receivers with maximum visibility in longer wavelengths, as well as the need to work in the "transparency windows" of the atmosphere. With low visibility, long-wavelength infrared thermal imaging devices are advantageous at long distances from objects. Devices with more shortwave sensitivity can be effective over short distances. Radar detection methods in many cases are not able to detect small objects with a low reflection coefficient, which is one of the reasons for the small effective scattering area. Known methods of observation using night vision devices and thermal imaging devices, having their advantages, are also ineffective in certain conditions. An attempt is made to combine the advantages of detecting aerial objects by the multi-position radar method with the advantages of thermal imaging devices. A method of illuminating air objects with infrared searchlights is proposed, which implies the presence of many geometrically spaced sources of infrared radiation with a working spectrum combined with areas of "windows of transparency" of the atmosphere and the maximum sensitivity of the thermal imager. In such conditions, the receiver registers the total signal of thermal radiation of the object and the reflected rays of searchlights. Remote control of the spotlight emission, on which the effective scattering area depends linearly, makes it possible to bring it to the level guaranteed with a high probability of detecting a small flying object.

1. Introduction

In recent years, observations of hostilities in various hotspots of the planet show that unmanned aerial vehicles are experiencing a real boom. High combat efficiency of unmanned aerial vehicles confirms it as one of the most important types of aircraft. As a result, the problem of airspace control has become more acute. Modern technologies used in small aircraft, such as unmanned aerial vehicles (UAVs), dramatically reduce their visibility. Special methods and surveillance systems are needed for timely detection of UAVs. UAVs are of small size and special geometry, have low engine power, are made of composite structural materials, which is expressed in small effective area of electromagnetic wave scattering (EWS), perform covert flights at medium, low and extremely low altitudes and are often invisible to radar systems (RS). In addition, RSs are quite structurally complex, cumbersome and at high power dangerous to humans, energy-intensive, and also unmasked. The use of traditional pulse RSs to increase the probability of detection of UAVs requires radical changes in their physical and technical parameters, and hence in the structural elements, which results in an increase in the cost of their production and maintenance. Multi-position radiolocation with a specific arrangement of radars and receivers is more effective for detecting small targets [1-6]. However, spaced multi-position systems have a major drawback – loss of functionality in the absence of radio communication. To reliably detect a UAV, it is necessary to maintain a continuous radar field over the controlled area at low and extremely low altitudes.

Content from this work may be used under the terms of the Creative Commons Attribution 3.0 licence. Any further distribution of this work must maintain attribution to the author(s) and the title of the work, journal citation and DOI.
Published under licence by IOP Publishing Ltd

The low visibility of UAVs and the limited possibilities to detect radars encourage the search for the possibility of detection by other unmasking features.

2. Literature analysis

Still new models with various possibilities and characteristics regularly appear on the UAV market, which requires immediate development of the theory and practice of combating them [7-8].

Trends in the development of small aircraft, ways to increase the probability of detection by the main types of visibility are analyzed [9-12]. Professional analysis of methods to increase the efficiency of UAV detection by traditional radio equipment showed that they are associated with an increase in the number of RSs, increasing power consumption [13-15]. Thus, to increase the effective scattering surface, the methods of single-position and multi-position radar are combined, for which the appropriate channels of synchronized processing of own signals and signals from third-party transmitters are formed in the RS. However, the developed recommendations for the optimal use of these systems do not reduce the financial costs of their production and operation [16-18].

One of the unmasking features of a UAV is thermal infrared (IR) radiation, which is mainly emitted by the power plant of the flying object. The possibility to be detected depends on the temperature contrast between the object and the background, the radiation area, the emissivity of materials. The targets in the IR region of the EMS are observed using night vision devices (NVD) and thermal imaging devices (TID). Each of them has its advantages and disadvantages. The use of active-pulse optoelectronic devices increases the detection range, but unmasks the observer. Such modern devices are complex in design and consume more energy [19-21].

Passive night vision devices are characterized by the covertness of reconnaissance surveillance and by their characteristics reach the theoretical limit of sensitivity [22-24]. Thermal imagers are inferior to night vision devices in resolution, but they can operate in complete darkness. Passive two-channel iconic optoelectronic systems are the most common for round-the-clock surveillance in poor visibility conditions [25-27].

Based on the problems of UAV detection, there is a need for a detailed study of the possibilities of increasing the detection efficiency in the IR range of EMS, the possibility of combining the methods of multi-position radar with the advantages of thermal imaging equipment.

2.1. Purpose of the paper.

Analyze possible mechanisms to increase the efficiency of detection of small aircraft TID.

3. Statement of the basic material

3.1. Optimization via the choice of the spectral section of the receiver sensitivity.

The efficiency of the system is its generalized functional property necessary to perform a certain task, which objectively expresses the degree of success in achieving the goal.

The signal-to-noise ratio at the output of the heat-sensitive radiation receiver is the physical and technical criterion of the efficiency of the TID observation.

The contrast of the thermal image of the object with temperature T_0 against the background with temperature T_ϕ is determined by difference signal ΔU [27-29]:

$$\Delta U_T = \frac{A_3 \cdot A_0}{\pi \cdot L^2} \cdot \int_0^{\infty} \tau_{oc}(\lambda) \cdot e^{-\alpha(\lambda) \cdot L} \cdot S_a(\lambda) \cdot \left[\varepsilon_0(\lambda) M(\lambda, T_0) - \varepsilon_\phi(\lambda) \cdot M(\lambda, T_\phi)\right] d\lambda, \quad (1)$$

where $A_з$ – the area of the entrance pupil of the optical system;

A_0 – the area of the object of observation;

λ – EMS length;

τ_{oc} – spectral transmittance of the optical system;

$\alpha(\lambda)$ – the natural indicator of the attenuation of the atmosphere;

$S_a(\lambda)$ – absolute voltage sensitivity of the radiation receiver;

L – the distance between the object and the optical system;

$\varepsilon_0(\lambda), \varepsilon_\phi(\lambda)$ – the coefficients of the radiation of the object and the background, respectively;

$M(\lambda, T_0), M(\lambda, T_\phi)$ – the spectral density of the radiation of an absolutely black body at the object temperature T_0 and background temperature T_ϕ respectively;

$$M(\lambda,T) = c_1\lambda^{-5}\left[\exp\left(\frac{c_2}{\lambda T}\right)-1\right]^{-1}$$, where $c_1=3.74\cdot10^{-16}$W·m^2; $c_2=1.439\cdot10^{-2}$m·°K.

Taking into account that $S_a(\lambda) = Д_a^*(\lambda)\cdot\frac{U_ш}{\sqrt{A_e\cdot\Delta f}} = Д_{\max}^*\cdot Д(\lambda)\cdot\frac{U_ш}{\sqrt{A_e\cdot\Delta f}}$, expression (1) can be rewritten as follows:

$$\Delta U_T = \frac{A_з\cdot A_0\cdot U_ш\cdot Д_{\max}^*}{\pi\cdot L^2\sqrt{A_e\cdot\Delta f}}\cdot\int_0^\infty \tau_{oc}(\lambda)\cdot e^{-\alpha(\lambda)L}\cdot Д(\lambda)\cdot\left[\varepsilon_0(\lambda)\cdot M(\lambda,T_0)-\varepsilon_\phi(\lambda)\cdot M(\lambda,T_\phi)\right]d\lambda, \quad (2)$$

where $Д_a^*(\lambda)$ – reduced detection ability, $Д_{\max}^*$ – its maximum value;

$U_ш$ – root-mean-square value of the noise;

Δf – equivalent noise bandwidth of the radiation receiver;

A_e – the area of the sensitive element of the receiver;

$Д(\lambda) = \frac{Д_a^*(\lambda)}{Д_{\max}^*}$ – relative spectral detection ability.

To simplify the calculations, we make the following assumptions:

– integration $\tau_{oc}(\lambda)$ in expression (2) is substituted by an approximate value of the optical system transmittance integration coefficient τ_o;

– assume $\varepsilon_0(\lambda) = \varepsilon_\phi(\lambda) = 1$;

– relative spectral detection ability $Д(\lambda)$ is restricted by section $\lambda\in[\lambda_1,\lambda_2]$ within the limits of which $Д(\lambda)=1$;

– at low differences between the object temperature T_0 and background temperature T_ϕ the difference in expression (2) can be changed by temperature differential [37-38]:

$$d_T M(\lambda,T) = \frac{\partial M(\lambda,T)}{\partial T}dT \approx \frac{\partial M(\lambda,T)}{\partial T}\Delta T.$$

Then signal-to-noise relation $q_1 = \frac{\Delta U}{U_ш}$ taking into account (2) will be of the form:

$$q_1(\lambda) = \frac{A_з\cdot A_0\cdot Д_{\max}^*\cdot\tau_0}{\pi\cdot L^2\cdot\sqrt{A_e\cdot\Delta f}}\Delta T\cdot\int_{\lambda_1}^{\lambda_2}\frac{\partial M(\lambda,T)}{\partial T}\cdot e^{-\alpha(\lambda)L}\cdot d\lambda. \quad (3)$$

The dependence of the attenuation natural indicator $\alpha(\lambda)$ on EMS length at the molecular or aerosol mechanisms of scattering and absorption is interpreted by G. Mie's known theory of radiation scattering in the turbid environment, which is based on the classical Maxwell's equations and is represented by rather complex mathematical expressions containing a number of special functions

[30]. In this form, non-numerical integration in (3) is impossible. In practice, a somewhat approximate semi-empirical formula is used [31-33]:

$$\alpha(\lambda)=A\lambda^{-\beta}, \tag{4}$$

where $A=\frac{3.91}{S_M}(0.555)^{\beta}$,

S_M – meteorological visibility range, km.

In the general case the value of indicator β and depends on the state of the atmosphere. Value β=4 corresponds to molecular (Rayleigh) scattering and in EMS IR-range is neglected small. Value β<4 corresponds to the attenuation of radiation due to absorption and scattering on aerosol particles of the atmosphere.

Value β can be roughly compared with meteorological visibility S_M according to the characteristics of the atmosphere in terms of the International Classification [34, 35]:

– the atmosphere is cloudless, very clear: $S_M > 20km$, $\beta \in (3.4)$;
– cloudless, clear: $10km \le S_M \le 20km$, $\beta \in (2.3)$;
– noticeable haze, strong haze: $2km < S_M \le 6km$, $\beta \in (1.5)$;
– light fog: $0.5km < S_M < 1km$, $\beta = 1$;
– thick, continuous fog: $0.05km < S_M < 0.2km$, $\beta \approx 0$.

In many cases dependence $\beta = 0.585 \cdot S_M^{1/3}$ is used, and under the conditions of moderate haze for $\lambda \in (0.61 \div 11.48)$ μm indicator $\beta \in 0.7$ [36, 37]. At operation sections $\lambda \in [8 \div 14]$ μm, where most TID function, dependence of β on λ is weak, therefore, without a great error, value $\alpha(\lambda)$ can be characterized by average value $\overline{\alpha}$ at interval $\Delta\lambda=\lambda_2-\lambda_1$, taking into account (4) (Figure 1):

$$\overline{\alpha}(\lambda_1,\Delta\lambda)=\frac{1}{\lambda_2-\lambda_1}\int_{\lambda_1}^{\lambda_2}\alpha(\lambda)d\lambda=\frac{A}{(1-\beta)\cdot\Delta\lambda}\cdot\left[(\lambda_1+\Delta\lambda)^{1-\beta}-\lambda_1^{1-\beta}\right],\ \beta\neq 1. \tag{5}$$

Then expression (3) will be of the form:

$$q_1(\lambda_1)=C_0\cdot e^{-\overline{\alpha}(\lambda_1)L}\cdot\frac{\Delta T}{L^2}\cdot\int_{\lambda_1}^{\lambda_2}\frac{\partial M(\lambda,T)}{\partial T}d\lambda, \tag{6}$$

where $C_0=\frac{A_з \cdot A_0 \cdot Д^{*}_{max} \cdot \tau_0}{\pi \cdot \sqrt{A_e \cdot \Delta f}}$.

In the range of temperatures (250÷350)°K and wave lengths (1÷15) μm $\exp\left(\frac{C_2}{\lambda T}\right)$ takes the value within limits (15÷10^{25}). At lower temperatures, the exponent grows even more significantly. That is, in the range of functional lengths of EMS of thermal imagers of different generations, the spectral energy luminosity of the object and the background can be considered under the condition $\exp\left(\frac{C_2}{\lambda T}\right)>>1$, which approximates Planck's formula by Wine's law and derivative in (6) $\frac{\partial M(\lambda,T)}{\partial T}\cong\frac{C_1C_2}{T^2\lambda^6}\cdot\exp\left(-\frac{C_2}{\lambda T}\right)$ (Figure 2).

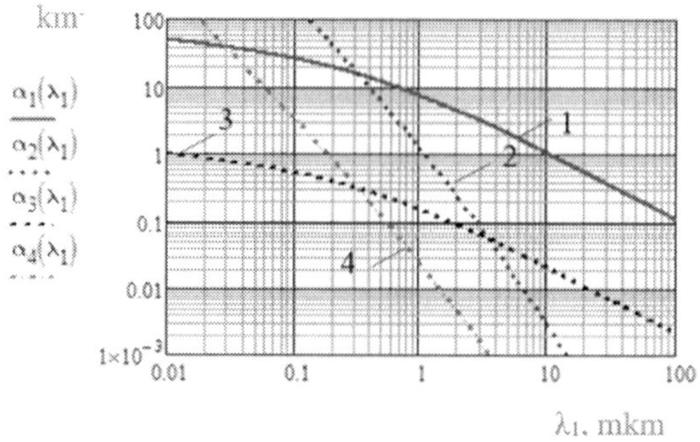

Figure 1. Dependence of the average natural index of atmospheric attenuation on fixed beginning λ1 of the section of TID thermal sensitivity (curve 1 – SM=0.2 km, β=1; curve 2 – SM=0.2 km, β=3; curve 3 – SM=10 km, β=1; curve 4 – SM=10 km, β=3).

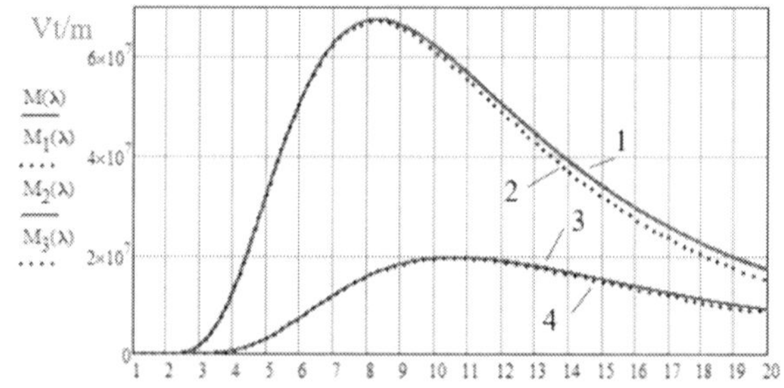

Figure 2. Spectral density of energy luminosity according to Planck's law (curve 1, 3 – solid) according to Wine's law (curves 2, 4 – dotted) at T=350°K (curves 1, 2) and T =273°K (curves 3, 4).

Taking into account the above said, relation (6) is of the form:

$$q_1(\lambda_1,T,\Delta T,\Delta\lambda)=\frac{C_0\cdot C_1\cdot T^3\cdot\Delta T}{C_2^4\cdot L^2}\cdot e^{-\bar{\alpha}(\lambda_1)L}\cdot\sum_{n=1}^{2}(-1)^n\cdot e^{-x_n}\cdot\left(x_n^4+4x_n^3+12x_n^2+24x_n+24\right), \quad (7)$$

where $x_n=\dfrac{C_2}{\lambda_n T}$, $\lambda_2=\lambda_1+\Delta\lambda$.

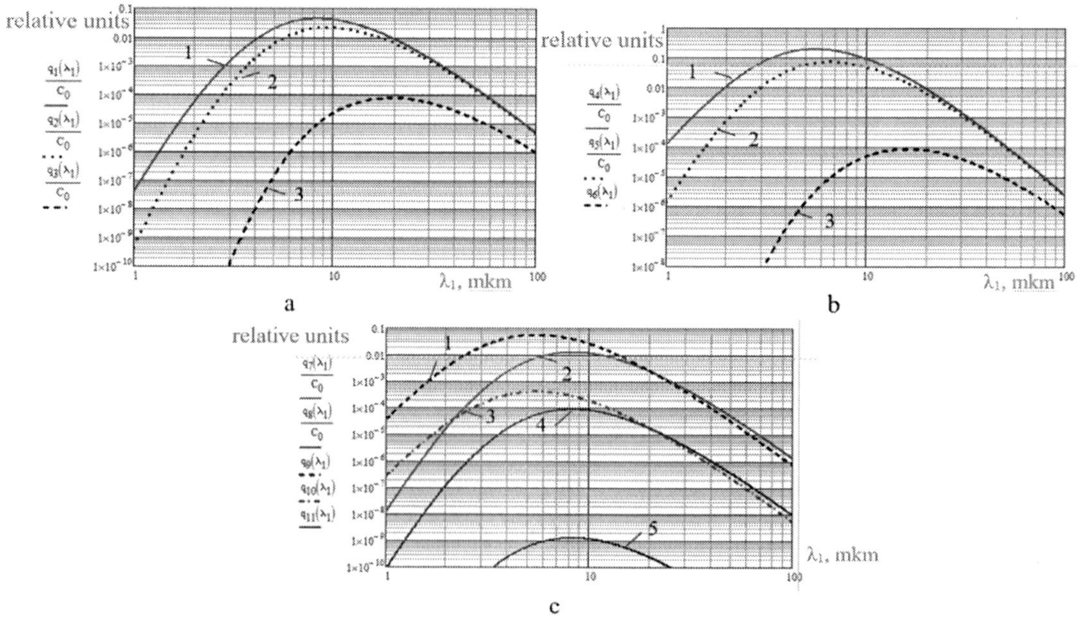

Figure 3. relation signal/noise at the output of TID receiver:
a) T=273°K; ΔT=1°K; L=3 km
(curve 1 – SM=30 km, β=3.5; curve 2 – SM=1 km, β=1; curve 3 – SM=0.2 km, β=0.7)
b) curves 1, 2, 3 at T=400°K
c) β=0; ΔT=1°K; SM=0.2 km
(T=273°K: curve 1 – L=0.3 km; curve 3 – L=0.5 km; curve 5 – L=1 km
T=400°K: curve 2 – L=0.3 km; curve 4 – L=0.5 km).

As can be seen from Figure 3, the maximum signal-to-noise ratio at the output of the receiver is shifted to the long-wavelength section of the IR range with decreasing β, that is, with impaired visibility. At β=0 relation q/C_0 does not depend on λ_1, but its value at a distance of L=3 km is very small and it becomes significant only at distances L≤0.2 km. The higher the temperature of the object and the background is the more max $q(\lambda_1)$ shifts into the short wavelength range. When temperature T changes within the limits of 273°K÷400°K, and S_M from 0.2 km to 30 km, max $q(\lambda_1)$ is within the limits ~ (5÷12) μm. That is, at high T (and available ΔT) objects can be detected by TID with receiver sensitivity in the range of λ_1~5 μm. At low T greater detection efficiency will be observed at longer operating waves. Fig. 3 illustrates the fact that with deterioration of visibility from S_M≥20 km to S_M≤0.2 km the probability of detecting objects for devices with short-wave sensitivity decreases and detection may not be possible at all, while for devices with long-wave sensitivity it may be most effective.

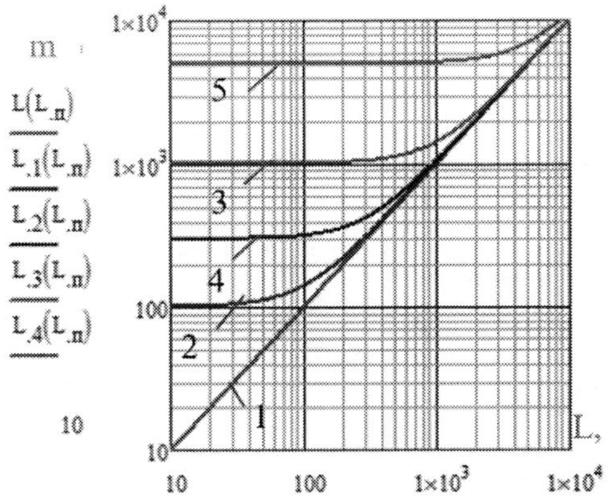

Figure 4. Dependence of the detection range of the object by the thermal imager on the distance caused by the illumination curve 1 – L_T=0; curve 2 – L_T=0.1 km; curve 3 – L_T=0.3 km; curve 4 – L_T=1 km; curve 5 – L_T=5 km).

3.2. Optimization via the choice of the section of the atmosphere transmittance spectrum.
When the TID operates with the sensitivity of the receivers, parts of the spectrum of which are combined with the "windows" of the transparency of the atmosphere, indicator $\alpha(\lambda)$ is small and can be neglected [17, 18]. In the case inequality $\bar{\alpha}(\lambda_1) << L^{-1}$ is true and $\exp(-\bar{\alpha}(\lambda_1)\cdot L) \cong 1$, signal-to-noise relation q_2 at the output of the receiver increases by $\exp(\bar{\alpha}(\lambda_1)\cdot L)$ times.

3.3. "Illumination" of the observation zone by artificial sources of IR radiation.
Assume there are N IR ray diodes, whose operating wave lengths λ_i are on the section of TID thermal sensitivity. Then signal-to-noise relation at the receiver output will increase by value of $q_{1nio}(r_i,\theta_i)$:

$$q_{1nio}(r_i,\theta_i) = C_0 \cdot \sum_{i=1}^{N} \cdot \int_{\lambda_1}^{\lambda_2} \frac{I_i(\lambda,\theta_i)\cdot\cos\theta_i\cdot\rho_i(\theta_i)}{L^2 r_i^2} d\lambda \cdot e^{-\alpha(\lambda_i)(L+r_i)}, \tag{8}$$

where $I_i(\lambda,\theta_i)$ – the spectral density of the radiant intensity of the i-th source;
$\rho_i(\theta_i)$ – the coefficient of infrared rays reflection from the object;
θ_i – the polar distance of the beam direction from the i-th source;
r_i – the distance between the object of observation and the i-th radiation source.

In the general case, taking into account the "illumination", relation q_1^* signal-to-noise is:

$$q_1^*(\lambda_1,T,\Delta T,\Delta\lambda,r_i,\theta_i) = q_1(\lambda_1,T,\Delta T,\Delta\lambda) + q_{1nio}(r_i,\theta_i). \tag{9}$$

When combining the spectral sensitivity of the receiver, the "transparency windows" of the atmosphere and the emission spectra of the ray diodes $\left(\overline{\alpha}(\lambda_1) << L^{-1}, r_i^{-1}\right)$ signal-to-noise relation q_2^* is:

$$q_2^*(\lambda_1, T, \Delta T, \Delta\lambda, r_i, \theta_i) = q_2(\lambda_1, T, \Delta T, \Delta\lambda) + q_{2nid}(r_i, \theta_i), \qquad (10)$$

where $q_{2nid}(r_i, \theta_i) = C_0 \cdot \sum_{i=1}^{N} \cdot \int_{\lambda_1}^{\lambda_2} \frac{I_{oi}(\lambda) f_i(\theta_i) \cdot \cos\theta_i \cdot \rho_i(\theta_i)}{L^2 r_i^2} d\lambda$

Without detailed calculation of q_{1nid} and q_{2nid}, which is a separate problem, we can estimate the contribution of illumination to the detection range of an aircraft object. Solving equation (10) with respect to L at $q_2^* = 1$, we have:

$$L = \sqrt{L_T^2 + L_{nid}^2}, \qquad (11)$$

where $L_T = \left[\frac{C_0 C_1 \cdot T^3 \cdot \Delta T}{C_2^4} \sum_{n=1}^{2} (-1)^n \cdot e^{-x_n} \left(x_n^4 + 4x_n^3 + 12x_n^2 + 24x_n + 24\right)\right]^{1/2}$ – range of detection of TID object due to its thermal radiation without "illumination";

$L_{nid} = \left(C_0 \cdot \sum_{i=1}^{N} \cdot \int_{\lambda_1}^{\lambda_2} \frac{I_i(\lambda, \theta_i) \cdot \cos\theta_i \cdot \rho_i(\theta_i)}{r_i^2} d\lambda\right)^{1/2}$ – increase in the detection range of the TID object due to additional "illumination" of the object (Figure 4).

It is easily seen that L_{nid} can vary widely, reaching a value predominant over L_T and can be the only factor in the detection of UAVs with low intrinsic thermal radiation. The detection range of the object can be increased by adjusting intensity $I_i(\lambda)$. For the continuity of the detection zone, the searchlights are selected to the required power and pattern. Distance d between them is calculated based on the specified minimum height of the field overlap $h_{\min}$:

$$d = 2h_{\min} \cdot tg\gamma; \qquad (12)$$

where γ – the aperture angle of searchlight emission.

The active work of EMS sources is a factor in unmasking their positions, however, the placement of a significant number of infrared searchlights in a large area is at the same time a disorientation for enemy air reconnaissance.

The use of a large number of IR searchlights to illuminate UAVs operating in areas of the thermal sensitivity spectrum of TID and located over a large area is nothing more than an analogue of multi-position multistatic radar [36-44].

4. Conclusions.

1. With decreasing meteorological visibility S_M the maximum signal from the TID receiver is shifted to the long-wavelength part of the IR spectrum.

2. At low visibility $(S_M \leq 0.2km)$ TIDs with long-wavelength sensitivity in the IR range have advantages in object detection at long distances.

Surveillance devices with short-wave sensitivity in the IR range can be effective at short distances.

3. The use of infrared searchlights to illuminate the air surveillance zone, the combination of searchlight emission spectra, thermal sensitivity of TID and "windows of transparency" of the atmosphere can significantly increase the efficiency of UAV detection.

References

[1] Surianinov M and Krutii Y 2018 To the solution of the problem of bending of a cylindrical shell by the boundary elements method *MATEC Web of Conferences* **230** 02032 doi:10.1051/matecconf/201823002032

[2] Krutii Y, Suriyaninov M and Vandynskyi V 2018 Analytic formulas for the natural frequencies of hinged structures with taking into account the dead weight *MATEC Web of Conferences* **230** doi:10.1051/matecconf/201823002016

[3] Bast J, Gorbatyuk S M and Kryukov I Yu 2011 Horizontal hcc-12000 unit for the continuous casting of semifinished products *Metallurgist* **55 (1-2)** pp 116–118 doi:10.1007/s11015-011-9399-1

[4] Romanchenko I S, Zahorka O M, Butenko O V and Deineha O V 2011 Theory and practice of combating small low-flying objects *Assessment of capabilities, trends in the development of air defense* monograph (Zhytomyr: Polissia) 344

[5] Shapoval A, Savchenko I and Markov O 2021 Determination coefficient of stress concentration using a conformed display on a circle of a single radius *Solid State Phenomena* **316** pp 928–935

[6] Vladov S, Shmelov Y and Shmelova T 2020 Modeling of the TV3-117 aircraft engine technical state as part of the helicopter power plant in the form of the markov process of death and reproduction *CEUR Workshop Proceedings* **2740** pp 400–407

[7] Danyk Yu H and Buhaiov M V 2015 Analysis of the effectiveness of detecting tactical unmanned aerial vehicles by passive and active means of surveillance *Collection of papers of ZhVI DUT* **10** pp 5–20

[8] Markov O E, Gerasimenko O V, Shapoval A A, Abdulov O R and Zhytnikov R U 2019 Computerized simulation of shortened ingots with a controlled crystallization for manufacturing of high-quality forgings *International Journal of Advanced Manufacturing Technology* **103(5-8)** pp 3057–3065

[9] Osadchiy V A, Albul S V, Kuprienko N S and Kirillova N L 2020 Future developments of a roll forming mill design algorithm *IOP Conference Series: Materials Science and Engineering* **709 (4)** 044079 doi:10.1088/1757-899X/709/4/044079

[10] Kukhar V V, Grushko A V and Vishtak I V 2018 Shape indexes for dieless forming of the elongated forgings with sharpened end by tensile drawing with rupture *Solid State Phenomena* **284** pp 408–415 doi:10.4028/www.scientific.net/SSP.284.408

[11] Khudov H V, Misiuk H V, Oleksenko O O, Raikov R Yu, Bezklubenko O O and Dobrev V Yu 2018 Analysis of existing ways to increase the efficiency of detection of small air objects by radar means of airspace control *Control, navigation and communication systems* **6(52)** pp 38–42

[12] Naumova M G, Morozova I G and Borisov P V 2020 Study of metal surface with color image obtained with laser marking *Solid State Phenomena* **299** pp 943–948 doi:10.4028/www.scientific.net/SSP.299.943

[13] Grushko A V, Kukhar V V and Slobodyanyuk Y O 2017 Phenomenological model of low-carbon steels hardening during multistage drawing *Solid State Phenomena* **265** pp 114–123 doi:10.4028/www.scientific.net/SSP.265.114

[14] Shapoval A, Drahobetskyi V, Savchenko I, Gurenko A and Markov O 2020 Profitability of production of stainless steel + zirconium metals combination adapters *Key Engineering Materials* **864** pp 285–291

[15] Shmelova T, Shmelov Y and Vladov S 2020 Concept of Building Intelligent Control Systems for Aircraft, Unmanned Aerial Vehicles and Aircraft Engines *IEEE 6th International Conference on Methods and Systems of Navigation and Motion Control, MSNMC 2020 – Proceedings* pp 14–19

[16] Kovalevskyi S M, Tiutiunnik V O and Khudov H V 2015 Method for calculating the scattering surface efficiency of small air objects with single-position and spaced signal reception in survey radar stations *Collection of papers of Kharkiv University of Aviation* **2(43)** pp 28–31
[17] Lukianova T, Donenko V and Klein R 2020 Implementing Building Information Modelling for the Reconstruction Process of Unfinished Building Projects *IEEE European Technology and Engineering Management Summit, E-TEMS 2020* 9111794
[18] Kobelev O A, Albul S V and Kirillova N L 2020 Research and development of broaching methods on mandrel of large-sized pipe forgings IOP Conference Series: Materials Science and Engineering **709** (4) 044104 doi:10.1088/1757-899X/709/4/044104
[19] Zarapin A Yu, Shur A I and Chichenev N A 1999 Improvement of the unit for rolling aluminum strip clad with corrosion-resistant steel *Steel in Translation* **29** (10) pp 69–71
[20] Savchenko I, Shapoval A and Gurenko A 2020 Modeling dynamic parameters of hard alloys during shock wave regeneration *IOP Conference Series: Materials Science and Engineering* **969(1)** 012079
[21] Artiukh V, Mazur V and Adamtsevich A 2017 Priority influence of horizontal forces at rolling on operation of main sheet rolling equipment *MATEC Web of Conferences* **106** 04001 doi:10.1051/matecconf/201710604001
[22] Volkov VH 2006 Aviation night vision devices *Special equipment* **4** pp 2–12
[23] Starodubov A V, Serdobintsev A A, Kozhevnikov I O, Pavlov A M, Galushka V V, Torgashov R A, Bessonov D A, Galkin A, Ryskin N M and Sakharov V K 2020 Comparison of nanoseconds and picoseconds laser ablation for microfabrication of planar slow-wave structures for D-band vacuum electronic devices with sheet electron beam *Progress in Biomedical Optics and Imaging - Proceedings of SPIE* **11458** 1145803 doi:10.1117/12.2564423
[24] Shmelov Y, Vladov S, Klimova Y and Kirukhina M 2018 Expert system for identification of the technical state of the aircraft engine TV3-117 in flight modes *IEEE 1st International Conference on System Analysis and Intelligent Computing, SAIC 2018 – Proceedings* **8516864**
[25] Volkov V H 2006 Ship night vision devices *Special equipment* **1** pp 2–8
[26] Sereda B, Sheyko S, Belokon Y and Sereda D 2011 The influence of modification on structure and properties of rapid steel *Paper presented at the AIST Steel Properties and Applications Conference Proceedings – Combined with MS and T'11* (Mater. Sci. and Tech. 2011) pp 457–460
[27] Hrudkina N, Aliieva L, Markov O, Marchenko I, Shapoval A, Abhari P and Kordenko M 2020 Predicting the shape formation of hollow parts with a flange in the process of combined radial-reverse extrusion *Eastern-European Journal of Enterprise Technologies* **4/1** (106) pp 55–62
[28] Shvab'yuk V I, Krutii Y S and Sur'yaninov M G 2016 Investigation of the Free Vibrations of Bar Elements with Variable Parameters Using the Direct Integration Method *Strength of Materials* **48** (3) pp 384–393 doi:10.1007/s11223-016-9776-x
[29] Volkov VH 2001 New generation ship night vision devices *Special equipment* **5** pp 2–8
[30] Dragobetskii V V, Shapoval A A, Mospan D V, Trotsko O V and Lotous V V 2015 Excavator bucket teeth strengthening using a plastic explosive deformation *Metallurgical and Mining Industry* **7** (4) pp 363–368
[31] Markov O, Gerasimenko O, Aliieva L and Shapoval A 2019 Development of the metal rheology model of high-temperature deformation for modeling by finite element method *EUREKA, Physics and Engineering* **2** pp 52–60
[32] Petrunenko I V, Belei S I, Petchenko M V, Bodnar O A and Maslak N G 2020 Organizational and financial principles for the development of euroregions *International Journal of Economics and Business Administration* **8(3)** pp 150–160

[33] Norton Y, Cambell S Horn D end Reago P 2000 Third generation imaging *Proc. SPJE* **4130** pp 226–235

[34] Keropyan A M 2016 Features of interaction of traction wheels of an electric locomotive and a diesel locomotive with rails in the conditions of open mountain works *Journal of Friction and Wear* **37** (1) pp 78–82 doi:10.3103/S1068366616010074

[35] Kondratenko V V, Sedykh L V, Mirzakarimov A and Aleksakhin A 2020 Static analysis and strength calculation of drive shaft of large-scale cone crusher *E3S Web of Conferences* **193** 01038 doi:10.1051/e3sconf/202019301038

[36] Malinov L S, Malysheva I E, Klimov E S, Kukhar V V and Balalayeva E Yu 2019 Effect of Particular Combinations of Quenching, Tempering and Carburization on Abrasive Wear of Low-Carbon Manganese Steels with Metastable Austenite *Materials Science Forum* **945** pp 574–578 doi.org/10.4028/www.scientific.net/MSF.945.574

[37] Shapoval A, Kantemyrova R, Markov O, Chernysh A, Vakulenko R and Savchenko I 2020 Technology of production of refractory composites for plasma technologies *Proceedings of the 25th IEEE Int. Conf. on Problems of Automated Electric Drive Theory and Practice* **9240830** 2020 doi:10.1109/PAEP49887.2020.9240830.

[38] Mykytenko V I and Kotovskyi V I 2008 Remote observations in the fog using passive two-channel iconic optoelectronic systems Methods and systems of optoelectronic and digital signal processing **35** pp 29–37

[39] Zagirnyak M, Zagirnyak V, Moloshtan D, Drahobetskyi V and Shapoval A 2019 A search for technologies implementing a high fighting efficiency of the multilayered elements of military equipment *Eastern-European Journal of Enterprise Technologies* **6** (1-102) pp 33–40 doi:10.15587/1729-4061.2019.183269

[40] Larin S N, Platonov V I and Solomonov K N 2017 Approach to assessment of microdamages accumulated during the constrained molding of shells made of the material subject to energy theory of creep and damage *Journal of Chemical Technology and Metallurgy* **52** (4) pp 679–684

[41] Oliinyk A, Fedorchenko I, Donenko V, Korniienko S and Kharchenko A 2020 Development of an Evolutionary Optimization Method for Financial Indicators of Pharmacies *International Journal of Computing* **19(3)** pp 449–463

[42] Khrebtova O A, Shapoval A A, Mos'pan D V, Dragobetsky V V, Gorbatyuk S M and Markov O E 2021 Automatic temperature control system for electrocontact annealing of steel welding wire *Metallurgist* **65** (3–4)

[43] Markov O E, Kukhar V V, Zlygoriev V N, Shapoval A A, Khvashchynskyi A S and Zhytnikov R U 2020 Improvement of Upsetting Process of Four-Beam Workpieces Based on Computerized and Physical Modeling *FME Transactions* **48** (4) pp 946–953 doi:10.5937/fme2004946M

[44] Kolobrodov V H and Lykholit M I 2007 Design of thermal imaging and television surveillance systems *Textbook* 364

To the problem of transverse anisotropy effect on the distribution of contact stresses in a beam-strip loaded with a rigid stamp

Shvabiuk V.I, D.Eng.Sc, prof., Rotko S.V., Cand.Eng.Sc., Assoc. Prof., Svabiuk V.V., Assoc. Prof., Hryshkova A.V. Asp.

Lutsk national technical university, Lutsk

Abstract. The contact interaction of an orthotropic bar-beam loaded with a rigid stamp. The nature of contact stresses distribution along contact surface of beam and the size of the contact region depending on the beam material orthotropy degree is studied. Analytical expressions for contact pressure under the stamp, dimensionless force, size of contact region and stamp settlement depending on contact region size are derived. Numerical results of calculations are given.

1. Introduction. The problem of forced bending of a thin rod over a predetermined rigid surface and the determination of maximum stresses in it were first considered by S.P. Timoshenko [1] in the formulation of hypotheses of the classical Bernoulli-Euler bending theory of rods. In a refined formulation, taking into account transverse shear deformations, the problem was first solved by M.M. Filonenko-Borodich [2] (1949 г.). On the example of thin strip which contacts the rigid template it is shown that at relatively large contact region the contact pressure is concentrated practically on the ends of contact region and gets the highest values there.

For the first time, most consistently and quite thoroughly, the periodic problem of bending an infinite strip by stamps based on the equations of the plane theory of elasticity was considered by Cyrus and Silva [3], where the magnitude of the contact region could be both large and relatively small. The solution was constructed in trigonometric series and was reduced to paired integral equations, which were solved numerically. The authors [3] showed that for small areas of contact the distribution of contact pressure under the stamp is close to the distribution according to G. Hertz, where the maximum pressure is concentrated in the center of the strip, and for large - close to its edge. The same problem, but already in the formulation of the cylindrical bend of the plate, was considered by B.L. Pelekh and V.I. Shvabiuk [4], but according to a refined theory, the equation of which took into account the transverse compression. E.I. Grigolyuk and V.M. Tolkachev [5], as well as S.O. Sargsyan [6] in the formulation of a plane deformation reduced this problem to the solution of the

Content from this work may be used under the terms of the Creative Commons Attribution 3.0 licence. Any further distribution of this work must maintain attribution to the author(s) and the title of the work, journal citation and DOI.
Published under licence by IOP Publishing Ltd

integral equation of Fredholm of the second kind, which was solved numerically. The reduction of the contact problem for an isotropic strip to a singular integral equation, with the subsequent use of the method of mechanical quadratures, was also carried out by the authors [7].

A similar problem of contact stress distribution arose in the mixed problem of bending an orthotropic beam lying on a rigid base [2,8]. In this case, the nature of the contact pressure distribution also very much depends on the calculation equations used to solve this problem. As well as for the parabolic stamp, found by V.I. Feodosiev [2], the contact pressure is expressed through hyperbolic functions with a maximum at the boundary of the contact of the plate with the rigid base, but a new fictitious concentrated force appears at the left end of half of the plate. This distribution of contact pressure corresponds to the solutions of contact problems for beams and plates, the bending equations of which take into account only the deformations of the transverse shear. Therefore, most of the works, the solutions of which are found on the basis of "shear" theories such as S.P. Tymoshenko [2], lead to the fact that the contact pressure at the boundary of the contact region is different from zero, even for hard surfaces without angular points. This fact contradicts the physical content of the problem, as well as the results obtained on the basis of the equations of the plane problem of the theory of elasticity of beam theories that take into account the deformations of the transverse shear and compression [3,5,8]. With the use of such models, these inconsistencies disappear and the distribution of contact pressure becomes similar to the distribution as in the case of a parabolic stamp.

To eliminate this shortcoming, in solving this problem, the authors [9,10] previously used refined theories, which in addition to the deformations of the transverse shear also took into account the transverse compression. Therefore, it is necessary to investigate in more detail the effect of this clarification on the nature of the contact stresses under the stamp.

2. Problem formulation. Let the strip of rectangular section is under the influence of rigid parabolic stamp, which is pressed to the upper surface of strip $(z=-h)$ by force P, directed downwards. On its ends $x=\pm l$ the strip is simply supported. Assume that the surface of the base of the stamp is perfectly smooth and is described by the equation $z=-\omega\cdot f(x)$. Contact pressure occurs between the strip and the stamp $q^{-}(x)=q(x)$. We will neglect the forces of friction. The basic bending equation of the system is reduced to the form:

$$EIw^{IV}=q(x)-\varepsilon_1 h^2 q^{II}-\varepsilon_2 h^4 q^{IV} E/4E', \quad (|x|\le a); \tag{1}$$
$$w^{IV}(x)=0, \quad (|x|>a),$$

where $\varepsilon_1 = 0,1\left(4\dfrac{E}{G'} - 3\nu''\right)$; $\varepsilon_2 = 0,2\left(1 - \nu''\dfrac{G'}{E}\right)$; $2a$ – contact region width, outside which the pressure on the upper face of the strip is zero $\left(q^-(x) = 0\right)$.

Let's find the solution of equation (1) separately in the contact region $\left(|x| \le a\right)$, where there is an unknown contact pressure, and in the region, which is free of contact interaction $\left(a \le |x| \le l\right)$. Assume that when $z = -h$ the top surface of the strip $\left(|x| \le a\right)$ tightly adjacent to the lower base of the stamp, i.e. —

$$W(x,-h) = \delta - \omega f(x), \tag{2}$$

where δ – stamp settlement; ω– arbitrary constant, which can be taken as a multiple of the curvature of the base of the stamp.
The left side of the equation for $W(x, -h)$ we obtain from the corresponding ratio for W.

Then, we get another relationship between the pressure under the stamp $q(x)$ and the deflection of the middle line of the strip $w(x)$:

$$A_0 q(x) = \delta - \omega f(x) - w(x) - \frac{6}{13}\nu'' h^2 \frac{d^2 w}{dx^2}. \tag{3}$$

Substituting expression (3) in equation (1), accepting $f(x) = x^2$, we obtain the following differential equation with respect to the deflection of the midline of the strip:

$$w^{IV} - 2g^2 w^{II} + \lambda^4 w = \lambda^4\left(\delta + 2\omega\varepsilon_1 h^2 - \omega x^2\right), \tag{4}$$

where $\lambda^4 = \left\{\left[5,9\dfrac{E}{E'} + 0,3\nu''\left(\dfrac{E}{G'} + 2\dfrac{G'}{E'}\right)\right]\dfrac{h^4}{12}\right\}^{-1}$; $g^2 = 0,2\left(\dfrac{E}{G'} - 1,9\nu''\right)\cdot h^2\lambda^4$,

$\alpha = \sqrt{\left(\lambda^2 + g^2\right)/2}$; $\beta = \sqrt{\left(\lambda^2 - g^2\right)/2}$.

The choice of one or another solution of equation (4) depends on the relationship E/G', E/E', values of Poisson's ratios ν and ν'', that is, of the characteristics of the strip material.

Для даної задачі співвідношення $g^2 \le \lambda^2$ означають наступні залежності:

$$\frac{E}{G'} - 2\nu'' \le 2,5\sqrt{\frac{1}{3}\left(5,9 - 0,3\nu'\nu'' + 0,3\nu'\left(\frac{E}{G'} + 2\frac{G'}{E'}\right)\right)\frac{E}{E'}}.$$

We will accept further, without loss of generality that $g^2 < \lambda^2$. Then, given the symmetry of the problem, the solution can be reduced to the form:

$$w^i = A_1 K_1(x) + A_2 K_2(x) + w^*, \tag{5}$$

where $K_1(x) = ch\alpha x \cdot \cos\beta x$; $K_2(x) = sh\alpha x \cdot \sin\beta x$; $K_3(x) = sh\alpha x \cdot \cos\beta x$;

$K_4(x) = ch\alpha x \cdot \sin\beta x -$ O.M. Krylov – V.Z. Vlasov fundamental functions.

3. Solution of the problem. Expression for the contact pressure under the stamp, the relationship between the dimensionless force $\overline{P}$, the size of the contact region and the stamp settlement can be written as follows:

$$q(x) = -A_1\left(C_1K_1(x) - C_2K_2(x)\right), \ \frac{1}{\overline{P}} = (1-\theta) + \frac{C_3K_1(a) - C_4K_2(a)}{l\left(C_5K_3(a) - C_6K_4(a)\right)}, \tag{6}$$

$$\overline{\delta} = \theta(2-\theta) + \frac{2}{3}\overline{P}(1-\theta)\left[(1-\theta)^2 + 3(\varepsilon_1' - \nu'')\right] - \frac{(1-\theta)P}{2\omega l} \cdot \frac{C_7K_3(a) - C_8K_4(a)}{C_5K_3(a) - C_6K_4(a)},$$

where $\overline{P} = \frac{Pl}{4\omega EI}$, $\overline{\delta} = \frac{\delta}{\beta\omega l^2}$, $\theta = \frac{a}{l}$; C_i – known constants.

The solution of a similar problem in the exact formulation of the plane problem of the theory of elasticity for an isotropic material was obtained in [3,5,8].

On the basis of the given formulas it is easy to show that at the decision of this type of contact problems it is necessary to consider cross compression even in case of very thin rods $(l/h = 200)$ and at large contact regions. So, M.M. Filonenko-Borodych [2] gives a numerical example for a steel strip $\left(E = 2{,}1\cdot 10^6 \text{кг}/\text{см}^2, \nu = 1/3\right)$ of length $2l = 20\text{см}$, width $t = 1\text{см}$ and thickness $2h = 0{,}1\text{см}$, which is pressed against the absolutely rigid base of the pattern by concentrated forces P', attached to its ends. The radius of curvature of the base of the pattern was equal to $R = 100\text{см}$. It was necessary to determine the magnitude of the forces applied to the strip to bend it along a given pattern, as well as to investigate the nature of the pressure distribution on the strip. M.M. Filonenko-Borodych finds the solution of the problem with the help of a theory that takes into account only the influence of the deformation of the transverse shear. According to the solution given in the paper, the magnitude of these forces is equal to $P' = 33{,}9\text{кг}$ each. The maximum pressure is reached at the ends of the strip and is equal to . The results obtained using the corresponding formulas (6) give the following values for maximum forces and maximum pressure:

$$P' = 0{,}5P = 25{,}2\text{кг}., \quad q_{\max} = 237\text{кг}/\text{см}.$$

4. Analysis of numerical calculation results. In fig. In Fig. 1 shows the graphs of the distribution of contact pressure in the mentioned thin steel strip under the stamp. The dashed line on it corresponds to the solution of M. Filonenko-Borodych [2] according to the theory of Tymoshenko-type rods. The solid line is constructed by formula (6).

Analysis of the nature of the curves in Fig.1 shows that the reaction of the pattern to the strip is concentrated at its ends in regions the size of 2*см*. The other part of the pattern is almost free of load. Also, from the numerical data for P' and $q_{\max}$ it is seen that the solution of MM Filonenko-Borodych gives inflated (by 34.5%) values both for the forces required to keep the strip in a given state, and for the contact pressure (almost twice) under the pattern. These results once again confirm the conclusion that solutions obtained on the basis of equations that do not take into account the compression over the thickness of the rod for this type of contact problems can lead to significant errors.

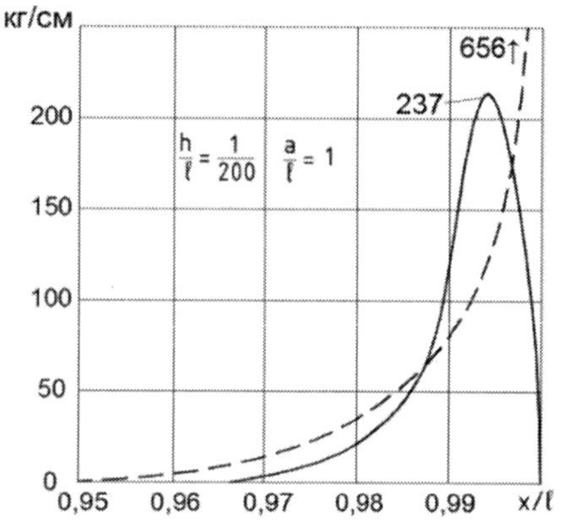

Fig.1. Distribution of contact pressure in a long rod

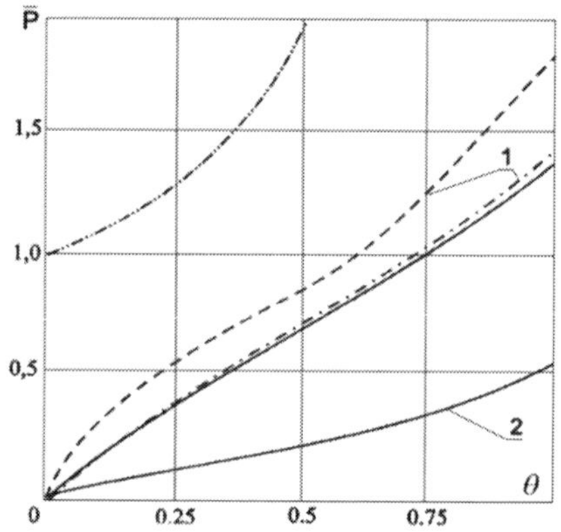

Fig.2 Dependence of the force applied to the stamp on the size of the contact region

In more detail, on the basis of the developed formulas (6), the study of the stress-strain state of the strip supported on two supports and loaded with a stamp of parabolic shape. The basis of the stamp is described by the function $f(x) = \delta + x^2 / 2R$. The supports are placed symmetrically relative to the stamp at a distance from each other. Here it is marked δ – stamp settlement, R – stamp base curvature radius.

Fig. 2 shows graphs characterizing the change in force $\bar{P} = Pl / 4\omega EI$ from the size of the contact area for the short beam ($h/l = 0,5$) according to the formula (6). The dashed curve is found according to the solution of M. Filonenko-Borodych [2] according to the formulas of the theory of Tymoshenko's type, and the dashed curve - the exact solution [3]. The double dashed line indicates the curve corresponding to the Bernoulli-Euler model of thin rods.

Curves 1 correspond to the isotropic case, and curve 2 to the transversely isotropic case (fiberglass 27-63C), with the following anisotropy parameters: $E/E'=5$, $G/G'=3$, $\nu=0,25$, $\nu''=0$.

From the analysis of the graphical data presented in Figure 2, it is seen that the relationship (6) between the magnitude of the contact area and the force applied to the stamp in the case of isotropic material is very close to exact solutions [3,8] and in the case of thin strips the curves almost coincide. At the same time, the double dash-dotted curve shown in Fig. 2 shows that the Bernoulli-Euler theory in these dependences leads to a number of physical inconsistencies. For example, according to this theory to make full contact of the rod with the stamp $(\theta=1)$ it is necessary to apply infinitely great force P to the latter, which even at zero contact $(\theta=0)$ and the weightless stamp must take values other than zero.

Analysis of the nature of the curves in Fig. 2 also shows that to achieve the same value of the contact region to the isotropic band it is necessary to apply a much greater (approximately three times) force P, than to the corresponding orthotropic strip with relations E/E', $G/G'>1$.

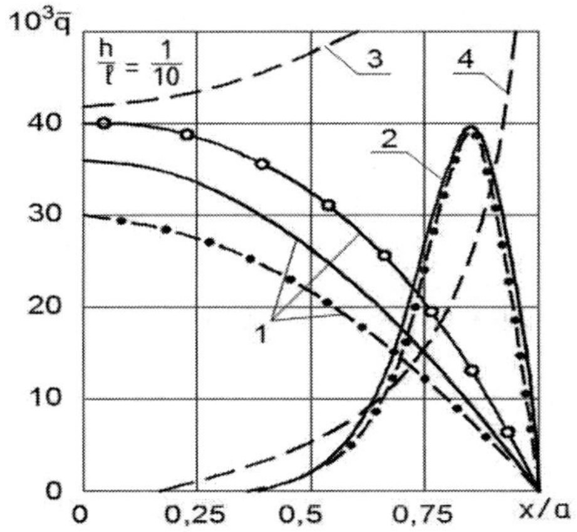

Fig.3. Distribution of contact pressure in the isotropic band

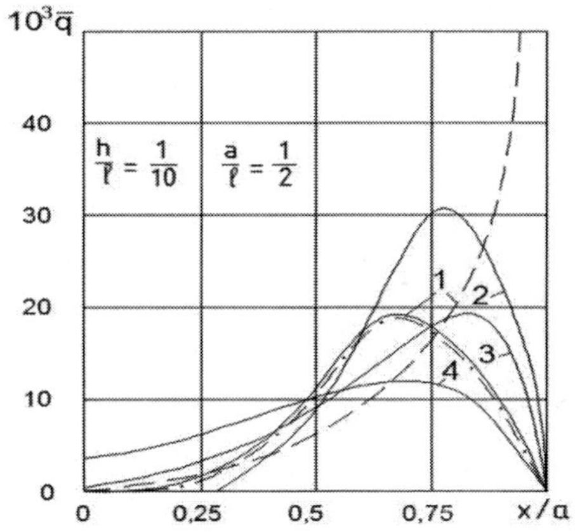

Fig.4. Influence of material anisotropy on the distribution of contact pressure in an orthotropic beam

Figures 3 and 4 show graphs of changes in contact pressure $\bar{q}=q/(Gl\omega)$ for isotropic and transtropic strips, respectively, for different values of the contact area and material, as well as the calculation model of the problem. In Fig. 3 curves 1, 3 are constructed for the size of the contact region $\theta = a/l = 0{,}1$, and curves 2,4 for $\theta = a/l = 0{,}8$. In particular, curves 1,2 (solid) are constructed by formula (6), and curves 3,4 (dashed) are found according to the solution of the shift theory of Tymoshenko-type beams [2]. The dashed-dotted curves are taken from [3], which are constructed according to the formulas of the exact solution of the plane problem of the theory of elasticity.. Curve 1, with circles, corresponds to the previous solution of the authors when the parameter $\varepsilon_2 \equiv 0$. From the graphs of changes in the reduced contact pressure $\overline{q}$ (curves 1,2) it is seen that the proposed model of bending of beams gives results quite close to the results of the plane problem of the theory of elasticity: for small regions of contact ($\theta < 0{,}2$) the nature of the distribution of contact stresses is close to the Hertz distribution - the value $q_{\max}$ is in the center of the contact region (curve 1); for large regions - , the maximum contact stress approaches the edge of the contact region (curves 2).

At the same time, solutions of the problem with the help of theories that take into account only the influence of the deformation of the transverse shear (dashed curves 3,4), give results that do not correspond to the real distribution of contact pressure under the stamp.

In fig. 4 it is studied the influence of material anisotropy on the distribution of contact pressure in the transtropic beam-strip depending on the values of the contact region and the characteristics of the material. In particular, curves 1 (solid, dashed and dashed) are constructed for isotropic material. Curves 2,3,4 are constructed for transverse-isotropic material with characteristics: $\frac{E}{E'} = 1$, $\frac{G}{G'} = 0$; $\frac{E}{E'} = 2$, $\frac{G}{G'} = 1$, $\nu = \nu'' = 0{,}25$ i $\frac{E}{E'} = 5$, $\frac{G}{G'} = 3$, $\nu = 0{,}25$ and $\nu'' = 0$ respectively. The dashed-dotted curve is taken from [3], and the dashed one is constructed by the formulas of the shear theory of Tymoshenko-type beams [2].

From the analysis of the nature of the contact pressure distribution in Fig. 3,4 we can conclude that the ratio of the characteristics of the transverse anisotropy - E/E' and G/G' differently affect the magnitude and distribution of contact pressure in the beam. Thus, the increase in the transverse shear modulus G' пleads to an increase in contact pressure (curves 1,2) and vice versa - a decrease in the shear modulus G' leads to a decrease in contact pressure (curves 1, 4). At the same time, increase the ratio E/E' leads to the spread of the magnitude of the contact pressure region towards its center.

5. Conclusion. The character of the contact pressure distribution under the rigid stamp along the surface of the contact region of the transverse isotropic beam is investigated. Based on the analysis of graphical data presented in Fig. 1-4, it is concluded that obtaining the real distribution of contact pressure on the surface of the beam-strip is impossible without the use of computational equations of nonclassical models (or the plane problem of the theory of elasticity), which take into account the deformation of transverse compression. It is shown that the values of the transverse shear modulus G' and modulus of elasticity E' differently affect the value (modulo) of the contact pressure and its distribution relative to the contact region: increase the transverse shear modulus G' leads to an increase in contact pressure and a decrease in the shear modulus G' leads to a decrease in contact pressure. Modulus of elasticity E' change has a greater effect on the very nature of the distribution curves and its reduction leads to a decrease in its maximum and the spread of the magnitude of the contact pressure region towards its center.

References

1. Timoshenko S.P., Lessels Dzh. Prikladnaya teoriya uprugosti. M.-L.: Gos. nauchno-tehn. izd.-vo, 1931. 391 s.
2. Filonenko-Borodich M.M. Izgib tonkogo sterzhnya po za - dannoy krivoy // Trudyi Mosk.el.-meh. in-ta inzh. transporta, 1949. V.58. S. 3-10.
3. Keer L.M., Silva M.A.G. Bending of a cantilever brought gradually into contact with a cylindrical supporting surface. – Int.J. Mech. Sci. Pergamon Press. 1970. Vol.12, №9, p.751-760.
4. Peleh B.L., Shvabyuk V.I. Ob odnom obobschenii teorii uprugih transversalno-izotropnyih plit primenitelno k nekotoryim kontaktnyim zadacham //Soprotivlenie materialov i teoriya sooruzheniy. K.: Budlvelnik, 1975. V.26. S.40-45.
5. Grigolyuk E.I., Tolkachev V.M. Kontaktnyie zadachi teorii plastin i obolochek. - M.: Mashinostroenie, 1980. 416 s.
6. Sarkisyan S.O. O tsilindricheskom izgibe plastinki zhestkimi shtampami // Dokladyi AN Arm.SSR, 1977. T.69. №4. S.216-223.
7. Shvab'yuk V.I., Maksimovich Ya.V. Kontaktna zadacha teorIYi pruzhnostYi dlya smugi, scho rozmIschena na oporah // VIsnik LvIv. un-tu. SerIya meh.- mat., 2000. Vip. 57. S.195-198.
8. Shvabyuk V.I., Pasternak Y.M., Rotko S.V. Refined solution of the Timoshenko problem for an orthotropic beam on a rigid base // Materials Science. – 2010. – Vol. 46. – No. 1. – P. 56–63.
9. Shvabyuk V.I., Rotko S.V., Uzhegova O.A. Bending of a Composite Beam with a Longitudinal Section . // Strenght of Materials. – 2014. – Vol. 46. – No. 4. – P. 558–566.
10. Goryk A.V. and S.B. Koval'chuk Solution of a Transverse Plane Bending Problem of a Laminated Cantilever Beam Under the Action of a Normal Uniform Load // Strength of Materials. – 2018. – Vol. 50, Iss. 3. – pp.406-418.

Corrosion-resistant epoxy coating for protection of structural glued-laminated timber in high humidity conditions

A Skripinets[1,*], N Saienko[1], Y Danchenko[1] and A Cherkashina[2]

[1]Kharkiv National University of Civil Engineering and Architecture, Sumska Str. 40, 61002, Kharkiv, Ukraine

[2]National Technical University «Kharkiv Polytechnic Institute» Kyrpychova str., 2, 61002, Kharkiv, Ukraine

[*]E-mail: a.skripinits87@gmail.com

Abstract. The article presents a solution to practical problem - the development of epoxy coating of low combustibility with corrosion-resistant properties for the protection of glued-laminated timber in conditions of high humidity. Based on the test results obtained, it was found that when using the developed epoxy polymer protective coating, the resistance of wooden beams to distilled and seawater increases by more than two orders of magnitude, ensuring the integrity of wood products and structures, protecting the ends and not causing warping of the material. In conditions of medium aggressiveness of the environment i.e. 10% H_2SO_4, 10% NaOH, 10% NaCl, the application of the developed epoxy polymer coating to glued-laminated timber provides their high indicators of bio- and chemical corrosion protection (water resistance, resistance to sea water and aggressive aqueous solutions). The developed protective coating has fungicidal properties, protecting structures from the appearance of blue stains, mold and wood-destroying fungi. Revealed a good correlation between the concentration of biocide and fungicidal coating properties. A comprehensive assessment showed that the developed epoxy coatings have increased biocorrosion-resistant properties and can be used as protective coatings for glued-laminated timber structures exposed to mold fungi, high temperatures and high humidity.

1. Introduction

Innovative technologies in the woodworking industry make it possible to obtain wood-based structural composite materials that exceed solid wood in terms of physical and mechanical characteristics. In modern practice, glued timber is widely used, obtained by gluing sheets of peeled coniferous veneer with phenol-formaldehyde glue. The technology for the production of glued timber of this type allows reducing the negative impact of natural deficiencies of wood, significantly increases the levels of its strength indicators [1].

Such glued materials make it possible to create new structural systems and architectural forms, expand the scope of wood use - low-rise residential buildings, public - buildings and some industrial and agricultural structures, where wood is used in load-bearing and enclosing structures. Compared to similar reinforced concrete structures, the use of glued wood structures allows to reduce the weight of the structure by 4-5 times, labor intensity of manufacturing and installation by more than 2 times [2].

The greatest economic effect from the use of glued structures is achieved when they overlap large spans (18-36 m) - such spans have cinemas, indoor markets, swimming pools, exhibition halls, athletics arenas, equestrian facilities, and they are also used in buildings and structures exposed to chemicals aggressive environment. The complex application of load-bearing timber glued structures together with lightweight enclosing structures is also practiced. In Europe, water parks and indoor pools are very popular, the vaults of which are usually made of glued wooden structures [3]. Also, the trends of recent years have shown an increase in interest in glued wooden structures in the countries of the equatorial belt, as well as in countries on the coast of the seas in other belts, where there is high humidity.

Content from this work may be used under the terms of the Creative Commons Attribution 3.0 licence. Any further distribution of this work must maintain attribution to the author(s) and the title of the work, journal citation and DOI.
Published under licence by IOP Publishing Ltd

However, significant disadvantages of such glued wooden structures in conditions of high humidity are low resistance to water and aggressive environmental action, aqueous solutions and biological damage.

Wooden products and structures that are constantly (regularly) exposed to salt (sea) water according to BS EN 335:2013 «Durability of wood and wood-based products. Use classes: definitions, application to solid wood and wood-based products» belong to the class of wood exploitation Hazard Class 5.

Under conditions simulating external environmental influences of medium aggressiveness (high humidity over 60-90% and temperatures over 20 °C, as well as aggressiveness of the environment), it is imperative to provide for constructive and chemical methods of protecting structures and elements based on glued wood.

The chemical measures to protect such wooden structures from bio- and chemical corrosion in conditions of high humidity and medium aggressiveness of the environment i.e. 10% H_2SO_4, 10% NaOH, 10% NaCl provide application of moisture-proof and chemically resistant moisture-proof coatings on wooden structures, such as acrylic, acrylic-alkyd, alkyd, urethane-alkyd, epoxy [4-6] according to SR 28.13330.2012 «Protection of building structures from corrosion».

The modern market for such protective coatings for glued timber is diverse and has a number of advantages and disadvantages when providing one or another protective function. So, as antiseptics, impregnating primers Belinka Impragnan, Belinka base, Pinotex Wood Primer, Fentak are used, intended for the main preventive protection of wood from biological pests (insects, blue fungus and rotting). Protective and decorative (moisture-proof) coatings such as Belinka exterier, yacht varnish Protex Yat Vernik, MeritYahti 30, Poli-R, Miranol paint, enamel PF-115, Enamel, Master 30 provide moisture resistance of structures, while the application of such coatings is carried out on already primed wooden surfaces.

However, the selection of such protective coatings for glued wood is complicated by the fact that the technology of its manufacture includes gluing wood elements with polymer adhesives of various chemical natures.

It is noted [7] that the presence of a polymer adhesive interlayer leads to a change in the structure of wood and its properties, which must be taken into account when selecting bio- and chemical protective agents. The phenol-formaldehyde adhesive used in Ultralam™ is not resistant to water and aqueous solutions, which eliminates the use of water-based protection. Taking into account all these factors significantly narrows the range of products that can be used for bio- and chemical protection of glued-laminated timber Ultralam™.

In modern construction, methods of applying polymer coatings to the surface of wooden structures and materials are increasingly used, which at the same time can protect wood from the effects of high air humidity, corrosive gases, fire and fungi, increasing the service life of wooden structures [8].

A promising direction for a comprehensive solution to these problems is the development of an epoxy polymer coating of low combustibility with corrosion-resistant properties for the protection of glued-laminated timber in conditions of high humidity. The advantage of epoxy polymer compositions over others is the ability to combine all the necessary properties in one material, they have high adhesion to wood and metal, and the technology of their preparation and application can take into account external climatic conditions [9-12].

According to [13] Ultralam™ belongs to the following fire hazard groups: E - highly flammable, C_{fl} - with a moderately spreading flame, s_3 - with a high smoke-generating ability, with high toxicity during combustion according to EN 13501. To improve the fire hazard properties of Ultralam™, as well as to expand the area of use, treatment with protective polymer coatings is necessary.

Earlier, in our work [14], it was found that the developed two-component composition based on epoxy oligomers, inorganic flame retardants and amine hardeners for fire protection of glued wood Ultralam™ provides a transfer according to the flammability group of group E building materials to A2, according to EN 13501.

However, it is necessary to solve the following tasks for the use of this composition as a protective coating with increased corrosion-resistant properties for glued-laminated timber in conditions of high humidity:
1. To assess the effect of a fire retardant on water and chemical resistance, as well as a biocide on fungal resistance of epoxy composites.
2. Determine the main physical and mechanical properties of the developed coating and carry out its comprehensive assessment to protect glued timber in conditions of high humidity.

2. Materials and methods

To obtain a protective coating used epoxy oligomer ED-20 with content of epoxy groups of 20.5%; modifying additive Laprolat-803 with the content of cyclocarbonate groups - 27.1% and epoxy groups - 2.17%; hardener PEPA; fire retardant - Exflam APP-201 with a particle size of 8-12 microns; modifying additives PMS-10, as well as biocide (guanidine phosphate).

The content of the fire retardant EXFLAME APP-201 in epoxy amine compositions is selected in the amount of 7-14 wt. %, which made it possible to increase the oxygen index of the coverage to 41% [15]. The content of biocide (guanidine phosphate) in epoxy amine compositions is selected in the amount of 0.2-5 wt. %.

To determine the resistance to water and aqueous solutions, samples of Ultralam™ beams (30x30x70 mm.) with and without coating were kept at a temperature of 18-25 °C in distilled water and in a solution whose salt composition imitated sea water. At the same time, attention was paid to the fact that the beams were completely immersed in solutions throughout the exposure time. Water resistance and resistance to sea water (w, %) was calculated as the relative weight gain over a certain time.

The tests of fungus resistance carried out in two ways. According to method «A», the samples were infected with an aqueous spore suspension of Trichoderma Sp. and exposed for 56 days with an intermediate view after 28 days. This method was used to determine the fungal resistance of materials by the degree of use of the latter as a food source for microscopic fungi.

Method «B» was used to determine the fungicidal properties of new polymeric materials and fungal resistance of the material by the degree of suppression and (or) suppression of the growth of those fungi on it, which are assimilated by other, besides the material itself, life-giving substrates. Experimental samples were placed in Petri dishes on agar nutrient medium, infected with a spore suspension of the fungus Trichoderma Sp in the same medium, diluted with water, and exposed for 28 to 84 days.

Some indicators of physical and mechanical characteristics of glued-laminated timber structures were determined according to standard methods.

3. Experimental

Results of water resistance and resistance to sea water (w, %) are shown in Figure 1-3. From the data obtained (Figure 1) it is seen that most intensively samples of glued-laminated timber Ultralam™ without a protective coating absorb water during 5 hours (0.167 days) exposure in distilled and sea water. The samples collect the maximum amount of water (up to 80% in distilled water and up to 70% in seawater) within 4 days. With further keeping the samples in water for the next day, the water absorption by the samples fluctuates within 1-2%.

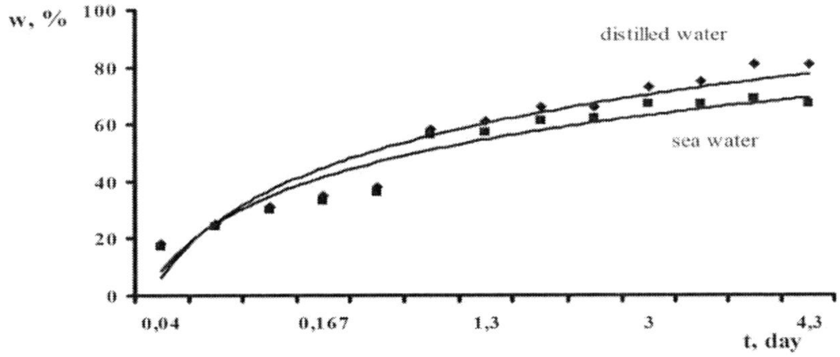

Figure 1. Water resistance and resistance in sea water of uncoated glued-laminated timber.

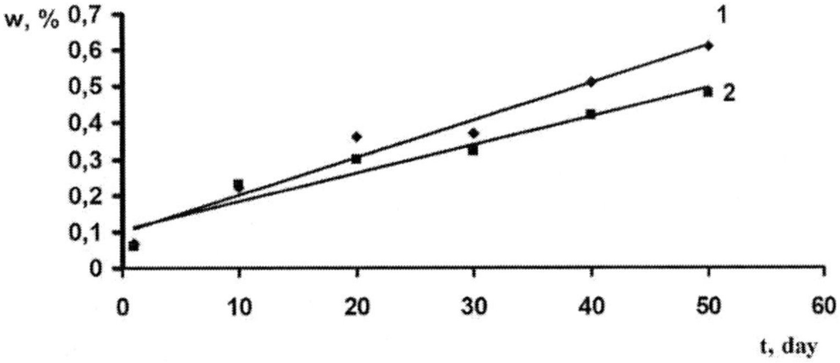

Figure 2. Water resistance of glued-laminated timber coated of epoxy composition:
1 - 7 wt. % and 2 - 14 wt. % fire retardant.

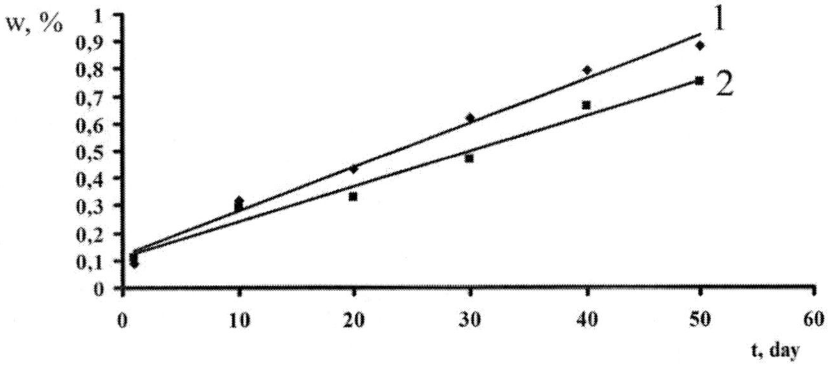

Figure 3. Sea water resistance of glued-laminated timber coated of epoxy composition:
1 - 10.5 wt. % and 2 - 14 wt. % fire retardant.

According to earlier studies [16] on strength, it was found that under exposure to moisture (soaking and boiling), the strength of adhesive joints decreases by 46% and 55% for wet and dry Ultralam^{TM} samples, respectively.

The decrease in strength under such temperature and humidity effects is largely due to the fact that wood and glue are characterized by unequal swelling coefficients. Thus, with a change in humidity in the area of the adhesive joint, the amount of deformation of the adhesive joint occurs, which entails the occurrence of stresses that form concentrations in the areas of joint of materials with different physical and mechanical properties. The stress concentration zone can be the places of the beginning of destruction under the action of moisture stresses, as well as when they are combined with stresses from an external load.

Humidification adversely affects the sides of the ends, as a result of which uneven stress occurs, which leads to delamination of the material. Humidification of the planes causes warping of the material, which is the reason for a decrease in the operational reliability of structures.

In Figure 1 and 2 show that when using a protective coating, the resistance of the beams to distilled and seawater increases by more than two orders of magnitude. Increasing the amount of fire retardant to 14 wt. % does not impair the protective properties of the coating. Also, the protective coating ensures the integrity of the product and structures of Ultralam™, protecting the ends and not causing warping of the material.

Research samples in aggressive environments i.e. 10% H_2SO_4, 10% NaOH, 10% NaCl were carried out for 2000 hours in accordance with GOST 12020-72. The results are shown in Table 1.

Table 1. The results of testing the resistance of samples to aggressive environments.

Indicator name	Indicator value
Resistance to 10% H_2SO_4 at a temperature of 20 ± 5 °C, %	1.85-2.05
Resistance to 10% NaOH at a temperature of 20 ± 5 °C, %	0.54
Resistance to 10% NaCl at a temperature of 20 ± 5 °C, %	1.08-1.20
Resistance to H_2O at a temperature of 20 ± 5 °C, %	0.2

As can be seen from Table 2, under conditions of medium aggressiveness of the environment i.e. 10% H_2SO_4, 10% NaOH, 10% NaCl, the application of the developed epoxy polymer coatings to glued-laminated timber structures ensures their high indicators of chemical corrosion protection.

Under conditions of operation with high or variable humidity, the developed protective coating must have fungicidal properties in order to protect structures from the appearance of blue stains, mold and wood-destroying fungi. Therefore, fungi resistance tests of epoxy polymers were carried out. The essence of the method lies in the exposure of materials contaminated with mold spores in conditions that are optimal for the development of mold.

The growth rate of microorganisms on the material is the opposite of the fungus resistance of the material. The highest score for fungal resistance is «0» (no growth of test cultures on the material when observed under a microscope). The lowest - score «5» (visual inspection clearly shows the development of fungi, covering 25% of the tested surface). The results of testing epoxy polymers are shown in Table 2.

Established based on the presented data Table 2:

• according to method «A», the development of fungi was observed under a microscope and only in sample No. 1, into which no biocide was added. In all other variants, with different quantitative biocide ratios, the development of fungi was not detected. According to the test results according to method «A», all investigated samples are fungus-resistant;

• according to method «B», the control samples were overgrown with fungal spores on the 84th day, which was accompanied by intensive sporulation of biotests (5-point degree of development of micromycetes). On the studied samples, the development of the fungus was noted only after 28 days of exposure. It is shown that samples № 3, 4 have an expression of fungicidal properties.

Table 2. Assessment of the resistance of the developed epoxy polymer coating to the effects of mold fungi.

№ composition	Biocide content, wt. %	Degree sample contamination						
		Method «A», 24 hours			Method «B», 24 hours			
		5	28	56	5	28	56	84
1	0	0/0	2/1	4/1	20/1	30/5	50/5	50/5
2	0.2	0/0	0/0	0/0	0/0	10/4	20/4	20/4
3	2	0/0	0/0	0/0	0/0	5/3	10/3	10/3
4	5	0/0	0/0	0/0	0/0	2/1	2/1	5/1

Note: the numerator shows the affected area of the samples, %, the denominator - the assessment of fungal resistance, points.

As a result of the research a stable correlation was found between the concentration of the biocide and the fungicidal properties of the polymer.

The main physical and mechanical properties of the developed coating have been determined and a comparative characteristic of glued-laminated timber structures with and without coating has been carried out (Table 3).

Table 3. Basic physical and mechanical characteristics of glued-laminated timber structures with and without coating.

Indicator name	Indicator value: without epoxy coating	Indicator value: with epoxy coating	Test Method
Flammability group	E	A2	EN 13501
Flame spread	C_{fl} -	B_{fl} -	EN 11925
Smoke generating ability	s_3	s_2	EN 9239-1
Adhesion strength, MPa	-	13.2	GOST 14760-69
Abrasion index, cm^3/m	0.36	0.19	ISO 9352
Impact strength, kJ/m^2	6.6	9.4	ISO 8256:2004
Resistant to mold fungi	Not fungus resistant	Fungus-resistant	GOST 9.48-9.049-75
Smoke production coefficient, m^2/kg	974	480	EN 13823:2010
Water resistance,%	20	0.2	GOST 12020-72

As a result of comprehensive studies, it was shown that the developed coating has a complex of improved physical and mechanical properties: high adhesive strength, increased wear resistance, impact strength, which indicates its high operational capabilities and the possibility of using it as a protective coating for glued wood structures exposed to mold. mushrooms, high temperatures and high humidity.

4. Conclusions

As a result of the conducted studies on water and chemical resistance, it was found that when using a protective coating, the resistance of glued-laminated timber to distilled and sea water increases by more than two orders of magnitude, ensuring the integrity of wooden products and structures, protecting the ends, and not causing warping of the material.

In conditions of medium aggressiveness of the environment (in conditions of 10% H_2SO_4, 10% NaOH, 10% NaCl), the application of the developed epoxy polymer coating to wooden structures provides their high indicators of bio- and chemical corrosion protection (water resistance, resistance to sea water and aggressive aqueous solutions).

The developed protective coating has fungicidal properties, protecting structures from the appearance of blue stains, mold, and wood-destroying fungi. Revealed a good correlation between the concentration of biocide and fungicidal coating properties.

A comprehensive assessment showed that the use of two-component polymer composite materials based on epoxy oligomers, fire retardants, biocide and hardeners for glued-laminated timber provides a coating with increased corrosion-resistant properties, as well as high adhesion to the processed surface. therefore, the developed epoxy polymer composition can be used as a protective coating for glued-laminated timber exposed to mold, high temperatures and high humidity.

References

[1] Chernykh A, Hryhor'ev K, Koval' P and Danylov E 2012 K voprosu opredelenyya nesushchey sposobnosty nahel'nykh soedynenyy v konstruktsyyakh yz brusa, kleenoho yz odnonapravlennoho shpona (LVL) *Suchasni problemy nauky ta osvity* **4** pp 109-118

[2] Weinig Group - Weinig, Wako, Grecon, Dimter, Rainmann, Concept 2016 *Wooden house building, demonstration materials*

[3] Verbitskiy I and Verbitskaya Ye 2017 Predposylki primeneniya kleyenykh derevyannykh konstruktsiy v sovremennom stroitel'stve *Polzunovskiy al'manakh* **2** pp 75-78

[4] Saienko N and Demidov D 2016 Primary assessment of fire-retardant properties of water-dispersion acrylic coatings for thermal insulation purposes *Scientific Bulletin of Construction* **86** pp 154-157 http://nbuv.gov.ua/UJRN/Nvb_2016_86_4_33

[5] Hryhorenko O, Saienko N, Lypovyi V and Harbuz S 2020 Research of Effectiveness of Wood Fire Protection by Modified Epoxy Polymers *In International Scientific Conference on Woods & Fire Safety* pp 125-128 doi:10.1007/978-3-030-41235-7_19

[6] Hryhorenko O, Zolkina Y, Saienko N, Popov Y and Bikov R 2020 Investigation of adhesive-strength characteristics of fire-retardant epoxy polymers modified with metal-containing additives *In IOP Conference Series: Materials Science and Engineering* **907** 1 012060 doi:10.1088/1757-899X/907/1/012060

[7] Yartsev V, Antipov D and Kiseleva O 2009 Vliyaniye temperatury i UF-oblucheniya na nesushchuyu i deformatsionnuyu sposobnost' kleyenoy drevesiny *Nauch. vestn. Voronezhskogo gosud. arkh.- stroit. un-ta* **4** pp 7-13

[8] Fire protection of materials and structures *Handbook* (M.: Special equipment, 2002) 240

[9] Skripinets A V, Saienko N V, Hryhorenko O N and BerezovskiyA I 2020 Development and Evaluation of the Possibility of Using Epoxyurethane Mastic in Railway Transport *In Materials Science Forum. Trans Tech Publications Ltd* **1006** pp 273-281 doi:10.4028/www.scientific.net/MSF.1006.273

[10] Danchenko Yu M, Skripinets A V and Kabus A V 2015 Issledovaniye tekhnologicheskikh i fiziko-khimicheskikh zakonomernostey izgotovleniya vibropogloshchayushchikh izdeliy na osnove epoksiuretanovykh polimernykh kompozitsiy *Eastern European Journal advanced technology* **3/11** (75) pp 4-8

[11] Andronov V A, Danchenko Yu M, Skripinets A V and Bukhman O M 2013 Efficiency of the use of vibration-absorbing polymer coating to reduce local vibration *Naukovyi Visnyk Natsionalnoho Hirnychoho Universytetu* **6** (144) pp 85-91: http://nbuv.gov.ua/UJRN/Nvngu_2013_6_16

[12] Berezovsky A I, Skripinets A V and Popov Yu V 2012 Doslidzhennya dynamichnykh mekhanichnykh i vibropohlynayuchykh vlastyvostey epoksyuretanovykh skladiv dlya vohnevibrozakhystu metalevykh vyrobiv *Pozhezhna bezpeka: teoriya i praktyka. Zbirnyk naukovykh prats'* (Cherkasy: APB im. Heroyiv Chornobylya) **10** pp 18 - 27

[13] Wooden structures. Ultralam laminated veneer. General technical requirements: STO 36554601-021-2010 [Valid from 2010-02-17] (M.: JSC «Research Center Construction) 13

[14] Andronov V, Danchenko Y and Saienko N 2014 Evaluation of the effectiveness of the use of epoxy polymer compositions for fire protection of glued wood *Problems of fire safety* **36** pp 10-16 http://repositsc.nuczu.edu.ua/handle/123456789/469
[15] Mostovoy A 2014 Development of compositions, technology and determination of the properties of micro- and nanofilled epoxy composites for functional purposes (PhD diss., University of Saratov)
[16] Scientific and technical report «Research of the multilayer glued veneer material Ultralam with substantiation of the provisions of the organization standard. Wooden structures. Multilayer glued veneer Ultralam», *General technical requirements»* (2010)

Limit state of a rectangular reinforced section using elastoplastic material strain diagrams

Mykola Soroka

Odesa State Academy of Civil Engineering and Architecture
Didrikhsona str., 4, Odesa, 65029, Ukraine

E-mail: soroka@ogasa.org.ua

Abstract. The paper considers the methodology of constructing the strength area of a rectangular reinforced section. The notion of "section strength area" is used in the calculations of structures according to the limit equilibrium. The peculiarity of the strength region is that inside the strength region the section works in the elastic stage, and at its boundary it transits to the limit state with the possibility of unrestricted plastic deformation. The equations describing the boundary of the section's strength region are often called yield conditions. In this paper, it is assumed that the material of which the section is made and the reinforcing material deform according to the ideal elastoplastic body law. Thus, the deformation diagrams of the materials are described by the Prandtl diagram. The material, of which the section is made, has different tensile and compressive yield strengths. The reinforcing material has the same tensile and compressive yield strengths. In deriving the equations describing the boundary of the section strength area, it was assumed that the bending moment and the longitudinal force applied in the center of the rectangle act in the section. Given that the section may have asymmetrical reinforcement, different equations are used to describe the upper and lower boundaries of the strength area. In order to construct the strength area, it is necessary to solve the optimization problem of finding the extreme value of the moment taking into account the constraints (equations and inequalities) for a given value of the longitudinal force. The analysis of results obtained in this way for a symmetrically reinforced cross-section made it possible to propose a simpler technique for constructing the strength area of a rectangular reinforced cross-section without solving the optimization problem.

1. Analysis of recent research

The problem of finding the destructive load for a structure was first formulated by Galileo Galilei. But only in the first half of the twentieth century was such a problem solved in the works of Gvozdev A. A. [1], Rzhanitsyn A. R. [2], Chiras A. A. [3], Pikovsky A. A. [4] and other researchers. In most of the works it was considered that the longitudinal force has little effect on the value of ultimate load and only bending moment was taken into account in the calculation. This is true for beam systems and, to some extent, for frame systems. As for the arch systems, the failure to consider the longitudinal forces leads to errors in determining the ultimate load [3], [5], [6].

The strength areas of unreinforced cross-sections of different shapes are presented in [3], [4], [5], [6]. In [7] the strength area of reinforced concrete rectangular section with symmetrical reinforcement is presented, in [8] the construction of strength area for rectangular section with single reinforcement is shown.

The equations describing the boundaries of the strength domain were used to compose an extreme problem in order to find the ultimate load. In [5], [6] the direct method [9] was proposed to find the ultimate load. This method has some advantages over other methods and makes it easy to compose a program for calculating the ultimate load.

2. The purpose of this work is to develop a methodology for constructing the strength domain of a reinforced rectangular section made of materials whose physical properties are described by a Prandtl diagram.

Content from this work may be used under the terms of the Creative Commons Attribution 3.0 licence. Any further distribution of this work must maintain attribution to the author(s) and the title of the work, journal citation and DOI.
Published under licence by IOP Publishing Ltd

3. Materials and methods

A rectangular reinforced section made of elastoplastic materials is considered. The strength area of the section is constructed by solving an extreme problem using standard programs.

4. Research results

Consider an asymmetrically reinforced rectangular cross-section in the limiting state. In this study, it is assumed that the physical characteristics of the material of which the section is made and of the reinforcing material correspond to the characteristics of an ideal elastic-plastic body (Figure 1).

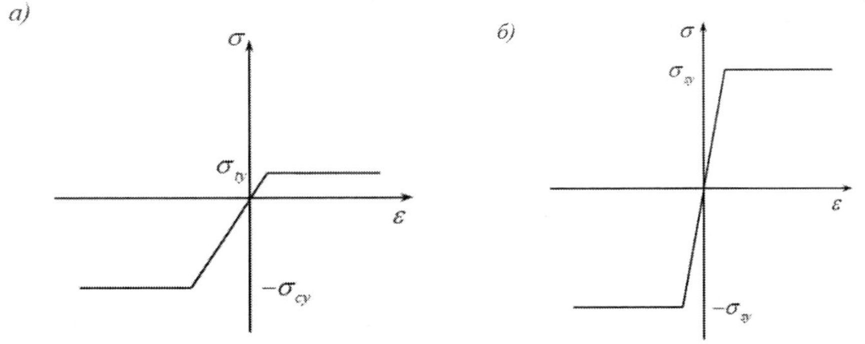

Figure 1. Deformation diagrams: a) – sectional material, б) – reinforcement.

The section is in equilibrium under the action of boundary forces M_b, N_b and forces arising in the reinforcement and the material of the section (Figure 2). For this section, it is required to obtain the dependences describing the boundary of the strength area and to plot the strength area in the coordinates "longitudinal force - bending moment".

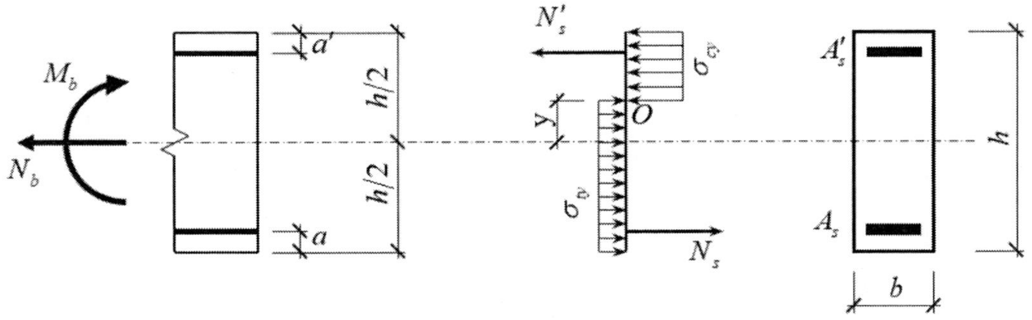

Figure 2. Scheme of forces in a rectangular cross-section under the action of moment and longitudinal force.

Write equilibrium equations:

$$\sum X = -b\sigma_{cy}(h/2-y)+b\sigma_{ty}(h/2+y)-N_s'+N_s-N_b=0; \qquad (1)$$

$$\sum m_o = M_b-0.5b\sigma_{cy}(h/2-y)^2-0.5b\sigma_{ty}(h/2+y)^2-N_s'(h/2-y-a') \\ -N_s(h/2+y-a)+N_b y=0. \qquad (2)$$

From equation (1) we obtain the expressions for y and for N_b:

$$y=\left[0.5h(\sigma_{cy}-\sigma_{ty})+\frac{1}{b}(N_s'-N_s+N_b)\right](\sigma_{cy}+\sigma_{ty})^{-1}; \qquad (3)$$

$$N_b = -b\sigma_{cy}\left(h/2 - y\right) + b\sigma_{ty}\left(h/2 + y\right) + N_s - N_s'. \tag{4}$$

Using equation (2), we write down the expression for the boundary moment:

$$M_b = 0.5b\sigma_{cy}\left(h/2 - y\right)^2 + 0.5b\sigma_{ty}\left(h/2 + y\right)^2 + N_s'\left(h/2 - y - a'\right) + N_s\left(h/2 + y - a\right) - N_b y. \tag{5}$$

The presented dependences (3) - (5) are used to construct the upper boundary of the section strength area. To construct the lower boundary of the section strength area, the dependences were obtained:

$$y = -\left[0.5h\left(\sigma_{cy} - \sigma_{ty}\right) + \frac{1}{b}\left(N_s' - N_s + N_b\right)\right]\left(\sigma_{cy} + \sigma_{ty}\right)^{-1}; \tag{6}$$

$$N_b = -b\sigma_{cy}\left(h/2 + y\right) + b\sigma_{ty}\left(h/2 - y\right) + N_s - N_s'; \tag{7}$$

$$M_b = 0.5b\sigma_{cy}\left(h/2 + y\right)^2 + 0.5b\sigma_{ty}\left(h/2 - y\right)^2 - N_s'\left(h/2 - y - a'\right) - N_s\left(h/2 + y - a\right) + N_b y. \tag{8}$$

In expressions (3) - (8) the unknowns are N_b, σ_s, σ_s'. To determine the boundary torque M_b at a given value of the longitudinal force N_b we form an optimization problem (9), (10), where the target function is the moment M_b, variable of design - stresses in the reinforcement σ_s and σ_s'. Since the upper and lower bounds of the strength area are described by different expressions, we use the solution of problem (9) to construct the upper bound, and the solution of problem (10) to construct the lower bound.

$$\begin{cases} M_b \to \max; \\ y - \left[0.5h\left(\sigma_{cy} - \sigma_{ty}\right) + \frac{1}{b}\left(N_s' - N_s + N_b\right)\right]\left(\sigma_{cy} + \sigma_{ty}\right)^{-1} = 0; \\ |y| \le 0.5h; \\ |\sigma_s| \le \sigma_{sy}; \\ |\sigma_s'| \le \sigma_{sy}, \end{cases} \tag{9}$$

where M_b is determined in accordance with (5).

$$\begin{cases} M_b \to \min; \\ y + \left[\frac{h}{2}\left(\sigma_{cy} - \sigma_{ty}\right) + \frac{1}{b}\left(N_s' - N_s + N_b\right)\right]\left(\sigma_{cy} + \sigma_{ty}\right)^{-1} = 0; \\ |y| \le 0.5h; \\ |\sigma_s| \le \sigma_{sy}; \\ |\sigma_s'| \le \sigma_{sy}, \end{cases} \tag{10}$$

where M_b is determined in accordance with (8).

Before solving problems (9), (10) it is necessary to determine the limits of change in the longitudinal force N_b.

The rightmost point of the strength region corresponds to the state when the entire section is stretched. Since the reinforcement is asymmetrical, a moment must be applied to the section to achieve this state (Figure 3).

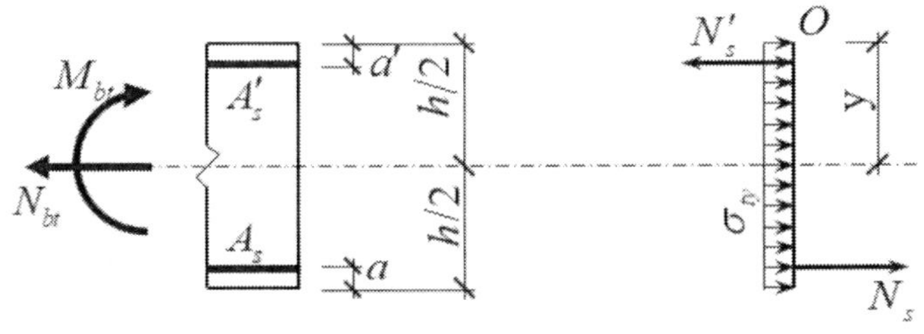

Figure 3. Scheme of the action of forces in a fully stretched section.

Write equilibrium equations:

$$\sum X = bh\sigma_{ty} + N_s - N_s' - N_{bt} = 0;$$
$$\sum m_o = M_{bt} - 0.5bh^2\sigma_{ty} - N_s\left(h-a\right) + N_s'a' + 0.5hN_{bt} = 0 .$$

From the equilibrium equations we find:

$$N_{bt} = bh\sigma_{ty} + N_s - N_s' ;$$
$$M_{bt} = 0.5bh^2\sigma_{ty} + N_s\left(h-a\right) - N_s'a' - 0.5hN_{bt} .$$

At the limit state $N_s = A_s\sigma_{sy}$; $N_s' = -A_s'\sigma_{sy}$. Because of this:

$$N_{bt} = bh\sigma_{ty} + \sigma_{sy}\left(A_s + A_s'\right); \tag{11}$$
$$M_{bt} = 0.5bh^2\sigma_{ty} + \sigma_{sy}\left[A_s\left(h-a\right) + A_s'\left(h-a'\right)\right] - 0.5hN_{bt} . \tag{12}$$

The leftmost point of the strength region corresponds to the state when the entire section is compressed. (Figure 4).

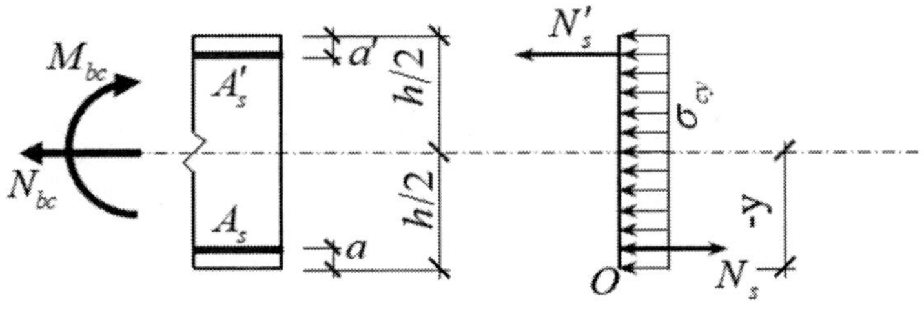

Figure 4. Scheme of forces at fully compressed cross-section.

Write equilibrium equations:

$$\sum X = -bh\sigma_{cy} + N_s - N_s' - N_{bc} = 0;$$
$$\sum m_o = M_{bc} - 0.5bh^2\sigma_{cy} + N_s a - N_s'\left(h-a'\right) - 0.5hN_{bc} = 0 .$$

From the equilibrium equations we find:

$$N_{bc} = -bh\sigma_{cy} + N_s - N_s' ;$$
$$M_{bc} = 0.5bh^2\sigma_{cy} - N_s a + N_s'\left(h-a'\right) + 0.5hN_{bc} .$$

At the limit state $N_s = -A_s\sigma_{sy}$; $N_s' = A_s'\sigma_{sy}$. Because of this:

$$N_{bc} = -bh\sigma_{cy} - \sigma_{sy}\left(A_s + A_s'\right); \tag{13}$$
$$M_{bc} = 0.5bh^2\sigma_{cy} + \sigma_{sy}\left[A_s\left(h-a\right) + A_s'\left(h-a'\right)\right] - 0.5hN_{bc} . \tag{14}$$

The upper and lower extremum points of the boundary of the strength area are defined at $y=0$. From expressions (4) and (7) we find:

$$N_b = -0.5bh\left(\sigma_{cy} - \sigma_{ty}\right) - N_s + N_s'.$$

From expressions (5) and (8) we find:

$$M_{b,\max} = 0.125bh^2\left(\sigma_{cy} + \sigma_{ty}\right) + N_s\left(h/2 - a\right) + N_s'\left(h/2 - a'\right).$$

The coordinates of the point of the upper extremum are determined at $\sigma_s = \sigma_{sy}$; $\sigma_s' = \sigma_{sy}$:

$$N_b = -0.5bh\left(\sigma_{cy} - \sigma_{ty}\right) - \sigma_{sy}\left(A_s - A_s'\right); \tag{15}$$

$$M_{b,\max} = 0.125bh^2\left(\sigma_{cy} + \sigma_{ty}\right) + \sigma_{sy}\left[A_s\left(h/2 - a\right) + A_s'\left(h/2 - a'\right)\right]. \tag{16}$$

The coordinates of the point of the lower extremum are determined at $\sigma_s = -\sigma_{sy}$; $\sigma_s' = -\sigma_{sy}$:

$$N_b = -0.5bh\left(\sigma_{cy} - \sigma_{ty}\right) + \sigma_{sy}\left(A_s - A_s'\right); \tag{17}$$

$$M_{b,\min} = 0.125bh^2\left(\sigma_{cy} + \sigma_{ty}\right) - \sigma_{sy}\left[A_s\left(h/2 - a\right) + A_s'\left(h/2 - a'\right)\right]. \tag{18}$$

4.1. Example 1

Construction of the Strength Area for a Rectangular Symmetrically Reinforced Section.

Consider a rectangular section with the characteristics: $b \times h = 0.2 \times 1.0м$, $E_c = 2.3 \cdot 10^7\, кН/м^2$, $\sigma_{cy} = 14500кН/м^2$, $\sigma_{ty} = 1300кН/м^2$, $E_{sy} = 2.1 \cdot 10^8\, кН/м^2$, $\sigma_{sy} = 365000кН/м^2$, $A_s' = A_s = 6.283см^2$ (2Ø20), $a' = a = 0.03м$.

Limits of variation of longitudinal force:

$$N_{bc} = -\left(bh\sigma_{cy} + 2A_s\sigma_{sy}\right) = -3358.67кН,\quad N_{bt} = bh\sigma_{ty} + 2A_s\sigma_{sy} = 718.67кН.$$

Coordinates of the extreme points of the section strength area are determined using the dependencies (15) – (18):

$$N_b = -0.5bh\left(\sigma_{cy} - \sigma_{ty}\right) = -1320.0кН;$$

$$M_{b,\max} = 0.125bh^2\left(\sigma_{cy} + \sigma_{ty}\right) + A_s\sigma_{sy}\left(h - a - a'\right) = 610.576кНм;$$

$$M_{b,\min} = 0.125bh^2\left(\sigma_{cy} + \sigma_{ty}\right) - A_s\sigma_{sy}\left(h - a - a'\right) = -610.576кНм.$$

To solve optimization problems (9) and (10) we use EXCEL. The boundary of the strength area is shown in Figure 1. Moreover, the areas EG, BD, EP, RD are described by linear functions, and the areas GJB and PQR are described by quadratic functions.

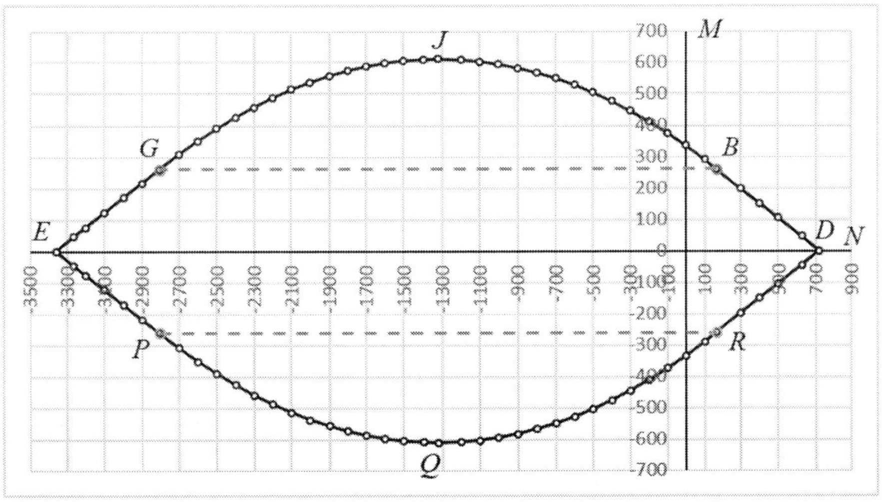

Figure 5. Strength area of a rectangular section with symmetrical reinforcement.

Plots of stresses in the reinforcement and coordinates of the bending center of the section at $M_b > 0$ are presented in Figures 6, 7. It should be noted that these graphs are broken lines, the break points of which correspond to the same values of N_b.

Analyzing the graphs shown in Figures 6, 7, we can note their characteristic features, namely:

- on segment EF $\sigma_s = -\sigma_{sy}$, $\sigma_s' = \sigma_{sy}$, $-(h/2-a) \le y \le h/2$;
- on segment FG $-\sigma_{sy} < \sigma_s < \sigma_{sy}$, $\sigma_s' = \sigma_{sy}$, $y = -(h/2-a)$;
- on segment GB $\sigma_s = \sigma_s' = \sigma_{sy}$, $-(h/2-a) < y < (h/2-a')$;
- on segment BC $\sigma_s = \sigma_{sy}$, $-\sigma_{sy} < \sigma_s' < \sigma_{sy}$, $y = h/2-a'$;
- on segment CD $\sigma_s = \sigma_{sy}$, $\sigma_s' = -\sigma_{sy}$, $(h/2-a') \le y \le h/2$.

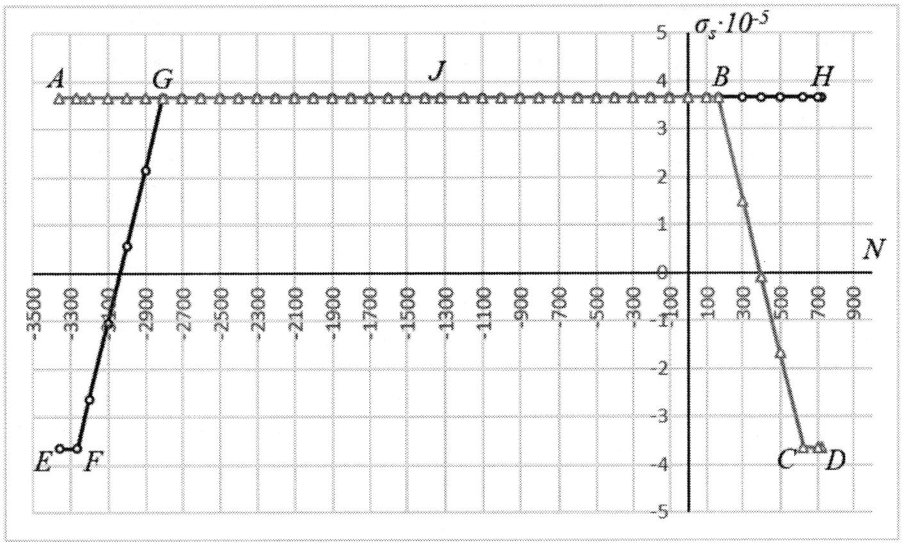

Figure 6. Stresses in the reinforcement for the upper boundary of the section strength area with symmetrical reinforcement – ─o─ – A_s, ─△─ – A_s'.

The coordinates of the break points of the graphs (Figures 6, 7), calculated by (4), (5) are presented in Table 1.

For the case where $M_b < 0$ (the lower boundary of the section strength area) the graphs of force changes in the reinforcement and the graph of bending center coordinate changes will be symmetrical to the graphs presented in Figures 6, 7 relative to the horizontal axis.

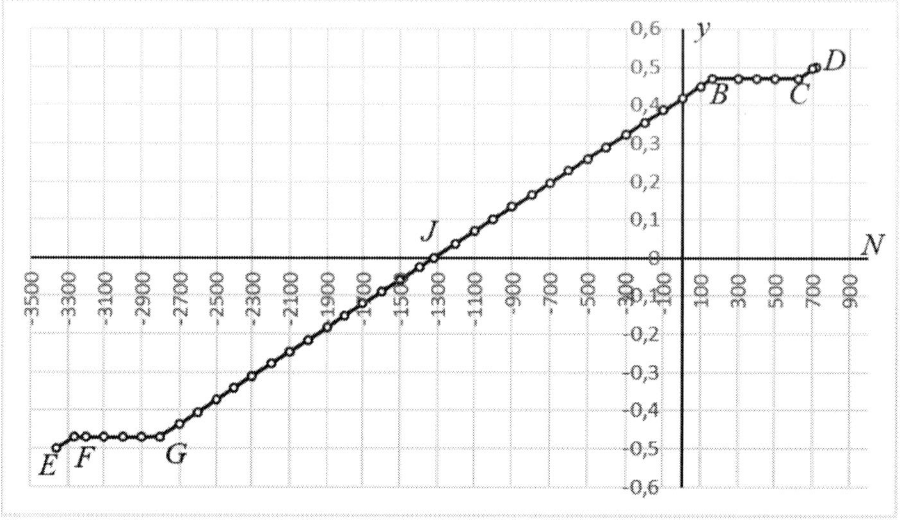

Figure 7. Change in the coordinate of the bending center of a section with symmetrical reinforcement for the upper boundary of the section strength area.

Table 1. The values of the limit forces at the break points of the graphs in Figures 6 and 7.

Point	y	σ_s	σ_s'	N_b, *kN*	M_b, *kNm*
E	$-h/2$	$-\sigma_{sy}$	σ_{sy}	-3358.67	0
F	$-(h/2-a)$	$-\sigma_{sy}$	σ_{sy}	-3263.873	45.978
G	$-(h/2-a)$	σ_{sy}	σ_{sy}	-2805.200	261.554
J	0	σ_{sy}	σ_{sy}	-1320.0	610.576
B	$h/2-a'$	σ_{sy}	σ_{sy}	165.200	261.554
C	$h/2-a'$	σ_{sy}	$-\sigma_{sy}$	623.873	45.978
D	$h/2$	σ_{sy}	$-\sigma_{sy}$	718.67	0

Taking into account that the coordinate of the bending center and the stresses in the reinforcement vary linearly, the boundary of the section strength area can also be constructed without solving problems (9), (10). The order of constructing the upper boundary of the section strength area is as follows:

– at a given value of the longitudinal force, we calculate the values of y, σ_s, σ_s';

– using expression (5), we calculate the boundary moment.

The lower boundary of the section strength area with symmetrical reinforcement is constructed by symmetrically mapping the upper boundary relative to the horizontal axis.

4.2. Example 2

Construction of the Strength Area for a Rectangular Asymmetrically Reinforced Section.

Consider the section from example 1 with reinforcement $A_s = 12.322 см^2$ (2Ø28), $A'_s = 6.283 см^2$ (2Ø20).

Dependencies (3) - (5) are used to construct the upper limit of the section strength area. First of all, we determine the coordinates of characteristic points on the diagrams of stresses change in reinforcement and change of coordinate of bending center of section (Table 2), (Figure 8, 9).

Table 2. The values of the limit forces at the break points of the graphs in Figures 8 and 9.

Point	y	σ_s	σ'_s	N_b, kN	M_b, kNm
E	$-h/2$	$-\sigma_{sy}$	σ_{sy}	-3578.835	-103.477
F	$-(h/2-a)$	$-\sigma_{sy}$	σ_{sy}	-3484.04	-57.4985
G	$-(h/2-a)$	σ_{sy}	σ_{sy}	-2585.04	365.0306
J	0	σ_{sy}	σ_{sy}	-1099.84	714.0526
B	$h/2-a'$	σ_{sy}	σ_{sy}	385.3628	365.0306
C	$h/2-a'$	σ_{sy}	$-\sigma_{sy}$	844.0353	149.4545
D	$h/2$	σ_{sy}	$-\sigma_{sy}$	938.835	103.477

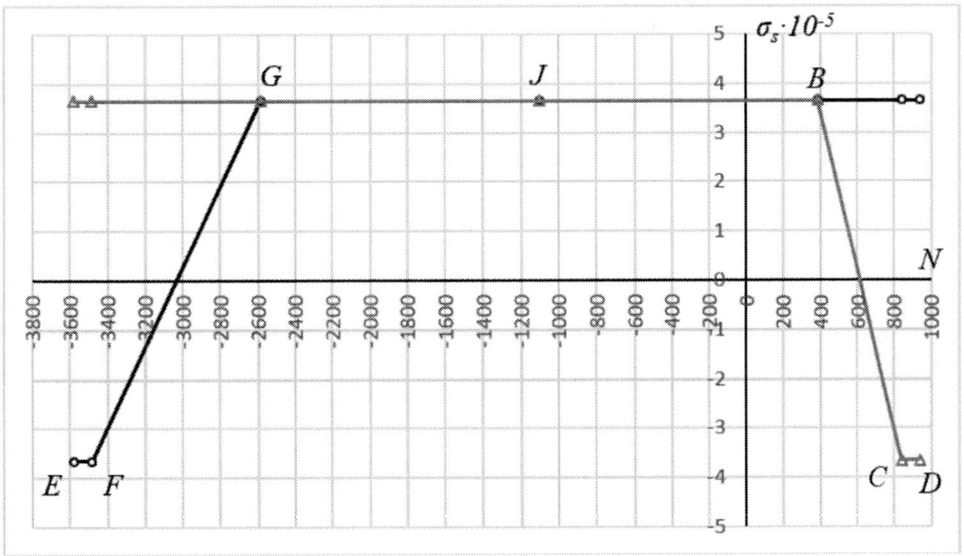

Figure 8. Stresses in the reinforcement for the upper boundary of the strength area of an asymmetrically reinforced section – —o— – A_s, —△— – A'_s.

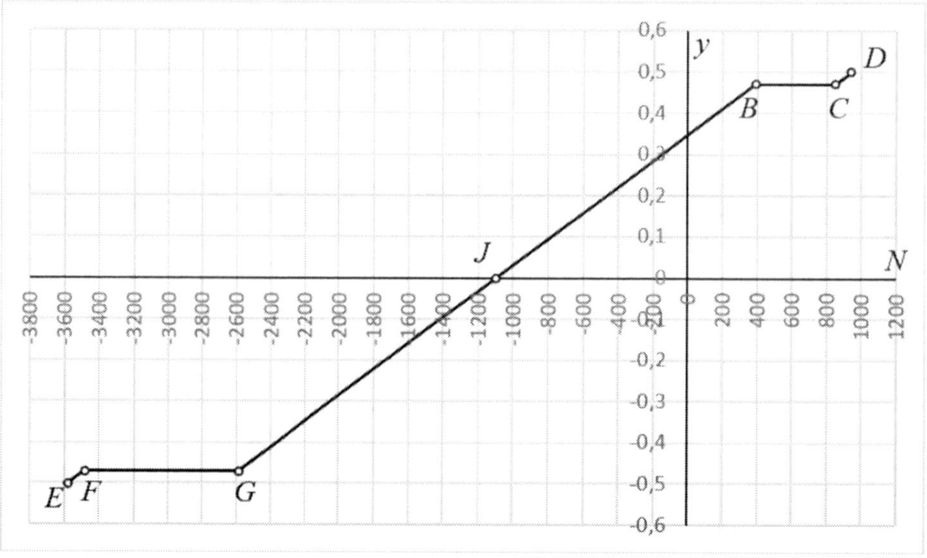

Figure 9. Change in the coordinate of the bending center of an asymmetrically reinforced section for the upper boundary of the strength region.

To construct the lower boundary of the section strength area, using dependencies (6) - (8) we calculate the coordinates of characteristic points on the graphs σ_s, σ_s', y (Table 3) and construct the corresponding graphs (Figures 10, 11).

Table 3. The values of the limit forces at the break points of the graphs shown in Figures 10 and 11.

Point	y	σ_s	σ_s'	N_b, *kN*	M_b, *kNm*
K	$h/2$	$-\sigma_{sy}$	σ_{sy}	-3578.835	-103.477
L	$h/2-a$	$-\sigma_{sy}$	σ_{sy}	-3484.035	-149.455
P	$h/2-a$	$-\sigma_{sy}$	$-\sigma_{sy}$	-3025.363	-365.031
Q	0	$-\sigma_{sy}$	$-\sigma_{sy}$	-1540.163	-714.053
R	$-(h/2-a')$	$-\sigma_{sy}$	$-\sigma_{sy}$	-54.963	-365.031
S	$-(h/2-a')$	σ_{sy}	$-\sigma_{sy}$	844.035	57.499
T	$-h/2$	σ_{sy}	$-\sigma_{sy}$	938.835	103.477

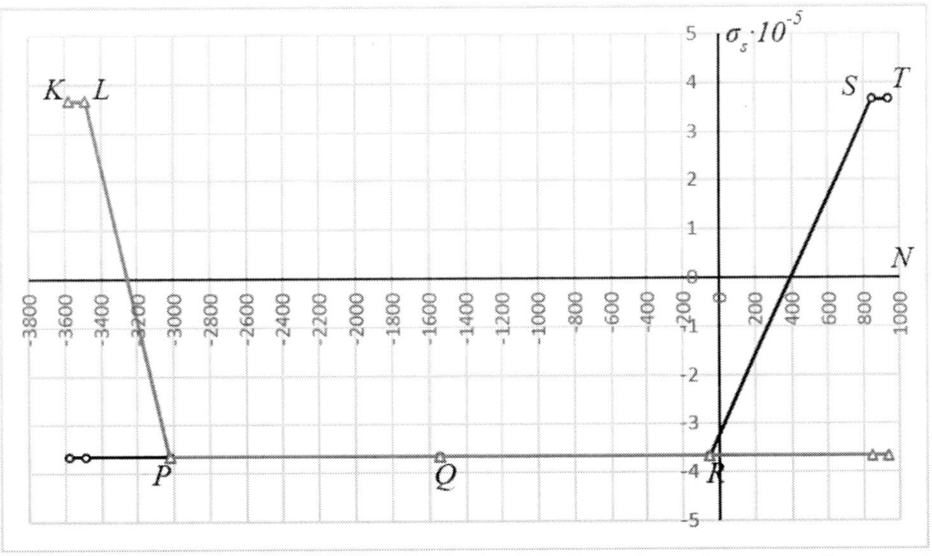

Figure 10. Stresses in the reinforcement for the lower boundary of the strength area with asymmetrical reinforcement – —o— – A_s, —△— – A_s'.

Having the graphs shown in Figures 8 - 11, you can easily build the boundary of the section strength area (Figure 12). For this purpose, using the available graphs (Figures 8 - 11) and Table 2, 3 for a given value of N_b we determine σ_s, σ_s', y and use the appropriate formulas to calculate M_b.

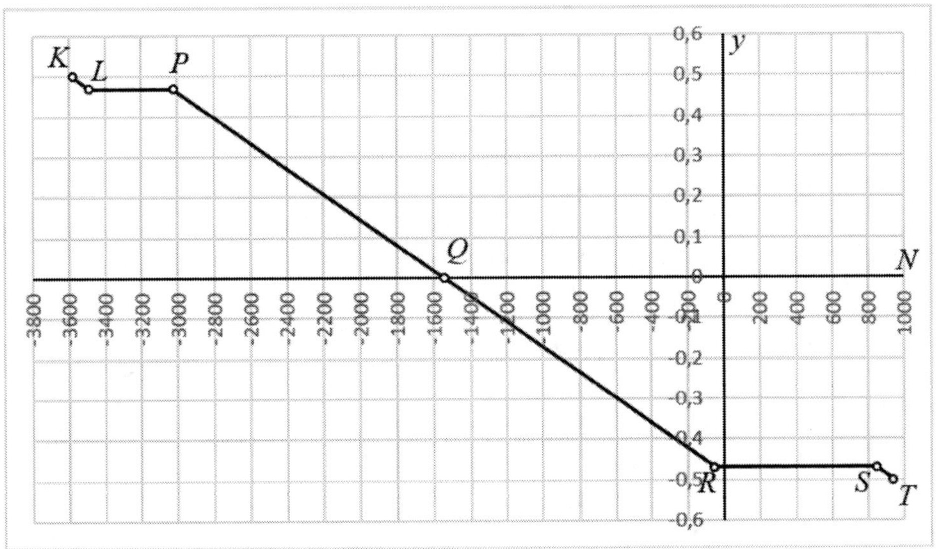

Figure 11. Change in coordinate of bending center of section with asymmetrical reinforcement for the lower boundary of the strength area.

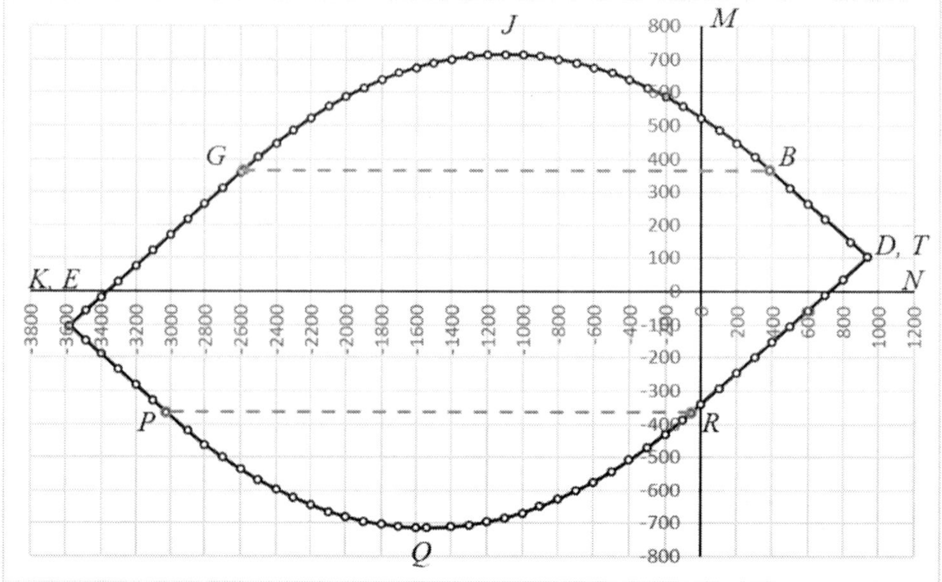

Figure 12. Strength area of asymmetrically reinforced rectangular section.

Similarly to the symmetrical reinforcement at the EG, KP, BD, RT sections, the boundary of the strength area is described by linear functions, and the upper and lower parts by quadratic functions.

The presence of equations describing the boundary of the section strength area allows us to solve the problem of finding the ultimate load for the structure using the method of limiting equilibrium [6].

5. Conclusions.

1. The boundary of the strength area of a rectangular reinforced section is described by linear and quadratic functions.

2. Extreme values of the boundary moments $M_{b,\max}$, $M_{b,min}$ and the corresponding values of longitudinal forces N_b both for symmetrical and asymmetrical reinforcement are determined by $y=0$, i.e., when the bending center of the rectangular section is on the symmetry axis of the rectangle.

3. The reinforced section reaches the limit state when the whole section and the reinforcement (or one of the reinforcements) enter the plastic state.

4. Linear graphs of changes in the coordinate of the bending center of the section and stresses in the reinforcement allow us to construct the boundary of the area of strength of a rectangular reinforced section without solving optimization problems (9), (10).

References

[1] Gvozdev A A 1949 *Raschet nesuschej sposobnosti konstrykcij po metody predelnogo ravnovesiya* (Moscow: Stroyizdat) 248

[2] Rzhanitsyn A R 1954 *Raschet soorygenij s ychetom plasticheskih svoistv materialov*. (Moscow: The state publishing house of literature on construction and architecture) 287

[3] Chiras A A 1989 *Stroitelnaya mekhanika, teorija i algoritmi* (Moscow: Stroyizdat) 256

[4] Pikovsky A A 1961 *Statika sterzhnevih system so szhatimy elementamy* (Moscow: State publishing house of physical and mathematical literature) 394

[5] Soroka Mykola *The limit state of non-hinged arch with a cross-section in the form of an idealized I-beam* http://www.enggjournals.com/ijet/vol10issue6.html

[6] Soroka Mykola 2019 Bearing capacity of structures made of materials with different tensile and compression strengths *6th Int. Conf. on Actual Problems of Engineering Mechanics (Odessa,*

Ukraine) vol **968** ed Mykola Surianinov (Switzerland Trans Tegh Publications Ltd) pp 200-208

[7] Russell Bridge and Andrew Wheeler 2000 *Cross-section Strength of Columns* (Part 1: AS 3600 Design) OneSteel Reinforcing Pty Ltd CAN 004 148 289

[8] Vrouwenvelder A C W M and Witteveen J 2003 *Plasticity Ct 4150. The plastic behaviour and the calculation of the plates subjected to bending* Technical University Delft Faculty of Civil Engineering and Geosciences 121

[9] Darkov A V and Shaposhnikov N N 1986 *Stroitelnaya mechanika* (Moscow: Higher school) 607

Expanded clay concrete mixtures and concretes on carbonate sand propertie improvement

I A Stolevich[1], A I Kostyuk[2,*], S A Kravchenko[2] and O O Posternak[2]

[1] Department of strength of materials, Odessa State Academy of Civil Engineering and Architecture Didrihsona str., 4, Odessa, Ukraine, 65029
[2] Department of Reinforced Concrete Structures and Transport Structures, Odessa State Academy of Civil Engineering and Architecture Didrihsona str., 4, Odessa, Ukraine, 65029

[*]E-mail: isi@ogasa.org.ua

Abstract. Results of researches of properties of expanded clay concrete on carbonate sand, mixtures and concrete, as well as a technique for selecting optimal concrete compositions for monolithic load-bearing and enclosing structures of residential and public buildings and structures are given.

1. Introduction

Today one of the main directions of building development is reduction of labour intensity and material consumption at building materials and structures manufacturing, reduction of their energy intensity and use of local materials.

Lightweight concrete on porous aggregates is a universal building material that allows, with its rational use, to solve many actual tasks of modern construction and simultaneously solve environmental, resource-saving and economic problems through technological and man-made waste when using and manufacturing local porous aggregates.

Additional economic effect can be reached at lightweight concretes on local aggregates of south of Ukraine (expanded clay, keralite, carbonate sand) and structures on their base use for buildings and structures construction.

Deficiency of fine aggregates for concrete in many regions of the country can be replenished by a wide application for their production of carbonate rock waste (porous limestone and shell limestone). Previous studies have shown the technical feasibility and economic feasibility of using limestone (carbonate) sand in concrete on artificial porous aggregates, and first of all in expanded clay aggregate, which makes up about 70% of the total volume of lightweight concrete.

2. Analysis of previous studies

Over the past years, in the laboratory of the Department of Reinforced Concrete and Stone OSACEA, research has been conducted on the use of rock-stone waste from various deposits in the south of Ukraine as a filler in the manufacture of concrete and reinforced concrete structures, including prestressed ones. These studies are ongoing.

3. Materials

The study of expanded clay concrete on carbonate sand and structures based on it was carried out in laboratory and production conditions.

In researches were used:

- Portland cement grade 400;
- expanded clay gravel of fractions 5…10 and 10…20 in the ratio by volume $V_{5...10}/V_{10...20}=1.5$;
- carbonate sand obtained from waste of shell limestone sawing.

Physic and mechanical characteristics of expanded clay gravel are shown in table 1.

Content from this work may be used under the terms of the Creative Commons Attribution 3.0 licence. Any further distribution of this work must maintain attribution to the author(s) and the title of the work, journal citation and DOI.
Published under licence by IOP Publishing Ltd

Table 1. Physic and mechanical properties of expanded clay gravel.

Fraction dimensions, mm	Strength (at squeezing in cylinder) $R_{сд}$, MPa	Bulk density, $\rho_{нас}$, kg/m³	Ratio of structure quality, kkk	Specific density, $\rho_{уд}$, g/sm³	The density in the cement paste, $\rho_{к}^{ц.т.}$, kg/m³	Water absorption w, %	Volume of intergranular voids, $V^{к}_{мп}$, %	The content of chopped grains,% by weight	Gravel grades by bulk density (DSTU B.V.2.7-17-95)
					Expanded clay				
5 – 10	2.8-4.67	504-583	559-801	2.31-2.48	955-1507	16.2-22.7	30-38.2	6.2	550-600
10 – 20	2.4-2.91	462-541	519-538	2.31-2.48	890-1493	20.8-24.9	38-41.1	8.7	500-550
Mixture	2.5-3.96	474-569	429-696	2.31-2.48	946-1502	18.2-24.1	33-40.1	7.9	500–600

As a fine aggregate carbonate sand was used. I was obtained as a result of crushing and reasoning of waste of stone-cutting and pieces of low-strength shell stone. Applying of such sand as a fine aggregate for different concretes, including lightweight concretes, is regulated by RST USSR 5014-82 [1] and confirmed in numerous studies.

The chemical composition of limestone-shells of this deposit and averaged data on the deposits of Ukraine are given in Table 2.

According to the chemical composition used in the study, limestones-shells of the Orel deposit can be attributed to pure limestones. The carbonate sands obtained from them do not contain harmful impurities: they are pure and suitable for use.

Table 2. Chemical composition of limestone-shells, % by mass.

Deposit	SiO_2	Al_2O_3	Fe_2O_3	CaO	MgO	SO_3	Loss on ignition
Orel	3.36	2.68	0.82	51.0	1.34	0.07	40.7
Glavan	4.2	2.82	0.9	50.2	1.0	0.06	41.8
Averaged by Ukraine	1.7	1.21	1.21	53.3	0.92	0.67	42.4

Comparison of the chemical composition of limestone-shell rock deposits of Ukraine shows their insignificant difference. Taking into account this circumstance in combination with the available data on the capacity of the Orel deposit, as well as the real possibilities of the industrial supply of carbonate sand from this field, justifies its use in the production of expanded clay based on it.

As studies [2] had shown, the carbonate aggregate is not an inert material, but enters into an active physicochemical interaction with clinker cement. This positive property of carbonate aggregates is complemented by their ability to create a self-priming effect, which leads to an increase in strength in concrete, both the sand itself and the contact layer with cement stone. In addition, it should be noted that the carbonate dust of the fraction (<0.14 mm) plays the role of a microfiller of cement, since it has not only a chemical similarity with it, but it is also close in particle size.

Taking into account this circumstance, it is possible to reduce cement consumption in concretes of equal strength, improve the workability of the concrete mix and increase the waterproofness of concrete.

Main physical and mechanical properties of carbonate sand are determined by DSTU B V.2.7-232:2010 [3]. Before the test, a fraction of more than 5 mm, whose content averaged 4% of the mass, was eliminated.

The average results of tests of carbonate sand batches, given in Table 3, showed that such sand meets the requirements of standards [1, 3] and can be used to produce expanded clay concrete with strength up to 30 MPa.

4. Properties study

Experimental studies of the basic properties of expanded clay-concrete on carbonate sand mixtures and concrete were carried out according to the procedure of the planned experiment.

Experimental studies consisted of two stages.

On the first stage the influence of prescriptive (consumption of cement and aggregate-structural factor) and technological (workability, mixing time of the mixture in the mixer and the time of its vibrocompaction) factors on the basic physical and mechanical properties of expanded clay mixture (porosity, delamination, density) and concrete (strength, density, strength uniformity and uniformity in density).

Table 3. General physical and mechanical properties of Orel deposit carbonate sand.

Grading: Private residues on control sieves, %						Strength of the original rock Ro.p., MPa	Strength of sand when squeezed in a cylinder Rc, MPa	Bulk density ρb, kg/m³	Density in a cement test ρc.t., kg/m³	Hollowness H, %	Water Absorption W, %	The size module Мк	Specific Gravity ρsp, g/cm³	The content of separately clay particles, %
2,5	1,25	0,63	0,315	0,14	less than 0,14									
17.4	12.4	17.0	15.7	17.4	19.8	0.9	1.6	1160	2390	42	5.4	2.38	2.99	0.48

As controlled parameters were chosen:

1. The volume of intergranular voids $V_п$, the index of delamination of the $П_р$ and the density of the expanded clay-concrete mixture ρ_0.
2. The water flow rate to achieve the desired workability of the mixture
3. Cubic strength R (28) and density of expanded clay concrete dried to a constant mass in a state of ρ.
4. The dispersion of strength S^2_R and density S^2_ρ of expanded clay concrete.

In addition, at the first stage, studies were made of the effect of prescription factors (cement consumption and aggregate-structural factor) on the mobility of the mixture over time. As the controlled parameter, the mobility of the mixture is accepted at predetermined intervals after its manufacture OK_t.

At the second stage, the strength and deformation characteristics of expanded clay concrete on carbonate sand under loads of different duration and intensity, adhesion of reinforcement to concrete, as well as the work of compressed concrete and bent reinforced concrete structures, including prestressed ones, were studied.

In the experimental studies conducted using the technique of a planned experiment, the following were chosen as the main controlled parameters:

1. Cubic $R_{(28)}$ and prism $R_{вt}$ strength in the age of 7, 28, 115, 300, 500 days.
2. Modulus of elasticity $E_в(t)$ in the same ages.
3. Ultimate compressive deformation $\varepsilon_{вu}$ of expanded clay concrete at short-term load loading.
4. Parametric levels of microcracks R^o_{crc} and R^v_{crc} at short-term load loading.
5. Relative shrinkage strain $\varepsilon_{sc,u}$ (t, tw). Start time countdown t_w corresponded to the end of the concrete setting time.
6. Relative creep strains $\varepsilon_{cc,u}(\sigma, t, t_0)$ of expanded clay concrete loaded at ages t_0 = 7, 28, 115 days with load equal to $0.2R_вS$; $0.6R_вS$; $0.8R_вS$.

5. Optimum compositions selection

Selection of the optimal composition of structural lightweight concrete, was carried out by the calculation and experimental method. Unlike conventional concrete, for lightweight concrete, in addition to the requirements for the strength of concrete and the workability of the mixture, it is necessary to ensure the requirement for a given density of concrete [2,4].

As a result of implementation of first stage of experimental studies (test and test results analysis) on developed technology there are achieved optimal by given criteria technological parameters for each studied composition of expanded clay concrete on carbonate sand, caused by prescription factors and the level of their variation.

The results of our preliminary experiments allowed us to select the following factors and assign the levels of their variation:

1. Cement consumption Cc, kg/m^3 – X_1.

Cement consumption (O) at main level is accepted as Cc = 375 kg/m^3. Interval of variation ±175 kg/m^3.

2. Aggregate-structural factor r – X_2.

The value of the aggregate-structural factor r at the main (O) level was chosen equal to 0.65. The variation interval is ±0.35.

3. Mixture workability – X_3.

The value of the workability index of the mixture at the main (O) level was chosen to be G = 35 s. Variation interval ±31 s.

4. Mixing time of keralite concrete mixture $t_п$, min. – X_4.

The value of the mixing time at the main (O) level was chosen to be $t_п$ = 5 min. Variation interval ±3 min.

5. Time of vibrocompaction of keralite concrete mixture $t_в$, s – X_5.

Time of vibrocompaction at the main (O) level was chosen to be $t_в$ = 105 s. Variation interval ±75s.

The optimal composition of structural expanded clay concrete should provide the specified strength, minimum possible density and the lowest possible cost. Obviously, the last two requirements cannot be ensured simultaneously, since a reduction in the density of expanded clay concrete leads to an increase in its cost, and vice versa.

To optimize expanded clay concrete, taking into account the economic criterion based on the experimental data on the actual compositions, their costs were calculated, processing the results of these calculations allowed to obtain with 95% reliability, the quadratic regression equation linking the cost of 1 m^3 expanded clay concrete with the costs of its components.

The resulting equation, taking into account only statistically significant coefficients, has the form:

$$C = 128,6 + 13,2X_1 - 37.6X_2 + 5.2X_1X_2 - 3,1X_2^2 \quad (1)$$

As seen from this equation, aggregate-structural factor r (X_2) has a somewhat greater influence on cost than cement flow (X_1).

In the presence of a priori information on the nature and power of the influence of composition factors on the main operational characteristics of concrete, it is most expedient to select the optimal compositions for the isolines constructed for each of these characteristics.

Isolines for main characteristics of expanded clay concrete $R_{(28)}$, ρ, C were built in axes "Cc – r" using according equations of regression.

$$R_{(28)} = 26.9 + 14.3X_1 - 7.2X_2 + 3.7X_3 - 2.2X_1^2 - 3.3X_3^2 - 1.3X_1X_2 + 1.7X_4X_5, \quad (2)$$

$$\rho = 1740 + 57X_1 + 156X_2 + 11X_4 - 26X_2^2 - 23X_1X_2 + 8X_4X_5. \quad (3)$$

At the same time, the values of factors X_3 (workability of the mixture) were taken at the lower level corresponding to OK = 6 cm, and the values of factors X_4 ($t_п$) and X_5 ($t_в$) were fixed at zero levels.

The integrated nomogram (Figure 1) allows for the given mobility of the mixture (OK = 6 cm) to obtain optimal compositions of expanded clay concrete (except for water flow).

The procedure for selecting the optimal compositions includes the following steps:

1. Calculation of strength values of expanded clay concrete $R_{(28)}$, corresponding to a given class of concrete for compressive strength, taking into account specific production conditions (C_{VR}), according to formula

$$R_{(28)} = B/(1-1.64C_{VR}) \tag{4}$$

2. According to the complex nomogram, a group of compositions (cement consumption Cc and value r) that satisfy the strength $R_{(28)}$ is established.
3. According to the complex nomogram for each of the compositions satisfying the strength $R_{(28)}$, the density of expanded clay concrete ρ is determined.
4. For each of composites on achieved cement consumptions Cc and values r it is calculated preliminary the water flow by equation:

$$B = 259 + 5X_1 + 14X_2 - 19X_3 + 14X_2^2 + 9X_1X_2 + 9X_2X_3 \tag{5}$$

5. For each composite on accepted cement consumption Cc and value r it is determined the concrete cost C.
6. If the density of concrete is not specified, then the optimal composition is selected by choosing from the established group the composition that provides the minimum cost of concrete. Otherwise, from the established group of compositions, a composition satisfying the given strength and density is selected. The assigned composition is adopted as the initial optimal composition.

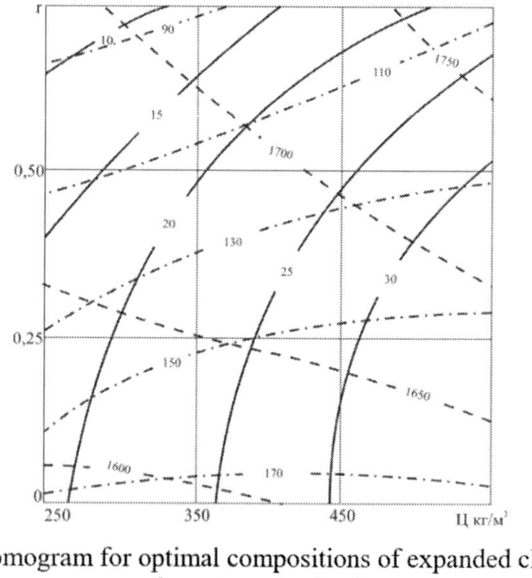

Figure 1. Nomogram for optimal compositions of expanded clay concrete on carbonate sand selection.
------ strength isolines $R_{(28)}$, MPa;
- - - - density isolines ρ, kg/m³;
-.-.-.- cost isolines Cc, c.hr./m³.

Based on the obtained dependencies and graphs, and also based on the requirements for lightweight concrete, the optimal composition of expanded clay concrete on carbonate sand was assigned.

The optimal composition of structural claydite concrete should provide the specified strength, minimum possible density and the lowest possible cost.

Optimal compositions of expanded clay concrete on carbonate sand with strength of 10…30 MPa are given in Table 4.

Table 4. Optimal compositions of expanded clay concrete on carbonate sand.

Designed strength, MPa	r	Materials consumptions on 1 m³ of concrete				Concrete density ρ, kg/m³
		Cement Cc, kg/l	Sand M, kg/l	Expanded clay K, kg/l	Water B, l	
10	1.85	250 / 80.6	1194 / 506	211 / 468	202	1685
15	0.75	280 / 90.3	1025 / 434	342 / 760	178	1675
20	0.75	360 / 116.1	975 / 413	325 / 722	192	1690
25	0.75	455 / 146.7	919 / 389	306 / 680	203	1710
30	0.75	540 / 174.2	870 / 368	290 / 644	210	1730

6. Use in elements and structures

From expanded clay concrete on carbonate sand, experimental batches of wall stones, wall blocks, slabs and coatings were made and tested.

Test results:

1. Average strength of SKU-1 stones was 3.01 MPa. Average strength of SKU-2 stones was 3.08 MPa. Average strength of control cubes was 3.12 MPa. Average strength of control prisms was 3.0 MPa. Frost resistance – F25.

2. Average value of blocks strength is 8.3 MPa, of cubes and prisms are respectively 10.6 MPa and 9.7 MPa, of modulus of elasticity is 7650 MPa, Poisson's ratio – 0.23. Cracks appears at load 0,92 N_{destr}. Density in a dry state is 1140 kg/m³.

3. Floor slabs and coatings are prestressed on series 1.141-1, issue 63, mark P63.15-8At-Vl, of continuous section from keralite concrete, class on compressive strength is B12.5 and density D1600. Average results of tests: cube strength is R=16.3 MPa, prism strength R_b=14.2 MPa, E_b=13630 MPa, ρ=1570 kg/m³. Control destructive load is q_p^{κ}=1310 kg/m², experimental destructive load on the nine floor was q_{destr}^{exp}=1616 kg/m², control deflection at control load q_f^{κ}=593 kg/m² is f_{κ}=14.8 mm. Actual deflection at control load q_f^{κ} is f_{κ}^{exp}=7.95 mm. First cracks appeared at load q^{exp}=1422 kg/m².

The experience of numerous domestic and foreign studies, practical experience shows that lightweight concretes on porous aggregates of various types are still an effective material of structural, structural heat-insulating and heat-insulating purpose.

The results of the research have proved the possibility of applying expanded clay concrete on carbonate sand as well as in monolithic load-bearing and enclosing structures.

7. Conclusions

1. The economic efficiency of using expanded clay concrete on carbonate sand is obvious, since the raw material for concrete is the waste from the stone-cutting of limestone shells.

2. Lightweight concretes on local porous aggregates of the south of Ukraine, both large and small, can be recommended for the manufacture of concrete and reinforced concrete structures with a concrete strength of 5 ... 30 MPa.

3. The conducted studies confirmed the high efficiency of lightweight concretes on porous aggregates, considerable savings in Portland cement, the expediency of using previously listed lightweight concretes for prefabricated and monolithic structures of residential and public buildings.

4. To optimize the composition of keralit concrete on carbonate sand, it is recommended to use the developed method of complex approach, which allows to obtain cost-effective compositions.

5. Researches proved the possibility of applying of expanded clay concrete on carbonate sand in monolithic load-bearing and enclosing structures and in reinforced concrete elements and structures.

References:

[1] DSTU B V.2.7-27-95 Sand from shell limestones for construction works. Specifications. Kyiv Derzhkommistobuduvannja Ukrai'ny 8

[2] Kostyuk A I, Stolevich A S and Makarov S V 1991 Influence of the structure on the strength and deformation properties of expanded clay concrete on carbonate sand *Stroitelnie konstrukcii: Resp. mezhvedomstv. nauch.-tekh. sb.* pp 104-107

[3] DSTU B V.2.7-232:2010. Sand for construction works. Test methods. Kyiv Minregionbud Ukrai'ny 28

[4] Stolevich A S, Makarov S V, Stolevich I A, Madii K M and Kravchenko S A 2005 Structural lightweight concrete *Visnyk Odeskoyi derzhavnoyi akademii budivnyctva ta arhitektury* **17** pp 246-255

Indentation of an elastic half-space by the cylindrical flat punch with a rounded edge under non-monotonic loading

Yu Streliaiev, O Titova

Mathematics Department, Zaporizhzhya National University, Zhukovsky str., 66, 69600 Zaporizhzhya, Ukraine

e-mail: strelkiny@gmail.com

Abstract. The paper presents a novel approach to solving the contact problem for interaction with friction between a rigid cylindrical flat punch with rounded corners and a linearly elastic half-space. Increasing and decreasing normal loads act on the punch sequentially. Coulomb's law is used to account for friction. The contact problem is reduced to nonlinear operator equations that must be solved at each step of the discrete loading process. The numerical method is applied. This method includes regularization of the equation, discretization of the regularized equations and the usage of an iterative process to obtain the approximate solution. The calculations were performed for different values of the relative radius of the flat area of the punch base. Distributions of contact traction and configuration of the stick and slip zones are studied at each monotonic stage of the loading process.

1. Introduction

The study of the contact of elastic bodies is important problem of mechanics and is widely used in mechanical engineering, transport, construction mechanics and other industries. When determining contact stresses in interacting elements of various mechanical systems, it is often necessary to take into account the friction between the contacting surfaces of these elements. To account for friction in contact problems of the elasticity theory, as a rule, Coulomb law is used [1].

The complexity of such problems is due to the fact that the contact surface, areas of adhesion and slippage are unknown in advance and can be complex and unpredictable. For the first time, approximate analytical solutions of the partial slip contact problems were independently obtained in papers [2] and [3]. The modern researches of the stick-slip contact problem are considered in articles [4–10] and books [11–15].

The numerical methods based on the variation statement of the problem are considered in researches [16-18]. The authors of papers [19-21] use methods based on reducing the problem to integral equations. Difficulties caused by the numerical realization of the variation methods for solving such complex contact problems are absent when we use the methods [22–24] based on nonlinear operator equations to simulate the contact interaction.

In this paper, nonlinear boundary integral equations method [24] was applied to obtain a numerical solution to the quasi-static contact problem of the interaction with Coulomb friction between the rigid cylindrical flat punch with rounded edges and a linearly elastic half-space when a non-monotonic normal load is acting on the punch.

Content from this work may be used under the terms of the Creative Commons Attribution 3.0 licence. Any further distribution of this work must maintain attribution to the author(s) and the title of the work, journal citation and DOI.
Published under licence by IOP Publishing Ltd

2. Operator equation of the contact problem

Consider the three-dimensional quasi-static contact problem on interaction of two linearly elastic bodies in the presence of Coulomb friction [1]. This problem is reduced to the solution of the nonlinear operator equations [24]:

$$\boldsymbol{p}_i = G_{\mu p_{1i-1}}\left(\boldsymbol{p}_i - E \cdot \left(A(\boldsymbol{p}_i) - \tilde{\boldsymbol{f}}_i\right)\right), i = \overline{1,l}, \tag{1}$$

where i is denotes step number of the discrete loading process; l is a total number of loading steps; $\boldsymbol{p}_i = \left(p_{1i}(s), p_{2i}(s), p_{3i}(s)\right) \in L_2^3(\Omega)$ [22] are unknown vector-functions from Hilbert space $L_2^3(\Omega)$ [22], where $p_{1i}(s) \in L_2(\Omega)$ [25] are normal contact tractions, $p_{2i}(s), p_{3i}(s) \in L_2(\Omega)$ are tangential contact tractions at a point $s \in \Omega$ at the ith step of loading, where Ω is bounded flat region, containing an unknown contact area at all loading steps; $\tilde{\boldsymbol{f}}_i$ is known element of $L_2^3(\Omega)$, that determines the configuration of bodies and the conditions for their loading; μ is a friction coefficient; E is a positive const.

The continuous nonlinear operator $G_{\mu g}: L_2^3(\Omega) \to L_2^3(\Omega)$ is given by the equalities

$$\begin{cases} \boldsymbol{x} = (x_1, x_2, x_3), \boldsymbol{y} = (y_1, y_2, y_3) \in L_2^3(\Omega); \\ \boldsymbol{y} = G_{\mu g}(\boldsymbol{x}), \\ y_1(s) = h\left(x_1(s)\right), \\ y_2(s) = q\left(x_2(s), x_3(s), \mu g(s)\right), \\ y_3(s) = q\left(x_3(s), x_2(s), \mu g(s)\right), \quad s \in \Omega, \end{cases}, \tag{2}$$

where $g(s) \in L_2(\Omega)$ is given positive function, the functions $h(x)$ and $q(x, y, z)$ have the form:

$$h(x) = \begin{cases} x, \text{ if } x \geq 0; \\ 0, \text{ if } x < 0; \end{cases} \tag{3}$$

$$q(x, y, z) = \begin{cases} x, & \text{if } \sqrt{x^2 + y^2} \leq z; \\ x \cdot \frac{z}{\sqrt{x^2+y^2}}, & \text{if } \sqrt{x^2 + y^2} > z. \end{cases} \tag{4}$$

The linear bounded operator of influence $A: L_2^3(\Omega) \to L_2^3(\Omega)$ is defined by the relations

$$\begin{cases} \boldsymbol{x} = (x_1, x_2, x_3), \boldsymbol{y} = (y_1, y_2, y_3) \in L_2^3(\Omega); \\ \boldsymbol{y} = A(\boldsymbol{x}), \ y_k = \sum_{j=1}^{3} A_{kj}(x_j) \ \forall k = \overline{1,3}. \end{cases} \tag{5}$$

where $A_{kj}: L_2(\Omega) \to L_2(\Omega)$ have the form:

$$A_{kj}(x)_s = \int_\Omega K_{kj}(s, s') \cdot x(s')ds', \ \left(k, j = \overline{1,3}\right). \tag{6}$$

According to the Boussinesq-Cerruti solution [13] the functions $K_{kj}(s, s')$ have the form:

$$\begin{cases} K_{11}(s, s') = \frac{c_1}{r}; \ K_{12}(s, s') = -K_{21}(s, s') = \frac{c_2(x-x')}{r^2}; \\ K_{13}(s, s') = -K_{31}(s, s') = \frac{c_2(y-y')}{r^2}; \\ K_{22}(s, s') = \frac{c_1}{r} + \frac{c_3(x-x')^2}{r^3}; \\ K_{23}(s, s') = -K_{32}(s, s') = \frac{c_3(x-x')(y-y')}{r^3}; \\ K_{33}(s, s') = \frac{c_1}{r} + \frac{c_3(y-y')^2}{r^3}; \end{cases} \tag{7}$$

$$\begin{cases} c_1 = \frac{1-\nu_1^2}{\pi E_1} + \frac{1-\nu_2^2}{\pi E_2}; \\ c_2 = \frac{(1+\nu_2)(1-2\nu_2)}{2\pi E_2} - \frac{(1+\nu_1)(1-2\nu_1)}{2\pi E_1}; \\ c_3 = \frac{\nu_1(1+\nu_1)}{\pi E_1} + \frac{\nu_2(1+\nu_2)}{\pi E_2}; \\ r = |s - s'| = \sqrt{(x-x')^2 - (y-y')^2}. \end{cases} \tag{8}$$

where E_1, E_2 and ν_1, ν_2 are Young's modulus's and Poisson's ratios for first and second bodies; x, x' and y, y' are coordinates of points s i s' in region Ω. If the first body is rigid then Young's modulus E_1 is equal infinity.

The vector-functions $\tilde{\boldsymbol{f}}_i$, have the form:

$$\tilde{\boldsymbol{f}}_i(s) = \left(-\delta_0(s) - \Delta_{1i}, -\tilde{\Delta}_{2i}(s), -\tilde{\Delta}_{3i}(s)\right), \ i = \overline{1,l}, \tag{9}$$

where $\delta_0(s) \in L_2(\Omega)$ defines the initial gap between the bodies; $\tilde{\Delta}_{2i}(s) = \Delta_{2i} - \sum_{j=1}^{3} A_{2j}\left(p_{ji-1}\right)_s - \Delta_{2i-1}$, $\tilde{\Delta}_{3i}(s) = \Delta_{3i} - \sum_{j=1}^{3} A_{3j}\left(p_{ji-1}\right)_s - \Delta_{3i-1}$; Δ_{ki} are components of the vector of relative displacement of remote points of bodies at the ith step of loading.

Taking into account the formulas (6), we can conclude that the operator equation (1) represent a system of three nonlinear integral equations with unknown functions $p_{1i}(s), p_{2i}(s), p_{3i}(s) \in L_2(\Omega)$.

3. Method of the numerical solution

The method of the numerical solution of the operator equation (1) is considered in paper [24]. It consists in the regularization of the equation (1), the discretization of the regularized equation and the use of an iterative process for solving the discrete analogue of the regularized equation. To obtain such a discrete analogue, we define the region Ω in the form of an open square. We split Ω into n^2 square regions $\omega_1, \omega_2, \ldots, \omega_{n^2}$. Let unknown functions $p_{1i}(s), p_{2i}(s), p_{3i}(s)$ is constants $x_{3k-2\,i}, x_{3k-1\,i}, x_{3k\,i}$ on each boundary element ω_k. Then, obtaining an approximate solution of the equation (1) at the ith step of loading is reduced to obtaining unknown numerical parameters $x_{1\,i}, x_{2\,i}, \ldots, x_{3n^2\,i}$ from the following system of $3n^2$ scalar equations:

$$\begin{cases} x_{3k-2\,i} = h\left(x_{3k-2\,i} - E \cdot \left(\sum_{j=1}^{3n^2} a_{3k-2\,j} \cdot x_{j\,i} - b_{3k-2\,i}\right)\right); \\ x_{3k-1\,i} = q\left(x_{3k-1\,i} - E \cdot \left(\sum_{j=1}^{3n^2} a_{3k-1\,j} \cdot x_{j\,i} - b_{3k-1\,i}\right), x_{3k\,i} - \right. \\ \qquad \left. -E \cdot \left(\sum_{j=1}^{3n^2} a_{3k\,j} \cdot x_{j\,i} - b_{3k\,i}\right), \mu \cdot x_{3k-2\,i-1}\right); \\ x_{3k\,i} = q\left(x_{3k\,i} - E \cdot \left(\sum_{j=1}^{3n^2} a_{3k\,j} \cdot x_{j\,i} - b_{3k\,i}\right), x_{3k-1\,i} - \right. \\ \qquad \left. - E \cdot \left(\sum_{j=1}^{3n^2} a_{3k-1\,j} \cdot x_{j\,i} - b_{3k-1\,i}\right), \mu \cdot x_{3k-2\,i-1}\right); \end{cases} \tag{10}$$

where $k = \overline{1,n^2}$, $i = \overline{1,l}$.

Numerical parameters a_{kj}, b_{ki} in system (10) are determined by relations

$$a_{3k-d\,3k-d} = \varepsilon + \int_{\omega_k} K_{3-j\,3-j}(s_k, s)ds;$$

$$a_{3k-2\,3k-1} = a_{3k-2\,3k} = a_{3k-1\,3k-2} = a_{3k-1\,3k} = a_{3k\,3k-2} = a_{3k\,3k-1} = 0; \tag{11}$$

$$a_{3k-d\,3j-e} = mes(\omega_j) \cdot K_{3-d\,3-e}(s_k, s_j);\ b_{3k-2\,i} = -\delta_0(s_k) - \Delta_{1i},$$

$$b_{3k-1\,i} = -\tilde{\Delta}_{2i}(s_k),\ b_{3k\,i} = -\tilde{\Delta}_{3i}(s_k);\ \forall k,j = \overline{1,n^2}\ (k \neq j);\ \forall d,e = \overline{0,2},$$

where $\varepsilon > 0$ is regularization parameter; s_j is the center of the square ω_j; $mes(\omega_j)$ is the area of the square ω_j.

The approximate solution of the system (10) can be found by the iterative process:

$$\begin{cases} \left(x_{1i}^{(0)}, x_{2i}^{(0)}, \dots, x_{3n^2 i}^{(0)}\right) \in R^{3n^2}; \\ x_{3k-2\,i}^{(m+1)} = h\left(x_{3k-2\,i}^{(m)} - E\cdot\left(\sum_{j=1}^{3n^2} a_{3k-2\,j}\cdot x_{j\,i}^{(m)} - b_{3k-2\,i}\right)\right); \\ x_{3k-1\,i}^{(m+1)} = q\left(x_{3k-1\,i}^{(m)} - E\cdot\left(\sum_{j=1}^{3n^2} a_{3k-1\,j}\cdot x_{j\,i}^{(m)} - b_{3k-1\,i}\right), x_{3k\,i}^{(m)} - \right. \\ \qquad \left. -E\cdot\left(\sum_{j=1}^{3n^2} a_{3k\,j}\cdot x_{j\,i}^{(m)} - b_{3k\,i}\right), \mu\cdot x_{3k-2\,i-1}\right); \\ x_{3k\,i}^{(m+1)} = q\left(x_{3k\,i}^{(m)} - E\cdot\left(\sum_{j=1}^{3n^2} a_{3k\,j}\cdot x_{j\,i}^{(m)} - b_{3k\,i}\right), x_{3k-1\,i}^{(m)} - \right. \\ \qquad \left. -E\cdot\left(\sum_{j=1}^{3n^2} a_{3k-1\,j}\cdot x_{j\,i}^{(m)} - b_{3k-1\,i}\right), \mu\cdot x_{3k-2\,i-1}\right); \end{cases} \tag{12}$$

where $i = \overline{1,l}$; $k = \overline{1,n^2}$; $m = 0,1,2,\dots$ is number of iteration step.

The iterative process (12) converges to the solution of the system (10) regardless of the choice of the initial approximation $\left(x_{1\,i}^{(0)}, x_{2\,i}^{(0)}, \dots, x_{3n^2\,i}^{(0)}\right) \in R^{3n^2}$ assuming a positive constant E satisfies inequality $E < \left(\max_{1\le i\le 3n^2} \sum_{j=1}^{3n^2} |a_{ij}|\right)^{-1}$ [26].

4. Numerical results and discussion

The proposed approach was applied to obtain a numerical solution of the contact problem of indentation of an elastic half-space by a rigid cylindrical flat punch with rounded edges (Figure 1).

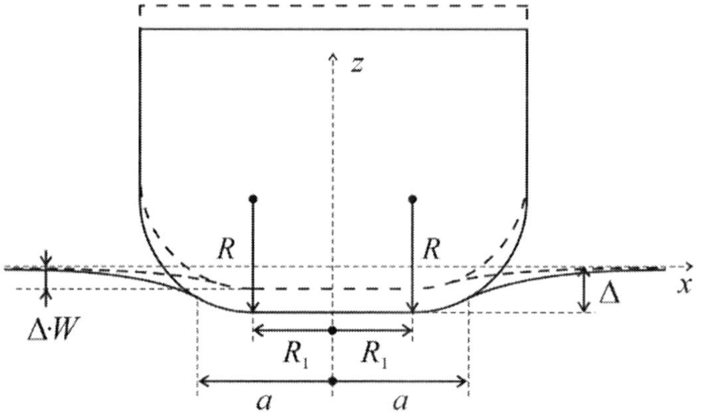

Figure 1. Diagram of contact.

The indenter profile has the shape

$$\delta_0(r) = \begin{cases} 0, & r \le R_1, \\ \frac{(r-R_1)^2}{2R}, & r > R_1, \end{cases} \tag{13}$$

where $r = \sqrt{x^2 + y^2}$; R is the radius of curvature of the rounded corner; R_1 is the radius of the flat punch surface.

The process of loading consisted of $3l$ steps in accordance with the following law of changing of the normal displacement of the punch:

$$\Delta_{1i} = \begin{cases} -\frac{\Delta}{l}\cdot i, & 1\le i \le l, \\ -\Delta + \frac{\Delta\cdot(1-W)}{l}\cdot(i-l), & l+1\le i \le 2l, \\ -\Delta\cdot W - \frac{\Delta\cdot(1-W)}{l}\cdot(i-2l), & 2l+1\le i \le 3l, \end{cases} \tag{14}$$

where Δ is the maximum indentation depth; $0 < W < 1$ is constant. The tangential displacements Δ_{2i} and Δ_{3i} are equal zero for all steps of loading.

Formula (14) defines non-monotonic normal loading, which consists of three stages. At the first stage $(1 \le i \le l)$ the punch is pressed into the half-space to the maximum indentation depth Δ, in the second stage $(l+1 \le i \le 2l)$ the punch is moved in the opposite direction to the indentation depth ΔW, in the third stage $(2l + 1 \le i \le 3l)$ the punch is again pressed into the half-space to the maximum depth Δ. The solution of the problem for the case of monotonic normal load was obtained in our earlier paper [27].

In the calculations, we used the following parameters: Young's modulus $E = 3.5 \cdot 10^4$MPa; Poisson's ratio $\nu = 0.44$; friction coefficient $\mu \approx 0.07$; $\mu/\gamma = 0.665$, where $\gamma = (1 - 2\nu)/(2 - 2\nu)$ [16]. In the formulas (14) we used the following parameters: $l = 20$, $\Delta = 10^{-4}$m, $W = 0.547$.

Numerical solutions were obtained for different values of parameter $\alpha = R_1/a$, where a is a maximum contact radius. The radius of curvature R in each case was selected from the condition of the constancy of the ratio $\Delta/a = 0.002$.

The calculation results are illustrated by the graphs (Figure 2-10). Here $\bar{q} = \frac{p_{2i}(x,0)}{\mu \cdot p_{1l}(0,0)}$ is normalized tangential traction acting on the punch, $\bar{p} = \frac{p_{1i}(x,0)}{p_{1l}(0,0)}$ is normalized contact pressure (normal traction), $\bar{x} = \frac{x}{a}$ is dimensionless distance to the center of the punch base.

In these figures, dotted lines with triangles correspond to the state after the first stage of loading ($i = l$ in formula (14)), with squares - after the second stage ($i = 2l$), with circles - after the third stage ($i = 3l$). Solid and dashed lines in Figure 8 correspond to the known solution [16] for a flat punch after the first and after the second stages of loading, respectively.

Analysis of the results shows that after the first stage of loading the contact pressure and tangential traction reach their maximum values at the points belonging to the rounded edges of the punch. On the flat punch surface, the contact traction increase monotonically with increasing distance from the center of the punch base. With a decrease in the radius of the flat section R_1 from $0.8a$ to $0.2a$, the maxima of contact traction shift closer to the center of the contact area (Figure 2-7).

After the second stage of loading (monotonic unloading to the indentation depth 0.547Δ), the contact pressures decrease (Figure 2, 4, 6) and their distribution (dotted line with squares) is similar to the first stage of loading (dotted line with triangles). The distribution of tangential traction (dotted line with squares in Figure 3, 5, 7) differs from their distribution after the first stage. At points near the boundary of the contact area, these stresses change sign to the opposite. The radii of the contact area and the central stick zone decrease; a second stick zone is formed at the boundary, while slipping occurs at the points on the contact area boundary (Figure 8-10). With a decrease in the radius of the flat region ($\alpha = 0.8$ 0.5 0.2) we see a decrease in the radius of the central adhesion zone, while the annular stick zone at the boundary has approximately the same width. Comparison of the results with the known solution [16] of this problem for a flat punch (Figure 8) shows that after the first stage of loading, the radius of the stick zone for all considered values of the parameter α is $0.68a$, which is approximately equal to its value for a flat punch. After the second stage of loading, the radius of the central stick zone for a flat punch is close to the value of $0.56a$, obtained at $\alpha = 0.8$. For the cases of α equal to 0.5 *and* 0.2 (Figure 9, 10) this radius is equal to $0.48a$ and $0.44a$, respectively.

After the third stage of loading, when the punch is again pressed into the half-space to the maximum indentation depth Δ, the contact area is again divided into a circular adhesion zone in the center and a slip zone at the boundary. As seen from Figure 2, 4, 6 the distribution of contact pressure (dotted line with circles) in this case coincides with the distribution obtained after the first stage (dotted line with triangles). In the graphs Figure 3, 5, 7 decrease in the values of tangential tractions in the slip zone formed after the second stage of loading is observed. This leads to an increase in the radius of the adhesion zone in comparison with its value achieved after the first stage of loading (by24% for the case of $\alpha = 0.8$ and by12% for the cases of $\alpha = 0.5$ *and* 0.2) (Figure 8-10).

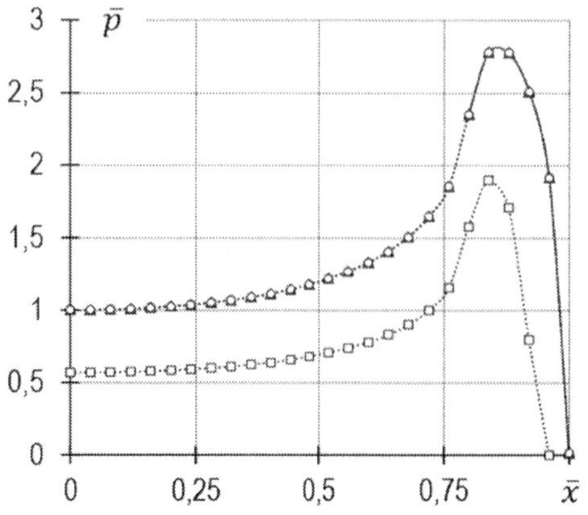

Figure 2. Contact pressure distribution, $\alpha = 0{,}8$.

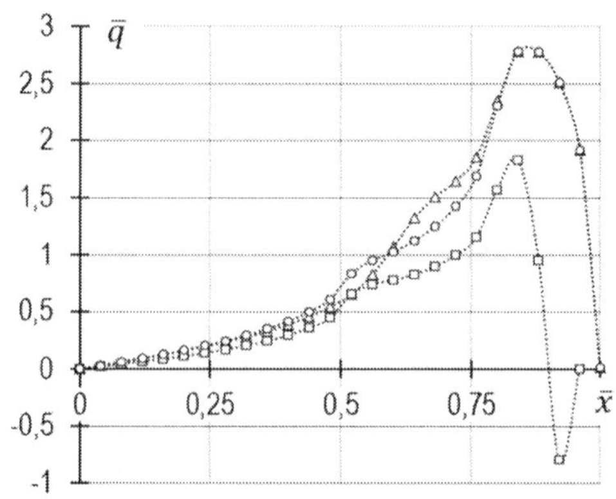

Figure 3. Tangential traction distribution, $\alpha = 0{,}8$.

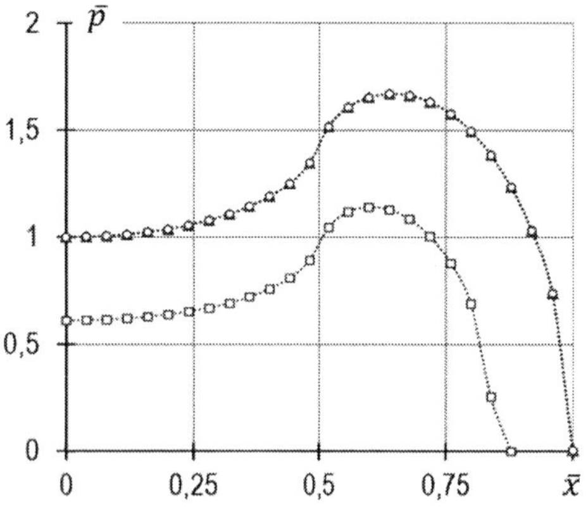

Figure 4. Contact pressure distribution, $\alpha = 0{,}5$.

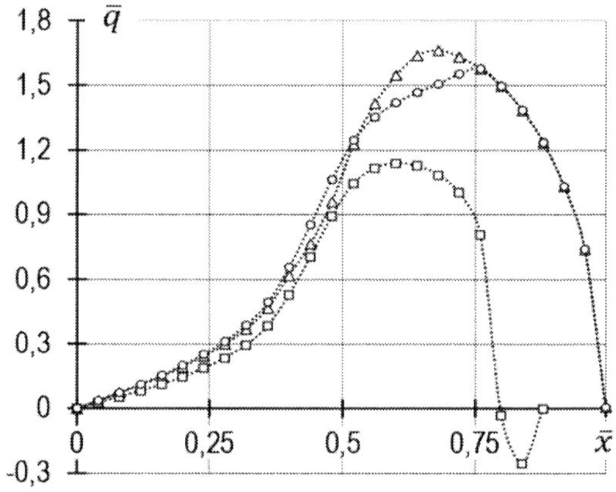

Figure 5. Tangential traction distribution, $\alpha = 0{,}5$.

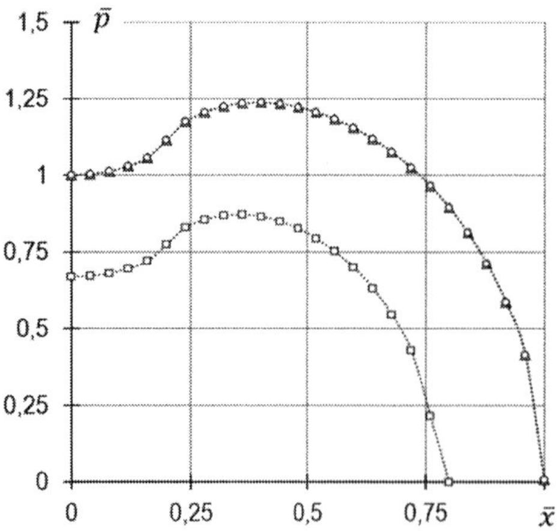

Figure 6. Contact pressure distribution, $\alpha = 0{,}2$.

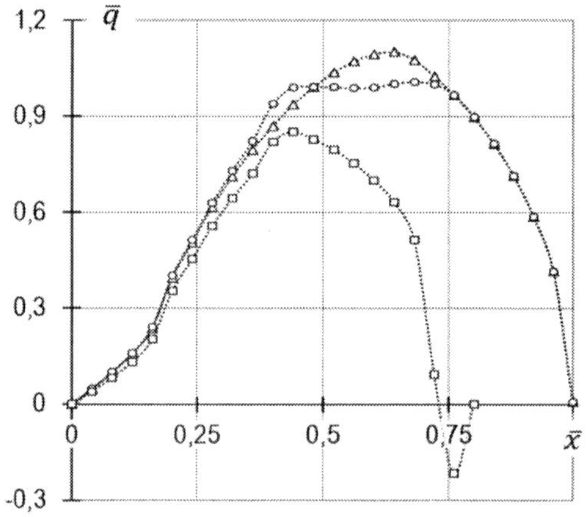

Figure 7. Tangential traction distribution, $\alpha = 0{,}2$.

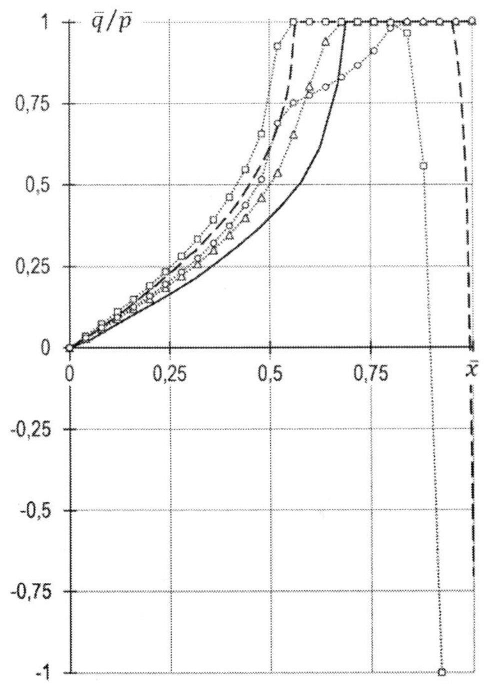

Figure 8. Solution to the problem, $\alpha = 0{,}8$.

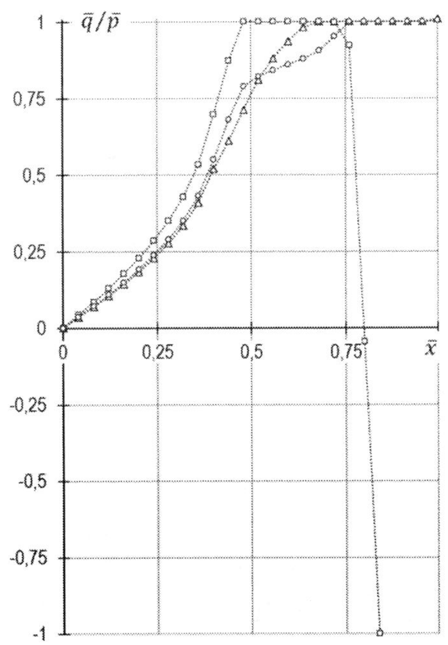

Figure 9. Solution to the problem, $\alpha = 0{,}5$.

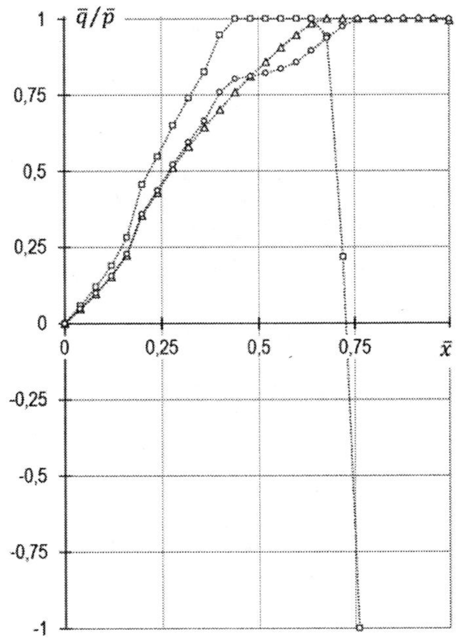

Figure 10. Solution to the problem, $\alpha = 0{,}2$.

5. Conclusions

The new approach to solving three-dimensional quasi-static contact problems of the interaction of elastic bodies with Coulomb friction is proposed. The contact problems are formulated using nonlinear operator equations for the unknown components of the contact traction. Each equation corresponds to a step in the discrete loading process. For the numerical solution of these equations, they were discretized and an iterative process was used to obtain an approximate solution to their discrete analogs.

The proposed approach is applied to solve the contact problem of interaction of a rigid cylindrical flat punch with rounded edge and an elastic half-space under non-monotonic normal loading, including the stages of loading, incomplete unloading and reloading. The distributions of normal and tangential contact tractions corresponding to each stage of loading were obtained for different values of the radius of the flat region of the punch base. Analysis of the obtained numerical results showed that the non-monotonic loads can lead to a change in the distribution of contact tractions and the configurations of the stick and slip zones formed at the initial stage of loading.

References

[1] Kalker J J 1977 A survey of the mechanics of contact between solid bodies *Zeitschrift für Angewandte Mathematik und Mechanik* **57(6)** T3–T17

[2] Cattaneo C 1938 Sul contatto di due corpi elastici: distribuzione locale degli stozzi *Rend. Dell'Academia nazionale dei Lincei* **27(6)** 342–348, 434–436, 474–478

[3] Mindlin R D 1949 Compliance of elastic bodies in contact *J. Appl. Mech.* **16(3)** 259–268

[4] Hills D A, Urriolagoitia Sosa G 1999 Origins of partial slip in fretting – a review of known and potential solutions *The Journal of Strain Analysis for Engineering Design* **34(3)** 175–181

[5] Barber J R, Ciavarella M 2000 Contact mechanics *Int. J. Solids Struct* .**37(1–2)** 29–43

[6] Ciavarella M 1999 Indentation by nominally flat or conical indenters with rounded edges. *Int. J. Solids Struct.* **36(27)** 4149–4181

[7] Zhupanska O I 2008 On the analytical approach to Galin's stick-slip problem. A survey *Journal of Elasticity* **90(3)** 315-333.

[8] Ostryk V I 2013 Axisymmetric contact of a punch of polynomial profile with an elastic half-space when there is friction and adhesion *J. Appl. Math. Mech.* **77(5)** 433-444

[9] Goryacheva I G, Martynyak R M 2014 Contact problems for textured surfaces involving frictional effects *Proc. Inst. of Mechanical Engineers Part J: Journal of Engineering Tribology* **228(7)** 707–716

[10] Andresen H, Hills D A. 2020 A review of partial slip solutions for contacts represented by half-planes including bulk tension and moments. *Tribology International* **143** 106050

[11] Johnson K L 1985 *Contact Mechanics* (Cambridge: Cambridge University Press) p 452

[12] Ostryk V I, Ulitko A F 2006 *The Wiener–Hopf Method in Contact Problems of Elasticity Theory* (Kyiv: Naukova Dumka) p 328

[13] Barber J R 2018 *Contact mechanics. Solid Mechanics and Its Applications* (Springer) p 585

[14] Ostryk V.I. 2018 *Factorization method and its generalization in mixed problems of elasticity theory* (Kyiv: VPC «Kyivskyi universytet») p 480

[15] Popov V L, Heß M, Willert E 2019 *Handbook of contact mechanics: exact solutions of axisymmetric contact problems* (Berlin: Springer Verlag) p 347

[16] Turner J R 1979 The frictional unloading problem on linear elastic half-space IMA J. Appl. Math. **24(5)** 439–469

[17] Kravchuk A S 2009 The variational method in contact problems. The present state of the problem and trends in its development *J. Appl. Math. Mech.* **73(4)** 351-357

[18] Reina S, Dini D, Hills D A, Iida Y 2011 A quadratic programming formulation for the solution of layered elastic contact problems: Example applications and experimental validation *European Journal of Mechanics - A/Solids* **30(3)** 236-247

[19] Galanov B A 1985 The method of boundary equations of the Hammerstein-type for contact problems of the theory of elasticity when the regions of contact are not known *J. Appl. Math.*

Mech. **49(6)** 634-640

[20] Aleksandrov V M, Pozharskii D A 2004 Three-dimensional contact problems taking friction and non-linear roughness into account *J. Appl. Math. Mech.* **68(4)** 463-472

[21] Narayan Sundaram, Farris T N 2010 Mechanics of advancing pin-loaded contacts with friction. *J. Mech. and Phys. of Solids* **58(11)** 1819-1833

[22] Aleksandrov A I 2015 A method for the solution of a three-dimensional contact problem of interaction of two elastic bodies in the presence of friction *J. Math. Sci.* **205(5)** 518-534

[23] Aleksandrov A I, Streliaiev Yu M 2014 Nonlinear boundary integral equations method for contact problems of the elasticity theory *Eastern-European J. of Enterprise Technologies* **3/7(69)** 36-40

[24] Streliaiev Yu M 2016 A nonlinear boundary integral equations method for the solving of quasistatic elastic contact problem with Coulomb friction *Vestn. Samar. Gos. Tekhn. Univ., Ser. Fiz.-Mat. Nauki [J. Samara State Tech. Univ., Ser. Phys. Math. Sci.*] **20(2)** 306-327

[25] Akhiezer N I, Glazman I M 1981 *Theory of linear operators in Hilbert space vol. I, II, Monographs and Studies in Mathematics vol. 9, 10.* (Edinburgh: Scottish Academic Press)

[26] Streliaiev Y M 2014 Solution of the quasistatic contact problem of elasticity with friction *Visnyk of Zaporizhzhya National University. Physical and Mathematical Sciences* **2** 161–172

[27] Streliaiev Yu M., Titova O O 2020 Stick-slip contact of the cylindrical flat punch with a rounded edge and elastic half-space *Visnyk of Zaporizhzhia National University. Physical and Mathematical Sciences* **1** 86-92

IOP Conf. Series: Materials Science and Engineering 1164 (2021) 012077 doi:10.1088/1757-899X/1164/1/012077

Accounting for dissipation of the energy of the supple base

L P Telipko[*], **V Yu Solod and A D Romaniuk**

Dniprovsky State Technical University, Kamenskoe, Ukraine

[*]E-mail: leo46din@gmail.com

Annotation. When calculating the foundations and structures on a compliant basis for the action of dynamic effects, the mathematical model of the base should reflect the real conditions for deformation of the base, due to their lamination, limited by the distribution property, weight and others. This is more, in our opinion, the axisymmetric columnar model of a compliant base is responsible, which takes into account the important properties of the ground base at a dynamic effect, as the extremity of the oscillation distribution zone and its inertia. In this case, the model allows us to determine not only the movement of the boundary plane, but also the stress-strain state of the base. With resonant phenomena, the accounting of energy dissipation during oscillations of the base is essential. The paper discusses the adaptation of the axisymmetric columnar model on a dissipative base, in which the dispersion of energy is carried out according to the complex hypothesis by E.S. Sorokin. The model in question is a cylindrical body in the form of a rod height H and radius R, large enough to believe that the amplitudes of displacements on the cylindrical surface of the rod are zero, when the rod on the outer end is loading within the centrally located circular area with a radius $\bar{r}$, much smaller R. The basis, obeying the three-axis linear physical law, is in the conditions of axial symmetry, therefore, only vertical u and radial w movements are observed, which are functions of variables x, r, t. The solution of the dynamic Liame equations described by the deformation of the cylindrical body was carried out by the discrete method of L.V. Vinokurov, according to which the solution was found in an analytical form in the variable x, and in the variable r - in the finite-difference. The use of a method of separation of variables by Fourier, a system of discrete differential equations in private derivatives was reduced to a system of differential equations in ordinary derivatives, the integration of which was carried out using Euler substitutions. Permanent integration was determined from the boundary conditions at the end of the cylindrical rod. The solutions obtained show that energy dissipation accounting causes a phase shift between the perturbing load and the movement of the base and to a significant decrease in the amplitudes of oscillations in the resonance zone.

In the calculation of foundations on an elastic basis, the reliability of the theoretical calculation results largely depends on the adopted mathematical model of the ground base, which, from our point of view, should meet two basic requirements: adequacy of its real conditions; mathematical simplicity, allowing us to use it in engineering practice.

This, in our opinion, with sufficient completeness corresponds to an axisymmetric columnar model of a compliant base [1, 2, 3], which takes into account the important properties of the ground base at a dynamic effect, as the extremity of the distribution zone of oscillations and its inertia. In this case, the model allows us to determine not only the movement of the boundary plane, but also the stress-strain state of the base.

With dynamic calculations of the foundations of industrial structures and machines with a large spectrum of frequencies, the accounting of irreversible losses of the oscillation energy at the base due to the presence of internal inelastic resistance is very significant, especially for the occurrence of resonance. Therefore, there is a generalization of an axisymmetric column model on a dissipative base.

In most cases, accounting the dissipation of energy is made according to the Foyht hypothesis of viscous friction, according to which the forces of inelastic resistance is the linear function of the deformation rate. Foyht hypothesis of viscous friction as a physical constant uses attenuation coefficient. It is convenient in mathematical terms, but contradicts experimental data. So according to

Content from this work may be used under the terms of the Creative Commons Attribution 3.0 licence. Any further distribution of this work must maintain attribution to the author(s) and the title of the work, journal citation and DOI.
Published under licence by IOP Publishing Ltd

the hypothesis of viscous friction, the attenuation coefficient and the attenuation decrement are proportional to the frequency, but in reality they do not depend on the frequency of oscillations. The disadvantage of this hypothesis is also that it does not reflect the dependence of inelastic resistance in the material from the amplitude of deformation.

In our opinion, E.S. Sorokin hypothesis of the hysteresis friction more corresponds to the reality, in which the forces of inelastic resistance are proportional to the movement and as a physical constant uses the coefficient of inelastic resistance. E.S. Sorokin hypothesis provides that the forces of inelastic resistance are proportional to the elastic ones, but they shifted relative to the elastic phase at the angle $\pi/2$. According to this hypothesis, internal resistance forces are presented in a comprehensive form.

$$S^* = (1+i)S, \qquad (*)$$

where: S^* – Total (elastic + elastic) internal force;

S – elastic force;

$\gamma = \dfrac{\psi}{2\pi}$ – the coefficient of inelastic resistance;

ψ – the absorption coefficient, which depends on the physical properties of the material;

the imaginary number $i = \sqrt{-1}$ – provides a phase shift by 90°.

The coefficient γ is the functions of the amplitude of movements, but in most cases you can take $\gamma = \text{const}$ for a wide range of movement amplitudes.

When determining the precipitate of the base and its stress-strain state, according to the axisymmetric columnar model of the base, the zone in which the localization of movements and stresses from a uniformly distributed load occurs $q(t) = F(t)/\pi \cdot \bar{r}^2$ within the centrally located platform, the radius, simulating a circular cylindrical pillar H high and a finite radius $R > \bar{r}$ in cross section (picture 1), which coincides with the radius of the attenuation of movements on the boundary plane of the base.

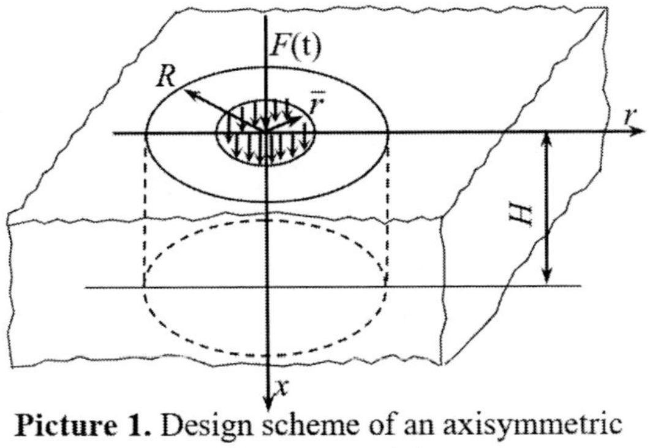

Picture 1. Design scheme of an axisymmetric columnar model.

Connect the beginning of the cylindrical coordinate system x, y, θ with an application point of force $F(t)$, and the vertical axis x will be directed according to force $F(t)$ to the lower end of the cylinder. The considered cylindrical body is in the conditions of axial symmetry, therefore only radial w and vertical u movements will be observed (tangential $v = 0$), which are functions of variables x, r and t.

The problem of movements and stresses of the compliant layer from dynamic loads will be solving in the absence of displacements u and w on the outer cylindrical column surface of the radius R isolated from the base. The area of the cargo platform of the $\bar{r}$ radius is consistent with the area of sections, which is divided by the foundation base when the distribution of reactive force will possibly be represented as a step-by-step diagram.

The solution of Liame equations, which describe the deformation of the cylindrical body we will carry out by L.V. Vinokurov discrete method [4]. According to the variable x, the solution is tried in an analytical form, and according to the variable r - in the finite-difference. We approximate the volume of the cylindrical body under consideration by the prisms dividing by its longitudinal sections passing through the vertical axis x and the components of the angles θ, and concentric circles. The movement of the prism edges is accepted as independent indeterminates. Since the cylindrical body under consideration is in axial symmetry, then in differential equations of the discrete method, movements on vertical lines-edges (lines resulting from the intersection of a longitudinal cross section with concentric circles) located in a single diametrical plane. We divide R into three equal parts (picture 2).

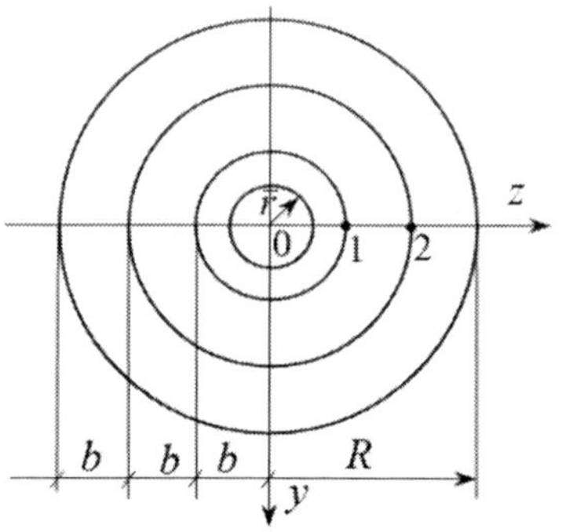

Picture 2. Dividing scheme of radius *R*.

Dynamic equilibrium of the compliant base represented by an axisymmetric columnar model, when taking into account the dissipation of energy according to the E.S. Sorokin hypothesis, is described by the following system of differential equations recorded for edges 0, j = 1.2

$$\left.\begin{aligned}
&(1+i\gamma)\left[\alpha_\mu \frac{\partial^2 u_0^*}{\partial \xi^2} + 2\beta_\mu \frac{\partial w_1^*}{\partial \xi} + 2(u_1^* - u_0^*)\right] - \frac{qb^2}{E}\rho\frac{\partial^2 u_0^*}{\partial t^2} = 0,\\
&(1+i\gamma)\left\{ j^2 \frac{\partial^2 w_j^*}{\partial \xi^2} + 2\alpha_\mu j\left[w_{j+1}^*\left(j+\frac{1}{2}\right) + w_{j-1}^*\left(j-\frac{1}{2}\right) - w_j^*\left(2j+\frac{1}{j}\right)\right] + \right.\\
&\left. + j^2\beta_\mu \frac{\partial}{\partial \xi}(u_{j+1}^* - u_{j-1}^*\right\} - \frac{2qj^2b^2}{E}\rho\frac{\partial^2 w_j}{\partial t^2} = 0,\\
&(1+i\gamma)\left\{ 2\alpha_\mu \frac{\partial^2 u_j^*}{\partial \xi^2} + \frac{1}{2}\left[u_{j+1}^*\left(2+\frac{1}{j}\right) + \left(2-\frac{1}{j}\right) - 4u_j^*\right] + \right.\\
&\left. + 2j\beta_\mu \frac{\partial}{\partial \xi}\left[\frac{1}{j^2} w_j^* + \frac{1}{2j}(w_{j+1}^* - w_{j-1}^*)\right]\right\} - \frac{2qb^2}{E}\rho\frac{\partial^2 u_j^*}{\partial t^2} = 0,
\end{aligned}\right\} \quad (1)$$

where: u^* and w^* – complex movements; $\xi = \frac{x}{b}$; $q = (1+\mu)$; $\alpha_\mu = \frac{1-\mu}{1-2\mu}$; $\beta_\mu = \frac{1}{2(1-2\mu)}$;

$\rho_\mu = \frac{\mu}{1-2\mu}$; μ – Poisson's ratio; ρ – mass density of base.

Stresses at the points of edges 0 and j, respectively, representing the complex harmonious stress by E.S. Sorokin in the form

$$\sigma^* = E(1+i\gamma)\varepsilon^*, \quad (2)$$

Are defined by expressions

$$\left.\begin{aligned}
\sigma^*_{x_0} &= \frac{E(1+i\gamma)}{1+\mu}\left(\alpha_\mu \frac{\partial u^*_0}{\partial x} + 2\rho_\mu \frac{w^*_1}{r_1}\right),\\
\sigma^*_{r_0} = \sigma^*_{\theta_0} &= \frac{E(1+i\gamma)}{1+\mu}\left[\frac{\alpha_\mu}{r_1} w^*_1 + \rho_\mu\left(\frac{1}{r_1} w^*_1 + \frac{\partial u^*_0}{\partial x}\right)\right],\\
\sigma^*_{x_j} &= \frac{E(1+i\gamma)}{2(1+\mu)}\left\{\frac{\rho_\mu}{r_j}\left[2w^*_j + \alpha_j(w^*_{j+1} - w^*_{j-1}\right] + 2\alpha_\mu \frac{\partial u^*_j}{\partial x}\right\},\\
\sigma^*_{r_j} &= \frac{E(1+i\gamma)}{2(1+\mu)}\left[\frac{2\rho_\mu}{r_j} w^*_j + \frac{\alpha_\mu \alpha_j}{r_j}(w^*_{j+1} - w^*_{j-1}) + 2\rho_\mu \frac{\partial u^*_j}{\partial x}\right],\\
\sigma^*_{\theta_j} &= \frac{E(1+i\gamma)}{2(1+\mu)}\left[\frac{2\alpha_\mu}{r_j} w^*{}_j + \frac{\alpha_j \rho_\mu}{r_j}(w^*_{j+1} - w^*_{j-1}) + 2\rho_\mu \frac{\partial u^*_j}{\partial x}\right],\\
\tau^*{}_{rx_j} &= \frac{E(1+i\gamma)}{2(1+\mu)}\left[\frac{\alpha_j}{2r_j}(u^*_{j+1} - u^*_{j-1}) + \frac{\partial w^*_j}{\partial x}\right],
\end{aligned}\right\} \quad (3)$$

where E – the deformation module of the base.

Accounting for the internal dispersion of energy with oscillations on the complex theory by E.S.Sorokin requires the formation of a complex excitation force *F(t)* for a given valid excitatory force *F(t)*. In the case when the excitation force varies by the harmonic law *cosωt* complex force is written as

$$F^*(t) = F_0 \cdot e^{i\omega t}. \quad (4)$$

At the same time, the imaginary part *F(t)* is

$$I_m\left[F^*(t)\right] = F_0 \cdot \sin\omega t. \quad (5)$$

In the future, the specified working strength is adopted for the real part of the complex force (4).

Solution of the system (1) of differential equations in private derivatives for the case of established oscillations, according to the Fourier method, we find in the form of

$$\left.\begin{aligned}
u^*_0(\xi,t) &= u_0(\xi)\cdot e^{i\omega t},\\
u^*_j(\xi,t) &= u_j(\xi)\cdot e^{i\omega t},\\
w^*_j(\xi,t) &= w_j(\xi)\cdot e^{i\omega t},
\end{aligned}\right\} \quad (6)$$

where: $u_0(\xi), u_j(\xi), w_j(\xi)$ – real function variable ξ.

Substitution (6) in the system (1) allows you to obtain a system of ordinary differential equations

$$\left.\begin{aligned}
&\alpha_\mu \frac{d^2u_0}{d\xi^2} + 2\beta_\mu \frac{dw_1}{d\xi} + 2(u_1 - u_0) + qk_*^2 u_0 = 0,\\
&j^2 \frac{d^2w_j}{d\xi^2} 2\alpha_\mu j \left[w_{j+1}\left(j+\frac{1}{2}\right) + w_{j-1}\left(j-\frac{1}{2}\right) - w_j\left(2j+\frac{1}{2}\right)\right] +\\
&+ j^2\beta_\mu \frac{d}{d\xi}(u_{j+1} - u_{j-1}) + 2qj^2k_*^2 w_j = 0,\\
&2\alpha_\mu \frac{d^2u_j}{d\xi^2} + \frac{1}{2}\left[u_{j+1}(2+\frac{1}{j}) + u_{j-1}(2-\frac{1}{j}) - 4u_j \right] +\\
&+ 2j\beta_\mu \frac{d}{d\xi}\left[\frac{1}{j^2} w_j + \frac{1}{2j}(w_{j+1} - w_{j-1}) \right] + 2qk_*^2 u_j = 0,
\end{aligned}\right\} \quad (7)$$

where
$$k_*^2 = \frac{\rho b^2 \omega^2}{(1+i\gamma)E}. \quad (8)$$

The frequency parameter k_*^2 is a comprehensive value.

We will integrate this system using Euler substitutions

$$\left.\begin{aligned}
u_0(\xi) &= A_0 \cdot e^{s\xi},\\
u_j(\xi) &= A_j \cdot e^{s\xi},\\
w_j(\xi) &= B_j \cdot e^{s\xi}.
\end{aligned}\right\} \quad (9)$$

After substitution (9) in (7), we obtain a system of algebraic equations, the determinant of which contains a complex frequency parameter. To solve this determinant is impossible. Therefore, from the expression (8) it is necessary to separate the real component

$$\mathrm{Re}(k_*^2) = \frac{\rho b^2 \omega^2}{(1+\gamma^2)E}. \quad (10)$$

The solution of the system of homogeneous differential equations (7) is written as

$$\left.\begin{aligned}
u_0^*(\xi,t) &= \left(\Sigma D_m^* e^{s_m \xi}\right) e^{i\omega t},\\
u_j^*(\xi,t) &= (\Sigma D_m^* a_{jm} e^{s_m \xi}) e^{i\omega t},\\
w_j(\xi,t) &= (\Sigma D_m^* b_{jm} e^{s_m \xi}) e^{i\omega t}.
\end{aligned}\right\} \quad (11)$$

Stress expressions (3), according to the Fourier method, is written in the form

$$\left.\begin{aligned}
&\sigma_{u_0}^*(\xi,t) = (1+i\gamma)\sigma_{x_0}(\xi)e^{i\omega t},\\
&\sigma_{r_0}^*(\xi,t) = \sigma_{\theta_0}^*(\xi,t) = (1+i\gamma)\sigma_{r_0}(\xi)e^{i\omega t},\\
&\sigma_{xj}(\xi,t) = (1+i\gamma)\sigma_{xj}(\xi)e^{i\omega t},\\
&\sigma_{rj}(\xi,t) = (1+i\gamma)\sigma_{rj}(\xi)e^{i\omega t},\\
&\sigma_{\theta j}(\xi,t) = (1+i\gamma)\sigma_{\theta j}(\xi)e^{i\omega t},\\
&\tau_{rj}(\xi,t) = (1+i\gamma)\tau_{rj}(\xi)e^{i\omega t},
\end{aligned}\right\} \quad (12)$$

where: σ_{x0}, σ_{r0} and $\sigma_{xj}, \sigma_{rj}, \sigma_{\theta j}, \tau_{rxj}$ are defined by expressions

$$\sigma_{x_0} = \frac{E}{1+\mu}\left(\alpha_\mu \frac{\partial u_0}{\partial x} + 2\rho_\mu \frac{w_1}{r_1}\right),$$

$$\sigma_{r_0} = \sigma_{\theta_0} = \frac{E}{1+\mu}\left[\frac{\alpha_\mu}{r_1} w_1 + \rho_\mu \left(\frac{1}{r_1} w_1 + \frac{\partial u_0}{\partial x}\right)\right],$$

$$\sigma^*_{x_j} = \frac{E}{2(1+\mu)}\left\{\frac{\rho_\mu}{r_j}\left[2w_j + \alpha_j (w_{j+1} - w_{j-1}\right] + 2\alpha_\mu \frac{\partial u_j}{\partial x}\right\},$$

$$\sigma_{r_j} = \frac{E}{2(1+\mu)}\left[\frac{2\rho_\mu}{r_j} w_j + \frac{\alpha_\mu \alpha_j}{r_j}(w_{j+1} - w_{j-1}) + 2\rho_\mu \frac{\partial u_j}{\partial x}\right],$$

$$\sigma_{\theta_j} = \frac{E}{2(1+\mu)}\left[\frac{2\alpha_\mu}{r_j} w_j + \frac{\alpha_j \rho_\mu}{r_j}(w_{j+1} - w_{j-1}) + 2\rho_\mu \frac{\partial u_j}{\partial x}\right],$$

$$\tau_{rx_j} = \frac{E}{2(1+\mu)}\left[\frac{\alpha_j}{2r_j}(u_{j+1} - u_{j-1}) + \frac{\partial w_j}{\partial x}\right].$$

Integration constant in expressions (11) are determined from the boundary conditions that, after the separation of variables, are the following

for $\xi = 0$, and $(1+i\gamma)\tau_{rx} = 0$, $(1+i\gamma)\sigma_x = \left\{-\frac{F}{\pi \bar{r}^2}\right.$ at $0 \le r \le \bar{r}$,

for $\xi = \frac{H}{b}$, $u = 0$, $w = 0$ at $\bar{r} \le r \le R$.

From the boundary conditions, a system of algebraic equations is followed, which in the matrix form is the following

$$CD^* = \overline{F}, \tag{13}$$

where: $C = \{c_{in}\}$ – matrix of coefficients with indeterminate integration constants D^*;

$\overline{F} = \left\{-\frac{qb}{E(1+i\gamma)}\frac{F}{\pi \bar{r}^2}, 0, 0, \ldots\ldots 0\right\}$ – column matrix of right parts.

After solving the system of equations (12), integration constants are determined by expressions

$$D^*_m = \frac{\overline{D_m} F b}{E(1+i\gamma)\pi \bar{r}^2},$$

where $\overline{D_m}$ – numeric coefficient obtained as a result of system solution (13).

Representing an integrated expression for D^* in an exponential form and substituting it into the solution (11), we get

$$\left.\begin{aligned}
u_0^*(\xi,t) &= \left(\frac{Fb}{(1+\gamma^2)^{1/2}E\pi\overline{\overline{r}}^2}\overline{\Sigma D_m e^{s_m\xi}}\right)e^{i(\omega t+\nu)},\\
u_j^*(\xi,t) &= \left(\frac{Fb}{(1+\gamma^2)^{1/2}E\pi\overline{\overline{r}}^2}\overline{\Sigma D_m a_{jm} e^{s_m\xi}}\right)e^{i(\omega t+\nu)},\\
w_j^*(\xi,t) &= \left(\frac{Fb}{(1+\gamma^2)^{1/2}E\pi\overline{\overline{r}}^2}\overline{\Sigma D_m b_{jm} e^{s_m\xi}}\right)e^{i(\omega t+\nu)}
\end{aligned}\right\} \tag{14}$$

where $\nu = arctg\gamma$.

Since the force $F\cdot\cos\omega t$ applied to the boundary plane represents the real part of the complex force (4), the solution of differential equations (1) in the real form will be introduced by the real part of the integrated solution (14).

Applying Euler identity to the latest one, we will write

$$\left.\begin{aligned}
u_0(\xi,t) &= \left(\frac{Fb}{(1+\gamma^2)^{1/2}E\pi\overline{\overline{r}}^2}\overline{\Sigma D_m\cdot e^{s_m\xi}}\right)\cos(\omega t+\nu),\\
u_j(\xi,t) &= \left(\frac{Fb}{(1+\gamma^2)^{1/2}E\pi\overline{\overline{r}}^2}\overline{\Sigma D\cdot a_{jm} e^{s_m\xi}}\right)\cos(\omega t+\nu),\\
w_j(\xi,t) &= \left(\frac{Fb}{(1+\gamma^2)^{1/2}E\pi\overline{\overline{r}}^2}\overline{\Sigma D_m b_{jm} e^{s_m\xi}}\right)\cos(\omega t+\nu).
\end{aligned}\right\} \tag{15}$$

Accounting of the energy dissipation, as follows from expressions (15), leads to the phase shift between the perturbing load and the base movement.

For forced oscillations picture 3 shows the curves of changes in the amplitudes of vertical longitudinal movements $u_0(0)$ of the base surface in the center of the cargo platform of radius $\bar{r}$ depending on the change in the frequency parameter k^2, taking into account and without taking into account the energy dissipation at $R=6\bar{r}$, $H=\overline{8}$ and $\mu=0{,}3$. Change $u_0(0)$ in the interval $k^2=0{,}34\div0{,}39$ is not given.

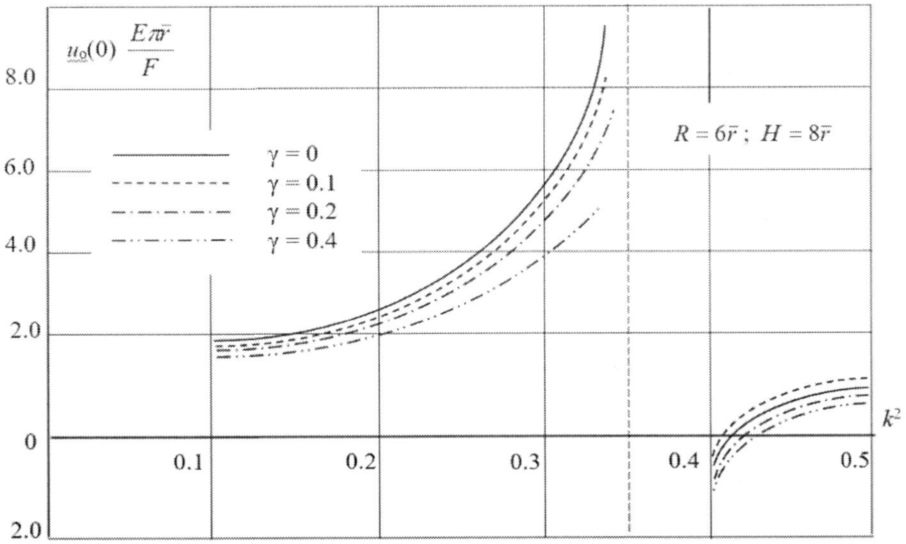

Picture 3. Changes in the amplitudes of the movements of $u_0(0)$ depending on the coefficient of inelastic resistance γ.

From picture 3 it follows that accounting the energy dissipation for values k^2 far from the resonance slightly affects the amplitudes of the base movements, while in the resonant zone ($k^2 = 0{,}3 \div 0{,}4$), the accounting of energy dissipation significantly reduces the movements of the base.

References

[1] Vinokurov L P and Telipko L P 1977 Solution of the dynamic contact problem for beams on a weighty compliant layer with the involvement of an axismetric column model *Construction and architecture* **2** pp 31-6

[2] Vinokurov L P and Telipko L P 1978 The solution of the contact problem for statically and dynamically loaded foundations on the compliant base, represented by the axisymmetric column model *Resistance of materials and theory of structures* **XXXII** pp 31-8

[3] Telipko L P, Mamaiev L M and Raksha S V 2018 Accounting for the heterogeneity of the foundation of the railway track when determining its stress-strain state *Science is the progress of transport Visnik Dnipropetrovsk National University of Railway Transport after Acad. V. Lazarian* **6 (78)** pp 101-17

[4] Sorokin Ye S 1960 On the theory of internal friction at oscillations of elastic systems (*Moscow: Gosstroyisdat)* 131

[5] Vinokurov L P 1956 Direct methods for solving spatial and contact problems for solid mass and foundations *(Kharkov: Publishing House of Kharkov University)* 280

Aggregation of identical mechanical systems with oscillations

I N Barabanov and V N Tkhai

Trapeznikov Institute of Control Sciences, 65 Profsoyuznaya Street, GSP-7, Moscow 117997, Russia

Abstract. We consider identical mechanical systems described by Lagrange's equations of the second kind and subject to the action of positional forces. It is assumed that a separate system allows for single-frequency oscillation. The problem of aggregating the set of systems into a coupled system with an attractive cycle close to the oscillation of uncoupled systems is solved. For this purpose, weak universal coupling controls are found.

1. Introduction

Aggregation of individual systems is performed to obtain a coupled system with a given property. This procedure should be accompanied by the study of the coupled system.

Usually, aggregation is performed after decomposition. Methods for aggregating complex systems by the Lyapunov method are given in [1]. It is proposed to study the set of systems in the framework of a model containing weakly coupled subsystems [2]. The idea of [2] was implemented for autonomous and periodic coupled systems described by general differential equations in [3-6]. In this paper, we consider identical mechanical systems.

Coupled systems appear in various fields of knowledge. A classic example in mechanics is the Sommerfeld sympathetic pendulum. For more examples, see [7-11].

2. Problem statement

Consider k n-degree-of-freedom identical mechanical systems described by Lagrange's equation of the second kind under the action of positional forces. Each system is assumed to allow for a single-frequency oscillation.

The phase portrait of an individual mechanical system is symmetrical with respect to the fixed space $M_l = \{q_l, \dot{q}_l : \dot{q}_l = 0\}$, where q_l is the generalized coordinate, $l = 1,\ldots,k$. The generalized velocity $\dot{q}_l$ turns into zero on M_l, while the periodic motion is symmetrical with respect to M_l and is called the symmetrical periodic motion (SPM). The set of SPMs form h-parametric families $\Sigma_l(h)$. We consider families $\Sigma_l(h)$ that contain only non-degenerate SPMs, i.e. SPMs with the period depending on h strictly monotonically.

SPMs cannot be stabilized in the framework of the considered mechanical model. In order to stabilize an SPM, it is necessary to apply a force that violates the symmetry of the phase portrait. If the applied force is ε-small, then the oscillation of the controlled mechanical system, which is stabilized, is ε-close to the SPM. The appropriate controlled mechanical system that possesses an orbitally asymptotically stable cycle is constructed in [12,13]. A small force becomes a weak coupling control for a set of mechanical systems.

Stated is the problem of aggregation of considered identical mechanical systems such that the resulting system allows for an attracting cycle, which is close to the oscillations of uncoupled systems. At that, universal coupling controls (i.e. suitable for any mechanical systems) are found.

3. Aggregation technique

The proposed technique interprets a set of uncoupled identical systems as a single mechanical system. This system, according to the problem statement, allows for a family $\Sigma(h) = \bigcup \Sigma_l(h)$ of nondegenerate SPMs. A controlled mechanical system is constructed, where an ε-small control is

Content from this work may be used under the terms of the Creative Commons Attribution 3.0 licence. Any further distribution of this work must maintain attribution to the author(s) and the title of the work, journal citation and DOI.
Published under licence by IOP Publishing Ltd

applied. The existence of such a control is justified in [12,13]. The further investigations are carried out in the framework of the controlled mechanical system

$$\frac{\mathrm{d}}{\mathrm{d}t}\frac{\partial T}{\partial \dot{q}_s}-\frac{\partial T}{\partial q_s}=Q_s(q)+\varepsilon u_s(q,\dot{q}),\quad s=1,\ldots,m, \tag{1}$$

where T is kinetic energy, $Q_s(q)$ is a positional force, $m=kn$, and $u=(u_1,\ldots,u_m)$ is chosen in the general form

$$u_s=\sigma\left[1-K\left(h^*\right)b(q,\dot{q})\right]\sum_{j=1}^{m}r_{sj}\dot{q}_j,\quad b>0,\quad r_{sj}=\text{const},\quad s=1,\ldots,m. \tag{2}$$

For k identical mechanical systems with n degrees of freedom, the dimension of system (1) is $m = kn$. The value of h^* corresponds to the attractive cycle; the choice of σ is described below.

Function $K(h)$ is the characteristics of the family of SPMs. It is defined by the identity

$$\int_0^{\tau(h)}\left[1-K(h)b\left(\varphi(h,t),\dot{\varphi}(h,t)\right)\right]\sum_{s,j=1}^{m}r_{sj}\dot{\varphi}_j(h,t)\psi_s(h,t)\mathrm{d}t\equiv 0.$$

As a result, we obtain

$$K(h)=\left(\int_0^{\tau(h)}\theta(h,t)\mathrm{d}t\right)\left(\int_0^{\tau(h)}b\left(\varphi(h,t),\dot{\varphi}(h,t)\right)\theta(h,t)\mathrm{d}t\right)^{-1},\quad \theta=\sum_{s,j=1}^{n}r_{sj}\dot{\varphi}_j(h,t)\psi_s(h,t).$$

In [13] it is proved that an orbitally asymptotically stable cycle always occurs in system (1),(2) at the proper choice of matrix $\left\|r_{sj}\right\|$. In (2), multiplier

$$\sigma=\begin{cases}1,\text{ if }\chi<0\\-1,\text{ if }\chi>0\end{cases},$$

where $\chi=\mathrm{d}K\left(h^*\right)/\mathrm{d}h$ is applied. Therefore, the problem of stabilization of k mechanical systems can be solved in principle, provided that appropriate function b and matrix $\left\|r_{sj}\right\|$ in (2) are found.

4. Cycle of the coupled mechanical system

Necessary and sufficient conditions for a τ-periodic SPM to exist are expressed as follows

$$\dot{q}_s\left(q_1^0,\ldots,q_n^0,t\right)=0,\quad t=0,\ \tau/2;\quad s=1,\ldots,n, \tag{3}$$

where q^0 is the initial point at $t=0$. This means that SPMs constitute a τ-parametric family.

Recall that we consider the family $\Sigma(h)$ of nondegenerate SPMs, so that the period depends on the numerical parameter strictly monotonously. Denote the trajectory on $\Sigma(h)$ by $q=\varphi(h,t+\gamma)$. Here h is the family parameter, and γ is the shift of the initial point along the trajectory. The solution is assumed to be in fixed set M at t=0, so put $\gamma=0$. Family Σ fills the invariant two-dimensional manifold $\hat{\Sigma}$ in the phase space. This fact, as well as the monotonous behaviour of function $\tau=\tau(h)$ is established in [14].

Since (1) is satisfied on $\hat{\Sigma}$, the following lemma holds.

Lemma. *A family of nondegenerate SPMs is described by a one-degree-of-freedom conservative system.*

Proof. Equation (3) implies that the following linear equation hold.

$$d\xi_s \equiv b_{s1}\left(q^0,t\right)dq_1^0 + \ldots + b_{sn}\left(q^0,t\right)dq_n^0 + c_s\left(q^0,t\right)dt = 0,$$
$$B = \left\|b_{sj}\left(q^0,t\right)\right\|,\quad b_{sj}\left(q^0,t\right) = \partial\dot{q}_s\left(q^0,t\right)/\partial q_j^0, \tag{4}$$
$$C = \left\|c_s\left(q^0,t\right)\right\|,\quad c_s = \partial\dot{q}_s\left(q^0,t\right)/\partial t = \ddot{q}_s\left(q^0,t\right);\quad s,j = 1,\ldots,n.$$

The above equations become identity on $\hat{\Sigma}$. Since the SPMs are nondegenerate, conditions $\det B = n$, $C = (c_1,\ldots,c_n) \neq 0$ are satisfied, therefore there exists a linear transform $\eta = P\xi$, $\xi = (\xi_1,\ldots,\xi_n)$, where $P = \left\|p_{sj}\right\|$ is a constant matrix, such that $\eta_2 = 0,\ldots,\eta_n = 0$. Since the transform holds for any point $\left(q^0,t\right)$, the first coordinate η_1 is chosen in $\hat{\Sigma}$, and (4) can be rearranged as follows.

$$d\eta_1 \equiv \sum_{j=1}^{n}\tilde{b}_{1j}\left(q^0,t\right)dq_j^0 + \tilde{c}_1\left(q^0,t\right)dt = 0,\quad d\eta_k\left(q^0,t\right) = 0,\quad k = 2,\ldots,n,\quad \tilde{c}_1 \equiv \sum_{j=1}^{n} p_{1j}c_j\left(q^0,t\right). \tag{5}$$

For Σ, the existence of the linear transform mentioned above means that there exist coordinates $w_1,\ldots,w_n$ such that $w_2 = 0,\ldots,w_n = 0$ on Σ. The initial point q^0 on Σ is the function of a single parameter. Take w_1^0 as this parameter to obtain the following equation from the first equation of (5).

$$\ddot{w}_1 + \tilde{c}_1 w_1 = 0.$$

Therefore, the dynamics on $\hat{\Sigma}$ is described by a one-degree-of-freedom conservative system, q.e.d.

According to Lemma, a one-degree-of-freedom conservative system is a key subsystem of the multi-degree-of freedom system that implement the family $\Sigma(h)$. The whole system can be described in the neighborhood of $\hat{\Sigma}$ by the generalized coordinates (x,y), where $x = w_1$, $y = (w_2,\ldots,w_n)$. Since $y = 0$ is satisfied on $\hat{\Sigma}$, we obtain $x^2 = \sum_{j=1}^{m} q_j^2$. Following [13], implement the universal control $\left(1 - Kx^2\right)\dot{x}$ (K is a constant value) on $\hat{\Sigma}$. Thus, function b in (2) can be found on $\hat{\Sigma}$.

Control $\varepsilon\sigma\left(1 - Kx^2\right)\dot{w}$ in the transformed mechanical system is given by positive definite form $\sum_{s=1}^{m} w_s^2$. The linear transform in Lemma is implemented with the constant nondegenerate matrix, so it does not change the signature of the quadratic form. So the positive definite form

$$R = \frac{1}{2}\sum_{s,j=1}^{m} r_{sj}\dot{q}_s\dot{q}_j > 0,\quad r_{sj} = \text{const}$$

of initial coordinates is implemented to establish control (2).

Let us recall that $m = kn$ in the case of k identical n-degree-of-freedom mechanical systems. Besides, the characteristics is calculated as $K(h)/k$, where $K(h)$ is the characteristics of the individual system. As a result, we obtain the following theorem that gives the solution to the problem of aggregation.

Theorem *Consider the set of* k *identical* n*-degree-of-freedom mechanical systems under the action of positional forces. Let the systems have a family of nondegenerate symmetrical periodic motions. Let this set be coupled by the following coupling controls*

$$u_s = \sigma\left[1 - \frac{K\left(h^*\right)\rho}{k}\right]\sum_{j=1}^{kn} r_{sj}\dot{q}_j, \quad \rho = \sum_{j=1}^{kn} q_j^2, \quad r_{sj} = \text{const}, \quad s = 1,\ldots,kn, \tag{6}$$

where r_{sj} are coefficients of a positive definite quadratic form. Then an orbitally asymptotically stable cycle close to the oscillation of uncoupled systems at $h = h^*$ occurs.

Remark 1 Coupling controls (6) ensure the natural stabilization of the cycle in the coupled mechanical system.

Remark 2 Given Theorem, oscillations of conservative systems become synchronized in frequency and phase.

Remark 3 For conservative mechanical systems, matrix $\left\|r_{sj}\right\|$ is block-diagonal with k blocks.

5. Example

Consider the set of k identical mathematical pendulums $\ddot{x} + \sin x = 0$. Take total mechanical energy as h. Characteristics $K(h)$ of a single pendulum is shown in figure 1.

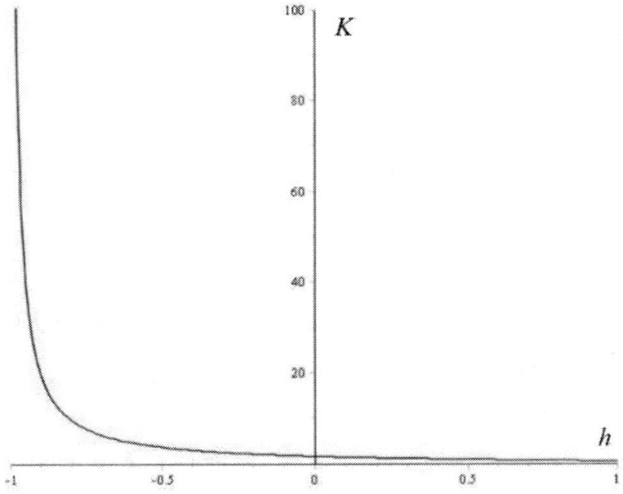

Figure 1. Characteristics of a single pendulum.

As it can be seen, $\mathrm{d}K(h)/\mathrm{d}h < 0$ for all $h \in (-1;1)$. Calculate the characteristics for the set of k pendulums as $\bar{K}(h) = K(h)/k$. So $\mathrm{d}\bar{K}(h)/\mathrm{d}h < 0$, and $\sigma = 1$. By the aggregation with the aid of coupling controls we obtain the controlled system

$$\ddot{x}_s + \sin x_s = \varepsilon\left[1 - \bar{K}\left(h^*\right)\sum_{j=1}^{k} x_j^2\right]\sum_{j=1}^{k} r_{sj}\dot{x}_j, \quad s = 1,\ldots,k, \tag{7}$$

where r_{sj} are coefficients of the positive definite quadratic form. Note that, according to Remark 3, the matrix of this quadratic form is block-diagonal, so equation (7) is rearranged as

$$\ddot{x}_s + \sin x_s = \varepsilon \left[1 - \bar{K}\left(h^*\right) \sum_{j=1}^{k} x_j^2 \right] \dot{x}_s, \quad s = 1, \ldots, k. \tag{8}$$

System (8) has an attracting cycle, which is ε -close to the oscillation of uncoupled pendulums with energy $h = h^*$.

6. Conclusion

Single-frequency oscillations of mechanical systems with positional forces are symmetric and form families. They cannot be stabilized in the framework of this model. The aggregation problem is stated in order to obtain a coupled system with an attracting cycle. The problem is solved by implementing appropriate weak coupling controls.

References

[1] Aleksandrov A Yu and Platonov A V 2012 *Comparison method and stability of motion of the nonlinear systems* (St. Petersburg: St. Petersburg University Press)

[2] Tkhai V N 2016 Stabilizing the oscillations of an autonomous system *Automation and Remote Control* **77(6)** pp 972–79

[3] Barabanov I N and Tkhai V N 2016 Oscillation family in weakly coupled identical systems *Automation and Remote Control* **77(4)** pp 561–8

[4] Barabanov I N and Tkhai V N 2017 Designing a stable cycle in weakly coupled identical systems *Automation and Remote Control* **78(2)** pp 217–23

[5] Barabanov I N and Tkhai V N 2018 Stabilization of oscillations in a periodic system by choosing appropriate Couplings *Automation and Remote Control* **79(12)** pp 2128–35

[6] Barabanov I N and Tkhai V N 2018 Solution to the natural stabilization problem for the oscillation of weakly coupled systems *Proc. 2018 14th International Conference 'Stability and Oscillations of Nonlinear Control Systems' (Pyatnitskiy's Conference)* (Moscow: IEEE Xplore) NSPEC Accession Number: 17899330 DOI: 10.1109/STAB.2018.8408348

[7] Morozov N F and Tovstik P E 2013 Transverse rod vibrations under a short-term longitudinal impact *Doklady Phys.* **58(9)** pp 387–91

[8] Kovaleva A and Manevitch L I 2016 Autoresonance versus localization in weakly coupled oscillators *Physica D: Nonlinear Phenomena* **320** pp 1–8

[9] Rompala K, Rand R and Howland H 2007 Dynamics of three coupled van der Pol oscillators with application to circadian rhythms *Communications in Nonlinear Science and Numerical Simulation* **12(5)** pp 794–803

[10] Yakushevich L V, Gapa S and Awrejcewicz J 2009 Mechanical analogs of the DNA base pair oscillations *Proc. 10th Conf. on Dynamical Systems Theory and Applications* (Lodz: Left Grupa) pp 879–86

[11] Kawamura Y 2014 Collective phase dynamics of globally coupled oscillators: noise-induced anti-phase synchronization *Physica D: Nonlinear Phenomena* **270** pp 20–29

[12] Tkhai V N 2019 Stabilizing the oscillations of a controlled mechanical system *Automation and Remote Control* **80(11)** pp 1996–2004

[13] Tkhai V N 2020 Stabilizing the oscillations of a controlled mechanical system with n degrees of freedom *Automation and Remote Control* **81(9)** pp 1637–46

[14] Tkhai V N 2012 The behaviour of the period of symmetrical periodic motions *J. Appl. Mathem. Mechan.* **76(4)** pp 446–450

Acknowledgments

This work is financially supported in part by the Russian Foundation for Basic Research, project No. 19-01-00146.

IOP Conf. Series: Materials Science and Engineering 1164 (2021) 012079 doi:10.1088/1757-899X/1164/1/012079

Topological approach to studying mechanochemical effects in the processes of corrosion of metals under stress

L E Trofimova

Odessa State Academy of Civil Engineering and Architecture, 4 Didrihsona Street, 65029, Odessa, Ukraine

E-mail: lara.reverberator119@gmail.com

Abstract. It is hereby suggested to use a topological approach based on the principles of synergetics and the theory of catastrophes to describe some aspects of the mechanochemical effect of the sign of deformation in stress corrosion phenomena. It was found that a steel plate being located in an aggressive environment and subjected to additional force impact may be informatively considered as a synergistic system. It has been established that the concept of sensitivity to imperfection can be applied to the interpretation of the deformation sign effect. It has been demonstrated that the violation of the symmetry of the sides of a bent plate is made provision for by the structure of standard models - catastrophes of the «ruffle» type. Possible scenarios for the development of mechanochemical effects are considered, taking into account Maxwell's principles and maximum delay principles. Topological models which describe both the above-mentioned mechanochemical effect of the deformation sign at small times (within 50 s) and its reversal with time are constructed and analyzed. Further possible directions of investigation of mechanochemical effects in the processes of corrosion of metals under stress are formulated.

1. Introduction

As you know, stress corrosion is a process which occurs under conditions when a material experiences external loads and deformation. It should be noted [1] that from the standpoint of mechanochemistry with making allowance for presence of the initial stress in the surface layer of any hard body in the form of surface tension, any naturally occurring corrosion is stress corrosion. As a result of the application of external forces to a body and its subsequent deformation, an additional surface stress is formed, which significantly affects the speed of the physicochemical dissolution process and, therefore, the rate of destruction of the material (in particular, metal structures). In this regard, finding of the laws of this phenomenon is not only of scientific, but also of practical interest.

It was experimentally established [2] that the applied mechanical stress accelerates corrosion. In order for the effect to be clearly expressed, sufficiently large stresses are needed. This result is most easily achieved through bending samples, in particular metal plates. In the thesis [3], it was for the first time experimentally discovered that the corrosion rate for the concave side of the plate is higher than for the convex side. Since the opposite sides of the curved plate differed in the sign of deformation, this phenomenon was called the mechanochemical effect of the sign of deformation in the phenomena of corrosion. It should be noted that for the first time the mechanochemical effect of the sign of deformation was recorded [4] in experiments on the dissolution of curved plates of single-crystal calcium chloride. In [3], it was also established that over time, the deformation sign effect reverses.

As follows from [1], the mechanochemical effect of the sign of deformation is inherent not only in stress corrosion processes, but also to any solid-body surface reactions. Thus, the effects discovered in [3] can be interpreted as universal and obeying a limited number of laws of nonlinear development of complex systems.

In connection with the above, the purpose of this research is to establish the regular patterns of the development of the processes of metals corrosion under stress. The research program provides for

Content from this work may be used under the terms of the Creative Commons Attribution 3.0 licence. Any further distribution of this work must maintain attribution to the author(s) and the title of the work, journal citation and DOI.
Published under licence by IOP Publishing Ltd

solving problems associated with the description and analysis of the phenomena, in which an increase in the influence leads to a qualitatively different behavior of the system.

2. Statement of the basic material

Some aspects of the mechanochemical effect of the sign of deformation in the phenomena of stress corrosion are proposed [5] to be interpreted from the standpoint of synergetics [6, 7] and catastrophe theory [8, 9]. In particular, it is assumed that a steel plate placed in an aggressive environment and subjected to additional force is recommended to be interpreted as a synergistic system, since within the framework of this approach, much attention is paid [6, 7] to the understanding of the fundamental significance of symmetry and violation of symmetry. In this aspect, the violation of symmetry (i.e., the manifestation of internal differentiation between different parts of the system or between the system and its environment) embodies one of the first preconditions of complex behavior and is accompanied by the emergence of new properties.

In turn, various symmetry violations (including spatial symmetry) are taken into account [6, 8] in the structure of standard models – catastrophes: canonical equations contain a term playing the role of a «symmetry violator» or the imperfection parameter, which is geometrically represented by the asymmetric bifurcation diagram. According to catastrophe theory, a «ruffle» type catastrophe corresponds to such an evolutionary picture. In accordance with this theory, whatever character the system imperfections may have, while it in the absence of imperfections is described in canonical form by means of a function:

$$V_c(x)=\frac{1}{4}x^4+\frac{1}{2}c_2x^2, \tag{1}$$

then, if imperfections are available, by the function:

$$V_c(x)=\frac{1}{4}x^4+\frac{1}{2}c_2x^2+c_1x. \tag{2}$$

A «trident»-type graph which is of frequent occurrence in bifurcation theory corresponds to the «perfect» system (1). The geometry of this diagram regulates the absence of catastrophic jumps, while in system (2) they necessarily occur when moving from one stable branch to another with increasing control parameter c_2. In this case, parameter c_1 is interpreted as initial imperfection. It should be noted that not only various defects (structural, geometric, etc.) are permissible to be considered as imperfections, but also the effects of external fields.

In light of the foregoing, it seems possible to use the above approaches in describing and analyzing both the previously indicated effect of deformation sign (Figure1) and the case of its reversal over time (Figure2).

For a more complete picture of the representation of the process under study at short time periods (Figure 1), two variables (control variables in terms of catastrophe theory) are additionally introduced: the load P on the sample and deformation ε of the sample (in [3] these parameters are constant). Accordingly, the mass loss of the sample (%) is interpreted in accepted terminology as a state parameter.

Catastrophe theory rigorously proves that the only type of model surface in such systems with one state parameter and two control parameters is «ruffle» (Figure 3). Qualitatively different system behavior is determined by various combinations of control parameters. In the case under consideration, the parameter P is called splitting parameter, since when its critical value is exceeded, the model surface splits into two sheets, i.e. its variation regulates the very probability of ambiguity of the dependence (%) on the sign ε and the occurrence of jumps. The normal parameter ε is directed along the normal (perpendicular) to the splitting factor.

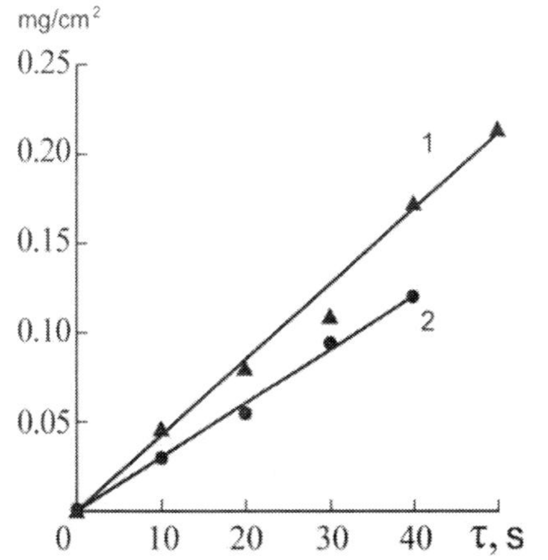

Figure 1. Dependence of sample mass decrease per a surface unit (in mg/cm^2) on the time τ for the concave (1) and convex (2) sides of a curved steel plate in a 35% hydrochloric acid solution. The slope of the lines represents the corrosion rate.

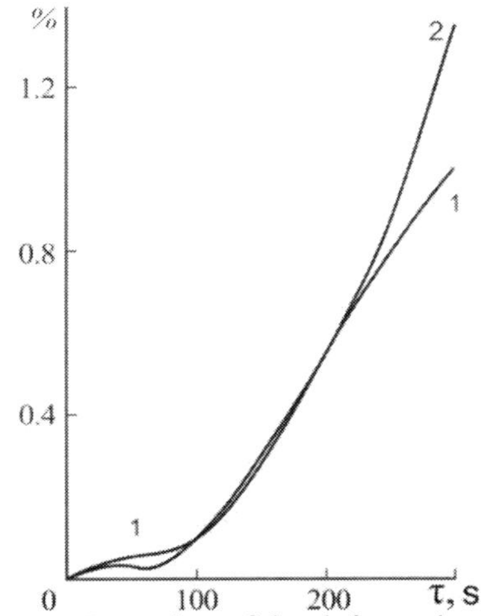

Figure 2. Inversion of the deformation sign. Curves of dependence of the sample mass loss (%) on the time τ for the concave (1) and convex (2) sides of a curved steel plate in a 35% hydrochloric acid solution.

An analysis of the scheme of distribution of deformations of a curved steel plate and curves of the dependence of the weight loss of the sample for its concave and convex sides in a 35% solution of hydrochloric acid at short time periods revealed [3], that the effect under consideration may be satisfactorily described by a model which geometry obeys the Maxwell's principle. In this case, a situation [6, 8] occurs, similar to the formation of a shock wave (gap) or phase transition of the first order in the areas of coexistence of different phases. In the absence of an external load ($P = 0$), the mass loss (%) is the same for the opposite sides of the sample. After passing through some value P_{cr} a manifestation of the deformation sign effect begins, which increases as the load increases. The structure of the model is designed with consideration to the fact that the dissolution rate is higher for the concave side than for the convex side. The upper sheet of the assembly corresponds to the mechanochemical dissolution effect for the compressed side of the curved plate, and the lower sheet – for the stretched side.

The position of the topological model in Figure 3 (a) at an angle to the axis (%) illustrates the well-known fact of acceleration of corrosion under stress, regardless of the sign of deformation [3]. Thus, all the points of the model surface are above the value of the parameter of state in the absence of external influence. For better visualization of the image of the quality situation being observed, the bottom sheet of the assembly is conditionally torn off.

The projection of the flat sector of manifold of the catastrophe «ruffle» onto the plane of control parameters is the so-called Maxwell's set (dashed line in Figure 3). According to this bifurcation scheme, movement from one sheet to another occurs whenever the control parameter ε crosses the set (structure of the shock wave), i.e. changes its sign to the opposite.

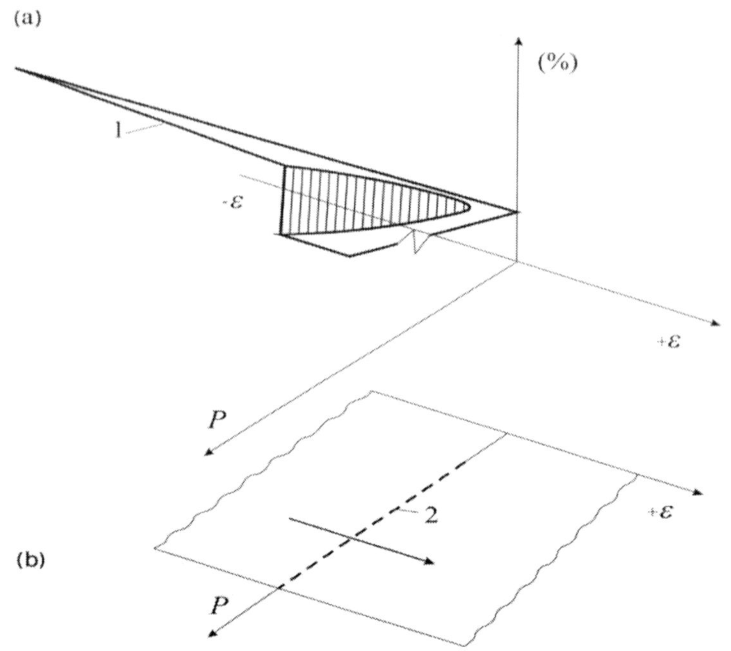

Figure 3. Topological model of the mechanochemical effect of deformation sign in the phenomena of corrosion under stress (catastrophe «ruffle», Maxwell's principle): (a) – three-dimensional surface in coordinates (%), ε and P (1 – cross section of the model); (b)–projection of the model onto the plane of the control parameters ε and P (2 – Maxwell's set).

Thus, deformation of the sample ε in this case should be interpreted as the initial imperfectionwith making allowance for presence in the initial (undistorted) state of surface tension which is added to the applied stress when the plate is stretched and subtracted from it when the plate is compressed. In other words, this symmetry-breaking single control parameter of geometric imperfection is adequate to the balance of all stresses in a bent sample. Consequently, the effect of the sign of deformation in the phenomena of corrosion under stress may be interpreted (in the terminology of the involved theory) as sensitivity to imperfection.

To simulate the reversal of the deformation sign effect (Figure 2) it also seems appropriate to use a catastrophe of the «ruffle» type, but under condition of observing the principle of maximum delay (Figure 4). According to this scenario, the system jumps into another state only when it has no other choice.

The investigated phenomenon is proposed to be described using two control parameters: P (as in the first case) and the time τ. As the state parameter, the difference in mass loss Δ (%) of the concave and convex sides of the curved sample was chosen:

$$\text{if } (\%)_{concave} > (\%)_{convex}, \quad \text{then } (\%)_{concave} - (\%)_{convex} = +\Delta(\%);$$
$$\text{if } (\%)_{concave} < (\%)_{convex}, \quad \text{then } (\%)_{concave} - (\%)_{convex} = -\Delta(\%).$$

Thus, the top sheet of the surface corresponds to the realization of the effect of the sign of deformation, and the bottom sheet corresponds to the fact of its reversal. Since in the absence of external influence the symmetry of the sides of the plate is not broken, i.e. $\Delta(\%) = 0$, thus the corresponding section of the assembly coincides with the axis τ.

The most interesting property of this surface is the presence of two fold lines starting at the so-called assembly point B which form a bifurcation curve with a tip at point B_1 on the plane of the control parameters τ and P (Figure 4). These points correspond to the P_{cr} value (in presence of the time coordinate), which, if reached, after some time [3] causes occurrence of cracks, which initiate the phenomenon of reversal of the deformation sign effect. It should be noted that when constructing the topological model, the acceleration of the onset of crack formation according to increase in load (i.e., a decrease in withstanding) is to be taken into account.

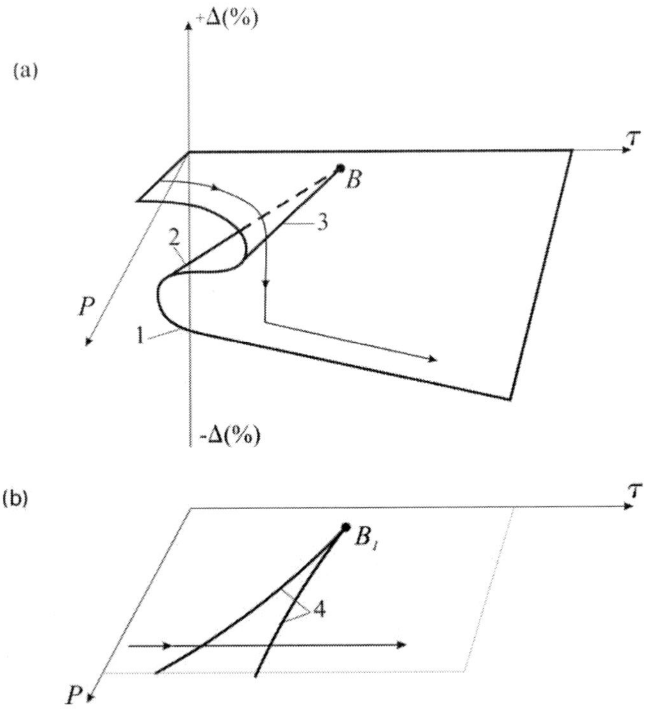

Figure 4. Topological model of reversing the effect of the deformation sign (catastrophe «ruffle», the principle of maximum delay): (a) – three-dimensional surface in the coordinates Δ(%), τ and P (1 – cross section of the model; 2 and 3 – fold lines; B – assembly point); (b) – projection of the model onto the plane of the control parameters τ and P (4 – bifurcation curve, i.e. bifurcation set; B_1 – bifurcation point).

In the general case, the bifurcation curve divides the control space into areas which are adequate to the various modes of the system functioning. Qualitative changes in its behavior occur only when the «trajectory», formed by a combination of control parameters, leaves the region inside this curve. It is in such a situation the state parameter Δ (%) jumps.

3. Conclusions and propositions

Qualitative features of behavior (signs of catastrophe) of the under study make it possible to simulate the general picture of the ongoing mechanochemical processes. If the sample mass loss (%) is presented as a function of two properly selected control parameters, then gap changes in the state of the system under study are described quite clearly with the help of a catastrophe of the «ruffle» type. In particular, the interpretation of the effect of the sign of deformation as sensitivity to imperfection does not contradict the provisions of mechanochemistry. In addition, expanding the scope of the experiment (due to the directed variation of the qualitative and quantitative composition of the aggressive environment, the type and value of external influences, etc.) will make it possible to use synergetic ideas for further research of these nontrivial effects, which have been first discovered in laboratory systems [3].

According to Poston and Stewart, the catastrophe theory approach should be seen as predicting what would be useful to look for, rather than what should certainly be seen. At present, only

IOP Conf. Series: Materials Science and Engineering 1164 (2021) 012079 doi:10.1088/1757-899X/1164/1/012079

experiment may serve for determining which aspects of the considered phenomena may be effectively described using this approach, and which may not [8].

References

[1] Rusanov A I 2006 *Thermodynamic Foundations of Mechanochemistry* (St.Petersburg: Nauka) 221
[2] Hutman E M 1974 *Mechanochemistry of Metals and Corrosion Protection* (Moscow: Metallurgiia) 230
[3] Rusanov A I, Uriev N B, Eryukin P V, Movchan T G and Esipova N E 2004 Discovery of effect of the sign of deformation in stress corrosion phenomena *Dokl. Russian Akad. of Sciences* **395** 3 pp 364–66
[4] Berenshtein G V, Diachenko A M and Rusanov A I 1988 Mechanochemical effect of dissolution *Dokl.Akad. Nauk SSSR* **298** 6 pp 1402–04
[5] Trofimova L 2020 Modeling and optimization of technology and physics-mechanical properties of composite materials *Key Engineering Materials* **864** pp 59–65
[6] Nicolis G and Prigogine I 1979 *Self-Organization in Non-Equilibrium Systems* [Russian translation] (Moscow: Mir) 512
[7] Nicolis G and Prigogine I 1990 *Exploring Complexity* [Russian translation] (Moscow: Mir) 344
[8] Poston T and Stewart I 1980 *Catastrophe Theory and Its Applications* [Russian translation] (Moscow: Mir) 608
[9] Arnold V I 1990 *Theory of Catastrophes* (Moscow: Nauka) 128

Study of resistance of thermomodified wood to the influence of natural conditions

Y Tsapko[1], O Horbachova[2], S Mazurchuk[3] and O Bondarenko[4]

[1]Scientific Research Institute for Binders and Materials, Kyiv National University of Construction and Architecture, Povitroflotskyj ave., 31, Kyiv, Ukraine, 03037 and Department of Technology and Design of Wood Products, National University of Life and Environmental Sciences of Ukraine, Heroiv Oborony str., 15, Kiyv, Ukraine, 03041
[2]Department of Technology and Design of Wood Products, National University of Life and Environmental Sciences of Ukraine, Heroiv Oborony str., 15, Kiyv, Ukraine, 03041
[3]Department of Technology and Design of Wood Products, National University of Life and Environmental Sciences of Ukraine, Heroiv Oborony str., 15, Kiyv, Ukraine, 03041
[4]Department of Building Materials, Scientific Research Institute for Binders and Materials, Kyiv National University of Construction and Architecture, Povitroflotskyj ave., 31, Kyiv, Ukraine, 03037

E-mail: mazurchuk.s.m@ukr.net

Abstract. It is established that thermal modification provides a decrease in the hydrophobicity of wood. Additional wood surface treatment reduces water absorption factor of 2 and modified sample at a temperature of 220°C – 3 times. Geometric and dimensional stability is improved by 2 times. The effect of heat treatment on the water absorption of wood is slightly smaller - for samples modified at 190 and 220°C for more than 10 hours – 20 %. Additionally, a 20 % deterioration in water absorption can be achieved by applying oil-wax or azure. With increasing temperature modification improves color stability to UV-rays. If the wood after termomodyfikuvannya 160°C color difference ΔE 2 times larger, 190 and 220°C – does not change. Protection after applying azure to the surface is slightly better compared to oil-wax. Thermal modification and finishing paint and varnish materials increases resistance to surface weathering. Therefore, such treatment is appropriate for the protection and aesthetic appeal of wood products.

1. Introduction

Unique technical properties (durability, low water absorption and dimensional stability) Thermomodified wood enable its application in various areas [1].

The material is often used by designers for interior decoration due to the acquired color under the action of high temperature. The stability of the geometric dimensions and resistance to environmental facilitates its use in furniture, window frames, doors, wood flooring. It is possible to use as external facing of facades, in the form of a terrace board and garden paths. The material is excellent for cladding saunas and baths [2].

The resistance of heat-treated wood to atmospheric influences (UV rays, wetting) is slightly better in comparison with untreated wood, but additional surface treatment with coatings is required. There are many substances on the market for external wood treatment [3-6]. For horizontal surfaces (deck, terrace cover, garden paths) recommended penetrating finish, which can provide oil [7]. In the market there are many oils that keep the "natural" look and resistant to atmospheric influences. In this work we wanted to study the effectiveness of such oils.

Therefore, the aim of this study was to study the effectiveness of such oils in comparison with film-forming substances to protect the surface of thermally modified and unmodified wood during weathering. These results allow to provide reliable indicator of efficiency heat treated wood in environmental conditions and develop measures to extend the life of the products.

Content from this work may be used under the terms of the Creative Commons Attribution 3.0 licence. Any further distribution of this work must maintain attribution to the author(s) and the title of the work, journal citation and DOI.
Published under licence by IOP Publishing Ltd

2. Analysis of recent research and publications

Wood treated at high temperatures eventually turns gray. It is established that the brown color of thermally modified wood is not resistant to light. Transparent coatings and oils do not protect the surface from discoloration during weathering, and they are only recommended for the treatment of thermally modified wood products that are operated away from direct sunlight and rain [8]. Recently, more and more attention is paid to the effects of sunlight on the surface of heat-treated wood. A new approach to stabilizing the dark color of thermodified wood, combining visible light screens with organic UV light absorbers, has been developed.

In [9] determined the discoloration is not covered and transparent coatings thermally modified beech, ash and hornbeam and found that the modified wood samples slowly fade in comparison with the unmodified samples. Accordingly, the FTIR spectra of samples of wood ash, beech and hornbeam exposed to UV light, showed similar chemical changes as unmodified wood samples that were exposed to UV light, but less pronounced.

Finishing heat-treated wood is similar to processing dried wood, but you should take into account changes in some of its properties. Thermally treated wood is more hydrophobic; absorbs moisture gradually; less prone to swelling and drying, but still requires elastic coatings; has a dark color, which is prone to fading due to the influence of light [10, 11].

Notice [12], which is a result of improved dimensional stability of wood cracking Thermomodified much less compared to natural. Cracking spruce and pine wood thermodified after entering the external conditions without protective coating cracking was equivalent to untreated wood with uncoated surfaces. The application of oil did not prevent the appearance of cracks on thermodified wood after exposure to natural conditions [13].

In [7], it was reported that samples of three wood species (oak, ash and beech) without a protective surface coating had fewer surface cracks than heat-treated samples also without coating during accelerated weathering. Cracking of samples of heat-treated wood was reduced by oil treatment.

Heat treated beech wood also showed better color stability compared to untreated samples after three months of natural weathering winter. Natural weathering effect on thermally modified beech wood samples less than unmodified in terms of loss of gloss and surface roughness. In addition, higher temperature and duration of thermal modifications give better surface properties after natural weathering [14].

Compatibility studied nine different coatings of thermally modified wood pine scottish [15]. Coating applied to the thermally modified timber showed a better performance than the unmodified coating on wood. This can be explained by the changed characteristics of thermally modified timber, such as a lower equilibrium moisture content lower permeability, increased dimensional stability, better resistance to UV radiation and resistance to fungal blue stain compared with unmodified wood. It was also demonstrated better penetration of modified wood coatings and better wetting of thermally modified wood protective substances. In addition, solvent-based paints showed better results after one year of external weathering compared to water-based coatings.

Drying oils provide the best protection for wood products, but their use is limited by their long-term polymerization. During the study [16] of the drying behavior of linseed oil, tung oil and their mixture 1:1 repeated measurements of the angle of contact with water coatings applied to beech and oak wood, showed clear differences in wetting. Tung oil gave hydrophobicity to all wood samples soon after application, even when used without a dryer. Flaxseed oil required a longer drying time and was more susceptible to substrate exposure, but eventually reached the highest contact angles after forced drying.

Unfortunately, the problem of discoloration of wood in the room under a transparent coating exists today and most paint coatings with UV protection work at an insufficient level. The market constantly appear new products that position as a defense of color of wood in the interior and to prevent graying, yellowing and darkening surfaces. To date, the amount of scientific information on the effectiveness of these products is insignificant.

Thus, from the literature it is established that thermal modification of wood has a positive effect on resistance to temperature and humidity factors. Additional surface treatment with protective substances contributes to the stability of the size of the product and protects against rapid weathering of the surface in open air conditions. This indicates the feasibility of this study to determine the effectiveness of protection of wood products with paints and varnishes, which will increase the service life of building structures in the environment.

3. The purpose of this work

The purpose of this work is to evaluate the effectiveness of surface treatment with paints and varnishes of different types on the resistance of thermo-modified wood products to temperature, humidity and light.

4. Materials and methods of research

To determine the moisture absorption were used randomly selected 90 samples of hornbeam wood of radial cross section, pre-dried. The samples had dimensions of 20×10×10 mm. Next, the samples were thermodified at temperatures of 160, 190 and 220°C, exposure in the chamber for 1, 10, 20 hours. According to the mode parameters of thermomodification, all samples were divided into 9 groups and a group of control samples (unmodified wood) in Figure 1.

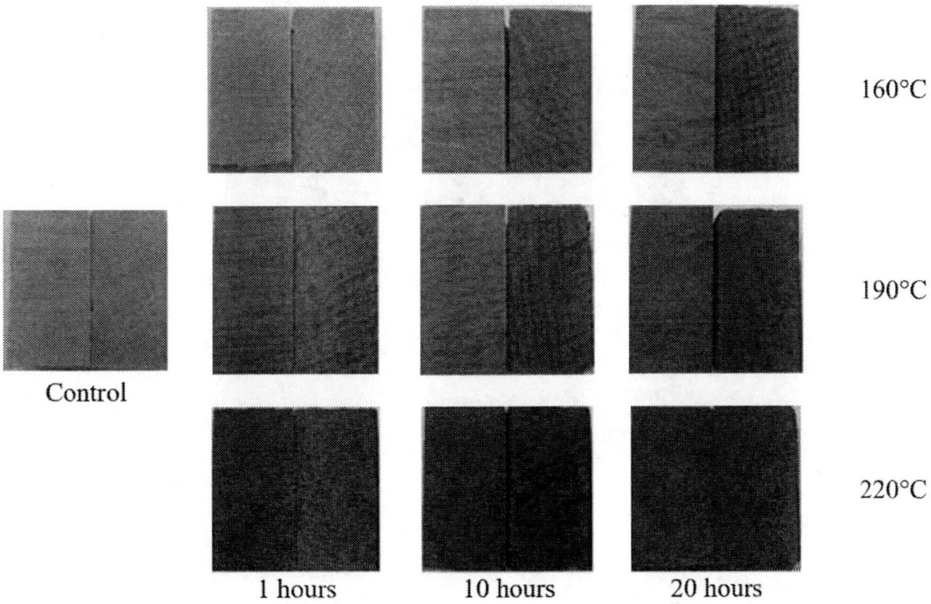

Figure 1. Prototypes of untreated and thermodified wood.

The samples were then treated with oil-wax and hydrophobic substance (azure) by immersion with re-treatment after drying the first layer for 24 hours. After drying, the samples were placed in a desiccator for 1 day. Then the dimensions and weight of each sample were recorded. The procedure was repeated for 2, 3, 6, 9, 13, 20, 30, 40 days.

To determine the water absorption, 90 samples similar to the previous experiment were used. After 2 hours in water, the samples were weighed and measured. The procedure was repeated for 1, 2, 3, 6, 9, 13, 20, 30, 40, 50 days.

The value and volume swelling αV amount of moisture absorbed W determined according to the equations:

$$\alpha_V = \frac{L_{r\max} \cdot L_{t\max} \cdot L_{a\max} - L_{r\min} \cdot L_{t\min} \cdot L_{a\min}}{L_{r\min} \cdot L_{t\min} \cdot L_{a\min}} \cdot 100, \quad (1)$$

where $L_{r\ max}$, $L_{t\ max}$, $L_{a\ max}$ – size samples after aging in a humid environment in the radial, plain areas and along the fiber, mm; $L_{r\ min}$, $L_{t\ min}$, $L_{a\ min}$ – size samples in a completely dry radially, and plain areas along the fiber, mm;

$$W = \frac{m_n - m}{m} \cdot 100, \quad (2)$$

where m_n – the mass of the sample on the n-th day of aging in a humid environment, g; m – is the mass of the sample in the dry state, g.

Also aware that the surface treatment of oil-wax product helps preserve the color and improves resistance to UV light. To establish the validity of this statement, studies were conducted to determine the effect of sunlight on similarly treated groups of samples of thermodified hornbeam wood size 20×20×10mm in vivo (Figure 2).

Figure 2. Exposure of samples in natural conditions.

Color was determined before and after incubation under natural conditions on the end and formation surfaces of the samples in three-dimensional color space CIEL* a* b*. In this system, L* (luminosity) characterizes the color from black to white (L* = 0 for black, L* = 100 for white), a* – from red (positive values) to green (negative values) and b* – from yellow to blue.

The change in the corresponding color parameter ΔL*, Δa* and Δb* was calculated relative to the control raw wood samples from the equations:

$$\Delta L^* = L^* \text{ treated} - L^*\text{control}, \quad (3)$$

$$\Delta b^* = b^* \text{ treated} - b^*\text{control}, \quad (4)$$

$$\Delta a^* = a^* \text{ treated} - a^*\text{control}. \quad (5)$$

The total color difference ΔE * was determined for each group of samples as follows:

$$\Delta E^* = \left[(\Delta a^*)^2 + (\Delta b^*)^2 + (\Delta L^*)^2\right]^{1/2}. \quad (6)$$

5. Research results

From the graph comparing the amount of moisture absorption (Figure 3) of thermodified wood in different modes (described above) it is seen that oil-wax and azure much better protect wood from moisture absorption from the air in all modes. In the first, the amount of moisture absorbed is 2 times less, and in the tenth mode, the difference was 2.7 times. In the first seven modes, the azure proved to be a little better, but from the eighth to the tenth mode, the wax proved to be better.

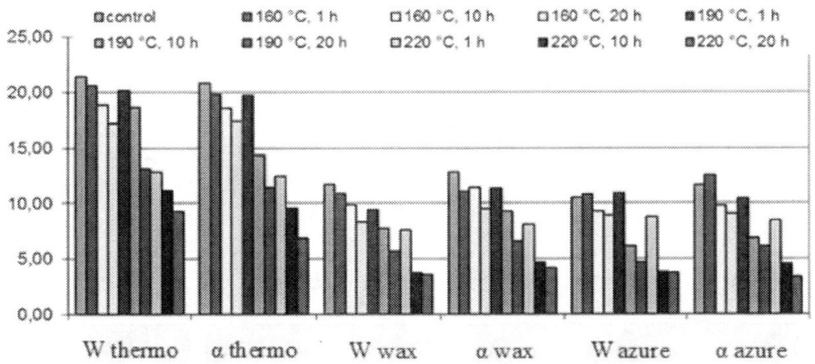

Figure 3. The results of determining the moisture absorption of samples of thermodified wood.

It was experimentally established that on the fortieth day of keeping the samples in a humid environment, the best result was shown by the samples of thermodified hornbeam wood additionally protected with oil-wax. They gained the lowest amount of moisture – 3.62 %, compared to thermodified wood under similar conditions (11.73 %) and additionally treated azure (3.76 %). However, the samples with additional azure treatment showed the best result in terms of volume edema, changing the size by 3.39 %. If we compare with untreated wood from this group of samples, the figure is 3.4 times higher and is 11.68 %.

Analyzing Figure 4 we see that in the conditions of exposure of samples in contact with water (water absorption) in terms of volume edema, the best result was shown by samples treated with azure – 11.85%. In comparison with untreated wood from this group, the value is 2 times higher and is 24.85%. In terms of the amount of water absorbed, the samples treated with azure performed better – 43.40%, compared to the samples treated with oil-wax – 45.99 %.

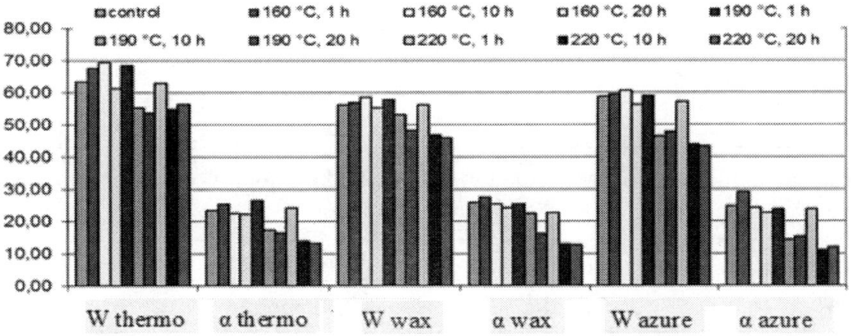

Figure 4. Water absorption and swelling of thermodified wood samples.

Also on the 6th day of keeping the samples in water, partial exfoliation was observed in the samples additionally treated with azure (Figure 5).

It was also found that the readings of water and moisture absorption are more influenced by the processing time than the temperature. Because in a short time in one hour the wood does not have time to evaporate extractives – terpenes, wax, phenol, fats. They prevent the adhesion of the paint material. The best result was shown by samples with a heat treatment time of 10 hours.

In the Table 1-6 show the results of measurement and calculation of color parameters of samples in the CIELAB coordinate system.

a b c d e f g h i j

Figure 5. Exfoliation of protective material (azure) on samples: a – untreated wood; b – modified at 160°C for 1 hours; c – modified at 160°C for 10 hours; d – modified at 160°C for 20 hours; e – modified at 190°C for 1 hours; f – modified at 190°C for 10 hours; g – modified at 190°C for 20 hours; h – modified at 220°C for 1 hours; i – modified at 220°C for 10 hours; j – modified at 220°C for 20 hours.

Table 1. Changing the color of the end surface of samples of thermodified wood without additional protection.

Thermomodification mode		Color options											
		Initial				In 1 month				In 5 month			
T, °C	τ, hours	L*	a*	b*	ΔE	L*	a*	b*	ΔE	L*	a*	b*	ΔE
Control		70	7	11	0	47	5	10	23.1	29	2	3	42.1
160	1	70	6	9	2.4	51	4	11	18.9	28	0	2	43.9
	10	51	10	13	19.3	54	7	13	16.1	31	2	3	40.6
	20	47	10	12	22.8	50	8	13	20.1	30	3	3	41.3
190	1	47	10	12	23.2	46	7	11	24.3	31	3	3	40.4
	10	25	5	2	45.7	32	7	7	38.6	28	3	2	43.2
	20	27	4	3	44.0	31	6	6	39.4	29	3	4	41.9
220	1	36	10	10	33.8	46	9	13	23.8	33	3	4	37.9
	10	25	3	2	45.8	27	3	3	44.3	26	1	1	45.9
	20	24	2	2	47.2	28	4	3	42.6	24	2	1	47.0

Table 2. Change in the color of the formation surface of the samples of thermodified wood without additional protection.

Thermomodification mode		Color options											
		Initial				In 1 month				In 5 month			
T, °C	τ, hours	L*	a*	b*	ΔE	L*	a*	b*	ΔE	L*	a*	b*	ΔE
Control		78	7	14	0	58	6	13	20.1	33	2	4	46.3
160	1	68	8	14	10.4	55	7	13	23.3	30	3	4	49.5
	10	55	10	15	23.8	50	7	12	28.0	35	3	23	44.4
	20	51	11	14	27.2	50	8	13	28.0	35	3	5	44.8
190	1	55	10	14	23.2	48	7	10	30.2	36	4	6	43.6
	10	35	9	7	44.2	38	7	9	40.6	30	4	5	49.0
	20	30	8	7	48.5	37	9	10	41.2	31	4	5	48.6
220	1	54	11	14	24.2	55	9	14	23.1	36	4	6	42.9
	10	30	3	3	49.9	31	5	4	48.0	27	3	2	52.7
	20	27	3	3	52.9	29	4	4	50.0	28	3	3	52.0

Research shows that wood modification significantly improves the resistance of wood to ultraviolet light. Thermally modified wood treated at a temperature of 190°C, 220°C for 1, 10, 20 hours is not additionally treated with protective substances and has a stable resistance to fading and color leaching.

Unmodified samples and samples modified at a temperature of 160°C lost half of the color in the first month of the experiment and for 5 months as much.

Table 3. Changing the color of the end surface of the samples of thermodified wood additionally treated with wax.

Thermomodification mode		Color options											
		Initial				In 1 month				In 5 month			
T, °C	τ, hours	L*	a*	b*	ΔE	L*	a*	b*	ΔE	L*	a*	b*	ΔE
Control		54	14	20	19.1	48	9	14	21.9	26	0	1	46.1
160	1	46	13	17	25.1	49	9	14	21.6	28	0	2	43.6
	10	27	9	6	43.4	37	8	9	33.5	27	1	2	44.1
	20	23	5	1	48.1	32	5	6	38.8	26	2	1	45.4
190	1	27	8	5	43.8	33	7	8	36.9	26	2	1	45.8
	10	23	1	0	48.3	26	3	2	45.2	25	1	0	46.8
	20	20	0	0	51.5	25	3	1	46.4	25	2	1	46.2
220	1	27	8	5	43.5	42	7	10	28.0	27	1	2	44.1
	10	21	0	0	50.9	23	1	0	48.8	24	0	0	47.9
	20	21	0	0	50.8	23	1	0	48.4	24	0	-1	47.7

Table 4. Changing the color of the end surface of samples of thermodified wood additionally treated with azure.

Thermomodification mode		Color options											
		Initial				In 1 month				In 5 month			
T, °C	τ, hours	L*	a*	b*	ΔE	L*	a*	b*	ΔE	L*	a*	b*	ΔE
Control		52	15	21	22.2	56	13	19	16.9	30	2	4	41,0
160	1	46	14	19	26.2	54	13	19	18.6	32	2	4	39.1
	10	25	10	8	44.9	35	12	12	35.0	33	2	4	37.8
	20	21	8	4	49.9	28	9	7	42.0	33	4	4	38.0
190	1	24	9	6	46.3	34	11	10	36.7	32	2	3	39.6
	10	18	1	0	53.6	20	1	0	51.3	26	1	0	46.1
	20	17	1	0	54.8	20	3	0	51.8	26	1	0	46.0
220	1	22	10	7	48.0	31	11	9	39.3	31	2	2	40.6
	10	18	0	0	54.1	21	0	0	51.1	24	1	0	47.6
	20	17	0	0	55.0	19	1	0	52.4	24	0	-1	48.3

The application of oil-wax and azure improved the color fastness by about 20% compared to untreated samples. Samples of thermodified wood have a strong resistance to ultraviolet light. Unmodified samples almost did not lose their color in the first month, but lost 50 % of the color in the next 4 months.

Layer samples treated with oil-wax and azure unmodified have less resistance to ultraviolet light than end samples. Exceptions are samples treated with azure at a temperature of 190°C for 10, 20 hours, which have the best resistance to ultraviolet light among all samples.

Table 5. Change in the color of the formation surface of the samples of thermodified wood additionally treated with wax.

Thermomodification mode		Color options											
		Initial				In 1 month				In 5 month			
T, °C	τ, hours	L*	a*	b*	ΔE	L*	a*	b*	ΔE	L*	a*	b*	ΔE
Control		68	10	19	11.9	58	10	17	21.0	27	1	2	52.6
	1	54	13	18	25.0	51	10	15	27.9	29	2	3	50.4
160	10	36	11	10	42.7	41	9	10	37.5	28	2	3	52.1
	20	34	10	8	44.4	39	8	9	40.0	30	4	4	49.4
	1	39	13	12	40.1	45	9	11	33.8	28	3	3	51.2
190	10	26	2	0	54.3	28	3	2	51.5	26	2	2	53.6
	20	23	2	1	57.0	28	4	3	51.8	27	2	2	53.2
	1	41	13	13	37.8	47	9	12	31.4	30	3	4	49.8
220	10	25	0	0	55.6	26	1	0	54.4	26	1	0	54.8
	20	22	0	0	58.1	24	0	-1	56.3	26	1	1	54.7

Table 6. Change in the color of the formation surface of samples of thermodified wood additionally treated with azure.

Thermomodification mode		Color options											
		Initial				In 1 month				In 5 month			
T, °C	τ, hours	L*	a*	b*	ΔE	L*	a*	b*	ΔE	L*	a*	b*	ΔE
Control		68	11	20	13.0	68	12	23	14.6	50	11	18	28.5
	1	53	14	21	27.3	57	15	22	23.9	39	11	15	39.2
160	10	31	14	14	47.5	37	16	15	41.9	31	8	8	47.4
	20	34	15	15	45.0	42	16	18	37.4	32	8	8	46.4
	1	43	15	18	36.2	48	15	19	31.4	38	10	12	40.5
190	10	20	2	1	60.3	21	3	0	58.7	25	2	1	55.4
	20	19	3	1	61.2	21	4	1	58.6	23	4	2	57.0
	1	37	16	18	42.4	43	15	17	36.7	33	9	9	45.6
220	10	18	-1	-1	62.8	19	0	-1	61.2	20	1	0	60.7
	20	17	0	0	62.9	20	0	-1	60.9	20	1	0	60.4

6. Conclusions

Thermally modified wood is increasingly used in outdoor conditions, but its surface is also subject to degradation due to weathering as well as unmodified wood. Paints and varnishes are largely used to protect the surface of wood in outdoor conditions, as they are easy to maintain in good condition and re-apply. They also improve the appearance of the wood surface. Protective substances for thermally modified wood provided similar protection of exposed thermally modified wood surfaces from water and ultraviolet radiation. The positive effect of thermal modification on reducing the level of water absorption has been established. Temperature has a significant influence on such changes. The geometric dimensional stability also improves. As the modification temperature increases, the color fastness to UV rays improves. The application of oil-wax and azure improved the color fastness by about 20 % compared to untreated samples. As the protective coating is also prone to photochemical degradation, timely restoration according to the manufacturer's instructions is very important for the pleasant appearance of wooden surfaces during outdoor use.

References

[1] Živković V, Pršа I, Turkulin H, Sinković T and Jirouš-Rajković V 2008 Dimensional stability of heat treated wood floorings *Drvna Ind* **59** pp 69–73

[2] Jirouš-Rajković V and Miklečić J 2019 Heat-treated wood as a substrate for coatings, weathering of heat-treated wood, and coating performance on heat-treated wood *Advances in Materials Science and Engineering* **8621486** p 9 https://doi.org/10.1155/2019/8621486
[3] Tsapko Yu, Tsapko A and Bondarenko O 2020 Research of conditions of removal of fire protection from building construction *Key Engineering Materials* vol **864** pp 141–148 doi:10.4028/www.scientific.net/KEM.864.141
[4] Tsapko Yu, Rogovskii I, Titova L, Bilko T, Tsapko A, Bondarenko O and Mazurchuk S 2020 Establishing regularities in the insulating capacity of a foaming agent for localizing flammable liquids *Eastern-European Journal of Enterprise Technologies* **5/10 (107)** pp 51–57 doi: 10.15587/1729-4061.2020.215130
[5] Tsapko Yu, Lomaha V, Bondarenko O and Sukhanevych M 2020 Research of mechanism of fire protection with wood lacquer *Materials Science Forum* **1006** pp 32–40 doi:10.4028/www.scientific.net/MSF.1006.32
[6] Tsapko Yu, Horbachova O, Tsapko A, Mazurchuk S, Zavialov D and Buiskykh N 2021 Establishing regularities in the propagation of phase transformation front during timber thermal modification *Eastern-European Journal of Enterprise Technologies* **1/10 (109)** pp 30–36 doi: 10.15587/1729-4061.2021. 225310
[7] Mikle˘circ J, Jirou˘s-Rajkovirc V, Pervan S and Grujirc S 2010 Oils usage in finishing of thermally modified wood in outdoor applications *Wood is Good-Transfer of Knowledge in Practice as a Way Out of the Crisis*
[8] Schaller C and Rogez D 2007 Light stabilisation of modified wood species *Proceedings of Dird European Conference on Wood Modification* pp 15–16
[9] Slabejová-Zuzana G and Šmidriaková V M 2019 Surface finishes for thermally modified beech wood *Acta facultatis xylologiae zvolen* **61 (2)** 41−50 doi: 10.17423/afx.2019.61.2.04
[10] Akyildiz M H and Kesik H I 2014 Effect of heat treatment on the adhesion strength of water based wood varnishes *Proceedings of the Seventh European Conference on Wood Modification*
[11] Humar M, Repič R, Kržišnik D and Lesar B 2020 Quality control of thermally modified timber using dynamic vapor sorption (dvs) analysis *Forests* vol **11 (6)** 666 doi: 10.3390/f11060666
[12] Vernois M 2001 Heat treatment of wood in France-State of the art *Proceedings of Special Seminar of Review on Heat Treatments of Wood*
[13] Jëamsëa S, Ahola P and Viitaniemi P 2000 Long-term natural weathering of coated thermowood *Pigment & Resin Technology* **29 (2)** pp 68–74
[14] Turkoglu T, Toker H, Baysal E, Kart S, Yuksel M, Emin M, Mugla E and Kocman S 2015 Some surface properties of heat treated and natural weathered oriental beech *Wood research* **60** pp 881–890
[15] Pavlič M, Petrič M and Zigon J 2021 Interactions of coating and wood flooring surface system properties *Coatings* **11 (1)** p 91 https://doi.org/10.3390/coatings11010091
[16] Arminger B, Jaxel J, Bacher M, Gindl-Altmutter W and Hansmann C 2020 On the drying behavior of natural oils used for solid wood finishing *Progress in Organic Coatings* **148** 105831

Acknowledgements
Authors express their gratitude to the Ministry of Education and Science of Ukraine for financial support of the research, that was performed in the framework of budget funding No. 0121U001007, as well as for the development of the theme of research according to the program of scientific cooperation COST Action FP 1407 "Understanding the modification of wood through an integrated scientific and environmental approach" of the European Union's framework program HORIZON 2020.

IOP Conf. Series: Materials Science and Engineering 1164 (2021) 012081 doi:10.1088/1757-899X/1164/1/012081

Design of the bracket's structures and their bracing for industrial climbers' service of high-rise buildings with facades of solid glazing

I Tvardovsky[1,*], A Chuchmai[1] and T Kalinina[1]

[1]Odesa State Academy of Civil Engineering and Architecture, 4, Didrihsona str., Odesa 65029, Ukraine

[*]E-mail: igortvardovsky@gmail.com

Abstract. The article presents one of the possible options for solving an urgent engineering problem of the development of special devices and their arrangement in high-rise buildings with continuous facade systems for the possibility of their service. The authors have developed a bracket design in two modifications made of metal plates, which is fastened to the end of a monolithic reinforced concrete slab at the level and along the perimeter of the eaves of the building with chemical anchors of the Hilti HVU series. The nature of the change in the stress-strain state in the plates, main pipes, and in anchor devices of the developed bracket construction was studied by calculation using the usual engineering method, as well as in the generated model in the SOFiSTiK computational and graphic complex. The results of short-term load testing of completed samples of the developed model before serial production are presented. The work of the elements of the developed structures of the brackets under the conditions of the action of an aggressive environment has been analyzed.

1. Introduction

During the construction of high-rise buildings in recent decades, the creation of facades with maximum usage of translucent constructions, including the entire contour, has become widespread in different countries of the world. Examples are the famous buildings: Commerzbank in Frankfurt am Main (Germany, 1997), City Hall in London (GB, 2002), Manitoba Hydro Place in Vinnipeg (Canada, 2009).

The usage of the translucent structures allows to diversify the building view and their shapes, which create the modern architectural style and its attractiveness – for example, the One Angel Square building, built in 2013 in Manchester (GB), has a peculiar configuration with a solid glass façade [1]. The construction of buildings with solid translucent facades has been developed in Ukraine too. The reason for the popularity of construction of buildings from translucent constructions is their practicality, reliability, durability, and, important for architects and designers, in a wide variety of types and forms. The translucent façade systems afford not only to insulate and decorate the façade effectively, but also give the building a stylish modern look. At the same time the cost of constructing the facade is considerably lower than for general construction works on its construction in the "standard" concrete form.

Alongside with the listed advantages, the facades of buildings made of solid translucent structures have a number of difficulties in their service, for example, when cleaning the plane of solid glazing from pollution. The solution of this problem by using temporary attachments with the formation of counterweights on the roof or on the upper operating sites of the building creates difficulties associated with the frequency of their arrangement and the cumbersome nature of the equipment used, including the often lack of access to these sites caused by the peculiarities of their operation.

The creation of special devices and technologies for their anchoring to the already existing facade systems in the constructed buildings (see figure 1) has become an urgent engineering task, the solution of which has the great practical importance.

Content from this work may be used under the terms of the Creative Commons Attribution 3.0 licence. Any further distribution of this work must maintain attribution to the author(s) and the title of the work, journal citation and DOI.
Published under licence by IOP Publishing Ltd

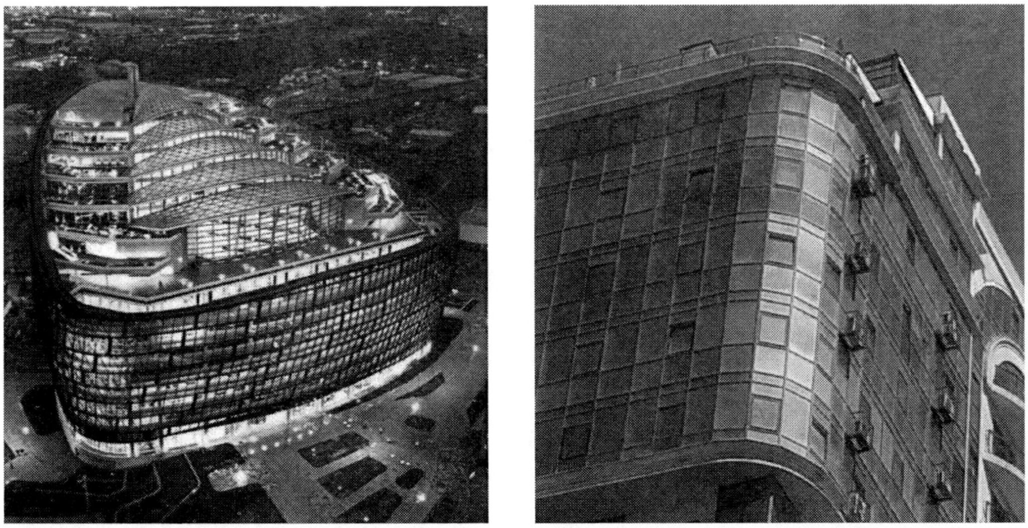

Figure 1. Examples of the buildings with the solid- glazing facades: the building One Angel Square (Manchester, UK) – on the left; the building the 7-th Zhemchuzhina (Odessa, Ukraine) – on the right.

2. The literature review

Engineering solutions for the formation of structures that provide the possibility of servicing facade systems by fixing industrial climbers to main pipes with brackets are based on calculations for the first and second groups of limiting condition, in accordance with the regulatory documents in force on the territory of Ukraine [2,3], and also European standards [4].

Many examples of metallic construction calculation are described in some detail in the designer's guide [5]

The joint operation of metal expanding anchors in a concrete environment is described in [6].

The calculation of plates with variable rigidity under the load in the transverse direction is described in the works [7,8], in which are also considered the plates pinched on one side. For such plates, the boundary conditions are used: the deflection and the angle of rotation at the pinched point are equal to zero, which leads to some simplification when writing the basic differential equation of bending of an orthotropic plate of variable rigidity [7].

Analysis of the stress-strain state of plates of variable stiffness, plates with holes, and other analytical methods becomes so time-consuming that in many cases it is not possible to obtain the result in analytical form [8]. The finite element method (FEM) is intended for such complex problems of plate calculation. On its basis, it is possible to solve practical problems of any complexity, including plate systems. In this case, there can be holes in the plates, on a part of the surface of the plate they can rest on an elastic foundation or have a nonlinear nature of deformation, caused by both changes in geometry and taking into account the inelastic properties of materials. Thus, the finite element method allows you to reflect the actual work of the constructions.

Calculation of plates by the finite element method (i.e. determination of the stress-strain state, in particular the calculation of the plate's deflections), same as by the finite difference method consists in determining the deflections of the finite number of plate's points -the nods of the discretization grid's, with the help of which the plate is divided into many smaller plates - finite elements.

3. The aim of the current work is creation of the design of brackets with main pipes to ensure the service of facades from translucent systems in high-rise buildings by industrial climbers; ensuring the reliability of fastening brackets in a concrete slab at the level of the cornice on the perimeter of the building; determination of the stress-strain state in the steel elements of the developed bracket under the transverse bend, shear forces and work to tear off the anchoring devices for attaching the brackets to the monolithic floor slab.

4. The materials and the methods

Metal plates and pipes, anchoring devices and a monolithic reinforced concrete slab were used. Determination of stresses and strains in steel elements and concrete of a monolithic reinforced concrete floor slab was carried out by generally accepted methods in accordance with the current regulatory documents [2,3] and using the SOFiSTiK calculation and graphic complex [9].

5. The statement of the main obtained results

The construction of the holding bracket is designed to allow industrial climbers to move in areas or around the entire perimeter of the building along metal pipes at the level of the cornice in order to service the surface of the facade made of solid glass of stained glass type.

The location of the brackets is provided at the same distance along the end of the monolithic reinforced concrete tube. The main parameters of the used metal plates of the bracket were determined initially based on the thickness of the monolithic reinforced concrete slab, as well as by calculation in accordance with the current regulatory documents [2,3].

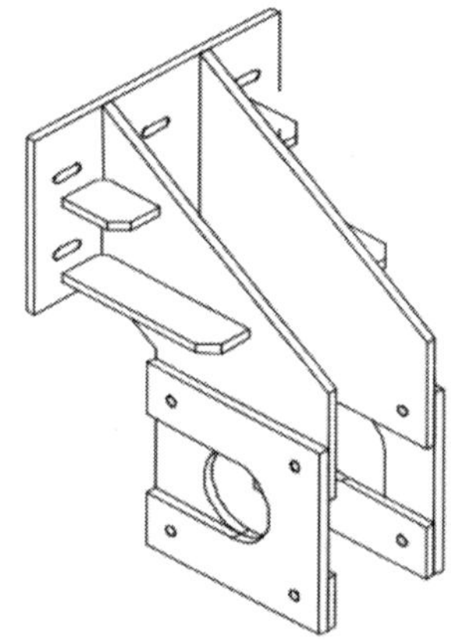

Figure 2. Construction of the designed bracket according to the assigned engineering task.

As the background information, the following was taken: the size of the plate should not exceed the thickness of the monolithic floor slab; the design load for one tube span is 0.2 kN; fastening of the pipe span with beading in the lugs of the plates is hinged.

The calculated span between the fixed brackets was determined by a calculation that takes into account the dynamic component of the vertical and horizontal load, the force of breaking the anchors from the concrete slab and the degree of possible metal corrosion during long-term operation under conditions of variable exposure to moisture and freezing and thawing.

Initially, the calculation determined the cross-section of the main tube which was broken along the edges with an estimated length of 1.0 m, the installation of which was provided using two intersecting plates fastened by mounting bolted joints (figure 2).

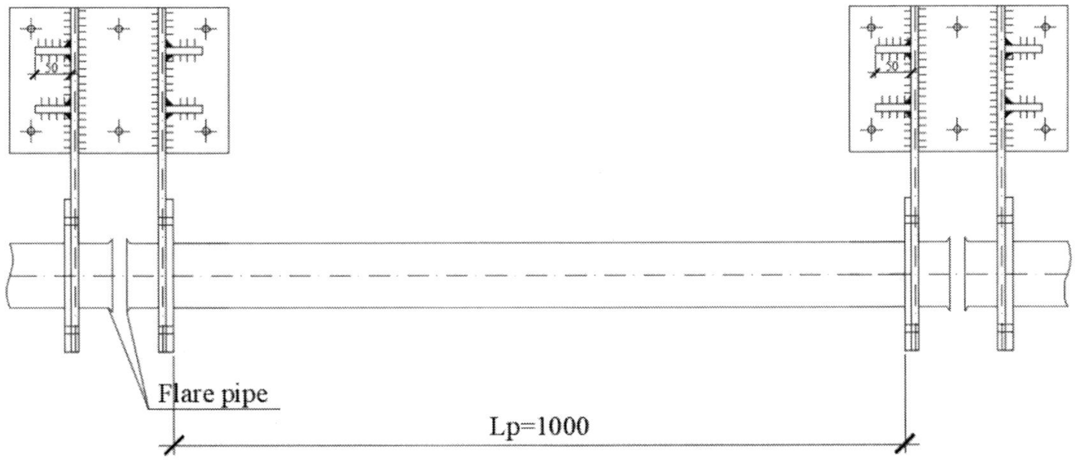

Figure 3. Bracing of the main pipe between brackets it the fixing plates.

The maximum bending moment perceived by the tube cross-section was determined by the following formula

$$M_{\max} = \frac{P \cdot n_{din} \cdot f_{cor}}{2} \cdot \frac{L_p}{2} = 50 \quad (1)$$

where: P – given load;

L_p – the calculated length;

n_{din} – dynamic coefficient;

f_{cor} – the time-dependent function of corrosive wear;

The required moment of resistance

$$W_{tr} = \frac{M_{\max}}{R_{u3}} = 12.5sm^3 \quad (2)$$

what determined the section of the main pipe on the calculated span Ø83x4 mm.

With the maximum deflection up to 1.1mm, according to the formula

$$f = \frac{5}{384} \cdot \frac{PL^3}{EI} \quad (3)$$

The maximum bending moment at the base of the support plate of the bracket was calculated by the formula:

$$M_{opr} = P \cdot n_{din} \cdot B_p . \quad (4)$$

The breakout force of the anchor from the reinforce concrete slab was calculated by formula:

$$N_{otr} = \frac{P \cdot n_{din} \cdot B_p}{b_{kr}} \quad (5)$$

where: B_p – the distance from the axis of the main tube to the bracket's side;

b_{kr} – the minimal indent between the brackets, which are forging the bending moment which supports the anchor from the toppling back.

The selection of the diameter and length of the bracket was carried out according to the condition $P_{crc} > N_{omp}$,

where:
$$P_{crc} = A_{пов} \cdot R_{bt} = 2\pi r \cdot L_{анк} \cdot R_{bt} \tag{6}$$

The salvaged elements obtained by the above formulas were compared with the results of the calculation in the software SOFiSTiK [9], the main results of which are given below in the figure 4.

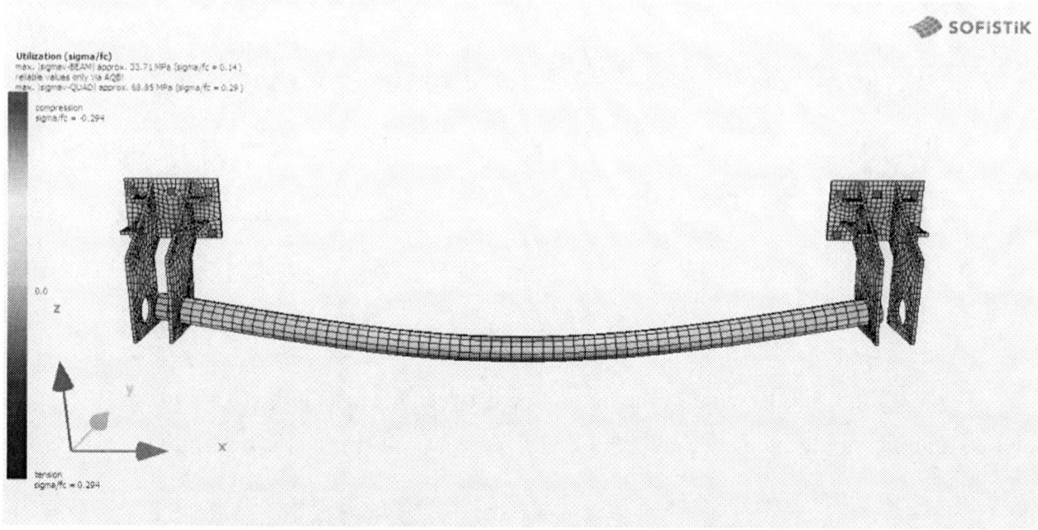

Figure 4. Stresses and strains in the anchor's elements.

Calculation of retaining vertical plates for the load from shear in the horizontal direction, taking into account the dynamic effect, is possible using the theory of calculation of plates with variable rigidity [7], and also via the calculation and graphical SOFiSTiK [9].
The main differential equation of the bend of the orthotropic plate with a variable rigidity is:

$$\begin{gathered} D_1\frac{\partial^4\omega}{\partial x^4}+2\frac{\partial D_1}{\partial x}\cdot\frac{\partial^3\omega}{\partial x^3}+\frac{\partial^2 D_1}{\partial x^2}\cdot\frac{\partial^2\omega}{\partial x^2}+\frac{\partial^2 D_2}{\partial y^2}\cdot\frac{\partial^2\omega}{\partial y^2}+2\frac{\partial D_2}{\partial y}\cdot\frac{\partial^3\omega}{\partial y^3}+D_2\frac{\partial^4\omega}{\partial y^4}+2\frac{\partial D_3}{\partial x}\cdot\frac{\partial^3\omega}{\partial x\partial y^3}+ \\ +2D_3\frac{\partial^4\omega}{\partial x^2\partial y^2}+2\frac{\partial D_3}{\partial y}\cdot\frac{\partial^3\omega}{\partial x^2\partial y}+\nu_2\frac{\partial^2 D_1}{\partial x^2}\cdot\frac{\partial^2\omega}{\partial y^2}+\nu_1\frac{\partial^2 D_2}{\partial y^2}\cdot\frac{\partial^2\omega}{\partial x^2}+4\frac{\partial^2 D_k}{\partial x\partial y}\cdot\frac{\partial^2\omega}{\partial x\partial y}=\bar{q}, \end{gathered} \tag{7}$$

where: D_1, D_2 - the bend rigidity of the plates in the main;

$D_3 = D_1\nu_2 + 2D_k$ - the main plate's rigidity;

D_k - torsional rigidity;

ν_1, ν_2 - Puasson's coefficients of the plate in the main directions;

$$\bar{q} = q\left(\frac{\partial m_x}{\partial x} + \frac{\partial m_y}{\partial y}\right).$$

The analytical solution is reduced to integrating the equation (7) subject to boundary and contact conditions which is well described in the work [7]. This approach is very complex and time-consuming for the engineering task to be performed, and the plates were calculated using a calculation and graphic complex [9], based on the finite elements method. The main calculation's result is given below.

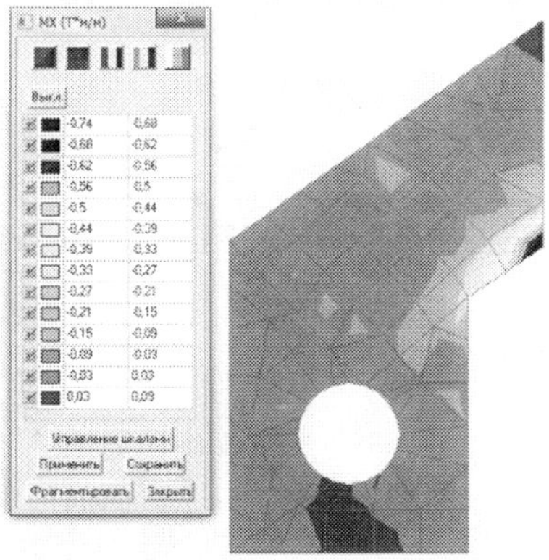

isofield of stresses M_x

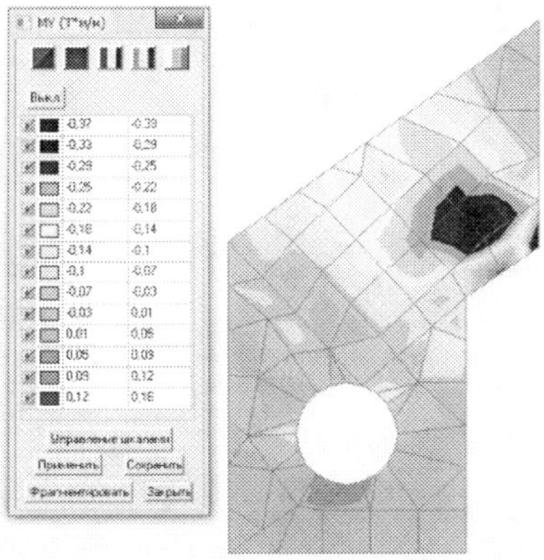

isofield of stresses M_y

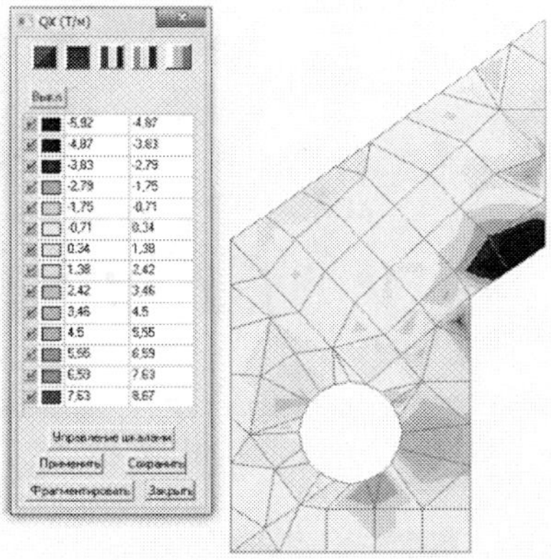

isofield of stresses Q_x

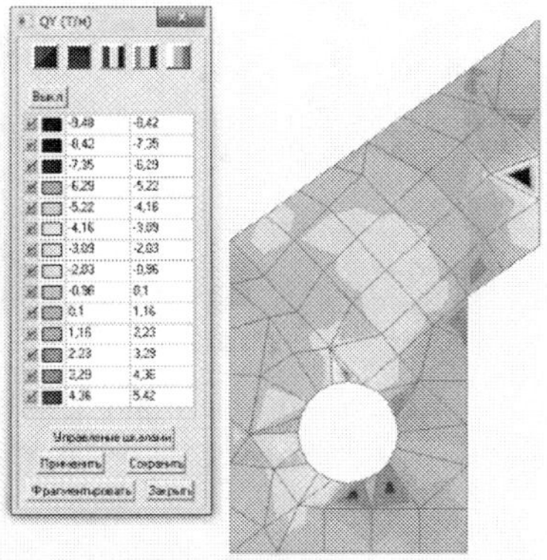

isofield of stresses Q_y

Figure 5. Stresses and strains in the anchor's elements.

In order to check the reliability of the selected bracket's elements and anchors of its bracing to the monolithic reinforced concrete slab and the possibility of further serial development, full-scale tests were performed using a load (figure 6) and via a special equipment (figure 7).

Figure 6. Testing under the load the fixing link consisting of the two prackets and the pipe.

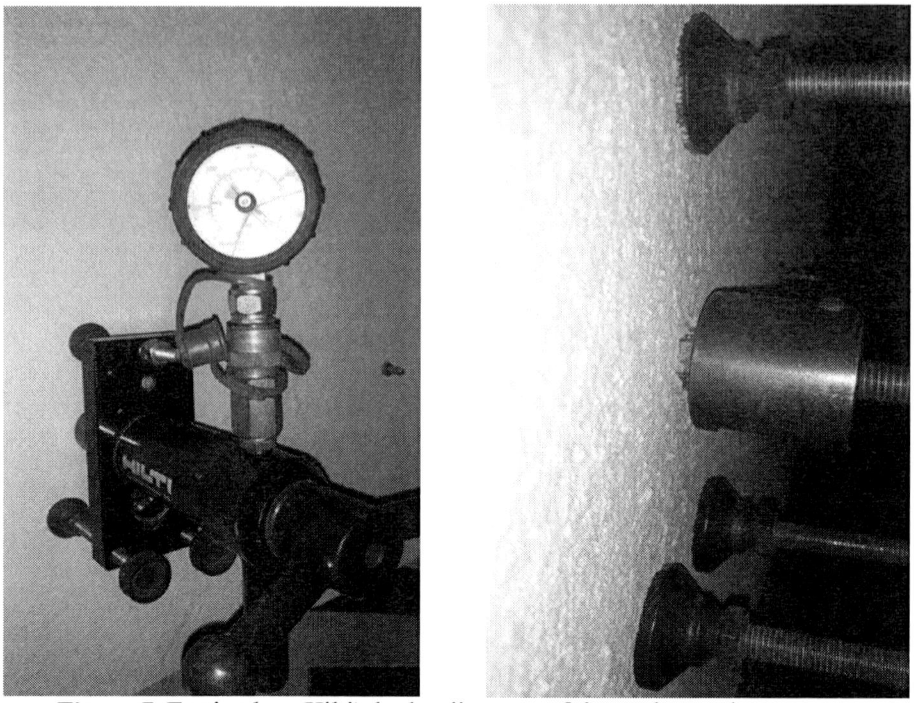

Figure 7. Testing by a Hilti's hydraulic press of the anchoress in concrete.

The final completion of the implemented design development and calculations is the installation of the anchors on the level of the high-rising's eaves and their use by industrial climbers for the service of the solid glazing facades (figure 8).

Figure 8. Installed brackets with main pipes for fixing and moving industrial climbers along them.

6. Conclusion

Designed construction of the bracket with the main pipes and the anchors allows solving the important problem of servicing the facades of high-rise buildings with solid stained-glass glazed.

The reliability of the structure is determined by comparing the calculations performed according to the engineering methodology and using the design and graphical software SOFiSTiK [9], as well as by the carried out full-scale bending and pull-off tests.

References

[1] *Inzhenernoe oborudovanie vyisotnyih zdaniy* pod red. M. M. Brodach. 2-e izd., ispr. i dop. (M.: AVOK-PRESS) 2011
[2] DBN V.2.6-198:2014 *Stalnyie konstruktsii Normyi proektirovaniya* Minregionbud Ukraine Kyiv 206
[3] DBN V.2.6-98:2009 *Betonnyie i zhelezobetonnyie konstruktsii Osnovnyie polozheniya* Minregionbud Ukraine, Kyiv 2011
[4] EN 1993-1.3 *Eurocode 3 Design of steel structures Part 1-1 General rules and rules for buildings*
[5] *Spravochnik proektirovschika Metallicheskie konstruktsii Tom 1 Obschaya chast* Pod obschey redaktsiey Kuznetsova V V 1998 (Moskva: ASV) 576
[6] *Spravochnik konstruktora Raschet i proektirovanie ankernyih uzlov Elementarnaya mehanika* (Kyiv: Elementa) 2017 155
[7] Vaynberg D V and Vaynberg E D 1970 *Raschet plastin* (K.: BudIvelnik) 436
[8] Monahov V A 2016 *Teoriya plastin i obolochek: ucheb. posobie po napravleniyu podgotovki 08.05.01 Stroitelstvo unikalnyih zdaniy i sooruzheniy* (Penza: PGUAS) 252
[9] Kuhtin V N, Bulaev I V and Baranov I S 2015 Primenenie raschetnogo kompleksa SOFiSTiK dlya rascheta mostovyih konstruktsiy Uchebnoe posobie (M.: MADI) 136

IOP Conf. Series: Materials Science and Engineering 1164 (2021) 012082 doi:10.1088/1757-899X/1164/1/012082

The impact of percentage content of basalt fiber-reinforced polymer and steel reinforcement on the strength indicators of experimental beams with hybrid reinforcement

Alexander Valovoi[1,*], Alexander Eremenko[1], Maksim Valovoi[1] and Sergii Stoianovych[2]

[1]Kryvyi Rih National University, Vitaly Matusevich Str. 11, Kryvyi Rih, Ukraine, 50036.
[2]National Academy of Fine Arts and Architecture, Smirnova-Lastochkina St. 20, Kyiv, Ukraine, 04053

[*]E-mail: valovojua@gmail.com

Abstract. The use of composite reinforcement in the structures subject to bending is limited due to their excessive deformation. One way to overcome this shortcoming is to use hybrid reinforcement where steel and composite reinforcement are used in combination. The effectiveness of this approach has been proven by previous studies. There are no studies on the establishment of the effective ratio between steel and composite elements. This paper presents the results of tests of prototype beams with hybrid reinforcement made of basalt fiber-reinforced polymer (BFRP) and steel. A total of 12 series of samples of beams, three beams in each series, were studied. The percentage of both types of reinforcement was variable; it varied from a ratio of 100% / 0% to 0% / 100% in increments of ≈ 20% (the numerator of the fraction corresponds to the percentage of steel reinforcement, and the denominator – to the percentage of BFRP reinforcement). According to the results of tests of beams by static loading, it was found that the use of hybrid reinforcement allowed to increase the bearing capacity of the samples by 33% – 74%, depending on the percentage of reinforcement in the cross section, compared with the beams of the control series. Based on the nature of breaks and load-bearing capacity, the most optimal ratio of steel and BFRP reinforcement was 60% – 40%. The optimal percentage of cross section reinforcement is 1.07%.

1. Introduction

The use of composite reinforcement in construction practice is increasing worldwide. One of the types of composite reinforcement is the one of basalt fiber-reinforced polymer. The advantages and disadvantages of the latter are widely known [1, 2, 3]. One of the main disadvantages of this type of reinforcement is the reduced modulus of elasticity [2], compared to steel reinforcement. In this case, structures reinforced with basalt fiber-reinforced polymer are more deformable, which does not allow, in certain cases, to ensure compliance of the moving structures to the requirements of the norms [1, 3]. The last remark is especially relevant for structures subject to bending.

At present, there are several areas for overcoming the excessive deformability of concrete structures reinforced with composite reinforcement, which are subject to bending. The latter include: prestressing the reinforcement, increasing the percentage of reinforcement in proportion to decreasing the modulus of elasticity of the reinforcement, hybrid reinforcement of the cross section [4, 5, 6]. The first two methods have their drawbacks. Due to the specifics of composite reinforcement, it is technologically difficult to generate prestress, and the cost of such a structure will be significantly higher than the structure without prestress. Increasing the percentage of reinforcement leads to increased spending of composite reinforcement and increased cost of the structure.

The use of hybrid reinforcement, when the working reinforcement of the cross section is composite and steel at the same time, is a promising method of increasing the stiffness parameters of concrete structures with composite reinforcement. This approach allows you to take full advantage of both

Content from this work may be used under the terms of the Creative Commons Attribution 3.0 licence. Any further distribution of this work must maintain attribution to the author(s) and the title of the work, journal citation and DOI.
Published under licence by IOP Publishing Ltd

IOP Conf. Series: Materials Science and Engineering 1164 (2021) 012082 doi:10.1088/1757-899X/1164/1/012082

types of reinforcement – the stiffness and fire resistance of steel reinforcement and the high strength and corrosion resistance of composite reinforcement. Studies of experimental samples of beams with hybrid reinforcement indicate its efficiency [7, 8, 9, 10, 11] and the ability to ensure compliance of the mobility of structures with the requirements of the norms. At the same time, there are virtually no studies to determine the effective ratio of steel and composite reinforcement in beams with hybrid reinforcement. In view of the above, the current issue is to determine the impact of changes in the percentage of composite and steel reinforcement on the load-bearing capacity of beams with hybrid reinforcement.

2. Research methodology

This paper is a continuation of the study described in [10, 11, 12]. The latter considered the performance of beams under load reinforced with basalt fiber-reinforced polymer reinforcement (BFRP), steel reinforcement (control series) and with hybrid reinforcement (which had both steel and BFRP reinforcement at the same time). To be able to correctly compare the results, the percentage of reinforcement was assumed to be the same for the beams of all series. According to the test results, it was found that the strength of beams with hybrid reinforcement increased compared to the beams of the control series and was at the level of beams reinforced with BFRP reinforcement [10]. At the same time, deflections and crack opening width decreased, and at 60% of the destructive load level did not exceed the maximum values allowable by the norms, while in beams reinforced exclusively with basalt fiber-reinforced polymer reinforcement they exceeded the limit values by 30… 120% [11, 12].

In order to clarify the impact of changes in the content of basalt fiber-reinforced polymer reinforcement in relation to steel on the performance of beams with hybrid reinforcement, 6 series of beam samples were designed, manufactured and studied.

Reinforcement of cross sections of beam samples of all series (new and previous studies [10, 11, 12]) is shown in Table 1.

To be able to compare them with each other and with the results of previous studies, the chosen class of concrete is the same as in [10, 11, 12], and the percentage of reinforcement is as close as possible to a similar figure in the mentioned studies. The percentage of beam reinforcement of new and previous studies was:

- BM, BMD, BB, BBD – 0.95%;
- BMB, BMBD – 0.84%;
- B 6/10, B 10/6 – 0.89%;
- B 8/10, B 10/8 – 1.07%;
- B 6/12, B 12/6 – 1.18%.

The length of all samples was the same – 2000 mm, cross sectional dimensions b x h = 120 x 220 mm. Reinforcement of the beams was performed with a spatial reinforcing frame. Transverse reinforcement is made of reinforcing bars Ø6 A240 with a step of 70 mm in the support zones. The accepted structural reinforcement in the compressed zone is 2Ø6 A240C.

Concreting of experimental beams was carried out in the inventory metal formwork at the reinforced concrete production plant, PJSC "Kryvorizhindustrbud" (Figure 1).

Table 1. Scheme of reinforcement of cross sections of beams.

Series no.	Marking of beams	Cross section of beams	Series no.	Marking of beams	Cross section of beams
1[1)]	BM-1 BM-2 BM-3	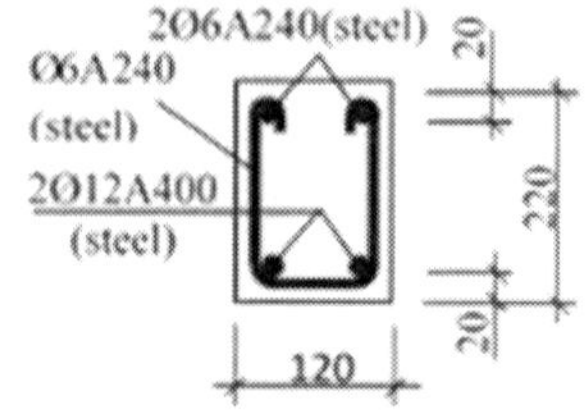	4[1)]	BMD-1 BMD-2 BMD-3	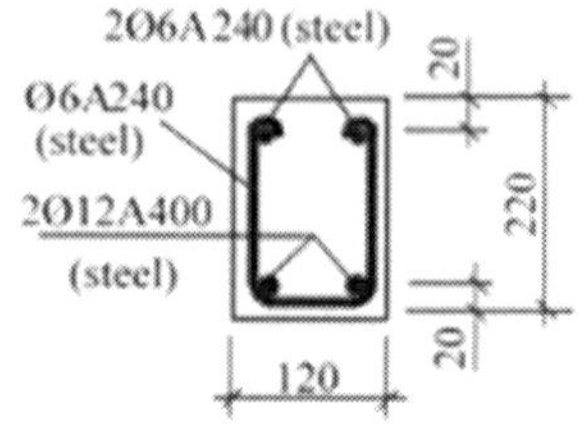
2[1)]	BB-1 BB-2 BB-3	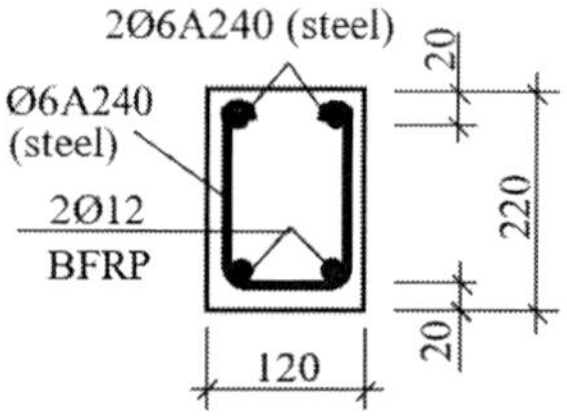	5[1)]	BBD-1 BBD-2 BBD-3	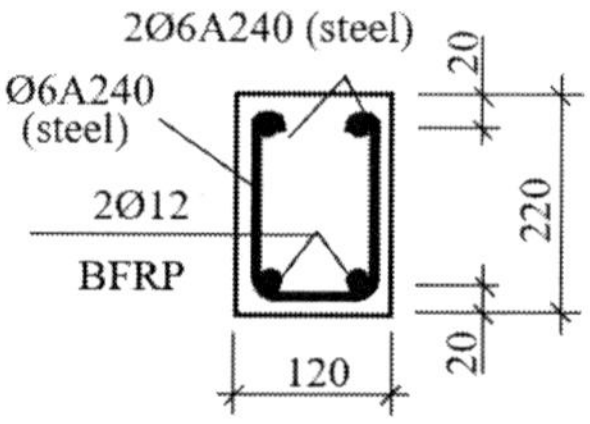
3[1)]	BMB-1 BMB-2 BMB-3	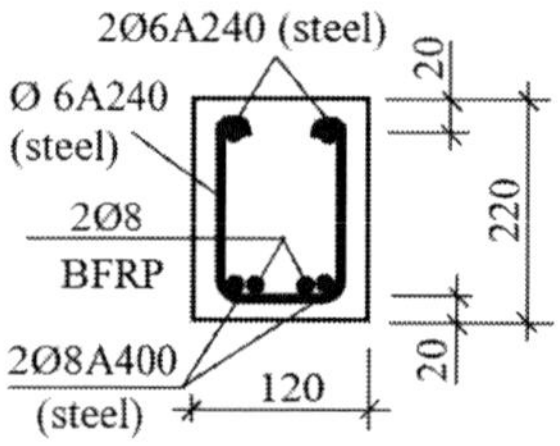	6[1)]	BMBD-1 BMBD-2 BMBD-3	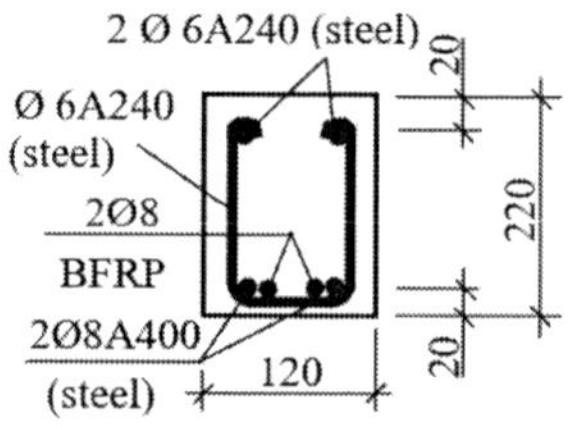
7[2)]	B 6/10-1 B 6/10-2 B 6/10-3	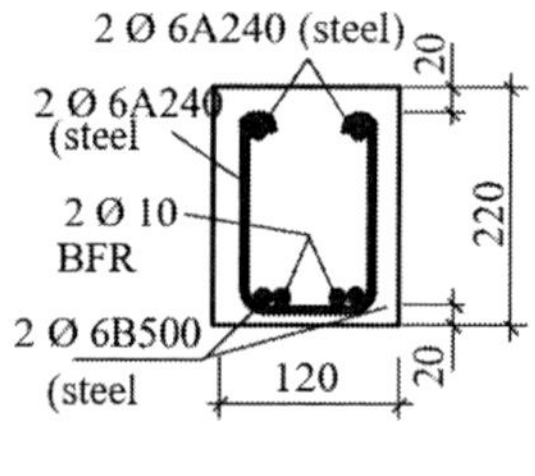	8[2)]	B 10/6-1 B 10/6-2 B 10/6-3	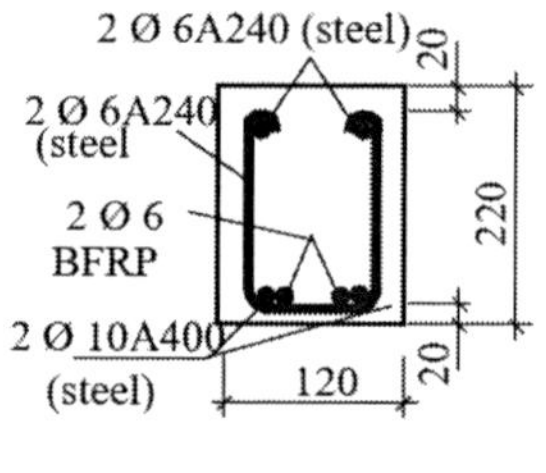
9[2)]	B 8/10-1 B 8/10-2 B 8/10-3	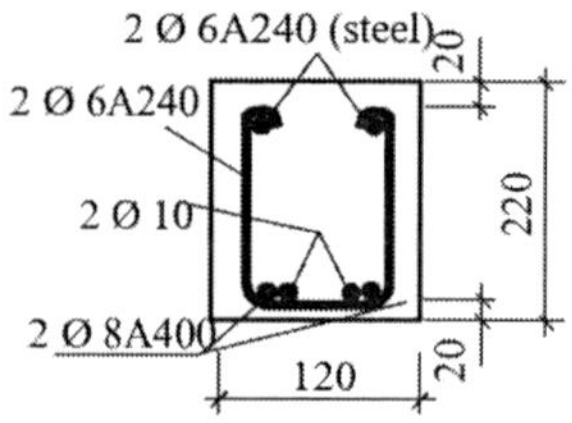	10[2)]	B 10/8-1 B 10/8-2 B 10/8-3	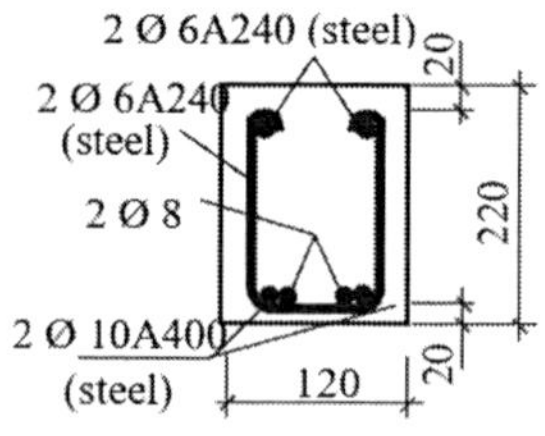

IOP Conf. Series: Materials Science and Engineering 1164 (2021) 012082 doi:10.1088/1757-899X/1164/1/012082

11[2)]	B 6/12-1 B 6/12-2 B 6/12-3	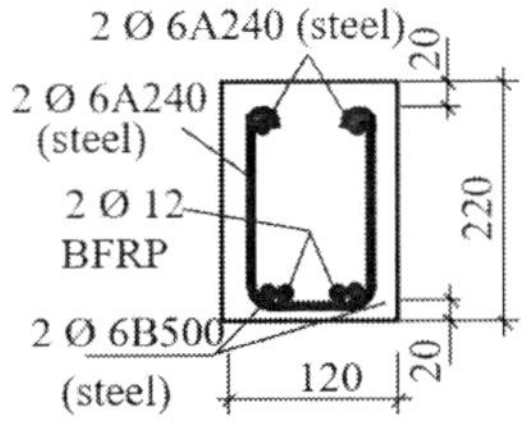	12[2)]	B 12/6-1 B 12/6-2 B 12/6-3	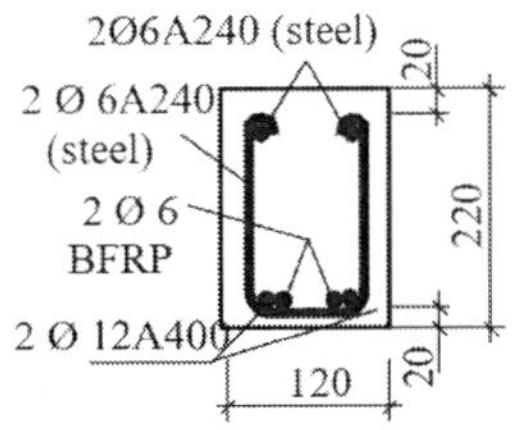

1 – cross sections of beams used in previous studies [10, 11, 12].

2 – accepted notation: the numerator of the fraction in the notation corresponds to the diameter of the steel reinforcement, and the denominator corresponds to the diameter of the BFRP reinforcement

Figure 1. Production of prototype beams. (a) Spatial reinforcement frames; (b) reinforcement frames installed in the inventory formwork; (c) formed beam samples.

To determine the physical and mechanical characteristics of concrete samples, the cube samples with a side of 100 mm and prisms with dimensions of 100x100x400 mm were made.

The tests of beams were carried out according to the scheme of a beam on two hinged supports loaded in the thirds of span. The load was created using a hydraulic press P-125 in steps of 0.1 from the estimated destructive load. Maksymov's deflectometer was used to measure the deflections; the deformation of the marginal fibers of the faces of the beam was measured using dial test indicators with a scale interval of 0.01 mm; the crack opening width was recorded using a MBP-2 microscope. A general view of the stand used for the testing of beam samples and placement of measuring instruments is shown in Figure 2.

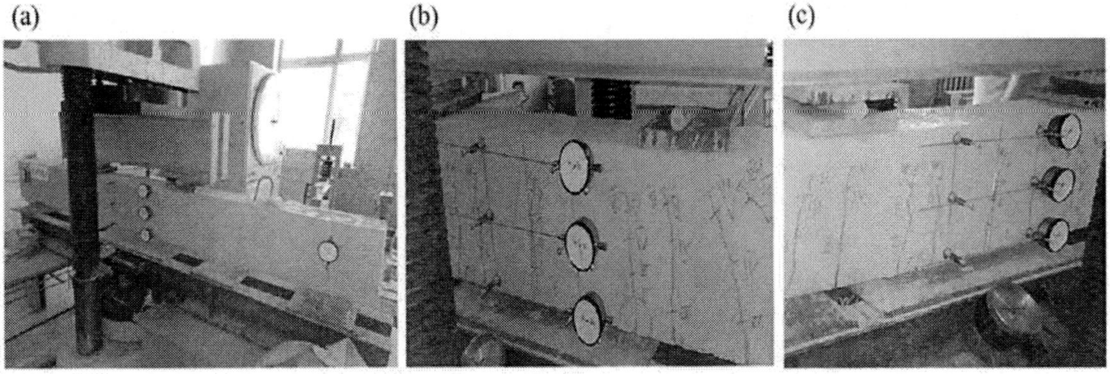

Figure 2. Stand for testing samples of beams with measuring instruments: (a) Test beam; (b) dial test indicators; (c) Maksymov's deflectometer.

3. Results and discussion

The testing of beams by short-term loads up until breaking began on the 151st day and ended on the 161st day after the production of samples.

The testing of beams was preceded by the study of strength, deformability of concrete and reinforcement. For the convenience of comparison in Table 2 and Table 3, the specified indicators of both concrete and reinforcement of "new" beams and beams of previous researches are given [10].

Table 2. Indicators of strength and deformability of concrete by its types.

Item no.	Type of concrete	Cube strength, R, MPa		Prism strength, R_b, MPa		Initial modulus of elasticity, $E_b \cdot 10^{-3}$, MPa		Tensile strength, R_{bt}, MPa		Boundary deformations during compression, $\varepsilon_{bR} \cdot 10^{-5}$
		Age of samples, days								
		28	151	28	151	28	151	28	151	151
1	Based on the waste material of the ore mining and processing plant	35.17	51.82	26.8	30.86	31.17	36.02	2.11	3.11	162
2*	Quartz sand concrete	32.09	46.13	23.86	27.49	29.72	34.7	1.93	2.77	204

* – used in the manufacture of samples of beams series BB, BM, BMB.

Table 3. Physical and mechanical characteristics of the reinforcement.

Reinforcement class	Yield limit, N/mm²	Ultimate tensile strength, N/mm²	Modulus of elasticity, N/mm²	Tensile elongation at break, %
A400C (∅ 8) steel	508	622	$198*10^3$	23.4
A400C (∅ 10) steel	519	629	$200*10^3$	21.6
A400C (∅ 12) steel	527	602	$201*10^3$	20.6
A240 (∅ 6) steel	298	373	$210*10^3$	17.8
B500 (∅ 6) steel	620 (conditional)	663	$192*10^3$	2.12
BFRP (∅ 6)	-	961	$57.3*10^3$	2.7
BFRP (∅ 8)	-	992	$59.52*10^3$	2.85
BFRP (∅ 10)	-	953	$46.1*10^3$	2.9
BFRP (∅ 12)	-	925	$53*10^3$	2.9

The nature of breaking of samples with hybrid reinforcement depended on the ratio of the content of steel and BFRP reinforcement. This content for beams of all series amounted to:

- BM, BMD – 100% / 0% [10];
- BB, BBD – 0% / 100% [10];
- BMB, BMBD – 50% / 50% [10];
- B 6/10 – 27% / 73%;
- B 10/6 – 73% / 27%;
- B 8/10 – 39% / 61%;
- B 10/8 – 61% / 39%;
- B 12/6 – 80% / 20%;
- B 6/12 – 20% / 80%.

In the dependences shown above, the numerator of the fraction corresponds to the percentage of steel reinforcement, and the denominator of the percentage – to the percentage of BFRP reinforcement.

The under load performance of beams with hybrid reinforcement can be divided into two stages – before the occurrence of yield in metal reinforcement and after it. The first normal cracks appeared at the load level of 0.2… 0.3 of the destructive one. Up to the level of loads 0.4… 0.5 of the destructive load the beams performed similarly to those reinforced with steel reinforcement – occurred in proportion to the increasing load the development of main cracks and the new cracks located in the middle third of the span occurred. The latter is due to the predominant influence of steel reinforcement. A further increase in load led to the occurrence of the reinforcement yield. At this stage, the main resistance to the load is provided by the basalt fiber-reinforced polymer reinforcement. At the same time the width of opening of existing cracks increases sharply, new ones are formed, deflections grow. This character of performance is typical for beams reinforced with composite reinforcement.

In general, the B 6/10, B 10/6, B 8/10, B 10/8, B 6/12, B 12/6 series beams showed a similar character of performance and destruction. The difference was that, depending on the percentage of steel reinforcement, the moment of crack formation in the beams with the predominant content of steel reinforcement occurred at load levels close to the similar indicator of the control series beams (BM series, see [10]). The width of the crack openings and their number by length was also similar to the beams of the BM series in the B 12/6, B 10/6, B 10/8 series beams, and close to the BB series beams in the B 6/12, B 6/10, B 8/10 series beams (see [12]).

Due to the fact that in the B 6/12, B 6/10 series beams the high-strength B500 cold-drawn reinforcement was used, the absence of a yield plateau in this reinforcement provided stable indicators of the beams' performance. The latter varied linearly, depending on the applied load. At loads ≈ 0.9 of the destructive load there occurred a break of the steel reinforcement, after which the resistance was continued by the basalt fiber-reinforced polymer reinforcement. Further increase in load led to the emergence of ultimate compressive stresses in the concrete of the compressed zone, followed by its crushing. No break of the basalt fiber-reinforced polymer reinforcement was recorded. In some areas, the fibers of the surface layer of this reinforcement were damaged (Figure 3).

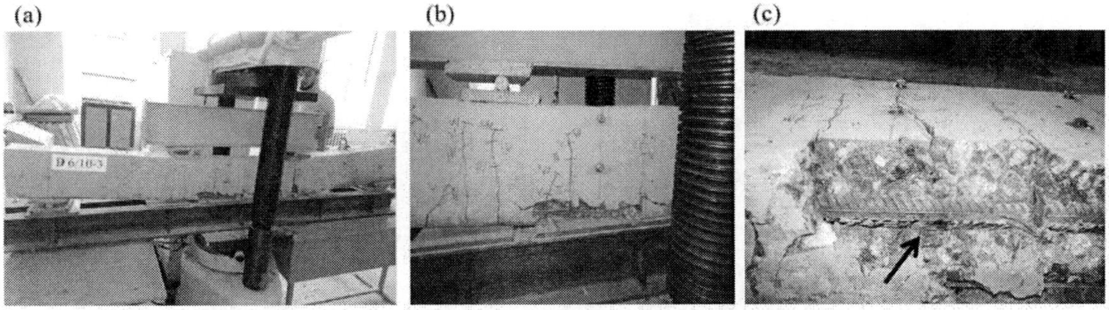

Figure 3. Sample beam series B 12-6-3 after the test. (a) Destroyed beam specimen; (b) destruction of the protective layer of the beam concrete; (c) breakage of part of the fibers of one of the rods used in the basalt fiber reinforced polymer reinforcement.

In B 12/6 series sample beams at a load level of 0.8 of the destructive load and in B 10/6 series samples at a load level of 0.6 of the destructive load, yield deformations occurred within the metal reinforcement. After that, the deformations of the samples increased as a result of which the deformations in the concrete of the compressed zone and stretched basalt fiber-reinforced polymer reinforcement reached the limit values. The destruction of the samples occurred due to the crushing of the concrete of the compressed zone with the simultaneous break of the BFRP reinforcement.

The destruction of the B 8/10, B 10/8 beam samples, as well as in the beams of other series, was caused by the fragmentation of the concrete of the compressed zone. The latter was preceded by the occurrence of the yield effect in steel reinforcement. The destruction of B 8/10 series beams was accompanied by the destruction (delamination) of the surface layer of the BFRP reinforcement fibers at loads equal to 0.9 – 0.95 of the destructive load, which led to a rapid increase in deformation and final crushing of the concrete in the compressed zone.

The only series of sample beams with hybrid reinforcement, the test of which was not accompanied by break of steel reinforcement or the occurrence of defects (partial break of fibers, fiber stratification) in BFRP reinforcement were B 10/8 series beams. The latter suggests that the ratio of metal reinforcement to the BFRP reinforcement in the proportion of 61% / 39% (≈ 60% / 40%) allows to perform reinforcement of the cross section that provides the sufficient level of structural reliability – the combined performance of both types of reinforcement at any load level, as well as the predictability of the nature of destruction of the beams due to crushing of the concrete in the compressed zone. The indicators of strength of experimental beams and beams used in previous studies [10], by series, are shown in Table 4.

Table 4. Indicators of strength of test sample beams.

Item no.	Series of beams	Breaking force, F_{ui}, kN	Relative value of strength, F_{ui}/F_u	Percentage of cross section reinforcement, μ, %	The ratio of the content of metal and BFRP reinforcement in cross section, % / %
1	BM[3)]	70.22[1)]	1	0.95	100 / 0
2	BMD[3)]	75.87[1)]	1.08	0.95	100 / 0
3	BB[3)]	100.88	1.43	0.95	0 / 100
4	BBD[3)]	101.17	1.44	0.95	0 / 100
5	BMB[3)]	96.1 (45.13[1)])	1.37	0.84	50 / 50
6	BMBD[3)]	99.24 (49.7[1)])	1.41	0.84	50 / 50
7	B 6/10[2)]	96.25	1.37	0.89	27 / 73
8	B 10/6	93.08 (63.77[1)])	1.33	0.89	73 / 27
9	B 8/10	119.67 (59.17[1)])	1.7	1.07	39 / 61
10	B 10/8	112.92 (69.53[1)])	1.61	1.07	61 / 39
11	B 6/12[2)]	106.25	1.51	1.18	20 / 80
12	B 12/6	122.05 (101.5[1)])	1.74	1.18	80 / 20

Notes:
1. F_u – breaking force of the BM series beams.
2. 1 – a force that corresponds to the beginning of the yield of the metal reinforcement.
3. 2 – high-strength B500 reinforcement B500 used.
4. 3 – the results of previous studies are given [10].

The data provided in Table 4 show that the replacement of metal reinforcement with BFRP reinforcement and the use of hybrid reinforcement increased the bearing capacity of the beam samples by 33% – 74% (see column 4, Table 4). The nature of destruction of the experimental beams indicates that the samples reinforced using the BFRP reinforcement collapsed due to the crushing of the concrete in the compressed zone. The latter is due to the high tensile strength of the BFRP reinforcement, as a result of which the beams had an overwhelming strength margin in the stretched area, and were "over-reinforced". That is, the strength of the beams was largely determined by the strength of the concrete in the compressed zone. According to the "traditional" method of beam design, "reinforcement" of the cross section should be avoided in order to save reinforcement materials. At the same time, due to the crumbling nature of breakage of the BFRP reinforcement, a more acceptable option is the destruction by crushing of the concrete in a compressed zone, because it does not occur suddenly, and BFRP reinforcement continues to perform resiliently without allowing the beam to collapse.

Analysis of the data given in columns 3, 5 of Table 4 indicates that, in addition to the strength of the concrete in the compressed zone, the strength of the beam samples was influenced by the percentage of reinforcement. The reinforcement of the stretched zone, when taking the tensile force, unloads the cross section, increasing the height of the compressed zone, thereby reducing the stress within it. Thus, the smallest increase in strength (33% – 41%) was shown by the BMB, BMBD, B 6/10, B 10/6 series beams with 0.84%, 0.89% percentage of reinforcement. Increase in the percentage of reinforcement has led to increase in strength. The optimal percentage of reinforcement should be considered 1.07%, which corresponded to the B 8/10, B 10/8 series beams. For those beams, the increase in strength was 61% – 70%, which is close to the maximum values.

Further increase in the percentage of reinforcement to 1.18% (B 6/12, B 12/6 series beams) virtually did not increase and, in case of the B 6/12 series, even decreased the value of the breaking force. The latter indicates that for this class of concrete in terms of the compressive strength, the optimal percentage of the BFRP reinforcement, or the combination of BFRP and steel reinforcement, is 1.07%. The latter statement is valid only in terms of strength and requires additional experimental studies in the case of the use of concrete of another class.

4. Conclusions

1. Replacement of steel reinforcement with BFRP reinforcement allows to increase the bearing capacity of beam samples by ≈ 44%.

2. The use of hybrid reinforcement allows to increase the bearing capacity of the beam samples by 33– 74% depending on the percentage of reinforcement of the cross section.

3. The samples of beams reinforced with BFRP reinforcement and beams with hybrid reinforcement are destroyed due to the crushing of the concrete in the compressed zone.

4. The determining factors of strength indicators of the beams reinforced with BFRP reinforcement and beams with hybrid reinforcement are the strength of concrete of the compressed zone and the percentage of reinforcement of the cross section.

5. The optimal percentage of reinforcement of beams with hybrid reinforcement is 1.07%, which allowed to achieve the maximum strength increase – 61–70% compared to the beams with steel reinforcement. The latter statement should be verified by conducting additional experimental tests on sample beams with different classes of concrete for compression.

6. For beams with hybrid reinforcement, from the point of view of reliability, the optimal percentage of the ratio of metal and BFRP reinforcement should be considered 60% / 40%. At this ratio, the joint work of both types of reinforcement took place at all levels of load without its physical destruction. In samples with a different percentage of metal and BFRP reinforcement, the destruction of one or another type of reinforcement at load levels close to destructive.

References

[1] Soldatchenko O S 2012 *Micnist', zhorstkist' ta trishhinostijkist' zginal'nih konstrukcij zi skloplastikovoju i bazal'toplastikovoju armaturoju*: dis. kand. tehn. nauk: 05.23.11 Soldatchenko Oleksandr Sergijovich Kyiv 196

[2] Valovoj O I, Eremenko O Ju and Valovoj M O 2017 Harakteristiki micnosti ta zhorstkosti balok armovanih bazal'tovoju armaturoju *Visnik Krivoriz'kogo nacional'nogo universitetu* **44** pp 142-146 http://nbuv.gov.ua/UJRN/Vktu_2017_44_30

[3] Pouya B 2011 *Experimental investigation of the mechanical and creep rupture properties of basalt fiber reinforced polymer* (bfrp) PhD thesis University of Akron (USA) 216

[4] Adhikari S 2009 *Mechanical Properties and Flexural Applications of Basalt Fiber Reinforced Polymer* (BFRP) Bars Thesis 385

[5] Eythor Jhorhallsson and Bjorgvin Smari Jonsson 2011 *Test of prestressed concrete beams with basalt fiber reinforcement polymer* (BFRP) tendons Thesis in Civil Engineering with specialization in structural design submitted to School of Science and Engineering at Reykjavik University (Iceland)

[6] Jónsson Björgvin Smári 2011 *Prestressed BFRP tendons in concrete beams* M S thesis Reykjavik University (Iceland)

[7] Ge W, Zhang J, Cao D and Tu Y 2015 Flexural behaviors of hybrid concrete beams reinforced with BFRP bars and steel bars. *Construction and Building Materials* **87** pp 28-37

[8] Elsayed T A, Eldaly A, El-Hefnawy A and Ghanem G 2011 Behavior of Concrete Beams Reinforced with Hybrid Fiber Reinforced Bars. *Advanced Composite Materials* **20(3)** pp 245-259

[9] Lau D and Pam H J 2010 Experimental study of hybrid FRP reinforced concrete beams. *Engineering Structures* **32(12)** pp 3857-3865

[10] Valovoi A, Koval P, Eremenko A, Valovoi M and Volkov S 2018 Durability of beams with hybrid reinforcement from metal and basalt fiber reinforced polymer (BFRP) armature *MATEC Web of Conferences* **230** URL: https://doi.org/10.1051/matecconf/201823002035 (Last accessed: 12.01.2021)

[11] Valovoi A, Eremenko A, Valovoi M and Volkov S 2019 Research of the Deflections of Beams Reinforced with BFRP Armature and Hybrid Reinforcement Using Metal and BFRP Armature *Actual Problems of Engineering Mechanics: Materials Science Forum. Trans Tech Publications Ltd.* (Switzerland) **968** pp 301–308 DOI:https://doi.org/10.4028/www.scientific.net/MSF.968.301 (Last accessed: 12.01.2021)

[12] Valovoi A, Eremenko A, Valovoi M and Volkov S 2020 Crack resistance and width of crack opening of beams with hybrid reinforcement using BFRP and metal armature. *Actual Problems of Engineering Mechanics: Materials Science Forum. Trans Tech Publications Ltd.* (Switzerland) **864** pp 149–157 DOI:https://doi.org/10.4028/www.scientific.net/KEM.864.149 (Last accessed: 12.01.2021)

IOP Conf. Series: Materials Science and Engineering 1164 (2021) 012083 doi:10.1088/1757-899X/1164/1/012083

Features of passing a shock wave in a long communication passageway with walls of different rigidity

A Vasilchenko[1,*], M Surianinov[2], Y Otrosh[1] and V Nikitin[1]

[1]National University of Civil Defence of Ukraine, 94, Chernishevska str., Kharkov 61023, Ukraine
[2]Odessa State Academy of Civil Engineering and Architecture, 4, Didrihsona str., Odessa 65029, Ukraine

*E-mail: avas2006@ukr.net

Abstract. The problem of attenuation of a shock wave during its propagation in long communication channels of industrial buildings or mining workings is considered. It is shown that an explosion in a channel produces a head shock wave with a plane wavefront, where the dynamic pressure significantly exceeds the pressure at fronts forming by the incident and reflected shock waves. A physical model of the formation and propagation of a shock wave in a channel with walls of different rigidity is proposed. It is shown that if one of the channel walls is movable or easily deformed, this leads to a violation of geometry of plane wavefront of the head shock wave and its weakening. The necessity of arranging pressure-relief structure in communication channels of buildings with increased explosive hazard and mine workings is grounded.

1. Introduction and literature review

Emergency explosions in industrial buildings related to highly explosion hazard facilities always are resonant events [1, 2]. Despite the improvement of safety methods in the implementation of technological processes, explosions due to the leakage of explosive liquids and gases lead to human casualties, damage equipment, destroy labor results, and cause significant losses [3, 4, 5]. Also resonant events are emergency explosions in mine workings of mines and tunnels.

A feature of explosions in mine workings is that shock waves (SW) propagate in narrow channels (drifts, adits, crosscuts) and, without the use of special anti-explosive measures, retain destructive force for a long distance [4, 5, 6].

To weaken the action of a shock wave in mines, dissipation methods are used, which provide at least partial absorption of the shock wave energy. There are many ways to extinguish an explosive shock wave during blasting in a mine [7–10]. All these methods are used precisely during blasting operations, which exclude presence of people and expensive equipment in the danger zone.

However, safety equipment for blasting operations cannot be used for other types of work, as they will interfere with the production process, hindering communications on drifts, tunnels, crosscuts (in general they can be represented as communication passageways or channels). Since emergency explosions cannot be excluded, when operating the communication channels of the mine, it is necessary, on the one hand, to ensure the free movement of people and equipment, and on the other, to reduce the negative consequences of an accidental explosion. The relevance of this problem is obvious [11].

The problem of weakening the blast wave is also relevant in the protection of industrial buildings. When inspecting construction sites damaged by accidental explosions and the fires caused by them, it is often necessary to explain the causes of visible damage to building structures [12–14]. In industrial buildings with increased explosion and fire hazards, as a rule, the main focus of the design is on manufacturing facilities. They are equipped by explosion-relief structures, which most often are as window and door openings. In the event of an emergency explosion, the shock wave destroys the explosion-relief structures, and overpressure in the room decreases rapidly. Structures in such rooms

Content from this work may be used under the terms of the Creative Commons Attribution 3.0 licence. Any further distribution of this work must maintain attribution to the author(s) and the title of the work, journal citation and DOI.
Published under licence by IOP Publishing Ltd

also count on the possible impact of an explosion. And in the communication passageways, which can be corridors or tunnels, reaching considerable length in industrial buildings, and where, after overcoming of the explosion-relief structure, shock wave propagates, it acts on the building envelope. At the same time, the state of building envelope is unpredictable: some walls on the path are destroyed, while others, similar, remain practically intact.

Based on these observations, it makes sense to consider the problem of attenuation of shock waves during their propagation in long communication passageways of industrial buildings and mining channels, assuming their fundamental similarity.

The communication passageway can be represented as a semi-closed space (channel), in which, according to [15, 16], the energy of the shock wave dissipates more slowly than in a large room or in open space. At the same time, energy losses for air heating and friction are increased when shock wave interact with the channel walls. The walls of the channel as a surface of reflection also influence the process of formation and propagation of shock wave, determining its intensity. For these reasons, the study of the interaction of the walls and partitions of the channel with the shock wave is of interest for predicting the behavior of both the shock wave itself and building structures in an explosion. In publications [17–19] problems of accounting for the behavior of shock wave in channels were considered, but it was assumed that the walls were equally rigid. However, in real construction projects, walls and partitions of communication channels often differ both structurally and in materials.

2. Construction of the model and its analysis

The objective of this work is to formulate recommendations for improving safety, both for building structures of communication passageways of buildings with increased explosion hazard, and for communication channels of mine workings based on the formation of a physical model of the formation and propagation of shock wave in channel with walls of different stiffness, i.e. when one of the walls can noticeably deform in comparison with the other.

A specific feature of the propagation of a shock wave in a semi-closed space is its interaction with the surface of the obstacle (wall) on which reflected shock waves are formed [18, 19]. For a shock wave formed during an explosion with overpressure at its front of less than 700 kPa, when air can still be considered an ideal gas, the reduced overpressure of reflected wave P^*_R is calculated from [20]:

$$P_R^* = 2P_I^* + \frac{(\gamma+1)P_I^*}{2\gamma+(\gamma-1)P_I^*}, \tag{1}$$

where P^*_I is the reduced overpressure of the incident wave; γ is the ratio of gas specific heats.

$$P_R^* = \frac{P_R}{P_0}; \qquad P_I^* = \frac{P_I}{P_0}, \tag{2}$$

where P_R is the overpressure of reflected wave; P_I is the overpressure of incident wave; P_0 is the initial pressure.

The load on the wall surface depends on the parameters of shock wave, the orientation of the shock front relative to the surface, and the surface rigidity. When the shock wave falls at an angle to the surface ($0 < \alpha_I < \pi/2$), the pressure in the contact zone will be greater than the pressure at the front of the incident shock wave ($P > P_I$), since when air is braked behind the shock front, its kinetic energy is converted into potential compression energy. The surface will experience additional load (P_u) caused by the action of impact air pressure, which depends on the air flow rate (u) and its density (ρ_B):

$$P_u = \frac{\rho_B u^2}{2g}, \tag{3}$$

where g is the acceleration of gravity.

Such reflection of shock waves is non-stationary. It depends on the shock wave intensity, the properties and the nature of the distribution of the parameters of the gaseous medium behind its front, the geometry of the wave and the surface with which the interaction occurs [19–21]. The ratio of the incident (I_I) and reflected (I_R) wave pulses is estimated from considerations of similarity:

$$\frac{I_R}{I_I} \approx \frac{P_R}{P_I}. \tag{4}$$

At small angles of incidence of the shock wave (α_I), the so-called regular reflection occurs, and the angle of the reflected shock wave is greater than the incident one $\alpha_R > \alpha_I$. In [15], for the case of regular reflection, a formula is proposed that allows calculating the overpressure at the front of the reflected shock wave:

$$P_R = 2P_I + \frac{(\gamma+1)}{2}\rho_B u^2 \cos^2 \alpha_I. \tag{5}$$

With an increase in the angle of incidence of the shock wave to a critical value ($\alpha_{Ikr} = \alpha_I$), a moment comes when regular reflection becomes impossible and only irregular (or Mach) reflection is possible [22, 23]. It is characterized by the fact that as a result of the interaction of the incident and reflected shock wave when $\alpha_{Ikr} > \alpha_I$ a third (Mach) wave is formed. The junction area of all three waves is called the triple point.

During the transition from regular to irregular reflection at near-critical angles of the shock wave incidence ($\alpha_{Ikr} = \alpha_I$), the trajectory of the triple point will be parallel to the rigid wall. If the angle of the shock wave incidence with an irregular reflection exceeds a critical value, then the triple point will move along the trajectory at an angle to the surface of the rigid wall. As a result, a continuously growing "Mach stem" is formed, which is a plane wave with a front normal to the surface.

Mach reflection is a rather complicated phenomenon. And if the pressure (and, correspondingly, the pulse) at the front of the reflected shock wave can be calculated for the zone of regular reflection of shock wave (see (5)), for irregular reflection this is difficult to do by simple methods, and only empirical methods remain. According to empirical estimates, the angle of movement of the triple point is 4...5 times smaller than the Mach transition angle α_{Ikr}. The Mach transition angle for the incident wave depends on the adiabatic index of detonation products and the Mach number of the shock wave (this angle lies within $\alpha_M = 39°...47°$) [20, 22, 23].

As mentioned above, internal emergency explosions of gas-air mixtures with subsequent propagation of shock wave in the channels are likely to occur in hazardous operations industrial facilities and in mine workings. The most important parameters of such explosions are geometric factors (the shape and cross-section of the channel, the presence of pressure-relief structures), the heat of gas combustion and the ratio of mixture components.

During the propagation of detonation products of explosion and/or shock wave in the channel, the limiting surfaces (walls), which are reflection surfaces, influence this process. The energy stored by the shock wave is dissipated much slower than in the atmosphere, but it is spent on air heating and friction when the shock wave interacts with the walls. The shock wave intensity in this case is determined not only by the explosion energy transmitted to the air, but also by the reflection processes.

The effect of the shock wave on the channel walls can be divided into two stages. On the first – the wall is loaded when the shock wave initially falls on it, and on the second – it is affected by several subsequent pressure pulses. This interaction can be very complex due to multiple reflections of shock waves and/or the presence of discharge openings.

A simplified model of how a shock wave is formed and propagated in a channel with rigid walls is shown in figure 1. This model can be represented in the following sequence:

1. The formation of curved surface of the front of shock wave.
2. Because the incident shock wave propagates in the channel in an unperturbed air medium, and the reflected shock waves propagate in the medium that has been compressed and heated by the transmitted incident shock wave, then the reflected shock waves propagate at higher speeds than the

incident shock wave and they can therefore catch up with the incident shock wave and merge with it [21].

3. The formation of a head shock wave (or bow shock wave) with a plane wavefront as a result of the confluence of the incident shock wave and the reflected shock waves. In the resulting head shock wave, the dynamic pressure significantly exceeds the pressures at the incident shock wave and reflected shock waves fronts.

The formation of a plane wavefront of the head shock wave is realized with oblique incidence of waves. If the angle of incidence is large enough, then Mach waves can form near the channel walls. Considering the angle of movement of the triple point trajectory, it can be expected that a continuous front of a plane head shock wave will form in a zone with a length of up to 8 characteristic dimensions of the channel cross section [22, 24].

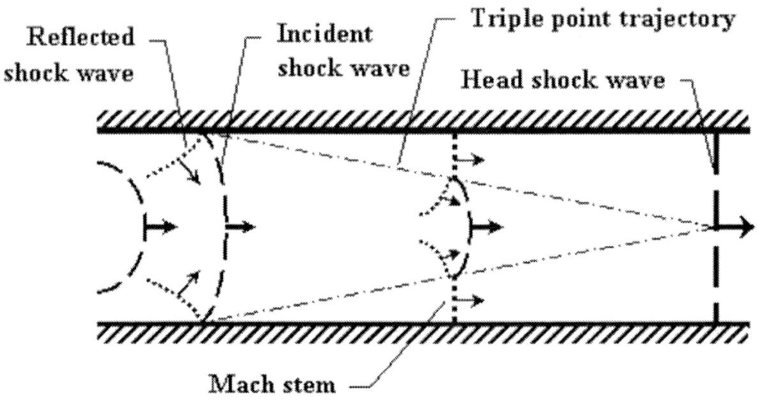

Figure 1. Scheme of interaction of shock waves in the channel with walls of the same rigid.

Formulas are known that describe the empirical dependence of overpressure (P_b) and specific impulse of head shock wave (I_b) during the explosion of a condensed explosive in a channel with rigid walls on the mass of explosive, the cross-sectional area, and the channel length [20]:

$$P_b = \left(a\frac{fm}{xS} + b\sqrt{\frac{fm}{xS}} \right) e^{-\beta x}; \tag{6}$$

$$I_b = c\frac{fm}{S} e^{-\frac{\beta x}{2d}}, \tag{7}$$

where P_b – overpressure, MPa; I_b – specific impulse, Pa·s; a, b, c – empirical coefficients; f – efficiency ratio of explosive as compared to trotyl; m – mass of explosive, kg; x – distance covered by the shock wave, m; S – channel cross-sectional area, m^2; β - reduced coefficient of aerodynamic resistance of the channel; d – reduced diameter of the channel, m.

Since, both in hazardous operations industrial facilities and in mine workings, accidental explosions of gas-air mixture are more likely, and when airborne shock wave propagate in the channel there is no fundamental difference in the nature of explosion, then formulas (6, 7) can be converted for the case of gas-air mixture:

$$P_b = \left(a\frac{\delta M Q_g}{xSQ_t} + b\sqrt{\frac{\delta M Q_g}{xSQ_t}} \right) e^{-\beta x}; \tag{8}$$

$$I_b = c\frac{\delta M Q_g}{SQ_t} e^{-\frac{\beta x}{2d}}. \tag{9}$$

where δ is the transition coefficient; M is the mass of explosive; Q is the calorific value of explosive; Q_t is the heat of TNT explosion.

The load that acts on channel walls during the passage of shock wave is characterized by the values of pressure P^*_1 and pulse I^*_1:

$$P_1^* = \frac{P_R F \sqrt{E_1}}{A_1 \sqrt{R_1^3}} \tag{10}$$

$$I_1^* = \frac{I_R h \sqrt{E_1}}{A_1 R_1 \sqrt{\rho_1}}, \tag{11}$$

where F – impact area of shock wave, m^2; E – modulus of elasticity, MPa; A – wall section area, m^2; R – tensile strength of the wall material, MPa; ρ – density of the wall material, N/m^3.

From the analysis of formulas (10, 11) it follows that if opposite walls of the channel are made of different materials, but with close values of tensile strengths and elastic modulus ($R_2 \approx R_1$; $E_2 \approx E_1$) that do not allow deformation under the influence of shock waves, then shock waves in the channel will be distributed according to the scheme shown in figure 1.

In the case when the strength and elastic modulus of one of the walls is much lower ($R_2 < R_1$; $E_2 < E_1$), and it is easily deformed, the reduced impulse required for its destruction decreases. This causes a decrease in the velocity of the reflected shock waves and an increase in the zone of formation of plane front of head shock wave. In this case, the front of shock wave unfolds towards a non-rigid wall, as shown in figure 2, the pressure increases on this wall, which can lead to its destruction.

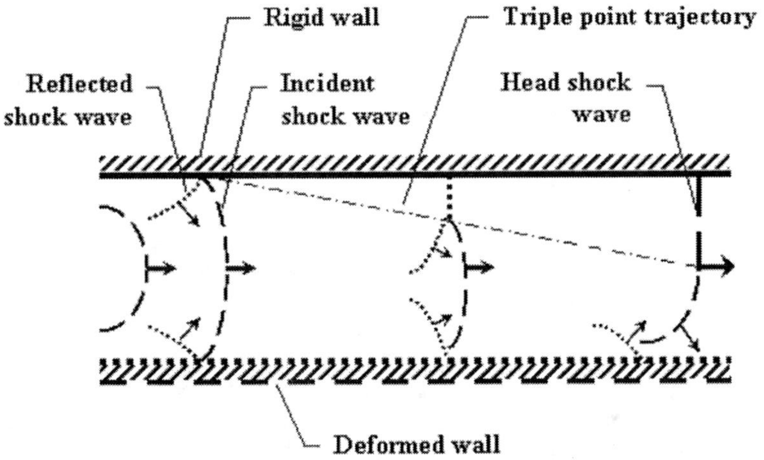

Figure 2. Scheme of the interaction of shock waves in the channel with one deformable wall.

Reversal of the front of shock wave means a violation of the geometry of plane front of head shock wave and a decrease in pressure at the front of shock wave. A new plane front will be able to form again only in zone with a length of 4…8 characteristic dimensions of channel section. For destruction of the wall and formation of a new plane front of head shock wave, energy must also be expended. For example, according to [25], even the creation of open dead end branch from the main channel with equal cross-section can reduce the load by 1.2...2 times.

Thus, if one wall is strong enough in the communication channel of large length and the other wall is easily deformed, then when a shock wave passes, a non-rigid wall can be fail fragmented in zones of formation of plane front of the head shock wave.

The proposed model allows us to justify a way to increase safety in long communication passageways of buildings with increased explosion hazard or mine workings. For this, it is necessary to provide easily breakable structures along the communication channel with an interval of 4-8 sizes of its width in the form of inserts (partitions) covering the expanders with a width and depth of not less than characteristic width of the channel. Such device of communication channels will prevent the formation of head shock wave and help to reduce the overpressure at the front of shock wave and its attenuation.

3. Conclusions

1. A physical model of formation and propagation of a shock wave in long communication channels is proposed, when one of the walls is easily deformed.

2. The proposed physical model explains the fragmentary damage of structures in long communication channels of communication passageways of industrial buildings and mine workings in case of emergency explosions.

3. The presented model shows that in the communication passageways of industrial buildings or the communication channels of mining, in which emergency internal explosions can occur, safety can be improved by reducing overpressure at the front of the shock wave, which contributes to its attenuation. For this, easily breakable structures in the form of partitions covering the expanders should be provided. Easily breakable structures should be located along the communication channel with an interval of 4-8 sizes of the width of this passageway.

4. The proposed physical model can be used to construct an algorithm for numerically calculating propagation of shock wave in a channel with walls of different rigid [26].

References

[1] Skob Y A, Ugryumov M L and Granovskiy E A 2020 Numerical assessment of hydrogen explosion consequences in a mine tunnel *International Journal of Hydrogen Energy* (In Press)

[2] Skob Y, Ugryumov M and Dreval Y 2020 Numerical Modelling of Gas Explosion Overpressure Mitigation Effects *Materials Science Forum* **1006** pp 117-122

[3] Orlov G G and Korolchenko A D 2016 Nagruzki, razrushayushchiye stroitelnyye konstruktsii v rezultate avariynykh vzryvov *Pozharovzryvobezopasnost – Fire and Explosion Safety* **25(3)** pp 45-56

[4] Mishuev A V, Kazennov V V, Khasnutdinov D Z, Gromov N V, Lukyanov I A and Prozorovskiy D V 2011 Explosion safety of extended road tunnels and ways of improvement *Vestnik MGSU* **8** pp 331-334

[5] Kolesnichenko I E, Artmiev V B, Kolesnichenko E A, Cherechukin V G and Lyubomishchenko E I 2017 Prevention of coal dust and methane explosions in mineworkings: theory and practice *Mining Industry Journal* **4(134)** 26

[6] Krainov A Yu and Moiseeva K M 2017 Mathematical modelling of sparkplug ignition of a coal-dust monodisperse suspension in a methane-air mixture *MATEC Web of Conferences* **115** 03017

[7] Lipnickij Yu M and Panasenko AV 2010 Formirovanie detonacionnyh voln v kanalah i ih vzaimodejstvie s pronicaemymi peregorodkami *Zhurnal tekhnicheskoj fiziki – J. Technical Physics* **80(8)** pp 53-58

[8] Sin A F and Chernyh A V 2010 *Sposob gasheniya vzryvnoj udarnoj volny* Patent № RU2404365S2 (Rossiya)

[9] Orlov G G, Korolchenko D A and Korolchenko A Ya 2015 Determination of the value and character of the exposive loads in case of using of the inertial safety constructions *Pozharovzryvobezopasnost – Fire and Explosion Safety* **24(4)** pp 47-55

[10] Gorev V and Medvedev G 2017 The flexible perforated screens are a protection of buildings against external explosions *MATEC Web of Conferences* **106** 02028

[11] Otrosh Y, Rybka Y, Danilin O and Zhuravskyi M 2019 Assessment of the technical state and the possibility of its control for the further safe operation of building structures of mining facilities *EDP Sciences. In E3S Web of Conferences* **123** 01012
[12] Otrosh Y, Kovaliov A, Semkiv O, Rudeshko I and Diven V 2018 Methodology remaining lifetime determination of the building structures *MATEC Web of Conferences* **230** 02023
[13] Surianinov M, Andronov V, Otrosh Y, Makovkina T and Vasiukov S 2020 Concrete and Fiber Concrete Impact Strength *In Materials Science Forum Trans Tech Publications Ltd.* **1006** pp 101-106
[14] Vasilchenko A, Doronin E, Ivanov B and Konoval V 2019 Effect of residual deformation of a steel column on its fire resistance under combined exposure "explosion-fire" *Materials Science Forum Trans Tech Publications Ltd.* **968** pp 288-293
[15] Clavin P 2017 Nonlinear dynamics of shock and detonation waves in gases *Combustion Science and Technology* **189(5)** pp 747-775
[16] Nalisko N N 2015 The estimation of shock air waves propagation in lengthy buildings subject to dissipation of gas stream internal energy *Construction, materials science, mechanical engineering* **83** pp 154-160
[17] Polandov Yu Kh, Dobrikov S A and Korolchenko A Ya 2016 Gas explosion in a cylindrical tube with a hole on the lateral surface *Pozharovzryvobezopasnost – Fire and Explosion Safety* **25(11)** pp 17-26
[18] Chernyshov M and Tyapko A 2018 Optimal regular reflection of shock and blast waves *MATEC Web of Conferences* **245** 12005
[19] Sui Yaguang, Zhang Dezhi, Tang Shiying, and Chen Bo 2017 *Experimental and Numerical Research on Cylindrical Tubes under Outer Cylindrical Explosive Waves, Shock and Vibration* ID 6150193.
[20] Bejker U, Koks P, Uestrajn P i dr. 1986 *Vzryvnye yavleniya: ocenka i posledstviya* Kn. 1. Per. s angl. Redaction Ya B Zel'dovich, B E Gel'fand Explosive phenomena: assessment and consequences. Book 1 (Moskow: Mir)
[21] Semenov A N, Berezkina M K and Krasovskaya IV 2009 *Classification of shock wave reflections from a wedge* Part 1: Boundaries and domains of existence for different types of reflections, Technical Physics **54(4)** pp 491-496
[22] Semenov A N, Berezkina M K and Krasovskaya I V 2009 *Classification of shock wave reflections from a wedge* Part 2: Experimental and numerical simulations of different types of Mach reflections, Technical Physics April **54(4)** pp 497-503
[23] Sandeman R J 2015 A simple physical theory of weak Mach reflection over plane surfaces (Australia/Oceania: Springer)
[24] Selivanov V V, Solov'ev V S and Sysoev N N 1990 *Udarnye i detonacionnye volny. Metody issledovaniya* (Moskva)
[25] Polandov Yu Kh, Babankov V A and Dobrikov S A 2016 Osobennosti razvitiya gazovogo vzryva v pomeshchenii pri nalichii smezhnoy komnaty *Pozharovzryvobezopasnost Fire and Explosion Safety* **25(1)** pp 38-46
[26] Krutii Y, Kovrov A, Otrosh Y and Surianinov M 2020 Analysis of Forced Longitudinal Vibrations of Columns Taking into Account Internal Resistance in Resonance Zones *In Materials Science Forum. Trans Tech Publications Ltd. Volume* **1006** pp 79-86

Improvement of technology of application of wood as a floor covering

Y Tsapko[1], R Vasylyshyn[2], O Horbachova[3,*] and O Bondarenko[4]

[1]Scientific Research Institute for Binders and Materials, Kyiv National University of Construction and Architecture, Povitroflotskyj ave., 31, Kyiv, Ukraine, 03037 and National University of Life and Environmental Sciences of Ukraine, Heroiv Oborony str., 15, Kiyv, Ukraine, 03041
[2]National University of Life and Environmental Sciences of Ukraine, Heroiv Oborony str., 15, Kiyv, Ukraine, 03041
[3]National University of Life and Environmental Sciences of Ukraine, Heroiv Oborony str., 15, Kiyv, Ukraine, 03041
[4]Kyiv National University of Construction and Architecture, Povitroflotskyj ave., 31, Kyiv, Ukraine, 03037

*E-mail: gorbachova.sasha@ukr.net

Abstract. The analysis of wood flooring has been made and it has been found that the manufacture of wood flooring poses certain difficulties in ensuring resistance to destruction. Thermally modified wood is one of the promising uses of flooring. Therefore, knowledge of the physico-mechanical featutes of wood composites, their quality indicators, adhesion with organic adhesives, allows to make a choice taking into account economic indicators, duration and safety of application. Experimental studies have found out that a floor covering with an upper layer of thermally modified oak wood is less resistant to abrasion. The additional moisture effect showed a better result for the thermally modified oak samples than for the ash and oak samples. The test pieces with an upper layer of thermally modified oak are found to have the lowest measure of power. The results of the research will enable the further development of new composites of thermally modified wood and the conditions under which they are operated at different sites.

1. Introduction

Today, wood is a major floor covering. Wooden floors have high wear resistance, durability, impact resistance, low heat conductivity and noiseless while walking. The main species for floor production are hard woods like: oak, beech, maple, ash [1, 2].

One of the main disadvantages of the use of wooden floor is its deterioration and knowledge of the physical and mechanical properties of such construction products, their quality indicators, allowing for the choice taking into account economic indicators, duration and safety of application, environmental aspects, etc. [3-6].

Therefore, the development of technological measures for the manufacture of wooden flooring and the research on the impact on the strength of different flooring of different wood species is an unresolved component of the sustainability of composite products and determines the need for research.

2. Analysis of recent research and publications

In the use of wooden floors it is assumed that the characteristics of application are significantly influenced by the conditions [7-9]. In this connection, a study has been carried out and it has been established that environmental certification and guarantee, especially in relation to the production of wood-based composites, have been the most important factor in determining preferences. The focus was on estimation methods of effects on wood characteristics. However, it is not specified the

Content from this work may be used under the terms of the Creative Commons Attribution 3.0 licence. Any further distribution of this work must maintain attribution to the author(s) and the title of the work, journal citation and DOI.
Published under licence by IOP Publishing Ltd

influence of differences among different wood species, as well as the wide variety of ways of composites creation.

During operation, the parquet board shall be exposed to various external factors affecting the stability of its shape [10], these include temperature, humidity and the load which the parquet board gets during operation. It has been established that each of these factors affects the stability of the shape differently, but that a long-term or joint action can alter its features or even destroy the parquet board. However, it is not identified how the exertion changes when the rocks in the composition article change according to the temperature change.

Despite increased biological stability and changes in size stability at temperature fluctuations which calls for durability tests that are the most sensitive to decomposition, but neither strength tests nor fungus identification which is responsible for decomposition are not included in the above-ground strength standards for field tests [11]. However, the necessary methods for characterizing the change in wood quality in terms of improved target surface properties are not specified either.

Changes caused by bacterial action that may affect surface quality have been measured and compared to the test samples, as bacterial destruction is closely related to surface wear [12]. After wood treatment, surface properties and aesthetic changes were assessed; and the performance of the coverings was evaluated. The antiseptic covering retains resistance to wood destruction, while without antiseptic increases surface destruction. However, there are no methods for estimating bacterial surface degradation.

There are a number of test methods simulating loads on wooden floors [13], where adhesion, impact strength, elasticity, wear resistance and resistance to various liquids aren all significant indicators of the quality of wooden floors. However, some of them do not specify minimum requirements or reference values for the specific technical properties of the wood surface.

The quality of oak and chestnut wood for floor coverings has already been confirmed [14] and it has been established that shrinkage and swelling, favourable to both species of wood, are encouraging the slight swelling of chestnut in the tangential direction; the hardness of oak was higher due to higher density. However, the characteristics aren't specified when making composite products from them.

Many wood species are used as flooring, but their characteristics may vary greatly depending on the properties of the wood and the type of application [15]. Industrial wooden floors have been subjected to tests on the falling ball, rolled loads, ground dents through small loads, and static and dynamic friction. In accordance with the characteristics of the wooden flooring, were established the resistant classes in the simulation tests. However, it is not specified how wood species are graded according to each class.

The study in [16] carried out dynamic hardness tests on solid and engineering wooden floors. The results showed similar behaviour towards traditional hardwood species applied to wooden floors, and this may be a promising growth in flooring industry. However, the use of pulp wood remains unresolved.

Preliminary tests were carried out to determine the geometric stabilization of wooden floor samples [17-19]. As a result, it has been determined that the flow of channels for hardwood parquet is no better for improving geometric stability under different climatic conditions. However, the method of estimating the surface after wood treatment is not specified.

The difference in adhesive properties through the different chemical composition of the fibre surface has been evaluated and it has been established that it is replaced by the penetration of adhesive into the volumetric fibre of the wood [20]. These results show more lignin or hydrophobic extractive substances on the fiber surface. With regard to the characteristics of adhesion, these results indicate the formation of a less polar surface of thermally modified fibres. However, the role of adhesive compositions on the properties of wood composites is not disclosed.

Thus, it has been established from literary sources that the application of wooden floors under different climatic conditions requires appropriate research to counteract the destruction. This suggests that a study to determine the parameters that ensure its durability would be useful. Research in this area is therefore an unfinished part of the sustainability of building structures and has necessitated researching.

3. The purpose of this work

The purpose of this work is to study the efficiency of wooden flooring for wood products and to substantiate its strength in operation.

4. Materials and methods of research

The physical-mechanical and operational properties of the parquet board were investigated in order to determine the prospects for the production of floor coverings depending on the conditions of use. The first group of samples consisted of two layers of ash wood and an upper layer of thermally modified oak wood; the second group consisted of a base of water-bearing plywood, a middle layer of XDF from the outside of ash wood; the third group consisted of pine veneer as a lower layer, in the middle of XDF, the wood of oak as the top layer is Figure 1.

a b c

Figure 1. Test samples of parquet board: a – top layer from thermally modified oak wood; b – top layer from ash wood; c – top layer from oak wood.

The durability of a material is defined as the wear on the top layer of a parquet board when used in the operation of friction forces. The essence of the method is to determine the loss of mass when the surface of the upper layer is abraded by an abrasive material with the specified load on the sample. The studies were carried out on samples measuring 50x150x14 mm, which were pre-stored 72 h in a room with an air temperature of 23±2 °C and relative humidity of 65%. Grinding material with a grain capacity of 240, the speed of movement of 1.55 m/min, the time of grinding 2 min, the load on the sample 2 kgf. At the end of the abrasion, the sample was weighted to an accuracy of 0.01 g and determined the abrasion by the formula:

$$t = h \cdot \frac{m_1 - m_2}{m_1} \quad (1)$$

where h – is the height of the test sample, mm; m_1 – is the mass of the sample before the test, r; m_2 – is the mass of the sample after the test, r.

During the operation of the flooring the contact with water is possible, so it is worth investigating the resistance of the upper layer to the effects of moisture. The moisture absorption experience was carried out in a sodium-saturated desiccator on the three parquet board types described above, measuring 40x50x14 mm. The test samples shall be preconditioned at a temperature of 23±2 °C and a relative humidity of 65% for three days for stabilization. The test samples were weighed with a daviation of not more than 0.01 g and placed in the desiccator with a lateral surface on the insert so that they do not touch each other and the walls of the desiccator. The soak was carried out in the air at a temperature of 20±2 °C. The samples were periodically weighed with a deviation of not more than 0.01 g the first weighing was carried out 24 hours after the sample was placed in the desiccator, followed by 2, 3, 6, 9, 13, 20 days later.

In the study of the glue compound, samples of three types of parquet board with dimensions of 40x50x14 mm were used after stabilization within three days. The essence of the strength limit method is to determine the breaking load vertically down at the glue junction (Figure 2).

Figure 2. Load test for the measure of power.

The sample was loaded continuously at the loading speed of the test vehicle head 0.60±0.15 mm/min. The tests continued until the fracture of the sample and determined the measure of power by the formula:

$$\sigma = \frac{P}{b \cdot l} \tag{2}$$

where P – is the destructive load, H; b – is the width of the glue, m; l – is the length of the glue, m.
The result was rounded to 0.1 MPa. The surface fracture rate was observed during the tests.

5. Research results

In Figure 3-5 show the abrasion results. In determining the resistance to abrasion, samples with an upper layer of oak wood showed the best results – less traces were left on them and on the polished skin as well (Figure 3). This suggests that this type of floor can be used in areas with large traffic.

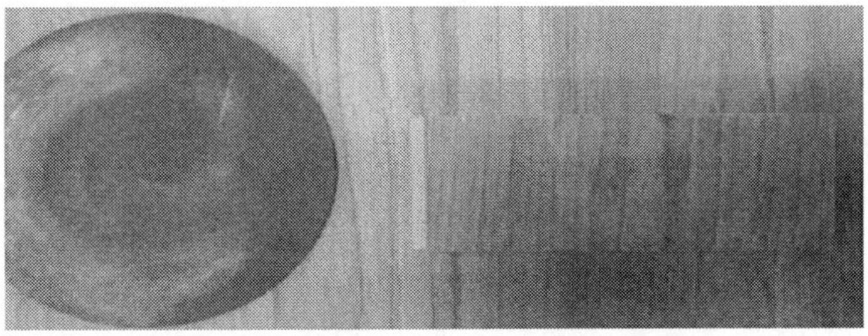

Figure 3. Results of the assessment of the abrasion resistance of the top layers of oak wood parquet board.

Under the same experience conditions, the grit paper removed a slightly larger layer of material – 0.2 mm. There are already visible friction marks in the form of rings and rubs in Figure 4. The samples with an upper layer of thermally modified oak did not perform as well as the previous ones (Figure 5). During grinding, black dust with a specific odour was released. Friction marks are clearly visible. The thickness of the layer removed is 0.36 mm.

After weighing the samples, it is established that the floor covering with an upper layer of thermally modified oak wood is less resistant to abrasion.

As a result of experience, it has been established that the most weight was lost in the sample with the upper layer of thermally modified wood – 1.01 g. The abrasion rate was 0.30. The most marks from the grint paper remain on the surface of the samples. The least weight – 0.38 g – has lost the sample with surface of oak, the abrasion rate is 0.07. There were less grinding marks on the ash wood sample.

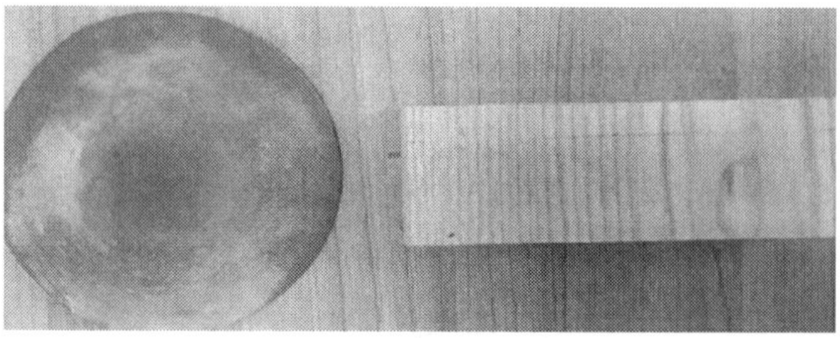

Figure 4. Results of the assessment of the abrasion resistance of a hardwood parquet board with top layers of ash wood.

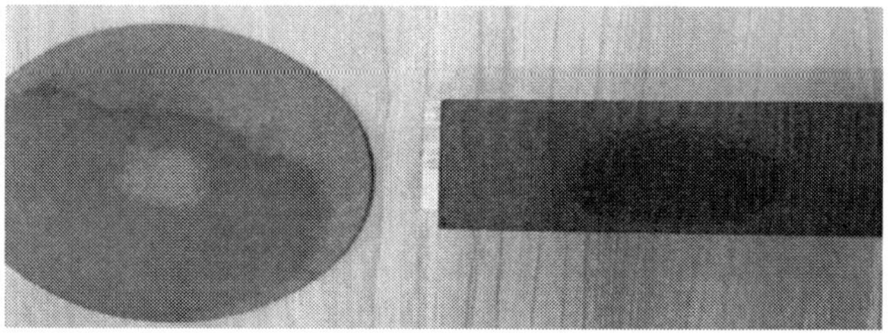

Figure 5. Results of the assessment of the abrasion resistance of the different top layers of parquet board with thermally modified oak.

Samples with an upper layer of ash wood showed a slightly worse result (Table 1).

Table 1. Results of the assessment of abrasion resistance of different materials.

Specimen (top layer)	Specimen thickness, mm	Specimen mass, g		Mass loss, g	Result	
		before abrasion	after abrasion		abrasion rate, t	visual assessment
Thermally modified oak	14.10	46.9	45.89	1.01	0.30	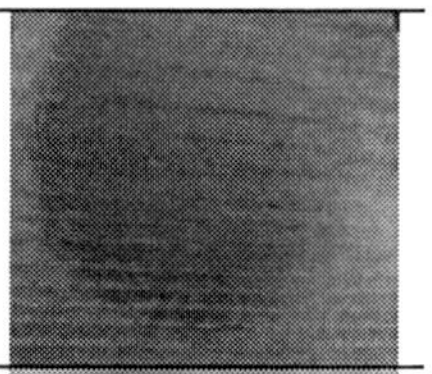
Ash	14.05	73.57	73.02	0.55	0.11	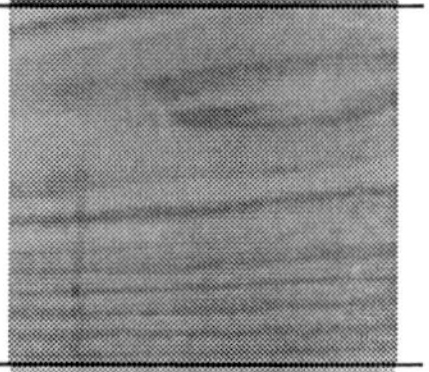
Oak	14.00	75.86	75.48	0.38	0.07	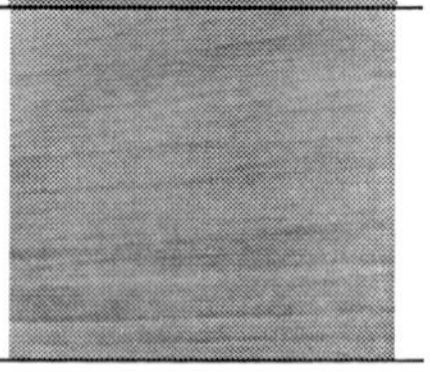

As a result of the experience with the determination of moisture absorption, the samples were found to have absorbed most of the moisture at the beginning of the experiment (Table 2).

Table 2. Change in mass of the floorboard samples in the wet environment*.

Specimen (top layer)	Change in mass of samples during keeping in a humid environment, g								Mass of absorbed moisture, g
	Duration of keeping in the environment with high humidity, days								
	0	1	2	3	6	9	13	20	
Thermally modified oak	15.40	17.62	17.95	18.50	19.99	21.99	23.57	24.54	9.14
Ash	19.21	21.82	23.60	24.23	26.42	27.90	29.55	31.37	12.15
Oak	23.87	26.54	27.21	28.29	30.60	32.93	35.45	37.80	13.93

*Note. Specimen initial moisture content 8 %

The first group of samples absorbed the least moisture in 24 hours – 2.22 g, and the most with the top layer of oak – 2.66 g. The difference is that the percentage is 19.17%, which is 2% less than the other two groups.

In 20 days, the humidity was determined to have changed by 71.28 % in samples with the top covering of ash wood. The samples with the covering of thermally modified oak performed better, the humidity increased by 65.7%. These samples also showed high size and shape resistance, unlike the others (Figure 6 a). The samples with the top layer of oak wood have buckled up (Figure 6 b) and the samples with the wood ash covering have peeled off in Figure 6 c.

a b c

Figure 6. Test samples of parquet board after soaking in a moist environment: a – top layer with thermo-modified oak wood; b – top layer of ash wood; c – top layer of oak wood.

Figure 7 shows the strength of the adhesive connection between the different groups of parquet board samples.

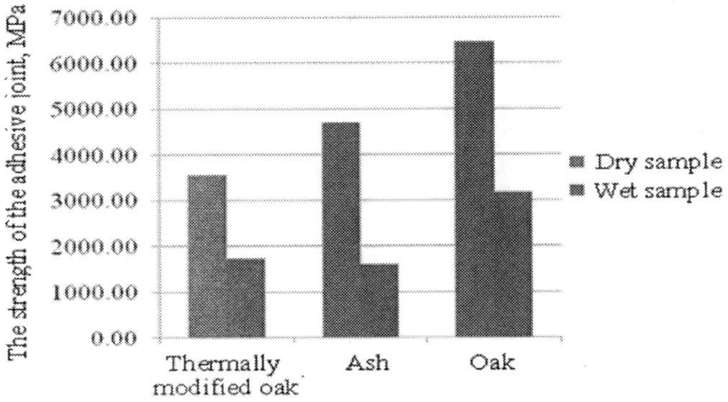

Figure 7. Measure of power of adhesive connections of different groups of parquet board samples.

Samples with an upper layer of thermally modified oak – 3557.1 MPa (Figure 7) have been determined to have the lowest strength.

A slightly better result in samples with an upper layer of ash – 4701.4 MPa; the hardwood parquet board with wood oak – 6458.1 MPa – showed the greatest strength. Similar samples were also examined to test the adhesive strength of the adhesive compound after being held in water. All the samples showed the twice worst result.

6. Conclusions

Thus, experimental studies have shown that floor covering with an upper layer of thermally modified oak wood is less resistant to abrasion. The additional moisture effect showed a better result for thermally modified oak-coated samples than for ash and oak samples. The test pieces with an upper layer of thermally modified oak are found to have the lowest strength. The results of the research will enable the further development of new composites and their operating conditions at different facilities.

References

[1] Abyzov V A, Pushkarova K K, Kochevykh M O, Honchar O A and Bazeliuk N L 2020 Innovative building materials in creation an architectural environment *IOP Conf. Series: Materials Science and Engineering Innovative Technology in Architecture and Design* **907** 012035 doi:10.1088/1757-899X/907/1/012035

[2] Tsapko Yu, Lomaha V, Tsapko O, Mazurchuk S, Horbachova O and Zavialov D 2020 Determination of regularities of heat resistance under flame action on wood wall with fire-retardant varnish *Eastern-European Journal Enterprise Technologies* **4/10 (106)** pp 55–60 doi: 10.15587/1729-4061.2020.210009

[3] Tsapko Yu, Tsapko A and Bondarenko O 2020 Research of conditions of removal of fire protection from building construction *Key Engineering Materials* **864** pp 141–148 doi:10.4028/www.scientific.net/KEM.864.141

[4] Pushkareva K K, Gonchar O A and Kaverin K O 2019 The role of the crystallo-chemical factor in the evaluation and improvement of the nanomodification efficiency of mortar and concrete *IOP Conf. Series: Materials Science and Engineering* **708** 012102 doi:10.1088/1757-899X/708/1/012102

[5] Tsapko Yu, Tsapko A and Bondarenko O 2020 Determination of the laws of thermal resistance of wood in application of fire-retardant fabric coatings *Eastern-European Journal of Enterprise Technologies* **2/10 (104)** pp 13–18 doi: 10.15587/1729-4061.2020.200467

[6] Tsapko Yu, Rogovskii I, Titova L, Bilko T, Tsapko A, Bondarenko O and Mazurchuk S 2020 Establishing regularities in the insulating capacity of a foaming agent for localizing flammable liquids *Eastern-European Journal of Enterprise Technologies* **5/10 (107)** pp 51–57 doi: 10.15587/1729-4061.2020.215130

[7] Tsapko Yu, Tsapko A and Bondarenko O 2020 Modeling of thermal conductivity of reed products *IOP Conf. Series: Materials Science and Engineering. Innovative Technology in Architecture and Design* **907** 012057 doi:10.1088/1757-899X/907/1/012057

[8] Berdnyk O Yu, Lastivka O V, Maystrenko A A and Amelina N O 2020 Processes of structure formation and neoformation of basalt fiber in an alkaline environment *IOP Conf. Series: Materials Science and Engineering. Innovative Technology in Architecture and Design (ITAD)* **907** 012036 doi:10.1088/1757-899X/907/1/012036

[9] Roos A and Hugosson M Consumer preferences for wooden and laminate flooring 2008 *Wood Material Science & Engineering* **3** pp 29–37

[10] Myakush B M and Kshyvetskyy B Y 2016 The effect of temperature on shape stability of three-layer parquet boards *Problems of Tribology* **81 (3)** pp 38–42

[11] Råberg U, Edlund M-L, Terziev N and Johan Land C 2005 Testing and evaluation of natural durability of wood in above ground conditions in Europe – an overview *Journal of Wood Science* **51** pp 429–440

[12] Chong J, Yang Z, Juqing C and Lu G 2019 The antibacterial properties and safety of a nanoparticle-coated parquet floor *Coatings* **9 (6)** 403
[13] Živković V, Turkulin H and Valdec N 2011 Surface of finished wood floorings – characteristics and testing pp 201–211
[14] Gorišek Z and Straže A 2011 Comparative analysis of relevant properties of oak- and chestnut wood for solid parquet production pp 43–49
[15] Oliveira M B, Silva J R M, Hein P R G and Lima J T 2019 Establishment of quality classes for hardwood floorings by simulated use *Cerne* **25 (1)** pp 105–109
[16] Acuña L, Sepliarsky F, Spavento E, Martínez1 R D and Balmori J-A 2020 Modelling of impact falling ball test response on solid and engineered wood flooring of two eucalyptus species *Forests* **11 (933)** doi:10.3390/f11090933
[17] Özdöl Koca G, Nihal A and As N 2013 Effect of some factors on solid wood parquet's geometrical stability *Forests* **63** pp 33–42 doi: 10.17099/jffiu.91429
[18] Adachi T 2018 Photocatalytic, superhydrophilic, self-cleaning TiO_2 coating on cheap, light-weight, flexible polycarbonate substrates *Applied Surface Science* **458** pp 917–923.
[19] Foliente G C, Leicester R H, Wang C H, Mackenzie C and Cole I 2002 Durability design of wood construction *For. Prod. Journal* **52** pp 10–19
[20] Herzele S, Herwijnen H, Griesser T, Gindl-Altmutter W, Rößler C and Konnerth J 2020 Differences in adhesion between 1C-PUR and MUF wood adhesives to (ligno)cellulosic surfaces revealed by nanoindentation *International Journal of Adhesion and Adhesives* **98** 102507

Acknowledgements
Authors express their gratitude to the Ministry of Education and Science of Ukraine for financial support of the research, that was performed in the framework of budget funding No. 0121U001007, as well as for the development of the theme of research according to the program of scientific cooperation COST Action FP 1407 "Understanding the modification of wood through an integrated scientific and environmental approach" of the European Union's framework program HORIZON 2020.

Physicochemical and physicomechanical studies of dolomite binder

V I Vinnichenko[1], A N Riazanov[2], R Z Rakhimov[3], A A Riazanov[2] and O V Vinnychenko[4,*]

[1]Cool Clean Researches & Technologies, 37 Rue Sainte Catherine, Le Cannet, 06110, France
[2]Ufa State Petroleum Technological University, 1 Kosmonavtov st., Ufa, 450064, Republic of Bashkortostan, Russia
[3]Kazan State University of Architecture and Engineering, 1 Zelenaya st., Kazan, 420043, Tatarstan, Russia
[4]Kharkiv National University of Civil Engineering and Architecture, 40 Sumska st., Kharkiv, 61002, Ukraine

*E-mail: vvinnichenko@ukr.net

Abstract. To obtain dolomite cement, it is proposed to use screenings of dolomite stone and coal waste. Both physicochemical and physicomechanical studies of cement samples are carried out. It is shown that, in contrast to the existing dolomite binders, the proposed cement has the ability to harden when mixed with water.

1. Relevance

According to literature analysis the magnesia binders have a number of significant advantages, i.e. quite fast setting, quick hardening, effective protection against the harmful effects of electromagnetic and high frequency radiation. There are significant reserves of dolomite stocks in Ukraine, with technological off-quality waste of dolomitic stone screenings being largely accumulating at mining combines, refractory making plants and metallurgical enterprises. Amount of waste in the piles is approaching millions of tons, with the quantity thereof constantly increasing, thereby generally resulting in environmental pollution and ecological problems [1]. On the other hand, the technology for the production of binders from dolomite waste has advantages over the Portland cement technology: lower fuel consumption for heat treatment; reduction in energy and operating costs, since crushing of raw materials and the use of expensive crushing equipment is no longer required; release of land plots used for landfills; the production process uses only one-component raw materials, such as dolomite, which does not require corrective additives; the firing temperature of dolomite cement is lower than that of Portland cement, therefore, the furnaces should not be lined with expensive refractory materials. All this makes it possible to reduce the production cost of dolomite cement at least twice as compared to Portland cement. In addition, the thermal processing of raw materials for the production of dolomite binders produces less greenhouse gases.

2. Problem setting

Currently, there is no unified and generally accepted theory of cement hardening [2,3]. The first theoretical concept of cement hardening was proposed by Le Chatelier. In accordance with his theory (crystallization theory), the cement hydration occurs through the solution [4]. Due to higher solubility of clinker minerals as compared with the formed crystalline hydrates, the latter, forming supersaturated solutions, are released from the cement-water suspension thereby forming a crystalline aggregate. The main position of Le Chatelier's theory of the fact that cement hydration occurs through the mortar is the starting point in current concepts for the hardening of mineral binders. According to the V. Michaelis theory (1893 - colloid theory), water directly interacts with the surface of cement

Content from this work may be used under the terms of the Creative Commons Attribution 3.0 licence. Any further distribution of this work must maintain attribution to the author(s) and the title of the work, journal citation and DOI.
Published under licence by IOP Publishing Ltd

minerals, and the hydration processes proceed without dissolution thereof (topochemical mechanism) [5]. Michaelis did not deny the occurrence of crystalline hydrates formation in the process of cement hardening, but paid less attention thereto. In 1923 A.A. Baikov tried to combine these theories (unified colloid chemical theory) [6]. He identified three periods: dissolution (before the start of setting); colloidation (hydration) and crystallization with the formation of a crystalline aggregate. The main position of Baikov's theory, current theoretical concepts are adhered to, is the ability of a significant part of the binder to hydrate according to the topochemical scheme.

P.A. Rehbinder divides the hardening process into three stages: dissolution of unstable clinker phases in water and separation of crystals; the formation of a coagulation structure; growth and aggregation of crystals [7].

As far as the magnesia binder is concerned, the forming $Mg(OH)_2$ [2] layer is known to prevent from the diffusion of water into MgO grains. The process is strongly accelerated, if salt electrolyte is dissolved in water. For caustic magnesite to form an artificial stone, it is tempered with $FeSO_4$, $MgSO_4$, H_2SO_4, $NaHSO_4$, $FeCl_2$, $ZnCl_2$ and mainly $MgCl_2 \bullet 6H_2O$ solutions [8-18, 46].

Regarding the process of curing magnesia binder M. Sorel [13], and then other researchers [14-18] found that when the magnesia binder hardened with an aqueous solution of magnesium chloride, an artificial stone is formed with different properties depending on the compounds formed. In theory, magnesium chloride reacts with MgO to form oxychlorides like nMgO · MgCl2 mH20. According to various scientists, n ranges from 3 to 7, and m - from 6 to 17 [14-29]. When magnesium oxide is mixed with highly concentrated solutions of MgCl2, the structure of magnesia stone is usually formed mainly by 5- and 3-hydroxychlorides, since the formation and appearance of magnesium hydroxide in it becomes energetically unfavorable.

The composition of the end products in magnesia cement is determined by ratio of initial components, since under insufficient solution content in conditions of high density of structure and under a significant change in MgCl2 concentration due to crystalline hydrates formation, the phase transitions of metastable compounds into stable ones may be stopped at one of the stages, and only $3MgO{\cdot}MgCl_2{\cdot}11H_2O$ or $5MgO{\cdot}MgCl_2{\cdot}13H_2O$ may be the end products, as well as mixtures thereof or mixtures of these oxychlorides with $Mg(OH)_2$ or $MgCl_2$.

In the majority of scientific works devoted to magnesian binder based on dolomite and materials based on it, much attention is paid to the problems of improving the quality of binders and materials [30-52].

The use of saline solutions negates the advantages of dolomite binder production due to the high price of this salt compared to water.

Modern theoretical and practical data suggest research on obtaining a magnesia binder that can harden in water.

3. Research hypothesis

Under aqueous conditions, magnesium oxide hydrates so slowly that this method has not found wide application, but when it is mixed with saline solutions (most often with a solution of MgC12 6H2O), the process is accelerated. An increase in the solubility of the Mg (OH) 2 layer appearing on the MgO surface and the involvement of magnesium oxide in the hydration process can be achieved due to the formation of minerals in the binder during its heat treatment, obtained by joint firing of dolomites and coal waste. The minerals formed during firing play the role of a salt mixture, and the interaction reactions proceed through the formation of calcium and magnesium hydrosilicates, which take part in the hardening process.

4. Theoretical

We have investigated the thermodynamic conditions for the decomposition of dolomite in a mixture with silicon dioxide - SiO2. The changes in the Gibbs free energy (ΔG) in the temperature range from 900 to 1500K are calculated.

The possibilities of forming the following connections are thermodynamically considered: calcium and magnesium oxides (reaction 1); calcium carbonate and magnesium oxide (reaction 2); β-wollastanite and magnesium oxide (reaction 3); α-wollastanite and magnesium oxide (reaction 4); magnesium silicate and calcium oxide (reaction 5); 2Ca SiO2β and magnesium and silicon oxides (reaction 6); 2Ca SiO2γ and magnesium and silicon oxides (reaction 7); clinoestatite (reaction 8); diopside (reaction 9); ackermanite (reaction 10); mervenite (reaction 11); monticellite (reaction 12).

Possible reactions in the mixture of dolomite and coal enrichment wastes are presented below:

1. $CaCO_3 + MgCO_3 + SiO_2 = CaO + MgO + SiO_2 + 2CO_2$
2. $CaCO_3 + MgCO_3 + SiO_2 = CaCO_3 + MgO + SiO_2 + CO_2$
3. $CaCO_3 + MgCO_3 + SiO_2 = \beta CaO\ SiO_2 + MgO + 2CO_2$
4. $CaCO_3 + MgCO_3 + SiO_2 = \alpha CaO\ SiO_2 + MgO + 2CO_2$
5. $CaCO_3 + MgCO_3 + SiO_2 = MgO\ SiO_2 + CaO + 2CO_2$
6. $CaCO_3 + MgCO_3 + SiO_2 = 1/2(2CaO\ SiO_2\beta) + MgO + 1/2SiO_2 + 2CO_2$
7. $CaCO_3 + MgCO_3 + SiO_2 = 1/2(2CaO\ SiO_2\gamma) + MgO + 1/2SiO_2 + 2CO_2$
8. $CaCO_3 + MgCO_3 + SiO_2 = 1/2(2MgO\ SiO_2) + CaO + 1/2SiO_2 + 2CO_2$
9. $CaCO_3 + MgCO_3 + SiO_2 = 1/2(CaO\ MgO\ 2SiO_2) + 1/2CaO + 1/2MgO + 2CO_2$
10. $CaCO_3 + MgCO_3 + SiO_2 = 1/2(2CaO\ MgO\ 2SiO_2) + 1/2MgO + 2CO_2$
11. $CaCO_3 + MgCO_3 + SiO_2 = 1/3(3CaO\ MgO\ 2SiO_2) + 2/3MgO + 2CO_2 + 1/3SiO_2$
12. $CaCO_3 + MgCO_3 + SiO_2 = CaO\ MgO\ SiO_2 + 2CO_2$

Table 1 shows the results of the thermodynamic identification of the Gibbs energy of these reactions.

Table 1. Gibbs energy values (ΔG) of the reactions.

Reaction No	Δ G, kCal/Mol at, K						
	900	1000	1100	1200	1300	1400	1500
1	-3117.40	-10808.88	-18378.13	-25824.91	-33149.16	-40350.89	-47430.16
2	-9470.24	-13368.47	-17189.48	-20931.18	-24591.86	-28170.08	-31664.62
3	-200670.00	-203732.10	-206650.40	-209421.80	-212043.90	-214514.50	-216831.90
4	-201508.40	-204777.40	-207899.00	-210870.70	-213690.30	-216356.20	-218866.80
5	-187658.60	-190663.00	-193522.70	-196235.00	-198797.90	-201209.60	-203468.60
6	-107364.40	-112886.10	-118286.00	-123563.20	-128717.10	-133747.30	-138653.40
7	-111561.80	-117501.50	-123331.10	-129053.00	-134669.40	-140182.40	-145593.80
8	-103807.00	-111089.00	-118537.00	-126149.40	-133924.60	-141886.60	-149959.30
9	-194460.30	-197254.90	-199909.60	-202421.30	-204787.20	-207005.00	-209072.80
10	-31607.10	-39722.50	-47821.10	-55920.30	-64002.70	-72074.90	-80137.50
11	-26189.05	-34010.30	-41812.61	-49598.91	-57385.21	-65149.51	-72876.81
12	-224951.00	-230422.50	-235781.00	-241021.90	-246141.70	-251137.90	-256008.70

The dependence of the Gibbs function change in the temperature range of (900 - 1500) K is represented by a diagram in Figure 1.

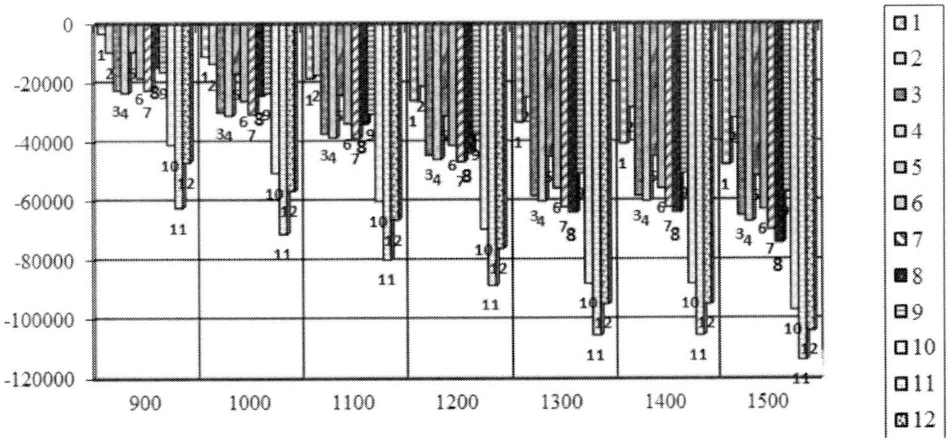

Figure 1. Dependence of Gibbs energy on temperature in 1-12 reactions.

An analysis of the results obtained shows that there is a thermodynamic probability of the formation of compounds during firing, capable of solidifying when mixed in water.

5. Characteristics of original materials

Dokuchaevsky dolomitic stone screenings (DSS) and waste coal (WAC) from Belorechensk processing plant are used as raw stock, with the chemical composition of DSS (mass %) being CaO – 28.89; MgO – 19.02; SiO_2 – 6.37; Fe_2O_3. – 0.85; Al_2O_3 – 1.21; ignition loss – 42.81, and that of WAC (mass %) CaO – 3.8; MgO – 1.3; SiO_2 – 55.3; Fe_2O_3. – 10.9; Al_2O_3 – 20.6; SO_3 – 2.8; K_2O – 2.6; Na_2O – 1.0.

X-ray patterns of dolomite screening and waste coal are shown in Figure 2 and Figure 3, respectively.

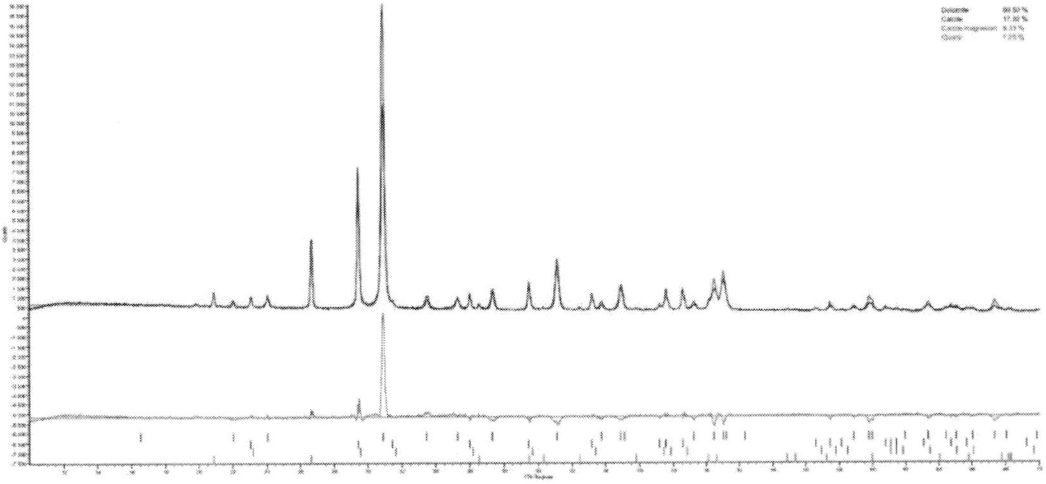

Figure 2. X-ray pattern of dolomitic stone screening.

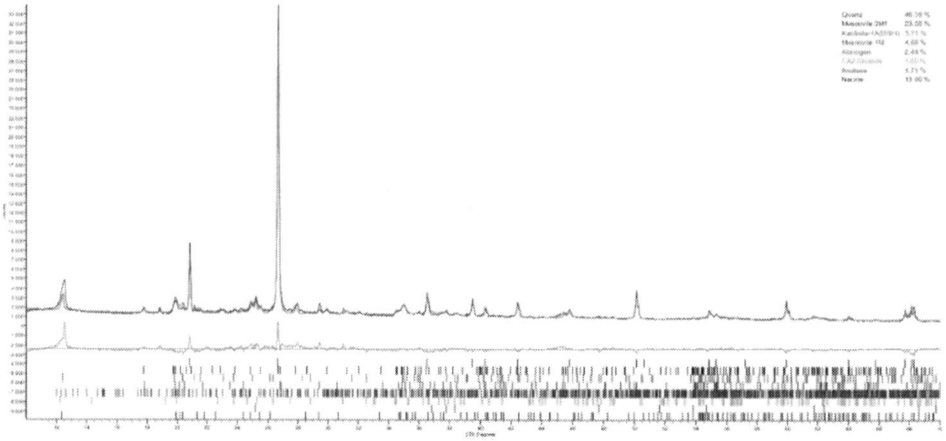

Figure 3. X-ray pattern of waste coal.

6. Experimental results

The prepared composition of DSS to WAC being as 1:1; 1:2, 1:3 was isothermally fired in a muffle furnace at 1000°C for 1 hour. After being fired the resulting material was cooled down in the muffle to 150-180°C for 6-8 hours. Afterwards the material was grounded in a ball mill up to a specific surface area of 3200-3300 cm^2/g.

The X-ray diagram of the obtained dolomite binder is shown in Figure 4.

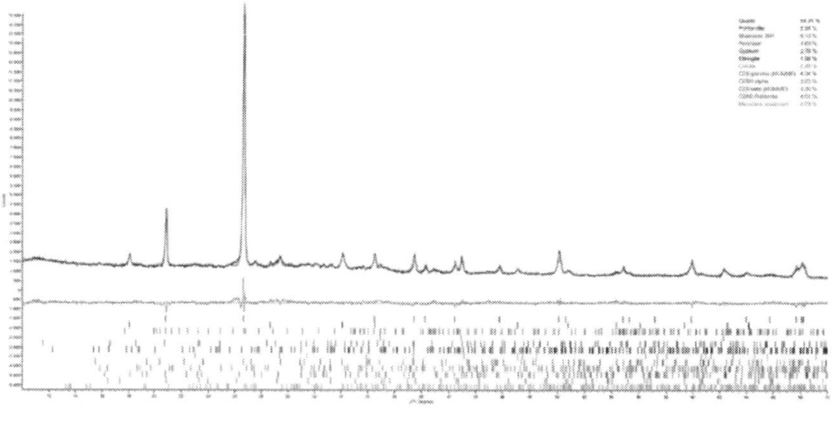

Figure 4. X-ray pattern of dolomite binder.

The results of X-ray analysis show that the binder contains compounds capable of hydrating when tempered with water, and having hydraulic properties. Then the binder and sand mixture was mixed with water. The water-cement ratio was taken considering cone flow of 109-110 mm on the shaker table.

The strength properties were determined with 40x40x160 mm and 31.6x31.6x31.6 mm samples with a binder-to-sand composition as 1:3. The samples were heat- and moisture-treated at 95°C according to 2+8+2h steam curing regime, with the prior exposition thereof to air for 2-3 hours. The physical and mechanical properties of the binder are given in table 2.

Dolomite binder samples with DSS-WAC composition of 1:3 are seen from table 2 to have the best strength properties characterized by strength of 4.0MPa after 28 days of leaving in moist room, and that of 21.0MPa after heat- and moisture-treated.

Table 2. Physical and mechanical properties of dolomite binder.

Mixture composition		True density, kg/m^3	Normal consistency of paste,%	Setting time		Distance b/n Le Chatelier mold indicator stems, mm	Water / cement	Compressive strength, MPa	
DSS	WAC			initial	final			bending	compression
50	50	2520	38	0-38	3-40	3.5	0.40	6.82	14.0
34	66	2480	33	0-40	3-35	2.7	0.45	6.91	16.0
25	75	2390	30	0-46	3-47	2.0	0.40	9.13	21.0

Therefore, the usage of DSS - WAC mixture under the heat treatment enabled to get a binder capable of hardening when mixed with water.

7. Results

- 1. Principles of binding properties found in the dolomite binder when mixed with water are theoretically verified and experimentally confirmed.
- 2. The research in making a dolomite binder by mutual firing of dolomitic stone screenings and waste coal with further grinding thereof to a specific surface area of 3200-3300 cm^2/g. The minerals capable of interacting with water and acting as a saline grout are found in the products of firing.
- 3. The dolomite binder was mixed with water at a water-cement ratio of 0.4-0.45.
- 4. Both physical and mechanical and physical and chemical studies have been carried out. The compressive strength of a dolomite binder in the cement to sand composition as 1:3 after being heat- and moisture- treated is 14-21 MPa.

References

[1] Yermakova E V 2000 Environmental assessment in the area of the Dokuchaev Flux-Dolomite Combine and possible ways to improve it *Donetsk DonNTU* pp 1-11

[2] Pashchenko O O, Serbin V P and Starchevskaya O O 1995 *Binding materials* Kyiv, Higher school p 440

[3] Mchedlov-Petrosyan O P 1998 *Chemistry of inorganic construction materials* (Moscow: Stroyizdat) 304

[4] Le Chatelier H 1919 Cristalloids against colloids in the theory of cements *Trans. Faraday Sos* **14 8**

[5] Michaelis W 1893 Uber Den Portland Cement *J fur Prakt Chemie Chem Zeitung* **17** p 982

[6] Baikov A A 1923 Portland cement and water-hardening cement theory *Technical and Engineering Bulletin* **3 6-7**

[7] Rehbinder P A 1958 Physical-chemical mechanics *Moscow Znanie* pp 321-322

[8] Vinnichenko V I and Ryazanov A N 2013 Cement production from dolomite wastes *Ecology and Industry* **2** pp 111-114

[9] Ryazanov A N, Vinnichenko V I, Shcheblykin S V and Telyatnikova N I 2013 The study of cement from dolomite seed *Visnyk NTU "HPI"* No **64** (1037)

[10] Borisov I N, Vinnichenko V I and Ryazanov A N 2013 Energy efficient construction materials of dolomite and waste coal *Energy efficient technology complexes and equipment for making construction materials. Interuniversity collection of papers* BSTU named after V. G. Shukov Belgorod Issue XX pp 114-117

[11] Ryazanov A N, Vinnichenko V I and Plugin A A 2013 Theoretical arguments for comprehensive application of dolomite and waste coal for the production of construction materials *Collection of scientific papers* Issue **138** Kharkiv, Ukr press pp 77-85

[12] Vinnichenko V and Ryazanov A. 2017 Ecological indices of manufacture of Portland cement

clinker and production of the dolomite clinker *MATEC Web of Conferences* 116 01020

[13] Sorel S 1866 Improved composition to be used as a cement and as a plastic material for molding Various articles *United States Patent Office Patent 53/092* Paris France

[14] Baikov A A 1913 Caustic magnesite, its properties and hardening *Russian Metallurgical Society J* No **3 1** pp 311-334

[15] De Wolff P M and Walter-Levy M L 1953 Hydratations prozesse und Erhartungs eigenschaften in Systemen MgO-MgCl2 *Zement-Kalk-Gips* II **4** pp 125-137

[16] Kasai J, Ichiba M and Nakanara M 1956 Mechanism of the Hydration of Magnesia Cement. *J. of Chem. Soc. of Japan* **63 7** pp 1182–1184

[17] Vyrodov I P and Bergman A G 1959 Problem of magnesia cement hardening *J of Applied Chemistry* **32 4** pp 716–723

[18] Vyrodov I P 1960 Magnesia cement structure formation *J of Applied Chemistry* **33 11** pp 2399-2404

[19] Smirnov B I, Smirnova Ye S and Segalova Ye Ye 1967 Investigation of chemical interaction of magnesium oxide with solutions of magnesium chloride of various concentrations *J of Applied Chemistry* **7 3** pp 505-514

[20] Solovyova Ye S, Smirnov B I, Segalova Ye Ye and Rehbinder P A 1968 Physical and chemical properties of magnesia cement hardening *J of Applied Chemistry* **30 3** pp 754-759

[21] Belyankin D S 1956 *Petrography of technical stone* Moscow Nedra 780

[22] Vayvad A Ya 1971 *Magnesia binders* Riga Science 315

[23] Kaminaskas A Yu 1987 *Construction materials technology based on magnesia raw stock* Riga Mokslas Press 342

[24] Vereshchagin V I, Zyryanova M A, Savinkina M A, Zyryanova V N and Filins S V 1991 Making of a waterproof magnesia binder based on magnesium-containing silicates and Sorel cement *Proc. Int. Conf on Cement Chemistry* Moscow p 76

[25] Zyryanova V N, Berdov G I and Tyuleneva N I 2007 Composite magnesia binders *Current problems in the construction and architecture. Education. Science, Practice: Proc. All-Russian Science and Technology Conf* Samara p 189

[26] Zyryanova V N, Berdov G I and Vereshchagin V I 2009 Physical and chemical processes in hydration hardening of composite magnesia binders *Current problems of production and application of composite building materials: Proc. All-Russian Conf.* Novosibirsk pp 50-52

[27] Berdov G I. Zyryanova V N, Mashkin A N and Khritankov V F 2009 Nanoprocesses in the technology of composite construction materials *Current problems of production and application of composite construction materials: Proc. All-Russian Conf* Novosibirsk pp 10-12.

[28] Zyryanova V N, Berdov G I and Vereshchagin V I 2009 Magnesia binders from sludge of magnesium chloride brines *Current problems of production and application of composite construction materials: Proc. All-Russian Conf.* Novosibirsk pp 168-170

[29] Zyryanova V N, Berdov G I and Vereshchagin V I 2009 Water-resistant composite magnesia binders with the application of natural and technological raw stock *Proc. III (XI) Int. Meeting on Cement Chemistry* Moscow pp 97-100

[30] Kozlova V K, Sweet T F, Dushevina A M, Chelyshev A S and Pimenov A T 2004 Comprehensive application of Taenzinsky deposit dolomites *Construction materials* **1** pp 29-31

[31] Kozlova V K, Dushevina A M and Chelyshev A S 2003 Dolomite-based construction materials of the Taenzinsky deposit *Reliability and durability of construction materials and structures:Proc. III Int. Sci-Tech Conf.* Volgograd: VolgaGASA Part 3 pp 108-110

[32] Kozlova V K, Dushevina A M and Pimenov A T 2001 Features of hydration and hardening of caustic dolomite *Current Problems of Construction Material Science. Seventh Academic Readings of the RAASN* Belgorod pp 223-227

[33] Piven N I 1972 Production and analysis of water-resistant magnesia cement *Author's abstract by*

PhD in Engineering Kharkiv p 20
[34] Rogacheva I N 1975 Research and development of water-resisting magnesia cement technology *Author's abstract by PhD in Engineering* Kharkiv p 22
[35] Ved Ye I, Bludov B F, Zharov Ye F and Piven N I 1972 Production of water-resistant magnesia cement *Proc. Belgorod TISM. Chemistry and chemical technology* Issue 2 pp 38-41
[36] A S 338505 Ved Ye I, Bludov B F and Piven N I 1972 *Magnesia cement*, kl S 04 V 9/04.
[37] Biryuleva D K, Shelikhov N S and Rakhimov R Z 1997 Mechanism of dolomite cement hardening *Abstracts of three academic readings "Current problems of construction materials*" Saransk p 119
[38] Biryuleva D K, Shelikhov N S and Rakhimov R Z 1999 Dolomite cement modified with silicate additives *Abstracts of the Anniversary Int. Sci-Pract. Conf. Construction 99* Rostov-on-Don p 49
[39] Komlev P G, Ivanov L K and Kurkina I D 1992 Production of Melekhopsky deposit dolomite-based magnesia binder *Construction materials waste utilization Abstracts* Penza
[40] Komlev P G and Ivanov L K 1993 Melekhopsky deposit dolomite-based magnesia binder *Resource-saving technologies of construction materials, items and structures.* Abstracts: Belgorod pp 53-54
[41] Erdman S V 1996 Natural magnesium-containing silicates in production of binders and ways to increase the water resistance of magnesia binders: *abstract of Ph.D. thesis in Engineering* Tomsk p 22
[42] Kramar L Ya 2006 The requirements of the standard for magnesia binder to be applied in construction *Construction materials* **1** pp 52-54
[43] Kramar L Ya, Chernykh T N and Trofimov B Ya 2006 Peculiarities of magnesia binder hardening *Cement and application thereof* **09** pp 58-61
[44] Zimich V V and L Ya Kramar 2009 Modified chlorine-magnesium binder for the production of heat-insulating materials *Problems of the construction complex of Russia: proceedings XIII Int. sci-tech Conf* Ufa, USPTU pp 107-109
[45] Shabanova G M, Taranenkova V V, Smal G L and Kuzmenkov E D 2012 Research of hydration products of caustic dolomite-based magnesia binder *Bulletin NTU* Kharkiv: NTU press **32** pp 184-188
[46] Vinnichenko V and Ryazanov A 2018 Energy Efficiency of Binder Application in Concrete *Int. J of Engineering Technology* pp 335-338 **7 4.3**
[47] A A Riazanov et al 2020 Resource saving and energy saving at the simultaneous production of two types of cemen *IOP Conf. Ser.: Mater. Sci. Eng.* **907** 012034
[48] Shinkevich E and Lutskin Y 2007 The Influence of Structure Modification of Silicate Materials after Hardening in Non-autoclave Conditions on Their Coefficient of Heat Conductivity. *Proceeding of International Conference "Alkali Activated Materials – Research, Production and Utilization"* pp 621–635
[49] Shinkevich E S and Zaytsev Y V 2010 Structural durability, deformation properties and fracture mechanics parameters of advanced silicate materials *18th european conference on fracture: fracture of materials and structures from micro to macro scale* Dresden Germany
[50] Shinkevich E 2011 Kinetic-mathematical model of hydration of lime-silica binder, which activated together with a fine-grained filler *Proceeding of 13th International Congress on the Chemistry of Cement* Madrid Spain **351**
[51] Lutskin E and Shinkevich E 2015 Analysis of the relationship between microstructure and properties of activated lime-silica composites on the basis of experimentally-statistical modelling *Tehnički glasnik* **9 1** pp 27-32
[52] Lyashenko T, Barabash I, Shinkevich E, Shcherbina S and Voznesensky V 1998 Experimental Statistical Modelling the Effect of Multi-Fractional Filler on Rheological Indices of Compositions *Progress and Trends in Rheology V* pp 104-105

IOP Conf. Series: Materials Science and Engineering 1164 (2021) 012086 doi:10.1088/1757-899X/1164/1/012086

The transportation of a particle by a vertical auger with a coaxial cylinder which rotate together around the common axis

Tatiana Volina[1,*], Sergiy Pylypaka[1], Olexandr Pavlenko[2], Oleksii Klochko[3] and Iryna Hryshchenko[1]

[1] National University of Life and Environmental sciences of Ukraine, 15 Heroyiv Oborony St., Kyiv 03041, Ukraine
[2] Bohdan Khmelnytsky Melitopol State Pedagogical University, 20 Hetmanska St., Melitopol 72300, Ukraine
[3] Sumy State Pedagogical University named after AS Makarenko, 87 Romenskaya St., Sumy 40002, Ukraine

*E-mail: t.n.zaharova@ukr.net

Abstract. Differential equations of relative movement of a particle on the periphery of a vertical auger bounded by a movable coaxial cylinder are compiled. Both surfaces form a single whole and rotate around a common axis. A partial case, when the surfaces are stationary, is considered. A qualitative analysis of the obtained equations is made and on this basis, the regularities of the particle movement along the helical line – the curve of the intersection of the auger with the cylinder are found. Structural and kinematic parameters in which the particle moves upwards during sliding along the helical line, or falls downwards are found. The relative and absolute trajectories of the particle movement are constructed.

1. Introduction

To begin with, blades are used for the ordered movement of particles on the surface of rotation, which performs a rotational movement around the vertical axis. If a cone without blades, which rotates around its vertical axis, is taken, the particle will slide on it, rising at the same time up. Juicers, in which the cone is used as a sieve, work on this principle [1].

If the particle needs to be accelerated and to be given the desired direction of its leaving the cone, flat vertical blades are used. Actually, centrifugal scattering devices [2] work on this principle (figure 1,a). In this case, the trajectory of the relative movement of the particle is a straight line – the line of intersection of the cone and the blade.

In addition, there are juicers with a cylindrical sieve [3], but the crushed particles must be removed manually after squeezing the juice because they do not rise upward. Clearly, a possible variant for lifting the particle is to use a curved blade in the form of a strip of helical conoid – auger (figure 1,b). Lifting of technological material by augers inside a cylindrical casing is widely applied in various mechanisms, however, the cylindrical casing thus is motionless.

Based on this foregoing, the aim of this research was to investigate the laws of movement of material particles inside the construction, which is a single whole of the cylinder and the coaxial strip of the helical conoid, and which rotates around a common vertical axis.

2. Literature review

It is a well-known fact, that the movement of the technological material along the surface depends on the shape of this surface. To accurately describe the movement of a material along a surface, it is necessary to take into account inertia forces from its rotation, which is extremely difficult to do. Therefore, in many cases, these forces are neglected if body sizes or angular velocities of its rotation are small [4]. A great deal is being written and said about the fact, that particles of material in contact with

Content from this work may be used under the terms of the Creative Commons Attribution 3.0 licence. Any further distribution of this work must maintain attribution to the author(s) and the title of the work, journal citation and DOI.
Published under licence by IOP Publishing Ltd

the surface may have a different origin: particles in a moving stream, mechanical particles, particles of a liquid or a gas [5, 6]. The movement of particles that are in contact with moving spiral working bodies, considered in the articles [7, 8]. Furthermore, the movement of particles on the surface of a rotating vertical helicoid is studied in [9]. In many cases, to simplify calculations, it is customary to consider the body as a material particle. The articles [10, 11] are the closest to the research topic. The first article [10] considers the movement of a particle along a gutter in the form of a helical surface under the action of its own weight. The second article [11] – the transportation of bulk material on the example of a single particle inside a vertical stationary cylinder by means of an auger rotating in this cylinder.

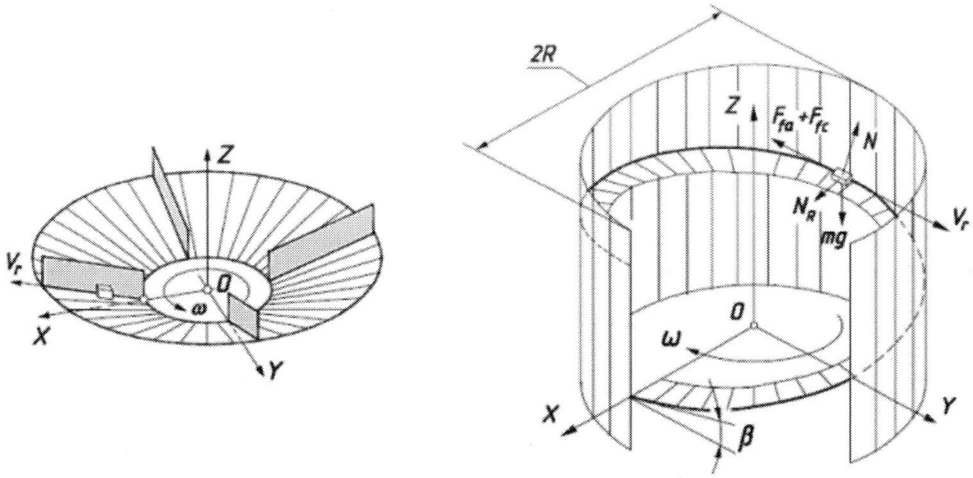

(a) (b)

Figure 1. Designs of devices for transportation of the particles, consisting of a combination of two surfaces: a) conical surface with flat vertical blades; b) cylindrical surface with a curved blade in the form of a strip of the helical conoid.

3. Research methodology

In the article [12] it is shown that when a particle moves under its own weight on the surface of a stationary vertical helical conoid, it moves away from its axis. This is due to the fact that the particle moves along a curved trajectory, resulting in a centrifugal force that causes the particle to move to the periphery. Thus, the particle will meet the surface of the cylinder. The common line of the conoid and the cylinder (the line of their intersection) is a helical line (in figure 1 it is shown in a thickened line). The particle will slide along this helical line, having simultaneous contact with the surface of the conoid (auger) and cylinder. Compilation of a mathematical model of particle movement begins with the parametrical equations of the helix – the trajectory of the relative movement of the particle. Due to the fact that in the absence of rotation of the surfaces the particle can move down under the action of its own weight, let us write the parametrical equations of the helix, the direction of the construction of which corresponds to the direction of the descent of the particle:

$$x = R\cos\alpha; \quad y = R\sin\alpha; \quad z = -R\alpha tg\beta, \tag{1}$$

where R is the radius of the limiting cylinder (a constant value);
β is the angle of the raising of the helix (a constant value);
α is the angle of rotation of the point of the helix around its axis (an independent variable).

When the particles move along the helical line (1), there are reaction forces N and N_R, directed along the normal to the conoid and the cylinder (figure 1,b). To determine the direction of this normal, the equations of these surfaces should be known.

The surface of the auger is formed by a set of rectilinear generatrices which are parallel to the horizontal plane, one end of which passes through the helical line (1), and the other is directed to the axis of the auger. Based on the method of formation of the auger, let us write its parametrical equations:

$$X=(R-u)\cos\alpha;\quad Y=(R-u)\sin\alpha;\quad Z=-R\alpha tg\beta, \qquad (2)$$

where u is the length of the rectilinear generatrix of the auger (the second independent variable of the surface).

Counting the length of the generatrix starts from the helix. When $u=0$ the parametrical equations of the helix (1) are obtained. To make the surface equations different from the line equations, lowercase letters are used for the line and uppercase letters – for the surface.

The vertical auger rotates around its axis with an angular velocity ω. The particle is located on the helical line (1) and is simultaneously in contact with the moving surfaces of the auger and cylinder (figure 1,b). The following forces are applied to it: the force of weight mg (m is the mass of a particle, g=9.81 m/s^2 is an acceleration of gravity), reaction N of a conoid, which is directed along normal to its surface, reaction N_R, which is directed along normal to a surface of the cylinder, and the friction force F_f (figure 1,b). Assume that at $\omega=0$ and at a sufficient value of the angle β the particle slides down, then the friction forces on the surface of the auger F_{fa} and on the surface of the cylinder F_{fc} are directed in the opposite direction of the sliding V_R (figure 1,b). The vectors of these forces will be tangent to the trajectory of relative movement (the helical line along which the particle slides), so $F_f=F_{fa}+F_{fc}$.

The equation of movement should be composed in the form $m\vec{a}=\vec{F}$ where m is the mass of the particle, $\vec{a}$ is the vector of absolute acceleration, and $\vec{F}$ is the resulting vector of the forces applied to the particle, which were listed earlier. The vector equation is described in projections on the coordinate axis $OXYZ$.

If in equations (1) the variable α is taken as time-dependent t ($\alpha=\alpha(t)$) then this internal dependence will specify the law of sliding of the particle along the helix (the law of relative movement). Therefore, under this condition, equations (1) are equations of relative movement.

The auger with the cylinder rotates around its axis with an angular velocity ω. During time t, they rotate at the angle $\alpha_a=-\omega t$ and move along the helix by a certain distance according to equations (1). The direction of rotation of the auger with the cylinder is selected so that the particle during sliding along the helix could rise in absolute movement. Let us rotate the helix (1) around the axis OZ at the angle $\alpha_a=-\omega t$ according to the known formulas of rotation:

$$x_a=R\cos\alpha\cos(-\omega t)-R\sin\alpha\sin(-\omega t)=R\cos(\omega t-\alpha);\quad y_a=R\cos\alpha\sin(-\omega t)+R\sin\alpha\cos(-\omega t)=-R\sin(\omega t-\alpha);\quad z_a=-R\alpha tg\beta, \qquad (3)$$

The parametrical equations (3) take into account two rotations: at the angle $\alpha=\alpha(t)$ in relative movement and at the angle $\alpha_a=-\omega t$ in transient motion, so they are the equations of absolute movement of the particle.

The equations (3) should be differentiated by the time t and projections of the absolute velocity vector can be obtained:

$$\dot{x}_a=-R(\omega-\dot{\alpha})\sin(\omega t-\alpha);\quad \dot{y}_a=-R(\omega-\dot{\alpha})\cos(\omega t-\alpha);\quad \dot{z}_a=-R\dot{\alpha}tg\beta. \qquad (4)$$

By differentiating equations (4) the projections of the absolute acceleration w of the particle can be received:

$$\begin{aligned} \ddot{x}_a &= R\ddot{\alpha}\sin(\omega t-\alpha)-R(\omega-\dot{\alpha})\cos(\omega t-\alpha); \\ \ddot{y}_a &= R\ddot{\alpha}\cos(\omega t-\alpha)+R(\omega-\dot{\alpha})\sin(\omega t-\alpha); \\ \ddot{z}_a &= -R\ddot{\alpha}tg\beta, \end{aligned} \tag{5}$$

To decompose the vector equation $m\vec{a}=\vec{F}$ in the projections on the axis of the coordinate system *OXYZ*, the direction of the forces $F_{fa}+F_{fc}$ should be determined (figure 1, b). They are directed in the opposite direction of the vector of relative velocity V_R. Projections of this vector are found by differentiation equations (1):

$$\begin{aligned} \dot{x} &= -R\dot{\alpha}\sin\alpha; \\ \dot{y} &= R\dot{\alpha}\cos\alpha; \\ \dot{z} &= -R\dot{\alpha}tg\beta. \end{aligned} \tag{6}$$

The relative velocity V_R of the movement of the particle is found:

$$V_R=\sqrt{\dot{x}^2+\dot{y}^2+\dot{z}^2}=\frac{R\dot{\alpha}}{\cos\beta}. \tag{7}$$

The projections of the unit directing vector of the relative velocity are found by dividing the projections (6) of this velocity by its absolute value (7):

$$V_{Rx}=-\cos\beta\sin\alpha;\ V_{Ry}=\cos\beta\cos\alpha;\ V_{Rz}=-\sin\beta; \tag{8}$$

In fact, the projections of the vector (7) are written without taking into account the rotational movement of the helix. All forces are projected on the fixed coordinate system *OXYZ*. In order for the force $F_{fa}+F_{fc}$ to be applied at the point of location of the particle, the projections (8) must be rotated at an angle $(-\omega t)$ around the axis Oz. After that, they take the form:

$$\{\cos\beta\sin(\omega t-\alpha);\quad \cos\beta\cos(\omega t-\alpha);\quad -\sin\beta\}. \tag{9}$$

Let us find the direction of action of the reaction forces of the surfaces of the auger and cylinder.

The direction of the normal to the surface is determined by the vector product of two vectors passing through a point on the surface (the point where the particle is located) and tangent to the coordinate lines passing through this point. The vector of the normal to the surface of the auger should be found. The vectors tangent to the coordinate lines of the surface are partial derivatives of equations (2) of the auger:

$$\begin{aligned} &\frac{dX}{d\alpha}=-(R-u)\sin\alpha; && \frac{dX}{du}=-\cos\alpha; \\ &\frac{dY}{d\alpha}=(R-u)\cos\alpha; && \frac{dY}{du}=-\sin\alpha; \\ &\frac{dZ}{d\alpha}=-Rtg\beta; && \frac{dZ}{du}=0. \end{aligned} \tag{10}$$

Therefore, the vector product of vectors (10) is:

$$\begin{vmatrix} X & Y & Z \\ -(R-u)\sin\alpha & (R-u)\cos\alpha & -Rtg\beta \\ -\cos\alpha & -\sin\alpha & 0 \end{vmatrix}=\{Rtg\beta\sin\alpha;\ Rtg\beta\cos\alpha;\ R-u\}. \tag{11}$$

The normal vector (11) is not a unit one. Its position on the surface is determined by two internal coordinates: u and α. The normal to the surface of the auger on the helical line at the location of the particle should be known, i.e. at $u=0$. Obviously, it is necessary to substitute $u=0$ in the expression of the vector (11) and to reduce it to a unit one:

$$\{-\sin\beta\sin\alpha;\quad \sin\beta\cos\alpha;\quad \cos\beta\}. \tag{12}$$

Similarly, the normal to the cylindrical casing can be found. However, its projections can be found much easier. From figure 1, it would be seen that the normal to the cylinder is parallel to the horizontal plane and directed to its axis. The result can be obtained:

$$\{-\cos\alpha;\quad -\sin\alpha;\quad 0\}. \tag{13}$$

Due to the above reasons, the unit vectors (12), (13) should be rotated at the angle $(-\omega t)$. After that, the unit vector of the normal to the surface of the auger at the location of the particle will be written:

$$\{\sin\beta\sin(\omega t-\alpha);\quad \sin\beta\cos(\omega t-\alpha);\quad \cos\beta\}. \tag{14}$$

The unit vector of the normal to the cylindrical casing (13) after rotation at an angle $(-\omega t)$ takes the form:

$$\{-\cos(\omega t-\alpha);\quad \sin(\omega t-\alpha);\quad 0\} \tag{15}$$

Doubtless, the directions of action of the applied forces are defined: for the friction force $F_{fa}+F_{fc}$ – by a unit vector (8), taken with the opposite sign; reactions of the surfaces N of the auger and N_R of the cylindrical casing – by unit vectors (14) and (15), respectively. The last force (the force of gravity of the particle mg) is directed downwards, so the unit vector is given by projections:

$$\{0;\quad 0;\quad -1\} \tag{16}$$

The value of the friction forces F_{fa} and F_{fc} is determined by the value of the surface reaction and the corresponding coefficient of friction: $F_{fa}=fN$, $F_{fc}=f_RN_R$, where f and f_R are the coefficients of friction of the particle on the surface of the auger and of the cylindrical casing, respectively. Therefore, the friction force $F_f=F_{fa}+F_{fc}=f_RN_R+fN$.

Hence, the vector equation $m\vec{a}=\vec{F}$ should be decomposed in the projections on the axis of the coordinate system $OXYZ$ taking into account the directions (9), (14), (15), (16) and the corresponding forces $F_f=f_RN_R+fN$, N, N_R and the force of weight mg:

$$\begin{aligned} m\ddot{x}_a &= -(f_RN_R+fN)\cos\beta\sin(\omega t-\alpha)+N\sin\beta\sin(\omega t-\alpha)-N_R\cos(\omega t-\alpha);\\ m\ddot{y}_a &= -(f_RN_R+fN)\cos\beta\cos(\omega t-\alpha)+N\sin\beta\cos(\omega t-\alpha)+N_R\sin(\omega t-\alpha);\\ m\ddot{z}_a &= (f_RN_R+fN)\sin\beta+N\cos\beta-mg. \end{aligned} \tag{17}$$

By substitution in (17) the expressions of absolute acceleration from (5) a system of three differential equations with three unknown functions: $\alpha=\alpha(t)$, $N=N(t)$ and $N_R=N_R(t)$ can be obtained. Let us solve the system regarding $\ddot{\alpha}, N, N_R$:

$$\ddot{\alpha}=\frac{g\cos\beta}{R}(\sin\beta-f\cos\beta)-f_R(\omega-\dot{\alpha})^2\cos\beta \tag{18}$$

$$N=m\cos\beta. \tag{19}$$

$$N_R=mR(\omega-\dot{\alpha})^2. \tag{20}$$

The differential equation (18) does not depend on (19) and (20), so it can be solved separately, after which reactions (19) and (20) can be found. Furthermore, equation (18) has an analytical solution, but even without a solution, some important conclusions can be done based on its qualitative analysis.

Generally, it can be assumed that over time the movement of the particle stabilizes, so the angular sliding velocity $\dot{\alpha}$ will be constant and the acceleration is $\ddot{\alpha}=0$. After substitution in (18) $\ddot{\alpha}=0$ and solving regarding $\dot{\alpha}$ the expression can be obtained:

$$\dot{\alpha} = \omega \pm \left[\frac{g}{Rf_R} (\sin\beta - f\cos\beta) \right]^{1/2} \quad (21)$$

4. Results
It can be concluded that if the surfaces are stationary (ω=0), then the sliding of the particle is possible, provided that the subroot expression is greater than zero. It becomes equal to zero at f=tgβ. This means that the angle of elevation of the helix is equal to the angular friction of the particle on the surface of the conoid. If an angle β is less than the angle of friction, the particle does not slide on the surfaces of both stationary and moving. During the rotation of the surfaces with an angular velocity ω, the particle is pressed under the action of centrifugal force to the cylinder and rotates with it, i.e. "sticks". The turning point occurs when the angle β is greater than the angle of friction. In this case, the particle will move down on stationary surfaces with an angular velocity, the value of which is determined by the square root of expression (21).

Indeed, if the surfaces have an angular velocity ω, then the angular velocity of the relative movement $\dot{\alpha}$ of the particle will be determined by the formula (21) with the sign "–" before the root, i.e. the difference between the angular velocity of rotation of the surfaces and the angular velocity of the particle sliding along the stationary surfaces. Let us denote $\omega_l = \left[\frac{g(\sin\beta - f\cos\beta)}{Rf_R} \right]^{1/2}$ and then write: $\dot{\alpha} = \omega - \omega_0$ where ω_0 is the velocity of sliding of the particle on stationary surfaces. For $\omega < \omega_0$ $\dot{\alpha}$ will be negative and the particle will fall down. As the angular velocity ω of the surface rotation increases, the lowering velocity will decrease. When the angular velocities become equal (ω=ω_o), the particle "hangs". Besides, with a further increase of the angular velocity ω (i.e., at ω>ω_o), the particle will rise, and the velocity of the raising will be directly proportional to the difference ω–ω_o. Thus, the transportation of the particle up is impossible under three conditions: 1) the angle β of the raising of the helical line is less than the angle of friction – the particle "sticks"; 2) the angle β is greater than the angle of friction, ω<ω_o – the particle is descending; 3) the angle β is greater than the friction angle, ω=ω_o – the particle "hangs", i.e. the angular velocities of the rotation of the surfaces and the sliding of the particle are equal in value and opposite in sign and the absolute velocity of rotation is equal to zero.

As a consequence, note some differences in the movement of the particle on an inclined plane and on the described surfaces, when f=tgβ, i.e. the angle of inclination of the plane and the angle of the raising of the helix are equal to the angle of friction. In this case, the particle will move on the plane with a constant velocity V_R, the value of which is set at the beginning of the movement. The particle on stationary surfaces will not move, because a centrifugal force arises, which causes the appearance of friction force. However, if the surface of the cylinder is absolutely smooth (i.e. f_R=0), then according to the differential equation (18) the movement of the particle on the surfaces is also possible with a given constant velocity V_R. If f_R≠0, then the particle on a stationary surface will remain at rest, and if you give it the initial velocity V_R, then over time it will stop. Moreover, if the surfaces are given the angular velocity ω, the particle can be either in the "sticking" mode or in the "hanging" mode. The "sticking" mode corresponds to the initial velocity V_R=0. The "hang" mode corresponds to such initial velocity V_R, which will provide an angular velocity of the sliding of the particle is equal to the angular velocity of the rotating of the surfaces. Then the absolute angular velocity of rotation of the particle is equal to zero and according to (20) the reaction N_R of the cylinder is absent, therefore, there is no corresponding friction force in equation (18).

As was already mentioned, the case (ω=ω_o) is the boundary between the lowering and raising of the particle. Moreover, at ω<ω_o the particle is moving down, at ω>ω_o the particle is moving up. When ω=ω_o the height of location of the particle does not change, and it can be both in the mode of "sticking" and in the mode of "hanging".

It should be noted that the relative and absolute trajectories of the particle movement were constructed by equations (1) and (3). Let the coefficient of friction f=0.3. It corresponds to the angle of friction β=arctg$f$=16.7°. In order to avoid sticking, the lifting angle of the helix must be greater than the angle of friction. Assume β=20°. At R=0.1 m, f_R=0.3 it can be found: $\omega_0 = \left[\frac{g(\sin\beta - f\cos\beta)}{Rf_R}\right]^{1/2}$=4.43 s^{-1}. It is the angular velocity of the descent of the particle on fixed surfaces (ω=0). It can be seen, that the relative and absolute trajectories coincide (figure 2,a). In figure 2,b the relative and absolute trajectories at ω=2 s^{-1} are constructed. The particle descends, but the velocity of descent has decreased. It is clear, that at $\omega=\omega_o$ the movement of the particle in the vertical direction is absent at all. In figure 2,c the relative and absolute trajectories at ω=10 s^{-1} are constructed. The particle rises and with increasing ω the velocity of the raising increases proportionally.

Accordingly, in figure 2,d relative and absolute trajectories with the previous parameters and a reduced coefficient f_R=0.2 are constructed. The velocity of the particle raising decreased as the value of ω_o was changed. With a further decrease in f_R, the raising may change to lowering, and at f_R=0 the particle will move down uniformly accelerated.

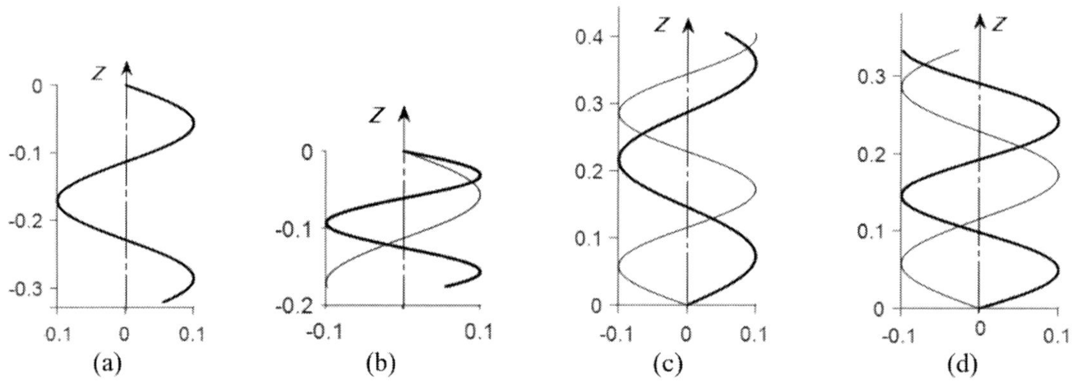

Figure 2. Absolute (thickened line) and relative trajectories of the particle movement during *2 s* period on surfaces with the following parameters: $R=0.1$ *m*, $\beta=20°$, $f=0.3$: a) $f_R=0.3$, $\omega=0$; b) $f_R=0.3$, $\omega=2$ s^{-1}; c) $f_R=0.3$, $\omega=10$ s^{-1}; d) $f_R=0.2$, $\omega=10$ s^{-1}.

The reaction of the surface of the auger (19) and the cylinder (20) are constant. By substituting in (20) $\dot{\alpha} = \omega - \omega_0$ one can obtain: $N_R = mR\omega_0^2$ Therefore, the value of the reaction of the cylinder does not depend on the angular velocity of its rotation and is determined by the angular velocity of the particle sliding on fixed surfaces.

5. Conclusions

If the angle of the raising of the helical line (the curve of the intersection of the cylinder with the helical conoid) is less than the angle of friction of the particle on the conoid, then the transportation of the particle is impossible both up and down. This applies to both stationary and movable surfaces that rotate around a common axis. The particle rotates together with the surfaces, i.e. "sticks". If the angle of the raising of the helical line is equal to the angle of friction, then transportation is also impossible and the particle can either "stick" or "hang". The last variant means that the particle slides along the helical line with an angular velocity of the rotation of the surfaces, but the angular velocities have the opposite sign. In absolute movement, the particle remains stationary. The mode of "sticking" or "hanging" depends on the initial conditions.

If the angle of the raising of the helix is greater than the angle of friction, the transportation of the particle is possible both up and down. In this case, the angular velocity ω_o of the particle during its descent on fixed surfaces is important. If the angular velocity ω of surface rotation is less than ω_o, then the particle will move downwards. As ω increases, the velocity of the descent of the particle will decrease. When $\omega=\omega_o$ the particle "hangs", when $\omega>\omega_o$ it moves upwards. The velocity of the raising of the particle is directly proportional to the increase of the angular velocity ω of the rotation of the surfaces. Prospects of further researches are the experimental verification of obtained theoretical results.

References

[1] Changzhong Wu, Fan Ge, Guangchao Shang, Guitao Wang, Mingpeng Zhao, Liang Wu and Hengshuai Guo 2021 Design of screw type automatic apple juicer *Journal of Physics: Conference Series* **1750** 012042

[2] Zhilenko D, Krivonosova O and Gritsevich M 2019 New type of centrifugal instability in a thin rotating spherical layer *Journal of Physics: Conference Series* **1163** 012011-1-012011-5

[3] Aremu, Ademola K and Ogunlade Clement A 2016 Development and evaluation of a multipurpose juice extractor *New York Science Journal* **9(6)** pp 7–14

[4] Liaposchenko O, Pavlenko I and Nastenko O 2017 The model of crossed movement and gas-liquid flow interaction with captured liquid film in the inertial-filtering separation channels *Separation and Purification Technology* **173(1)** pp 240–243

[5] Golub G, Szalay K, Kukharets S and Marus O 2017 Energy efficiency of rotary digesters *Progress in Agricultural Engineering Sciences* **13(1)** pp 35–49

[6] Kobets A, Ponomarenko N and Kharytonov M 2017 Construction of centrifugal working device for mineral fertilizers spreading *INMATEH – Agricultural Engineering* **51(1)** pp 5–14

[7] Trokhaniak O, Hevko R, Lyashuk O, Dovbush T, Pohrishchuk B and Dobizha N 2020 Research of the of bulk material movement process in the inactive zone between screw sections *INMATEH–Agricultural Engineering* **60(1)** pp 261–268

[8] Hevko R, Zalutskyi S, Hladyo Y, Tkachenko I, Lyashuk O, Pavlov O, Pohrishchuk B, Trokhaniak O and Dobizha N 2019 Determination of interaction parameters and grain material flow motion on screw conveyor elastic section surface *INMATEH–Agricultural Engineering* **57(1)** pp 123–134

[9] Pavlenko I, Liaposhenko A, Ochowiak M and Demyanenko M 2018 Solving the stationary hydroaeroelasticity problem for dynamic deflection elements of separation devices *Vibrations in Physical Systems* 29(2018026)

[10] Pylypaka S, Nesvidomin V, Zaharova T, Pavlenko O and Klendiy M 2020 The investigation of particle movement on a helical surface *Lecture Notes in Mechanical Engineering* pp 671–681

[11] Nnamdi U, Onyejiuwa C and Ogbuke C 2020 Review of orange juice extractor machines *Advances in Science, Technology and Engineering Systems Journal* **5(5)** pp 485–492

[12] Pylypaka S F, Klendii M B and Klendii O M 2017 Particle motion over the surface of a rotary vertical axis helicoid *INMATEH–Agricultural Engineering* **51(1)** pp 15–28

IOP Conf. Series: Materials Science and Engineering 1164 (2021) 012087 doi:10.1088/1757-899X/1164/1/012087

Characteristics of the limit state of the rope

P Vovk[1,*] and I Chaiun[1]

[1]Odessa National Polytechnic University, 1, Shevchenko Avenue, Odessa 65044, Ukraine

*E-mail: vovk.pavel.1995@gmail.com

Abstract. According to the Rules, in the calculation of ropes for strength, one characteristic is used, and it is the same regardless of the loading scheme. In the calculation method for permissible stresses, two strength characteristics are used: the yield strength or the tensile strength of the material of the part. It's proposed the method for the theoretical determination of the characteristics of the ultimate elastic P_e and ultimate elastic-plastic state P_u under four loading schemes, and taking into account the operational breaks of the wires (two characteristics of tension with winding on a drum with a weight in guides and with a free suspension and two when stretching a straight rope.) The ultimate elastic state is determined by the longitudinal force in the section of the rope, at which the longitudinal deformation of one of the wires reaches the yield point ε_T according to the schematized wire tension diagram $\sigma - \varepsilon$. The criterion for the ultimate elastic-plastic state of the rope is the achievement of one of the wires of the ultimate uniform deformation according to the diagram $\sigma - \varepsilon$. It is assumed that on the basis of the characteristics P_e of the stretching of the rope according to the ultimate elastic state, it is possible to construct a method for calculating the strength of the ropes, which will have advantages over the existing one. Advantage arguments:
– strength characteristics P_e are directly related to the rope loading scheme
– the values P_e take into account all geometric parameters of single and double lay, as well as deformation properties of wires
– characteristics P_e allow to take into account the effect of wire breaks during the operation of the ropes.

1. Formulation of the problem.

The general essence of strength calculation methods is a comparison of two states of an element of the structure: ultimate and nominal, interconnected by a standard safety factor $[n]\geq 1$. The guarantee of strength consists in the fact that the actual strength state of an element of the structure, changing during operation, will not become ultimate (will not exceed it). This is ensured by the standard safety factor.

The existing methods of strength calculations use different criteria for the nominal and ultimate states. In the method of calculation for "permissible stresses", the ultimate state is the state in which the stress in the element reaches the strength characteristic of its material: the yield point σ_T or ultimate strength σ_B. With regard to normal stress, the strength condition has the form.

$$\max\sigma \leq [\sigma] = \frac{\sigma_L}{[n]}, \qquad (1)$$

where $[\sigma]$ is the so-called allowable stress;

$\sigma_L = \sigma_T$ or $\sigma_L = \sigma_B$ is ultimate stress.

Content from this work may be used under the terms of the Creative Commons Attribution 3.0 licence. Any further distribution of this work must maintain attribution to the author(s) and the title of the work, journal citation and DOI.
Published under licence by IOP Publishing Ltd

On the diagram $\sigma - \varepsilon$ (Figure 1), the meaning of condition (1) is determined by three stresses: allowable $[\sigma]$ ultimate σ_L, and the third stress – current $0 \le \max\sigma \le \sigma_L$, which, under the specified operating conditions, can approach the ultimate one σ_L but will not exceed it. This is guaranteed by the standard $[n]$ safety factor.

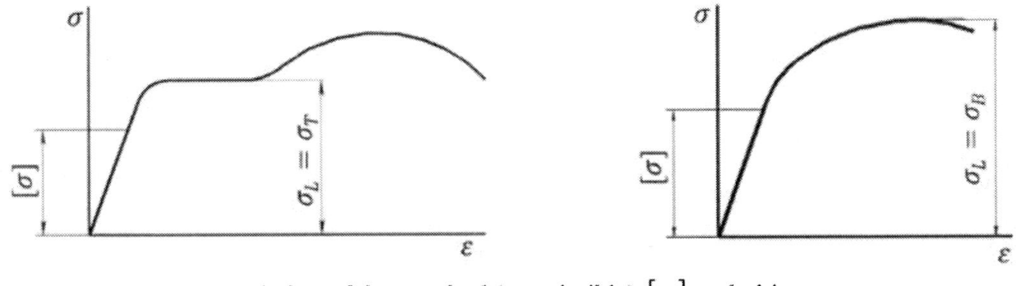

Figure. 1 Relation of the standard (permissible) $[\sigma]$ and ultimate stresses σ_L.

The term "allowable stress" does not quite accurately reflect the essence of the method. More precisely, the term "nominal stress" would answer the essence.

In the method of calculating the loading capacity the nominal state is assumed to correspond to the passport data, for example, the lifting capacity of the crane. The ultimate state is the load capacity of a part or structure as a whole (this is a load, the excess of which leads to destruction).

In work [1] the strength calculation of ropes is mistakenly called the "method of permissible stresses". In fact, this is a calculation of the loading capacity.

In the regulated *Rules for the calculation of ropes*, one characteristic is used, which in [2] is called the “breaking force of the rope as a whole”, in [3] - *Mindest bruchkraft (minimum breaking force of the rope*). Generalized essence of these characteristics

$$P_H = KP_{C,} \tag{2}$$

where $K = 0{,}83$ according to [2] and $K = 0{,}75 - 0{,}90$ according to [3] (the lower value refers to a three-layer double-lay rope, the larger value to single-layer structures);

P_c is the total breaking force of all wires of the rope (where $P_c = F_c\sigma_в$ (F_c is the total cross-sectional area of the wires, $\sigma_в$ is their ultimate strength).

In essence $P_н$, roughly represents the loading capacity of the rope under tensile in the guides. It is known that such a characteristic does not take into account the loading scheme of the rope, its structure and the plastic properties of the wires. At the same time reliability is ensured due to large values of standard safety factors (from 3 to 10 and more [1-3]).

We suggest, by analogy with the method of "permissible stresses" in the calculation of ropes, to use two characteristics of the ultimate state: the tensile force of the rope, corresponding to the ultimate elastic state P_e and the tensile force, equal to the bearing capacity of the rope $P_н$.

2. Presentation of the main material.

The purpose of this work is the analytical determination of the force of the ultimate elastic state P_e, deformations of the rope and stresses in the wires in this state, as well as the loading capacity $P_н$ of the rope. These studies refer to both the first loading period and the operational state associated with the appearance of wire breaks.

Four schemes of rope loading are considered, two of which are tension with coiling onto a drum (Figure 1, a, b). The other two are stretching of a straight rope. They can be considered as special case of the first two (formally, they are obtained with a drum diameter $D_б \to \infty$).

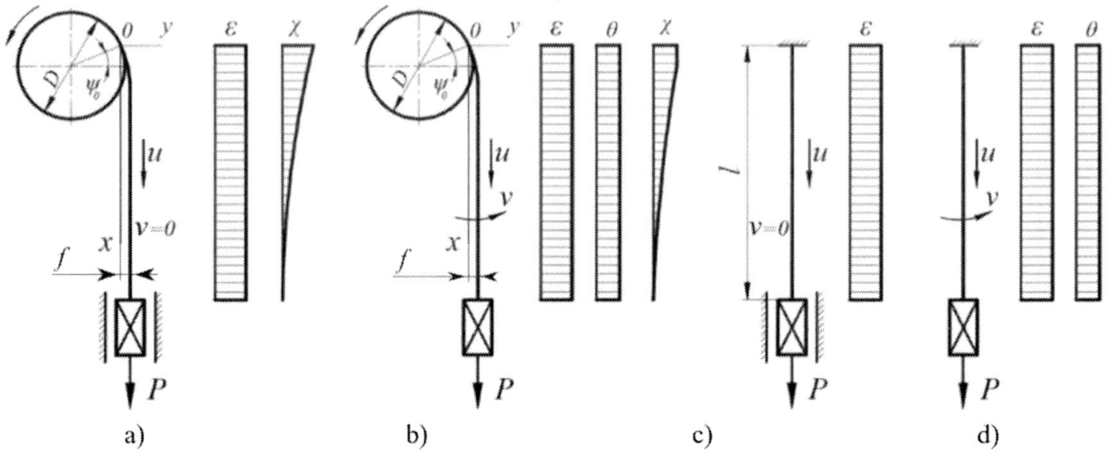

Figure 2. Rope loading schemes
a) stretching with bending in the guides; b) stretching with bending with free suspension of the load; c) pure stretching; d) free stretching.

2.1. Parameters of the curved axis of the rope. Curved axis equation (Figure 2, a)

$$y = \varphi(1 - e^{-\kappa x}); \ f = G_{33}/PR; \tag{3}$$

functions of curvature and rotation angle of rope sections

$$\chi = R^{-1}e^{-\kappa x}; \quad \psi = (G_{33} / PR^2)^{0.5} e^{-\kappa x}, \tag{4}$$

where G_{33} – is the bending stiffness of the rope section in the winding plane;

$\kappa = (P / G_{33})^{0.5}$; $R = 2(D + d_к)$; $d_к$ – rope diametr.

2.2. Deformed state of the rope. It is determined by 4 deformations [5]:

$$|DK| = |\varepsilon\ \theta\ \chi \zeta| = |G|^{-1}|F|. \tag{5}$$

Here the stiffness matrix of the rope [4]:

$$|G| = \begin{vmatrix} G_{11} & G_{12} & G_{13} & G_{14} \\ G_{21} & G_{22} & G_{23} & G_{24} \\ G_{31} & G_{32} & G_{33} & G_{34} \\ G_{41} & G_{42} & G_{43} & G_{44} \end{vmatrix} = \sum_1^S sc\,\alpha_i \cdot sc\,\beta_i |K|_i \begin{vmatrix} \Phi_{P_i} & & & 0 \\ & \Phi_{t_i} & & \\ & & \Phi_{u_i} & \\ 0 & & & \Phi_{u_i} \end{vmatrix} \cdot |K|_i^T \tag{6}$$

where $G_{11}, G_{22}, G_{33}, G_{44}$ are the main stiffness of the rope section (longitudinal, torsional, bending);

$G_{12} = G_{21}; G_{13} = G_{31}$ etc. are the stiffnesses of the influence;

$\Phi_{P_i} = E\pi\delta^2 / 4; \Phi_{t_i} = E\pi\delta^4 / 80; \Phi_{u_i} = E\pi\delta^4 / 64$ are the stiffness of the wire section (longitudinal, torsional, bending; E is elastic modulus; δ is wire diameter).

The matrix $|K|_i$ of specific deformations of the i-th wire, the elements of which are obtained on the basis of the geometric equations of the deformation of wires and strands

$$|K|_i = \begin{vmatrix} K_{e\varepsilon} & K_{t\varepsilon} & K_{b\varepsilon} & K_{n\varepsilon} \\ K_{e\theta} & K_{t\theta} & K_{b\theta} & K_{n\theta} \\ K_{e\chi} & K_{t\chi} & K_{b\chi} & K_{n\chi} \\ K_{e\varsigma} & K_{t\varsigma} & K_{b\varsigma} & K_{n\varsigma} \end{vmatrix}$$

$$= \begin{vmatrix} \overline{K}_{E\varepsilon} & \overline{K}_{T\varepsilon} & \overline{K}_{B\varepsilon} & 0 \\ \overline{K}_{E\theta} & \overline{K}_{T\theta} & \overline{K}_{B\theta} & 0 \\ \overline{K}_{E\chi} & \overline{K}_{T\chi} & \overline{K}_{B\chi} & \overline{K}_{N\chi} \\ \overline{K}_{E\varsigma} & \overline{K}_{T\varsigma} & \overline{K}_{B\varsigma} & \overline{K}_{N\varsigma} \end{vmatrix} \cdot \begin{vmatrix} K_{eE} & K_{tE} & K_{bE} & 0 \\ K_{eT} & K_{tT} & K_{bT} & 0 \\ K_{eB} & K_{tB} & K_{bB} & K_{nB} \\ K_{eN} & K_{tN} & K_{bN} & K_{nN} \end{vmatrix}; \qquad (7)$$

where $|KT|_i$ is the matrix of specific strains in the rope, which includes the i-th wire;

$|KF|_i$ is the matrix of specific deformations of the i-th wire in a strand.

In the elements of the matrix $|K|_i$, the first index indicates the deformation of the wires in the rope (E - longitudinal; T - torsion; B and N - bending), and the second indicates the deformation of the rope, from which this deformation of the wire occurs.

In the elements of the matrix $|KF|_i$, the first index indicates the deformation of the wire in the strand (e - longitudinal; t - torsion; b and n- bending), and the second - the deformation of the strand, with which this wire deformation is directly related.

Elements of matrices $|KF|_i$ and $|KT|_i$ are obtained on the basis of geometric equations of deformation of wires in a single-lay rope (strand) and strands in a double-lay rope, taking into account transverse narrowing and friction [6].

The vector of internal force factors in the sections of the rope

$$|F| = |T\,M\,L\,\Gamma|^T, \qquad (8)$$

where T – is the longitudinal force; M is the torque;

$L = P f e^{-\kappa x}; \Gamma = -PA_{12} / A_{22} \sin\psi$ – are bending moments.

2.3. Characteristic of the ultimate elastic state of the rope.

It is determined by the internal force N, at which the longitudinal deformation of one of its elements reaches the elastic limit (the yield point ε_T according to the schematized $\sigma - \varepsilon$ wire diagram):

$$P_e = N = \frac{\varepsilon_T \lambda}{\max \overline{e}_i}; \quad i = 1, 2, ..., s. \qquad (9)$$

where $\lambda \le 1$ – is the coefficient of the influence of torsional deformation and bending of the wire in the rope on its ultimate elastic state (in the first approximation we accept $\lambda = 1$);

s is the number of elements (in the general case, elements are each wire of the rope);

$\max \overline{e}_i$ is the highest (among all the wires of the rope) value of the specific tensile deformation from the action of the end weight $\overline{P} = 1$, which depends on the stiffness characteristics of the rope, the loading scheme, the uniformity of the twisting tension, the presence of wire breaks [6]

$$\max \overline{e}_i = \overline{\varepsilon} K_{e\varepsilon_i} + \overline{\theta} K_{e\theta_i} + \overline{\chi} K_{e\chi_i} + \overline{\xi} K_{e\xi_i}, \qquad (10)$$

$\bar{\varepsilon}, \bar{\theta}, \bar{\chi}, \bar{\xi}$ – are specific deformations of tension, torsion, bending of the rope from $\bar{P} = 1$ for specific loading schemes (Figure 2).

2.3.1. Stretching of the rope with coiling onto a drum with a weight in the guides

$$\bar{\varepsilon} = (A_{11}\cos\psi - A_{12}(A_{12}\cos\psi + A_{14}\sin\psi)/A_{22} + A_{13}fe^{-kx})/|D| \quad (11)$$

$$\bar{\theta} = (A_{23}fe^{-kx} - A_{24}\frac{A_{12}}{A_{22}}\sin\psi)/|D| \quad (12)$$

$$\bar{\chi} = (A_{13} - A_{12}(A_{23}\cos\psi + A_{34}\sin\psi)/A_{22} + A_{33}fe^{-kx})/|D| \quad (13)$$

$$\bar{\zeta} = (A_{14}\cos\psi - A_{12}(A_{24}\cos\psi + A_{44}\sin\psi)/A_{22} + A_{34}fe^{-kx})/|D| \quad (14)$$

where $|D|$ and $A_{11}, A_{12}, ..., A_{34}$ – are the determinant and the algebraic complements of the rope stiffness matrix.

2.3.2 Stretching the rope with coiling on the drum with free suspension of the weight

$$\bar{\varepsilon} = (A_{11}\cos\psi + A_{13}fe^{-\kappa x})/|D|; \; \bar{\theta} = (A_{12}\cos\psi + A_{23}fe^{-\kappa x})/|D| \quad (15)$$

$$\bar{\chi} = (A_{13}\cos\psi + A_{33}fe^{-\kappa x})/|D|; \bar{\zeta} = (A_{14}\cos\psi + A_{34}fe^{-\kappa x})/|D|. \quad (16)$$

Specific deformation (10) *max* $\bar{e}_i$ depends on the corresponding specific deformations of the *i-th* wire and on the bending parameters (3), (4), and they are determined by the level of the end weight *P*. Therefore, calculate the tensile force P_e (9) of the limiting elastic state for these two loading schemes is possible only by consecutive approximations. As a first approximation, one should take P_e the one corresponding to the stretching pattern of a straight rope.

2.3.3 Stretching of a straight rope with a weight in the guides

$$\bar{\varepsilon} = (A_{11} - A_{12}^2/A_{22})/|D|; \; \bar{\chi} = (A_{13} - A_{12}A_{23}/A_{22})/|D|; \; \bar{\zeta} = (A_{14} - A_{12}A_{24}/A_{22})/|D|. \quad (17)$$

2.3.4 Stretching of a straight rope with a freely suspended weight

$$\bar{\varepsilon} = A_{11}/|D|; \quad \bar{\theta} = A_{12}/|D|; \quad \bar{\chi} = A_{13}/|D|; \quad \bar{\xi} = A_{14}/|D|. \quad (18)$$

2.4. Rigidity of the rope section taking into account wire breaks. The basis for determining the rigidity of the rope is the formula (6). In this case, the stiffness of wires with breaks depends on their number and location relative to the calculated section of the rope, which is taken into account on the basis of [6] in this way:

$$\Phi_p = \frac{\pi\delta^2}{4}E(1-e^{-fb}); \; b = \frac{2\pi l \sin\alpha}{h\cos\beta}, \quad (19)$$

where f – is the coefficient of friction; l – is the distance from the break point to the calculated section of the rope;

h – is the wire lay step; α and β – are wire and strand lay angles.

According to the proposed method, calculations were performed for several single and double lay ropes.

As an example, calculations are given for two ropes, the cross-sections of which with wire numbering are shown in Figure 3 and 4, where wires with breaks are indicated by shading.

Figure 3. Rope 1+6+12

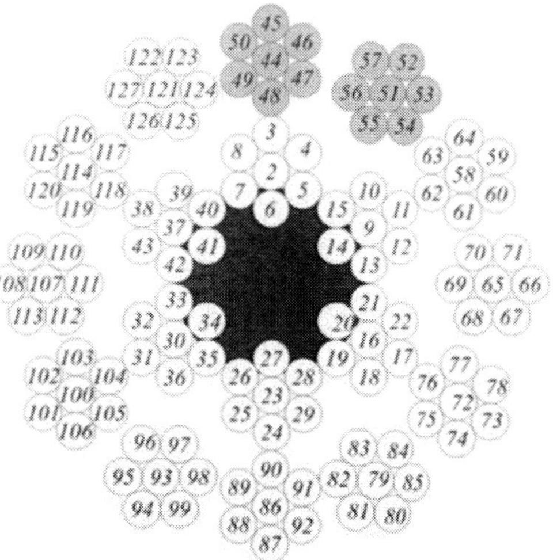

Figure 4. Rope 12(1– 6) – (1+6) +o.c.

Single lay rope (figure 3) construction $1/1.15+6/1+12/1$; diameter 5.15 *mm*; $F_c = 15.175 mm^2$; the total breaking force of all wires $P_c = 27315H$; the pitch of the wires of the outer layer $h =$ 45 *mm*; lay angles $\alpha_2 = 16.75^\circ$; $\alpha_3 = 16.16^\circ$ deformation of the yield point of the wire $\varepsilon_T = 0.0073$. Double lay rope (Figure 4) of the structure $12(1-6)-6(1+6)+o.c.$; diameter 9.3 *mm*; $F_c = 35.625$ *mm*²; lay pitch of the outer layer of strands $H = 256$ *mm*; wire lay spacing in strands $h = 112$ *mm*; wire diameters $\delta = 0.6$ *mm*; strand angles $\beta_2 = 17.5^\circ$; $\beta_3 = 19.5^\circ$ total cross-sectional area of wires $F_c = 35.625$ *mm*²; $P_c = 5700 N$, deformation of the yield strength of the wire $\varepsilon_T = 0.00665$.

Table 1 shows the value of the cross-sectional stiffness of these ropes for three situations: 1 - without wire breaks; 2 - cliffs are evenly spaced within one step; 3 - breaks in one section (this means in formula (19) $b = 0$).

Table 1. Stiffness of the section of the ropes x 10⁻⁴

Situation	G_{11}, H	G_{22}, Nmm^2	G_{33}, Nmm^2	G_{44}, Nmm^2	G_{12}, Nmm	G_{13}, Nmm	G_{14}, Nmm	G_{23}, Nmm^2	G_{24}, Nmm^2	G_{34}, Nmm^2
	Rope 1+6+12; diameter 5,15 *mm*; шаг *h=45mm*									
1	270	85.9	24.4	24.5	127	~0	~0	~0	~0	~0
2	233	72.5	23.3	23.9	104.6	-6.2	-3.2	-3.7	-1.9	-0.43
3	229	68.3	20.5	20.8	101.8	-6.7	-3.9	-4.2	-3.4	0.063
	Rope 12(1-6) - 6(1+6) + o.c.; diameter 9,3 *mm*; *h=112mm*; *H=256mm*									
1	473	631	17.9	17.9	337	~0	~0	~0	~0	~0
2	436	565	17.6	17.8	287	-2.6	0.42	-3.5	0.54	0.039
3	419	535	16.1	15.9	266	-3.7	1.03	-5.08	0.95	0.45

Tables 2 and 3 shows the calculated values for 3 situations of tensile forces P_e (9) of the limiting elastic state of the indicated ropes during deformations ε_T. The forces are presented in relative dimensions: $\bar{P}_e = P_e / P_c$. Next to the values P_e in brackets are the numbers of the wires in which the ultimate elastic tensile deformation is reached first, equal to ε_T. Further in table. 2 and 3 is shown the deformation of the ropes $\varepsilon, \theta, \chi, \zeta$ in the ultimate elastic state; experimental values of tensile stresses of wires $\max\sigma$ and $\min\sigma$ (in brackets wire numbers are indicated according to Figure 2 and 3); parameter $\bar{\Delta} = \min\sigma / max\sigma$, finally, the average tensile stresses are indicated: $\sigma = P_e / F_c$ (this stress is conditional, but it is convenient for an integral assessment of the stress state of the ropes [1]).

Table 2. Parameters of the ultimate elastic state of the rope 1+6+12.

Calculation parameters	Stretching schemes			
	1	2	3	4
1. Nominal rope condition				
$\bar{P}_e = P_e/P_C$ (№)	0.522 (2)	0.223 (1)	0.723 (1)	0.222 (1)
$\varepsilon \cdot 10^4$	52.7	73	73	73
$\theta \cdot 10^4$, *rad/mm*	~0	-107	0	-108
$\chi \cdot 10^4$, *rad/mm*	125	125	~0	~0
$\zeta \cdot 10^4$, *rad/mm*	14.2	~0	~0	~0
$\max\sigma$, N/mm^2 (№)	1460 (2)	1459 (1)	1459 (1)	1460 (1)
$\max\sigma$, N/mm^2 (№)	472 (5)	-328 (14)	1338 (5)	151 (11)
$\bar{\Delta}$	0.230	-0.225	0.911	0.103
σ, N/mm^2	845	401	1301	400
2. Wire breaks № 8÷10 within a step $h = 45mm$				
$\bar{P}_e$(№)	0.389 (1)	0.218 (1)	0.618 (1)	0.221 (1)
$\varepsilon \cdot 10^4$	47.3	73	73	73
$\theta \cdot 10^4$, *rad/mm*	4.96	98.7	0	-105
$\chi \cdot 10^4$, *rad/mm*	139	129	19.6	2.64
$\zeta \cdot 10^4$, *rad/mm*	18	4.19	10.2	1.37
$\max\sigma$, N/mm^2(№)	1445 (19)	1463(1)	1459 (1)	1459 (1)
$\max\sigma$, N/mm^2(№)	348 (5)	-239 (16)	1261 (5)	169 (15)
$\bar{\Delta}$	0.24	-0.163	0.864	0.116
σ, N/mm^2	701	392	1112	397
3. Wire breaks № 8÷10 in one section				
$\bar{P}_e$.(№)	0.369 (1)	0.204 (1)	0.603 (1)	0.205 (1)
$\varepsilon \cdot 10^4$	45.07	73	73	73
$\theta \cdot 10^4$, *rad/mm*	7.79	-101	0	-109
$\chi \cdot 10^4$, *rad/mm*	141	128	23.7	1.4
$\zeta \cdot 10^4$, *rad/mm*	2.49	-3.16	13.6	-4.1
$\max\sigma$, N/mm^2(№)	1430 (2)	1460 (1)	1460 (1)	1459 (1)
$\max\sigma$, N/mm^2(№)	314 (5)	-266 (14)	1242 (15)	121 (12)
$\bar{\Delta} = \min\sigma/\max\sigma$	0.220	-0.182	0.851	0.0829
σ, N/mm^2	665	367	1086	369

Table 3. Parameters of the ultimate elastic state of the rope 12(1–6) – 6(1+6) +o.c.

Calculation parameters	Stretching schemes			
	1	2	3	4
Nominal rope condition				
$\overline{P}_e = P_e / P_C$(№)	0.612 (123)	0.301[a] (3)	0.669 (44)	0.308 (2)
$\varepsilon \cdot 10^4$	73.7	58.6	80.6	60
$\theta \cdot 10^4$, *rad/mm*	~0	-31,3	0	-32.1
$\chi \cdot 10^4$, *rad/mm*	69.4	69.4	~0	~0
$\zeta \cdot 10^4$, *rad/mm*	21.8	~0	~0	~0
$\max \sigma$, *N/mm²* (№)	1330 (117)	1333 (3)	1305 (2)	1305 (2)
$\min \sigma$, *N/mm²* (№)	974 (123)	100 (87)	1230 (3)	232 (44)
$\overline{\Delta}$	0.73	0.075	0.94	0.174
σ, *N/mm²*	979	482	1071	493
Wire breaks № 44–57 within a step $h = 66.5mm$				
$\overline{P}_e$.(№)	0.586 (123)	0.301 (3)	0.649 (44)	0.308 (2)
$\varepsilon \cdot 10^4$	72.9	58.8	80.7	61.2
$\theta \cdot 10^4$, *rad/mm*	0.14	-30.7	0	-30.4
$\chi \cdot 10^4$, *rad/mm*	73.9	70.4	18.9	4.9
$\zeta \cdot 10^4$, *rad/mm*	22.3	-0.6	-5.8	-2.3
$\max \sigma$, *N/mm²* (№)	1329 (117)	1333 (3)	1329 (58)	1329 (2)
$\min \sigma$, *N/mm²* (№)	950 (123)	112 (87)	1189 (24)	291 (58)
$\overline{\Delta}$	0.71	0.084	0.93	0.219
σ, *N/mm²*	938	481	948	492
Wire breaks № 44-57 in one section				
$\overline{P}_e$.(№)	0.582 (123)	0.301 (5)	0.630 (44)	0.308 (3)
$\varepsilon \cdot 10^4$	72.6	58.8	80.7	60.7
$\theta \cdot 10^4$, *rad/mm*	0.15	-30.7	0	-30.9
$\chi \cdot 10^4$, *rad/mm*	74.5	70.5	12.0	2.96
$\zeta \cdot 10^4$, *rad/mm*	24.1	1.3	-1.94	-0.5
$\max \sigma$, *N/mm²* (№)	1329 (123)	1333 (3)	1329 (14)	1329 (2)
$\min \sigma$, *N/mm²* (№)	944 (24)	113 (87)	1204 (44)	272 (122)
$\overline{\Delta}$	0.71	0.085	0.906	0.205
σ, *N/mm²*	931	480	986	493

[a] The value $\overline{P}_e = 0.301^*$ with a discrepancy of 3 - 5% (depending on ε_T in formula (9)) coincides with the experimental data of [7].

The loading capacity is determined on the basis of a study of the deformation of the rope beyond the elastic limit, which occurs when the stiffness of its section changes. Taking a discrete change in stiffness, we divide the rope deformation process into intervals with constant stiffnesses within each of them. The loading capacity of the rope is determined by the sum of the increments of the longitudinal force in its section for each interval. The first interval represents elastic deformation. Subsequent - intervals of elastic-plastic deformation

$$P_н = \sum_{j=1}^{K} \Delta N_j, \tag{20}$$

where ΔN_j is the longitudinal force in the section of the rope for the j interval.

Calculation methods of ΔN_j for various rope loading schemes are similar to determining the force P_e.

3. Conclusions.

The strength characteristics P_e of the ultimate elastic state of the ropes very significantly depend on the loading scheme, the design of the rope and the presence of wire breaks, as well as on the unevenness of the twisting tension of the wires and strands. For the convenience of the analysis, they are presented in a relative dimension $\bar{P}_e = P_e / P_c$, in fact, adequately to the coefficient K in formula (2), which is approximately equal to the third loading scheme of the rope.

The main reason for the low values of the characteristics $\bar{P}_e$ is the uneven deformations and tensile stresses of the wires in the rope. This is primarily due to the deformation θ of twisting of the rope, and secondly to the deformation χ of bending. The values $\bar{P}_e$ for the 2nd and 4th tension schemes significantly depend on the stiffness of the influence G_{12}, with which the twisting deformation of the rope is associated. So for a rope of construction 1 + 6 + 12 $G_{12} = 1270 kNmm$ and $\bar{P}_e = 0.222$. In the case of 1–6 + 12 construction with opposite direction of layers $G_{12} = 736 kNmm$ and $\bar{P}_e = 0.47$.

The values of the average conditional stress $\sigma = P_e / F_c$ show that their level in the ultimate elastic state of the ropes is less than the advising parameter at working (operational) loads [1]. It is useful to take this into account when analyzing the standard safety factors for specific rope loading schemes.

The presence of wire breaks leads to the appearance of the stiffness of influence: longitudinal-bending G_{13}, G_{14}; torsionally-bending G_{23}, G_{24}, as well as bending-bending G_{34}. When the ropes are stretched in a state with these stiffnesses, additional deformations of bending and twisting of the rope appear, and therefore deformations of the wires, both upward compared to the nominal state, and downward.

The characteristics $\bar{P}_e$ in the states with wire breaks in the 1st and 3rd schemes are reduced. With the 2nd and 4th, they remain practically unchanged, because torsional deformations θ change slightly

For the 3rd tension scheme, the ratio between the characteristic P_e of the ultimate elastic state of the rope and the characteristic of its bearing capacity is $P_e(0.7 - 0.75)P_н$. For other stretching schemes associated with torsion and bending, the ratio is less stable (depends on the stiffness of the influence and the ratio of the drum diameter to the rope diameter).

We assume that, based on the characteristics P_e of the stretching of the rope according to the ultimate elastic state, it is possible to construct a method for calculating the strength of ropes, which will have advantages over the existing one, based on the characteristic (1). Advantage arguments

– strength characteristics P_e are directly related to the rope loading scheme;

– the value P_e takes into account all geometric parameters of single and double lay, as well as defmation properties of wires;

– characteristics P_e allow taking into account the effect of wire breaks during the operation of ropes both within the rejection norm and in the direction of optimizing the norms.

4. References

[1] Berezhinskiy V and Shatilo A 2015 *Mine hoist ropes* University book (Moscow, Russia) 232
[2] *Rules for the construction and safe operation of cranes* (Kharkiv: Fort) 2007 268
[3] *DIN-Taschenbuch. Drahtseile: Normen* (Berlin-Koln: Beuth) 1990 360

[4] Chaiun I 2013 Substantiation of increasing the accuracy of the calculated strength scheme of hoisting ropes *Journal Handling equipment* **4** pp 36 – 44
[5] Chaiun I 2003 *Load-bearing capacity of lifting ropes and belts* (Odessa: Astroprint) 233
[6] Chaiun I and Vovk P 2019 Friction of a thread located on a screw line during a cylindrical surface *Journal Handling equipment* **(2)** pp 75 – 81
[7] Glushko M 2013 *Steel hoisting ropes* (Odessa: Astroprint) 326

IOP Conf. Series: Materials Science and Engineering 1164 (2021) 012088 doi:10.1088/1757-899X/1164/1/012088

Structural organization and operation of the structure-system

Valerii Vyrovoy[1,a], **Oksana Korobko**[1,b], **Oleg Pishchev**[1,c] **and Hanna Varych**[1,d]

Odessa State Academy Civil Engineering and Architecture, Department of Architectural Structures, Didrihsona st. 4, 65029 Odessa, Ukraine

E-mail: [a]vyrovoy@ukr.net, [b]okskorobko71@gmail.com, [c]tapi74@ukr.net, [d]varich09@gmail.com

Abstract. The main factor is presentation of the building construction as open complex self-organizing system. The initial theoretical provisions on processes of self-organization of the construction both during its production period and during operational loads action period are formulated. The active elements of structure material are selected. Representation of the construction as system allows determining the conditions for safe functioning of the construction-system in the system of similar constructions.

1. Introduction

Building objects of various purposes, individual products, constructions and materials can be considered as certain type systems [1, 2]. Presentation of individual building objects as specific systems [3, 4] allows determining the level of intra-system connections, to assess the reliability of the object-system under the action of various external loads and thus to solve problems of object safe operation. Of basic importance is a choice of the object in the building complex, which to be represented as system. This allows systems of other hierarchical levels to be analyzed through the dedicated system object. Representation of the building construction as system makes it possible to integrative examines the construction itself and analyzes its role in the system of construction-systems (buildings, structures). This approach allows to reasonably including the material in the analysis of the state of building object-systems.

The system of construction-systems puts forward specific conditions to the individual construction-system. The construction-system to satisfy these conditions must have the strictly defined set of properties. The material is the end element of the future system, technological solutions of which will determine a future safe operation of construction-systems and, as a result, the system of construction-systems. Representing the construction as system assumes an existence of certain structural organization, which allows acquiring the set of specified properties. Constructions lose their functional abilities under action of a complex of external loads and influences. This is due to a change in properties of the material from which they are made. In turn, a change in properties is determined by corresponding changes in the structure of material, and, therefore, construction. Based on this, the task was set to analyze a role of the construction in the general building hierarchy, taking into account its structural organization and features of functioning as open complex self-organizing system.

2. The foundation for a representation of the building construction as certain type system

The building construction as open complex self-organizing system.

The material as independent object does not exist and manifests its properties in samples, products, constructions. Moreover, material properties level depends on type of the construction in which it is applied. Therefore, to present the material as basic object-system is senselessly. This suggests that the construction should be represented as system in which the material properties should be manifested

Content from this work may be used under the terms of the Creative Commons Attribution 3.0 licence. Any further distribution of this work must maintain attribution to the author(s) and the title of the work, journal citation and DOI.
Published under licence by IOP Publishing Ltd

and realized. In turn, the construction-system is part of the system of construction-systems, forming buildings and structures – different level systems.

All systems, regardless of their nature, perceive external influences and react accordingly to them. Building constructions, regardless of their purpose and location in buildings and structures, are constantly forced to repel environmental attacks. This means that building constructions is open systems and implies continuous interactions of constructions with their environment.

The construction can be considered as specially figuration material. In turn, the material consists of different in nature, amount and dimensions of initial components, which through rather complex physical-chemical processes of hydration and physical-mechanical processes of structure formation provide the required set of mechanical characteristics of the end product. The complexity of any building construction as open system is related not only to complex mineralogical compositions of mineral binders, but also to the multifaceted phenomena and processes that occur both during structure formation and during operation. Processes of force stationary and dynamic action are superimposed on complexity of material structure that leads to development of local and general deformations in the material. This, in turn, provokes redistribution of forces between components, local changes of structure parameters, diffusion migration of pore fluid, etc. Due to the fact that the material is poly-structural, consisting of structural heterogeneities that are different in mechanisms of structure organization, each structural level has the individual complexity and characteristic properties. All of the above allows analyzing building constructions as complex systems.

Building constructions as open complex systems have the ability to transform their structure under influence of external factors in such a way as to preserve properties of the system under operating conditions. Such spontaneous structural changes are self-organization processes. Self-organization processes in the construction material can be divided into two types: - self-organization processes in the technological period of construction production; - processes of spontaneous structure restructuring in the period of operating constructions.

Spontaneous processes of structure organization during its production of the construction cause spontaneous generation and development of diverse set of structural elements, which can be classified into active, metastable and conservative [6]. A distinctive feature of active elements of the material structure and construction is their adequate response to almost any external and internal influences. Cracks and inner surfaces of partition, which exist in the material at different levels of structural heterogeneities, are active elements. In addition, local and integral fields of technological (initial, residual, hereditary) deformations are formed at the construction level. The construction as open complex system enters an active phase of functioning with a certain set of structural elements that spontaneously originated in the technological period of its production.

Changing parameters of active elements causes change in system structure parameters. An adequate reacting to external and internal influences implies that active elements, which «sensitive» to certain effects, change their parameters in a timely manner with influences. The term «sensitivity» here means ability and speed of response to certain influences. One of the ways to ensure safe functioning of the construction-system in the system of construction-systems at a negative action of operation environment is to preserve networked relations of active elements at all levels of subsystems. This can be achieved by directed organization of the structure taking into account phenomena of self-organization in the technological period of constructions production.

The building construction reacts to whole complex of external and internal loads that arise as a result of environment actions. Multi-component material and complexity of structure organization, complex of physical-chemical processes of binder hydration and physical-mechanical processes of structure formation at all hierarchical levels of the system, constant structural transformations under influence by external and internal factors during operation period indicate complexity of the construction. Self-organization processes of structure determine periods of creation and operation of the construction. Thus, the building construction can be represented as open complex self-organizing system.

The interrelation of material structure and construction structure.
The material of construction is always present as changing factor, regardless of structure hierarchical levels of the building. This is particularly emphasized in scientific papers [5, 6], in which the construction is considered as specially figuration material.

Representation of the building construction as especially figuration material in certain geometric shapes implies the following: - construction properties are determined by material properties; - structural features of the material are automatically included in construction structure; - geometrical construction parameters create local and integral fields of residual deformations in the material in the technological period of material production; - formation nature of residual fields of deformations leads to fluctuations in a density of concrete mixture, change of kinetics of physical-chemical hydration processes, change of mechanical and deformation characteristics of concrete in various volumes of the construction; - change of concrete properties in various parts of a particular construction suggests that the properties of concrete of the same composition should differ in the product with different geometric characteristics; - existence of material properties gradients in the construction leads to an uneven development of humidity and thermal deformations, diffusion mass transfer processes of different intensities, etc.; - permanent change of material structure and therefore its properties during the period of operation loads results in permanent non-equilibrium state of the construction; - concrete quality indicators in operated construction should be determined by a special method depending on type of the product; - average characteristics of material properties cannot give reliable information about the product condition considering that all catastrophic processes start from localized zones.

All these factors, as well as many others that have been left without due attention, indicate, in our opinion, at least three fundamental points. The first point is due to the fact that it is irrational to consider technical and economic tasks of designing optimal concrete compositions without taking into account type of the construction with possible fluctuation of properties by their volume. The second point is that methods should be developed that allow, at the stage of designing construction, to take into account not only average characteristics of strength and deformation properties of the material, but also its structural features related with this construction. The third point is based on the fact that it is not technically and economically advantageous to design concrete compositions for construction production without taking into account possible influences of operating environment.

The emphasis on structural organization constructions, taking into account the material structure, gives opportunity to study its structure, to determine the structural components of the material, which can remove the construction from conditions of safe functioning, to assess its role in ensuring cooperation with its environment.

The construction as system in system of construction-systems.
Construction-systems, depending on their individual structural figuration, determine creation and functioning of buildings, structures and their complexes as systems of construction-system, in turn, imposing requirements on their own basic elements, including the material from which constructions are made. This approach makes it possible to evaluate contribution of material to organization of structural levels of hierarchical systems of any kind of complexity, to identify structural elements that ensure the preservation of construction properties under influence of operational loads, to present its structure as a function of movement, which determines through change of structural parameters safety of functioning of the object-system in the system of object-systems.

The selection of the construction-system as basic integral element in the general building hierarchy implies the following: - the construction-system, being an element of the system of construction-system, creates continuous structural series of any type systems complexity; - presentation of the construction-system in the system of constructions-system and in mega-systems allows considering issues of the rational use of material and technical resources from single technical and economic positions taking into account environmental requirements both at the construction stage, and during operation of individual constructions, structures and town-planning complexes; - the construction-system, being a knot of inter-structural interactions in the system of constructions-system, determines the safe functioning of the whole system by own state; - the structure and therefore properties of the material are realized in the construction-system, that allowing entering a material into all structural levels of hierarchical systems.

The role of the construction-system in the general building hierarchy is demonstrated by the scheme shown in figure 1.

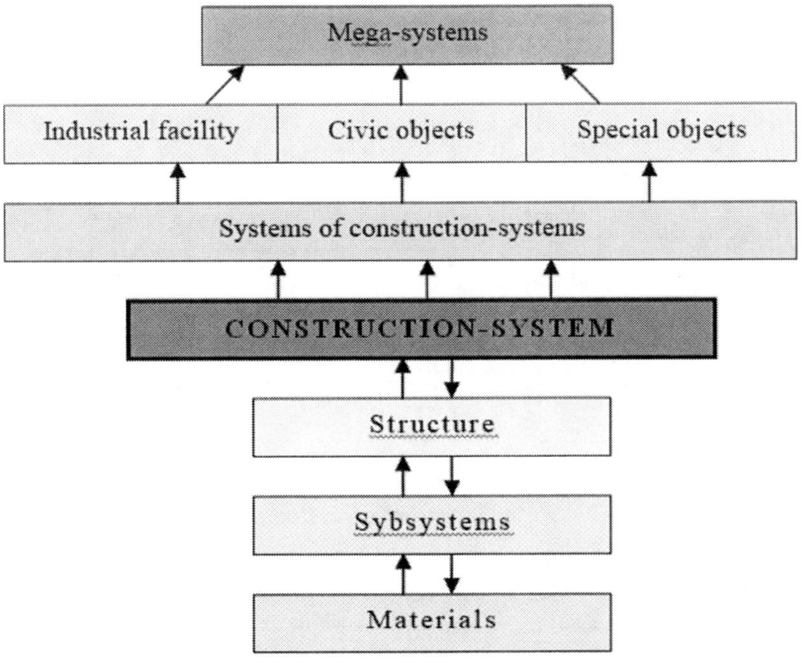

Figure 1. The role of the construction-system in the general building hierarchy.

Mega-systems are complex-organized formations with multi-purpose functioning. Components of mega-systems are self-contained systems, which include different for the purpose buildings and structures (systems of construction-systems).

The building (structure) can be reasonably presented as complex open system. This implies that the building is in the certain way structurally hierarchical formed and perceives all influences of the environment. Structural figuration allows distinguishing individual subsystems of the building: subsystems in the form of electric power supply, water supply, ventilation and heating (sewage and communication systems are not included in our tasks). Load-bearing walls or a frame can be selected as subsystems depending on the design features of buildings. Building bases and foundations, overlapping and flat roofs, stair landings and flights, elevator shafts, etc, can be are individual subsystems. The principle of subsystems selection is based on their role in the building, types of perceived loads, design features and required properties. At the same time, an interaction of subsystems is necessarily taken into account that implies manifestation of emergent in the building, and rather complicated organization of the subsystems themselves. We will accept that the building is built of concrete and reinforced concrete of appropriate types and purpose. Therefore, concrete can be included as the subsystem in the previously selected subsystems. In turn, concrete is a poly-structural material with a rather complex structural organization at each level of structural heterogeneities [7]. Structural heterogeneities can be interpreted as subsystems of subsystem at the concrete level, which in turn is the subsystem of subsystem at construction level or structural element level.

The selection of individual building construction as research object makes it possible to decompose the construction itself and analyze its role in individual structural schemes of the building and the building itself (Figure 2).

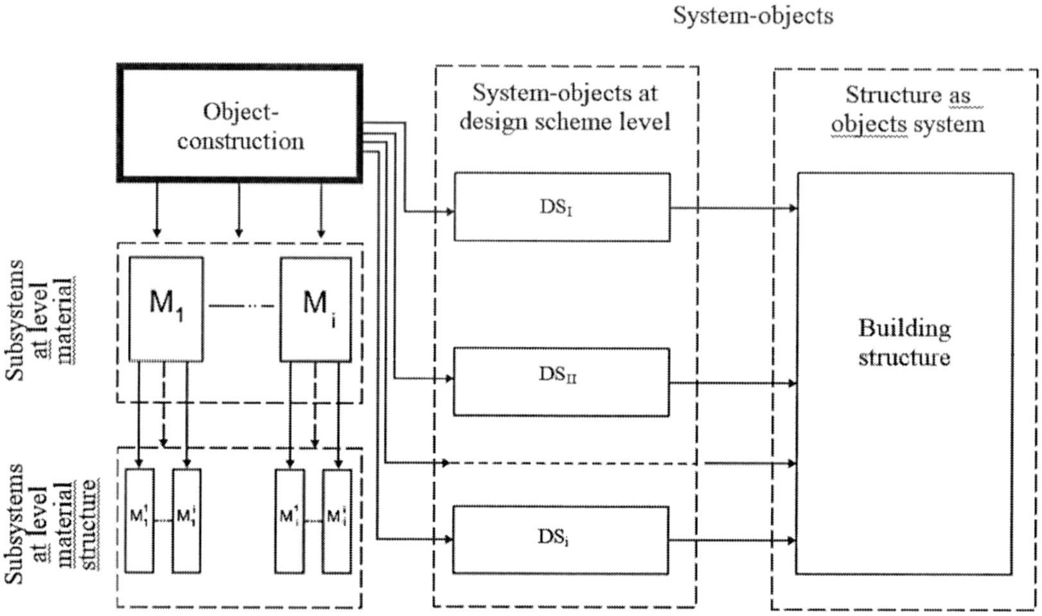

Figure 2. The analysis of a role of structure organization of the construction-system in the system of construction-systems.

This allows evaluating properties of construction; to determine main factors that define them, and to assess a role of the construction in the system of objects, both at individual structural schemes level, and at the entire object level as system with a certain structural figuration of structural elements. This approach seems to be the most rational, because it allows studying an influence of structural organization of the material on construction properties and determining its role in individual structural schemes and in a building itself. The representation of the construction as system allows determining its structural organization, which ensures the realization of its functions, and analyzing a impact of changes in the functional construction properties on working state of individual structural schemes and whole building-system.

3. Functioning of the construction-system.

The construction-system is for realization individual purposes, so special attention should be paid to the period of functioning. It is during this period that the effectiveness of design and technological solutions is determined. At the beginning of realization of main functions (τ_0) the construction-system (CS) is structurally figurate in a certain way, which determines its state CS-I (Figure 3).

During operation, the construction perceives external influences, and the material of construction is subject to internal factors action. The response to external influences is the reaction of the structure material. Material response is compensatory structural changes (CI), which prevent required properties from being out of value tolerance. Internal factors action is to processes that cause structure changes (SI).

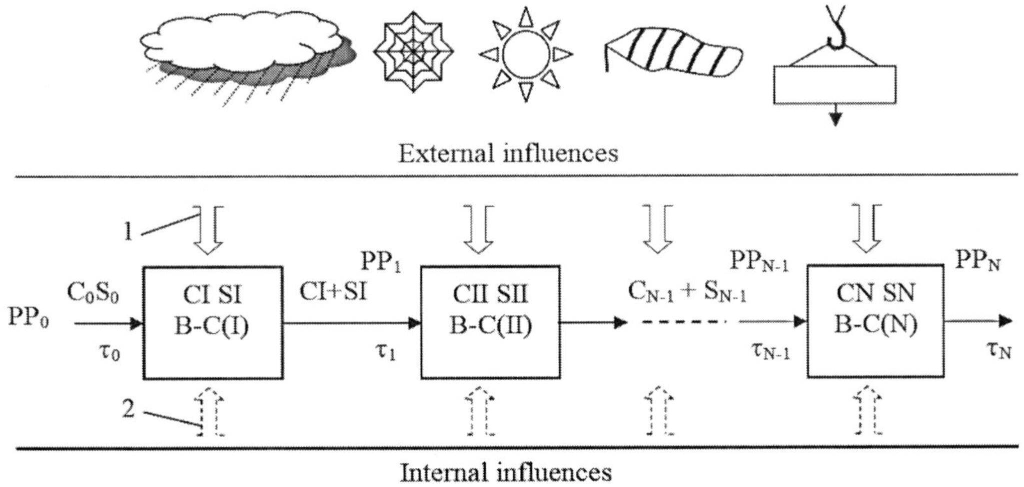

Figure 3. Functioning of the construction-system: 1 – the complex of external influences; 2 – the complex of internal influences; K-C(I)...K-C(N) – the functional state of construction-systems; CI...CN – the complex of structural changes under external influences; SI...SN – the complex of structural changes under internal factors influences; $PP_0...PP_N$ – properties parameters.

Integrated changes of structure (CI+SI) after τ_1 time transfer the construction to the state of CS-II with the structure parameters CII and SII.

For the entire period of active functioning of τ_N, structural parameters of construction-system change proportionally to the periods of action of external and internal factors to C_N and S_N values, which determines the functional state of construction-system CS-N. The functional state of construction-system meets the requirements if material properties are within the specified limits.

There is the peculiar history of properties parameters (PP) changes during the active functioning period of construction-system. Each story begins with a certain event or point in time. The history of changing material properties begins from the time the construction was operated.

The parameters of material properties are changed to the level of PP, remaining in allowable limits, under external influences, which are perceived by material during the construction operation period τ_1. The further predicted safe functioning demands clarification of the reasons of properties change and a description of the history of construction behaviour during the previous operational period.

In process of extension of terms of operation to τ_N values in the conditions of permanent influence of operating environment, the next stage of changing properties to PP_N critical level occurs at the end of the existence material history satisfying to the level of requirements. A new history of completion of the stage of construction functioning begins with transition to the stage of irreversible degradation.

Presentation of the construction as system *a priori* assumes its structural figuration. This gives rise to different interpretation of the history of structural parameter changes and, therefore, properties of the construction-system. For different interpretation, functioning of the construction-system should be presented as continuous structural transitions in which the system homeostasis is preserved in conditions of external and internal factors influences. The safe limits for changing properties of the construction-system are set by the range of changes in construction parameters.

The very fact of continuous change of structure indicators allows representing the structure as a function of movement, which sets a certain range of change of material properties and determines functional state of the construction-system. Structural transformations intensity of the material under acting of operational loads on the construction depends on initial qualitative composition and quantitative ratios of structural elements (CoSo). Initial structure parameters are defined as a certain set of structural elements at the time of construction operating. At this point, the structure of the construction-system has already passed a certain history of formation and development. The beginning of this story falls during the system formation period. Disparate models prior to this begin to merge into the construction-system model.

Due to the fact that functioning of construction-systems is considered as continuous processes of structural transformations, which depend on the history of structural figuration, it is necessary to solve the following tasks: - a choice of construction-system structure model taking into account possible options of its interaction with its environment during the operation period; - identification of dominant elements of the structure, which should ensure safe operation of the construction-system under influences of external and internal factors; - design of concrete compositions and assignment of technological modes for concrete mix production taking into account geometrical parameters of the construction to create required set of structural elements according to their qualitative composition and quantitative ratios; - purpose of production technology of the construction-system with required set of physical-mechanical characteristics.

The adoption of rational technological decisions to obtain the construction-system with required structural characteristics at this stage of technology development of production of building materials and products is based primarily on determination of structural elements, which provide the appropriate level of required properties and which are able to maintain this level in adverse operating conditions. It should be noted that the original structure is an instant «portrait» of the structure, which entered the «family album» of the construction-system not by some exceptional internal characteristics, but by the fixed date. From this point on, the history of functioning of the construction-system in the system of construction-systems begins.

The motivated source of initiation and development of self-organization processes can be considered active elements in the form of cracks and inner surfaces of partition, which are inherent components of individual subsystems. A developed network of cracks and inner surfaces of partition connects all hierarchical levels of material into a single dissipative system. Therefore, the structure of complexly organized material can be represented as the network of active elements, which organized according to the principle «network within networks» at individual levels of heterogeneities.

The «Large» network of cracks and inner surfaces of partition includes the network of cracks and inner surfaces of partition of different scale heterogeneities at the level of material structure.

Each individual network in the integrated network is a certain set of active elements at a certain level of structural heterogeneities (subsystems). With this model of material representation it is possible to distinguish interaction zones of different-level network-components, in which they manifest themselves through internal and inter-structural connections. A different order in transition zones is formed in case of structural changes caused by internal or external factors. Structures automatically emerge that are new to each individual network and the only ones possible to preserve the interconnected network of active elements in the product material as complex system. If it is not possible to incorporate individual networks into the global network, the interconnectedness of structural elements is violated. At the same time an uneven load on individual network elements arises, the transmission of information about the state of individual subsystems ceases and, as a result, the principle of auto-support is violated. This leads to formation of the fundamentally different material structure. The new structure is active due to the fact that its organization has manifested itself as a response to the formation of active elements with critical parameters. The new structure turns the basic system into a scleronomous one through its desire for self-organization. The base system loses its ability to perform its functions.

One of ways to ensure safe functioning of the construction-system in the system of construction-systems at negative action of the operational environment is to preserve interconnected relations of active elements at all levels of subsystems. This can be achieved by directed organization of the structure taking into account the phenomena of self-organization during the technological period of construction production. A certain set of active elements at all levels of structural heterogeneities should be created in the material by the time the construction starts working in the system with similar ones. The individual set involves designing an interconnected and interdependent network of active elements, taking into account their interaction and self-support during the period of action of external and internal loads on the system. Thus, the principle of system self-organization will be realize throughout the entire period of its functioning that makes it possible become apparent whole complex of adaptation mechanisms and thereby avoid crisis situations for systems.

4. Summary

The analysis leads to the conclusion that the building construction can be represented as open self-organizing system. Such presentation will make it possible to realize more fully joint efforts of designers and technologists aimed at further revealing potential possibilities of realizing material properties formed in the construction. Special attention should be paid to active elements, as they are able to change their own parameters under influence of external and internal factors. Changing parameters of active elements causes change in the system structure. When critical parameters of self-developing active elements are reached, catastrophic loss of functional properties with early failure of the construction-system may occur. Favourable structural changes associated with self-support and self-development of active elements network make it possible to realize adaptation effects (structural self-organization), which allows the construction-system to function during the standard period. Due to the individual structural organization, which provides manifestation of required properties, the individual building construction should be identified as independent research object. Representation of the construction as system in system of construction-systems allows determine structural components of the material, which provide the required construction properties and preserve them in predetermined values during the period of operation loads action. This makes it possible to work together in the system of similar construction-systems through spontaneous manifestation of internal and external safety effects (homeostasis phenomena).

References

[1] Travush V I and Fedorova N V 2018 Survivability of structural systems of buildings with special effects *Magazine of Civil Engineering* Vol. **81** Issue 5 pp 73–80

[2] Hewitt E, Oberg A, Coronado C and Andrews C 2019 Assessing "green" and "resilient" building features using a purposeful systems approach Sustainable Cities and Society (Elsevier) Volume **48** https://doi.org/10.1016/j.scs.2019.101546

[3] Artyukhov V V 2009 *General systems theory: Self-organization, sustainability, diversity, crises* (Moscow: Librocom) 224

[4] Soroko E M 2006 *Golden ratios, processes of self-organization and evolution of systems: Introduction to the general theory of harmony systems* (Moscow: KomKniga) 264

[5] Sukhanov V G, Vyrovoy V N and Korobko O A 2016 *Material structure in construction structure* (Odessa: POLIGFAF) 244

[6] Vyrovoy V N, Dorofeev V S and Sukhanov V G 2010 *Composite Building Materials and Constructions. Structure, Self-organization, Properties* (Odessa: TES) 152

[7] Cherniavsky V L 2008 *Adaptation of Abiotic Systems: Concrete and Reinforced Concrete* (Dnepropetrovsk: DNURT) 4125

IOP Conf. Series: Materials Science and Engineering 1164 (2021) 012089 doi:10.1088/1757-899X/1164/1/012089

Experimental studies of the deformation properties of concrete based on sulfate-resistant slag Portland cement as a material for hard coatings of agricultural airfields

O Yanin[1, a], T Yemelianova[1, b] and S Novikova[1, c]

[1]Kherson State Agrarian and Economic University, Stritenskaya, 23, Ukraine

E-mail: [a]yanin_a@ukr.net, [b]e.tatyana.2014@ukr.net, [c]novikova_svetla@ukr.net

Abstract. The paper presents the results of experimental studies of the deformation properties of concretes based on sulfate-resistant slag Portland cement intended for hard coatings of agricultural tracks and airfields. The effect of concrete hardening conditions on the value of its relative shrinkage deformation has been studied. The role of surfactants (plasticizers) as an additional ingredient to concrete is also investigated. Based on these experimental tests, the regularities of the development of plastic deformations formed in concrete in time at a constant load are revealed. The results of the two series of experiments have made it possible to demonstrate the influence of concrete hardening conditions on the values of creep deformation that are also described and analyzed. This article analyzes the kinetics of concrete shrinkage deformations without additives, comparing the results with those for concrete with the addition of a lignosulfonate superplasticizer and also taking into account various modes of heat and humidity treatment of concrete. A comparison of the deformation properties in concrete with sulfate-resistant Portland slag cement with the deformation properties of concrete based on ordinary Portland cement is carried out. The conclusion is made about suitability of concrete with Portland slag cement for use in construction of hard coating for agricultural tracks and airfield runways.

1. Introduction

To improve the efficiency and reduce the cost of rigid concrete coatings of agricultural tracks and airfields, it becomes necessary to develop new solutions in the field of materials science. In particular, the reduction in the cost of cement, which is part of concrete, is promising direction. The corrosion resistance of the coating is also of great importance. In this case, to determination of the coating deformation it seems important, since this is associated with the safety of aviation and automobile transport.

2. Formulation of the Problem

Since slag Portland cement is a new material in the construction of pavements of agricultural tracks and airfields, determination of the deformation modulus, and to study the creep and shrinkage of concrete based on this material is the scientific and practical interest.

Studies of the deformation properties of concrete are devoted to the work of many scientists [1-10]. A A Gaikovich and R V Ovchinnikov substantiated the possibility of obtaining concretes of different strengths with given properties based on a cement-ash binder. This made it possible to achieve a reduction in the cost of production by 8.5%, which is equivalent to saving almost 100 kg of cement per 1 m3 of concrete mixture [1-3].

The influence of the mineral-chemical additive of the boiler slag on the hardening of cement is considered in the works of Korovkin M O, Eroshkina N A , Zamchalin M N. It has been proven that replacement of the 8–15% of cement by crushed slag reduces the mobility of the mixture, but with an increase its consumption to 20%, workability remains the same as for compositions without additives [4]. To eliminate the negative effect of slowing down the rate of cement hardening with the introduction of slag, it is proposed to introduce a hardening accelerator into the complex additive

Content from this work may be used under the terms of the Creative Commons Attribution 3.0 licence. Any further distribution of this work must maintain attribution to the author(s) and the title of the work, journal citation and DOI.
Published under licence by IOP Publishing Ltd

[4-6].

The works of Belov V V, Subbotin S L, Kulyaev P V are devoted to the study of the deformative properties of concretes with a finely dispersed limestone component. The use of this additive made it possible to reduce the creep and shrinkage of concrete in comparison with concretes without additives [7].

The possibility of using in construction of fine-grained slag concrete on the basis of sifting out from of crushing cast slag crushed stone is substantiated in the works of such scientists as B A Bondarev, N N Chernousov, R N Chernousov, V A Sturov. Due to the introduction into the composition of the cinder-concrete mixture the steel fiber, which has high tensile strength and increased modulus of elasticity, it has become possible to abandon the rod reinforcement in some structures [8].

Studies [9] have shown that the use of high-strength concretes can increase the service life of the structures of cement-concrete road surfaces or achieve significant material savings due to the thinner slab thickness. The use of complex plasticizing additives made it possible to obtain road concretes of C35 / 40 and higher classes [9-10].

The work [11] is devoted to the study of the strength properties and frost resistance of concretes based on slag Portland cement. It substantiates the possibility and feasibility of using sulfate-resistant slag Portland cement for the construction of rigid coatings of agricultural tracks and airfields. At the same time, it is important to determine the deflection of the coating from the load and volumetric deformations of concrete. Therefore, it became necessary to study the deformation characteristics of such concrete.

The aim of the work is to experimentally determine the deformation modulus of concrete on sulfate-resistant slag Portland cement under different hardening conditions; revealing the influence of concrete hardening conditions on the value of its relative shrinkage and creep deformation; study of the effect of additives of surfactants - plasticizers in the form of lignosulfonate on the deformation characteristics of concrete.

3. Materials and methods

Samples in the form of a prism with dimensions of 10 × 10 × 30 cm were made for the experimental determination of the modulus of deformation of concrete on sulfate-resistant slag Portland cement binder [1, 12, 13]. They were tested at the age of 28 days.

For testing, samples of grades CK 5, CK 3 and CK 9 were selected, which after thermal and moisture treatment were in normal hardening conditions and had the composition shown in table 1 [13]. Determination of the concrete deformation modulus for these samples was carried out by loading them with a load equal to 0.3 - 0.5 of the prismatic strength (fcd) and measuring the corresponding deformations. For this, metal corners were glued to the samples on four sides before testing in order to install dial indicators, figure 1.

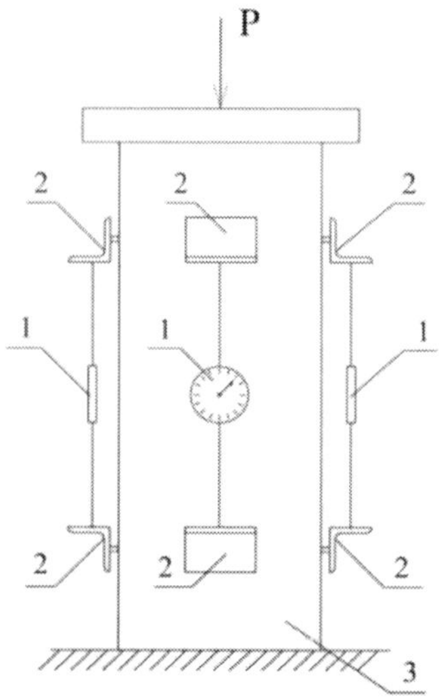

Figure 1. Installation scheme for determining the modulus of concrete deformation and for examining it for shrinkage and creep: 1 – dial indicator; 2 – corners; 3 – sample.

The deformation modulus was determined by dividing the vertical stress in the sample by the corresponding averaged relative deformation.

Samples in the form of a prism with dimensions of 10×10×30 cm were made to study the shrinkage deformations of concrete on a slag Portland cement binder. They began to be tested at the age of one day. As in the determination of the deformation modulus, metal corners were glued to the samples before the study to establish dial indicators, figure 1.

The samples were regularly monitored and the readings of indicators were periodically taken. It is made it possible to determine the relative deformations of concrete shrinkage at various points in time. The maximum duration of the experiments was 250 days. For the study samples of grades CK 1, CK 3, CK 5, CK 9, CK 11, CK 13 and CK 17 [11] were selected, the hardening they took place under normal conditions at a humidity of 90 - 95% [11].

In total, several series of experiments were carried out, which made it possible to reveal the influence of concrete hardening conditions on the value of its relative shrinkage deformation. In addition, the role of surfactant additives - plasticizers introduced into concrete was investigated.

As you know, concrete creep is the phenomenon of a gradual increase of deformation under a constant load. Creep is not only a negative phenomenon, since it indicates the relaxation of stresses in time [14].

Typically, in a creep study, concrete specimens are loaded with a load that is about 30% of the breaking value. Taking into account the importance of this characteristic, concrete on slag Portland cement was subjected to tests, which made it possible to reveal how its plastic deformations develop over time under constant load.

Samples were made of concrete on slag Portland cement with dimensions of 10×10×30 cm to carry out these tests, which began to be tested at the age of 7 days. As in the previous cases, corners were glued to the samples to establish dial indicators, figure 1.

Samples of grades CK 3 and CK 9 were selected (table 1 [11]) to test concrete for creep. They hardened after thermal and moisture treatment under normal conditions. After a constant load equal to 30% of the destructive load was applied to the samples, they were regularly monitored and readings of indicators were periodically taken.

Two series of experiments were carried out, which made it possible to reveal the influence of concrete hardening conditions on the values of creep deformation. The maximum duration of the experiments was 220 days.

doi:10.1088/1757-899X/1164/1/012089

4. Research results

The results of determining the modulus of concrete deformation are given in table 1.

Table 1. Results of determining the deformation modulus of concrete on sulfate-resistant slag Portland cement grade 400.

Sample group number	Sample brand	Curing conditions	Modulus of concrete deformation, MPa, under load	
			$0.3f_{cd}$	$0.5f_{cd}$
1	СК5	М	$3.04 \cdot 10^4$	$2.65 \cdot 10^4$
2	СК3	М1	$2.5 \cdot 10^4$	$2.21 \cdot 10^4$
3	СК9	Ж1	$2.74 \cdot 10^4$	$2.35 \cdot 10^4$

As can be seen from table 1, the values of the modulus of deformation of concrete on slag Portland cement coincide practically with the data given in the literature for heavy concrete on Portland cement [15, 16, 17].

The study results of the concrete shrinkage deformations are presented in figures 2 and 3.

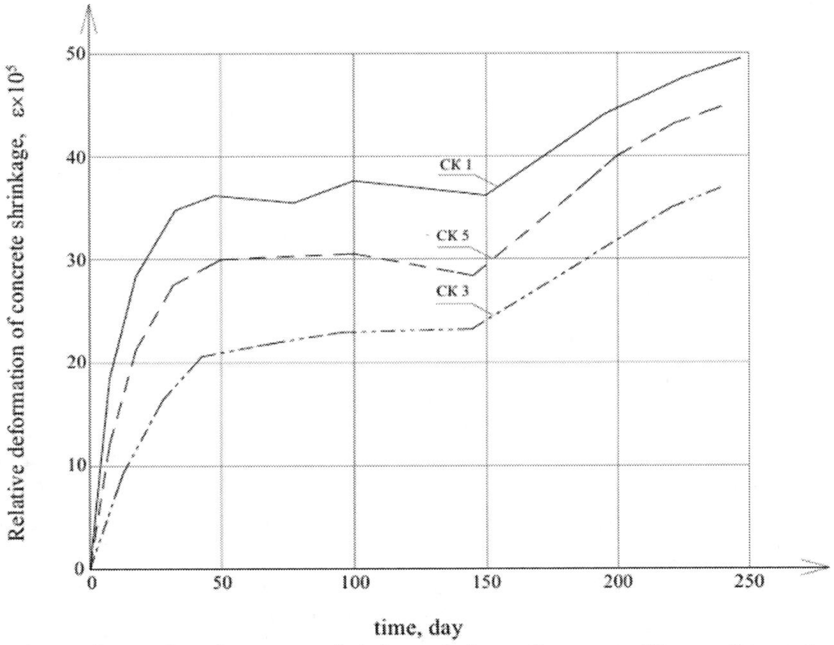

Figure 2. The study results of concrete shrinkage deformations on sulfate-resistant slag Portland cement without plasticizing additives.

The greatest amount of shrinkage is observed in samples hardened under normal conditions without preliminary thermal and moisture treatment (СК 1). Below there is a curve corresponding to a softer mode of thermal and moisture treatment (СК 5). The smallest shrinkage was observed in specimens hardened in a more severe mode (СК 3).

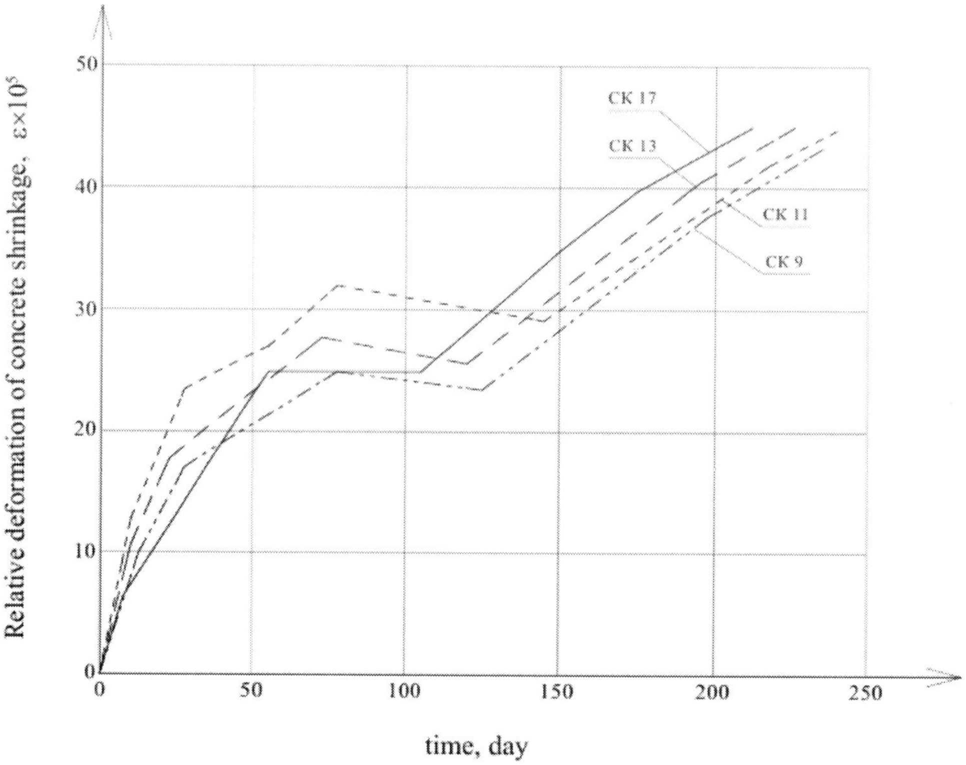

Figure 3. The study results of concrete shrinkage deformations on sulfate-resistant slag Portland cement with the introduction of a superplasticizer in the form of lignosulfonate (LCT).

As can be seen from the graph of the kinetics of shrinkage deformations of concrete without additives, figure 2, the curves are characterized by reaching a plateau parallel to the abscissa axis by 40-50 days. Subsequently, shrinkage deformations begin to increase from the moment of time 150 days, and second part of the curve has a damping character by 250 days. The noted regularities also take place for the graphs of the kinetics of concrete shrinkage deformations of with additives, figure 3, but they are less pronounced.

It is known that concrete shrinkage deformations arise from a reduction in the volume of the cement-water system during cement hydration [18-21]. The values of the concrete relative deformation of shrinkage on ordinary Portland cement by 180 days at a water-cement ratio of 0.4 are $4\text{-}8 \cdot 10^{-4}$, depending on the ratio between the aggregate and the cement [22].

The data obtained as a result of experimental studies indicate that, in general, the relative shrinkage deformations of concretes on a slag Portland cement binder, both with and without additives, are in close to the values of the concrete relative shrinkage deformations on ordinary Portland cement in order of magnitude.

The study results of concrete on sulfate-resistant slag Portland cement on creep are presented in the form of a graph, figure 4.

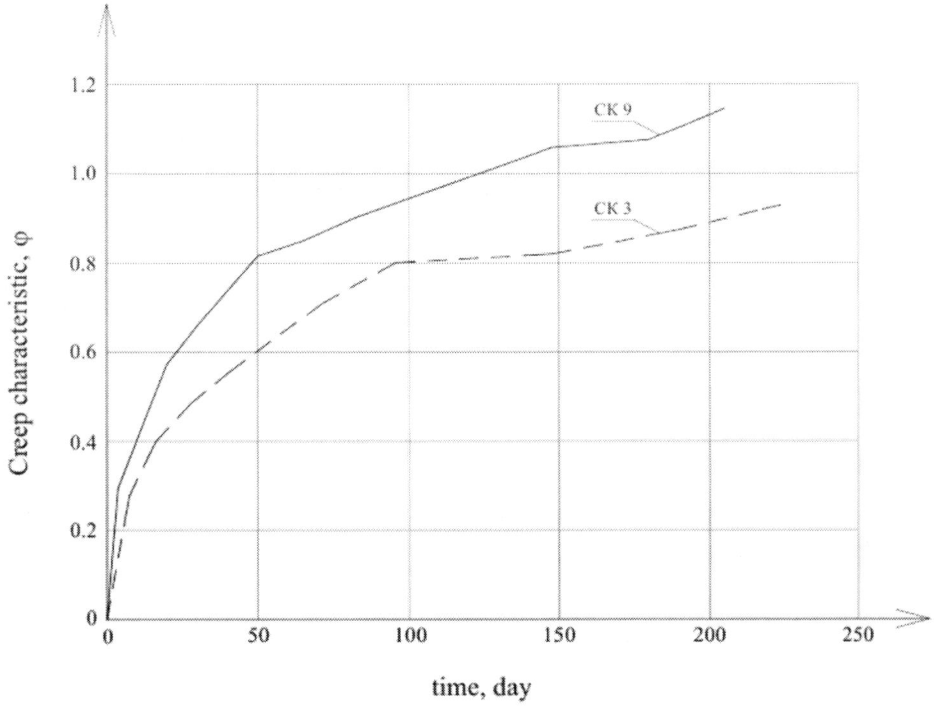

Figure 4. The study results of concrete on sulfate-resistant slag Portland cement on creep.

Time was plotted along the horizontal axis on the graph, and the creep characteristic φ was plotted along the vertical axis. It is the ratio of the amount of plastic deformation at a given time to the amount of elastic deformation at the time of loading. The obtained data indicate that, in all experiments, creep deformations are damped. They are lower in samples that have undergone thermal and moisture treatment in a milder mode (lower curve).

In accordance with the data available in the literature [14-22], the value of the creep relative deformation of concrete on Portland cement when using gravel aggregate is about $6 \cdot 10^{-4}$ 0,5 years after the start of loading.

From the graph shown in Figure 4, it can be seen that the greatest value of the creep characteristic for concretes based on slag Portland cement is 0,9 – 1,1. It corresponds to the value of the relative creep deformation, equal to $(2{,}7 - 3{,}2) \cdot 10^{-4}$ also half a year after the start of loading. It follows from this that the relative creep deformations for concretes both on slag Portland cement and ordinary Portland cement are of the same order.

5. Conclusions

Experimental studies of the deformation properties of concretes on sulfate-resistant slag Portland cement have shown that it has:

1) shrinkage deformations of the same order as that of concrete on Portland cement;

2) the same order of magnitude of creep deformation as for concretes on Portland cement. This magnitude has a damping character at loads of 30% of the destructive value.

The following conclusions can be drawn based on the analysis of the kinetics of the development of concrete shrinkage deformations on a slag Portland cement binder:

1) the nature of the curves for all, without exception, samples hardened under different conditions with and without additives of a superplasticizer in the form of lignosulfonate is the same;

2) for concretes without additives, the kinetic curves of the development of shrinkage deformations have virtually the same shape, and the quantitative differences between them are relatively small;

3) the curves of the kinetics of shrinkage deformations of concretes with additives under various

conditions of hardening are more closely spaced. The influence of the modes of thermal and moisture treatment on shrinkage deformations for concretes with additives is less pronounced.

The deformation moduli of concrete on sulfate-resistant slag Portland cement under different hardening conditions practically coincide with the values of the deformation modulus for heavy concretes on Portland cement.

The creep characteristic for concretes based on slag Portland cement is close in order of magnitude to the corresponding characteristic of concrete on ordinary Portland cement.

It was found that elastic deformations and shrinkage and creep deformations are among the structure-sensitive characteristics. This can explain the influence of the steaming mode on their values.

Thus, the studies carried out have shown that concrete based on slag Portland cement binder under use plasticizing additives, in all basic deformation characteristics is at the level of concrete on Portland cement binder and can be used for the construction of prefabricated coverings for agricultural airfields.

References

[1] Ovchinnikov R V and Avakyan A G 2014 Modifikatsiya portlandtsementa zoloshlakovymi otkhodami *Novyye tekhnologii* 02

[2] Cherepanov A A and Kardash V T 2009 Kompleksnaya pererabotka zoloshlakovykh otkhodov TETS (rezul'taty laboratornykh i polupromyshlennykh ispytaniy) *Geologiya i poleznyye iskopayemyye Mirovogo okeana* **2** 98

[3] Fedynin N I 1998 Rol' chastits nesgorevshego topliva v formirovanii svoystv yacheistogo zolobetona *Stroitel'nyye materialy* **9** pp 26-29

[4] Korovkin M O, Yeroshkina N A and Zamchalin M N 2015 Vliyaniye mineral'no-khimicheskoy dobavki na osnove kotel'nogo shlaka na tverdeniye tsementa *Sovremennyye problemy nauki i obrazovaniya*

[5] Tkach Ye V, Rakhimov M A, Toimbayeva B M and Rakhimova G M 2012 Vliyaniye organomineral'nogo modifikatora na fiziko-mekhanicheskiye i deformativnyye svoystva betona *Fundamental'nyye issledovaniya* **3-2** pp 428–431

[6] Anisimov S N, Kononova O V, Leshkanov A Yu and Smirnov A O 2014 Issledovaniye vliyaniya kompleksa modifikatorov na kinetiku tverdeniya betonov *Sovremennyye problemy nauki i obrazovaniya* **4**

[7] Belova V V, Subbotina S L and Kulyayeva P V 2015 Prochnostnyye i deformativnyye svoystva betonov s karbonatnymi mikronapolnitelyami *Stroitel'nyye materialy* **3** pp 25–29

[8] Bondarev B A, Chernousov N N, Chernousov R N and Sturova V A 2017 Issledovaniye deformativnykh svoystv stale-fibroshlakobetona pri osevom rastyazhenii i szhatii s uchetom yego vozrasta *Vestnik Permskogo natsional'nogo issledovatel'skogo politekhnicheskogo universiteta. Stroitel'stvo i arkhitektura* **8(1)** pp 18–31 DOI: 10.15593/2224-9826/2017.1.02

[9] Hamelyak I P 2011 Pro efektyvnist' vykorystannya vysokomitsnoho tsementobetonu dlya budivnytstva zhorstkykh pokryttiv *Avtomobil'ni dorohy i dorozhnye budivnytstvo* (Kyiv: NTU) **81** 30

[10] Chistyakov V V, Shurgaya A G, Doroshenko Yu M and Chizhenko N P 2012 Modifitsirovannyye tsementobetony dlya pokrytiya dorog *Stroitel'nyye materialy, izdeliya i sanitarnaya tekhnika* **43** 212

[11] Yanin O, Yemelianov T and Novikova S 2020 Experimental investigations of concrete on slag portland cement as a coating material for agricultural aerodromes *Key Engineering Materials* **864** pp 19–26

[12] Weia Y and Hansen W 2013 Early-age strain–stress relationship and cracking behavior of slagcement mixtures subject to constant uniaxial restraint *Construction and Building Materials* **49** pp 635–642

[13] Said Kenai, Walid Yahiaoui and Belkacem Menadi Effect of granulated blast furnace slag on the durability of self-compacting concrete in hot environment Service *Life of Cement-Based Materials and Structures* **79**

[14] Pane I and Hansen W 2002 Early age creep and stress relaxation of concrete containing blended cements *Materials and Structures* **35(2)** pp 92-96

[15] Eurocode-2 2004 *Design of Concrete Structures* Part 1-1: General Rules and Rules for Building: EN 1992-1-1 Brussels: CEN 225

[16] DBN V.2.6-98:2009 2011 *Betonni ta zalizobetonni konstruktsiy* (Osnovni polozhennya) Ministerstvo rehional'noho rozvytku ta budivnytstva Ukrayiny Kyiv 71

[17] *Proyektirovaniye zhelezobetonnykh konstruktsiy: Spravochnoye posobiye* 1985 Pod red. A B Golysheva (Kyiv: Budivel'nik) 496

[18] Lu T, Li Z and Breugel K 2020 Modelling of autogenous shrinkage of hardening cement paste *Construction and Building Materials* **264** 120708

[19] Malbois M, Socie A, Darquennes A, De Sa C and Benboudjema F 2016 Experimental and numerical analysis of drying shrinkage on cement-based materials *Int. RILEM Conf. on Materials, Systems and Structures in Civil Engineering* **1** (Denmark: Lyngby/Technical University) 325

[20] Wei Y, Hansen W, Biernacki J J and Schlangen E 2011 Unified shrinkage model for concrete from autogenous shrinkage test on paste with and without ground-granulated blast-furnace slag ACI *Mater. J.* **108** pp 12-20

[21] Li Z, Lu T, Liang X, Dong H and Ye G 2020 Mechanisms of autogenous shrinkage of alkali-activated slag and fly ash pastes *Cement and Concrete Research* **135** 106107

[22] Nevill' A M 1972 *Svoystva betona*: Sokr. per. s angl. (Moskva) 344

Application of improved gradient projection method to parametric optimization of steel lattice portal frame

V V Yurchenko[1], I D Peleshko[2] and Nikita Biliaiev[3]

[1]Department of Steel and Timber Structures, Kyiv National University of Construction and Architecture, 31, Povitroflotsky av., Kyiv, 03680, Ukraine

[2]Building Production Department, Lviv Polytechnic National University, 6, Karpins'kogo str., ap. 419, 79013, Lviv, 79013

[3]BM Prefab Engineering, 14a, Nikolya Ushakova str., Kyiv, 03179, Ukraine

Corresponding author's e-mail address: iurchenko.vv@knuba.edu.ua

Abstract. The paper has proposed a mathematical model for parametric optimization problem of the steel lattice portal frame. The design variable vector includes geometrical parameters of the structure (node coordinates), as well as cross-sectional dimensions of the structural members. The system of constraints covers load-carrying capacities constraints formulated for all design sections of structural members of the steel structure subjected to all ultimate load case combinations. The displacements constraints formulated for the specified nodes of the steel structure subjected to all serviceability load case combinations have been also included into the system of constraints. Additional requirements in the form of constraints on lower and upper values of the design variables, constraints on permissible minimal thicknesses, constraints on permissible maximum diameter-to-thickness ratio for the structural members with circle hollow sections, as well as the conditions for designing gusset-less welded joints between structural members with circle hollow sections have been also considered in the scope of the mathematical model. The method of the objective function gradient projection onto the active constraints surface with simultaneous correction of the constraints violations has been used to solve the formulated parametric optimization problem. New optimal layouts of the steel lattice portal frame by the criterion of the minimum weight, as well as minimum costs on manufacturing and erection have been presented.

1. Introduction

Over the past 50 years, numerical optimization and the finite element method have individually made significant advances and have together been developed to make possible the emergence of structural optimization as a potential design tool. In recent years, great efforts have been also devoted to integrate optimization procedures into the CAD facilities. With these new developments, lots of computer packages are now able to solve relatively complicated industrial design problems using different structural optimization techniques [1].

Applied optimum design problems for bar structures in some cases are formulated as parametric optimization problems, namely as searching problems for unknown structural parameters, which provide an extreme value of the specified purpose function in the feasible region defined by the specified constraints [2]. In this case, structural optimization is performed by variation of the structural

Content from this work may be used under the terms of the Creative Commons Attribution 3.0 licence. Any further distribution of this work must maintain attribution to the author(s) and the title of the work, journal citation and DOI.
Published under licence by IOP Publishing Ltd

parameters when the structural topology, cross-section types and node type connections of the bars, the support conditions of the bar system, as well as loading patterns and load design values are prescribed and constants.

Kibkalo et al. in the paper [3] formulated a parametric optimization problem for thin-walled bar structures and considered methods to solve them. The searching for the optimum solution has been performed by varying the structural parameters providing the required load-carrying capacity of structural members and the minimum value of manufacturing costs.

Alekseytsev has described the process of developing a parametrical-optimization algorithm for steel trusses in the paper [4]. Parametric optimization has been performed taking into account strength, stability and stiffness constraints formulated for all truss members. The objective function has been formulated depending on the specific manufacturing of the truss panel joints in term of the manufacturing cost calculated based on the labor costs and materials used.

Serpik et al. in the paper [5] developed an algorithm for parametric optimization of steel flat rod systems. The optimization problem has been formulated as a structural weight minimization problem taking into account strength and displacement constraints, as well as overall stability constraints. The cross-sectional dimensions of the truss members and the coordinates of the truss panel joints have been considered as design variables. The structural analysis of internal forces and displacements for considered structures has been performed using the finite element method. An iterative procedure for searching for optimum solution has been proposed in [6].

Sergeyev et al. in the paper [7] formulated a parametric optimization problem with constraints on faultless operation probability of bar structures with random defects. The weight of the bar structures has been considered as the objective function. Initial global imperfections have been considered as small independent random variables distributed according to normal distribution law, as well as buckling load value has been also considered as a random variable.

The mathematical model of the parametric optimization problem of structures includes a set of design variables, an objective function, as well as constraints, which reflect generally non-linear dependences between them [8]. If the purpose function and constraints of the mathematical model are continuously differentiable functions, as well as the search space is smooth, then the parametric optimization problems are successfully solved using gradient projection non-linear methods [9]. The gradient projection methods operate with the first derivatives or gradients only both of the objective function and constraints. The methods are based on the iterative construction of such a sequence of the approximations of design variables that provides convergence to the optimum solution (optimum values of the structural parameters).

Additionally, a sensitivity analysis is a useful optional feature that could be used in scope of the numerical algorithms developed based on the gradients methods [10, 11]. Thus, in the paper [12] Sergeyev et al. formulated a parametric optimization problem of linearly elastic space frame structures taking into account the stress and multiple natural frequency constraints. The cross-sectional parameters of structural members as well as node positions of the considered bar structures has been considered as design variables. The sensitivity analysis of multiple frequencies has been performed using analytic differentiation with respect to the design variables. The optimal design of the structure has been obtained by solving a sequence of quadratic programming problems.

In this paper, steel lattice portal frame is considered as research object, which investigated for the searching for optimum parameters of the structural form. The following research tasks are formulated: to develop a mathematical model for parametric optimization of the considered steel structures taking into account load-carrying capacities and stiffness constraints; to propose a numerical algorithm for parametric optimization of the steel structures based on the gradient projection method; to confirm the validity of the optimum solutions obtained using the proposed methodology based on numerical examples.

2. Problem formulation for parametric optimization of steel structures

Let us consider a parametric optimization problem of a structure consisting of bar members. The problem statement can be performed taking into account the following assumptions widely used in structural mechanic problems: the material of the structure is ideal elastic; the bar structure is deformable linearly; external loadings applied to the structure are quasi-static.

Let us also formulate the following pre-conditions for calculation: cross-section types and dimensions of structural members are constant along member lengths; external loadings are applied to the structural members without eccentricities relating to the center of mass and shear center of its cross-sections; an additional restraining by stiffeners are provided in the design sections where point loads (reactions) applied with the exception of cross-section warping and local buckling of the cross-section elements; load-carrying capacity of the structural joints, splices and connections are provided by additional structural parameters do not covered by the considered parametric optimization problem.

A parametric optimization problem of the structure can be formulated as presented below: *to find optimum values for geometrical parameters of the structure and member's cross-section dimensions, which provide the extreme value of the determined optimality criterion and satisfy all load-carrying capacities and stiffness requirements*. We assume, that the structural topology, cross-section types and node type connections of the bars, the support conditions of the bar system, as well as loading patterns are prescribed and constants.

The formulated parametric optimization problem can be considered integrally using the mathematical model in the form of the non-linear programming task including an objective function, a set of independent design variables and constraints, which reflect generally non-linear dependences between them. The validity of the mathematical model can be estimated by the compliance of its structure with the design code requirements.

The parametric optimization problem of steel structures can be stated in the following mathematical terms: to find unknown structural parameters $\vec{X}=\{X_\iota\}^T$, $\iota=\overline{1,N_X}$ (N_X is the total number of the design variables), providing the least value of the determined objective function:

$$f^*=f\left(\vec{X}^*\right)=\min_{\vec{X}\in\Im_N} f\left(\vec{X}\right); \qquad (2.1)$$

in a feasible region (search space) $\Im$ defined by the following system of constraints:

$$\boldsymbol{\psi}\left(\vec{X}\right)=\left\{\psi_\kappa\left(\vec{X}\right)=0 \mid \kappa=\overline{1,N_{EC}}\right\}; \qquad (2.2)$$

$$\boldsymbol{\varphi}\left(\vec{X}\right)=\left\{\phi_\eta\left(\vec{X}\right)\le 0 \mid \eta=\overline{N_{EC}+1,N_{IC}}\right\}; \qquad (2.3)$$

where $\vec{X}$ is the vector of the design variables (unknown structural parameters); f, ψ_κ, φ_η are the continuous functions of the vector argument; $\vec{X}^*$ is the optimum solution or optimum point (the vector of optimum values of the structural parameters); f^* is the optimum value of the optimum criterion (objective function); N_{EC} is the number of constraints-equalities $\psi_\kappa\left(\vec{X}\right)$, which define hyperplanes of the feasible solutions; N_{IC} is the number of constraints-inequalities $\varphi_\eta\left(\vec{X}\right)$, which define a feasible region in the design space $\Im$.

The vector of the design variables comprises of unknown geometrical parameters of the structure $\vec{X}_G=\{X_{G.\chi}\}^T$, $\chi=\overline{1,N_{X.G}}$, and unknown cross-sectional dimensions of the structural members $\vec{X}_{CS}=\{X_{CS.\alpha}\}^T$, $\alpha=\overline{1,N_{X.CS}}$:

$$\vec{X}=\left\{\vec{X}_G,\vec{X}_{CS}\right\}^T=\left\{\{X_{G.\chi}\},\{X_{CS.\alpha}\}\right\}^T; \qquad (2.4)$$

where $N_{X,G}$ is the total number of unknown node coordinates of the steel structure; $N_{X,CS}$ is the total number of unknown cross-sectional dimensions of the structural members, $N_{X,G}+N_{X,CS}=N_X$.

The specific technical-and-economic index (material weight, material cost, construction cost etc.) or another determined indicator can be considered as the objective function (2.1) taking into account the ability to formulate its analytical expression as a function of design variables $\vec{X}$.

Load-carrying capacities constraints (strength and stability inequalities) for all design sections of the structural members subjected to all design load combinations at the ultimate limit state as well as displacements constraints (stiffness inequalities) for the specified nodes of the bar system subjected to all design load combinations at the serviceability limit state should be included into the system of constraints (2.2) – (2.3). Additional requirements which describe structural, technological and serviceability particularities of the considered structure can be also included into the system of equations (2.2) – (2.3).

The design internal forces in the structural members used in the strength and stability inequalities of the system of equations (2.2) – (2.3) are considered as state variables depending on design variables $\vec{X}$ and can be calculated from the following linear equations system of the finite element method [13]:

$$\mathbf{K}\left(\vec{X}_G,\vec{X}_{CS}\right)\times\vec{z}_{ULS,k}=\vec{p}_{ULS,k}\left(\vec{X}_G,\vec{X}_{PS}\right),\ k=\overline{1,N_{LC}^{ULS}}\ ; \quad (2.5)$$

where $\mathbf{K}\left(\vec{X}_G,\vec{X}_{CS}\right)$ is the stiffness matrix of the finite element model of the bar system, which should be formed depending on the unknown (variable) cross-sectional dimensions of the structural members $\vec{X}_{CS}$, as well as unknown (variable) node coordinates of the structure $\vec{X}_G$; $\vec{p}_{ULS,k}\left(\vec{X}_G,\vec{X}_{PS}\right)$ is the column-vector of the node's loads for k th design load combination of the ultimate limit state, which should be formed depending on unknown (variable) initial pre-stressing forces $\vec{X}_{PS}$, as well as unknown (variable) node coordinates of the structure $\vec{X}_G$; $\vec{z}_{ULS,k}$ is the result column-vector of the node displacements for k th design load combination of the ultimate limit state, $\vec{z}_{ULS,k}=\mathbf{Z}_{\mathbf{FEM},k}^{ULS}\left(\vec{X}_G,\vec{X}_{CS},\vec{X}_{PS}\right)=\mathbf{Z}_{\mathbf{FEM},k}^{ULS}\left(\vec{X}\right)$; N_{LC}^{ULS} is the number of the design ultimate load combinations. For each i th design section of j th structural member subjected to k th ultimate design load combination the design internal forces (axial force, bending moments and shear forces) can be calculated depending on node displacement column-vector $\vec{z}_{ULS,k}$.

The node displacement of the bar system used in stiffness inequalities of the system of equations (2.2) – (2.3) are also considered as state variables depending on design variables $\vec{X}$ and can be calculated from the following linear equations system of the finite element method [13]:

$$\mathbf{K}\left(\vec{X}_G,\vec{X}_{CS}\right)\times\vec{z}_{SLS,k}=\vec{p}_{SLS,k}\left(\vec{X}_G,\vec{X}_{PS}\right),\ k=\overline{1,N_{LC}^{SLS}}\ ; \quad (2.6)$$

where $\vec{p}_{SLS,k}\left(\vec{X}_{PS}\right)$ is the column-vector of the node's loads for k th design load combination of the serviceability limit state, which should be formed depending on unknown (variable) initial pre-stressing forces $\vec{X}_{PS}$, as well as unknown (variable) node coordinates of the structure $\vec{X}_G$; $\vec{z}_{ULS,k}$ is the result column-vector of the node displacements for k th design load combination of the serviceability limit state, $\vec{z}_{SLS,k}=\mathbf{Z}_{\mathbf{FEM},k}^{SLS}\left(\vec{X}_G,\vec{X}_{CS},\vec{X}_{PS}\right)=\mathbf{Z}_{\mathbf{FEM},k}^{SLS}\left(\vec{X}\right)$; N_{LC}^{SLS} is the number of the design serviceability load combinations. For each m th node of the finite element model subjected to k th serviceability design load combination the design vertical and horizontal displacements can be calculated depending on node displacement column-vector $\vec{z}_{SLS,k}$.

3. An improved gradient projection method for solving the formulated parametric optimization problem

The parametric optimization problem stated as non-linear programming task by equations (2.1) – (2.3) can be solved using a gradient projection method. The method of objective function gradient projection onto the active constraints surface with simultaneous correction of the constraints violations ensures effective searching for solution of the non-linear programming tasks occurred when optimum designing of the building structures [15, 16].

The gradient projection method operates with the first derivatives or gradients only of both the objective function (2.1) and constraints (2.2) – (2.3). The method is based on the iterative construction of such sequence presented by equation (3.1) of the approximations of the design variables $\vec{X} = \{X_i\}^T$, $i = \overline{1, N_X}$, that provides the convergence to the optimum solution (optimum values of the structural parameters):

$$\vec{X}_{t+1} = \vec{X}_t + \Delta\vec{X}_t; \tag{3.1}$$

where $\vec{X}_t = \{X_i\}^T$, $i = \overline{1, N_X}$ is the current approximation to the optimum solution $\vec{X}^*$ that satisfies both constraints-equalities (2.2) and constraints-inequalities (2.3) with the extreme value of the objective function (2.1); $\Delta\vec{X}_t = \{\Delta X_i\}^T$, $i = \overline{1, N_X}$, is the increment vector for the current values of the design variables $\vec{X}_t$; t is the iteration's index. The start point of the iterative searching process $\vec{X}_{t=0}$ can be assigned as engineering estimation of the admissible design of the structure.

The active constraints only of the constraints system (2.2) – (2.3) should be considered at each iteration. A set of active constraints numbers $\mathbf{A}$ calculated for the current approximation $\vec{X}_t$ to the optimum solution (current design of the structure) is determined as:

$$\mathbf{A} = \boldsymbol{\kappa} \cup \boldsymbol{\eta},\ \boldsymbol{\kappa} = \left\{\kappa \,\middle|\, \left|\psi_\kappa\left(\vec{X}_t\right)\right| \geq -\varepsilon\right\},\ \boldsymbol{\eta} = \left\{N_{EC} + \eta \,\middle|\, \phi_\eta\left(\vec{X}_t\right) \geq -\varepsilon\right\}; \tag{3.2}$$

where ε is a small positive number introduced here in order to diminish the oscillations on movement alongside of the active constraints surface.

The increment vector $\Delta\vec{X}_t$ for the current values of the design variables $\vec{X}_t$ can be determined by the following equation:

$$\Delta\vec{X}_t = \Delta\vec{X}_\perp^t + \Delta\vec{X}_\parallel^t; \tag{3.3}$$

where $\Delta\vec{X}_\perp^t$ is the vector calculated subject to the condition of elimination the constraint's violations; $\Delta\vec{X}_\parallel^t$ is the vector determined taking into consideration the improvement of the objective function value. Vectors $\Delta\vec{X}_\parallel^t$ and $\Delta\vec{X}_\perp^t$ are directed parallel and perpendicularly accordingly to the subspace with the vectors basis of the linear-independent constraint's gradients, such that:

$$\left(\Delta\vec{X}_\perp^t\right)^T \Delta\vec{X}_\parallel^t = 0. \tag{3.4}$$

The values of the constraint's violations for the current approximation $\vec{X}_t$ of the design variables are accumulated into the following vector:

$$\mathbf{V} = \left(\psi_\kappa\left(\vec{X}\right) \forall \kappa \in \boldsymbol{\kappa};\ \phi_\eta\left(\vec{X}\right) \forall \eta \in \boldsymbol{\eta}\right).$$

Let us introduce a set $\mathbf{L}$, $\mathbf{L} \subseteq \mathbf{A}$, of the constraint's numbers, such that the gradients of the constraints at the current approximation $\vec{X}_t$ to the optimum solution are linear-independent.

Component $\Delta\vec{X}'_{\perp}$ is calculated from the equation presented below:

$$\Delta\vec{X}'_{\perp} = [\nabla\varphi]\vec{\mu}_{\perp}; \tag{3.5}$$

where $[\nabla\varphi]$ is the matrix that consists of components $\frac{\partial\psi_{\kappa}}{\partial X_{\iota}}$ and $\frac{\partial\phi_{\eta}}{\partial X_{\iota}}$, here $\iota = \overline{1, N_X}$, $\kappa \in \mathbf{L}$, $\eta \in \mathbf{L}$; $\vec{\mu}_{\perp}$ is the column-vector that defines the design variables increment subject to the condition of elimination the constraint's violations. Vector $\vec{\mu}_{\perp}$ can be calculated as presented below.

In order to correct constraint's violations $\mathbf{V}$, vector $\Delta\vec{X}'_{\perp}$ to a first approximation should also satisfy Taylor's theorem for the continuously differentiable multivariable function in the vicinity of point $\vec{X}_{t}$ for each constraint from set $\mathbf{L}$, namely:

$$-\mathbf{V} = [\nabla\varphi]^T \Delta\vec{X}'_{\perp}. \tag{3.6}$$

With substitution of equation (3.5) into equation (3.6) we obtain the system of equations to determine column-vector $\vec{\mu}_{\perp}$:

$$[\nabla\varphi]^T [\nabla\varphi]\vec{\mu}_{\perp} = -\mathbf{V}. \tag{3.7}$$

Component $\Delta\vec{X}'_{\parallel}$ is determined using the following equation:

$$\Delta\vec{X}'_{\parallel} = \xi \times \vec{p}_{\nabla f} = \xi\left(\nabla\vec{f} - [\nabla\varphi]\vec{\mu}_{\parallel}\right), \tag{3.8}$$

where $\nabla\vec{f}$ is the vector of the objective function gradient in the current point (current approximation of the design variables) $\vec{X}_{t}$; $\vec{p}_{\nabla f}$ is the projection of the objective function gradient vector onto the active constraints surface in the current point $\vec{X}_{t}$; $\vec{\mu}_{\parallel}$ is the column-vector that defines the design variable's increment subject to the improvement of the objective function value. Column-vector $\vec{\mu}_{\parallel}$ can be calculated approximately using the least-square method by the following equation:

$$[\nabla\varphi]\vec{\mu}_{\parallel} \approx \nabla\vec{f}, \tag{3.9}$$

or from the equation presented below:

$$[\nabla\varphi]^T [\nabla\varphi]\vec{\mu}_{\parallel} = [\nabla\varphi]^T \nabla\vec{f}, \tag{3.10}$$

where ξ is the step parameter, which can be calculated subject to the desired increment Δf of the purpose function on movement along the direction of the purpose function anti-gradient. The increment Δf can be assign as 5...25% from the current value of the objective function $f(\vec{X}_{t})$:

$$\Delta f = \xi\left(\nabla\vec{f}\right)^T \nabla\vec{f},\ \xi = \frac{\Delta f}{\left(\nabla\vec{f}\right)^T \nabla\vec{f}}, \tag{3.11}$$

where in case of minimization the function (2.1) Δf and ξ accordingly have negative values. The parameter ξ can be also calculated using the dependency presented below:

$$\xi = \frac{\Delta f}{\left(\vec{p}_{\nabla f}\right)^T \nabla\vec{f}}, \tag{3.12}$$

that follows from the condition of attainment the desired increment of the objective function Δf on the movement along the direction of the objective function anti-gradient projection onto the active constraints surface. Step parameter ξ can be also selected as a result of numerical experiments performed for each type of the structure individually [17, 18].

Using equations (3.5) and (3.8), equation (3.3) can be rewritten as presented below:

$$\Delta\vec{X}_t = [\nabla\varphi]\,\vec{\mu}_\perp + \xi\left(\nabla\vec{f} - [\nabla\varphi]\,\vec{\mu}_\parallel\right), \tag{3.13}$$

or

$$\Delta\vec{X}_t = \xi\,\nabla\vec{f} + [\nabla\varphi]\left(\vec{\mu}_\perp - \xi\,\vec{\mu}_\parallel\right), \tag{3.14}$$

where column-vectors $\vec{\mu}_\perp$ and $\vec{\mu}_\parallel$ are calculated using equation (3.7) and equation (3.9) or equation (3.10), respectively.

The linear-independent constraints of the system (2.2) – (2.3) should be detected when constructing the matrix of the active constraints gradients $[\nabla\varphi]$ used by equation (3.7) and equation (3.9) or equation (3.10). Selection of the linear-independent constraints can be performed based on the equivalent transformations of the resolving equations of the gradient projection method using the non-degenerate transformation matrix $\mathbf{H}$, such that the sub-diagonal elements of the matrix $\mathbf{H}[\nabla\varphi]$ equal to zero. An orthogonal matrix of the elementary mapping (Householder's transformation) [19] has been used to select linear-independent constraints of the system (2.2) – (2.3), as well as to form triangular structure of the nonzero elements of matrix $\mathbf{H}[\nabla\varphi]$ [15].

Using Householder's transformations described above triangular structure of the nonzero elements of matrix $\mathbf{H}[\nabla\varphi]$ is formed step-by-step. Besides, equation (3.7) and equation (3.9) can be rewritten as follow:

$$\left([\nabla\varphi]^T\,\mathbf{H}^T\right)\left(\mathbf{H}[\nabla\varphi]\right)\vec{\mu}_\perp = -\mathbf{V}; \tag{3.15}$$

$$\mathbf{H}[\nabla\varphi]\vec{\mu}_\parallel \approx \mathbf{H}\nabla\vec{f}\,. \tag{3.16}$$

Equivalent Householder transformations of the resolving equations (2.15), (2.16) have been proposed by the paper [15]. They increase numerical efficiency of the algorithm developed based on the considered method.

In order to calculate column-vectors $\vec{\mu}_\perp$ and $\vec{\mu}_\parallel$, it is required only to perform forward and backward substitutions in equation (3.15) and equation (3.16).

To accelerate the convergence of the minimization algorithm presented above, h^{th} columns should be excluded from matrix $\mathbf{H}[\nabla\varphi]$. These columns correspond to those constraints from equation (2.3), for which the following inequality satisfies:

$$\mu_{\perp h} - \xi \times \mu_{\parallel h} > 0\,. \tag{3.17}$$

As presented by the papers [15, 16], when $\mu_{\perp h} - \xi \times \mu_{\parallel h} > 0$, then the return onto the active constraints surface from the feasible region $\Im$ is performed with simultaneous degradation of the objective function value. At the same time, in case of $\mu_{\perp h} - \xi \times \mu_{\parallel h} < 0$, both the improvement of the objective function value and the return from the inadmissible region onto the active constraints surface are performed.

When excluding h^{th} columns from matrix $\mathbf{H}[\nabla\varphi]$ corresponded to those constraints for which equation (3.17) is satisfied, the matrix $\left(\mathbf{H}[\nabla\varphi]\right)_{red}$ with a broken (non-triangular) structure of the nonzero elements is obtained. The set $\mathbf{L}$ of the linear-independent active constraints numbers transforms into the set $\mathbf{L}_{red}$ respectively. At the same time, the vector of the constraint's violations $\mathbf{V}$ reduced into the vector $\mathbf{V}_{red}$ accordingly. In order to restore the triangular structure of the matrix $\left(\mathbf{H}[\nabla\varphi]\right)_{red}$ with zero sub-diagonal elements, Givens transformations (Givens rotations) [19] can be used.

Considering Givens transformations, equation (3.15) and equation (3.16) for column-vectors $\left(\vec{\mu}_{\perp}\right)_{red}$ and $\left(\vec{\mu}_{\|}\right)_{red}$ can be rewritten as:

$$\left([\nabla\varphi]^T \mathbf{H}^T\right)_{red} \mathbf{G}^T\mathbf{G}\left(\mathbf{H}[\nabla\varphi]\right)_{red}\left(\vec{\mu}_{\perp}\right)_{red} = -\mathbf{V}_{red}; \tag{3.18}$$

$$\mathbf{G}\left(\mathbf{H}[\nabla\varphi]\right)_{red}\left(\vec{\mu}_{\|}\right)_{red} \approx \mathbf{GH}\nabla\vec{f}. \tag{3.19}$$

Equivalent transformations of the resolving equations (3.18), (3.19) using Givens rotations (transformations with matrix $\mathbf{G}$) ensure acceleration of the iterative searching process presented by equation (3.1) in those cases when equation 3.17) is considered due to decreasing the amount of calculations [15].

The main resolving equations (3.13) and (3.14) of the gradient method can be rewritten as presented below:

$$\Delta\vec{X}_t = \left(\mathbf{H}[\nabla\varphi]\right)_{red}\left(\vec{\mu}_{\perp}\right)_{red} + \xi\left(\nabla\vec{f} - \left(\mathbf{H}[\nabla\varphi]\right)_{red}\left(\vec{\mu}_{\|}\right)_{red}\right); \tag{3.20}$$

or

$$\Delta\vec{X}_t = \xi\,\nabla\vec{f} + \left(\mathbf{H}[\nabla\varphi]\right)_{red}\left(\left(\vec{\mu}_{\perp}\right)_{red} - \xi\left(\vec{\mu}_{\|}\right)_{red}\right). \tag{3.21}$$

It should be noted that the lengths of the gradient vectors for the objective function (2.1), as well as for constraints (2.2) – (2.3), remain as they were in scope of the proposed equivalent transformations ensuring the dependability of the optimization algorithm [15].

The determination the convergence criterion is the final question when using the iterative searching for the optimum point presented by equation (3.1) described above. Considering the geometrical content of the gradient steepest descent method, we can assume that at the permissible point $\vec{X}_t$ the component of the increment vector $\Delta\vec{X}_{\|}^t$ for the design variables should be vanish, $\Delta\vec{X}_{\|}^t \to 0$, in case of approximation to the optimum solution of the non-linear programming task presented by equations (2.1) – (2.4). So, the following convergence criterion of the iterative procedure presented by equation (3.1) can be assigned:

$$\left\|\Delta\vec{X}_{\|}^k\right\| = \sqrt{\sum_{t=1}^{N_X}\left(\Delta X_{\|,t}^k\right)^2} < \varepsilon_1, \tag{3.22}$$

where ε_1 is a small positive number. In the paper [15] the convergence criteria for the iterative procedure presented by equation (3.1) has been presented in detail.

4. Mathematical model for parametric optimization of the lattice portal frame

A parametric optimization task for lattice portal frame of the steel warehouse framework designed as repository for the granulated sulfur has been considered. Building object locates in seaport Ust-Luga of Russian Federation The general building sizes are length 247,25 m and width 69,0 m. Steel framework of the building consists of portal frames with span 69,0 m positioned along building length with bay 7,5 m.

Steel portal frames consist of the lattice structural members fabricated from pipes with steel grade St20 according to design code [14]. Joints in the lattice structural members were designed using welded connections without gussets.

There is a service platform at the level +28,25 m provided for supporting the crane-loader and conveyor (see figure 4.1). The load-bearing structures of the platform are suspended to the structural members of the portal frames. Welded I-beams of this platform were manufactured from the universal steel sheets of grade S245.

Figure 4.1. Assembling process on job site (Photo has been provided by V. Shymanovsky Ukrainian Research and Design Institute of Steel Construction)

Design scheme of the steel lattice portal frame was assumed as a hinged-bar structure with hinged column bases. Geometrical scheme of the portal frame was described using the set of nodes and bars with orientation on implementation of the finite element method for linear static analysis. Node coordinates of the design scheme were determined in Cartesian coordinate system and presented as expressions in dependence of geometrical design variables of the optimization task.

Dead loads included self weight of the portal frame, roof purlins, roof bracings, fire escape staircase and mezzanines, profiled panels which is used as non-warmth-keeping walling as well as service loads on fire escape staircase and mezzanines. Safety factors for the design loads and effects as well as safety factor for the building responsibility were defined according to [14]. Live loads (or technological loads) were accepted according to the target specification. Calculation the design values for climate loadings has been performed according to the requirements [14]. So, tree types of snow loads and two types of wind load have been considered when optimum designing of the steel portal frames.

Design loads and effects have been combined in 16 design load case combinations taken into account the combination factors according to requirements of design code [14]. All loads and effects on the structure were presented as concentrated loads at the nodes and were determined analytically depending on the variable parameters of the geometrical scheme.

Mathematical model of the parametric optimization task for the steel portal frame with lattice structural members has been formulated as nonlinear programming task including the set of design variables, system of constraints as well as specified purpose function.

4.1. Design variables.

Parameters of the geometric scheme of the portal frames have been considered as design variables. Variable parameters of the geometrical scheme were building height at the eave node H_{zdk} and at the ridge H_{zd}, distance between upper and lower chords of the lattice rafter at the eave node h_{op} and at the erection joints h_2, h_4, h_6 i h_7, distance between chords of the lattice column at the eave node b_{op} and parameter b_7 (see figure 4.2). Start values for the design variables were accepted according to the design decision of the steel framework developed by Open Join-Stock Company "V. Shymanovsky Ukrainian Research and Design Institute of Steel Construction", namely: $H_{zd} = 39,58$ m, $H_{zdk} = 10,63$ m, $h_2 = h_4 = h_6 = h_7 = h_{op} = b_{op} = 2,6$ m, $b_7 = 0,56$ m. Additionally, cross-sectional sizes of the lattice structural members for each stiffness type were considered as design variables (see table 4.1).

Table 4.1. Variable cross-sectional sizes of lattice structural elements of transverse frames

Destination and location of structural member	Stiffness type number	Design variables name, diameter× thickness	Start values, mm×mm	Stiffness type number	Design variables name, diameter× thickness	Start values, mm×mm
Chords of the lattice structural members	1	$d_1 \times t_1$	299×25	7	$d_6 \times t_7$	299×16
	2	$d_1 \times t_2$	299×14	8	$d_8 \times t_8$	299×10
	3	$d_1 \times t_3$	299×10	9	$d_9 \times t_9$	299×10
	4	$d_1 \times t_4$	299×14	10	$d_{10} \times t_{10}$	299×14
	5	$d_1 \times t_5$	299×14	11	$d_{11} \times t_{11}$	180×12
	6	$d_6 \times t_6$	299×10	–	–	–
Elements of the lattice rafters	12	$d_{12} \times t_{12}$	152×8	14	$d_{14} \times t_{14}$	102×5
	13	$d_{13} \times t_{13}$	121×8	–	–	–
Frame ridge	15	$d_{15} \times t_{15}$	152×8	17	$d_{17} \times t_{17}$	180×12
	16	$d_{16} \times t_{16}$	102×5	18	$d_{18} \times t_{18}$	180×12
Elements of the lattice columns	19	$d_{19} \times t_{19}$	299×10	21	$d_{21} \times t_{21}$	299×25
	20	$d_{19} \times t_{20}$	299×25	22	$d_{22} \times t_{22}$	102×5
Suspension arm of the service platform	23	$d_{23} \times t_{23}$	180×12	–	–	–

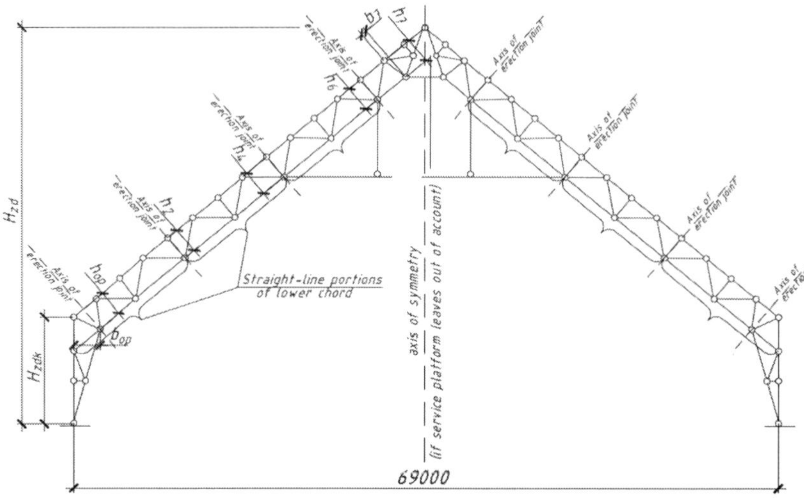

Figure 4.2. The design scheme of the portal frame with specification of the variable geometrical parameters.

4.2. System of constraints.
The system of constraints (2.2) – (2.3) should cover strength and stability constraints formulated for all design sections of all structural members of the considered steel structure subjected to all design load combinations at the ultimate limit state. The following *strength constraints* have been included in the system of constraints (2.2) – (2.3), formulated for all design sections, $\forall i = \overline{1, N_{DS}}$ (N_{DS} is the total number of the design sections in structural members), of all structural members, $\forall j = \overline{1, N_B}$ (N_B is the total number of the structural members), subjected to all ultimate load case combination, $\forall k = \overline{1, N_{LC}^{ULS}}$, namely normal stresses verifications:

$$\frac{N_{ijk}\left(\vec{X}\right)}{A_{n,j}\left(\vec{X}_{CS}\right)R_{y,j}\gamma_c} - 1 \le 0; \tag{4.1}$$

where $\sigma_{ijk}\left(\vec{X}\right) = \frac{N_{ijk}\left(\vec{X}\right)}{A_{n,j}\left(\vec{X}_{CS}\right)}$ is the value of the normal stresses caused by axial force $N_{ijk}\left(\vec{X}\right)$ acting in i^{th} design section of jth structural member subjected to k^{th} ultimate load case combination calculated from the linear equations system of the finite element method presented by equation (2.5); $A_{n,j}\left(\vec{X}_{CS}\right)$ is the net cross-sectional area of jth structural member calculated depending on the variable cross-sectional dimensions of the structural members $\vec{X}_{CS}$; γ_c is the safety factor [14]; $R_{y,j}$ is the design strength for steel member subjected to tension, bending and compression; $R_y\gamma_c$ are allowable value for normal stresses [14]; $\sigma_{x,ijk}\left(\vec{X}\right)$ are normal stresses at the specified cross-section point caused by internal forces acting in i^{th} design section of jth structural member subjected to k^{th} ultimate load case combination calculated from the linear equations system of the finite element method presented by equation (2.5). The value of the normal stresses $\sigma_{x,ijk}\left(\vec{X}\right)$ at the specified cross-section point has been calculated depending on the variable geometrical parameters of the structure $\vec{X}_G$ and variable cross-sectional dimensions of the structural members $\vec{X}_{CS}$.

The following *constraints on slenderness* of the structural members have been included in the system of constraints (2.2) – (2.3), $\forall j = \overline{1, N_B}$:

$$\frac{l_{ef,y,j}\left(\vec{X}_G\right)}{i_{y,j}\left(\vec{X}_{CS}\right)\lambda_{uy,j}} - 1 \le 0; \tag{4.2}$$

$$\frac{l_{ef,z,j}\left(\vec{X}_G\right)}{i_{z,j}\left(\vec{X}_{CS}\right)\lambda_{uz,j}} - 1{,}0 \le 0; \tag{4.3}$$

where $i_{y,j}\left(\vec{X}_{CS}\right)$ and $i_{z,j}\left(\vec{X}_{CS}\right)$ are radiuses of inertia for j^{th} structural member's design cross-section relative to the main axis of inertia and calculated depending on the variable cross-sectional dimensions of the structural members $\vec{X}_{CS}$; $l_{ef,y,j}\left(\vec{X}_G\right)$ and $l_{ef,z,j}\left(\vec{X}_G\right)$ are design lengths for j^{th} structural member in the main planes of inertia calculated depending on the variable geometrical parameters of the structure $\vec{X}_G$; $\lambda_{uy,j}$ and $\lambda_{uz,j}$ are the ultimate slenderness for j^{th} structural member. Design lengths of the structural members $l_{ef,y,j}\left(\vec{X}_G\right)$ and $l_{ef,z,j}\left(\vec{X}_G\right)$ were defined according to [5] as: for chords, support diagonals and support columns of the lattice structural members – $l_{ef,y,j}\left(\vec{X}_G\right) = l_j\left(\vec{X}_G\right)$; $l_{ef,z,j}\left(\vec{X}_G\right) = l_{1,j}$; for other elements of the lattice structural members – $l_{ef,y,j}\left(\vec{X}_G\right) = 0{,}85 l_j\left(\vec{X}_G\right)$; $l_{ef,z,j}\left(\vec{X}_G\right) = 0{,}85 l_{1,j}$; here l_j is the geometrical length for j^{th} bar of lattice structural member; $l_{1,j}$ is the distance between out-of-plane restraints of the member from the horizontal displacements in out-of-plane direction. Ultimate values for the slenderness of the lattice structural members were specified according to [5] as: $\lambda_{uy,j} = \lambda_{uz,j} = 400$ – for all tensioned members; $\lambda_{uy,j} = \lambda_{uz,j} = 150$ – for all compressed members.

The following *stability constraints* have been included in the system of constraints (2.2) – (2.3), formulated for all design sections, $\forall i = \overline{1, N_{DS}}$, of the structural members subjected to all ultimate load case combination, $\forall k = \overline{1, N_{LC}^{ULS}}$, namely flexural buckling verifications for all structural members subjected to axial compression force $N_{ijk}\left(\vec{X}\right)$, $\forall j = \overline{1, N_B}$:

$$\frac{N_{ijk}\left(\vec{X}\right)}{\varphi_{y,j}\left(\vec{X}_G, \vec{X}_{CS}\right) A_j\left(\vec{X}_{CS}\right) R_{y,j}\gamma_c} - 1 \le 0; \tag{4.4}$$

$$\frac{N_{ijk}\left(\vec{X}\right)}{\varphi_{z,j}\left(\vec{X}_G, \vec{X}_{CS}\right) A_j\left(\vec{X}_{CS}\right) R_{y,j}\gamma_c} - 1 \le 0; \tag{4.5}$$

where $A_j\left(\vec{X}_{CS}\right)$ is the gross cross-sectional area of jth structural member calculated depending on the variable cross-sectional dimensions of the structural members $\vec{X}_{CS}$; $\varphi_{y,j}\left(\vec{X}_G, \vec{X}_{CS}\right)$ and $\varphi_{z,j}\left(\vec{X}_G, \vec{X}_{CS}\right)$ are column's stability factors corresponded to flexural buckling relative to the main axes of inertia and calculated depending on the design lengths $l_{ef,y,j}\left(\vec{X}_G\right)$, $l_{ef,z,j}\left(\vec{X}_G\right)$, cross-section type and cross-section geometrical properties for the jth structural member [14]. The flexural buckling factors $\varphi_{y,j}\left(\vec{X}_G,\vec{X}_{CS}\right)$ and $\varphi_{z,j}\left(\vec{X}_G,\vec{X}_{CS}\right)$ calculated depending on the variable geometrical parameters of the structure $\vec{X}_G$ and variable cross-sectional dimensions of the structural members $\vec{X}_{CS}$.

The following *local buckling constraints* have been also included into the system of constraints:

$$\frac{\bar{\lambda}_{w,j}\left(\vec{X}_{CS}\right)}{\bar{\lambda}_{uw,j}\left(\vec{X}\right)}-1\leq 0; \tag{4.6}$$

$$\frac{\bar{\lambda}_{f,j}\left(\vec{X}_{CS}\right)}{\bar{\lambda}_{uf,j}\left(\vec{X}\right)}-1\leq 0; \tag{4.7}$$

where $\bar{\lambda}_{w,j}\left(\vec{X}_{CS}\right)$ and $\bar{\lambda}_{f,j}\left(\vec{X}_{CS}\right)$ are the non-dimensional slenderness of the web and flange respectively of the cross-section for j^{th} structural member; $\bar{\lambda}_{uw,j}\left(\vec{X}\right)$ and $\bar{\lambda}_{uf,j}\left(\vec{X}\right)$ are the maximum values for corresponded non-dimensional slenderness for column structural members calculated depending on the internal forces (ration of the bending moment to the axial force), as well as depending on the design lengths $l_{ef,y,j}$, $l_{ef,z,j}$, cross-section type and cross-section geometrical properties for the jth structural member [14]. The non-dimensional slenderness $\bar{\lambda}_{w,j}\left(\vec{X}_{CS}\right)$ and $\bar{\lambda}_{f,j}\left(\vec{X}_{CS}\right)$ calculated depending on the variable cross-sectional dimensions of the structural members $\vec{X}_{CS}$ only. At the same time, the maximum values for corresponded non-dimensional slenderness $\bar{\lambda}_{uw,j}\left(\vec{X}\right)$ and $\bar{\lambda}_{uf,j}\left(\vec{X}\right)$ calculated depending on the variable geometrical parameters of the structure $\vec{X}_G$ and variable cross-sectional dimensions of the structural members $\vec{X}_{CS}$.

The system of constraints (2.2) – (2.3) has been also covered the *displacements constraints* (stiffness inequalities) for the specified nodes of the considered steel structure subjected to all design load combinations at the serviceability limit state. The following horizontal and vertical displacements constraints have been included into the system of constraints (2.2) – (2.3), formulated for all nodes, $\forall m=\overline{1,N_N}$ (N_N is the total number of nodes in the considered steel structure), of the steel structure subjected to all serviceability load case combination, $\forall k=\overline{1,N_{LC}^{SLS}}$, namely:

$$\frac{\delta_{x,mk}\left(\vec{X}\right)}{\delta_{ux,m}}-1\leq 0; \tag{4.8}$$

$$\frac{\delta_{z,mk}\left(\vec{X}\right)}{\delta_{uz,m}}-1\leq 0; \tag{4.9}$$

where $\delta_{x,mk}\left(\vec{X}\right)$ and $\delta_{z,mk}\left(\vec{X}\right)$ are the horizontal and vertical displacements respectively for l^{th} node of the steel structure subjected to k^{th} serviceability load case combination calculated from the linear equations system of the finite element method presented by equation (2.6); $\delta_{ux,m}$ and $\delta_{uz,m}$ are the allowable horizontal and vertical displacements for m^{th} structural node. Ultimate values for linear node displacements of the steel lattice portal frame were calculated according to the Table 4 [6] considering item 4 of this Table as $\delta_{ux,m}=H_{zd}/210$ and $\delta_{uz,m}=L/300=230$ mm.

Additional requirements that describe structural, technological and serviceability particularities of the considered structure, as well as constraints on the building functional volume (see figure 4.3) can be also included into the system (2.2) – (2.3). In particular these requirements can be presented in the form of constraints on lower and upper values of the design variables, $\forall i=\overline{1,N_X}$:

$$1-\frac{X_i}{X_i^L}\le 0; \tag{4.10}$$

$$\frac{X_i}{X_i^U}-1\le 0; \tag{4.11}$$

where X_i^L and X_i^U are the lower and upper bounds for the design variable X_i ; N_X is the total number of the design variables.

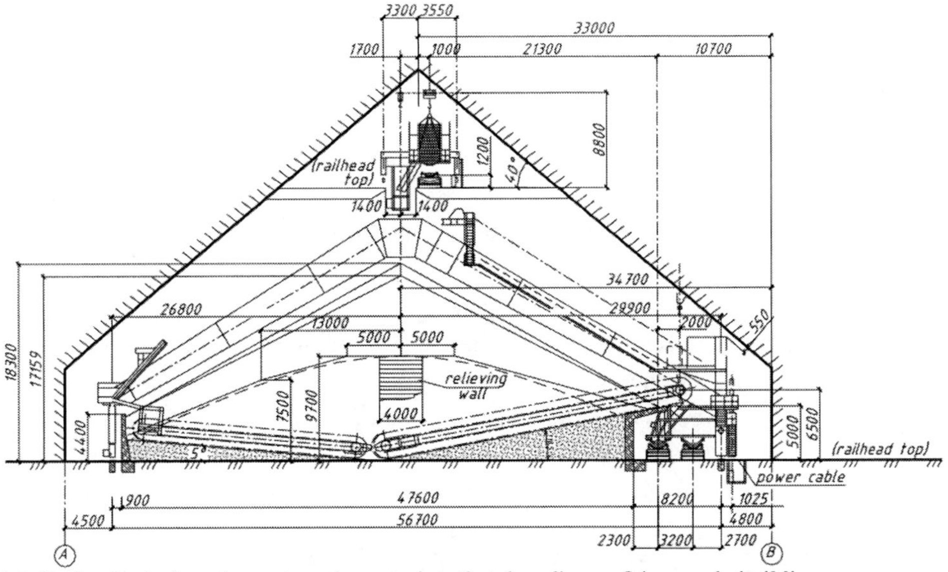

Figure 4.3. Technological equipments and constraints that describe useful space in building

Additional constraints on cross-sectional sizes of the lattice structural members with circle hollow sections type have been formulated according to the requirements of [14]. There were constraints of permissible minimal thickness and permissible maximum diameter-to-thickness ratio for the structural members, namely, $\forall l=\overline{1,N_{ST}}$:

$$1{,}0-\frac{t_l}{t_{\min,l}}\le 0; \tag{4.12}$$

$$\frac{D_l}{t_l\delta_{\max,l}}-1{,}0\le 0; \tag{4.13}$$

where l is the number of the stiffness type; N_{ST} – the overall quantity of the stiffness types in the considered steel lattice portal frame; t_l and D_l are thickness and diameter of the circle hollow section for l^{th} stiffness type (see Table 4.1) respectively; $t_{\min,l}$ – the minimum thickness of the circle hollow section in accordance with design code[14] as $t_{\min,l}=3\,\text{mm}$ for chords, support diagonals and support columns of lattice structural members and $t_{\min,l} = 2.5$ mm for other lattice elements; $\delta_{\max,l}$ – maximum diameter-to-thickness ratio for the structural member with circle hollow section in accordance with design code [14] depending on the yield stress value $R_{yn} = 245$ MPa < 295 MPa as $\delta_{\max,l} = 30$ for chords elements of the lattice structural members and $\delta_{\max,l} = 90$ for other lattice elements.

The following constraints that describe *conditions for designing gusset-less welded joints* between structural members with circle hollow sections formulated according to the requirements [14] have been also included in the system (2.2) – (2.3), $q, p = 1...N_{ST}$:

$$0{,}3d_p \le d_q \le d_p; \tag{4.14}$$

where p and q are the numbers of the stiffness types of structural members connected in the joint, here p is the number of the chord's stiffness type; q is the number of the lattice stiffness type.

4.3. Objective function.
Minimum weight as well as minimum construction budget has been considered as purpose function. Analytical expression for the structural weight depending on the variable cross-sectional sizes of the members can be written by the following formula:

$$\mathbf{M}\left(\vec{X}_G, \vec{X}_{CS}\right) = \varphi\rho\sum_{j=1}^{N_E} A_j l_j = 4\pi\varphi\rho\sum_{j=1}^{N_E} t_j\left(D_j - t_j\right)l_j \to \min; \tag{4.15}$$

where A_j and l_j – gross cross-sectional area and geometrical length for j^{th} structural member respectively; ρ – steel density, $\rho = 7850\,\text{kg/m}^3$; φ – factor that takes into account the increment of structural weight due to the present of the adjunct elements in the structural members and joints (stiffeners, ribs, end-plates, gussets etc.), φ was defined according to the steel specification mentioned in the source project for the warehouse framework, $\varphi \simeq 1{,}1$.

Construction budget of the steel portal frame with lattice structural members taken into account construction budget of mezzanines erected at the level +28,25 can be presented as follow:

$$K = C_{mnf}^{lt} + C_{mnf}^{sl} + C_{asm} + C_{asm}^{wl} + C_{c.p.} + C_{q.c.} + C_{mat};$$

where C_{mnf}^{lt} – manufacturing cost of lattice structural members; C_{mnf}^{sl} – manufacturing cost of mezzanine's structural members; C_{asm} – assembly cost for steel portal frame; C_{asm}^{wl} – assembly cost for walling; $C_{c.p.}$ – cost on the work package for corrosion protection of the steel framework; $C_{q.c.}$ – cost for the quality control of welded connections; C_{mat} – material cost for structural members of the steel portal frame. Analytical expression of the construction budget for manufacturing and erection of the steel lattice portal frame depending on the design variables have been presented by the following, UAH:

$$K = 20670\mathbf{M}\left(\vec{X}_G, \vec{X}_{CS}\right) + 1525D_{ov}\left(\vec{X}_G\right) + 1300H_{zdk} + 110\mathbf{A}_s\left(\vec{X}_G, \vec{X}_{CS}\right) + 21120 \to \min; \tag{4.16}$$

where D_{ov} – half-rafter length, $D_{ov} = \left(\left(0{,}5L\right)^2 + \left(H_{zd} - H_{zdk}\right)^2\right)^{\frac{1}{2}}$; L – portal frame span, $L = 69$ m.

5. A parametric optimization algorithm based on the gradient projection method

Let present the following numerical algorithm to solve the parametric optimization problem for steel structures formulated above.

Step 1. Describing an initial design (a set of design variables) and initial data for structural optimization.

The design variable vector $\vec{X}_k = \left(\vec{X}_G, \vec{X}_{CS}\right)_k^T$ has been specified, where k is the iteration index, $k = 0$. The structural topology, cross-section types and node type connections of the bars, the support conditions of the bar system, as well as loading patterns, load case combinations and load design values are prescribed and constants.

Initial data for optimization of the considered steel structure are design strength for steel member R_y, safety factor γ_c, factors to define flexural design lengths $l_{ef,y,j}$, $l_{ef,z,j}$ for all column structural members; allowable values for horizontal and vertical displacements $\delta_{ux,l}$ and $\delta_{uz,l}$ of the specified nodes of the considered steel structure; lower $\vec{X}^L$ and upper $\vec{X}^U$ bounds for the design variables; as well as specified objective function $f(\vec{X}_k)$.

Step 2. Calculation of the geometrical and design lengths for all structural members.

The geometrical lengths l_j of all structural members are calculated based on the node coordinates of the considered steel structure. The latter depend on the unknown (variable) geometrical parameters of the structure $\vec{X}_G$. The design lengths $l_{ef,y,j}$, $l_{ef,z,j}$ of all column structural members are calculated using calculated geometrical lengths l_j and initial data relating to the design length factors. The latter are constant during the iteration process presented below. Variation of the geometrical lengths l_j and corresponded design lengths $l_{ef,y,j}$, $l_{ef,z,j}$ on the further iterations has been performed based on the current values of the variable (unknown) parameters $\vec{X}_G$ of the geometrical scheme.

Step 3. Calculation of the cross-section dimensions and geometrical properties for all design cross-sections.

Geometrical properties of the design cross-sections (areas, moments of inertia, elastic section moments, radiuses of inertia, etc.), as well as non-dimensional slenderness for cross-section elements (webs and flanges) $\bar{\lambda}_{w,j}\left(\vec{X}_{CS}\right)$ and $\bar{\lambda}_{f,j}\left(\vec{X}_{CS}\right)$ have been calculated depending on the current values of the unknown (variable) cross-section dimensions $\vec{X}_{CS}$.

Step 4. Linear structural analysis of the considered steel structure.

For each mth node of the finite element model subjected to k^{th} serviceability load case combination the displacements and rotations, as well as the design horizontal $\delta_{x,mk}\left(\vec{X}\right)$ and vertical $\delta_{z,lk}\left(\vec{X}\right)$ displacements can be calculated using the linear equations system of the finite element method presented by equation (2.6).

For each ith design section of jth structural member subjected to k^{th} ultimate load case combination the design internal forces can be calculated using the linear equations system of the finite element method presented by equation (2.5).

Step 5. Calculation of the state variables (stresses, buckling factors, allowable non-dimensional slenderness etc.).

The value of the normal $\sigma_{x,ijk}\left(\vec{X}\right)$ stresses at the specified cross-section point has been calculated depending on the axial force acting in i^{th} design section of jth structural member subjected to k^{th} ultimate load case combination as presented by the design code.

The flexural buckling factors $\varphi_{y,j}\left(\vec{X}_G,\vec{X}_{CS}\right)$, $\varphi_{z,j}\left(\vec{X}_G,\vec{X}_{CS}\right)$ have been calculated depending on the corresponded design lengths, cross-section type and cross-section geometrical properties for the structural members according to the design code [14].

The maximum values for corresponded non-dimensional slenderness $\bar{\lambda}_{ttw,j}(\vec{X})$ and $\bar{\lambda}_{tf,j}(\vec{X})$ for column structural members have been calculated depending on the design lengths $l_{ef,y,j}$, $l_{ef,z,j}$, cross-section type and cross-section geometrical properties for the jth structural member [14].

Step 6. Verifications of the constraints and construction the set of active constraints numbers $\mathbf{A}$.

Verification of the constraints (4.1), (4.4), (4.5) has been performed for all ultimate load case combinations and all design cross-sections of all structural members. Verification of the constraints (4.8), (4.9) have been also conducted for all serviceability load case combinations and all design structural nodes. Additional requirements in the form of constraints (4.10), (4.11) on lower and upper values of the design variables, local buckling constraints (4.6), (4.7), constraints on the member's slenderness (4.2), (4.3), constraint (4.12) on permissible minimal thickness, constraint (4.13) on permissible maximum diameter-to-thickness ratio for the structural members, as well as the conditions (4.14) for designing gusset-less welded joints between structural members with circle hollow sections have been also verified. Set of the active constraints numbers $\mathbf{A}$ calculated for the current approximation $\vec{X}_k$ has been constructed according to equation (3.2).

Step 7. Calculation of the current objective function value $f(\vec{X}_k)$, objective function gradient $\nabla f(\vec{X}_k)$ and determination of the desired decrement of the objective function value $\Delta f(\vec{X}_k)$.

The objective function gradient $\nabla f(\vec{X}_k)$ can be calculated by the numerical differentiation with respect to the design variables using the finite difference approximation. The desired decrement of the objective function value $\Delta f(\vec{X}_k)$ can be assigned as 5...25% from the current objective function value $f(\vec{X}_k)$.

Step 8. Construction of the constraint's violations vector $\mathbf{V}$ and the matrix of the active constraint's gradients $[\nabla\varphi]$. The vector of the values of the constraint's violations $\mathbf{V}$ and the matrix of the constraint's gradients $[\nabla\varphi]$ are constructed for active constraints only according to the set of active constraints numbers $\mathbf{A}$.

Step 9. Construction the matrix of active linear-independent constraint's gradients with triangular structure. The set of linear-independent constraint's numbers $\mathbf{L}$ and the matrix of active linear-independent constraint's gradients $\mathbf{H}[\nabla\varphi]$ with triangular structure are constructed according to the algorithm presented by the paper [15].

Step 10. Step parameter ξ calculation. Step parameter ξ has been calculated according to equation (3.11) or equation (3.12) and can be modified on the further iterations depending on convergence of the iterative process presented by equation (3.1).

Step 11. Calculation the column-vectors $\vec{\mu}_{\perp}$ and $\vec{\mu}_{\parallel}$ which define the design variables increment subject to the condition of elimination the constraint's violations and subject to the improvement of the objective function value. The vectors $\vec{\mu}_{\perp}$ and $\vec{\mu}_{\parallel}$ can be calculated using equation (3.18) and equation (3.19) respectively.

If some h^{th} component of the column-vectors $\vec{\mu}_{\perp}$ and $\vec{\mu}_{\parallel}$ satisfies equation (3.17), the corresponded constraint gradient $\nabla\varphi_h$ has been excluded from the matrix $[\nabla\varphi]$, and corresponded violations V_h has been excluded from the vector $\mathbf{V}$, as well as the return to step 9 has to be conducted. In contrary case transition to the step 11 has been performed.

Step 12. Calculation the increment vector for the current design variables and determination the improved approximation to the optimum solution. The increment vector $\Delta\vec{X}_k$ for the current design variables values $\vec{X}_k$ has been calculated according to equation (3.20) or equation (3.21). The improved approximation $\vec{X}_{k+1}$ to the optimum solution has been determined according to equation (3.1).

Step 13. Stop criteria verification of iterative searching for the optimum solution. If all constraints (4.1) – (4.14) are satisfied with appropriate accuracy, as well as inequality (3.22) or one of the stop criteria described by the paper [15] is also satisfied, then transition to the step 13 has been performed. In contrary case return to the step 1 has been conducted with $k \leftarrow k+1$.

Step 14. Discretization the optimum solution $\vec{X}_k$ obtained in the continuum space of the design variables.

Step 15. Optimum parameters of the structure is $\vec{X}_k$ with optimum value of the objective function $f(\vec{X}_k)$.

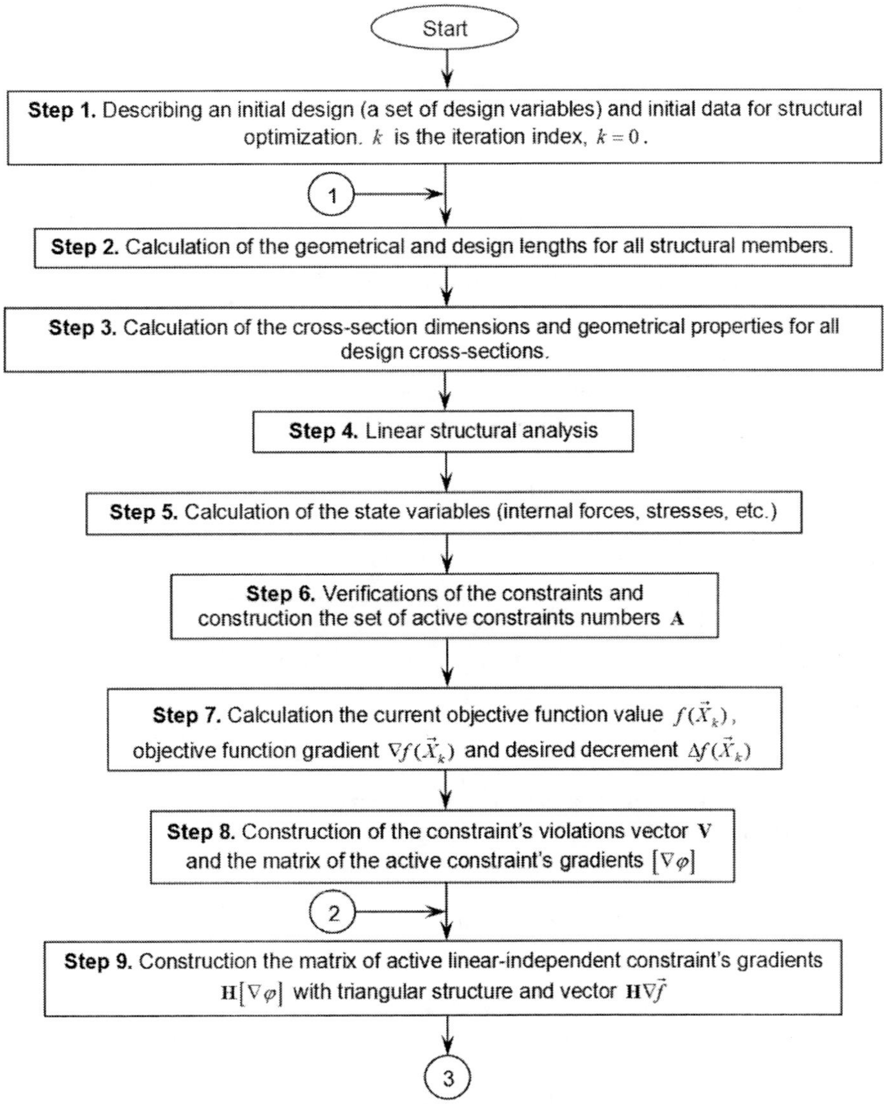

Figure 5.1. The flow chart for structural optimization according to the searching technique based on the gradient projection method.

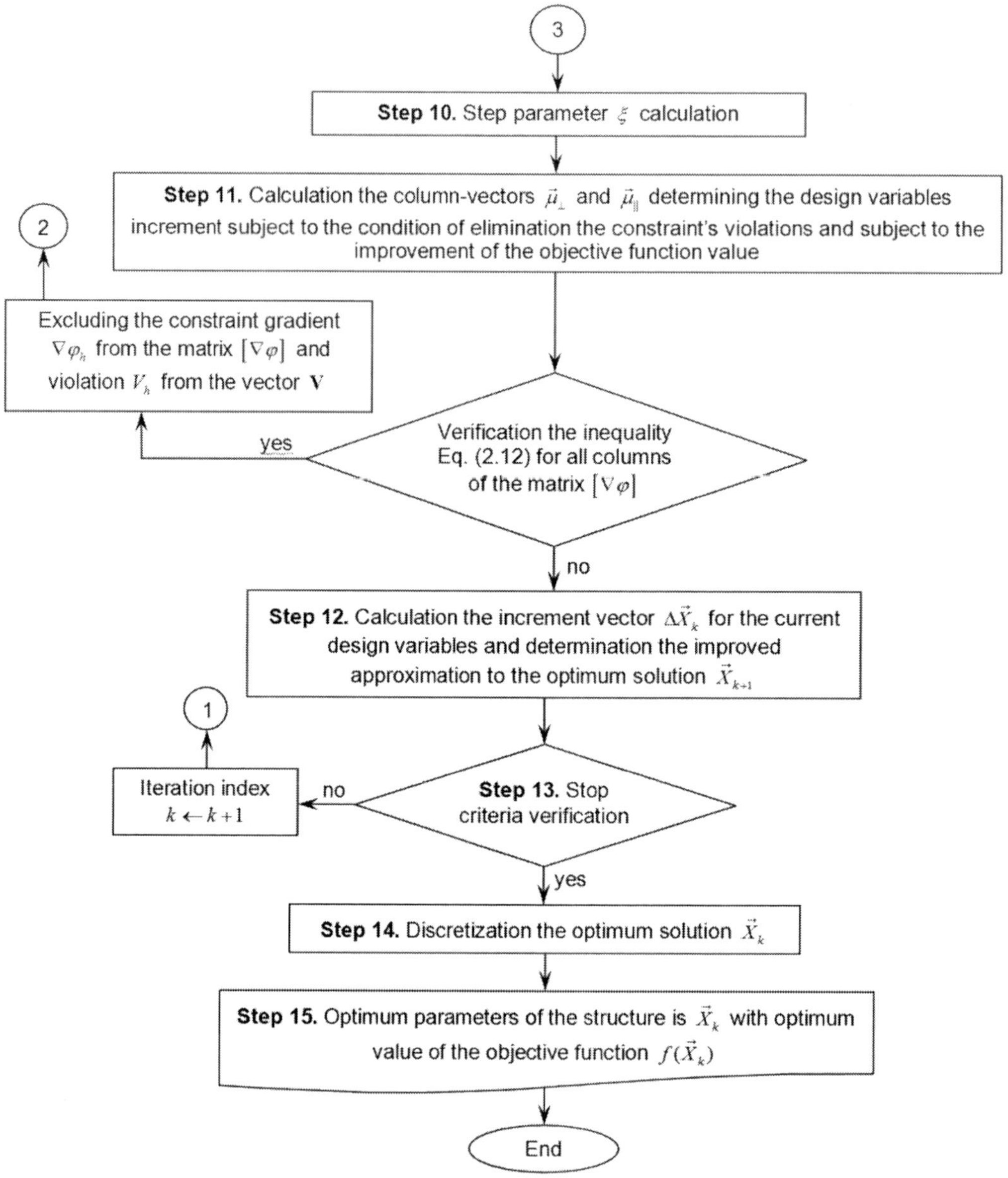

Figure 5.1 (continuation). The flow chart for structural optimization according to the searching technique based on the gradient projection method.

Figure 5.1 presents the flow chart for structural optimization according to the searching technique describing by the gradient projection method considered above.

6. Parametric optimization results for the steel lattice portal frames

A parametric optimization methodology presented above has been realized in software OptCAD [20]. This software provides solutions to a wide range of problems, namely: (i) linear static analysis of bar structures; (ii) verification of the load-bearing capacity of the structural members according to specified design code; (iii) searching for values of the structural parameters when structure complies with design code requirements and designer's criterions; (iv) parametric optimization of the steel bar structures by the determined criterion.

Formulated parametric optimization problem for the steel lattice portal frame has been solved using software OptCAD. Mathematical apparatus of the software combines the finite element method to perform linear static analysis of the bar system, as well as improved gradient projection method to solve parametric optimization problems formulated as non-linear programming tasks [20]. Task dimensions are: account of design variables is 50, account of problem constraints is 14000.

Optimization results received using software OptCAD are presented by the tables 6.1 and 6.2. Figure 6.1 show optimal design decision by the criterion of minimum structural weight of the steel lattice portal frame (project with start values of the design variables is indicated by the red colour, optimum project is indicated by the blue colour).

Table 6.1. Optimal values for variable geometrical parameters of the portal frame's design scheme

Design variable	Start value, m	Optimum values, m, by the criterion of minimum			
		weight		costs on fabrication and erection	
		when lower chord of lattice rafter is			
		straight-line	polygonal	straight-line	polygonal
H_{zd}	39,58	38,82	39,55	38,74	39,44
H_{zdk}	10,63	11,92	11,84	12,08	12,03
h_2	2,60	–	3,23	–	3,31
h_4	2,60	–	3,21	–	3,23
h_6	2,60	–	2,68	–	2,70
h_7	2,60	2,18	2,68	2,13	2,61
b_7	0,55773	1,00	0,56	1,05	0,63
h_{op}	2,60	3,61	3,72	3,73	3,87
b_{op}	2,60	3,08	3,02	3,02	2,95
Weight, ×10³,kg	**30,78**	**18,98**	**18,45**		
Costs, UAH	**786681**			**534647**	**524257**

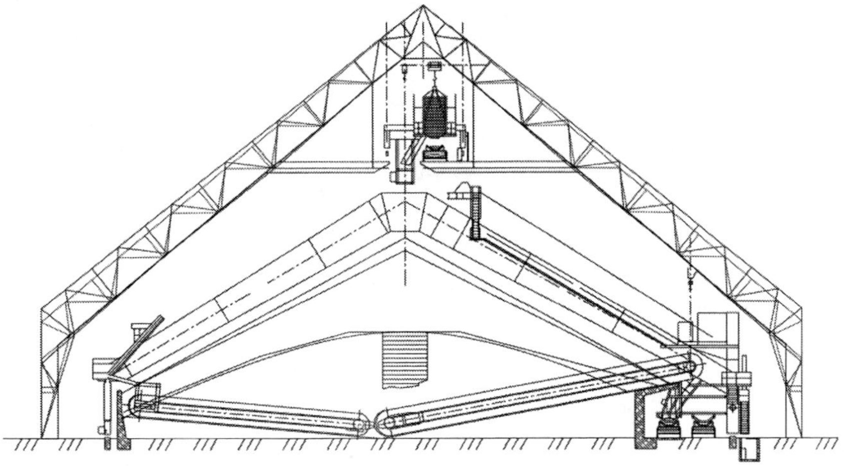

Figure 6.1. The optimum design decision by the criterion of minimum structural weight when lower chord of the lattice rafter is straight-line

IOP Conf. Series: Materials Science and Engineering 1164 (2021) 012090 doi:10.1088/1757-899X/1164/1/012090

Table 6.2. Optimal values for variable cross-sectional sizes of the lattice structural members of the portal frame

Design variable	Start value, m	Optimum values, mm×mm, by the criterion of minimum					
		weight			costs on fabrication and erection		
		when lower chord of the lattice rafter is					
		straight-line	polygonal		straight-line	polygonal	
		in the search space					
		continuous		discrete	continuous		discrete
$d_1 \times t_1$	299×25	183,4×8,4	192,3×8,6	194×9,0	173,9×9,3	172,9×8,6	168×11,0
$d_1 \times t_2$	299×14	183,4×6,1	192,3×6,4	194×7,0	173,9×6,9	172,9×6,3	168×8,0
$d_1 \times t_3$	299×10	183,4×6,9	192,3×6,5	194×6,5	173,9×7,3	172,9×7,4	168×8,5
$d_1 \times t_4$	299×14	183,4×8,5	192,3×9,3	194×9,0	173,9×10,7	172,9×9,3	168×11,0
$d_1 \times t_5$	299×14	183,4×6,1	192,3×6,4	194×6,5	173,9×5,8	172,9×5,8	168×6,0
$d_6 \times t_6$	299×10	297,4×9,9	303,1×10,1	299×10,0	298,1×9,9	292,3×9,7	299×10,0
$d_6 \times t_7$	299×16	297,4×11,3	303,1×11,4	299×12,0	298,1×11,6	292,3×11,5	299×12,0
$d_8 \times t_8$	299×10	264,4×8,8	269,4×9,0	299×10,0	269,0×9,0	263,6×8,8	273×9,5
$d_9 \times t_9$	299×10	203,0×6,8	210,7×7,0	194×8,5	211,5×7,0	202,8×6,8	219×7,5
$d_{10} \times t_{10}$	299×14	143,5×4,8	147,1×4,9	152×5,5	148,1×4,9	143,3×4,8	146×5,5
$d_{11} \times t_{11}$	180×12	183,4×9,7	192,3×9,6	194×9,5	173,9×11,5	172,9×11,1	168×13,0
$d_{12} \times t_{12}$	152×8	161,2×4,5	165,6×4,6	152×5,5	164,9×4,6	160,8×4,5	168×5,0
$d_{13} \times t_{13}$	121×8	110,3×3,9	107,4×3,6	108×4,0	80,7×4,9	79,1×5,6	83×5,0
$d_{14} \times t_{14}$	102×5	89,2×3,5	90,9×3,5	95×3,5	89,4×3,5	87,7×3,5	95×3,5
$d_{15} \times t_{15}$	152×8	143,5×5,2	147,1×5,1	152×5,0	133,5×5,9	135,1×5,6	133×6,0
$d_{16} \times t_{16}$	102×5	55,0×3,5	57,9×3,5	60×3,5	52,2×3,5	51,9×3,5	54×3,5
$d_{17} \times t_{17}$	180×12	112,8×4,9	116,3×4,7	108×5,0	90,4×6,1	101,8×5,4	95×6,0
$d_{18} \times t_{18}$	180×12	143,5×6,3	147,1×6,6	152×6,5	148,1×6,6	143,3×6,4	146×7,5
$d_{19} \times t_{19}$	299×10	297,4×9,9	303,2×10,1	299×10,0	298,1×9,9	292,3×9,7	299×10,0
$d_{19} \times t_{20}$	299×25	297,4×9,9	303,2×10,1	299×10,0	298,1×9,9	292,3×9,7	299×10,0
$d_{21} \times t_{21}$	299×25	297,4×21,0	303,2×21,1	299×22,0	298,1×22,3	292,3×22,3	299×24,0
$d_{22} \times t_{22}$	102×5	89,2×3,5	90,9×3,5	95×3,5	89,4×3,5	87,7×3,5	95×3,5
$d_{23} \times t_{23}$	180×12	148,9×5,3	151,6×5,3	152×5,5	79,7×11,4	81,8×10,6	83×1,2
Weight, ×10³,kg	**30,78**	**18,45**	**18,98**	**19,70**	–	–	–
Costs, UAH	**786681**	–	–	–	**534647**	**524257**	**552368**

Conclusion

The paper has proposed a mathematical model for parametric optimization problem of the steel lattice portal frame. The design variable vector includes geometrical parameters of the structure (node coordinates), as well as cross-sectional dimensions of the structural members. The system of constraints covers load-carrying capacities constraints formulated for all design sections of structural members of the steel structure subjected to all ultimate load case combinations. The displacements constraints formulated for the specified nodes of the steel structure subjected to all serviceability load case combinations have been also included into the system of constraints. Additional requirements in the form of constraints on lower and upper values of the design variables, constraints on permissible minimal thicknesses, constraints on permissible maximum diameter-to-thickness ratio for the structural members with circle hollow sections, as well as the conditions for designing gusset-less welded joints between structural members with circle hollow sections have been also considered in the scope of the mathematical model. The method of the objective function gradient projection onto the active constraints surface with simultaneous correction of the constraints violations has been used for solving the parametric optimization problem. A numerical algorithm for solving the formulated parametric optimization problems of steel structures has been presented in the paper. New optimal layouts of the steel lattice portal frame by the criterion of the minimum weight, as well as minimum costs on manufacturing and erection have been presented.

References

[1] Degertekin, S. O., Lamberti, L. Sizing, layout and topology design optimization of truss structures using the Jaya algorithm. Applied soft computing. 2018. Vol. 70. Pp. 903–928. DOI: 10.1016/j.asoc.2017.10.001

[2] Alpatov, V. The task of multi-criteria optimization of metal frame structures. MATEC Web of Conferences. 2017. No. 117. No. 00007. DOI: 10.1051/matecconf/201711700007

[3] Kibkalo, A., Lebedeva, M., Volkov, M. Methods of parametric optimization of thin-walled structures and parameters which influence on it. MATEC Web of Conferences. 2016. No. 53. No. 01051. DOI: 10.1051/matecconf/20165301051

[4] Alekseytsev, A. V. Evolutionary optimization of steel trusses with the nodal joints of rods. Magazine of Civil Engineering. 2013. No. 40(5). Pp. 28–37. (rus). DOI: 10.5862/MCE.40.3.

[5] Serpik, I. N., Alekseytsev, A. V., Balabin, P. Yu., Kurchenko, N. S. Flat rod systems: optimization with overall stability control. Magazine of Civil Engineering. 2017. No. 76(8). Pp. 181–192. DOI: 10.18720/MCE.76.16.

[6] Serpik, I. N., Alekseytsev, A. V. Optimization of frame structures with possibility of emergency actions. Magazine of Civil Engineering. 2013. No. 44(9). Pp. 23–29. (rus). DOI: 10.5862/MCE.44.3.

[7] Sergeyev, O. A., Kiselev, V. G., Sergeyeva, S. A. Overall instability and optimization of bar structures with random defects in case of constraints on faultless operation probability. Magazine of Civil Engineering. 2013. No. 44(9). Pp. 30–41. (rus). DOI: 10.5862/MCE.44.4.

[8] Permyakov, V. O., Yurchenko, V. V., Peleshko, I. D. An optimum structural computer-aided design using hybrid genetic algorithm. Proceeding of the International Conference "Progress in Steel, Composite and Aluminium Structures". Taylor & Francis Group. London, 2006. Pp. 819–826.

[9] Peleshko I., Yurchenko V. Parametric optimization of steel structures based on gradient projection method // Strength of Materials and Theory of Structures: Scientific-and-technical collected articles. – Kyiv: KNUBA, 2020. – Issue 105. – P. 192-220. DOI: 10.32347/2410-2547.2020.105.192-220.

[10] Yurchenko V.V., Peleshko I. D. Searching for optimal pre-stressing of steel bar structures based on sensitivity analysis // Archives of Civil Engineering, Vol. 66, No. 3, 2020. – P. 525-540. DOI: 10.24425/ACE.2020.134411

[11] Kuci, E., Henrotte, F., Duysinx, P., Geuzaine, C. Design sensitivity analysis for shape optimization based on the Lie derivative. Computer methods in applied mechanics and engineering. 2017. Vol. 317. Pp. 702–722. DOI: 10.1016/j.cma.2016.12.036

[12] Sergeyev, O. A., Kiselev, V. G., Sergeyeva, S. A. Optimal design of 3D frame structures taking into account the stress and multiple natural frequency constraints. Magazine of Civil Engineering. 2016. No. 61(1). Pp. 74–81. DOI: 10.5862/MCE.61.7

[13] Huebner, K. H., Dewhirst, D. L., Smith, D. E., Byrom, T. G. The finite element method for engineers. 4th ed. John Wiley & Sons, 2001. 744 p.

[14] SP 16.13330.2017, Steel structures. Moscow: Minstroy, 2017. 140 p. (rus)

[15] Peleshko, I. D., Yurchenko, V. V. Improved gradient projection method for parametric optimization of bar structures. Magazine of Civil Engineering. 2020. № 98(6). Article No. 9812. DOI: 10.18720/MCE.98.12.

[16] Peleshko I., Yurchenko V. An improved gradient-based method to solve parametric optimisation problems of the bar structures // Strength of Materials and Theory of Structures:

Scientific-and-technical collected articles. – Kyiv: KNUBA, 2020. – Issue 104. – P. 265-288. DOI: 10.32347/2410-2547.2020.104.265-288.

[17] Haug, E. J., Arora, J. S. Applied optimal design: mechanical and structural systems. John Wiley & Sons, 1979. 520 p.

[18] Reklaitis, G. V., Ravindran, A., Ragsdell, K. M. Engineering optimization. Methods and applications. Wiley, 2006. 688 p.

[19] Wilkinson J. H., Reinsch C. Handbook for Automatic Computation. Volume II: Linear Algebra. Heidelberg New York Springer-Verlag Berlin. 1971. 441 p. DOI: 10.1137/1014116

[20] Peleshko I. D., Yurchenko V. V., Beliaev N. A. Computer-aided design and optimization of steel structural systems // Zeszyty naukowe Politechniki Rzeszowskiej "Budownictwo i inżynieria środowiska", Nr. 264. – 2009. – z. 52. – Rzeszów, Oficyna Wydawnicza Politechniki Rzeszowkiej, 2009. – P. 145-154

A method for solving boundary value problems of mathematical theory of thick transversely isotropic plates

A G Zelensky[1,*], **A G Demianenko**[2]

[1]State Higher Educational Institution Pridneprovsk State Academy of Civil Engineering and Architecture, 24 A, Chernyshevsky str., Dnipro 49600, Ukraine
[2]Dniprovsk State Agrarian and Economic University, 25, Efremova str., Dnipro 49027, Ukraine

[*] E-mail: a.zelensky@ukr.net

Abstract. A method for solving the boundary value problems of a variant of the mathematical theory of thick transversely isotropic plates has been developed. Systems of high-order equilibrium differential equations are reduced by this method to second-order differential equations. The constructed version of the theory makes it possible to solve three-dimensional boundary value problems with any high accuracy. All components of the stress-strain state and boundary conditions are considered functions of three coordinates. They are represented by infinite mathematical series in transverse coordinate with the help of Legendre polynomials. The three-dimensional boundary-value problem is reduced to two-dimensional using the Reisner variational principle. The transverse normal and tangential stresses are displayed in such a way that the boundary conditions on the horizontal faces of the plate are exactly satisfied. Boundary value problems for semifinite transversely isotropic plates of arbitrary thickness with different boundary conditions are considered. Boundary effects are taken into account. The analysis of the stress-strain state is performed.

1. Introduction and literature review

The classical theories of plates and shells, the Timoshenko-Reisner theory [1, 2], their modifications, theories based on certain models of deformation [3-5], and other theories using various hypotheses, as a rule, cannot describe the state of these elements for the most diverse tasks with the necessary accuracy. The main disadvantage of model theories is the inability to refine the results within the same theories, since the accuracy is determined by a set of accepted hypotheses. In mathematical terms, such theories are reduced to systems of differential equations (DE) of relatively low order. At the same time, obtaining analytical solutions of boundary value problems for plates and shells based on three-dimensional elasticity theory is fraught with great mathematical difficulties [6-9]. A variant of mathematical theory (MT) [10-17] is based on the representation of all components of the stress-strain state (SSS) in the form of an infinite mathematical series with three coordinates. It can be different mathematical series: tensor, power, using Legendre polynomials. In all variants of the MT, three-dimensional problems of the theory of elasticity are reduced to two-dimensional by various methods: projection, variational, and others. Depending on the method of obtaining the basic relations, equilibrium equations, boundary conditions, the accuracy and efficiency of each theory are different. From this point of view, a MT variant, started in [17] for skew-symmetric deformation of plates in the first two approximations, and developed for plates and shells of arbitrary constant thickness in [18-22] for arbitrary approximations and transverse loads, gives high accuracy already at low approximations. As the number of terms in the mathematical series increases for the SSS, the order of the systems of equations increases, but also the accuracy of solving boundary-value problems in determining the internal SSS and boundary effects increases. At present, theories of the Tymoshenko-Reissner type and their refinement variants are usually used in Western studies. These theories are, as a rule, partial cases of a variant of the MT [17-22]. They can be obtained as a first approximation of a variant of the MT with some modifications and refinements. An extensive bibliography of Western works is given in

Content from this work may be used under the terms of the Creative Commons Attribution 3.0 licence. Any further distribution of this work must maintain attribution to the author(s) and the title of the work, journal citation and DOI.
Published under licence by IOP Publishing Ltd

[22]. The development of the theory of calculation of plates and shells can be found, in particular, in [3, 4, 6, 8, 9, 16].

2. Method of solving boundary value problems for plates

The method of solving the boundary value problems of the constructed variant of the theory is as follows: 1. All components of the SSS develop into mathematical series with the help of Legendre polynomials. Boundary conditions and systems of differential equilibrium equations in the components of displacements are obtained from the Reisner variational equation. 2. The initial system of equations is reduced to two defining subsystems by algebraic, differential and operator transformations. One subsystem (homogeneous) describes the vortex boundary effect and is reduced to Helmholtz differential equations. The other subsystem is heterogeneous and describes the internal SSS and the potential boundary effect, which are further separated. This subsystem depends on the new functions, its left parts are the same, and the right ones are structurally the same and contain Helmholtz operators. By the developed method of order reduction the last system is reduced to inhomogeneous differential equations of the second order. 3. Next, we find the general solutions of the defining system of differential equations, and then find the general solutions of the initial systems of equilibrium equations and the components of the SSS by inverse transformations.

3. Variant of mathematical theory of plates of arbitrary thickness

Here are the main provisions and equations of a variant of the MT of plates of arbitrary thickness [18-20, 22] (hereinafter referred to as a variant of the MT). The variant takes into account all SSS components, which are considered to be functions of three variable coordinates x, y, z. The axes x, y belong to the middle plane, which coincides with the plane of isotropy. The z axis is perpendicular and directed upward, $z \in [-h/2, h/2]$, where h is the plate thickness. The plate is subjected to the downward distributed loads $q_1(x,y)$ and $q_2(x,y)$. The $q_1(x,y)$ load acts on the upper face of the plate and $q_2(x,y)$ acts on the lower face.

The boundary conditions (for $z = \pm h/2$) are as follows:

$$\sigma_z(z = \pm h/2) = (\mp q(x,y) - p(x,y))/2; \; \sigma_{xz}(z = \pm h/2) = \sigma_{yz}(z = \pm h/2) = 0, \tag{1}$$

where $p(x,y)$ and $q(x,y)$ are symmetrical and obliquely symmetrical loads relative to the median plane.

The components of displacements and stresses are represented as infinite mathematical series in transverse coordinate by means of Legendre polynomials.

Components of displacements:

$$U(x,y,z) = \sum_{k=0}^{\infty} P_k(2z/h) u_k(x,y) \; (U,V;\; u,v); \; W(x,y,z) = \sum_{k=1}^{\infty} P_{k-1}(2z/h) w_k(x,y), \tag{2}$$

where P_k is Legendre polynomials; u_k, v_k, w_k–are unknown components of displacement.

If in the mathematical seriess (2) for the components of the tangent displacements we consider only terms with indices $k = 0,1,\ldots,N$, where N is an odd number, then we call this approximation K0-N an approximation; if we consider the terms with the $k = 1,3$ indices, then this is an approximation of K13. We write down the partial sums of a series of displacements and stresses in the K0-N approximation.

The displacement components according to (2) are as follows:

$$U^{(N)}(x,y,z) = \sum_{k=0}^{N} P_k(2z/h) u_k(x,y), (U,V; u_k, v_k); \; W^{(N)}(x,y,z) = \sum_{k=1}^{N} P_{k-1}(2z/h) w_k(x,y), \tag{3}$$

where the index (N) at the top indicates the approximation number.

The stress components are determined from the Reisner variation equation [23], taking into account (3):

$$\sigma_{xz}^{(N)}(x,y,z)=\sum_{n=0}^{N+1}P_n t_{xn};\ \sigma_{yz}^{(N)}(x,y,z)=\sum_{n=0}^{N+1}P_n t_{yn};\sigma_{z}^{(N)}(x,y,z)=\sum_{n=0}^{N+2}P_n s_{zn};$$
$$\sigma_{x}^{(N)}(x,y,z)=\sum_{n=0}^{N+2}P_n s_{xn};\ \sigma_{y}^{(N)}(x,y,z)=\sum_{n=0}^{N+2}P_n s_{yn};\ \sigma_{yx}^{(N)}(x,y,z)=\sum_{n=0}^{N}P_n t_{yxn}, \tag{4}$$

where

$$t_{xn}(x,y)=\sum_{j=1,3,}^{N}(h_{0nj}w_{j,x}+l_{0nj}u_j),\ (n=0,2,...,N+1);$$
$$t_{xn}(x,y)=\sum_{j=2,4,}^{N-1}(h_{0nj}w_{j,x}+l_{0nj}u_j),\ (n=1,3,...,N);\ (x,y;u,v);$$
$$s_{zn}(x,y)=\sum_{j=2,4}^{N-1}(p_{nj}w_j+g_{nj}\phi_j)+g_{np}p\ ,\ (n=0,2,...,N+1);\quad \varphi_j=\varphi_j(x,y)=u_{j,x}+v_{j,y};$$
$$s_{zn}(x,y)=\sum_{j=3,5}^{N}(p_{nj}w_j+g_{nj}\phi_j)+g_{nq}q\ ,\ (n=1,3,...,2n+1); \tag{5}$$
$$s_{xn}(x,y)=\sum_{j=0,2}^{N-1}(a_{nuj}u_{j,x}+a_{nvj}v_{j,y})+\sum_{j=2,4}^{N-1}a_{nwj}w_j+a_{np}p,\ (n=0,2....,N+1);$$
$$s_{xn}(x,y)=\sum_{j=1,3}^{N}(a_{nuj}u_{j,x}+a_{nvj}v_{j,y})+\sum_{j=1,3}^{N}a_{nwj}w_j+a_{nq}q,\ (n=1,3....,N+2);$$
$$t_{yxn}(x,y)=G(u_{n,y}+v_{n,x}),(n=0,1,...,N).$$

In formulas (5) h, l, p, g, a with the lower indices depend on the mechanical and geometric parameters (MGP) of the plate. For the approximations K0-3, K0-5, they are given in [19, 20]. The components of the transverse stresses (4), (5) satisfy the boundary conditions (1).

The system of equilibrium equations is obtained on the basis of the Reisner variation principle. In the K0-N approximation, the system looks like:

$$\sum_{j=0}^{N}(L_{i,uj}u_j+L_{i,vj}v_j)+\sum_{j=1}^{N}L_{i,wj}w_j=L_{ipq}(x,y),\ (i=1,2,...,3N+2), \tag{6}$$

where $L_{i,uj}$, $L_{i,vj}$, $L_{i,wj}$ - differential operators not higher than the 2nd order; $L_{ipq}(x,y)$- functions that depend on the intensity of the external load $p(x,y)$ and $q(x,y)$. The integration constants of the system of DE (6) are determined from the boundary conditions on the lateral surface.

The boundary conditions follow from the Reisner variational equation and are given in [18–20].

4. Problem statement

A semi-infinite transversal isotropic plate of arbitrary constant thickness ($x\in[0;\infty)$, $y\in(-\infty;\infty)$) is considered. The plate has a transverse load, which disappears at $x\to\infty$:

$$q(x,y)=q_0\exp(-\alpha x)\cos\beta y\ ,\ (\alpha>0;\ \beta>0;\ \alpha\neq\beta),(q_0\text{–const}). \tag{7}$$

On the basis of the MT variant, we investigate the analytically SSS in this plate under skew-symmetric deformation and different boundary conditions at the edge $x=0$. Consider the approximation of K13. In this approximation, a theory variant MT describes with high precision the internal SSS and boundary effects under smooth loads. System (6) is reduced by algebraic and differential transformations to two independent systems of DE: the first – inhomogeneous system of

DE describing the internal SSS and potential boundary effect, the second – homogeneous system of equations of vortex boundary effect [19, 20].

5. Transformation of equations. General solutions

The system of equations of the internal SSS and the potential limit effect is reduced to two defining inhomogeneous differential equations of the eighth order with respect to new functions $\Phi_1(x,y), \Phi_3(x,y)$ by the operator method.

$$D_0 D_0 D_1 D_2 \Phi_k(x,y) = a_{k0} D_{k0}\, q(x,y),\ (k=1,3), \tag{8}$$

where $D_0 = \nabla^2$, $D_i = (\nabla^2 - s_i)$ $(i=1,2)$, $D_{k0} = (\nabla^2 - s_{k0})$; a and s with indexes – MGP.

General solution for components of transverse displacements [20]:

$$\begin{aligned} w_1(x,y) &= \Pi_{33}\Phi_1(x,y) - \Pi_{13}\Phi_3(x,y); \\ w_3(x,y) &= -\Pi_{31}\Phi_1(x,y) + \Pi_{11}\Phi_3(x,y), \end{aligned} \tag{9}$$

where $\Pi_{33}(x,y),...,\Pi_{11}(x,y)$–differential operators of the fourth order, $\Phi_1(x,y)$ and $\Phi_3(x,y)$ are the general solutions of equations (8), and

$$\Phi_1 = \Phi_{1B} + \Phi_{1\Pi} + \Phi_{1r};\ \Phi_3 = \Phi_{3r}\,. \tag{10}$$

In Equations (10) $\Phi_{1B}(x,y)$ and $\Phi_{1\Pi}(x,y)$ is the general solution of the equation

$$\nabla^4 \Phi_{1B}(x,y) = 0;\ \ (\nabla^2 - s_1)(\nabla^2 - s_2)\Phi_{1\Pi}(x,y) = 0, \tag{11}$$

$\Phi_{1r}(x,y), \Phi_{3r}(x,y)$ - partial solutions of inhomogeneous equations (8).

The total solution of $\Phi_{1\Pi}(x,y)$ equation (11) is equal to the sum of $\Phi_{11}(x,y)$ and $\Phi_{12}(x,y)$:

$$\Phi_{1\Pi}(x,y) = \Phi_{11}(x,y) + \Phi_{12}(x,y), \tag{12}$$

where $\Phi_{1i}(x,y)$ $(i=1,2)$ is the general solutions of the Helmholtz equations:

$$(\nabla^2 - s_i)\Phi_{1i}(x,y) = 0\,. \tag{13}$$

The second equation (11) with respect to the function $\Phi_{1\Pi}(x,y)$ describes the potential boundary effect, and the biharmonic equation with respect to the function $\Phi_{1B}(x,y)$ together with the partial solutions $\Phi_{1r}(x,y)$ and $\Phi_{3r}(x,y)$ of equations (8) describe the internal SSS.

The method of reducing a non-homogeneous high-order DE to a low-order inhomogeneous equations [22] has proved that partial solutions $\Phi_{kr}(x,y)$ $(k=1,3)$ of a non-uniforms DE (8) are represented in the following form:

$$\Phi_{kr}(x,y) = \frac{a_{k0}\,D_{k0}}{s_1 s_2}\left(\frac{s_2}{s_1(s_1-s_2)}(f_{1r} - f_{0r}) + \frac{s_1}{s_2(s_2-s_1)}(f_{2r} - f_{0r}) + f_{00r}\right). \tag{14}$$

In dependencies (14): $f_{0r}(x,y)$, $f_{1r}(x,y)$, $f_{2r}(x,y)$, $f_{00r}(x,y)$ – partial solutions of the following equations, respectively:

$$D_0 f_0(x,y) = q(x,y);\ D_i f_i(x,y) = q(x,y), (i=1,2);\ D_0 D_0 f_{00}(x,y) = q(x,y)\,. \tag{15}$$

The vortex functions $\psi_1(x,y), \psi_3(x,y)$ are expressed by the operator method through the new function $\psi(x,y)$:

$$\psi_1(x,y)=(\beta_{333}+\beta_{332}\nabla^2)\psi(x,y);\ \ \psi_3(x,y)=-\beta_{133}\psi(x,y). \tag{16}$$

Function $\psi(x,y)$ satisfies DE

$$(\nabla^2-r_1)(\nabla^2-r_2)\psi(x,y)=0, \tag{17}$$

where r_i is the plate MGP ($r_i \rangle 0$).

The general solution of equation (17) is shown in the form

$$\psi(x,y)=\psi_{10}(x,y)+\psi_{20}(x,y), \tag{18}$$

where $\psi_{i0}(x,y)$ is the general solutions of the Helmholtz equations $(\nabla^2-r_i)\psi_{i0}(x,y)=0$.

After determining the transverse displacements of (9) and the vortex functions of (16), we find the tangential displacements of $u_k(x,y)$, $v_k(x,y)$:

$$\begin{aligned} u_k(x,y)&=\lambda_{k\phi1}\phi_{1,x}+\lambda_{k\phi3}\phi_{3,x}+\lambda_{k\psi1}\psi_{1,y}+\lambda_{k\psi3}\psi_{3,y}+\lambda_{kw1}w_{1,x}+\lambda_{kw3}w_{3,x}+\lambda_{kq}q_{,x};\\ v_k(x,y)&=\lambda_{k\phi1}\phi_{1,y}+\lambda_{k\phi3}\phi_{3,y}-\lambda_{k\psi1}\psi_{1,x}-\lambda_{k\psi3}\psi_{3,x}+\lambda_{kw1}w_{1,y}+\lambda_{kw3}w_{3,y}+\lambda_{kq}q_{,y}, \end{aligned} \tag{19}$$

where $\phi_k=\lambda_{k1}\nabla^2 w_1+\lambda_{k2}w_3+\lambda_{k3}\nabla^2 w_3+\lambda_{k4}q$ $(k=1,3)$; λ with indexes – MGP.

6. Solving boundary value problems. General solutions for the components of transverse displacements and vortex functions

Consider the boundary value problem formulated in Section 4 for different boundary conditions on the side surface.

Given (7), (9)–(19) and the disappearance of all functions at infinity, we obtain:

$$\begin{gathered}
\Phi_{1B}(x,y)=(C_1+C_2x)\exp(-\beta x)\cos\beta y;\\
\Phi_{11}(x,y)=C_5\exp(-\beta\omega_1 x)\cos\beta y, (\omega_1=\sqrt{1+s_1/\beta^2}\rangle 1);\\
\Phi_{12}(x,y)=C_7\exp(-\beta\omega_2 x)\cos\beta y, (\omega_2=\sqrt{1+s_2/\beta^2}\rangle 1);\\
\Phi_{1\Pi}(x,y)=(C_5\exp(-\beta\omega_1 x)+C_7\exp(-\beta\omega_2 x))\cos\beta y;\\
\Phi_{10}(x,y)=((C_1+C_2x)\exp(-\beta x)+C_5\exp(-\beta\omega_1 x)+C_7\exp(-\beta\omega_2 x))\cos\beta y;\\
\Phi_{30}(x,y)=0;\ f_{0r}(x,y)=A_0\exp(-\alpha x)\cos\beta y;\ A_0=q_0/(\alpha^2-\beta^2);\\
f_{ir}(x,y)=A_i\exp(-\alpha x)\cos\beta y;\ A_i=q_0/(\alpha^2-\beta^2-s_i);\\
f_{00r}(x,y)=A_3\exp(-\alpha x)\cos\beta y;\ A_3=q_0/(\alpha^2-\beta^2)^2;\\
\Phi_{kr}(x,y)=A_{kr}\exp(-\alpha x)\cos\beta y,\ (k=1,3);\\
\Phi_1(x,y)=((C_1+C_2x)\exp(-\beta x)+A_{1r}\exp(-\alpha x)\\
+C_5\exp(-\beta\omega_1 x)+C_7\exp(-\beta\omega_2 x))\cos\beta y;\\
\Phi_3(x,y)=\Phi_{3r}(x,y)=A_{3r}\exp(-\alpha x))\cos\beta y;\\
w_1(x,y)=((C_1b_1+C_2b_2(x))\exp(-\beta x)+c_{1r}\exp(-\alpha x))\cos\beta y\\
+(C_5b_5\exp(-\beta\omega_1 x)+C_7b_7\exp(-\beta\omega_2 x))\cos\beta y;\\
w_3(x,y)=(C_2d_2\exp(-\beta x)+c_{3r}\exp(-\alpha x))\cos\beta y
\end{gathered} \tag{20}$$

$$+(C_5 d_5 \exp(-\beta\omega_1 x)+C_7 d_7 \exp(-\beta\omega_2 x))\cos\beta y,$$

where here and further C with indices – constant integrations,

$$A_{kr}=\frac{a_{k0}}{s_1 s_2}(\frac{s_2}{s_1(s_1-s_2)}(A_1-A_0)+\frac{s_1}{s_2(s_2-s_1)}(A_2-A_0)+A_3)(\alpha^2-\beta^2-s_{k0});$$
$$b_1=\mu_{330};b_2(x)=\mu_{332}a_2+\mu_{330}x;\ c_{1r}=A_{1r}b_{1r}+A_{3r}b_{3r};\ b_5=\mu_{334}a_5^2+\mu_{332}a_5+\mu_{330};$$
$$b_7=\mu_{334}a_7^2+\mu_{332}a_7+\mu_{330};\ d_2=-\mu_{312}a_2;\ c_{3r}=A_{1r}d_{1r}+A_{3r}d_{3r};\ d_5=-(\mu_{314}a_5^2+\mu_{312}a_5);$$
$$d_7=-(\mu_{314}a_7^2+\mu_{312}a_7);\ d_{1r}=-(\mu_{314}a_{1r}^2+\mu_{312}a_{1r});\ d_{3r}=\mu_{114}a_{1r}^2;$$
$$b_{1r}=\mu_{334}a_{1r}^2+\mu_{332}a_{1r}+\mu_{330};\ b_{3r}=-(\mu_{134}a_{1r}^2+\mu_{132}a_{1r});\ a_2=-2\beta;\ a_5=\beta^2(\omega_1^2-1);$$
$$a_7=\beta^2(\omega_2^2-1);\ a_{1r}=\alpha^2-\beta^2.$$

Vortex functions $\psi_1(x,y)$ and $\psi_3(x,y)$ are determined from the following dependences:

$$\psi_1(x,y)=(C_9 k_{1r}\exp(-\beta\omega_3 x)+C_{11}k_{2r}\exp(-\beta\omega_4 x))\sin\beta y;$$
$$\psi_3(x,y)=(C_9 k_{3r}\exp(-\beta\omega_3 x)+C_{11}k_{3r}\exp(-\beta\omega_4 x))\sin\beta y, \quad (21)$$

where $k_{1r}=\beta_{332}r_1+\beta_{333};\ k_{2r}=\beta_{332}r_2+\beta_{333};\ k_{3r}=-\beta_{133}$.

7. Stresses and tangential displacements

Given (4), (5), (19)–(21), we find the stresses and tangential displacements:

Components of tangential displacements:

$$u_k(x,y)=((C_1 a_{k1\beta}\beta+C_2 a_{k2\beta})\exp(-\beta x)-a_{k\alpha}\alpha\exp(-\alpha x))\cos\beta y$$
$$+(C_5 a_{k5\beta}(-\beta\omega_1)\exp(-\beta\omega_1 x)+C_7 a_{k7\beta}(-\beta\omega_2)\exp(-\beta\omega_2 x))\cos\beta y$$
$$+(C_9 a_{k9\beta}\beta\exp(-\beta\omega_3 x)+C_{11}a_{k11\beta}\beta\exp(-\beta\omega_4 x))\cos\beta y;$$
$$(22)$$
$$v_k(x,y)=((C_1 a_{k1\beta}\beta+C_2 b_{k2\beta})\exp(-\beta x)-b_{k\alpha}\beta\exp(-\alpha x))\sin\beta y$$
$$+(C_5 a_{k5\beta}(-\beta)\exp(-\beta\omega_1 x)+C_7 a_{k7\beta}(-\beta)\exp(-\beta\omega_2 x))\sin\beta y$$
$$+(C_9 a_{k9\beta}\beta\omega_3\exp(-\beta\omega_3 x)+C_{11}a_{k11\beta}\beta\omega_4\exp(-\beta\omega_4 x))\sin\beta y.\ (k=1,3),$$

where

$$a_{k1\beta}=-b_1\lambda_{kw1};\ a_{k2\beta}=\lambda_{kw1}(\mu_{330}-\beta b_2(x))-\beta(\lambda_{k\phi1}t_{12}+\lambda_{k\phi3}t_{32})-d_2\beta\lambda_{kw3};$$
$$a_{k\alpha}=\lambda_{k\phi1}(t_{10}+\lambda_{14}q_0)+\lambda_{k\phi3}(t_{30}+\lambda_{34}q_0)+(\lambda_{kw1}c_{1r}+\lambda_{kw3}c_{3r})+\lambda_{kqx}q_0;$$
$$a_{k5\beta}=\lambda_{k\phi1}t_{15}+\lambda_{k\phi3}t_{35}+\lambda_{kw1}b_5+\lambda_{kw3}d_5;\ a_{k7\beta}=\lambda_{k\phi1}t_{17}+\lambda_{k\phi3}t_{37}+\lambda_{kw1}b_7+\lambda_{kw3}d_7;$$
$$a_{k9\beta}=\lambda_{k\psi1}k_{1r}+\lambda_{k\psi3}k_{3r};\ a_{k11\beta}=\lambda_{k\psi1}k_{2r}+\lambda_{k\psi3}k_{3r};\ b_{k2\beta}=a_{k2\beta}-\lambda_{kw1}\mu_{330};$$
$$b_{k\alpha}=\lambda_{k\phi1}(t_{10}+\lambda_{14}q_0)+\lambda_{k\phi3}(t_{30}+\lambda_{34}q_0)+(\lambda_{kw1}c_{1r}+\lambda_{kw3}c_{3r})+\lambda_{kqy}q_0;$$
$$t_{k0}=\lambda_{k1}a_{1r}c_{1r}+\lambda_{k2}c_{3r}+\lambda_{k3}c_{3r}(\alpha^2-\beta^2);\ t_{k2}=\lambda_{k2}d_2-2\lambda_{k1}\beta\mu_{330};$$
$$t_{k5}=\lambda_{k1}a_5b_5+\lambda_{k2}d_5+\lambda_{k3}a_5d_5;\ t_{k7}=\lambda_{k1}a_7b_7+\lambda_{k2}d_7+\lambda_{k3}a_7d_7.$$

Stress components;

$$\sigma_{xz}^{(3)}(x,y,z)=\sum_{n=0,2}^{4}P_n t_{xn};\ \sigma_{yz}^{(3)}(x,y,z)=\sum_{n=0,2}^{4}P_n t_{yn};\sigma_{z}^{(3)}(x,y,z)=\sum_{n=1,3}^{5}P_n s_{zn};$$
$$\sigma_{x}^{(3)}(x,y,z)=\sum_{n=1,3}^{5}P_n s_{xn};\ \sigma_{y}^{(3)}(x,y,z)=\sum_{n=1,3}^{5}P_n s_{yn};\ \sigma_{yx}^{(3)}(x,y,z)=\sum_{n=1,3}^{3}P_n t_{yxn}. \quad (23)$$

The components in stresses $\sigma_x(x,y,z),\sigma_y(x,y,z)$ have the form:

$$s_{xn}=s_{xn}(x,y)=\sum_{k=1,3}^{3}((C_1c_{1xkn}+C_2c_{2xkn})\exp(-\beta x)+c_{\alpha xkn}\exp(-\alpha x)$$
$$+C_5c_{5xkn}\exp(-\beta\omega_1 x)+C_7c_{7xkn}\exp(-\beta\omega_2 x) \quad (24)$$
$$+C_9c_{9xkn}\exp(-\beta\omega_3 x)+C_{11}c_{11xkn}\exp(-\beta\omega_4 x))\cos\beta y,\ (n=1,3,5),$$

where

$$c_{1xkn}=a_{k1\beta}\beta^2(a_{nvk}-a_{nuk});\ c_{2xkn}=a_{nvk}\beta(a_{k2\beta}-\lambda_{kw1}\mu_{330})-a_{nuk}a_{k2\beta}\beta+a_{nw3}d_2;$$
$$c_{\alpha xkn}=a_{nuk}a_{k\alpha}\alpha^2-a_{nvk}b_{k\alpha}\beta^2+a_{nw3}c_{3r}+a_{nq}q_0;$$
$$c_{5xkn}=a_{k5\beta}\beta^2(a_{nuk}\omega_1^2-a_{nvk})+a_{nw3}d_5;c_{7xkn}=a_{k7\beta}\beta^2(a_{nuk}\omega_2^2-a_{nvk})+a_{nw3}d_7; \quad (25)$$
$$c_{9xkn}=a_{k9\beta}\beta^2\omega_3(a_{nvk}-a_{nuk});\ c_{11xkn}=a_{k11\beta}\beta^2\omega_4(a_{nvk}-a_{nuk}).$$

In order to obtain $S_{yn}(x,y)$, it is necessary to replace $S_{xn}(x,y)$ with $S_{yn}(x,y)$ $(s_{xn}\to s_{xn})$, in formulas (24) and $a_{nuk}\to a_{nvk}$; $a_{nvk}\to a_{nuk}$ in formulas (25).

Formulas for components in other stresses:

$$t_{yxn}=t_{yxn}(x,y)=((C_1c_{1tn}+C_2c_{2tn})\exp(-\beta x)+c_{\alpha tn}\exp(-\alpha x)+C_5c_{5tn}\exp(-\beta\omega_1 x)$$
$$+C_7c_{7tn}\exp(-\beta\omega_2 x)+C_9c_{9tn}\exp(-\beta\omega_3 x)+C_{11}c_{11tn}\exp(-\beta\omega_4 x))\sin\beta y,\ (n=1,3);$$
$$t_{xn}=t_{xn}(x,y)=((C_1d_{1xn}+C_2d_{2xn})\exp(-\beta x)+d_{\alpha xn}\exp(-\alpha x)+C_5d_{5xn}\exp(-\beta\omega_1 x)$$
$$+C_7d_{7xn}\exp(-\beta\omega_2 x)+C_9d_{9xn}\exp(-\beta\omega_3 x)+C_{11}d_{11xn}\exp(-\beta\omega_4 x))\cos\beta y,\ (n=0,2,4),$$
$$(t_{xn}\to t_{yn},d_{ixn}\to d_{iyn},d_{\alpha xn}\to d_{\alpha yn},\ \cos\beta y\to\sin\beta y); \quad (26)$$
$$s_{zn}=s_{zn}(x,y)=((C_1d_{1zn}+C_2d_{2zn})\exp(-\beta x)+d_{\alpha zn}\exp(-\alpha x)+C_5d_{5zn}\exp(-\beta\omega_1 x)$$
$$+C_7d_{7zn}\exp(-\beta\omega_2 x)+C_9d_{9zn}\exp(-\beta\omega_3 x)+C_{11}d_{11zn}\exp(-\beta\omega_4 x))\cos\beta y,\ (n=1,3,5),$$

where c and d with the indices depend on α,β,n and the plate MGP.

Thus, the components in displacements and stresses are as follows:

$$w_1(x,y)=f_{w1}(x,C_1,C_2,C_5,C_7,c_{1r})\cos\beta y;\ w_3(x,y)=f_{w3}(x,C_2,C_5,C_7,c_{3r})\cos\beta y;$$
$$u_n(x,y)=f_{un}(x,C_1,C_2,C_5,C_7,C_9,C_{11},a_{n\alpha})\cos\beta y;$$
$$v_n(x,y)=f_{vn}(x,C_1,C_2,C_5,C_7,C_9,C_{11},b_{n\alpha})\sin\beta y;\ (n=1,3);$$
$$s_{xn}(x,y)=f_{xn}(x,C_1,C_2,C_5,C_7,C_9,C_{11},c_{\alpha xkn})\cos\beta y;$$
$$s_{yn}(x,y)=f_{yn}(x,C_1,C_2,C_5,C_7,C_9,C_{11},c_{\alpha ykn})\cos\beta y,\ (n=1,3,5); \quad (27)$$
$$t_{yxn}(x,y)=f_{yxn}(x,C_1,C_2,C_5,C_7,C_9,C_{11},c_{\alpha tn})\sin\beta y,\ (n=1,3);$$
$$t_{xn}(x,y)=f_{xn}(x,C_1,C_2,C_5,C_7,C_9,C_{11},d_{\alpha xn})\cos\beta y;$$
$$t_{yn}(x,y)=f_{yn}(x,C_1,C_2,C_5,C_7,C_9,C_{11},d_{\alpha yn})\sin\beta y,\ (n=0,2,4);$$

$$s_{zn}(x,y) = f_{zn}(x, C_1, C_2, C_5, C_7, C_9, C_{11}, d_{\alpha zn})\cos\beta y,\ (n=1,3,5).$$

The coefficients c, a, b, d with the indexes depend on α, β, n and the MGP of the plate.

8. Boundary problems and their analysis

8.1. Boundary problems. Let us consider some boundary value problems for a half-infinite plate under the action of transverse loading (7).

Boundary problem 1. Consider the following boundary conditions for $x = o$.

$$U(x=0,y,z)=0,\ V(x=0,y,z)=0,\ W(x=0,y,z)=0.$$

These conditions, given (3), are as follows:

$$u_n(x=0,y) = v_n(x=0,y) = w_n(x=0,y) = 0,\ (n=1,3). \tag{28}$$

Based on (20), (22), (27), (28) we have six linear algebraic equations with respect to constant integrations $C_1, C_2, C_5, C_7, C_9, C_{11}$:

$$u_n(x=0,y) = f_{un}(x=0, C_1, C_2, C_5, C_7, C_9, C_{11}, a_{n\alpha}) = 0;$$
$$v_n(x=0,y) = f_{vn}(x=0, C_1, C_2, C_5, C_7, C_9, C_{11}, b_{n\alpha}) = 0,$$
$$w_n(x=0,y) = f_{wn}(x=0, C_1, C_2, C_5, C_7, c_{nr}) = 0,\ (n=1,3).$$

The right-hands sides of these equations depend on $c_{1r}, c_{3r}, a_{n\alpha}, b_{n\alpha}$ respectively.

After determining $C_1, C_2, C_5, C_7, C_9, C_{11}$, all displacements and stresses according to (3), (20), (22–26) are found.

Boundary problem 2. Let us now consider the following boundary conditions:

$$U(x=0,y,z) = (A_1^0 P_1(2z/h) + A_3^0 P_3(2z/h))\cos\beta y;$$
$$V(x=0,y,z) = (B_1^0 P_1(2z/h) + B_3^0 P_3(2z/h))\sin\beta y;$$
$$W(x=0,y,z) = (C_1^0 P_0(2z/h) + C_3^0 P_2(2z/h))\cos\beta y,\ (A_1^0, ..., C_3^0 - const).$$

We also obtain a system of six linear algebraic equations with respect to $C_1, C_2, C_5, C_7, C_9, C_{11}$:

$$u_n(x=0,y) = f_{un}(x=0, C_1, C_2, C_5, C_7, C_9, C_{11}, a_{n\alpha}) = A_n^0;$$
$$v_n(x=0,y) = f_{vn}(x=0, C_1, C_2, C_5, C_7, C_9, C_{11}, b_{n\alpha}) = B_n^0;$$
$$w_n(x=0,y) = f_{wn}(x=0, C_1, C_2, C_5, C_7, c_{nr}) = C_n^0,\ (n=1,3).$$

Boundary problem 3. Consider the following boundary conditions:

$$X_v(x=0,y,z) = (A_{1x} P_1(2z/h) + A_{3x} P_3(2z/h))\cos\beta y,\ A_{1x}, A_{3x} - const;$$
$$V(x=0,y,z) = 0;\ W(x=0,y,z) = 0.$$

Continuous integrations will be determined from the following equations:

$$s_{xn}(x=0,y) = f_{xn}(x=0, C_1, C_2, C_5, C_7, C_9, C_{11}, a_{n\alpha}) = x_{sn} \equiv A_{nx};$$
$$v_n(x=0,y) = f_{vn}(x=0, C_1, C_2, C_5, C_7, C_9, C_{11}, b_{n\alpha}) = 0;$$

$$w_n(x=0,y)=f_{wn}(x=0,C_1,C_2,C_5,C_7,c_{nr})=0,\ (n=1,3).$$

Boundary problem 4.

$$X_v(x=0,y,z)=(A_{1x}P_1(2z/h)+A_{3x}P_3(2z/h))\cos\beta y;$$

$$Y_v(x=0,y,z)=(A_{1y}P_1(2z/h)+A_{3y}P_3(2z/h))\sin\beta y;\ W(x=0,y,z)=0,\ (A_{1y},A_{3y}-const).$$

The system of linear algebraic equations for determining the integration constants will look like:

$$s_{xn}(x=0,y)=f_{xn}(x=0,C_1,C_2,C_5,C_7,C_9,C_{11},a_{n\alpha})=x_{sn}\equiv A_{nx};$$

$$t_{yxn}(x=0,y)=f_{yxn}(x=0,C_1,C_2,C_5,C_7,C_9,C_{11},c_{\alpha tn})=y_{sn}\equiv A_{ny};$$

$$w_n(x=0,y)=f_{wn}(x=0,C_1,C_2,C_5,C_7,c_{nr})=0,\ (n=1,3).$$

The sequence of determining stresses and displacements in boundary value problems 2, 3 and 4 is similar to boundary value problem 1.

8.2. Analysis of boundary value problems.

In the considered boundary value problems for a semi-infinite plate subjected to loading (7), we have six conditions for determining six constant integrations of systems of equations.

The internal SSS is described by terms that do not contain integration constants, together with terms containing C_1, C_2 constants. Terms with C_5, C_7 constants determine the potential boundary effect, and terms with constants C_9, C_{11}-vortex boundary effect.

In [18], boundary effects were investigated based on a variant MT. The vortex boundary effect was characterized by decaying exponential solutions containing functions $\exp(-\tilde{\lambda}_1 x/h)$ and $\exp(-\tilde{\lambda}_3 x/h)$, ($\tilde{\lambda}_1=\sqrt{r_1}$, $\tilde{\lambda}_3=\sqrt{r_3}$). Parameters $\tilde{\lambda}_1$ and $\tilde{\lambda}_3$ are indicators of the variability of the eddy effect of the first and second boundary layers. The potential boundary effect was characterized by the disappearing solutions of the exponential form ($\tilde{\alpha}_1=\sqrt{s_1}$, $\tilde{\alpha}_3=\sqrt{s_3}$ variability indicators) and the disappearing solutions of the oscillating form ($\tilde{\alpha}_0$ variability indicator).

The table 1 shows the variability indicators for isotropic (ν=0,3) and transtropic plates ($\nu=\nu'=0,3$; $E'=E$); $\tilde{\lambda}_k$ is variability indicator of the edge effect according to Reisner theory. The depth of penetration of eddy boundary layers is greater than the potential boundary layers for $G'/G\langle 1$. At $G'/G=0.7708$ for the potential boundary effect $\tilde{\alpha}_1=\tilde{\alpha}_3=\tilde{\alpha}_0=7.9478$. At G'/G <0.7708, the attenuation is exponential and at G'/G>0.7708 it is oscillatory. According to the exact theory of isotropic Lurie plates [24] $\tilde{\lambda}_1=3.142$; $\tilde{\lambda}_3=9.423$; $\tilde{\alpha}_1$=7.498; $\tilde{\alpha}_3$=13.90.

Table 1. Variability indicators of the vortex and potential boundary layers in the K13 approximation.

G'/G	$\tilde{\lambda}_1$	$\tilde{\lambda}_3$	$\tilde{\lambda}_k$ ($\psi_3=0$)	$\tilde{\lambda}$ [25]	$\tilde{\alpha}_1$	$\tilde{\alpha}_3$	$\tilde{\alpha}_0$
0.01	0.3142	1.011	0.3347	0.2876	0.5622	112.4	-
0.1	0.9935	3.196	1.058	0.9094	1.805	35.00	-
1.0	3.142	10.11	3.347	2.876	-	-	7.331
10.0	9.935	31.96	10.58	9.094	-	-	5.025
100.0	31.42	101.1	33.47	28.76	-	-	4.733

A variant of the theory describes with high accuracy the vortex edge effects in the K13 approximation.

The potential boundary effect needs to be clarified. Other theories describe boundary effects with less accuracy.

A variant of mathematical theory and the developed method can be generalized to solve boundary value problems of dynamics [26].

9. Conclusions

A method for solving the boundary value problems of a variant of the mathematical theory of thick transversely isotropic plates under arbitrary transverse loads has been developed. The proposed method makes it possible to significantly simplify the solution of systems of differential equations of high-order equilibrium by reducing them to second-order differential equations.

The method is used to solve the boundary value problem for a semi-infinite transverse-isotropic plate of arbitrary thickness at a transverse load $q(x, y) = q_0 \exp(-\alpha x)\cos\beta y$ in the K13 approximation. The general solutions of the problem for the components of displacements and stresses are deduced and their research is carried out.

Four different classes of boundary conditions are considered, covering the boundary-value problem in displacements and the mixed problem. Boundary conditions and algebraic equations for determination of constant integrations are obtained. The terms that describe the internal SSS and boundary effects are highlighted. All components of the state in the variable x, which is perpendicular to the edge of the plate, have a decaying exponential character.

The transverse displacements depend on the internal SSS and the potential boundary effect. Other components depend on internal SSS, potential and vortex boundary effects. The internal SSS on the variable x depends exponentially on α and β. Potential boundary effect is characterized by $\beta\omega_1$ and $\beta\omega_2$, and vortex by $\beta\omega_3$ and $\beta\omega_4$.

As parameter β increases, the permeation of the vortex and potential boundary layers for $G'/G \langle 1$ decreases, with the decrease of β, the permeability of the vortex and potential boundary layers increases, with $\beta \to 0$ the variability indicators go to the values given in the table.

A high-precision theory variant describes the internal SSS (at smooth loads) and eddy boundary effects in the K13 approximation [18-20, 22]. The potential marginal effect needs clarification

Similarly to the K13 approximation, analytic solutions can be found in the K0-3, K135, K0-5, and others approximations, using the basic equations previously obtained [19] for these approximations. Quantitative results are determined on the basis of analytical solutions.

References

[1] Timoshenko S P 1921 On the correction for shear of the differential equation for transverse vibrations of prismatic bars *Philosophical Magazine and J. of sciene* **41** Ser. **6** **№ 245** pp 744-746

[2] Reissner E On the theory of bending of elastic plates *J. of Math. and Phys.* **33** (1944) 184–191

[3] Lo K H, Christensen R M and Wu E M 1977 A high-order theory of plate deformation Part 1 Homo-geneous plates *J. Appl. Mech.* **44** pp 663–668

[4] Ambarcumyan S A 1987 *Theory of Anisotropic Plates* (Moskva: Nauka)

[5] Altenbach H and Eremeyev V A 2009 On the linear theory of micropolar plates *J. of applied mathematics and mechanics* **89** pp 242–256

[6] Jemielita G 1991 Plate *Theory Meanders* (in Polish) Publisher of the Warsaw University of Technology (Warsaw)

[7] Nemish Yu N 2000 Development of analytical methods in three-dimensional static problems of anisotropic bodies *Prikladnaya mehanika* **36** **№ 2** pp 3–38

[8] Grigorenko Ya M Vasilenko A T and Pankratova N D 1991 *Problems of the theory of elasticity of inhomogeneous bodies* (Kiev: Naukova dumka)

[9] Nemish Yu N Homa I Yu 1991 The stress-strain state of non-thin shells and plates. Generalized theory *Prikladnaya mehanika* **29** **№11** pp 3–27
[10] Vekua I N 1955 On a method of calculating prismatic shells *Trudy Tbilisskogomatematicheskogo instituta* **21** pp 191–293
[11] Cicala P 1959 Sulla teria elastica della plate sottile *Giorn genio Civile* **97** № **4** pp 238-256
[12] Ponyatovskij V V 1962 To the theory of plates of average thickness *Prikladnaya matematika i mehanika* T **24** **№2** pp 335-341
[13] Kilchevskij N A 1963 *Fundamentals of Analytical Shell Mechanics* (Kiev: AN USSR)
[14] Homa I Yu 1973 On the general solution of the system of equations of equilibrium for bending plates of constant thickness *Doklady AN SSSR* **1** pp 59–62
[15] Plehanov A V and Prusakov A P 1976 On an asymptotic method for constructing a theory of bending of plates of average thickness *Mehanika tverdogo tela* **3** pp 84–90
[16] Gulyaev V I Bazhenov V A and Lizunov P P 1978 *Non-classical theory of shells and its application to solving engineering problems* (Lvov: Izd-vo Lvovskogo universiteta)
[17] Prusakov A P 1993 On the construction of twelve-order bend equations for a transversely isotrop-ic plate *Prikladnaya mehanika* **29** № **12** pp 51-58
[18] Zelenskyi A H 2005 Edge effects in non-thin plates *Visnyk Dnipropetrovskoho universytetu* № **10/2** Mekhanika v **9** t 2 pp 51–58
[19] Zelenskyi A H and Serebrianska P A 2007 The method of interconnected equations in the analytic theory of transistor plates, taking into account higher approximations *Visnyk Dnipropetrovskoho universytetu* № **2/2** *Mekhanika* v **11** t **2** pp 84–94
[20] Zelenskyi A H 2009 Models of analytical theory of transversely isotropic plates *Visnyk Dnipropetrovskoho universytetu* T **17** **№5** *Mekhanika* v **13** t **2** pp 54–62
[21] Zelensky A G 2019 Mathematical theory of transversally isotropic shells of arbitrary thickness at static load Home Materials Science Forum *Actual Problems of Engineering Mechanics Switzerland Trans Tech Publications Ltd* V **968** pp 496–510
[22] Zelensky A and Privarnikov A 2020 The method of integrating systems of high-order equilibrium equations of the mathematical theory of thick plates under intermittent loads (part 1) Innovative scientific researches: *European development trends and regional aspect Monograph Izdevniecība "Baltija Publishing"* (Riga Latvia) pp 221–255
[23] Reissner E 1950 On a variotional theoremin elasticiti, *J. Math. and Phys.* **33** pp 90–95
[24] Lur'e A I 1942 To the theory of thick plates *Prikladnaja matematika i mehanika* T **6** № **2** pp 151–169
[25] Piskunov V G Burygina A V and Rasskazov A A 1998 The effect of shear effects on the stress state in transversely isotropic plates *Problemy prochnosti* № **1** pp 56–71
[26] Demianenko A G 2019 Dynamics of Elastic Objects under Movable Inertial Loading Materials Science Forum *Actual Problems of Engineering Mechanics Switzerland Trans Tech Publications Ltd* v **968** pp 427–436

Stress-deformed state of vertical cylindrical metal shell under temperature climate impact

A A Zhdanov[1,*] **and V N Petrov**[1]

[1]Odesa State Academy of Civil Engineering and Architecture, 4, Didrihsona str., Odesa 65029, Ukraine

*E-mail: zhdanov@ogasa.org.ua

Abstract. There are continued the previous executed researches of the tense-deformed condition of a vertical metal thin cylindrical shell under temperature climatic influences in this work. So, in [1] the temperature climatic influences on cylindrical metal shells are investigated, in [2] the parameters of the elastic base simulating the wheat repulse are investigated, in [3] the tense-deformed condition of a shell on the elastic base with piecewise-constant is investigated, and in [4] with a piecewise-linear Winkler coefficient for axisymmetric cooling of the shell. The tense-deformed condition of a vertical cylindrical thin-wall shell lies on a piecewise-constant elastic base under non-axisymmetric temperature climatic influences has been studied in the present study on the basis of membrane theory of shells and a simple edge effect. The decision is received in the Fourier series, containing in the general case of heating-cooling the shell, ten members of the Fourier series.

1. Introduction

Storages of bulk materials in the form of vertical cylindrical metal thin-walled shells, during operation, are empty and filled with bulk material and experience the temperature climatic effects described in [1]. A change in the temperature of the outside air, the intensity of direct and scattered solar radiation induces a flat temperature field in the shell - the temperature of the shell is constant along the generatrix and variable in the circumferential direction, approximated by a Fourier series containing 5 terms of the cosine expansion series for the shell fully illuminated by the sun and 10 members of the series expansion in cosines and sines - for a shell half of which is covered by a screen at its full height. The temperature drop across the thickness of the shell can be neglected due to the small thickness of the shell (3 ... 10 mm) and a significant coefficient of thermal diffusivity of metals (steel, aluminum). has a direct effect on the stress-strain state (SSS) of the shell. In this case, it is necessary to take into account the variability in time of the position of the sun in the sky and the corresponding change in the position of the normal of the incidence of sunlight on the shell. The stress-strain state of the walls of bulk materials storages under temperature climatic influences is necessary to know when designing new bulk materials storages and assessing the strength of existing ones.

2. Analysis of recent research and publications

In [1], the temperature climatic effects on cylindrical metal shells were investigated, in [2, 3], the stress-strain state of the shell on an elastic foundation with a piecewise constant was investigated, and in [4] with a piecewise linear bed coefficient with axisymmetric cooling of the shell.

The small thickness of the shells, the absence of significant gradients of changes in the shell temperature led to the application in [2-4] of the momentless theory of shells, supplemented, to eliminate the residuals of the momentless solution, the theory of a simple edge effect. In this study, we use a similar proven approach.

3. Goals and objectives

The purpose of this work is to study the stress-strain state of an empty vertical cylindrical metal thin-walled shell of piecewise constant (in height) thickness, (Figure 1, a), caused by temperature climatic

Content from this work may be used under the terms of the Creative Commons Attribution 3.0 licence. Any further distribution of this work must maintain attribution to the author(s) and the title of the work, journal citation and DOI.
Published under licence by IOP Publishing Ltd

IOP Conf. Series: Materials Science and Engineering 1164 (2021) 012092 doi:10.1088/1757-899X/1164/1/012092

influences - non-axisymmetric heating-cooling of the shell to the state of the same temperature in all of its points.

4. Methods and Results

Let us write down the basic equations of the momentless theory of thin cylindrical shells [5, 6]:
Geometric Equations:

$$\varepsilon_1 = \frac{1}{R}\frac{\partial u}{\partial \alpha} + \alpha_t t\,,\ \varepsilon_2 \frac{1}{R}\left(\frac{\partial v}{\partial \beta} + w\right) + \alpha_t t\,, \tag{1}$$

where u, v, w - components of displacements of points of the middle surface of the shell (positive directions are shown in Figure 1, a);

$\varepsilon_1, \varepsilon_2$ - deformation of the middle surface, respectively, in the axial and circumferential directions;

α, β - dimensionless coordinates of the points of the middle surface: $\alpha = x/R$, β - angular coordinate;

t - shell temperature is a given function of coordinates α and β;

R - radius of the middle surface of the shell;

h - shell thickness;

E, μ, α_t - Young's modulus, Poisson's ratio and coefficient of linear thermal expansion of the shell material.

Positive directions of efforts and displacements are shown in Figure 1.

Physical equations:

$$\left.\begin{aligned} N_1 &= \frac{Eh}{\left(1-\mu^2\right)R}\left[\frac{\partial u}{\partial \alpha} + \mu\left(\frac{\partial v}{\partial \beta} + w\right) - (1+\mu)\cdot\alpha_t\cdot R\cdot t\right], \\ N_2 &= \frac{Eh}{\left(1-\mu^2\right)R}\left[\frac{\partial v}{\partial \beta} + w + \mu\frac{\partial u}{\partial \alpha} - (1+\mu)\cdot\alpha_t\cdot R\cdot t\right], \\ S &= \frac{Eh}{2(1+\mu)R}\left(\frac{\partial u}{\partial \beta} + \frac{\partial v}{\partial \alpha}\right). \end{aligned}\right\} \tag{2}$$

Equilibrium equations in displacements have the form:

$$\left.\begin{aligned} &\frac{\partial^2 u}{\partial \alpha^2} + \frac{1-\mu}{2}\frac{\partial^2 u}{\partial \beta^2} + \frac{1+\mu}{2}\frac{\partial^2 v}{\partial \alpha \partial \beta} + \mu\frac{\partial w}{\partial \alpha} = (1+\mu)\cdot\alpha_t\cdot R\cdot\frac{\partial t}{\partial \alpha}, \\ &\frac{1+\mu}{2}\frac{\partial^2 u}{\partial \alpha \partial \beta} + \frac{1-\mu}{2}\frac{\partial^2 v}{\partial \alpha^2} + \frac{\partial^2 v}{\partial \beta^2} + \frac{\partial w}{\partial \beta} = (1+\mu)\cdot\alpha_t\cdot R\cdot\frac{\partial t}{\partial \beta}, \\ &\mu\frac{\partial u}{\partial \alpha} + \frac{\partial v}{\partial \beta} + w = (1+\mu)\cdot\alpha_t\cdot R\cdot t. \end{aligned}\right\} \tag{3}$$

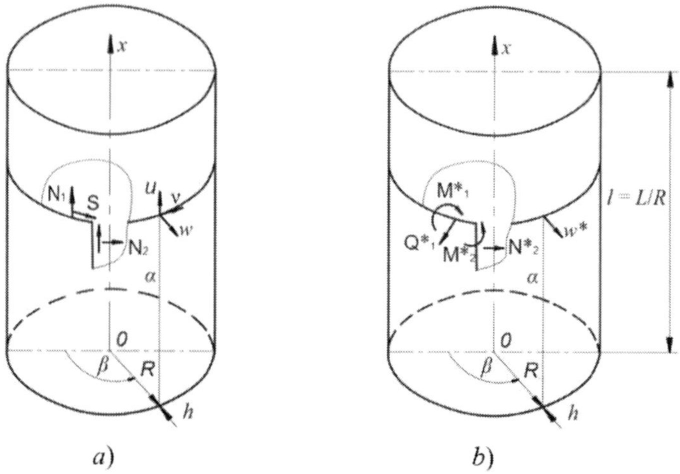

Figure 1. Positive directions of effort and displacement:
a - momentless stress-strain state; b - the state of the edge effect.

The external influence is the flat temperature field of the shell of the form:

$$t(\beta) = t_o + \sum_{n=1}^{10} \left(t_{nc} \cos n\beta + t_{ns} \sin n\beta \right). \tag{4}$$

The amplitudes of harmonics in (4) can be calculated according to the recommendations [1].

For stitching the belts and on the supporting edges of the shell, we use the theory of a simple edge effect, the main dependences of which are presented below [5]:

The resolving equation of the simple edge effect for the ith belt has the form [5,6]:

$$\frac{d^4 w_i^*}{d\alpha^4} + 4 g_i^4 w_i^* = 0, \tag{5}$$

where $g_i^4 = 3\left(1-\mu^2\right)\frac{R^2}{h_i^2}$, $D_i = \frac{E h_i^3}{12\left(1-\mu^2\right)}$.

Geometric Equations:

$$\varepsilon_2^* = \frac{w^*}{R}, \; \gamma^* = \frac{1}{R}\frac{dw^*}{d\alpha}, \; \varepsilon_1^* = 0. \tag{6}$$

Physical equations

$$\left. \begin{aligned} N_1^* = 0, \; N_2^* = \frac{Eh}{R} w^*, \; M_1^* = -\frac{D}{R^2}\frac{d^2 w^*}{d\alpha^2}, \\ M_2^* = \mu \cdot M_1^*, \; Q_1^* = \frac{D}{R^3}\frac{d^3 w^*}{d\alpha^3}. \end{aligned} \right\} \tag{7}$$

The positive directions of forces, bending moments and displacements of the edge effect are shown in Figure 1, b.

Taking into account the form of the temperature field (4), the equilibrium equations (3) for each belt of the shell take the form:

$$\left.\begin{aligned}
&\frac{\partial^2 u}{\partial\alpha^2}+\frac{1-\mu}{2}\frac{\partial^2 u}{\partial\beta^2}+\frac{1+\mu}{2}\frac{\partial^2 v}{\partial\alpha\partial\beta}+\mu\frac{\partial w}{\partial\alpha}=0,\\
&\frac{1+\mu}{2}\frac{\partial^2 u}{\partial\alpha\partial\beta}+\frac{1-\mu}{2}\frac{\partial^2 v}{\partial\alpha^2}+\frac{\partial^2 v}{\partial\beta^2}+\frac{\partial w}{\partial\beta}=(1+\mu)\cdot\alpha_t\cdot R\cdot\frac{\partial t}{\partial\beta},\\
&\mu\frac{\partial u}{\partial\alpha}+\frac{\partial v}{\partial\beta}+w=(1+\mu)\cdot\alpha_t\cdot R\cdot t.
\end{aligned}\right\} \tag{8}$$

In expressions (6-8), the index of the belt, for which the system of equations is written, is omitted.

4.1. Axisymmetric temperature action $(n=0)$

In this case, $v=0$, all derivatives with respect to β are equal to zero, and the system of equations (8) takes the form:

$$\left.\begin{aligned}
&\frac{d^2u_o}{d\alpha^2}+\mu\frac{dw_o}{d\alpha}=0,\\
&\mu\frac{du_o}{d\alpha}+w_o=(1+\mu)\alpha_t Rt_o.
\end{aligned}\right\} \tag{9}$$

Integrating the first equation (9) and substituting it into the second, we find:

$$\frac{du_o}{d\alpha^2}=-\mu w_o+C_1, \tag{10}$$

$$w_o=\frac{\alpha_t Rt_o}{1-\mu}-\frac{\mu}{1-\mu^2}C_1. \tag{11}$$

Substituting (11) into (10) and performing the integration, we obtain:

$$u_o(\alpha)=-\mu\frac{\alpha_t Rt_o}{1-\mu}\alpha+\frac{C_1}{1-\mu^2}\alpha+C_2. \tag{12}$$

Arbitrary constants C_1 and C_2 included in (11) and (12) must be determined from the tangential boundary conditions:

$$u_{1o}(0)=0,\ N_{1m}(l)=N_1^H-\vartheta u_m(l) \tag{13}$$

and tangential conditions for joining the belts:

$$\left.\begin{aligned}
&u_i(l)=u_{i+1}(0),\\
&N_{1,i}(l)=N_{1,i+1}(0),
\end{aligned}\right\}\ (i=1,2,\dots,m-1), \tag{14}$$

where m - the number of belts of the shell of different thickness, the numbering of the belts starts from the fixed edge of the shell;

N_1^H - pre-tensioning of the suspension shell in the axial direction;

ϑ - stiffness coefficient of elastic fastening of the shell at the edge $\alpha_m=l_m$.

Taking into account [2], tangential conditions (13) and (14) explicitly look like this:

$$C_{21}=0,$$

$$\left.\begin{aligned}&\frac{l_i}{1-\mu^2}C_{1i}+1\cdot C_{2i}+0\cdot C_{1,i+1}-1\cdot C_{2,i+1}=\frac{\mu\cdot l_i}{1-\mu}\alpha_t Rt_0,\\&h_i\cdot C_{1i}+0\cdot C_{2i}-h_{i+1}\cdot C_{1,i+1}+0\cdot C_{2,i+1}=(1+\mu)\alpha_t Rt_0(h_i-h_{i+1}),\end{aligned}\right\}(i=1,2,...,m-1) \quad (15)$$

$$\left[\frac{Eh_m}{(1-\mu^2)R}-\vartheta\frac{l_m}{(1-\mu^2)}\right]C_{1m}-\vartheta\cdot C_{2m}=N_1^H+\vartheta\frac{\mu l_m}{1-\mu}\alpha_t Rt_0+\frac{Eh_m}{(1-\mu)}\alpha_t t_0.$$

Discrepancies of the momentless solution arise due to different thicknesses of adjacent belts.

To satisfy the nontangential boundary conditions and the conditions for the continuity of radial displacements and angles of rotation at the joints of the chords, we will use the theory of a simple edge effect, with success used earlier [2,3,4]. In the corresponding formulas [4], it is necessary to set the bedding coefficient of the base $k(\alpha)\equiv 0$.

4.2. *Impact of non-axisymmetric components of the temperature field* $(n\geq 1)$

We seek the solution of the system of inhomogeneous equations (8) in the form of the sum of the general solution of the corresponding homogeneous and particular solution of the inhomogeneous system. Let's get a solution for an arbitrary belt. We omit the belt index in the calculations.

We represent the sought solutions in terms of two scalar functions $\Phi(\alpha,\beta)$ and $\Phi t(\alpha,\beta)$. In this case, the function $\Phi(\alpha,\beta)$ determines the solution of the homogeneous system of equations and is introduced in accordance with [6] as follows:

$$\left.\begin{aligned}&u_o=\frac{\partial}{\partial\alpha}\left(\frac{\partial^2\Phi}{\partial\beta^2}-\mu\frac{\partial^2\Phi}{\partial\alpha^2}\right),\\&v_o=-\frac{\partial}{\partial\beta}\left(\frac{\partial^2\Phi}{\partial\beta^2}+(2+\mu)\frac{\partial^2\Phi}{\partial\alpha^2}\right),\\&w_o=\nabla^2\nabla^2\Phi,\end{aligned}\right\} \quad (16)$$

where $\nabla^2=\frac{\partial^2}{\partial\alpha^2}+\frac{\partial^2}{\partial\beta^2}$ - Laplace operator.

By direct substitution, one can make sure that the introduced function satisfies the first two equations of the system. After substitution in the third equation (8), we obtain the resolving equation for the function $\Phi(\alpha,\beta)$:

$$\mu\frac{\partial^4\Phi}{\partial\alpha^2\partial\beta^2}-\mu^2\frac{\partial^4\Phi}{\partial\alpha^4}-\frac{\partial^4\Phi}{\partial\beta^4}-(2+\mu)\frac{\partial^4\Phi}{\partial\alpha^2\partial\beta^2}+\nabla^2\nabla^2\Phi=0. \quad (17)$$

Expanding the double Laplace operator and reducing similar terms in (17), we obtain the resolving equation for the function $\Phi(\alpha,\beta)$ in the form:

$$\frac{\partial^4\Phi}{\partial\alpha^4}=0. \quad (18)$$

A particular solution is determined by the function $\Phi t(\alpha,\beta)$, which, taking into account [7], we introduce so that through it the displacements ut, vt and wt are expressed as follows:

$$\left.\begin{aligned} u_t &= (1-\mu)\frac{\partial^3 \Phi_t}{\partial \alpha^3}, \\ v_t &= -(1-\mu)\frac{\partial^3 \Phi_t}{\partial \alpha^2 \partial \beta}, \\ w_t &= (1-\mu)\frac{\partial^2}{\partial \alpha^2}\nabla^2 \Phi_t, \end{aligned}\right\} \tag{19}$$

Taking into account the form of the temperature field of the shell, the functions $\Phi(\alpha,\beta)$ and $\Phi_t(\alpha,\beta)$ can be represented as a product of two functions, one of which depends only on α, the other only on β:

$$\Phi(\alpha,\beta)=\sum_{n=1}^{10}\left[\Phi_c^n \cos n\beta + \Phi_s^n \sin n\beta\right], \tag{20}$$

$$\Phi_t(\alpha,\beta)=\sum_{n=1}^{10}\left[\Phi_{tc}^n \cos n\beta + \Phi_{ts}^n \sin n\beta\right]. \tag{21}$$

Further calculations, without loss of generality of transformations, are carried out for the shell completely illuminated by the sun, i.e. containing 5 terms of the cosine expansion series. Substitution of (19) into (3) leads to the fact that all three equations coincide and the resolving equation for the function $\Phi_t(\alpha,\beta)$ can be written in the form:

$$\frac{\partial^4 \Phi_t}{\partial \alpha^4}=\frac{\alpha_t R t}{1-\mu}. \tag{22}$$

Taking into account (20) and (21), the resolving equations (18) and (22) for the nth term of the series are transformed into ordinary differential equations of the fourth order:

$$\frac{d^4 \Phi^n}{d\alpha^4}=0, \tag{23}$$

$$\frac{d^4 \Phi_t^n}{d\alpha^4}=\frac{\alpha_t R t}{1-\mu}. \tag{24}$$

The solutions to these equations are written in the form:

$$\Phi_c^n(\alpha)=C_{1c}^n\alpha^3+C_{2c}^n\alpha^2+C_{3c}^n\alpha+C_{4c}^n, \tag{25}$$

$$\Phi_{tc}^n(\alpha)=\frac{\alpha_t R t_{nc}}{1-\mu}\cdot\frac{\alpha^4}{24}. \tag{26}$$

Taking into account (25) and (26) n - e terms of the series of displacements of the momentless solution look like this:

$$\left.\begin{aligned} u_c^n(\alpha) &= \alpha_t R t_{nc}\alpha-\left(3n^2\alpha^2+6\mu\right)C_{1c}^n-2n^2\alpha C_{2c}^n-n^2C_{3c}^n, \\ v_c^n(\alpha) &= \alpha_t R t_{nc}\frac{n\alpha^2}{2}-\left[n^3\alpha^3-(2+\mu)6n\alpha\right]C_{1c}^n-\left[n^3\alpha^3-(2+\mu)2n\right]C_{2c}^n-n^3\alpha C_{3c}^n-n^3C_{4c}^n, \\ w_c^n(\alpha) &= \alpha_t R t_{nc}\left(1-\frac{n^2\alpha^2}{2}\right)+\left(n^4\alpha^3-12n^2\alpha\right)C_{1c}^n+\left(n^4\alpha^2-4n^2\right)C_{2c}^n+n^4\alpha C_{3c}^n+n^4C_{4c}^n. \end{aligned}\right\} \tag{27}$$

Changing the index "c" to "s" in (27), we obtain formulas for calculating the amplitudes of the harmonics of the corresponding displacements at $\sin n\beta$. The first terms in formulas (27) are particular solutions.

Substituting (27) into physical equations (2) after appropriate transformations, we obtain expressions for tangential forces:

$$\left.\begin{aligned} N_{10c}(\alpha) &= \sum_{n=1}^{10} N_{10c}^{n}(\alpha)\cos n\beta, \\ N_{20c}(\alpha) &= \sum_{n=1}^{10} N_{20c}^{n}(\alpha)\cos n\beta, \\ S_{0c}(\alpha) &= \sum_{n=1}^{10} S_{0c}^{n}(\alpha)\sin n\beta, \end{aligned}\right\} \tag{28}$$

where

$$\left.\begin{aligned} N_{10c}^{n}(\alpha) &= -\frac{Eh}{R}2n^2\left(3\alpha C_{1c}^{n} + C_{2c}^{n}\right), \\ N_{20c}^{n}(\alpha) &= 0, \\ S_{0c}^{n}(\alpha) &= \frac{Eh}{R}6nC_{1c}^{n}. \end{aligned}\right\} \tag{29}$$

The formulas for the amplitudes of the harmonics of tangential forces (29) do not include particular solutions - they are identically equal to zero. It is characteristic that the circumferential force identically equal to zero, this is a consequence of the free movement of the shell in the radial direction

Arbitrary constants in the expressions for the displacements (27) and effort (29), are determined from the tangential boundary conditions and docking conditions zones recorded for the amplitudes of the n-th term of a similar (15).

Supplementing the obtained solutions with the edge effect written for the nth harmonic similarly to [2, 3], we obtain a complete solution of the problem posed.

An empty vertical metal shell far from the fixed edges freely deforms in the radial direction, which explains the vanishing of the circumferential force of the momentless state N_2 in the shell.

A feature of the temperature climatic effect on a vertical circular cylindrical shell is that the temperature field of the shell seems to follow the sun, enveloping the shell, while in the shell without a screen, the temperature field differs slightly from the one that is symmetric relative to the normal incidence of sunlight.

5. Conclusions

1. For a vertical empty cylindrical shell, with one-sided heating-cooling of the shell, a momentless solution in the Fourier series for the components of the stress-strain state (SSS) is obtained.
2. A simple edge effect in the Fourier series eliminates the residuals of the momentless solution.
3. Taking into account [1], five terms of the cosine expansion series are sufficient to describe the SSS components during cooling of a shell heated without a screen.
4. To describe the components of the stress-strain state of a shell during cooling of a shell heated in the presence of a screen covering half of the shell along its entire height [1], it is necessary to take into account ten terms of the expansion series in the circumferential direction in sines and cosines.
5. The circumferential force $N_2(\alpha)$ of the momentless state is equal to zero as a consequence of the absence in the empty shell of restrictions on the shell displacement in the radial direction.

References

[1] Zhdanov A A 2014 Vliyanie solnechnoj radiacii na temperaturnye polya vertikal'nyh cilindricheskij hranilishch sypuchih materialov *Visnik ODABA* **56** (Odesa:

Zovnishreklamservis) pp 249-255
[2] Zhdanov A A 2015 Cilindricheskaya metallicheskaya obolochka na uprugom osnovanii pri temperaturnom klimaticheskom vozdejstvii *Visnik ODABA* **57** (Odesa: Zovnishreklamservis) pp 152-160
[3] Zhdanov A A 2017 Napryazhenno-deformirovannoe sostoyanie cilindricheskoj metallicheskoj obolochki na kusochno-postoyannom uprugom osnovanii pri temperaturnom klimaticheskom vozdejstvii *Visnik ODABA* **68** (Odesa: ODABA) pp 9-19
[4] Zhdanov A A 2016 Cilindricheskaya metallicheskaya obolochka na uprugom osnovanii pri osesimmetrichnom ohlazhdenii *Tezisi dokladov III Mezhdunarodnoy nauchno-prakticheskoy konferentsii Aktualnye problemy inzhenernoy mehaniki* (Odesa: Vnishreklamservis) pp 73-78
[5] Goldenveyzer A L 1953 *Teoriya uprugih tonkih obolochek* (Moscow: Gostehteorizdat) 544
[6] Vlasov V Z 1949 *Obshchaya teoriya obolochek i ee prilozheniya v tekhnike* (Moscow: Gostehteorizdat) 784
[7] Fridman L I 1956 Temperaturnye napryazheniya v kozhuhe koltsevoy kamery sgoraniy Vibratsionnaya prochnosct i nadezhnost aviatsionnyh dvigateley *Trudy Kujbyshevskogo aviacionnogo instituta* (Kuybishev: Izd-vo KuAI) **19** pp 299-306

Cracks and their role in the life cycle of building materials

V N Vyrovoy[1], V G Sukhanov[1], O A Korobko[1], S V Sukhanova[1]

[1] Odessa State Academy of Civil Engineering and Architecture,
Didrikhsona Str. 4, 65029 Odessa, Ukraine

E-mail: vyrovoy@ukr.net, svetlanasukhanova@rambler.ru

Abstract. Cracks as active elements are presented in all building structures. Cracks define a spontaneous internal restructuring of the structure of the construction material. This is a necessary condition for the adaptation of the system to the influence of unfavorable factors. The presence of active elements in the material of the structure-system contributes to its structural development, which indicates the creative role of cracks. Upon reaching the maximum variety of the structure, cracks-creators are able to develop to cracks of destruction. Cracks-destroyers should be considered as open complex self-organizing systems with their own life cycle. The predominance of the goal of the new crack-system over the goal of the existence of the structure-system leads to the failure of the structure as a basic system.
Key words: structure-system, structure-forming cracks, destruction cracks, life cycle, structural changes, active elements.

1. Introduction

The life cycle is understood as the period of time of the existence of a structure-system - from the manufacturing and performance of its inherent functions to its withdrawal from the operating mode. This assumes that during this period, a set of interrelated and interdependent processes and stages of structural transformations are implemented, ensuring the safe functioning of the structure. The life story of the system is formed under the influence of the particular life stories of individual structural elements. The structural elements, which by their very presence contribute to the course of the life story of the structure-system, include technological cracks (TC) and internal interface surfaces (IIS). Cracks and interface surfaces are active structural elements capable of responding in the same tempo-rhythm to external and internal influences and, thus, to participate in the processes of structural "restructuring" of the system, ensuring its external and internal safety [1,2]. Such participation of active elements in spontaneous structural rearrangements should be regarded as the only acceptable one. Otherwise, as a result of similar phenomena of self-organization, active elements can change the life story of the structure-system up to its degradation and death. Therefore, the analysis of the role of cracks as active elements in the life cycle of building structures should be considered an important task.

2. The role of cracks in the destruction of structural material

Failure of structures is associated with the irreversible growth of a crack, which, as a rule, divides the structure into separate parts. Experts note certain stages in the "life" of a crack - from the moment of its initiation and growth to its transformation into a main one with the exit of the crack front to the surface of a specimen or structure. The specific role of a crack is its ability to concentrate the arising deformations and stresses at its mouth, which greatly facilitates the process of its growth in the material environment. Methods for the quantitative assessment of the fracture toughness (fracture

Content from this work may be used under the terms of the Creative Commons Attribution 3.0 licence. Any further distribution of this work must maintain attribution to the author(s) and the title of the work, journal citation and DOI.
Published under licence by IOP Publishing Ltd

toughness) of a material by determining the stress concentration factors are proposed. Measures have been developed to improve the crack resistance of various materials, including construction materials. Attempts have been made to use critical stress intensity factors in assessing the durability of building materials and in calculating structures. The data presented are quite convincing evidence of a crack as the main factor determining the destruction of materials and structures made of them.

Our analysis showed the inevitability of the appearance of cracks at various levels of structural inhomogeneities for polystructural materials and made it possible to assess the role of cracks in its structural development as elements capable of stabilizing the properties of a structure and thereby ensuring its functionality. In addition, it is shown that the presence of cracks causes a permanent nonequilibrium state of individual subsystems and the system as a whole, and this contributes to the structural development of the system, an increase in its structural diversity and, as a result, a transition to a more equilibrium and stable state. This made it possible to draw a conclusion about the creative role of cracks in the structural development of open complex dynamic systems, which include building structures. There is no alternative to the cause of the destruction of structures as through the irreversible development of cracks. Therefore, the task was determined to study the conditions for the transition from cracks that play a creative role to cracks that cause the destruction of composite building materials and structures.

For the analysis, one should consider the full cycle of the structure's existence, including the active phase of its functioning as an open complex dynamic system.

Due to the fact that cracks as active structural elements of each subsystem and system are included in the model of the structural material and they play an essential role both in the processes of reorganization of the structure in order to stabilize the properties of the system and in irreversible processes of destruction, the main attention in the life cycle of the structure is given to cracks.

The main periods of the life of a structure as a system and the processes that occur at each stage of its existence are presented in table 1.

Table 1

№	Stages of the system's existence	The main processes characteristic of each stage of the life of the system
1	The origin and formation of the system	Initial structure (product geometry; distribution of aggregates, cement, water) ↓ Complex of physical-chemical and physical and mechanical phenomena and processes ↓ Organization of the structure of various levels of structural heterogeneities ↓ Manifestation of equifinality effects ↓ Technological cracks
2	System operation	Technological cracks ↓ Cracks-relaxers; Cracks-triggers; Cracks-transducers of the structure; Cracks-dissipators ↓ Structural changes (increased degree of structural diversity) ↓ Ultimate structural changes ↓ Nucleation of fracture cracks - trunk cracks
3	System death	Trunk cracks ↓ Irreversible growth of the trunk crack ↓ Destruction of a structure as a system

The origin and development of technological cracks in the structure of the material occurs at the stage of the birth and formation of the system. In this case, various mechanisms for organizing the structure of each level of inhomogeneities lead to one result - the appearance of technological cracks. The number and orientation of the initial cracks of various subsystems and the entire system is determined by the initial composition and intensity of the occurrence of physical-chemical and physical-mechanical phenomena and processes that are involved in organizing the structure of the material. The beginning of the system's functioning should be considered as the end of the period of its formation. The structure enters the working life cycle with a certain set of technological cracks at all levels of structural inhomogeneities.

During the period of operation, the material of the structure perceives the entire complex of operational loads, to which active elements react first of all. At this time, the following properties of "cracks-creators" are manifested: - relaxation of deformations and stresses in the material adjacent to the crack edges (cracks-relaxers); - the formation of new areas of the coastal surface, which contributes to the inclusion in the work of metastable elements (trigger cracks); - transformation of cracks into internal interfaces and vice versa, which leads to the formation of new structural elements - blocks (cracks-transducers of the structure); - dissipation of excess energy through the formation of new surfaces (dissipator cracks).

It should be noted that these properties are not possessed by individual cracks. Each crack, depending on the specific situation, exhibits one or another property. This reveals the creative role of cracks at each level of inhomogeneities (subsystems) and in the entire system. Constant restructuring is taking place, which contributes to increasing the diversity of structure in hierarchical systems. According to R. Ashby's figurative expression, only diversity can destroy diversity. However, in our opinion, the change in structural diversity in order to maintain the required parameters of properties has its limits.

In a functioning system, it can happen that the "cracks-creators" degenerate into "cracks-destroyers", which is a harbinger of a decrease in the properties parameters and the beginning of the period of the system's death. A situation arises in the system in which one of the structure elements takes on the function of the system in which it arose. He closes the goal of creating and the existence of the system on himself. Such a structural element primarily perceives the entire range of operational loads, which contributes to its own growth. With a size comparable to the size of individual subsystems (structural inhomogeneities), he no longer "feels" their structural features. This drastically reduces the structural diversity of the system. The processes that inevitably take place in separate subsystems, structural blocks, etc. are insignificant for the dominant structural element. Such a structural element, in our case a fracture crack, itself becomes a system that can be regarded as an open complex dynamic system.

Openness assumes that the crack, with its elements, is capable of perceiving, transmitting and redistributing deformations and stresses in the environment - the original system.

The complexity in this case is associated not only with the complex structure of the crack itself, but also with its rather complex behavior. Being an unstable system itself, a fracture crack, reacting to external influences, is capable of stopping, bending the outlines of the front, absorbing other cracks, and responding to local and integral fields of residual deformations.

The dynamism of a crack as a system assumes that each subsequent state is determined by the previous one. In this case, the new state may differ from the previous one in many parameters and indicators. This presupposes the irreversibility of the development of a dynamic crack-system in the process of reaching the target setting.

The target setting of the resulting system, for which the original system becomes a subsystem, is to ensure a complete life cycle.

A crack as a system goes through the full path of its "life" - birth, active functioning and death. The death of a crack as a system occurs when its main attributes - the mouth and front - disappear. In our opinion, the difference between “cracks-destroyers” and “cracks-creators” lies in the fact that the death of the former is associated with the disappearance of the front when it reaches the surface of a sample of a product, structure, etc. The death of "cracks-creators" occurs when their front reaches the banks of other cracks and interfaces. In this case, it is not so much the death of the "cracks-creators" that occurs, but their spontaneous transformation into a qualitatively different structural element - the inner interface.

During the period of active life, in order to ensure the priority of its own development, the fracture crack as a system uses all the potential capabilities of the base system and its structural features. The characteristic features of the "crack-system" that have arisen include: - the size (volume), which includes several subsystems of the base system and thus turns it into an independent extremely unstable system; - the ability to "draw" into itself on the way of its own growth other active elements of the structure (the multiplicity of attractors gives rise to the emergence of one attractor); - the ability of directed concentration of energy in the zones of its development, which practically excludes the influence of the structural diversity of the basic system on the growth conditions of the new system.

Upon reaching a certain stage of growth, the new system is capable of developing irreversibly using its own resources. It begins to exist at its own pace, striving to complete its development. This leads to the death of two systems at once - the new system due to the achievement of the goal of its existence and the basic system due to the fact that the new system, having died, led to the loss of its main functions. This completes the life of the original system due to the completion of the growth of the fracture crack.

3. Conclusion

The studies carried out allow us to draw the following conclusions:

Fracture crack should be considered as an open complex dynamic system. The purpose of the existence of such system is its own development. The appearance of fracture cracks (main cracks) simplifies the structural design of the base system. The activity of the new system presupposes that the goals of the new system prevail over the goal of the functioning of the original system. Achieving the goal of crack development as a system leads to its death and to the completion of the life cycle of the structure as a system.

References

[1] Broek D 1980 *Elementary engineering fracture mechanics* (Moscow: High School) p 368
[2] Zaitsev Yu 2010 Strength *and durability of cracked structural materials* (Minsk: BNTU) p 362
[3] Vyrovoy V, Sukhanov V and Korobko O 2016 *Material structure in structure of construction* (Odessa: Poligraf) p 244

AUTHOR INDEX